Oxford University Press
Digital Course Materials
for

Immunology

FIRST EDITION

Stephen J. Juris

Carefully scratch off the silver coating to see your personal redemption code.

This code can be redeemed only once.

Once the code has been revealed, this access card cannot be returned to the publisher.

Access can also be purchased online during the registration process.

The code on this card is valid for two years from the date of first purchase. Complete terms and conditions are available at learninglink.oup.com

Access Length: 6 months from redemption of the code.

OXFORD
UNIVERSITY PRESS

Directions for accessing your
Oxford University Press Digital Course Materials

Your OUP digital course materials can be delivered several different ways, depending on how your instructor has elected to incorporate them into his or her course.

BEFORE REGISTERING FOR ACCESS, be sure to check with your instructor to ensure that you register using the proper method.

VIA YOUR SCHOOL'S LEARNING MANAGEMENT SYSTEM

Use this method if your instructor has integrated these resources into your school's Learning Management System (LMS)—Blackboard, Canvas, Brightspace, Moodle, or other.

- Log in to your instructor's course within your school's LMS.
- When you click a link to a resource that is access-protected, you will be prompted to register for access.
- Follow the on-screen instructions.
- Enter your personal redemption code (or purchase access) when prompted.

VIA OXFORD learning link

Use this method if you are using the resources for self-study only. **NOTE**: *Scores for any quizzes you take on the OUP site will not report to your instructor's gradebook.*

- **Visit www.oup.com/he/juris1e**
- Select the edition you are using, then select student resources for that edition.
- Click the link to upgrade your access to the student resources.
- Follow the on-screen instructions.
- Enter your personal redemption code (or purchase access) when prompted.

VIA OXFORD learning cloud

Use this method only if your instructor has specifically instructed you to enroll in an Oxford Learning Cloud course. **NOTE**: *If your instructor is using these resources within your school's LMS, use the Learning Management System instructions.*

- Visit the course invitation URL provided by your instructor.
- If you already have an oup.instructure.com account you will be added to the course automatically; if not, create an account by providing your name and email.
- When you click a link to a resource in the course that is access-protected, you will be prompted to register.
- Follow the on-screen instructions, entering your personal redemption code where prompted.

For assistance with code redemption, Oxford Learning Cloud registration, or if you redeemed your code using the wrong method for your course, please contact our customer support team at **learninglinkdirect.support@oup.com** or 855-281-8749.

Immunology

IMMUNOLOGY

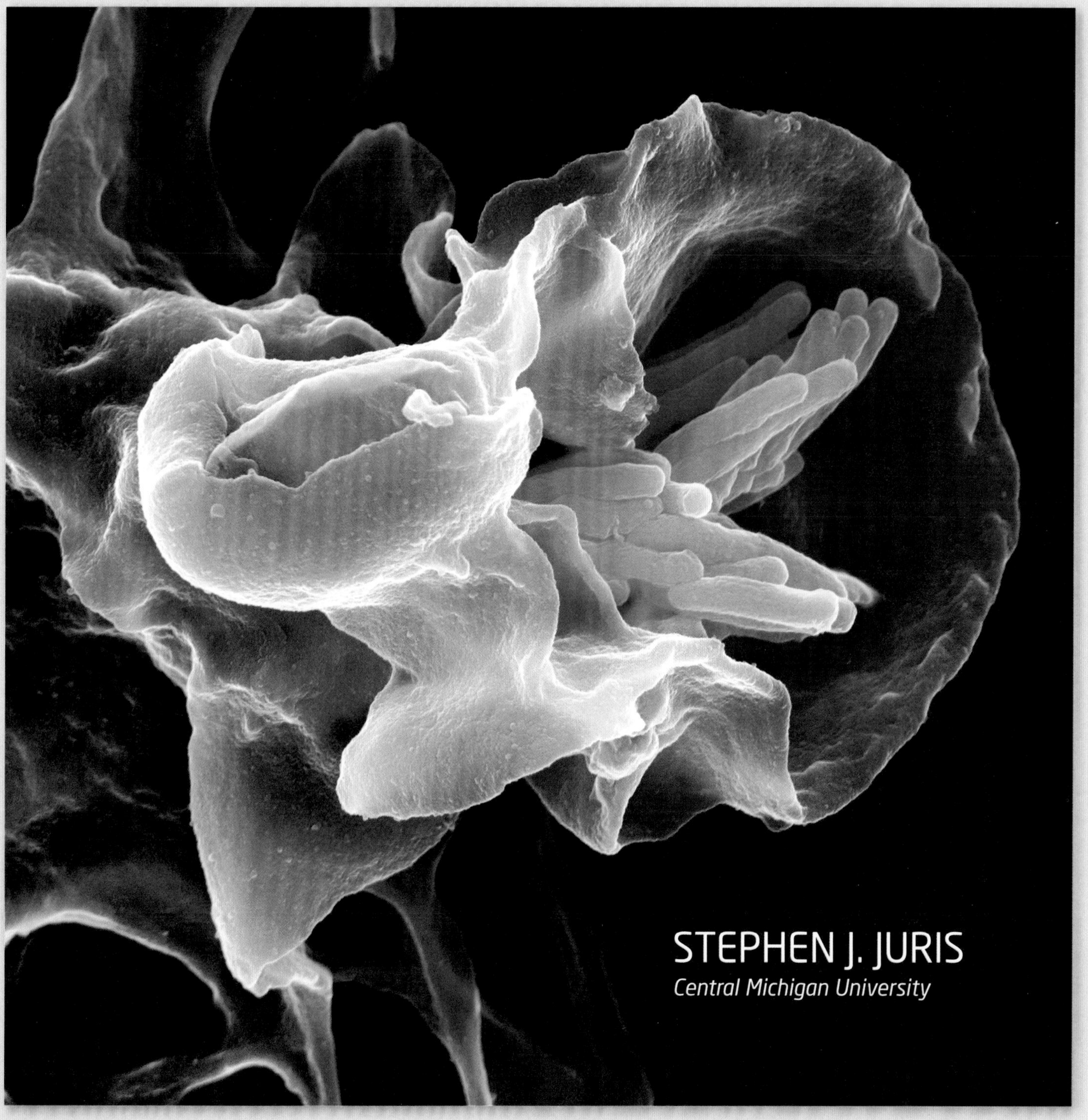

STEPHEN J. JURIS
Central Michigan University

OXFORD
UNIVERSITY PRESS

NEW YORK   OXFORD
OXFORD UNIVERSITY PRESS

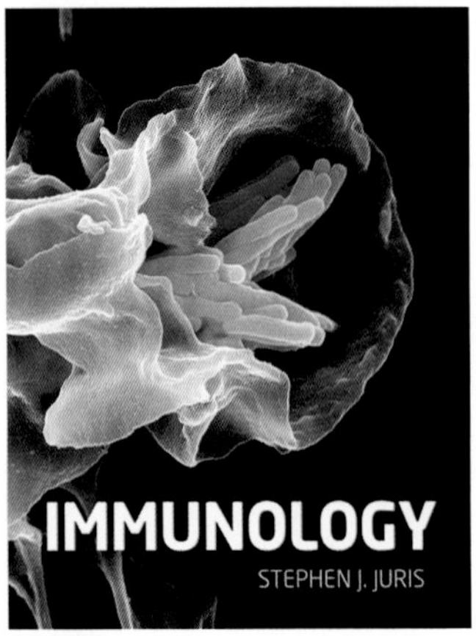

© SPL/Science Source

On the cover:

Macrophage engulfing TB bacteria. Colored scanning electron micrograph (SEM) of a macrophage (brown) engulfing tuberculosis (*Mycobacterium tuberculosis*) bacteria (green). This process is called phagocytosis. Macrophages are white blood cells of the body's immune system. One role of macrophages in the immune system is to phagocytose and destroy pathogens, dead cells, and cellular debris. Magnification: 750× at 10 cm on the long side.

Oxford University Press is a department of the University of Oxford. It furthers the University's objective of excellence in research, scholarship, and education by publishing worldwide. Oxford is a registered trade mark of Oxford University Press in the UK and certain other countries.

Published in the United States of America by Oxford University Press
198 Madison Avenue, New York, NY 10016, United States of America.

For titles covered by Section 112 of the US Higher Education Opportunity Act, please visit www.oup.com/us/he for the latest information about pricing and alternate formats.

ACCESSIBLE COLOR CONTENT Every opportunity has been taken to ensure that the content herein is fully accessible to those who have difficulty perceiving color. Exceptions are cases where the colors provided are expressly required because of the purpose of the illustration.

NOTICE OF TRADEMARKS Throughout this book trademark names have been used, and in some instances, depicted. In lieu of appending the trademark symbol to each occurrence, the authors and publisher state that these trademarks are used in an editorial fashion, to the benefit of the trademark owners, and with no intent to infringe upon the trademarks.

Library of Congress Cataloging-in-Publication Data

Names: Juris, Stephen, author.

Title: Immunology / Stephen Juris, Department of Biology, Central Michigan University.

Description: First edition. | New York : Oxford University Press, [2022] | Includes bibliographical references and index. | Summary: "Immunology offers the most contemporary perspective on the science available, providing a clear, easy-to-follow introduction to the discipline suitable for undergraduate students. In a course where students often get lost in vast amounts of detail and the sheer complexity of the immune response, Immunology helps students see "the big picture" with an approachable narrative that presents the exquisite details of immunology while emphasizing the connections between key themes that students so often lose sight of when learning the material. Immunology features an exceptional illustration program and includes simple, clear explanations, abundant examples, and features that unravel the mysteries of immunology through accounts of classical discoveries and recent, cutting-edge research. Since many students in the course are preparing to enter careers in research, medicine, and other health professions, an appropriate amount of applied knowledge and clinical content is included in the narrative, features, and engaging case studies. Students will easily be able to make connections, moving beyond memorizing just what we know to truly understanding how we know what we know-and why"-- Provided by publisher.

Identifiers: LCCN 2021020440 (print) | LCCN 2021020441 (ebook) | ISBN 9780190200312 (paperback) | ISBN 9780190926878 | ISBN 9780197631393 (ebook)

Subjects: LCSH: Immunology. | Immunity.

Classification: LCC QR181 .J87 2022 (print) | LCC QR181 (ebook) | DDC 616.07/9--dc23

LC record available at https://lccn.loc.gov/2021020440

LC ebook record available at https://lccn.loc.gov/2021020441

Printing number: 9 8 7 6 5 4 3 2 1

Printed by Quad, Printed in Mexico.

Brief Contents

To my wife, Michelle,
whose support, wisdom, and tolerance were essential for this journey;
and to my sons, Evan and Joshua,
who reminded me along the way to take breaks to play

Contents

1 Introduction to Immunology and the Immune System 1

2 Innate Immunity 34

5 Development of T Lymphocytes 123

6 Antigen Recognition by T Lymphocytes 151

7 T-Cell-Mediated Adaptive Immunity 180

Preface

My starting point: A passion for immunology

My love for immunology began when I was a student, perhaps not much older than you. As a graduate student researching how pathogens manipulate our bodies' lines of defense, I was excited to unravel the story of a human pathogen and how it causes disease. However, I soon realized that how a pathogen causes disease was only half the story: I needed to understand our immune system's response to the pathogen in order to understand the "big picture" of immunology. That was when I came to appreciate the connection between how our immune system operates and how pathogens evolve to evade our defense mechanisms.

My academic career since then has focused on the connection between host–pathogen interactions and the role of our immune responses to those pathogens. I have been able to share this appreciation as an instructor teaching immunology to undergraduate and graduate students. I have made it my goal to provide students like you with the tools they need to understand the overarching mechanisms of our immune systems so that they can apply this knowledge as they march through their own academic careers, whether in applied fields in the health professions or in research.

The overarching goal: A comprehensive text that doesn't lose sight of the big picture

My goals as an immunology author mirror my goals as an immunology instructor. This new text offers the most contemporary perspective on the science available. It is designed to provide undergraduate students who have little or no prior exposure to the field with a straightforward, easy-to-understand introduction to the subject. More importantly, though, it is designed to guide students like you through vast amounts of detail and terminology while emphasizing connections so that you will be able to see the big picture of the immune response that came as such a revelation to me as a graduate student.

All immunology texts contain vast amounts of detail and unfamiliar terms. These details are central to understanding the subject. However, immunology is an advanced undergraduate course typically taken in second or third year, when students are often expected to be thinking critically and applying knowledge rather than simply memorizing information. So, throughout the text I have endeavored to connect foundational knowledge of immunology to stories about the pioneering discoveries and contemporary cutting-edge science that drive our understanding of the field. By taking this approach, I hope to help you see the big picture of immunology while enabling you to apply your knowledge through a variety of problem-solving exercises. Instructors can easily adapt these exercises to an active-learning classroom, where they can be used to solidify fundamental concepts. I have also highlighted many of the real-world applications and evolutionary connections of immunology. These features are the foundation of a comprehensive textbook that is learner-centered and targeted at the student in search of relevant application examples.

A logical organization and student-friendly approach

I have observed, and my students have confirmed, that immunology is easier to understand when the material is organized starting from the moment the immune system responds to a pathogen. For this reason, *Immunology* begins with an introduction to innate immunity and then follows the human body's response to a pathogen and the various defenses humans use throughout the course of an infection. This starts with barriers at surfaces and tissues devoted to immune system function and continues with innate immune responses and then adaptive immune responses. The text concludes with a discussion of immune system dysfunction.

Students often become overwhelmed when learning adaptive immunity concepts, as the connections between cellular and humoral immunity can make the ordering of this material confusing. To alleviate misconceptions, *Immunology* incorporates commonalities in a unique overview chapter of the adaptive immune system (Chapter 4). Subsequent chapters focus on T-cell development and function prior to B-cell development and function, as T cells are vital in B-cell activation. After covering the fundamental concepts of adaptive immunity, *Immunology* focuses on the application of adaptive immune system memory and its medical application in vaccination (Chapter 11). Later chapters focus on connections between the immune system and health, including chapters on allergies (Chapter 15), transplantation (Chapter 17), and cancer (Chapter 18).

Since you may be taking this course in preparation to enter a career in research, medicine, or another health profession, I have integrated within the narrative an appropriate amount of applied knowledge and clinical content to keep you engaged. Supported by an exceptional illustration program that

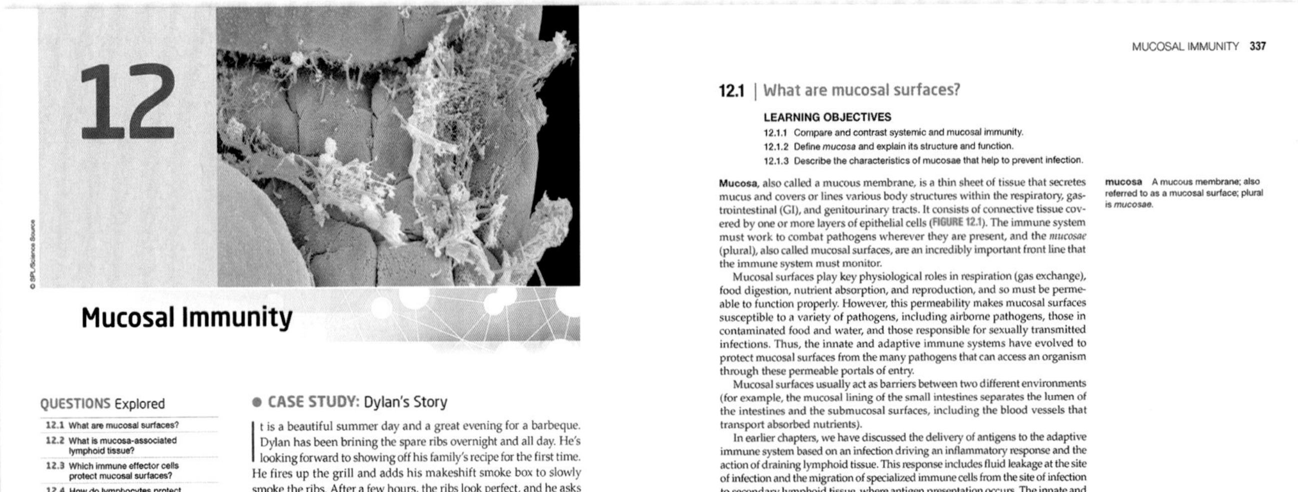

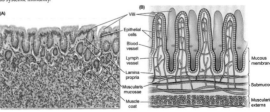

shows processes and structures in exquisite detail, the text includes many features that unravel the mysteries of immunology through accounts of classical discoveries and the latest advances in research. These features make it easier for you, the student, to make connections between key mechanisms of the human immune response and move beyond memorization to truly understand *how* we know what we know.

Illustrations that help you visualize features and processes in clear, vivid detail

Immunology features a striking art program comprising beautifully illustrated figures that are vivid, simple, and clear so that you can *see* the material you're reading about. Carefully thought-out and precisely rendered process diagrams walk readers through each step of key sequences in the immune response. Exquisitely drawn illustrations give students insight into cellular and molecular details you can otherwise only imagine.

These illustrations come together in a signature feature of the book, **Making Connections**, which relates material you are reading about in one chapter to key concepts and mechanisms you have learned about in previous chapters. You will gain a big picture understanding of immunology by visualizing how, for example, genetic recombination events that produce T-cell and B-cell receptors (Chapter 4) relate to T-cell and B-cell development (Chapters 5 and 8). One Making Connections figure in each chapter is expanded to include videos, interactive figures, vocabulary quizzing, and assessments so you can work through problems on your own and master these topics.

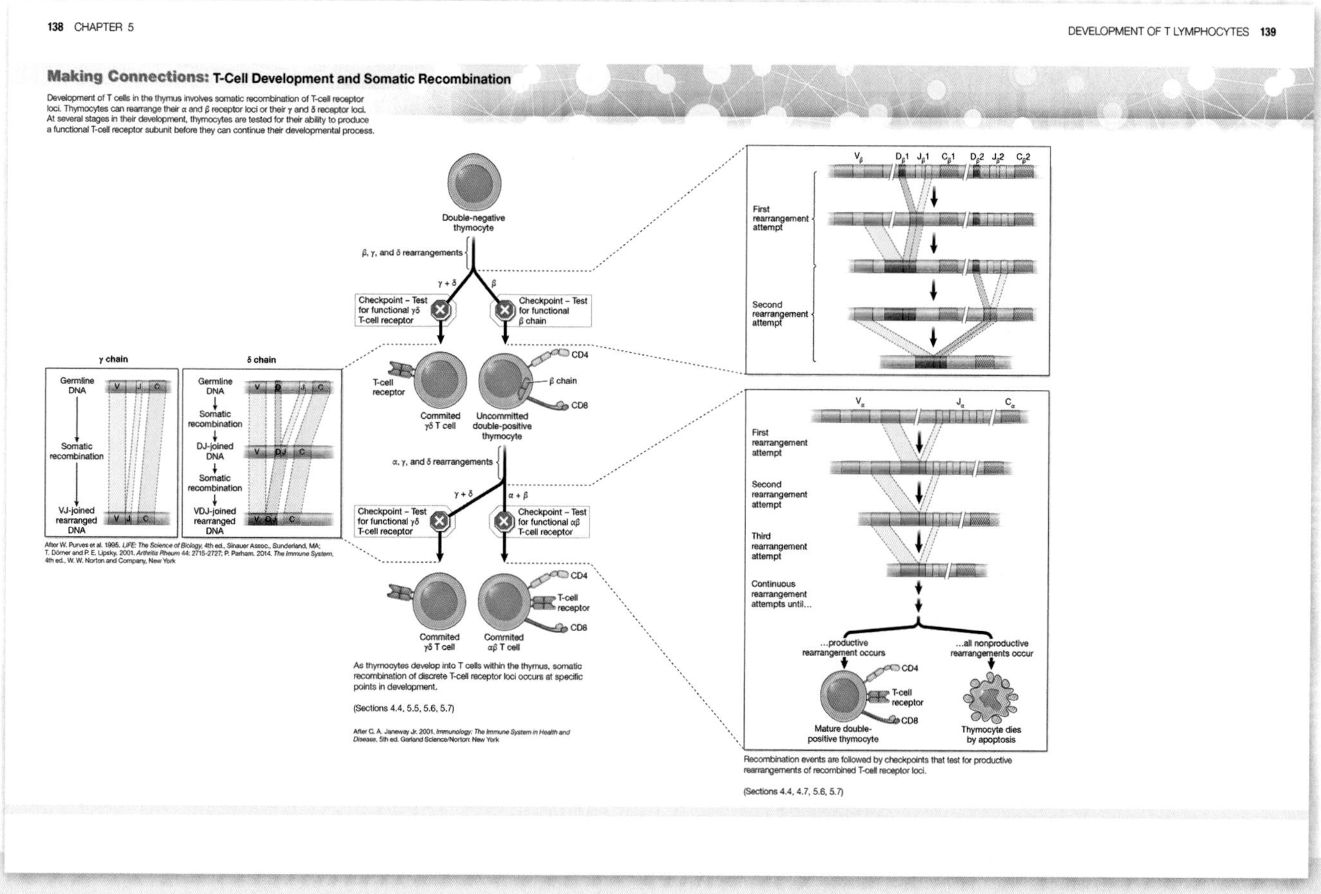

Classic cases and recent research highlight applications of immunology

Immunology unravels the mysteries of the discipline through accounts of classic discoveries and recent, cutting-edge research.

- **Case Studies:** Undergraduate immunology courses tend to draw students who aspire to careers in health-related professions, as well as those who wish to pursue research. Since both of these audiences are interested in clinical connections to their course content, this textbook uses *Case Studies* featuring real-life clinical examples to contextualize the material of each chapter. *Case Studies* introduced at the start of each chapter are revisited at the end of the chapter with questions for you to tackle on your own and in groups within an active-learning environment.
- **Emerging Science:** Examples of modern research in immunology give students both a connection to cutting-edge technology and a means to practice understanding of primary literature. *Think About…* questions provide opportunities for classroom discussion, active learning exercises, journal clubs, and practice for individual student conceptual understanding.
- **Key Discoveries:** This text rests on the bedrock of research presented in primary literature. To give you an appreciation for *how we know what we know*, this feature summarizes pioneering research connected to the chapter material. Students will learn about "what's next" and how certain processes were understood before a fundamental finding opened new avenues for further scientific inquiry.
- **Evolution and Immunity:** This feature helps students see the bigger picture of the immune system by exploring the evolutionary connections of immune systems from diverse organisms. Evolutionary conservation and evolutionary themes observed in immune responses of bacterial and eukaryotic organisms are just two of the fascinating topics you will encounter.
- **Controversial Topics:** This feature of later chapters explores our different societal connections to immunology. The questions included at the end of the feature promote critical thinking about the topic and will prepare you to address challenging questions in your future careers when you encounter alternative viewpoints.

Teaching and learning features that engage students and support instructors

Immunology presents the molecular details at the heart of the discipline within a pedagogical structure that ensures you don't lose sight of the big picture. Key questions—like *How does the immune response combat the pathogens that cause disease?*—frame the contents of each chapter. Guideposts—including learning objectives, review questions, and summaries—lead you through the clear, approachable narrative and keep you oriented to each chapter's key questions. All of these tools are designed to help students like you make connections between the concepts, the research, and the applications of immunology.

- **Learning Objectives:** *Learning Objectives* guide your learning of each chapter's key questions. Crafted using action verbs taken from Bloom's taxonomy, *Learning Objectives* stress the application of knowledge and are tied to *Checkpoint Questions* at the end of each section and *Review Questions* at the end of each chapter.

KEY DISCOVERIES

How was vaccination developed against smallpox?

Article

Jenner, E. 1798. An inquiry into the causes and effects of the variole vaccinae, a disease discovered in some of the western counties of England, particularly Gloucestershire and known by the name of the cow-pox. London: Sampson Low, 1798.

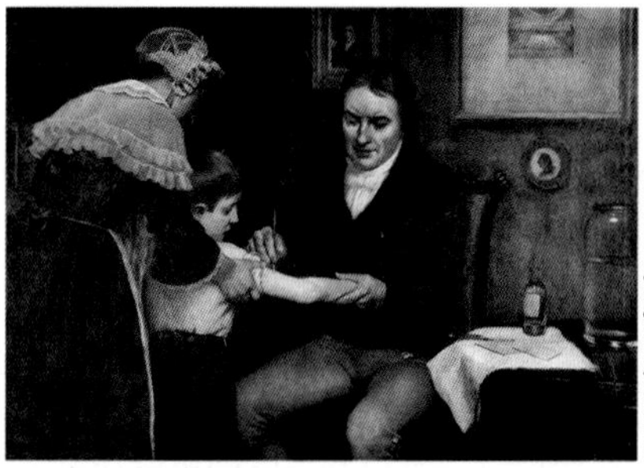

© Science History Images/Alamy Stock Photo

Early twentieth-century painting by Ernest Board depicts Dr. Edward Jenner performing his first vaccination in 1796.

Background

Smallpox was having a profound effect on the global population in the 1700s, infecting approximately 60% of the world, with an associated mortality rate of 30%. The massive impact of the disease warranted attempts to combat the infection and increase survival by protecting the population by any means necessary. However, at the time, pharmaceuticals did not yet exist and medical knowledge was limited. The pressing question was: How can the global population be protected from such a devastating disease?

Early Research

One of the standard means of protection against pathogens of the time was the practice of variolation: inoculation with a small amount of the disease-causing agent. Observations had been made that this could provide protection against subsequent infection from the same pathogen. However, this practice came with major risk—approximately 1% of those inoculated with the pathogen (by scratching the skin surface and applying infected material) would subsequently be infected with the pathogen. Although the practice of variolation brought with it the potential for protection against a pathogen, scientists and doctors of the time scrambled to find a safer alternative.

Think About...

1. Was variolation worth the risk of infection? How might a medical doctor of today wrestle with the conundrum of that time—deliberately infecting an individual with a pathogen to attempt to provide future protection? What are some of the ethical and moral questions associated with this situation?
2. If you lived in the eighteenth century and had to design a vaccine for smallpox (or any other pathogen), what questions might need to be answered? (Remember, scientists at the time had very limited technology compared to today!)
3. In the eighteenth century, what means might be available to test your vaccine?

Article Summary

Edward Jenner, ofte[…] utilized current know[…] attempt to design a [...] that bypassed variol[…] with the live virus. A [...] who had been expos[…] milkmaids) did not s[…] variolation, suggesti[…] similarly to smallpox [...] provided resistance [...] lated that cowpox bl[…] might have a protect[…] smallpox infection.

Jenner inoculated [...] had isolated from a [...] rienced pain in his a[…] developed a fever, h[…] Over the course of s[…] with smallpox sever[…] the smallpox virus. J[…] this child, and the ot[…] nal paper, gave birth [...]

EMERGING SCIENCE

Can messenger RNA be used for vaccination to a pathogen?

Article

Sahin, U., A. Muik, E. Derhovanessian, I. Vogler, L. M. Kranz, M. Vormehr, A. Baum, K. Pascal, J. Quandt, D. Maurus, S. Brachtendorf, V. Lorks, J. Sikorski, R. Hilker, D. Becker, A.-K. Eller, J. Grutzner, C. Boesler, C. Rosenbaum, M.-C. Kuhnle, U. Luxemburger, A. Kemmer-Bruck, D. Langer, M. Bexon, S. Bolte, K. Kariko, T. Palanche, B. Fischer, A. Schultz, P.-Y. Shi, C. Fontes-Garfias, J. L. Perez, K.A. Swanson, J. Loschko, I. L. Scully, M. Cutler, W. Kalina, C. A. Kyratsous, D. Cooper, P. R. Dormitzer, K. U. Jansen, and O. Tureci. 2020. COVID-19 vaccine BNT162b1 elicits human antibody and T_H1 T cell responses. *Nature* 586: 594–599.

Background

The COVID-19 pandemic caused by the SARS-CoV-2 virus took the world by storm, first identified in Wuhan, China, in December 2019 and rapidly infecting individuals worldwide. By the end of 2020, Johns Hopkins University reported more than 81 million COVID-19 cases globally, resulting in more than 1.7 million deaths. While mitigating factors, including lockdowns, travel restrictions, and requirements for social distancing and face coverings, were instituted in certain areas to lower exposure rates, it became increasingly apparent that true protection might not occur until the development of a safe and effective vaccine. Researchers raced to develop a vaccine, using both tried-and-true technology and the newer mRNA technology.

The Study

Researchers developed a vaccine that included an mRNA molecule that encodes for the receptor-binding domain of the viral spike protein, which is predicted to be a target for the generation of neutralizing antibodies. The mRNA molecule was made in vitro using modified nucleotides to protect the mRNA and minimize activation of an innate immune response and was packaged in lipid nanoparticles to aid in [...]

vaccinated 60 individuals with varying doses of the mRNA vaccine to test for the generation of a protective immune response. The vaccine induced a dose-dependent generation of immunoglobulins that recognized the spike protein receptor-binding domain that was capable of neutralizing virus infection (FIGURE ES 11.1). The researchers further demonstrated the activation of both CD4 and CD8 T cells due to the mRNA vaccine, with CD4 T cells preferentially differentiating into T_H1 helper T cells.

The researchers were able to successfully demonstrate the use of an mRNA vaccine to express the receptor-binding domain of the spike protein of SARS-CoV-2. The vaccine was shown to promote the production of immunoglobulins that bind to the spike protein and act as neutralizing antibodies. This research further paved the path for the development of the first mRNA vaccine produced to combat the COVID-19 pandemic.

Think About...

1. Why did the researchers choose to generate the mRNA to express the virus spike protein? After looking into the biology of coronaviruses, are there any other potential targets the researchers could have chosen?
2. With your knowledge of induction of the adaptive immune response, which cells would be most beneficial in receiving the mRNA molecule for activation of the adaptive immune response and CD4 and CD8 T cells?
3. How would CD8 T cells be activated by an mRNA vaccine? How would CD4 T cells be activated by an mRNA vaccine?

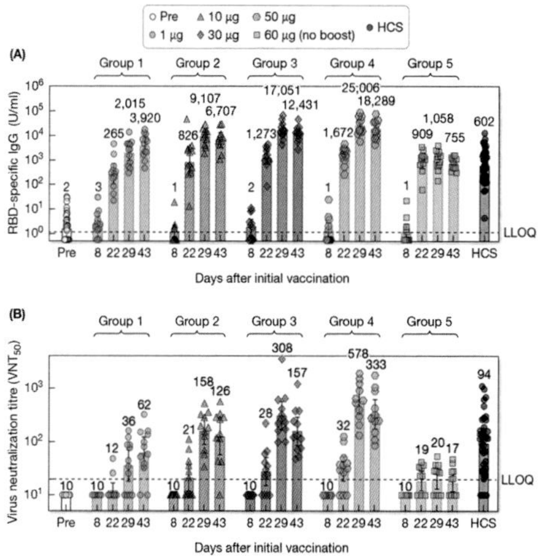

CONTROVERSIAL TOPICS

Do we really need all those gluten-free foods found in grocery stores?

The answer to this question may be more complex than you think. Over the past decade, there has been an increasing preponderance of food products labeled "gluten-free," including foods that a reasonable person would not expect to contain gluten (such as peanut butter, dried cranberries, and canned mushrooms) since gluten is a protein specific to wheat, rye, and barley. These gluten-free labels have increased in number largely due to the fact that a higher percentage of the population now chooses to eat a gluten-free diet. According to a 2017 study published in *Mayo Clinic Proceedings*, the population of individuals eating a gluten-free diet tripled from 0.5% to 1.7% of the U.S. population between 2009 and 2014, whereas the number of confirmed celiac disease cases (the autoimmune disease known to be caused by a reaction to gluten), has remained unchanged at approximately 0.7% of the population.

Individuals without celiac disease who avoid gluten in their diets may be perceived as following a fad diet since they don't suffer from the autoimmune disease or a wheat allergy. While some who claim that they experience fewer intestinal issues after switching to a gluten-free diet may be experiencing a placebo effect, there may be a significant portion of the population who are actually sensitive to gluten or to some other antigen present in gluten-containing foods.

Individuals with non-celiac gluten sensitivity (NCGS) suffer from intestinal symptoms associated with the ingestion of gluten-containing foods yet lack a positive celiac disease diagnosis. The pathogenesis of NCGS is not well understood, and it may be caused by gluten or by another protein present in gluten-containing foods. Because the actual cause of NCGS has yet to be discovered, it remains a controversial issue, one that some researchers question as a true clinical diagnosis. Until a confirmed causative agent of NCGS is discovered, the controversy about gluten sensitivity is likely to remain a topic of debate.

Think About...

1. How might a scientist attempt to discover a cause of NCGS?
2. The estimated number of individuals affected by NCGS worldwide is between 0.5% and 13%. Why do you think there is such a wide disparity between the lower and upper estimates? Do you think this wide range further strengthens or weakens the case of NCGS being an accepted clinical diagnosis?
3. Do you think that NCGS is truly a clinical illness or is related to the popularity of the gluten-free diet? What evidence from an opposing viewpoint might sway your opinion?

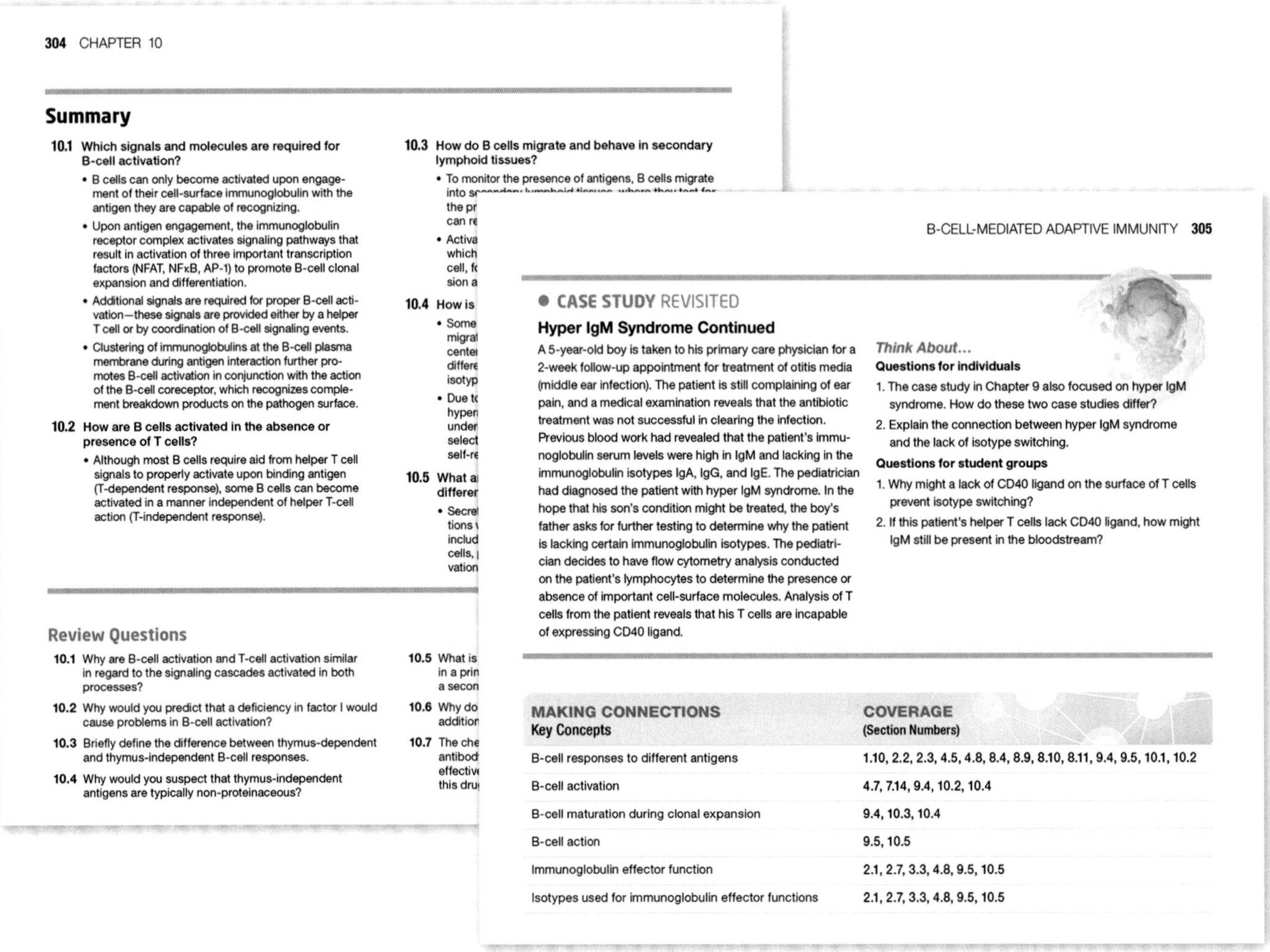

304 CHAPTER 10

Summary

10.1 Which signals and molecules are required for B-cell activation?
- B cells can only become activated upon engagement of their cell-surface immunoglobulin with the antigen they are capable of recognizing.
- Upon antigen engagement, the immunoglobulin receptor complex activates signaling pathways that result in activation of three important transcription factors (NFAT, NFκB, AP-1) to promote B-cell clonal expansion and differentiation.
- Additional signals are required for proper B-cell activation—these signals are provided either by a helper T cell or by coordination of B-cell signaling events.
- Clustering of immunoglobulins at the B-cell plasma membrane during antigen interaction further promotes B-cell activation in conjunction with the action of the B-cell coreceptor, which recognizes complement breakdown products on the pathogen surface.

10.2 How are B cells activated in the absence or presence of T cells?
- Although most B cells require aid from helper T cell signals to properly activate upon binding antigen (T-dependent response), some B cells can become activated in a manner independent of helper T-cell action (T-independent response).

10.3 How do B cells migrate and behave in secondary lymphoid tissues?
- To monitor the presence of antigens, B cells migrate into se... the pr... can re...
- Activa... which... cell, fc... sion a...

10.4 How is...
- Some... migrat... center... differe... isotyp...
- Due to... hyperr... under... select... self-re...

10.5 What a... differer...
- Secret... tions... includ... cells,... vation...

Review Questions

10.1 Why are B-cell activation and T-cell activation similar in regard to the signaling cascades activated in both processes?

10.2 Why would you predict that a deficiency in factor I would cause problems in B-cell activation?

10.3 Briefly define the difference between thymus-dependent and thymus-independent B-cell responses.

10.4 Why would you suspect that thymus-independent antigens are typically non-proteinaceous?

10.5 What is... in a prin... a secon...

10.6 Why do... addition...

10.7 The che... antibod... effective... this dru...

B-CELL-MEDIATED ADAPTIVE IMMUNITY 305

CASE STUDY REVISITED

Hyper IgM Syndrome Continued

A 5-year-old boy is taken to his primary care physician for a 2-week follow-up appointment for treatment of otitis media (middle ear infection). The patient is still complaining of ear pain, and a medical examination reveals that the antibiotic treatment was not successful in clearing the infection. Previous blood work had revealed that the patient's immunoglobulin serum levels were high in IgM and lacking in the immunoglobulin isotypes IgA, IgG, and IgE. The pediatrician had diagnosed the patient with hyper IgM syndrome. In the hope that his son's condition might be treated, the boy's father asks for further testing to determine why the patient is lacking certain immunoglobulin isotypes. The pediatrician decides to have flow cytometry analysis conducted on the patient's lymphocytes to determine the presence or absence of important cell-surface molecules. Analysis of T cells from the patient reveals that his T cells are incapable of expressing CD40 ligand.

Think About...

Questions for individuals
1. The case study in Chapter 9 also focused on hyper IgM syndrome. How do these two case studies differ?
2. Explain the connection between hyper IgM syndrome and the lack of isotype switching.

Questions for student groups
1. Why might a lack of CD40 ligand on the surface of T cells prevent isotype switching?
2. If this patient's helper T cells lack CD40 ligand, how might IgM still be present in the bloodstream?

MAKING CONNECTIONS Key Concepts	COVERAGE (Section Numbers)
B-cell responses to different antigens	1.10, 2.2, 2.3, 4.5, 4.8, 8.4, 8.9, 8.10, 8.11, 9.4, 9.5, 10.1, 10.2
B-cell activation	4.7, 7.14, 9.4, 10.2, 10.4
B-cell maturation during clonal expansion	9.4, 10.3, 10.4
B-cell action	9.5, 10.5
Immunoglobulin effector function	2.1, 2.7, 3.3, 4.8, 9.5, 10.5
Isotypes used for immunoglobulin effector functions	2.1, 2.7, 3.3, 4.8, 9.5, 10.5

- **Checkpoint Questions:** End-of-section *Checkpoint Questions* reinforce your learning by focusing on the main concepts from each section and providing you an opportunity to review what you have just read. You can answer these questions as you read through the chapter or use them as you study for summative assessments in your class.
- **Chapter Summaries:** Each chapter concludes with a summary of essential points covered in the exploration of each key question. You can use these essential points as an outline to guide your review of the material when preparing for quizzes and exams.
- **Review Questions:** Students typically learn best when applying knowledge. You can practice applying the knowledge you gain from the text by solving the end-of-chapter *Review Questions,* which not only test your ability to recall details from the chapter but also challenge you to think critically about situations based on material drawn from the chapter.

Oxford Insight: A homework platform to optimize student success

Immunology is available with Oxford Insight, a data-driven courseware platform designed to optimize student success. Oxford Insight delivers the entire text of *Immunology* along with thousands of gradable quiz and homework problems tailored to individual student needs. Topic-specific review questions are assigned to you based on how well you have performed in earlier practice quizzes.

In this way, Oxford Insight delivers a customized learning experience that actively engages you with assigned reading and additional review materials adapted to your level of mastery.

Other support for students

Immunology comes with a number of additional resources to support you along your unique learning pathway. These resources include:

- **Flashcards** and **vocabulary quizzes** to help you master the language of immunology
- **Quizzes** that help you review the content of every section and every chapter
- **Personalized practice** sections that help you review the material you need the most help with
- **Case Study quizzes** that help you review the stories that open each chapter
- **Interactive figures** that test your ability to label illustrations from the text using drag-and-drop exercises
- **Making Connections modules**, featuring my own overview and summary videos and review exercises.

Support for instructors

Oxford University Press offers a comprehensive ancillary package for instructors who adopt this new edition of *Immunology*. Oxford Learning Link (OLL) (www.oup.com/he/juris1e) contains the following teaching tools:

- **PowerPoint Images for each chapter:** All figures from each chapter are available in PowerPoint slides.
- **Lecture notes for each chapter:** Editable lecture notes in PowerPoint format make preparing lectures faster and easier than ever. The notes include a succinct outline of key concepts and featured research studies, key figures from the chapter, and 2–3 clicker questions, to increase student engagement.
- **Test bank:** The test bank includes over 60 multiple-choice and short-answer exam questions for each chapter—over a thousand questions in all.

Contact your local OUP sales representative or visit the Oxford Learning Link instructors page (learninglink.oup.com/instructors) to learn more and gain access to these instructor resources.

Acknowledgments and thanks

The idiom "it takes a village" comes to mind as I reflect on the fruition of this project. Given that my love for immunology was born during my graduate career, I would be remiss not to thank my Ph.D. advisor, Jack Dixon, for guiding and training me as a scientist. He gave me freedom to explore and the support I needed as I developed my passion for immunology. Thank you to my friends and colleagues at Central Michigan University who supported me in the development of this textbook.

I deeply appreciate the team at Oxford University Press who all came together to make my vision a reality. I'd first like to thank Jason Noe, Senior Acquisitions Editor at Oxford. I still remember the day you knocked on my office door when you came to campus and the conversation we had: the timing of your inquiry about authoring an immunology textbook was truly an aligning of the stars. If I hadn't been starting a sabbatical shortly after that fateful meeting, I'm not sure this project would have ever started. Thank you for your belief in me and the project. Thank you, also, to Joan Kalkut, Senior Editor, who helped the project reach the finish line and Sarah D'Arienzo, Editorial Assistant. I owe thanks to the team of development editors I worked with on this project: Tanya Martin, who worked closely with me from the beginning of the project; and Janna Green, Kerry O'Neill, and Eric Sinkins, who helped me along the way as I honed my craft and ensured we put out a product of which I am extremely proud. Thank you to the production team that brought my vision to life so vividly on the page and on the screen: Production Manager Joan Gemme, Senior Production Editors Martha Lorantos and Johannah Walkowicz, Book Designer and Production Specialist Meg Britton Clark, Photo Researcher Mark Siddall, Permissions Manager Michele Beckta, and much appreciation to Dragonfly Media Group for their painstaking work in creating the elegant illustrations that so clearly highlight the content. I am especially proud of the digital resources we have assembled to support student learning, and so I would like to thank Media Editor Lauren Elfers and Digital Resource Development Editor Zan Carter. In addition, I express my sincere gratitude to the sales and marketing groups, particularly Joan Lewis-Milne, Marketing Manager, and Ashendri Wickremasinghe, Marketing Assistant.

I would not have been able to see this project to completion without the continued and unwavering support of my family. To Michelle, my partner in science and in life, and our sons, Evan and Joshua: your enthusiasm, support, and patience were inspirational as I worked through many deadlines and pushed this project to completion. Through many late nights and long weekends, you continued to be my cheering section and lifted me up to complete this monumental task. I love you all very much.

About the author

Stephen Juris is Associate Professor at Central Michigan University. He received his Ph.D. from the University of Michigan and completed postdoctoral research at Harvard Medical School. Stephen's research focuses on the role of bacterial toxins in bacterial pathogenesis. His dissertation work was conducted on *Yersinia pestis* (causative agent of the bubonic plague) and its role in shutting down the immune response. His postdoctoral work examined the biochemical and biophysical action of anthrax toxin, including transport across endosomal membranes and action within cells. His current research at CMU focuses on a toxin made by *Vibrio cholerae* (causative agent of cholera) and the role the toxin plays in disruption of the cytoskeleton in intestinal epithelial cells. During his tenure at CMU, he has taught courses in immunology, bacterial pathogenesis, biochemistry, cell biology, and other topics. When he isn't teaching or in the lab, Stephen enjoys running, playing golf, and spending time with his family and friends.

Reviewers

Igor Alemeida, *University of Texas at El Paso*
Camilla Ambivero, *University of Central Florida*
Janet Andersen, *Stony Brook University*
Michael Angell, *Eastern Michigan University*
Rustom Antia, *Emory University*
Ivica Arsov, *The City University of New York*
Ali Azghani, *University of Texas at Tyler*
Fengwei Bai, *University of Southern Mississippi*
Kenneth Balazovich, *University of Michigan*
Simran Banga, *Western Kentucky University*
Lisa Banner, *California State University, Northridge*
Amorette Barber, *Longwood University*
Brianne Barker, *Drew University*
Harry Bernheim, *Tufts University*
Elliott Blumenthal, *Purdue University Fort Wayne*
James Bottesch, *Eastern Florida State College*
Catherine Brennan, *California State University, Fullerton*
Heather Bruns, *Ball State University*
Jerry Brunson, *Northwestern State University of Louisiana*
Gerald Buldak, *Loyola University Chicago*
Richard Bungiro, *Brown University*
David Kim Burnham, *Oklahoma State University*
Todd Camenisch, *University of Arizona*
Jean Cardinale, *Alfred University*
Lynne Cassimeris, *Lehigh University*
Maria Castillo, *New Mexico State University*
Tammy Castro, *Bloomfield College*
Matina Cetkovic-Cvrlje, *St. Cloud State University*
Yung Chang, *Arizona State University*
Stephen Chapes, *Kansas State University*
Thomas Comollo, *Kean University*
Danielle Condry, *University of North Dakota*
Jeff Copeland, *Eastern Mennonite University*
Kenneth Curr, *California State University, East Bay*
Candice Damiani, *University of Pittsburg*
Phillip Danielson, *University of Denver*
Beckley Davis, *Franklin & Marshall College*
Kelley Davis, *Nova Southeastern*
Anthony De Tomaso, *University of California, Santa Barbara*
Diane Dixon, *Southeastern Oklahoma State University*
Brian Doctor, *The University of Mississippi*
Megan Doczi, *Norwich University*
Robert Dotson, *Tulane University of Louisiana*
Jeannine Durdik, *University of Arkansas*
Jennifer Easterwood, *Queens University of Charlotte*
Michael Edidin, *Johns Hopkins University*
Anthony Ejiofor, *Tennessee State University*
Terri Ellis, *University of North Florida*
Uthayashanker Ezekiel, *Saint Louis University*
Anthony Farone, *Middle Tennessee State University*
Shawn Flanagan, *The University of Iowa*
Laura Flatow, *Berry College*
Sherry Fleming, *Kansas State University*
Teresa Foley, *University of Colorado at Boulder*
Thomas Fondy, *Syracuse University*
Clifton Franklund, *Ferris State University*
Kenneth Frauwirth, *University of Maryland, College Park*
Mihaela Gadjeva, *Harvard University*
Chiara Gamberi, *Concordia University*
Julie Torruellas Garcia, *Nova Southeaster University*
Brittany Gasper, *Florida Southern College*
Raffaella Ghittoni, *University of Southern California*
Swapan Ghosh, *Indiana State University*
Lindsay Gielda, *Purdue University Northwest*
Alan Goodman, *Washington State University*
Brian Gray, *York College of Pennsylvania*
Jennifer Gubbels, *Augustana University*
Severina Haddad, *University of New Haven*
Michael Hanna, *University of Texas at San Antonio*
Andrea Henle, *Carthage College*
Caroline Hennigan, *McNeese State University*
Charmaine Henry, *Baker University*
Kathleen Hoag, *Michigan State University*
Christopher Hughes, *University of California, Irvine*
Mette Ibba, *The Ohio State University*
Janaki Iyer, *Northeastern State University*
Liesl Jeffers-Francis, *North Carolina A&T State University*
Douglas Johnston, *Louisiana State University*
Louis Justement, *University of Alabama at Birmingham*
Richard Karp, *University of California, Berkeley*
Jinsil Kim, *Biola University*
Kevin Kinney, *DePauw University*
David Kittlesen, *University of Virginia*
Gary Koski, *Kent State University*
Ashwini Kucknoor, *Lamar University*
James Kumi-Diaka, *Florida Atlantic University*
Alexandra Kurtz, *Georgia Gwinnett College*
Allison Land, *Minnesota State University*
Joseph Larkin, *University of Florida*
Alan Levine, *Case Western Reserve University*
Nancy Liu-Sullivan, *College of Staten Island*
Li-Fan Lu, *University of California at San Diego*
Lisa Lyford, *University of the Cumberlands*
Colin Martin, *University of St. Thomas*
Nathaniel Martinez, *California Polytechnic State University*
Andrea Mastro, *The Pennsylvania State University*
Clinton Mathias, *Western New England University*
Matthew Maurer, *Robert Morris University*
Victoria McCurdy, *Mississippi State University*
Dennis McGee, *Binghamton University*
Thomas McGuire, *Pennsylvania State University, Abington*
Jennifer Metzler, *Ball State University*

Jeffrey Mital, *Quinnipiac University*
Fernando Monroy, *Northern Arizona University*
Brigette Morin, *Michigan Technological University*
Robert Moss, *Wofford College*
Andrea Moyer, *Oakland University*
Rita Moyes, *Texas A&M University*
Mustafa Mujtaba, *Florida Gulf Coast University*
Hao Nguyen, *California State University, Sacramento*
Shanna Nifoussi, *University of Wisconsin, Superior*
Mary Jane Niles, *University of San Francisco*
Erin Norcross, *Mississippi College*
Fran Norflus, *Clayton State University*
Cynthia Norrgran, *Colorado School of Mines*
Amanda Norvell, *College of New Jersey*
Amy Obringer, *University of Saint Francis*
Robert O'Donnell, *The State University of New York College at Geneseo*
Michael Opata, *Appalachian State University*
John Palisano, *Sewanee University*
Kirstin Park, *Michigan State University*
Robert Paulson, *The Pennsylvania State University*
Marcia Pierce, *Eastern Kentucky University*
Roberta Pollock, *Occidental College*
Marianne Poxleitner, *Gonzaga University*
Madhura Pradhan, *The Ohio State University*
Dorothy Pumo, *Albany College of Pharmaceutical Sciences*
Shira Rabin, *University of Louisville*
Kathryn Rafferty, *University of Nevada, Las Vegas*
Sarah Redmond, *Radford University*
Carol Reis, *New York University*
Clifford Renk, *Florida Gulf Coast*
Paulson Robert, *The Pennsylvania State University*
Ellen Robey, *University of California, Berkeley*
Michael Roner, *University of Texas Arlington*
Ken Roth, *James Madison University*
Shereen Sabet, *California State University, Long Beach*
Tobili Sam-Yellowe, *Cleveland State University*
Jerry Sanders, *University of Michigan, Flint*
Abhay Satoskar, *The Ohio State University*
Elizabeth Schwartz, *Auburn University*
Laurie Shornick, *St. Louis University*
Trevor Siggers, *Boston University*
Michael Sikes, *North Carolina State University*
David Singleton, *York College of Pennsylvania*
Denise Slayback-Barry, *Indiana University–Purdue University Indianapolis*
L. Courtney Smith, *George Washington University*
Vladislav Snitsarev, *Montclair State University*
Janice Speschock, *Tarleton State University*
Henry Spratt, *University of Tennessee at Chattanooga*
Kurt Spurgin, *University of California, Riverside*
David Stachura, *California State University, Chico*
Paul Storer, *Coe College*
Jennifer Stueckle, *West Virginia University*
Kevin Suh, *High Point University*
Julie Swartzendruber, *Midwestern University*
Joe Taube, *Baylor University*
Venkataswarup Tiriveedhi, *Tennessee State University*
Clara Toth, *St. Thomas Aquinas College*
Adrianne Vasey, *The Pennsylvania State University*
Hannah Venit, *Chestnut Hill College*
Calli Versagli, *Saint Mary's College*
Laura Vogel, *Illinois State University*
Debby Walser-Kuntz, *Carleton College*
Helen Walter, *Mills College*
Cornelius Watson, *Roosevelt University*
Yanzhang Wei, *Clemson University*
Gregory Weigel, *University of Central Florida*
Robert Wheeler, *University of Maine*
Denise Wingett, *Boise State University*
Candace Winstead, *California Polytechnic State University*
Kelly Woytek, *Texas State University*
Eric Yager, *Albany College of Pharmacy and Health Sciences*
Qian Yin, *Florida State University*
Michael Zimmer, *Purdue University Northwest*

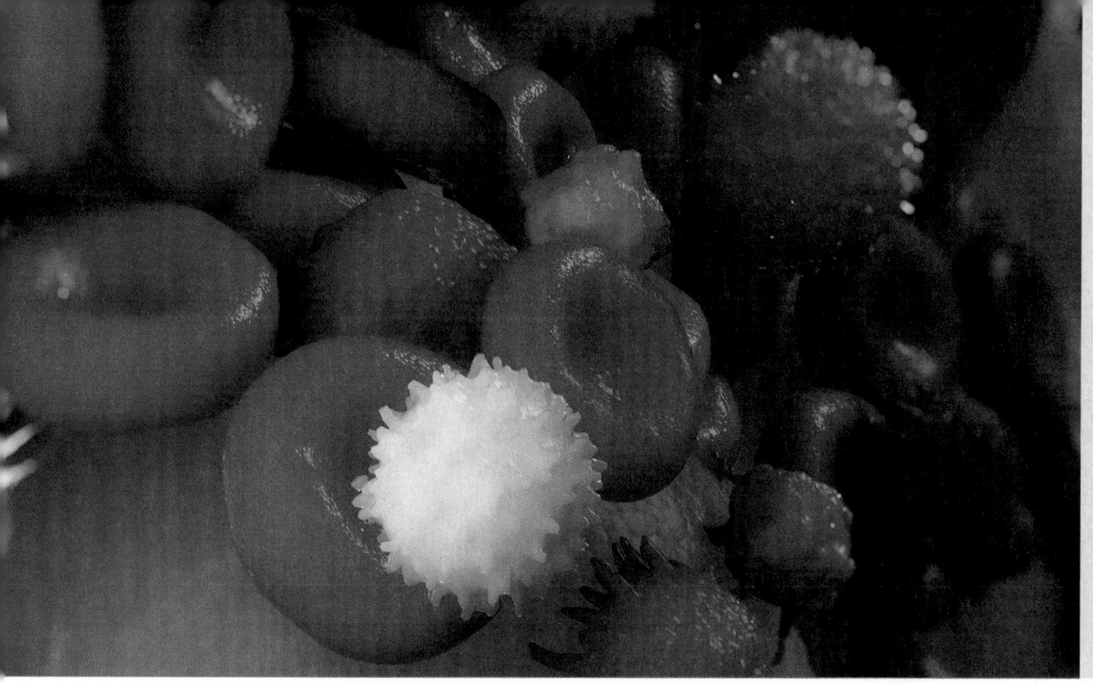

© SciencePicture.Co/Medical Images

1

Introduction to Immunology and the Immune System

● CASE STUDY: John's Story

John is driving home from an uneventful day at work. He is looking forward to a tasty dinner with his wife and son, and watching the football game that evening. The traffic is typical for rush hour, so he doesn't anticipate any problems making it home with plenty of time for both dinner and the game. As John merges onto the highway, he is suddenly greeted by a loud honking horn and screeching tires...

John wakes up to find nurses and a doctor bustling near his hospital bed. Seeing that John is awake, the doctor begins to explain what had happened—John had been in a terrible car accident yesterday evening. While everyone involved had survived, John had suffered a puncture wound to his left side from some of the car shrapnel. The injury had forced doctors to perform a splenectomy (a surgical procedure whereby a surgeon removes the spleen) to prevent internal bleeding. The doctor reassures John that he will pull through and be able to go home with his family.

Six months later, John wakes up feeling under the weather. Maybe he ate something that didn't agree with him, or maybe he was coming down with the flu. Unfortunately, as the day progresses, John begins to feel much worse. He is fevered and nauseated and cannot keep any food down. John tells his wife that he thinks it would be a good idea for him to go to the hospital.

In order to understand what John is experiencing, and to answer the case study questions at the end of this chapter, we must first understand how the body keeps infections at bay. Our body employs an intricate system known as the immune system that protects us

QUESTIONS Explored

1.1 How did the scientific field of immunology emerge?

1.2 What are the types of pathogens?

1.3 How does the immune response combat the pathogens that cause disease?

1.4 What is the body's first line of defense against pathogens?

1.5 Which cell lineages are associated with innate and adaptive immunity?

1.6 How does the innate immune system limit the spread of most infections?

1.7 How does the adaptive immune system target and destroy pathogens that have evaded the innate immune system?

1.8 How do primary and secondary lymphoid tissues differ?

1.9 What are the specialized types of secondary lymphoid tissues and how are they organized?

1.10 What happens when the immune system malfunctions?

from the causative agents of infectious disease. As you progress through this textbook, you will gain an appreciation of the vast array of mechanisms our body relies on to keep us healthy in spite of all the challenges it faces daily.

Our immune system is a result of the evolutionary "arms race" against various microorganisms. Over thousands of years, these microorganisms promoted the development of tools inside our bodies that are designed to detect and destroy pathogens with the potential to cause us harm. Our understanding of the immune system has not only revealed the complex mechanisms we use to fight infectious disease, but it has also led to medical breakthroughs that promote the well-being of the human race. Vaccine development, organ and tissue transplants, and targeted cancer therapies all owe their success to our knowledge of our immune response.

In this chapter, we will explore the two major branches of the immune system—the innate immune system and the adaptive immune system—and the cells, molecules, and tissues that drive their responses to protect us from infectious disease.

1.1 How did the scientific field of immunology emerge?

LEARNING OBJECTIVE

1.1.1 Explain the focus of the science of immunology and how this field originated.

Immunology is the branch of biomedical science that studies resistance to infection, as well as the mechanisms by which organisms defend themselves against foreign particles or microorganisms, known as the **immune response**. Modern immunology was born in the nineteenth century when advances in medicine and epidemiology led to an initial understanding of the processes of **immunity**, a state of being resistant to infection by a specific pathogen. The discovery of microorganisms led to the first true understanding of the cause of infectious disease. This discovery also provided insight into the function and evolution of the immune system and its role in fighting disease. **FIGURE 1.1** highlights significant events in immunology from the earliest discoveries to the present day.

Early observations and immunization

As early as 430 BCE, it was observed that when a population suffered from a deadly disease such as the plague, some individuals survived. These individuals were then able to care for the sick during a second bout of the disease without getting sick themselves, but no one understood why. These initial observations and practices helped to establish a framework for understanding the connection between disease onset and progression, exposure, and protection from infection.

In the sixteenth century, scientists and physicians began to connect the dots between the causative agents of disease and the process of protecting populations. In 1546 (over 100 years before the first direct observation of microorganisms), physician Girolamo Fracastoro wrote that diseases are caused by seed-like entities capable of being transmitted by direct or indirect contact. Fracastoro's idea contributed to the initial hypothesis of the germ theory of disease.

Germ theory of disease

Our modern-day definition of the germ theory of disease recognizes that diseases are caused by **pathogens**, which are microorganisms with the potential to

immunology Branch of biomedical science that studies immunity and the mechanisms of the immune response.

immune response The body's response to the presence of foreign particles or microorganisms (antigens).

immunity State of being resistant to infection by a specific pathogen.

pathogen Organism with the potential to cause disease.

430 BCE Thucydides	Eyewitness account of plague of Athens to promote survival upon future disease occurrences
10th Century Rhazes	Use of humoral theory to distinguish different contagious diseases
1000 AD in China	Immunization practiced using crusts of smallpox lesions
11th Century Avicenna	Refinement of the theory of acquired immunity
1546 Girolamo Fracastoro	Proposal that epidemics are caused by transferral of seed-like entities
1720 in Great Britain	Inoculation using smallpox to prevent smallpox infection
1796 Edward Jenner	Inoculation with cowpox to prevent smallpox infection
1862 Ernst Haeckel	Recognition of phagocytosis
1877 Paul Ehrlich	Recognition of mast cells
1880 Louis Pasteur and Emile Roux	Vaccination of chickens against cholera using an attenuated culture of *Vibrio cholerae*
1884 Ilya Mechnikov	Cellular theory of immunology
1890 Emil von Behring and Shibasaburo Kitasato	Theory of humoral immunity
1891 Robert Koch	Discovery of delayed-type hypersensitivity
1894 Jules Bordet	Discovery of the complement system
1897 Paul Ehrlich	Theory of antibody formation
1901 Karl Landsteiner	Discovery of A, B, O blood groups
1906 Clemens von Pirquet	The term “allergy” coined
1932 Alexander Glenny	Discovery of adjuvants
1944 Peter Medawar and Thomas Gibson	Discovery of the immunological basis of transplantation immunology
1948 Astrid Fagraeus	Discovery that antibody production occurs in plasma B cells
1958 Frank McFarlane Burnet	Development of the clonal selection theory
1959–1962 Gerald Edelman and Rodney Porter	Determination of the structure of antibodies
1961–1962 Jacques Miller	Discovery of the thymus involvement in cellular immunity
1974 Rolf Zinkernagel and Peter Doherty	Demonstration of MHC restriction of T cells in their antigen response
1976 Susumu Tonegawa	Identification of somatic recombination of immunoglobulin genes
1983 Ellis Reinherz, Phillipa Marrack, John Kappler, and James Allison	Discovery of the T-cell antigen receptor
1990 David Baltimore	Discovery of the role of RAG genes in V(D)J recombination
1994 Polly Matzinger	Development of the “danger” model of immunological tolerance, whereby self-reactive lymphocytes are deleted from the population

FIGURE 1.1 **Timeline of significant discoveries in immunology**

cause disease. During the nineteenth century, the germ theory of disease was solidified among the scientific community as observations were made that established an association between the presence of microorganisms and the onset of disease. The first observations in support of microorganisms as the causative agent for disease were made by Agostino Bassi. Beginning in 1807, Bassi demonstrated that the muscardine disease in silkworms (worms are covered with a powdery material, and the disease kills silkworm larvae) was caused by a parasitic fungus. Bassi's observations of microorganisms causing disease were further supported by the observations of Ignaz Semmelweis and Louis Pasteur.

In the 1840s, Semmelweis noticed a high rate of mortality in mothers from fever when the mothers were treated by doctors and medical students during delivery. By contrast, midwife deliveries seemed to result in a lower incidence of mortality. Semmelweis hypothesized that the higher incidence of postpartum fever was due to contact with doctors who had also conducted autopsies and that the fever was caused by a transmissible disease. Based on his hypothesis, Semmelweis suggested that doctors wash their hands prior to assisting with deliveries, and this practice resulted in a sharp decline in postpartum fever deaths. In the early 1860s, Louis Pasteur connected a microorganism to the transmission of postpartum fever, the first reported confirmation of the germ theory of disease.

In the late nineteenth century, Robert Koch developed four criteria for determining whether a specific microorganism was responsible for causing a specific disease through his experiments on anthrax (an infectious disease caused by bacteria). Koch's criteria sought to definitively identify a microorganism as the causative agent of a disease by requiring:

1. Isolation of a microorganism from a diseased host
2. Infection of a healthy host with the isolated microorganism
3. Subsequent onset of the same disease found in the diseased host
4. Isolation of an identical microorganism from the infected host

While Koch's ideas have been modified based on our current understanding of microbial pathogenesis, as sometimes it is difficult to isolate a pathogen from a diseased host, his work was extremely important in support of the germ theory of disease.

Early vaccination

Much earlier than the development of the germ theory of disease, various cultures acted on their observations that individuals who survived exposure to a deadly disease were protected from subsequent infection with that disease. In 1000 AD, individuals in China were protected against smallpox infection by inhaling pulverized crusts from smallpox lesions.

The earliest accounts of injection of individuals with smallpox come from sixteenth-century China. In a process referred to as *variolation*, smallpox lesion material was injected under the skin of healthy individuals in the hope of promoting a protective response. Variolation was also used widely in Africa and was introduced to America in the early eighteenth century when Cotton Mather learned about the process from his African slave. During the same era, Western Europe also began to use the process of variolation. While variolation was able to generate immunity against smallpox in many cases, because the process involves the introduction of a pathogenic organism into individuals, the disease rate was high (approximately 1% of individuals treated through variolation were ultimately afflicted with the disease).

The advent of vaccination, which hypothesized that exposure to a nonpathogenic form of the infectious agent might promote a protective response, was rooted in phenomena witnessed during variolation. The observations of Edward

Jenner, an English physician, in the late eighteenth century played a critical role in the development of a vaccine for smallpox. The use of variolation to protect against smallpox had spread throughout Europe. However, physicians observed that some farmers seemed to be unresponsive to this process. This unresponsiveness in farmers seemed to be due to their exposure to cowpox (a disease that affects a cow's udder that can also be transmitted to humans), suggesting that the farmers were already protected against smallpox and the variolation process did not provide any further protection. As a practicing apprentice, Jenner heard the stories of the unsuccessful variolation of farmers exposed to cowpox and hypothesized that there might be a link between the two. To test this hypothesis, he inoculated an 8-year-old boy with cowpox pus from a pustule on an infected milkmaid and found that the boy was protected against smallpox infection.

One hundred years later, Louis Pasteur demonstrated that a weakened strain of a pathogenic bacteria could provide protection against infection by that bacteria. Pasteur's inoculation work began in his studies of the bacteria that cause chicken cholera. He observed that a spoiled culture of the causative agent or a culture that he had weakened caused mild symptoms in the chickens but did not kill them. More importantly, he observed that the chickens were protected from infection by the causative bacteria. Pasteur expanded his observations to the protection of cattle from anthrax using an injection of a weakened strain of the bacteria that causes anthrax. Pasteur coined the term *vaccine* (*vaccinus* is Latin for "of the cow") to honor Jenner's initial discovery.

Advent of the field of immunology

While the late eighteenth and early nineteenth centuries gave us the advent of vaccination and the development of the germ theory of disease, the field of immunology was born in the mid- to late nineteenth century with several key discoveries. German physician Paul Ehrlich identified a *granulocyte* (a type of white blood cell) as a component of the immune system and named these cells *mast cells*. Shortly after this discovery, the Russian biologist Ilya Mechnikov identified white blood cells capable of *phagocytosis* (the process of a cell engulfing a foreign cell) and presented the first *cellular theory of immunology*. This theory provides the foundation for the cellular component of the immune response, describing how certain white blood cells use phagocytosis to engulf and destroy foreign microorganisms.

While Mechnikov's observations provide us with a historical window to the identification of phagocytes, we will see that the immune system employs both cellular and soluble components to protect against infection. The first identification of soluble molecules capable of protecting an organism from infection came from observations made by Emil von Behring and Shibasaburo Kitasato, who identified an antitoxin to diphtheria and tetanus in the blood of individuals who had been infected with those diseases. This discovery led to the *humoral theory of immunology*, which states that a soluble molecule within the blood can react with and inactivate a toxin. Building on this theory, Ehrlich coined the term *antibody* and developed the *side-chain theory*, which held that certain cells had receptors (side chains) capable of binding to toxins and that these cells, when bound to toxins, would produce antibodies that halted toxin action. The findings of von Behring, Kitasato, and Ehrlich sparked the study of the humoral arm of the immune system.

• CHECKPOINT QUESTIONS

1. Why were Edward Jenner's observations critical to our understanding of immunology?
2. What key observations led to the discovery of soluble molecules that protect us from infection?

1.2 | What are the types of pathogens?

LEARNING OBJECTIVE

1.2.1 List the five types of pathogens and the two classifications for pathogens based on their location.

opportunistic pathogen Organism with the potential to cause disease in an immunocompromised host.

extracellular pathogen Pathogen that resides within an organism without being enclosed by a host's cellular membrane.

intracellular pathogen Pathogen that resides within a cell of the infected organism.

To understand the mechanisms the immune system uses to combat pathogens that cause disease, we must understand how pathogens interact with the body. Some pathogens can cause disease in all hosts, regardless of whether the organism is healthy and robust or sick and weak. Others, known as **opportunistic pathogens**, cause disease only when the immune system is not functioning properly or when the pathogen accidently enters a site within the body and can grow uncontrollably in that location. Pathogens can reside both in and on the body, so any surface and any tissue is a potential breeding ground.

Types of pathogens

Pathogenic microorganisms come in a huge variety of shapes, sizes, and types (**FIGURE 1.2** and **TABLE 1.1**). Disease may be caused by bacteria, viruses, fungi, or protozoa. Parasites are another type of pathogen. This group includes parasitic animals such as worms or helminths, and some protozoa can be parasitic as well.

Pathogen types by location

Pathogens may also be categorized as extracellular or intracellular. **Extracellular pathogens** reside outside of the cells of organisms they have invaded. In contrast, the life cycle of **intracellular pathogens** includes a stage in which they reside within the organism's cells (although most intracellular pathogens have an extracellular component to their infection cycle at some point in order to infect a new host). Because pathogens have life cycles that differ in their location (extracellular vs. intracellular) and even in how they interact with the host's environment (e.g., extracellular pathogens can remain soluble or can interact with cell surfaces), our immune system has a variety of mechanisms to combat these different types.

CHECKPOINT QUESTION

1. What are the five types of pathogens?

TABLE 1.1 | Pathogen Types and Related Diseases

Type of Pathogenic Organism	Example	Diseases Caused
Bacteria	*Staphylococcus aureus*	Skin infection, meningitis, toxic shock
	Haemophilus influenzae	Pneumonia
	Salmonella typhimurium	Food poisoning
	Vibrio cholerae	Cholera
Viruses	Influenza A	Influenza (flu)
	Hepatitis B	Hepatitis
	Epstein–Barr	Mononucleosis
	Ebolavirus	Hemorrhagic fever
Fungi	*Candida albicans*	Yeast infection
	Cryptococcus neoformans	Meningitis
	Aspergillus flavus	Aspergillosis
Parasites	*Plasmodium falciparum*[a]	Malaria
	Toxoplasma gondii	Toxoplasmosis
	Trypanosoma brucei	Sleeping sickness
Protozoa	*Giardia intestinalis*	Giardiasis
	Leishmania	Leishmaniasis

[a] Also a protozoan
Source: www.cdc.gov

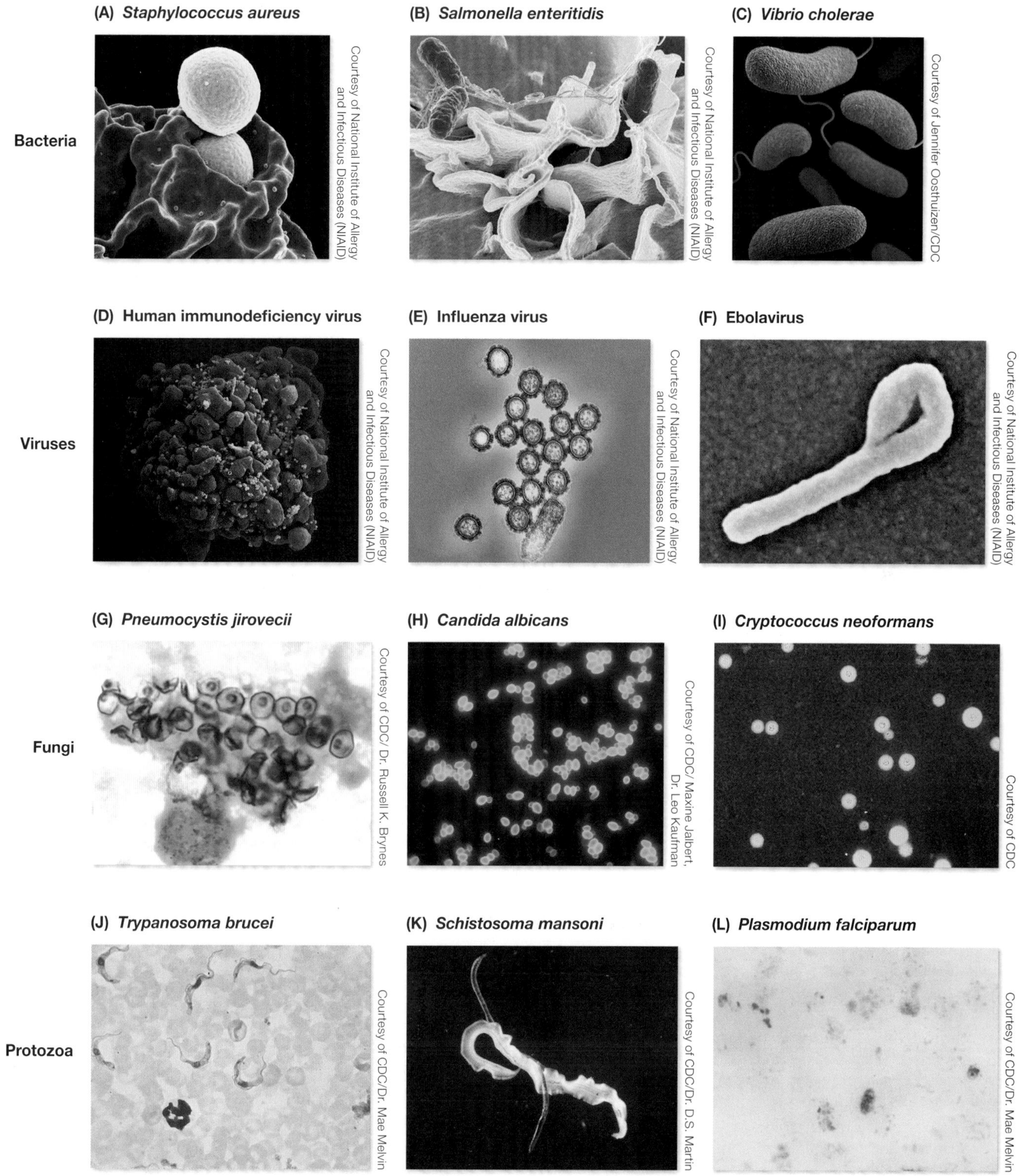

FIGURE 1.2 Examples of various types of human pathogens (A) *Staphylococcus aureus*, a common organism of the microbiota within the upper respiratory tract and skin, can cause skin infections, respiratory infections, and food poisoning. (B) *Salmonella enteritidis*, a common cause of food poisoning. (C) *Vibrio cholerae*, a waterborne bacteria that causes the gastrointestinal disease cholera. (D) Human immunodeficiency virus (HIV) is the causative agent of acquired immunodeficiency syndrome (AIDS). (E) Influenza virus, the causative agent of the flu. (F) Ebolavirus causes hemorrhagic fever in humans. (G) *Pneumocystis jirovecii*, the causative agent of pneumocystis pneumonia. (H) *Candida albicans*, a common fungus of the gut microbiota, causes candidiasis. (I) *Cryptococcus neoformans* can infect the lungs and the central nervous system, where it causes cryptococcal meningitis. (J) *Trypanosoma brucei*, a parasite carried by tsetse flies, causes African sleeping sickness. (K) *Schistosoma mansoni*, a waterborne blood fluke that causes the intestinal infection schistosomiasis. (L) *Plasmodium falciparum*, a red blood cell parasite that causes malaria.

1.3 How does the immune response combat the pathogens that cause disease?

LEARNING OBJECTIVES

1.3.1 Describe the body's three lines of defense against infection.

1.3.2 Differentiate between innate and adaptive immunity and humoral and cell-mediated immunity.

Humans regularly come in contact with a huge number of microbes, some of which are pathogens with the potential to cause disease. How then do we not succumb to infection early in life? You will learn the amazing story of how the body accomplishes this feat as you progress through this text.

Lines of defense

The immune system has three main lines of defense against invaders. The first line of defense consists of physical barriers. These "walls" make it difficult for pathogens to enter the body, take hold, and thrive. The primary physical barriers are the skin and mucosa. The second line of defense consists of cells, chemicals, and processes that work to immobilize or kill pathogens. These two initial lines of defense are collectively known as **innate immunity** because we are equipped with these resources from birth. Innate immunity is also called *nonspecific immunity*, because the innate immune system recognizes pathogens by relying on commonalities among different pathogens.

What happens if these initial lines of defense cannot adequately protect us from a disease-causing pathogen? The body's third line of defense is **adaptive immunity**, also known as *specific immunity*, because the recognition system used by the adaptive immune system is highly specific to one molecule. Adaptive immunity is "learned" immunity that develops and is enhanced after contact with a specific pathogen and protects the organism upon subsequent exposure to that particular pathogen.

Initial observations made in the early days of immunology contributed to the understanding that there are two types of adaptive immune response that target specific pathogens: cell-mediated (cellular) and humoral. The term *humoral* is based on the Latin word for moisture and means "relating to body fluids." This is rooted in the old use of the word *humor* to refer to a body fluid (such as the aqueous and vitreous humors of the eyeball). Both of these immune responses function to identify **antigens**, which are substances the body recognizes as foreign, thus stimulating an immune response. Both humoral and cell-mediated immunity employ specific proteins that identify antigens by recognizing **epitopes** on the surface of antigenic molecules.

HUMORAL IMMUNITY Humoral immunity is driven by the action of **B cells**, which are immune cells that use a cell-surface protein known as an **immunoglobulin** to bind to and recognize an antigen (FIGURE 1.3). Once a B cell has recognized an antigen using its cell-surface immunoglobulin, the B cell activates and differentiates into a cell capable of producing a soluble form of its immunoglobulin, also called an **antibody**. B cells commonly protect against extracellular pathogens through the action of soluble immunoglobulins. Antibodies are secreted by B cells and are a major protein component of blood plasma, where they function to bind to specific antigens and neutralize them using mechanisms discussed in later chapters.

CELL-MEDIATED IMMUNITY Cell-mediated immunity is driven by the action of **T cells**, which are immune cells that use a protein known as the T-cell

innate immunity Mechanisms used by the immune system that target pathogens based on common components and in a relatively nonspecific manner.

adaptive immunity Mechanisms used by the immune system that target specific products of pathogens through the use of T cells and B cells.

antigen Molecule recognized by a T-cell receptor, immunoglobulin, or antibody.

epitope Region of an antigen that is recognized by an immunoglobulin, antibody, or T-cell receptor.

B cell Lymphocyte that is the primary cell involved in the humoral adaptive immune response; recognizes specific antigens using cell surface immunoglobulins and differentiates into plasma cells or centrocytes when activated.

immunoglobulins Proteins made by B cells that act as specific antigen receptors for B-cell recognition, activation, and effector function.

antibodies Soluble immunoglobulins synthesized by B cells to bind to antigens; act to neutralize pathogens and toxins, aid in complement activation, and aid as a tag to promote phagocytosis.

T cell Lymphocyte that is the primary cell involved in the cell-mediated adaptive immune response; once activated, functions to activate other immune cells or target intracellular infections.

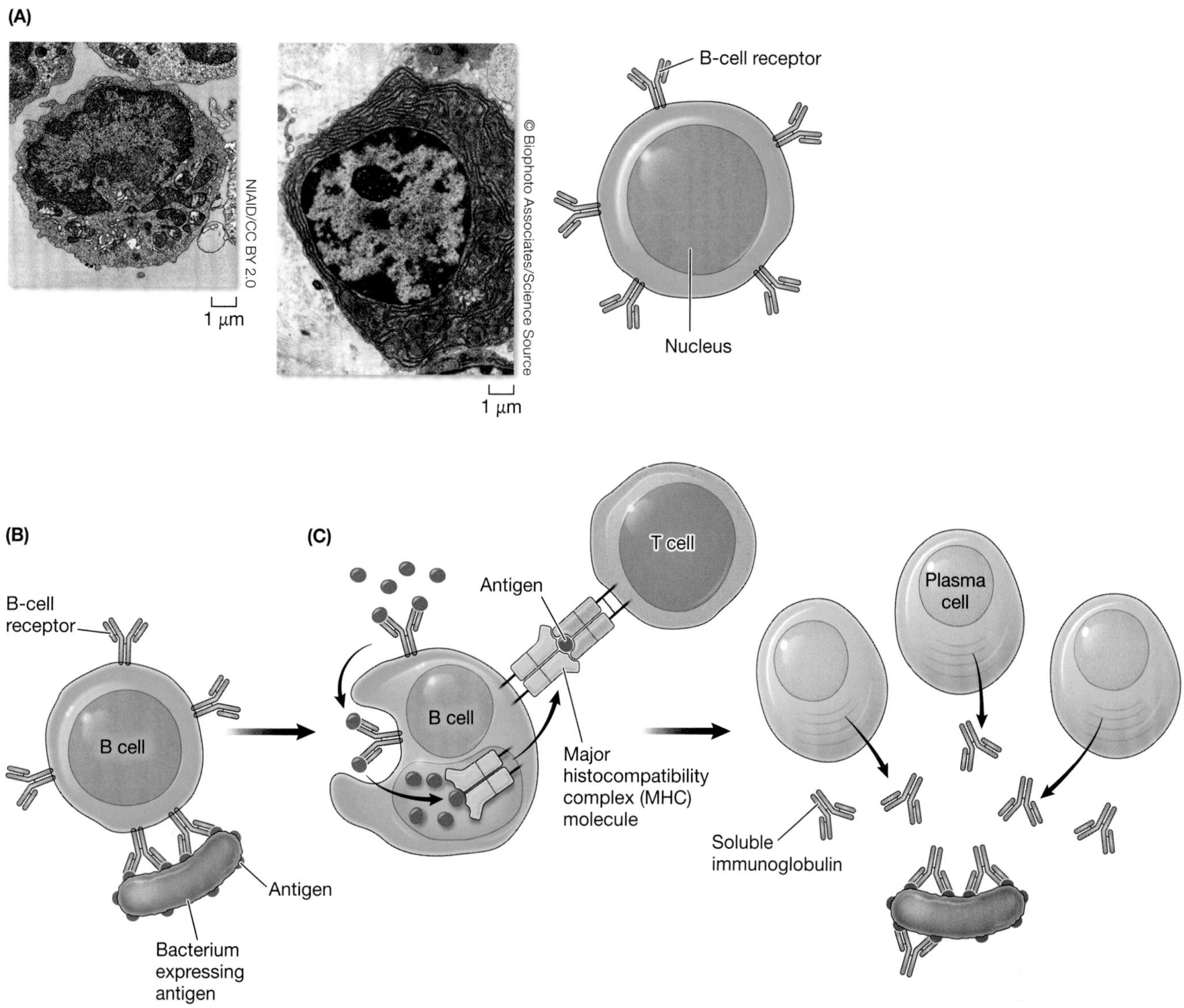

FIGURE 1.3 B cells, the lymphocytes of humoral immunity (A) Electron micrograph showing a naïve B cell (left) and an activated B cell, known as a plasma cell (middle). Right: Structure of a B cell, illustrating the presence of membrane-bound immunoglobulins at the cell surface, which serve as the B-cell receptor. (B) The B-cell receptor (membrane-bound immunoglobulin) is responsible for binding a specific foreign antigen. (C) Left: Upon binding of the B-cell receptor to its specific antigen, the B cell engulfs the antigen, digests the internalized material, and presents it via a major histocompatibility complex (MHC) molecule on its cell surface. There, T cells recognize the MHC and antigen at the surface of the B cell and facilitate B-cell activation. Right: Activated B cells become plasma cells, which secrete soluble immunoglobulins capable of binding to the same antigen recognized by the receptor of the activated B cell.

receptor to recognize a foreign antigen (**FIGURE 1.4**). Upon antigen recognition, T cells become activated and acquire different mechanisms to aid in clearing the pathogen. Some T cells work in support of other immune system cells to combat extracellular pathogens, and others act directly to target and destroy pathogen-infected cells.

Tolerance

The humoral and cell-mediated immune responses are powerful in their ability to recognize antigens. Immunoglobulins and T-cell receptors that work to recognize antigens are extremely diverse in nature due to processes unique to lymphocytes (B cells and T cells of the lymphatic system) that cause gene recombination.

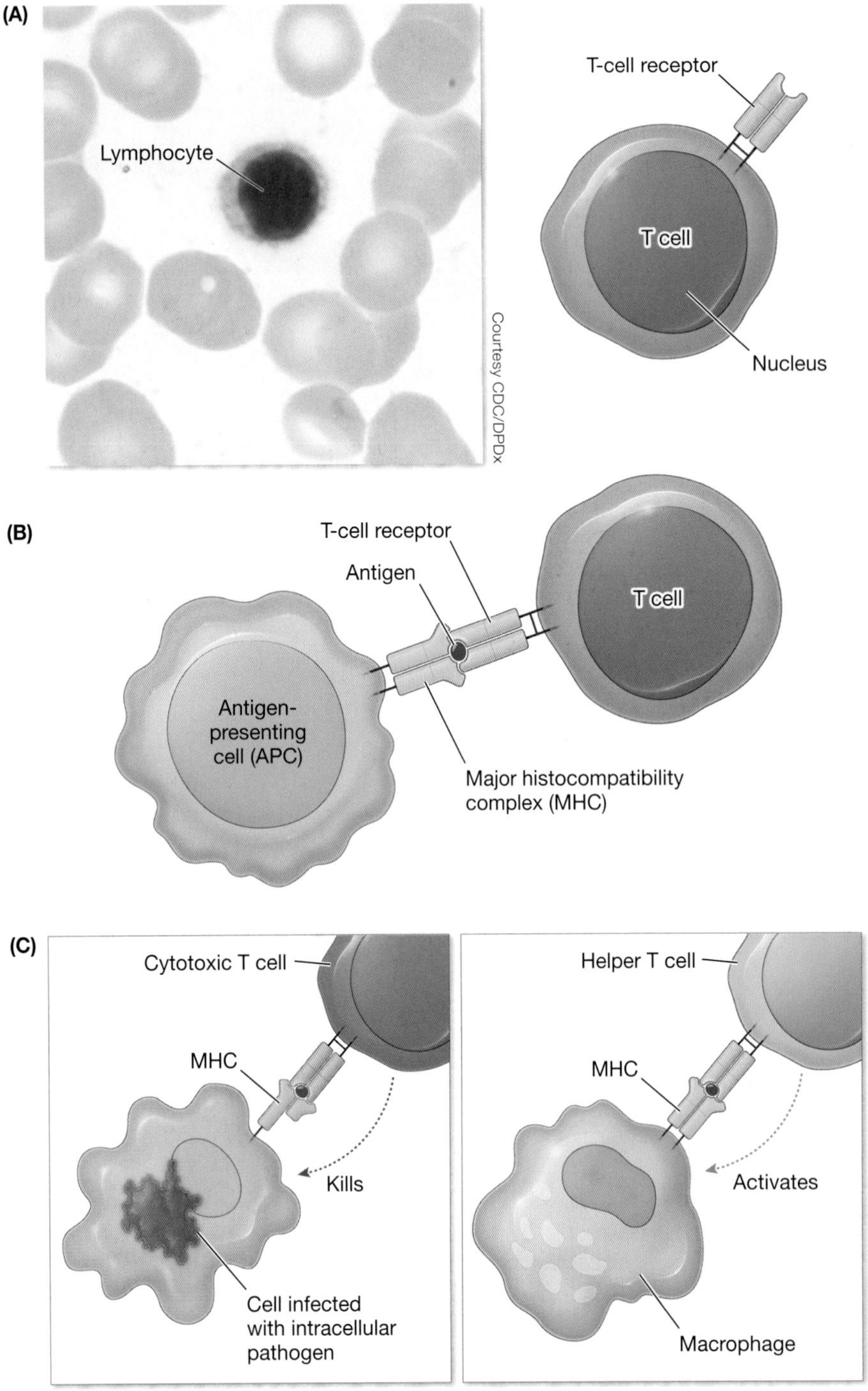

FIGURE 1.4 Overview of the cellular arm of the adaptive immune system
(A) Left: Micrograph of a lymphocyte. Right: Each T cell in circulation expresses a unique receptor. (B) The T-cell receptor recognizes a specific antigen presented via a major histocompatibility complex (MHC) cell-surface complex from an antigen-presenting cell (APC). (C) Left: When a cytotoxic T cell recognizes an antigen presented via an MHC molecule, it targets the APC for destruction. Right: When a helper T cell recognizes an antigen presented via an MHC molecule, it activates the APC.

Each B cell has one specific type of immunoglobulin, and each T cell has a specific T-cell receptor, but due to recombination events (discussed in later chapters), there are a variety of types of immunoglobulins and T-cell receptors. The mechanism that drives diverse antigen recognition, however, can also cause the development of immunoglobulins and T-cell receptors that recognize self-antigens. Also known as autoantigens, **self-antigens** are components of normal tissue that stimulate an immune response.

Since the goal of the immune response is to combat foreign antigens, not attack the body's own components, there has to be a mechanism to provide **tolerance**, or inactivation of immune responses to self. Tolerance to self-antigens occurs at multiple times and locations during the lifespan of B cells and T cells, including during their development and at peripheral locations throughout the body. Mechanisms of tolerance will be discussed in detail as we explore B-cell and T-cell development and the functions of activated B cells and T cells.

Real-time view of the immune response

Imagine you cut your hand—your wound is of concern not just because it is bleeding but also because it is a potential entry point for pathogens. Your body employs a wide range of mechanisms to prevent disease that could be caused by these pathogens. Some of these mechanisms are present immediately and work relatively quickly (within hours), while others require more time for proper activation (days or weeks).

In learning the mechanisms of the immune response, it is generally easiest to address events in the order that they occur in the body. That said, certain concepts need to be introduced at particular points to help you understand and appreciate the complexity of the immune system. For the most part, in this text, immune system mechanisms will be introduced and discussed in an order that reflects their timing as part of the immune response (**FIGURE 1.5**). Selected parts of the overall timeline will be enhanced and expanded in later chapters to highlight the significant events and processes being discussed (**TABLE 1.2**).

self-antigen Constituent of normal tissue that stimulates an immune response. Also known as *autoantigen*.

tolerance Inactivation of immune response against self-antigens.

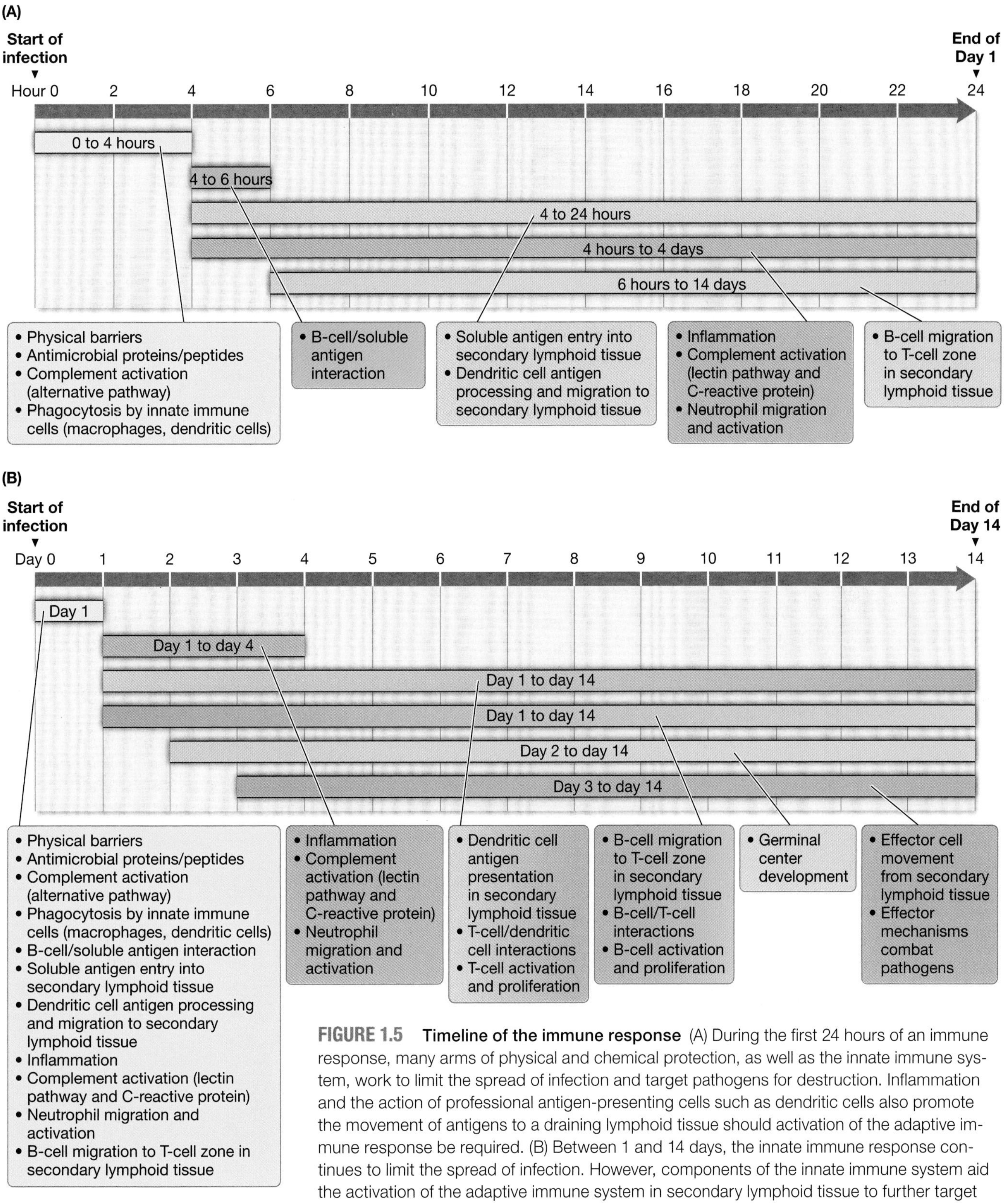

FIGURE 1.5 Timeline of the immune response (A) During the first 24 hours of an immune response, many arms of physical and chemical protection, as well as the innate immune system, work to limit the spread of infection and target pathogens for destruction. Inflammation and the action of professional antigen-presenting cells such as dendritic cells also promote the movement of antigens to a draining lymphoid tissue should activation of the adaptive immune response be required. (B) Between 1 and 14 days, the innate immune response continues to limit the spread of infection. However, components of the innate immune system aid the activation of the adaptive immune system in secondary lymphoid tissue to further target pathogens for destruction.

CHECKPOINT QUESTIONS

1. What are the three lines of defense against pathogen infection?
2. What are the two major cell types that function in the adaptive immune system? In which arm of the adaptive immune system does each cell type function?

TABLE 1.2 Overview of Events of the Immune Response

Component/Event of the Immune Response	Time Frame in the Body	Chapter Where Topic is Covered
Physical barrier	Immediate	1
Antimicrobial agents (defensins, etc.)	Immediate	2
Protease inhibitors	Immediate	2
Alternative complement pathway	Immediate	3
Macrophage phagocytosis	Immediate	2
Neutrophil/dendritic cell migration/phagocytosis	4 hours–4 days	2
Inflammation/fever	4 hours–4 days	2
Lectin/classical complement pathway	4 hours–4 days	3
NK cell action (if viral infection)	4 hours–4 days	2
Dendritic cell migration to secondary lymphoid tissue	1–14 days	7
T-cell activation (clonal selection/expansion)	1–14 days	7
Effector T-cell function (CD4, CD8 T cells)	1–14 days	7
B-cell activation (clonal selection/expansion)	1–14 days	10
Effector B-cell function (antibody production/action)	1–14 days	10
Immunological memory generation	1–14 days	11

1.4 What is the body's first line of defense against pathogens?

LEARNING OBJECTIVE

1.4.1 Describe the body's first line of defense against infection.

Our first line of defense against pathogenic organisms is a physical barrier between the organism and our tissues (**FIGURE 1.6**). Thus, the skin and mucosa play an important protective role. In addition, trillions of microbes in our body prevent potential pathogens from colonizing various locations, also serving as a first line of defense. They do this by competing for physical space and nutrients or by actively limiting pathogen growth.

Skin and mucosal surfaces

The epithelium of the skin (**FIGURE 1.7**) prevents entry of pathogenic organisms into underlying tissue. Before the advent of antiseptic use in surgery, a patient's life was often at risk during a surgical procedure due to infection caused by breaching the skin.

Mucous membranes line body cavities and passages such as the respiratory, urogenital, and gastrointestinal tracts. Although mucous membranes are potential breeding grounds for various pathogens, they resist infection through mucus production. Mucus contains proteoglycans, enzymes, and other glycoproteins that limit infection. Mucus is also constantly flushed and replenished in the body, and so any microorganisms present are flushed away with the removed mucus.

All epithelial cells secrete antimicrobial substances, including fatty acids and lactic acid, which inhibit bacterial growth. They also secrete antimicrobial peptides called **defensins** (see **EVOLUTION AND IMMUNITY**) and proteins such as lysozyme. Defensins disrupt pathogen membranes, and the enzyme lysozyme damages bacterial cell walls. The low pH environment (pH 1.5–5.5) of the skin, stomach, and vagina also deter microorganism growth (**FIGURE 1.8**).

defensin Antimicrobial peptide that is capable of disrupting cellular membranes by inserting into them.

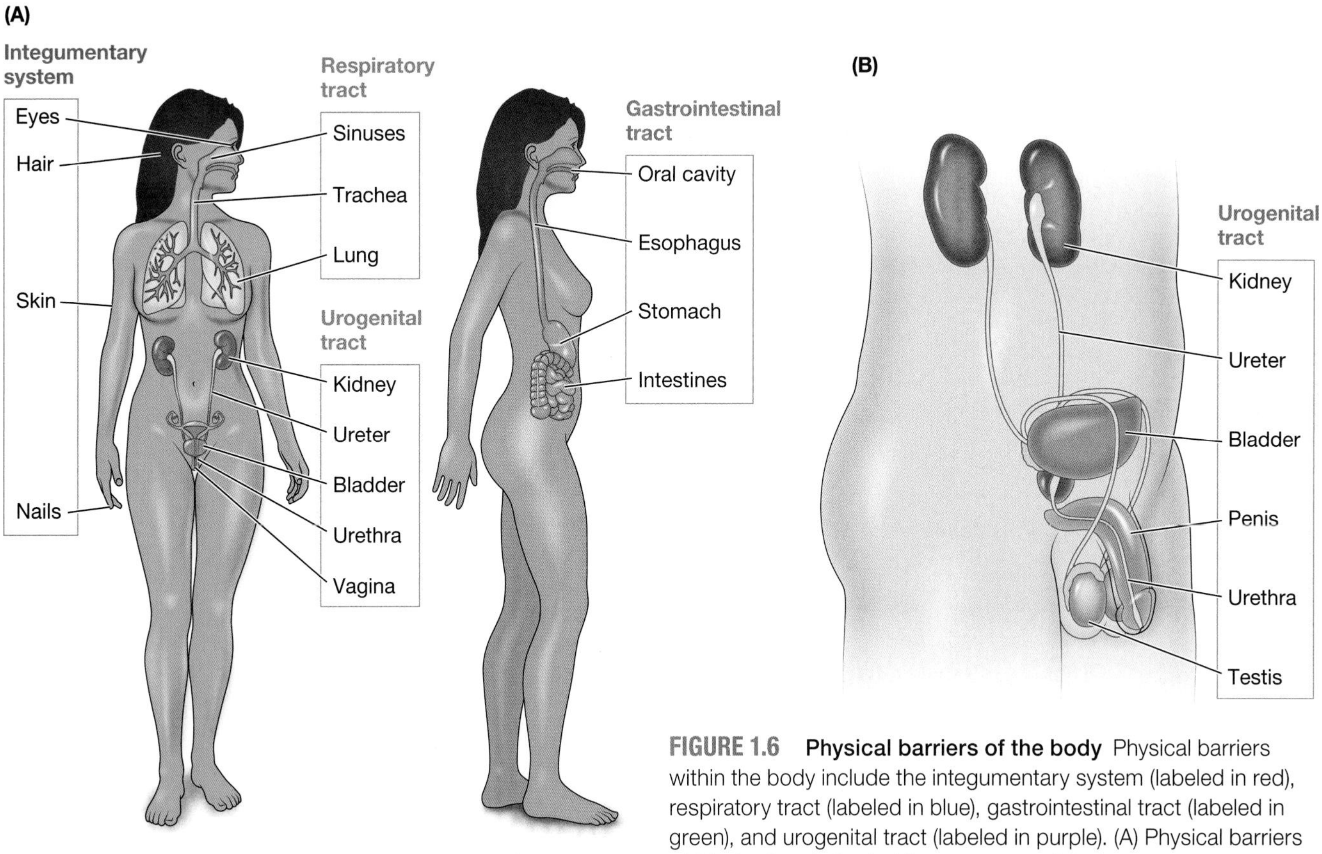

FIGURE 1.6 Physical barriers of the body Physical barriers within the body include the integumentary system (labeled in red), respiratory tract (labeled in blue), gastrointestinal tract (labeled in green), and urogenital tract (labeled in purple). (A) Physical barriers within the female body. Key structures of the integumentary system include eyes, hair, skin, and nails. Key tissues of the respiratory tract include the oral cavity, esophagus, stomach, and intestines. Key structures of the female urogenital tract include the kidneys, bladder, urethra, and vagina. (B) Physical barriers of the male urogenital tract include the ureter, bladder, urethra, penis, and testes.

Commensal organisms

While the skin and mucosa serve as physical barriers, because these surfaces come in contact with the environment they are typically colonized with a diverse array of microorganisms. The human body is a very rich environment in terms of resources, nutrients,

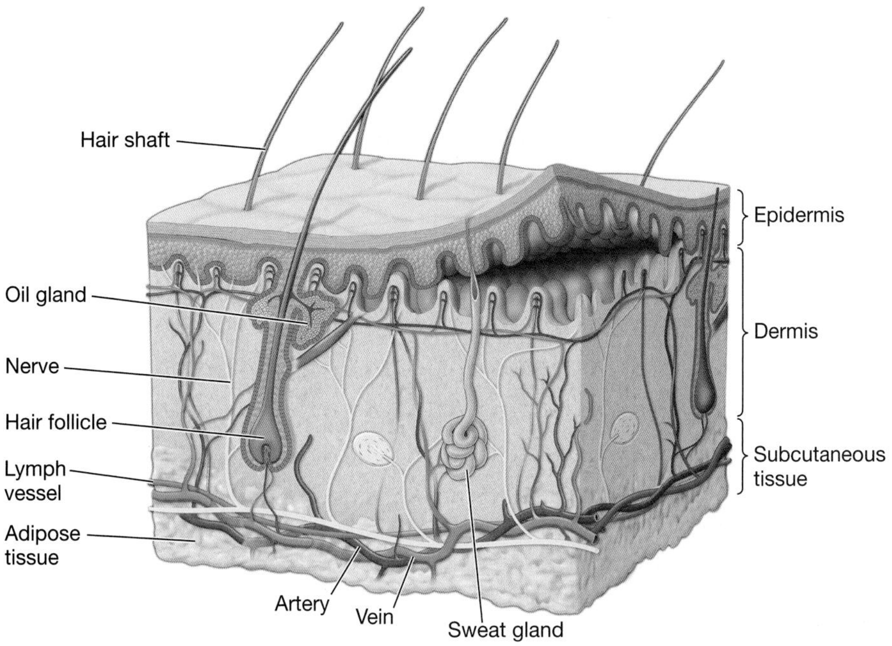

FIGURE 1.7 Structure of epithelial tissue The structure of epithelial tissue allows it to serve as a physical barrier to potential pathogens. Its vascularization promotes efficient delivery of cells and components of the immune system to the site of infection and from the site of infection to a draining lymphoid tissue.

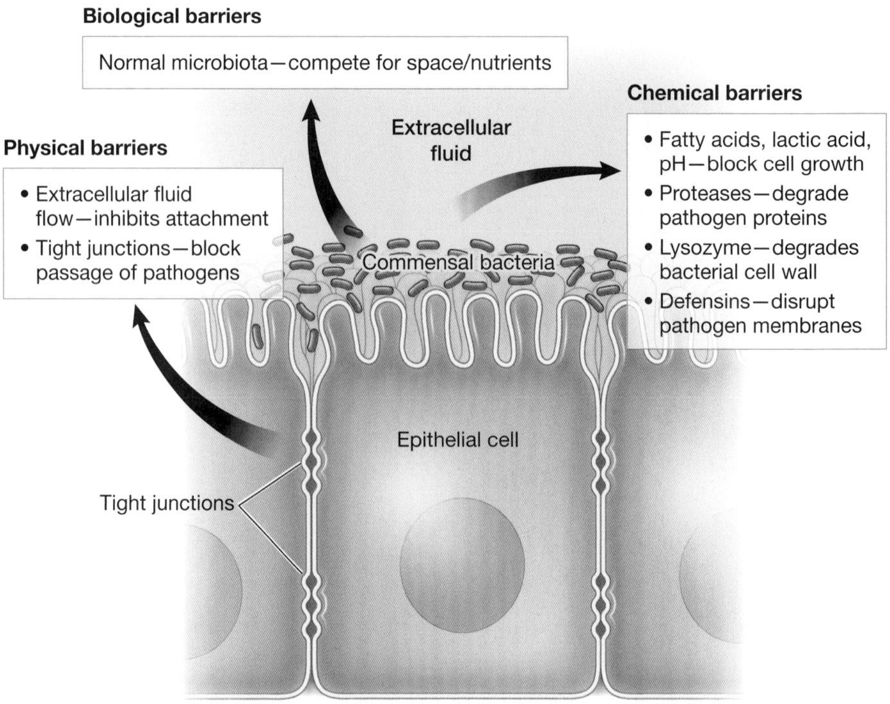

FIGURE 1.8 Physical, chemical, and biological methods of protection of epithelial tissues lining surfaces Physical barriers that protect epithelial tissues include tight junctions, which prevent passage of pathogens between epithelial cells, and extracellular fluid, which prevents pathogens from attaching to nearby cells. Chemical methods of protection include the presence of antimicrobial agents (such as fatty acids and lactic acid), the pH of the environment, which may be inhospitable to pathogens, enzymes that cleave proteins and bacterial cell walls, and defensins. Normal microbiota can serve as a means of protection by competing with pathogens for space and nutrients.

microbiota Microbial species that reside in symbiosis with an organism on or in tissues such as the skin and intestines.

water, and the proper temperature to nurture most microorganisms. Our gut alone provides a nutrient-rich environment in which trillions of microbes from more than 500 different species thrive. This collection of microorganisms that normally inhabit the body is known as **microbiota** (formerly called microbial flora). These microbes do not typically cause disease within us. They are an example of symbiotic organisms, which are organisms of different species that live in close association without causing harm to each other. In fact, one organism often benefits from the presence of the other (**FIGURE 1.9**).

The organisms of our microbiota benefit from their nutrient-rich environment, and they also serve us by aiding in digestion and preventing colonization by pathogenic organisms. Symbiotic microorganisms colonize locations that pathogenic organisms might seek out, including the skin, mouth, and gut.

EVOLUTION AND IMMUNITY

Antimicrobial Peptides

In addition to our examination of the fascinating molecular mechanisms of the immune system, it's valuable to examine the evolutionary history of both innate and adaptive immunity. Exploring the evolution of immune responses allows us to understand the unique ways that organisms have developed and diversified in response to pathogen exposure. We will see how different organisms have developed diversity in their immune systems by taking different evolutionary routes that achieve the same goal (convergent evolution).

One common host defense mechanism that is present in both plants and animals (and thus predates the separation of these two kingdom lineages on the evolutionary tree of life) is *antimicrobial peptides*. Antimicrobial peptides are also produced by bacteria, demonstrating their preservation (at least in terms of function) throughout evolutionary history.

One of the widely conserved classes of antimicrobial peptides in multicellular organisms is the *defensin* family, which disrupts cellular membranes. Multicellular organisms make a variety of defensins that target different microorganisms, including gram-positive and gram-negative bacteria and fungi. Some plants produce as many as 13 different defensins. The fruit fly *Drosophila melanogaster* produces 15, and humans produce 21. The conservation of these and other antimicrobial peptides throughout evolutionary history provides strong evidence of their longstanding role in host defense.

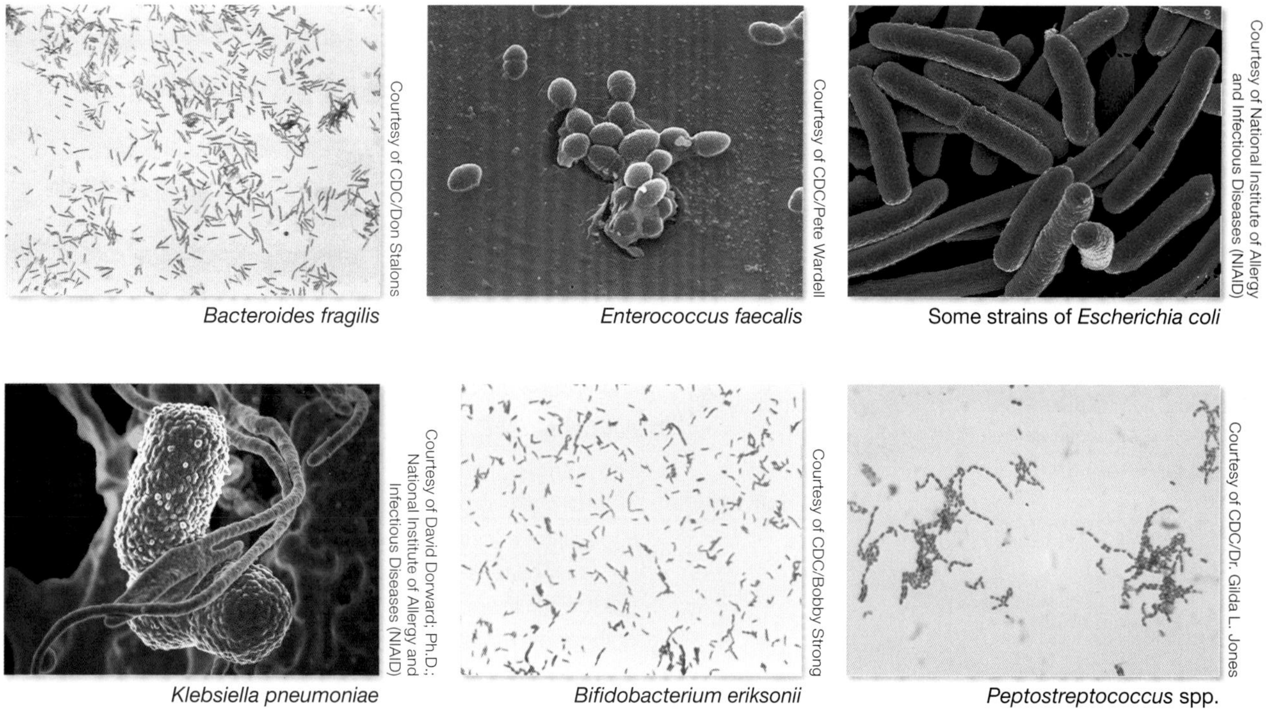

FIGURE 1.9 Beneficial gut microbes The microbiota of the gut can serve as a biological method of protection of gut epithelia. Common gut microbes include *Bacteroides fragilis*, *Enterococcus faecalis*, some strains of *Escherichia coli*, *Klebsiella pneumoniae*, *Bifidobacterium eriksonii*, and species of *Peptostreptococcus*.

Besides simply competing for space, symbiotic microorganisms compete for nutrients in the same ecological niche to a potential pathogen. This competition has the effect of limiting nutrient availability to pathogens (thus limiting pathogen growth) but can also affect how virulent a pathogen is (thus limiting pathogen colonization, growth, and invasion), lowering the potential to cause disease. Furthermore, some of the metabolic products of these microorganisms can lower virulence gene expression in pathogens, preventing the pathogen from being able to cause disease within the host. Symbiotic microorganisms also promote the formation of an environment less favorable to pathogen growth—for example, lactobacilli can aid in lowering the pH of the vaginal environment, thus limiting the ability of some pathogens (that cannot grow in this lower pH) to gain hold within that environment. Finally, some microorganisms secrete antimicrobial substances known as *bacteriocins* that prevent other microorganisms from gaining a foothold in that particular niche (see **EMERGING SCIENCE**).

Interestingly, in addition to helping us actively combat pathogens through competition, inhibition of virulence of pathogens, alteration of the local environment, and production of antimicrobial substances, the microbiota play a critical role in the development of our immune system, both during fetal development and after birth. The developing immune system is tolerant to symbiotic microorganisms shortly after birth to ensure establishment of the microbiota. These microorganisms can also positively affect the development of *secondary lymphoid tissue* (the sites of lymphocyte activation in response to pathogens) and physical mucosal barriers.

CHECKPOINT QUESTION

1. How do symbiotic microorganisms protect us from pathogen infection?

EMERGING SCIENCE

Can probiotic treatment successfully combat gastrointestinal infections?

Article

E. Deriu, J. Z. Liu, M. Pezeshki, R. A. Edwards, R. J. Ochoa, H. Contreras, S. J. Libby, F. C. Feng, and M. Raffatellu. 2013. Probiotic bacteria reduce typhimurium intestinal colonization by competing for iron. *Cell Host Microbe* 14: 26–37.

Background

Physical barriers and competition for space and nutrients are important initial mechanisms used by the body to fight off pathogens. This phenomenon led scientists to postulate the advantage of using probiotics to aid in combating gastrointestinal infections.

The Study

Researchers focused their attention on iron usage in the gut. This is because the inflammatory response in the gut that results from an infection reduces iron availability to inhibit microbial growth. Previous studies had demonstrated that the gut pathogen *Salmonella enterica* serovar Typhimurium overcomes this iron depletion by scavenging for iron using a family of proteins known as siderophores, which are capable of binding iron with a high affinity.

The authors hypothesized that administering a nonpathogenic bacteria that can acquire iron in a similar manner would prevent pathogen colonization by competing for iron with potential pathogens. To test this hypothesis, they used a particular mouse strain that develops *Salmonella* colitis and a persistent infection.

First, the antibiotic streptomycin was administered to kill off natural gut flora. The mice were then inoculated with *Salmonella*, which caused a persistent infection and inflammatory response. However, a single dose of a nonpathogenic strain of bacteria, *Escherichia coli* Nissle, given 72 hours after *Salmonella* infection was enough to lower *Salmonella* presence in feces by two orders of magnitude and reduce the inflammatory response (**TABLES**).

Conclusion

The authors of this study demonstrate that this inhibition of *Salmonella* colonization and inflammatory response by *E. coli* Nissle is dependent on the probiotic's ability to take up iron. These findings support the hypothesis that treatment with a commensal organism that can behave like a pathogenic counterpart can limit infection by the pathogen.

Think About...

1. Given the nature of this study, what might be some of the limitations of *E. coli* Nissle treatment?
2. How would you be able to predict whether probiotic treatment with *E. coli* Nissle would behave similarly in cases of infection with a different pathogen?
3. Since microorganisms can evolve at a rapid rate, what are some potential evolutionary advantages gut pathogens may evolve to overcome *E. coli* Nissle probiotic treatment? What could be done to minimize these evolutionary changes?

Effect of probiotic *E. coli* Nissle treatment on bacterial infection

	Average CFU/mg bacteria isolated 22 days post infection
Mock infected	1.33×10^6 *S. typhimurium* ($n = 4$)
E. coli Nissle infected	5.41×10^5 *E. coli* ($n = 4$)
E. coli Nissle infected	1.6×10^4 *S. typhimurium* ($n = 6$)

Effect of probiotic *E. coli* Nissle treatment on cecal inflammation

	Inflammation score geometric mean ($n = 7$)
Mock infected	9.8
Wild-type *E. coli* Nissle infected	7.5
tonB mutant *E. coli* Nissle infected	13

Effect of probiotic *E. coli* Nissle treatment on intestinal colonization

Infection ($n = 5$)	*S. typhimurium* CFU/mg		*E. coli* CFU/mg	
	48 hpi	96 hpi	48 hpi	96 hpi
S. typhimurium	1×10^6	1×10^6	—	—
E. coli	—	—	7.8×10^5	5×10^5
S. typhimurium + wild-type *E. coli*	5×10^4	2.2×10^3	1×10^6	5.5×10^5
S. typhimurium + *tonB E. coli*	5×10^5	5.6×10^5	1×10^6	5.5×10^3

hpi = hours post infection
Source: Data from E. Deriu et al. 2013. *Cell Host Microbe* 14: 26–37.

1.5 Which cell lineages are associated with innate and adaptive immunity?

LEARNING OBJECTIVE

1.5.1 Outline the cell lineages of hematopoiesis and describe the functions of immune cells produced through this process.

A wide variety of white blood cells (WBCs), also called **leukocytes**, are involved in innate and adaptive immunity. Leukocytes lack hemoglobin, so they are colorless (white). There are two major categories of WBCs: granulocytes and agranulocytes.

Granulocytes

Granulocytes (FIGURE 1.10) are so named because they contain granules within their cytoplasm that package important proteins and chemicals that work to destroy pathogens and increase inflammation. The predominant granulocyte is the **neutrophil**, which makes up 50% or more of the total WBC population in the body. Neutrophils specialize in **phagocytosis** (cell eating). They are rapidly mobilized to sites of infection to engulf and destroy pathogens using chemicals within their granules. While neutrophils are fast-acting, they are short-lived and die at the site of infection, ultimately forming pus. Because of the rapid death of these cells, neutrophils may cause localized tissue damage if they release the toxic products contained within their granules into the infection site. Thus, *macrophages* play an important role in removing rapidly dying neutrophils at a site of infection.

The other granulocytes are eosinophils, basophils, and mast cells. **Eosinophils** make up 2% to 4% of total WBCs and play a role in defending against parasitic worms. **Basophils** make up less than 1% of total WBCs. They also work in fighting off worm infections and recently have been shown to play a role in the activation of T cells necessary for combating parasitic infection. **Mast cells** are granular cells found in connective tissue. Activation and degranulation (release of granule content to an extracellular location) of these three types of granulocytes causes release of inflammatory mediators from their granules, and these inflammatory mediators play a major role in the start of an inflammatory response. Mast cells also secrete molecules that aid in wound repair and angiogenesis (blood vessel growth).

Agranulocytes

Lymphocytes and monocytes are the two types of **agranulocytes** (FIGURE 1.11). **Lymphocytes** are the second most prevalent leukocyte, making up 25% or more of the total WBC population. Lymphocytes are primarily found in the lymphoid tissues; they include B cells and T cells of the adaptive immune

leukocyte White blood cell. Includes granulocytes, agranulocytes, and cells involved in blood clotting.

granulocyte White blood cell that contain granules. Refers to neutrophils, basophils, eosinophils, and mast cells.

neutrophil Granulocyte that migrates to sites of infection and phagocytoses pathogens located there.

phagocytosis Process of engulfing and killing cells through receptor binding and endocytosis.

eosinophil Granulocyte that specializes in clearance of parasitic infections.

basophil Granulocyte responsible for the clearance of parasites.

mast cell Granulocyte that specializes in induction of an inflammatory response; often closely associated with IgE isotype antibodies.

agranulocyte White blood cell that is devoid of granules. Refers to lymphocytes (B cells, T cells, NK cells, innate lymphoid cells) and monocytes (macrophages, dendritic cells).

lymphocyte Cells of the lymphoid arm of hematopoiesis, including the cells of the adaptive immune system (T cells and B cells) and some cells of the innate immune system (innate lymphoid cells and NK cells).

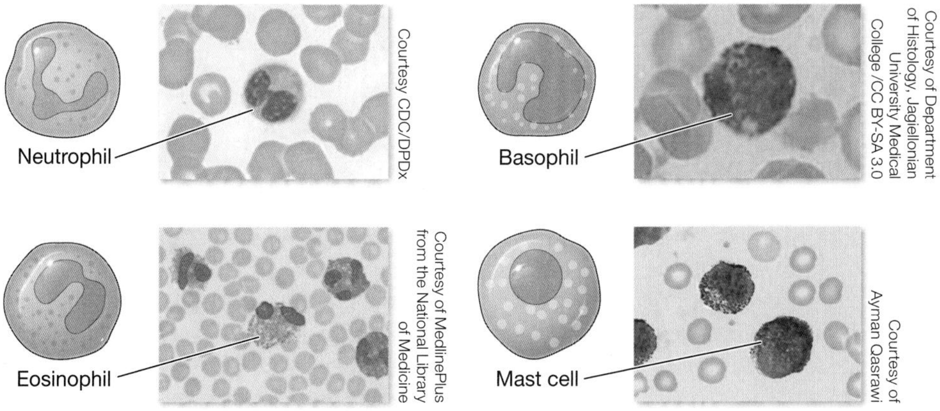

FIGURE 1.10 Granulocytes of the immune system The innate immune response uses a variety of granulocytes to promote pathogen elimination through engulfment and destruction, as with the action of neutrophils (top left), or through the induction of an inflammatory response, as with the action of eosinophils, basophils, and mast cells.

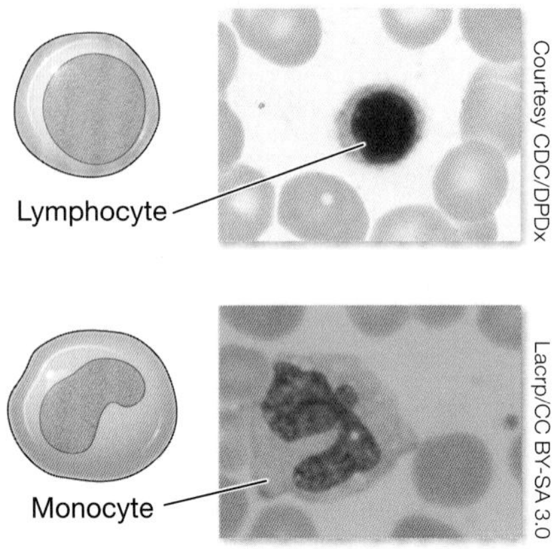

FIGURE 1.11 Agranulocytes of the immune system The immune response uses two types of agranulocytes to promote pathogen elimination. Lymphocytes (top) comprise T cells and B cells, the arm of the adaptive immune response. Monocytes (bottom) can differentiate into macrophages (professional phagocytic cells and antigen-presenting cells) and dendritic cells (professional antigen-presenting cells).

system and natural killer (NK) cells, which work to combat intracellular pathogens. **Monocytes** make up 3% to 8% of total WBCs. They are the largest of the leukocytes and differentiate into dendritic cells and macrophages (long-lived phagocytes).

Hematopoiesis

All cells within the blood are derived from a common stem cell precursor. The process of formation of blood cells is known as **hematopoiesis** (see **KEY DISCOVERIES** for an explanation of how scientists learned about this process). Hematopoiesis produces the **erythrocytes** (red blood cells) used for gas transport throughout the body, the **megakaryocytes** used for blood clotting and wound repair, and the leukocytes responsible for immunity (**FIGURE 1.12**).

Self-renewing hematopoietic stem cells begin their cell division and differentiation journey by becoming **progenitor cells** for one specific cell line: as an erythroid megakaryocyte progenitor, a myeloid progenitor, or a lymphoid progenitor.

ERYTHROID MEGAKARYOCYTE PROGENITOR Hematopoietic stem cells differentiate into erythroid megakaryocyte progenitor cells, which divide and differentiate into either erythrocytes or megakaryocytes. Erythrocytes are one of the formed elements in blood. The mature cells are nonnucleated (without nuclei) and contain hemoglobin; they specialize in delivery of oxygen from the lungs to peripheral tissues. Megakaryocytes are responsible for platelet production. **Platelets** are nonnucleated cell fragments that play a critical role in blood clotting and wound repair.

MYELOID PROGENITOR The myeloid progenitor gives rise to most cells involved in innate immunity and antigen presentation. **Antigen-presenting cells** are cells that process engulfed material into small peptides. They then present the processed peptides at their surface as markers for adaptive immune cells, namely T cells.

Hematopoietic stem cells differentiate into myeloid progenitor cells, which divide and differentiate into a number of different white blood cells associated with innate immunity. One branch of the myeloid lineage produces most of the granulocytes described earlier in this chapter (neutrophils, eosinophils, and basophils), and the other branch produces monocytes and mast cells. Monocytes further differentiate into macrophages and dendritic cells (see Figure 1.12).

Macrophages Like neutrophils, **macrophages** specialize in phagocytosis. Unlike neutrophils, macrophages are long-lived phagocytes. They inhabit specific organs and body areas, whereas monocytes (macrophage precursors) circulate in the blood. Macrophages also act as scavengers that rid the body of pathogens, dead cells, and cellular debris produced through **apoptosis** (programmed cell death). Macrophages secrete **cytokines**, which act as signaling molecules as part of the immune response. Some of the cytokines secreted by macrophages are those responsible for guiding neutrophils to the site of infection while others are important in inflammation.

monocyte Circulating white blood cell that migrates to tissue and differentiates into a macrophage or a dendritic cell.

hematopoiesis Development of cells of the circulatory system from a common stem cell ancestor.

erythrocyte Red blood cell responsible for transporting oxygen from the lungs to peripheral tissues.

megakaryocyte White blood cell that specializes in the production of platelets.

progenitor cell Precursor cell that can differentiate into several specialized blood cell types during hematopoiesis.

platelet Nonnucleated product of megakaryocytes that functions in blood clotting.

antigen-presenting cells Phagocytic cells of the innate immune system that process engulfed materials and present them to T cells and B cells.

macrophage Agranulocyte that specializes in phagocytosis of foreign antigens and apoptotic cells.

apoptosis Programmed cell death; results in the formation of cellular debris removed via phagocytic cells such as macrophages.

cytokine Soluble protein product secreted by cells to aid the immune system by increasing inflammation, chemotaxis to sites of infection, or immune cell signaling and differentiation.

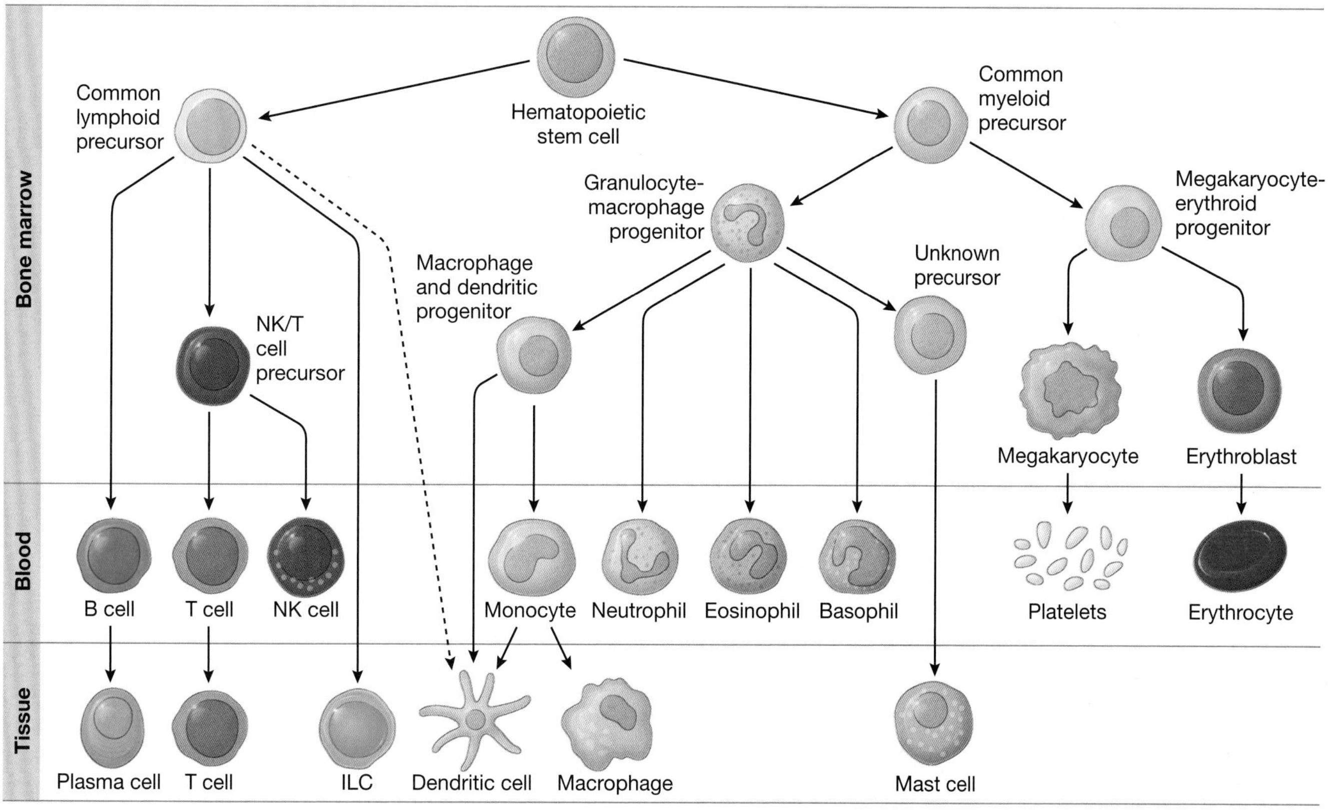

FIGURE 1.12 The process of hematopoiesis A hematopoietic stem cell possesses the potency to give rise to all cell types within the blood. The three cell lineages within the circulatory system are the erythroid, myeloid, and lymphoid. The erythroid progenitor gives rise to erythrocytes and platelets. The myeloid precursor gives rise to granulocytes (including neutrophils, eosinophils, basophils, and mast cells), as well as macrophages and dendritic cells. The lymphoid precursor gives rise to B cells, T cells, innate lymphoid cells (ILC), and natural killer (NK) cells.

Dendritic cells Like macrophages, **dendritic cells** take up residence in many tissues in the body. Dendritic cells play an important role in phagocytosis and pathogen destruction. These cells also carry out an extremely important job by linking the innate and adaptive immune response. While macrophages can present antigens and activate T cells, dendritic cells play a dominant role in antigen uptake and presentation during an innate immune response. This is due not only to the dendritic cell's efficiency in processing many different types of pathogens, but also to the dendritic cell's ability to leave a site of infection. Dendritic cells are able to migrate to a specialized tissue known as secondary lymphoid tissue, which serves as a center for antigen display and adaptive immunity activation.

LYMPHOID PROGENITOR The lymphoid progenitor gives rise to T cells, B cells, innate lymphoid cells, and natural killer cells (see Figure 1.12). Thus, hematopoietic stem cells that differentiate into lymphoid progenitor cells give rise to cells involved in both innate and adaptive immune responses. The innate immune response cells that are part of the lymphoid cell lineage are **innate lymphoid cells (ILCs)**, which include **natural killer (NK) cells**. ILCs secrete cytokines that activate other innate immune system cells at a site of infection, including dendritic cells and macrophages. NK cells play an important role in preventing viral infections. They use a variety of mechanisms to recognize and destroy virus-infected cells. In addition, they secrete cytokines that minimize viral replication within infected cells.

dendritic cell White blood cell that specializes in phagocytosis of foreign pathogens and presentation of antigen epitopes to the adaptive immune system.

innate lymphoid cell (ILC) White blood cell of the innate immune system that works to activate other innate immune cells, including macrophages and dendritic cells, at a site of infection.

natural killer (NK) cell White blood cell of the innate immune system that works to clear intracellular infections, including those caused by viruses.

KEY DISCOVERIES

How did we learn that there is a common stem cell precursor to all lineages of blood cells?

Article

A. J. Becker, E. A. McCulloch, and J. E. Till. 1963. Cytological demonstration of the clonal nature of spleen colonies derived from transplanted mouse marrow cells. *Nature* 197: 452–454.

Background

It is well established that blood cells (both red and white) originate from a common ancestral cell. But we can take a look back and explore the scientific discoveries that led to this knowledge. How was it determined that there is a common stem cell precursor to all lineages of blood cells?

Early Research

In 1908, Alexander Maksimov, a Russian histologist, hypothesized the existence of a stem cell as a common ancestor to all blood cells in his model of hematopoiesis. Maksimov is credited with coining the term "stem cell."

In the 1930s, Florence Sabin, an American medical researcher, and her colleagues made major strides in hematology research. They demonstrated that radiation of tissue in rabbits resulted in damage and atrophy of lymph nodes and bone marrow. Importantly, these researchers hypothesized that the radiation altered the chromatin (material in the cell nucleus from which chromosomes are formed) within the stem cell in bone marrow, preventing that cell from producing differentiated lymphocytes.

This research, as well as that of others, led to the development of bone marrow transplantation for individuals suffering from various bone marrow diseases, including lymphomas. While there was compelling evidence that a common ancestral stem cell was present in bone marrow, direct observation of this stem cell population had not been observed.

Think About...

1. In light of this summary of early research, what would your hypothesis be concerning hematopoietic stem cells?
2. What experiment(s) would you conduct to test your hypothesis? How could you test whether an isolated stem cell from bone marrow (or another tissue) is capable of giving rise to a number of differentiated cells?
3. How did the findings by Becker and colleagues support the hypothesis of a singular hematopoietic stem cell?

Article Summary

Till and McCulloch had demonstrated in 1961 that irradiated mice developed macroscopic colonies in the spleen when they were transplanted with cells from hematopoietic tissue from a donor mouse. They further observed that these colonies contained cells with all three lineages of blood cells (erythroid, myeloid, and lymphoid). The major question the scientists addressed was whether these lineages were derived from a single clonal ancestor.

To test the clonality of a hematopoietic stem cell, the researchers used techniques that were state of the art at the time. These included karyotyping (mapping the chromosomes of an organism) and observing the chromatin structure of cells within the colony. The authors speculated that if they observed that all cells within the colony had a

cytotoxic T cell CD8-positive T cell that has been activated by recognition of a specific epitope; targets cells with intracellular infections through the actions of cytokines and cytotoxins.

plasma cell B cell that has been activated by antigen and specializes in the secretion of antibodies into plasma.

The other cells produced from the lymphoid lineage are the T cells and B cells of the adaptive immune system. After activation, T cells can differentiate into a variety of effector T cells, including cytotoxic T cells and helper T cells, to promote their role in the cellular arm of the adaptive immune system. The primary function of **cytotoxic T cells** is to attack cells that have been invaded by a pathogen. Helper T cells assist in the activation of other cells, including macrophages, B cells, and cytotoxic T cells.

When B cells differentiate after activation, the main effector cell produced is the **plasma cell**. The plasma cell's main function is to produce soluble immunoglobulins (antibodies) that bind and neutralize pathogens and their toxins.

CHECKPOINT QUESTIONS

1. Which three lineages are derived from hematopoietic stem cells?
2. What are the different types of granulocytes of the immune system?
3. What are the different types of agranulocytes of the immune system?

single karyotype, the most likely explanation would be that all the cells came from a common ancestor cell.

One complication they faced in developing this experiment was finding a way to alter the karyotype of the donor bone marrow cells to make them distinctive and detectable. To "mark" the donor cells, Becker and colleagues transplanted bone marrow cells into heavily irradiated mice (to ensure that the mice did not have their own source of bone marrow cells). They then re-irradiated the mice at a level that was capable of altering but not destroying the karyotype of the donor cells. In this manner, the donor bone marrow cells could be detected.

The authors speculated that the radiation in the animal before transplantation would likely cause a different karyotype change than the radiation after transplantation. They would look at the karyotypes of cells within the colony with the expectation that they would observe either one karyotype (denoting ancestry from a single precursor) or multiple karyotypes (denoting ancestry from multiple precursors).

Conclusion

When the authors observed the karyotypes of dividing cells within four different colonies, they found that virtually all cells observed contained the same distinguishing abnormalities (**TABLE**). These findings support the hypothesis that all cells found within a singular colony are derived from a common stem cell ancestor. They are the source of the modern definition of hematopoiesis and stem cell biology.

Common abnormalities observed in donor bone marrow cells transplanted into irradiated mice

Colony with abnormal karyotype	Number of cells scored	Number of abnormalities present	Percentage of cells with all characteristic abnormalities	Percentage of cells with at least one characteristic abnormality
A	100	2	81	99
B	100	2	95	95
C	100	1	91	97
D	75	2	97	97

Source: Data from A. J. Becker et al. 1963. *Nature* 197: 452–454.

1.6 How does the innate immune system limit the spread of most infections?

LEARNING OBJECTIVE

1.6.1 Compare and contrast the roles played by components of the innate immune system in fighting infection.

Although the skin and mucosal surfaces provide an effective first line of defense, inevitably some pathogens will breach these barriers. This can happen when we consume contaminated food, breathe polluted air, or touch a contaminated surface or object and then touch our eyes, nose, or mouth. Furthermore, pathogens can gain access through the skin via wounds or insect bites. Most infections remain localized at the point of entry due to the actions of the innate immune system, which has two functions: *pathogen recognition* and *pathogen destruction*.

Pathogen recognition and destruction

Pathogen recognition involves the use of soluble plasma proteins of the *complement system* (described in Chapter 3) or receptors on the surface of innate immune system cells, including macrophages and neutrophils (**FIGURE 1.13**). The soluble plasma proteins tag foreign molecules so cells of the innate immune system can remove them, or they actively promote pathogen destruction by

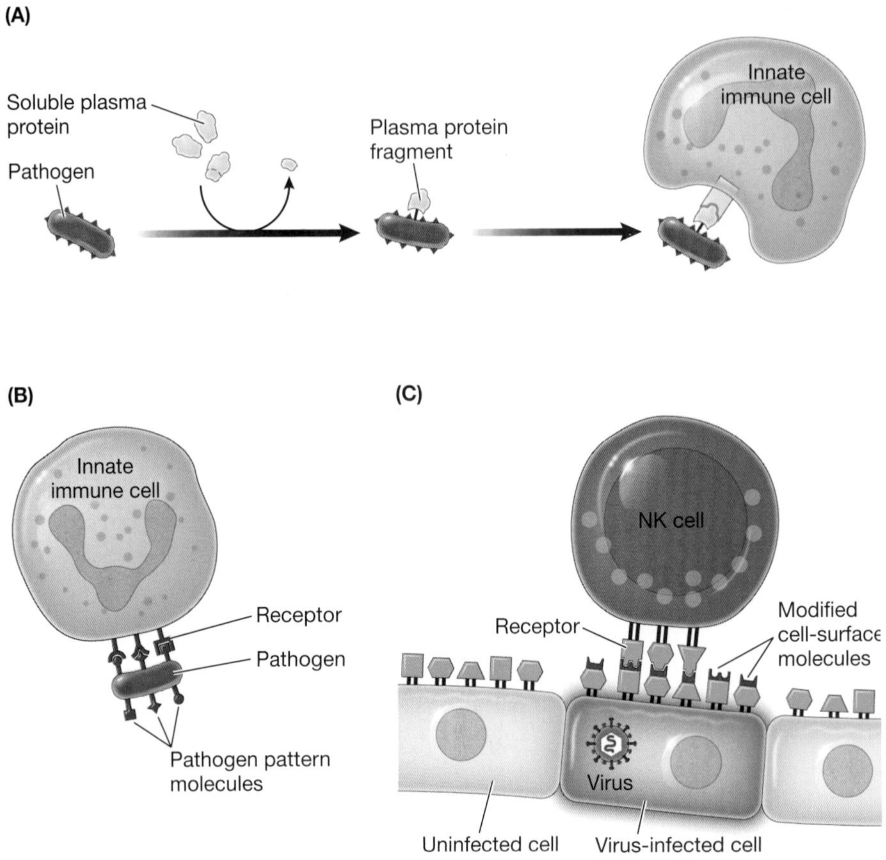

FIGURE 1.13 Pathogen recognition by soluble proteins and receptors of the innate immune system (A) Soluble proteins label foreign pathogen, allowing for recognition by innate immune cells. (B) Innate immune cell receptors recognize pathogen surface molecules. (C) Innate immune cell receptors recognize cell-surface changes in virus-infected cells.

destabilizing the membrane structure of a pathogen. Macrophages and neutrophils are relatively nonspecific, so they recognize patterns on the surfaces of multiple types of pathogens rather than just one specific type. Once recognized, cells of the innate immune system can engulf these pathogens via phagocytosis and kill the pathogens using one of several possible methods (described in Chapter 2). Furthermore, this recognition and destruction promotes the innate immune system cells to induce inflammation.

Inflammation

An important component of the innate immune system response is **inflammation** (**FIGURE 1.14**), which draws fluid to the site of an infection. The inflammatory response occurs when innate immune cells recognize an infection and release either cytokines or inflammatory mediators. These secreted molecules act locally at the site of infection to promote movement of immune system cells and soluble plasma proteins to the site of infection through vasodilation (increasing the diameter of a blood vessel). Furthermore, the inflammatory response can promote secretion of other plasma proteins of the complement system to further arm the innate immune response and can induce epithelial cells in the vicinity of the inflammatory response to express different surface proteins. This prompts the recruitment of leukocytes such as neutrophils to the site of infection. Finally, vasodilation caused by the inflammatory response promotes the movement of dendritic cells and soluble pathogen antigens to nearby lymphoid tissue, where they may activate an adaptive immune response.

Inflammation over a short duration of time (hours to a few days) is known as acute inflammation. A longer-lasting inflammatory response is termed chronic inflammation. The four cardinal signs of inflammation are pain, redness, warmth, and swelling. Temporary loss of function of the inflamed tissue may also occur.

inflammation Response triggered by cytokines that results in vasodilation, redness, and swelling; this response aids the innate immune response by promoting immune cell migration to a site of infection and immune cell action at an infection site.

(A)

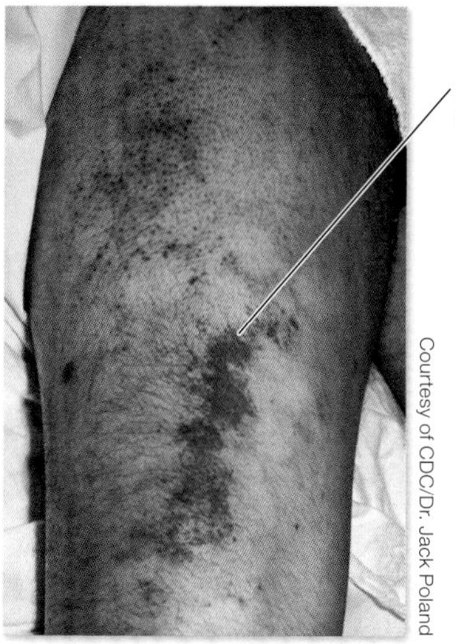

(B)

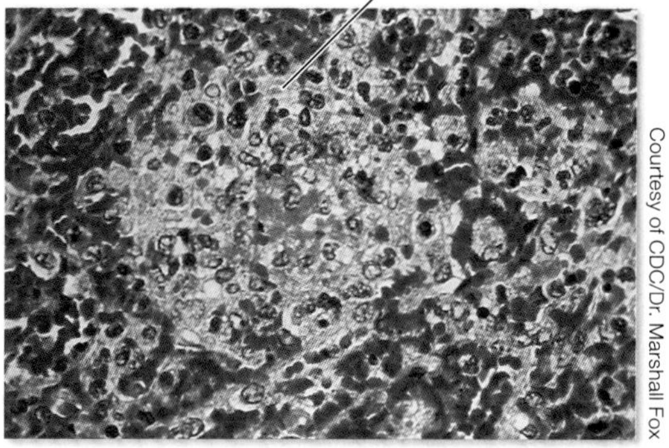

FIGURE 1.14 **Photo and micrograph of inflamed tissue** (A) Inflammation and necrosis associated with infection with *Yersinia pestis*, the bacteria that causes human plague. (B) Micrograph of spleen tissue from individual infected with this same bacteria (inflammation of the arterioles can be seen).

Fever

One byproduct of inflammation is a rise in body temperature, referred to as fever. The rise in temperature is caused by cytokines and inflammatory mediators that drive inflammation—these same molecules alter metabolism in tissues to increase heat output, thus increasing body temperature. This rise in temperature aids the immune system in two ways: it lowers the rate of replication of many pathogens and it increases the activity of the adaptive immune response.

CHECKPOINT QUESTIONS

1. What are the two main functions of an immune response?
2. What is the purpose of inflammation in an immune response?

1.7 How does the adaptive immune system target and destroy pathogens that have evaded the innate immune system?

LEARNING OBJECTIVE

1.7.1 Name and describe the cells of the adaptive immune system and explain the mechanisms they use to target and destroy pathogens that have evaded the innate immune system.

The T cells and B cells of the adaptive immune system provide powerful, long-term protection against pathogens. While the innate immune response manages to keep most infections at bay, some infections persist. Impaired immune function can occur for many reasons, including a congenital condition, illness, certain drugs, chronic alcoholism, chronic stress, or malnutrition.

Clonal selection and expansion

The adaptive immune system is the most powerful arm of the immune response. It specifically targets and destroys pathogens that have evaded the innate immune system. T cells and B cells contain highly specific **receptors**, made specific during the process of T-cell and B-cell development. Each specialized receptor of a T cell or B cell recognizes a specific antigen. This recognition process is known as **clonal selection**. Due to the diverse population of T cells and B cells in the body, the adaptive immune system is capable of recognizing a wide variety of foreign molecules.

receptor Molecule (typically a protein) that binds to another molecule to sense the environment, interact with other cells, and respond to stimuli.

clonal selection Process by which a specific T cell or B cell recognizes its antigen through action of its receptor.

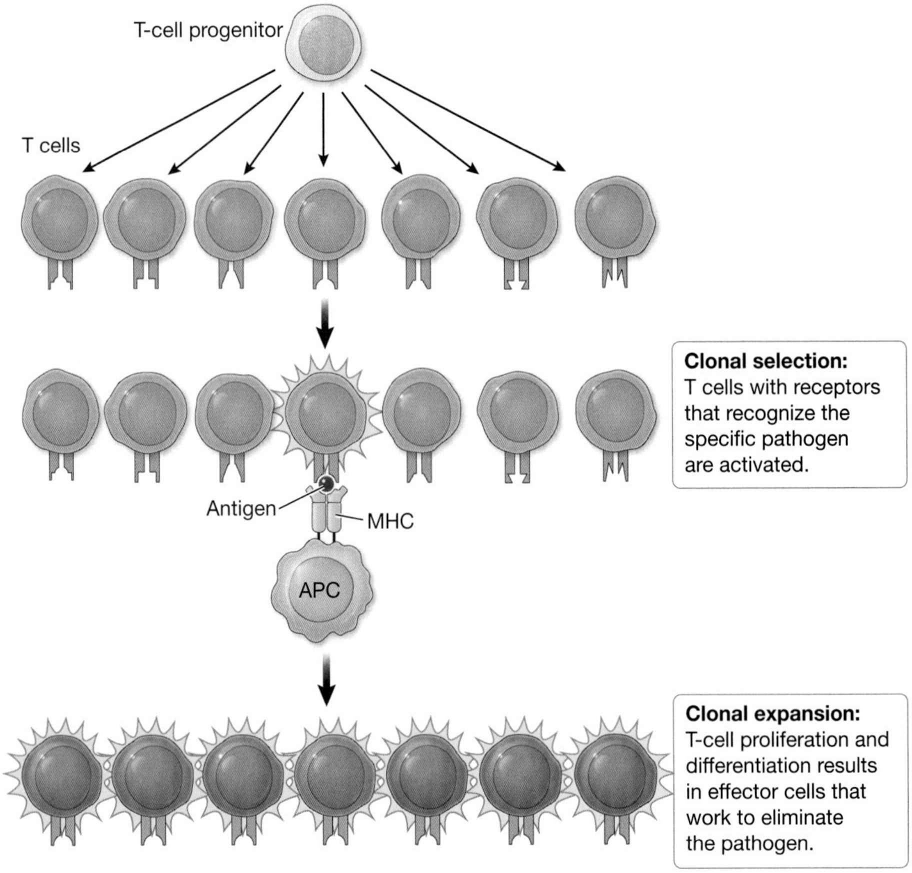

FIGURE 1.15 Clonal selection and expansion produces effector cells that target and eliminate the pathogen Each T cell produces a unique receptor (top). Recognition of a specific pathogen by a T-cell receptor (middle, clonal selection), promotes T-cell activation (bottom, clonal expansion), which includes rapid division of the selected T cell and differentiation into effector T cells.

clonal expansion Proliferation and differentiation of a T cell or B cell that has engaged its receptor with its antigen.

effector cell Immune system cell that has been activated and changes its function to specialize in pathogen recognition and destruction.

immunological memory Production of memory T cells and memory B cells during clonal expansion, which act in a faster, more robust adaptive immune response upon subsequent contact with the same antigen.

memory cells Differentiated long-lived T cells and B cells that drive secondary immune responses.

vaccine Nonpathogenic product from a pathogen formulated to be administered to stimulate a primary immune response and provide protection from infection by that specific pathogen.

immunization Process used to prevent disease by exposing an individual to a nonpathogenic product of a pathogenic organism to promote mounting of a primary immune response and immunological memory.

lymphoid tissue Specialized tissue of the lymphatic system that functions in the development and activation of T cells and B cells.

primary lymphoid tissue Lymphoid tissue that serves as a location for lymphocyte development.

bone marrow Primary lymphoid tissue that serves as the location of hematopoiesis and the maturation of B cells.

thymus Primary lymphoid tissue that serves as the location for T cell development.

secondary lymphoid tissue Lymphoid tissue that serves as a location for activating lymphocytes.

Once recognition occurs, the receptor activates the lymphocyte in a way that results in cell proliferation and differentiation, a process called **clonal expansion**. The daughter cells, known as **effector cells**, possess the same antigen-specific receptors as the parent cell (**FIGURE 1.15**). They go on to fight the infection using mechanisms we will explore in later chapters.

Immunological memory

A key characteristic of the adaptive immune response is the development of **immunological memory**. When an individual encounters a pathogen for the first time, a *primary immune response* occurs. This initial adaptive immune response uses T cells and B cells to destroy the pathogen.

During the primary immune response, some of the effector cells produced during clonal selection and clonal expansion are **memory cells**. These memory cells are long-lived cells that are primed to act upon future exposure to the same pathogen. Thus, when an individual encounters the same pathogen at a later time, the *secondary immune response* is stronger and faster (2 to 3 days vs. approximately 2 weeks for the primary response).

The phenomenon of immunological memory allows for the development of vaccines. **Vaccines** work by priming the adaptive immune system with a nonpathogenic form of a potentially deadly microorganism or a nontoxic product of a pathogen. After **immunization**, contact with the same microorganism stimulates adaptive immunity very quickly.

CHECKPOINT QUESTIONS

1. What are the differences between clonal selection and clonal expansion?
2. What is the importance of immunological memory in the immune system?

1.8 How do primary and secondary lymphoid tissues differ?

LEARNING OBJECTIVE

1.8.1 Describe the tissues and organs that make up primary and secondary lymphoid tissues and explain their role in immune system function.

lymphatic system Circulatory system of lymph and lymphoid tissue that allows for lymphocyte migration to and from these tissues.

lymph Fluid of the lymphatic system that allows for plasma and lymphocyte movement to and from lymphoid tissue.

The diversity of B cells and T cells—with each of these lymphocytes expressing a receptor capable of recognizing a different antigen—complicates the means by which a proper adaptive immune response can be raised. Pathogens at a site of infection may be easily recognized by a subset of lymphocytes with the proper cell-surface receptor, but this would require movement of the lymphocytes to the site of infection. To concentrate the efforts of pathogen recognition by the adaptive immune response, most lymphocytes in the body reside in specialized tissue known as **lymphoid tissue**. The major lymphoid organs are the lymph nodes. Lymphoid tissue is also found in the bone marrow, thymus, spleen, tonsils, appendix, and the Peyer's patches of the small intestine (**FIGURE 1.16**). The gut and mucosal surfaces of the body have their own organized lymphoid tissue known as mucosa-associated lymphoid tissue, as these surfaces are constantly in contact with potential pathogens.

Primary lymphoid tissues are sites where lymphocytes develop and mature. There are two types of primary lymphoid tissue: bone marrow and the thymus. **Bone marrow** is where B cells and T cells (along with other hematopoietic cells) are produced and where B cells mature. T cells migrate from the bone marrow to develop and mature in the **thymus**. T cells are named for this connection to the thymus.

All other lymphoid tissue in the body is referred to as **secondary lymphoid tissue**. These tissues are responsible for presenting antigens from pathogens and are the sites of lymphocyte activation in response to pathogens.

Lymph nodes are an important secondary lymphoid tissue. They play a role in lymphocyte activation and also form a connection between the circulatory and lymphatic systems. The **lymphatic system** is a network made up of lymphatic vessels and lymphoid tissues and organs. Lymphatic vessels drain extracellular fluid, known as **lymph**, from the tissues and return it to the blood, where it recirculates. The movement of lymph is driven by pressure from localized muscle contractions and body movement (unlike blood driven by pressure in the circulatory system). Lymph moves in one direction due to one-way valves that direct lymph flow toward the upper ducts of the body, especially the thoracic duct. The edema (excess fluid in tissues that results in swelling) seen in patients restricted to bed rest is caused by fluid accumulation due to lack of movement.

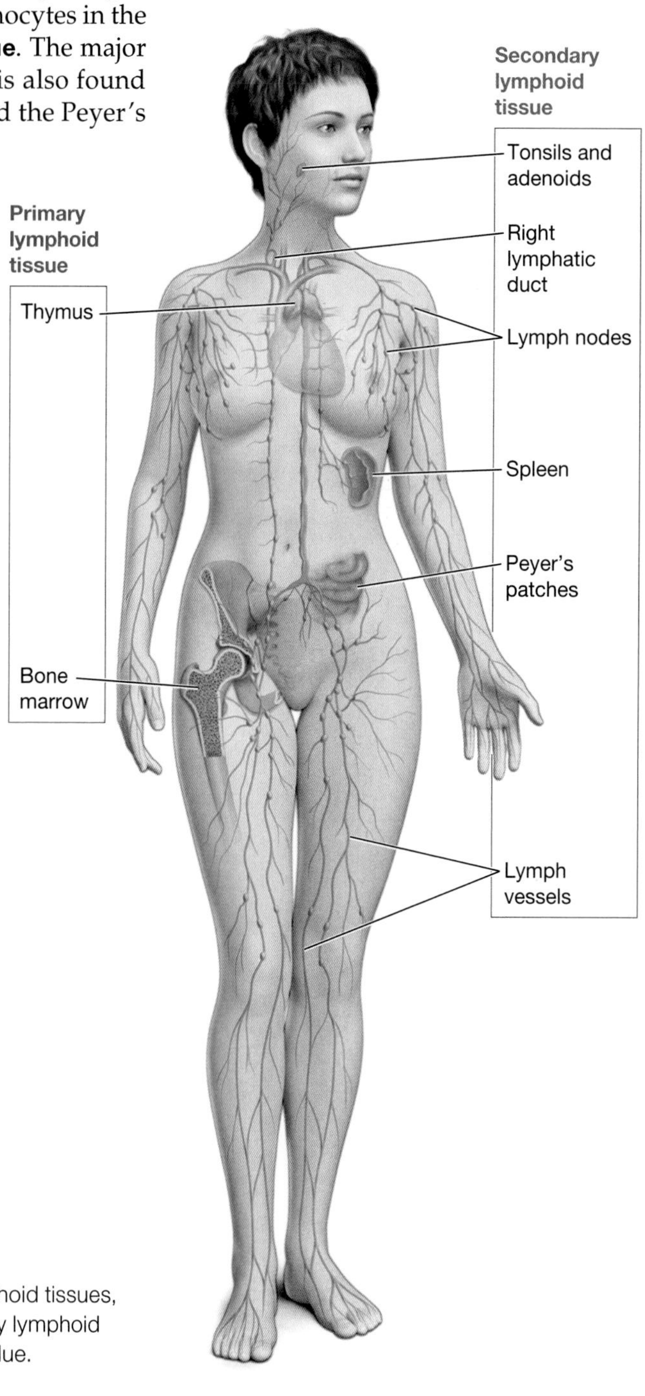

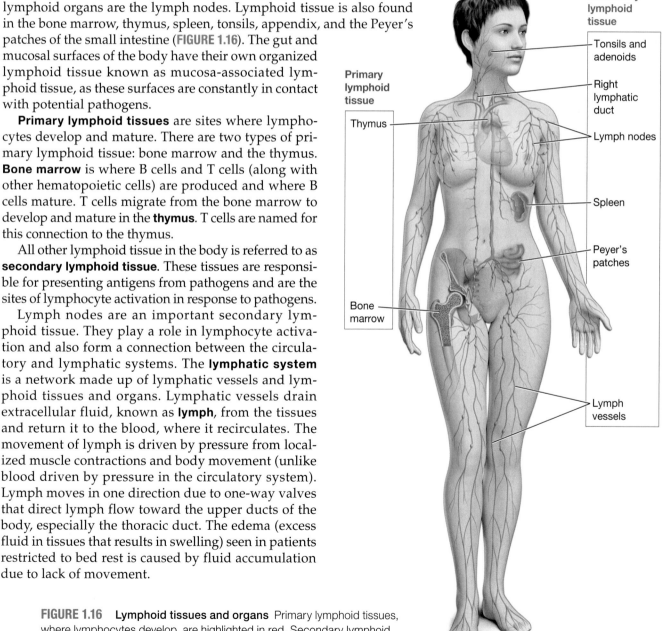

FIGURE 1.16 Lymphoid tissues and organs Primary lymphoid tissues, where lymphocytes develop, are highlighted in red. Secondary lymphoid tissues, where lymphocytes are activated, are highlighted in blue.

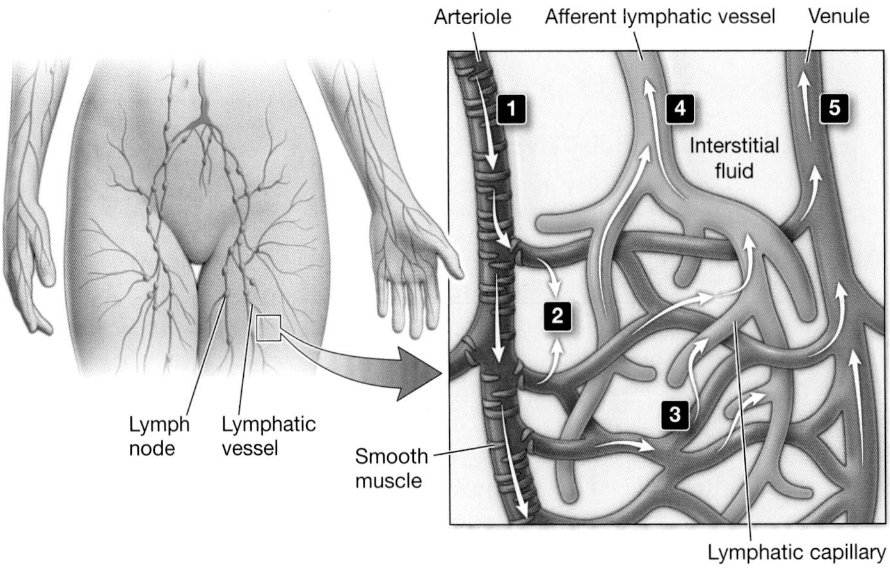

FIGURE 1.17 Circulation of lymphocytes in the lymphatic and circulatory systems Lymphocytes can migrate through the body through both the circulatory and lymphatic systems. (1) Lymphocytes move with red blood cells through arteries via heart muscle contractions. (2) Lymphocytes can migrate between endothelial cells of the arterioles of capillaries and enter the interstitial fluid. (3) Porous lymphatic vessels can allow entry of interstitial fluid, soluble proteins, and lymphocytes into the lymphatic system. (4) Lymphocytes can circulate via the lymphatic system to secondary lymphoid tissues, or (5) lymphocytes that never exited blood capillaries can continue their migration through the circulatory system.

Mature T cells and B cells have the ability to circulate through both the lymphatic and circulatory systems (**FIGURE 1.17**). Both types of cells leave the primary lymphoid tissue where they originated and enter the circulatory system. When these cells reach secondary lymphoid tissue, such as a lymph node, they can leave the blood and enter this tissue in search of antigens. If the B cells or T cells recognize an antigen presented in the secondary lymphoid tissue, they remain and go through activation and expansion. If the lymphocytes do not encounter an antigen that their receptor recognizes, the cells leave via the efferent (exiting) lymph and eventually recirculate back into the bloodstream.

CHECKPOINT QUESTIONS

1. What is the difference between primary and secondary lymphoid tissue?
2. What are the two types of primary lymphoid tissue in humans?

1.9 What are the specialized types of secondary lymphoid tissues and how are they organized?

LEARNING OBJECTIVES

1.9.1 Describe the organization of lymph nodes, spleen, and mucosa-associated lymphoid tissue.

1.9.2 Explain the role of secondary lymphoid tissue in the adaptive immune response.

As previously noted, secondary lymphoid tissue is important in presenting antigens to circulating T cells and B cells. During an infection, fluid at the infection site contains dendritic cells that have encountered and processed pathogens. These cells leave the site of infection through the normal flow of lymph and drain into a secondary lymphoid tissue, referred to as a *draining secondary lymphoid tissue*. The following sections focus on several specialized secondary lymphoid tissues: lymph nodes, the spleen, and gut/mucosa-associated secondary lymphoid tissue.

Lymph nodes

Along with other secondary lymphoid tissues, the **lymph nodes** (**FIGURE 1.18**) are responsible for facilitating antigen presentation and activation of the adaptive immune response. The anatomy of lymph nodes is ideal for this process.

lymph node Secondary lymphoid tissue that contains anatomy specialized in antigen presentation to T cells and B cells and lymphocyte activation and proliferation.

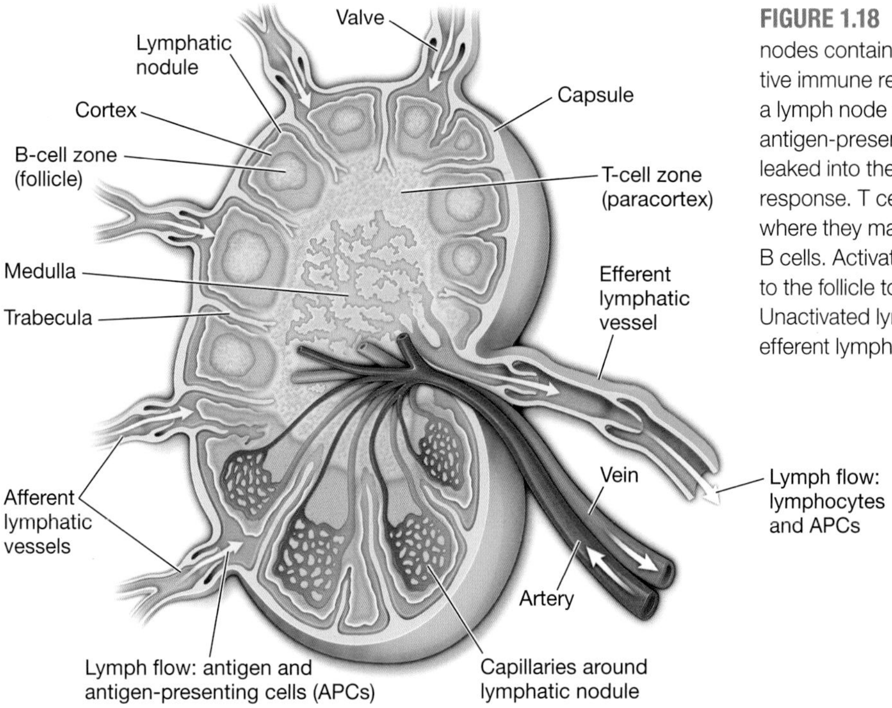

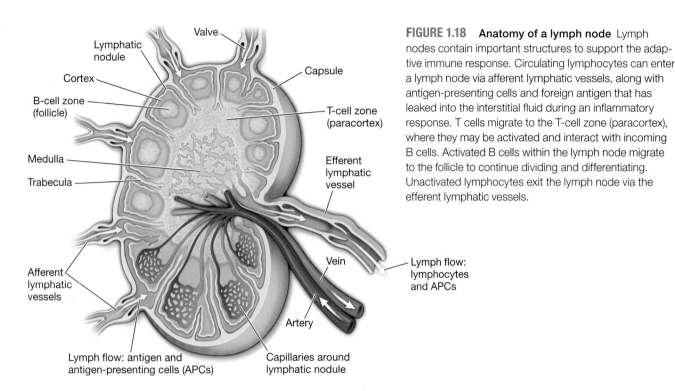

FIGURE 1.18 Anatomy of a lymph node Lymph nodes contain important structures to support the adaptive immune response. Circulating lymphocytes can enter a lymph node via afferent lymphatic vessels, along with antigen-presenting cells and foreign antigen that has leaked into the interstitial fluid during an inflammatory response. T cells migrate to the T-cell zone (paracortex), where they may be activated and interact with incoming B cells. Activated B cells within the lymph node migrate to the follicle to continue dividing and differentiating. Unactivated lymphocytes exit the lymph node via the efferent lymphatic vessels.

As fluid drains into the lymph node, T cells and B cells enter and migrate to different regions. T cells migrate to T-cell areas, and B cells migrate to **lymphoid follicles** (**FIGURE 1.19**). Dendritic cells present antigen to T cells in the T-cell area, where, if activated, they become effector cells. These effector cells flow out via the lymphatic vessels or migrate to lymphoid follicles to help in the process of B-cell activation. B cells that have been activated in a lymphoid follicle proliferate and form a germinal center within the follicle. Some of the cells produced and differentiated during B-cell activation will produce antibodies that will also leave the lymph node via the lymphatic vessels.

Spleen

Pathogens can enter the bloodstream in a variety of ways, including being carried by vectors that feed off blood (such as mosquitos) or if an infection has not been properly cleared. Macrophages and dendritic cells that reside in the **spleen** can engulf these bloodborne pathogens and activate T cells and B cells circulating through the blood into the spleen. Antigen presentation and lymphocyte activation occurs in the white pulp of the spleen (**FIGURE 1.20**).

Individuals who lack a spleen due to a genetic disorder or surgical removal are more prone to bacterial infections in the bloodstream, especially those caused by encapsulated bacteria such as *Streptococcus pneumoniae*. Because of this susceptibility to infection, it is often advised that individuals without a spleen be vaccinated against encapsulated bacteria to prevent potential infections of the blood.

Mucosa-associated lymphoid tissue

Mucosa-associated lymphoid tissues (**MALT**) are located within the mucous membranes of the digestive, respiratory, and urogenital tracts. They make up a large surface area that can potentially come in contact with pathogens. Two main types of MALT are the gut-associated lymphoid tissue (GALT) of the digestive system and bronchial-associated lymphoid tissue (BALT) of the

lymphoid follicle Location within a lymph node where activated T cells migrate to promote activation of B cells during an infection.

spleen Secondary lymphoid tissue important in red blood cell filtration and monitoring pathogen presence in the bloodstream.

mucosa-associated lymphoid tissue (MALT) Secondary lymphoid tissue that specializes in the uptake and clearance of pathogens at mucosal surfaces.

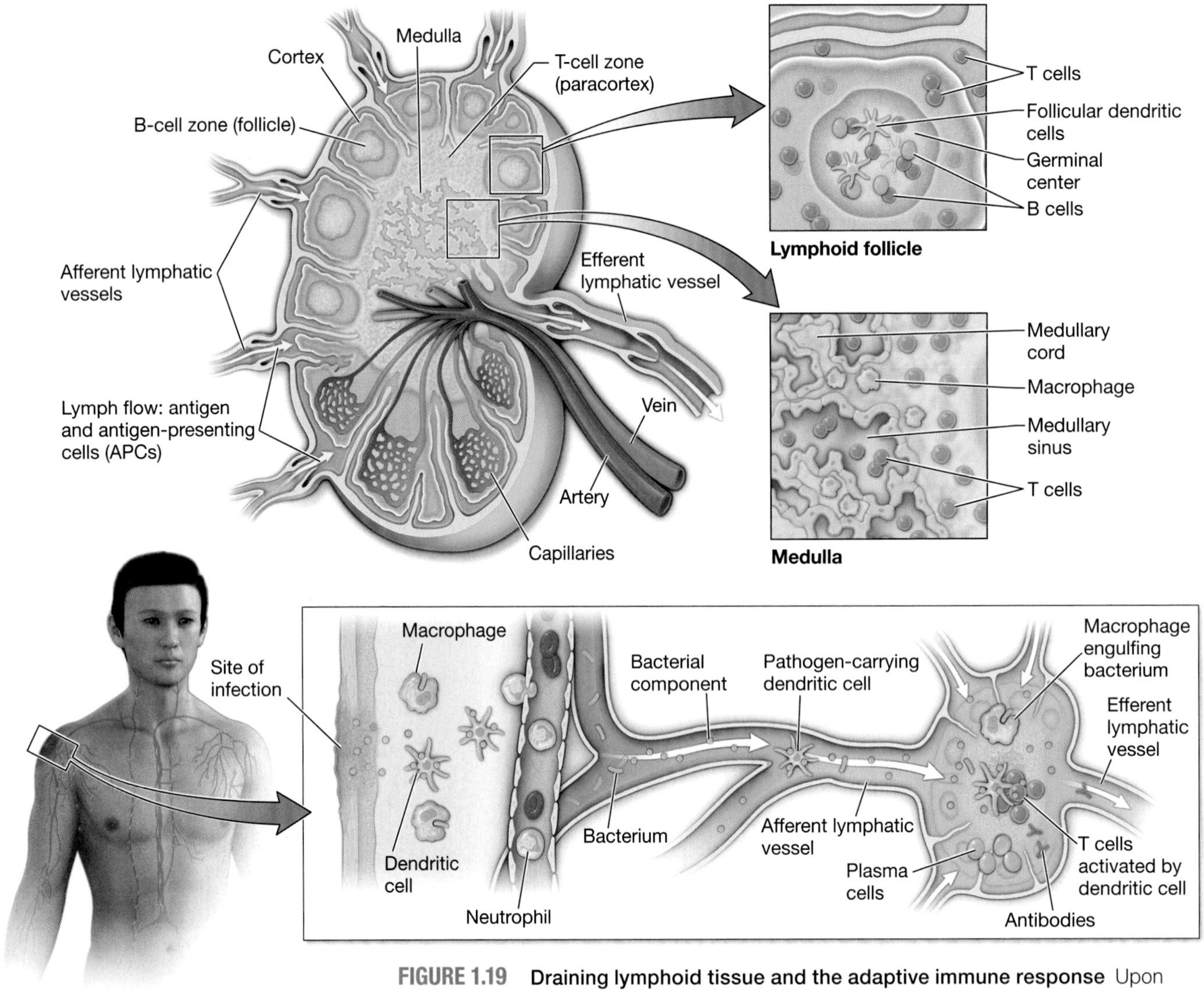

FIGURE 1.19 Draining lymphoid tissue and the adaptive immune response Upon infection, such as via a cut at the skin surface, pathogens gain entry to underlying tissue. Bottom: Pathogens at the site of infection are recognized and engulfed by resident macrophages and dendritic cells, and activation of an inflammatory response triggers the exit of neutrophils from the bloodstream and their migration to the site of infection. The inflammatory response promotes drainage of material, pathogens, and dendritic cells from the site of infection to a secondary lymphoid tissue such as a lymph node. The draining lymph node serves as a centralized site for the activation of T cells and B cells. Upper right: T cells are activated in the T-cell zone of the medulla. The majority of B cells are activated within the lymphoid follicle.

respiratory system. These secondary lymphoid tissues bear anatomic similarities to lymph nodes and the spleen (**FIGURE 1.21**).

Although lymph nodes, spleen, and MALT bear anatomic similarities and function similarly in the activation of an adaptive immune response, a major difference is the mechanism of antigen delivery to these tissues. While many secondary lymphoid tissues gain antigen via the inflammatory response and act as draining secondary lymphoid tissue, it is not desirable to induce an inflammatory response in mucosal tissue, since these tissues may contain molecules that can cause damage to the underlying tissue, such as the digestive enzymes seen in the gastrointestinal tract. Thus, MALT gains antigens for presentation

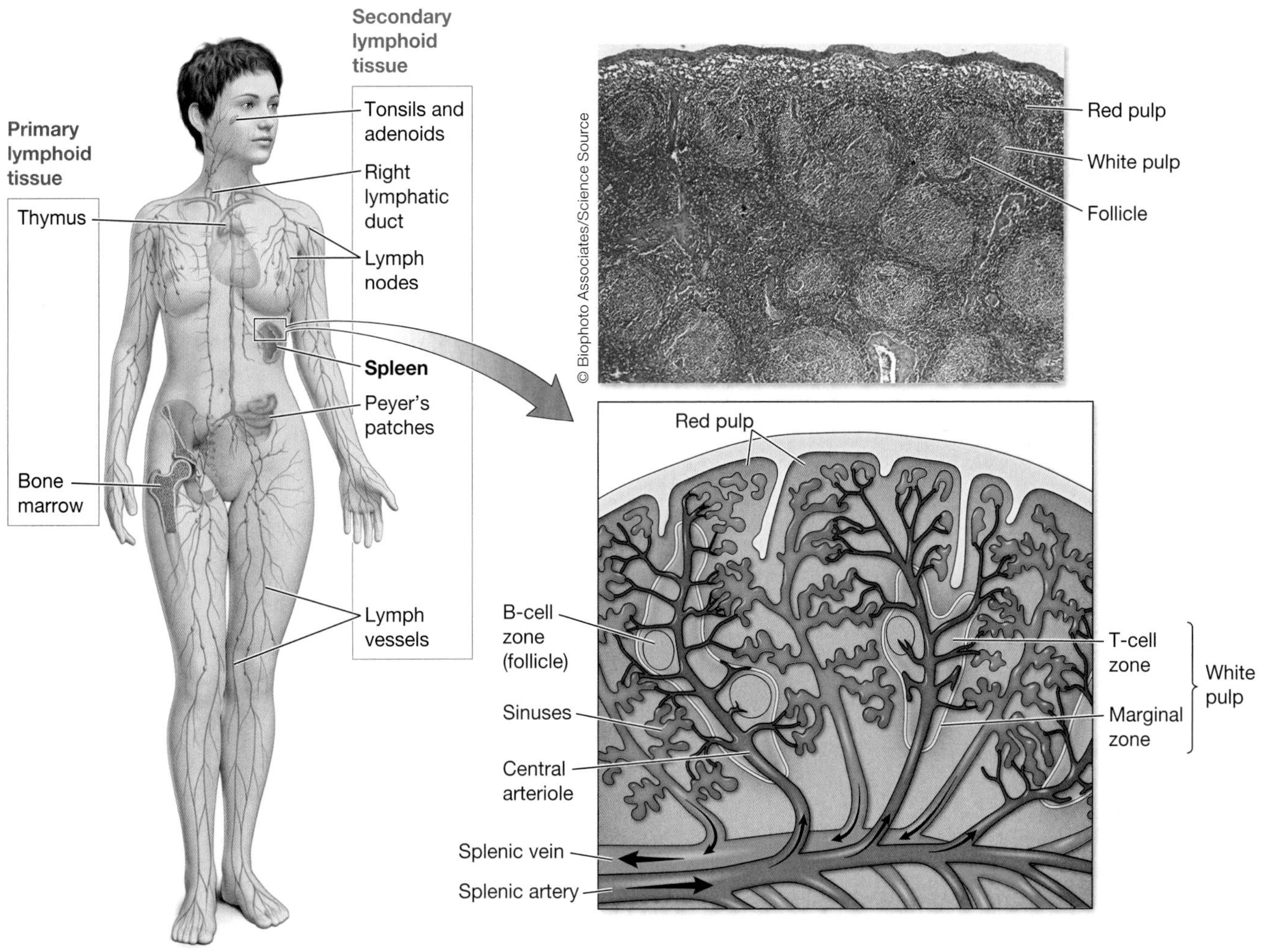

FIGURE 1.20 Spleen anatomy showing key structures involved in the adaptive immune response Upper left: Primary (red) and secondary (blue) lymphoid tissue. Upper right: Micrograph of the spleen. Lower right: Key structures within the spleen, including white pulp, which contains secondary lymphoid tissue structures used for T-cell activation (T-cell zone), B-cell maturation (marginal zone), and B-cell differentiation (follicle).

to lymphocytes differently by using **M cells** as the means by which antigens enter. Pathogens enter the MALT via M cells and come into contact with resident dendritic cells and macrophages. They are then processed and presented to lymphocytes.

Lymphocytes enter the MALT from the bloodstream and exit via lymphatic vessels. Lymphocytes that have been activated within MALT tend to remain in these tissues, where they await subsequent attack by their target pathogen.

M cell Cells associated with mucosa-associated lymphoid tissue that specialize in pathogen and antigen delivery to those tissues.

CHECKPOINT QUESTIONS

1. At what locations within lymph nodes are T cells and B cells activated?
2. Most secondary lymphoid tissue relies on inflammation for the delivery of antigens, whereas mucosa-associated lymphoid tissue (MALT) does not. How are antigens delivered to MALT?

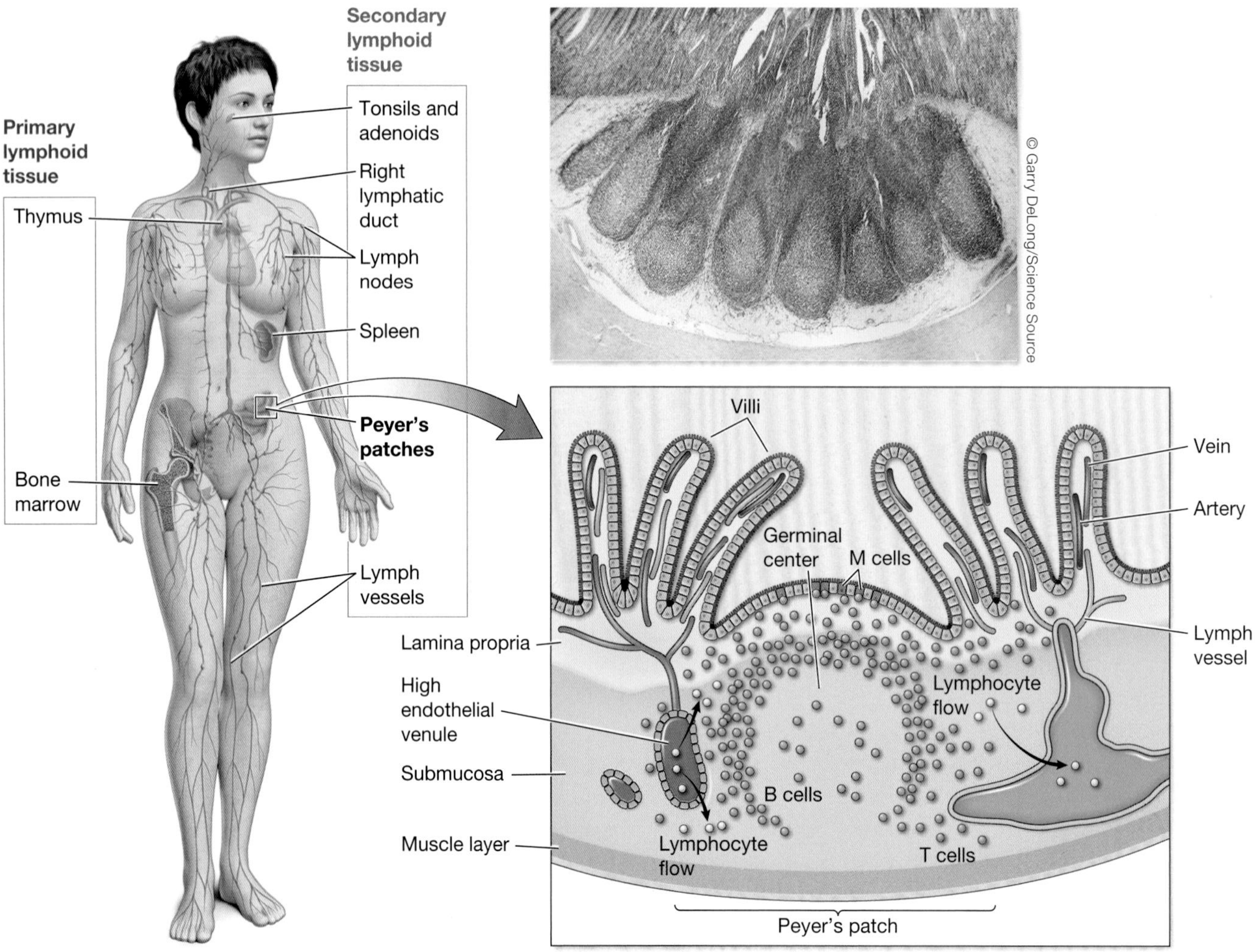

FIGURE 1.21 Mucosa-associated lymphoid tissue (MALT) showing key structures involved in the adaptive immune response Upper left: Primary (red) and secondary (blue) lymphoid tissue. Upper right: Micrograph of MALT. Lower right: Key structures within MALT, including M cells, Peyer's patch, and lymphocytes that have exited a high endothelial venule.

1.10 What happens when the immune system malfunctions?

LEARNING OBJECTIVE

1.10.1 Explain the four major ways the immune system can malfunction and the likely outcomes of each type of malfunction.

Although the immune system works hard to prevent pathogen infection and disease, certain genetic disorders and environmental causes can lead to immune system malfunction. These malfunctions can result in disorders in which individuals raise immune reactions to innocuous materials, raise immune reactions to self-antigens (losing tolerance), or do not efficiently mount any type of immune response. Consequences of immune system malfunction can be as simple as sneezing to as complex as immune system failure and ultimately death, if untreated. The major types of immune system malfunction are:

- *Hypersensitivity reactions*: Recognition of a foreign, but innocuous, material can result in immune responses discussed earlier in this chapter. These types of reactions are referred to as hypersensitivity reactions or, more commonly, **allergies**. The foreign material responsible for inducing a hypersensitivity reaction is referred to as an **allergen**. Most commonly, hypersensitivity reactions activate granulocytes such as mast cells, basophils, and eosinophils and induce localized symptoms, including coughing, sneezing, vomiting, and diarrhea, as a means to expel the foreign molecule. However, some allergens can induce a more severe hypersensitivity reaction that is capable of inducing anaphylaxis (an acute reaction that can cause difficulty breathing and shock), which can lead to death if not treated properly.
- *Autoimmune diseases*: The power of the adaptive immune response is driven by the diversity of B cells and T cells. During cell development, if everything is functioning properly, the immune system gains tolerance to self-molecules. However, if the adaptive immune system malfunctions and B cells and T cells that recognize self-molecules enter the circulatory and lymphatic systems, then there is potential to mount an adaptive immune response to tissues within the body. These autoimmune diseases (diseases driven by an immune response against self-tissues) can result in inflammation at the sites of recognition (e.g., rheumatoid arthritis) or can result in destruction of the tissue recognized by the immune system (e.g., type 1 diabetes).
- *Immunodeficiencies*: While the previous two malfunctions result from abnormal antigen recognition, another immune system malfunction lies on the other extreme—lack of immune system function. When immune system cells are incapable of mounting a proper immune response, there is a higher likelihood of pathogen infection. There are two common causes: inherited immunodeficiencies, which result from genetic mutations preventing immune cell development or function, and acquired immunodeficiencies, caused by environmental factors or infection by pathogens that can inhibit the immune system.
- *Cancer*: The adaptive immune response is fine-tuned to recognize many different foreign antigens. As mentioned earlier in this section, B cells and T cells gain tolerance to self-molecules during the development process. This tolerance is highly specific to each individual's self-molecules, but over the course of the lifespan of an individual, random mutations in genes may occur. These mutations may alter self-molecules enough to elicit an adaptive immune response to the mutated cell, resulting in its destruction. However, if the immune response fails to recognize the mutated cell, eventually the mutations could result in changes in genes responsible for controlling cell division. If these mutated genes can no longer function properly in controlling cell division, the cells will grow uncontrollably, resulting in cancer. Thus, while the immune system functions to protect us from foreign pathogens, it is also capable of recognizing our own cells that have mutated, and if it does not recognize these mutations, cancerous growth might occur.

allergy Hypersensitivity reaction to an innocuous foreign antigen.

allergen Innocuous foreign molecule that can elicit a hypersensitivity response.

CHECKPOINT QUESTIONS

1. What are the four major types of immune system malfunction?
2. What causes each type?

Summary

1.1 How did the scientific field of immunology emerge?

- Immunology is the study of immunity from diseases and the mechanisms used by the immune system to provide protection against pathogens.

1.2 What are the types of pathogens?

- Pathogens have the ability to cause disease. Pathogens can be bacteria, viruses, fungi, protozoa, or parasites. Extracellular pathogens largely reside outside of host cells, and intracellular pathogens largely reside inside host cells.

1.3 How does the immune response combat the pathogens that cause disease?

- The two arms of the immune system are the innate immune system and the adaptive immune system (which works to provide humoral immunity and cell-mediated immunity).

1.4 What is the body's first line of defense against pathogens?

- The body's first lines of defense against pathogens include the physical barriers of the skin and mucosal surfaces, the presence and action of commensal organisms, and antimicrobial agents.

1.5 Which cell lineages are associated with innate and adaptive immunity?

- All blood cells are derived from a common stem cell precursor through a process known as hematopoiesis. Hematopoiesis gives rise to three blood cell progenitor cells (the erythroid megakaryocyte progenitor, the myeloid progenitor, and the lymphoid progenitor); each of these differentiate into specialized cells.

1.6 How does the innate immune system limit the spread of most infections?

- The innate immune system uses several mechanisms to recognize and destroy pathogens. These include pathogen recognition and destruction, inflammation, and fever.

1.7 How does the adaptive immune system target and destroy pathogens that have evaded the innate immune system?

- The adaptive immune system is driven by the action of B cells and T cells, which are capable of recognizing specific antigens (clonal selection) and developing into specialized cells (clonal expansion).
- The adaptive immune system develops immunological memory, which allows for faster and stronger response to a pathogen upon subsequent exposure.

1.8 How do primary and secondary lymphoid tissues differ?

- Primary lymphoid tissue is the location of lymphocyte development, and secondary lymphoid tissue is the location of lymphocyte activation and proliferation.

1.9 What are the specialized types of secondary lymphoid tissue and how are they organized?

- Secondary lymphoid tissues, including the lymph nodes, spleen, and MALT, have evolved anatomically to promote pathogen sensing and clearance based on the tissues and systems with which they are associated.

1.10 What happens when the immune system malfunctions?

- Immune system malfunction can be classified into four major categories: immunodeficiencies, hypersensitivity reactions, autoimmune diseases, and cancer. These can result in conditions that range from mild allergies to serious and life-threatening infections or complications.

Review Questions

1.1 What is the purpose of immunology?

1.2 What are the five types of pathogens that can cause disease? Provide an example of each type.

1.3 **a.** Compare/contrast humoral and cell-mediated immunity

b. *Escherichia coli* colonizes our small intestine and aids in food digestion. This bacterium is an example of a(n) ______ organism.

A. pathogenic
B. opportunistic
C. commensal
D. parasitic
E. intracellular

1.4 How do defensins function as antimicrobial peptides?

1.5 **a.** The mechanism used by neutrophils and macrophages to uptake foreign pathogens is

A. phagocytosis.
B. inflammation.
C. cytokine production.
D. defensin production.
E. antigen presentation.

b. Explain why phagocytic cells are important in activating an adaptive immune response.

c. Name the three progenitor lines involved in hematopoiesis and the differentiated cells produced by each line.

1.6 Explain the role of inflammation in the immune response.

1.7 Explain the functions of clonal selection and clonal expansion as they pertain to mounting an adaptive immune response.

1.8 Primary lymphoid tissues are the locations where lymphocytes ______, whereas secondary lymphoid tissues are the locations where lymphocytes _______.

A. undergo apoptosis; develop and mature

B. are activated; develop and mature

C. develop and mature; undergo apoptosis

D. develop and mature; are activated

E. are activated; undergo apoptosis

1.9 Briefly describe the anatomy of a lymph node and how this tissue interacts with adaptive immune cells.

1.10 Describe three ways in which immune system malfunction can lead to disease.

CASE STUDY REVISITED

Pneumococcal Sepsis

A male, age 31, is admitted to the hospital with symptoms of fever, chills, nausea, vomiting, and diarrhea. Food poisoning is ruled out, as the medical history reveals that the family ate the same meal and are not suffering from any of these symptoms. A blood culture reveals that the patient has an invasive *Streptococcus pneumoniae* infection in his bloodstream. Pneumococcal sepsis occurs in people with weakened immune systems. The patient's medical records reveal that he underwent a splenectomy 6 months ago following an automobile accident.

Think About...

Questions for individuals

1. Define sepsis.
2. What are some mechanisms that *Streptococcus pneumoniae* uses to evade the immune response?

Questions for student groups

1. Is the patient's history consistent with the bacterial infection seen? Why or why not?
2. What is the likely treatment for this infection?
3. What actions could/should have been taken after the splenectomy to prevent subsequent bacterial infection?

2

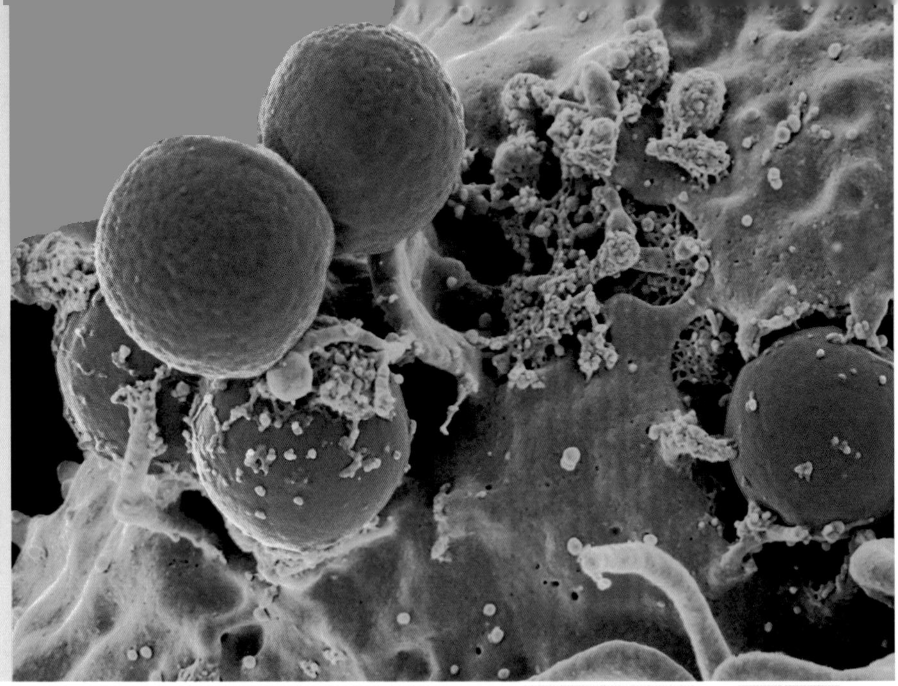

© BSIP SA/Alamy Stock Photo

Innate Immunity

QUESTIONS Explored

- **2.1** What are the primary defenses of the innate immune system?
- **2.2** How are pathogens recognized by cellular receptors?
- **2.3** Once PAMP recognition occurs, what innate immune responses are activated?
- **2.4** How do the Toll-like receptors function in pathogen recognition and cellular activation?
- **2.5** What are the functions of macrophages in innate immunity?
- **2.6** What are the functions of neutrophils in innate immunity?
- **2.7** What are the functions of natural killer cells in innate immunity?
- **2.8** How does the innate immune system initiate and help regulate the adaptive immune response?
- **2.9** What happens when the innate immune system malfunctions?

CASE STUDY: Diane's Story

Diane and her husband are preparing a new recipe for dinner that they have never tried before: chicken cacciatore. To save a bit of money, Diane decides to buy a whole chicken and cut it into pieces herself, which is a trick she's done plenty of times before. While her husband works on the sauce for dinner, Diane begins slicing the chicken; however, while slicing off one of the wings, Diane slices her finger as well.

Diane's husband drives her to the emergency room, where they apply six stitches and bandage her up. The doctor also prescribes an antibiotic to prevent an infection. Diane and her husband return home a bit shaken up and resign themselves to the fact that the chicken cacciatore will have to wait and it will be pizza for dinner.

The next day, Diane wakes up with the chills, and the cut on her finger seems to be swollen. Diane's husband takes her temperature and finds that she has a fever. Diane's ankles are also very swollen, making it difficult for her to walk or stand. She has a hard time catching her breath, and she feels like her heart is racing. Diane's husband decides to take her to the emergency room again to find out what is wrong.

In order to understand what Diane is experiencing, and to answer the case study questions at the end of this chapter, we must first explore the many mechanisms that our innate immune system employs to combat pathogens. In this chapter, we will explore the ways in which the innate immune system uses physical and chemical barriers to stave off infection and how it employs innate immune cells to seek out and destroy pathogens. The innate immune system also uses a variety of effector functions (mechanisms) to efficiently clear infections. These include the action of innate immune cells and the use of soluble signaling molecules that function to increase the effectiveness of the innate immune response and facilitate pathogen destruction.

2.1 What are the primary defenses of the innate immune system?

LEARNING OBJECTIVES

2.1.1 Describe in real time the means by which the innate immune system functions to combat a pathogen.

2.1.2 Distinguish the different mechanisms used by the innate immune system to fight infection, including the action of antimicrobial proteins and peptides and the process of phagocytosis.

Recall from Chapter 1 that innate immune system defenses consist of physical barriers, chemical barriers, and responses of several different innate immune cells. As a brief review of these defenses, let's consider a *Salmonella* infection, a common cause of food poisoning. How does *Salmonella* travel within the body from a contaminated food source to ultimately colonize the small intestine, and what defense mechanisms are employed to combat colonization?

The contaminated food enters the mouth, passes through the esophagus, is processed in the stomach, and reaches the small intestine. During this transit, *Salmonella* bacteria come into contact with several physical and chemical barriers, including saliva, the mucus of the digestive tract, and the low pH of the stomach (**FIGURE 2.1**). Saliva contains chemicals and enzymes that begin the digestion process and also destroy potential pathogens in food. The digestive tract is coated in mucus, which is removed and replenished constantly and contains chemicals aimed at pathogen destruction. The pH in the stomach is low, mainly to aid in digestion, but this low pH also serves as a barrier to pathogens incapable of surviving the low pH. Thus, the body sets up nonspecific physical and chemical barriers as its initial defense when pathogens are first encountered.

If *Salmonella* bacteria escape these initial physical and chemical barriers and reach the small intestine, different defense mechanisms take over as the body attempts to prevent pathogen colonization. The small intestine contains a large number of bacterial organisms as part of the normal microbiota that compete for nutrients and space with any other organisms that enter. These microorganisms may actively alter the environment by changing the pH or releasing antimicrobial agents to prevent infection by other pathogens in the area. Furthermore, the lumen of the small intestine contains proteins such as lysozyme that aid in combating pathogens. And cells of the innate immune system (closely associated with the small intestine) specialize in pathogen recognition and removal (see Figure 2.1).

We have seen *Salmonella* as an example of an ingested pathogen that the immune system fights using a variety of physical and chemical barriers. Physical barriers in other areas of the body protect against pathogens using a different mode of entry such as via a cut that punctures the skin or through inhalation.

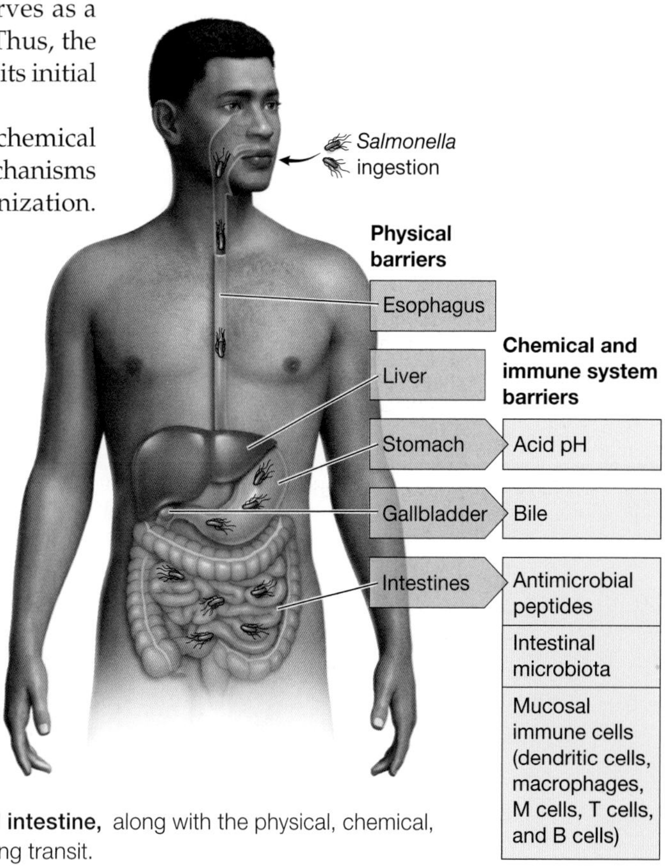

FIGURE 2.1 **Transit mechanism of *Salmonella* to small intestine,** along with the physical, chemical, and immune system barriers the pathogen encounters during transit.

lysozyme Enzyme present in saliva, tears, and respiratory tract fluid that cleaves the polysaccharide of bacterial cell walls.

protease inhibitor Molecule that is capable of preventing the enzymatic activity of proteases.

Our skin, the largest organ of the body, provides a physical barrier to prevent pathogens from entering. Our mucous membranes secrete fluids such as mucus and tears to prevent microorganism colonization by ongoing clearance of these fluids. Tissues lined with epithelial barriers employ tight junctions between cells to prevent pathogens from migrating from one epithelial cell to another.

While physical barriers are an important first line of defense, the cells lining the tissues that create physical barriers secrete molecules that target nearby microorganisms. These antimicrobial agents act to limit colonization and growth at the point of contact. Another important component of the immune system is the *complement system*, which is composed of soluble plasma proteins that aid in pathogen detection and elimination. Finally, phagocytosis of microorganisms at a site of infection also plays an important role in triggering an adaptive immune response, if one is required (**FIGURE 2.2**).

Antimicrobial proteins

Many of the epithelial cells lining tissues that form physical barriers also secrete antimicrobial proteins. These proteins either directly target microorganisms for destruction or inhibit their growth by preventing them from obtaining vital nutrients.

Lysozyme, an enzyme secreted in saliva, tears, and respiratory tract mucus, cleaves (splits apart) the polysaccharide component of bacterial cell walls that normally provides structural support to bacteria. Loss of this structural support can cause bacterial lysis due to the internal pressure of the bacterial cytoplasm.

Other antimicrobial proteins that destroy or inhibit the growth of microorganisms include *psoriasin*, *calprotectin*, and *lactoferrin*. Psoriasin is found in the skin and binds to calcium, and can, like lysozyme, promote bacterial lysis. It has antimicrobial activity against *Escherichia coli* but not against other common skin bacteria. Calprotectin and lactoferrin bind to metal ions such as calcium and iron. This binding inhibits bacterial and fungal growth by limiting the access of potential pathogens to these vital nutrients.

Some antimicrobial proteins act as **protease inhibitors**, blocking the action of specific proteases (enzymes that cleave proteins) produced by pathogens that can help pathogens migrate into surrounding tissues. One example of a secreted protease inhibitor is α2-macroglobulin, which acts as a decoy for secreted proteases. A region of α2-macroglobulin called the bait region is cleaved by a variety of different proteases, which causes the α2-macroglobulin to collapse around the protease and prevent it from further action.

Antimicrobial peptides

Antimicrobial peptides are another family of antimicrobial agents secreted by many epithelial cells. They are conserved through evolutionary history, suggesting that they represent a primitive component of the immune system of multicellular organisms. Many of these peptides are amphipathic in nature (the molecules contain both polar and nonpolar regions). Antimicrobial peptides are positively charged, which allows them to interact with lipid bilayers, and the nonpolar regions of these peptides allow the peptides to insert into pathogen membranes, causing membrane damage. *Defensins* are an important group of peptides that target a wide array of pathogens, including *E. coli*, *Staphylococcus aureus*, *Pseudomonas aeruginosa*, *Haemophilus influenzae*, and herpesvirus. *Histatins* are another family of peptides that defend the body; they are secreted in saliva and work by disrupting microbial and fungal membranes.

The complement system

While antimicrobial proteins and peptides work to limit pathogen colonization and growth at the point of contact, an important component of the plasma

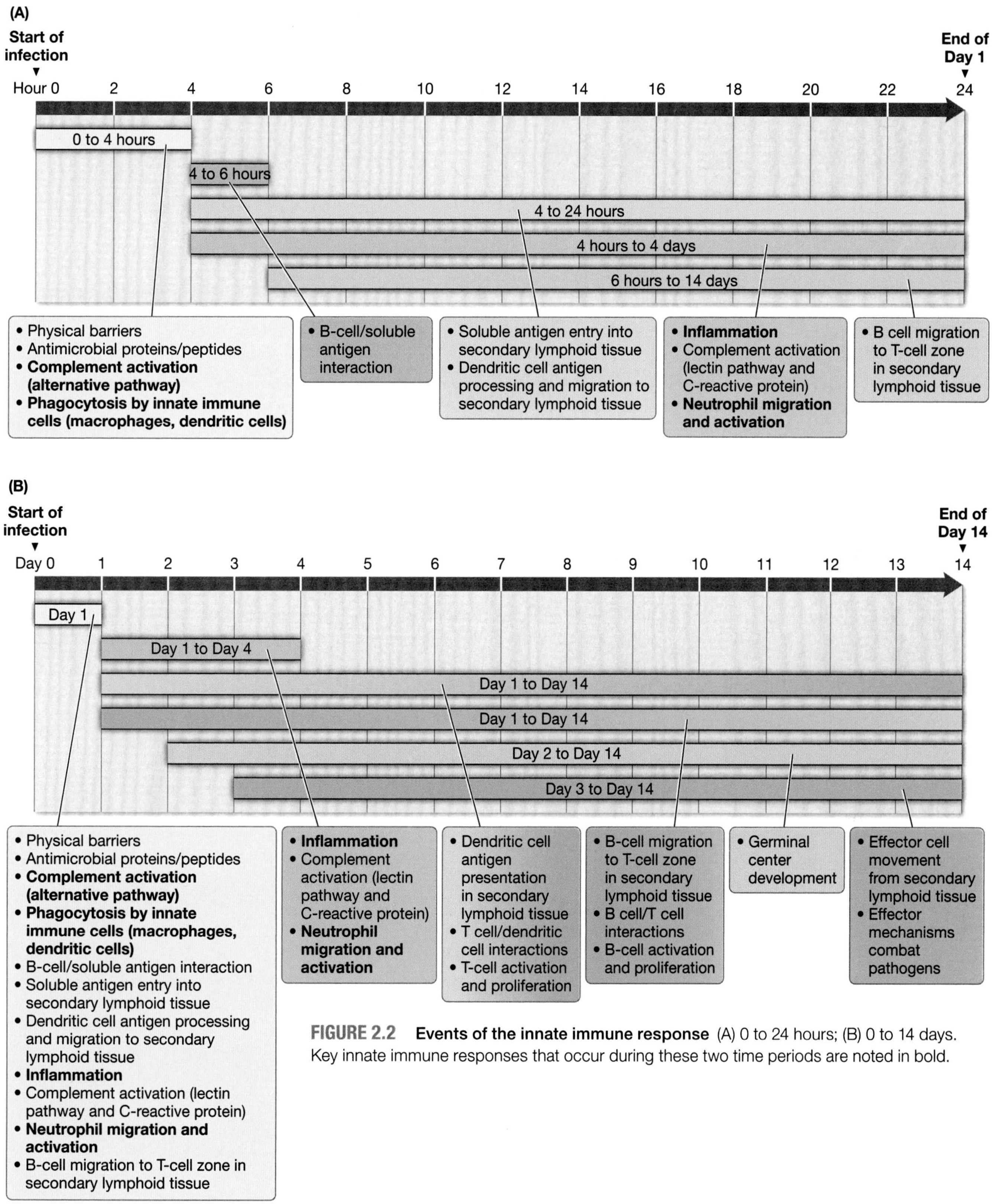

FIGURE 2.2 Events of the innate immune response (A) 0 to 24 hours; (B) 0 to 14 days. Key innate immune responses that occur during these two time periods are noted in bold.

that aids in pathogen recognition and destruction is the soluble proteins of the **complement** system. Many of these proteins are expressed as *zymogens*, or inactive precursors, that require activation through proteolytic cleavage (cleavage by a protease) to function. For example, once activated, the protein

complement System of soluble plasma proteins that act to opsonize and lyse pathogens; also known as *complement system*.

complement component 3 (C3) Protein component of the complement system that is cleaved during complement activation to produce C3a (a chemokine) and C3b (an opsonin).

membrane-attack complex (MAC) Complex of complement proteins C5, C6, C7, C8, and C9 that work in concert to form holes in bacterial and eukaryotic membranes.

opsonin Protein product on a pathogen surface that marks the pathogen for phagocytosis by neutrophils and macrophages.

phagosome Membrane-bound vesicle created during the internalization of a pathogen during phagocytosis.

lysosome Organelle in eukaryotic cells that facilitates digestion of internalized material due to the presence of digestive enzymes and toxic substances.

phagolysosome Membrane-bound organelle that results from fusion of a lysosome with a phagosome; specializes in digestion of phagocytosed material.

complement component 3 (**C3**) can function as a tag, marking something as foreign to be removed by innate immune cells. Other components of the complement system form a **membrane-attack complex** (**MAC**), which can create pores in pathogen membranes, causing lysis of the pathogen. The complement system will be discussed in detail in Chapter 3.

Phagocytosis

Phagocytosis is a major effector mechanism employed by innate immune cells to destroy pathogens. This process of phagocytosis is initiated through pathogen recognition by a receptor on the surface of macrophages and neutrophils (**FIGURE 2.3**). These receptors can recognize either an **opsonin** (a molecule that tags a foreign substance for removal) or a pattern on the surface of the pathogen. This recognition triggers binding of the pathogen and, since most pathogens are small in size compared to macrophages and neutrophils, typically results in clustering of cell-surface receptors engaged with molecules on the pathogen surface (Figure 2.3, Step 1). Receptor clustering causes the pathogen to be ingested as a portion of the phagocyte cell membrane encloses it (Figure 2.3, Step 2). The membrane pinches off, forming a **phagosome**, a membrane-bound intracellular vesicle containing the pathogen (Figure 2.3, Step 3). The phagosome then matures and fuses with an organelle called a **lysosome** to become a **phagolysosome** (Figure 2.3, Step 4). This fusion brings the pathogen into contact with a low-pH environment, digestive enzymes, and other toxic substances within the lysosome that result in its destruction and digestion by the phagocytic cell (Figure 2.3, Step 5).

CHECKPOINT QUESTIONS

1. Which soluble protein families function in the innate immune system?
2. How do phagocytes recognize pathogens that need to be destroyed by phagocytosis?

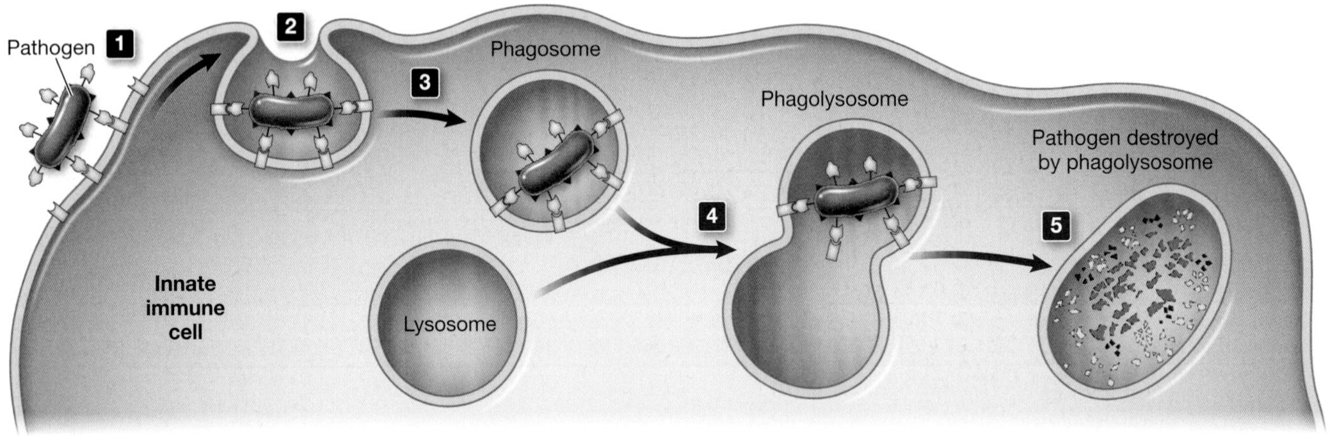

FIGURE 2.3 **Phagocytosis by a professional phagocytic cell** Step 1: Binding of a pathogen to a professional phagocyte occurs via receptors at the surface of the phagocyte with cell-surface components of the pathogen. Step 2: Clustering of receptors upon pathogen binding causes invagination of the phagocyte cell membrane. Step 3: The pathogen is internalized into a vesicle known as a phagosome. Step 4: The phagosome will fuse with a lysosome to form a phagolysosome. Step 5: The pathogen is digested and destroyed in the phagolysosome.

2.2 How are pathogens recognized by cellular receptors?

LEARNING OBJECTIVE

2.2.1 Name and describe the various types of innate immune cell receptors that recognize foreign pathogens.

The innate immune system employs a number of means to recognize *self* versus *nonself*, including a variety of cellular receptors known as **pattern recognition receptors** (**PRRs**). PRRs identify foreign molecules that commonly appear on microorganisms but that are lacking on host cells (**FIGURE 2.4**). These foreign molecules are referred to as **pathogen-associated molecular patterns** (**PAMPs**). Innate immune cells also employ receptors that bind to pathogens with opsonins on their surface.

PAMPs are typically carbohydrates and lipids on the surface of microorganisms. Families of receptors that recognize PAMPs include:

- *Toll-like receptor.* The **Toll-like receptor** (**TLR**) family includes various transmembrane proteins capable of recognizing a variety of PAMPs, including lipopolysaccharide from gram-negative bacteria, lipoteichoic acid from gram-positive bacteria, single- and double-stranded RNA, and other bacterial proteins. Upon recognition of their cognate PAMP, TLR signaling induces activation of responses directly connected with the targeting and destruction of the recognized pathogen (discussed in detail in section 2.5).
- *Lectin.* The **lectin** family of PRRs encompasses receptors that bind to carbohydrates. An example is the mannose receptor—also known as CD206. The CD nomenclature used here and throughout this text refers to **cluster of differentiation** (**CD**), a mechanism for the identification of cell-surface molecules in white blood cells. CD206 recognizes sulfated sugars and polysaccharides ending in D-mannose, L-fucose, or *N*-acetylglucosamine (common sugars in pathogen polysaccharides). Binding of lectin PRRs

pattern recognition receptor (PRR) Proteins on the surface of innate immune cells that are capable of recognizing molecules known as PAMPs.

pathogen-associated molecular pattern (PAMP) Molecules commonly seen on the surfaces of microorganisms that have been conserved evolutionarily but are absent on the surface of eukaryotic cells; used by innate immune cells to recognize foreign microorganisms.

Toll-like receptor (TLR) Innate immune cell receptor family capable of recognizing a variety of different microbial products and activating key signaling cascades to induce an innate immune response.

lectin Molecule capable of binding to carbohydrates.

cluster of differentiation (CD) System used in the nomenclature of different cell-surface molecules in white blood cells.

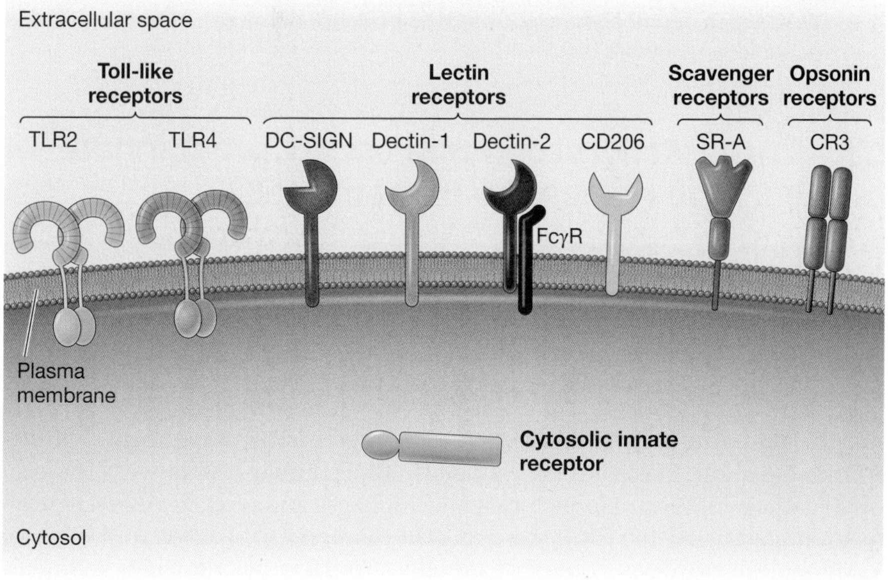

FIGURE 2.4 Common innate immune cell receptors that recognize PAMPs and opsonins Innate immune cells use four major families of receptors (Toll-like receptors, lectin receptors, scavenger receptors, and cytosolic innate receptors) to recognize a variety of PAMPs and one family (opsonin receptors) to recognize tagged pathogens. Binding of ligand to these receptors induces effector mechanisms of the innate immune system. (After M. S. Lionakis and M. G. Netea. 2013. *PLOS Pathog* 9: e1003079.)

scavenger receptor Innate immune cell receptor that binds to negatively charged molecules at the surface of microorganisms.

lipopolysaccharide (LPS) Endotoxin that is a major component of cell walls in gram-negative bacteria.

cytosolic innate receptor Intracellular receptor capable of recognizing PAMPs that may be present in the cytosol of cells.

opsonin receptor Protein capable of recognizing an opsonin on a foreign pathogen or molecule.

Fc receptor Receptor on the surface of an innate immune cell capable of recognizing the Fc component (tail region) of an immunoglobulin.

to their cognate PAMP prompts the white blood cell to phagocytose the bound PAMP-containing microorganism.

- *Scavenger receptor.* The **scavenger receptor** family includes receptors on macrophages such as SR-A and SR-B, which have affinity for many different ligands (binding partners for a receptor) that are negatively charged, including sulfated polysaccharides and lipoteichoic acid present in gram-positive bacterial cell walls. Some scavenger receptors also bind to **lipopolysaccharide** (**LPS**), a major component of the cell wall of gram-negative organisms, and to other cell-surface structures on bacterial and eukaryotic pathogens. Scavenger receptors function in a manner similar to that of lectin receptors. Binding of scavenger receptors to their cognate PAMPs induces phagocytosis of the microorganism.
- *Cytosolic innate receptor.* The **cytosolic innate receptor** family is capable of recognizing intracellular PAMPs, such as cytosolic nucleic acids—like DNA and double-stranded RNA, which are indicative of an intracellular bacterial or viral infection—and cytosolic signaling molecules produced by bacteria such as cyclic dinucleotides. One example of a cytosolic innate receptor that recognizes cytosolic nucleic acids is RIG-I, which recognizes the presence of viral RNA. Another family of cytosolic innate receptors includes the nucleotide-binding oligomerization domain (NOD)-like receptor family, which recognizes cytosolic cell wall components of intracellular bacteria. Recognition of intracellular PAMPs by this family of receptors activates pathways that inhibit the growth of intracellular pathogens and recruit white blood cells to destroy the infected host cells.
- *Opsonin receptor.* The **opsonin receptor** family is responsible for recognizing pathogens or foreign molecules that have been tagged with opsonins in order to facilitate phagocytosis of that pathogen. These types of receptors include the complement receptors CR3 and CR4, and **Fc receptors**, which are capable of recognizing immunoglobulins bound to the surface of a pathogen or soluble foreign molecule.

CHECKPOINT QUESTIONS

1. What are the major classes of pattern recognition receptors?
2. What types of pathogen-associated molecular patterns can pattern recognition receptors bind?

2.3 Once PAMP recognition occurs, what innate immune responses are activated?

LEARNING OBJECTIVES

2.3.1 Explain the mechanism by which innate immune cells recognize and respond to pathogens.

2.3.2 Explain the important functions of cytokines in inducing innate immune responses.

Imagine that you've cut your finger, allowing entry of a pathogen via the wound. We know that when this happens, our immune system will kick into action. Innate immune cells, including macrophages and neutrophils, recognize the pathogen using the PRRs described in the previous section and bind to PAMP molecules. Many of the PRRs on the surface of macrophages and neutrophils are capable of both recognizing pathogens and inducing their destruction through phagocytosis. Typically, PRRs recognize repeated molecules on the pathogen surface and bind to these ligands, which results in clustering of the receptors on the phagocytic cell surface. Pathogen recognition at the site of infection also drives signaling through these PRRs to further promote activation of an effective immune response, including activating an inflammatory response and recruiting white blood cells to the site of infection.

cytokine Soluble protein product secreted by cells to aid the immune system by increasing inflammation, chemotaxis to sites of infection, or immune cell signaling and differentiation.

interleukin (IL) Cytokine secreted by white blood cells.

Receptor signaling

The clustering of PRRs on the surface of an innate immune cell signals the cell to initiate phagocytosis and also triggers signaling cascades and an inflammatory response. Engagement of the PRR causes activation of signaling pathways within the binding cell, along with activation of transcription factors. The activation of transcription factors in turn induces changes in gene expression that result in the production of signaling molecules and the recruitment of innate immune cells from the bloodstream.

Cytokines

Activation of signaling pathways upon engagement of PRRs in innate immune cells triggers the production and release of signaling molecules known as **cytokines**. Cytokines promote immune cell activation and function. These secreted molecules interact with cell-surface receptors on cells of the immune system and throughout the body. They help activate the pathways needed to promote both innate and adaptive immunity, as well as pathways involved in the inflammatory response. Some cytokines work to recruit other immune cells to the site of infection.

Cytokines can produce a variety of physiological effects, but they commonly induce cellular changes that lead to inflammation, fever, cell proliferation, or cellular activation (**TABLE 2.1**). Many cytokines are members of the **interleukin** (**IL**) family of proteins, which are involved in activating a variety of immune responses. A majority of immune cells in the body require an input of specific cytokines to elicit the proper immune response from that cell.

TABLE 2.1 | Classes of Cytokines

Cytokine	Secreting Cells	Function
Inflammatory Cytokines		
TNF-α	Macrophages	Inflammation, acute-phase response
IL-1	Macrophages Dendritic cells Fibroblasts	Inflammation, acute-phase response
Signaling Cytokines		
IL-2	T cells	T-cell activation
IL-4	Dendritic cells	B-cell activation
IL-10	Monocytes	Anti-inflammatory
IL-12	Dendritic cells	T_H1 helper T-cell activation
IL-17	T cells	Neutrophil activation
IFN-γ	T cells Macrophages NK cells	Macrophage activation
IFN-α/IFN-β	Macrophages Virally infected cells	Activation of NK cells, prevents viral replication
TGF-β	Regulatory T cells	Peripheral tolerance, T-cell inactivation

Sources: Data from W. P. Arend et al. 2008. *Immunol Rev* 223: 20–38; H. Fickenscher et al. 2002. *Trends Immunol* 23: 89–96; K. Gee et al. 2009. *Inflamm Allergy Drug Targets* 8: 40–52; L. L. Jones et al. 2011. *Immunol Res* 51: 5–14; M. Michaeu et al. 2003. *Cell* 114: 181–190; A. J. Sadler et al. 2008. *Nat Rev Immunol* 8: 559–556.

INFLAMMATORY CYTOKINES Inflammatory cytokines stimulate inflammation at the site of infection. This group is important to the innate immune response and includes IL-1 and tumor necrosis factor-α (TNF-α). Inflammatory cytokines act to increase local fluid accumulation in tissue, which causes swelling and pain due to cellular changes in capillaries that increase capillary diameter (dilation). Dilation of the blood vessel, along with greater permeability to cells and plasma fluids, brings innate immune cells from the circulation into the tissue at the site of infection. Thus, when we come into contact with a pathogen, our innate immune system induces an inflammatory response to allow circulating innate immune cells to migrate to the site of infection.

Inflammatory cytokines also promote changes in gene expression. These changes alter cell-surface molecule expression on local blood vessels, producing receptors on blood vessel endothelial cells that bind tightly to innate immune cells. As a result, immune cells in circulation can recognize a site of inflammation, exit the bloodstream, and migrate toward the infecting pathogen. Innate immune cells that migrate to the site of an inflammatory response include monocytes, which can differentiate into macrophages and dendritic cells at the site of infection, and neutrophils.

chemokine Soluble secreted protein that acts as an attractant molecule to promote cell migration to a specific area.

interferon (IFN) Cytokine that activates cells important in targeting viral infection.

Inflammation also promotes fluid leakage into a draining secondary lymphoid tissue, where activation of an adaptive immune response occurs, if required. Thus, the inflammatory response serves a role during both innate and adaptive immunity. If our innate immune response is not adequate to fight off the invading pathogen, the inflammatory response promotes delivery of pathogen components to the adaptive immune system for a targeted response. Finally, inflammatory cytokines can play a role in activating other arms of the innate immune system, including fever and the complement system (see Chapter 3).

CHEMOKINES Another family of cytokines produced by our innate immune system are **chemokines**, which are cytokines that recruit white blood cells such as neutrophils to the site of infection. Chemokines function as soluble chemoattractants, meaning that migrating cells move toward them. Chemokine release during an inflammatory response signals the location of the infection and drives the recruitment of white blood cells to that site to combat pathogens more effectively.

TABLE 2.2 | Toll-Like Receptors of the Immune System

Receptor	Ligand	Expressing Cells
TLR1	Bacterial lipoprotein	Macrophages Dendritic cells B cells
TLR2	Lipoteichoic acid Bacterial cell wall components	Macrophages Neutrophils Dendritic cells Mast cells
TLR3	Double-stranded RNA	Dendritic cells B cells
TLR4	LPS	Macrophages Neutrophils Dendritic cells Mast cells B cells
TLR5	Flagellin	Macrophages Dendritic cells
TLR6	Diacyl lipopeptides	Macrophages Mast cells B cells
TLR7	Single-stranded RNA	Macrophages Dendritic cells B cells
TLR8	Single-stranded RNA Phagocytosed bacterial RNA	Macrophages Dendritic cells Mast cells
TLR9	CpG oligodeoxynucleotide DNA	Macrophages Dendritic cells B cells
TLR10	Unknown	Unknown
TLR11/12	Profilin (from *Toxoplasma*)	Macrophages Dendritic cells
TLR13	Bacterial ribosomal RNA	Macrophages Dendritic cells

Source: K. Vijay. 2018. *Int Immunopharmacol* 59: 391–412.

ANTIVIRAL CYTOKINES Another family of cytokines is involved in the activation of cells that are important in combating viral infections. Viral infections are challenging to the immune system since they are typically an intracellular pathogen. Only cells with the capacity to patrol the body looking for intracellular pathogens can eliminate virally infected cells.

One innate immune cell that plays a key role in combating viral infections is the natural killer (NK) cell. NK cells are activated by a cytokine, IL-12, produced during an inflammatory response, and by a family of secreted molecules known as **interferons** (**IFN**), especially IFN-α and IFN-β. Interferons are secreted to activate NK cells when a viral infection is detected. Once activated, NK cells seek out the source of infection and target the infected cells for destruction.

CHECKPOINT QUESTION

1. What are the major actions of cytokines secreted during an innate immune response?

2.4 How do the Toll-like receptors function in pathogen recognition and cellular activation?

LEARNING OBJECTIVE

2.4.1 Illustrate the major signaling pathways used by Toll-like receptors that result in activation of transcription factors NFκB and IRF3.

TLRs play a vital role in the innate immune response, inducing key signaling events necessary for mounting a robust defense against pathogens. This family of receptors has been evolutionarily conserved (see **EVOLUTION AND IMMUNITY**).

TLRs are expressed on a variety of cells and are activated upon recognizing their microbial ligand (**TABLE 2.2**). These ligands are specific to microorganisms and are not expressed on human cells.

Making Connections: Pathogen Pattern Recognition and Innate Immune Cell Function

Innate immune cells employ a variety of pattern recognition receptors (PRRs) to detect different pathogens. Specific effector functions of these cells are determined by the detected pathogen.

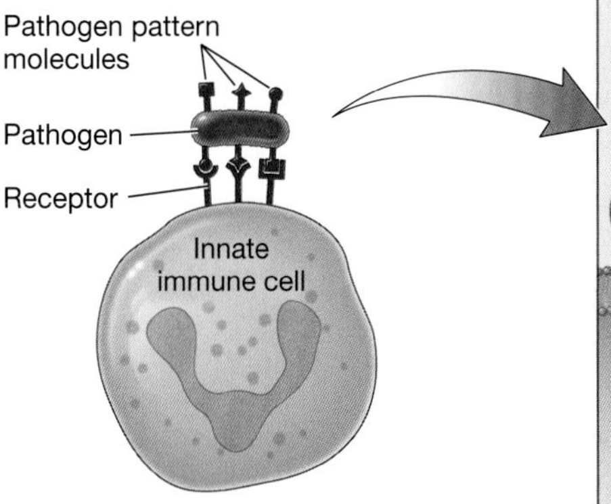

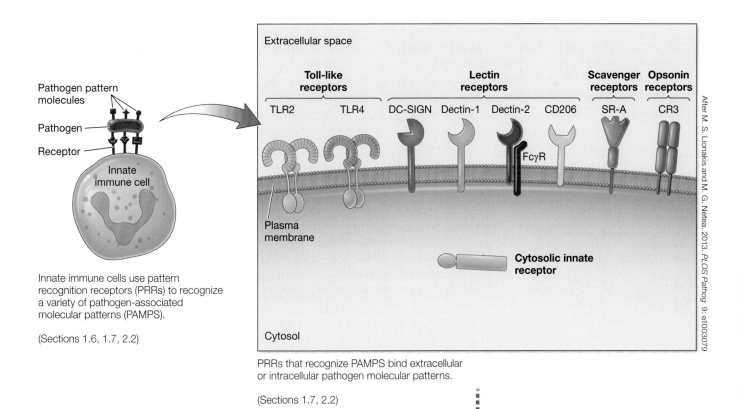

Innate immune cells use pattern recognition receptors (PRRs) to recognize a variety of pathogen-associated molecular patterns (PAMPS).

(Sections 1.6, 1.7, 2.2)

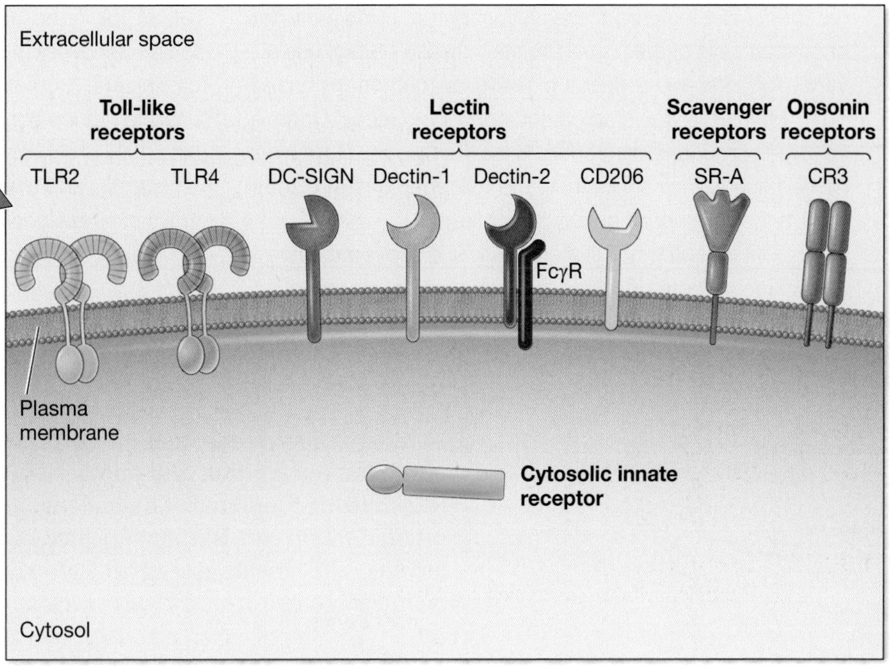

After M. S. Lionakis and M. G. Netea. 2013. *PLOS Pathog* 9: e1003079

PRRs that recognize PAMPS bind extracellular or intracellular pathogen molecular patterns.

(Sections 1.7, 2.2)

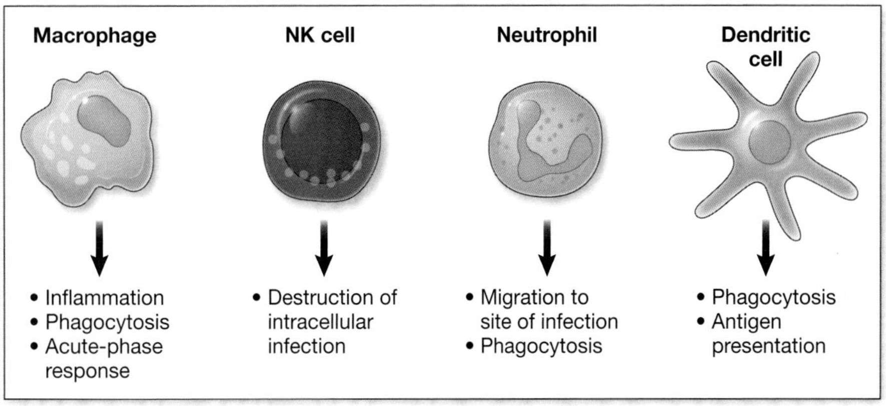

Upon recognition of PAMPs by PRRs, a variety of innate immune responses may be activated, including the activation of macrophages and an inflammatory response, activation of NK cells, recruitment and activation of neutrophils, and activation of dendritic cells, which bridge the innate and adaptive immune responses.

(Sections 1.6, 1.7, 2.2)

EVOLUTION AND IMMUNITY

Toll-Like Receptors

Since antimicrobial peptides are likely the most ancient host defense mechanism in evolutionary history, we can imagine that mechanisms used to drive antimicrobial peptide production (including recognizing microbes via cell surface receptors) also entered the evolutionary tree of life early. We know that cells of the human innate immune system use a variety of different receptors to recognize pathogens based on PAMPs. One well-studied and conserved receptor family capable of recognizing PAMPs is the TLR family. TLRs were named due to their similarity to the Toll receptor first discovered in the *Drosophila* genus of fruit flies.

The Toll receptor in fruit flies was first discovered as an important molecule involved in embryonic development and dorsal-ventral patterning. Later studies discovered that Toll mutations in adult flies led to an inability to produce the antimicrobial peptide drosomycin, which increased their susceptibility to infection. This discovery highlighted the role of Toll in the innate immune response in flies and led to the discovery of similar receptors, called TLRs, in other organisms. Studies on the signal transduction pathway used to induce gene expression upon ligand binding have also highlighted similarities between the Toll pathway and *Drosophila* and the TLRs in other organisms. A total of 13 TLRs have been identified in mammals (10 in humans) that are responsible for recognizing microorganisms, including bacteria, fungi, and viruses.

Most TLRs are transmembrane proteins. They contain a domain (region) for pathogen recognition and another on the opposite side of the membrane for transmitting the signal of ligand engagement (**FIGURE 2.5**). The pathogen recognition domain consists of a repeated motif of 20 to 30 amino acids that is rich in leucine. The repeating segments are known as leucine-rich repeats (LRRs). The overall domain forms a C or horseshoe shape.

Some TLRs recognize extracellular ligands and thus are located at the plasma membrane of human cells. Others recognize nucleic acids of pathogens; these reside in endosomal membranes, where they can detect released nucleic acids. Ligand recognition leads to changes in gene expression that result in the production and release of cytokines. TLRs sometimes require a cofactor (called a coreceptor) to assist with ligand recognition.

Toll-like receptor 4

The best-studied TLR is TLR4. TLR4 uses the coreceptor CD14 to recognize the gram-negative endotoxin LPS. Soluble LPS released from gram-negative organisms can also activate TLR4. This type of recognition requires another molecule, MD2, to promote binding (**FIGURE 2.6**). This interaction at the surface of cells is the initial step in a signal transduction cascade (signal transmission across the plasma membrane) that ultimately results in the production of cytokines. (See **KEY DISCOVERIES** to learn how scientists discovered the link between TLR4 and LPS.)

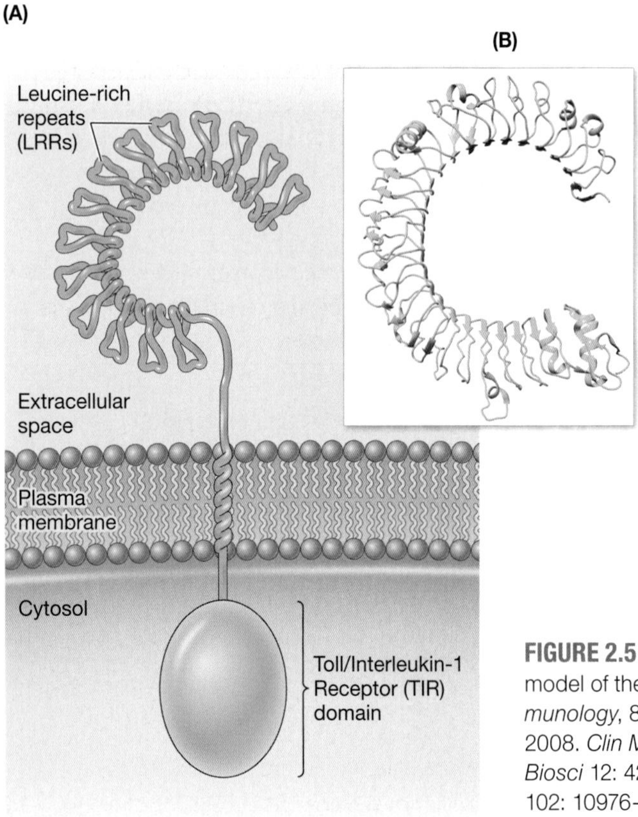

FIGURE 2.5 Toll-like receptor (TLR) (A) Illustration; (B) Three-dimensional model of the extracellular domain of TLR3. (A, after J. Punt et al. 2018. *Kuby Immunology*, 8th ed., W. H. Freeman and Company: New York; E. Vercammen et al. 2008. *Clin Microbiol Rev* 21: 13–25; B. M. Tesar and D. R. Goldstein. 2007. *Front Biosci* 12: 4221–4238; B, data from J. K. Bell et al. 2005. *Proc Natl Acad Sci USA* 102: 10976–10980. © 2005 National Academy of Sciences, U.S.A., redrawn using Chimera and PDB file 2A0Z.)

How was the receptor for the endotoxin lipopolysaccharide discovered?

Article

A. Potlorak, X. He, I. Smirnova, M. Y. Liu, C. V. Huffle, X. Du, D. Birdwell, E. Alejos, M. Silva, C. Galanos, M. Freudenberg, P. Ricciardi-Castagnoli, B. Layton, and B. Beutler. 1998. Defective LPS signaling in C3H/HeJ and C57BL/10ScCr mice: Mutations in *Tlr4* gene. *Science* 282: 2085–2088.

Background

One classic example of mechanisms of action of innate immune receptors is that of TLRs, which function to recognize different molecular patterns that pathogens express. TLR4 is responsible for recognizing the gram-negative endotoxin LPS. How was it determined that TLR4 is responsible for recognizing and responding to LPS?

Early Research

In the 1960s, mice of the strain C3H/HeJ were found to lack the ability to respond to the bacterial endotoxin LPS. The only known protein that recognized endotoxin at the time was CD14, which is a soluble protein that cannot interact with the cytoplasm. The authors hypothesized that for cells to properly respond, they must activate signaling pathways to alter cellular function in the presence of endotoxin. Genetic analysis determined that a codominant mutation at a single locus (named *Lps*) within the mice was responsible for the defect in endotoxin response.

Another mutation preventing an endotoxin response was identified in a second mouse strain, C57BL/10ScCr. This allele was demonstrated to be recessive, as crossing this strain with a control C57BL strain produced offspring that were able to respond to endotoxin. A cross between the two strains produced offspring also unresponsive to the endotoxin, indicating that a single allele was responsible for the lack of endotoxin response in both mutant strains.

Think About...

1. Considering the background information and early research, what would your hypothesis be on the function of the *Lps* gene and its protein product?
2. What experiment(s) would you need to conduct to test your hypothesis? Consider that at the time of this experiment, the only knowledge was that there was a single allele responsible for the phenotype. This allele was only mapped to a region on chromosome 4 of the mouse.
3. If you sequenced the *Lps* gene, what differences (if any) would you expect to see in the sequence within the mutant strains compared to a wild-type counterpart?
4. What experiments could you conduct to determine if the protein product of the *Lps* gene is responsible for recognizing endotoxin? Would information you obtained in experiments conducted to sequence the *Lps* gene be useful?

Article Summary

Through shotgun cloning (a method used to duplicate DNA) and sequencing of the region known to map to the *Lps* allele, the authors identified two genes (a portion of the *Pappa* locus and the entire *Tlr4* locus). Further analysis of the *Pappa* locus demonstrated that it is not expressed in macrophages or macrophage cell lines. This suggests that it is not expressed in cells responsible for responding to endotoxin. Thus, the authors focused their attention on *Tlr4*.

Further analysis of *Tlr4* demonstrated that there is a single-point mutation in the *Tlr4* gene in the C3H/HeJ mouse strain. Interestingly, the authors also found that no *Tlr4* mRNA is produced in the C57BL/10ScCr mice. These two findings led the authors to conclude that *Tlr4* and *Lps* were indeed the same allele and also explain the nature of the phenotypes with the two *Lps* alleles. The single-point mutation in C3H/HeJ mice must act in a dominant negative fashion to allow for the codominant phenotype seen in these mice. The mutation in C57BL/10ScCr mice must prevent the production of *Tlr4* mRNA (and, thus, protein). When crossed with a wild-type strain, the mutation would be masked by the other gene, thus requiring both mutations to see the phenotype.

Finally, the authors demonstrated that macrophages alter *Tlr4* mRNA production in response to LPS (**FIGURE KD 2.1**). Since *Tlr4* was mutated in these two mice strains incapable of responding to LPS, it can be inferred that normal *Tlr4* is responsible for LPS recognition.

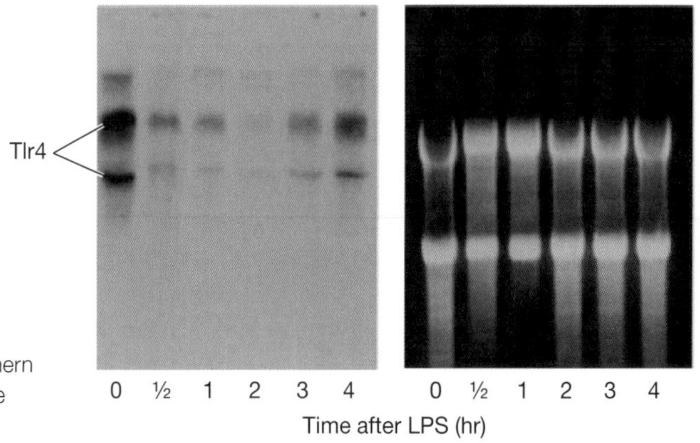

FIGURE KD 2.1 Lipopolysaccharide (LPS) treatment in a macrophage cell line lowers expression of TLR4 in a time-dependent manner. TLR4 mRNA levels were measured via Northern blot (left). Total mRNA levels were visualized via ethidium bromide staining (right).

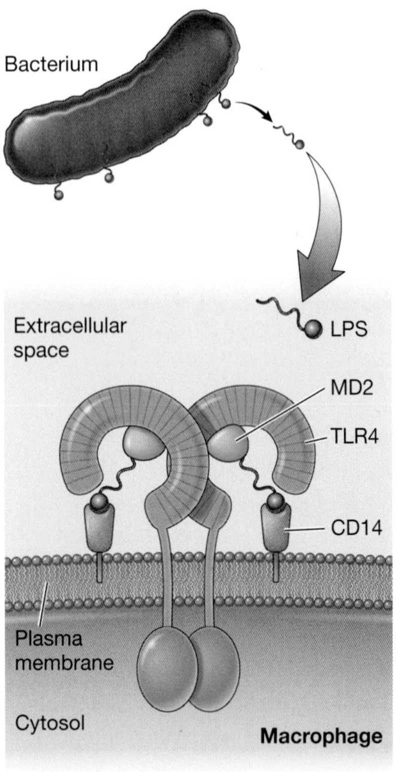

FIGURE 2.6 Toll-like receptor 4 (TLR4) recognition of lipopolysaccharide (LPS) TLR4 at the surface of macrophages can bind to LPS from gram-negative bacteria with the aid of the coreceptors MD2 and CD14.

TLR signaling pathway initiated by MyD88

Many TLRs signal through a pathway initiated by the adaptor protein MyD88. The cascade of events occurs primarily in the cytoplasm and results in the activation of the transcription factor **nuclear factor κB (NFκB)**, which plays a role in both innate and adaptive immunity. Activation of NFκB ultimately induces transcription of cytokine genes necessary for inflammation and immune response activation. Recognition of the gram-negative bacterial PAMP LPS by TLR4 drives signaling events to produce and release cytokines that stimulate an immune response against the gram-negative pathogen. This initial interaction at the white blood cell surface initiates a kinase cascade that ultimately results in destruction of an inhibitor of NFκB, allowing the transcription factor to be activated, inducing cytokine gene expression.

The activation of NFκB during TLR4 signaling begins with the binding of the extracellular TLR4 domain to LPS and the coreceptors MD2 and CD14 (**FIGURE 2.7**, Step 1). This interaction induces the cytoplasmic domain of TLR4, known as a Toll-interleukin receptor (TIR) domain, to interact with the TIR domain of the adaptor molecule MyD88 (Figure 2.7, Step 2). MyD88 functions as a bridge by interacting with a protein kinase called IRAK4, which phosphorylates (attaches a phosphate group) and activates another kinase, IRAK1 (Figure 2.7, Step 3).

Activation of protein kinases within cells is a common mechanism used to amplify a signaling response within the cell. These kinases can trigger altered activity in downstream targets. The kinase IRAK1 activates another protein, TRAF6 (Figure 2.7, Step 4), which results in the marking and proteolytic degradation of two other signaling molecules, NEMO and TAB (Figure 2.7, Step 5). Degradation of NEMO and TAB ultimately results in the activation of another protein kinase, TAK1 (Figure 2.7, Step 6).

Active TAK1 further activates another kinase, IκB kinase (IKK) (Figure 2.7, Step 7). IKK is responsible for phosphorylating the inhibitory IκB subunit of the IκB/NFκB complex, which is present in the cytoplasm (Figure 2.7, Step 8). When linked with IκB, NFκB is prevented from entering the nucleus and thus cannot function as a transcription factor. Phosphorylation of IκB targets it for destruction, freeing NFκB to move to the nucleus, where it can function in transcription to activate important cytokine genes (Figure 2.7, Step 9). These genes encode for inflammatory cytokines that activate cells at the site of infection and promote phagocyte migration.

TLR signaling pathway initiated by TRIF and TRAM

While some pathogens such as gram-negative bacteria induce an inflammatory response by activating NFκB, others require the secretion of different cytokines to promote an effective immune response. For instance, intracellular pathogens such as viruses are recognized by different PRRs and are managed by our immune system using different cytokines. Thus, signaling events through TLRs recognizing intracellular pathogens require activation of different transcription factors to drive expression of cytokines required to fight intracellular infections. Some TLRs that recognize intracellular pathogens signal through a pathway that culminates in the activation of transcription factors IRF3 and IRF7 (**FIGURE 2.8**). It is this pathway that comes into play in response to recognition of viral infections and results in the production of interferons. In this cascade, upon ligand association (Figure 2.8, Step 1), the cytoplasmic TIR domain of the TLR interacts with two adaptor proteins: Toll receptor-associated activator of interferon (TRIF) and Toll receptor-associated molecule (TRAM) (Figure 2.8, Step 2).

TRIF and TRAM activate TRAF3 (Figure 2.8, Step 3), a kinase that phosphorylates IRF3 and IRF7 (Figure 2.8, Step 4). Upon phosphorylation, the two transcription factors dimerize (join to form one compound) and move into the

nuclear factor κB (NFκB) Transcription factor that activates gene expression during both innate and adaptive immune responses; inactivated by binding to IκB in the cytoplasm.

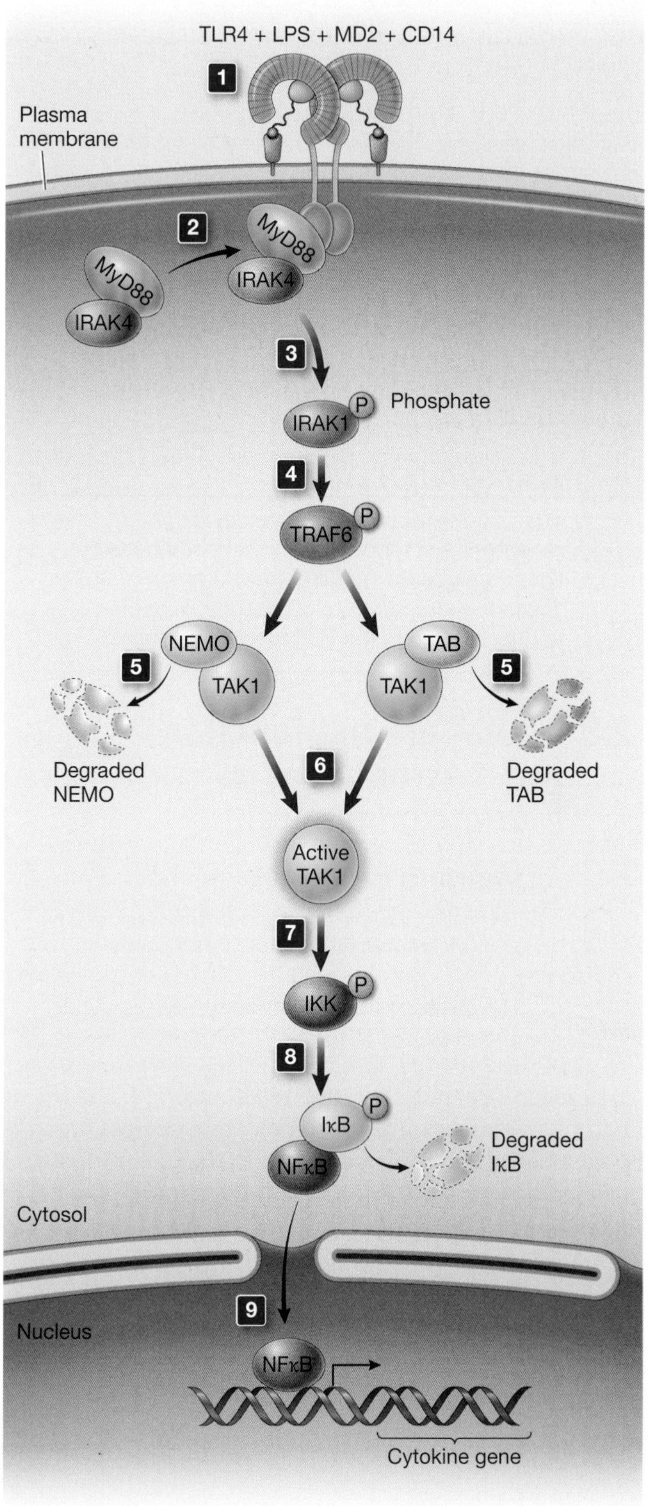

FIGURE 2.7 The TLR4 signaling pathway Step 1: TLR4 binds to LPS, MD2, and CD14. Step 2: Binding of TLR4 to LPS recruits MyD88 to the TLR4 TIR domain. Step 3: IRAK4 is recruited to MyD88 and phosphorylates and activates IRAK1. Step 4: Active IRAK1 activates TRAF6. Step 5: Active TRAF6 results in degradation of two proteins, TAB and NEMO. Step 6: Degradation of TAB and NEMO causes activation of another protein kinase, TAK1. Step 7: TAK1 phosphorylates and activates the IκB kinase (IKK) complex. Step 8: Active IKK phosphorylates IκB, which marks it for degradation. Step 9: With IκB degraded, the transcription factor NFκB moves from the cytosol to the nucleus, where it activates transcription of genes involved in inflammation.

nucleus (Figure 2.8, Step 5). Within the nucleus, they activate gene expression of antiviral cytokines, typically interferon-α (IFN-α) and interferon-β (IFN-β) (Figure 2.8, Step 6). These two cytokines activate cells involved in a viral response. TLRs that signal through the IRF3 and IRF7 transcription factors are activated,

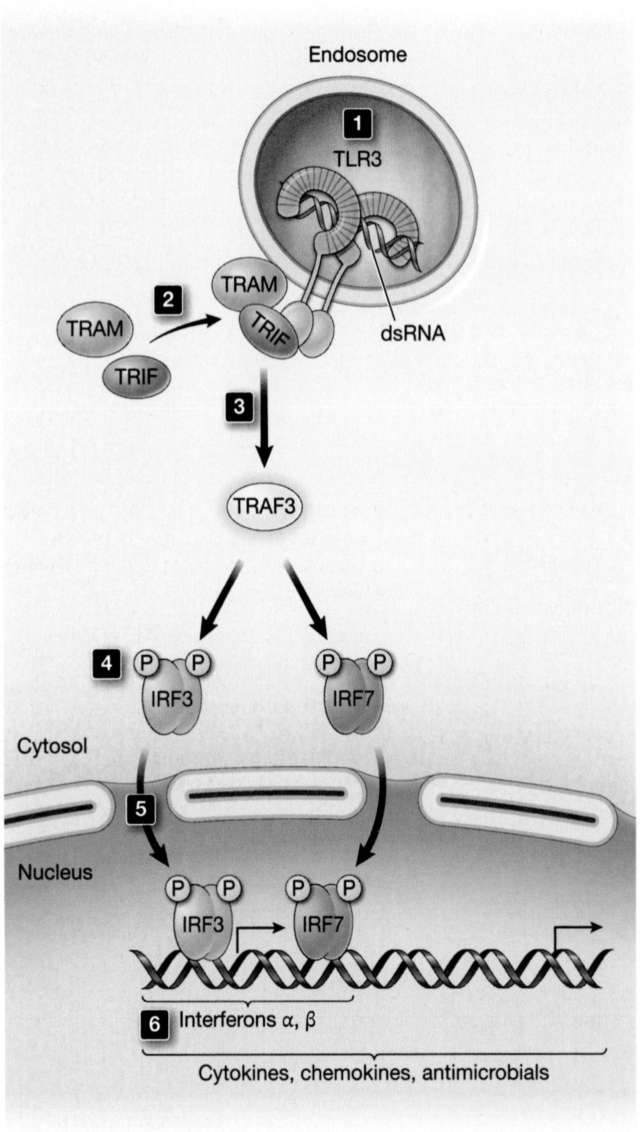

FIGURE 2.8 Illustration of TLR signaling to produce interferons Step 1: The TLR receptor TLR3 binds to viral double-stranded (ds) RNA. Step 2: Binding of the ligand allows the recruitment of TRIF and TRAM to TLR3. Step 3: TRIF and TRAM activate the protein kinase TRAF3. Step 4: TRAF3 phosphorylates and activates the transcription factors IRF3 and IRF7. Step 5: When phosphorylated, IRF3 and IRF7 localize to the nucleus. Step 6: Within the nucleus, IRF3 and IRF7 activate genes involved in defense against viral infections, including the genes that encode interferons.

in part, based on their location within a cell. TLRs that signal through this pathway are typically located at the endosomal membrane, where they are more likely to interact with viral particles.

CHECKPOINT QUESTION

1. The Toll-like receptor (TLR) family contains receptors that recognize a variety of different ligands, but each member has common features within the structure of the receptor. What features are shared within the TLR family of proteins?

2.5 What are the functions of macrophages in innate immunity?

LEARNING OBJECTIVE

2.5.1 Describe the effector mechanisms used by macrophages to combat pathogens.

Macrophages engulf and digest pathogens at a site of infection through the process of phagocytosis, as previously described. These cells possess a variety of surface receptors that allow for recognition of PAMPs on pathogen surfaces. They also possess receptors for components of the complement system that act as opsonins after activation. In addition, macrophages express TLRs that enable changes in gene expression, through activation of NFκB.

Key inflammatory cytokines expressed by macrophages

Upon activation of PRRs that drive cytokine production, macrophages secrete inflammatory cytokines that play a vital role in promoting migration of innate immune cells toward the site of infection. Key cytokines expressed by macrophages upon activation of TLR signaling pathways include IL-1, IL-6, IL-12, TNF-α, and CXCL8. These all play key roles in enhancing the innate immune response (**FIGURE 2.9**).

IL-1 and TNF-α induce both fever and an inflammatory response. The fever helps to inhibit bacterial and viral growth and promotes activation of the adaptive immune response, if needed. The inflammatory response increases vascular permeability, plasma proteins, and innate immune cells at the site of infection. It also drives the expression of cell-surface molecules on endothelial cells of blood vessels that aid in neutrophil migration to the site of infection. Vascular permeability allows for fluid drainage into nearby secondary lymphoid tissue as a means to activate an adaptive immune response, if needed.

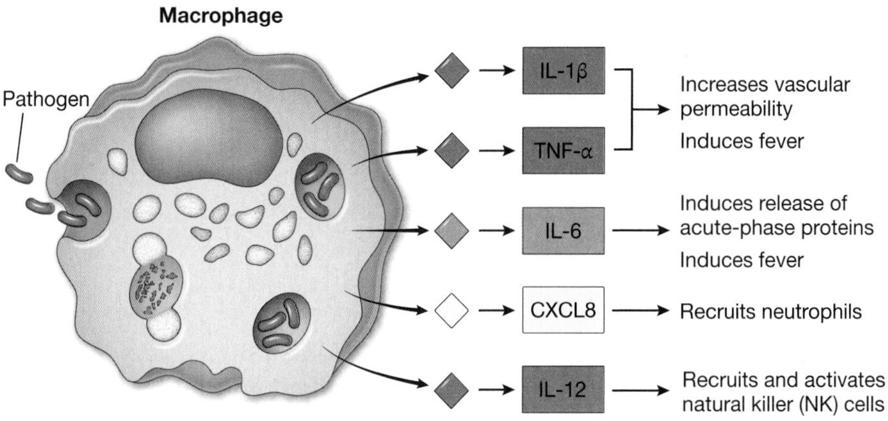

FIGURE 2.9 Cytokines produced by activated macrophages and their actions Signaling events within a macrophage that occur after pathogen recognition result in the production and secretion of a variety of cytokines and chemoattractants. Each cytokine or chemoattractant acts on a specific set of cells within the immune system and will ultimately drive an inflammatory response at the site of pathogen recognition.

IL-6 is also capable of inducing fever, inhibiting pathogen growth, and aiding in the adaptive immune response in a manner similar to IL-1 and TNF-α. IL-6 activates hepatocytes (liver cells) to produce acute-phase response proteins responsible for the activation of complement. Macrophage secretion of IL-6 thus enhances the innate immune response by increasing the levels of important complement proteins within the plasma.

Finally, macrophages also secrete CXCL8, a chemokine, which is a type of cytokine that attracts neutrophils and basophils to the site of infection.

inflammasome Complex of proteins that interact to produce a means to efficiently activate caspase-1 and allow efficient activation of IL-1.

Activation of IL-1 by the inflammasome

The inflammatory cytokine IL-1 requires activation before it can induce fever and an inflammatory response since it is produced as an inactive precursor, proIL-1. Thus, a protease inside macrophages must cleave the inactive proIL-1. The protease responsible for cleavage of proIL-1 is caspase-1, which is also produced as an inactive precursor.

The activation of IL-1 through cleavage by caspase-1 is greatly enhanced by the formation of an inflammatory complex known as the **inflammasome**. An inflammasome is composed of a cytosolic innate receptor of the NOD-like receptor (NLR) family (most commonly NLRP3) and a protein known as PYCARD, or ASC (for *a*poptosis-associated *s*peck-like protein containing a CARD; CARD stands for *c*aspase *a*ctivation and *r*ecruitment *d*omain). Assembly of the inflammasome is facilitated by intracellular potassium ions, intracellular viruses, or intracellular bacteria (**FIGURE 2.10**, Step 1). The cytosolic innate receptor component of the inflammasome is capable of recognizing these intracellular changes. The inflammasome can recruit procaspase-1 to activate it and produce active caspase-1 (Figure 2.10, Step 2). Active caspase-1 then activates proIL-1 (Figure 2.10, Step 3).

CHECKPOINT QUESTIONS

1. What key signaling molecules responsible for aiding the innate immune response are secreted by activated macrophages?
2. What are the main functions of each of these signaling molecules?

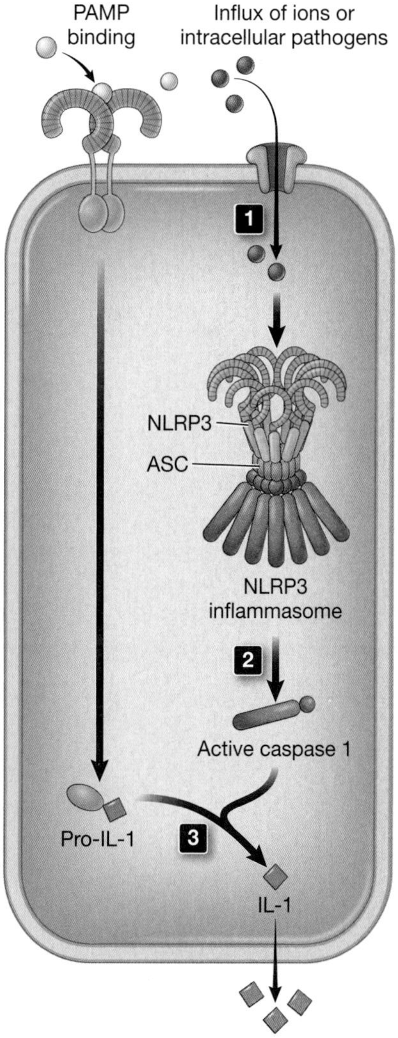

FIGURE 2.10 Inflammasome and the activation of IL-1 PAMP binding to the appropriate pattern recognition receptor results in the production of pro-IL-1, which must be cleaved by caspase-1 to be active. Step 1: The presence of intracellular pathogens or an influx of ions prompts formation of the inflammasome. Step 2: The inflammasome activates caspase-1. Step 3: Active caspase-1 cleaves pro-IL-1 to produce active IL-1. (After K. Schroder and J. Tschopp. 2010. *Cell* 140: 821–832.)

2.6 What are the functions of neutrophils in innate immunity?

LEARNING OBJECTIVES

2.6.1 Explain the process of neutrophil migration to a site of infection.

2.6.2 Describe the effector mechanisms used by neutrophils to combat pathogens.

adhesion molecule Receptor on the surfaces of two interacting cells that aids in cellular contact.

selectin Cell-adhesion molecule that binds to specific carbohydrates on another cell-adhesion molecule.

glycoprotein Cell-adhesion molecule containing specific carbohydrates that binds to selectin.

integrin Cell-surface glycoprotein receptor important in mediating cell-to-cell adhesion interactions.

intercellular adhesion molecule (ICAM) Cell-adhesion molecule that mediates cell-to-cell contact, commonly with integrins.

Neutrophils play a vital role in pathogen uptake and destruction. Like macrophages, they possess receptors that recognize PAMPs on pathogen surfaces. Neutrophils express CR3 and CR4, two receptors that can recognize pathogens marked with complement on their surface. Upon binding of any of these receptors, neutrophils initiate phagocytosis.

In comparison to macrophages, neutrophils are more mobile; they make up about 50% of the circulating white blood cells. Neutrophils are very short-lived. Upon targeting a pathogen, they act quickly under the anaerobic conditions of damaged tissue and then die within a few hours. Dead neutrophils are a primary component of the pus formed by an infected wound.

Neutrophil migration

The movement of neutrophils to a site of infection requires the action of **adhesion molecules**, proteins that hold cells together. These molecules are a pair of complementary receptors: one is present on the neutrophil, and its "partner" is located on various types of tissue, including endothelial cells. Four types of adhesion molecules (**FIGURE 2.11**) mediate cell-to-cell contact:

- **Selectins** bind to specific carbohydrate groups present on the neutrophil.
- **Glycoproteins** (proteins with attached carbohydrates) bind to a complementary selectin.
- **Integrins** bind to other proteins; they are involved in cell-to-cell and cell-to-substrate adhesion.
- **Intercellular adhesion molecules** (**ICAMs**) bind mainly with integrins.

During a macrophage response, CXCL8 is produced in response to receptor signaling and NFκB activation. This chemoattractant binds to receptors on the surface of neutrophils to promote migration to the secreted chemokines. Neutrophils express the receptors CXCR1 and CXCR2, which recognize the CXCL8 chemokine. Recognition of CXCL8 produces changes in the neutrophil that allow it to better adhere to blood vessels and to exit the bloodstream and travel to the site of infection.

Upon binding to the chemokines, neutrophils induce changes at their cell surfaces that increase the number of adhesion molecules. The inflammatory response increases the number of adhesion molecules on vascular endothelial cells to promote interaction with neutrophils.

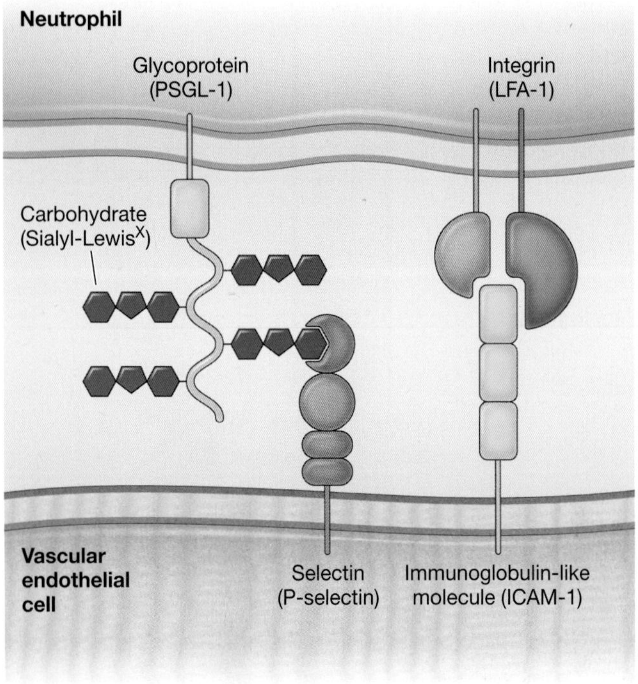

FIGURE 2.11 Cell–cell adhesion molecules facilitate binding of neutrophils to vascular endothelial cells Vascular endothelial cells express selectins such as P-selectin at their surface, which bind to glycoproteins such as PSGL-1 containing sialyl-Lewis^X. Likewise, immunoglobulin-like molecules such as ICAM-1 at the endothelial cell surface bind to the neutrophil cell-surface integrin LFA-1. (After P. Parham. 2014. *The Immune System*, 4th ed., W. W. Norton and Company.)

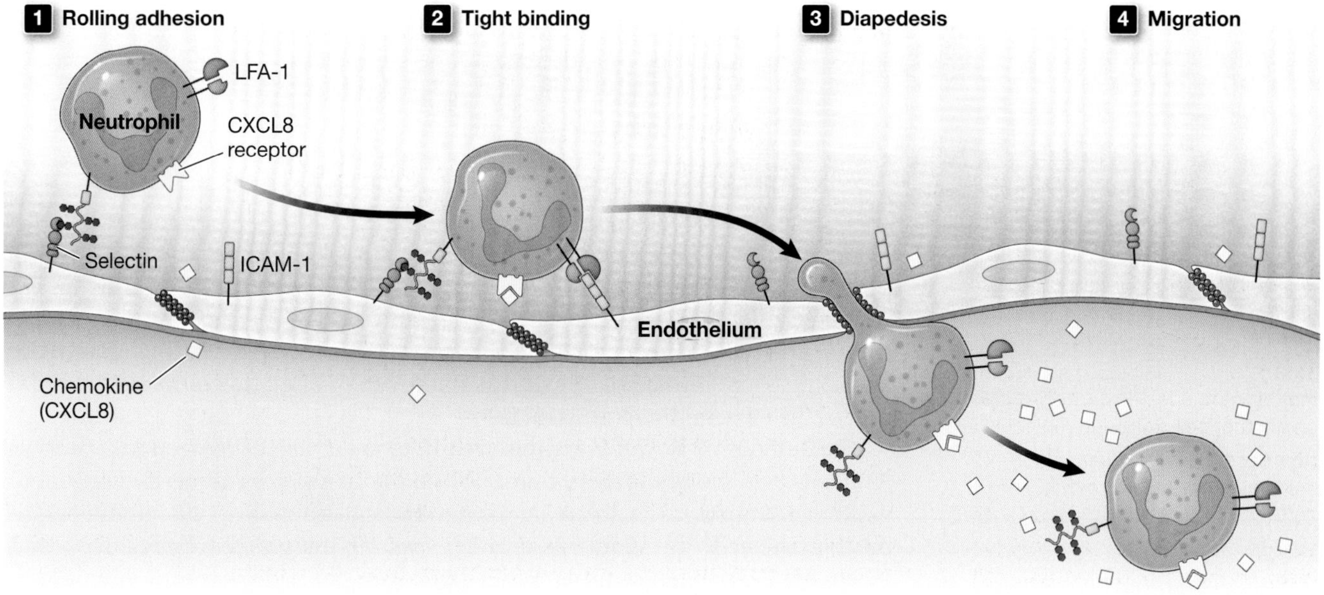

FIGURE 2.12 Extravasation of neutrophils Neutrophils within the bloodstream must be able to recognize a potential site of infection and exit the circulatory system to migrate to that site. The inflammatory response caused by pathogen recognition prompts neutrophil extravasation to the site of infection. Neutrophils first use surface carbohydrates to loosely associate with selectins on vascular endothelial cells and roll along the endothelial cell layer until they are close to the site of inflammation. Changes to the vascular endothelial cells allow for neutrophils to tightly bind through an interaction between LFA-1 and ICAM-1. Neutrophils then move through the vascular endothelial cells (diapedesis). Neutrophils that have crossed the endothelial cell layer migrate to a site of inflammation, where chemoattractants such as CXCL8 are being produced.

Migration of neutrophils to the site of infection is a process called **extravasation**, which occurs in four steps (**FIGURE 2.12**):

1. *Rolling adhesion* is mediated by selectins on the endothelial cells and carbohydrate chains in glycoproteins and glycolipids on the neutrophil surface. The carbohydrate in both these macromolecules contains an abundance of sialic acid and is referred to as sialyl-Lewis^{X}. These interactions are relatively weak, which allows the neutrophil to adhere to the lining of a blood vessel and roll slowly along its surface.
2. *Tight binding* is induced by the presence of CXCL8 and the inflammatory mediator TNF-α. The inflammatory response induces the presence of adhesion molecules such as ICAM-1 on the surface of vascular endothelial cells. These, in turn, interact with adhesion molecules on the neutrophil surface, especially LFA-1 and CR3. In the presence of CXCL8, LFA-1 and CR3 increase their adhesion power so they can tightly bind with ICAM-1.
3. *Diapedesis* is the third step in neutrophil migration. **Diapedesis** involves the movement of the neutrophil between two vascular endothelial cells. The process is mediated by LFA-1 and CR3. As the neutrophil moves across the endothelial wall, it eventually comes in contact with extracellular matrix proteins of the basement membrane, which it degrades by secreting proteases into the extracellular matrix.
4. *Migration* occurs as the gradient of CXCL8 directs the movement of neutrophils via chemotaxis toward the site of infection.

extravasation Process of migration of neutrophils from the circulation to a site of infection.

diapedesis Mechanism by which a leukocyte squeezes between two endothelial cells.

Effector mechanisms

Upon phagocytosis of a pathogen, the phagosome fuses with granules present in the neutrophil. These granules are filled with substances that target the

respiratory burst Process of oxygen consumption due to the action of NADPH oxidase.

pathogen for destruction, including proteases and antimicrobial proteins and peptides such as lysozyme, defensins, and lactoferrin.

A key protein contained in the granules is NADPH oxidase, which is present at the plasma membrane and in phagosomes of neutrophils. NADPH oxidase is an enzyme consisting of five oxidase subunits (gp91^{phox}, p22^{phox}, p40^{phox}, p47^{phox}, and p67^{phox}) that produces superoxide radicals when oxygen is present. These superoxide radicals are converted to hydrogen peroxide by superoxide dismutase. These reactions raise the pH of phagosomes as they consume protons. The pH level rises to between 7.8 and 8.0, which is the optimal pH at which other antimicrobial proteins and peptides act. The phagosome matures further and fuses with a lysosome to become a phagolysosome. Enzymes of the lysosome come into contact with the pathogen, leading to complete degradation of the engulfed microorganism.

The activity of NADPH oxidase requires rapid use of oxygen in a process referred to as the **respiratory burst**. While the respiratory burst is essential to neutrophil function, it can create oxygen radicals capable of damaging neighboring tissue. To prevent this damage, certain enzymes are produced that inactivate the oxygen radicals. While hydrogen peroxide is commonly used in solution as an antiseptic and bleaching agent, high concentrations can cause tissue damage. Following the respiratory burst, the enzyme catalase effectively clears the hydrogen peroxide produced by the earlier reactions by converting it to water and oxygen.

Another effector mechanism used by neutrophils to combat pathogens is the secretion of neutrophil extracellular traps (NETs). NETs are extracellular fibers consisting of neutrophil DNA and components of antimicrobial proteins from neutrophil granules, including neutrophil elastase. The function of NETs is to increase localized destruction of pathogens at the site of infection without requiring the neutrophils to phagocytose the pathogen.

Once a neutrophil has "used up" its granules, it dies by apoptosis and is devoured by macrophages in the vicinity.

CHECKPOINT QUESTIONS

1. What are the four steps of neutrophil migration to a site of infection?
2. What effector mechanisms are used by neutrophils to kill phagocytosed pathogens?

2.7 What are the functions of natural killer cells in innate immunity?

LEARNING OBJECTIVE

2.7.1 Describe the effector mechanisms used by natural killer cells to combat intracellular pathogens.

While macrophages and neutrophils are efficient in recognizing and clearing extracellular pathogens, they are not effective against intracellular organisms such as viruses. How then does the immune system prevent viral growth and spread? In this case, NK cells, large lymphocytes filled with cytotoxic granules, step up and play a central role.

The process begins when TLRs or cytosolic innate receptors such as RIG-I sense viral molecules. Subsequent signal transduction results in the expression of interferons, especially IFN-α and IFN-β. These cytokines are potent inducers of NK cell proliferation and also function to prevent viral replication (**FIGURE 2.13**). They activate interferon-stimulated genes that encode:

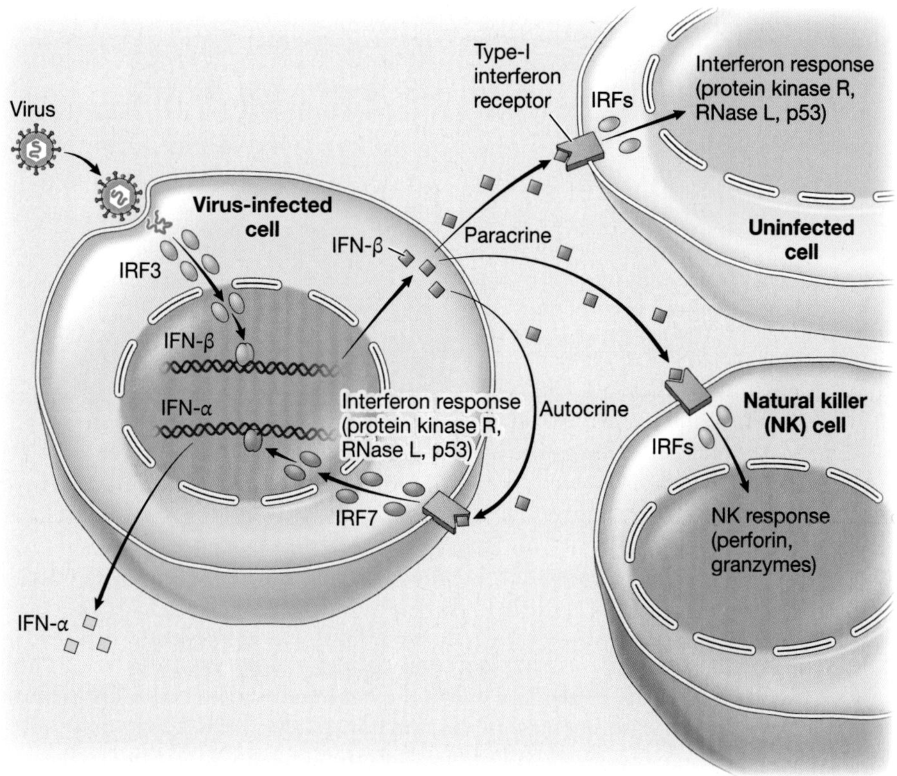

FIGURE 2.13 Actions of type-I interferons Recognition of viral particles within an infected cell activates the expression and release of type-I interferons such as IFN-α and IFN-β. These cytokines can bind to a receptor on the same cell (autocrine signaling) to limit the viral infection. Type-I interferons can also bind to receptors on adjacent cells (paracrine signaling), including uninfected cells and natural killer (NK) cells. Type-I interferon signaling in uninfected cells triggers protective mechanisms against viral spread. Type-I interferon signaling in NK cells activates the cells, which then destroy infected cells via the actions of perforin and granzymes.

- Protein kinase R—which inhibits protein synthesis
- RNAse L—which destroys RNA in the cell; RNA is a typical component of many viral genomes.
- p53—which can induce apoptosis within cells stimulated by IFN-α and IFN-β

In addition, IFN-α and IFN-β can induce the expression of receptors that can signal to NK cells the presence of an intracellular virus. NK cells are also activated by the inflammatory cytokines IL-12 and TNF-α. Upon activation, NK cells target infected cells, resulting in their death.

Recognizing intracellular infection

NK cells employ a unique method to recognize an intracellular infection. Unlike macrophages and neutrophils, which bear receptors that bind to PAMPS, NK cells monitor the exterior of host cells to learn what's going on inside. To do this, NK cells rely on two broad families of receptors, *inhibitory* and *activating*. Both types sense the cell-surface environment, looking for markers of stress or infection. Within these two larger families, NK cell receptors can be further divided into two structurally distinct types: killer-cell immunoglobulin-like receptors (KIRs) and lectin-like receptors (TABLE 2.3).

TABLE 2.3 | NK Cell Receptor Types and Structural Classes

Receptor Type	Structural Class
Activating receptors	
Ly49[a]	Lectin-like receptor
CD94/NKG2[b]	Lectin-like receptor
NCR	Immunoglobulin-like receptor
CD16/FcγIIIA	Immunoglobulin-like receptor
Inhibitory receptors	
Killer-cell immunoglobulin-like receptors	Immunoglobulin-like receptor
CD94/NKG2[b]	Lectin-like receptor
Leukocyte inhibitory receptors (LIR)	Immunoglobulin-like receptor
Ly49[a]	Lectin-like receptor

[a] Ly49 can act as both an activating and inhibitory receptor, depending on the isoform.

[b] CD94/NKG2 can act as both an activating and inhibitory receptor, depending on the isoform.

Source: H. J. Pegram et al. 2011. *Immunol Cell Biol* 89: 216–224.

perforin Protein expressed by cytotoxic T cells that functions to perforate small holes in target cell membranes in order to deliver cytotoxins and induce apoptosis.

granzyme Enzymes produced by cytotoxic innate and adaptive immune cells that promote the induction of apoptosis to limit the presence of an intracellular infection.

INHIBITORY RECEPTORS Inhibitory receptors on NK cells bind to all cell types and recognize a variety of molecules. Most often, inhibitory receptors detect cell-surface molecules that are normally present. These molecules play an important role in processing and presenting intracellular proteins at the cell surface. Many viruses alter the presentation pathway and decrease the amount of these proteins. When fewer cell-surface proteins are present, the inhibitory receptor can't interact with the cell it is monitoring. Thus, lack of interaction indicates an abnormal cell. Healthy cells generally have higher numbers of these cell-surface proteins, allowing for interaction with the inhibitory receptors. Recognition of cells through inhibitory receptors prevents NK cells from targeting normal cells for destruction, promoting self-tolerance (**FIGURE 2.14**).

ACTIVATING RECEPTORS To fully activate the effector functions of an NK cell, the cell must engage with a target cell through an activating receptor. A common activating receptor is the protein NKG2D, which binds to two ligands on the surface of target cells: MIC-A and MIC-B. These proteins are only expressed by cells in response to stress (see Figure 2.14).

Going in for the kill

When the NKG2D of an NK cell interacts with an MIC protein on a target cell, the NK cell activates, releasing the cytotoxic contents of the granules toward the infected cell (see Figure 2.14). The cytotoxic contents within granules include molecules such as **perforin**, which can form small holes within the plasma membrane of the target cell, and **granzymes**, which can be transported into the target cell to initiate apoptosis. The inhibitory and activating receptors work synergistically to ensure that only stressed and altered cells are targeted for destruction. The altered cells may be infected with an intracellular pathogen, or they have been mutated in some way (as occurs with certain cancers). NK cells have actually been tested as a therapy for targeting tumors (see **EMERGING SCIENCE**).

CHECKPOINT QUESTION

1. Describe how natural killer cells can detect an intracellular infection.

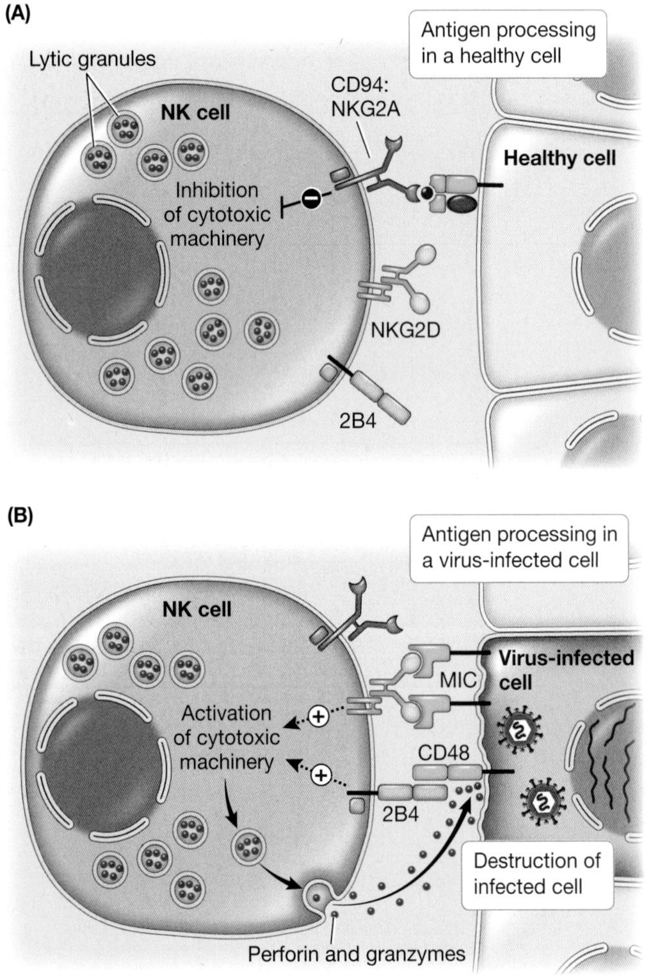

FIGURE 2.14 Inhibitory and activating receptors of NK cells and target cells (A) In a healthy cell, antigen processing is not affected. When an NK cell monitors the surface of a healthy cell, inhibitory receptors such as CD94:NKG2A bind to cell-surface molecules that are present because of proper antigen processing. This interaction shuts down NK activation and prevents destruction of healthy cells. (B) Virus-infected cells commonly alter antigen processing to hide from the immune system. This altered antigen processing can be recognized by a monitoring NK cell through activating receptors such as NKG2D and 2B4. Binding of these activating receptors to the virus-infected cell activates the NK cell, resulting in release of perforin and granzymes and destruction of the infected cell. (After P. Parham. 2014. *The Immune System*, 4th ed., W. W. Norton and Company.)

EMERGING SCIENCE

Can NK cells be used as an effective immunotherapy to fight cancer?

Article

T. Tonn, D. Schwabe, H. G. Klingemann, S. Becker, R. Esser, U. Koehl, M. Suttorp, E. Seifried, O. G. Ottmann, and G. Bug. 2013. Treatment of patients with advanced cancer with the natural killer cell line NK-92. *Cytotherapy* 15: 1563–1570.

Background

NK cells are potent innate immune cells that act by targeting cells with altered surface molecules. Cancer cells are altered in regard to the types and amounts of cell-surface molecules present. Scientists postulated that cancer cells that were unresponsive to other treatments might respond to NK cell therapy.

The Study

Since NK cells have the ability to monitor differences between normal cells and malignant cancerous cells, researchers hypothesized that NK cell therapy could be an effective treatment for patients with malignant tumors. They used an isolated NK cell line called NK-92, which was known to have predictable cytotoxic activity. It is also easily grown in culture conditions. The authors employed NK cell therapy on 15 patients with advanced treatment-resistant malignancies (that is, they were unresponsive to other tested treatments). Patients were divided into three groups. Each group received a different number of infused NK cells as treatment. The authors found that NK cell treatment in 75% of the patients with lung cancer produced antitumor responses (TABLE).

Conclusion

The authors observed that NK cell treatment, which did not appear to produce any side effects, resulted in positive responses in individuals suffering from treatment-resistant lung cancer. These findings support the hypothesis that NK cell treatment may be a viable cancer therapy option.

Think About...

1. Given the nature of this study, what might be some limitations of NK cell therapy in individuals with metastatic cancer?
2. What might be some limitations of using the NK cell line NK-92? Consider the role of innate and adaptive immune cells and their actions to monitor self.
3. Although there were no noticeable side effects, what might you predict as potential side effects of a cell-based therapy as described in this study? How could these side effects be minimized?

Effect of NK Cell Infusion in Patients with Different Forms of Lung Cancer

Disease	NK-92 starting cell dose ($10^9/m^2$)	NK-92 total cell dose (10^9)	Response to NK cell therapy	Overall survival in days
Primitive neuroectodermal tumor	1	2.3	Progressive disease	44
Soft tissue sarcoma	0.85	2.4	Progressive disease	262
Rhabdomyosarcoma	1	4.0	Progressive disease	42
Osteosarcoma	1	2.6	Progressive disease	99
Chronic lymphocytic leukemia-transformed	1	2.0	Progressive disease	163
Adrenal carcinoma	1	2.0	Progressive disease	801
Small-cell lung cancer	1	3.6	Mixed response	388
Soft tissue sarcoma	3	9.4	Progressive disease	13
Medulloblastoma	3	8.6	Progressive disease	593
Medulloblastoma	3	6.6	Progressive disease	19
Colorectal cancer	3	12	Progressive disease	310
Small-cell lung cancer	3	5.8	Progressive disease	218
Small-cell lung cancer	3	10	Mixed response	242
Non-small-cell lung cancer	10	42.4	Stable disease	707
Bone-non-Hodgkin lymphoma	10	18	Progressive disease	296

Source: T. Tonn et al. 2013. *Cytotherapy* 12: 1563–1570.

2.8 How does the innate immune system initiate and help regulate the adaptive immune response?

LEARNING OBJECTIVE

2.8.1 Describe the means by which the innate immune system can activate an adaptive immune response.

While the innate immune system employs mechanisms that are vital to ensuring ongoing health, sometimes the innate immune response is not sufficient to completely thwart pathogen colonization and growth. In these cases, the adaptive immune response, driven by the action of B cells and T cells, must be engaged to limit the infection. Innate immune cells perform certain functions that "wake up" the antigen-specific adaptive immune response. We can appreciate this connectivity between the two systems, as it is designed to ensure that pathogens that escape the body's initial defenses will face attack by mechanisms of the adaptive immune system.

Dendritic cells are messengers

Dendritic cells (discussed in detail in Chapter 6) are phagocytes that work at the site of infection. They function as antigen-presenting cells, acting as messengers between the innate and adaptive immune systems. They engulf organisms and transport them to a secondary lymphoid tissue (such as lymph nodes and spleen), where they present fragments of the material to T cells. They contain a variety of PAMP receptors and induce changes in gene expression to secrete the specific cytokine required to activate the appropriate adaptive immune cell (**FIGURE 2.15**).

Dendritic cells that sense intracellular bacteria or viruses secrete the cytokines IL-12 and IFN-γ to promote T-cell activation toward helper T cells that activate macrophages and cytotoxic T cells. Dendritic cells that sense a pathogen requiring a B-cell response secrete IL-4 and IL-10 to promote T-cell activation toward helper T cells that activate B cells. Dendritic cells that bind extracellular bacteria and fungi that require a neutrophil response secrete IL-6 and TGF-β to promote T-cell activation toward helper T cells that activate neutrophils. Thus, each cytokine released by dendritic cells triggers activation of the exact T-cell effector mechanism needed to fight a specific infection.

NK cells activate cytotoxic T cells

NK cells also play an important role in bridging innate and adaptive immunity. While their main function is to target intracellularly infected cells, NK cells also secrete the cytokine IFN-γ, which recruits cytotoxic T cells. The effector function of cytotoxic T cells mirrors that of NK cells, but cytotoxic T cells are much more efficient in monitoring for the presence of intracellular infection.

CHECKPOINT QUESTION

1. Why are dendritic cells so important in linking the innate and adaptive immune responses?

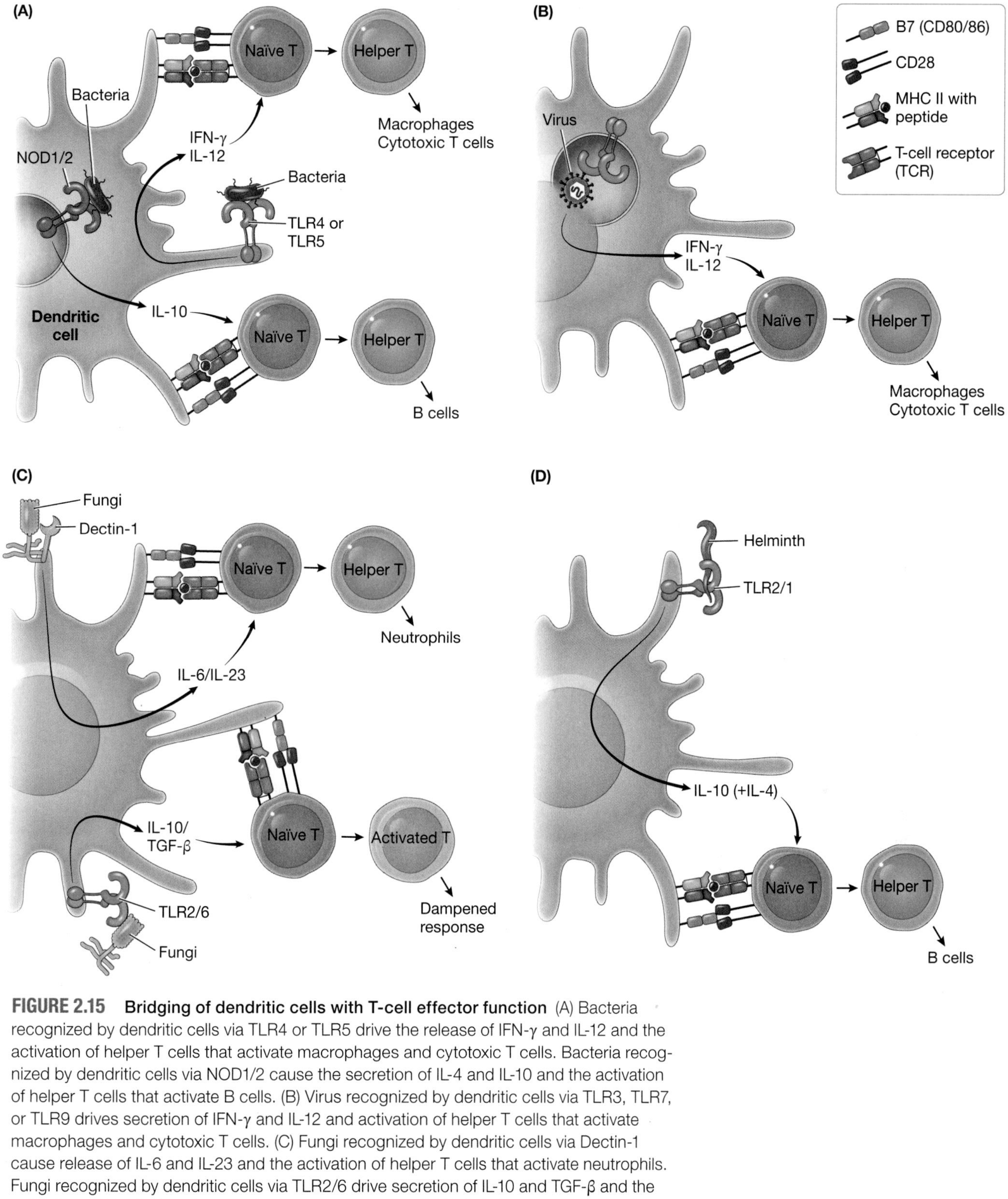

FIGURE 2.15 Bridging of dendritic cells with T-cell effector function (A) Bacteria recognized by dendritic cells via TLR4 or TLR5 drive the release of IFN-γ and IL-12 and the activation of helper T cells that activate macrophages and cytotoxic T cells. Bacteria recognized by dendritic cells via NOD1/2 cause the secretion of IL-4 and IL-10 and the activation of helper T cells that activate B cells. (B) Virus recognized by dendritic cells via TLR3, TLR7, or TLR9 drives secretion of IFN-γ and IL-12 and activation of helper T cells that activate macrophages and cytotoxic T cells. (C) Fungi recognized by dendritic cells via Dectin-1 cause release of IL-6 and IL-23 and the activation of helper T cells that activate neutrophils. Fungi recognized by dendritic cells via TLR2/6 drive secretion of IL-10 and TGF-β and the activation of T cells that dampen the immune response. (D) Helminths recognized by dendritic cells via TLR2/1 drive secretion of IL-4 and IL-10 and the activation of helper T cells that activate B cells.

Making Connections: Innate Immune Cell Functions Targeting Intracellular and Extracellular Pathogens

Innate immune cells employ a variety of pattern recognition receptors (PRRs) to detect different pathogens, including mechanisms designed to recognize pathogens where they reside (extracellularly or intracellularly). Specific effector functions of these cells are determined by the detected pathogen.

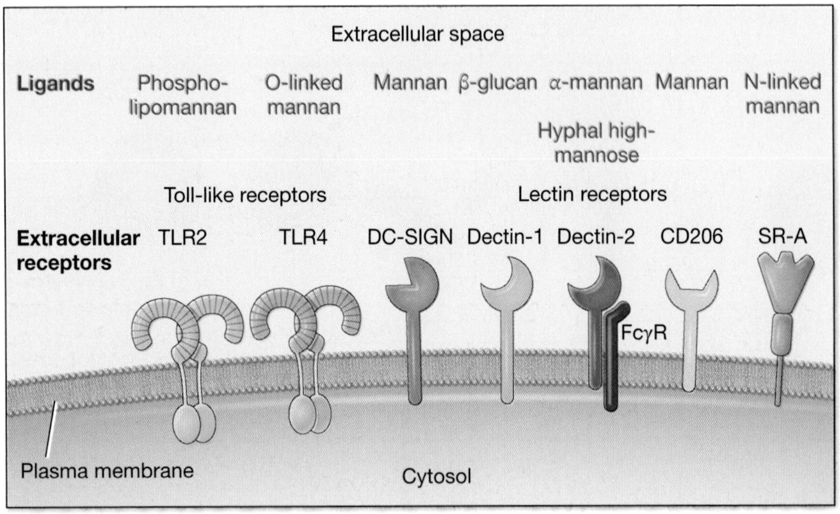

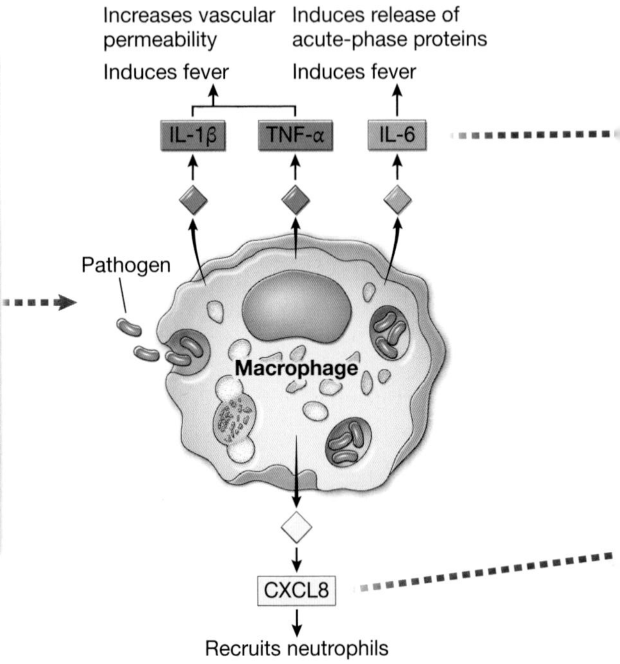

Extracellular pathogens are commonly recognized by PRRs of innate immune cells present at the cell surface.

(Sections 1.6, 1.7, 2.2, 2.3)

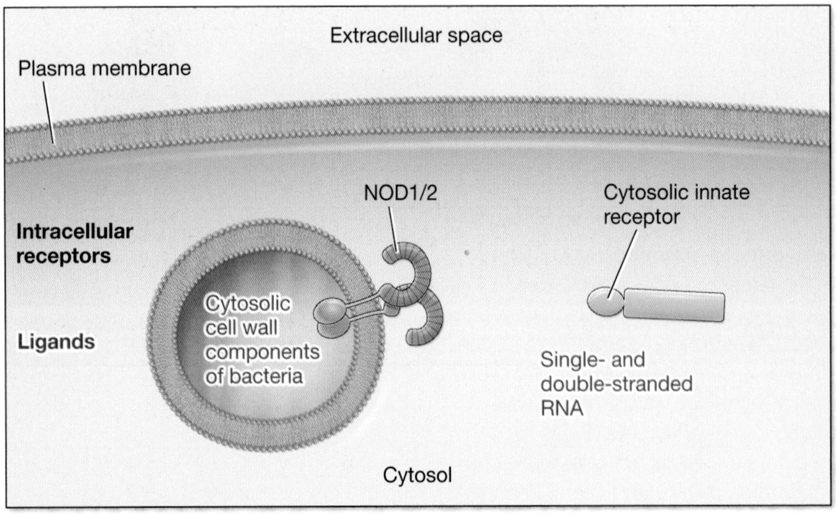

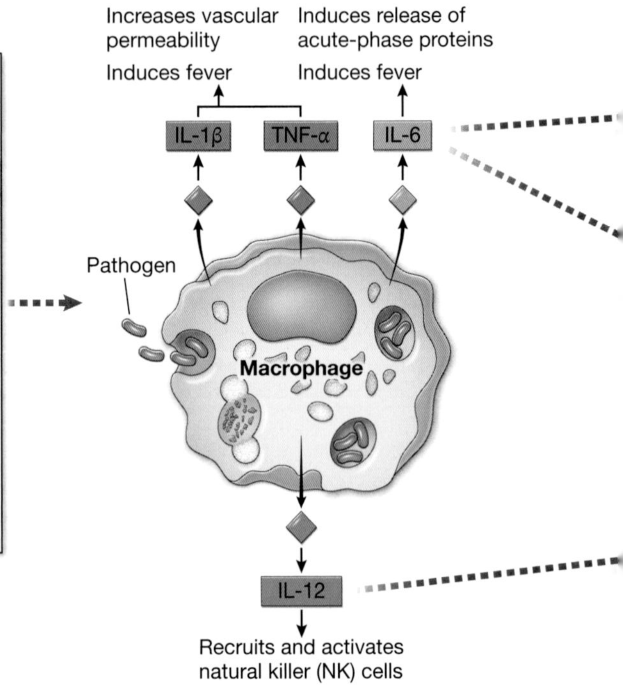

Intracellular pathogens are commonly recognized by PRRs of innate immune cells present within the cytoplasm, such as NOD-like receptors (NLRs), or within the lysosome.

(Sections 1.6, 1.7, 2.2, 2.3)

After M. S. Lionakis and M. G. Netea. 2013. *PLOS Pathog* 9: e1003079

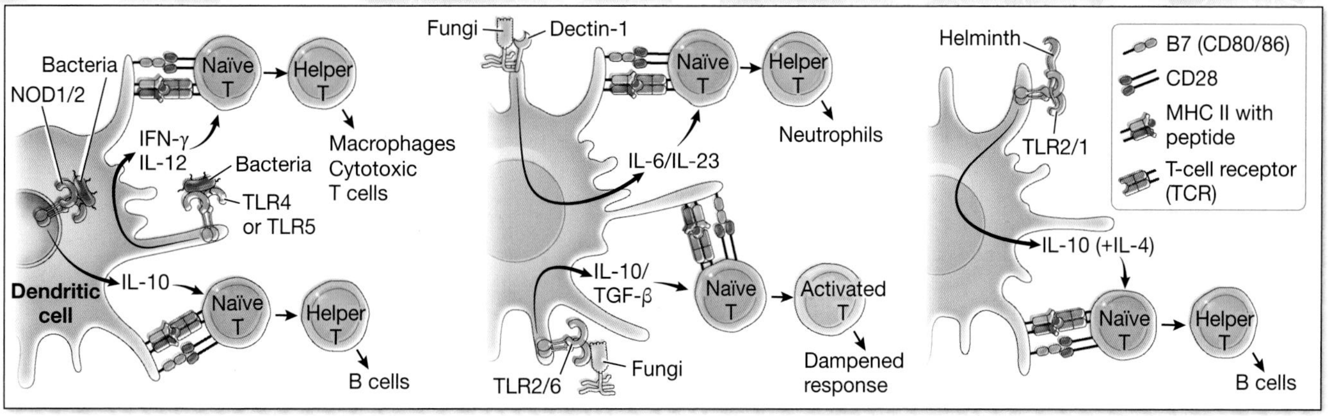

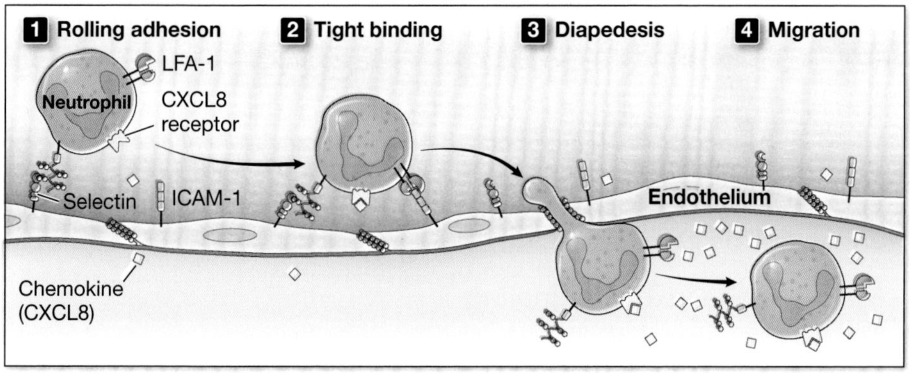

Upon recognition of extracellular PAMPs by PRRs, a variety of innate immune responses may be activated, including the activation of macrophages and an inflammatory response, recruitment and activation of neutrophils, and activation of dendritic cells, which bridge the innate and adaptive immune responses.

(Sections 1.6, 1.7, 2.5, 2.6, 2.8)

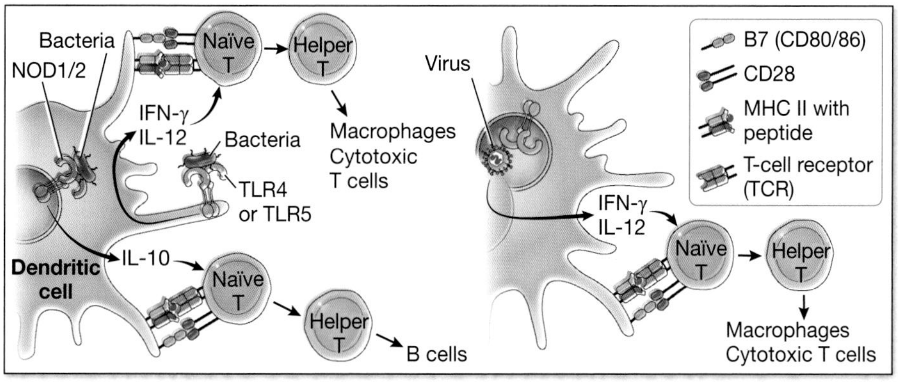

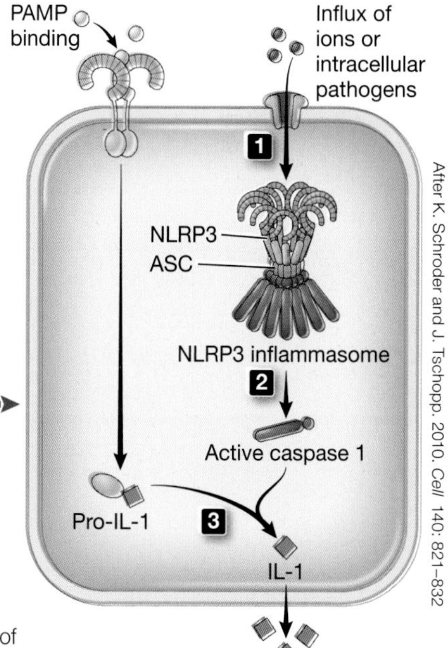

After K. Schroder and J. Tschopp. 2010. *Cell* 140: 821–832

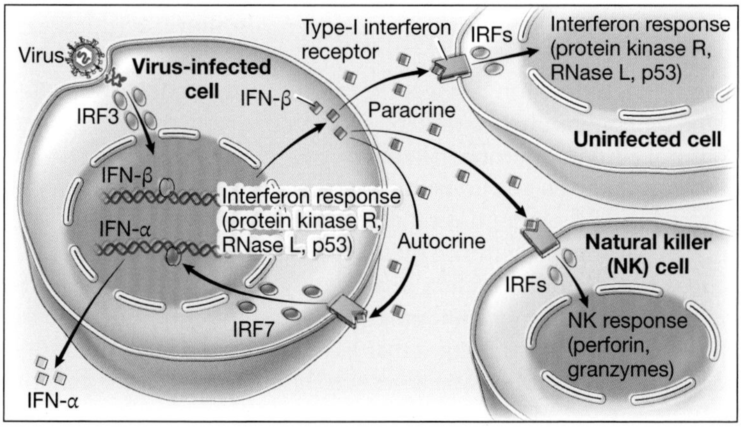

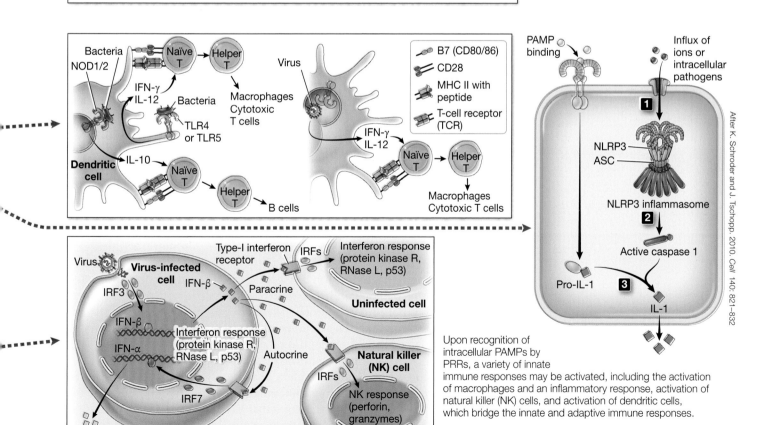

Upon recognition of intracellular PAMPs by PRRs, a variety of innate immune responses may be activated, including the activation of macrophages and an inflammatory response, activation of natural killer (NK) cells, and activation of dendritic cells, which bridge the innate and adaptive immune responses.

(Sections 1.6, 1.7, 2.5, 2.8)

2.9 What happens when the innate immune system malfunctions?

LEARNING OBJECTIVE

2.9.1 Explain how genetic deficiencies that affect innate immune system function cause specific disorders.

systemic infection Infection that is present in the bloodstream and widespread throughout the body.

granuloma Inflammatory mass that contains fused macrophages that have engulfed more neutrophils or antigen than they are capable of digesting.

The innate immune system plays a significant role in limiting the effects of pathogens and fighting infection. It also serves as a bridge to activate an adaptive immune response when T-cell and B-cell action is required. The importance of a healthy innate immune system is highlighted by problems that can occur when the immune system is activated in a systemic manner or completely malfunctions.

Systemic infection

The inflammatory response is critical in activating innate immune cells. As you learned, this response is triggered by inflammatory and signaling cytokines. Inflammatory cytokines cause vasodilation to increase plasma proteins and prompt cell migration to a site of infection. But induction of an inflammatory response at the wrong time can have dire consequences.

During a **systemic infection**, an infection in the bloodstream and throughout the body, secretion of the inflammatory cytokine TNF-α can cause body-wide vasodilation, which can result in massive fluid leakage from blood vessels into surrounding tissue. This fluid loss causes a rapid drop in blood pressure and can lead to *septic shock*, a life-threatening condition that can cause failure of vital organs, including the kidneys, liver, heart, and lungs.

Malfunctions in TLR signaling

TLR recognition of specific PAMP molecules and subsequent signaling is important in the production of inflammatory and signaling cytokines. Malfunction in the TLR signaling cascade can lead to an inadequate innate immune response due to a lack of production of essential cytokines. Mutations in the signaling cascades discussed in this chapter lead to profound limitations in mounting an effective immune response. For example, genetic deficiencies in MyD88 or IRAK4 in the TLR4 signaling pathway increase susceptibility to infection by pyogenic bacteria due to lack of production of inflammatory cytokines. Deficiency in TRAF3 or TRIF can lead to increased susceptibility to viral infections due to lack of expression of interferons. These deficiencies demonstrate the importance of the TLR signaling pathways in inducing a proper and robust innate immune response.

Malfunctions in innate immune cell action

The innate immune cells all play vital roles in pathogen clearance and limiting the spread of infection by activating inflammation, phagocytosing pathogens, and sensing viral infections. Certain genetic deficiencies alter the function of the innate immune cells.

Improper macrophage function can lead to the persistence of various infections. In fact, some pathogens alter normal phagocytic function of macrophages to survive in an intracellular environment. For instance, *Mycobacterium tuberculosis* is efficiently engulfed into an endosome in a macrophage but after uptake, the bacteria halt fusion of the phagosome with the lysosome, a process that is essential for normal pathogen clearance by macrophages.

The importance of neutrophil function in an innate immune response is also highlighted by the formation of **granulomas**, which are inflammatory masses containing accumulations of macrophages that have engulfed a large amount of neutrophils or antigen without efficiently digesting the engulfed cells or material.

Granulomas form due to the neutrophil's inability to efficiently destroy the engulfed pathogens. A deficiency in the NADPH oxidase enzyme, which is essential in producing reactive oxygen species required for pathogen destruction, is one disorder that can lead to granuloma formation. Neutrophils lacking NADPH oxidase are capable of phagocytosis but cannot destroy the pathogen, and thus these neutrophils become filled with intracellular infections. Macrophages attempt to clear the infected neutrophils and eventually fuse to form granulomas.

Because NK cells play an essential role in sensing viruses and removing virally infected cells, NK-cell malfunction can lead to recurring and chronic viral infections. Individuals with an NK cell deficiency commonly suffer from infections by herpesvirus, cytomegalovirus (CMV), and even Epstein-Barr virus (EBV), which is typically only seen as a threat to individuals with severe immunodeficiency. Since NK cells sense changes in cell surfaces and can detect altered or cancerous cells, NK cell deficiency can result in death due to the body's inability to fight cancer.

CHECKPOINT QUESTION

1. Why is release of TNF-α in the bloodstream problematic?

Summary

2.1 What are the primary defenses of the innate immune system?

- The innate immune system employs a variety of mechanisms to combat pathogens at the site of an infection, including antimicrobial peptides and proteins and the complement system.

2.2 How are pathogens recognized by cellular receptors?

- Cells of the innate immune system recognize pathogens using cellular receptors called PRRs (pattern recognition receptors) to interact with PAMPs (pathogen-associated molecular patterns). This process induces changes in gene expression through the activation of transcription factors in order to express cytokines and can induce phagocytosis.

2.3 Once PAMP recognition occurs, what innate immune responses are activated?

- Inflammatory and antiviral cytokines enhance the innate immune response by stimulating the action of important immune cells, the secretion of plasma proteins, and the migration of immune cells to the site of infection.

2.4 How do the Toll-like receptors function in pathogen recognition and cellular activation?

- Signaling through Toll-like receptors (TLRs) triggers an intracellular response that activates transcription factors to produce an inflammatory response or secretion of antiviral cytokines.

2.5 What are the functions of macrophages in innate immunity?

- Macrophages are innate immune cells that function to phagocytose foreign pathogens and secrete inflammatory cytokines.

2.6 What are the functions of neutrophils in innate immunity?

- Neutrophils are innate immune cells. They phagocytose foreign pathogens and destroy pathogens through the action of granules that can digest engulfed particles and use a respiratory burst to target pathogens.
- Neutrophils can migrate from the bloodstream to a site of infection through a process known as extravasation, which occurs through rolling adhesion, tight binding, diapedesis, and migration. This process is mediated by inflammatory cytokines secreted by activated macrophages.

2.7 What are the functions of natural killer cells in innate immunity?

- NK cells are innate immune cells that monitor host cells for intracellular infection, especially viral infections, by using activating and inhibitory receptors to sense the status of host cell surfaces.

2.8 How does the innate immune system initiate and help regulate the adaptive immune response?

- Innate immune cells, especially dendritic cells and NK cells, work to promote activation of the adaptive immune system, should it be required.

2.9 What happens when the innate immune system malfunctions?

- Genetic deficiencies that affect innate immune system function can cause recurring bacterial infections, improper soluble immune complex clearance, and autoimmune diseases.

Review Questions

2.1 Explain the process of phagocytosis.

2.2 What term is given to molecules recognized by innate immune cells on the surfaces of multiple types of pathogens?

2.3 **a.** Why does PAMP recognition promote efficient phagocytosis of a microorganism?

b. Which cytokine family is responsible for signaling viral infection?

A. Interleukins

B. Chemokines

C. Integrins

D. Interferons

E. Defensins

2.4 Summarize the signaling pathway used by TLR4 to induce an inflammatory response in response to LPS.

2.5 Name the important inflammatory cytokines that macrophages secrete in response to pathogens.

2.6 Describe the four steps that occur during neutrophil extravasation. Be sure to include important events and molecules involved in the process.

2.7 How do NK cells recognize intracellular infections without targeting uninfected cells for destruction?

2.8 Why are dendritic cells an important bridge between innate and adaptive immunity?

2.9 What types of innate immune system malfunction can occur, and how do these malfunctions affect the body?

CASE STUDY REVISITED

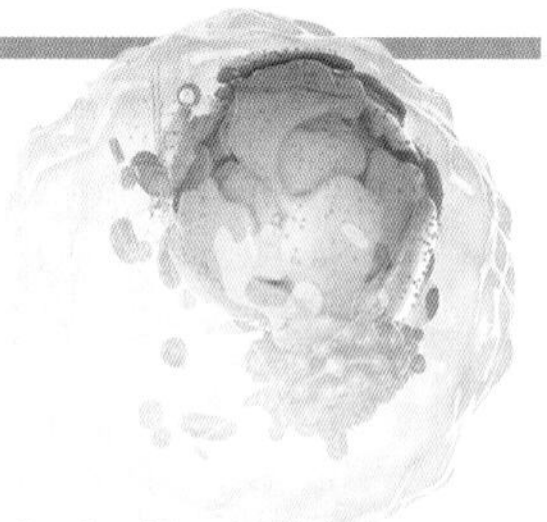

Septic Shock

A female, age 25, is rushed to the emergency room suffering from kidney failure and extreme low blood pressure. There is a bandage on her left index finger, and the patient's husband reveals that she received stitches from a kitchen injury yesterday. There is observable edema in multiple locations on her body. The patient has a rapid breathing rate of 25 breaths per minute, an elevated heart rate of 95 beats per minute, and a fever of 101°F. Blood analysis detects *Staphylococcus aureus* infection in her bloodstream. The kidney failure and sepsis lead to the doctor's diagnosis of septic shock.

Think About...

Questions for individuals

1. Define septic shock.
2. Describe how a bacterial infection in the bloodstream can lead to septic shock.

Questions for student groups

1. Is the patient's history and blood analysis consistent with septic shock? Why or why not?
2. What molecule may be responsible for inducing septic shock as seen in this patient? Which molecule would you expect to find at elevated levels within the patient's bloodstream?

MAKING CONNECTIONS Key Concepts	**COVERAGE** (Section Numbers)
Mechanisms of pathogen pattern recognition	1.6, 1.7, 2.2
Pattern recognition receptor signaling	1.7, 2.2, 2.3
Effector mechanisms activated upon PAMP recognition	1.7, 2.1, 2.3, 2.5, 2.6, 2.7
Recognition of extracellular infection by the innate immune system and subsequent activated effector mechanisms	1.6, 1.7, 2.2, 2.3, 2.5, 2.6, 2.8
Recognition of intracellular infection by the innate immune system and subsequent activated effector mechanisms	1.6, 1.7, 2.2, 2.3, 2.5, 2.8

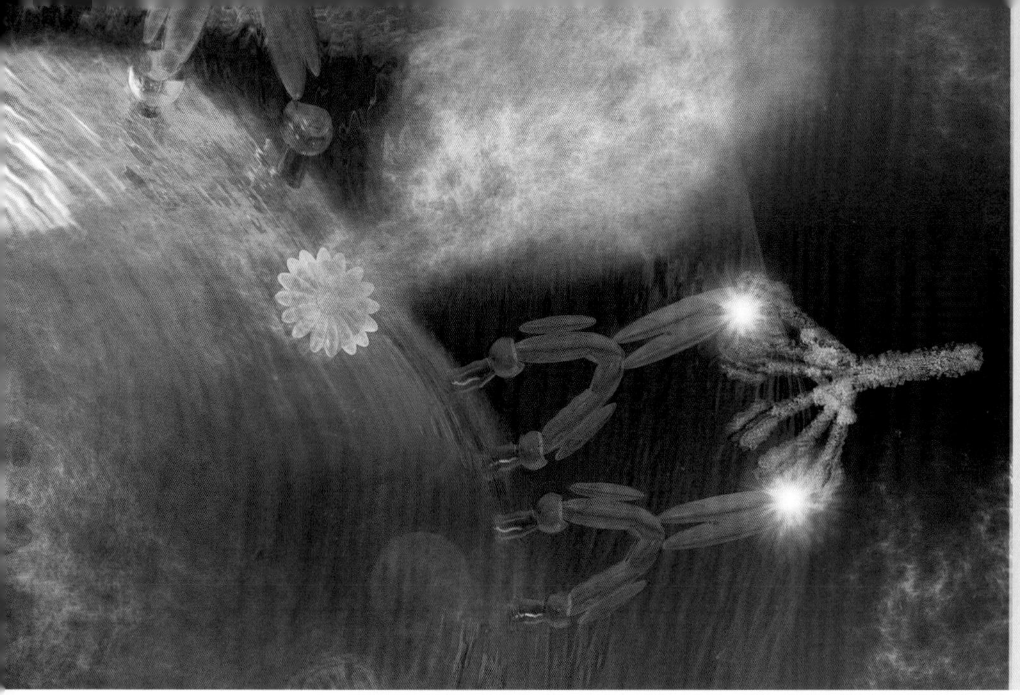

© Carol and Mike Werner/Science Source

3

The Complement System

CASE STUDY: Chase's Story

It is another typical Monday morning in the Harris household. Shondra is busy getting Chase ready so that she can drop him off at daycare before heading to work. Chase, as usual, is not happy about getting dressed or eating breakfast in a hurry, so he is his typical crabby self. Shondra thinks nothing of Chase's attitude and continues to run around the house, preparing her lunch and getting Chase dressed. In a flash, the two of them are out the door and driving to Chase's daycare. Chase continues to whine for the entire ride and during drop off.

About an hour later, Shondra's cellphone rings. It is Chase's daycare calling. The daycare supervisor says that Chase has been crying for the past 45 minutes. Shondra knows this isn't like Chase and tells the supervisor that she will be there as soon as she can get off of work.

Shondra arrives at Chase's daycare 30 minutes later. She talks to the supervisor, who tells her that Chase has been crying and grabbing his left ear and that he is running a fever. Suspecting yet another ear infection, Shondra decides to go directly to the emergency room. Unfortunately, Shondra is familiar with these types of trips and thinks the emergency room triage nurses are going to be on a first-name basis with her before long.

In order to understand what Chase is experiencing, and to answer the case study questions at the end of this chapter, we must understand some of the important components of our innate immune system that act to keep pathogens at bay. In Chapter 2, we explored various innate immune cell effector mechanisms designed to recognize and destroy pathogens. In this chapter, we will explore a key component of the innate immune system involving soluble plasma proteins: the complement system.

QUESTIONS Explored

3.1 What is the complement system and how is it activated?

3.2 When and how is the alternative pathway initiated?

3.3 When and how are the lectin and classical pathways initiated?

3.4 Which plasma proteins work with complement to fight infection?

3.5 What happens when the complement system malfunctions?

The complement system constitutes a variety of proteins involved in recognizing and tagging foreign molecules and pathogens. This family of proteins coats the surface of foreign molecules and cells, tagging them to mark them as candidates for destruction by innate immune cells. The complement system bears a remarkable similarity to the body's blood-clotting system—it involves a cascade of different proteins, each of which must be activated by cleavage via a protease to induce the next step in the process. Complement activation culminates in the activation of two key complement proteins—C3 and C5—that function in the tagging and removal of pathogens.

3.1 What is the complement system and how is it activated?

LEARNING OBJECTIVES

3.1.1 Describe the function and components of the complement system.

3.1.2 Compare and contrast the three pathways of complement activation.

Once a pathogen enters the body, the innate immune system gears up to fight it. Innate immune cells such as macrophages and neutrophils recognize molecular patterns on the pathogen through pattern recognition receptors (PRRs). Recognition of a pathogen by macrophages and neutrophils by PRRs results in phagocytosis of the pathogen and the production and release of inflammatory cytokines. A key line of defense in the detection and destruction of pathogens is a group of soluble proteins synthesized mainly by the liver that are present in blood, the lymphatic system, and extracellular fluid. These soluble proteins are referred to as *complement* or the **complement system**. More than 30 proteins function in a working complement system; they can be classified into 7 major types (**TABLE 3.1**).

Many components of the complement system act as proteases, enzymes that catalyze the breakdown of proteins into peptides or amino acids. These proteases are commonly zymogens, or inactive precursors. They work to ensure that the complement system is regulated properly and only activated when necessary. **Complement activation** occurs via a cascade of enzymatic reactions in which one component activates the next.

Complement in pathogen recognition

An important process in the immune response is the differentiation between self and nonself. The complement system plays an important role in marking foreign molecules and pathogens as nonself to allow detection by the immune system. Through a process referred to as **complement fixation**, activation of complement proteins promotes the attachment of specific proteins to the pathogen surface that then serve as an opsonin for recognition by phagocytic cells such as macrophages and neutrophils.

Complement in pathogen destruction

Another important process in the immune response is pathogen destruction and preventing further growth or infiltration into other tissues of the body. While complement fixed on the surface of the pathogen facilitates its

complement system System of soluble plasma proteins that acts to opsonize and lyse pathogens. Also known as *complement*.

complement activation Proteolytic processing of complement proteins to promote function in opsonization and pathogen destruction.

complement fixation Covalent attachment of a complement protein product to the surface of a pathogen.

TABLE 3.1 | Key Components of the Complement System

Complement Protein	Function	Complement Pathway
C1	Protease involved in cleaving C2 and C4	Classical
C2	Cleaved by either C1 or MASP-2 to form C2b fragment of classical C3 convertase	Lectin Classical
C3	Cleaved to form C3a (anaphylatoxin) and C3b (opsonin); also functions in forming alternative C3 convertase	Alternative Lectin Classical
C4	Cleaved by either C1 or MASP-2 to form C4b fragment of classical C3 convertase	Lectin Classical
C5	Cleaved by C3 convertase to form C5a (anaphylatoxin) and C5b (initiates formation of the membrane-attack complex by interacting with C6 and C7)	Alternative Lectin Classical
C6	Interacts with C5b and C7 to initiate formation of the membrane-attack complex	Alternative Lectin Classical
C7	Interacts with C5b and C6 to initiate formation of the membrane-attack complex	Alternative Lectin Classical
C8	Binds C5b67 to serve as a docking site for C9	Alternative Lectin Classical
C9	Binds C8; forms transmembrane pores in pathogen surfaces	Alternative Lectin Classical

destruction through the action of phagocytic cells, some components of the complement system are capable of directly damaging and destroying pathogens. Insertion of complement components into the pathogen membrane and formation of pores within the membrane disrupt the osmotic balance across the membrane and promote pathogen death.

Complement component 3

Complement component 3 (**C3**) is arguably one of the most important factors of the complement system. Patients deficient in C3 suffer from severe immunodeficiency and experience repeated and sometimes dangerous infections. Complement activation leads to the cleavage of C3 into two fragments: C3a and C3b (**FIGURE 3.1**). This breakdown occurs through the action of enzymes known as *C3 convertases* (explained further in Section 3.2)

The small C3a fragment serves as an **anaphylatoxin**, a soluble factor that aids in the induction of an inflammatory response. The C3a fragment induces degranulation (release of active substances) of immune system cells, including mast cells and phagocytes. When produced during complement activation, C3a can also recruit phagocytes to the site of infection. The C3b fragment, when produced, may be covalently attached (complement fixation) to the surface of the pathogen (see Figure 3.1). The pathogen tagged with C3b is now marked for destruction. In marking the pathogen, C3b functions as an opsonin, rendering the pathogen more susceptible to phagocytosis.

C3b fixation at the proper time during complement activation is assisted by the presence of a high-energy thioester bond (bond between a cysteine side chain and a hydroxyl group) within the structure of C3 (see Figure 3.1). In the zymogen form of C3, the reactive thioester bond is hidden within the interior of the protein. Once C3 is cleaved, the thioester bond is exposed, and the result is C3b attaching to the surface of the pathogen through the formation of an intermolecular thioester bond. The thioester also reacts with water in the body, creating an inactive C3b soluble fragment, ensuring that cleaved C3b can only form thioester bonds in the immediate vicinity of its activation.

Complement activation pathways

To ensure that complement activation occurs only when necessary, it is driven by proteolytic events. Three different proteolytic pathways can lead to complement activation (**FIGURE 3.2**). The very first time we come into contact with a pathogen, the first pathway to activate is the **alternative pathway**. The **lectin pathway** is the second pathway to come into play, and the **classical pathway** activates last.

complement component 3 (C3) Protein component of the complement system that is cleaved during complement activation to produce C3a (a chemokine) and C3b (an opsonin).

anaphylatoxin Molecule capable of activating an inflammatory response by triggering degranulation of cells capable of inducing inflammation.

alternative pathway First pathway of complement activation to be activated that relies on complement system proteins always present in the plasma for the activation of C3.

lectin pathway Second pathway of complement activation to be activated that relies on the use of mannose-binding lectin to initiate complement activation.

classical pathway Third pathway of complement activation to be activated that relies on the use of C-reactive protein or an antibody to initiate complement activation.

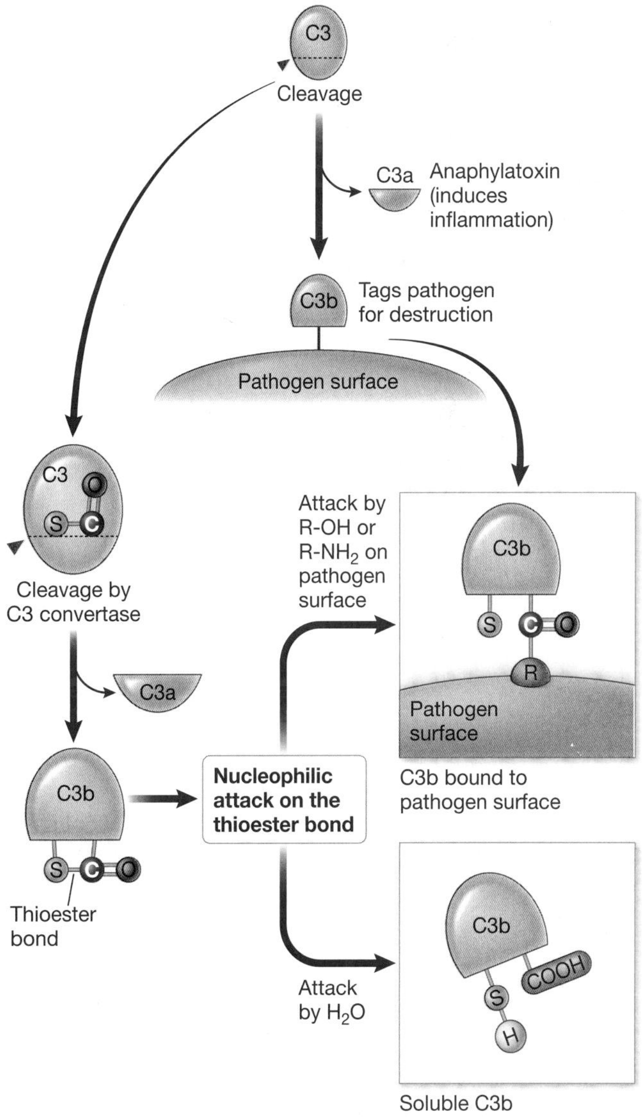

FIGURE 3.1 Proteolytic processing of complement component 3 (C3) Top: The complement system uses C3 as an inactive precursor. During complement activation, C3 is proteolytically cleaved into two smaller fragments, C3a and C3b. C3a is a soluble fragment that acts as an anaphylatoxin to induce inflammation. C3b becomes covalently bound to the pathogen surface to tag it for destruction. Bottom: Cleavage of C3 by a C3 convertase exposes a thioester bond in C3b, which can either remain soluble when it reacts with water or form a covalent bond when it reacts with a functional group on the surface of a pathogen.

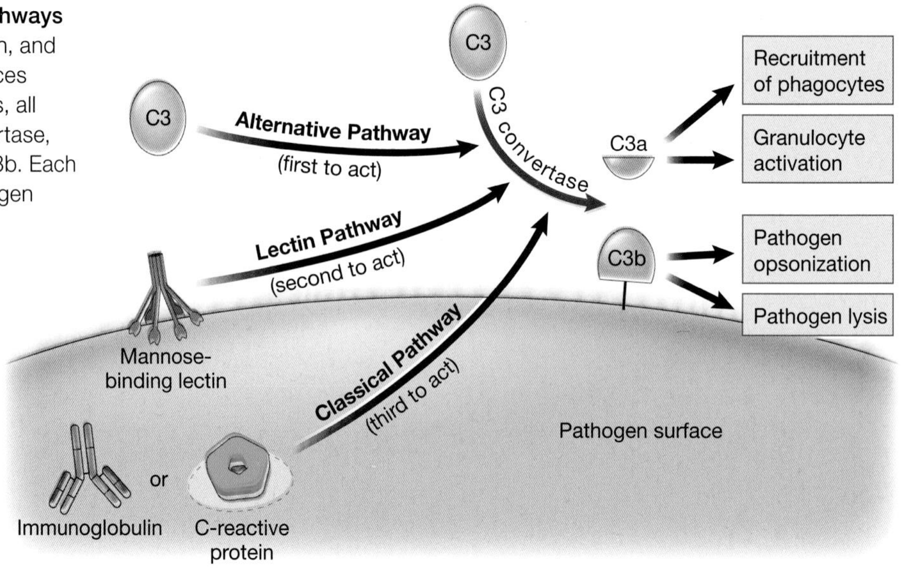

FIGURE 3.2 Complement activation pathways The three pathways are the alternative, lectin, and classical pathways. While each initially induces complement activation via a different means, all three result in the production of a C3 convertase, responsible for cleaving C3 into C3a and C3b. Each of these fragments functions to aid in pathogen recognition and destruction.

CHECKPOINT QUESTIONS

1. Which complement protein is the most important in directly targeting pathogens for destruction?
2. What are the cleavage products of that protein, and what does each cleavage product do to facilitate pathogen elimination?
3. What are the three pathways of complement activation?

3.2 When and how is the alternative pathway initiated?

LEARNING OBJECTIVES

3.2.1 Describe the molecular mechanisms of the alternative pathway of complement activation.

3.2.2 Explain the role of complement receptors in recognizing fixed complement.

3.2.3 Describe the regulation of complement activation.

3.2.4 Illustrate the molecular mechanisms of formation of the membrane-attack complex.

3.2.5 Describe the negative regulation of the membrane-attack complex.

Imagine again that you have gotten a cut and a pathogen has entered your body. In Chapter 2, we examined the action of a variety of different innate immune cells to recognize and destroy this pathogen. The complement system is another component of the innate immune system that functions to facilitate pathogen recognition and destruction by these innate immune cells. How then does the complement system work? The alternative pathway is the first to act during an infection and relies on soluble proteins, including C3, that are always present within the plasma. As previously noted, C3 possesses an intramolecular thioester bond that plays a role in the fixation of C3 on the surface of a foreign molecule. When this protein enters the plasma, the thioester bond is naturally exposed to hydrolysis (breaking down due to the presence of water). Hydrolysis of the exposed thioester bond produces a modified form of C3 known as iC3. This process is sometimes referred to as *tickover* because C3 spontaneously "ticks over" (changes) to the iC3 form (FIGURE 3.3, Step 1). The production of iC3 can occur throughout the body, but the process is accelerated in the presence of a pathogen.

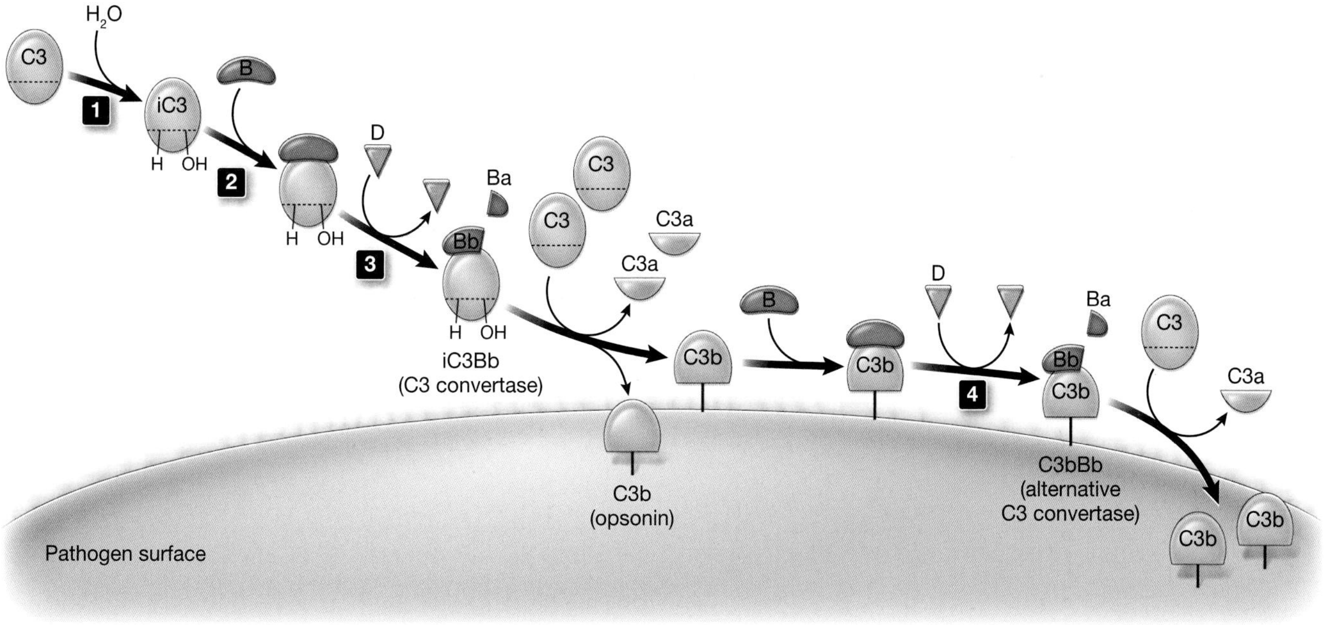

FIGURE 3.3 Alternative pathway of complement activation The alternative pathway, the first to act, begins by spontaneous "ticking over" (changing) of soluble C3 in the plasma by water, producing iC3 (Step 1). iC3 binds to factor B present in the plasma (Step 2). The iC3B complex is cleaved by factor D to produce soluble Ba (which does not function in the pathway) and iC3Bb, a C3 convertase (Step 3). This C3 convertase cleaves C3 to C3a (an anaphylatoxin) and C3b (an opsonin). C3b can serve as a tag for pathogen destruction or bind to factor B (similar to Step 2). The C3bB complex can be cleaved by factor D (Step 4) to produce the alternative C3 convertase C3bBb, which has the same activity as iC3Bb, namely to cleave C3 to C3a and C3b.

Alternative C3 convertase and C3b fixation

iC3 is capable of binding with an inactive complement protein known as **factor B** (Figure 3.3, Step 2). Once factor B is bound to iC3, it becomes susceptible to cleavage by a protease, **factor D**. Cleavage of factor B produces a small inactive fragment (Ba) and fragment Bb, which is a protease that remains bound to iC3. The product formed by factor D cleavage is known as iC3Bb, referred to as a **C3 convertase** (Figure 3.3, Step 3). The identification of factor D as the protease responsible for cleaving factor B was first described in the early 1970s (see **KEY DISCOVERIES**).

A C3 convertase is a protease capable of converting soluble C3 into C3a and C3b. The C3a produced works to recruit neutrophils and macrophages to the site of infection by stimulating an inflammatory response. The C3b produced has an exposed thioester that allows it to react with the surface of the pathogen. This creates a covalently attached C3b, which marks the pathogen as a target for phagocytosis.

To further enhance complement activation, the attached C3b can bind more factor B, which is again activated by factor D to produce Ba and Bb. The Bb produced remains bound to the C3b to form C3bBb, which acts as an **alternative C3 convertase** (Figure 3.3, Step 4). The alternative C3 convertase works to cleave more C3 into C3a and C3b, and the cycle continues. Thus, a positive feedback loop is established that promotes the activation of C3 for pathogen tagging and destruction. The alternative C3 convertase is stabilized on the pathogen surface by the plasma protein **properdin**, also known as **factor P**.

Factor P stabilizes the alternative C3 convertase at the pathogen surface and also acts as an initiator of the alternative pathway. It can bind to certain pathogens, where it serves as a docking site for C3b and factor B on the pathogen surface. When factor P binds C3b and factor B, factor B can be cleaved by factor D (as previously described) and form the alternative C3 convertase.

One product of an inflammatory response is the activation of the blood coagulation cascade of proteases. This system ensures that blood clotting occurs

factor B Protease zymogen that functions in the complement cascade; cleavage of factor B by factor D causes its activation and ability to function as a subunit of C3 convertase.

factor D Protease of the complement system that functions to activate factor B through proteolytic cleavage.

C3 convertase Protease produced during activation of complement that is capable of cleaving the complement protein C3 into C3a and C3b.

alternative C3 convertase C3 convertase produced when the opsonin C3b binds to factor B and bound factor B is cleaved by factor D to produce C3bBb at the pathogen surface.

properdin Plasma protein that enhances the activity of the alternative C3 convertase to aid in complement activation and fixation. Also known as *factor P*.

factor P Plasma protein that enhances the activity of the alternative C3 convertase to aid in complement activation and fixation. Also known as *properdin*.

efficiently to prevent pathogen spread into the bloodstream. The coagulation system has also been shown to activate the alternative pathway. In inflammation studies in mice in which the C3 gene was disrupted, activation of C5 (a member of the membrane-attack complex, discussed later in Section 3.2) was observed. C3 serves as the major component of the protease involved in activating C5; however, in these mice, the absence of C3 suggests that another protease can serve the

KEY DISCOVERIES

How was factor D discovered as the protease responsible for factor B cleavage?

Article

H. J. Muller-Eberhard and O. Gotze. 1972. C3 proactivator convertase and its mode of action. *Journal of Experimental Medicine* 135: 1003–1008.

Background

The alternative pathway of complement activation requires the action of factor D, a protease, to cleave factor B in the formulation of the C3 convertase active in the alternative pathway. How was factor D identified as a protease responsible for cleavage of factor B?

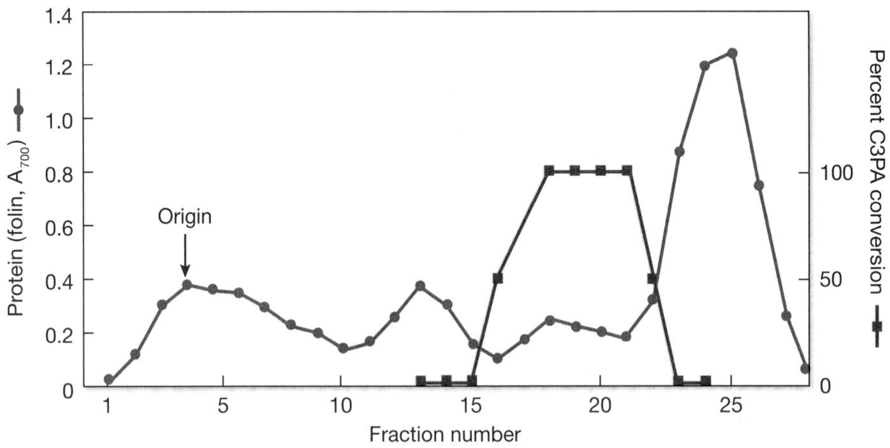

FIGURE KD 3.1 C3PAse activity can be found in the α-globulin region of serum separated by block electrophoresis Serum proteins were separated by Pevikon block electrophoresis (circles). Collected fractions were assayed for C3PAse activity, and fractions corresponding to the α-globulin region (fractions 18–21) were highest in C3PAse activity (squares). (© 1972 H. J. Müller-Eberhard and O. Gotze originally published in *J Exp Med* doi.org/10.1084/jem.135.4.1003.)

Early Research

Factor B (originally named the C3 proactivator, or C3PA) was shown to be an enzyme that was cleaved and capable of activating C3 when serum was incubated with a variety of different initiating signals, including yeast cell walls, inulin, endotoxin, and immunoglobulins. It was hypothesized that another serum enzyme, a C3PA convertase (or C3PAse) was capable of activating factor B (C3PA).

Think About...

1. Considering this brief summary, what would your hypothesis be on the activity of the C3PAse, with the assumption that cleavage of C3PA was required for activation?
2. What experiment(s) would you need to conduct to test your hypothesis?
3. In this study, the researchers identified that C3b was required for the activity associated with C3PAse and the activation of C3PA. Why would you predict this need?

Article Summary

The authors first identified a hydrazine-sensitive factor in serum that prevented cleavage of C3PA (factor B) in the presence of inulin (a polysaccharide). By biochemically separating components of serum, the researchers were able to reconstitute the hydrazine-sensitive factor with isolated C3, suggesting that the hydrazine-sensitive factor and C3 were one and the same molecule. In conducting these studies, the researchers found another hydrazine-sensitive factor (hydrazine-sensitive factor a) that was able to activate C3PA in the absence of inulin. Subsequent characterization of this second hydrazine-sensitive factor led the researchers to hypothesize that this hydrazine-sensitive factor was a fragment of C3, namely C3b.

Using their purified hydrazine-sensitive factor a (C3b) and their purified C3PA (factor B), researchers were able to study the activity and characterize C3PAse (factor D) from serum. Their purification led them to identify that C3PAse was a 3S α-globulin requiring a fragment of C3 for activity (**FIGURE KD 3.1**). This study was the first to characterize the activity of factor D, its ability to cleave factor B, and the requirement of the C3b fragment in this process. It was also the first to postulate the positive feedback loop of the action of the alternative C3 convertase, whereby C3b functions to aid in further activation of factor B.

role of C3 in C5 activation. These observations suggest that coagulation proteases activated during inflammation may play a role in complement activation.

Complement receptors

Up to this point, we have only described the activation of C3 and the fixation of C3b on the pathogen surface. How does this fixation of C3b on the surface of the pathogen function in pathogen recognition and destruction? The process of coating a pathogen with opsonins is known as **opsonization**. C3b serves as an opsonin, tagging pathogens for phagocytosis. C3b can carry out this role because it serves as a ligand.

Macrophages are phagocytes that are prevalent in the lungs, the liver, the connective tissue, and the lining of the respiratory and gastrointestinal tracts. They have the capacity to phagocytose bacteria and other microorganisms in a relatively nonspecific manner due to their possessing multiple types of receptors.

Complement receptors are cell-surface receptors that bind to complement proteins attached to the surface of pathogens. Macrophages express a surface receptor called **complement receptor 1** (**CR1**). Upon recognizing C3b on the pathogen surface, bound CR1 clusters on the macrophage surface, which facilitates phagocytosis (**FIGURE 3.4**).

Efficient phagocytosis requires the interaction of multiple C3b molecules with CR1 molecules. However, the serine protease **factor I** inactivates C3b by further cleavage, which produces a smaller fragment known as iC3b. The resulting iC3b fragment cannot assemble into an active C3 convertase. This inactivation process prevents efficient fixation of C3b on macrophages near the location of complement activation, which means that macrophages at the site of infection serve as a poor target.

The production of iC3b by factor I can occur at the pathogen surface during complement activation. In this case, the iC3b produced can serve as a ligand for two other macrophage receptors: **complement receptor 3** (**CR3**) and **complement receptor 4** (**CR4**). These two receptors are members of the *integrin* family of receptors, which are cell-surface glycoproteins that function in cell-to-cell adhesion.

The action of the three cell-surface receptors (CR1, CR3, and CR4) on macrophages allows for efficient attachment to pathogens and enhanced phagocytosis. Thus, complement activation, which generates products that act as ligands for

opsonization Process of marking a pathogen or foreign molecule as one that needs to be targeted for removal or destruction by phagocytosis.

complement receptor Cell-surface receptor family that binds to complement proteins fixed on the surface of a pathogen, allowing for recognition and phagocytosis.

complement receptor 1 (CR1) Cell-surface receptor expressed by macrophages that binds to C3b fixed on the surface of a pathogen, allowing for enhanced recognition and phagocytosis.

factor I Serine protease that inactivates C3b through its cleavage into a smaller fragment known as iC3b, which cannot function as a component of a C3 convertase.

complement receptor 3 (CR3) Cell-surface receptor expressed by macrophages and neutrophils that binds to iC3b fixed on the surface of a pathogen, allowing for enhanced recognition and phagocytosis.

complement receptor 4 (CR4) Cell-surface receptor expressed by macrophages and neutrophils that binds to iC3b fixed on the surface of a pathogen, allowing for enhanced recognition and phagocytosis.

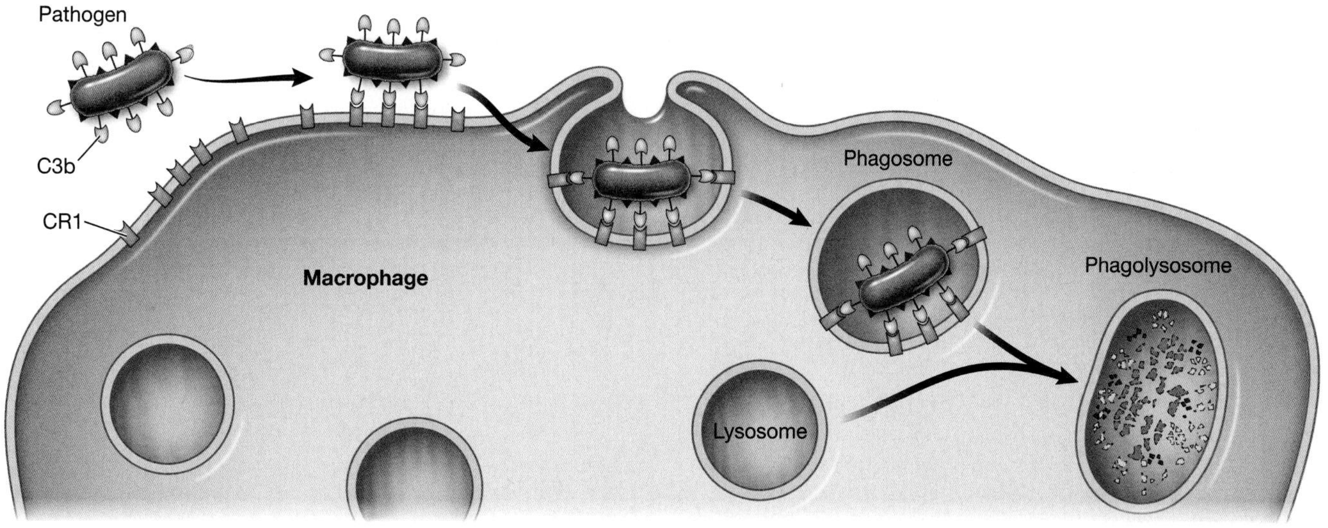

FIGURE 3.4 Complement fixation tags pathogens for phagocytosis Fixation of C3b on the surface of a pathogen allows phagocytes such as macrophages to recognize the pathogen as foreign. Binding and clustering of C3b on the pathogen surface by CR1 on the phagocyte surface triggers phagocytosis of the opsonized pathogen. The ingested pathogen is now enclosed within the phagosome, which fuses with lysosomes to form the phagolysosome. Within the phagolysosome, the pathogen is destroyed.

factor H Plasma protein that enhances the cleavage of C3b to iC3b by factor I.

the receptors, promotes clearance of pathogens by macrophages present at the site of infection.

Regulating complement activation

We have seen that complement activation provides a powerful mechanism for pathogen removal through phagocytosis. Later in this chapter, we will examine how complement activation can also result in the creation of transmembrane pores in pathogen membranes. Complement activation also produces an inflammatory response that assists in innate immune cell activation and migration. However, the fixation of C3b onto cells is nonspecific. C3 is normally present in plasma, and the exposed thioester of C3b during complement activation has the ability to interact with amines of any proteins and hydroxyl groups of any carbohydrates, including those on our own cell surfaces. Therefore, complement activation and fixation must be highly regulated to ensure that our own cells and tissues are not targeted as foreign invaders. The rapid proliferation of complement activation via the positive feedback loop must also be regulated to ensure that complement is not depleted from the bloodstream. Two families of regulatory proteins are involved in controlling the activation of complement: plasma proteins and membrane proteins.

PLASMA PROTEINS Plasma proteins interact with C3b covalently attached to cell surfaces. One plasma protein, properdin, also known as factor P, increases the efficiency of the alternative C3 convertase at the surface of a pathogen by protecting it from protease degradation. Conversely, another plasma protein, **factor H**, works to "apply the brakes" to complement activation. Factor H binds to C3b to facilitate its cleavage to the inactive iC3b form by the serine protease factor I (**FIGURE 3.5**).

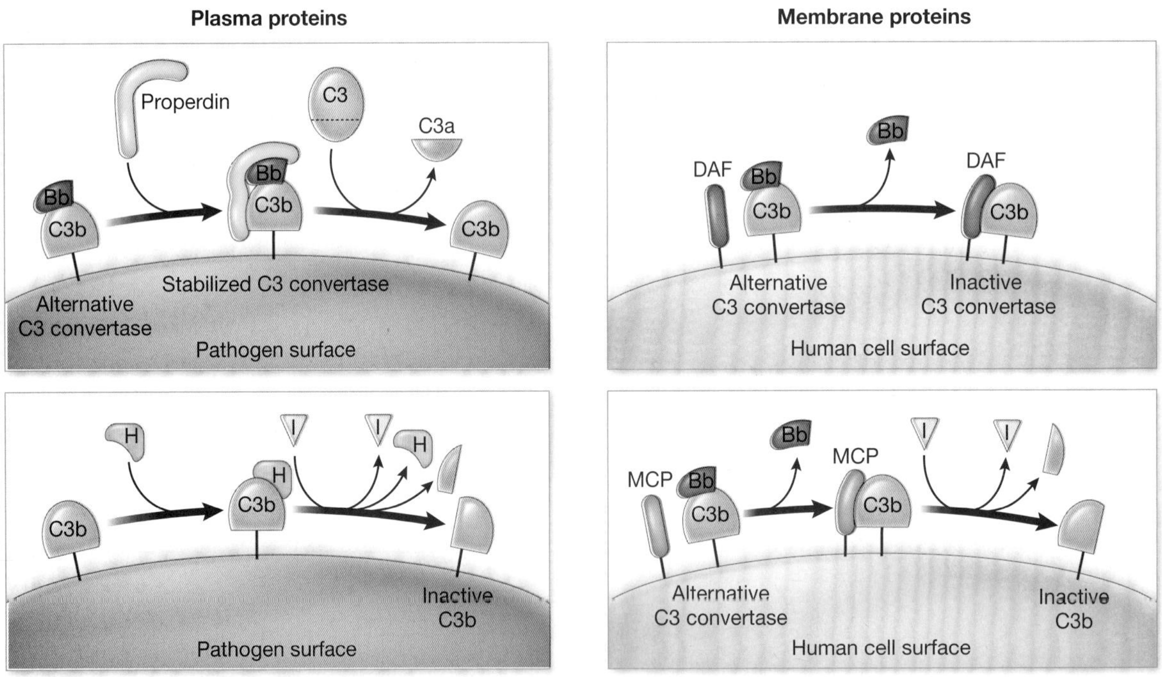

FIGURE 3.5 Regulation of the complement system The complement system is regulated through the action of both plasma proteins (left) and membrane proteins (right). Properdin (top left) is a positive regulator of the alternative C3 convertase, stabilizing it at the pathogen surface. Factors H and I (bottom left) prevent too much complement fixation by inactivating C3b on the pathogen surface. Decay-accelerating factor (DAF; top right) is a human membrane protein that breaks apart the subunits of the alternative C3 convertase. Membrane cofactor protein (MCP; bottom right) is a human membrane protein that breaks apart alternative C3 convertase subunits and further promotes inactivation of C3b at the cell surface by cleavage through the action of factor I to the inactive iC3b. (After P. Parham. 2014. *The Immune System*, 4th ed., W. W. Norton and Company.)

Factor H binds to cell membranes by interacting with sialic acid, a common carbohydrate on the surface of eukaryotic cells. Some bacteria, including *Staphylococcus aureus* and *Streptococcus pyogenes*, have evolved a defensive mechanism that works to limit complement activation on their surface by incorporating sialic acid on their surfaces. In a manner similar to inhibition on the surface of host cells, these bacteria bind to factor H, thus inhibiting complement activation.

MEMBRANE PROTEINS To ensure that C3b fixation on host cell surfaces does not result in targeting these cells for destruction, human cells express membrane proteins that inhibit complement activation. Two such proteins are **decay-accelerating factor** (**DAF**) and **membrane cofactor protein** (**MCP**). DAF inhibits complement activation by causing the breakdown of the alternative C3 convertase. MCP inhibits complement activation by binding to C3b and enhancing its cleavage to the inactive iC3b by factor I (see Figure 3.5).

Membrane-attack complex

The production of C3b and subsequent opsonization of foreign material is generally regarded as the most important function of complement activation. However, other complement proteins also function in pathogen destruction. Five complement proteins (C5, C6, C7, C8, and C9) work in concert to promote the formation of a **membrane-attack complex** (**MAC**). This complex of complement proteins works to create holes in bacterial and eukaryotic cell membranes, ultimately causing cellular lysis.

C3b produced during the alternative pathway of complement activation interacts with the alternative C3 convertase to form $C3b_2Bb$, also known as the **alternative C5 convertase**, which cleaves C5 into C5a and C5b fragments (**FIGURE 3.6**). C5b functions as an initiating factor in the formation of the MAC.

To form a MAC large enough to disrupt membrane activity and promote cellular lysis, complement proteins bind to each other. This binding exposes hydrophobic regions that facilitate interaction with the hydrophobic composition of the plasma membrane. Five proteins form a complex that results in the creation of larger and larger pores in the target membrane. The first three proteins of the MAC—C5b, C6, and C7—bind together to form C5b67. C7 contributes a hydrophobic region to allow for insertion of the complex into the target membrane. Once C5b67 inserts into the membrane, C8 can bind to the complex. C8 in the C5b678 complex serves as a docking site for C9, which is responsible for polymerization in the target membrane and formation of a pore large enough to cause loss of membrane integrity and cellular lysis (**FIGURE 3.7**).

REGULATING THE MEMBRANE-ATTACK COMPLEX Since the MAC can insert into eukaryotic membranes, it is imperative to ensure that this complex does not form in human cells. Several plasma proteins, including S protein, clusterin, and factor J, prevent formation of the MAC by inhibiting the

decay-accelerating factor (DAF) Cell-surface protein expressed by human cells that inhibits complement activation by inactivating the alternative C3 convertase on the surface of human cells.

membrane cofactor protein (MCP) Cell-surface protein that inhibits complement activation by enhancing C3b cleavage by factor I on the surface of human cells.

membrane-attack complex (MAC) Complex of complement proteins C5, C6, C7, C8, and C9 that work in concert to form holes in bacterial and eukaryotic membranes.

alternative C5 convertase C5 convertase produced when C3b binds to the alternative C3 convertase C3bBb; responsible for cleaving C5 into C5a and C5b.

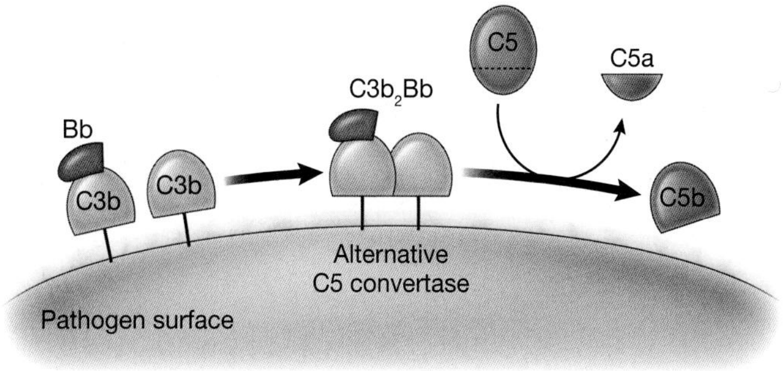

FIGURE 3.6 Formation and action of the alternative C5 convertase During the alternative pathway of complement activation, both the alternative C3 convertase (C3bBb) and the complement opsonin C3b are covalently attached to the pathogen surface. They interact to form the alternative C5 convertase, which cleaves the complement protein C5 into C5a and C5b. C5a acts as an anaphylatoxin, and C5b on the pathogen surface initiates the formation of the MAC. (After P. Parham. 2014. *The Immune System*, 4th ed., W. W. Norton and Company.)

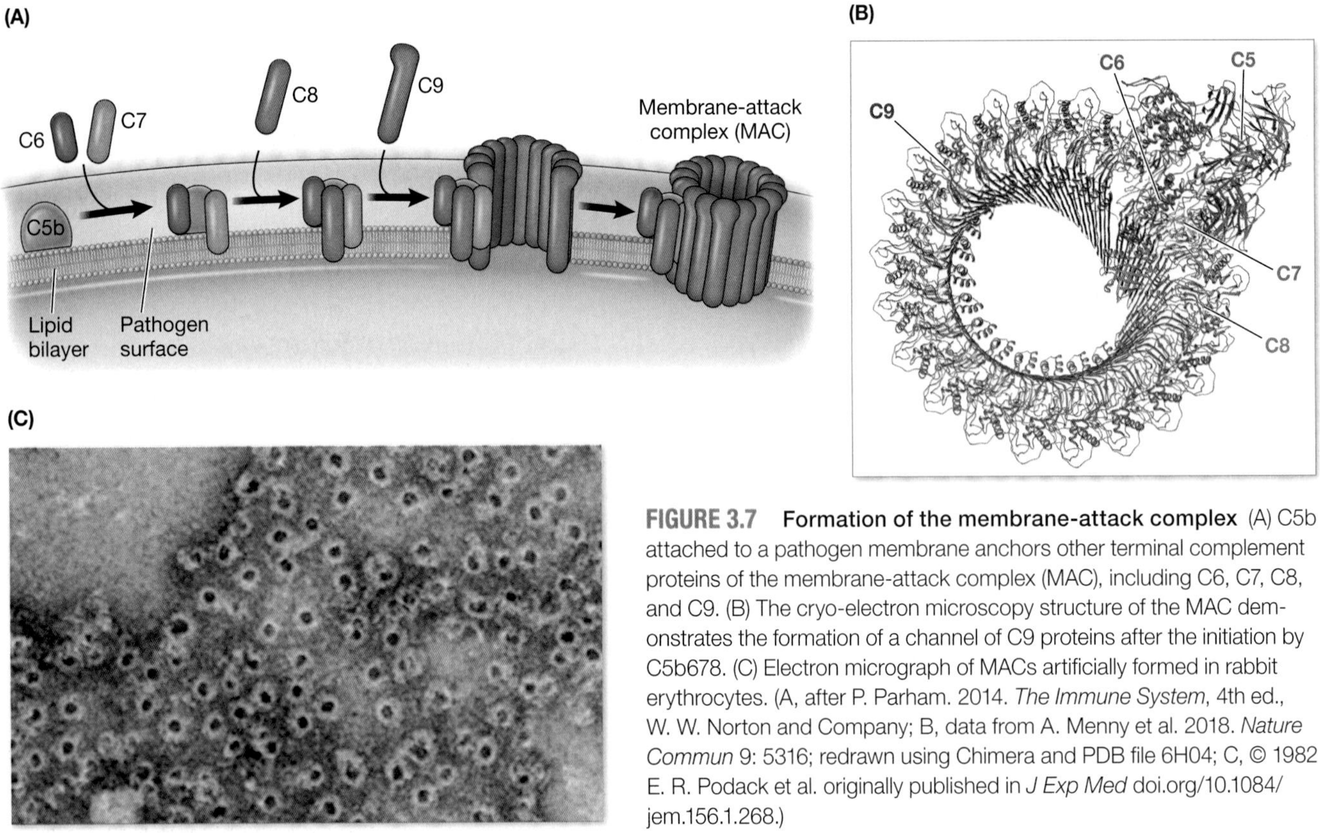

FIGURE 3.7 Formation of the membrane-attack complex (A) C5b attached to a pathogen membrane anchors other terminal complement proteins of the membrane-attack complex (MAC), including C6, C7, C8, and C9. (B) The cryo-electron microscopy structure of the MAC demonstrates the formation of a channel of C9 proteins after the initiation by C5b678. (C) Electron micrograph of MACs artificially formed in rabbit erythrocytes. (A, after P. Parham. 2014. *The Immune System*, 4th ed., W. W. Norton and Company; B, data from A. Menny et al. 2018. *Nature Commun* 9: 5316; redrawn using Chimera and PDB file 6H04; C, © 1982 E. R. Podack et al. originally published in *J Exp Med* doi.org/10.1084/jem.156.1.268.)

interaction of soluble C5b67 with human cell membranes (**FIGURE 3.8**). Human cells also express receptors at the plasma membrane that prevent the formation of MAC pores at the surface. Cell-surface proteins, including protectin (CD59) and homologous restriction factor (HRF), prevent C9 from being recruited to cell surfaces by the C5b678 complex.

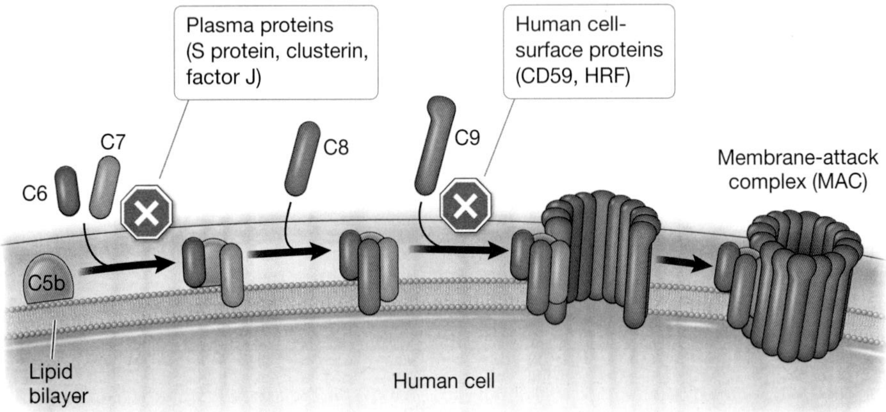

FIGURE 3.8 Regulation of formation of the membrane-attack complex Because complement fixation and production of C5b can occur in the absence of infection nonspecifically, MAC formation is inhibited by a variety of mechanisms so that human cells are not lysed. Soluble plasma proteins, including S protein, clusterin, and factor J, block the association of C5b67 with human plasma membranes, thus preventing MAC formation. Human cell surface proteins, including CD59 (protectin) and homologous restriction factor (HRF), block the recruitment of C9 to the plasma membrane of human cells. (After P. Parham. 2014. *The Immune System*, 4th ed., W. W. Norton and Company.)

Functions of C3a and C5a

During complement fixation and the formation of the MAC, C3 and C5 are cleaved into smaller fragments. The larger fragments (C3b and C5b) mark a pathogen for phagocytosis and promote the creation of pores in pathogen membranes. The smaller fragments (C3a and C5a) also play a role in the innate immune response.

C3a and C5a can induce an inflammatory response as anaphylatoxins. Both proteins bind to receptors on the surface of endothelial cells and granulocytes such as mast cells and basophils. C3a and C5a directly affect local blood vessels, increasing vascular permeability and blood flow. C3a and C5a also cause smooth muscle contraction and degranulation of mast cells and basophils (**FIGURE 3.9**). Degranulation leads to the secretion of histamine. Histamine increases vascular and capillary permeability, allowing plasma proteins and innate immune cells to exit the bloodstream. They migrate to the tissue in which complement activation occurred to fight the pathogen at the site of infection.

C3a and C5a also act as chemoattractants, chemical agents that cause a cell or organism to move toward them. They bind to phagocyte receptors, drawing the cells toward the site of infection. They also aid in recruiting phagocytes (especially neutrophils) by increasing their adherence to blood vessels close to the site of complement activation. C3a and C5a increase the potency of phagocytic cells by promoting an increase in expression of complement receptors such as CR1 and CR3, which work to bind pathogens opsonized by the complement system. Thus, C3a and C5a further enhance a robust innate immune response at the site of infection.

CHECKPOINT QUESTIONS

1. During the alternative pathway of complement activation, two C3 convertases are formed. Which proteins make up each of these convertases?
2. Which complement protein is responsible for forming pores in pathogen membranes within the MAC?

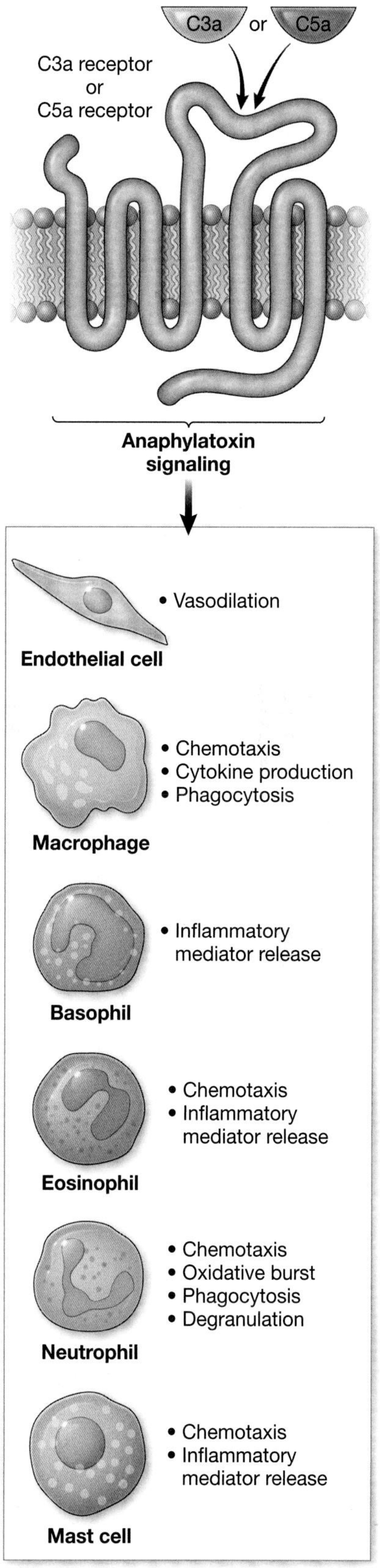

FIGURE 3.9 Action of C3a and C5a as anaphylatoxins C3a and C5a can bind to a cell-surface receptor on a variety of different cells, including white blood cells and endothelial cells. Signaling events triggered after C3a or C5a bind to their receptor result in actions that promote immune cell function, including inflammation, chemotaxis of cells to a site of infection, and effector mechanisms of innate immune cells. (After J. Punt et al. 2018. *Kuby Immunology*, 8th ed., W. H. Freeman and Company: New York.)

Making Connections: The Alternative Pathway of Complement Activation and Innate Immune Cell Action

The alternative pathway of complement activation uses plasma proteins to produce the opsonin C3b and anaphylatoxin C3a, which drive effector mechanisms of the innate immune system.

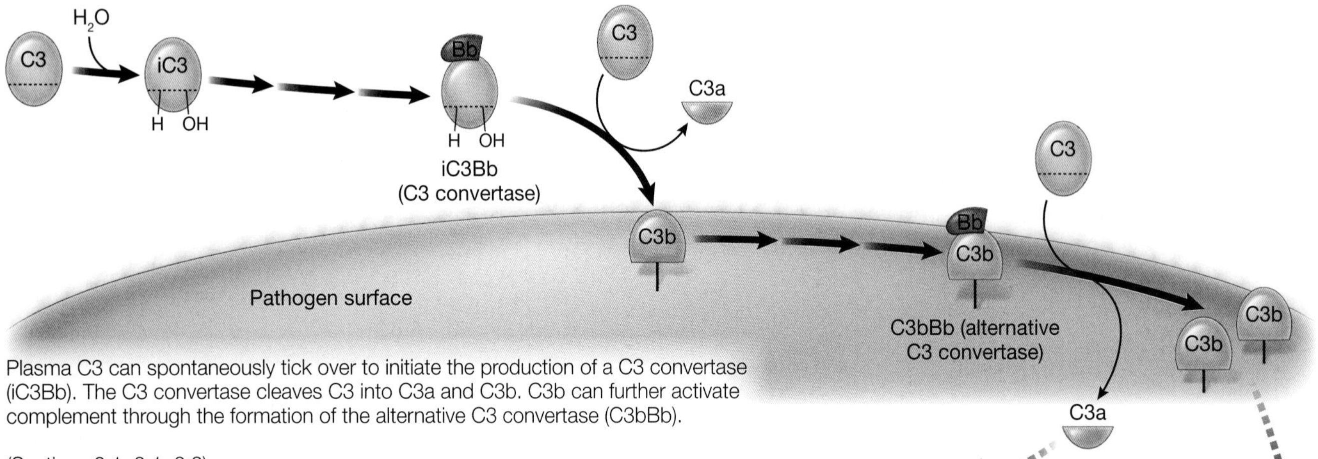

Plasma C3 can spontaneously tick over to initiate the production of a C3 convertase (iC3Bb). The C3 convertase cleaves C3 into C3a and C3b. C3b can further activate complement through the formation of the alternative C3 convertase (C3bBb).

(Sections 2.1, 3.1, 3.2)

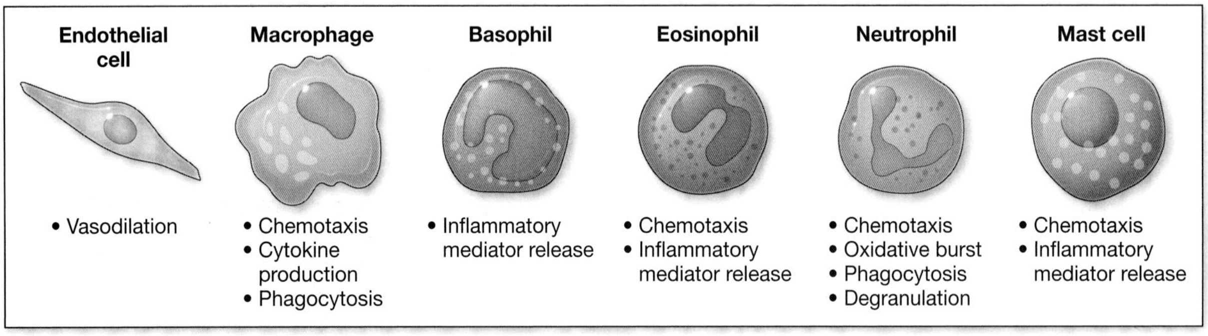

C3a, an anaphylatoxin, activates granulocytes, induces vasodilation, and acts as a chemoattractant.

(Sections 1.6, 1.7, 2.1, 2.6, 3.1, 3.2)

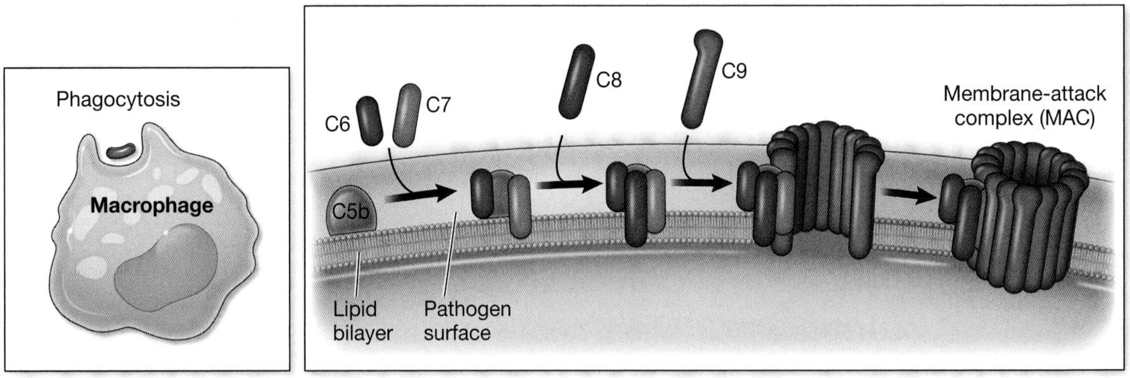

C3b, an opsonin, induces phagocytosis and facilitates formation of the membrane-attack complex.

(Sections 1.6, 1.7, 2.1, 2.2, 2.5, 3.1, 3.2)

3.3 When and how are the lectin and classical pathways initiated?

LEARNING OBJECTIVE

3.3.1 Explain the molecular mechanisms of the lectin and classical pathways of complement activation.

The alternative pathway is the first complement system to act due to the ubiquitous presence of factors involved in that pathway. However, inflammatory cytokines secreted during an innate immune response can trigger the lectin and classical pathways of complement activation. The lectin and classical pathways are means of activating the surfaces of specific microbes through the binding of specific molecules (sugars in the lectin pathway and phospholipids in the classical pathway). The molecules responsible for binding to microbes are also plasma proteins but are only released into the plasma when inflammatory response cytokines have been secreted during an infection. Systemic actions of these inflammatory cytokines such as IL-1, IL-6, and TNF-α cause hepatocytes of the liver to change the plasma proteins they are secreting and induce the acute-phase response.

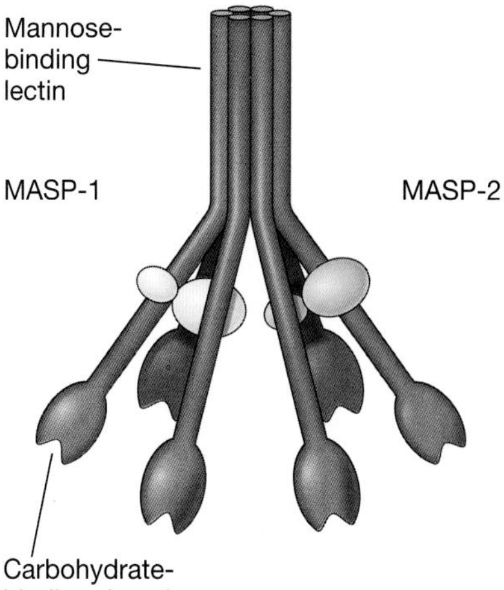

FIGURE 3.10 The opsonin mannose-binding lectin in complex with proteases MASP-1 and MASP-2 Mannose-binding lectin (purple) interacts with mannose-binding-lectin–associated serine proteases MASP-1 and MASP-2 (yellow), which are each present as a dimer in the interaction. (After J. Dobó et al. 2016. *Immunol Rev* 274: 98–111.)

Opsonins of the lectin and classical complement pathways

Two major binding proteins used in the lectin and classical pathways recognize their specific microbe surface molecule: mannose-binding lectin (MBL) in the lectin pathway and C-reactive protein in the classical pathway. Lectins are proteins that bind to specific carbohydrates. **Mannose-binding lectin** (**MBL**) is the key acute-phase protein involved in initiating the lectin pathway. MBL binds to mannose, a sugar on the pathogen surface, and its structure provides up to 18 sites for attachment. MBL is present in the circulation bound to two serine proteases, MASP-1 and MASP-2, which function during complement activation (**FIGURE 3.10**). MBL can also serve as an opsonin. Monocytes have receptors that can bind to MBL attached to the pathogen surface, thereby facilitating phagocytosis (**FIGURE 3.11**). While the surface of human cells contains the mannose carbohydrate, MBL cannot efficiently attach to human cells.

mannose-binding lectin (MBL) Acute-phase response protein that functions to initiate the lectin pathway of complement activation.

C-reactive protein An acute-phase response protein that functions to initiate activation of the classical complement pathway.

In addition to stimulating the expression of MBL by the liver, inflammatory cytokines also prompt the production of C-reactive protein, which can initiate the classical pathway of complement activation. **C-reactive protein** is made up of five identical subunits (**FIGURE 3.12**). It binds to phosphocholine within lipopolysaccharide in bacterial and fungal cell walls but not in human cell membranes. Like MBL, when C-reactive protein binds to a pathogen surface, it can serve as an opsonin and promote phagocytosis (**FIGURE 3.13**). Furthermore, the classical pathway is part of both the innate and adaptive immune responses, since the initiating component can be either an innate immune protein (C-reactive protein) or a soluble immunoglobulin. (To see how immunoglobulins activate complement, see Chapter 10.)

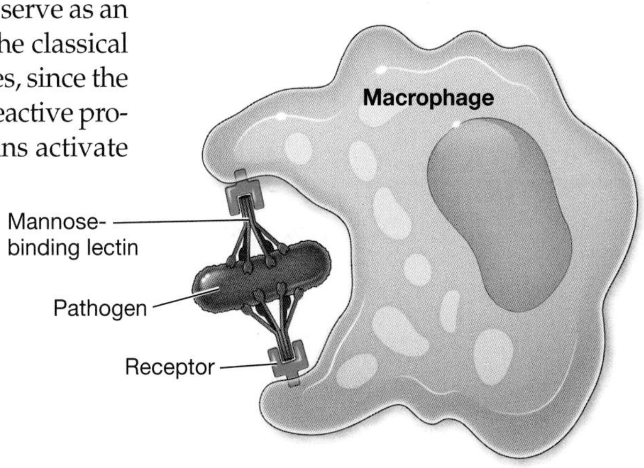

FIGURE 3.11 Action of mannose-binding lectin as an opsonin Macrophages produce receptors that bind to MBL on the pathogen surface. When the receptor binds to MBL, phagocytosis of the pathogen begins. (After P. Parham. 2014. *The Immune System*, 4th ed., W. W. Norton and Company.)

Activation of the lectin and classical complement pathways

When MBL binds to a pathogen surface, the associated protease dimer of MASP-1 and MASP-2 initiates the lectin complement pathway. MASP-1 autoactivates and triggers MASP-2 activation. The activated MASP-1/MASP-2 dimer has two important substrates it cleaves during the lectin pathway: C4, a protein that resembles C3, and C2, a serine protease similar to factor B.

Cleavage of C4 results in the formation of C4a and C4b fragments (**FIGURE 3.14**, Step 1). C4b is structurally similar to C3b, containing a thioester bond that is exposed upon

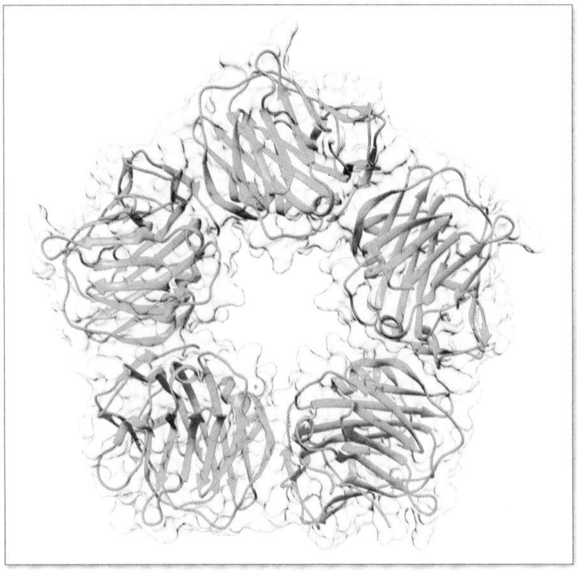

FIGURE 3.12 Structure of the opsonin C-reactive protein The acute-phase protein C-reactive protein binds to phosphocholine on the pathogen surface. (Data from M. A. Ramadan et al. 2002. *Acta Crystallogr D Biol Crystallogr* 58: 992–1001; redrawn using Chimera and PDB file 1LJ7.)

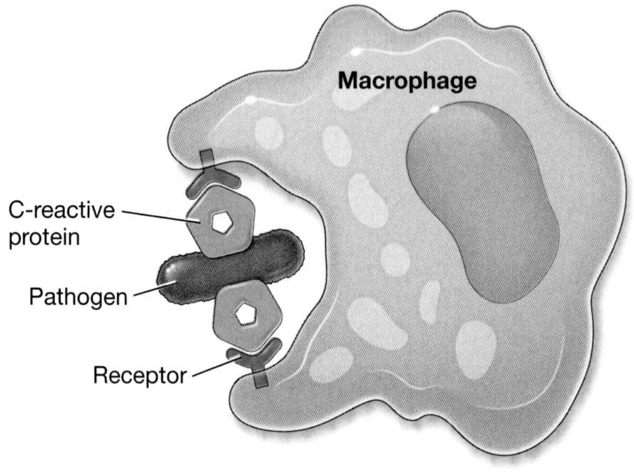

FIGURE 3.13 Action of C-reactive protein as an opsonin Macrophages produce receptors that bind to C-reactive protein on the pathogen surface. When the receptor binds to C-reactive protein, phagocytosis of the pathogen begins. (After P. Parham. 2014. *The Immune System*, 4th ed., W. W. Norton and Company.)

protease cleavage. This allows it to attach to the pathogen surface. C4a can function as a chemoattractant (similar to C3a and C5a), recruiting leukocytes to the site of C4 activation. However, its chemoattractant activity is weaker than that of C3a and C5a.

classical C3 convertase C3 convertase produced when C4b binds to C2b (both products of the lectin and classical pathways of complement activation).

C2 is also cleaved to form C2a and C2b fragments (Figure 3.14, Step 2). C4b and C2b can interact to form C4bC2b (Figure 3.14, Step 3), which acts as a C3 convertase known as the **classical C3 convertase**. (Note that most textbooks typically refer to the classical C3 convertase as C4bC2a. However, the convention of naming the complement cleavage products has moved toward naming the larger fragments produced during complement cleavage as the "b" component. The classical C3 convertase C4bC2b named here uses this convention of naming the complement cleavage products.) This enzyme functions as the C3 convertase in both the lectin and classical pathways (although its formation in the classical pathway differs). Because it is a C3 convertase, C4bC2b can cleave C3 into C3a and C3b (Figure 3.14, Step 4). As in the alternative pathway, the resultant C3b can act as an opsonin, and it can also interact with factor B to eventually produce the alternative C3 convertase C3bBb.

The classical pathway closely mirrors that of the lectin pathway, only differing in the initial binding molecule and the proteases involved in the cleavage of C4 and C2 (see Figure 3.14). Upon binding to a pathogen surface, C-reactive protein interacts with another complement protein, C1. C1 is similar in structure to the

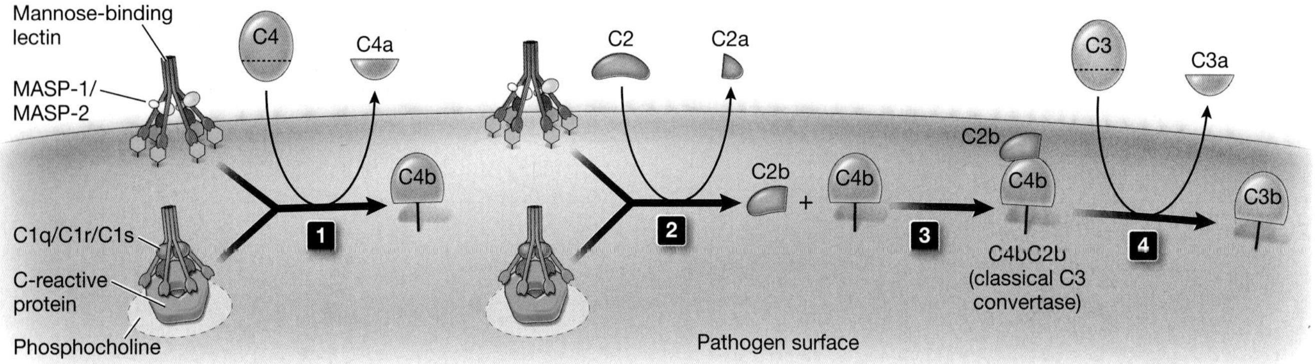

FIGURE 3.14 The lectin and classical pathways of complement activation The lectin pathway begins with the binding of MBL to carbohydrate on the pathogen surface. The associated protease MASP-1/MASP-2 can cleave both C4 (to form C4a and C4b) (Step 1) and C2 (to form C2a and C2b) (Step 2). C4b can be covalently attached to the surface of the pathogen and can interact with C2b to form C4bC2b (the classical C3 convertase) (Step 3). This C3 convertase can cleave C3 into C3a and C3b (Step 4). The classical pathway begins with the binding of C-reactive protein and its associated C1q/C1r/C1s to phosphocholine on the pathogen surface. The protease C1r/C1s can act in a similar manner fashion to MASP-1/MASP-2 to cleave both C4 (Step 1) and C2 (Step 2) and form the same classical C3 convertase (Step 3) formed by the lectin pathway.

MBL/MASP-1/MASP-2 complex. It contains three proteins: C1q, C1r, and C1s. C1q serves as the ligand of C-reactive protein. C1r and C1s are serine proteases that are cleaved and activated upon C1q binding to C-reactive protein. Active C1s cleaves both C4 (into C4a and C4b) and C2 (into C2a and C2b) in a manner identical to MASP-1/MASP-2 cleavage of C4 and C2 in the lectin pathway (Figure 3.14, Steps 1 and 2). The resultant C4b and C2b creates C4bC2b (Figure 3.14, Step 3), the same classical C3 convertase created in the lectin pathway. To appreciate the evolutionary conservation of the complement system and the different pathways of complement activation, see **EVOLUTION AND IMMUNITY**.

While in this chapter we are discussing complement activation in the context of soluble proteins of the innate immune system, in Chapter 10 we will discover that soluble immunoglobulins can also activate complement. Immunoglobulins can act similarly to C-reactive protein by interacting with their antigen on a pathogen surface and recruiting the same protease that C-reactive protein binds to activate complement fixation on foreign molecules and cells.

CHECKPOINT QUESTIONS

1. Which acute-phase proteins are the two initiating opsonins of the lectin and classical pathways of complement activation?
2. Which complement cleavage products constitute the classical C3 convertase?

EVOLUTION AND IMMUNITY

Complement Component 3 and Complement Pathways

As we have seen, the complement system functions to opsonize, or tag, foreign organisms for removal by phagocytes of the immune system. One extremely important protein in the complement system is C3, which is the tag that gets attached to foreign organisms. One can envision that an extremely ancient complement system would require the action of three molecules. The first molecule would be C3, which binds to the surface of the foreign organism. The second molecule would be another factor similar to a molecule known as factor B, which combines with C3 to form an enzyme known as C3 convertase (responsible for processing and amplifying the cleavage of C3). The final molecule would have to be a C3 receptor on the surface of professional phagocytes, which recognizes C3 on the surface of the foreign organism.

While we know these proteins function in higher eukaryotic organisms, the hypothesized vision of a primitive complement system has been brought to fruition through the discovery of these molecules in invertebrate organisms such as echinoderms. These organisms have been shown to express C3 and a factor B homolog (functional equivalent), which aid in initial tagging and amplification of the complement system. Although no formal C3 receptor has been identified to date, it has been established that echinoderm phagocytes take up C3-opsonized organisms more efficiently. Interestingly, this evolutionarily ancient complement system is induced in the presence of bacteria, showing its role in bacterial removal and its evolution of control. Indeed, other C3 homologs have been isolated from more distantly related invertebrates, including *Drosophila*, which has at least four C3 homologs, termed thioester-containing proteins. These proteins are thought to have a role in *Drosophila* immune function, as they are upregulated during bacterial infection.

The complement system is also evident in the genomes of chordates, where homologs of C3 and factor B are often seen. Furthermore, it is easier to establish potential C3 receptors (such as integrins) in the genomes of chordates. Finally, several regulatory proteins to the complement system that drive complement expression and secretion can be found in the genomes of chordates.

Evolution of the lectin and classical pathways of complement activation

Ficolins, which are related to MBLs, have been identified in both vertebrates and some invertebrates, including urochordates. C-reactive protein homologs have also been identified in urochordates. As with the proteolytic processing of C3, the lectin and classical pathways use cleavage events to initiate the pathway. The lectin pathway uses MASPs and the classical pathway uses C1q, C1r, and C1s. Homologs of both MASPs and the C1q/C1r/C1s system of complement activation have been identified in urochordates. It is apparent that while the minimalist complement system (C3, factor B, C3 receptor) was originally identified in echinoderms, the lectin and classical pathways are younger on the evolutionary tree, likely due to an added ability to mark and target pathogens with the complement system.

Making Connections: The Acute-Phase Response and Complement Activation

Cytokine activation occurs when a macrophage recognizes a pathogen-associated molecular pattern (PAMP) via a pattern recognition receptor (PRR). Some cytokines prompt the acute-phase response, releasing opsonins and inducing the lectin and classical pathways of complement activation.

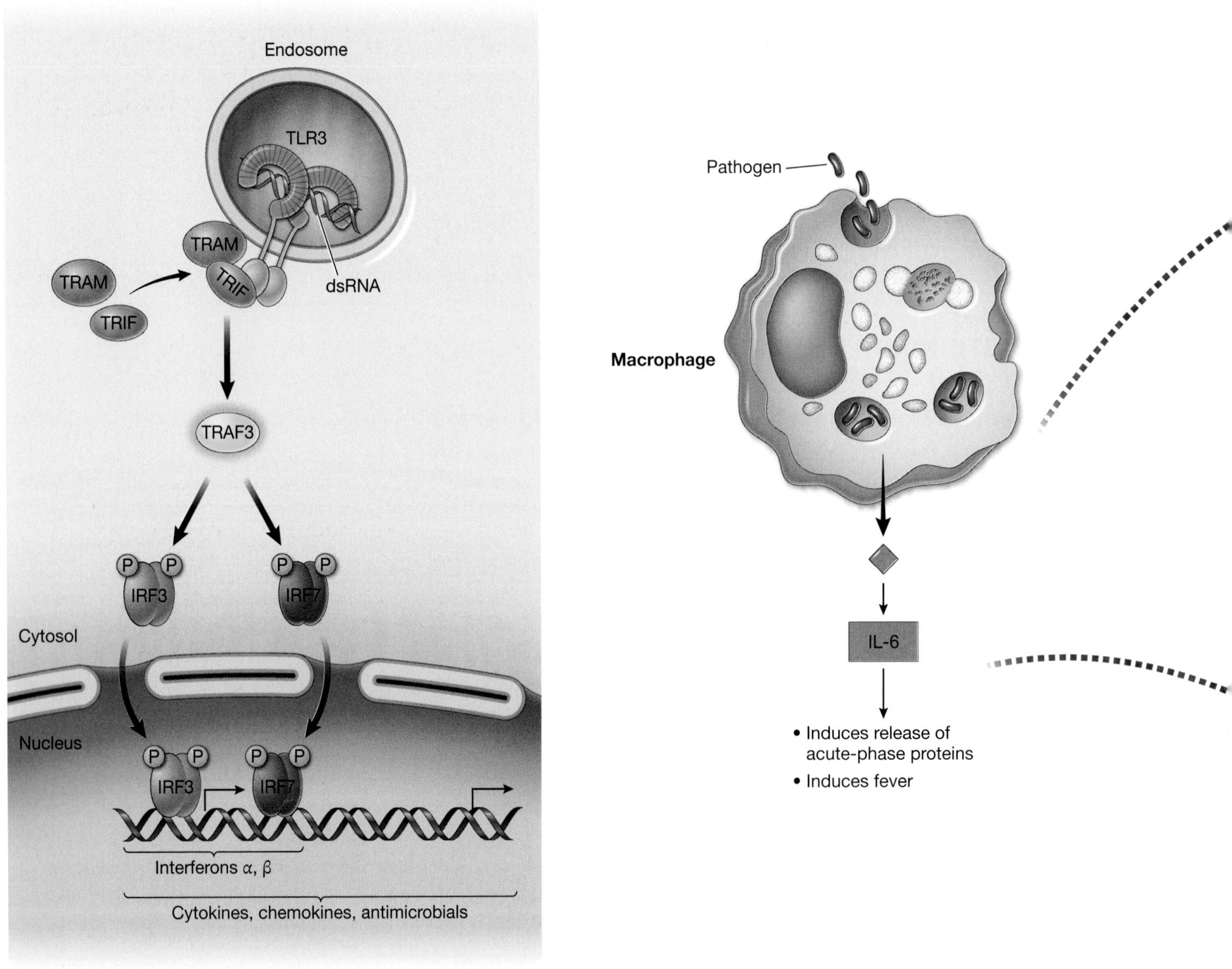

Macrophage activation upon PRR engagement causes cytokine release. The cytokine IL-6 drives release of mannose-binding lectin and C-reactive protein from the liver.

(Sections 1.6, 1.7, 2.5)

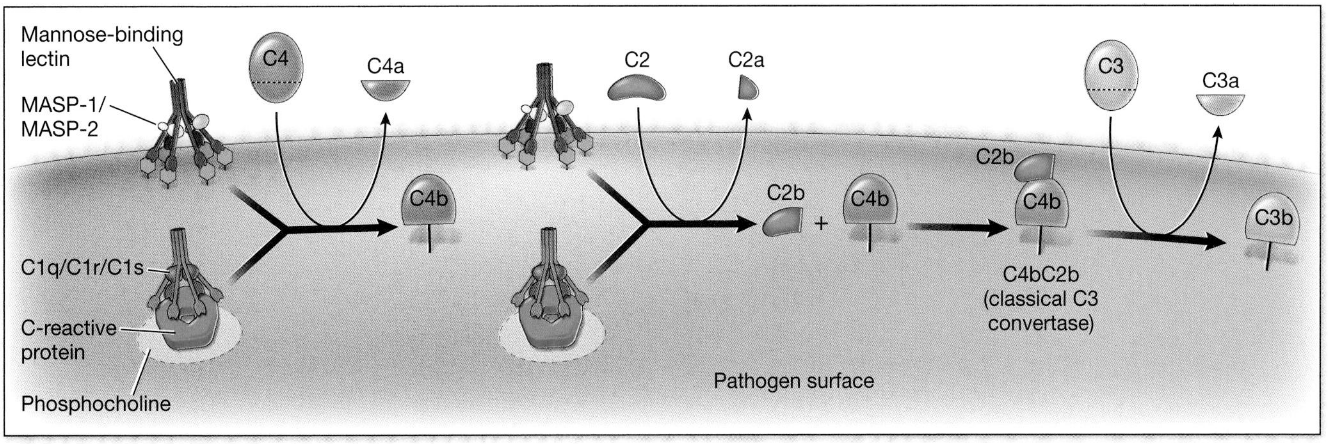

The acute-phase proteins mannose-binding lectin and C-reactive protein activate complement through the production of the classical C3 convertase.

(Sections 2.2, 2.3, 2.4, 2.5, 2.6, 3.3)

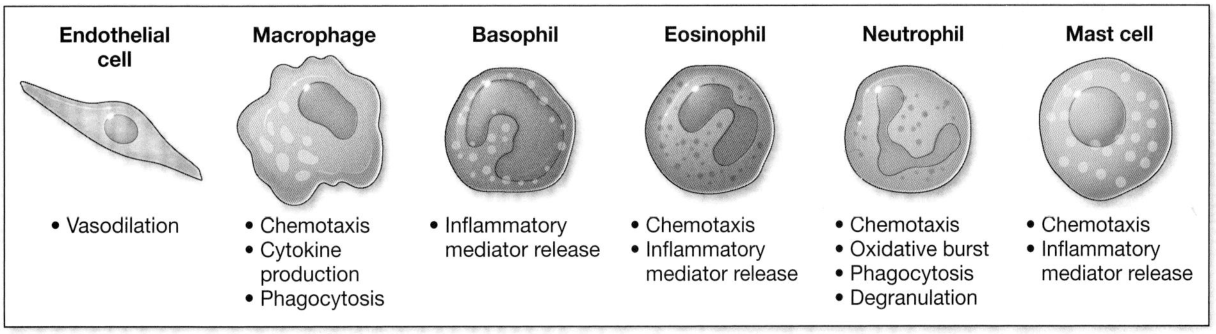

Activation of complement via the lectin and classical pathways mirrors the outcomes of the alternative pathway.

(Sections 1.6, 1.7, 2.1, 2.2, 2.5, 3.1, 3.2, 3.3)

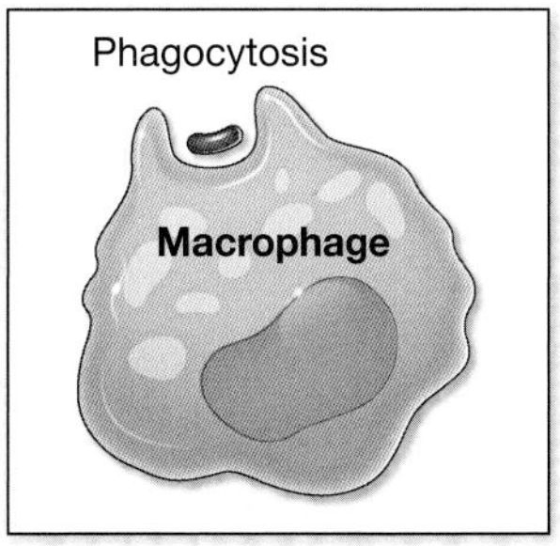

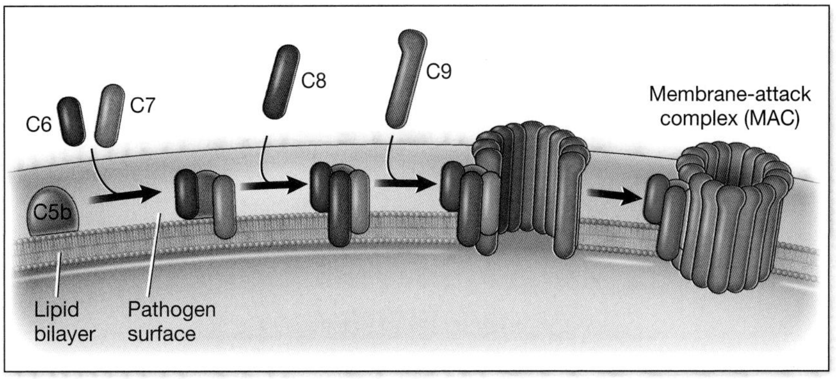

3.4 Which plasma proteins work with complement to fight infection?

LEARNING OBJECTIVE

3.4.1 Define the roles of other plasma proteins in the innate immune response.

Complement plays an important role in marking pathogens for destruction and initiating pathogen lysis via the MAC. Several other plasma proteins work to prevent the spread of microorganisms into neighboring tissues.

The *blood coagulation system* involves a cascade of proteolytic enzymes that play a vital role in preventing microorganisms from entering the circulatory system. When a wound occurs, this system kicks in to minimize blood loss, and clotting factors prevent the movement of pathogens into the bloodstream. *Platelets* play a key role in activating the coagulation cascade. They also secrete signaling molecules (such as prostaglandins), hydrolytic enzymes, and other factors that aid in both wound repair and the innate immune response. *Bradykinin* is a plasma protein that is activated when tissue damage occurs. It increases vasodilation and attracts important innate immune response mediators to the site of infection.

Protease inhibitors

Many pathogens express proteases to mediate digestion of extracellular matrix proteins and facilitate invasion into neighboring tissues. These enzymes break down proteins that resist pathogen migration and degrade antimicrobial peptides secreted during an innate immune response. To combat the actions of proteases, plasma contains protease inhibitors. Approximately 10% of the proteins in plasma are protease inhibitors.

An important protease inhibitor is the 180-kDa glycoprotein α_2-macroglobulin, which is structurally related to the complement protein C3 (both contain a thioester bond). The α_2-macroglobulin contains a region that serves as a substrate for microorganism proteases. However, when a pathogen's protease cleaves this region, it initiates a mechanism that inhibits the pathogen's protease activity (**FIGURE 3.15**). Cleavage exposes the protease to covalent linkage to the internal thioester, thereby locking the protease to α_2-macroglobulin. Cleavage of α_2-macroglobulin also causes it to wrap around the protease, preventing the protease from being able to reach and cleave any other substrates. Protease complexes bound to α_2-macroglobulin are cleared by macrophages, fibroblasts, and hepatocytes.

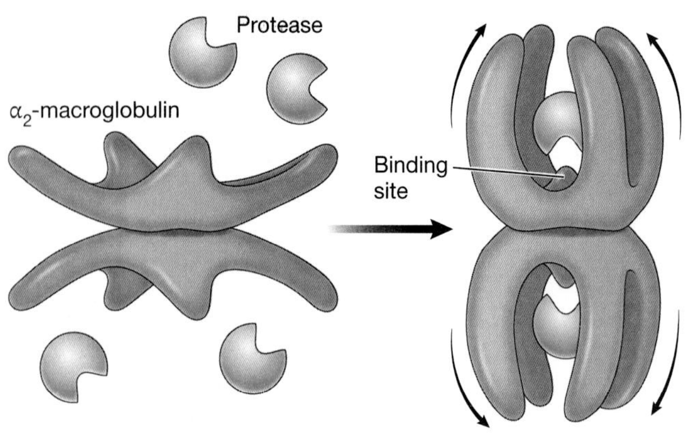

FIGURE 3.15 Inhibition of proteases by α_2-macroglobulin Proteases in close vicinity to α_2-macroglobulin will bind to a binding site on α_2-macroglobulin, which mimics a protease cleavage site. Binding of the protease causes α_2-macroglobulin to close around the protease and render it inactive.

Defensins

Defensins are a group of antimicrobial peptides that contain 35 to 40 amino acids. Defensins are amphipathic, with both hydrophobic and hydrophilic regions. The hydrophobic regions allow defensins to penetrate pathogen membranes and target pathogens for destruction.

There are two classes of defensins: α-defensins, with 6 subtypes, and β-defensins, with 32 different genes of β-defensins present in the human genome. The various subtypes differ in regard to the cells that express them and the tissues they defend (**TABLE 3.2**).

The α-defensins are expressed by both neutrophils and Paneth cells. Paneth cells are specialized epithelial cells of the small intestine. They are located in the crypts at the base of microvilli. In addition to defensins, Paneth cells also secrete lysozyme and other important innate immune response molecules.

β-defensins are expressed by many different types of epithelial cells throughout the skin and the respiratory and urogenital tracts. The physical environment provided by sweat, tears, mucus, and the gut lumen promotes efficient defensin action.

Defensins are among the oldest immune system mechanisms in evolutionary history. Pathogens have been repeatedly exposed to these peptides and have evolved mechanisms to escape their action. And in response, humans have evolved a variety of defensins to fend off many different pathogen types. As time moves forward, this evolutionary "arms race" continues.

TABLE 3.2 | Defensin Classes and Subtypes

Class	Gene Name	Expressing Cells
α-defensins	α-defensin 1 α-defensin 1B α-defensin 3 α-defensin 4	Neutrophils Natural killer cells T cells
	α-defensin 5 α-defensin 6	Paneth cells
β-defensins	β-defensin 1 β-defensin 2 β-defensin 3 β-defensin 103B β-defensin 106A β-defensin 106B β-defensin 107A	White blood cells Epithelial cells

Source: J. Jarczak, et al. 2013. *Hum Immunol* 74: 1069–1079.

CHECKPOINT QUESTION

1. What is the role of α_2-macroglobulin in the innate immune response?

3.5 | What happens when the complement system malfunctions?

LEARNING OBJECTIVE

3.5.1 Predict the onset of disease states given a deficiency in the complement system.

We have seen that the complement system plays an important role in pathogen recognition and destruction. Because of its importance in the innate immune system, deficiencies in any of the genes encoding for complement proteins can increase the susceptibility of an individual to various types of infections (**TABLE 3.3**). Interestingly, the complement system may play a role in suppressing neuronal growth and promoting neuronal loss (see **EMERGING SCIENCE**).

The immune system relies on complement activation to clear many gram-positive organisms that tend to evade phagocytic cells by masking PAMPs from PRRs. Thus, many complement-component deficiencies lead to abnormal clearance of gram-positive pyogenic (pus-forming) bacteria, including *Streptococcus pyogenes* and *Staphylococcus aureus*, because of an inability to efficiently opsonize the bacteria for destruction.

Deficiencies in early components of the complement system also cause improper clearance of soluble immune complexes, which are small foreign particles opsonized by either complement or antibodies. Accumulation of soluble immune complexes can lead to hypersensitivity responses and autoimmune disorders such as lupus.

Other mutations in early complement genes, especially C4, can lead to recurring infections of both gram-negative and gram-positive bacteria, including *Streptococcus pneumoniae*, *Neisseria meningitidis*, and *Haemophilus influenzae*.

As previously noted, C3 is one of the most important complement proteins. Individuals who lack C3 due to a genetic defect or the malfunction of a regulatory protein suffer from recurring bacterial infections and other diseases. Interestingly, the first patient diagnosed with a C3 deficiency had suffered so many severe recurring bacterial infections that the original diagnosis was a

TABLE 3.3 | Common Diseases Resulting from Complement Deficiencies

Deficient Complement Protein	Common Diseases
C1	Systemic lupus erythematosus Rheumatoid arthritis
C1-inhibitor (C1-inh)	Hereditary angioedema
C2	Recurring bacterial infections Systemic lupus erythematosus
C3	Recurring bacterial infections
C4	Systemic lupus erythematosus Rheumatoid arthritis
Factor H	Recurring bacterial infections Kidney disease Macular degeneration
Properdin (factor P)	*Neisseria* infection susceptibility (meningitis)

Source: K.R. Mayilyan. 2012. *Protein Cell* 3: 487–496.

EMERGING SCIENCE

Does the complement system play a role in regulating neuronal growth?

Article

S. L. Peterson, H. X. Nguyen, O. A. Mendez, and A. J. Anderson. 2017. Complement protein C3 suppresses axon growth and promotes neuron loss. *Scientific Reports* 7:12904.

Background

Much of this chapter focused on the role of the complement system in the innate and adaptive immune responses. However, recent research has suggested that the complement system is involved in other regulatory mechanisms within the body, including central nervous system development, cell proliferation and differentiation, and remodeling of neuronal synapses. Scientists wanted to analyze the potential mechanistic connection between the complement system and neuronal growth.

The Study

Researchers wanted to study the importance of the complement protein C3 in the regeneration of axons following spinal cord injury in mice and in vitro in neurite growth. In vivo studies in which a spinal cord injury was surgically induced were conducted in mice containing wild-type copies of C3 ($C3^{+/+}$) or mice knocked out for C3 ($C3^{-/-}$). The scientists found that $C3^{-/-}$ mice had much longer regenerating sensory axons in their surgical spinal cord injury compared to their wild-type littermates (**FIGURE ES 3.1**). The researchers correlated this finding to demonstrate that the presence of C3 in a culture used to observe neurite growth (neurons grown in vitro on myelin) caused a significant decrease in neurite growth and length and the number of neurons. Further characterization demonstrated that C3 cleavage was likely responsible for this inhibitory activity.

Conclusion

The scientists' findings suggest that C3 has a negative impact on axon growth and neuron number in vitro and in vivo. They correlated these findings to normal C3 processing, suggesting that proteases associated with myelin may be responsible for the aberrant processing of C3 following spinal cord injury.

Think About...

1. Would you suggest the development of a drug targeting C3 processing as a means to treat spinal cord injury? Why or why not?
2. If these data are correct, might you predict that individuals suffering from spinal cord injury are more susceptible to infections of the central nervous system? Why or why not?
3. How might you use this study as a means to prevent C3 processing and subsequent axon growth inhibition in patients?

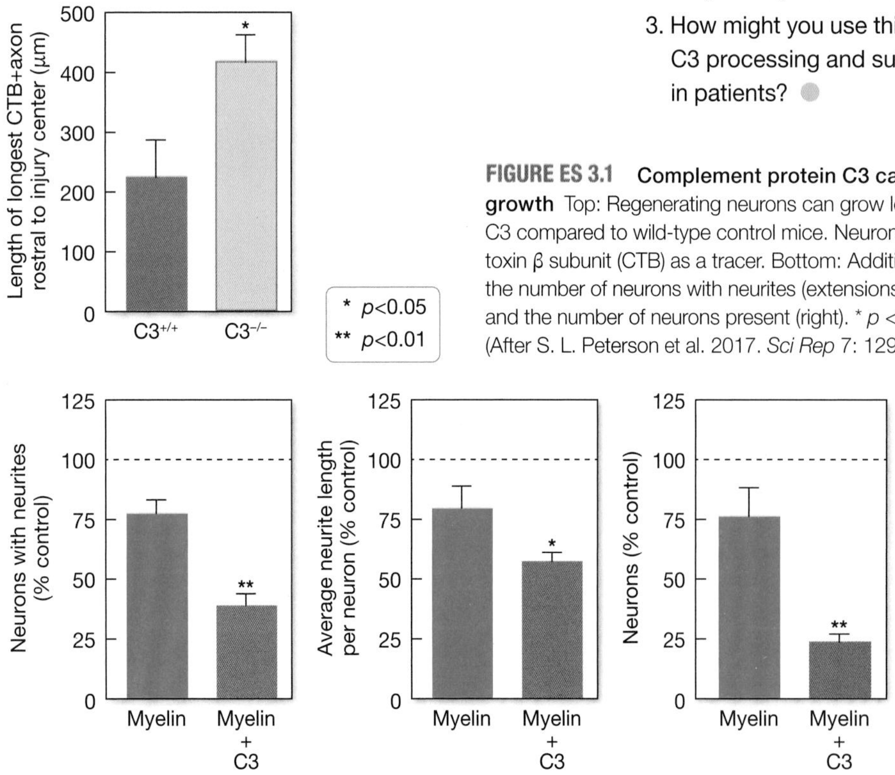

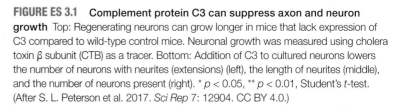

FIGURE ES 3.1 Complement protein C3 can suppress axon and neuron growth Top: Regenerating neurons can grow longer in mice that lack expression of C3 compared to wild-type control mice. Neuronal growth was measured using cholera toxin β subunit (CTB) as a tracer. Bottom: Addition of C3 to cultured neurons lowers the number of neurons with neurites (extensions) (left), the length of neurites (middle), and the number of neurons present (right). * $p < 0.05$, ** $p < 0.01$, Student's *t*-test. (After S. L. Peterson et al. 2017. *Sci Rep* 7: 12904. CC BY 4.0.)

complete deficiency in B-cell signaling. The patient's lack of C3 demonstrated the central role of C3 in innate immune function.

Malfunction of the MAC due to mutations in any of the MAC genes can lead to an increased infection rate, especially by *N. meningitidis*. This suggests that the MAC is especially important in the clearance of these bacteria.

CHECKPOINT QUESTION

1. What are some common complications associated with a deficiency in the complement system?

Summary

3.1 What is the complement system and how is it activated?

- The complement system works through a variety of plasma proteins to promote pathogen recognition and destruction, especially through the action of the complement protein C3.
- The three pathways of complement activation used by the innate immune system are the alternative, lectin, and classical pathways.

3.2 When and how is the alternative pathway initiated?

- The alternative pathway is initiated by the action of a C3 convertase to form C3a (a chemoatttractant) and C3b (an opsonin) to facilitate pathogen clearance.
- Activation of the alternative pathway promotes formation of the MAC, which uses other complement proteins to create pores in pathogen membranes, disrupting membrane integrity.

3.3 When and how are the lectin and classical pathways initiated?

- The lectin and classical pathways of complement activation use different initiator proteins (MBL and C-reactive protein, respectively) but result in the formation of C3a and C3b through the action of the same classical C3 convertase.
- The complement system is regulated to ensure that host cells are not actively marked by activation of the complement system.

3.4 Which plasma proteins work with complement to fight infection?

- Plasma proteins, including defensins, protease inhibitors, and proteins of the blood coagulation system, function to further target pathogens and prevent their spread.

3.5 What happens when the complement system malfunctions?

- Deficiencies in components of the complement system can result in a variety of immune disorders, including recurring infections and allergic responses.

Review Questions

3.1 What is the main output of all three pathways of complement activation? What is the role of the molecules made during complement activation?

3.2 What are the three pathways of complement activation in the order of their timing during an innate immune response?

3.3 Which complement protein is responsible for forming the transmembrane pore of the MAC?

A. C5
B. C6
C. C7
D. C8
E. C9

3.4 **a.** Which two complement proteins interact to form the classical C3 convertase?

b. What might you predict would be the effect on an individual lacking a functional C1 gene?

3.5 Which plasma protein acts as a protease inhibitor to limit pathogen invasion into neighboring tissue?

A. Defensins
B. α_2-macroglobulin
C. TNF-α
D. Bradykinin
E. C3

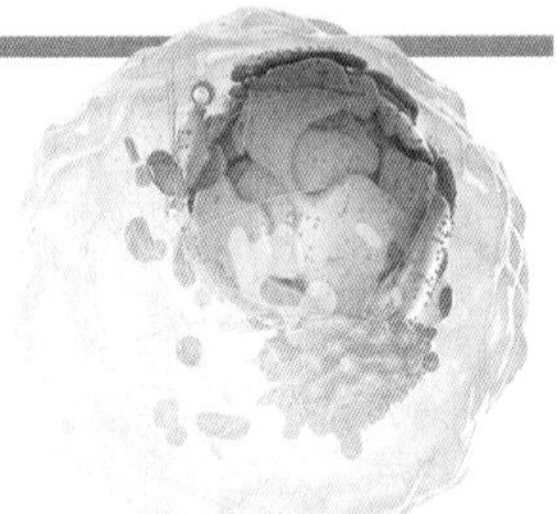

CASE STUDY REVISITED

Otitis Media

A boy, age 5, is admitted to the emergency room with symptoms of fever and an earache. His mother tells the attending physician that the boy often suffers from ear infections and she worries that this may be another one. Furthermore, she mentions that the child commonly suffers from skin abscesses that can be treated with topical antibiotic cream. Visual observation confirms the presence of an ear infection. The attending physician hypothesizes that the recurring skin abscesses may be a result of a pyogenic bacterial infection, consistent with the recurring ear infections. The physician decides to order a complete blood analysis to determine the underlying cause of the child's recurring skin and ear infections. A complete blood analysis shows that he completely lacks factor I in his plasma.

Think About...

Questions for individuals

1. Define pyogenic bacteria.
2. What is the role of factor I in regulation of complement?

Questions for student groups

1. What would you expect a specific analysis of plasma levels of C3 to reveal? Why?
2. Is the patient's history and blood analysis consistent with the bacterial infection seen? Why or why not?
3. What might be a proper treatment for this infection?

MAKING CONNECTIONS Key Concepts	COVERAGE (Section Numbers)
Action of the alternative pathway	2.1, 3.1, 3.2
Action of C3a and C3b, including innate immune responses	1.6, 1.7, 2.1, 2.2, 2.5, 2.6, 3.1, 3.2
Macrophage activation and the induction of the acute-phase response	1.6, 1.7, 2.5
Acute-phase protein action in complement activation	2.2, 2.3, 2.4, 2.5, 2.6, 3.3
Connection of three complement activation pathways	1.6, 1.7, 2.1, 2.2, 2.5, 3.1, 3.2, 3.3
Dendritic cell interaction with T cells	2.1, 2.8, 4.2, 4.5, 4.6, 6.4, 6.5, 6.6, 7.3, 7.6, 7.7

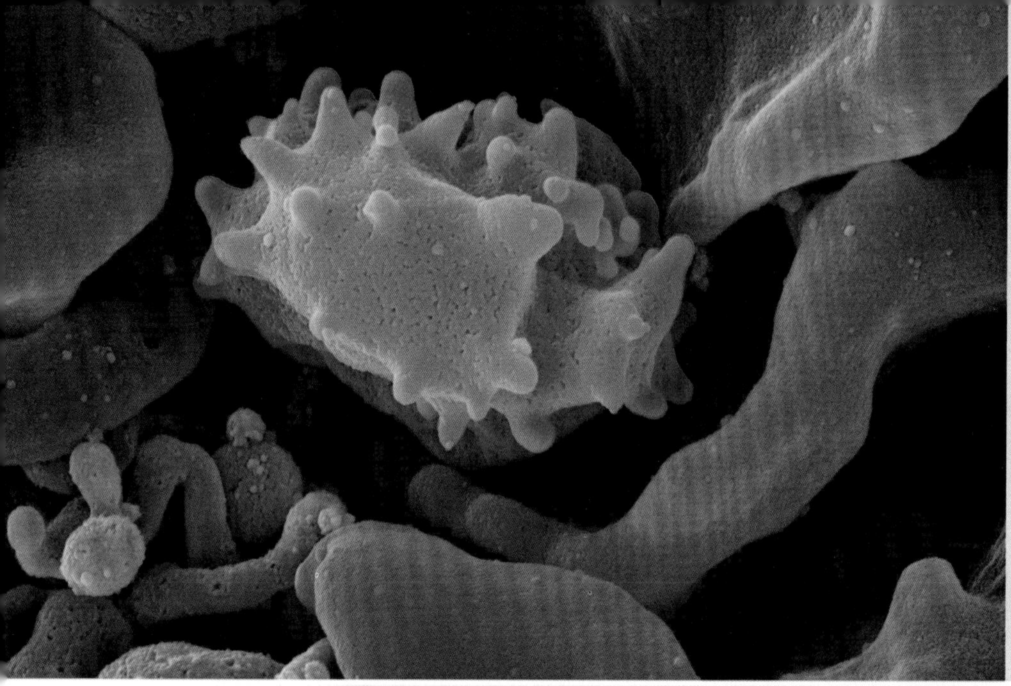

4

Overview of Adaptive Immunity

CASE STUDY: Haruto's Story

No one ever warns you about the first 6 months of parenthood, Mei thinks to herself as the baby monitor lights up again from the crying Haruto. Mei jabs Yuki in his side, waking him from the typical light sleep they have both been getting as new parents. "It's your turn," Mei says, and reluctantly, Yuki rolls out of bed and walks to the nursery down the hall. Suddenly, Mei hears Yuki yell, "Mei, you better come down here quick! This definitely isn't normal!" Mei hopes that this will just be another night battling diarrhea, but instead, the night takes a downward turn.

Mei enters the nursery, where Yuki has the lights on and is holding Haruto, trying to soothe him. Mei notices that Haruto has a rash all over his face, which was something she certainly had never seen before. Both Mei and Yuki worry for Haruto's health and decide to rush him to the emergency room.

In order to understand what Haruto is experiencing, and to answer the case study questions at the end of this chapter, we will examine the function of our adaptive immune system. Our adaptive immune system employs two key lymphocytes, B cells and T cells, that specialize in pathogen recognition and destruction. Each circulating B cell and T cell expresses a specific receptor involved in pathogen recognition. Circulating T cells that are activated when their receptors recognize a pathogen function in the cellular arm of our adaptive immune system, and circulating B cells that are activated when their receptors recognize a pathogen function in the humoral arm of our adaptive immune system. During development

QUESTIONS Explored

4.1 How does adaptive immunity differ from innate immunity?

4.2 How are pathogens recognized by cells of the adaptive immune system?

4.3 What is the timeline of an adaptive immune response?

4.4 What mechanism provides diversity of T-cell and B-cell receptors?

4.5 How does tolerance to self-molecules occur given the diversity of our adaptive immune system?

4.6 What is the importance of the major histocompatibility complexes in adaptive immunity?

4.7 What roles do T-cell receptors play in adaptive immunity?

4.8 What roles do immunoglobulins play in adaptive immunity?

4.9 What is immunological memory and how is it developed?

4.10 Why is immunological memory important to an adaptive immune response?

of these lymphocytes, elegant mechanisms of genetic recombination alter the genes of these receptors in a manner that allows them to be expressed. Since both B cells and T cells undergo these recombination events, we will explore this mechanism in detail in this chapter. This chapter will then serve as a "view from 30,000 feet," providing an overview of the adaptive immune system, which will be examined in detail in the following six chapters.

4.1 How does adaptive immunity differ from innate immunity?

LEARNING OBJECTIVE

4.1.1 Compare and contrast the antigen-recognition mechanisms of the innate and adaptive immune system.

As discussed in Chapters 2 and 3, the innate immune system plays a vital role in combating pathogens during the initial stages of an infection. Plasma proteins, especially those of the complement system, promote pathogen recognition and destruction. Innate immune cell receptors bind to complement attached to the pathogen surface or recognize pathogen-associated molecular patterns (PAMPs) at the pathogen surface. For example, infection with gram-negative bacteria such as *Escherichia coli* is likely to induce an inflammatory response due to the recognition of lipopolysaccharide (LPS), the major component of the outer membrane of the bacteria, by Toll-like receptor 4 (TLR4). The bacteria will also be targeted by complement proteins, leading to the fixation of C3b on its surface or the formation of the membrane-attack complex. Thus, the recognition and targeting of pathogens by the innate immune system serves to limit an infection.

However, because pathogen recognition relies on complement fixation on the pathogen surface or pattern recognition receptors that recognize common pathogen macromolecules, this process is limited and not capable of recognizing all types of diverse pathogens. So even though the innate immune system is equipped with a variety of infection-fighting mechanisms, due to the lack of diverse receptor types, there are times when these alone are not sufficient to halt an infection.

When the innate immune system defenses are not adequate, the adaptive immune system comes into play. The adaptive immune response is based on *specificity* and *diversity*—it functions by recognizing and combating specific antigens. T-cell receptors and immunoglobulins from B cells have the ability to recognize very specific foreign molecules. Due to the elegant process of lymphocyte development, T-cell receptors and immunoglobulins are almost limitless in the diversity of molecules they can recognize. When this recognition occurs, the T cell or B cell rapidly divides and produces differentiated cells that target the pathogen for destruction. Per our example of *E. coli* infection, only T cells and B cells that produce receptors that recognize *E. coli* antigens will become activated, specifically those with receptors to cell-surface *E. coli* adhesin molecules that allow the bacteria to interact with target cells.

A key aspect of adaptive immunity is the subset of differentiated cells produced during an initial adaptive immune response. These cells monitor for a subsequent occurrence of the same type of infection, serving as a *memory* of pathogen encounter and allowing the adaptive immune system to respond more quickly and more robustly than innate defenses. So while the end goal of both the innate and adaptive immune responses is the same—fighting infection—the cells of the adaptive immune system are capable of *recognizing specific pathogens and combating previously encountered pathogens more rigorously.*

CHECKPOINT QUESTION

1. Explain the concept of memory as it relates to the adaptive immune system.

4.2 How are pathogens recognized by cells of the adaptive immune system?

LEARNING OBJECTIVE

4.2.1 Name and describe molecules of the adaptive immune system that play a role in pathogen recognition.

Cells of the innate immune system rely on receptors to recognize molecular patterns on the pathogen surface or complement that has become fixed to the surface. In contrast, adaptive immune system cells use specialized receptors—T-cell receptors and membrane-bound immunoglobulins (B-cell receptors)—with both a high degree of specificity for the molecule they recognize and a high degree of diversity among lymphocytes. Their specificity and diversity are driven by antigen-binding sites at particular locations in their structure.

T-cell receptors and MHC/HLA molecules

The T-cell receptor is a transmembrane protein with an almost entirely extracellular structure. It contains two protein chains, most typically an α chain and a β chain, although some T-cell receptors incorporate γ and δ chains. Each chain of the T-cell receptor contains both a **constant region**, which is very similar in amino acid structure among all T-cell receptors, and a **variable region**, with variances in amino acids from receptor to receptor (**FIGURE 4.1**). The variable region contains the antigen-binding site.

Differences in the variable region allow receptors in different T cells to recognize different antigens, providing diversity in the T cells in circulation. However, each T-cell receptor is only capable of recognizing one specific antigen. Since each T cell expresses a unique type of receptor with a different antigen-binding site, each T cell offers potential protection against only one pathogen.

The vast majority of T-cell receptors can only recognize small peptides as antigens, and these peptides must be presented to them via cell-surface protein receptors of the **major histocompatibility complex** (**MHC**) family in mice

constant region Region of a T-cell receptor, immunoglobulin, or antibody where the amino acid sequence is similar among molecules.

variable region Region of a T-cell receptor, immunoglobulin, or antibody where the amino acid sequence is different among molecules; the antigen-binding site of T-cell receptors, immunoglobulins, and antibodies is created by the interaction of variable regions of different receptor subunits (e.g., the variable regions of the T-cell receptor α and β chains interact to form a domain containing the antigen-binding site).

major histocompatibility complex (MHC) Family of immune proteins in mice and other higher vertebrates that includes cell-surface proteins responsible for the presentation of peptide antigens to T cells. Referred to as HLA proteins in humans.

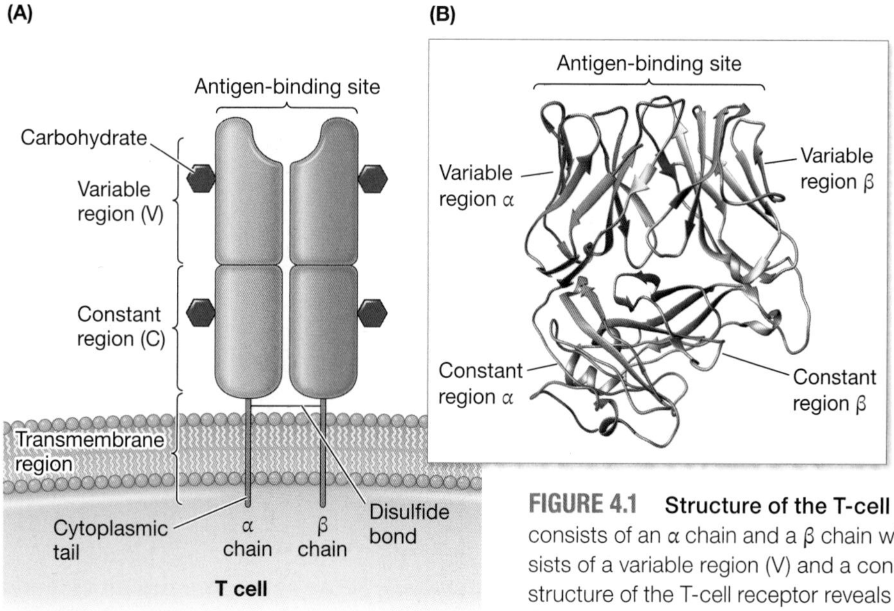

FIGURE 4.1 Structure of the T-cell receptor (A) The T-cell receptor typically consists of an α chain and a β chain within the cell membrane. Each chain consists of a variable region (V) and a constant region (C). (B) The three-dimensional structure of the T-cell receptor reveals the immunoglobulin-like domain folds of the variable and constant regions as well as the interaction between both chains. (B, data from K. E. J. Rödström et al. 2015. *PLOS One* 10: 01318; redrawn using Chimera and PDB file 4UDT.)

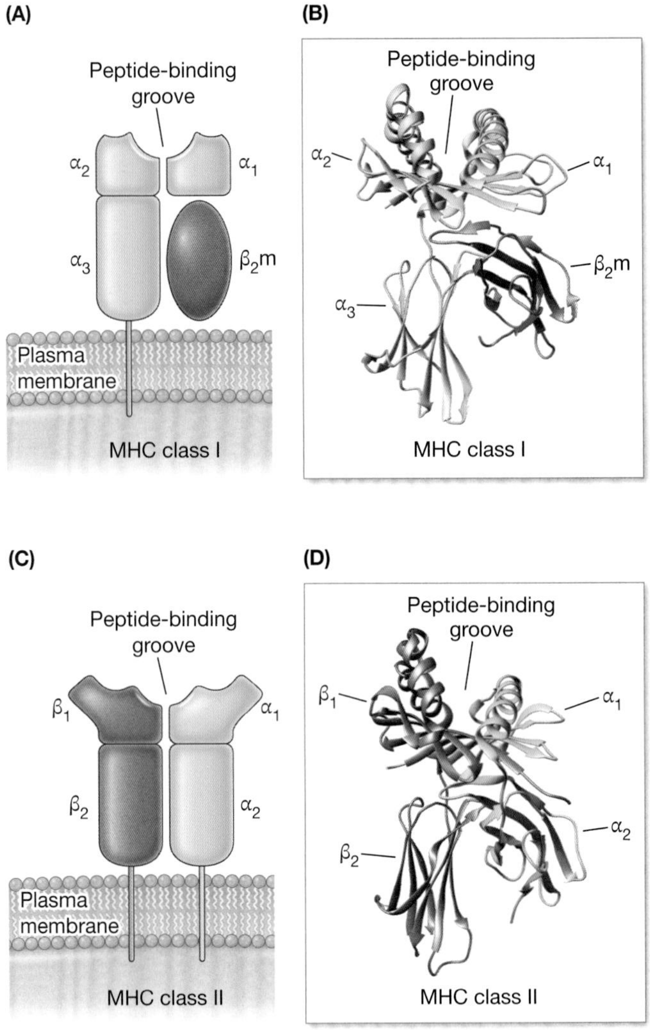

FIGURE 4.2 Structure of MHC class I and class II molecules (A) MHC class I molecules contain two different polypeptides, an α chain and β_2-microglobulin (β_2m). The α chain contains three motifs (α_1, α_2, and α_3). (B) The peptide-binding groove of MHC class I molecules is sandwiched between two alpha-helices, one from the α_1 motif and one from the α_2 motif. (C) MHC class II molecules also contain two different polypeptides, an α chain and a β chain. Each chain contains two different motifs (α_1, α_2 for the α chain and β_1, β_2 for the β chain). (D) The peptide-binding groove resembles that of MHC class I molecules, but it is sandwiched between the α and β chains. (B, data from N. L. La Gruta et al. 2008. *Proc Natl Acad Sci USA* 105: 2034–2039. ©2008 National Academy of Sciences, USA; redrawn using Chimera and PDB file 3BUY; D, data from L. J. Stern et al. 1994. *Nature* 368: 215–221; redrawn using Chimera and PDB file 1DLH.)

human leukocyte antigen (HLA) Family of immune proteins in humans that includes cell-surface proteins responsible for the presentation of peptide antigens of T cells. Referred to as MHC proteins in other vertebrates.

MHC class I A major histocompatibility protein consisting of two polypeptides, an α chain transmembrane protein and soluble β_2-microglobulin; displays peptides produced through intracellular processing of antigens.

MHC class II A major histocompatibility protein consisting of two transmembrane polypeptides, an α chain and a β chain; displays peptides produced through extracellular processing of antigens.

CD4 T-cell coreceptor involved in selection and binding to MHC class II molecules.

CD8 T-cell coreceptor involved in selection and binding to MHC class I molecules.

polymorphic Genes that encode for the same protein; represented in a large number of different alleles in the population.

allele Alternative form of a gene located at the same location on a chromosome.

and other higher vertebrates or the **human leukocyte antigen** (**HLA**) family in humans. MHC class I and class II glycoproteins bind to peptides that have been processed through proteolysis in the cytosol or lysosomal digestion and present them on the surface of cells. **MHC class I** molecules typically present intracellular antigens, and **MHC class II** molecules present extracellular antigens (**FIGURE 4.2**).

The MHC molecule presenting the antigen binds to both the T-cell receptor and a coreceptor expressed on the T-cell surface. T cells can express one of two coreceptors: CD4 and CD8. **CD4** receptors bind to MHC class II molecules, and **CD8** receptors bind to MHC class I molecules (**FIGURE 4.3**).

MHC molecules are expressed by a variety of genes in higher vertebrate chromosomes. These genes are highly **polymorphic**, which means they contain many different functional genes that, when expressed, differ in their protein sequences. An **allele** is a variant of a gene that encodes for a specific trait. The various MHC alleles within the population result in a diversity of peptides that can be displayed on the surface of cells.

Immunoglobulins

Immunoglobulins (also known as antibodies) act as both B-cell receptors of the adaptive immune system—when expressed on the surface of a B cell—and as

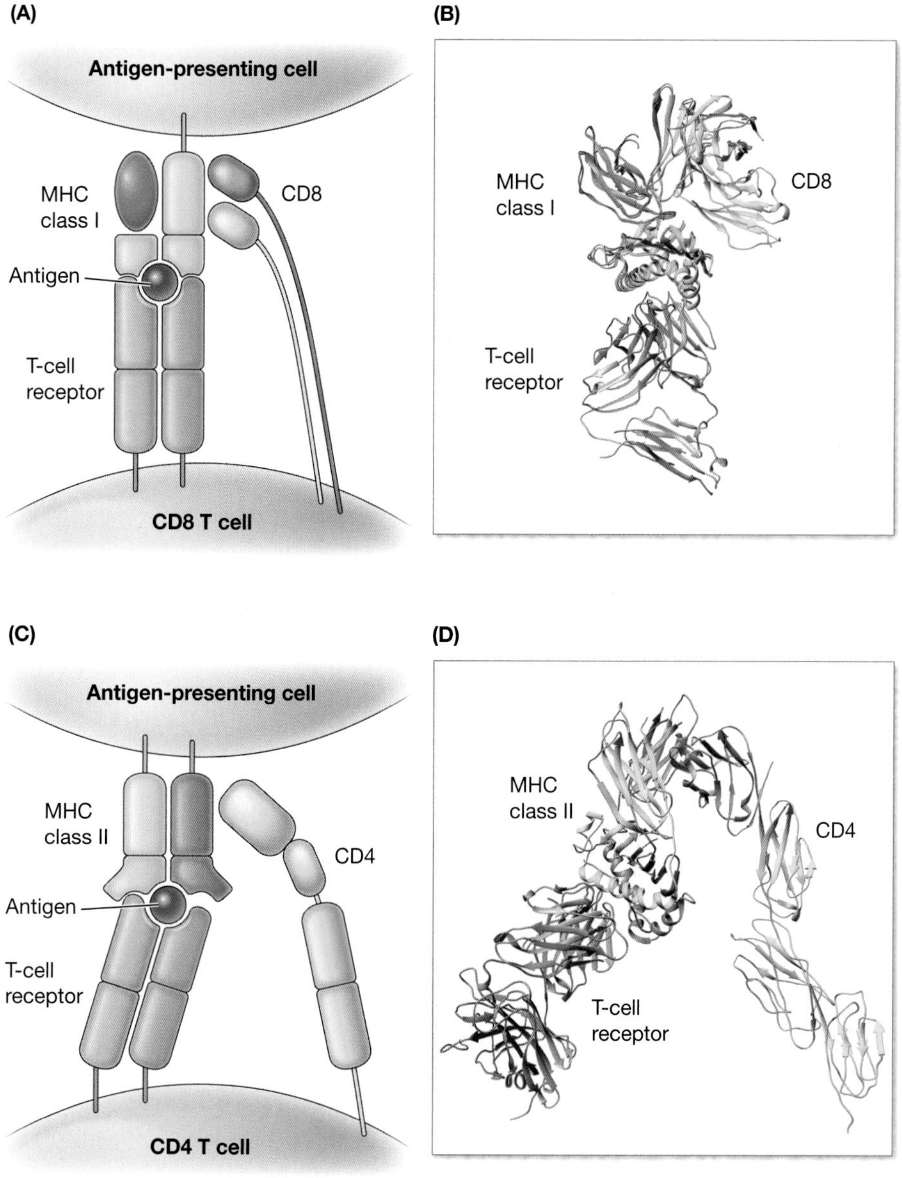

FIGURE 4.3 The T-cell receptor binds to both antigen and presenting MHC molecule (A) CD8 T cells express a receptor that recognizes antigen presented by an MHC class I molecule. The CD8 coreceptor binds to the α chain of MHC class I. (B) CD8 interacts with the α subunit of MHC class I molecules when the T-cell receptor engages with the MHC:peptide complex. Structures shown are superimposed images of the T-cell receptor-MHC complex and the MHC-CD8 complex. (C) CD4 T cells express a receptor that recognizes antigen presented by an MHC class II molecule. The CD4 coreceptor binds the β chain of an MHC class II molecule. (D) CD4 interacts with the β subunit of MHC class II molecules when the T-cell receptor engages with the MHC-peptide complex. (B, data from D. N. Garboczi et al. 1996. *Nature* 384: 134–141; redrawn using Chimera and PDB file 1AO7; D, data from Y. Yin et al. 2012. *Proc Natl Acad Sci USA* 109: 5405–5410. © 2012 National Academy of Sciences, USA, redrawn using Chimera and PDB file 3T0E.)

soluble effector molecules of B cells when secreted by *plasma cells* (differentiated and activated B cells). The B-cell receptor resembles the T-cell receptor in that the protein is a transmembrane protein, with the bulk of the protein located on the extracellular side of the membrane.

Immunoglobulins are composed of four polypeptides: two identical heavy chains and two identical light chains. Like T-cell receptors, both chains contain constant regions and variable regions. The variable regions come together to make an immunoglobulin antigen-binding site. Because of their double-chain structure, immunoglobulins contain two antigen-binding sites (**FIGURE 4.4**).

Like T cells, B cells have a diverse array of immunoglobulins, but each B cell produces a single type of immunoglobulin that recognizes one specific antigen. Unlike T-cell receptors, immunoglobulins can bind antigen without it being presented by an MHC molecule and can recognize any biological macromolecule (proteins, carbohydrates, lipids, and nucleic acids).

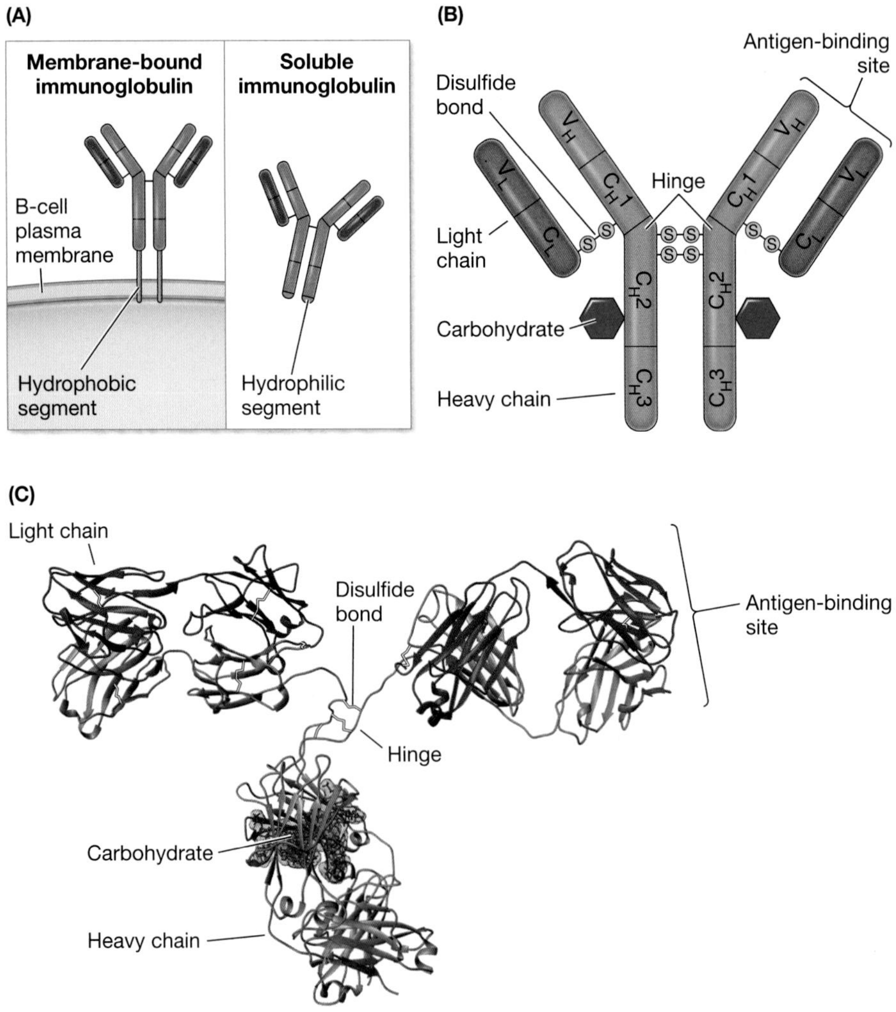

FIGURE 4.4 Structure of immunoglobulins (A) B cells synthesize immunoglobulins that remain associated with the plasma membrane (membrane-bound immunoglobulins) and immunoglobulins that are secreted (soluble immunoglobulins). (B) Each immunoglobulin contains two identical heavy chains and two identical light chains. Each heavy chain has a variable region (V) and 3–5 constant regions (C). Each light chain contains a single variable region and a single constant region. The variable regions of the heavy and light chains constitute the antigen-binding site. (C) The three-dimensional structure of an immunoglobulin reveals the immunoglobulin folds that make up the variable and constant regions and the interaction between heavy and light chains. (C, data from L. J. Harris et al. 1997. *Biochem* 36: 1581–1597. © 1997 American Chemical Society; redrawn using Chimera and PDB file 1IGT.)

When B cells are activated during an adaptive immune response, some cells divide and differentiate into plasma cells. These specialized effector B cells secrete soluble immunoglobulins. Plasma cells can make one of five **isotypes** (varieties) of soluble immunoglobulins, depending on the specific response that needs to be generated.

Soluble immunoglobulins (see Figure 4.4) recognize the same antigen that the cell-surface immunoglobulin of the parent B cell recognized. However, due to the mechanisms of differentiation into plasma cells, some soluble immunoglobulins can have a higher affinity for antigen.

isotypes Different protein versions of a gene family that have related functions.

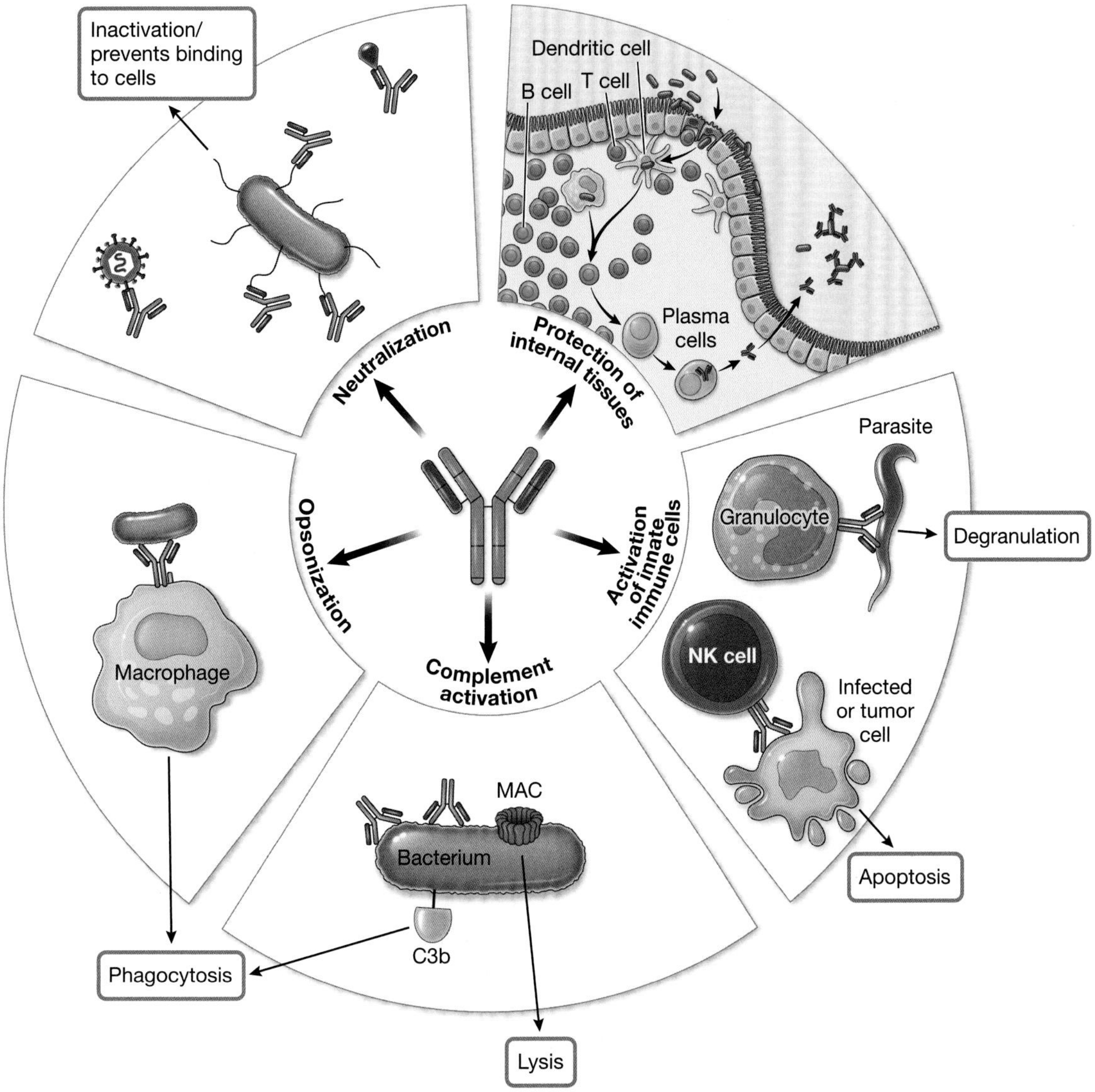

FIGURE 4.5 Effector functions of soluble immunoglobulins Secreted immunoglobulins have a variety of effector functions when made by plasma cells, including neutralizing foreign particles or pathogens, opsonizing foreign particles or pathogens, activating complement, activating innate immune cells (including granulocytes and natural killer cells), and protecting internal tissues (including mucosal surfaces).

Soluble immunoglobulins produced during an adaptive immune response have five major functions (**FIGURE 4.5**):

- Neutralization of a foreign particle or pathogen
- Opsonization of a foreign particle or pathogen
- Complement activation
- Activation of innate immune cells such as mast cells and natural killer cells
- Protection of internal/mucosal tissues

CHECKPOINT QUESTIONS

1. What is the importance of an MHC molecule in the adaptive immune response?
2. What are the five major functions of soluble immunoglobulins?

Making Connections: Comparing and Contrasting the Structure and Function of Innate and Adaptive Immune System Receptors

The innate and adaptive immune systems employ receptors to detect foreign antigen. While receptors from both systems are important in recognizing pathogens, their structures and modes of diversity differ. Ultimately, the goal of a receptor binding to an antigen is to invoke an intracellular response needed to combat the pathogen.

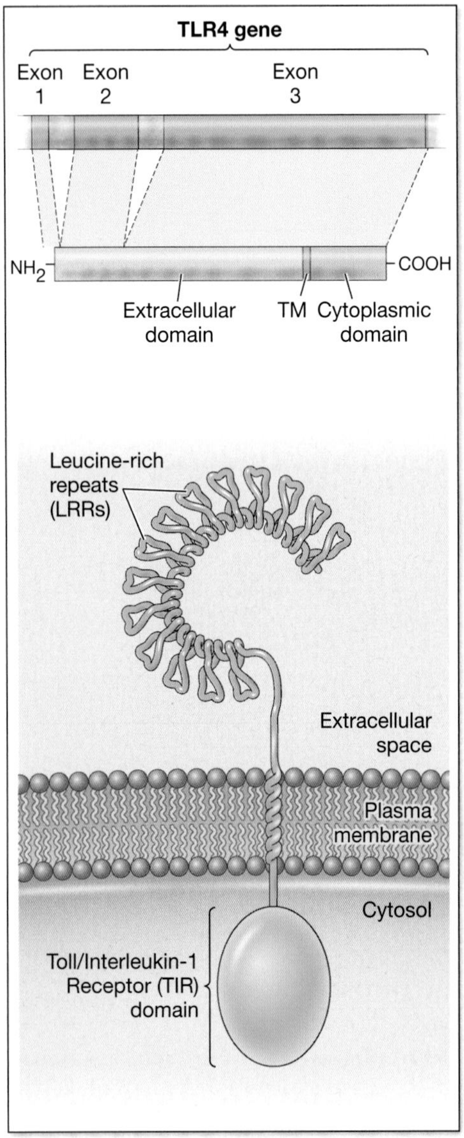

After J. Punt et al. 2018. *Kuby Immunology*, 8th ed., W. H. Freeman and Company: New York; E. Vercammen et al. 2008. *Clin Microbiol Rev* 21: 13–25; B. M. Tesar and D. R. Goldstein. 2007. *Front Biosci* 12: 4221–4238

Innate and adaptive immune system receptors vary in their DNA structure—genes of pattern recognition receptors (PRRs) do not recombine, whereas lymphocyte receptors do. PRR diversity is driven by presence of a variety of different receptors in the genome. Diversity of lymphocyte receptors occurs because of recombination.

(Sections 2.2, 2.4, 4.2, 4.4, 5.5, 6.3, 8.5, 9.2)

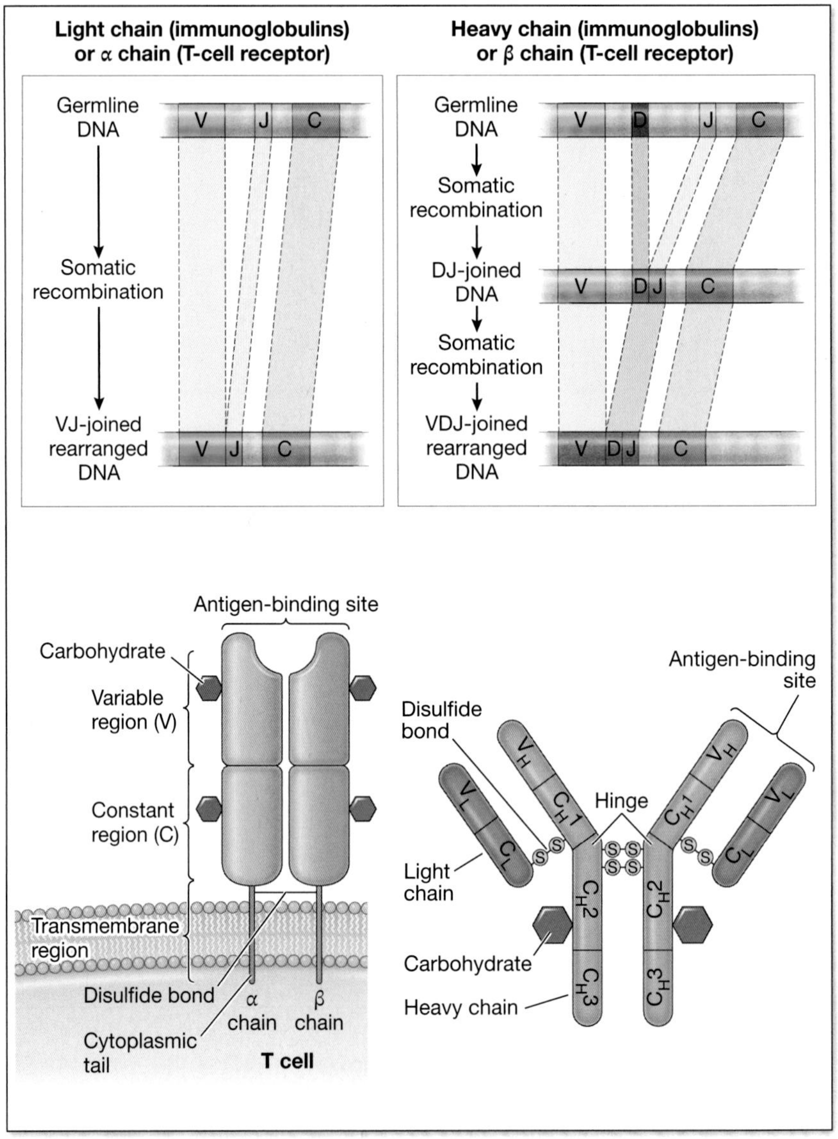

After W. Purves et al. 1995. *LIFE: The Science of Biology*, 4th ed., Sinauer Assoc., Sunderland, MA; T. Dörner and P. E. Lipsky. 2001. *Arthritis Rheum* 44: 2715–2727; P. Parham. 2014. *The Immune System*, 4th ed., W. W. Norton and Company, New York

Innate and adaptive immune response receptor protein structure promotes the ability to bind a single receptor. PRR diversity is driven by expression of different receptors that bind different ligands, while diversity of adaptive immune response receptors is driven by somatic recombination and different antigen-binding sites.

(Sections 2.2, 2.4, 4.2, 4.7, 4.8, 6.2, 9.1)

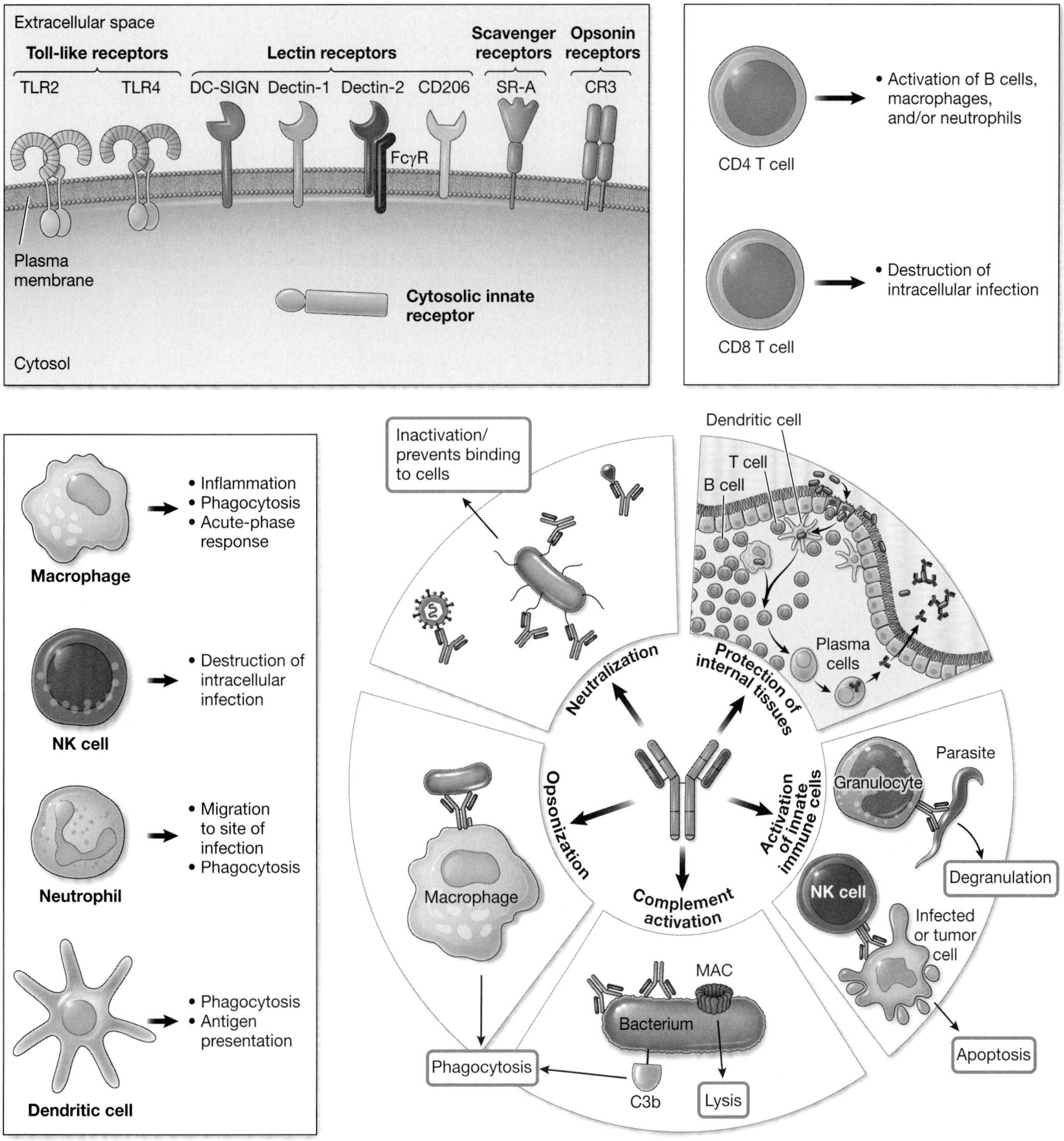

After M. S. Lionakis and M. G. Netea. 2013. *PLOS Pathog* 9: e1003079

Although the structure and mechanism of diversity of innate and adaptive immune receptors differ, many of the effector mechanisms driven by receptors binding to their respective ligand overlap or complement each other to promote pathogen destruction.

(Sections 1.7, 1.8, 2.5, 2.6, 2.7, 3.1, 4.7, 4.8, 7.13, 7.14, 7.16, 10.5)

4.3 What is the timeline of an adaptive immune response?

LEARNING OBJECTIVE

4.3.1 Outline the steps involved in the development and activation of an adaptive immune response.

Unlike the innate immune response, which is activated within minutes to hours of encountering a pathogen, the adaptive immune response requires a longer period of time, usually a few days to 2 weeks. The longer response time is due to two key factors: the immense diversity of T-cell receptors and immunoglobulins and the requirement to deliver and present antigen in secondary lymphoid tissues to lymphocytes that have never been activated. The diverse populations of lymphocytes can't monitor every potential location of infection to determine if their receptors can recognize a pathogen there. Thus, fully developed lymphocytes that have never interacted with a corresponding antigen enter and exit secondary lymphoid tissues to monitor for antigens. If none are found, they leave the tissue and reenter the circulation.

Antigen-presenting cells (APCs) present antigen to circulating lymphocytes that have entered secondary lymphoid tissue to determine if the lymphocyte recognizes the specific antigen. Upon antigen recognition, T cells and B cells become activated and differentiate into effector cells to begin fighting the pathogen. The steps involved in the adaptive immune response are shown in **FIGURE 4.6** and described in the subsections that follow.

Step 1: T-cell and B-cell development

The adaptive immune response depends on the action of T cells and B cells. T cells originate in the bone marrow and mature in the thymus. During their development, T cells engage in *somatic recombination* to rearrange T-cell receptor genes, test the function of the rearranged T-cell receptor genes, and undergo positive and negative selection. The development process also ensures that the right coreceptor (CD4 or CD8) is present on the surface of the T cell. Each coreceptor has specific responsibilities in the recognition of MHC molecules.

Like T cells, B cells must develop in primary lymphoid tissue to allow for genetic recombination of loci associated with immunoglobulins. For B cells, this development occurs in the bone marrow and provides a continuous supply of B cells. As B cells develop and as recombination occurs at the heavy and light chain loci, these recombination events are tested to ensure that each has resulted in a functional gene capable of producing a functional protein product.

Step 2: Antigen processing and presentation

As discussed in Chapter 2, dendritic cells play a role in both the innate and adaptive immune responses. They function at the site of infection to phagocytose extracellular pathogens and to process viral antigens. Dendritic cells then migrate to a draining lymphoid tissue, where they process foreign antigens by one of two different mechanisms. Intracellular antigens, including those from intracellular pathogens such as viruses and *Shigella* bacteria, are digested in the cytoplasm. Digested peptides are transported into the endoplasmic reticulum, where they are loaded onto MHC class I molecules that are then transported to the cell surface. Extracellular antigens, including those from extracellular pathogens such as *Vibrio cholerae* bacteria, are digested within a phagolysosome, which then fuses with a vesicle containing MHC class II molecules to load peptides on the class II molecule before transport to the cell surface. Other innate immune system cells process pathogens in a similar manner.

Thus, the function of class I molecules is to present cytoplasmic peptides (and potentially intracellular pathogen antigens), and the function of class II

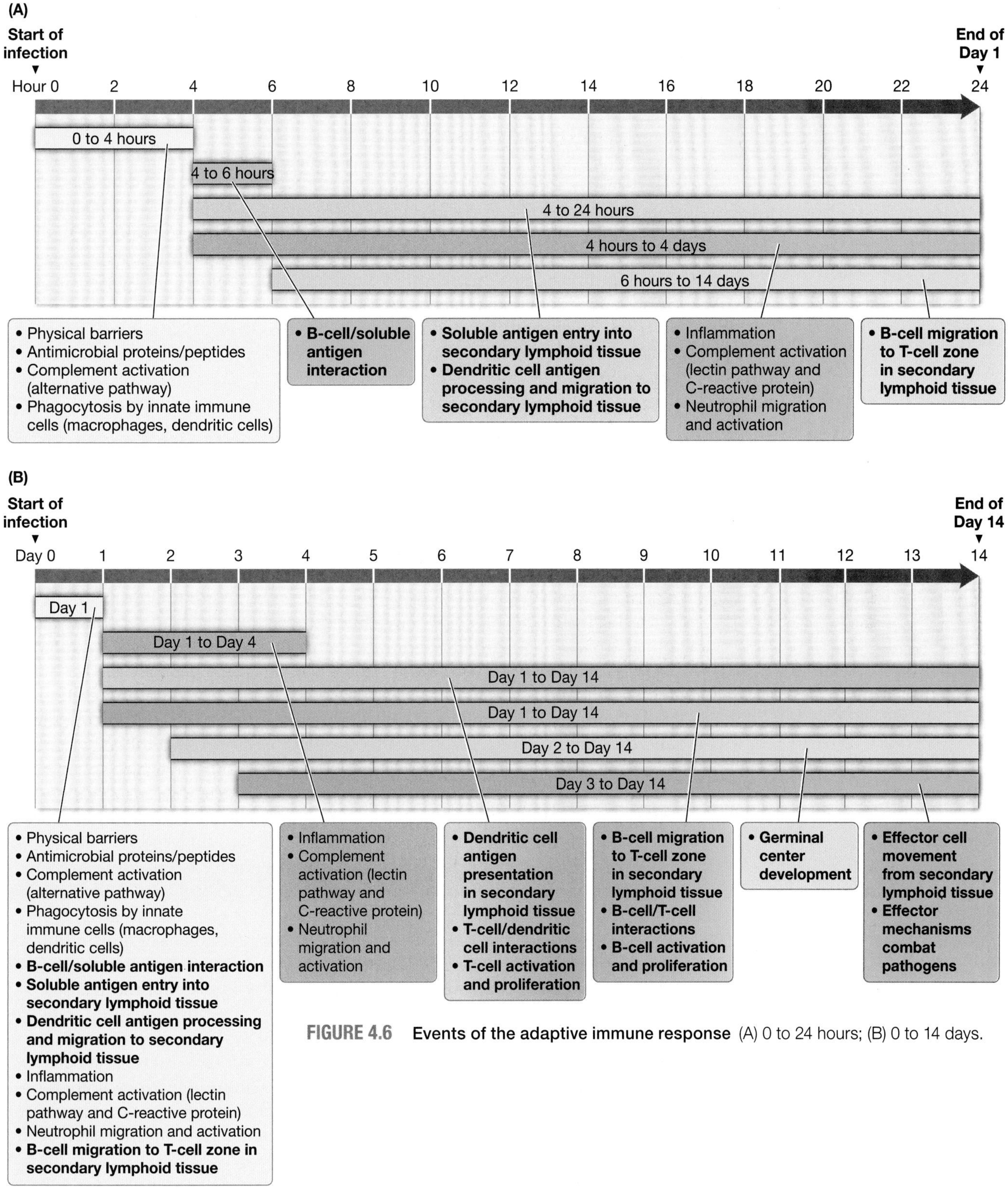

FIGURE 4.6 **Events of the adaptive immune response** (A) 0 to 24 hours; (B) 0 to 14 days.

molecules is to present phagocytosed peptides (and potentially extracellular pathogen antigens). APCs that have migrated from a site of infection to a draining lymphoid tissue wait there to interact with circulating T cells to determine if the T-cell receptor recognizes the presented antigen.

Step 3: T-cell and B-cell clonal selection/expansion

Mature T cells leave the thymus and enter the circulatory and lymphatic system to migrate to secondary lymphoid tissues. Here, they interact with dendritic cells that are presenting antigens via MHC molecules on the cell surface. Molecules on the surface of T cells such as LFA-1 bind to ICAM-1 and ICAM-2 on the dendritic cell. The T-cell receptors scan MHC molecules on the dendritic cell for potential recognition and interaction.

If the T-cell receptor does not detect its coinciding antigen, it disengages from the dendritic cell, exits the secondary lymphoid tissue, and continues circulating. If the T-cell receptor and coreceptor engage with an MHC molecule containing the correct antigen, the receptor remains tightly associated, activating further signaling pathways. This activation drives proliferation of the engaged T cell and promotes the action of cytokines that prompt differentiation into various types of effector T cells.

If the T cell has engaged with an MHC class I molecule, that T cell will differentiate into a cytotoxic T cell. Cytotoxic T cells monitor the presence of intracellular infections, including viruses and intracellular bacteria and parasites. If it recognizes an MHC class I molecule presenting a foreign antigen on the surface of a cell via its T-cell receptor, the cytotoxic T cell will use effector mechanisms to target that cell for destruction. However, if the T cell engages with an MHC class II molecule, that T cell will differentiate into a helper T cell. Helper T cells activate other cells that are important in combating extracellular infections, including B cells and macrophages, which both work to neutralize and remove the extracellular pathogens.

Upon activation and differentiation, some helper T cells migrate to a different location in the secondary lymphoid tissue where they wait to interact with B cells. B cells travel through the lymphatic system to secondary lymphoid tissues in a manner similar to that of T cells. Those B cells migrate to a location known as the T-cell zone where they test whether they recognize the same pathogen as the helper T cell. If the B cell and T cell recognize the same pathogen, the B cell engages with the helper T cell and B-cell proliferation and differentiation begins.

Signals that drive the process come from T-cell cytokines and cell-surface ligands on the T cell that interact with other B-cell surface molecules (the specific mechanism is discussed in Chapter 10). Upon receiving the right signals from the T cell, the B cell begins dividing and produces effector B cells known as plasma cells that secrete soluble immunoglobulins. Activated B cells induce transcription of immunoglobulin genes to produce the secreted antibodies.

Some of the activated B cells undergo further refinement of their immunoglobulin genes to allow for better antigen recognition and to promote specialized functions of these immunoglobulins before differentiating into plasma cells. The genes progress through two important modifications that enhance B-cell effectiveness: *affinity maturation* and *isotype switching*.

Affinity maturation is the process whereby the immunoglobulin gene is randomly altered in a subset of activated B cells to promote the production of B cells containing immunoglobulins with higher affinity for antigen. This maturation is driven by **somatic hypermutation**, which is random mutation at the variable regions of the immunoglobulin gene to drive altered affinity of that immunoglobulin for its antigen.

Incorporation of different constant regions of the heavy chain also occurs in this subset of activated B cells. The heavy chain constant region determines the immunoglobulin isotype. The events that occur in this subset of B cells promote recombination of the constant regions of the immunoglobulin heavy chain locus to switch the isotype of the expressed immunoglobulin.

During the clonal expansion and differentiation of T cells and B cells, a subset of lymphocytes known as *memory cells* are produced. These cells function to promote a more rapid adaptive immune response if the same antigen that promoted their differentiation is encountered again.

somatic hypermutation Preferential action of activation-induced cytidine deaminase (AID) at the variable regions of immunoglobulin genes to further enhance affinity for its epitope.

- **CHECKPOINT QUESTION**

1. What general steps must occur for a proper adaptive immune response to take place?

4.4 What mechanism provides diversity of T-cell and B-cell receptors?

LEARNING OBJECTIVE

4.4.1 Describe the mechanisms used to generate diversity in T-cell receptors and immunoglobulins.

Cells of the adaptive immune system are extremely specific in the antigens they recognize due to expressing only one type of receptor per lymphocyte. How then is the adaptive immune system capable of recognizing an incredibly vast number of pathogens? Specific pathogens are recognized by lymphocytes with variable regions in their receptors that recognize a particular antigen.

During lymphocyte development, several mechanisms are involved in generating functional T-cell or B-cell receptors and promoting the development of a vast number of diverse lymphocytes, each with a receptor specific to a particular antigen. These mechanisms include:

- Gene rearrangement of random variable segments of T-cell and B-cell receptor genes
- Addition or removal of nucleotides at junction sites of variable segments
- Combination of different receptor subunits, each of which have undergone different gene rearrangement and junction site changes

gene rearrangement Recombination event in a gene that changes the original genetic allele and promotes alternate function of the gene.

somatic recombination DNA recombination (gene rearrangement) that occurs in developing T cells and B cells, also known as V(D)J recombination.

Gene rearrangement and its role in receptor diversity

The diversity of antigens recognized by T cells and B cells is primarily due to **gene rearrangement**. The process by which recombination occurs at specific genes is known as **somatic recombination**. The genes for both the T-cell receptor and B-cell receptor are present in all cells but cannot encode for a protein product when gene rearrangement has not occurred in germline DNA (the DNA present in most somatic cells in our body) (see **KEY DISCOVERIES**).

The variable regions of the genes are separated on the chromosome into different segments, referred to as V (for *variable*), J (for *joining*), and D (for *diversity*). Note that only certain receptor subunits contain D segments. During T-cell and B-cell development, specific enzymes change the loci that encode for T-cell receptors or immunoglobulins. This results in joining of the V, J, and sometimes D segments to produce a functional variable region of the receptor, which can then be connected with the constant region to make a functional protein product (**FIGURE 4.7**).

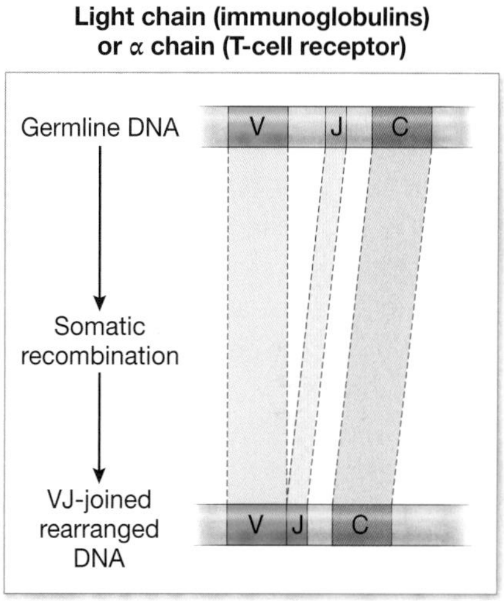

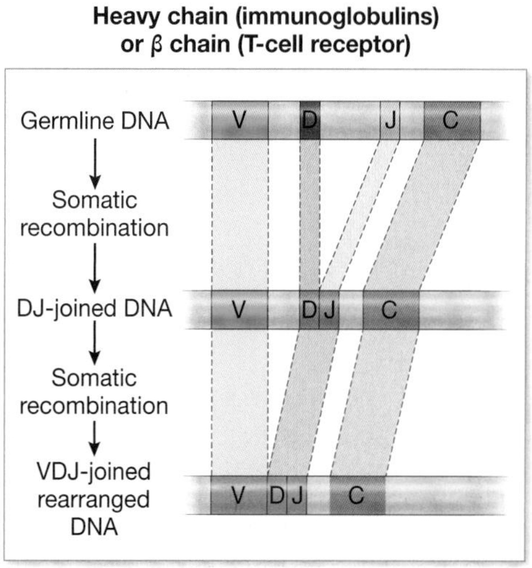

FIGURE 4.7 Somatic recombination in lymphocytes During B-cell and T-cell development, immunoglobulin and T-cell receptor genes are recombined from the germline configuration to create a functional gene within these lymphocytes. Recombination within lymphocytes combines different gene segments to produce a variable region of the receptor gene. Light chain immunoglobulin and α-chain rearrangement combine V and J segments to form their variable regions. Heavy chain immunoglobulin and β-chain rearrangement combine V, D, and J segments to form their variable regions. (After W. Purves et al. 1995. *LIFE: The Science of Biology*, 4th ed., Sinauer Assoc., Sunderland, MA; T. Dörner and P. E. Lipsky. 2001. *Arthritis Rheum* 44: 2715–2727; P. Parham. 2014. *The Immune System*, 4th ed., W. W. Norton and Company, New York.)

KEY DISCOVERIES

How did we learn that immunoglobulin genes undergo recombination in B cells?

Article

N. Hozumi and S. Tonegawa. 1976. Evidence for somatic rearrangement of immunoglobulin genes coding for variable and constant regions. *Proceedings of the National Academy of Sciences USA* 73: 3628–3632.

Background

It has been well established that T-cell receptors and immunoglobulins undergo recombination of their chromosomal DNA to generate variable regions. These regions differ in each cell, thus allowing for the production of a population of T cells and B cells with the ability to recognize a diverse number of antigens. The recombination events that occur in developing lymphocytes do not typically occur in somatic cells, so how did scientists uncover this mechanism of the adaptive immune system?

Early Research

In the 1960s, it was established that the heavy and light chains of immunoglobulins consist of a variable region connected to a constant region. Nucleic acid studies conducted in the 1970s confirmed the generation of V regions from germline DNA. These studies established the paradigm of variable and constant regions composing protein products of lymphocytes, but the question arose: How are closely related variable regions incorporated into final products? Since genomic sequencing had not been established at the time, it was unclear how many variable regions would have to be part of the genome to generate the diversity seen in lymphocytes. Two models had been proposed: either the V and C regions were linked as a result of RNA processing mechanisms (as a post-transcriptional event, such as by alternative splicing of the RNA), or integration of V and C regions occurred at the DNA level.

Think About...

1. Considering the research summary above, state the proposed models as testable hypotheses.
2. What experiment(s) would you conduct to test each hypothesis? How could you differentiate among your hypotheses with these methods?
3. How did the findings of Hozumi and Tonegawa support one of the models suggested as a means to drive V and C integration? What key finding ruled out the other possible models?
4. What possible mechanisms did the authors suggest as means to drive V and C integration that were supported by their findings? How do these proposed mechanisms compare to our current understanding of somatic recombination within lymphocytes?

Article Summary

At the time of this study, Hozumi and Tonegawa were at the forefront of understanding the mechanism by which variable regions and constant regions were joined to produce functional immunoglobulins. A key question at the time was the mechanism by which these regions were joined together: Did joining occur at the RNA level or the DNA level?

To determine the answer, researchers studied the DNA from two different sources: early embryos of Balb/C mice and the plasmacytoma cell line MOPC 321, which was known to produce functioning κ light chains. They isolated the DNA from both sources, digested it with a restriction endonuclease, and separated fragments by electrophoresis. DNA fragments were allowed to hybridize with two different radioactive probes made from κ light chain mRNA isolated from MOPC 321 cells, either the full-length mRNA or the 3′ end of the mRNA. DNA fragments were analyzed

How does the joining of V, D, and J segments at the T-cell receptor and immunoglobulin loci promote the vast diversity of adaptive immune system receptors? One of the sources of diversity is based on the sheer numbers of possible segment combinations that can occur. Furthermore, the actions of the enzymes involved in recombination are imprecise, leading to further diversity. These events result in a seemingly infinite number of possible variable regions that can be formed during the recombination events associated with lymphocyte development.

T-cell receptor rearrangement and diversity

As previously discussed, the T-cell receptor commonly has two chains, an α chain and a β chain. Some T cells can have a T-cell receptor with a γ chain and a δ chain. While most of our consideration of T-cell receptor diversity will focus on $\alpha\beta$ T-cell receptors, rearrangement and diversity also occurs at the γ and δ T-cell receptor loci. We will see in Chapter 12 that $\gamma\delta$ T cells are involved in recognizing based on their size and their ability to hybridize with either or both of the κ light chain probes.

Conclusion

When the authors compared the DNA isolated from mouse embryos with that isolated from MOPC 321 cells, they made several critical observations.

- Mouse embryo DNA hybridized differently depending on the probe used—whole κ light chain mRNA hybridized to two distinct regions of DNA, and the 3′ end of the κ light chain mRNA only hybridized to one region of DNA.
- DNA from MOPC 321 cells only produced one hybridizing DNA no matter what mRNA probe was used.
- More interesting than hybridization patterns was the size of the DNA that hybridized to the probe. In embryo DNA, the prominent DNA containing both V and C regions was approximately 6 million base pairs, while in MOPC 321 DNA, the prominent DNA containing both V and C regions was approximately 2.4 million base pairs (**FIGURE KD 4.1**).

These data were interpreted by the authors to suggest that V and C regions of the light chain are far apart in germline DNA, and that recombination events occur to bring the DNA closer together and promote V and C integration.

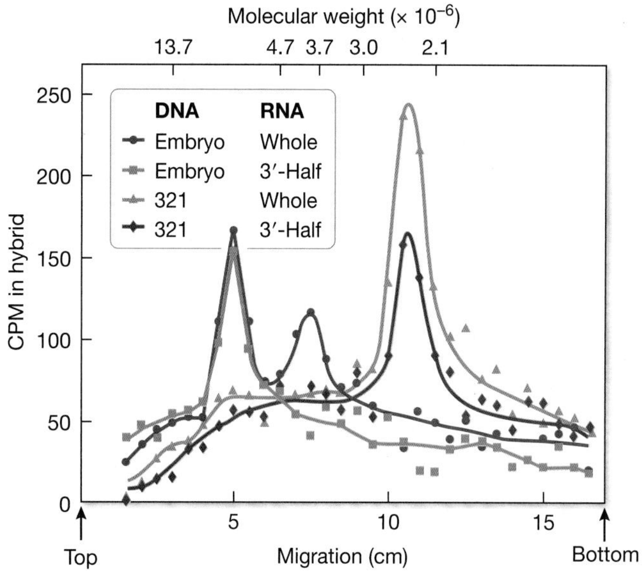

FIGURE KD 4.1 Light chain DNA isolated from light-chain–producing cells differs from embryonic light chain DNA DNA was isolated from either mouse embryos or from the cell line MOPC 321 (321) and was digested with restriction endonucleases. The DNA was separated by electrophoresis and was measured by hybridization with either whole light chain RNA or the 3′ half of light chain RNA and quantified by the amount of hybridized DNA in counts per minute (CPM). The light chain DNA detected in the light-chain–producing MOPC 321 cells is smaller than the light chain DNA detected in mouse embryos. (From N. Hozumi and S. Tonegawa. 1976. *Proc Natl Acad Sci USA* 73: 3628–3632.)

a subset of pathogen antigens, including lipid antigens, and function to help protect mucosal surfaces. When considering the α and β chains of the T-cell receptor, the α chain contains variable regions that link V and J segments (termed V_α and J_α), while the β chain contains variable regions that link V, D, and J segments (termed V_β, D_β, and J_β). During recombination of the β chain, a D segment and a J segment are recombined first before a V segment recombines with the DJ segment. The germline DNA contains a large variety of V, D, and J components (**FIGURE 4.8**), which will be explored in Chapter 5 as an additional mechanism of generating diversity.

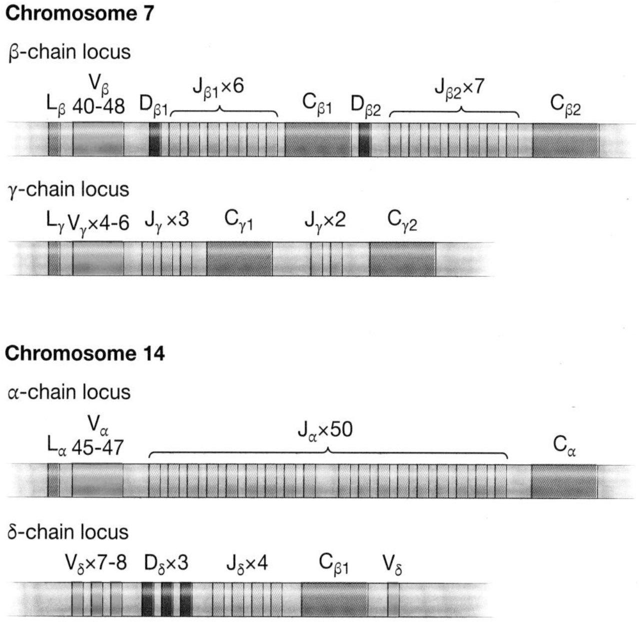

FIGURE 4.8 Germline gene organization of T-cell receptor loci The β-chain and γ-chain loci of the T-cell receptor are located on chromosome 7, whereas the α-chain and δ-chain loci are located on chromosome 14. Diversity of T-cell receptors in lymphocytes is driven in part by the number of V, D, and J segments present within the genome. (Data from T. Boehm and T. H. Rabbitts. 1989. *FASEB J* 3: 2344–2359.)

Immunoglobulin rearrangement and diversity

Immunoglobulins have two of each type of subunit (two heavy chains and two light chains). During B-cell development, immunoglobulin gene rearrangement also occurs via V(D)J recombination. Heavy chains contain V, D, and J segments (V_H, D_H, and J_H) that are recombined to make their variable regions. In a similar fashion to recombination of the β chain of the T-cell receptor, during heavy chain recombination a D segment and a J segment are recombined first before a V segment recombines with the DJ segment. Light chains contain only V and J segments; however, there are two different loci capable of expressing the light chain of immunoglobulins: κ and λ (variable segments known as V_κ, J_κ, V_λ, and J_λ). Similar to T-cell receptor genes, the germline DNA of immunoglobulin genes contains a large variety of V, D, and J segments of the variable regions at each of the loci (**FIGURE 4.9**), which will be explored in Chapter 8 as an additional mechanism of generating diversity.

Immunoglobulins produced by B cells that have come into contact with antigen are further refined during activation to bind to their antigen with higher affinity. This process, referred to as **affinity maturation**, is driven by mutation of the immunoglobulin genes that are actively producing soluble immunoglobulins. Affinity maturation (discussed further in Chapter 9) does not increase the number or type of antigens that can be recognized by immunoglobulins, but it does play an important role in ensuring that subsequent exposure to the same antigen is met with a faster, more robust adaptive immune response.

Immunoglobulins have five different constant regions, and incorporation of a constant region constitutes one of five different isotypes of immunoglobulins. These five constant regions (C_μ, C_δ, C_γ, C_ε, and C_α) can all be used as components of the heavy chain through a process known as **isotype switching** (also discussed further in Chapter 9), in which a different immunoglobulin is produced due to further rearrangement of the immunoglobulin locus and use of a different heavy chain constant region. These isotypes provide specialized proteins that work in different tissues of the body to promote an effective adaptive immune response.

V(D)J recombinase

Under normal circumstances, cells will not usually recombine genetic information unless their DNA has been damaged, as recombination events can lead to potentially abnormal or cancer-causing gene products. But developing lymphocytes activate a special set of genes to recombine their receptor loci.

Two primary genes activated in developing lymphocytes are **recombination activating gene 1** (***RAG1***) and **recombination activating gene 2** (***RAG2***). Their

affinity maturation Process by which immunoglobulin genes gain higher affinity for antigen through mutagenesis of the immunoglobulin gene during B-cell activation.

isotype switching Mechanism used by activated B cells to incorporate a different heavy chain constant region into the functioning immunoglobulin gene to refine the activity of the immunoglobulin based on need.

recombination activating gene 1 (*RAG1*) Gene activated in developing lymphocytes that functions in the recombination of variable regions in T-cell receptors and immunoglobulins as part of the V(D)J recombinase enzyme complex.

recombination activating gene 2 (*RAG2*) Gene activated in developing lymphocytes that functions in the recombination of variable regions in T-cell receptors and immunoglobulins as part of the V(D)J recombinase enzyme complex.

V(D)J recombinase Enzyme complex that functions in developing lymphocytes to recombine segments of the variable regions of T-cell receptors and immunoglobulins and promote the formation of a functional gene product.

recombination signal sequences (RSSs) Sequence motifs that flank variable region gene segments and serve as recognition sites of enzymes involved in V(D)J recombination.

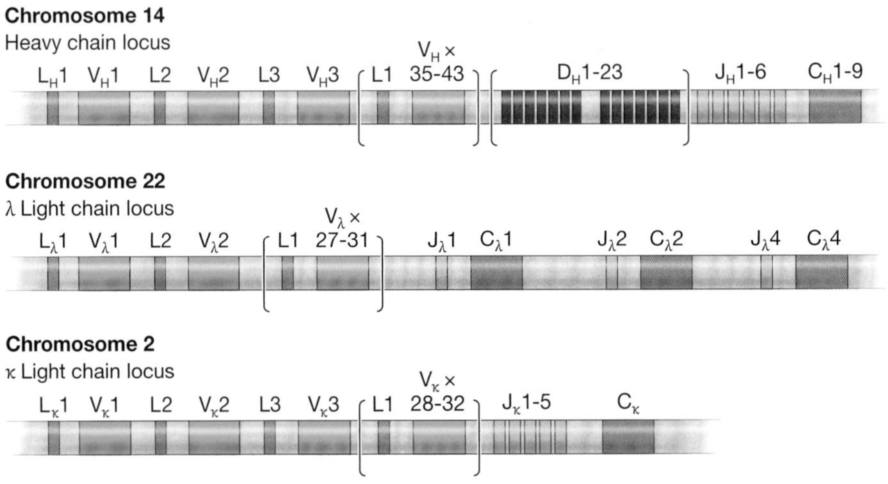

FIGURE 4.9 Germline gene organization of immunoglobulin loci The immunoglobulin heavy chain locus is located on chromosome 14, the λ light chain locus is located on chromosome 22, and the κ light chain locus is located on chromosome 2. Diversity of immunoglobulins in lymphocytes is driven in part by the number of V, D, and J segments present within the genome. (Data from S. Tonegawa. 1983. *Nature* 302: 575–581.)

two gene products are major players in the **V(D)J recombinase** enzyme complex, which recombines V, D, and J segments to produce a functional receptor (see **EVOLUTION AND IMMUNITY**). The variable region of the β-chain subunit of a T-cell receptor and heavy chain of an immunoglobulin contain all three segments (and thus recombine V, D, and J segments), while the α-chain subunit of a T-cell receptor and the light chain of an immunoglobulin contain only V and J segments (and recombine the two to make a functional subunit).

Although these gene products are essential for receptor diversity, another system found in all cells is also involved in the recombination events that occur during lymphocyte development. Gene products of the non-homologous end joining (NHEJ) DNA repair recombination pathway are required for efficient genetic rearrangement needed for lymphocyte receptor processing.

The elegant system of recombination that occurs during lymphocyte development is specific to genes of the receptors. This specificity is driven by the enzymes involved in V(D)J recombination that recognize these variable segments by binding to specific DNA sequence motifs known as **recombination signal sequences** (**RSSs**). RSSs are found at specific locations in relation to each of the three segments: at the downstream ends of V segments, both the upstream and downstream ends of D segments, and the upstream end of J segments (**FIGURE 4.10**). Properly recombined variable regions will always contain

coding joint Segment of DNA joined together during V(D)J recombination that encodes for the variable region of a T-cell receptor or immunoglobulin.

signal joint Circle of DNA originally located between two variable segments in the germline DNA that is removed upon V(D)J recombination of variable segments.

P nucleotides Nucleotides present in the overhangs left when V(D)J recombinase opens up the hairpin formed during V(D)J recombination.

exonuclease Enzyme capable of removing nucleotides from the ends of nucleic acids.

terminal deoxynucleotidyl transferase (TdT) Enzyme involved in the addition of nucleotides to the ends of recombining variable segments during lymphocyte development.

EVOLUTION AND IMMUNITY

Somatic Recombination Machinery

While it seems apparent that the innate immune system has evolved in an ordered manner, becoming more sophisticated in more complex organisms, the adaptive immune system is fascinating to observe in terms of evolutionary history. Originally, it appeared that adaptive immunity evolved at the same time as jawed vertebrates. This evolution seemed abrupt in part due to the insertion of a transposable element into an ancient immunoglobulin-like gene cluster, conferring an ability to undergo somatic genetic recombination events. However, there are examples of species outside the jawed vertebrates that evolved mechanisms of diversifying receptors involved in pathogen recognition. Some of these mechanisms are driven by alternative splicing, others by somatic mutation, and still others by somatic genetic recombination of genes related to TLRs.

Regardless of the mechanism of receptor diversity, adaptive immune systems seem to represent an example of convergent evolution in which different organisms evolved different mechanisms toward a common goal: diversification of pathogen recognition. One of the means by which diversity arises is genetic recombination using the enzyme V(D)J recombinase. The theory of V(D)J recombinase evolution in higher vertebrate species is related to the functional connection its action has with another family of enzymes, the transposases.

Jawed fish and all higher vertebrates have an adaptive immune system that uses somatic gene rearrangement to produce a diverse set of receptors involved in pathogen recognition. This rearrangement is driven by the action of a V(D)J recombinase, which through the function of several proteins (especially the RAG1/RAG2 proteins) recombines germline DNA to produce a variable receptor within the immune cell. It appears that evolution of this mechanism of receptor diversification abruptly entered evolutionary history through the insertion of a mobile genetic element (known as a transposon) containing an ancient V(D)J recombinase inserted into an ancient immunoglobulin gene or T-cell receptor gene. Upon its insertion into the ancient gene, the V(D)J recombinase must have lost its ability to excise DNA with its original excision sequences that it recognized. Instead, it must have targeted the RSSs that flank the recombined sequences of T-cell receptors and immunoglobulins.

RAG1/RAG2 can catalyze the insertion of a DNA fragment into DNA containing RSSs, behavior identical to that of transposons. Another interesting aspect of the action of transposons is that the very nature of transposition is prone to error within the genetic information being altered. This transposition could have harmful effects under normal circumstances. However, since the goal of recombination in the T-cell receptor and immunoglobulins is that of diversity, transposition furthers that goal due to the error-prone nature of its mechanism.

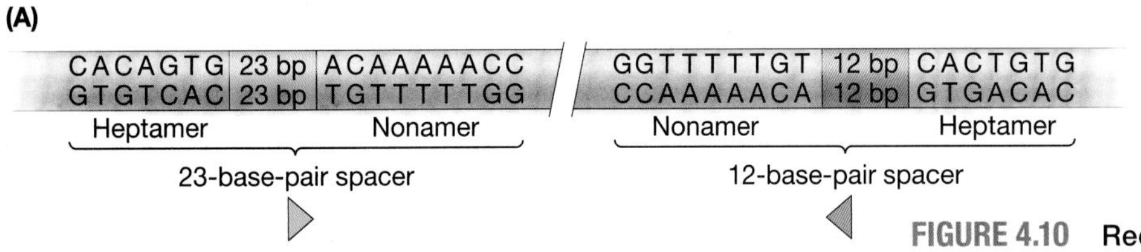

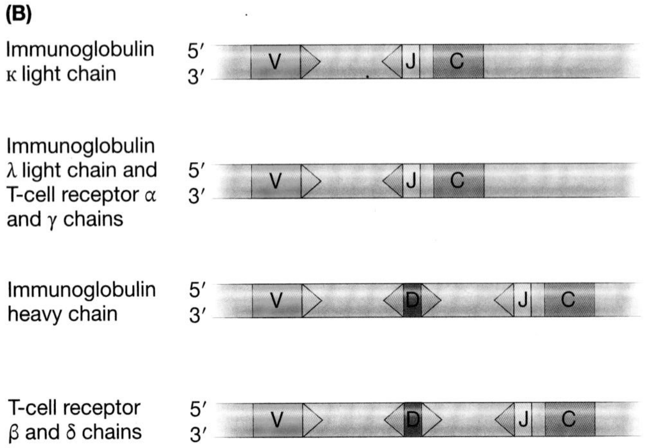

FIGURE 4.10 Recombination signal sequences (RSSs) and variable gene segment arrangement of immunoglobulin and T-cell receptor genes (A) RSSs in immunoglobulin and T-cell receptor gene segments contain a 7-base-pair heptamer sequence closest to the variable gene segment, either a 23- or 12-base-pair spacer, and a 9-base-pair nonamer sequence farthest from the variable gene segment. (B) RSS arrangement in both immunoglobulin and T-cell receptor genes ensures that segments being recombined will contain one segment with a 23-base-pair spacer and another segment with a 12-base-pair spacer. (Data from J. Punt et al. 2018. *Kuby Immunology*, 8th ed., W. H. Freeman and Company: New York.)

a recombined V, D, and J segment (as is the case of the T-cell receptor β chain or immunoglobulin heavy chain) or a recombined V and J segment (as is the case of the T-cell receptor α chain or immunoglobulin light chain).

Because DNA recombination can result in a variety of disease states, including the onset of cancer, the structure of the RSSs present in T-cell receptor and immunoglobulin genes is crucial in ensuring that recombination driven by V(D)J recombinase only occurs at these genes. RSSs contain three very important structural features that allow for V(D)J recombination events to occur specifically and efficiently at lymphocyte receptor genes:

- A conserved *heptamer* (7 base pair) sequence closest to the associated coding region of the segment
- A conserved *nonamer* (9 base pair) sequence furthest away from the associated region of the segment
- A *spacer* (either 23 base pairs or 12 base pairs).

The conserved sequences act as recognition sites for the V(D)J recombinase machinery. The spacers ensure that the RSSs are located on the same physical side of the DNA double helix (12 base pairs and 23 base pairs represent one or two full turns of the DNA double helix; see Figure 4.10).

Recognition of RSSs by V(D)J recombinase allows for random variable region segments to be brought close together and fused. If we consider the simple recombination of a V segment and a J segment, the first event that occurs is recognition of the RSS for the V segment and subsequent recognition of the RSS for the J segment (**FIGURE 4.11**, Step 1). After the two segments are brought close

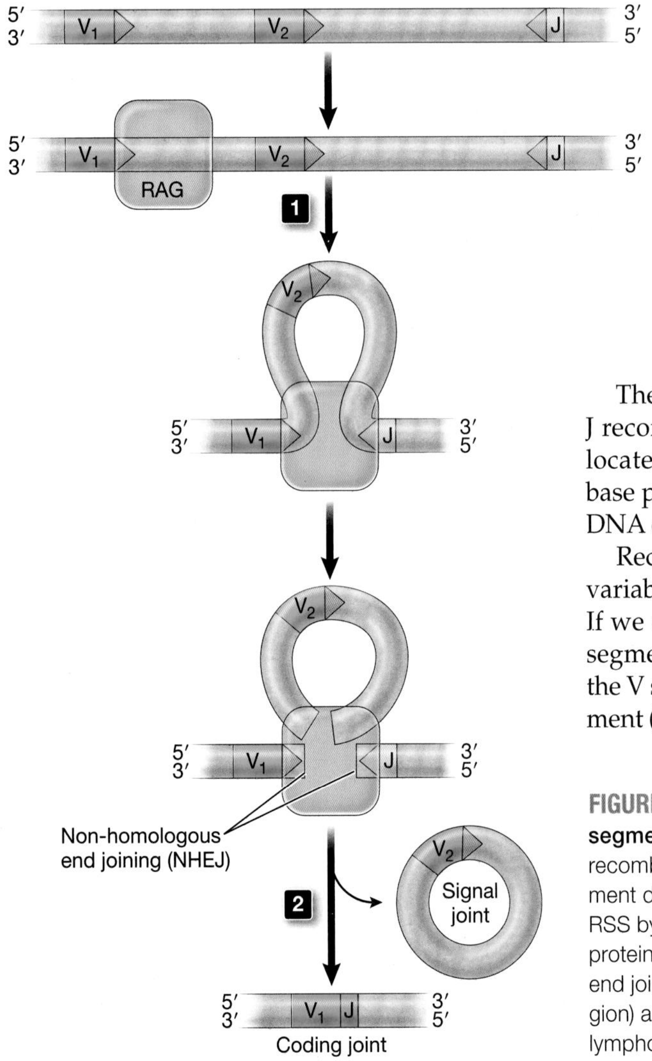

FIGURE 4.11 V(D)J recombination of lymphocyte receptor gene segments Recombination of lymphocyte genes requires the action of V(D)J recombinase recognizing RSSs at different gene segments. In VJ rearrangement depicted here, V(D)J recombinase (designated RAG) recognizes the RSS by a random V segment followed by a random J segment (Step 1). RAG proteins cleave the two RSSs that it bound and facilitates non-homologous end joining repair to produce a coding joint (representing a functional V region) and a circular piece of DNA known as a signal joint, which is lost by the lymphocyte (Step 2).

together, V(D)J recombinase cleaves the DNA to create ends that can be joined and releases the DNA between the segments, forming a circle (Figure 4.11, Step 2). The joined segment that codes for the variable region of the receptor is the **coding joint**, and the circle of DNA that has been removed is the **signal joint**.

N nucleotides Nucleotides added to the ends of variable segments involved in the formation of a coding joint; these are added to the ends by terminal deoxynucleotidyl transferase (TdT).

Diversity of the variable regions of T-cell receptors and immunoglobulins is also enhanced by another process. As RAG1 and RAG2 function to form the coding joint, the heptamer sequence of the RSS is removed from the two segments to be linked together (**FIGURE 4.12**, Step 1). This results in the formation of a hairpin in the two segments to be joined (Figure 4.12, Step 2). The hairpin is nicked by the DNA repair nuclease Artemis, creating palindromic overhangs (DNA sequences that would be identical when read properly on either strand of DNA) (Figure 4.12, Step 3). The nucleotides associated with the overhangs are referred to as **P nucleotides**. The overhangs can be filled in by NHEJ DNA repair enzymes and the two segments ligated together to form a potentially functional variable segment (Figure 4.12, Step 4).

Further diversity can arise by the action of **exonucleases**, which cleave nucleotides from the ends of DNA, and **terminal deoxynucleotidyl transferase** (**TdT**). TdT has the ability to add additional nucleotides known as **N nucleotides** before the segments are joined together through DNA repair (see Figure 4.12, Step 4). The addition of N nucleotides and removal of P nucleotides can alter the sequence at the junction between the two joined fragments. If this rearrangement does not occur in sets of three nucleotides, the result is a variable region that cannot be translated into protein.

CHECKPOINT QUESTIONS

1. What mechanisms result in T-cell and B-cell receptor diversity?
2. Define the role of V(D)J recombinase in the recombination events that occur at the T-cell and B-cell receptor loci during lymphocyte development.

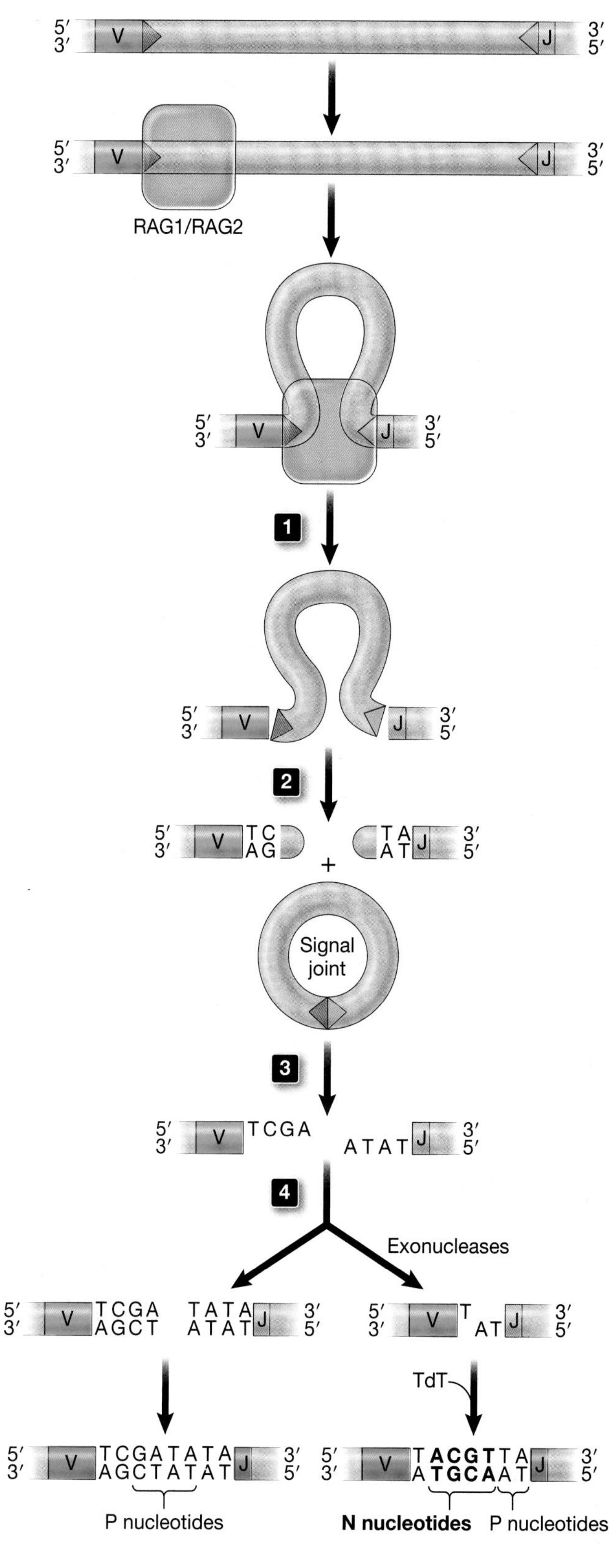

FIGURE 4.12 Mechanism of V(D)J recombination and junctional diversity Recombination begins when RAG1/RAG2 binds to RSSs on the target gene segments. RAG1/RAG2 cleaves the heptamer sequence of the two adjoining segment RSSs (Step 1). This cleavage results in the formation of a hairpin at the end of each segment to be joined and the removal of the internal DNA segment as a signal joint (Step 2). The hairpins are opened up to create palindromic overhangs (Step 3). There are several fates after creation of these overhangs, which include the overhangs remaining, nucleotides from the overhangs being removed by exonucleases, or nucleotides being added by terminal deoxynucleotidyl transferase (Step 4). Once the complementary strand of the overhangs is synthesized, the two segments are joined to create the coding strand. (After S. D. Fugmann. 2001. *Immunol Res* 23: 23–39.)

4.5 How does tolerance to self-molecules occur given the diversity of our adaptive immune system?

LEARNING OBJECTIVE

4.5.1 Explain how cellular receptors of the adaptive immune system are tested to ensure functionality and self-tolerance.

positive selection Process by which lymphocytes are tested to determine if their receptor is functioning properly.

negative selection Process by which lymphocytes are removed from the body if their receptors recognize self-molecules.

The mechanisms that drive T-cell receptor and immunoglobulin diversity are crucial in the development of T cells and B cells with the ability to recognize a wide variety of antigens. Without these systems in place, the adaptive immune system fails to function properly, which can lead to serious disorders such as severe combined immunodeficiencies.

Because these mechanisms are nonspecific, they produce T-cell receptors and immunoglobulins that can potentially recognize any molecule, including those produced by your own body (self-molecules). How does the adaptive immune system maintain a diverse set of T-cell receptors and immunoglobulins and simultaneously ensure tolerance of self-molecules? As T cells develop in the thymus and B cells develop in the bone marrow, T-cell receptors and immunoglobulins are tested for both functionality and specificity. **Positive selection** tests for a functioning T-cell receptor and surface immunoglobulin. **Negative selection** is used to test self versus nonself recognition.

Positive selection

Receptors must function properly for T cells and B cells to mount an effective adaptive immune response. While genetic recombination provides the required diversity, recombination also has the potential to render the receptor loci incapable of producing a functioning receptor subunit. During T-cell and B-cell development, receptor loci are rearranged in a deliberate and orderly manner. After each recombination event, the protein product is tested to determine whether the rearrangement events maintained the integrity of the gene or rendered it nonfunctional. If the protein functions, further development ensues. If the protein is nonfunctional, recombination occurs on the other locus. If all loci have been subjected to unproductive rearrangements, the developing lymphocyte undergoes apoptosis (**FIGURE 4.13**).

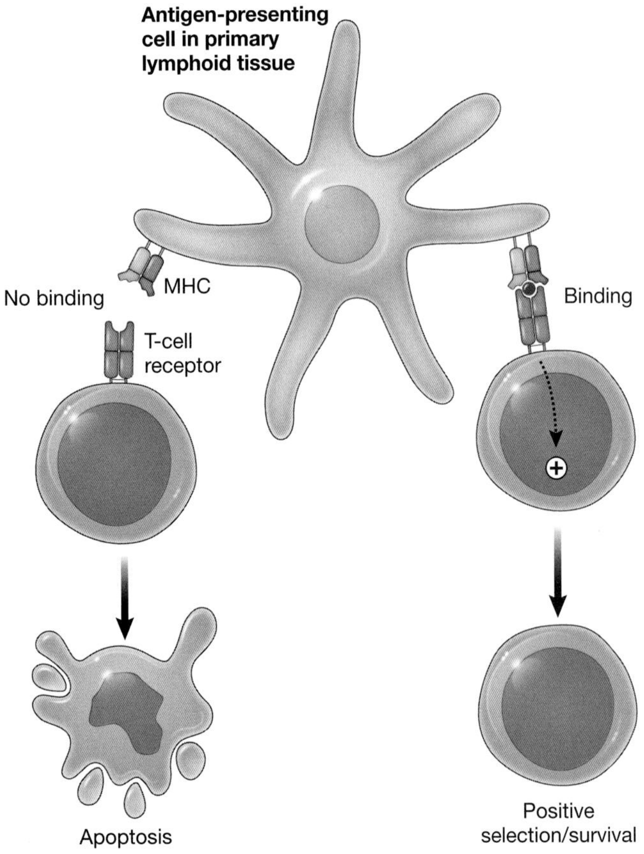

FIGURE 4.13 Positive selection of lymphocytes Developing lymphocytes are tested for the presence and binding capacity of their receptor. When a lymphocyte lacks a membrane receptor that can recognize antigen, it dies by apoptosis. Lymphocytes with functioning receptors are positively selected to continue their development.

Negative selection

Because positive selection tests only for receptor functionality, the developing lymphocyte may produce a functional receptor that can recognize epitopes naturally expressed by the body. Cells with these receptors would be extremely harmful because they could stimulate an immune response targeted at the body's own tissues. To ensure that this doesn't happen, developing lymphocytes are tested to determine if their receptor recognizes a self-molecule.

MHC molecules on the surface of cells in the thymus display a wide variety of self peptides; they assess the ability of functioning T-cell receptors to recognize each antigen. Any T cell that expresses a receptor that tightly engages with an MHC molecule during development is retained in the thymus and eventually removed by apoptosis. Any T cell that expresses a receptor that only moderately engages with an MHC molecule is permitted to continue development and leave the thymus as a T cell looking for a foreign epitope (**FIGURE 4.14**).

CHECKPOINT QUESTION

1. Explain the importance of negative selection in lymphocyte development.

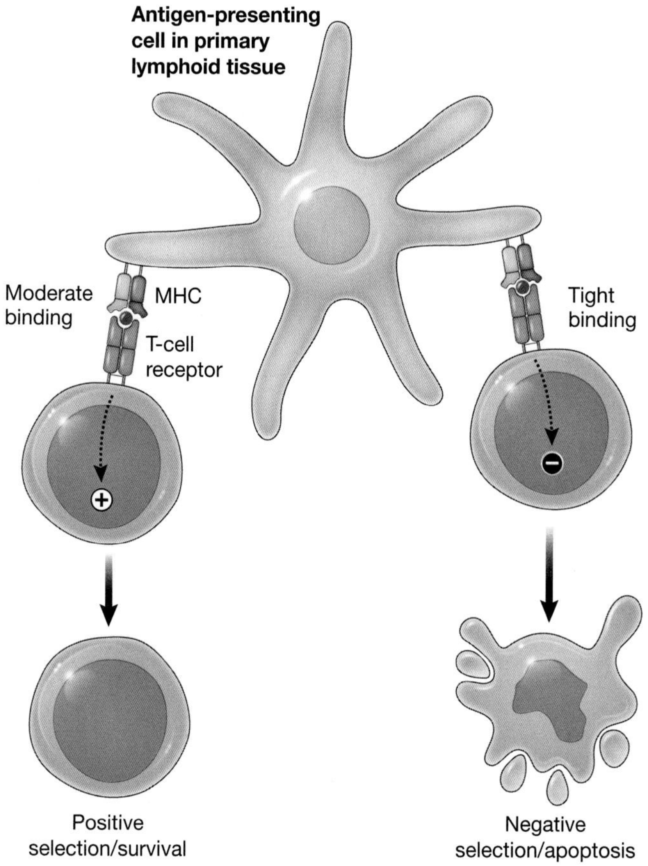

FIGURE 4.14 Negative selection of lymphocytes To prevent circulation of self-reactive lymphocytes, receptors are tested for binding affinity to self-antigen. Lymphocytes that are too likely to recognize self-antigen are negatively selected and die by apoptosis.

Making Connections: Lymphocyte Development in Primary Lymphoid Tissue Requires Recombination of Receptor Genes and Positive and Negative Selection

The adaptive immune system requires proper development and action of T cells and B cells. These developmental events take place in primary lymphoid tissues (bone marrow for B cells and thymus for T cells). During this process, somatic recombination occurs in developing lymphocytes to create a diverse array of cells capable of mounting an adaptive immune response to specific antigens. Before lymphocytes are released into circulation, they must be tested for both functioning receptors and potential for self-reactivity.

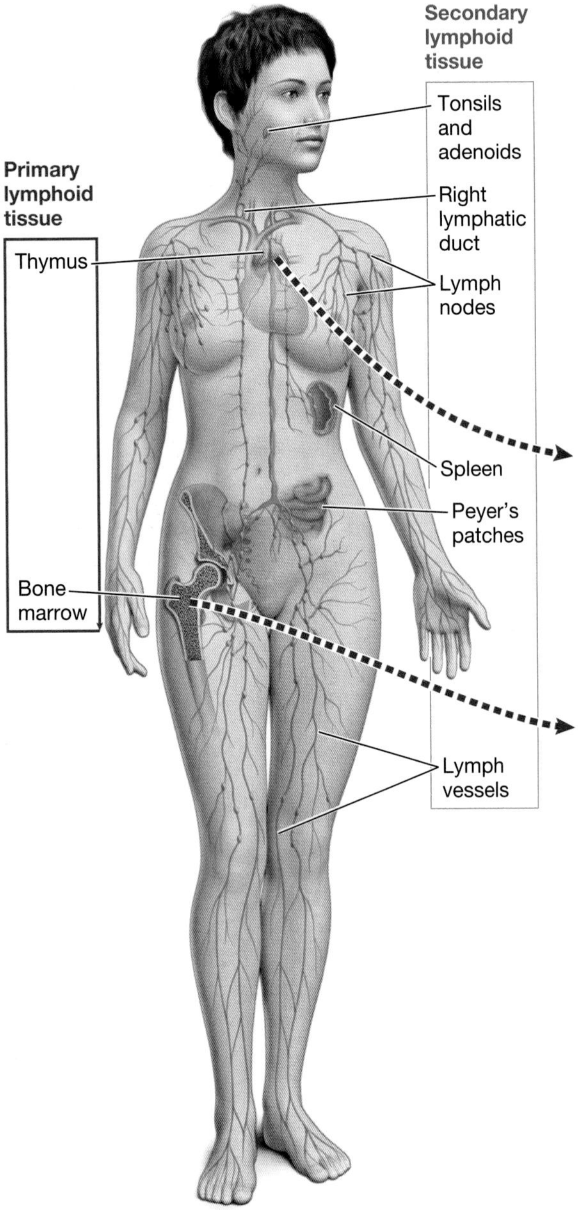

Lymphocytes begin their development during hematopoiesis in the bone marrow. Lymphocyte precursors destined to become B cells continue their development in the bone marrow, while lymphocyte precursors destined to become T cells migrate to the thymus and continue their development there.

(Sections 1.6, 1.9)

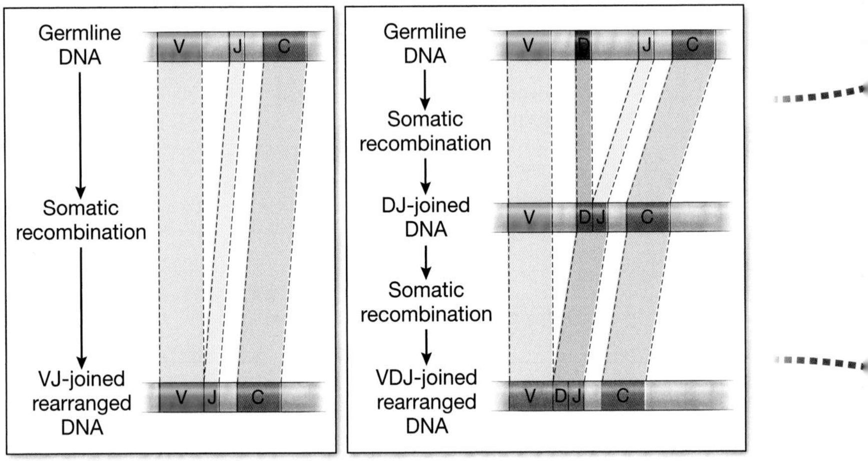

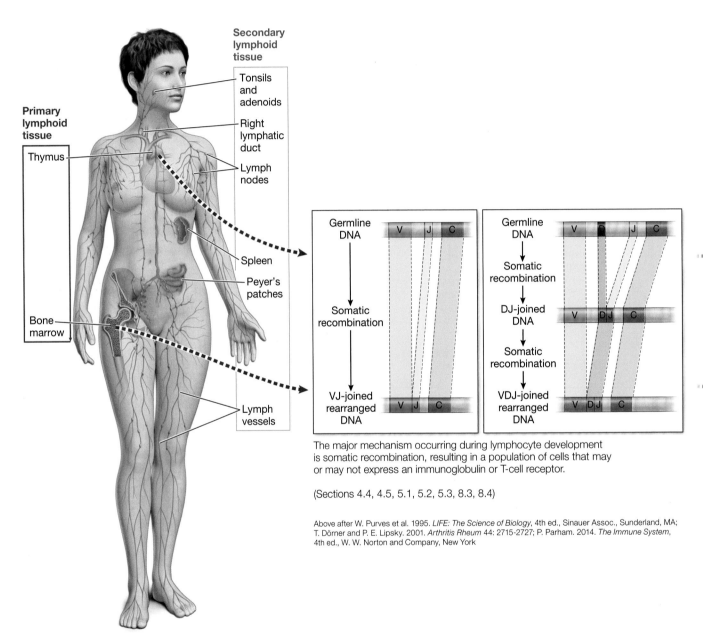

The major mechanism occurring during lymphocyte development is somatic recombination, resulting in a population of cells that may or may not express an immunoglobulin or T-cell receptor.

(Sections 4.4, 4.5, 5.1, 5.2, 5.3, 8.3, 8.4)

Above after W. Purves et al. 1995. *LIFE: The Science of Biology*, 4th ed., Sinauer Assoc., Sunderland, MA; T. Dörner and P. E. Lipsky. 2001. *Arthritis Rheum* 44: 2715-2727; P. Parham. 2014. *The Immune System*, 4th ed., W. W. Norton and Company, New York

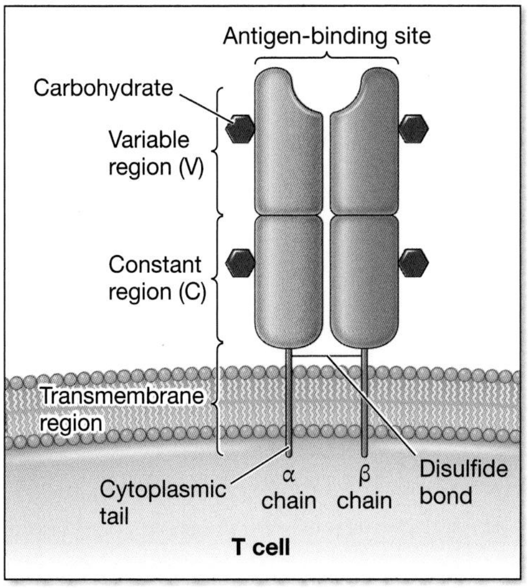

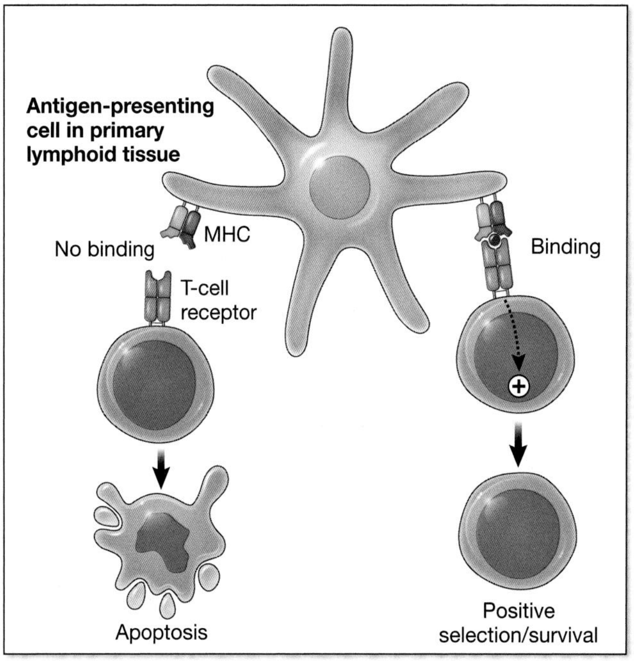

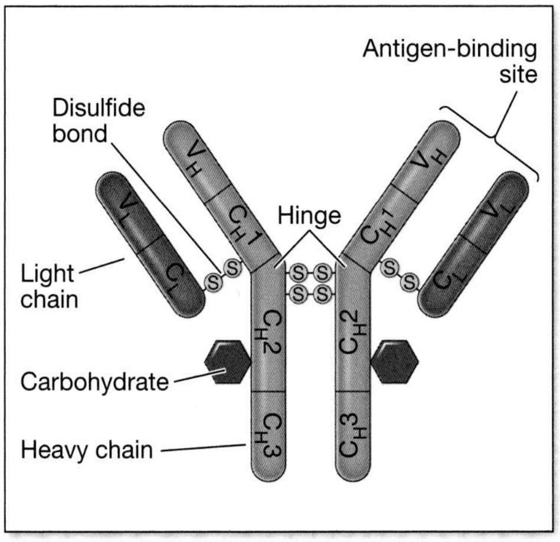

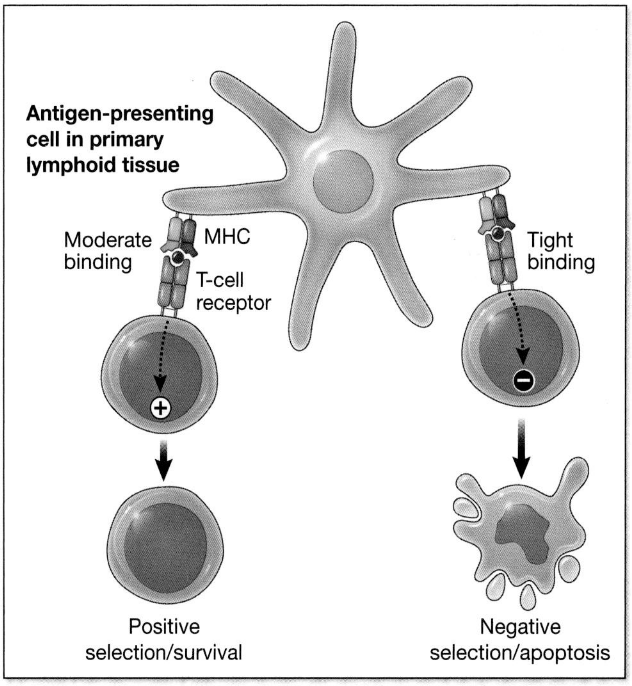

Developing lymphocytes test for both the presence of a functioning receptor (positive selection) and for receptor self-reactivity (negative selection). If a developing lymphocyte does not express a functioning receptor, or expresses a receptor that recognizes self-antigen, it is removed by apoptosis.

(Sections 2.2, 2.4, 4.2, 4.7, 4.8, 6.2, 9.1)

4.6 What is the importance of the major histocompatibility complexes in adaptive immunity?

LEARNING OBJECTIVE

4.6.1 Explain the role of MHC molecules in adaptive immunity and antigen presentation.

T-cell receptors recognize peptides displayed on cell surfaces via MHC molecules. Thus, the majority of genes associated with MHC molecules are those of the MHC molecules themselves or those involved in peptide processing (**FIGURE 4.15**). MHC molecules bind to protein peptides and move to the cell surface where they wait to engage a T-cell receptor. Most MHC molecules on the cell surface present self-molecules and are not recognized by any T cells in circulation (since self-recognizing T cells are usually eliminated by negative selection if they recognize a self-molecule in the thymus). During an infection, pathogens are processed to present peptides via MHC molecules on the cell surface, where they may bind to T-cell receptors.

MHC classes

MHC molecules are encoded by a gene family that is divided into three classes: class I, class II, and class III (see Figure 4.15). Each of these gene clusters is important in immune system function:

- *MHC class I*—These molecules function in the presentation of peptides generated from intracellular proteins. MHC class I molecules consist of two polypeptides: α, a transmembrane protein, and β_2-microglobulin, a soluble protein that associates with the α subunit (see Figure 4.2). While all of the α polypeptides are located on chromosome 6 in humans, the β_2-microglobulin gene is located on chromosome 15. MHC class I molecules are translated via ribosomes on the endoplasmic reticulum, where they are maintained until they bind to a peptide.
- *MHC class II*—These molecules function in the presentation of peptides generated from extracellular proteins. MHC class II molecules consist of two chains that are both transmembrane proteins: an α chain and a β chain (see Figure 4.2). MHC class II molecules are also translated via ribosomes at the endoplasmic reticulum; however, MHC class II molecules cannot initially associate with peptides in the endoplasmic reticulum as their peptide-binding site is blocked. Instead, they are loaded after being transported to secretory vesicles.

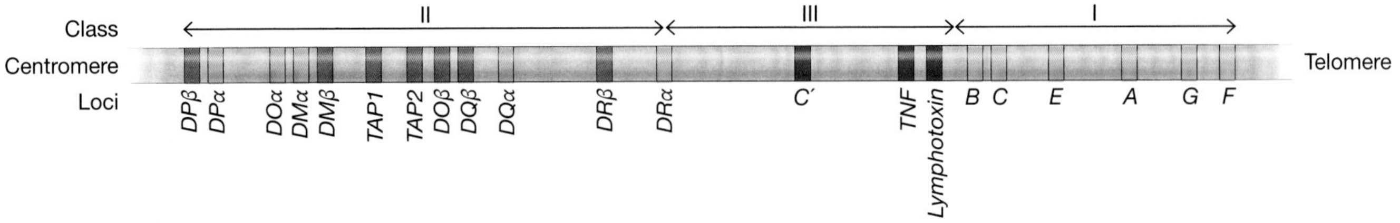

FIGURE 4.15 Structure of the major histocompatibility complex (MHC) gene clusters The MHC gene clusters, located on chromosome 6 in humans, are divided into three class locations, I, II, and III. The class I gene cluster contains all MHC class I α-chain genes (*A, B, C, E, F, G*), but the β_2-microglobulin gene is present on chromosome 15. The class II gene cluster contains all MHC class II α-chain genes (*DMα, DOα, DPα, DQα, DRα*), all MHC class II β-chain genes (*DMβ, DOβ, DPβ, DQβ, DRβ*), as well as genes for antigen processing and presentation such as *TAP1* and *TAP2*. The class III gene cluster contains some cytokine genes such as *TNF* and *lymphotoxin*, as well as some genes of the complement pathway (*C′*). (Data from The MHC sequencing consortium. 1999. *Nature* 401: 921–923.)

- *MHC class III*—MHC class III genes are part of the MHC cluster on the chromosome but do not directly function in peptide presentation. Genes that are part of this cluster still produce important immune response regulators, including inflammatory cytokines and components of the complement system (see Figure 4.15).

MHC class I peptide presentation

MHC class I molecules present peptides for cytoplasmic proteins, and most presented peptides are generated from self-molecules (**FIGURE 4.16**). Cytoplasmic proteins are digested by the proteasome, a protease responsible for degrading cellular proteins into small peptides (Figure 4.16, Step 1). Peptides generated by the proteasome are transported into the endoplasmic reticulum, where they bind to the peptide-binding site of MHC class I molecules (Figure 4.16, Step 2).

Once a peptide has been loaded onto the MHC class I molecule, the complex leaves the endoplasmic reticulum and travels through the secretory pathway, ultimately reaching the plasma membrane where peptides are presented to T cells (Figure 4.16, Step 3). T-cell receptors, in conjunction with the CD8 coreceptor, may recognize the MHC class I:peptide complex and prompt activation of the bound T cell. Since any cell can potentially be infected by an intracellular pathogen (a virus or bacteria such as *Listeria monocytogenes*), most cells express MHC class I molecules and present peptides on their surface.

MHC class II peptide presentation

MHC class II molecules present peptides for extracellular proteins (**FIGURE 4.17**). MHC class II molecules are first translated and transported into the endoplasmic reticulum. Once there, these complexes are maintained in an inactive state that prevents them from binding to peptides generated for MHC class I molecules (Figure 4.17, Step 1).

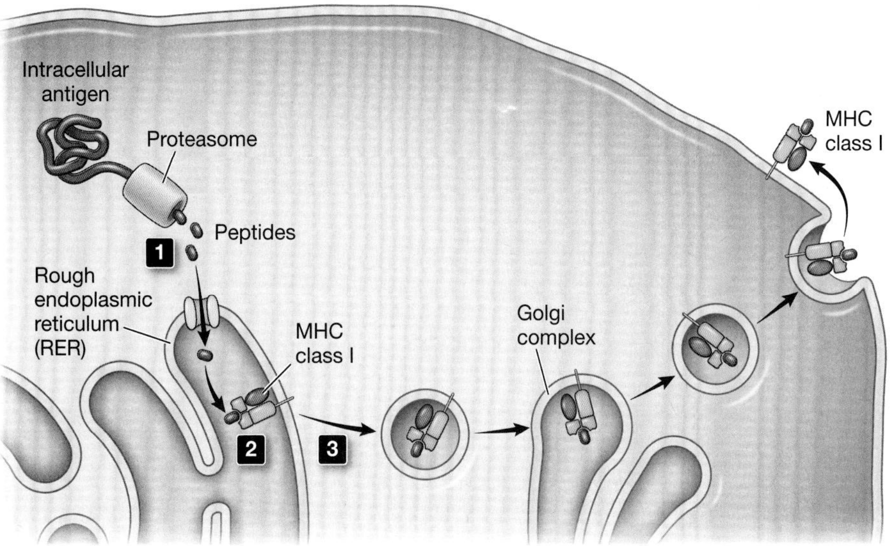

FIGURE 4.16 Overview of MHC class I antigen processing and presentation
Intracellular proteins present in the cytoplasm of all nucleated cells are eventually digested by the proteasome to create small peptides (Step 1). These peptides are transported into the rough endoplasmic reticulum (RER), where they bind to MHC class I molecules (Step 2). Once loaded with peptide, MHC class I molecules travel through the secretory pathway via vesicles to the Golgi complex and eventually to the plasma membrane, where they present peptides to CD8 T cells (Step 3).

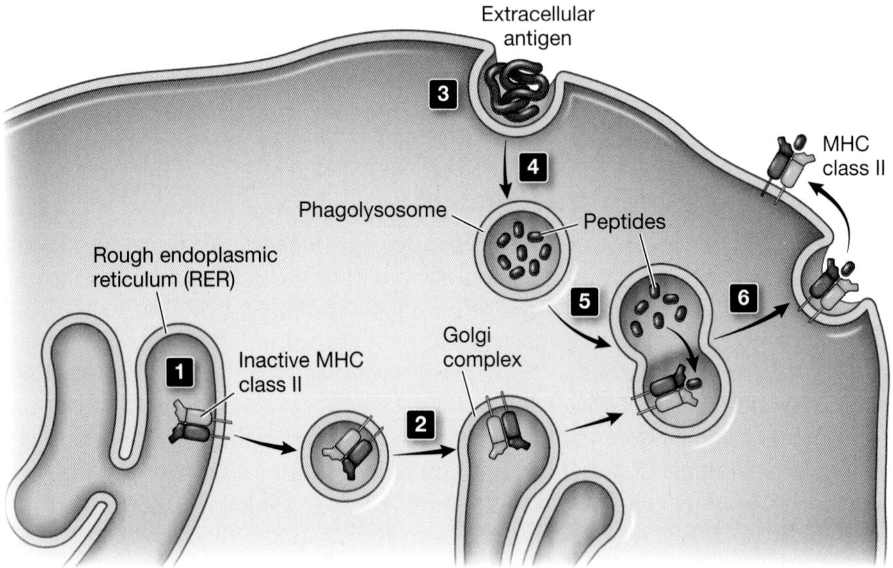

FIGURE 4.17 Overview of MHC class II antigen processing and presentation MHC class II molecules are synthesized co-translationally at the rough endoplasmic reticulum (RER) but remain inactive and unable to bind to class I peptides (Step 1). Once synthesized, MHC class II molecules travel through the secretory pathway via vesicles and wait for fusion with a phagolysosome (Step 2). Extracellular proteins are phagocytosed by professional antigen-presenting cells (APCs) (Step 3). After phagocytosis, the antigens are digested within a phagolysosome (Step 4). The phagolysosome fuses with an MHC class II-containing vesicle to load the MHC molecule with peptides generated from digestion of extracellular antigen (Step 5). Loaded MHC class II molecules travel to the cell surface, where they can present antigen to CD4 T cells (Step 6).

MHC class II molecules travel through the secretory pathway to a secretory vesicle, where they wait for further action driven by a phagolysosome (Figure 4.17, Step 2). Extracellular molecules and pathogens are endocytosed by professional APCs, including macrophages, dendritic cells, and B cells (Figure 4.17, Step 3).

The endosome containing the extracellular material matures into a phagolysosome, which results in the digestion of proteins by lysosomal proteases (Figure 4.17, Step 4). After digestion, phagolysosomes fuse with secretory vesicles containing MHC class II molecules, which can bind to peptides generated during lysosomal digestion (Figure 4.17, Step 5). These newly loaded MHC class II molecules then travel to the cell surface where they present peptides generated by the digestion of extracellular material to T cells (Figure 4.17, Step 6).

T-cell receptors, in conjunction with the CD4 coreceptor, may recognize the MHC class II:peptide complex and prompt activation of the bound T cell. Since only professional APCs process extracellular pathogens in this manner, they are the primary cells that express MHC class II molecules (although thymic epithelial cells involved in positive and negative selection during T-cell development also express class II molecules).

• CHECKPOINT QUESTION

1. What types of antigens do MHC class I and class II molecules typically present?

4.7 What roles do T-cell receptors play in adaptive immunity?

LEARNING OBJECTIVE

4.7.1 Describe the roles of T cells in antigen recognition and the adaptive immune response.

T cells play a major role in inducing an adaptive immune response. Activated T cells can directly target infected cells or assist in activating other components of the adaptive response. To recognize when foreign pathogens are being presented by class I or class II MHC molecules, these cells rely on the T-cell receptor.

T-cell receptor complex

The T-cell receptor consists of two different subunit chains: either α and β or γ and δ. The T-cell receptor α chain (and γ chain) contain a variable region generated by recombination of a V segment and J segment connected to a constant region, a transmembrane domain, and a small cytoplasmic domain. The β chain (and δ chain) contain a variable region generated by recombination of a V segment, D segment, and J segment connected to a constant region, a transmembrane domain, and a small cytoplasmic domain (**FIGURE 4.18**).

The variable and constant domains are folded in a common fold known as the *immunoglobulin superfamily domain*. The variable regions of the T-cell receptor subunits, generated by V(D)J recombination, are responsible for recognizing the MHC-peptide complex. Upon recognition of the MHC-peptide complex via the T-cell receptor, receptor proteins known as the CD3 complex (composed of four subunits) use their cytoplasmic domains to engage in signaling events. These events lead to T-cell activation and drive clonal expansion and the formation of effector T cells. However, T-cell receptor interaction with the MHC molecule and subsequent signaling events also require the action of the T-cell coreceptor.

T-cell coreceptors

To properly engage a cell presenting an MHC:peptide complex, T cells use both the T-cell receptor and a coreceptor. There are two coreceptors:

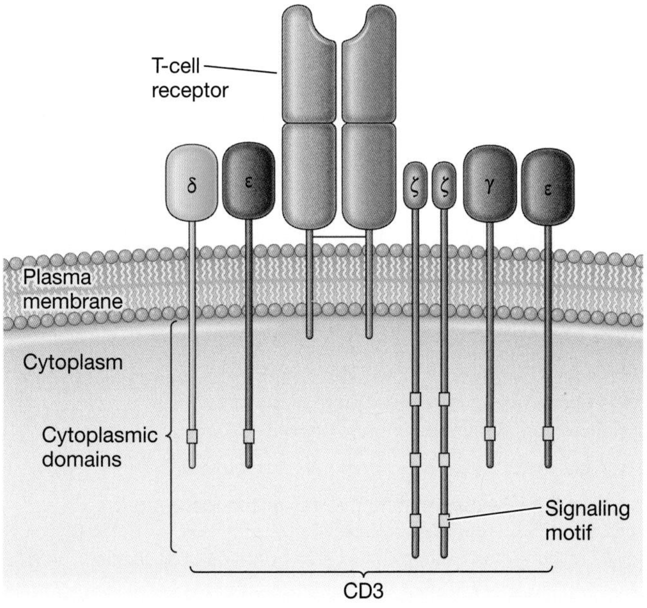

FIGURE 4.18 T-cell receptor complex While the T-cell receptor is responsible for recognizing MHC-peptide complexes, other membrane proteins of the CD3 complex are required for T-cell signaling once this interaction occurs. Proteins of the CD3 complex (δ, ε, γ, ζ) contain cytoplasmic domains with key signaling motifs that promote intracellular signaling events.

- The CD4 coreceptor recognizes MHC class II molecules. It is a single subunit protein that consists of four immunoglobulin domains and is anchored at the membrane through a transmembrane domain. It facilitates interaction with MHC class II molecules (and thus is on T cells that combat extracellular pathogens).
- The CD8 coreceptor recognizes MHC class I molecules. It is composed of two subunits (CD8α and CD8β), each containing a single immunoglobulin domain. The CD8 coreceptor facilitates interaction with MHC class I molecules (and thus is on T cells that combat intracellular infections).

The cytoplasmic domains of CD4 and CD8 also recruit important signaling molecules to the T-cell receptor complex to drive signaling events necessary for T-cell activation.

Effector T cells

Effector T cells specialize in promoting an adaptive immune response against a specific group of pathogens. The type of effector T cell produced upon clonal expansion and differentiation is a direct result of the type of pathogen recognized. The effector T cells are cytotoxic T cells; T_H1, T_H2, and T_H17 helper T cells; natural killer (NK) T cells; regulatory T cells; and memory T cells. **TABLE 4.1** summarizes the functions of each type, which will be discussed in detail in Chapter 7.

CHECKPOINT QUESTION

1. What subunits are required for a properly functioning T-cell receptor complex to both recognize its antigen and signal the process of clonal expansion?

TABLE 4.1 | Effector T Cells

Type of Effector T Cell	Description/Function(s)
Cytotoxic T cells	Recognize, target, and destroy cells infected with an intracellular pathogen Activated upon engagement of a T cell containing a CD8 coreceptor with an MHC class I APC
T_H1 helper T cells	Aid in combating extracellular pathogens by activating macrophages Activated upon engagement of a T cell containing a CD4 coreceptor with an MHC class II APC
T_H2 helper T cells	Work to combat extracellular pathogens by activating B cells (A subset known as follicular helper T cells activates within secondary lymphoid tissue to promote isotype switching and somatic hypermutation.) Also activated upon engagement of a T cell containing a CD4 coreceptor with an MHC class II APC
T_H17 helper T cells	Work to combat extracellular pathogens by activating neutrophils Also activated upon engagement of a T cell containing a CD4 coreceptor with an MHC class II APC
Natural killer (NK) T cells (also known as type I or classical NKT cells)	Unique subset of T cells that express an invariant αβ T-cell receptor capable of recognizing lipids and glycolipids Function as helper T cells through the secretion of inflammatory cytokines and as NK cells targeting intracellular infection Also function in a manner similar to that of cytotoxic T cells
Regulatory T cells	Inactivate self-reactive T cells to promote peripheral tolerance
Memory T cells	Long-lived T cells that promote an adaptive immune response to a pathogen upon subsequent exposure Can be rapidly activated and can monitor areas outside of secondary lymphoid tissues All activated T cells can produce memory T cells.

4.8 What roles do immunoglobulins play in adaptive immunity?

LEARNING OBJECTIVE

4.8.1 Describe the role of B cells in antigen recognition and the adaptive immune response.

B cells are activated with the assistance of helper T cells. These B cells then develop into specialized cells that secrete proteins that aid in targeting and destroying foreign pathogens. To recognize when foreign antigens are present, B cells use immunoglobulins that bind to specific antigens.

B-cell receptor complex

Recall that immunoglobulins are composed of two subunit chains: two identical heavy chains and two identical light chains. The light chain consists of a variable region generated by recombination of a V segment and J segment connected to a constant region. The heavy chain consists of a variable region generated by recombination of a V segment, D segment, and J segment connected to a constant region.

Immunoglobulins that act as cell-surface receptors also contain a transmembrane domain and a small cytoplasmic domain. The heavy and light chains are composed of immunoglobulin domains that support the structure of the protein, and disulfide bonds link the two chains.

The variable regions of one heavy chain and one light chain come together to form an antigen-binding site, so there are two antigen-binding sites per immunoglobulin. Most naturally produced immunoglobulins typically recognize the same antigen at the two antigen-binding sites.

Upon antigen recognition by a cell-surface immunoglobulin, associated proteins known as Igα and Igβ use their cytoplasmic domains to activate signaling pathways that promote B-cell activation, expansion, and differentiation (**FIGURE 4.19**). This process is similar to the way in which CD3 proteins of the T-cell receptor complex activate signaling pathways.

Effector B cells produce secreted antibodies that recognize the same antigen as the cell-surface immunoglobulin. These antibodies are further refined during the B-cell activation process based on different heavy chain constant regions that determine the antibody isotype.

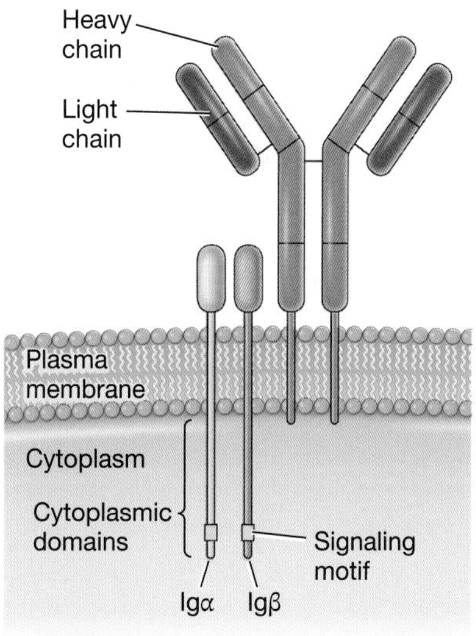

FIGURE 4.19 B-cell receptor complex While the B-cell receptor is responsible for recognizing extracellular antigen, other membrane proteins, including Igα and Igβ, are required for B-cell activation. Igα and Igβ contain cytoplasmic domains with key signaling motifs that promote intracellular signaling events.

Antibody isotypes

Upon B-cell selection, expansion, and activation, the immunoglobulin genes engage in somatic hypermutation and isotype switching to enhance both the affinity of the variable regions for their epitopes and the ability of soluble antibodies to function as part of the adaptive immune response. Both processes occur as an enzyme known as activation-induced cytidine deaminase (AID) stimulates changes in the immunoglobulin genes (discussed further in Chapter 9). Proper isotype switching is driven by the action of cytokines and the activation of transcription factors that enhance switching for certain heavy chain constant region isotypes to produce the IgG, IgE, and IgA isotypes.

Antibody isotypes specialize in their functions and actions. The five isotypes of immunoglobulins are IgM, IgD, IgG, IgE, and IgA. **TABLE 4.2** summarizes the functions of each type.

Effector B cells

Following clonal expansion and differentiation of an engaged B cell, the B cell is activated to become an effector B cell. Effector B cells function primarily to promote humoral immunity against extracellular pathogens, although some

TABLE 4.2 | Antibody Isotypes

Isotype	Description/Function(s)
IgM	Serves as one of the two types of B-cell receptors on the surface of naïve B cells Produced in activated B cells in conjunction with IgD through alternative splicing Neutralizing antibody that works primarily as the original antibody opsonin, activating the classical complement pathway in a manner similar to that of C-reactive protein Effector functions similar to those of IgG Can serve as a B-cell receptor in naïve B cells
IgD	Serves as one of the two types of B-cell receptors on the surface of naïve B cells Produced in activated B cells in conjunction with IgM through alternative splicing Plays a role in activating mast cells and basophils Can serve as a B-cell receptor in naïve B cells
IgG	Functions as both a neutralizing antibody and an opsonin to activate the classical complement pathway Also plays a role in the formation and clearance of soluble immune complexes and in activating NK cells in the process of antibody-dependent cell-mediated cytotoxicity
IgE	Normally associated with mast cells due to those cells having a tight-binding receptor for IgE Binding to antigen causes mast cell degranulation and the release of inflammatory mediators Activates mast cells, basophils, and eosinophils
IgA	Primarily found at mucosal surfaces Neutralizes pathogens at these surfaces and delivers pathogens to mucosa-associated lymphoid tissue

play a role in adaptive immune system memory and regulation of the immune system. The types of effector B cells are:

- *Plasma cells*—Activated B cells that specialize in secreting antibodies. Activated B cells known as centrocytes can produce a subset of plasma cells that have undergone affinity maturation and isotype switching to produce high-affinity, specialized antibodies of specific isotypes.
- *Regulatory B cells*—Work to suppress the immune system. Their primary role is to suppress inflammation by secreting the anti-inflammatory cytokine IL-10.
- *Memory B cells*—Long-lived B cells that promote an adaptive immune response to a pathogen upon subsequent exposure. These cells can be rapidly activated, can recognize antigen due to higher affinity produced during somatic hypermutation, and can monitor areas outside of secondary lymphoid tissues.

Immunoglobulins' variety of effector functions allows them to combat many extracellular pathogens and toxins, including viruses. One common strategy in treating viral infections is the use of neutralizing antibodies to block virus interaction with target cells. This strategy is being tested as a possible treatment for COVID-19, caused by SARS-CoV-2 (see **EMERGING SCIENCE**).

CHECKPOINT QUESTION

1. List the five major immunoglobulin isotypes and provide a brief description of their function.

EMERGING SCIENCE

Can neutralizing antibodies be used as a treatment for COVID-19?

Article

A. C. Wells, Y-J. Park, M. A. Tortorici, A. Wall, A. T. McGuire, and D. Vessler. 2020. Structure, function, and antigenicity of the SARS-CoV-2 spike glycoprotein. *Cell* 180: 1–12.

Background

Immunoglobulins contain two identical heavy chains and two identical light chains that associate during immunoglobulin synthesis and processing. The variable regions of the heavy and light chains are in close proximity to each other to create the antigen-binding site. Thus, immunoglobulins have two identical antigen-binding sites per molecule.

One major effector function of immunoglobulins is neutralizing foreign particles or pathogens by interacting with a component of the pathogen required for binding with target cells. If the pathogen component that binds with target cells is known, it might be possible to create immunoglobulins that can prevent the pathogen from infecting those target cells. The development of neutralizing immunoglobulins to combat viruses is an attractive therapeutic strategy since viruses must interact with target cells during their life cycle to hijack the target-cell machinery and replicate.

The Study

Researchers studied the major spike glycoprotein responsible for the interaction of SARS-CoV-2. To do this, they analyzed the target-cell protein that the spike glycoprotein of the coronavirus binds to. They also analyzed the three-dimensional structure of the spike glycoprotein using cryo-electron microscopy. The researchers were able to determine that the coronavirus spike glycoprotein binds to the human protein angiotensin-converting enzyme 2 (ACE2) to infect target cells. This binding to ACE2 is similar to the manner in which SARS-CoV (the causative agent of the SARS pandemic that occurred between 2002 and 2004) infects target cells.

Because of the similarities of the structure of the spike glycoproteins from SARS-CoV-2 and SARS-CoV, the researchers hypothesized that neutralizing antibodies of SARS-CoV might be capable of blocking SARS-CoV-2 infection of target cells. To test this, the researchers used plasma from mice that had been inoculated with SARS-CoV to attempt to block cells from being infected with SARS-CoV-2 in culture. Using several different mouse sera raised against SARS-CoV, they found that SARS-CoV-2 could be blocked from infecting cells in culture (**FIGURE ES 4.1**).

Conclusion

The ability to use SARS-CoV neutralizing antibodies to block SARS-CoV-2 infection is a promising step in the development of therapeutics against COVID-19 and a strategy that might aid in the development of a successful vaccine.

Think About...

1. Based on your knowledge of immunoglobulin structure and the immune system, why did the researchers compare the structures of the spike glycoproteins of SARS-CoV and SARS-CoV-2 before testing the mouse sera of SARS-CoV as neutralizing sera to SARS-CoV-2?
2. Given the similarities in structure of the two spike glycoproteins, what might you predict about the evolution of the two coronaviruses in the study?
3. Given the results of this study, would you predict that individuals who had been infected with SARS-CoV in 2002 would be protected against SARS-CoV-2 infection? How might you test this prediction?

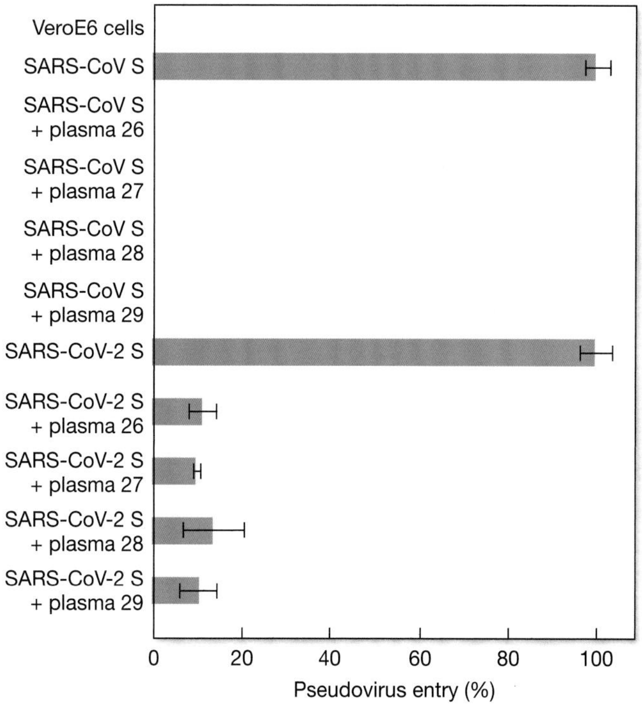

FIGURE ES 4.1 Murine plasma containing antibodies to SARS-CoV can neutralize SARS-CoV-2, the causative agent of COVID-19, from infecting cells Cells in culture are readily infected with SARS-CoV and SARS-CoV-2 pseudoviruses, which were created and purified through transfection. Plasma from mice immunized against SARS-CoV could completely block SARS-CoV infection into cells. The same plasma could lower the infectivity of SARS-CoV-2 in culture by 90%, suggesting there is an ability to produce neutralizing antibodies against SARS-CoV-2. (From A. C. Walls et al. 2020. *Cell* 180: 1–12.)

Making Connections: Secondary Lymphoid Tissue and the Adaptive Immune Response

Because of the diverse number of T cells and B cells with different lymphocyte receptors, it is important to concentrate activation of an adaptive immune response near a site of infection. To do so, secondary lymphoid tissue serves as a depository for antigen that is surveyed by the receptors of circulating T cells and B cells. If a T cell or B cell recognizes antigen via its cell-surface receptor, activation of that cell ensues.

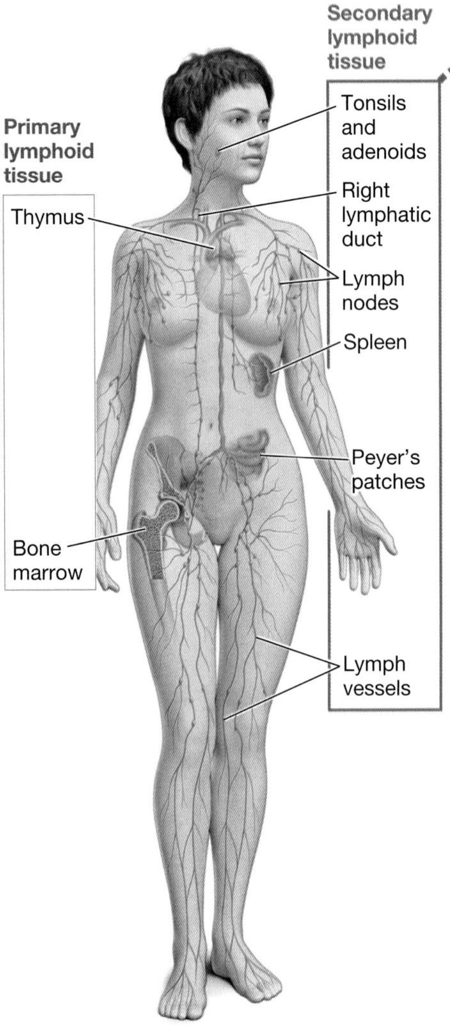

Secondary lymphoid tissue, including lymph nodes, the spleen, and mucosa-associated lymphoid tissue, serves as a depository for antigen in the event activation of an adaptive immune response becomes necessary.

(Sections 1.8, 1.9)

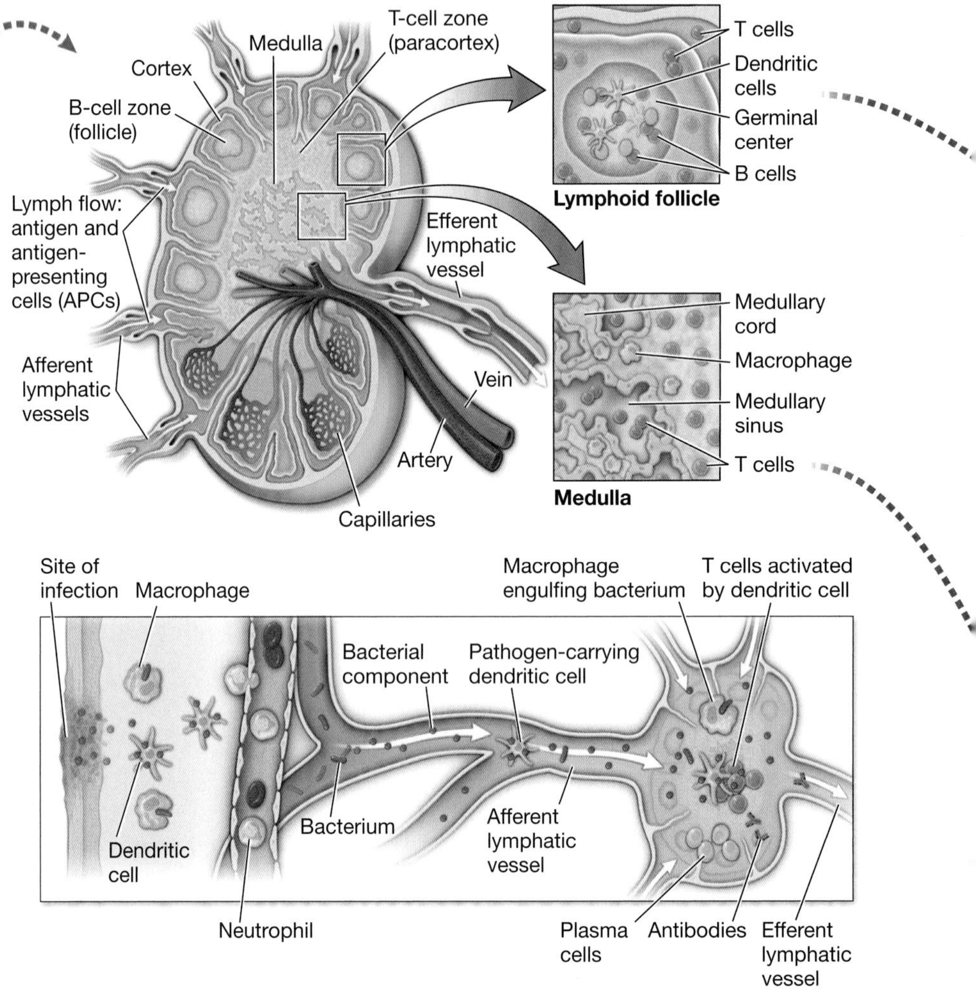

Delivery of antigen to secondary lymphoid tissue can result in the activation of T cells and B cells, clonally selected by the presence of a receptor that recognizes that specific antigen. Binding of the receptor induces clonal expansion and activation of effector mechanisms of the selected lymphocyte.

(Sections 1.8, 1.9, 1.10, 4.3, 4.6, 4.7, 4.8, 7.6, 7.12, 10.2, 10.3)

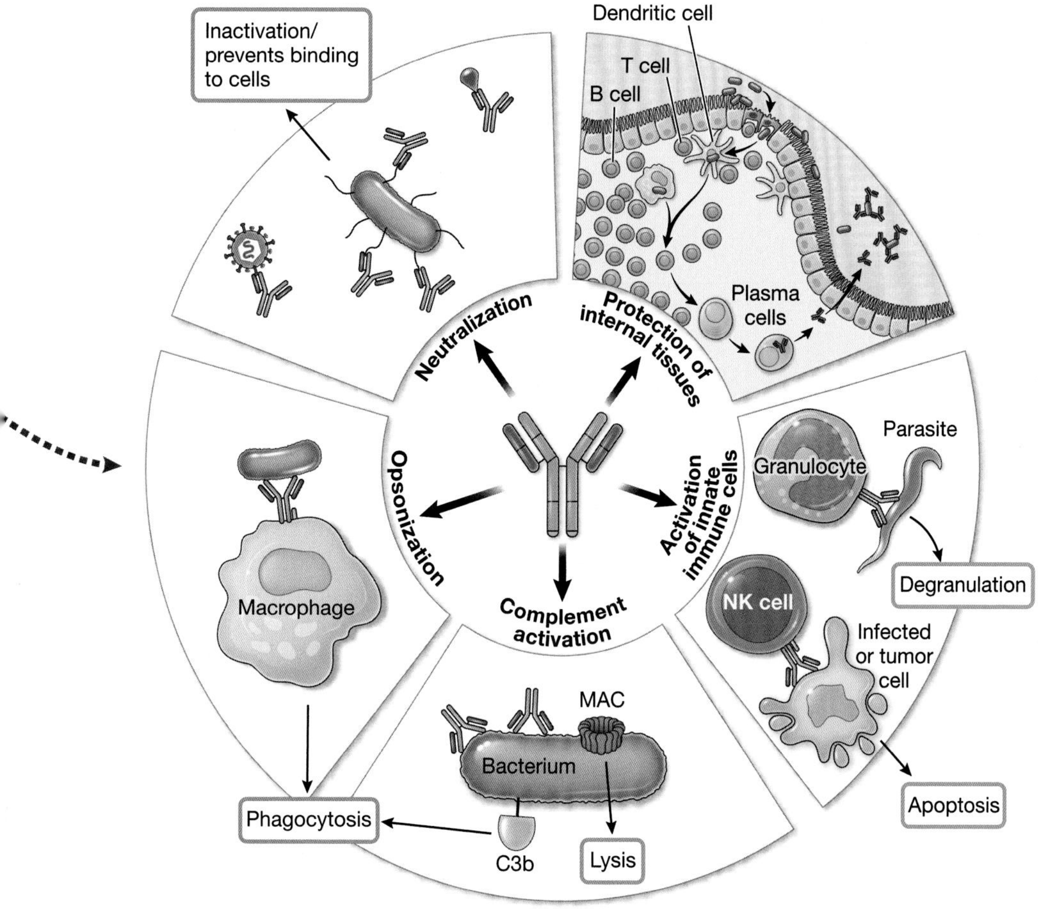
Inactivation/
prevents binding
to cells
Dendritic cell
T cell
B cell
Plasma
cells
Neutralization
Protection of
internal tissues
Opsonization
Activation
of innate
immune cells
Complement
activation
Parasite
Granulocyte
Degranulation
NK cell
Infected
or tumor
cell
Apoptosis
Macrophage
MAC
Bacterium
Phagocytosis
C3b
Lysis

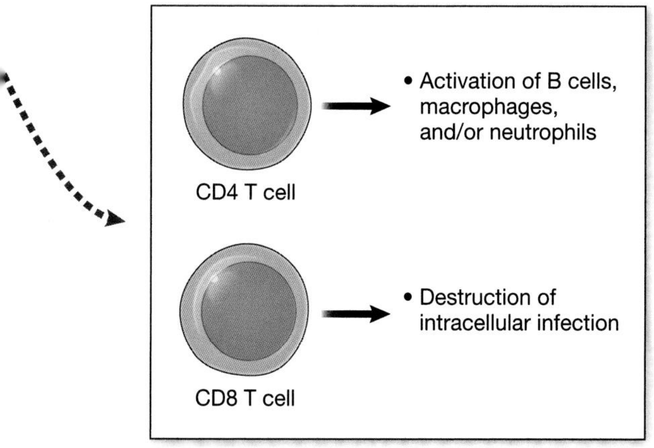
• Activation of B cells,
macrophages,
and/or neutrophils
CD4 T cell
• Destruction of
intracellular infection
CD8 T cell

4.9 What is immunological memory and how is it developed?

LEARNING OBJECTIVE

4.9.1 Explain the importance of immunological memory for the adaptive immune system.

primary immune response Adaptive immune response generated upon initial exposure to a pathogen.

secondary immune response Adaptive immune response generated upon second and subsequent exposure to a pathogen.

The first exposure to an antigen that activates the adaptive immune system is known as the **primary immune response**. This initial process requires time for the pathogen to be processed at the site of infection, for dendritic cells that have taken up pathogens to migrate to a draining lymphoid tissue, and for antigen presentation to circulating T cells. Expansion of activated lymphocytes also takes time before a robust adaptive immune response can properly target a foreign pathogen.

The power of the adaptive immune response is employed upon the first encounter with a specific pathogen, both to target the pathogen for destruction and to produce cells that recognize this pathogen, resulting in faster activation upon subsequent exposure. During the activation and expansion of T cells and B cells, memory T cells and memory B cells are produced that maintain memory of the pathogen that activated their initial production. These cells circulate in the body to monitor for a recurrence of the same infection or reside in tissues close to where the original activation occurred.

Memory B cells are also produced after somatic hypermutation and isotype switching have occurred, enabling them to recognize antigen faster and with a higher affinity than cells that were activated during the initial encounter. The adaptive immune response that occurs upon subsequent exposure to the same pathogen is known as the **secondary immune response**. Since immunological memory has been established, a stronger, more rapid defense is mounted against the invading pathogen.

CHECKPOINT QUESTION

1. Explain why a primary immune response usually takes longer than a secondary immune response.

4.10 Why is immunological memory important to an adaptive immune response?

LEARNING OBJECTIVE

4.10.1 Differentiate between a primary and secondary adaptive immune response.

In Chapter 2, we explored the importance of the innate immune system and the many mechanisms involved in preventing the spread of infection. While the innate immune response is efficient, pathogens are sometimes able to evade the innate immune defenses. When this happens, the adaptive immune system must step in to target and eliminate pathogens. Individuals who lack an adaptive immune system, due to mutations that prevent the proper development of lymphocytes or suppression by a virus such as HIV, have a very difficult time clearing infections (**FIGURE 4.20**).

Because of the many steps involved, the primary adaptive immune response can take weeks to properly develop and target an infecting pathogen. Memory cells produced as part of the primary immune response ensure that there are primed cells "at the ready" that were produced during a previous encounter with a pathogen and can monitor sites throughout the body and respond more

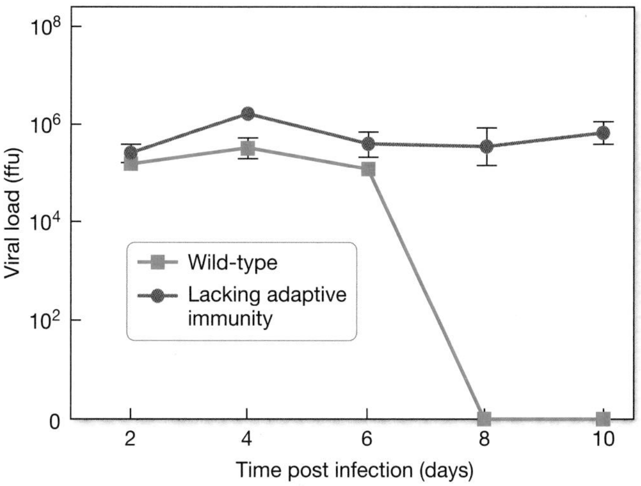

FIGURE 4.20 Loss of an adaptive immune response prevents clearance of an infection Mice were infected with influenza A and tested at regular intervals to determine the amount of virus present (focus-forming units [ffu]) in wild-type mice and mice deficient in adaptive immunity. Wild-type mice are able to clear the viral infection, evidenced by undetectable ffu, while mice lacking an adaptive immune system cannot clear the virus. (After H. Wu et al. 2010. *Virol J* 7: 172.)

robustly upon subsequent exposure. The means by which immunological memory is generated is the basis for the development of safe, protective vaccines.

The primary and secondary adaptive immune response both produce effector cells and soluble immunoglobulins that target the pathogen in a similar manner. For instance, exposure to an extracellular pathogen such as *Vibrio cholerae* results in the activation of phagocytes such as macrophages and promotes the production of antibodies that will aid in opsonization of the bacteria. The mechanisms employed upon first exposure to *V. cholerae* will closely parallel those of a secondary immune response to the same bacteria.

The important difference between a primary and secondary response is the *speed* at which the response occurs (primarily due to the presence of memory cells) and the *power* of the response. The secondary immune response is usually faster and more robust than a primary response (**FIGURE 4.21**). On average, a primary response can take 1 to 2 weeks to clear the pathogen, while a secondary response can take as little as 3 to 4 days. Thus, secondary immune responses can be extremely helpful in the removal of infectious agents, but this power does have a downside. In Chapters 15, 16, and 17, we will discuss ways in which the adaptive immune response can be detrimental, such as when it targets innocuous materials or transplanted tissues.

CHECKPOINT QUESTION

1. Compare the timeline of a primary immune response to a secondary immune response.

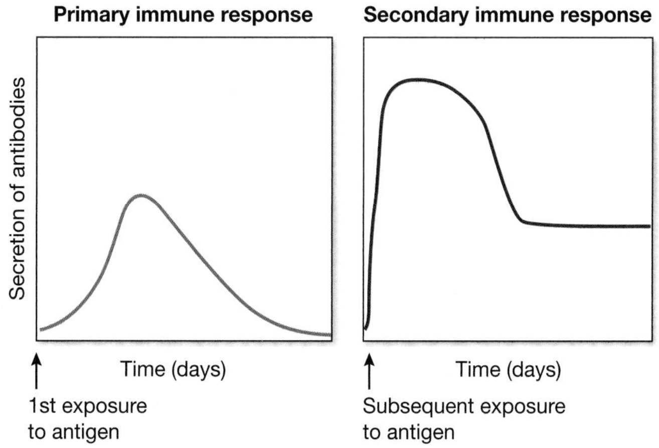

FIGURE 4.21 Adaptive immunity increases in efficiency and speed While a primary adaptive immune response results in maximal antibody secretion weeks after exposure to antigen, subsequent exposure to the same antigen produces a much faster and more robust antibody response.

Summary

4.1 How does adaptive immunity differ from innate immunity?

- The adaptive immune system employs specificity to target particular pathogens and memory to ensure that subsequent infection by the same pathogen is cleared faster and more robustly.

4.2 How are pathogens recognized by cells of the adaptive immune system?

- T-cell receptors recognize antigens presented by major histocompatibility complex (MHC) molecules in conjunction with the T-cell coreceptors CD4 and CD8.
- Immunoglobulins (B-cell receptors) recognize antigens at the surface of B cells or soluble antigens through the production of soluble immunoglobulins called antibodies.

4.3 What is the timeline of an adaptive immune response?

- A primary adaptive immune response requires a few days to 2 weeks. Immature B cells and T cells must develop in primary lymphoid tissue and then migrate to secondary lymphoid tissue, where they interact with antigen-presenting cells (APCs).

4.4 What mechanism provides diversity of T-cell and B-cell receptors?

- Gene rearrangement, or somatic recombination, is an essential process that drives T-cell receptor and immunoglobulin diversity.
- Somatic recombination occurs through the action of the RAG1/RAG2 proteins, which constitute the major activity of the V(D)J recombinase.
- V(D)J recombination occurs at very specific locations within the T-cell receptor and immunoglobulin loci known as recombination signal sequences (RSSs).
- Recombination events in B cells and T cells combine DNA at T-cell receptor and immunoglobulin loci and eliminate DNA located between the recombination sites.
- Random recombination events of different V, D, and J regions at T-cell receptor and immunoglobulin loci in germline DNA provide diversity in T-cell receptors and immunoglobulins.
- Junctional diversity, which is produced by the addition and removal of nucleotides and recombination sites, provides another level of diversity within T-cell receptors and immunoglobulins.
- A third level of diversity of T-cell receptors and immunoglobulins is the random combination of T-cell receptor and immunoglobulin subunits, which is required for a functional receptor.
- Immunoglobulin genes proceed through further recombination events during B-cell activation to produce isotypes with specialized roles in the adaptive immune response.

4.5 How does tolerance to self-molecules occur given the diversity of our adaptive immune system?

- Positive and negative selection provide a mechanism for testing T-cell receptors and immunoglobulins for proper function and prevention of autoreactivity.

4.6 What is the importance of the major histocompatibility complexes in adaptive immunity?

- MHC molecules are important in presenting antigen to T cells: MHC class I molecules present intracellular antigens, and MHC class II molecules present extracellular antigens.
- MHC class I loading occurs through digestion of cytosolic proteins, transport of the peptides into the endoplasmic reticulum, and loading of MHC class I molecules before they migrate to the cell surface and await $CD8^+$ T cells.
- MHC class II loading occurs through phagocytosis of extracellular material, digestion within the phagolysosome, and loading of MHC class II molecules within a vesicle before they migrate to the cell surface and await $CD4^+$ T cells.

4.7 What roles do T-cell receptors play in adaptive immunity?

- A functioning T-cell receptor complex includes the T-cell receptor, T-cell coreceptor, and the CD3 complex, which transmits the signal of a bound T-cell receptor to the cytosol of the engaged T cell.
- Effector T cells produced during T-cell activation function to either target an intracellular pathogen (cytotoxic T cells) or aid in the activation of other immune cells (helper T cells).

4.8 What roles do immunoglobulins play in adaptive immunity?

- A functioning immunoglobulin complex includes the immunoglobulin chains and Igα and Igβ, which transmit the signal of a bound immunoglobulin to the cytosol of the engaged B cell.
- The antibody isotypes (IgM, IgD, IgG, IgE, IgA) function at different times during an adaptive immune response and play different roles in the tissues they protect and immune responses they activate.

4.9 What is immunological memory and how is it developed?

- Effector B cells produced during B-cell activation produce soluble immunoglobulins of varying affinities and memory B cells for long-term memory.
- Memory T cells and B cells produced during a primary immune response stand "at the ready" to combat the same pathogen upon subsequent exposure.

4.10 Why is immunological memory important to an adaptive immune response?

- The secondary immune response (produced through activation of memory T cells and memory B cells) is faster and more robust than the primary immune response; it is important in effective clearance of pathogens upon subsequent exposure.

Review Questions

4.1 Compare and contrast the pathogen recognition systems of the innate and adaptive immune systems.

4.2 How are T-cell receptors and immunoglobulins similar in how they recognize antigen? How are they different?

4.3 What might the impact be on the adaptive immune response if an individual were unable to produce β-microglobulin? What types of pathogens (intracellular versus extracellular) would you predict this individual might be more susceptible to?

4.4 What are the major mechanisms used to diversify T-cell receptors and immunoglobulins?

4.5 A patient is admitted to the hospital due to recurring infections. Blood analysis shows a high concentration of IgM immunoglobulins compared to other types. Genetic evaluation determines that the patient has a mutation in activation-induced cytosine deaminase (AID). Why does this mutation cause production of only IgM immunoglobulins?

4.6 Explain why both positive and negative selection of developing lymphocytes is critical in the function of a normal adaptive immune system.

4.7 Which effector T cells use CD4 as a coreceptor? Which effector T cells use CD8 as a coreceptor? Name these effector cells and match them with their target and the epitopes they can recognize based on coreceptor function.

4.8 A patient is admitted to the hospital due to recurring gastrointestinal and respiratory infections with extracellular pathogens. The suspected cause is a lack of the adaptive immune response to protect the mucosal surfaces. Which immunoglobulin isotype do you think is lacking in this patient? Which primary immunoglobulin effector function(s) is/are eliminated at the mucosal surfaces because of lack of this particular immunoglobulin?

4.9 What is the major function of memory cells in the adaptive immune system?

4.10 Explain why development of memory cells is important in regard to the effectiveness of vaccines.

CASE STUDY REVISITED

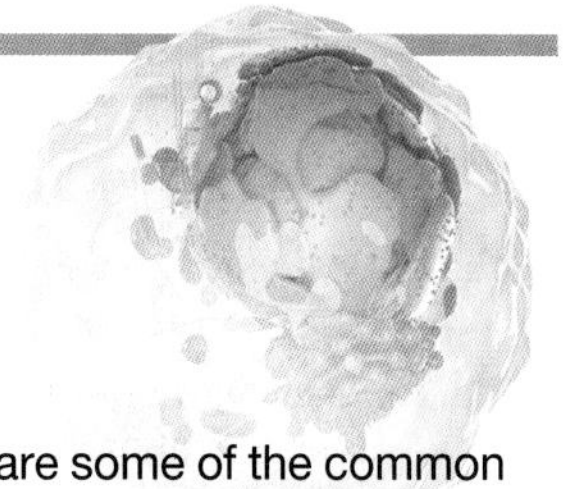

Omenn Syndrome

A male infant, age 6 months, is admitted to the emergency room with a red rash all over his body. The parents state that he has suffered from chronic diarrhea but that this is the first time they have seen a red rash all over his body. Physical examination reveals many swollen lymph nodes, suggesting a chronic infection. Because of the chronic diarrhea, the red rash, and the presence of a large number of swollen lymph nodes, the physician orders a complete blood panel to analyze the infant's white blood cell count. The complete blood panel shows that the baby lacks circulating B and T cells, suggesting a problem with the development of these lymphocytes in the infant. To further ascertain the issue associated with lymphocyte development, a sequence analysis is ordered. The sequence analysis demonstrates that the infant suffers from a missense mutation in the *RAG1* gene, and based on the sequence analysis, the absence of circulating B and T cells, and the infection, the physician diagnoses the infant with Omenn syndrome. An MHC crossmatch is ordered for bone marrow transplantation.

Think About...

Questions for individuals

1. Define Ommen syndrome. What are some of the common symptoms associated with Ommen syndrome?
2. Are there any causes of Ommen syndrome besides a mutation in *RAG1*?

Questions for student groups

1. Would you expect that the infant's innate immune system is functioning? Why or why not?
2. Do the child's symptoms support the diagnosis of Ommen syndrome? Are there any other diseases associated with mutation in *RAG1*?
3. Why would mutation in the *RAG1* gene prevent B cells and T cells from circulating in the bloodstream?
4. Why is a bone marrow transplant warranted in this case?

MAKING CONNECTIONS **Key Concepts**	**COVERAGE** **(Section Numbers)**
Comparison of innate and adaptive immune receptor gene and protein structure	2.2, 2.4, 4.2, 4.4, 4.7, 4.8, 5.5, 6.2, 6.3, 8.5, 9.1, 9.2
Comparison of signaling mechanisms used in innate and adaptive immune receptor signaling	2.4, 7.7, 7.8, 10.1
Comparison of effector mechanisms of the innate and adaptive immune systems	1.7, 1.8, 2.5, 2.6, 2.7, 3.1, 4.7, 4.8, 7.13, 7.14, 7.16, 10.5
Primary lymphoid tissue and its role in lymphocyte development	1.6, 1.9, 4.4, 4.5, 5.1, 5.2, 5.3, 8.3, 8.4
The processes of positive and negative selection in lymphocyte development	4.5, 5.8, 8.7, 8.8, 8.9
Secondary lymphoid tissue and its role in the adaptive immune response	1.8, 1.9, 1.10, 4.7, 4.8, 7.6, 10.3
Specialized areas of secondary lymphoid tissue	4.3, 4.6, 4.7, 4.8, 7.12, 10.2

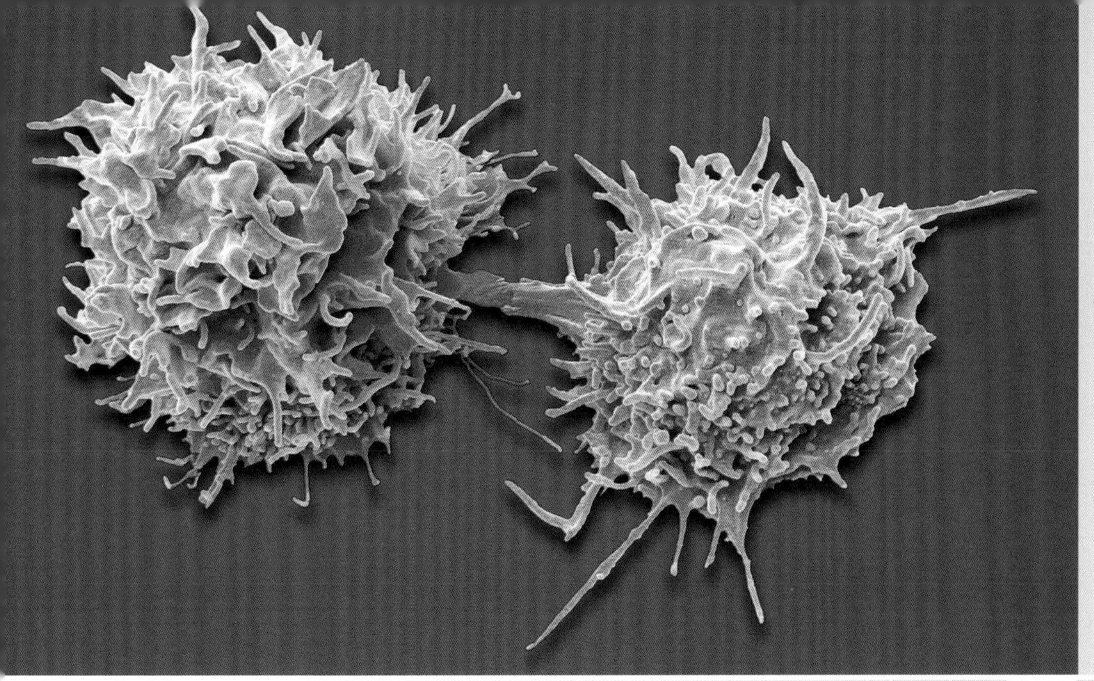

5

Development of T Lymphocytes

CASE STUDY: Deion's Story

Deion wakes up feeling exhausted. His mother enters the room to get him ready for school and notices that Deion is looking more tired than usual. Deion recently joined the football team, so his mother assumes that he is tired from all of the after-school football practices. Lately, she has noticed that Deion has been drinking more water than usual, but she again attributes this to football.

She asks Deion how he's feeling and if he slept alright. Deion tells his mother that he was up almost every hour needing to go to the bathroom and feeling really thirsty. Deion mentions that he had been running sprints at practice the day before, but he felt achy all over, including his arms and back. Since this is out of character for Deion—an athletic 11-year-old—his mother decides to make an appointment with a pediatrician. She remembers when her brother started having similar symptoms when he was about Deion's age; her brother's diagnosis was much worse than delayed-onset muscle soreness.

In order to understand what Deion is experiencing, and to answer the case study questions at the end of this chapter, we must first explore the process of T-cell development. Recall that lymphocytes use the elegant mechanism of somatic recombination (see Chapter 4) to produce a diverse set of receptors responsible for recognizing foreign antigens. While we have already explored this mechanism, we have yet to define when somatic recombination occurs during lymphocyte development, or discuss the means by which we test that these events produce a functioning lymphocyte receptor (i.e., one that can differentiate between foreign antigens and self molecules). T cells must produce a cell-surface T-cell receptor to function properly in an adaptive immune response. The T-cell receptor needs to engage with a peptide presented via an MHC molecule by an

QUESTIONS Explored

5.1 What is the role of the thymus in T-cell development?

5.2 Which cells are precursors to developing thymocytes?

5.3 What is the role of Notch1 in T-cell development?

5.4 What are the stages of T-cell development?

5.5 How does a developing thymocyte pass through the β-chain checkpoint?

5.6 How does a developing thymocyte pass through the α-chain checkpoint?

5.7 How does positive and negative selection of T cells occur?

5.8 How is T-cell receptor diversity achieved?

antigen-presenting cell. Thus, during T cell development, proper recombination of each subunit of the T-cell receptor must occur, and its ability to recognize MHC-peptide complexes must be tested. One major problem that could occur during these recombination events is the production of a T cell that is autoreactive, or capable of recognizing and combating our own tissue. In order to prevent T cells that can recognize self antigen from circulating in the body, T-cell receptors must be tested to ensure they cannot bind too tightly to MHC-self-peptide complexes.

5.1 What is the role of the thymus in T-cell development?

LEARNING OBJECTIVES

5.1.1 Describe the roles of the thymus and resident cells of the thymus in proper T-cell development.

5.1.2 Describe the roles and determinants of key stem cells as precursors to thymocytes and T cells.

Because T cells play such a critical role in the adaptive immune response and protection against infection, their proper development is crucial to adaptive immune system function. The events leading to the development of circulating T cells require coordination, accuracy, and a means to ensure proper cell development. These events occur in the *thymus*, a primary lymphoid organ (gland) located in the upper mediastinum above the heart and extending superiorly into the neck to the lower edge of the thyroid gland (see Figure 1.16). Undifferentiated lymphocyte precursor cells migrate from the bone marrow and blood to the thymus, where T cells develop and mature. While this chapter focuses on the development of T cells within the thymus, it should be noted that thymus involution (loss of thymic cells) begins at birth and continues at a rate of approximately 3% per year until age 35–45. At middle age, the rate of shrinkage slows to approximately 1% per year. While thymic cells are still present in middle-aged adults, transplant studies show that the ability of the thymus to function in the development of new circulating T cells is greatly diminished. These studies suggest that while the T-cell population is initially generated in infants in the thymus through the processes discussed in this chapter, it is maintained later in life not by new T-cell development but rather by division of circulating T cells.

Anatomy of the thymus

The thymus consists of two lobes bound by a connective tissue capsule. The gland is largest at puberty and gradually atrophies over time. Each lobule consists of an outer cortex and an inner medulla. Immature T cells known as **thymocytes** reside in the cortex, where they undergo key developmental steps to ensure proper T-cell receptor expression. The medulla is the site where mature thymocytes undergo negative selection and finish their development into naïve T cells (**FIGURE 5.1**).

Cells of the thymus

Thymocytes must receive the correct signals to promote their development, activate pathways required for the development process, test cells at designated checkpoints, and prevent self-recognition. Other resident cells of the thymus drive the development of thymocytes into naïve T cells; these include thymic epithelial cells, macrophages, and dendritic cells:

- *Thymic epithelial cells* play an important role in activating genes required for thymocyte development and in their maturation. They also play a role in several key developmental checkpoints of thymocyte differentiation, and they function in both positive and negative selection of developing thymocytes.

thymocyte Developing immature T cells present in the thymus.

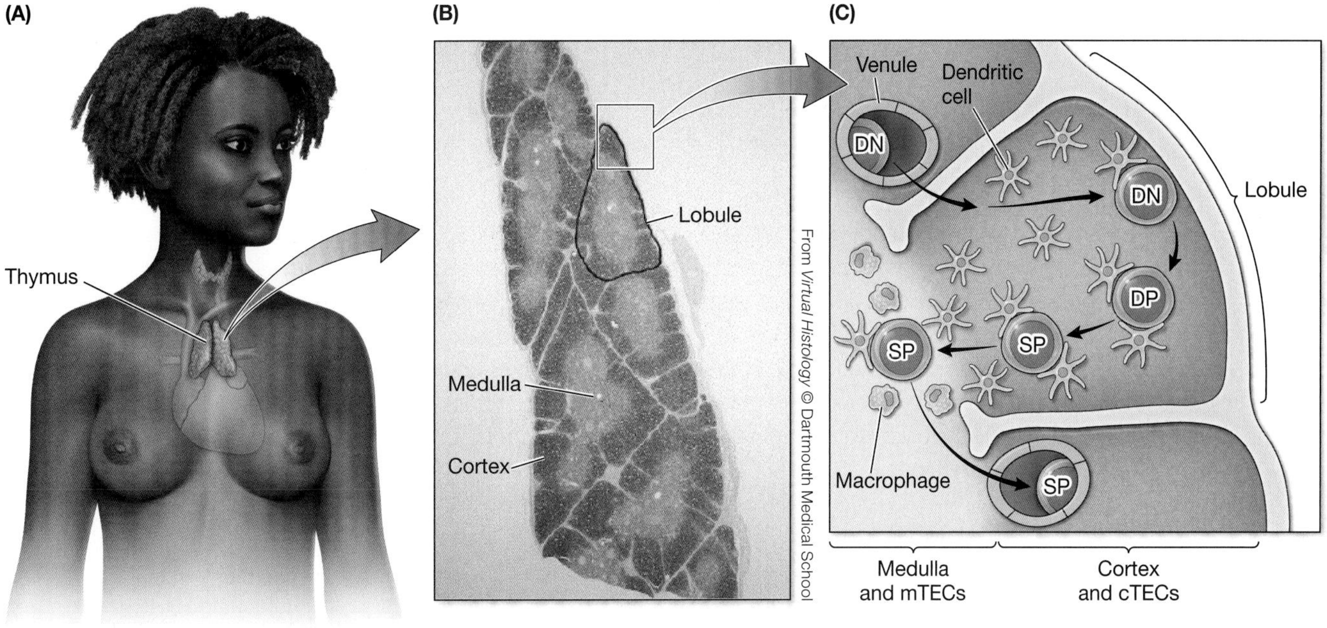

FIGURE 5.1 Anatomy of the thymus (A) The thymus, the primary lymphoid tissue for T-cell development, is located just above the heart and below the thyroid gland. (B) The thymus contains many lobules, each consisting of cortex and medulla, where thymocytes develop into T cells (used with permission of Dartmouth College). (C) Developing thymocytes enter the thymus as double-negative (DN) thymocytes and interact with cortical thymic epithelial cells (cTECs) to begin their development into T cells. DN thymocytes undergo cell-surface changes that transform them into double-positive (DP) thymocytes. Further maturation of DP thymocytes occurs in the medulla, where they are tested for self-reactivity by medullary thymic epithelial cells (mTECs), dendritic cells, and macrophages before they exit the thymus as single-positive (SP) T cells. (C, after L. Klein et al. 2009. *Nat Rev Immunol* 9: 833–844.)

- *Macrophages* remove thymocytes that have not developed properly and have undergone apoptosis (programmed cell death). The cellular debris that remains is phagocytosed by resident macrophages.
- *Dendritic cells* present self-antigens to developing thymocytes to test the functionality of the T-cell receptor and to define tolerance to self-molecules.

The medulla contains the Hassall's corpuscles (also known as thymic corpuscles), which contain epithelial cells that express an important protein, thymic stromal lymphopoietin (TSLP). The TSLP protein triggers thymic dendritic cells to activate a subset of thymocytes to develop into regulatory T cells by activating a transcription factor known as FOXP3 (forkhead box P3) within the engaged thymocytes.

Within the thymus, cells of hematopoietic origin receive signals directing them to continue development and become members of the T-cell lineage. The thymus plays an important role in promoting recombination events within the developing T cell, and thymic cells are essential to the selection events required to ensure tolerance and prevent T-cell autoreactivity. For example, an individual with a small or missing thymus, as in DiGeorge syndrome (DGS), may experience frequent infections. DGS is most commonly associated with the deletion of a small part of chromosome 22, which can result in abnormal development of the thymus and impaired T-cell production.

Before we explore the steps involved in T-cell development and selection, let's take a look at the origin of the thymus and its cells. Since it is involved primarily in T-cell development, it is not surprising to learn that the timeline of thymus evolution closely mirrors that of T-cell evolution (see **EVOLUTION AND IMMUNITY**).

CHECKPOINT QUESTION

1. What three major cell types found in the thymus play a role in T-cell development? Explain the function of each type in T-cell development.

EVOLUTION AND IMMUNITY

Thymus Organogenesis

In Chapter 4, we saw that T-cell receptor diversity arose abruptly in jawed fish. It is thus likely that T-cell development had to arise according to a similar evolutionary timeframe. We know that the thymus is the primary lymphoid organ responsible for T-cell development, so it stands to reason that thymus organogenesis (development of an organ at the embryonic stage) coevolved at the same time as T cells and T-cell receptor diversity.

A thymus can be found not only in humans and mice (as described in this chapter) but also in jawed fish, the most primitive surviving jawed vertebrates. The thymus in these animals is arranged in a very similar manner to that of humans and mice, suggesting similar biological function. However, some questions remain: How did the thymus evolve? Is there evidence of a primitive tissue precursor to the thymus?

A primitive thymus, known as the *thymoid*, has been described in lamprey, a jawless fish. Within lamprey, lymphocyte-like cells encode for multiple leucine-rich repeat (LRR) domains, termed variable-lymphocyte receptors (VLRs). Interestingly, the LRRs present in these lamprey lymphocyte-like cells are highly variable at the amino-acid level. Analysis of the genetic structure of lamprey VLRs has revealed that the VLR genes are reorganized from the germline structure via somatic gene rearrangement. The germline structure contains an incomplete VLR gene (a signal peptide, a partial N-terminal LRR, and a partial C-terminal LRR) but also contains flanking LRR domains in DNA cassettes (short segments of DNA containing a coding sequence and recombination site). Properly developed lamprey lymphocytes have complete VLR genes that have undergone recombination in which flanking LRRs have recombined with the germline VLR gene, resulting in a unique VLR in each lymphocyte.

Within the thymoid in lamprey, cells with improperly recombined VLR genes are associated with high apoptosis activity. This suggests that the thymoid is an active location of lymphocyte-like cell development in lamprey, just as the thymus is an active location of T-cell development in mice and humans.

5.2 Which cells are precursors to developing thymocytes?

LEARNING OBJECTIVE

5.2.1 Explain the role of lymphoid progenitor cells in T-cell development and their pluripotency in regard to lymphocyte development.

As discussed in Chapter 1, all cells of the hematopoietic system, including those of the immune system, arise from a hematopoietic stem cell. This stem cell is responsible for producing all immune system cells, including thymocytes.

Hematopoietic stem cells revisited

Hematopoietic stem cells are often characterized by cell-surface markers that are present on the stem cell precursors, along with the absence of lineage-specific markers (for example, a cell expressing CD8 has a T-cell lineage). Human and mouse hematopoietic stem cells differ in their cell-surface markers (**TABLE 5.1**).

Mouse hematopoietic stem cells have a low (or no) expression of CD34 and express both CD38 and the stem cell factor receptor c-kit. Human hematopoietic stem cells have an opposite expression profile; they express CD34 and have a low (or no) expression of CD38. While stem cells can give rise to any cell of the circulatory system, to develop into a T cell, a stem cell must first differentiate into a *lymphoid progenitor cell.*

TABLE 5.1 | Cell-Surface Markers: Mouse and Human Hematopoietic Stem Cells

Hematopoietic stem cell marker	Mouse	Human
CD34	+/–	++
CD38	++	+/–
c-kit (CD117)	++	++

Legend: +/– (low/no expression); ++ (high expression)

Lymphoid progenitor cells

We know that all T cells descend from common lymphoid progenitor cells. Progenitor cells also give rise to B cells, dendritic cells, and natural killer cells. Lymphoid progenitors migrating from the bone marrow to the thymus express cell-surface molecules that distinguish them from other cells.

Both mouse and human lymphoid progenitors express CD34 and CD38 at their cell surface (TABLE 5.2). Interestingly, mouse lymphoid progenitors express the interleukin 7 (IL-7) receptor, while human cells committed to becoming thymocytes do not begin to express the IL-7 receptor until they have been further differentiated. It is known that IL-7 is secreted by thymic epithelial cells as a signaling molecule, as individuals who are homozygous recessive for a nonfunctional mutant IL-7 receptor are immunocompromised due to a lack of T cells.

TABLE 5.2 | Cell-Surface Markers: Mouse and Human Lymphoid Progenitor Cells

Lymphoid progenitor cell marker	Mouse	Human
CD34	++	++
CD38	++	++
c-kit	++	++
IL-7 receptor	++	+/–

Legend: +/– (low/no expression); ++ (high expression)

CHECKPOINT QUESTIONS

1. Lymphoid progenitor cells are stem cells that can produce both T cells and B cells during hematopoiesis. Explain how lymphoid progenitor cells can differentiate into T cells based on their location.
2. Why are individuals who lack the IL-7 receptor unable to produce functioning T cells?

5.3 | What is the role of Notch1 in T-cell development?

LEARNING OBJECTIVE

5.3.1 Explain the role of Notch1 in T-cell development and its importance in lymphoid progenitor cell commitment to thymocyte development.

As noted previously, developing thymocytes require signals to continue their development into naïve T cells. These signals are provided by other resident cells, especially thymic epithelial cells. The epithelial cells directly contact lymphoid progenitor cells that have migrated from the bone marrow to the thymus through cell-to-cell receptor interactions.

Because lymphoid progenitors are precursors to several cell types, including B and T cells, they must receive the appropriate signal to commit to a differentiated cell type. The *Notch signaling pathway* regulates cell proliferation and differentiation. The protein Notch is a cell-surface receptor that interacts with transmembrane ligands on adjacent cells. In mammals, there are four different Notch receptors that play a role in a variety of processes, including neuronal development, growth of blood vessels, and cardiac function.

Notch1 has been shown to be vital in committing a lymphoid progenitor to the T-cell lineage, presumably by inhibiting B-cell development within the thymus. For instance, in individuals who have a mutation preventing proper Notch1 function, lymphoid progenitors in the thymus can develop into B cells rather than T cells. Conversely, overexpression of Notch1 in bone marrow can program lymphoid progenitors there to develop into T cells instead of driving normal B-cell development.

Thymic epithelial cells provide signals required for T-cell development, including secretion of IL-7, to promote proper thymocyte development and rearrangement of T-cell receptor genes. They also provide *Notch ligand* to drive commitment to a T-cell fate. The interaction between Notch1 on the lymphoid progenitor and Notch ligand on the thymic epithelial cells activates Notch1 within the progenitor cell. This activation occurs through the proteolytic cleavage of Notch1 at the plasma membrane of the lymphoid progenitor, which releases an active cytosolic domain. The cytosolic Notch1 domain translocates into the nucleus, where it activates gene expression required for T-cell differentiation (FIGURE 5.2).

FIGURE 5.2 Notch1 signaling commits thymocytes to T-cell development Binding of Notch1 on the surface of lymphoid progenitor cells to Notch ligand on thymic epithelial cells promotes the proteolytic cleavage of Notch1. This releases the intracellular domain of Notch1 from the membrane, allowing it to translocate to the nucleus to initiate transcription of genes necessary for development into T cells. (After C. Nowell and F. Radtke. 2017. *Nat Rev Cancer* 17: 145–159.)

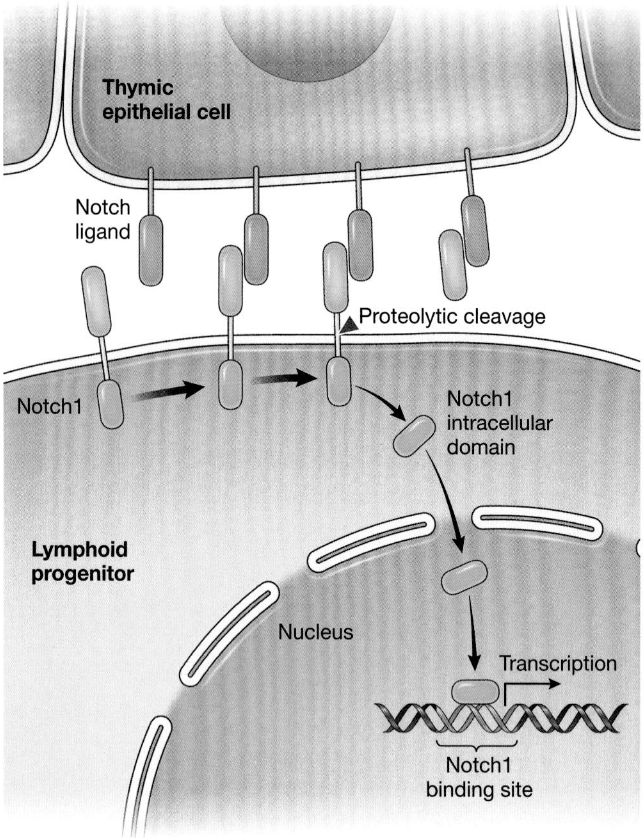

● **CHECKPOINT QUESTION**

1. What two pieces of evidence demonstrate the importance of Notch1 for committing lymphoid progenitor cells to T-cell development?

5.4 | What are the stages of T-cell development?

LEARNING OBJECTIVES

5.4.1 List the developmental stages of thymocytes as they progress to development into naïve T cells.

5.4.2 Describe the major checkpoints of thymocyte development as they pertain to proper T-cell receptor recombination.

We have learned that T-cell development begins through the production of a hematopoietic stem cell in the bone marrow, followed by differentiation to a lymphoid progenitor cell. The lymphoid progenitor migrates to the thymus, where its interaction with thymic epithelial cells (through Notch1 at the lymphoid cell surface) initiates commitment to maturation into a T cell. The committed thymocyte progresses through several developmental stages as it matures and activates somatic recombination of the T-cell receptor genes (see Chapter 4). Thymic epithelial cells further promote somatic recombination of T-cell receptor genes through secretion of the cytokine IL-7.

Developmental stages

A thymocyte progresses through several developmental stages (**FIGURE 5.3**). The primary goal of these developmental stages is to produce a population of T cells with a diverse array of T-cell receptors.

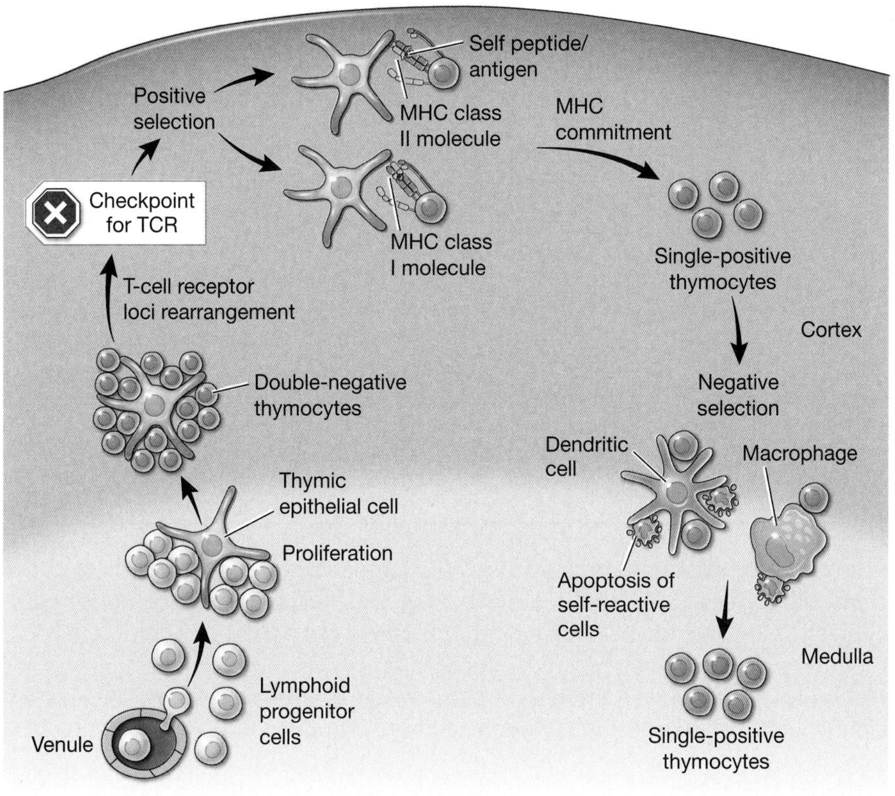

FIGURE 5.3 Key steps in T-cell development During the development process, lymphoid progenitor cells within the thymus receive signals from thymic epithelial cells to initiate proliferation and commitment to the T-cell lineage. Developing thymocytes undergo gene rearrangement at their T-cell receptor loci and are tested to ensure productive rearrangements occur before continuing their developmental process. Two checkpoints for productive rearrangement of T-cell receptor loci occur, one for rearrangement of the β chain and one for rearrangement of the α chain. Cells that fail to pass either checkpoint die by apoptosis. After positive selection of the T-cell receptor, T cells undergo MHC commitment based on the affinity of the T-cell receptor with either MHC class I or MHC class II self-peptide complexes to express either the coreceptor CD4 or CD8. These single-positive thymocytes undergo negative selection to promote central tolerance, and self-reactive cells are removed by apoptosis. The resulting single-positive thymocytes enter the circulation to function as circulating naive T cells. (After J. Parkin and B. Cohen, 2001. *Lancet* 357: 1777–1789.)

At each stage, the term used to describe the thymocyte is based on the presence or absence of T-cell receptor and coreceptor components on its surface. During initial development in the thymic cortex, the thymocyte is called a **double-negative thymocyte** because it does not express either coreceptor (CD4 or CD8). A double-negative thymocyte begins somatic recombination at the T-cell receptor loci, passing through checkpoints to ensure proper T-cell receptor recombination, until both subunits of the receptor have properly rearranged. At this point, both coreceptors (CD4 and CD8) are present on the surface of the developing T-cell, and the thymocyte becomes a **double-positive thymocyte**.

Double-positive thymocytes have several possible fates as they continue developing into T cells. Some develop into specialized T cells, including regulatory T cells and NK T cells. Others begin the process of positive and negative selection to test the T-cell receptor and select for a single coreceptor to ultimately become a **single-positive thymocyte**. During these developmental stages, important checkpoints test to ensure that proper recombination events have taken place at the T-cell receptor loci.

Differences in cell-surface markers

Several markers at the surface of the developing thymocyte change during the developmental process. Similar to the marker differences seen on the hematopoietic stem cell and lymphoid progenitor population in mice and human cells, developing thymocytes also differ in terms of their cell-surface markers.

In mice, lymphoid progenitor cells migrating from the bone marrow to the thymus express two important cell-surface molecules (c-kit/CD117 and CD44; see Table 5.2 and **TABLE 5.3**) and are referred to as **DN1** (double-negative 1) thymocytes. In mice, the expression of these surface molecules changes as the developing cells further mature toward becoming a double-positive thymocyte.

double-negative thymocyte Developing thymocyte within the thymus that does not express either T-cell coreceptor (CD4 or CD8).

double-positive thymocyte Developing thymocyte within the thymus that expresses both T-cell coreceptors (CD4 and CD8).

single-positive thymocyte Developing thymocyte that has undergone lineage commitment and expresses only a single coreceptor.

DN1 Double-negative 1 thymocytes; thymocytes that have entered the thymus and initiated somatic recombination at the T-cell receptor loci.

TABLE 5.3 | Cell-Surface Markers: Mouse and Human Thymocytes

Thymocyte development stage	Mouse cell surface markers	Human cell surface markers
DN1	c-kit^{++}, CD44	CD2^{+}, CD5^{+}, CD7^{+}, CD34^{+}, CD44^{+}
DN2	c-kit^{++}, CD44^{+}, CD25^{+}	CD2^{+}, CD5^{+}, CD7^{+}, CD34^{+}, CD44^{+}
DN3	c-kit^{+}, CD25^{+}	CD34^{+}, CD38^{+}, CD25^{+}, CD1^{+}
DN4	c-kit$^{+/-}$	CD34$^{+/-}$, CD38$^{+/-}$, CD1$^{+/-}$
DP	CD4^{+}, CD8^{+}	CD4^{+}, CD8^{+}

Legend: +/– (low/no expression); + (low expression); ++ (high expression)

DN2 Double-negative 2 thymocytes; thymocytes that have committed to T-cell development.

DN3 Double-negative 3 thymocytes; thymocytes that have finished β-chain rearrangement and are active in β-chain allelic exclusion and proliferation to engage in α-chain rearrangement.

DN4 Double-negative 4 thymocytes; thymocytes that are active in α-chain rearrangement.

Mouse thymocytes committed to development into T cells become **DN2** cells and begin to express the cell-surface molecule CD25 (see Table 5.3). As mouse thymocytes continue their development, they become **DN3** cells (cells that have gone through somatic recombination of their β chain and turned off expression of CD44). Once a mouse thymocyte has made it through the β-chain checkpoint, it continues its development into a **DN4** cell, where cell-surface molecules are further refined, ultimately expressing a low level of c-kit/CD117 and no longer expressing CD44 or CD25 at the surface.

Human thymocytes differ from mouse thymocytes in regard to the cell-surface molecules present during the developmental process. Human lymphoid progenitor cells that have migrated from the bone marrow to the thymus shut down expression of the stem cell marker CD34 and begin expressing the adhesion and signaling molecules CD2, CD5, and CD7 (see Table 5.3). As human thymocytes continue their development, they begin to express CD1 at their cell surface. In a manner similar to that of mouse thymocytes, they do not express CD4 or CD8 and are thus double-negative thymocytes early in development.

Critical checkpoints

There are three key checkpoints that a thymocyte must pass to continue development and become a functional naïve T cell (**FIGURE 5.4**):

1. *Checkpoint of the γδ subunit of the T-cell receptor*—A thymocyte that has properly gone through development can only express either an αβ T-cell receptor or a γδ T-cell receptor. Rearrangement of the T-cell receptor loci begins at the β, γ, and δ genes, and a T cell that has properly recombined both the γ and δ loci is destined to become a γδ T cell.
2. *Checkpoint of the β subunit of the T-cell receptor*—This checkpoint occurs during the somatic recombination events to test the β subunit of the T-cell receptor, which recombines at the same time as the γ and δ subunits. If the β subunit recombines properly, somatic recombination is inactivated to allow the thymocyte to ready itself for the next developmental stage.
3. *Checkpoint of the α subunit of the T-cell receptor*—The purpose of this checkpoint is to test the α subunit of the T-cell receptor. Once a developing thymocyte has properly recombined a β locus and progressed to the next developmental stage, recombination events continue at the α, γ, and δ T-cell receptor genes. If the γ and δ genes recombine properly, the thymocyte will be committed to becoming a γδ T cell. Proper recombination of the α subunit locus allows the thymocyte to continue development toward becoming an αβ T cell.

Development into one of two T-cell lineages

As developing thymocytes induce somatic recombination events to create a functional T-cell receptor, the race is on. The factor that determines whether a developing thymocyte becomes an αβ T cell or a γδ T cell is how fast the receptor genes recombine. Initially, recombination events target the β, γ, and δ loci of the T-cell receptor genes (see Figure 5.4). If a thymocyte properly recombines the γ and δ genes for the T-cell receptor, it will continue the developmental process as a γδ T cell. T cells that express γδ T-cell receptors typically do not express the coreceptors CD4 or CD8 (and therefore are double-negative T cells).

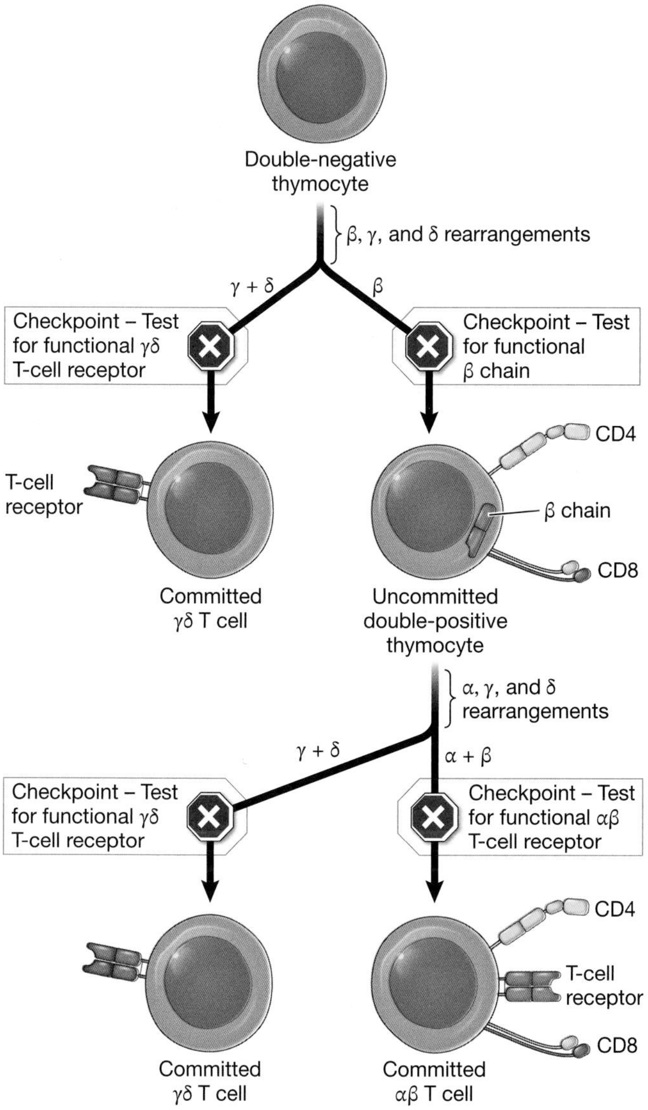

FIGURE 5.4 Timing of somatic recombination in T-cell development Thymocytes rearrange their T-cell receptor loci and test these rearrangements at defined checkpoints before continuing their developmental process. Developing thymocytes can commit to the γδ T-cell lineage if they first productively rearrange their γ and δ T-cell receptor loci, or they can commit to the αβ T-cell lineage if they first productively rearrange their α and β T-cell receptor loci. (After C. A. Janeway Jr. 2001. *Immunology: The Immune System in Health and Disease*, 5th ed. Garland Science/Norton: New York.)

If the β-chain locus recombines first, action is taken to test the β-chain recombination event and continue development. After passing through the β-chain checkpoint, recombination events continue and another race is on—one of recombination at the α, γ, and δ loci (see Figure 5.4). If proper recombination of the γ and δ loci occurs, the thymocyte is destined to become a γδ T cell. Interestingly, the loci for the δ chain is positioned within the loci for the α chain. Thus, if proper recombination of the α chain occurs, the δ chain is lost in a signal joint of somatic recombination, and the cell cannot express a proper γδ T-cell receptor.

Because γδ T cells require two successful recombination events at the T-cell receptor loci compared to one event in the race between β, γ, and δ loci, adults do not have a large quantity of γδ T cells. (They represent 0.5% of T cells in mice and 3.5% in humans.) However, γδ T cells are important during fetal development; they are the first T cells produced and work to protect the fetus. In adults, γδ T cells play an important role in protecting mucosal surfaces. These cells can recognize unconventional antigens, including lipids, using their γδ T-cell receptors.

CHECKPOINT QUESTION

1. What three checkpoints can occur as part of the T-cell development process? Why don't all three checkpoints need to be passed by a developing T cell?

Making Connections: Hematopoiesis and the Commitment to the T-cell Lineage

While hematopoiesis produces all cell types of the circulatory system from a common stem cell, T cells begin their life in the bone marrow as a common lymphoid progenitor. Migration of common lymphoid progenitors to the thymus along with signals given by cells commit the common lymphoid progenitors to the T-cell lineage.

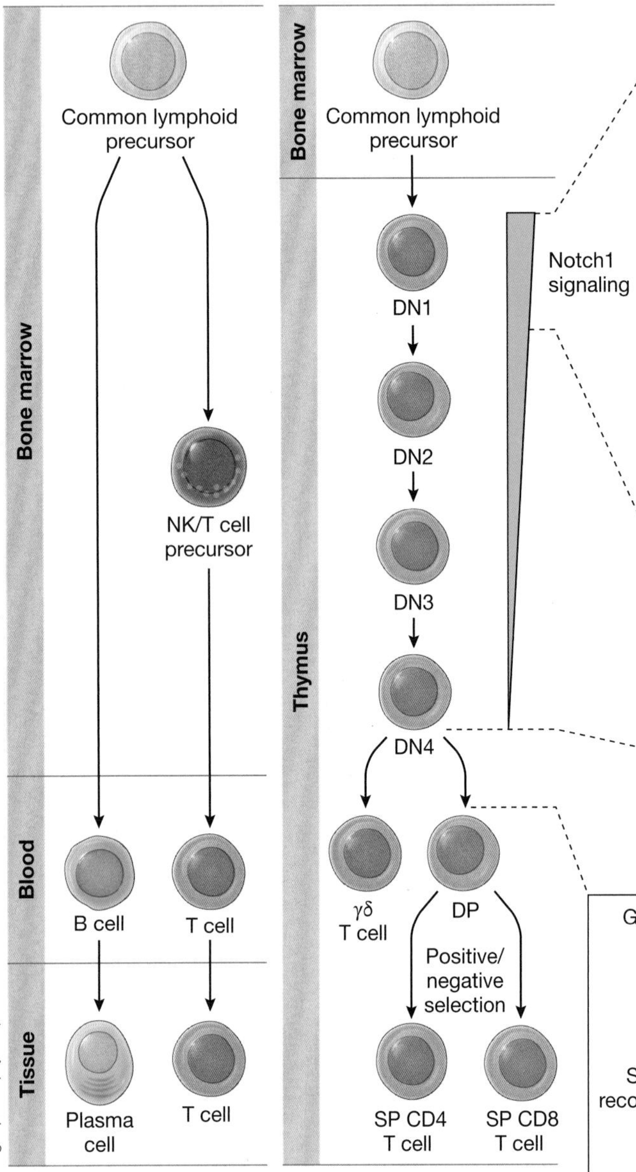

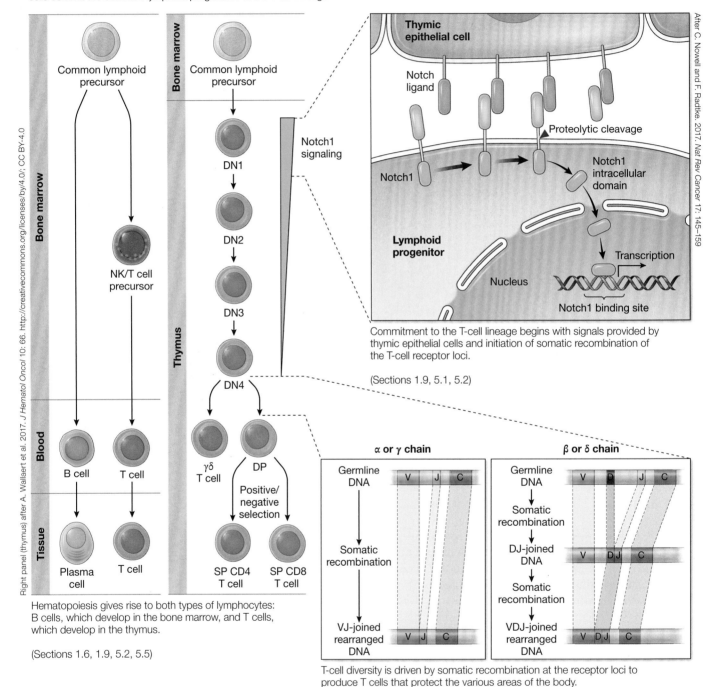

Hematopoiesis gives rise to both types of lymphocytes: B cells, which develop in the bone marrow, and T cells, which develop in the thymus.

(Sections 1.6, 1.9, 5.2, 5.5)

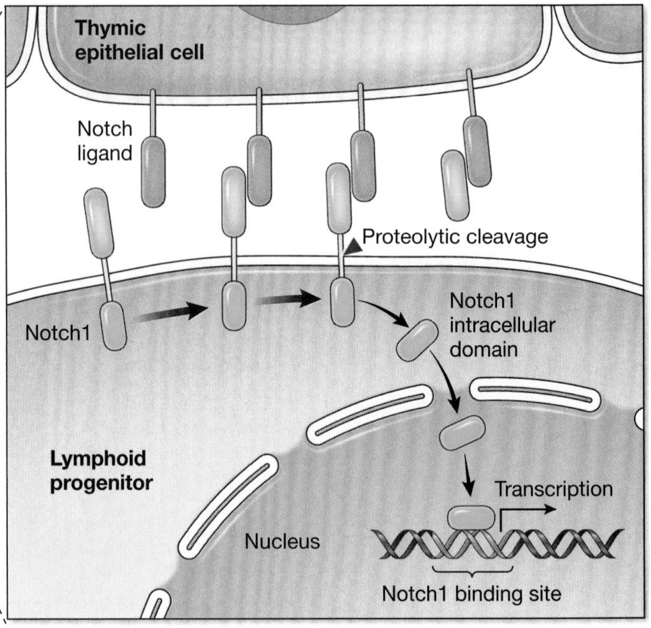

After C. Nowell and F. Radtke. 2017. *Nat Rev Cancer* 17: 145–159

Commitment to the T-cell lineage begins with signals provided by thymic epithelial cells and initiation of somatic recombination of the T-cell receptor loci.

(Sections 1.9, 5.1, 5.2)

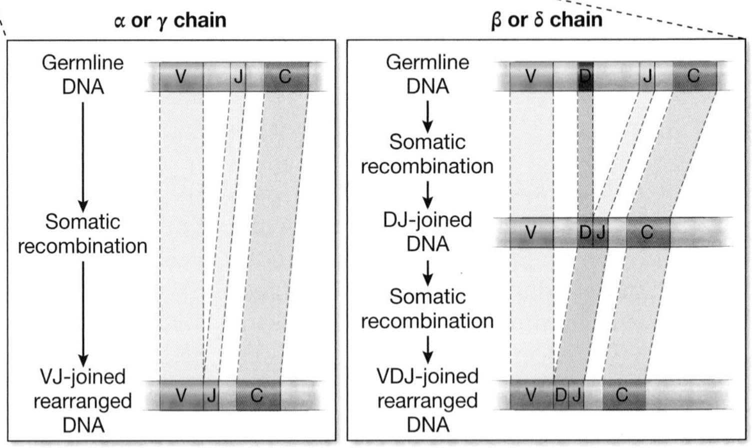

T-cell diversity is driven by somatic recombination at the receptor loci to produce T cells that protect the various areas of the body.

(Sections 4.4, 4.7, 5.5, 7.13, 7.14, 7.15, 7.16, 12.4)

Above after W. Purves et al. 1995. *LIFE: The Science of Biology*, 4th ed., Sinauer Assoc., Sunderland, MA; T. Dörner and P. E. Lipsky. 2001. *Arthritis Rheum* 44: 2715-2727; P. Parham. 2014. *The Immune System*, 4th ed., W. W. Norton and Company, New York

5.5 How does a developing thymocyte pass through the β-chain checkpoint?

LEARNING OBJECTIVES

5.5.1 Differentiate between development into an αβ thymocyte versus development into a γδ thymocyte.

5.5.2 Outline the process of passage through the β-chain checkpoint and explain the process and importance of allelic exclusion.

Once a thymocyte has begun the developmental process, somatic recombination is activated through the expression of the RAG1 and RAG2 proteins. Recombination events begin at the β, γ, and δ loci of the T-cell receptor (**FIGURE 5.5**, and see Chapter 4). If productive rearrangements occur at the γ and δ loci, the thymocyte is destined to become a γδ T cell. However, because only a single productive rearrangement of the β-chain locus is required, most thymocytes reach the β-chain checkpoint before rearrangement of the γ and δ loci occurs. Furthermore, there are both β1 and β2 D and J segments on each chromosome, providing an opportunity for a β-chain V segment to rearrange with either a β1 D and J segment or to rearrange with a β2 D and J segment on either chromosome, providing four possible opportunities for a productive β-chain rearrangement.

pre-T α chain (pTα) Protein expressed in thymocytes that have finished β-chain rearrangement; acts as a surrogate α chain to test proper expression and action of the recombined β chain.

pre-T cell Thymocyte that has successfully passed the β-chain checkpoint; pre-T cells engage in β-chain allelic exclusion and proliferate to produce cells that can further engage in T-cell development through α-chain rearrangement.

To test for a productive arrangement of the β chain, the developing thymocyte must assemble a surrogate T-cell receptor complex. Since recombination of the α-chain locus does not occur until the β-chain checkpoint has been passed, a surrogate α chain must be used to permit assembly of a receptor complex. This surrogate α chain is referred to as the **pre-T α chain (pTα)**, and it assembles with the rearranged β chain and the CD3 complex that plays a role in downstream signaling events upon T-cell receptor engagement. This complex ensures that a functioning T-cell receptor can be formed and used to signal via the same transduction pathways used to activate T cells. **Pre-T cells** are those that can form a complex and continue their development.

If a thymocyte does not properly rearrange the β-chain locus (and has also not rearranged the γ and δ chain loci), it is removed by apoptosis. The large number of thymocytes that do not pass the checkpoints are programmed for cellular death and removal by resident macrophages (**FIGURE 5.6**).

Four attempts at β-chain rearrangement

As discussed in Chapter 4, a productive β-chain rearrangement requires proper recombination of a V, D, and J segment within the β-chain locus. Another means by which the β-chain locus is productively recombined before the γ- and δ-chain loci is based on the presence of two different constant regions and their associated D and J segments ($D_\beta1$, $J_\beta1$, and $C_\beta1$ or $D_\beta2$, $J_\beta2$, and $C_\beta2$). Recombination at the β-chain locus can occur using a V segment recombining with either the $D_\beta1$ and $J_\beta1$ segments or the $D_\beta2$ and $J_\beta2$ segments. Since these two constant regions (and their corresponding D and J segments) are present on both chromosomes within developing thymocytes, four possible rearrangements can occur (**FIGURE 5.7**).

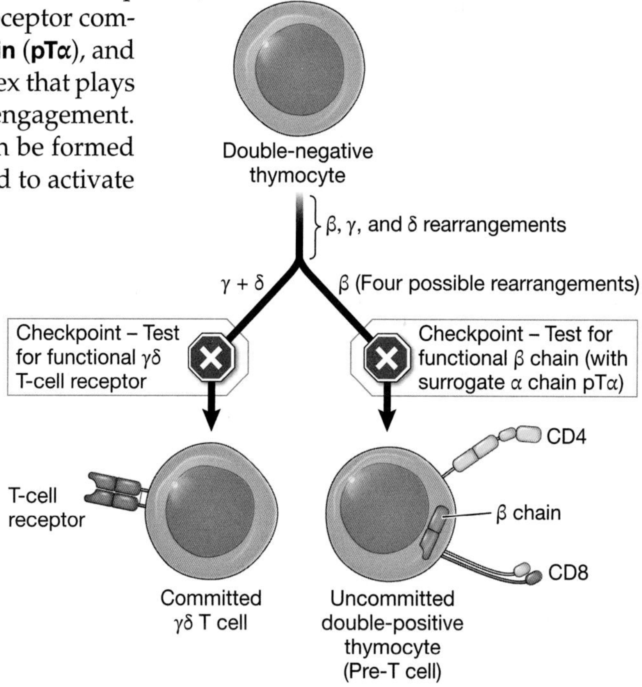

FIGURE 5.5 First series of recombination events during T-cell development Double-negative thymocytes first recombine the β, γ, and δ chain T-cell receptor loci. If both the γ- and δ-chain loci are productively recombined, the cell is committed to the γδ T-cell lineage. If the β-chain loci recombines first, it is tested with a surrogate α-chain pT α before becoming a pre-T cell. If all recombination events are unproductive, the thymocyte dies by apoptosis. (After C. A. Janeway Jr. 2001. Immunology: *The Immune System in Health and Disease*, 5th ed. Garland Science/Norton: New York.)

(A)

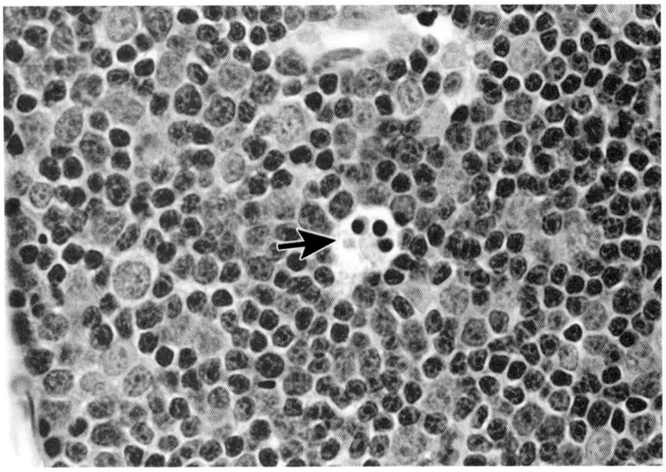

(B)

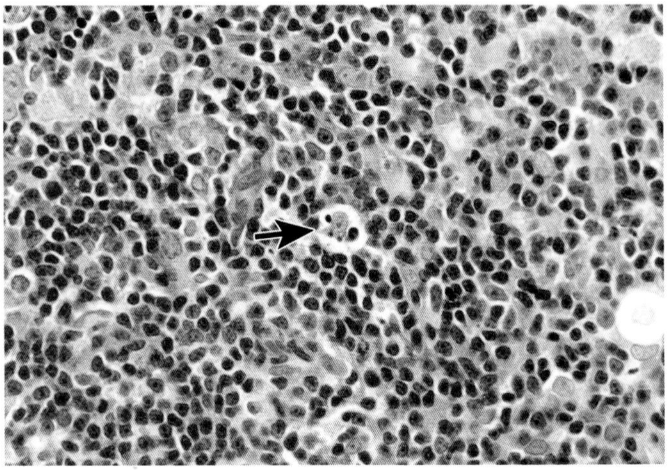

Both photos from S. A. Elmore et al. 2016. *Toxicol Pathol* 44:173–88. doi: 10.1177/0192623315625859

FIGURE 5.6 Macrophages phagocytose apoptotic thymocytes in the thymus (A) Thymocytes that lack functional receptors die by apoptosis during positive selection in the thymic cortex. (B) Thymocytes with functional receptors that are potentially self-reactive die by apoptosis during negative selection in the thymic medulla.

allelic exclusion Inactivation of somatic recombination and active chromatin remodeling in order to prevent further recombination of specific loci of adaptive immune receptors.

Allelic exclusion at the β-chain locus

Once a productive rearrangement at the β-chain locus has taken place (as tested by the formation of a functioning pre-T-cell receptor), events occur to prevent further recombination of the β chain. This recombination shutdown is referred to as **allelic exclusion**. This process is vital because it prevents the thymocyte from producing more than one functional β chain as it continues development.

Allelic exclusion occurs at the β-chain locus in the short term and also more permanently. Immediate allelic exclusion occurs due to signals activated by a properly functioning pre-T-cell receptor. Some of these signaling events create negative feedback to downregulate the expression of the recombination machinery. The proteins RAG1 and RAG2 are vital to somatic recombination as components of the V(D)J recombinase (see Chapter 4), so inhibiting their

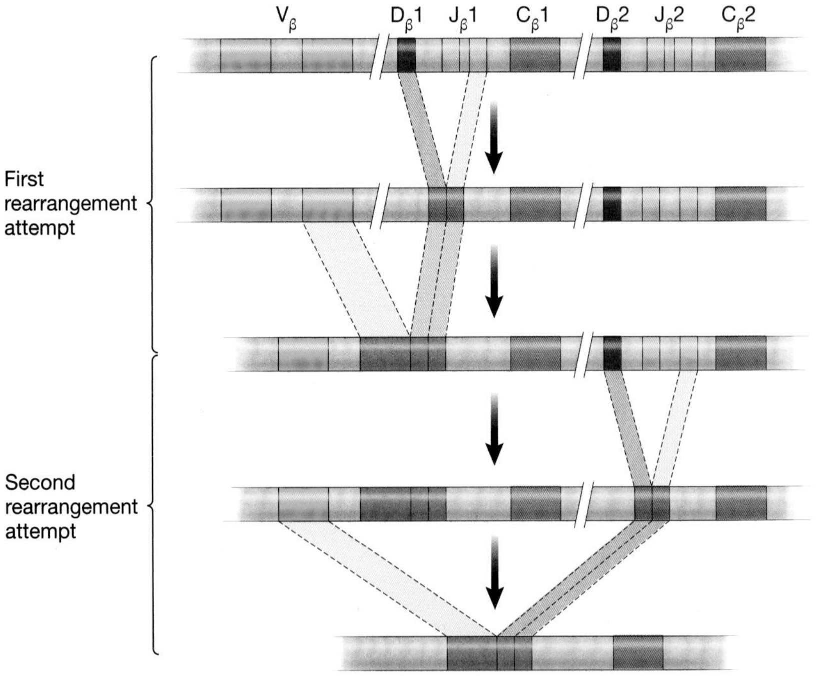

FIGURE 5.7 Four possible rearrangement attempts to produce a functional β chain Each β-chain locus has two constant regions preceded by two regions that contain D and J segments. Each constant region with its corresponding D and J segments may be recombined with a V segment to attempt to produce a functional β chain. All somatic cells have two chromosomes. Because there are two constant regions and two possible rearrangement attempts per chromosome, there are four possible rearrangement attempts of a β-chain locus.

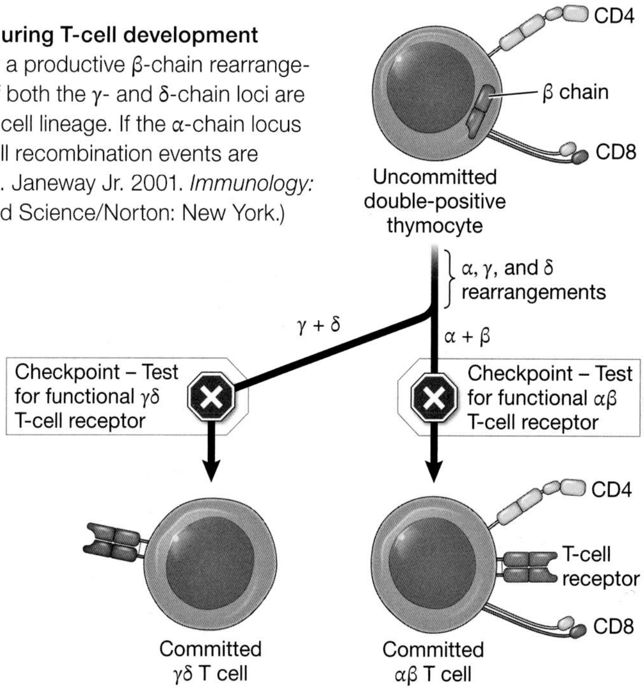

FIGURE 5.8 Second series of recombination events during T-cell development Immature double-negative thymocytes that have tested for a productive β-chain rearrangement recombine the α, γ, and δ chain T-cell receptor loci. If both the γ- and δ-chain loci are productively recombined, the cell is committed to the γδ T-cell lineage. If the α-chain locus recombines first, it is tested with the functional β chain. If all recombination events are unproductive, the thymocyte dies by apoptosis. (After C. A. Janeway Jr. 2001. *Immunology: The Immune System in Health and Disease*, 5th ed. Garland Science/Norton: New York.)

expression prevents further recombination from occurring. However, the thymocyte needs to continue to recombine the α chain to develop into a functioning αβ T cell. How, then, does the thymocyte ensure permanent allelic exclusion at the β chain when the recombination machinery will shortly be reactivated to promote α-chain recombination? The answer lies in chromatin remodeling; the β-chain locus is repackaged into tightly condensed chromatin to prevent the recombination machinery access to the other β-chain locus. This prevents further recombination at the remaining β-chain locus when the recombination machinery is reactivated.

Gene rearrangement after passage through the β-chain checkpoint

Once a pre-T cell has passed through the β-chain checkpoint, the cell begins to proliferate and produce a large number of thymocytes that can express the same β chain. These cells eventually reactivate expression of the recombination machinery, including RAG1 and RAG2, which begins to recombine at the other T-cell receptor loci (α, γ, and δ; **FIGURE 5.8**). As previously noted, recombination can no longer occur at the β chain because of allelic exclusion. A thymocyte that successfully rearranges the γ and δ loci of the T-cell receptor is destined to become a γδ T cell.

Recombination also occurs at the α-chain locus. To produce a productive rearrangement event of the α chain, a V_α and J_α gene segment must recombine. Because of the many V_α and J_α gene segments within the α-chain locus, many rearrangement attempts can occur to increase the probability of a productive rearrangement (**FIGURE 5.9**). This rearrangement event must also be tested (as the β-chain rearrangement was tested), leading to another developmental checkpoint.

CHECKPOINT QUESTION

1. Why do developing thymocytes need to use the pre-Tα chain to progress through the β-chain checkpoint?

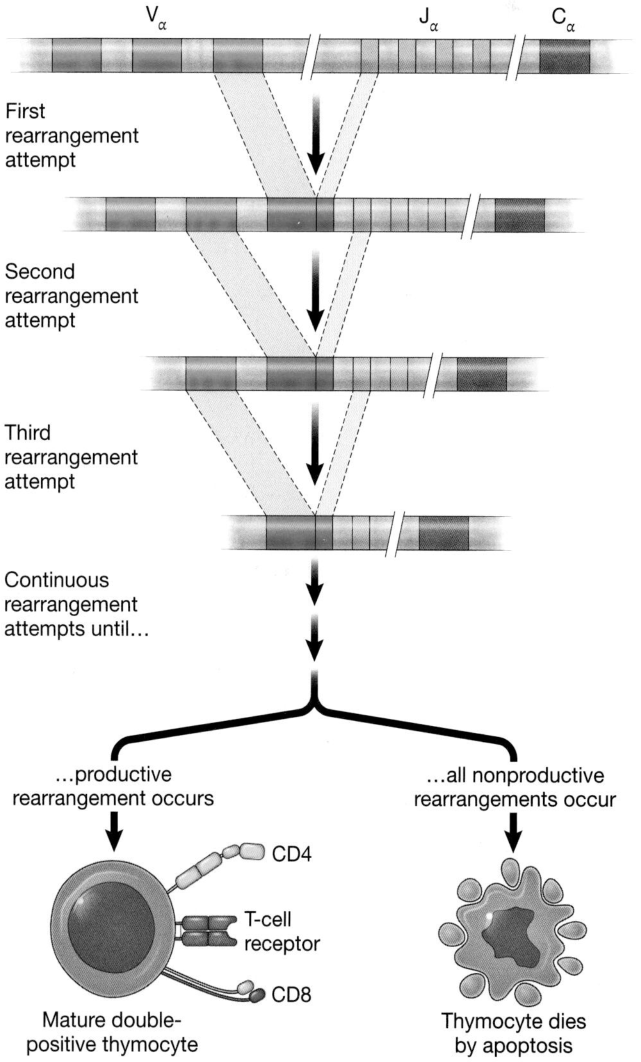

FIGURE 5.9 Possible rearrangement attempts to produce a functional α chain Each α-chain locus has a constant region preceded by many V and J segments, and each V and J segment can randomly recombine. Each recombined locus will be tested to determine whether it can produce a functional α chain capable of interacting with the recombined β chain to form a T-cell receptor.

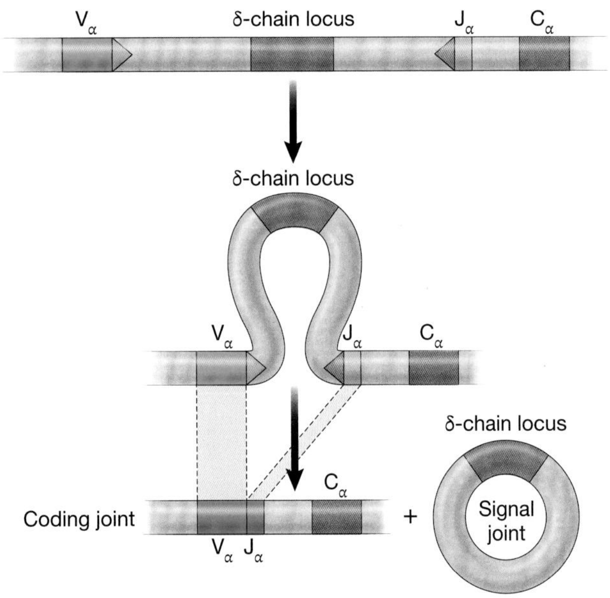

FIGURE 5.10 Structure of the α chain T-cell receptor loci After the T-cell receptor β chain is tested for productive rearrangement, recombination continues at the α, γ, and δ T-cell receptor loci. Productive rearrangement of the α-chain locus (tested at the αβ-chain checkpoint) results in removal of the δ-chain locus in the signal joint, thus preventing a T cell from being able to express both an αβ T-cell receptor and a γδ T-cell receptor.

5.6 How does a developing thymocyte pass through the α-chain checkpoint?

LEARNING OBJECTIVE

5.6.1 Outline the process of passage through the α-chain checkpoint.

Rearrangement of the α chain must be tested to ensure that the chain can function as a subunit of the T-cell receptor. As with all membrane-bound components of the T-cell receptor complex, upon activation of the expression of the α chain, it is translocated into the endoplasmic reticulum membrane. There, the α chain is tested. If the expressed α chain cannot support the needed role in the T-cell receptor complex, rearrangement of the α-chain loci continues (see Figure 5.9). This rearrangement and testing process continues until a functional α chain can be produced or until α-chain loci rearrangement can no longer occur.

If a functional α chain is produced, the developing thymocyte continues development into a naïve T cell. Proper α-chain rearrangement eliminates the possibility of producing a γδ T cell, as the δ chain locus is positioned in the middle of the recombining α-chain locus. Proper rearrangement of the α chain removes the δ-chain locus as part of a signal joint during the recombination process (**FIGURE 5.10**).

CHECKPOINT QUESTION

1. Why does productive rearrangement of the α-chain locus prevent developing T cells from becoming γδ T cells?

5.7 How does positive and negative selection of T cells occur?

LEARNING OBJECTIVE

5.7.1 Compare and contrast positive and negative selection of thymocytes and the role of MHC recognition by the T-cell receptor in both processes, including the role of T-cell receptor binding affinity to MHC molecules.

MHC restriction Selection process driven by a recombined and expressed cell-surface T-cell receptor that can only interact with a specific self-MHC molecule.

A thymocyte that has passed through the β-chain and α-chain checkpoints has the capacity to continue the developmental process. Cells with functional α and β chains within their receptor complexes also begin to express the coreceptors CD4 and CD8 and become double-positive thymocytes. They can now continue through the later stages of development: positive and negative selection.

Positive and negative selection promotes the selection of thymocytes that can bind to self-MHC molecules (**MHC restriction**) and prevents the release of thymocytes that can recognize MHC-self-peptide complexes with high-affinity (self-tolerance) (**FIGURE 5.11**). Most cells that have passed through the α and β checkpoints (about 98% of the double-positive thymocytes) die by apoptosis and are removed at this stage.

Recall from Chapter 4 that within the context of a true adaptive immune response, a T cell recognizes antigen–MHC complex via its receptor. The next

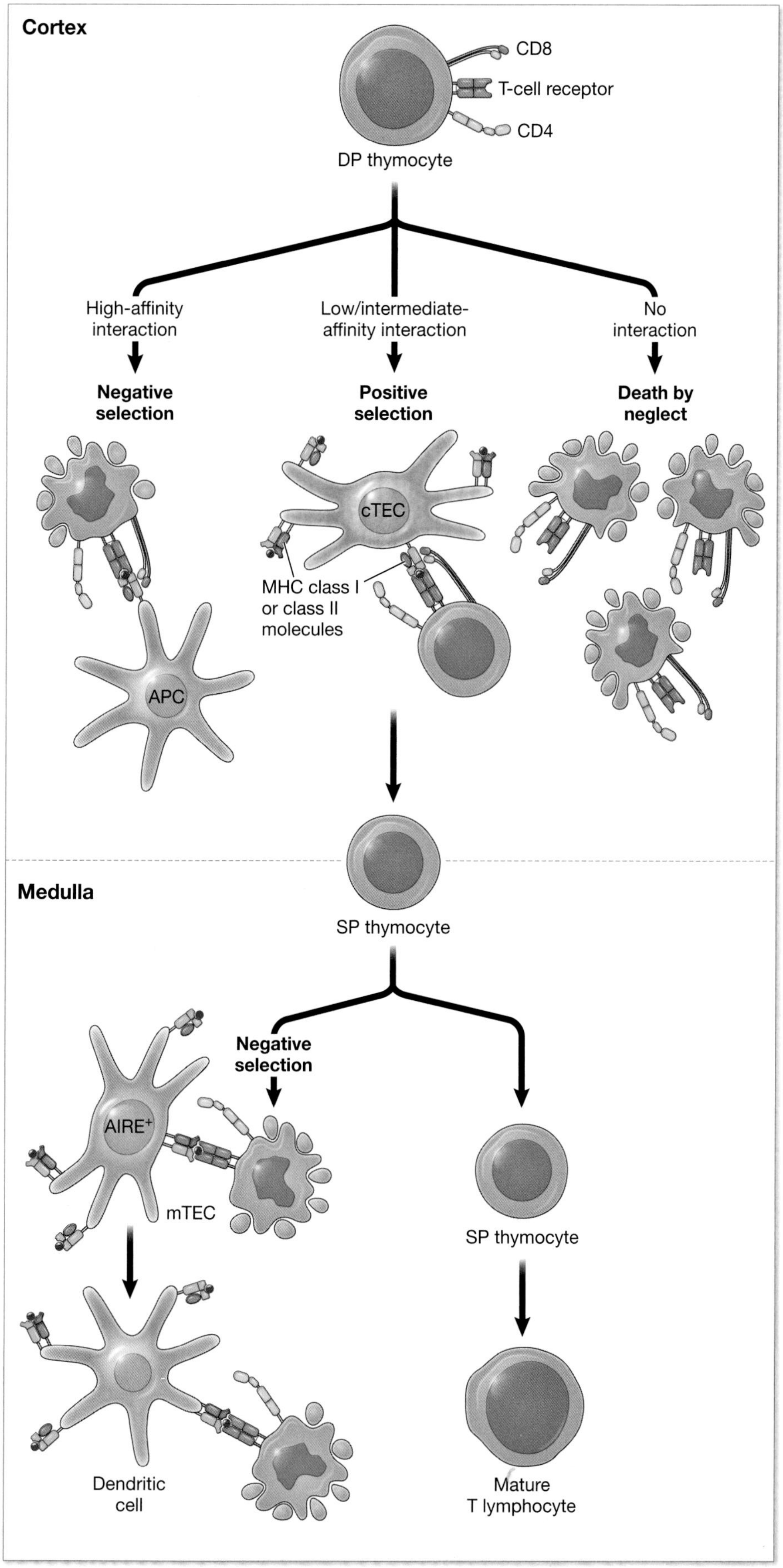

FIGURE 5.11 Positive and negative selection of developing thymocytes in the thymus Double-positive (DP) thymocytes that have passed the checkpoints that occur during somatic recombination are tested by antigen-presenting cells (APCs) or cortical thymic epithelial cells (cTECs) for the affinity of their T-cell receptor for MHC class I or II molecules presenting antigen. Thymocytes that cannot bind to MHC molecules or that bind with too high an affinity die by apoptosis. Those with a low or moderate affinity for MHC class I or II molecules presenting antigen first select for their coreceptor to become single-positive (SP) thymocytes and then undergo a second round of negative selection in the medulla. SP thymocytes in the medulla that bind to MHC molecules presenting antigen on the surface of either medullary thymic epithelial cells (mTECs) or dendritic cells die by apoptosis. Self antigens from endocrine tissues are expressed in mTECs due to the action of autoimmune regulator (AIRE). SP thymocytes that pass this round of negative selection exit the thymus as mature T lymphocytes.

Making Connections: T-Cell Development and Somatic Recombination

Development of T cells in the thymus involves somatic recombination of T-cell receptor loci. Thymocytes can rearrange their α and β receptor loci or their γ and δ receptor loci. At several stages in their development, thymocytes are tested for their ability to produce a functional T-cell receptor subunit before they can continue their developmental process.

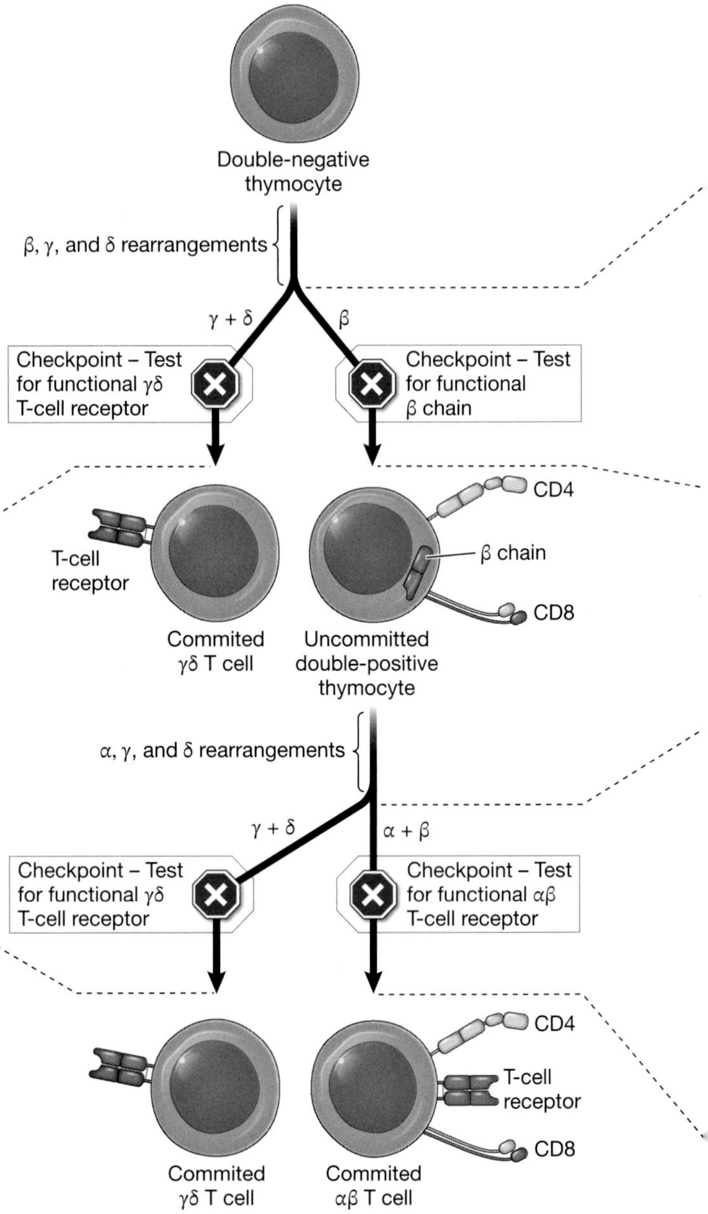

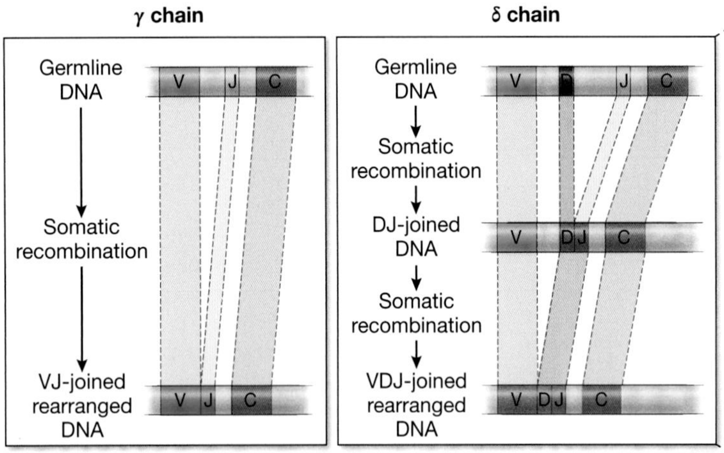

After W. Purves et al. 1995. *LIFE: The Science of Biology*, 4th ed., Sinauer Assoc., Sunderland, MA; T. Dörner and P. E. Lipsky. 2001. *Arthritis Rheum* 44: 2715-2727; P. Parham. 2014. *The Immune System*, 4th ed., W. W. Norton and Company, New York

As thymocytes develop into T cells within the thymus, somatic recombination of discrete T-cell receptor loci occurs at specific points in development.

(Sections 4.4, 5.5, 5.6, 5.7)

After C. A. Janeway Jr. 2001. *Immunology: The Immune System in Health and Disease*, 5th ed. Garland Science/Norton: New York

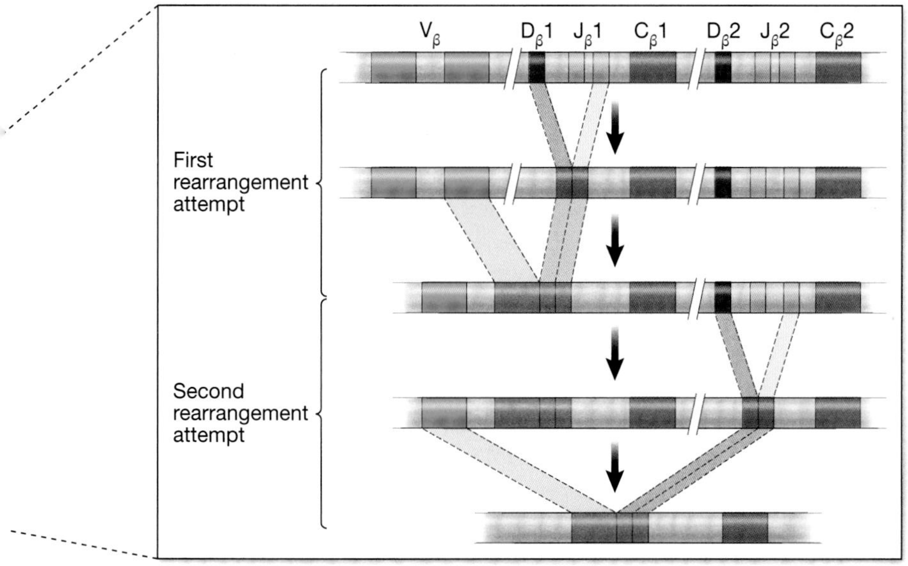

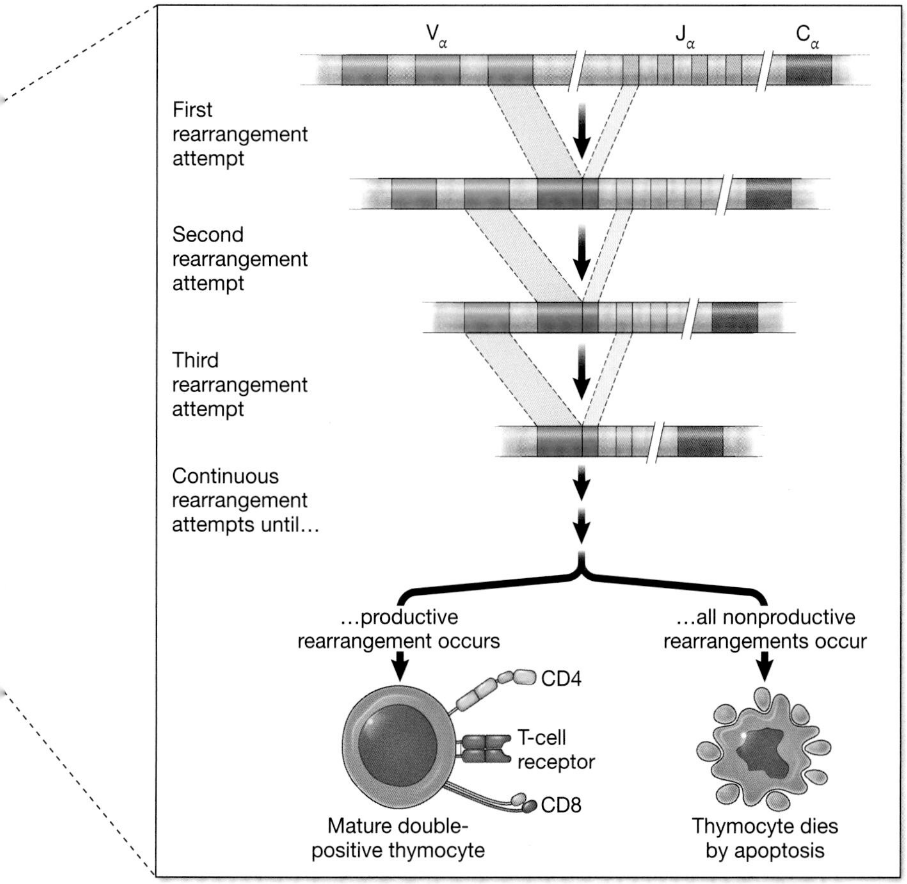

Recombination events are followed by checkpoints that test for productive rearrangements of recombined T-cell receptor loci.

(Sections 4.4, 4.7, 5.6, 5.7)

lineage commitment Process by which developing double-positive thymocytes switch to the expression of a single coreceptor and become single-positive thymocytes.

developmental stage within the thymus is an assessment of whether the thymocyte's receptor (now bearing functional α and β chains) is capable of recognizing an MHC molecule presenting a self-peptide (only self-peptides are present in the thymus). It must also be determined whether that recognition is cause for alarm.

Selection within the thymic cortex

Double-positive thymocytes interact with thymic epithelial cells within the cortex to test the affinity of the T-cell receptor with MHC–peptide complexes at the cell surface. The epithelial cells can express both MHC class I and class II molecules and present a variety of self-peptides to promote selection of double-positive thymocytes. The selection process is based on the affinity of the T-cell receptor for the MHC–peptide complexes. This view suggests three possible outcomes for a double-positive thymocyte that interacts with thymic epithelial cells (**FIGURE 5.12**):

- *Death by neglect*—A majority of double-positive thymocytes with a proper T-cell receptor cannot interact with any MHC–peptide complexes presented within the thymus. These cells cannot recognize self-MHC molecules and so are of no use to the organism in regard to an adaptive immune response. These cells do not receive survival signals from thymic epithelial cells because they cannot interact with those cells, and the lack of a survival signal causes them to die by apoptosis.
- *Negative selection*—Double-positive thymocytes with T-cell receptors that bind too tightly with an MHC–peptide complex on the surface of thymic epithelial cells are potentially dangerous to the organism. These cells could mount an adaptive immune response against self-molecules, so they are negatively selected and die by apoptosis (see **KEY DISCOVERIES**).
- *Positive selection*—Double-positive thymocytes with T-cell receptors that can interact with an MHC–peptide complex with a low or intermediate affinity (about three times lower than negatively selected cells) are positively selected. These cells survive and proliferate.

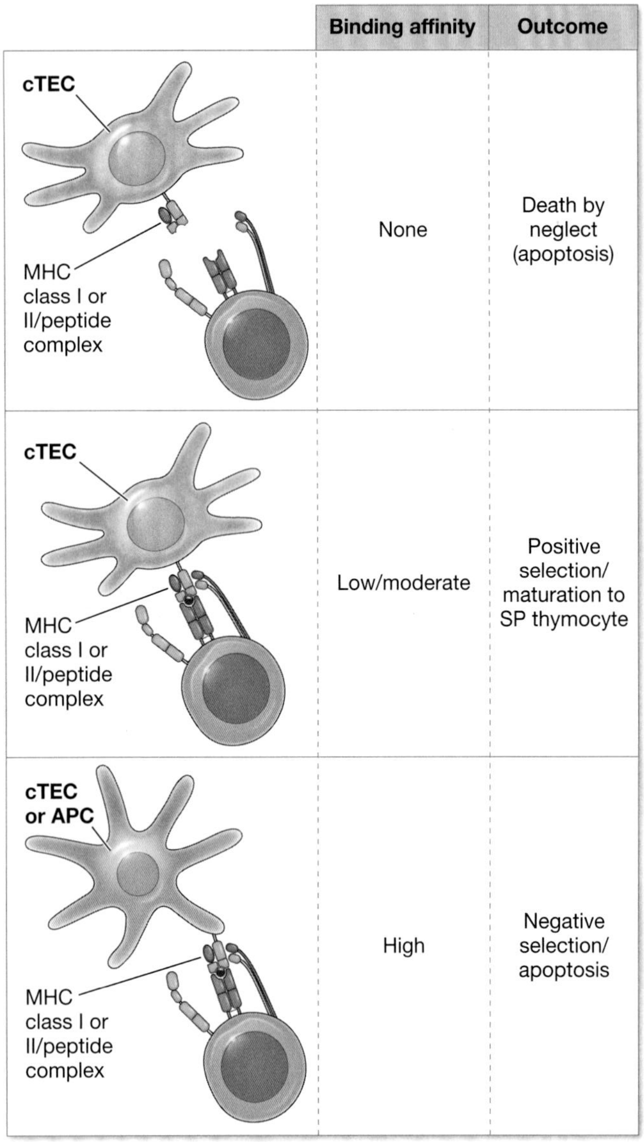

FIGURE 5.12 Possible fates of developing thymocytes in the thymic cortex Double-positive (DP) thymocytes that have passed the checkpoints during somatic recombination are tested by antigen-presenting cells (APCs) or cortical thymic epithelial cells (cTECs) for the affinity of their T-cell receptor for MHC class I or II molecules presenting antigen. Thymocytes that cannot bind to MHC molecules or that bind with too high an affinity die by apoptosis. Thymocytes with a low or moderate affinity for MHC class I or II molecules presenting antigen first select for their coreceptor to become single-positive (SP) thymocytes. (After J. Punt et al. 2018. *Kuby Immunology*, 8th ed., W. H. Freeman and Company: New York.)

MHC restriction and lineage commitment

In Chapter 4 we saw that T cells use the CD4 or CD8 coreceptor to fully engage an MHC–peptide complex. During the selection process within the thymic cortex, thymocytes are still positive for both CD4 and CD8. How, then, does a double-positive thymocyte further develop to become a single-positive thymocyte and express only a single coreceptor?

There is much debate about the mechanism of **lineage commitment** (commitment of a thymocyte to express a single coreceptor). Both transcriptional control and epigenetics (changing the phenotype by modifying gene expression rather than the DNA sequence itself) likely

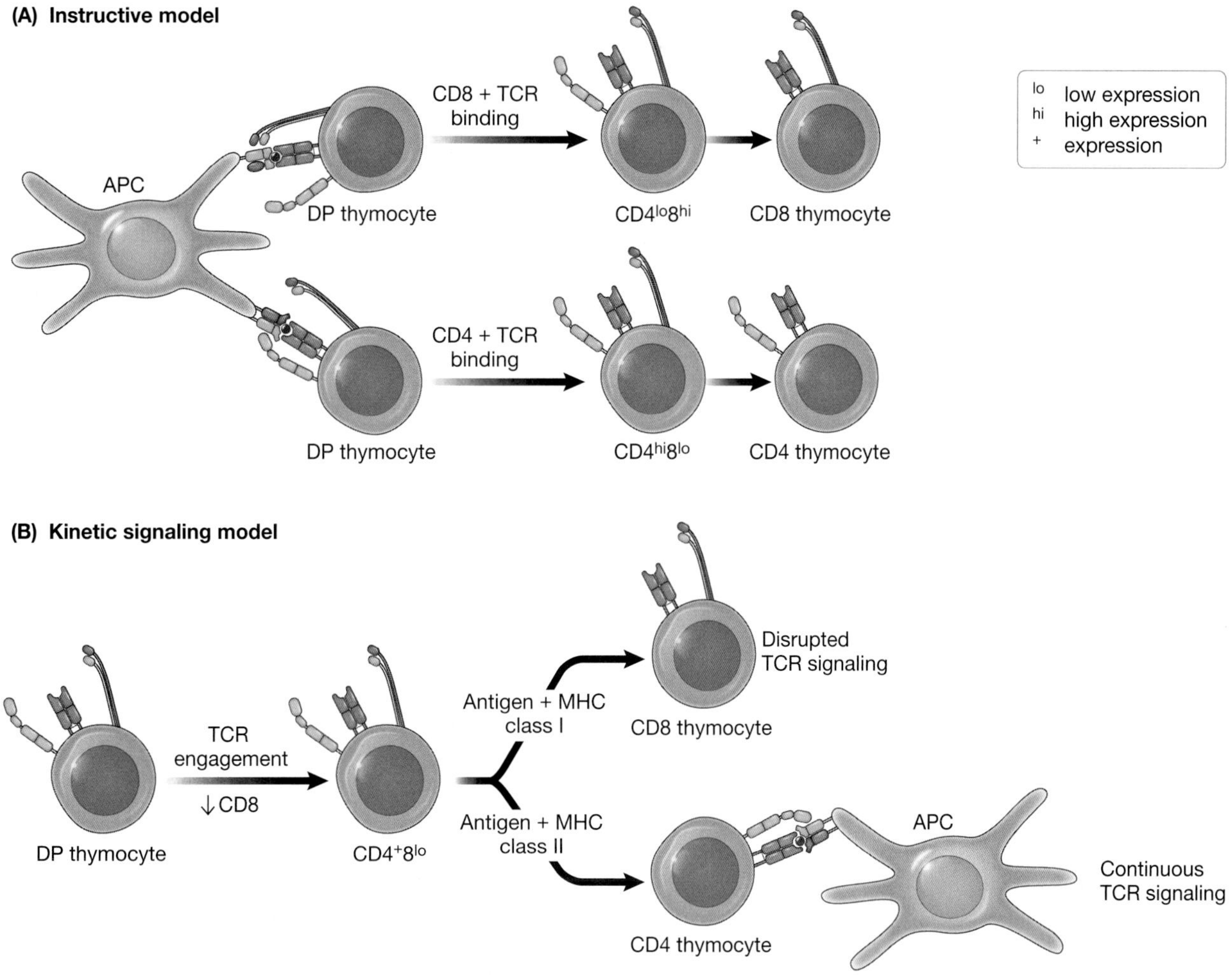

FIGURE 5.13 Two models of MHC restriction of coreceptors on developing thymocytes (A) The instructive model states that binding of the T-cell receptor and the correct coreceptor to an MHC molecule "instructs" the cell to express only the engaged coreceptor. (B) The kinetic signaling model states that all double-positive (DP) thymocytes first express high levels of CD4 coreceptor on the surface. Engagement of the CD4 coreceptor and T-cell receptor with MHC class II/antigen complexes promotes T-cell signaling and maintenance of CD4 as the coreceptor. If the T-cell receptor cannot signal (because it is only capable of binding to MHC class I/antigen complexes), the thymocyte switches coreceptor expression from CD4 to CD8. (After J. Punt et al. 2018. *Kuby Immunology*, 8th ed., W. H. Freeman and Company: New York.)

play roles in this process. The models that follow are simplified descriptions of key aspects of the process (**FIGURE 5.13**):

- *Instructive model*—This model of lineage commitment states that the interaction between the T-cell receptor on a double-positive thymocyte and the MHC–peptide complex on a thymic epithelial cell drives the selection of the coreceptor. Thus, if a T-cell receptor engages an MHC class I molecule presenting a peptide, the interaction will also promote CD8 interaction with the complex and shut down CD4 expression. Conversely, if a T-cell receptor engages an MHC class II molecule presenting a peptide, the CD4 coreceptor will engage in the interaction, and the thymocyte will receive a signal to prevent CD8 expression.
- *Kinetic signaling model*—Although this model is also simplistic and not the whole story, it takes into account data from experiments that demonstrated that $CD4^+$ or $CD8^+$ thymocytes could be reprogrammed to express the other coreceptor by modulating the signaling strength within

KEY DISCOVERIES

How did we learn that self-reactive thymocytes are eliminated from the thymocyte population via negative selection?

Article

P. Kisielow, H. Bluthmann, U. D. Staerz, M. Steinmetz, and H. von Boehmer. 1988. Tolerance in T-cell receptor transgenic mice involves deletion of nonmature $CD4^+8^+$ thymocytes. *Nature* 333: 742–746.

Background

It had been well documented that an $\alpha\beta$ T-cell receptor interacts with MHC–peptide complexes to activate a T cell. It was also known that mature T cells do not interact with self-MHC:self-peptide complexes within an organism. The lack of autoreactive T cells was hypothesized to be due to a deletion mechanism during T-cell development; however, demonstration of this process had never occurred due to the complex nature of thymocytes developing within the thymus of an organism. How was it determined that negative selection occurs—that is, how do we know that autoreactive T cells are deleted from the lymphocyte population during T-cell development?

Early Research

In the late 1980s, two research groups were able to create an important detection reagent to autoreactive thymocytes—monoclonal antibodies to certain β chains expressed at high frequency on T cells that can recognize a certain self-MHC:self-peptide complex. Using these antibodies, the research groups demonstrated that peripheral T cells and thymocytes within the thymic medulla do not contain an autoreactive β chain. However, they were able

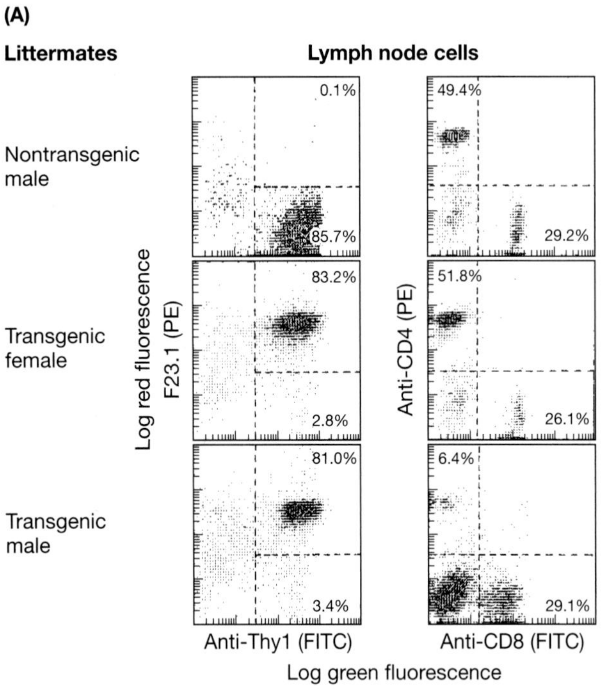

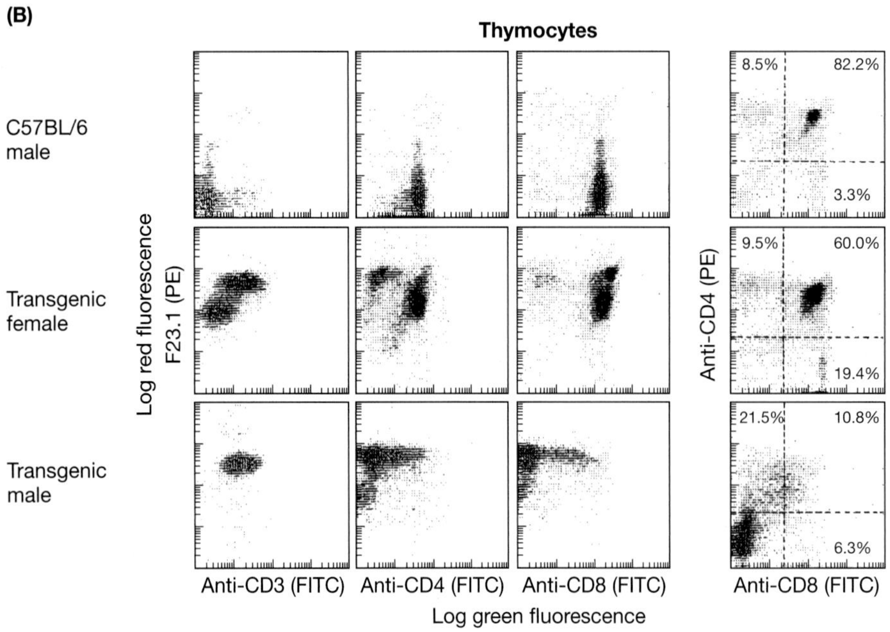

FIGURE KD 5.1 T cells expressing a T-cell receptor that recognizes the male-specific antigen H-Y are removed from the T-cell population in male mice (A) While female mice expressing a T-cell receptor transgene capable of recognizing a male-specific self-antigen (H-Y) contain similar populations of T cells isolated from lymph nodes as that of their nontransgenic littermates, T cells from male mice expressing the transgene are mainly double-negative with a small population of CD8 T cells that express the coreceptor at low levels. (B) Similar data to those observed in T cells isolated from lymph nodes were obtained when observing thymocytes developing in the thymus of male and female mice expressing the T-cell receptor transgene. Females contained thymocytes that were double-positive for the coreceptors, while males contained double-negative thymocytes, suggesting the removal of transgene-expressing thymocytes by negative selection. (From P. Kisielow et al. 1988. *Nature* 333: 742–746.)

to detect this autoreactive β chain within thymic cortex double-positive thymocytes. These findings suggested that a mechanism prevented the autoreactive β chain from progressing further than the double-positive thymocyte stage of development. However, two possibilities existed for how this β chain was no longer found in medullary thymocytes and peripheral T cells: 1) Thymocytes containing the autoreactive β chain were deleted; or 2) The autoreactive β chain was absent or masked within more advanced thymocytes and peripheral T cells.

Think About...

1. State the proposed models as testable hypotheses.
2. What experiment(s) would you conduct to test each hypothesis? How could you differentiate among your hypotheses with these methods?
3. How did the research findings support one of the models suggested for the absence of autoreactive peripheral T cells?
4. How do the findings compare to our current understanding of thymocyte development and autoreactive thymocytes?

Article Summary

To test the hypothesis that autoreactive T cells are removed by deletion during T-cell development, the authors created transgenic mice by injecting genomic DNA from a $CD8^+$ T-cell clone containing a rearranged αβ T-cell receptor that recognizes the minor histocompatibility antigen H-Y (only expressed in males when presented by the MHC class I allotype H-2Db). The researchers then tested for the presence of transgene-expressing thymocytes and peripheral T cells among male and female littermates.

Conclusion

The researchers found differences between male and female transgenic mice at the level of both peripheral T cells and developing thymocytes. Transgenic females expressed the β-chain transgene and possessed normal levels of $CD4^+$ and $CD8^+$ peripheral T cells within lymph nodes. However, transgenic males also expressed the β-chain transgene but primarily possessed double-negative peripheral T cells and a small number of low-level–expressing $CD8^+$ peripheral T cells (**FIGURE KD 5.1**).

Furthermore, the researchers found that the lack of $CD8^+$ peripheral T cells occurs because of removal of these cells during thymocyte development. Most thymocytes in transgenic males that expressed the β-chain transgene were double-negative thymocytes. In contrast, the β-chain transgene thymocytes in transgenic females were double-positive thymocytes (see Figure KD 5.1). These findings provided strong evidence that the removal of autoreactive T cells must occur by deletion within the thymus during T-cell development.

the thymocyte. Stronger signaling events can drive CD4 expression, and weaker signaling events can drive CD8 expression. Based on these and historical observations, the kinetic signaling model suggests that positively selected thymocytes will become $CD4^+$ if the T-cell receptor/coreceptor signal is continuous; they will become $CD8^+$ if this signal is interrupted. This model also suggests that interruption in this signal does not prevent thymocyte development and is likely rescued by the cytokine IL-7, which can drive lineage commitment to CD8.

Negative selection within the thymic medulla

We know that negative selection of double-positive thymocytes can occur if the T-cell receptor:MHC-peptide complex has too high an affinity. One extremely complicating factor in the process of **central tolerance** is that certain endocrine tissues within the body are the source of very specific proteins. For instance, insulin is only produced by the pancreas. To ensure central tolerance and minimize the presence of thymocytes that may be self-reactive to tissue-specific antigens, epithelial cells within the thymic medulla express tissue-specific antigens by utilizing the transcriptional activator AIRE (autoimmune regulator). This transcriptional activator allows these cells to express genes not normally expressed by epithelial cells of the thymus.

AIRE enables thymic epithelial cells to undergo promiscuous gene expression. It contains several protein domains that facilitate its role as a transcriptional activator, including a CARD domain (*ca*spase *r*ecruitment *d*omain),

central tolerance Negative selection processes that occur in primary lymphoid tissues that are responsible for the removal of self-reactive lymphocytes.

Making Connections: T-Cell Selection and Tolerance

T cells that have rearranged their T-cell receptor loci undergo both positive and negative selection in the thymus. T cells undergo positive selection to ensure that they have a functional receptor and negative selection to ensure that they are not self-reactive. A further means of tolerance is provided by the action of regulatory T cells, which can inactivate self-reactive T cells in the periphery.

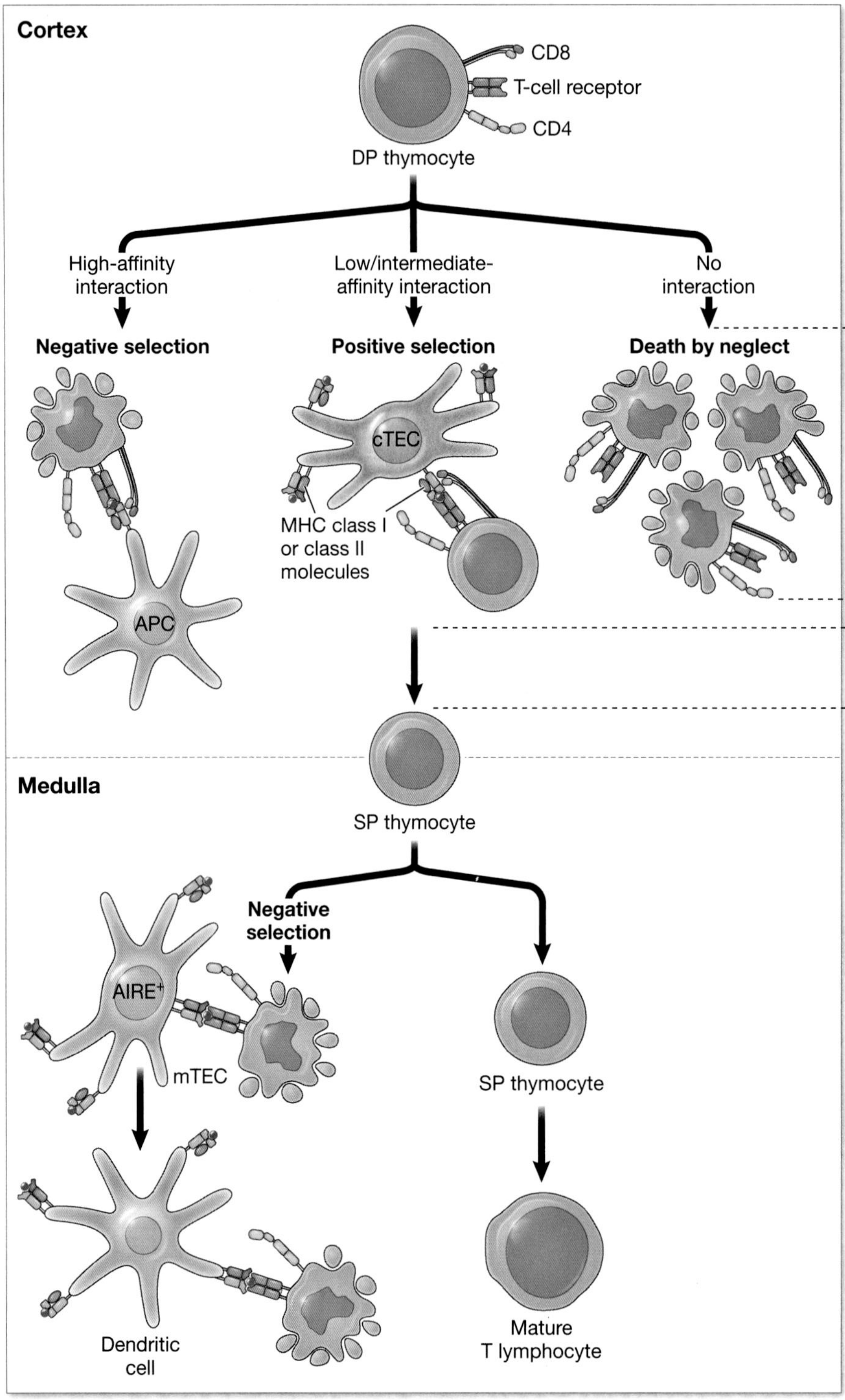

Double-positive thymocytes engage in both positive and negative selection to ensure that they have a functional T-cell receptor that cannot react with self-antigens.

(Sections 4.4, 4.5, 5.8, 7.15, 16.1)

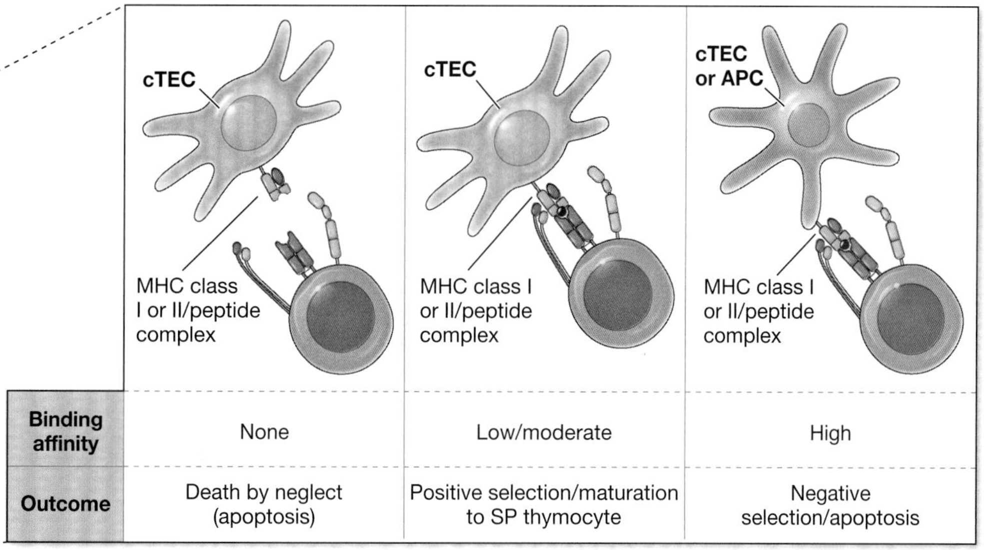

After J. Punt et al. 2018. *Kuby Immunology*, 8th ed., W. H. Freeman and Company: New York

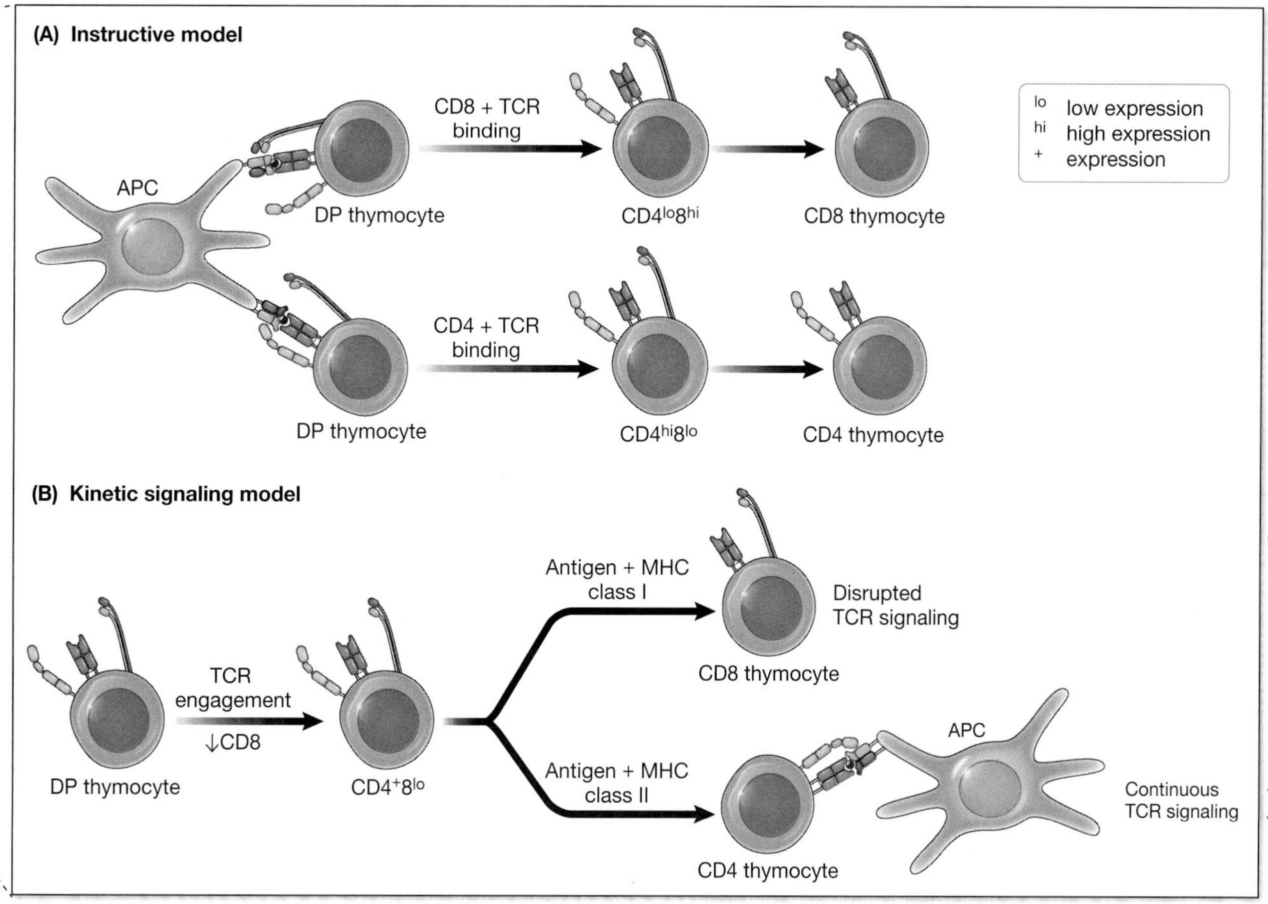

After J. Punt et al. 2018. *Kuby Immunology*, 8th ed., W. H. Freeman and Company: New York

Double-positive thymocytes that successfully pass the positive selection stage are committed to a single coreceptor lineage through MHC lineage commitment.

(Sections 4.4, 4.6, 4.7, 5.8, 6.4, 6.6)

peripheral tolerance Negative selection/inactivation processes that occur in peripheral tissues to prevent the activation of self-reactive lymphocytes that have escaped negative selection in their primary lymphoid tissues.

a SAND domain (*S*P100, *A*IRE1, *N*ucP41/PP75, and *D*EAF1), and two PHD domains (*p*lant *h*omeo*d*omain). These domains work together to allow AIRE to interact with histones within chromatin of thymic epithelial cells.

AIRE also promotes gene transcription by acting as a "gas pedal" for RNA polymerase II, which would normally not transcribe genes of tissue-specific antigens within thymic epithelial cells. In this manner, AIRE ensures that genes of tissue-specific antigens are activated in thymic epithelial cells to promote the expression and processing of these antigens within the thymus.

The protein products of genes activated by AIRE can be processed via MHC class I antigen presentation (described in detail in Chapter 6) to present tissue-specific self-peptides via MHC class I molecules on the surface. Any $CD8^+$ thymocyte within the thymic medulla that strongly engages such an MHC–tissue-specific peptide complex will be negatively selected and will die by apoptosis.

The thymic medulla also houses dendritic cells. These cells can engulf epithelial cells within the medulla and present tissue-specific antigens via MHC class II antigen presentation, allowing for negative selection of $CD4^+$ thymocytes that bear a T-cell receptor that interacts too strongly with an MHC class II–tissue-specific antigen peptide. This means of negative selection drives central tolerance and limits the circulation of self-reactive T cells within an organism. Failure of this negative selection process can lead to autoimmune disorders.

To further promote tolerance, a subset of CD4 T cells that express a self-reactive T-cell receptor (one with a high avidity to MHC class II proteins displaying self-antigen) as tested in the thymic medulla express the transcription factor FOXP3. These cells continue the developmental process to become natural regulatory T cells that are released into circulation to promote **peripheral tolerance** against cells that escaped the negative selection process (discussed further in Chapter 7).

CHECKPOINT QUESTION

1. What are the three possible fates of double-positive thymocytes continuing their development in the thymus?

5.8 | How is T-cell receptor diversity achieved?

LEARNING OBJECTIVE

5.8.1 Compare and contrast the three models driving coreceptor selection and lineage commitment.

We have seen that the cellular arm of the adaptive immune system employs a variety of T cells that either directly target pathogens for destruction or activate other immune system components for this purpose. Extensive T-cell receptor diversity exists due to somatic recombination, which controls gene rearrangement of T-cell receptor loci.

The majority of circulating T cells express an $\alpha\beta$ T-cell receptor that, when activated, can differentiate into effector T cells that activate a range of innate or adaptive immune responses (including macrophages, neutrophils, and B cells) or directly destroy an infected cell. These circulating T cells play a vital role in the adaptive immune response. Because of this, they may be a target for infection and destruction by SARS-CoV-2, the virus responsible for the COVID-19 pandemic (see **EMERGING SCIENCE**).

A small subset of T cells in the body express a $\gamma\delta$ T-cell receptor. These cells reside mainly within the gut mucosa as intraepithelial lymphocytes (explored

EMERGING SCIENCE

Does the causative agent of COVID-19 target the destruction of T cells?

Article

C. Qin, L. Zhou, Z. Hu, S. Zhang, S. Yang, Y. Tao, C. Xie, K. Ma, K. Shang, W. Wang, and D.-S. Tian. 2020. Dysregulation of immune response in patients with Coronavirus 2019 (COVID-19) in Wuhan, China. *Clinical Infectious Diseases* 71: 762–768.

Background

T-cell development and function are vital in the proper combat and clearance of many different pathogens. Because T-cell development within the thymus occurs early in childhood and later shifts to long-term maintenance through division of circulating T cells, ensuring the survival of existing T cells is crucial in mounting a proper adaptive immune response. Pathogens that target the circulating population of T cells may be at an advantage during an infection. The emergence of SARS-CoV-2, the causative agent of COVID-19, and its frequent severity of disease prompted scientists to ask the question: Are cells of the immune system a target for SARS-CoV-2 infection and destruction?

The Study

Researchers hypothesized that cells of the immune system might be susceptible to infection and destruction by SARS-CoV-2. To test their hypothesis, they analyzed lymphocyte and leukocyte numbers in 452 patients in Wuhan, China, who were infected with SARS-CoV-2. Analysis of the 286 severe cases from their patient pool demonstrated that individuals with severe infection, characterized by respiratory distress (respiratory rate of over 30 per minute), an oxygen saturation level of less than 93%, and arterial blood oxygen partial pressure to oxygen concentration ratio of less than 300 mm Hg, had fewer lymphocytes in circulation (14% compared to the normal range of 20%–50%). Further analysis of the lymphocyte population revealed a significant decrease in circulating CD4 (helper) T cells and regulatory T cells, as compared to relatively normal amounts of NK cells and B cells (**TABLE**).

Conclusion

The findings provide evidence that circulating T cells were decreased in more severe cases of COVID-19, suggesting that SARS-CoV-2 may cause more severe disease states by targeting T cells within infected individuals. The study also lays the foundation for future studies to test the ability for SARS-CoV-2 to directly infect and destroy CD4 T cells or to study whether SARS-CoV-2 may influence an infected individual's ability to maintain CD4 T cells in peripheral tissues.

Think About...

1. If you were part of this research team and wanted to determine whether SARS-CoV-2 could directly target CD4 T cells, what experiment(s) could you do to test this hypothesis?
2. The researchers evaluated the number of circulating lymphocytes in individuals infected with SARS-CoV-2. What other information might be important in light of the claim that SARS-CoV-2 directly affects the number of circulating lymphocytes?
3. Most viruses infect cells by binding to a surface protein on the target cell via a surface protein on the viral envelope. It is known that SARS-CoV-2 uses a spike glycoprotein in the viral envelope to bind to the angiotensin-converting enzyme 2 (ACE2) protein on target cells. Based on this knowledge, how could you test whether SARS-CoV-2 could directly target CD4 T cells?

Lymphocyte Counts in Patients with Mild to Severe Cases of COVID-19

		Mean (SD)			
	Normal range	All patients (n=44)	Non-severe (n=17)	Severe (n=27)	*P* value
T cells+B cells+NK cells/μl	1100.0–3200.0	852.9 (412.0)	1020.1 (396.5)	743.6 (384.4)	0.032
B cells/μl	90.0–560.0	179.7 (143.1)	196.1 (144.9)	169.0 (140.9)	0.559
T cells/μl	955.0–2860.0	541.5 (292.7)	663.8 (291.3)	461.6 (264.7)	0.027
NK cells/μl	150.0–1100.0	131.7 (83.1)	160.2 (90.8)	113.0 (71.8)	0.072
CD4 T cells/μl	550.0–1440.0	338.6 (196.3)	420.5 (207.8)	285.1 (168.0)	0.027
Regulatory T cells/μl	5.36–6.30	4.1 (1.2)	4.5 (0.9)	3.7 (1.3)	0.040

Source: Data from C. Qin et al. 2020. *Clin Infect Dis* 71: 762–768. doi.org/10.1093/cid/ciaa248.

TABLE 5.4 | Variable Segments of the T-Cell Receptor Loci

Variable segment	Functional segments
V_β	40–48
D_β	2
J_β	12–13
V_α	45–47
J_α	50
V_δ	7–8
D_δ	3
J_δ	4
V_γ	4–6
J_γ	5

Source: Data accessed from M.-P. Lefranc. 2001. IMGT®, the international ImMunoGeneTics database. *Nucl Acids Res* 29: 207–209. DOI:10.1093/nar/29.1.207; M.-P. Lefranc. 2014. *Front Immunol* 5: 22. DOI:10.3389/fimmu.2014.00022.

in Chapter 12), or they express a γδ T-cell receptor that behaves like a pattern recognition receptor of the innate immune system. The receptors that mimic pattern recognition receptors are produced through somatic recombination but are limited in their diversity. They can recognize similar molecules from different pathogens, such as the case of γδ T-cell receptors that recognize compounds of mevalonate synthesis of different microorganisms.

T-cell receptor diversity is driven in part by the sheer number of V, D, and J subunits present for all chains associated with the receptor (**TABLE 5.4**). Since recombination of variable segments is random, the diversity of variable regions is extensive, allowing for T-cell recognition of a wide array of antigens. Furthermore, junctional diversity by the addition and removal of P and N nucleotides during recombination adds to the variation of T-cell receptors present.

- There are approximately 45 different Vα segments and 50 different Jα segments, so about 2,150 different combinations are possible to create a functional T-cell receptor α chain.
- There are approximately 42 different Vβ segments, 2 different Dβ segments, and 13 different Jβ segments, so about 1,092 different combinations can occur to create a functional β chain.
- Any subunit of a T-cell receptor that has properly recombined needs to associate with another randomly recombined subunit. Random association of α and β chains can result in 2.3×10^6 different T-cell receptors.
- While random association of α and β chains results in a large number of possible different T-cell receptors, research has shown that in terminal deoxynucleotidyl transferase (TdT)-deficient mice, the number of circulating T cells with different T-cell receptors is approximately 10% of the number of T cells with different receptors in wild-type mice. This suggests that junctional diversity may increase the number of different T-cell receptors to 2.3×10^7!

CHECKPOINT QUESTION

1. What are the three major factors that contribute to T-cell receptor diversity?

Summary

5.1 What is the role of the thymus in T-cell development?

- The thymus is a primary lymphoid organ and the site of thymocyte development.
- Resident cells of the thymus, including thymic epithelial cells, macrophages, and dendritic cells, play critical roles in thymocyte development and the removal of apoptotic thymocytes.

5.2 Which cells are precursors to developing thymocytes?

- Hematopoietic stem cells serve as the stem cell precursors to lymphoid progenitor cells, which can undergo differentiation into several lines of lymphocyte origin.
- Lymphoid progenitor cells, which are pluripotent to several lines of hematopoietic origin, change their physical determinants (especially cell-surface markers) as they further commit to thymocyte development.

5.3 What is the role of Notch1 in T-cell development?

- Notch1 is a lymphoid progenitor cell transcription factor activated by engagement with thymic epithelial cells that drives the commitment of thymocyte development.

5.4 What are the stages of T-cell development?

- Developing T cells mature by progression through multiple stages, including several stages as double-negative thymocytes and as a double-positive thymocyte.
- A developing thymocyte is destined to express either an αβ T-cell receptor or a γδ T-cell receptor, if critical checkpoints are passed during the developmental process.

5.5 How does a developing thymocyte pass through the β chain checkpoint?

- As a thymocyte matures into a double-positive thymocyte that expresses an αβT-cell receptor, it first must have productive rearrangement of the β chain locus, then allelic exclusion of recombination of the other β chain, followed by productive rearrangement of the α chain locus.

5.6 How does a developing thymocyte pass through the α chain checkpoint?

- Progression through proper thymocyte development requires testing of the rearranged T-cell receptor; improper rearrangement can result in apoptosis of the developing thymocyte.

5.7 How does positive and negative selection of T cells occur?

- Upon proper rearrangement of an αβ T-cell receptor and production of a double-positive thymocyte, the T-cell receptor is tested for affinity for MHC molecules; lack of affinity or too strong an affinity for MHC molecules leads to apoptosis of the thymocyte through death by neglect or negative selection.
- Double-positive thymocytes become single-positive thymocytes through MHC engagement and coreceptor selection, although the mechanism driving coreceptor selection is still under debate.

5.8 How is T-cell receptor diversity achieved?

- The diversity of T-cell receptors is driven by the large number of V, D, and J segments present in the loci for the different T-cell receptor subunits.
- Randomness of which V, D, and J segments are used during somatic recombination of each T-cell receptor subunit, along with random association of differentially recombined subunits, is predicted to allow for 2 million possible T-cell receptor combinations.

Review Questions

5.1 List the major resident cells of the thymus and describe their function in T-cell development.

5.2 **a.** If you were conducting analysis of cell-surface molecules on cells isolated from the blood of a mouse, would you expect to see CD4 or CD8 expressed on the surface of hematopoietic stem cells? Why or why not?

b. A lymphoid progenitor is capable of differentiating into a number of lymphocytes, including T cells, B cells, natural killer cells, and dendritic cells. What signal(s) promote the differentiation of lymphoid progenitors into T cells?

5.3 What would you predict would be the outcome of thymocyte development in individuals who are unable to produce Notch ligand?

5.4 **a.** Compare and contrast double-negative and double-positive thymocytes.

b. Outline the recombination events that occur as a double-negative thymocyte progresses toward a double-positive thymocyte.

c. What is the main purpose of having the α, β, and γδ chain checkpoints during T-cell development?

5.5 **a.** Why is it important to initiate allelic exclusion at the β-chain locus once a thymocyte has passed through the β-chain checkpoint?

b. Why is it more likely for the α-chain locus to recombine productively before the γ- and δ-chain loci recombine?

5.6 **a.** Why is it impossible for a T cell to express both an αβ receptor and γδ T-cell receptor?

b. Why is it important for thymocytes that have just finished α- and β-chain recombination to express both coreceptors?

5.7 List the three major fates of double-positive thymocytes within the thymus and explain what actions are required to dictate each fate.

5.8 Why would you predict that there is less T-cell receptor diversity in cells expressing γδ T-cell receptors compared to cells expressing αβ T-cell receptors?

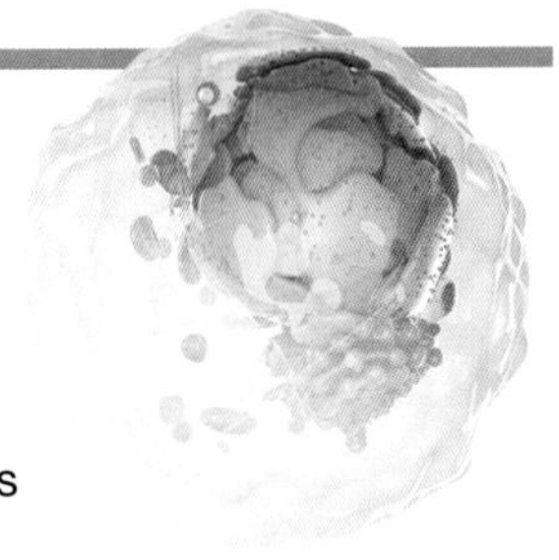

● CASE STUDY REVISITED

Type 1 Diabetes

An 11-year-old boy is seen by his primary care physician due to lethargy and flu-like symptoms. His mother mentions that the boy had recently taken up football and that practices had just begun, but that he is an athletic boy and had passed a sports physical earlier that summer. The medical history reveals that he has recently had bouts of increased thirst and urination. The primary care physician orders a blood test to see if the boy is suffering from a systemic infection. While there is no increase in white blood cell numbers (suggesting there is no apparent evidence of an infection), a blood test reveals elevated fasting blood sugar and glycated hemoglobin (A1C) levels. Further blood tests reveal the presence of anti-insulin autoantibodies. Analysis of peripheral lymphocytes reveals the presence of T cells that are autoreactive to insulin. The diagnosis is type 1 diabetes.

Think About...

Questions for individuals

1. Why is type 1 diabetes defined as an autoimmune disorder?
2. What types of autoantigens have been discovered in patients with type 1 diabetes?

Questions for student groups

1. What process within thymocyte development is likely faulty to allow the production of autoreactive T cells within the patient?
2. Insulin is produced in the islets of Langerhans in pancreatic β cells. Knowing the process of T-cell development within tissues and cells of the thymus, where in the thymus has the defect that led to the production of autoreactive T cells occurred?
3. Speculate on which cells and factors may be deficient in the patient, resulting in the production of autoreactive T cells.

MAKING CONNECTIONS **Key Concepts**	**COVERAGE** **(Section Numbers)**
Locations where hematopoiesis occurs	1.6, 1.9
T-cell arm of hematopoiesis and effector T cells	5.1, 5.2, 5.8, 7.13, 7.14, 7.15, 7.16, 12.4
Stages of T-cell development and recombination events that occur during these stages	4.4, 4.7, 5.5, 5.6, 5.7
Positive and negative selection of T cells	4.4, 4.5, 5.8, 16.1
Lineage commitment of T cells and connection between coreceptor and MHC class molecules	4.4, 4.6, 4.7, 5.8, 6.4, 6.6
Connection of negative selection, central tolerance, and peripheral tolerance	4.5, 5.8, 7.15

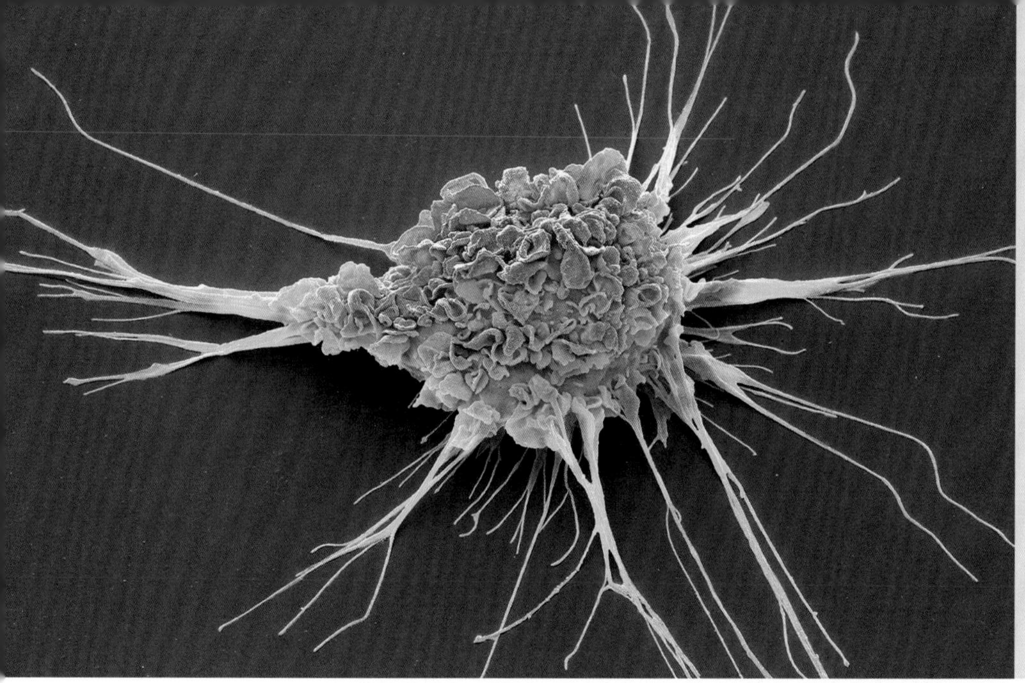

6 Antigen Recognition by T Lymphocytes

CASE STUDY: Amy's Story

It's the middle of the night when the quiet is broken by a loud, raspy cough. Amy's parents rush to her room to comfort her and take her temperature—104°F. Amy's father gives her some ibuprofen and runs a cool bath to try and bring her temperature down. After the bath, he turns the shower on high to try and build up steam in the bathroom so that Amy might breathe a bit better. Unfortunately, this seems like another typical night in the Wang household.

Over the past 6 months, Amy has been afflicted with four respiratory infections, all handled swiftly by antibiotic treatment. Tonight's latest bout seems worse; Amy hasn't had a fever this high in the past 6 months, and her cough seems much deeper and straining. Also, the skin lesions that Amy has been suffering from seem to be getting worse. The doctors had attributed the skin lesions to the antibiotic treatments, but Amy's parents aren't convinced, since Amy seems to get them all the time.

The ibuprofen, cool bath, and steam seem to be making Amy feel a bit better, but her parents are still frightened from the ordeal and decide that it would be worth the extra effort to take Amy to the emergency room.

In order to understand what Amy is experiencing, and to answer the case study questions at the end of the chapter, we must first explore the important interactions between T cells and antigen-presenting cells. Through these interactions, our adaptive immune system is able to recognize an infection. In Chapters 4 and 5, we explored the recombination of the T-cell receptor genes and precisely when, during development, these recombination events occur. We also learned

QUESTIONS Explored

6.1 What role do T cells play in antigen recognition?

6.2 What is the structure of the T-cell receptor?

6.3 How does somatic recombination account for T-cell receptor diversity?

6.4 What is the structure of MHC molecules?

6.5 How do T-cell receptors interact with MHC-peptide complexes?

6.6 How does peptide loading occur in MHC class I and class II molecules?

6.7 How does MHC diversity affect peptide presentation to T cells?

6.8 What benefits and problems are associated with MHC diversity?

about the importance of the interaction between the T-cell receptor and MHC-peptide complexes. However, what are the defining features of the T-cell receptor structure that allow it to be both diverse and to function in recognizing an MHC-peptide complex? What are the molecular mechanisms of antigen processing and presentation by MHC class I and class II molecules? This chapter will answer these questions and conclude by exploring the diversity in MHC molecules, which is a result of different MHC genes in the population. Maintenance of MHC diversity promotes organism survival in the evolutionary arms race against pathogens, but it also causes problems for medical breakthroughs, such as tissue and hematopoietic stem cell transplantation.

6.1 What role do T cells play in antigen recognition?

LEARNING OBJECTIVE

6.1.1 Explain the importance of T cells and their receptors in recognizing foreign antigens.

T cells play an important role in adaptive immune system function. Effector T cells generated during clonal selection and expansion activate other immune system cells, including macrophages, neutrophils, and B cells. Pathogen clearance that requires an adaptive immune response relies on the action of T cells.

Recall from Chapter 4 that each T cell has a surface receptor and coreceptor that interact with a specific antigenic peptide presented by a cell using a *major histocompatibility complex* (MHC) protein. This interaction plays a crucial role in determining the fate of the binding T cell. Antigens are presented differently depending on the pathogen's location within the organism, so the T cell must be able to first recognize and then combat the specific pathogen. Thus, a range of effector mechanisms designed to accomplish this task develop in various T cells. For example, T cells that recognize an antigen presented via an MHC class I molecule (typically from an intracellular pathogen) activate effector mechanisms that fight off an intracellular infection.

MHC molecules that present peptides differ in both their protein structure and the means by which they are loaded with peptide antigens. Antigen processing and presentation via MHC molecules is driven not only by the nature and location of the pathogen but also by how the peptide antigens are derived. The peptide-binding site of MHC molecules must be able to bind to many different peptides since these molecules present a wide variety of antigens, including the thousands of proteins naturally produced by cells and foreign antigens.

MHC molecule diversity is not generated through recombination events such as those described in Chapter 4. Instead, diversity of MHC molecules is due to the presence of MHC **gene families** and **genetic polymorphism**. Different yet related genes encode MHC class I and class II subunits. Multiple alleles of each gene encode each subunit, resulting in diversity due to genetic polymorphism within the population. The wide variety of MHC molecules work to bind and present the many antigens that must be displayed to T cells as part of the adaptive immune response.

gene family Genes present within an organism that express isotypes of the same biochemical function; these genes are related evolutionarily as they are formed through duplication events from a single original gene.

genetic polymorphism Presence of multiple alleles within a population that encode for a particular gene product/isotype.

CHECKPOINT QUESTION

1. How does the generation of diversity of MHC molecules differ from that of T-cell receptors?

6.2 | What is the structure of the T-cell receptor?

LEARNING OBJECTIVE

6.2.1 Describe the composition of the T-cell receptor and other proteins involved in T-cell signaling and activation.

The major function of the T-cell receptor is to recognize a specific MHC-peptide complex. The specificity of this recognition event is driven by somatic recombination of the variable region of the receptor gene loci during T-cell development. This event also involves recognition of the presenting MHC molecule, as the T cell's effector mechanisms must be geared specifically to the infection it needs to fight. This recognition occurs, in part, through a coreceptor located on the cell surface. Once a T-cell receptor:MHC-peptide complex forms, signaling pathways activate two types of genes: those required for proper division and differentiation of the T cell and those required to carry out the effector functions of the activated T cell. These signaling pathways are initiated by several other proteins that function in conjunction with the T-cell receptor/coreceptor complex.

hypervariable region Loops of the variable regions of the T-cell receptor subunits or the heavy and light chains of immunoglobulins that have the most diverse differences; these regions contribute significantly to the engagement of the T-cell receptor or an immunoglobulin with its antigen.

Exploring the structure of T-cell receptors

Recall that T-cell receptors are composed of two subunits, an α and a β subunit or a γ and a δ subunit. Each subunit has a variable domain and a constant domain, and each domain is folded into an immunoglobulin superfamily domain fold (**FIGURE 6.1**).

The immunoglobulin superfamily domain fold of each variable domain contains three **hypervariable regions**, also known as *complementarity-determining regions* (CDRs). These hypervariable regions are found at the junctions of V, D, and J segments—they represent the location where junctional diversity is manifested in the intact T-cell receptor. This fold allows the hypervariable regions of

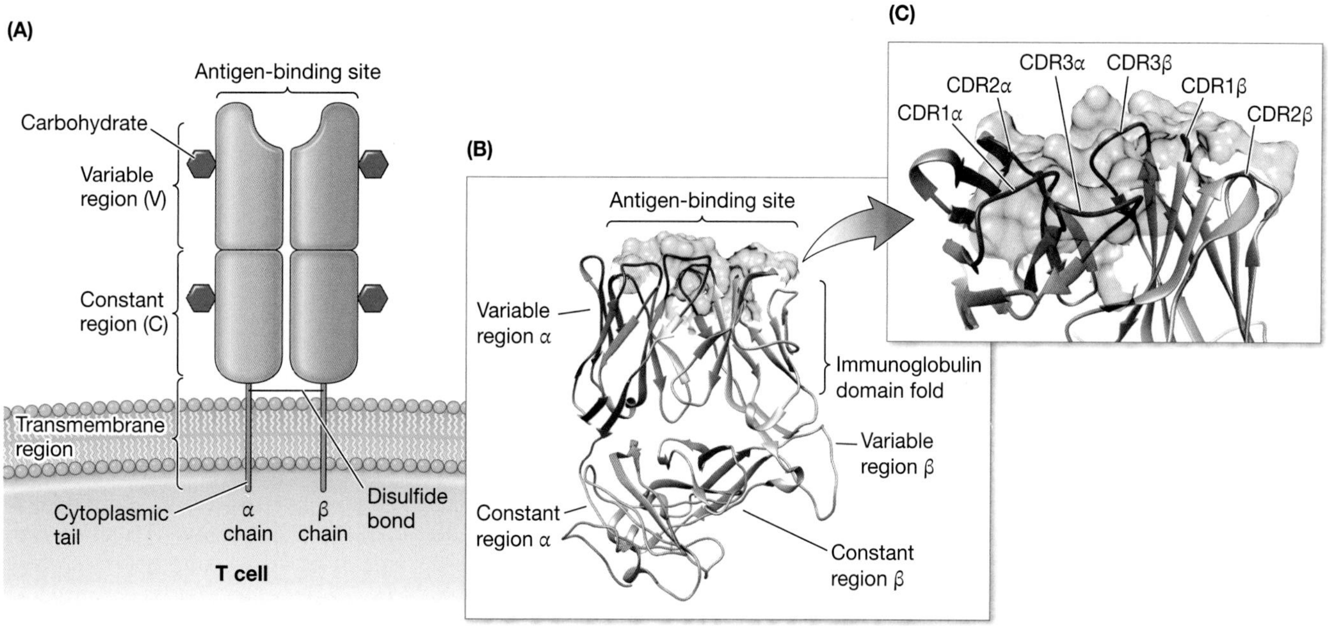

FIGURE 6.1 Structure of the T-cell receptor (A) The T-cell receptor always consists of two membrane-associated polypeptides, most commonly an α chain and a β chain. The two chains each contribute one variable region to an antigen-binding site. (B) The three-dimensional structure of the T-cell receptor shows the two immunoglobulin-like domains that make up each T-cell receptor subunit. (C) The variable region of each T-cell receptor subunit contributes three complementarity-determining regions (CDRs), or hypervariable regions, to the antigen-binding site. The surface representation of these loops demonstrates how close the loops are in three dimensions. (B,C after K. E. Rödström et al. 2015. *PLOS ONE* 10: e0131988/CC BY 4.0. doi:10.1371/journal.pone.0131988; redrawn using PDB file 4UDT.)

each receptor subunit (the sequences most affected by somatic recombination and junctional diversity) to be close together to form an antigen-binding site. This site can vary based on the mechanisms of somatic recombination (see Chapter 4).

There are two types of functional T-cell receptors: those that employ α and β subunits and those that employ γ and δ subunits (**FIGURE 6.2**). The γ and δ gene loci are also capable of undergoing somatic recombination and junctional diversity during T-cell development, as they contain several V, D, and J segments. However, there is much less diversity in γδ receptors as there are fewer V, D, and J regions associated with the γ and δ loci. A circulating T cell can have only one type of receptor (αβ or γδ), as somatic recombination of the α locus removes the δ locus in a signal joint (see Chapter 5).

All αβ receptors recognize peptide antigens presented by MHC molecules. Conversely, γδ receptors recognize their antigen without the presence of an MHC molecule, and they do not necessarily have to recognize a peptide antigen. For example, a subset of γδ receptors can recognize a lipid as an antigen. Because researchers know more about the αβ receptor (due to abundance and conservation of function across organisms), αβ T cells are the typical cell type described. (We will explore the effector mechanisms of γδ T cells and their recognition of lipid antigens in Chapters 7 and 12.)

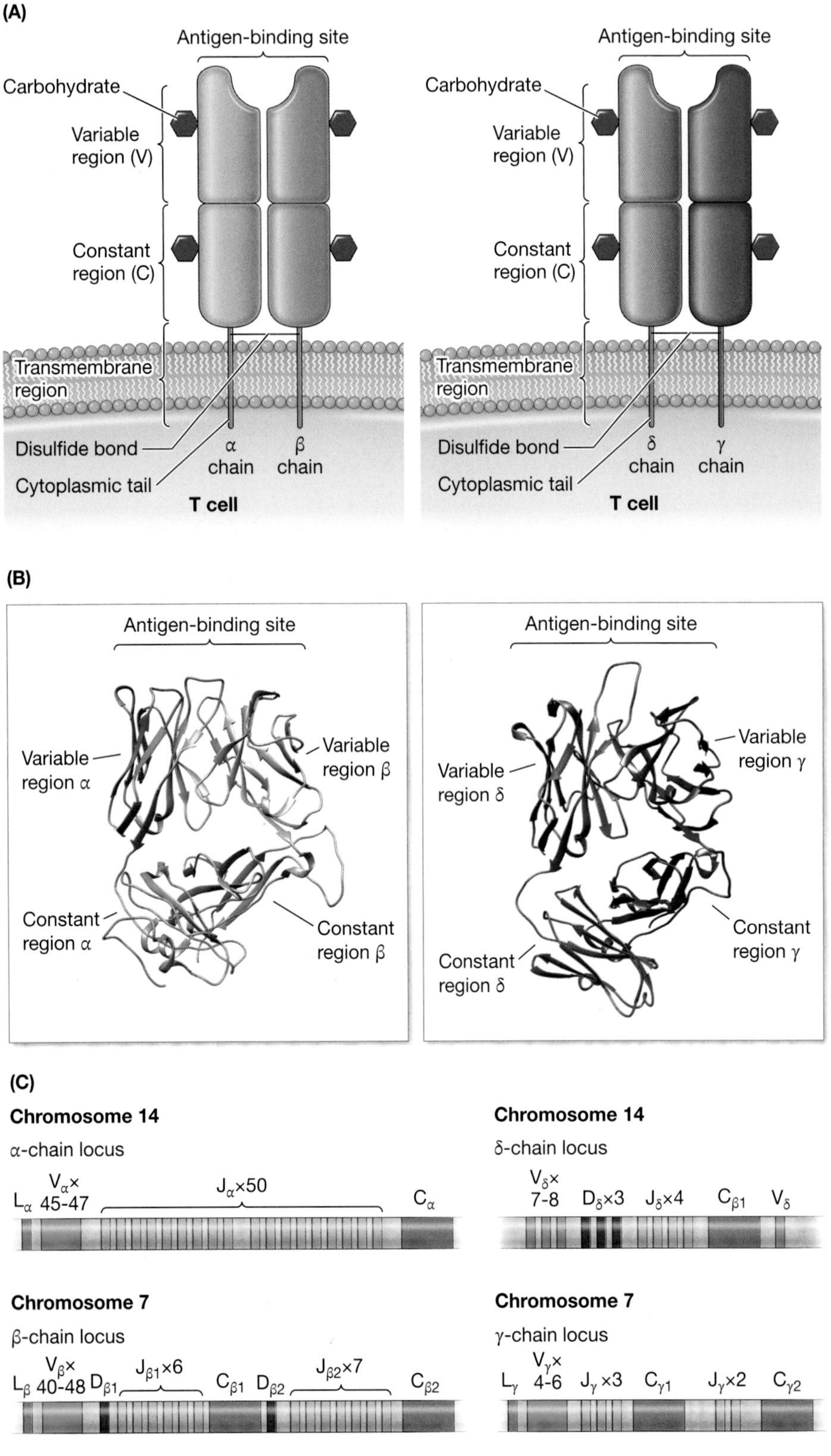

FIGURE 6.2 Comparing the structures of αβ and γδ T-cell receptors (A) Both αβ and γδ T-cell receptors have two subunits—the α and γ chains are similar, and the β and δ chains are similar. (B) While the three-dimensional structures of αβ and γδ T-cell receptors reveal that each chain has two immunoglobulin-like domains, there are noticeable differences in the structures. (C) The α- and γ-chain loci of T-cell receptors contain V and J segments to recombine and produce a variable region, whereas the β- and δ-chain loci contain V, D, and J segments. αβ T-cell receptors are more varied due to the number of V, D, and J segments present in the genome. (B, left after K. E. Rödström et al. 2015. *PLOS ONE* 10: e0131988. CC BY 4.0. doi:10.1371/journal.pone.0131988, redrawn using PDB file 4UDT; B, right after T. J. Allison et al. 2001. *Nature* 411: 820–824, redrawn using PDB file 1HXM; C after P. Parham. 2014. *The Immune System*, 4th ed., W. W. Norton and Company.)

Making Connections: Structure of the T-Cell Receptor and Connection to Diversity

The T-cell receptor is required for the recognition of foreign antigen by T cells. The protein contains two different subunits to make up either an $\alpha\beta$ receptor or $\gamma\delta$ receptor. Developing thymocytes (precursors to T cells) undergo somatic recombination of their T-cell receptor loci to produce a diverse population of T cells, each expressing a different receptor that can recognize a different antigen. The T-cell receptor's structure allows it to utilize this mode of generating diversity to recognize different antigens.

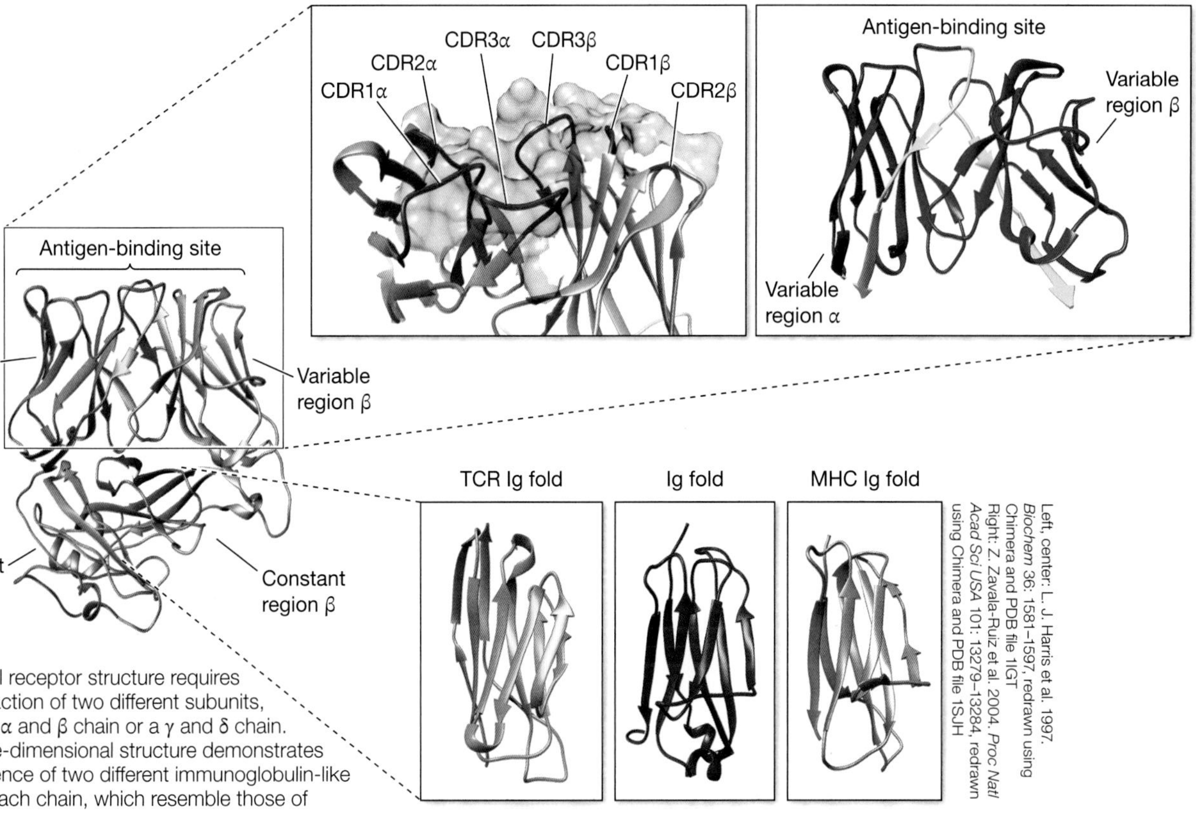

Data from K. E. J. Rödström et al. 2015. *PLOS One* 10: 01318; redrawn using Chimera and PDB file 4UDT

Both after K. E. Rödström et al. 2015. *PLOS ONE* 10: e0131988/CC BY 4.0. doi:10.1371/journal.pone.013198; redrawn using PDB file 4UDT

Left, center: L. J. Harris et al. 1997. *Biochem* 36: 1581–1597, redrawn using Chimera and PDB file 1IGT
Right: Z. Zavala-Ruiz et al. 2004. *Proc Natl Acad Sci USA* 101: 13279–13284, redrawn using Chimera and PDB file 1SJH

The T-cell receptor structure requires the interaction of two different subunits, either an α and β chain or a γ and δ chain. The three-dimensional structure demonstrates the presence of two different immunoglobulin-like folds in each chain, which resemble those of immunoglobulins and MHC molecules.

(Sections 4.2, 4.7, 4.8, 6.2, 6.4, 9.1)

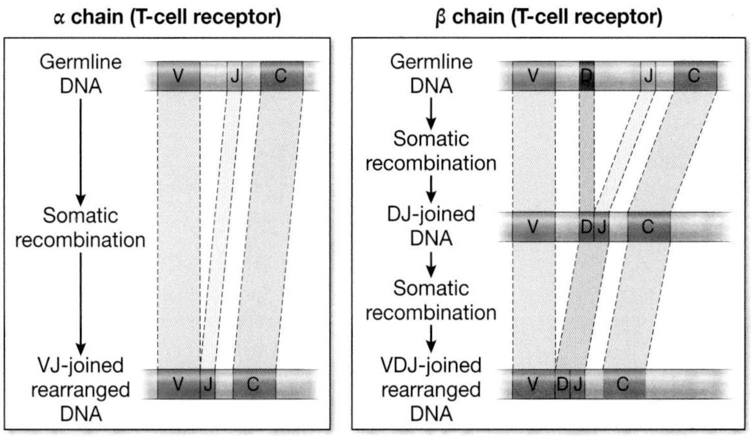

After W. Purves et al. 1995. *LIFE: The Science of Biology* 4th ed., Sinauer Assoc., Sunderland, MA; T. Dörner and P. E. Lipsky. 2001. *Arthritis Rheum* 44: 2715-2727; P. Parham. 2014. *The Immune System*, 4th ed., W. W. Norton and Company, New York

The antigen-binding site of a T-cell receptor is located at the interface of the variable regions of the two different receptor chains. Each subunit contributes a variable region, and each variable region has three hypervariable loops. The variability of this region is driven by somatic recombination.

(Sections 4.2, 4.3, 4.7, 5.4, 6.2, 6.3)

CD4 and CD8 coreceptors

Along with the T-cell receptor, T cells interact with MHC-peptide complexes using one of two coreceptors: CD4 or CD8. Recall that activated CD8 T cells become cytotoxic T cells that target intracellular pathogens, while activated CD4 T cells become helper T cells (T_H). Helper T cells activate cells of the innate and adaptive immune systems (including macrophages, neutrophils, and B cells) that combat extracellular pathogens.

CD4 and CD8 coreceptors have different structures. The CD4 coreceptor is a single transmembrane protein that contains four immunoglobulin-like domains, while the CD8 coreceptor consists of two transmembrane polypeptides, each with a single immunoglobulin-like fold (**FIGURE 6.3**). Both coreceptors are important in the recognition of the MHC-peptide complex on an antigen-presenting cell, but each type recognizes this complex in a location separate from the **peptide-binding groove** (**FIGURE 6.4**).

peptide-binding groove Domain within an MHC molecule that binds to a peptide to present the bound peptide to a T-cell receptor.

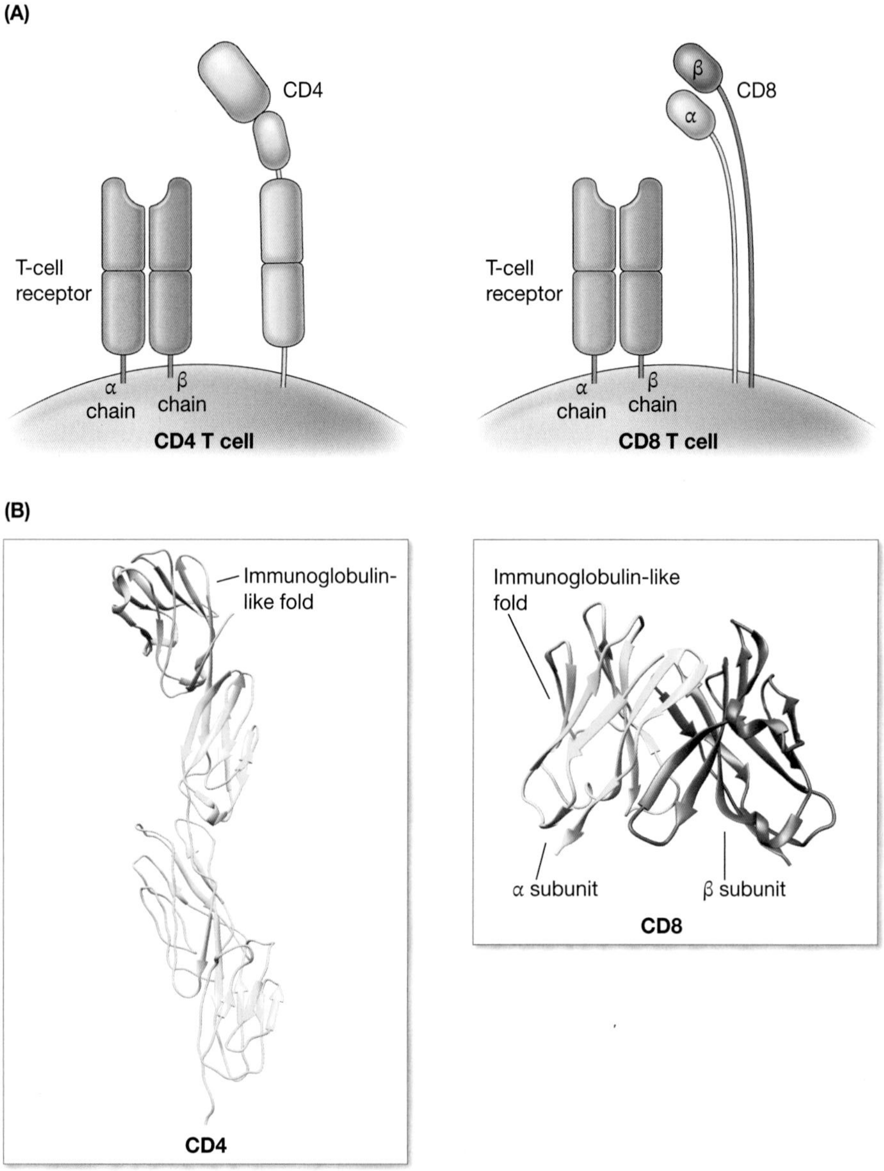

FIGURE 6.3 T-cell coreceptors CD4 and CD8 (A) Circulating T cells express either a CD4 or CD8 coreceptor after they have undergone selection in the thymus and lineage commitment. (B) The CD4 coreceptor contains four immunoglobulin-like folds, whereas each subunit of the CD8 coreceptor contains a single immunoglobulin-like fold. (B data from Y. Yin et al. 2012. *Proc Natl Acad Sci USA* 109: 5405–5410, redrawn using Chimera and PDB file 3T0E; and H. C. Chang, et al. 2005. *Immunity* 23: 661-671, redrawn using Chimera and PDB file 2ATP.)

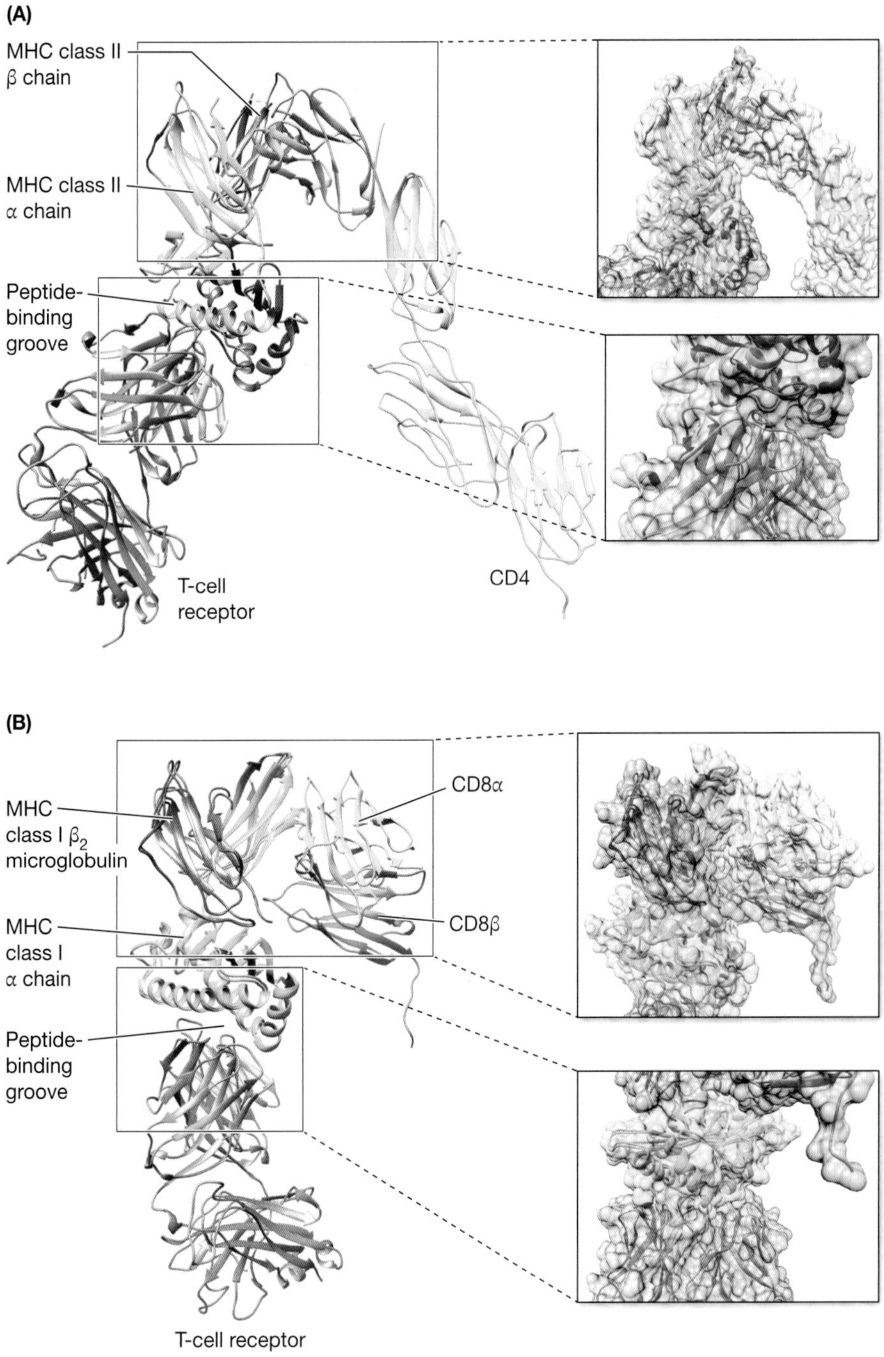

FIGURE 6.4 Three-dimensional structures of the T-cell-receptor/MHC-peptide/coreceptor complexes (A) CD4 T cells interact with antigen presented via an MHC class II molecule. The CD4 coreceptor interacts with the α and β chain of the MHC class II molecule distant from the peptide-binding groove. (B) CD8 T cells interact with antigen presented via an MHC class I molecule. The CD8 coreceptor α and β subunits interact with the α chain of the MHC class I molecule distant from the peptide-binding groove. (A data from Y. Yin et al. 2012. *Proc Natl Acad Sci USA* 109: 5405–5410, redrawn using Chimera and PDB file 3T0E; B data from K. Natarajan et al. 2017. *Nat Commun* 8: 15260, CC BY 4.0. doi.org/10.1038; redrawn using Chimera and PDB file 5IVX; and R. Wang et al. 2009. *J Immunol* 183: 2554–2564, redrawn using Chimera and overlaying PDB files 5IVX and 3DMM.)

The T-cell receptor signaling complex

Upon antigen presentation to the T cell with the correct receptor, the next step in the adaptive immune response requires rapid growth and differentiation (clonal expansion). A signal is relayed from the cell surface to the nucleus through a series of events that activate specific genes.

Typically, transmission of an extracellular signal to the nucleus is initiated by cytoplasmic domains of the cell-surface receptors recognizing the extracellular binding partner. However, the two polypeptide chains of the T-cell receptor lack a significant cytoplasmic domain and cannot initiate intracellular signaling events on their own. The receptor must interact with other cell-surface molecules to initiate these signaling events.

FIGURE 6.5 Proteins of the T-cell receptor signaling complex While the T-cell receptor is critical in recognizing antigen presented via MHC molecules, other plasma membrane proteins are required to stimulate an intracellular signal required for T-cell activation. These proteins include the T-cell coreceptor (CD4 or CD8), proteins of the CD3 complex (δ, ε, γ, ζ), the costimulatory modulator CD28 or CTLA4, and the receptor phosphatase CD45.

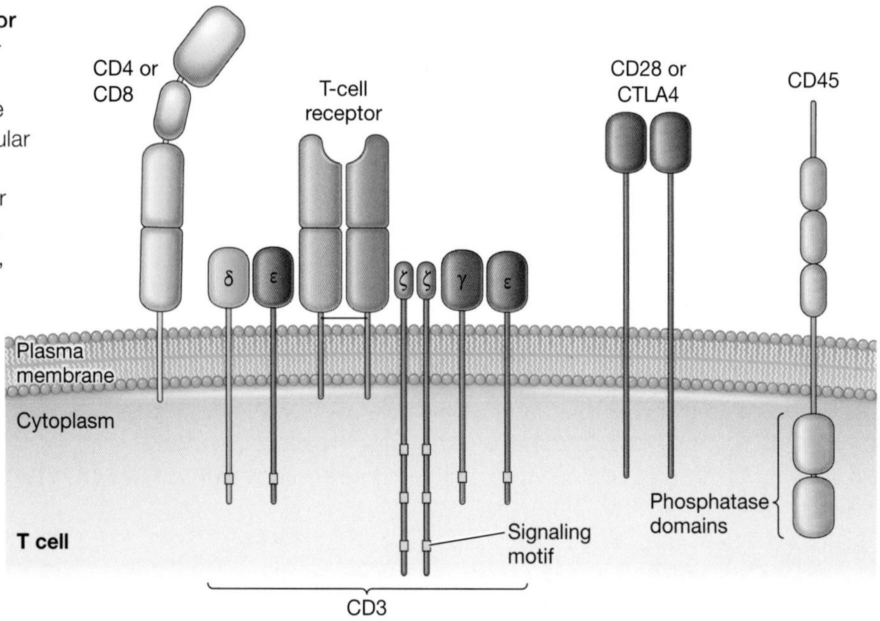

T cells use the CD4 or CD8 coreceptor to initiate signaling, along with the CD3 complex, which is composed of the δ, ε, and γ chains and two ζ chains (**FIGURE 6.5**). These polypeptides work to recruit signaling molecules that are activated upon T-cell receptor engagement (**TABLE 6.1**).

CD28 is also involved in this signaling complex, at least in naïve T cells. CD28 works to provide a costimulatory signal to the T cell. Conversely, CTLA4, which binds to the same molecule as CD28, works to downregulate T-cell activation and prevent unchecked T-cell activation and effector functions. CD45 also functions in signaling through the T-cell receptor complex. (The action of these additional signaling molecules will be explored in detail in Chapter 7.)

CHECKPOINT QUESTION

1. Why is the CD3 complex required in a functional T-cell receptor complex?

TABLE 6.1 | T-Cell Receptor Signaling Complex

T-cell receptor complex subunit	Function
αβ or δγ	Recognize antigen presented to T cell
CD4	Recognize MHC class II molecule; transduction of antigen-binding signal
CD8	Recognize MHC class I molecule; transduction of antigen-binding signal
CD3δ	Transduction of antigen-binding signal
CD3ε	Transduction of antigen-binding signal
CD3γ	Transduction of antigen-binding signal
CD3ζ	Transduction of antigen-binding signal
CD28	Required for co-stimulation of T-cell activation during primary immune response
CD45	Transduction of antigen-binding signal and preferentially activating memory T cells
CTLA4	Downregulate T-cell activation by dampening costimulatory signal

Sources: S. Hedrick et al. 1984. *Nature* 308: 149–153; M. S. Kuhns and and H. B. Badgandi. 2012. *Immunol Rev* 250: 120–143; W. E. Paul. 2012 *Fundamental Immunology*, 7th ed, Lippincott Williams & Wilkins, Philadelphia.; J. E. Smith-Garvin et al. 2009. *Ann Rev Immunol* 27: 591–619.

6.3 How does somatic recombination account for T-cell receptor diversity?

LEARNING OBJECTIVE

6.3.1 Summarize the role of recombination in the production of a functional T-cell receptor.

In Chapter 4, we explored the elegant mechanism by which immunoglobulins and T-cell receptors recombine to promote T-cell and B-cell diversity. The inclusion of a large number of V, D, and J regions at the loci for T-cell receptor subunits, the random combination of receptor subunits, and the junctional diversity driven by this recombination event all enhance the diversity of T-cell receptors.

Exonucleases can trim the P nucleotides left in the overhang region of the opened end of the gene segment. Terminal deoxynucleotidyl transferase (TdT) can add N nucleotides to the ends of each gene segment without the requirement of a template strand. Although the removal of P nucleotides and addition of N nucleotides is random—because protein translation occurs via encoding a single amino acid within a protein sequence using codons read in non-overlapping sequence—a proper reading frame must be maintained to produce a protein product that can function as a T-cell receptor subunit. However, the reading frame of the recombined locus can be altered through recombination events and the addition or removal of N and P nucleotides. These events result in **productive rearrangements**, which result in the production of a functional receptor subunit, or **unproductive rearrangements**, which alter the production of the subunit and result in a nonfunctional T-cell receptor (**FIGURE 6.6**).

During T-cell development, checkpoints test whether a productive rearrangement has occurred at the α and β loci. If an unproductive rearrangement prevents the production of a functional α or β subunit, checkpoint mechanisms promote further recombination at those loci to determine whether a productive rearrangement can occur. Because a T cell is nonfunctional without a receptor, exhaustive unproductive rearrangements that eliminate the α or β loci within the lymphocyte genes will cause that T cell to activate apoptosis.

Another complication that can arise during rearrangement events is the production of a functional but self-reactive T-cell receptor. Should T cells with a self-recognizing receptor escape the thymus, they may lead to the development of autoimmune disorders (see Chapter 16) and/or the destruction of healthy tissue. To prevent these potentially dangerous T cells from leaving the thymus, receptors are tested for the ability to recognize self-antigens through the positive and *negative selection* processes described in Chapter 5.

productive rearrangement Genetic recombination at the immunoglobulin and T-cell receptor loci that creates a gene that can produce a functional subunit.

unproductive rearrangement Genetic recombination at the immunoglobulin and T-cell receptor loci that creates a gene that cannot produce a functional subunit.

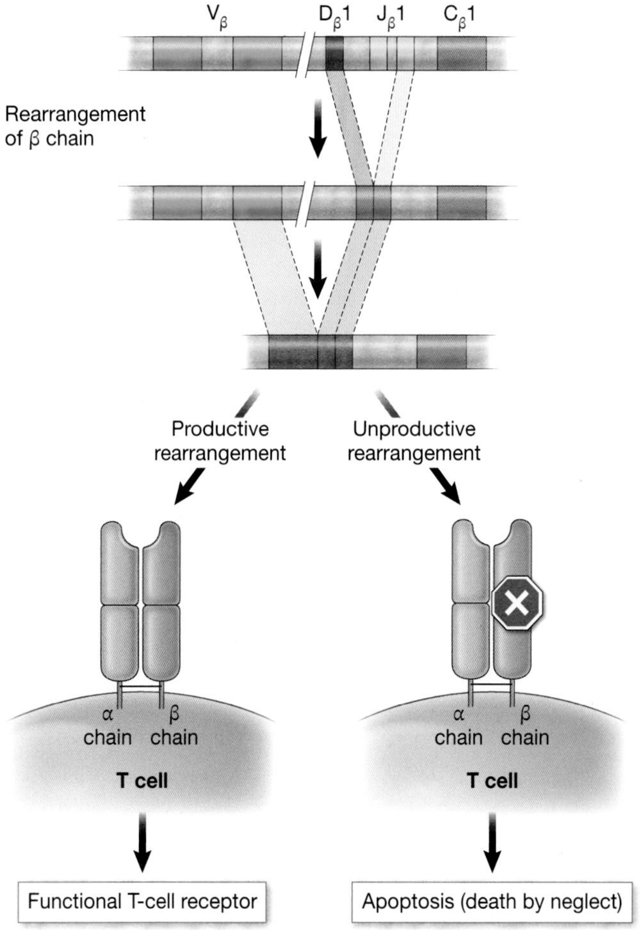

FIGURE 6.6 Productive versus unproductive rearrangements of lymphocyte receptors Because rearrangement and the addition and removal of P and N nucleotides is random, somatic recombination can result in either a productive or an unproductive rearrangement. For example, if a productive rearrangement of the β chain of the T-cell receptor occurs, the T cell can make a functional β-chain subunit. If all β-chain loci undergo unproductive rearrangements, no β chain can be made, and the developing T cell dies by neglect.

CHECKPOINT QUESTION

1. Contrast productive and unproductive rearrangements of the T-cell receptor loci.

6.4 | What is the structure of MHC molecules?

LEARNING OBJECTIVE

6.4.1 Describe the composition and structure of MHC class I and class II molecules.

promiscuous binding specificity Ability of a single MHC molecule to bind to a variety of peptides in its peptide-binding groove.

For a T cell to undergo clonal expansion and differentiation in the context of an infection, the cell must recognize its cognate antigen and undergo clonal selection. However, for clonal selection to occur, antigen for a T-cell receptor must be displayed to the T cell. This antigen display occurs through antigen processing and presentation on the surface of the antigen-presenting cell (APC) via proteins of the MHC family.

Peptides generated by the destruction of intracellular cytosolic proteins are presented via MHC class I molecules, and peptides derived through phagocytosis and destruction in the phagolysosome are typically presented via MHC class II molecules. These two classes of molecules have similarities in their three-dimensional structure but differ in their composition and the types of peptides they can bind. Both MHC class I and class II molecules possess **promiscuous binding specificity**, which allows a single MHC molecule to bind to a variety of peptides.

MHC class I

MHC class I molecules are made up of two subunits, the α chain, which anchors the MHC molecule to the plasma membrane (contains a transmembrane segment), and the soluble small protein β_2-microglobulin (**FIGURE 6.7**).

The α subunit has three domains—α_1, α_2, and α_3. The α_1 and α_2 domains provide the structural features of the peptide-binding groove, and α_3 (along with β_2-microglobulin) provides structural support using immunoglobulin-like domains (similar to the folds seen in immunoglobulin and T-cell receptor domains).

The peptide-binding groove of MHC class I molecules promotes the binding of small peptides 8 to 10 amino acids in length. Two pockets within the groove bind to the ends (N- and C-terminus) of the peptide. The molecule's main interactions are at the two ends of the bound peptide and the peptide backbone. The side chains of the amino acids of the bound peptide play little role in binding to the peptide-binding groove. The only interaction between an MHC class I molecule and the side chain occurs at the C-terminus of the bound peptide, which usually has a hydrophobic or basic (positively charged) amino acid side chain. The MHC class I molecule accommodates these side chains through hydrophobic or negatively charged amino acids located close to the site of the side chain of the C-terminal amino acid bound to the pocket (see Figure 6.7).

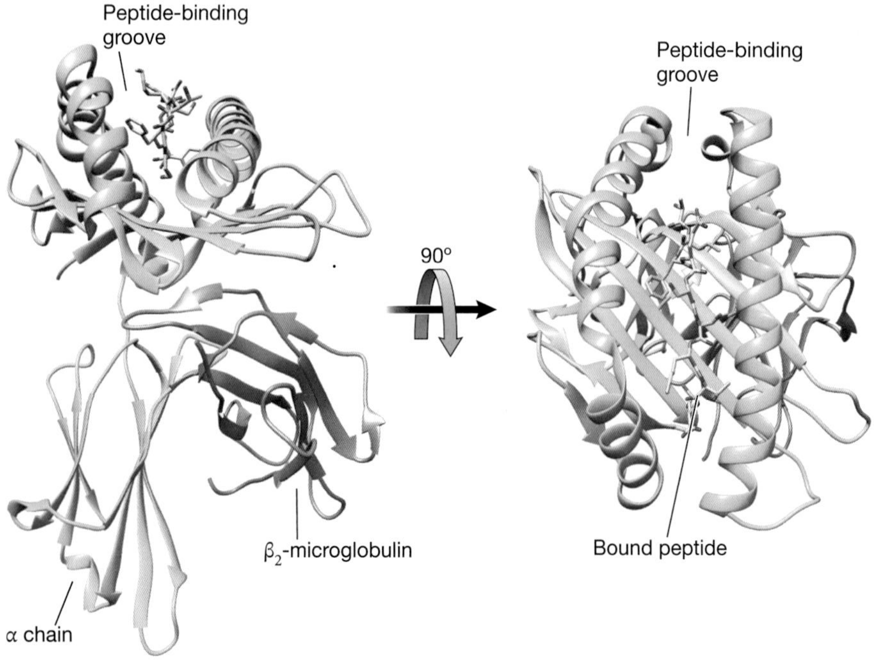

FIGURE 6.7 Three-dimensional structure of an MHC class I protein bound to peptide MHC class I molecules consist of a membrane-bound α chain and the soluble protein β_2-microglobulin. The peptide-binding groove of an MHC class I molecule is generated by folding of the α chain, as seen in the image of the class I molecule bound to an HIV Pol peptide. (Data from L. Niu et al. 2013. *Mol Immunol* 55: 381–392; redrawn using Chimera and PDB file 4HWZ.)

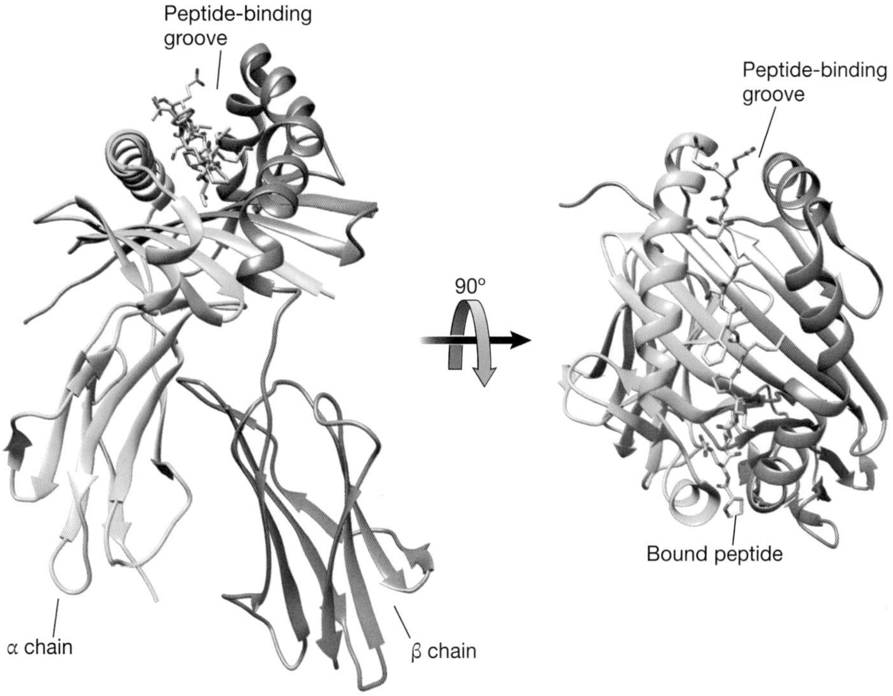

FIGURE 6.8 Three-dimensional structure of an MHC class II protein bound to peptide MHC class II molecules consist of two membrane-bound subunits: an α chain and a β chain. The peptide-binding groove of an MHC class II molecule is generated by the interaction of the folded α and β chains, as seen in the image of the class II molecule bound to an HIV GAG peptide. (Data from Z. Zavala-Ruiz et al. 2004. *Proc Natl Acad Sci USA* 101: 13279–13284. © 2004. National Academy of Sciences, U.S.A., redrawn using Chimera and PDB file 1SJH.)

MHC class II

MHC class II molecules are also made up of two subunits, α and β (**FIGURE 6.8**). Both subunits are transmembrane polypeptides (differing from the one-transmembrane polypeptide subunit of MHC class I molecules). Another major difference between MHC class I and class II molecules is that the peptide-binding groove of MHC class II molecules is formed through the interface of the α and β chains (while the α subunit of class I molecules is the only contributor of the peptide-binding groove).

Each of the two subunits of an MHC class II molecule folds to form a peptide-binding groove similar to that seen in class I molecules. But the class II peptide-binding groove can bind to much longer peptides. Often, peptides with lengths of 13 to 25 amino acids extend from each end of the peptide-binding groove of MHC class II molecules. Other portions of each MHC class II subunit fold to contain immunoglobulin-like domains that support the structure of the peptide-binding groove.

● CHECKPOINT QUESTION

1. Compare and contrast MHC class I and class II molecules in terms of their structure and the peptides they can bind.

6.5 How do T-cell receptors interact with MHC-peptide complexes?

LEARNING OBJECTIVE

6.5.1 Describe the interactions that occur when T-cell receptors and coreceptors bind to MHC molecules.

Once processing and presentation of intracellular or extracellular peptides has occurred and MHC molecules are present at the plasma membrane of the APC, the APC waits for a T cell with the proper receptor to bind. This receptor–peptide interaction requires the correct peptide to be presented and also requires recognition of the MHC molecule presenting the peptide (through interactions with the T-cell receptor and coreceptor).

T-cell receptor:MHC-peptide complex

The MHC molecules' peptide-binding groove allows the peptide to protrude as it sits between the two α-helices forming the groove (see Figures 6.7 and 6.8). This protrusion promotes the formation of a plane of interaction of the peptide and α-helices that can bind to the T-cell receptor. Thus, the interaction between the receptor and MHC-peptide complex involves not only the bound peptide but also a larger surface area composed of the MHC molecule and the bound peptide (see Figure 6.4). The variable regions of the two T-cell receptor subunits are involved, with the most highly variable loop of each subunit recognizing the bound peptide, and the other two loops interacting with the MHC peptide-binding groove α-helices.

Coreceptors and MHC molecules

The T-cell receptor:MHC-peptide complex is essential to the T cell's recognition of the correct antigen, and activation of the proper effector mechanisms is equally important. Extracellular pathogens are cleared by the activation of phagocytic cells such as macrophages and neutrophils, and by the activity of B cells and antibody production. Thus, T-cell effector mechanisms require a subset of T cells that prompt activation of macrophages, neutrophils, and B cells.

CD4 plays an important role in MHC class II recognition. The CD4 coreceptor on naïve T cells that is responsible for recognizing MHC class II-peptide complexes promotes the targeting of extracellular pathogens (see Figure 6.3). The CD4 molecule makes contact with both immunoglobulin-like domains of each subunit of an MHC class II molecule.

CD8 works similarly in recognizing MHC class I molecules to prompt the targeting of intracellular pathogens. The CD8 molecule makes contact with the immunoglobulin-like domain of the α-subunit of MHC class I molecules and has also been shown to interact with β_2-microglobulin in the MHC class I complex (see Figure 6.3).

T-cell receptor signaling molecules

While the T-cell receptor/MHC-peptide/coreceptor complex is necessary for clonal selection of the T cell that will work to combat a specific pathogen, it is not sufficient for T-cell activation and clonal expansion. Recall that other signaling molecules (including the CD3 complex, CD28, and CD45) are required to link the receptor–peptide interaction to signaling events within the bound T cell to promote activation of that cell. These signaling molecules ultimately activate gene transcription within the bound T cell to drive production of cytokines required for proper T-cell activation and differentiation. (The signaling cascade and the cytokines produced will be explored in detail in Chapter 7.)

CHECKPOINT QUESTION

1. Which proteins are required for a complete interaction between a T cell and an antigen-presenting cell?

6.6 How does peptide loading occur in MHC class I and class II molecules?

LEARNING OBJECTIVE

6.6.1 Compare and contrast the mechanisms of antigen processing and presentation via MHC class I and class II molecules.

T cells containing an αβ T-cell receptor must interact with an MHC-peptide complex to promote clonal expansion of the selected T cell. Thus, peptide loading onto MHC molecules is vital to the successful activation of T cells. Since the two major classes of MHC molecules present peptides from proteins of different origins (intracellular or extracellular), peptide loading must allow for presentation of peptides generated from the appropriate locations (**FIGURE 6.9**).

MHC class I molecules are typically loaded with peptides from intracellular proteins in the endoplasmic reticulum (ER). Upon proper loading, the MHC class I molecules continue through the secretory pathway to the plasma membrane, where they present those peptides to CD8 T cells.

MHC class II molecules also begin their transition to the plasma membrane in the ER. However, they cannot bind to peptides in the ER and instead travel to a vesicle, where they wait to interact with a phagolysosome that has generated peptides from extracellular proteins. Upon loading, MHC class II molecules travel to the plasma membrane, where they present peptides to CD4 T cells.

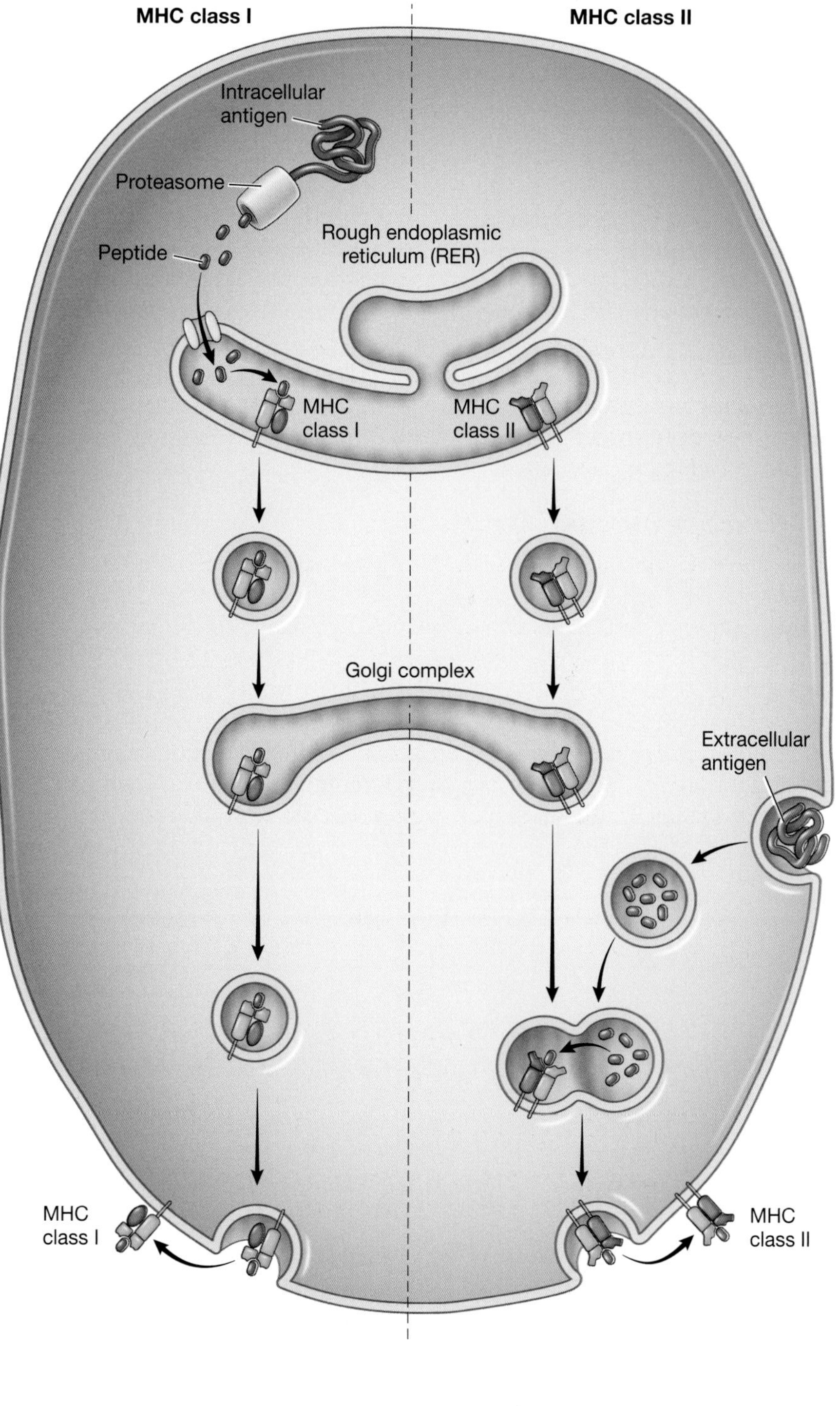

FIGURE 6.9 Overview of intracellular and extracellular antigen processing and presentation Intracellular proteins present in the cytoplasm of all nucleated cells are eventually digested by the proteasome to create small peptides. These peptides are transported into the rough endoplasmic reticulum (RER), where they bind to MHC class I molecules. Once loaded with peptide, MHC class I molecules travel through the secretory pathway to the plasma membrane, where they present peptides to CD8 T cells. MHC class II molecules are synthesized co-translationally at the RER but remain inactive and unable to bind to class I peptides. Once synthesized, MHC class II molecules travel through the secretory pathway via vesicles and wait for fusion with a phagolysosome. Extracellular proteins are phagocytosed by professional antigen-presenting cells (APCs). After phagocytosis, the antigens are digested within a phagolysosome. The phagolysosome fuses with an MHC class II-containing vesicle to load the MHC molecule with peptides generated from digestion of extracellular antigen. Loaded MHC class II molecules travel to the cell surface, where they can present antigen to CD4 T cells. (After J. Punt et al. 2018. *Kuby Immunology*, 8th ed., W. H. Freeman and Company: New York.)

MHC class I peptide loading

Because MHC class I molecules can present peptides generated due to an intracellular infection, it is imperative that nearly all cells of the body express MHC class I molecules (a few exceptions exist, such as erythrocytes, which are short-lived and lack a nucleus). Peptide loading of class I molecules generated from intracellular proteins occurs in the ER, and the molecules remain there until properly loaded. MHC class I peptide loading occurs in five steps (**FIGURE 6.10**):

calnexin Membrane-bound ER chaperone that aids in folding of proteins in the secretory pathway.

STEP 1: FORMATION OF THE PEPTIDE-LOADING COMPLEX Both the α chain and β_2-microglobulin contain signal sequences in their protein products driving their co-translational import via the ribosome into the ER. Upon entry, the α chain binds to the ER chaperone **calnexin**, which keeps MHC class I molecules

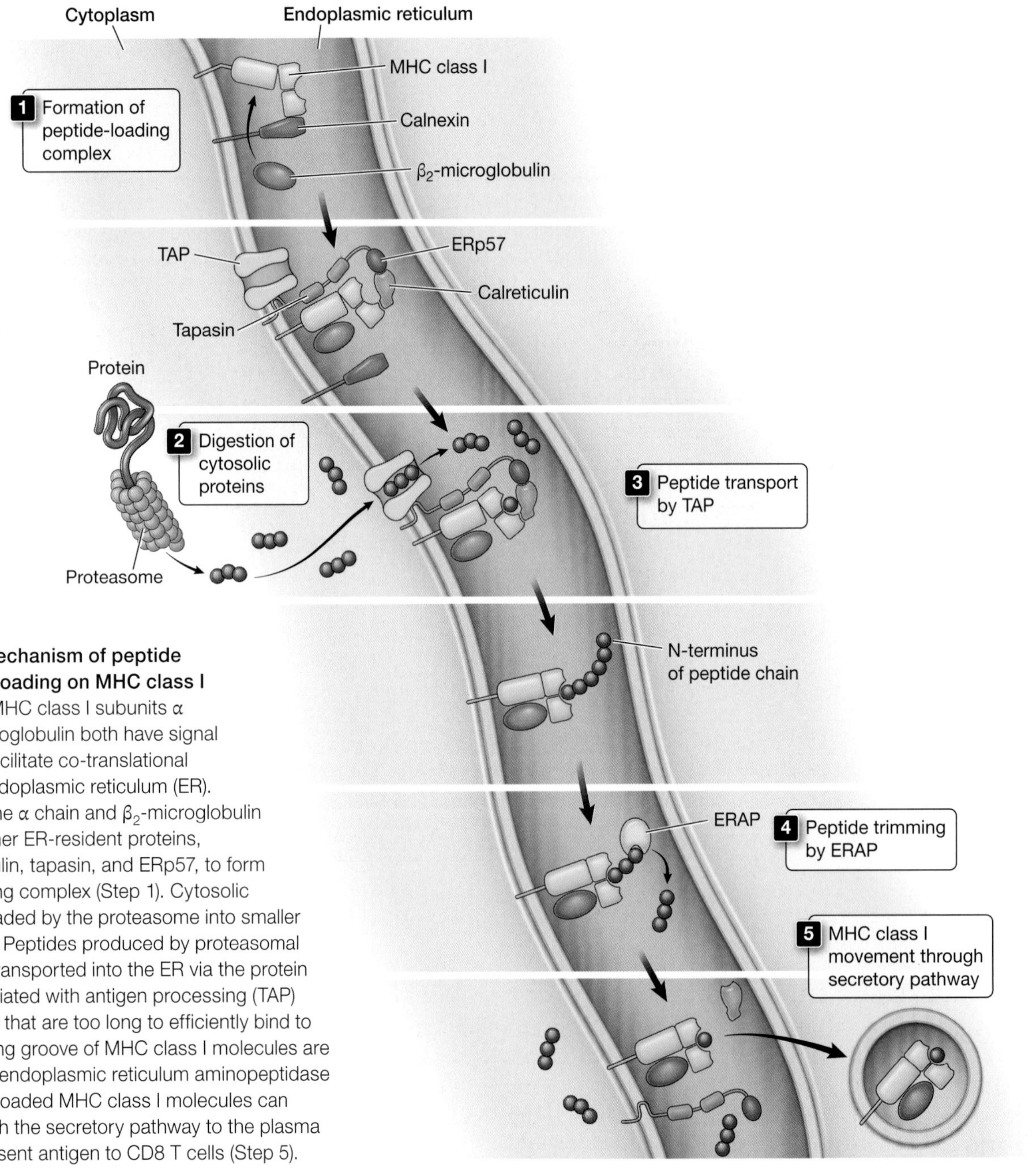

FIGURE 6.10 Mechanism of peptide processing and loading on MHC class I molecules The MHC class I subunits α chain and β_2-microglobulin both have signal sequences that facilitate co-translational import into the endoplasmic reticulum (ER). Once in the ER, the α chain and β_2-microglobulin associate with other ER-resident proteins, including calreticulin, tapasin, and ERp57, to form the peptide-loading complex (Step 1). Cytosolic proteins are degraded by the proteasome into smaller peptides (Step 2). Peptides produced by proteasomal degradation are transported into the ER via the protein transporter associated with antigen processing (TAP) (Step 3). Peptides that are too long to efficiently bind to the peptide-binding groove of MHC class I molecules are shortened by the endoplasmic reticulum aminopeptidase (ERAP) (Step 4). Loaded MHC class I molecules can then travel through the secretory pathway to the plasma membrane to present antigen to CD8 T cells (Step 5).

in the ER until they have become properly folded. Once in the ER and properly folded, the α chain and β_2-microglobulin associate with the **peptide-loading complex**. The protein **tapasin** is a major player in the peptide-loading complex; it works to allow peptides to be loaded more efficiently on MHC class I molecules. It is also important in promoting the association of a peptide that can bind tightly with MHC class I molecules (see EMERGING SCIENCE). Another important player is the ER chaperone **calreticulin**, which further promotes proper assembly of MHC class I molecules with the tightly bound peptide.

STEP 2: DIGESTION OF PROTEINS BY THE PROTEASOME Cells constantly degrade intracellular proteins due to the normal half-lives of proteins, the need to dispose of damaged proteins, and targeted regulation of pathways in the cell. To degrade intracellular proteins, cells use a proteolytic enzyme complex known as the **proteasome**. The proteasome cleaves intracellular proteins into small peptides of varying lengths. Interestingly, cytokines secreted during an inflammatory response induced by the innate immune system (including IFN-γ) can drive the expression of subunits of the proteasome, including the 11S subunit and different β subunits with altered specificity, to produce the **immunoproteasome**. The subunits of the immunoproteasome further promote MHC class I presentation at the surface of cells. The 11S subunit aids in processing MHC class I molecules, and the β subunits activated by IFN-γ change the specificity to enhance generation of peptides of 8 to 10 amino acids in length (the size needed to properly bind to the MHC class I peptide-binding groove).

STEP 3: PEPTIDE TRANSPORT INTO THE ER Peptides generated by the proteasome or immunoproteasome are released into the cytosol but must be transported into the ER to be in the same location as expressed MHC class I molecules. A peptide transporter in the ER membrane known as **transporter associated with antigen processing** (**TAP**) carries proteasome-derived peptides into the ER where they can interact with MHC class I molecules.

STEP 4: PEPTIDE TRIMMING Tapasin functions in the peptide-loading complex to ensure that a tightly bound peptide is engaged in the peptide-binding groove of an MHC class I molecule. Some peptides produced by the proteasome are too long to fit in the groove. In some cases, the C-terminal amino acid of a peptide can bind tightly within the groove but the length of the peptide stretches beyond it. In these cases, another protease that resides in the ER, **endoplasmic reticulum aminopeptidase** (**ERAP**), trims the N-terminal end of the overhanging peptide to the length needed (8 to 10 amino acids) for tight binding.

STEP 5: TRANSPORT TO THE PLASMA MEMBRANE Upon proper loading of a bound peptide, the MHC class I molecule is transported via vesicles through the secretory pathway. It travels through the Golgi apparatus before its cargo vesicle fuses with the plasma membrane. Since the peptide-binding groove is located in the lumen of the ER to promote peptide loading, the bound peptide ultimately resides on the extracellular surface of the plasma membrane. There, the loaded MHC class I molecule waits for its cognate CD8 T cell bearing a T-cell receptor that can recognize the MHC-peptide complex.

MHC class II peptide loading

MHC class II molecules present peptides generated from extracellular proteins on the surface of cells. Because extracellular pathogens are only taken up by phagocytic cells, MHC class II molecules are restricted to cells such as macrophages, dendritic cells, and B cells. Peptide loading of class II molecules occurs later in the secretory pathway than that of MHC class I molecules. MHC class II

peptide-loading complex Protein complex within the ER that assists in binding intracellular peptides to the peptide-binding groove of MHC class I molecules.

tapasin ER protein responsible for peptide loading and peptide exchange on MHC class I molecules.

calreticulin ER chaperone involved in the folding of proteins in the secretory pathway; functions in the peptide-loading complex to promote proper folding of MHC class I molecules loaded with peptide.

proteasome Multi-subunit enzyme complex that cleaves intracellular cytosolic proteins into peptides for presentation on MHC class I molecules or removal of damaged or abnormal proteins.

immunoproteasome Proteasome complexes with 11S subunit that promote digestion of proteins into peptides of 8 to 10 amino acids in length for better presentation on MHC class I molecules.

transporter associated with antigen processing (TAP) Protein complex embedded in the ER membrane; responsible for transporting peptides generated by the proteasome into the ER for MHC class I loading.

endoplasmic reticulum aminopeptidase (ERAP) Protease located in the ER lumen that trims the N-terminus of peptides bound to MHC class I molecules to bind peptides more tightly to the peptide-binding groove.

EMERGING SCIENCE

How does tapasin promote peptide exchange on MHC class I molecules?

Article

S.K. Saini, H. Schuster, V.R. Ramnayaran, H.G. Rammensee, S. Stevanovic, and S. Springer. 2015. Dipeptides catalyze rapid peptide exchange on MHC class I molecules. *Proceedings of the National Academy of Sciences U.S.A.* 112: 202–207.

Background

The presentation of peptides via MHC class I molecules is essential in mounting adaptive immune responses to viruses and cancer cells. MHC class I molecules are loaded with intracellular peptides processed by the proteasome and imported into the ER. Tapasin is a major component of peptide loading in the ER, but how this protein coordinates peptide loading and peptide exchange into the MHC class I peptide-binding groove is largely unknown. Studies demonstrated that movement of MHC class I molecules from the ER to the cell surface requires proper MHC class I molecule folding. This folding was shown, in vitro, to be supported by dipeptides binding to class I molecules. Although these dipeptides could assist in MHC class I folding, whether they could function in an analogous manner to tapasin (by promoting dipeptide exchange in the peptide-binding groove) was not known.

The Study

Researchers used their previous finding that dipeptides could promote proper MHC class I molecule folding to test whether they could promote exchange of peptides in the peptide-binding groove. To do this, they expressed recombinant MHC class I molecules and folded them in the presence of a peptide that could bind to the peptide-binding groove. They then tested to see whether dipeptides of different sequence could promote the exchange of the bound peptide with a fluorescently labeled similar peptide. The researchers found that dipeptides, especially those that ended with a hydrophobic amino acid, could efficiently exchange peptides within the peptide-binding groove of both recombinant and cell-surface preloaded MHC class I molecules (**FIGURE ES 6.1**).

Conclusion

The exchange of peptides on MHC class I molecules in the presence of dipeptides suggests a mechanism of tapasin action. It also provides a potential therapeutic target for loading dendritic cells with an appropriate peptide in vitro to drive antigen-specific lymphocyte activation as a means of vaccination or tumor targeting.

Think About...

1. What model is proposed to explain how the dipeptide can promote peptide exchange? Which portion of the bound peptide is likely to be important in this exchange reaction?
2. In light of what you know about the peptide-binding motifs of class I molecules, why do you think the dipeptides worked best with a hydrophobic amino acid at their C-terminus?
3. Knowing the dipeptide properties needed to promote exchange of peptide in the peptide-binding groove, how can this information be applied to determine how tapasin functions in peptide exchange?

invariant chain Protein in the secretory pathway that binds to MHC class II molecules and blocks the peptide-binding groove from binding to intracellular peptides transported into the ER via TAP; is proteolytically cleaved in the MHC compartment to CLIP.

molecules begin in the secretory pathway but are not loaded until they are part of a vesicle capable of fusing with a phagolysosome. MHC class II peptide loading occurs in four steps (**FIGURE 6.11**):

STEP 1: MHC CLASS II MOLECULE ASSEMBLY IN THE ER MHC class II molecules are composed of the transmembrane subunits α and β and also have signal sequences that drive co-translational import into the ER. Because both class I and class II molecules are present in the ER, and because intracellular peptides generated via the proteasome are also present, a mechanism must be in place to ensure that class II molecules are not loaded with intracellular peptides. To ensure this protection, MHC class II molecules assembled in the ER bind to another protein known as the **invariant chain**, which blocks the peptide-binding groove and prevents binding of proteasomal peptides.

STEP 2: CLIP PRODUCTION MHC class II molecules bound to the invariant chain are transported via vesicles through the secretory pathway. They first travel through the Golgi apparatus to a specialized endosome known as the

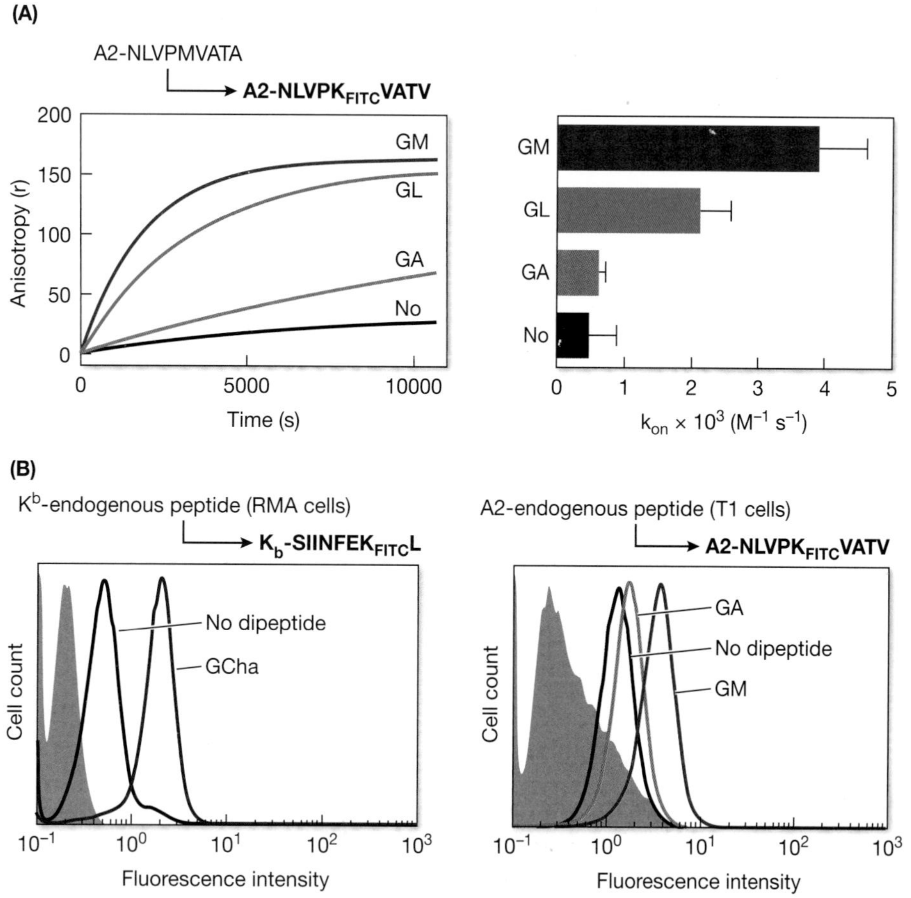

FIGURE ES 6.1 Dipeptides can facilitate peptide exchange on MHC class I molecules (A) MHC class I molecules with a bound peptide were tested for peptide exchange in vitro in the absence or presence of different dipeptides (GA, GL, or GM). Dipeptides facilitated peptide exchange, with better exchange occurring in the presence of dipeptides with larger hydrophobic amino acids at the C-terminus. (B) RMA or T1 cells with endogenous class I molecules were tested for peptide exchange in the absence or presence of different dipeptides (GCHa, GA, GM). The dipeptides facilitated peptide exchange on both cell lines as observed by flow cytometry. (From S. K. Saini et al. 2015. *Proc Natl Acad Sci*, 112: 202–207; doi.org/10.1073/pnas.1418690112.)

MHC compartment. Within this endosome, the protease cathepsin S cleaves the invariant chain but leaves a peptide called the **class II-associated invariant chain peptide** (**CLIP**) bound to the peptide-binding groove. This endosome remains in an intracellular pool until fusion with a phagolysosome.

class-II associated invariant chain peptide (CLIP) Peptide produced upon digestion of the invariant chain that remains bound to MHC class II molecules until exchange with extracellular peptides generated during phagocytosis.

HLA-DM MHC class II molecule that functions in promoting peptide exchange of CLIP with extracellular peptides generated by phagocytosis.

STEP 3: PHAGOCYTOSIS AND FUSION WITH THE MHC COMPARTMENT Phagocytic cells engulf pathogens through receptor-mediated endocytosis and internalize the pathogens in phagosomes. Each phagosome fuses with a lysosome, acidifies, and becomes a phagolysosome. Protease activity within the phagolysosome promotes the generation of peptides from ingested material. The phagolysosome then fuses with the MHC compartment.

STEP 4: PEPTIDE LOADING MHC class II molecules present in the fused vesicle are still bound to CLIP and need to release the peptide to enable exchange with peptides generated during lysosomal digestion. An MHC family member known as **HLA-DM** (human leukocyte antigen DM) promotes the exchange of CLIP for lysosomal peptides on the peptide-binding groove of MHC class

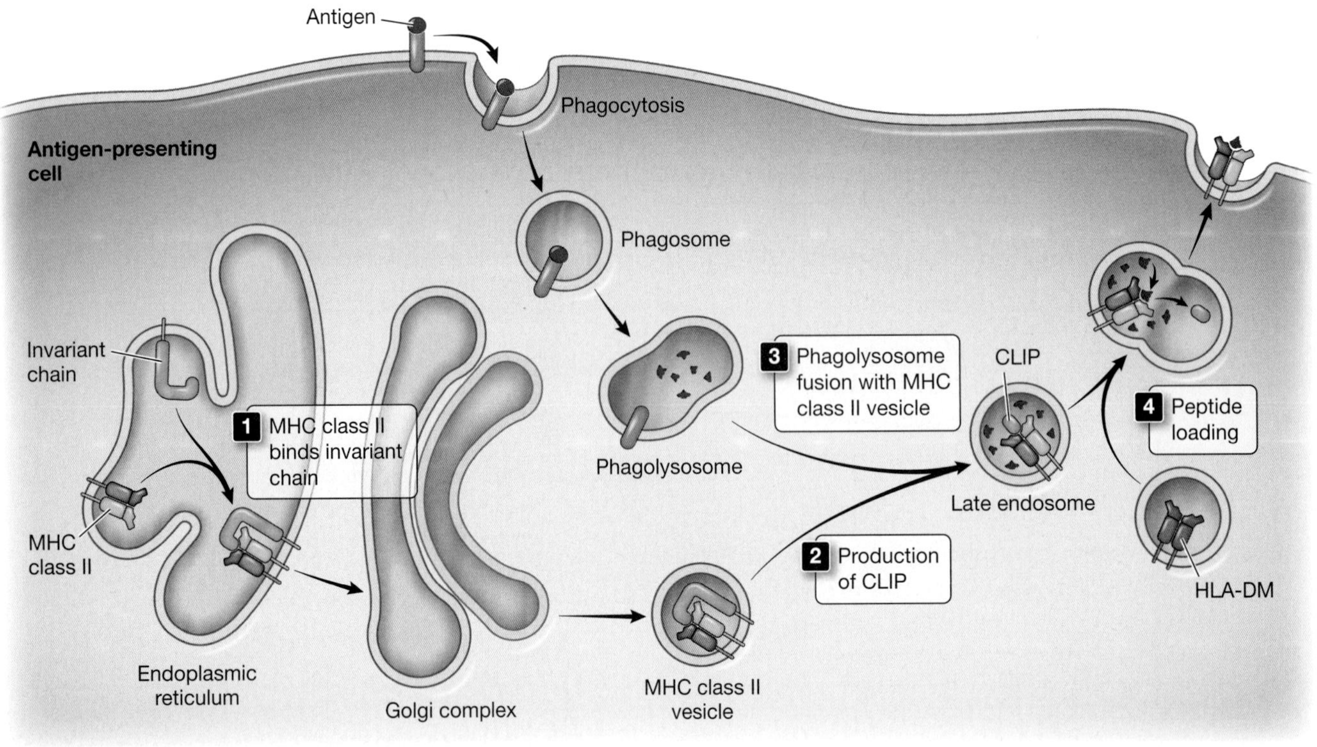

FIGURE 6.11 Mechanism of peptide processing and loading on MHC class II molecules The MHC class II subunits α and β chains both have signal sequences that facilitate co-translational import into the endoplasmic reticulum (ER). To prevent loading of MHC class II molecules with peptides generated for class I loading, the MHC class II molecules associate with the invariant chain in the ER (Step 1). The MHC class II bound to the invariant chain travels through the secretory pathway to a late endosome, where the invariant chain is digested and the peptide CLIP remains associated with MHC class II (Step 2). Phagocytic cells internalize and digest material through phagocytosis and the phagolysosome fuses with the MHC class II vesicle (Step 3). A vesicle containing HLA-DM fuses with the MHC class II vesicle and facilitates loading of peptides generated during phagocytosis in place of CLIP on MHC class II molecules (Step 4). Loaded MHC class II molecules can then travel to the plasma membrane to present antigen to CD4 T cells.

II molecules. Upon peptide exchange, the vesicle and MHC class II molecules travel to the plasma membrane. The loaded MHC class II molecules then wait for their cognate CD4 T cell bearing a T-cell receptor capable of recognizing the MHC-peptide complex.

MHC cross-presentation

We have explored the means by which peptides are processed and presented on MHC class I and class II molecules. These peptides reside on the cell surface awaiting recognition by a T cell bearing the appropriate receptor. One caveat to T-cell activation is the need for a naïve T cell to not only engage the MHC-peptide complex using its receptor and coreceptor but to also use a costimulatory signal for proper activation.

Although phagocytic cells such as dendritic cells bear the appropriate costimulatory receptor and can stimulate naïve CD4 and CD8 T cells at any time, most cells do not express a costimulatory molecule unless an inflammatory response has been activated. Cytokines produced during an innate immune response drive expression of these costimulatory molecules on a potential APC. However, all cells in the body are a possible haven for an intracellular infection and may not express the necessary costimulatory molecule on their surface. Furthermore, the adaptive immune response is centralized within the lymphatic system since it is virtually impossible for the diverse array of T cells to migrate to every area of the body to monitor for the specific antigen their receptor can recognize. How, then, can naïve CD8 T cells become activated unless a professional APC has become infected by an intracellular pathogen? And if an intracellular pathogen

cross-presentation Mechanism of peptide presentation whereby phagocytic cells transport endocytosed proteins into the cytosol for presentation via the MHC class I pathway.

cytosolic diversion Process by which endocytosed proteins are transported into the cytosol for cross-presentation.

is taken up through phagocytosis, how does the phagocytic cell ensure that peptides from the intracellular pathogen are presented by MHC class I molecules, since phagocytosed material is typically presented via MHC class II molecules?

Because naïve T cells use secondary lymphoid tissue and professional APCs to centralize their activation, phagocytic cells (especially dendritic cells, but also macrophages and B cells) that have taken up intracellular pathogens through phagocytosis can use **cross-presentation** to present phagocytosed material via MHC class I molecules. Cross-presentation activates the CD8 T cells responsible for combating the intracellular infection (**FIGURE 6.12**).

The mechanism of cross-presentation is unclear. It has been suggested that endocytosed material in a cross-presenting cell is transported to a specialized endosome capable of **cytosolic diversion** (transport of endocytosed material into the cytosol). Diverted material within the cytosol can then use the normal MHC class I antigen processing and presentation machinery to load endocytosed proteins onto class I molecules.

Cross-presentation is believed to play an important role in activating naïve CD8 T cells required to mount an adaptive immune response to an intracellular pathogen such as a virus. Cross-presentation can also aid in peripheral self-tolerance, ensuring that self-reactive CD8 T cells do not mount an autoimmune response. Dendritic cells that cross-present can present both foreign and self-antigens taken in through phagocytosis. Thus, CD8 T cells may be self-reactive to extracellular antigen presented via the cross-presentation pathway, as they may not have been exposed to this antigen during their development. To promote self-tolerance to these self-reactive CD8 T cells, it has been postulated that the dendritic cell must be *licensed* to properly cross-present antigen.

Licensing of cross-presentation in dendritic cells is postulated to require CD4 T cells using the following mechanism: Dendritic cells present phagocytosed extracellular antigens via the canonical MHC class II pathway, where they may activate a naïve CD4 T cell. Presumably, self-reactive CD4 T cells should have been negatively selected against during T-cell development, and thus only CD4 T cells that are not self-reactive should be activated via this process. Activation of the engaged CD4 T cells would promote cytokine release, signaling the dendritic cell to begin cross-presentation, potentially activating naïve CD8 T cells (see Figure 6.12). This mechanism would promote tolerance to self-reactive CD8 T cells since autoreactivity would require the presence and activation of two self-reactive T cells: a self-reactive CD4 T cell and a self-reactive CD8 T cell.

CHECKPOINT QUESTIONS

1. Explain the processes of antigen presentation via MHC class I and class II pathways.
2. What is the role of cross-presentation in dendritic cells?

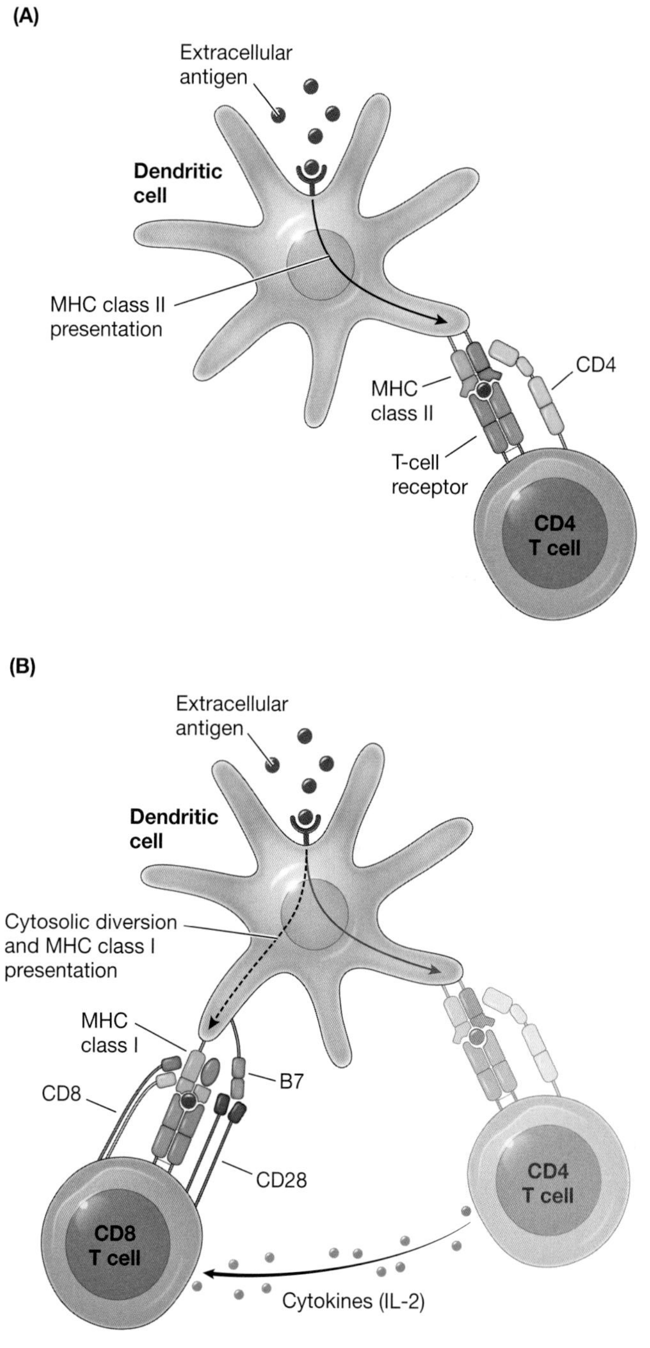

FIGURE 6.12 Cross-presentation by dendritic cells (A) In order for a dendritic cell to activate CD8 T cells, they must be licensed to cross-present antigen by a CD4 T cell. This event occurs by processing of extracellular antigen processed via the MHC class II processing and presentation pathway. (B) Licensing of a dendritic cell by a CD4 T cell promotes the dendritic cell to undergo cross-presentation, enabling extracellular antigen to be diverted from the phagocytic process through cytosolic diversion, where it can be processed via the MHC class I pathway and presented to CD8 T cells.

Making Connections: Action of Dendritic Cells in Draining Lymphoid Tissue

Recognition of pathogens via PAMPs induces an inflammatory response, which activates dendritic cells. This allows for antigen delivery by dendritic cells processing the antigen via draining into underlying lymphoid tissue. Dendritic cells function as professional antigen-presenting cells by allowing for antigen processing via both HLA class I and class II molecules and processing extracellular material through cross-presentation to efficiently activate CD8 T cells.

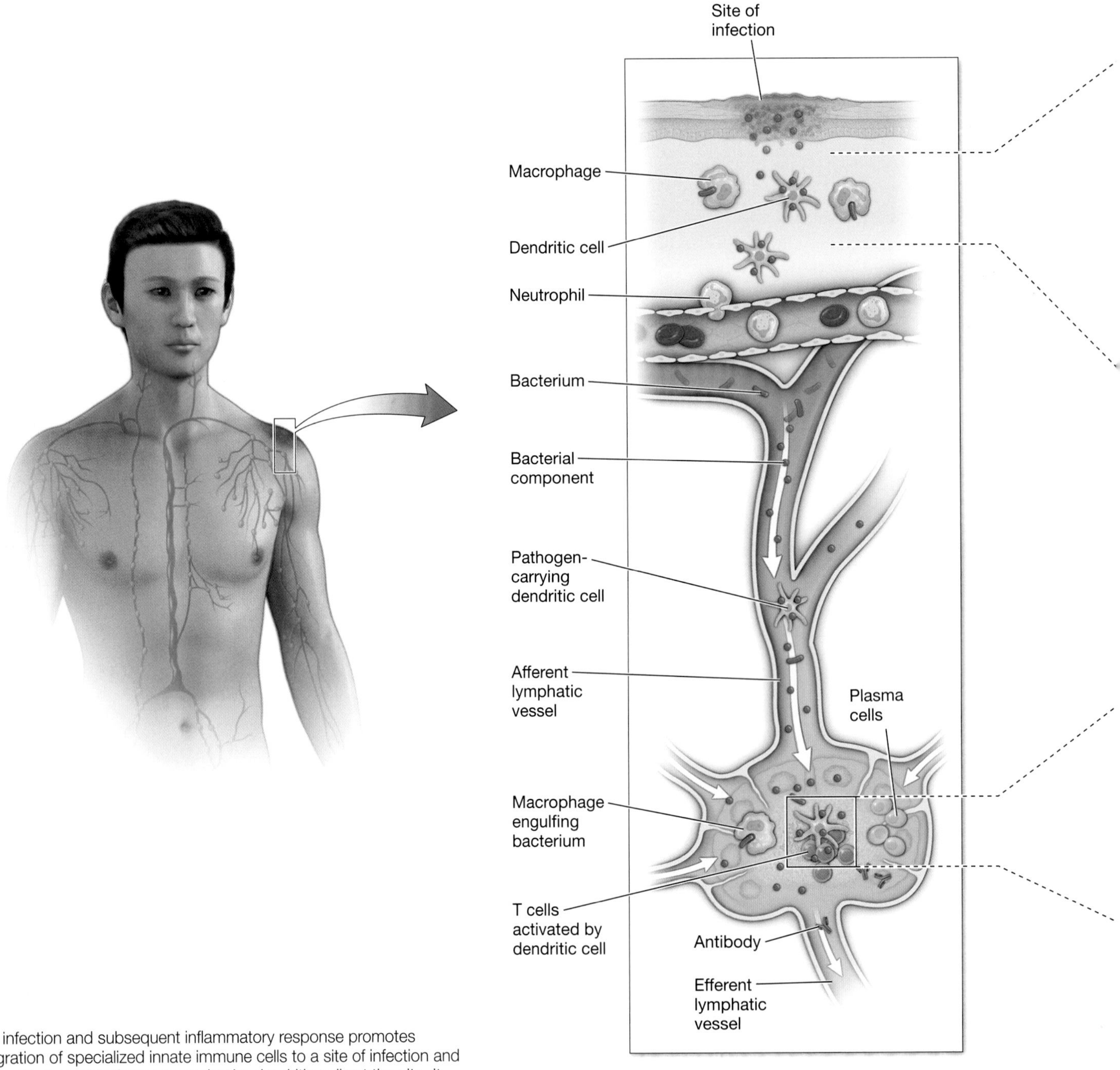

An infection and subsequent inflammatory response promotes migration of specialized innate immune cells to a site of infection and the activation of antigen processing by dendritic cells at the site. It also prompts dendritic cell migration to a draining lymphoid tissue.

(Sections 1.10, 2.1, 2.8. 4.2, 6.4, 6.5)

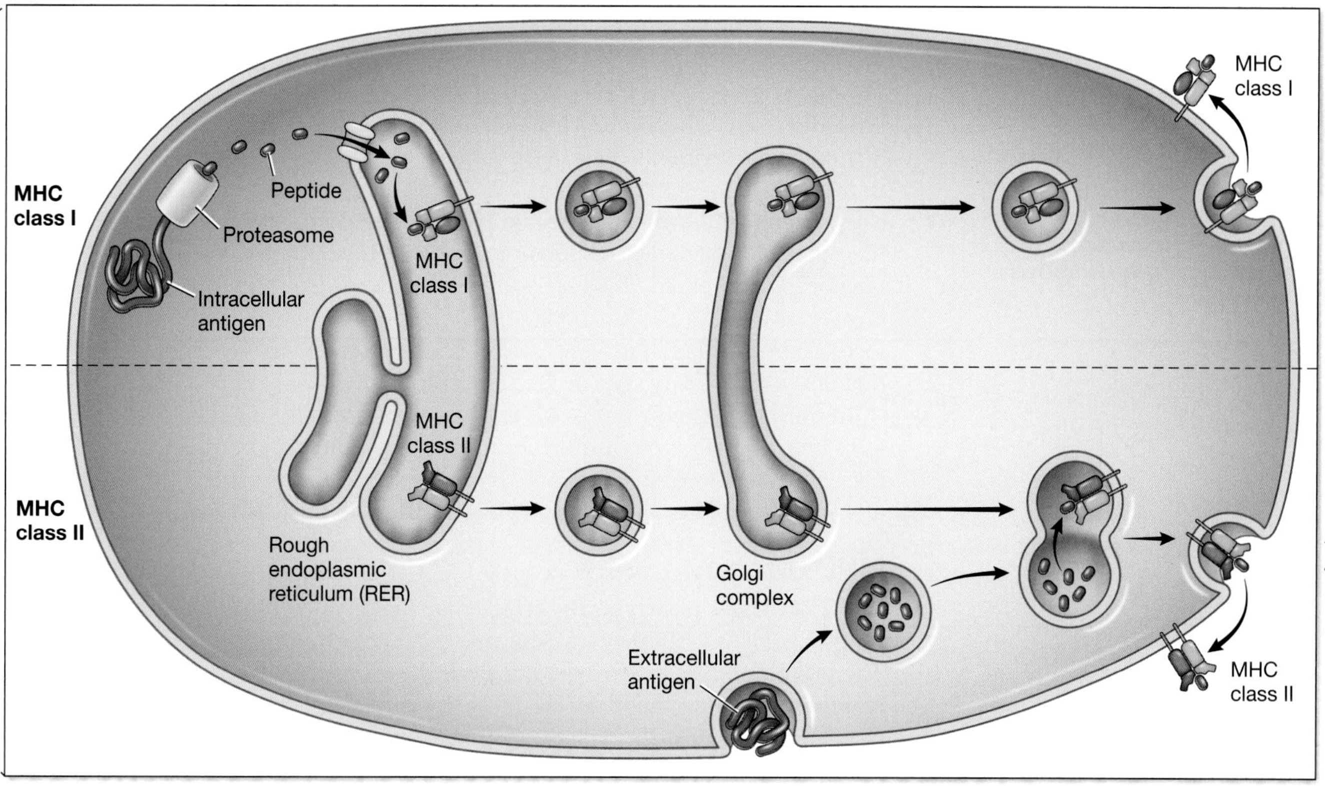

After J. Punt et al. 2018. *Kuby Immunology*, 8th ed., W. H. Freeman and Company: New York

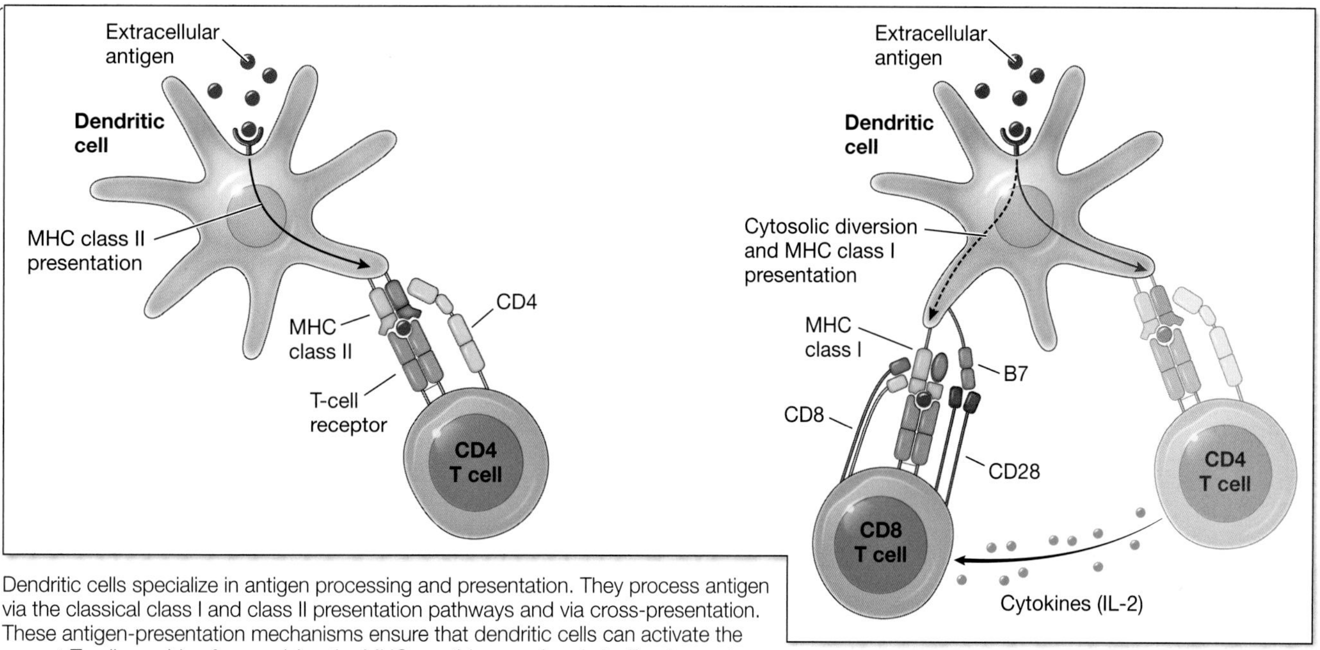

Dendritic cells specialize in antigen processing and presentation. They process antigen via the classical class I and class II presentation pathways and via cross-presentation. These antigen-presentation mechanisms ensure that dendritic cells can activate the correct T cell capable of recognizing the MHC-peptide complex via its T-cell receptor.

(Sections 2.1, 2.8, 4.2, 4.5, 4.6, 6.4, 6.5, 6.6, 7.3, 7.6, 7.7)

6.7 | How does MHC diversity affect peptide presentation to T cells?

LEARNING OBJECTIVE

6.7.1 Describe the diversity of MHC molecules and explain how polymorphisms of MHC molecules present an extensive array of antigens to T cells.

isotype Various protein products that have functional and structural similarities within an organism.

allotype Protein product of a particular allele of a gene.

haplotype Expressed isotypes of all alleles of a gene family within an individual.

The discovery of MHC molecules dates to the early twentieth century, when C. C. Little discovered that tumors transplanted between different strains of mice failed to grow in the recipient mice. During World War II it was discovered that self-skin transplants were better tolerated than those involving donor skin from a genetically different individual. In the late 1940s, George Snell and Peter Gorer identified the genetic locus in mice responsible for tissue transplant acceptance or rejection. The locus was named H-2, for histocompatibility 2 (see **KEY DISCOVERIES**) and is analogous to the *human leukocyte antigen* (HLA) locus. Further studies of the MHC locus demonstrated that MHC molecules represent a vastly diverse array, which parallels the diversity seen in T-cell receptors.

In the following sections we explore the diversity of MHC molecules and why this diversity is so important for peptide binding and presentation. It has been shown that MHC molecule diversity seems to mirror that of T-cell receptors in evolutionary history (see **EVOLUTION AND IMMUNITY**).

MHC diversity

MHC diversity is driven by genes that make up different but functionally related MHC molecules and the presence of different alleles for each gene. The proteins expressed by the gene families for each MHC class are known as **isotypes**, meaning that they are made by different genes, but each protein has a structural and functional connection.

Also contributing to MHC diversity within each individual is the presence of two copies of human chromosomes within somatic cells. Thus, any gene that encodes for an isotype subunit is physically present in two copies in each cell. Because these copies may differ due to mutations that have occurred over the course of human evolutionary history (polymorphism), each individual may have two different alleles for any given MHC subunit. Both alleles are active gene units, and if the alleles differ on the two chromosomes, both isotypes are expressed in cells and have an equal impact on the phenotype of the cell as it relates to the presence of MHC molecules. Thus, MHC alleles do not exhibit dominant or recessive trait characteristics; rather, they exhibit *codominance* in their phenotype (similar to how individuals containing A and B antigens in their blood cells are blood type AB because they make both antigens).

The proteins expressed by these alleles are **allotypes**. All the alleles of every MHC class I and class II gene in an individual make up that person's **haplotype**, which provides the overall MHC diversity (usually a good thing in terms of immune function). But this genetic aspect must be taken into consideration if tissue, organ, or bone marrow transplantation becomes necessary.

TABLE 6.2 | HLA Class I Genes and Proteins of the Human Genome

HLA locus	Number of alleles	Number of allotypes
A	6291	3896
B	7562	4803
C	6223	3681
E	256	110
F	45	6
G	82	22

Source: Data from J. Robinson et al. 2015. *Nuc Acid Res* 43: D423–431; J. Robinson et al. 2000. *Tissue Antigens* 55: 280–287; CC BY-ND 3.0.

CLASS I DIVERSITY In humans, there are six MHC class I isotypes: HLA-A, HLA-B, HLA-C, HLA-E, HLA-F, and HLA-G. These isotypes vary in regard to the number of allotypes in the population (**TABLE 6.2**). All isotypes function within the context of the immune system, but each has different capacities. HLA-A, HLA-B, and HLA-C function as previously described for MHC class I molecules—they present peptides generated from intracellular proteins to CD8 T cells. These class I molecules, along with HLA-E and HLA-G, can also act as ligands for receptors on the surface of NK cells, promoting targeting of damaged or abnormal cells or infections that need to be addressed by destroying

KEY DISCOVERIES

How were MHC loci discovered?

Article

G.D. Snell and G.F. Higgins. 1951. Alleles at the histocompatibility-2 locus in the mouse as determined by tumor transplantation. *Genetics* 36: 306–310.

Background

The importance of MHC molecules and their diversity is not only critical to normal T-cell function, but the diversity must be considered when attempting tissue transplantation. Thus, the discovery of MHC molecules is crucial in understanding the role of antigen-presenting cells in T-cell activation and in regard to the issues associated with organ and tissue transplantation. How were multiple loci of the MHC complex identified?

Early Research

Research in the early twentieth century demonstrated that alleles at the histocompatibility locus are important in susceptibility or resistance to transplanted tumors and determining blood group antigen. Observations suggested that there was a common gene locus associated with reactivity to a particular blood group after transplantation with a specific tumor: C57BL mouse sera will agglutinate blood type A after it has been inoculated with an A-strain tumor. Furthermore, this locus can be shown to be connected genetically as crosses between an A strain and C57BL strain always resulted in susceptibility to A-strain tumors and the red blood cell A antigen. These analyses led to the identification of three alleles associated with these phenotypes: *H-2*, *H-2*^{d}, and *h-2*. These alleles were also shown to be genetically linked to the *Fused* gene *Fu* (mutations in the *Fused* gene cause mice to have fused/kinked tails). This study aimed to further identify new alleles at the *H-2* locus.

Think About...

1. Design experiments to identify new alleles at the *H-2* locus. Consider the resistance/susceptibility to specific tumor strains based on this locus and linkage analysis as a starting point to identify these new alleles.
2. Why does a linked gene need to be used in linkage analysis? What is the purpose of a second gene in these experiments?
3. Explain how linkage analysis is affected by the nature of the gene present: Is the phenotype based on typical dominant/recessive characteristics, or is the phenotype based on codominance, such as in blood typing?

Article Summary

The authors used linkage analysis and different parental strains to test resistance or susceptibility to specific tumors connected to the same phenotype of the parental strains. They crossed different strains of inbred mice with the following pattern (M × F) × N and then implanted an M-strain tumor. If *Fused* showed linkage with resistance, then both M and N and M and F were carrying different alleles at the *H-2* locus (**TABLE**). The authors identified four

(Continued)

Linkage Analysis of *Fused* (*Fu*) to Tumor Susceptibility

Cross	++	+–	*Fu*+	*Fu*–	Conclusion
(A × *Fu*) × dba/2	12	1	2	11	dba/2 is not *H-2*, is *H-2*^{d}
(A × *Fu*) × Balb/c	17	0	0	11	Balb/c is not *H-2*
(dba/2 × *T*) × Balb/c	10	0	12	0	Balb/c is *H-2*^{d}
(Balb/c × *Fu*) × dba/2	13	0	5	0	Balb/c is *H-2*^{d}
(A × *Fu*) × C57BL/10	5	1	0	8	C57BL/10 is not *H-2*
(dba/2 × *T*) × C57BL/10	10	2	0	17	C57BL/10 is not *H-2*^{d}
(C57BL/6 × *Fu*/C) × A	10	0	1	10	C57BL/6 is *h-2*^{b}; *Fu*/C is not *h-2*^{b}
(C57BL/6 × *Fu*/C) × C57BL/6	4	19	13	0	*Fu*/C is *H-2*
(A × *Fu*/C) × C57BL/10	28	0	15	1	*Fu*/C is *H-2*
(A × *Fu*) × P	13	2	1	10	P is not *H-2*
(dba/2 × *Fu*) × P	35	3	5	33	P is not *H-2*^{d}
(C57BL/6 × *Fu*/C) × P	16	0	0	9	P is not *h-2*^{b}
(P × *Fu*/C) × C57BL/10	4	3	0	5	P is *h-2*^{p}

Legend: ++ (Tumor susceptible); +– (Tumor resistant); *Fu*+ (Linkage to *Fused*); *Fu*– (Absence of Linkage to *Fused*)

Source: From G. D. Snell and G. F. Higgins. 1951. *Genetics* 36: 306–310.

KEY DISCOVERIES *(continued)*

alleles at the histocompatibility-2 locus: *H-2*, *H-2^d*, *h-2^p*, and *h-2^b*. Not only was the identification of these alleles at the histocompatibility-2 locus essential in our understanding of the MHC family, but the researchers also observed that MHC molecule expression seemed to behave similarly to blood type antigens, an example of codominance.

Conclusion

The researchers' observations are important in understanding that each individual has a haplotype dictated by the summation of the alleles he or she possesses in the genome rather than a true dominant/recessive nature of genes at particular alleles. These key observations opened our understanding of the complexity of the MHC loci and the nature of their expression in mice, which translates to HLA expression in humans.

the problematic or infected cell. HLA-F plays a role in retrieving class I molecules that no longer have peptide bound in their peptide-binding groove.

If we consider the alleles of the HLA-A, HLA-B, and HLA-C isotypes (which present peptides to CD8 T cells) and consider the possible combinations on two chromosomes in somatic cells (multiplying the alleles together and multiplying that number by 2), there are over 5.1×10^{11} different haplotypes for MHC class I!

CLASS II DIVERSITY There are five MHC class II isotypes in humans: HLA-DM, HLA-DO, HLA-DP, HLA-DQ, and HLA-DR (**TABLE 6.3**). These isotypes not only vary in the allotypes present in the human population, but they can also vary in the number of genes present for each subunit in the genome. For instance, there are four different HLA-DR β-chain genes (*HLA-DRB1*, *HLA-DRB3*, *HLA-DRB4*, and *HLA-DRB5*).

As with class I isotypes, class II isotypes play different roles within the immune system. HLA-DP, HLA-DQ, and HLA-DR function as MHC class II molecules and present peptides to CD4 T cells. HLA-DM (discussed previously) and HLA-DO aid in peptide loading of MHC class II molecules. In considering the alleles of MHC class II molecules to determine the number of possible haplotypes, there are 1.6×10^{23} class II haplotypes. Since class I and class II are both present in the human genome, the total number of possible haplotypes is an astonishing 8.2×10^{34}!

TABLE 6.3 | HLA Class II Genes and Proteins of the Human Genome

HLA locus	Number of alleles	Number of allotypes
DMA	7	4
DMB	13	7
DOA	12	3
DOB	13	5
DPA1	216	80
DPA2	5	2
DPB1	1654	1064
DQA1	246	114
DQB1	1930	1273
DRA	29	2
DRB1	2838	1973
DRB3	363	272
DRB4	180	121
DRB5	142	110

Source: Data from J. Robinson et al. 2015. *Nuc Acid Res* 43: D423–431; J. Robinson et al. 2000. *Tissue Antigens* 55: 280–287; CC BY-ND 3.0.

EVOLUTION AND IMMUNITY

Major Histocompatibility Complex Diversity

In the Chapter 4 "Evolution and Immunity" feature, we learned that T-cell receptor diversity evolved abruptly in jawed fish. Thus, it would be expected that the cognate ligand of T-cell receptors, namely MHC molecules, should have evolved at the same time in evolutionary history. Indeed, MHC molecules can be found in all cartilaginous fish and higher vertebrates, but they are not found in agnathans (jawless fish) or invertebrates. The polymorphism in MHC molecules that aids in diversity in peptide production is also evident as early as the evolutionary history of MHC molecules. For instance, in some organisms there are more than 20 different MHC class I alleles. In sharks, all the MHC class II loci are polymorphic. So, as expected, a diversity of T-cell receptors and MHC molecules required to interact with those receptors evolved at the same time. However, the selective pressure of maintaining this mechanism of adaptive immunity is still not well understood.

MHC polymorphisms and peptide binding

As we have seen, the alleles associated with isotypes of class I and class II molecules are highly polymorphic. These polymorphisms are not random mutations; they represent changes within the peptide-binding groove and the α helices that interact with T-cell receptors (**FIGURE 6.13**). This concentration of mutations in polymorphic alleles provides yet another level of diversity, which ensures that a wide variety of peptides can be presented by class I and class II molecules.

Each isotype also contains a **peptide-binding motif** (sequence of amino acids) that drives peptide-binding specificity. The peptide-binding motif allows for flexibility regarding the presence of most amino acids within the bound peptide. However, there are typically two or three sites within the bound peptide where there is less flexibility in the type of amino acid that can be present at that specific location. These less-flexible amino acids represent **anchor residues** that support peptide binding to a specific isotype. For example, HLA-A class I molecules have a peptide-binding motif in which amino acids 2 and 9 are anchor residues (**FIGURE 6.14**).

Anchor residues allow for tight binding of a peptide to the peptide-binding groove of the class I molecule and also promote promiscuity in binding. This can be seen in the binding of different viral peptides from influenza A and human immunodeficiency virus (HIV) to HLA-A*0301, which all have a hydrophobic amino acid at position 2 and a positively charged amino acid at position 9 but differ at the other 7 positions. Thus, peptide-binding grooves are both flexible in the types of peptides they can bind and specific in regard to certain amino acids within the peptide. Each isotype with a bound peptide can bind only one T-cell receptor.

peptide-binding motif Sequence of amino acids of a peptide that are essential for interaction with an MHC molecule.

anchor residues Residues of a peptide bound to an MHC molecule that are essential in interacting with the peptide-binding groove; typically need to have specific properties in all peptides that can bind to that particular MHC molecule.

• CHECKPOINT QUESTION

1. Define *promiscuous binding site* as it relates to MHC molecules and the role of anchor residues in peptide binding to MHC molecules.

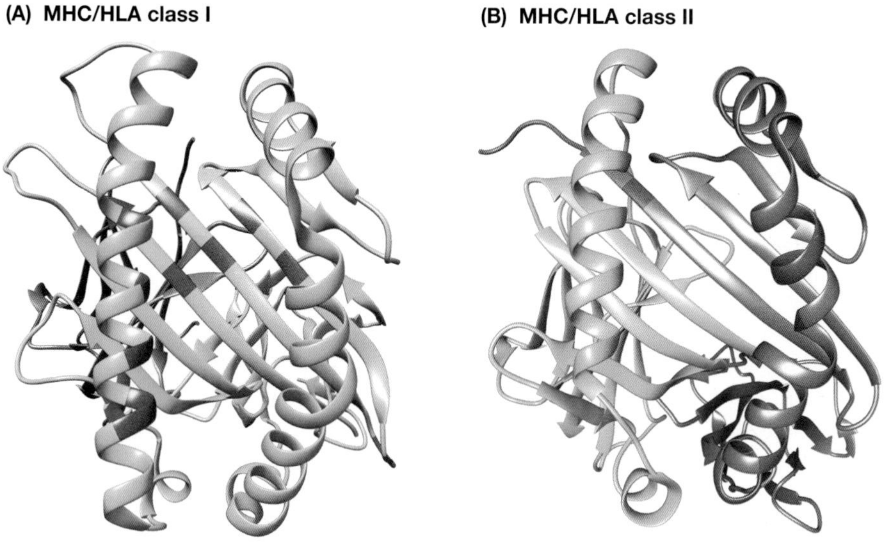

FIGURE 6.13 Polymorphisms of MHC molecules are located in the peptide-binding groove (A) Analysis of polymorphisms of the HLA class I molecule HLA-A show that variations among HLA-A molecules are focused in the peptide-binding groove (highlighted in orange). (B) Analysis of polymorphisms of the HLA class II molecule shows that variations among HLA class II molecules are also focused in the peptide-binding groove, mainly in the β chain (highlighted in orange). (A data from L. Niu et al. 2013. *Mol Immunol* 55: 381–392, redrawn using Chimera and PDB file 4HWZ; B data from Z. Zavala-Ruiz et al. 2004. *Proc Natl Acad Sci USA* 101: 13279–13284. © 2004. National Academy of Sciences, U.S.A., using Chimera and PDB file 1SJH.)

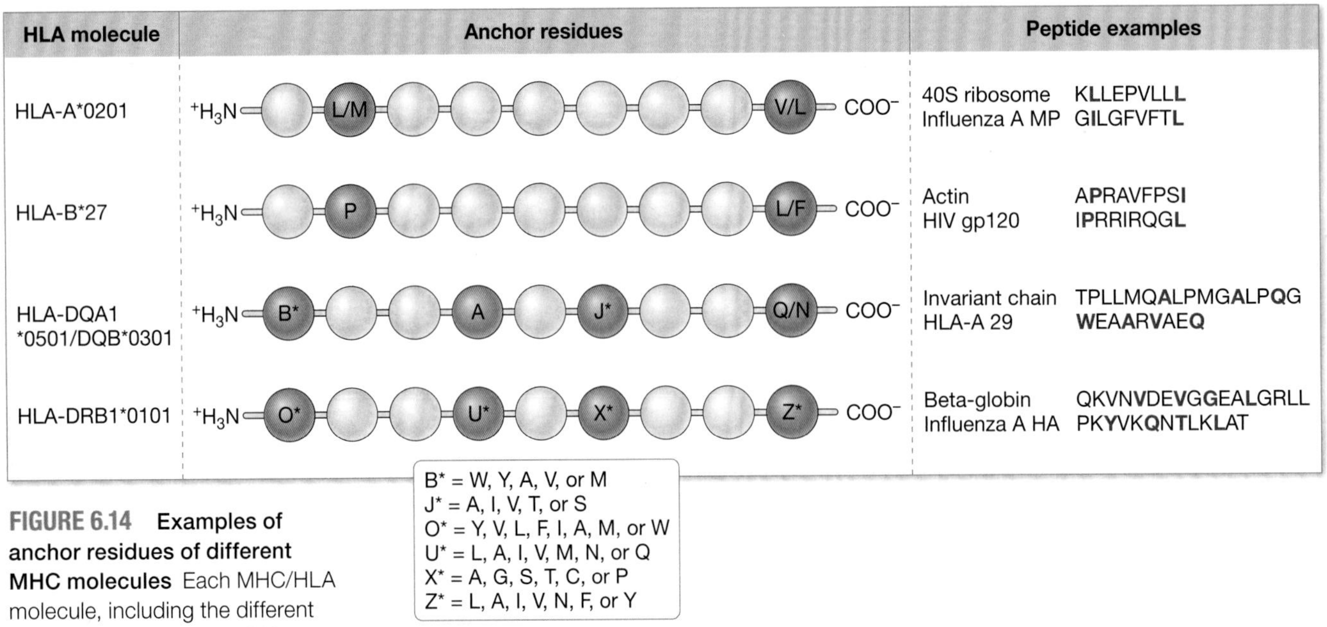

FIGURE 6.14 Examples of anchor residues of different MHC molecules Each MHC/HLA molecule, including the different polymorphism variants within the population, has promiscuous binding affinity dictated by several anchor residues that facilitate a peptide binding to the peptide-binding groove. Shown are several examples of both HLA class I and class II molecules that bear different peptide-binding motifs due to differences in the anchor residues, as well as possible peptides that could be presented by these variants, with anchor residues highlighted in red. (Data from SYPEITHI www.syfpeithi.de)

6.8 What benefits and problems are associated with MHC diversity?

LEARNING OBJECTIVE

6.8.1 Contrast the importance of MHC molecule diversity in immune system function with potential complications related to tissue transplantation.

Diversity of MHC molecules promotes protection from a wide range of pathogens. Indeed, infectious disease is thought to be a selective pressure on the maintenance of **heterozygosity** (variation of alleles) of MHC molecules within an individual and in a population. The heterozygosity of different MHC alleles allows the display of more peptides that can activate appropriate adaptive immune responses as needed.

Pathogens will likely evolve in a way that exploits an MHC polymorphism, making an organism more susceptible to infection. A diverse array of heterozygosity through many MHC polymorphisms limits the presence of a susceptible MHC allele and confers genetic herd immunity by reducing the number of individuals in the population who might have this specific allele. The downside is that this immense variety of MHC molecules can lead to complications during infectious disease outbreaks and in organ and tissue transplantation.

MHC heterozygosity: benefits

We have seen the importance of MHC diversity in displaying a large variety of peptides to CD4 and CD8 T cells. This diversity is based on selective maintenance of alleles within the human population and is likely due to pressure imposed by infectious pathogens. The advantage of heterozygosity is that more peptides from the pathogen can be displayed, providing a higher potential for T cells to be activated during infection. Because this heterozygosity is advantageous in the body's defense against any given pathogen, it is selected for through **balancing selection**. Individuals who are homozygous for a specific allele have less ability to present peptides in a diverse manner, so they have more difficulty clearing and surviving an infection.

heterozygosity The presence of two different alleles of a gene within an individual's genome or the genome of a population.

balancing selection Selective pressure that promotes heterozygosity of a gene within a population.

MHC heterozygosity: drawbacks

While balancing selection tends to be the major evolutionary selective force on polymorphisms in a population due to the benefits of MHC heterozygosity, exposure to an infectious disease will have a much different effect. A population exposed to an emerging infectious disease (such as COVID-19 caused by the SARS-CoV-2 virus) will drive selection of MHC molecule alleles that are most capable of presenting peptides. This contrasts with the normal exposure to multiple different pathogens that drives balancing selection. A shift in selective pressure is referred to as **directional selection**. It results in a biased population in which the best alleles are maintained at the expense of heterozygosity.

directional selection Selective pressure that promotes maintenance of a single allele within a population.

autologous Self-molecules within an individual that are not recognized by lymphocytes due to immune tolerance.

allogeneic Molecules from a source within the same species that are genetically different, as from a different individual.

MHC heterozygosity: dangers

During development, T cells that recognize self-molecules in context of MHC molecules are negatively selected and rarely make it out of the thymus, as they pose a threat to self-tolerance. However, T-cell development only occurs in context of the MHC molecules expressed in each individual. These self-MHC, or **autologous**, molecules work to prevent autoreactive T cells from leaving the thymus.

Introduction of **allogeneic** MHC molecules (from the same species but genetically different) can elicit an adaptive immune response. This phenomenon can result in rejection of transplanted tissues and organs. For example, in the case of a kidney transplant, the donor's kidney may display allogeneic MHC molecules that will activate the recipient's alloreactive T cells, leading to rejection. To minimize this risk, a donor must be selected who has a close match to the recipient's MHC allotypes. Allotype matching minimizes the risk of tissue rejection, but as we will see in Chapter 17, it only slows the rejection process and must be supplemented with immunosuppressant drugs. So, while the heterozygosity of MHC alleles protects most of us against infection, it can severely complicate medical conditions that require tissue or organ transplantation.

CHECKPOINT QUESTION

1. While MHC diversity provides a mechanism to protect against a wide variety of pathogens, when can MHC diversity cause problems?

Summary

6.1 What role do T cells play in antigen recognition?

- T cells recognize foreign antigens via the T-cell receptor, which recognizes an MHC-peptide complex, and a coreceptor, which recognizes the type of infection (intracellular or extracellular).

6.2 What is the structure of the T-cell receptor?

- The T-cell receptor has two subunits (α and β or δ and γ); each subunit consists of a constant region and a variable region.
- The variable regions of each T-cell receptor subunit are close in three-dimensional space to form an antigen-binding site.
- Most T-cell receptors can only recognize peptide antigens presented via MHC molecules.
- The T-cell coreceptor CD4 interacts with MHC class II molecules and recognizes extracellular antigens processed by the antigen-presenting cell.
- The T-cell coreceptor CD8 interacts with MHC class I molecules and recognizes intracellular antigens processed by the antigen-presenting cell.
- T-cell receptors use subunits of the CD3 complex, as well as other signaling molecules such as CD28 and CD45, to transmit the signal of antigen binding to the cytosol and drive T-cell activation.

6.3 How does somatic recombination account for T-cell receptor diversity?

- Somatic recombination of the T-cell receptor loci and junctional diversity promote diversity of the T-cell receptor; these events can result in the production of a T-cell receptor (productive rearrangements) or a nonfunctional T-cell receptor (unproductive rearrangements).

(Continued)

Summary *(continued)*

6.4 What is the structure of MHC molecules?

- MHC class I and class II molecules differ slightly in subunit structure but maintain the important structural feature of a peptide-binding groove, which is responsible for presenting different peptides to T-cell receptors.
- The peptide-binding grooves of MHC class I and class II molecules differ in their ability to bind to peptides of various lengths due to the types of antigens they present.

6.5 How do T-cell receptors interact with MHC-peptide complexes?

- The antigen-binding site of a T-cell receptor recognizes both a bound peptide and the peptide-binding groove of an MHC molecule.

6.6 How does peptide loading occur in MHC class I and class II molecules?

- MHC class I molecules are loaded with intracellular peptides that result from the digestion of cytosolic proteins by the proteasome that have been transported into the ER.
- Loading of MHC class I molecules is facilitated by the molecules TAP (transports proteins into the ER), tapasin (aids in peptide loading), calnexin and calreticulin (promote MHC folding), and ERAP (trims loaded peptides to the right size).
- MHC class II molecules are loaded with extracellular peptides that result from the digestion of extracellular proteins during phagocytosis in the phagolysosome.
- MHC class II molecules are kept inactive in the ER by the invariant chain, which is trimmed to CLIP, before peptide exchange with extracellular peptides occurs with the aid of HLA-DM.
- Professional antigen-presenting cells can also cross-present extracellular antigens via MHC class I molecules to aid in the activation of cytotoxic T cells needed for an intracellular response.

6.7 How does MHC diversity affect peptide presentation to T cells?

- MHC molecules are diverse in the population due to polymorphisms within individual MHC genes and multiple MHC genes that express different MHC isotypes.
- MHC diversity (heterozygosity) promotes a diverse array of molecules that can present a large number of different peptides to T-cell receptors.
- Most MHC polymorphisms are found at the peptide-binding groove, promoting diversity in bound peptides and diversity in T-cell receptors with which MHC molecules can interact.

6.8 What benefits and problems are associated with MHC diversity?

- While MHC heterozygosity promotes adaptive immune response diversity, associated complications can cause health issues. Directional selection is driven by newly emerging pathogens, so only a small subset of MHC molecules will be effective at eliciting a proper immune response. Mismatched MHC molecules during tissue transplantation can prompt an adaptive immune response to transplanted tissue and lead to rejection.

Review Questions

6.1 By what means do T cells recognize their specific antigen? Which molecules are important in this recognition event?

6.2 Which of the following is *not* a component of the T-cell receptor signaling complex?

A. ζ chain
B. CD3δ
C. CD45
D. CD3γ
E. CLIP

6.3 Explain why negative selection must occur when a productive rearrangement of both T-cell receptor subunits occurs.

6.4 Compare and contrast the structure of MHC class I and class II molecules.

6.5 MHC class I and class II molecules bind to T-cell receptors and the coreceptors CD4 or CD8 (depending on the class of the molecule). Which coreceptor binds to each class of MHC molecule? Why are each of these interactions important when considering the effector functions of the T cells containing each of these coreceptors?

6.6 An individual has a homozygous recessive mutation in the HLA-DM gene. Predict how this would affect MHC class I and class II presentation on the surface of cells.

6.7 Describe how diversity in MHC molecules is gained in a given population of organisms.

6.8 Why is MHC matching so important in regard to tissue and organ transplantation?

CASE STUDY REVISITED

Bare Lymphocyte Syndrome

A 3-year-old girl is taken to the emergency room due to a high fever and lethargy. A chest x-ray reveals that she is suffering from upper and lower respiratory tract infections. A medical history reveals that she has experienced an unusually high number of infections and a history of chronic skin lesions. The emergency room physician prescribes antibiotics but, hoping to get at the root cause of the patient's recurring infections, he also orders a white blood cell (WBC) analysis by flow cytometry. The analysis reveals a very low CD8 T-cell count and a lack of MHC class I molecules on the cell surface. The diagnosis is bare lymphocyte syndrome I.

Think About...

Questions for individuals

1. Define bare lymphocyte syndrome. What are the common symptoms associated with bare lymphocyte syndrome?
2. What are the different classes of bare lymphocyte syndrome and the cause of each class?

Questions for student groups

1. Why might there be a lack of MHC class I molecules on the surface of the analyzed cells? Speculate in regard to potential malfunctions or mutations that would cause this result.
2. How could the lack of MHC molecules be connected to a lack of CD8 T cells?
3. Would you predict that mutations leading to bare lymphocyte syndrome would affect MHC class II molecules on the WBC surface? Why or why not?

MAKING CONNECTIONS

Key Concepts	COVERAGE (Section Numbers)
T-cell receptor structure	4.2, 4.7, 4.8, 6.2, 6.4, 9.1
T-cell receptor junctional diversity	4.2, 4.7, 6.2, 6.3
T-cell receptor antigen-binding sites	4.3, 4.7, 5.4, 6.2
Draining lymphoid tissue and its role in presentation of infection	1.3, 1.4, 1.10, 4.2, 6.4, 6.5
Dendritic cells and their role in different antigen processing and presentation pathways	2.1, 2.8, 4.2, 4.5, 4.6, 6.4, 6.5, 7.3
Dendritic cell interaction with T cells	2.1, 2.8, 4.2, 4.5, 4.6, 6.4, 6.5, 6.6, 7.3, 7.6, 7.7

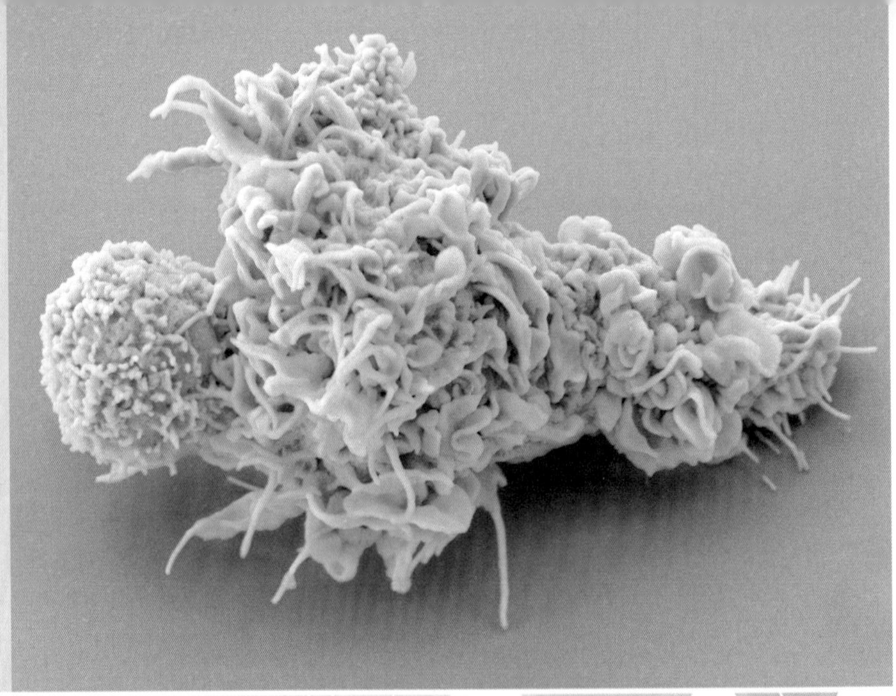

7

T-Cell-Mediated Adaptive Immunity

QUESTIONS Explored

7.1 How do naïve T cells become effector T cells?

7.2 Which cells present antigen to T cells in secondary lymphoid tissue?

7.3 What is the role of dendritic cells in antigen presentation and T-cell activation?

7.4 What is the role of macrophages in antigen presentation and T-cell activation?

7.5 What is the role of B cells in antigen presentation and T-cell activation?

7.6 How do T cells migrate into secondary lymphoid tissue?

7.7 How are T cells activated?

7.8 What signal transduction process must occur for T-cell activation?

7.9 What is the role of interleukin-2 in T-cell activation?

7.10 Why is costimulation important in T-cell activation?

7.11 How do effector T cells utilize cytokines as part of the immune response?

7.12 What signaling process drives T-cell differentiation into various types of effector T cells?

7.13 How do cytotoxic T cells destroy target cells?

7.14 What are the functions of helper T cells?

7.15 What are the functions of regulatory T cells?

7.16 What are the functions of natural killer T cells?

CASE STUDY: Meg's Story

Alicia and James wake up early Tuesday morning to a loud, hacking cough coming from their daughter's room. Alicia walks down the hall and asks, "Meg, are you OK? Do you need a drink of water?" Two-year-old Meg has a hard time catching her breath to answer, but eventually she asks her mother for a drink. Alicia sits in Meg's room while Meg catches her breath and calms down, and shortly after, Meg is able to fall back to sleep.

Later that morning, Meg is coughing again. Alicia hopes that Meg has a seasonal cold and not another pneumonia infection. Alicia checks Meg's temperature and is relieved to see that it is normal. The family gets ready to begin the day.

As the week progresses, Meg continues to cough and have trouble catching her breath. One evening, Meg wakes up and cries out. Alicia runs into Meg's room and sees that Meg has been sweating. Meg complains about feeling cold, but Alicia touches her forehead and can instantly tell that Meg has a fever. Alicia also notices a mix of green and red mucus on Meg's bedsheet. Alicia calls out to James, "James, get dressed! We need to take Meg to the emergency room again!"

In order to understand what Meg is experiencing, and to answer the case study questions at the end of the chapter, we must first examine the role of T cells in our adaptive immune system. Chapters 5 and 6 explored the mechanisms involved in T-cell development and the cell-surface molecule that allows recognition of a specific antigen. The questions we will address next are: How are T cells activated when they recognize an MHC-peptide complex via their receptor, and what are the effector functions of T cells in the cellular arm of the adaptive immune system?

During a primary adaptive immune response, T cells can migrate to secondary lymphoid tissue, where they assess various antigens presented by antigen-presenting cells via MHC class I and class II molecules. When a T cell recognizes a specific MHC-peptide complex and undergoes clonal selection, signaling ensues that promotes its activation. These signaling events drive clonal expansion via cell division and allow the expanded T cells to differentiate into effector T cells. The effector T cells produced during clonal expansion and differentiation will match what is required to combat the specific pathogen whose antigen was presented. Extracellular pathogens induce differentiation of helper T cells, which activate cells that promote neutralization and phagocytosis of the pathogen. Intracellular pathogens induce differentiation of cytotoxic T cells, which hunt down infected cells and target them for apoptosis.

T-cell priming Process whereby effector T cells are produced to drive the cellular arm of the adaptive immune system; also called *T-cell activation*.

7.1 How do naïve T cells become effector T cells?

LEARNING OBJECTIVE

7.1.1 Explain why naïve T cells require priming in secondary lymphoid tissue to become effector T cells.

In Chapter 5, we learned about T-cell development in the thymus, and in Chapter 6 we explored the nature of T-cell receptors and their interactions with MHC molecules as part of antigen recognition. We know that naïve T cells remain inactive until the right peptide in the right MHC is presented to the appropriate T-cell receptor.

Naïve T cells circulate in the blood and lymph and ultimately reach secondary lymphoid tissue. Here they interact with antigen-presenting cells (APCs), including dendritic cells and macrophages, which present a variety of peptides on MHC molecules. When the T-cell receptor and MHC-peptide complex interact, a signaling cascade results in T-cell activation. The T cell then differentiates into an effector T cell primed to fight pathogens in the secondary lymphoid tissue and at other sites of infection (FIGURE 7.1). This "priming" step is why T-cell activation is also known as **T-cell priming**.

CHECKPOINT QUESTION

1. What is the role of T-cell priming in an adaptive immune response?

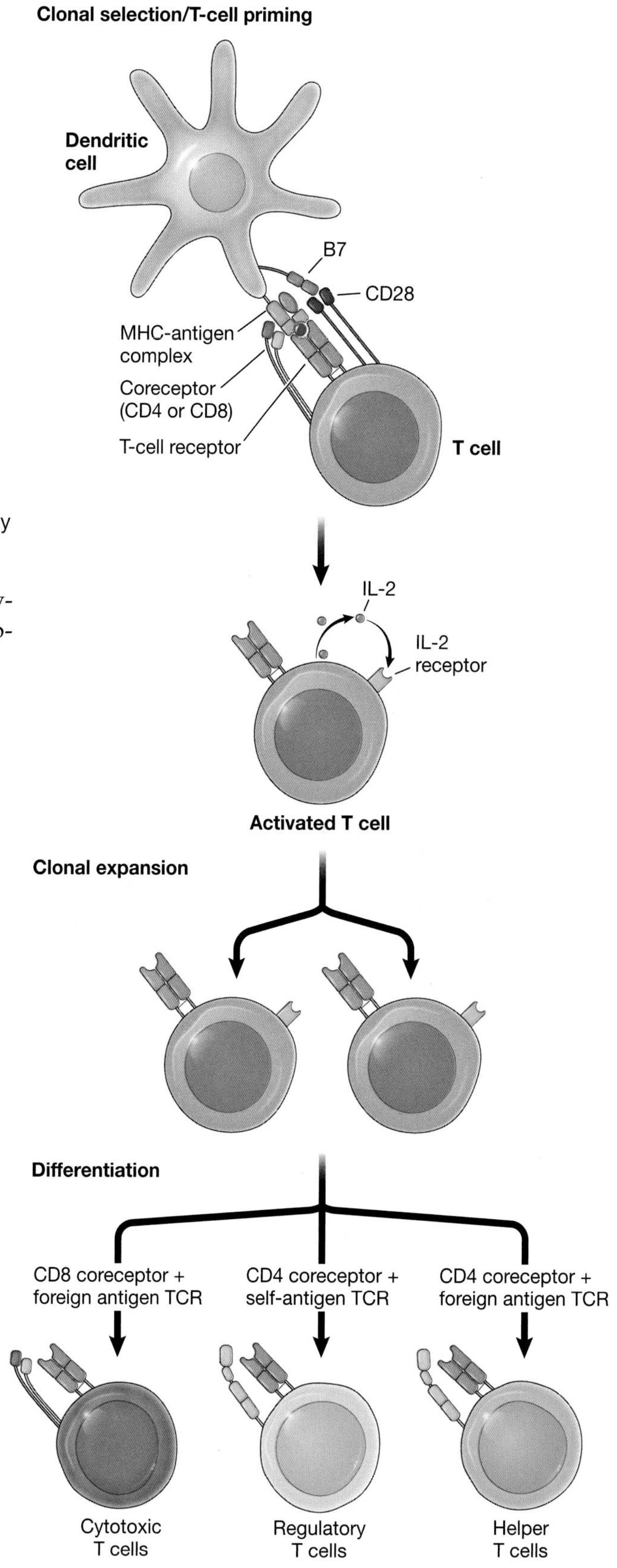

FIGURE 7.1 Priming, activation, and differentiation of T cells T cells require engagement of an MHC-antigen complex with their specific T-cell receptor (TCR) and a costimulatory signal through CD28 to be clonally selected and primed. This process prompts activation of signaling events within the T cell that result in changes in gene expression and the release of cytokines, including interleukin-2 (IL-2), which induces division and clonal expansion of the activated T cell. Differentiation of expanded T cells produces cytotoxic T cells, regulatory T cells, and helper T cells. The effector function of each type of cell depends on specific signals and the type of coreceptor. Each of these three cell types (and their subtypes) have specific effector functions that are required for a proper adaptive immune response.

7.2 Which cells present antigen to T cells in secondary lymphoid tissue?

LEARNING OBJECTIVE

7.2.1 List the three types of cells involved in presenting antigen to T cells.

Given the enormous surface area of the body and its tissues and organs, it would be virtually impossible for naïve T cells to circulate throughout the entire body in search of one particular antigen. How, then, are the appropriate T cells activated in a timely manner when needed?

Recall that the secondary lymphoid tissue plays an important role in activating an adaptive immune response. Phagocytes of the innate immune system engulf pathogens at the site of infection and transport them to secondary lymphoid tissue, where antigen presentation occurs. The cells that carry out the important role of presenting antigens to T cells are dendritic cells, macrophages, and B cells.

CHECKPOINT QUESTION

1. What three types of APCs play a role in activating naïve T cells in secondary lymphoid tissues?

7.3 What is the role of dendritic cells in antigen presentation and T-cell activation?

LEARNING OBJECTIVE

7.3.1 Explain the role of dendritic cells in antigen presentation to T cells.

During an infection, the innate immune response activates dendritic cells and macrophages to promote pathogen uptake and presentation of foreign antigens on surface MHC molecules. The dendritic cells are most important in T-cell activation due to their ability to phagocytose foreign pathogens and migrate to secondary lymphoid tissue. While the secondary lymphoid tissue is somewhat specialized, the mechanism by which dendritic cells migrate to these tissues and activate naïve T cells is essentially the same regardless of tissue type.

Immature dendritic cells (**FIGURE 7.2A**) function in peripheral tissues by phagocytosing foreign pathogens. They specialize in antigen uptake and processing. When these cells containing foreign antigens migrate into secondary lymphoid tissue, they become mature dendritic cells. Mature cells contain many cellular projections, which enhance their ability to interact with and activate naïve T cells (**FIGURE 7.2B**).

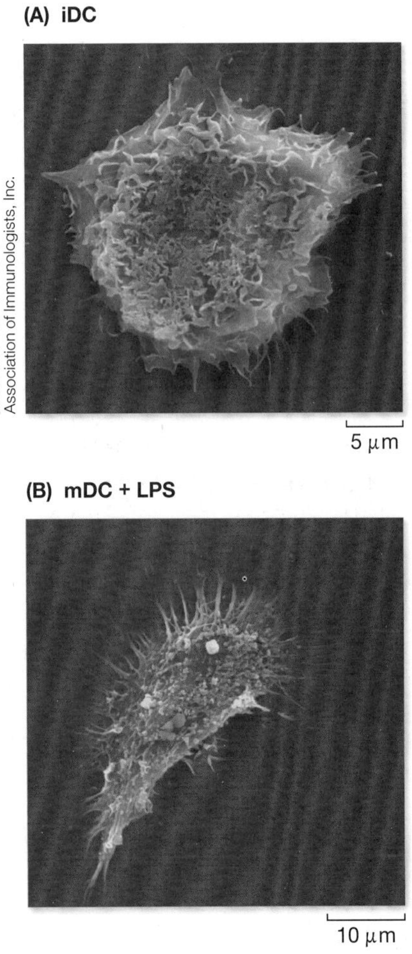

FIGURE 7.2 **Micrographs of immature and mature dendritic cells**
(A) Scanning electron micrograph of an immature dendritic cell (iDC). (B) Dendritic cell that was induced to mature (mDC) through lipopolysaccharide (LPS) treatment. Immature dendritic cells specialize in antigen uptake and processing, whereas mature dendritic cells specialize in antigen presentation via their characteristic dendrite extensions, as seen in (B).

TABLE 7.1 | Pathogen Presentation by Dendritic Cells

Pathogen Presented	MHC Molecule Presenting Peptides	Type of T Cell Activated
Extracellular pathogen (bacteria) Viral particles Toxins	MHC class II	CD4 T cells
Intracellular viruses	MHC class I	CD8 T cells
Phagocytosed viruses	MHC class I (cross-presentation)	CD8 T cells

upregulation Increase in production of proteins within a cell due to activation of gene expression.

Dendritic cells are equipped with a variety of receptors that enable them to recognize and engulf different types of pathogens (**TABLE 7.1**). Recall from Chapter 2 that pathogen-associated molecular patterns (PAMPs) are recognized by innate immune cell receptors. Dendritic cells use these receptors (including the mannose receptor and Toll-like receptors [TLRs]) to bind to a wide variety of pathogens and engulf and process them. As described in Chapter 6, antigens of extracellular pathogens phagocytosed in this manner are processed and presented on MHC class II molecules to naïve CD4 T cells in secondary lymphoid tissue. Alternately, the antigens are processed and presented on MHC class I molecules to naïve CD8 T cells via cross-presentation.

Dendritic cells can also process viruses. Cytosolic proteins (including viral antigens) are processed via the proteasome into short peptides, which are presented on the surface via MHC class I molecules. MHC class I-peptide complexes are then presented to naïve CD8 T cells in secondary lymphoid tissue.

Upon PAMP recognition, the cell-surface receptor is activated, causing signal transduction cascades and changes in gene regulation. These changes lead to **upregulation** of proteins required for antigen processing and migration of dendritic cells from infected tissue into secondary lymphoid tissue.

Upregulated proteins include those involved in MHC class II antigen processing and receptors needed for migration such as CCR7, which binds to the chemokine CCL21. CCL21 is synthesized in secondary lymphoid tissue, allowing dendritic cells to sense and migrate to those tissues during times of infection. Maturation of a dendritic cell from antigen processing to antigen presenting occurs due to the interaction between CCR7 and CCL21. This process ensures that when a dendritic cell enters a secondary lymphoid tissue, it is prepared to present antigen to naïve T cells.

Mature dendritic cells then express the coreceptor B7 (also known as CD80/CD86), which is required for costimulation of naïve T cells. They also begin to secrete the chemokine CCL18. CCL18 functions as a chemokine to both naïve T cells and immature dendritic cells. Homing of naïve T cells increases the number of possible T cells that can be activated in the secondary lymphoid tissue. Homing of immature dendritic cells increases the migration of dendritic cells acting at a site of infection to the secondary lymphoid tissue so they may act in antigen presentation to T cells.

CHECKPOINT QUESTION

1. Upon recognition of a pathogen through a PAMP–receptor interaction, dendritic cells are activated to increase antigen processing. Activated dendritic cells migrate into secondary lymphoid tissues and interact with T cells. What proteins do dendritic cells make, and what is the role of these proteins?

7.4 What is the role of macrophages in antigen presentation and T-cell activation?

LEARNING OBJECTIVE

7.4.1 Explain the role of macrophages in antigen presentation to T cells.

Like dendritic cells, macrophages possess a variety of receptors responsible for recognizing PAMPs. Activation of a PAMP receptor on a macrophage induces expression of the costimulatory receptor B7 at the cell surface and an increase in antigen processing in MHC pathways.

A significant difference between macrophages and dendritic cells is that dendritic cells specialize in both phagocytosis and T-cell activation (due to their ability to migrate to draining secondary lymphoid tissue). Macrophages mainly specialize in phagocytosis. While those in secondary lymphoid tissue are capable of activating naïve T cells that have been transported to their location, the majority of macrophages remain at the site of infection and are not involved in T-cell activation.

CHECKPOINT QUESTION

1. Both macrophages and dendritic cells are capable of phagocytosis and processing antigens. Why are macrophages not a major contributor in T-cell activation?

7.5 What is the role of B cells in antigen presentation and T-cell activation?

LEARNING OBJECTIVE

7.5.1 Explain the role of B cells in antigen presentation to T cells.

B cells, another specialized cell type of the adaptive immune system, present specific antigens to T cells in secondary lymphoid tissue. Recall from Chapter 4 that B cells use immunoglobulins to recognize antigens in a manner similar to that of T-cell receptor recognition. Unlike the T-cell receptor, immunoglobulins can recognize antigen in the absence of an MHC molecule.

B cells can express both cell-surface immunoglobulins (the B-cell receptor) and soluble immunoglobulins, produced by plasma cells (effector B cells). Immunoglobulins recognize a single antigen and, upon recognition, the immunoglobulin–antigen complex is taken into the B cell via receptor-mediated endocytosis. At this point, the antigen is degraded in a lysosome to display antigen peptides on the surface via the MHC class II pathway (**FIGURE 7.3**). Thus, B cells can act like APCs, but this activity is limited to the one type of antigen recognized by the immunoglobulin.

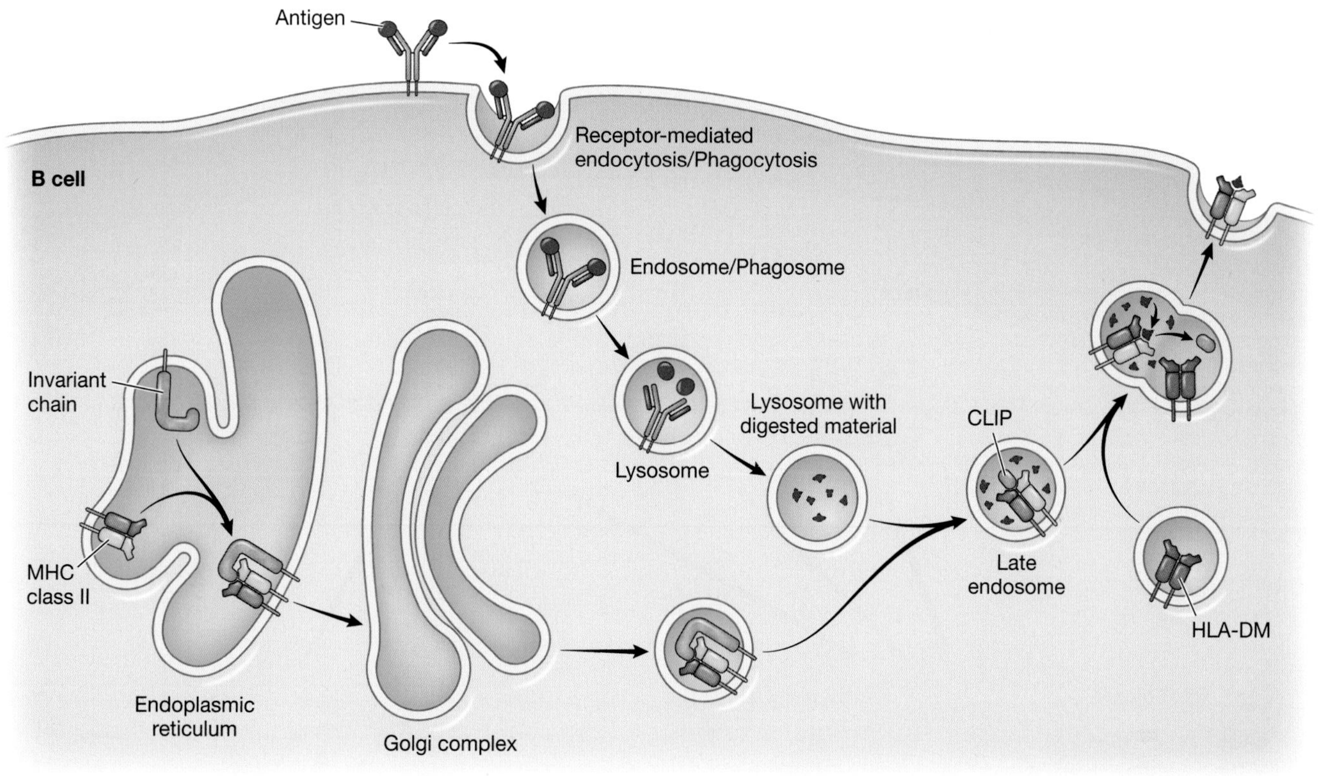

FIGURE 7.3 B-cell antigen processing and presentation B cells utilize cell-surface immunoglobulins to bind to antigen and internalize antigen through receptor-mediated endocytosis (small antigens) or phagocytosis (cellular antigens). Internalized antigen is digested in the lysosome. Through fusion of the lysosome with an MHC class II vesicle and the action of HLA-DM (human leukocyte antigen DM), antigen peptides can be exchanged for class II-associated invariant chain peptide (CLIP) onto the MHC class II molecule and presented at the surface of the B cell. This process of endocytosis/phagocytosis using the cell-surface immunoglobulin is specific to the antigen the immunoglobulin can bind.

B cells migrate through both the lymphatic and circulatory systems, but only B cells within a secondary lymphoid tissue (localized in the T-cell zone of that lymphoid tissue) are capable of activating naïve T cells if they recognize the same antigen. Under normal circumstances, B cells do not play a major role in T-cell activation. Instead, B cells are activated by **CD4 helper T cells**, which we will discuss in Section 7.14.

CD4 helper T cells Activated and differentiated T cells that function to activate either B cells or macrophages.

CHECKPOINT QUESTION

1. What is the major difference between extracellular pathogen processing and antigen presentation of B cells compared to that of macrophages and dendritic cells?

Making Connections: Professional Antigen-Presenting Cells in the Adaptive Immune Response

While all nucleated cells express MHC/HLA class I molecules at their surface, dendritic cells, macrophages, and B cells are also capable of processing and presenting antigen via the class II pathway. All three of these professional antigen-presenting cells are involved in the uptake of antigen or pathogen via receptor-mediated endocytosis or phagocytosis. Thus, all three cell types can activate CD4 and CD8 T cells in draining lymphoid tissue.

Dendritic cells, macrophages, and B cells are considered professional antigen-presenting cells because they process antigen via both class I and class II pathways. Class II processing involves endocytosis or phagocytosis via cell-surface receptors (or immunoglobulins in the case of B cells). Furthermore, dendritic cells can be licensed to cross-present extracellular antigens via class I presentation.

(Sections 2.2, 2.3, 2.5, 2.8, 4.2, 4.6, 6.5, 7.2, 7.3, 7.4)

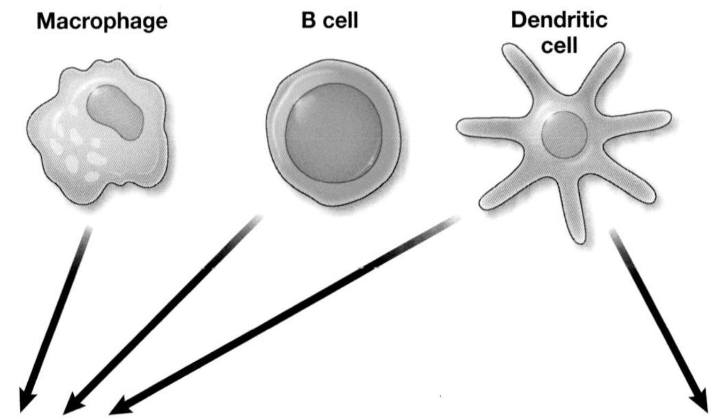

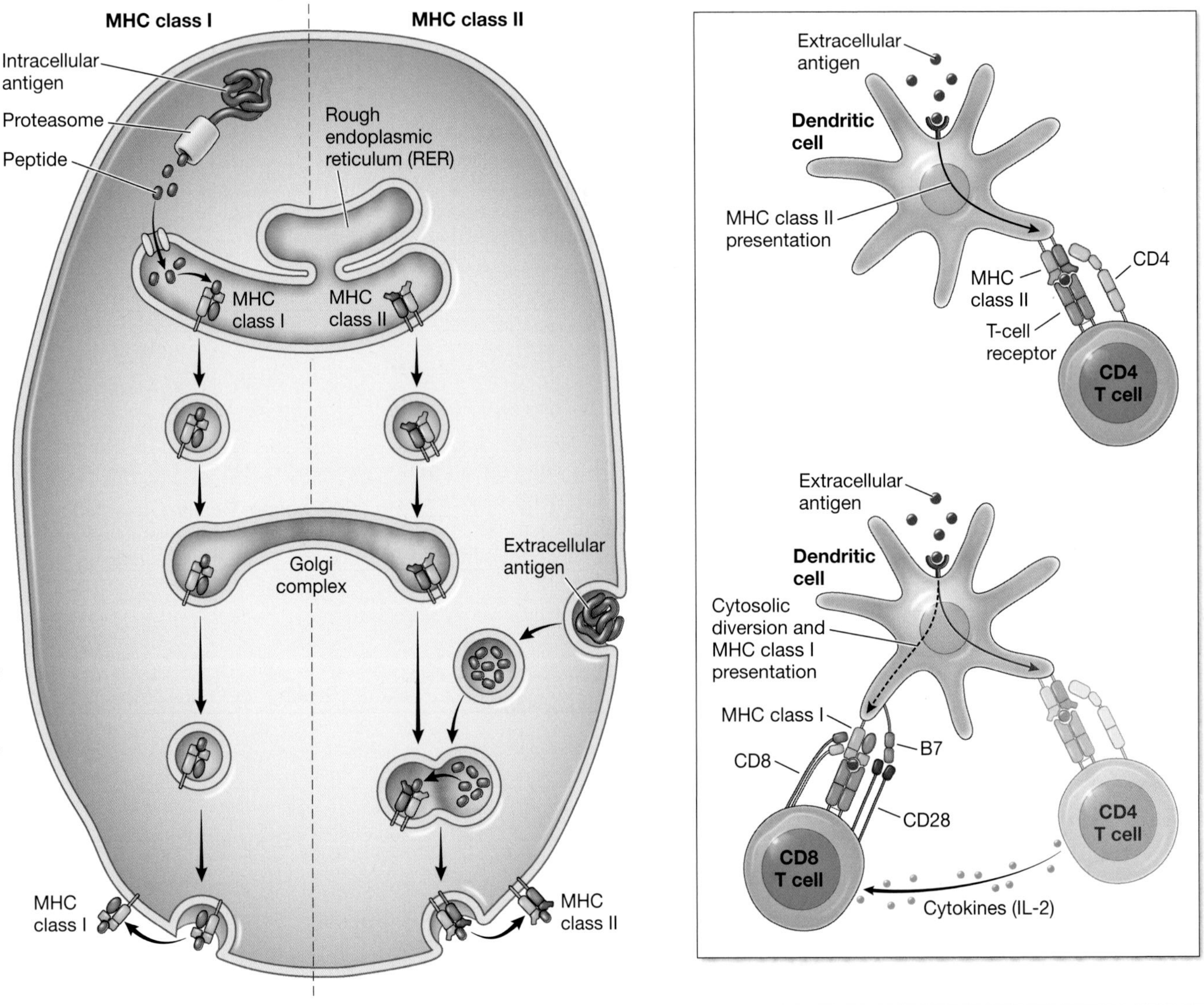

After J. Punt et al. 2018. *Kuby Immunology*, 8th ed., W. H. Freeman and Company: New York

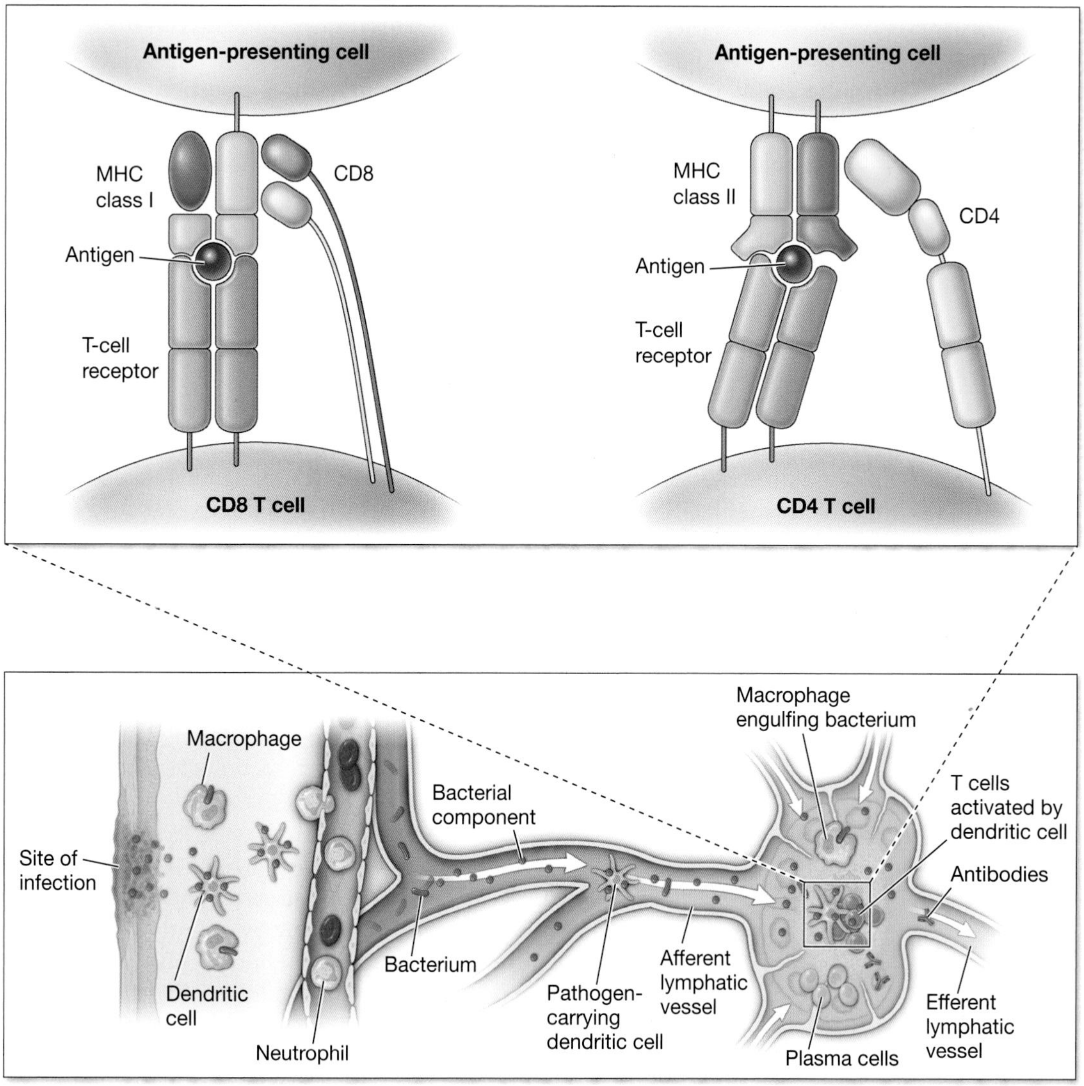

Professional antigen-presenting cells can either migrate to draining lymphoid tissue to present antigen to circulating T cells or take up and process antigen that has drained into nearby lymphoid tissue and present it to circulating T cells.

(Sections 1.9, 1.10, 2.8, 4.2, 4.7, 6.2, 6.4, 7.3, 7.4, 7.5)

7.6 | How do T cells migrate into secondary lymphoid tissue?

LEARNING OBJECTIVE

7.6.1 Describe the role of secondary lymphoid tissue in T-cell activation.

We have learned that naïve T-cell activation occurs in a draining secondary lymphoid tissue where professional APCs have migrated (as with dendritic cells) or are localized (as with macrophages and B cells). Thus, activation requires T-cell migration into the secondary lymphoid tissue. Migration to a specific organ or tissue is known as *T-cell homing*. This process depends on the action of chemokines and cell-adhesion molecules.

The ability of any immune cell to migrate to a site of infection or to any tissue requires:

- A signal to the cell that it must migrate to a particular location
- The action of adhesion molecules to allow the cell to attach to a specific tissue.

Cell signaling

Naïve T cells must be able to respond to the "call" to move out of the circulatory and lymphatic systems and bind to secondary lymphoid tissues to aid in migration. When a lymph node is involved, dendritic cells and stromal cells (connective tissue cells) within the node secrete chemokines CCL19 and CCL21. These chemokines bind to endothelial cells of the venule and establish a chemokine gradient that facilitates T-cell homing. T cells express the CCL19 and CCL21 chemokine receptor CCR7, which allows them to bind to endothelial cells and begin the migration process (FIGURE 7.4).

high endothelial venule (HEV) Specialized blood vessel in secondary lymphoid tissues containing large endothelial cells that express lymphocyte adhesion molecules.

Cell adhesion

Cell adhesion drives T cells' ability to migrate into secondary lymphoid tissue. Cell adhesion molecules are present on both the T cell and the secondary lymphoid tissue, specifically on endothelial cells of the **high endothelial venules** (**HEVs**) of the secondary lymphoid tissue. The adhesion of T cells and cells of the secondary lymphoid tissue mimics adhesion of neutrophils with endothelial cells during neutrophil migration. The T-cell adhesion molecule L-selectin binds to adhesion molecules on the endothelial cell surface, including CD34 and GlyCAM-1. This adhesion slows the movement of the T cell in the circulation (rolling adhesion, FIGURE 7.5, Step 1), allowing for attachment to the endothelial cell surface through the interaction of the T-cell integrin LFA-1 binding to integrin receptors such as ICAM-1 and ICAM-2 on the endothelial cell surface (tight binding, Figure 7.5, Step 2). The tight binding of the T cell and endothelial cell surfaces allows

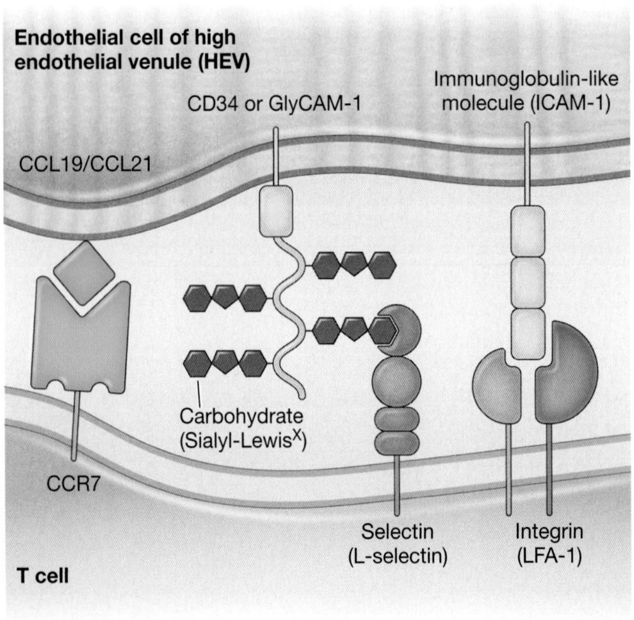

FIGURE 7.4 **T-cell homing adhesion molecules** T cells express two adhesion molecules that they use to bind to endothelial cells of high endothelial venules (HEVs): L-selectin, which binds to Sialyl-Lewis^X carbohydrate on either CD34 or GlyCAM-1, and LFA-1, an integrin that binds to the immunoglobulin-like molecule ICAM-1. The affinity of these interactions allows for T cells to undergo rolling adhesion on the HEV or tightly interact with the HEV. (After P. Parham. 2014. *The Immune System*, 4th ed., W. W. Norton and Company.)

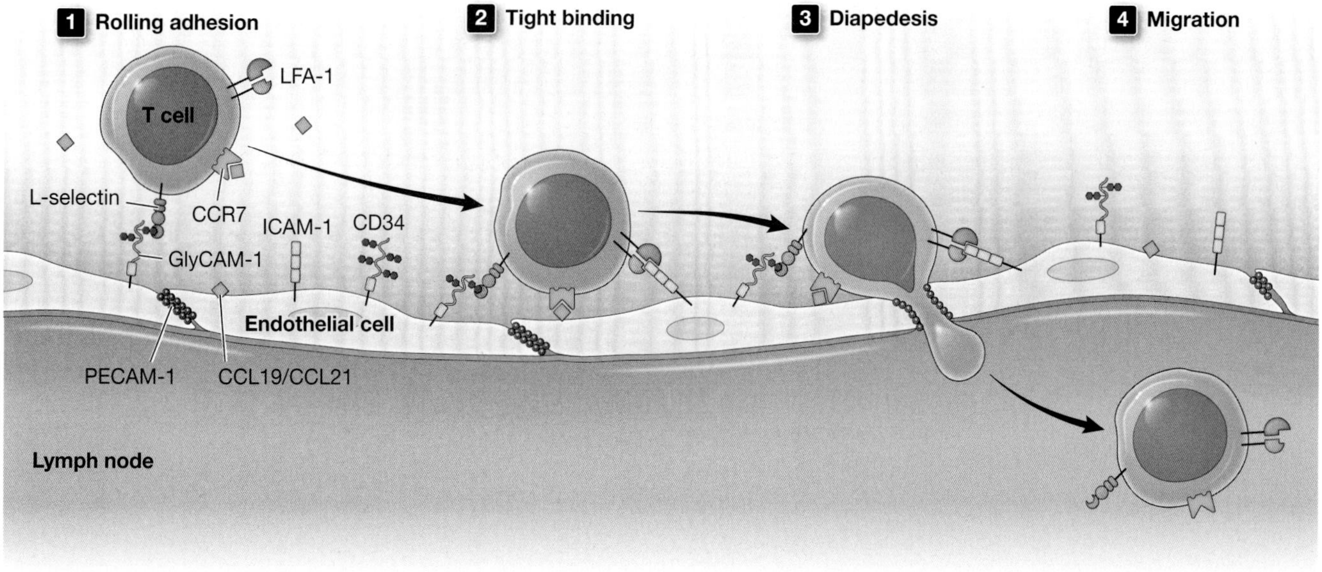

FIGURE 7.5 T-cell migration into lymphoid tissue In a manner similar to neutrophil migration to a site of infection, T cells can migrate into lymphoid tissue using cell–cell interaction. The interaction of L-selectin on the T cell with CD34 or GlyCAM-1 on the endothelial cell allows the T cells to undergo rolling adhesion (Step 1). Slowly rolling T cells can then tightly bind to ICAM-1 using their cell-surface molecule LFA-1 (Step 2). After tight binding, T cells squeeze through the endothelial wall of the high endothelial venule in a process known as diapedesis (Step 3). Finally, T cells migrate into the lymphoid tissue (Step 4). (After P. Parham. 2014. *The Immune System*, 4th ed., W. W. Norton and Company.)

the T cell to slowly move along the endothelial surface toward the chemokine gradient. It eventually allows the T cell to squeeze between individual endothelial cells (diapedesis, Figure 7.5, Step 3), exiting the lymphatic system and entering the draining secondary lymphoid tissue (migration, Figure 7.5, Step 4).

Upon entering the secondary lymphoid tissue, T cells interact with dendritic cells via the same adhesion molecules that were used for binding to the endothelial surface (LFA-1 on the T-cell surface and ICAM-1 and ICAM-2 on the dendritic cell surface). Other interactions between the dendritic cell and T cell strengthen the connection between the two cells. These interactions involve LFA-1, LFA-3, and DC-SIGN molecules on the dendritic cell surface and CD2 and the integrin receptor ICAM-3 on the T-cell surface (**FIGURE 7.6**). The strengthened connection allows the T cell to scan the MHC-peptide complexes on the surface of the dendritic cell.

T-cell migration

If a T cell does not have an activating interaction with an MHC-peptide complex on an APC, it exits the secondary lymphoid tissue and re-enters the lymphatic system through an efferent lymphatic vessel. Movement from the secondary lymphoid tissue back into circulation is mediated by the action of the lipid molecule sphingosine-1-phosphate. T cells expressing a sphingosine-1-phosphate receptor are attracted to the lipid molecule gradient, which facilitates their travel through the lymphoid tissue and back into circulation.

Activated cells suppress sphingosine-1-phosphate receptor expression for several days, providing time for clonal expansion and differentiation into effector T cells. Once expansion and differentiation are complete, the receptor is "switched on" again, allowing effector T cells to exit the secondary lymphoid tissue and return to the circulation. Activated T cells exit the secondary lymphoid tissue through an efferent lymphatic vessel and migrate to the site of infection.

CHECKPOINT QUESTION

1. Explain the process of T-cell migration into a secondary lymphoid tissue.

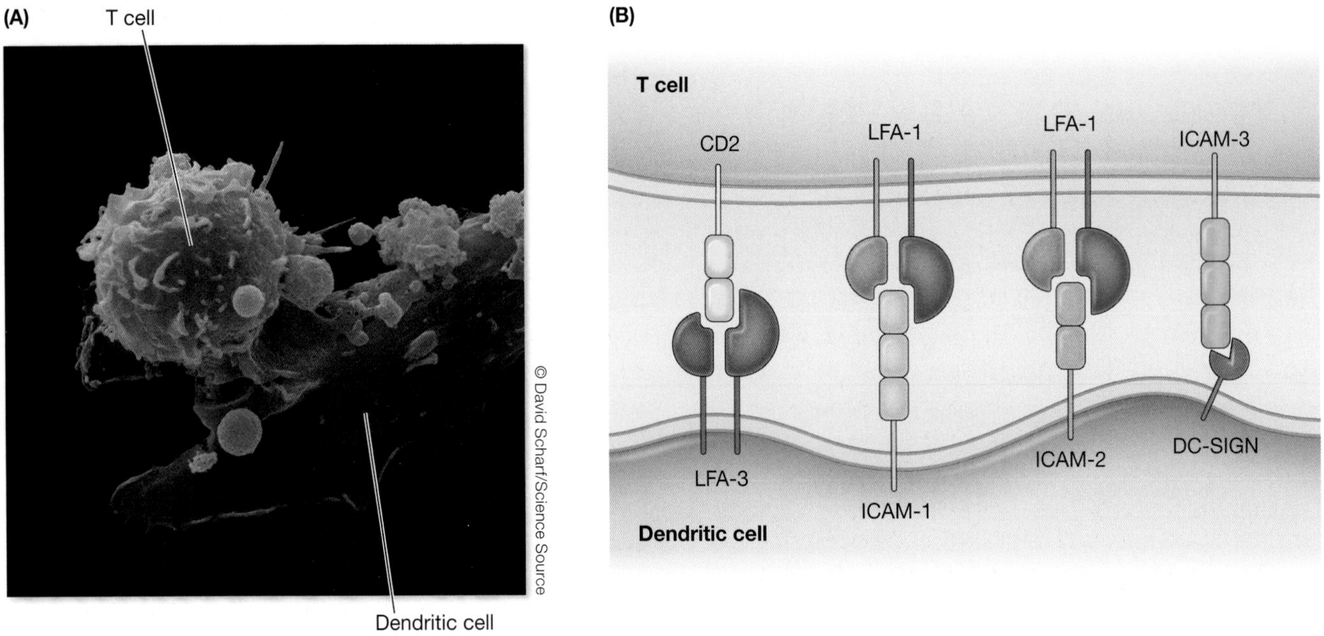

FIGURE 7.6 Interaction of dendritic cells and T cells (A) Electron micrograph of a T cell interacting with a dendritic cell. (B) Dendritic cells and T cells interact via cell–cell adhesion molecules, including CD2, LFA-1, and ICAM-3 on the surface of T cells with LFA-3, ICAM-1, ICAM-2, and DC-SIGN on the surface of dendritic cells. These adhesion molecules allow T cells to test whether their receptor is capable of interacting with MHC-peptide complexes on the dendritic cell surface.

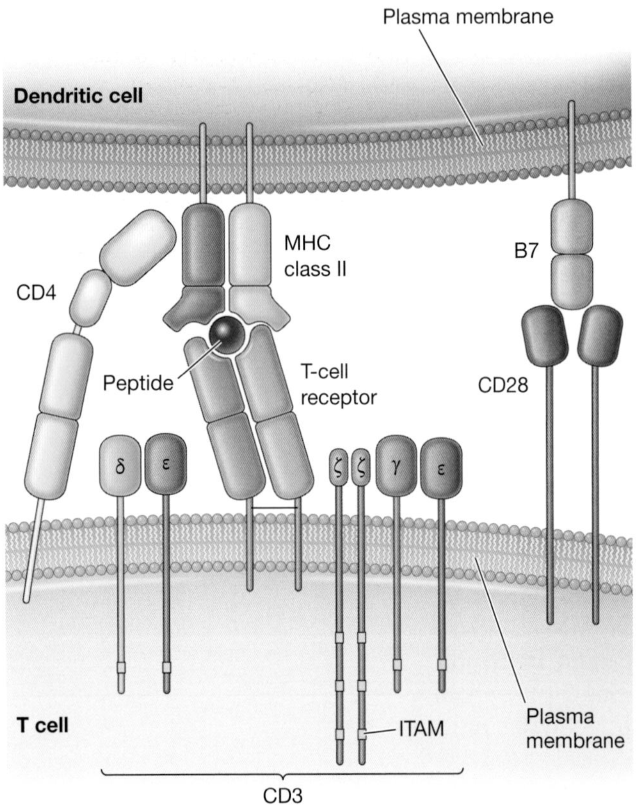

FIGURE 7.7 T-cell activation requires binding of the T-cell receptor and CD28 at an immunological synapse The T-cell receptor signaling complex contains the receptor subunits as well as the coreceptor (CD4 shown) and the CD3 complex. T cells require signaling through the receptor signaling complex and a costimulatory signal provided by CD28, which binds to B7 on a dendritic cell.

7.7 | How are T cells activated?

LEARNING OBJECTIVE

7.7.1 Name the cell-surface molecules required to form a cell–cell interaction between a T cell and an antigen-presenting cell, and explain how this process occurs.

Chapter 6 explored the interaction between the T-cell receptor and MHC-peptide complex in detail. This interaction signals the naïve T cell that foreign antigen is present and requires a response. However, the interaction between the T-cell receptor and MHC-peptide complex alone is not sufficient for T-cell activation: a **costimulatory signal** from an APC is also required. This costimulatory signal must come from the same cell that is interacting with the T-cell receptor through an MHC-peptide complex.

T cells express a cell-surface protein called CD28, which is required for T-cell activation. CD28 binds to the cell-surface protein B7 on an APC. There are actually two related B7 molecules (B7.1, also called CD80, and B7.2, also called CD86); collectively, they are referred to as B7 molecules.

To activate a naïve T cell, two interactions are required: the one between the T-cell receptor and the MHC-peptide complex on the APC and the interaction between CD28 on the T cell and a B7 molecule on the APC. Upon binding of these molecules, signals are initiated that begin the process of T-cell expansion and differentiation (**FIGURE 7.7**). Expression of B7 molecules on the surface of professional APCs only occurs when infection is present. PAMP receptors, including TLRs, are responsible for inducing B7 expression.

EVOLUTION AND IMMUNITY

T-Cell Costimulatory Molecules

Recall that both T-cell receptors and MHC class I and class II molecules entered the evolutionary tree of life abruptly with the evolution of jawed fish. We also know that proper T-cell activation requires the action of a costimulatory signal with CD28 on the T cell. Thus, it would be reasonable to predict that costimulatory molecules evolved in a similar manner as the T-cell receptor and MHC molecules.

In a 2007 study, Bernard and colleagues identified both CD28 homologs and homologs to the inhibitory receptor CTLA4 in birds, amphibians, and jawed fish (Bernard et al., 2007). Using sequence similarity to search for homologs, the authors were able to identify CD28 homologs in six different fish species and in the chicken and *Xenopus* (African clawed frog) genomes. Interestingly, a key cytoplasmic signaling motif, which contains a critical tyrosine residue that is phosphorylated during signaling, was conserved in all homologs. This key finding suggests that CD28 homologs likely play similar signaling roles in all species in which they have been identified.

The authors also found homologs of the inhibitory receptor CTLA4 in four different fish species and in the chicken and *Xenopus* genomes. Given that T-cell receptors and MHC molecules also entered the evolutionary tree of life in jawed fish, it may be speculated that the functional roles of CD28 and CTLA4 homologs in these species are conserved. Indeed, studies done by the same authors demonstrated that there is functional conservation in the cytoplasmic domains of the CD28 and CTLA4 homologs from rainbow trout when connected to a human CD28 extracellular domain. These studies suggest that the signaling molecules associated with T-cell activation likely evolved at the same time.

Naïve T cells express CD28 as a receptor for B7 molecules. However, activated T cells express CTLA4, an inhibitory receptor, instead of CD28. When CTLA4 binds to a B7 molecule, the adhesion is 20 times stronger than the connection between B7 and CD28. This provides an internal mechanism for controlling the proliferation of activated T cells. (See **EVOLUTION AND IMMUNITY** to learn more about CD28 and CTLA4.)

costimulatory signal Signal required during activation of lymphocytes in addition to antigen engagement with its receptor.

immunological synapse Connection between a T cell and an antigen-presenting cell that aids in the clustering of T-cell receptor signaling molecules and activation of the T cell.

CHECKPOINT QUESTION

1. What costimulatory interaction is required between T cells and APCs?

7.8 What signal transduction process must occur for T-cell activation?

LEARNING OBJECTIVE

7.8.1 Outline the steps involved in T-cell activation upon T-cell receptor stimulation.

The interaction between the T-cell receptor and an MHC-peptide complex, coupled with the proper costimulatory signaling, promotes the formation of an **immunological synapse** between the T cell and APC (see Figure 7.7).

The immunological synapse contains interactions between:

- The T-cell receptor and the MHC-peptide complex
- The costimulatory signaling molecules
- The T-cell coreceptor (CD4 or CD8) and the MHC-peptide complex
- Other cell-adhesion molecules, which allow a seal to form between the two cells.

Because the α and β chains of the T-cell receptors have small cytoplasmic domains, the associated CD3 proteins play a key role in relaying the signal

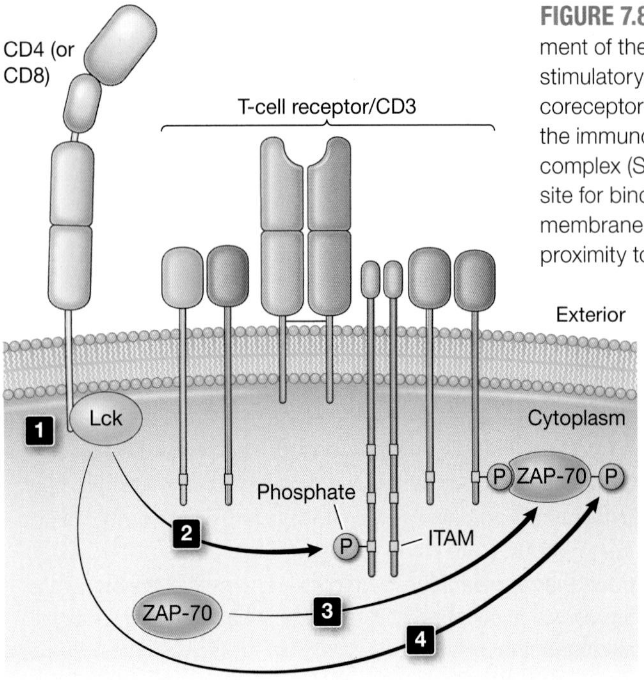

FIGURE 7.8 Signal transduction at the T-cell receptor complex Engagement of the T-cell receptor with an MHC-peptide complex (along with a co-stimulatory signal) prompts intracellular events to drive T-cell activation. The coreceptor (CD4 or CD8) interacts with the kinase Lck (Step 1). Formation of the immunological synapse allows Lck to phosphorylate ITAMs in the CD3 complex (Step 2). ITAMs with phosphate covalently bound serve as a docking site for binding of the soluble kinase ZAP-70, which is recruited to the plasma membrane (Step 3). Recruitment of ZAP-70 to the plasma membrane brings it in proximity to be phosphorylated by Lck (Step 4).

to the cytoplasm of the T cell that an immunological synapse has formed. CD3 proteins contain amino acid sequences called **immunoreceptor tyrosine-based activation motifs (ITAMs)**. ITAMs are important in signal transduction because, when phosphorylated, they form docking sites for other proteins involved in the signaling process. ITAMs mediate the interaction with cytoplasmic protein-tyrosine kinases. Remember that kinases are enzymes that catalyze the transfer of a phosphate group from a donor to an acceptor. Upon activation, kinases phosphorylate tyrosine residues in the ITAMs to form the docking sites.

immunoreceptor tyrosine-based activation motifs (ITAMs) Cytoplasmic amino acid sequences containing tyrosine that are targets of phosphorylation and subsequent activation of signaling events.

The T-cell receptor and CD4 or CD8 coreceptor are important in initiating the signal transduction cascade required for T-cell activation. The CD4/CD8 coreceptor cytoplasmic domain associates with the protein-tyrosine kinase Lck (**FIGURE 7.8**, Step 1). Upon formation of an immunological synapse, ITAMs are phosphorylated (Figure 7.8, Step 2), resulting in the recruitment of ZAP-70 (Figure 7.8, Step 3). Upon recruitment of ZAP-70 to the T-cell cytoplasmic face of the immunological synapse, it is phosphorylated and activated by the coreceptor-associated kinase Lck (Figure 7.8, Step 4). This event is key in initiating the signal transduction cascade. (To learn more about the identification of ZAP-70 and its role in T-cell receptor signaling, see **KEY DISCOVERIES**).

Upon activation, ZAP-70 induces three signal transduction pathways that lead to changes in gene expression. These signaling pathways are common in many types of cells, but their activation in naïve T cells ultimately drives clonal expansion and differentiation of the T cell. ZAP-70 initiates the three pathways through the activation of the enzyme phospholipase C-γ (PLC-γ) (**FIGURE 7.9**, Step 1), which catalyzes the cleavage of phosphatidylinositol bisphosphate to inositol triphosphate (IP_3) and diacylglycerol (DAG) (Figure 7.9, Step 2).

As in most cells, IP_3 signals the opening of calcium channels, promoting an influx of calcium ions into the cytoplasm of the T cell from extracellular and intracellular stores (Figure 7.9, Step 3a). Calcium ions activate a phosphatase called calcineurin (Figure 7.9, Step 4a), which in turn dephosphorylates the transcriptional activator nuclear factor of activated T cells (NFAT) (Figure 7.9, Step 5a). This event allows cytoplasmic NFAT to relocate to the nucleus, where it can activate genes involved in T-cell expansion and differentiation (Figure 7.9, Step 6a).

DAG, the other product of the PLC-γ reaction, is capable of initiating the other two signal transduction pathways. DAG activates protein kinase C-θ (PKC-θ) (Figure 7.9, Step 3b), which then activates the transcription factor NFκB through destruction of the associated IκB protein (Figure 7.9, Step 4b), in a mechanism similar to TLR signaling. Localization of NFκB to the nucleus of the T cell further promotes T-cell expansion and differentiation (Figure 7.9,

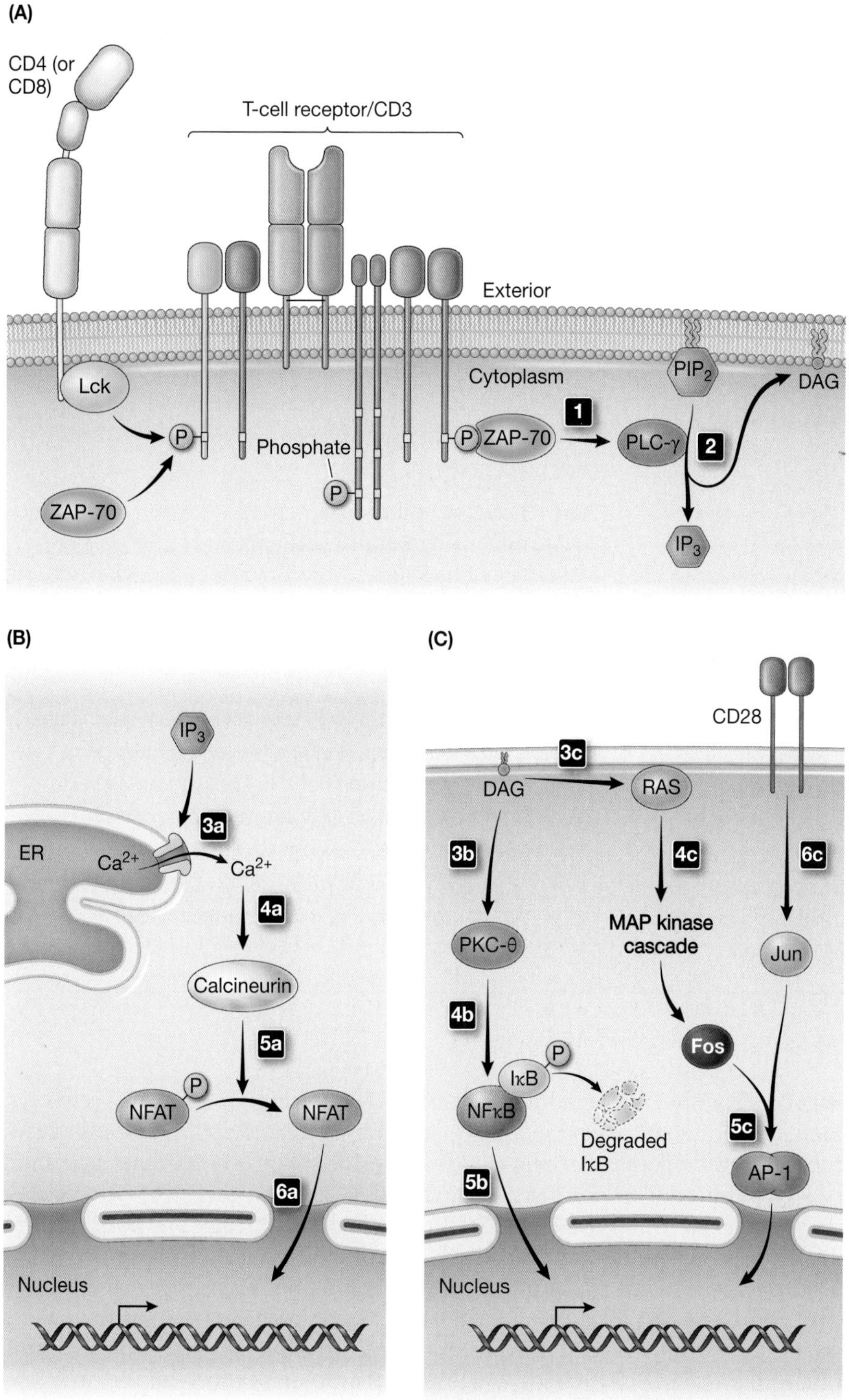

FIGURE 7.9 Signaling events that induce T-cell activation (A) Active ZAP-70, a kinase, phosphorylates and activates phospholipase C-γ (PLC-γ) (Step 1). PLC-γ is a lipase that cleaves the lipid phosphatidylinositol bisphosphate (PIP_2) into diacylglycerol (DAG) and inositol triphosphate (IP_3) (Step 2). (B) IP_3 activates a transcription factor involved in T-cell activation. As a second messenger, IP_3 also activates calcium channels in the endoplasmic reticulum (ER) membrane, resulting of movement of calcium ions (Ca^{2+}) from the ER into the cytosol (Step 3a). Ca^{2+} activates the phosphatase calcineurin (Step 4a), which removes phosphates from the transcription factor NFAT (Step 5a). Dephosphorylated NFAT translocates to the nucleus and activates genes involved in T-cell activation (Step 6a). (C) DAG also activates two transcription factors involved in T-cell activation. It activates the protein kinase PKC-θ (Step 3b), which phosphorylates IκB in order to target its destruction (Step 4b). Degradation of IκB releases the transcription factor NFκB within the cytosol, allowing it to translocate into the nucleus and transcribe genes involved in T-cell activation (Step 5b). DAG also activates the small GTPase Ras (Step 3c), triggering signaling through the MAP kinase cascade (Step 4c) and activating Fos. Fos interacts with another protein, Jun, to form the transcription factor AP-1 (Step 5c). Jun is activated through the costimulatory signal provided by CD28 (Step 6c).

KEY DISCOVERIES

How was ZAP-70 identified as a pivotal protein-tyrosine kinase in T-cell activation?

Article

A. C. Chan, M. Iwashima, C. W. Turck, and A. Weiss. 1992. ZAP-70: A 70 kd protein-tyrosine kinase that associates with the TCR zeta chain. *Cell* 71: 649–662.

Background

We have explored the mechanism of signal transduction used by T cells to signal activation once its T-cell receptor has become engaged with an MHC-peptide complex. At the crux of this activation is tyrosine phosphorylation and the activity of ZAP-70, a protein-tyrosine kinase. ZAP-70 activates phospholipase C and ultimately alters gene expression through the activation of several transcription factors. How was it determined that ZAP-70 is involved in this process and that it is a protein-tyrosine kinase?

Early Research

At the time of the 1992 study, it was known that the T-cell receptor was composed of a multi-subunit complex, including the heterodimer α and β subunits responsible for MHC-peptide recognition, as well as multiple subunits of the CD3 complex (γ, δ, ε, and ζ). It was proposed that the CD3 subunits were important in intracellular signal transduction, as the α and β subunits of the T-cell receptor only had five amino acids located on the cytoplasmic side of the membrane. It was also known that signaling through the T-cell receptor resulted in an increase in tyrosine phosphorylation in the cell, notably on both the ζ chain of CD3 and phospholipase C. It had also been shown that cross-linking of the ζ chain could bypass the activity of the T-cell receptor and result in signals related to T-cell activation. This suggested that the ζ chain plays an important role in T-cell activation signaling. Before this study, the authors had isolated a 70-kd phosphoprotein that could associate with phosphorylated ζ and named it ZAP-70 (for zeta chain associated protein of 70 kd).

Think About...

1. If you were a researcher on the cutting edge of studying ZAP-70 function, what would your hypothesis be concerning ZAP-70 and its role in T-cell signaling?
2. What experiments would be important to conduct to test your hypothesis?
3. Since the function of ZAP-70 was unknown, what critical experiments would you do to help infer the function of ZAP-70? What would you need to know about ZAP-70 to determine its function? For example, how could you predict that ZAP-70 is a protein-tyrosine kinase?
4. What would be some critical experiments for determining that ZAP-70 indeed functions as a protein-tyrosine kinase?

Step 5b). Finally, DAG activates the small GTP-binding protein Ras (Figure 7.9, Step 3c), which in turn activates the mitogen-activated protein kinase (MAP kinase) cascade (Figure 7.9, Step 4c). This cascade culminates with the activation of Fos, a component of transcription factor AP-1 that associates with another component, Jun (Figure 7.9, Step 5c). Jun is activated through the costimulatory signal received through CD28 on the T cell (Figure 7.9, Step 6c).

The actions of the three transcription factors, NFAT, NFκB, and AP-1, all promote T-cell proliferation and differentiation into effector T cells. One of the most important genes activated through this signaling process is the cytokine interleukin-2 (IL-2), which we will explore further in Section 7.9.

CHECKPOINT QUESTIONS

1. What is the importance of activating protein-tyrosine kinases at the initiation of the formation of an immunological synapse between T cells and APCs?
2. Clonal expansion of T cells requires the activation of transcription factors to promote gene activation. What three transcription factors are activated during T-cell activation?

Article Summary

After the authors had identified ZAP-70 as a ζ chain interacting protein, they used peptide sequencing and classical cloning experiments to isolate a cDNA clone of ZAP-70. Sequence analysis of the clone suggested that ZAP-70 could be a protein-tyrosine kinase. Their study went on to demonstrate that ZAP-70 has protein-tyrosine kinase activity and interacts with the ζ chain in a phosphorylation-dependent manner (**FIGURE KD 7.1**). These findings were paramount in determining that ZAP-70 plays a pivotal role in T-cell receptor signaling and activation.

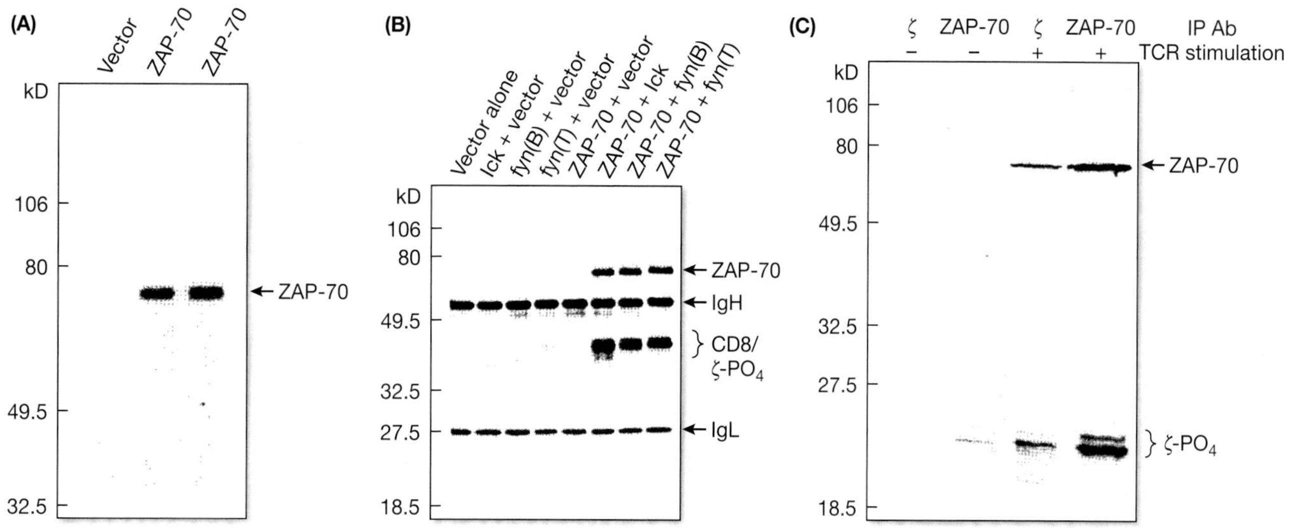

FIGURE KD 7.1 ZAP-70 is a protein-tyrosine kinase involved in T-cell signaling and activation (A) ZAP-70 expressed in COS-1 cells is capable of autophosphorylation in an in vitro kinase assay. Two identical transfections of a ZAP-70 expression vector demonstrate purification and phosphorylation of ZAP-70 in the kinase assay. (B) Phosphorylated ZAP-70 interacts with CD8-ζ chain when other kinases involved in T-cell signaling are present. COS-18 cells expressing the kinases Lck and Fyn allow phosphorylated ZAP-70 to associate with the phosphorylated ζ chain as observed by immunoprecipitation. (C) Phosphorylated ZAP-70 associates with phosphorylated ζ chain in the Jurkat T-cell line. When the T-cell receptor is stimulated in the Jurkat T-cell line, ZAP-70 can be shown to associate with phosphorylated ζ chain by immunoprecipitation. (From A. C. Chan. 1992. *Cell* 71: 649–662.)

7.9 What is the role of interleukin-2 in T-cell activation?

LEARNING OBJECTIVE

7.9.1 Define the role of interleukin-2 in T-cell activation.

How do the signal transduction cascade and subsequent changes in gene expression drive T-cell expansion and differentiation? At the center of this process is the cytokine IL-2, which is secreted by the activated T cell. The IL-2 binds to a surface receptor to drive mitosis, and thus clonal expansion, of the activated cell (**FIGURE 7.10**). While IL-2 is not the only cytokine produced by the activated T cell, it is regarded as the key player in this process.

Production of IL-2 is induced by the formation of an immunological synapse, so it requires a costimulatory signal via CD28. Cytokines, including IL-2, are typically regulated through destabilization of the mRNA responsible for

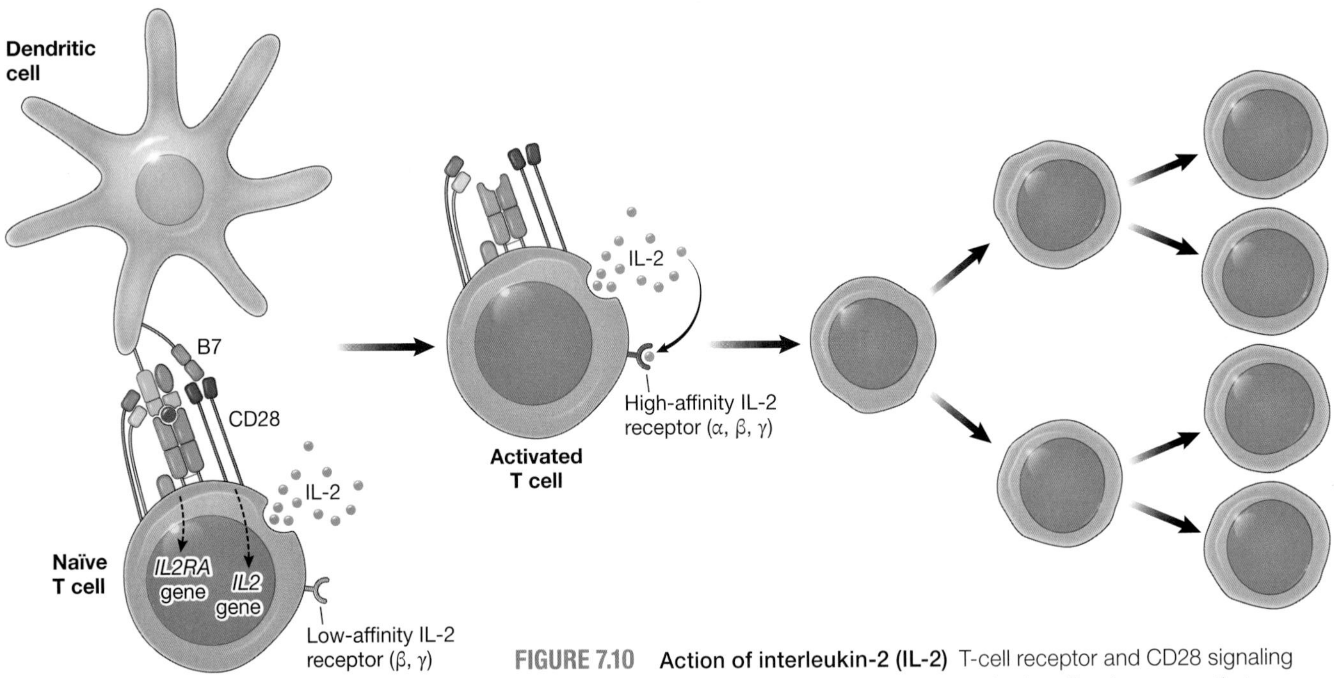

FIGURE 7.10 Action of interleukin-2 (IL-2) T-cell receptor and CD28 signaling induces expression of genes involved in T-cell activation. Two key genes that are activated are the interleukin-2 (IL-2) gene and the α subunit of the IL-2 receptor. Naïve T cells express only the low-affinity IL-2 receptor because they only produce the β and γ subunits. Activated T cells that make the high-affinity IL-2 receptor bind to IL-2, which promotes their proliferation and differentiation.

protein synthesis. The signal transduction activated through CD28 employs mechanisms that work to stabilize the IL-2 mRNA. This signal also enhances transcriptional activation of the IL-2 gene. The signal prolongs the half-life of IL-2 mRNA and enhances IL-2 mRNA transcription, resulting in a large increase in IL-2 production by the activated T cell. Conversely, the downregulatory signaling molecule CTLA4 prevents CD28 from acting as a costimulatory signal in an immunological synapse, so IL-2 mRNA is not stabilized in cells when CTLA4 engages with B7 on the APC.

Naïve T cells express an IL-2 receptor, but it has low affinity for IL-2. The receptor is made up of two protein subunits (β and γ). Upon T-cell activation, a third protein component of the IL-2 receptor (α) is synthesized. This protein is then transported to the cell surface, where it interacts with the β and γ subunits to form a high-affinity IL-2 receptor. This increase in IL-2 affinity ensures that only activated T cells will bind and respond to the IL-2 that is produced and that clonal expansion driven by IL-2 is restricted to those activated T cells. Upon binding, IL-2 stimulates T-cell proliferation, ultimately producing thousands of antigen-specific T cells (see Figure 7.10).

CHECKPOINT QUESTION

1. Why does IL-2 produced by activated T cells induce clonal expansion of activated T cells but not naïve T cells?

7.10 Why is costimulation important in T-cell activation?

LEARNING OBJECTIVE

7.10.1 Contrast the effects of the presence and absence of a costimulatory signal on the T-cell response.

We have seen that T-cell activation requires a functional T-cell receptor/MHC-peptide interaction coupled with the interaction of CD28 and B7 molecules. Why is this costimulatory signal so important?

We know that naïve T cells that recognize self-antigens can be produced during development (see Chapter 5). This is likely due to the thymus not presenting that particular antigen to the T cell. How, then, do we prevent potentially self-reacting T cells from mounting what could be a life-threatening adaptive immune response? Therein lies the importance of the costimulatory signal.

Recall that professional APCs will only produce B7 molecules when an infection is present. In the absence of B7 production, APCs still process and present antigen, but they cannot properly activate a T cell since B7 is not produced. If a naïve T cell comes into contact with an APC that is presenting self-peptides and binds tightly but does not receive the costimulatory signal, the cell is not activated. Instead, the T cell enters a state known as **anergy**, a condition in which the cell is nonresponsive and incapable of expansion and differentiation.

This mechanism of downregulating activation of a T cell that recognizes self-peptides is an alternate means of peripheral tolerance. Self-reactive T cells that escape the negative selection process of the thymus undergo negative selection in peripheral tissues by undergoing anergy, thereby preventing an autoimmune response.

anergy Cellular state of non-responsiveness in which the lymphocyte is incapable of expansion and differentiation.

autocrine Signaling mechanism whereby the same cell producing a signaling molecule responds to that signaling molecule.

paracrine Signaling mechanism whereby cells adjacent to a cell producing a signaling molecule respond to that signaling molecule.

endocrine Signaling mechanism whereby cells in one area respond to a signaling molecule produced in a different area; typically these signaling molecules are transported via the bloodstream to the responding cell.

CHECKPOINT QUESTION

1. What is the purpose of induction of anergy in T cells in circulation?

7.11 How do effector T cells utilize cytokines as part of the immune response?

LEARNING OBJECTIVE

7.11.1 Outline the steps involved in cytokine signaling that result in transcriptional activation within target cells.

Effector T cells either alter cell behavior or destroy damaged or infected cells. These cells carry out their functions through the actions of cytokines, which are produced during both the innate and adaptive immune responses.

Recall that cytokines are secreted or membrane-bound proteins that bind to cell-surface receptors to initiate signal transduction cascades. These cascades ultimately result in changes in gene expression. Cytokines can act in an **autocrine** manner, which means that they act on the cell that produced them. Alternatively, they can act in a **paracrine** manner, affecting adjacent cells. Finally, some cytokines have the potential to act in an **endocrine** manner, affecting cells throughout the body, if those cytokines are released into the bloodstream.

Making Connections: T-Cell Activation and Signal Transduction

The interaction of a T cell with an antigen-presenting cell involves an MHC/HLA-T-cell receptor interaction, binding of the coreceptor CD4 or CD8, and a costimulatory signal. These interactions drive the formation of an immunological synapse, which activates T cells through signaling events inside the cytosol. Signal transduction in the engaged T cell promotes clonal expansion, and cytokines secreted from the antigen-presenting cell promote differentiation into effector T cells.

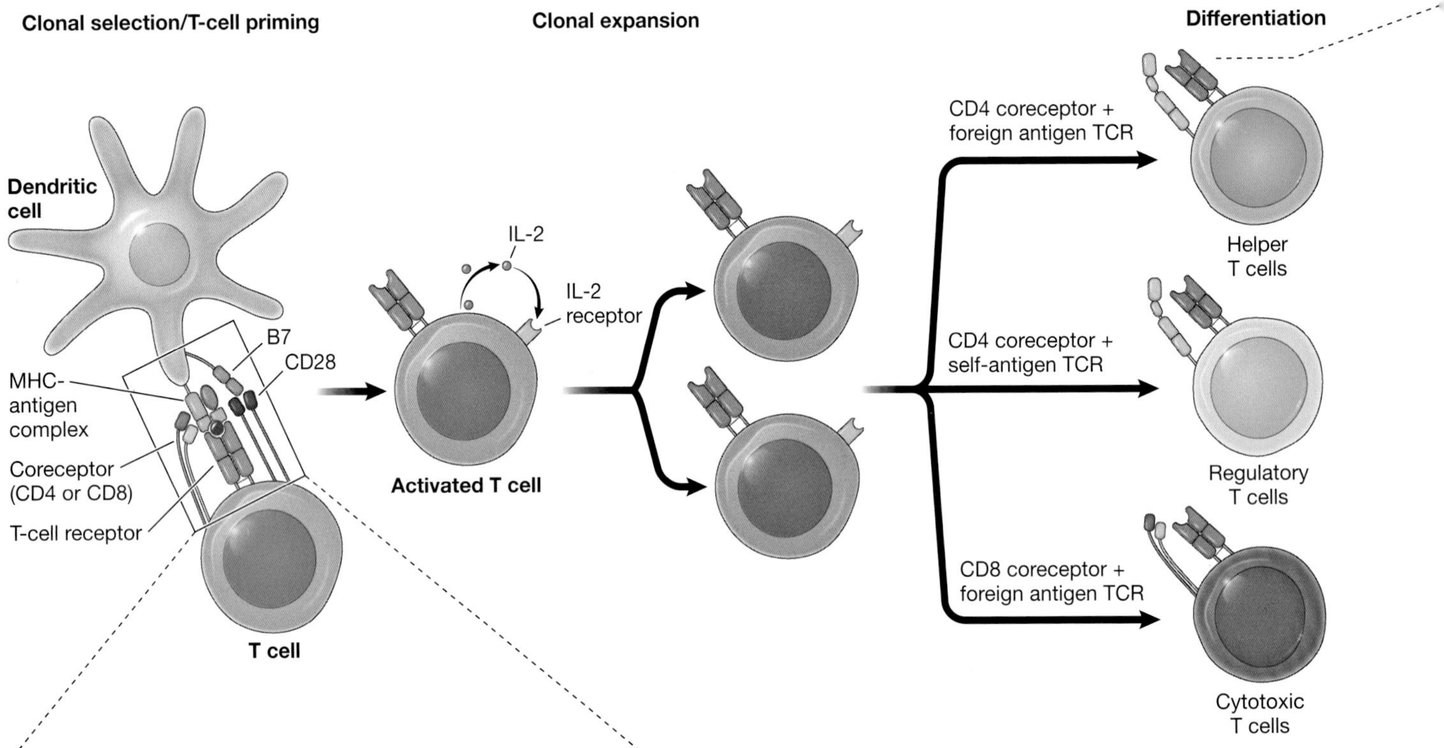

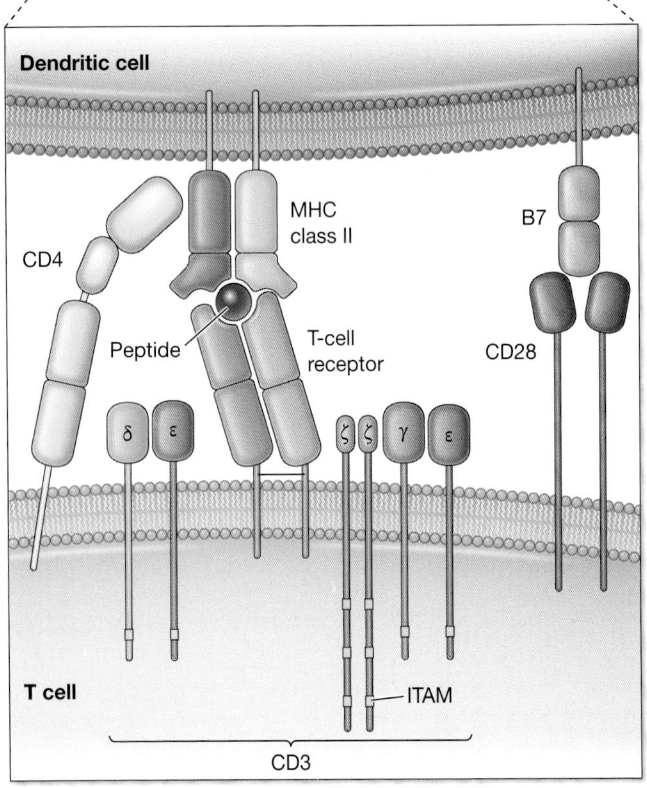

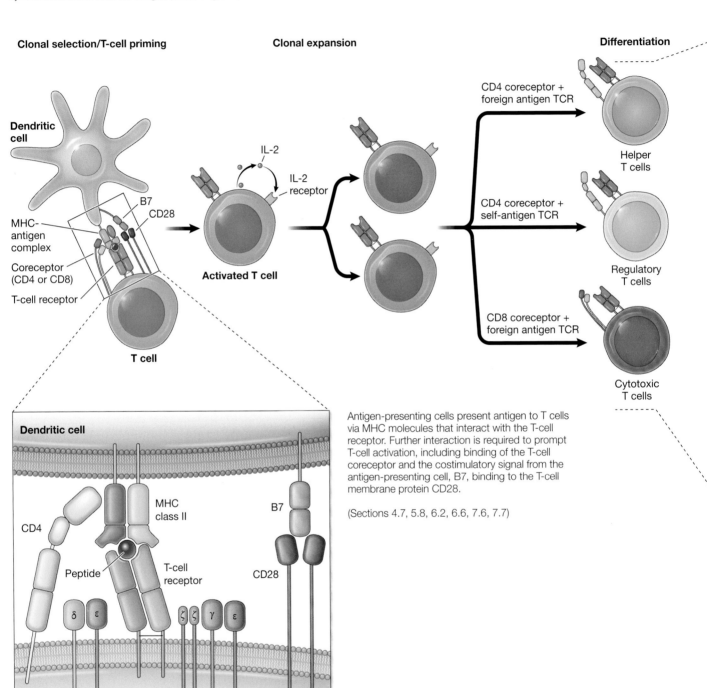

Antigen-presenting cells present antigen to T cells via MHC molecules that interact with the T-cell receptor. Further interaction is required to prompt T-cell activation, including binding of the T-cell coreceptor and the costimulatory signal from the antigen-presenting cell, B7, binding to the T-cell membrane protein CD28.

(Sections 4.7, 5.8, 6.2, 6.6, 7.6, 7.7)

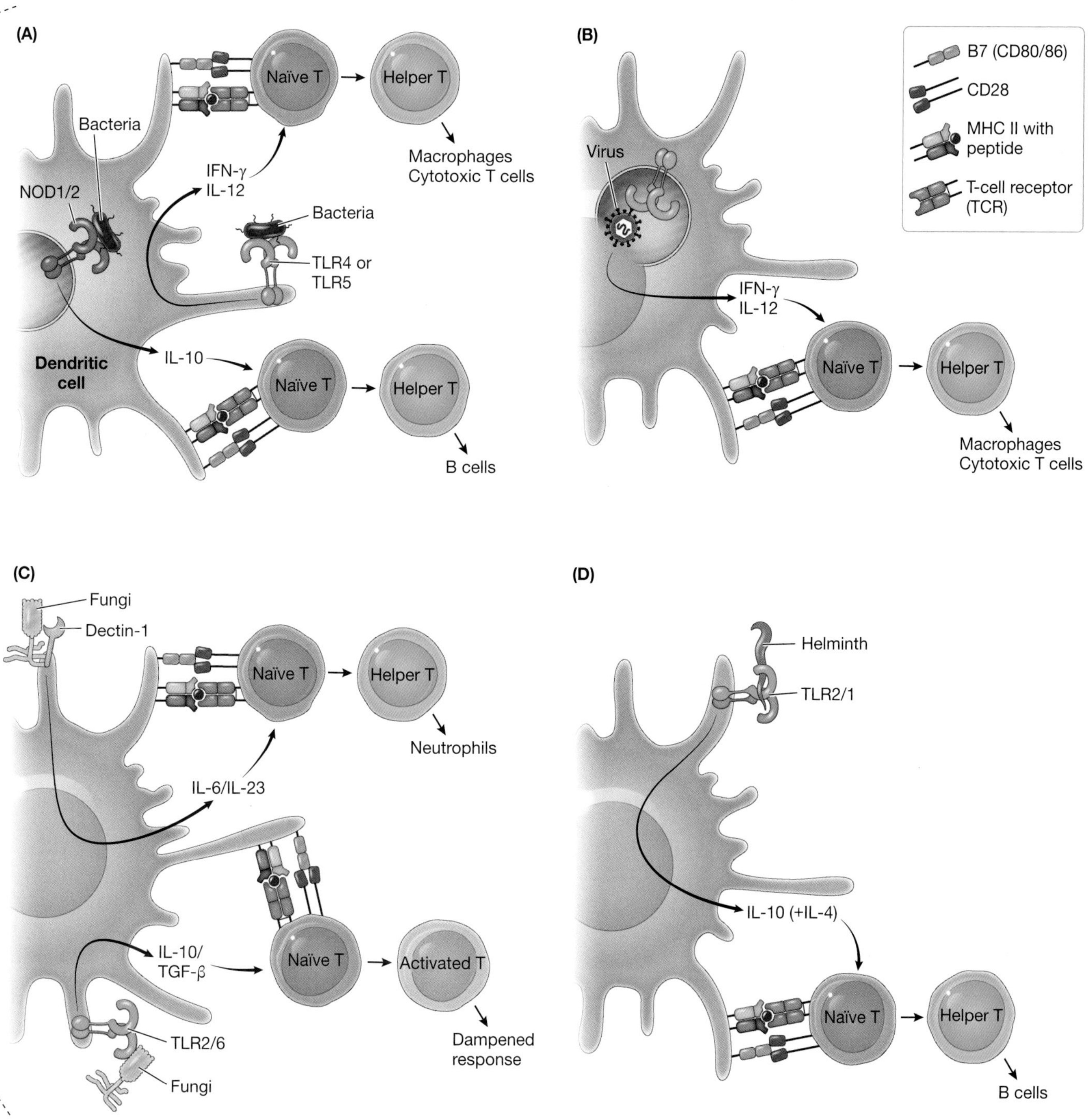

Signal transduction in activated T cells drives expression of IL-2, which is responsible for clonal expansion. Cytokines secreted from the antigen-presenting cell, such as those secreted from a dendritic cell, drive differentiation into effector T cells.

(Sections 4.7, 7.6, 7.7, 7.8, 7.9, 7.12)

FIGURE 7.11 Cytokine signaling pathway Cytokine receptors are transmembrane proteins with a cytosolic domain that interacts with the kinase JAK. Cytokine binding to the receptor causes the cytokine receptors to dimerize and JAKs to phosphorylate each other and the cytokine receptors (Step 1). The transcription factor STAT is recruited to the phosphorylated cytokine receptor, and JAK phosphorylates the recruited STAT (Step 2). Phosphorylated STATs dimerize and translocate to the nucleus, where they activate transcription (Step 3).

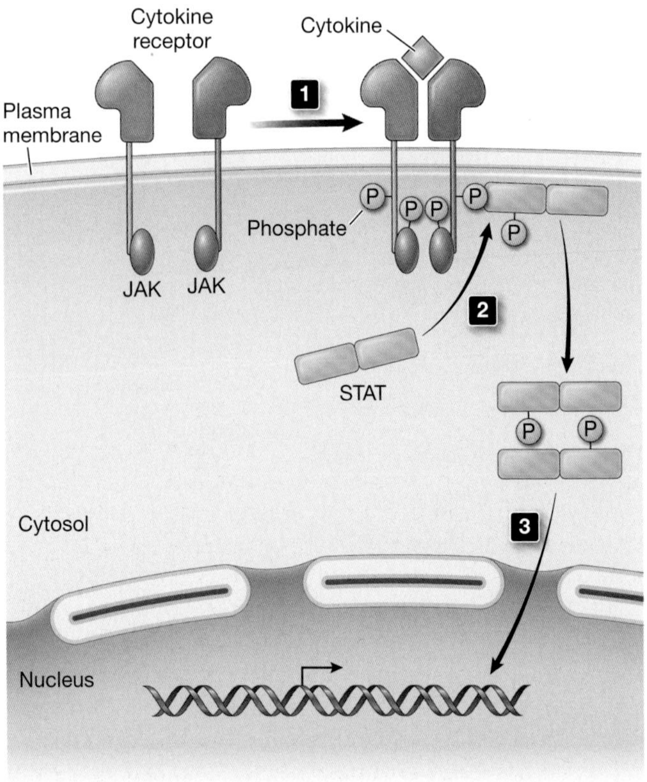

T_H1 helper T cells Effector T cells that function to aid in activation of macrophages.

T_H2 helper T cells Effector T cells that function to aid in activation of B cells.

T_H17 helper T cells Effector T cells that function to aid in neutrophil activation.

Most cytokine receptors have a cytoplasmic tail that associates with kinases known as Janus kinases (JAKs). Cytokine binding to its cognate receptor causes receptor dimerization and kinase activation (**FIGURE 7.11**, Step 1). Activated JAKs phosphorylate proteins known as signal transducers and activators of transcription (STATs), which function as transcription factors (Figure 7.11, Step 2). Phosphorylated STATs dimerize and relocate from the cytoplasm to the nucleus, where they act as transcription factors to induce gene expression (Figure 7.11,Step 3).

Effector T cells differ in the cytokines that they produce. This difference is due to the type of immune response that each type of effector cell is responsible for inducing. The cytokines produced by **T_H1 helper T cells** act primarily to activate macrophages, those produced by **T_H2 helper T cells** work to activate B cells and granulocytes in response to parasitic infections, those produced by T_{FH} helper T cells work to activate B cells for immunoglobulin production, and those produced by **T_H17 helper T cells** work to activate neutrophils. Cytotoxic T cells primarily work to destroy infected cells, but they can produce cytokines to activate other cells of the immune system.

During the process of T-cell activation, a subset of cells produced during clonal expansion of activated T cells will promote *memory* of the interaction with the activating antigen. These *memory T cells* typically reside near where they were first produced and monitor for the presence of the activating antigen. Upon subsequent exposure, they work to produce a faster and more robust adaptive immune response. We'll explore the production and function of memory T cells in Chapter 11.

CHECKPOINT QUESTION

1. Summarize the signal transduction events involved in cytokine signaling.

7.12 What signaling process drives T-cell differentiation into various types of effector T cells?

LEARNING OBJECTIVE

7.12.1 List the cytokines required for the differentiation of activated T cells into helper T cells or cytotoxic T cells.

We have seen that T cells differentiate into effector cells, either helper T cells or cytotoxic T cells, that carry out specific duties. So what signaling process drives this differentiation into the proper type of effector T cell?

Production of CD4 effector T cells

T_H1 helper T cells primarily activate macrophages. Activated CD4 T cells in the presence of the cytokines IL-12 and IFN-γ (produced by the innate immune system during an infection) respond and differentiate into T_H1 helper T cells. These cells secrete a large quantity of IFN-γ, which is required for macrophage activation and also promotes differentiation of more T_H1 helper T cells.

T_H2 helper T cells primarily activate B cells to express immunoglobulins needed for the activation of granulocytes (including basophils, mast cells, and eosinophils) in response to multicellular parasites. Activated CD4 T cells in the presence of the cytokine IL-4 respond by secreting IL-4 (to stimulate further B-cell activation to combat parasites by activating granulocytes) and IL-5.

T_{FH} helper T cells primarily activate B cells to produce immunoglobulins required to combat extracellular pathogens. Some activated CD4 T cells in the T-cell zone of the secondary lymphoid tissue will migrate to a primary follicle within the secondary lymphoid tissue and differentiate into T_{FH} cells. While in the follicle they secrete IL-21 and IL-4 to promote B-cell activation and isotype switching within the activated B cells.

T_H17 helper T cells primarily activate neutrophils so they can target extracellular pathogens. Activated CD4 T cells in the presence of cytokines including TGF-β, IL-6, and IL-21 respond by secreting IL-17, a cytokine that aids in neutrophil activation.

Production of CD8 effector T cells

CD8 T cells require a much stronger costimulatory signal to become cytotoxic T cells. Typically this signal can only be provided by dendritic cells, which synthesize the B7 molecule. However, all cells must be able to activate CD8 T cells to potentially thwart an intracellular infection. How, then, can the naïve CD8 T cell be activated?

The answer lies in the action of CD4 T cells. CD4 T cells either can interact simultaneously with APCs as CD8 T cells or can interact successively. CD4 T cells that recognize antigen can first interact with an APC and activate it to produce more B7 molecules on the cell surface. This enhanced B7 molecule expression then provides the costimulatory signal needed to activate naïve CD8 T cells. If the B7 molecule production is at a low level on the APC surface, naïve CD4 T cells recognizing the antigen presented by MHC class II can activate to produce IL-2. If a naïve CD8 T cell interacts with the same APC at the same time, the IL-2 produced by the activated CD4 T cell can stimulate proliferation and differentiation of the adjacent CD8 T cell.

The stringent requirements of CD8 T-cell activation are related to their cytotoxic activity. As we will see shortly, cytotoxic T cells act to destroy infected cells but can also cause tissue damage. The CD8 T-cell activation process ensures that a cytotoxic T-cell response is indeed necessary before risking damage to infected tissues.

CD8 cytotoxic T cells Activated and differentiated T cells that function to target and destroy cells infected intracellularly.

cytotoxins Protein products produced by immune cells that act to target and kill intracellularly infected cells.

Effector T cells do not require costimulation

We learned earlier that naïve T cells require a costimulatory signal to become effector T cells. However, effector T cells undergo cellular changes that eliminate the need for costimulation. This change allows effector T cells to mount an effective immune response.

Cytotoxic T cells have the ability to kill any cell infected with an intracellular pathogen, but most of these cells do not express B7 molecules. Since cytotoxic T cells do not require costimulation to respond to an infection, they can be activated in a secondary lymphoid tissue and go right to work fighting the infection.

Similarly, helper T cells interact with both macrophages and B cells, and these cells have varying amounts of B7 molecules on their surface. Since there is no need for costimulation of helper T cells, the macrophages and B cells at a site of infection can activate a larger number of helper T cells within the secondary lymphoid tissue during the initial stages of infection.

CHECKPOINT QUESTION

1. Describe the overall roles of the effector T cells T_H1, T_H2, T_{FH}, T_H17, and cytotoxic T cells.

7.13 | How do cytotoxic T cells destroy target cells?

LEARNING OBJECTIVE

7.13.1 Explain the mechanism utilized by cytotoxic T cells to target and destroy cells infected with intracellular pathogens.

Pathogens that take up residence inside cells are protected from antibodies and other components of the immune system such as complement. To eliminate an intracellular pathogen, the infected cell must destroy the pathogen, or the body's immune defenses must destroy the infected cell. Cytotoxic T cells attack and destroy infected cells that cannot otherwise control or eliminate an intracellular pathogen. This self-sacrifice of the infected cell minimizes the spread of infection into neighboring cells. The action of **CD8 cytotoxic T cells** is so potent that it is being exploited to destroy tumor cells in patients (see **EMERGING SCIENCE**).

Cytotoxic T cells contain lytic granules, and the granules contain specialized proteins called cytotoxins. **Cytotoxins** are substances that kill infected cells. The lytic granules begin to form and are filled with cytotoxins as soon as a naïve CD8 T cell is activated in a secondary lymphoid tissue. Upon activation, cytotoxic T cells leave the secondary lymphoid tissue and migrate to sites of infection. Upon arrival, they survey for infected cells presenting foreign peptides via MHC class I molecules.

When a cytotoxic T cell binds to the proper MHC-peptide complex, cytotoxins are released. To minimize damage to healthy tissue, cytotoxic T cells focus granule secretion at the location of interaction with the target cell (**FIGURE 7.12**).

FIGURE 7.12 Cytotoxic T cells polarize to secrete cytotoxins toward target cells Upon interaction with the target cell, the cytotoxic T cell releases the lytic granules (stained in red) toward the bound target cell, eventually causing the death of the target cell (and loss of the green marker peptide) through the action of the lytic granule components.

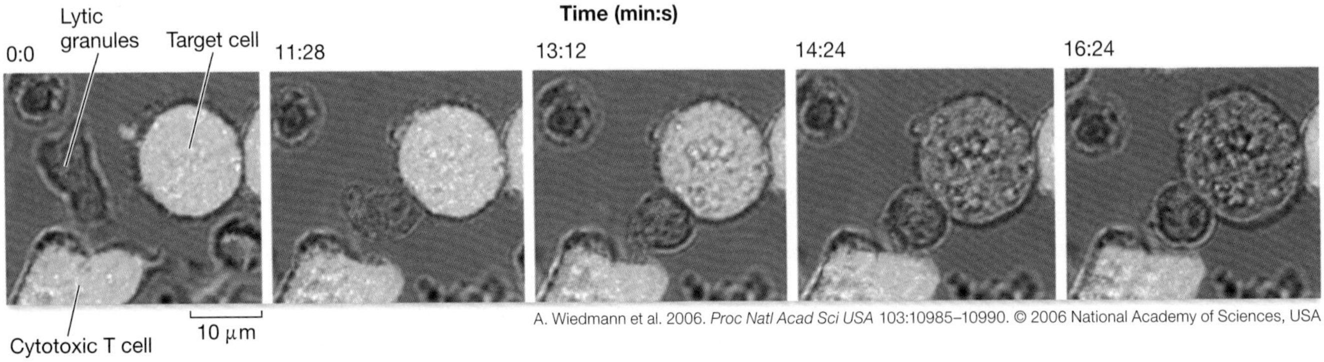

A. Wiedmann et al. 2006. *Proc Natl Acad Sci USA* 103:10985–10990. © 2006 National Academy of Sciences, USA

Can the immune system be trained to destroy tumor cells?

Article

M. Heibesen, L. Baitsch, D. Presotto, P. Baumgaertner, P. Romero, O. Michielin, D. E. Speiser, and N. Rufer. 2013. SHP-1 phosphatase activity counteracts increased T-cell receptor affinity. *Journal of Clinical Investigation* 123: 1044–1056.

Background

We have seen that T-cell activation is dependent on T-cell receptor engagement with the MHC-peptide complex coordinated with costimulatory signals. We also know that T-cell development in the thymus depends on T-cell receptor affinity to MHC-peptide complexes and that extremely high affinity results in negative selection of those cells. An avenue currently being explored by researchers is adoptive immune cell transfer in the treatment of tumors. Using this process, CD8 cells are developed that will selectively bind to tumor cells to target them for destruction. The challenge lies in developing a T-cell line that can do two things: bind selectively to tumor cells and thereby minimize damage to healthy tissue, and "hide from" regulatory T cells, whose job is to anergize self-reactive T cells.

The Study

Researchers developed a series of T-cell lines that contained receptors with varying degrees of affinities to a specific tumor antigen HLA-A2/NY-ESO-1, from K_ds of 15 nM to 5 μM. They found that although T-cell receptors that had affinities to the tumor antigen ranging between 1 and 5 μM maintained optimal function, higher affinities (K_ds less than 1 μM) actually caused the T cells to behave as if they hadn't been activated (exhibiting lower gene expression and signaling). The researchers went on to demonstrate that the phosphatase SHP-1 (for Src homology 2 containing phosphatase 1) was responsible for negatively regulating CD8 T cells with high-affinity T-cell receptors and that inhibition of SHP-1 could reverse this negative regulatory effect (**FIGURE ES 7.1**). The findings of this study could lead to new pharmacological treatments of individuals going through adoptive immune cell transfer to further inhibit tumor growth.

Think About...

1. Based on your knowledge of T-cell signaling, what might some of the substrates of SHP-1 be?
2. How would you design an in vivo mouse experiment to test the role of SHP-1 in negatively regulating adoptive-transfer-designed T cells with varying affinities for tumor cells? What would be your negative control in this experiment?
3. What are some possible negative impacts of adoptive immune transfer used as a type of chemotherapy? How could these negative effects be minimized?

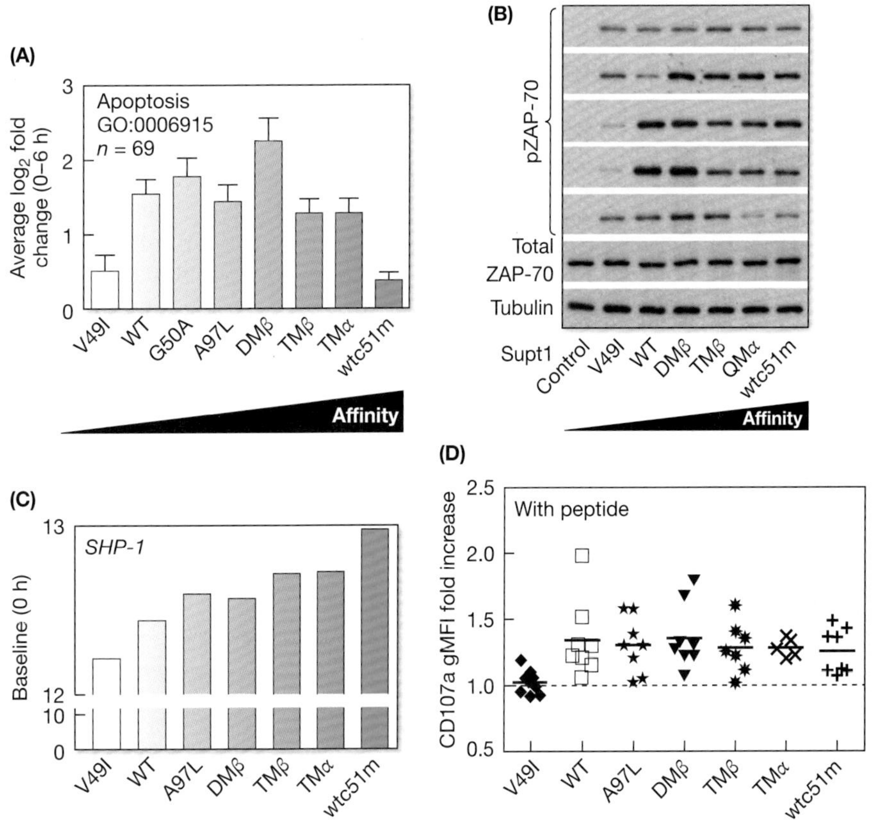

FIGURE ES 7.1 T-cell receptors with high affinity are inhibited by the action of the phosphatase SHP-1 (A) CD8 T cells expressing high-affinity receptors were less capable of inducing apoptosis in target cells compared to non-mutated T-cell receptors (WT) or those with lower affinities. (B) Activated ZAP-70 (pZAP-70) was lower in T cells expressing T-cell receptors with high affinity compared to cells expressing T-cell receptors with moderate affinity, suggesting an impairment in proper T-cell receptor signaling in these cells. (C) Cells expressing higher-affinity T-cell receptors also expressed higher amounts of the phosphatase SHP-1, suggesting that SHP-1 is capable of negatively impacting receptor signaling in T cells expressing high-affinity receptors. (D) The negative effect SHP-1 has on signaling in cells expressing high-affinity T-cell receptors could be blocked using a peptide inhibitor of SHP-1. All T-cell receptors of differing affinities had the same level of activity when an inhibitor of SHP-1 was used to treat the cells with different receptors, showing that inhibition of SHP-1 could reverse the negative impact of signaling in cells with high-affinity receptors. (From M. Hebeisen et al. 2013. *J Clin Invest* 123: 1044–1056. doi.org/10.1172/JCI65325. © 2013 The American Society for Clinical Investigation.)

Upon destruction of the infected cell, the cytotoxic T cell is released and moves on to target and destroy neighboring infected cells (**FIGURE 7.13**).

As cytotoxic T cells work to cap the spread of an intracellular infection, they must also ensure that contents of the target cell are not released, as release can exacerbate an immune response. To do this, they induce **apoptosis** (programmed cell death). With apoptosis, the target cell undergoes a series of reactions that destroy its genetic information but maintain the integrity of the cellular membrane. Apoptotic cells ultimately shed membrane vesicles containing cellular content, which are removed by phagocytes.

Cytotoxic T cells initiate apoptosis using two different mechanisms (**FIGURE 7.14**):

- *Creating pores in the target cell membrane.* The cytotoxins released from lytic granules include **perforin** (a protein that creates pores in membranes), **granulysin**, and **granzymes** (serine proteases). Upon their release, perforin, granulysin, and granzymes are endocytosed by the neighboring target cell. Perforin and granulysin create small pores in the endosomal membrane to deposit granzymes into the target cell. Granzymes then induce a proteolytic cascade that results in activation of the protease caspase-3, which cleaves key proteins involved in the induction of apoptosis.
- *Expressing the protein Fas ligand.* **Fas ligand** (**FasL**) binds to Fas on the target cell surface. Binding of FasL to Fas on the target cell causes receptor clustering of Fas on the target cell and the recruitment of a protein known as Fas associated protein with death domain (FADD) to the plasma membrane. FADD movement to the plasma membrane attracts the inactive apoptosis protease pro-caspase-8 to the membrane. Here, it is proteolytically activated to caspase-8, which activates caspase-3 and induces apoptosis in the target cell.

CHECKPOINT QUESTION

1. Summarize the two mechanisms of apoptosis induction that cytotoxic T cells use on target cells.

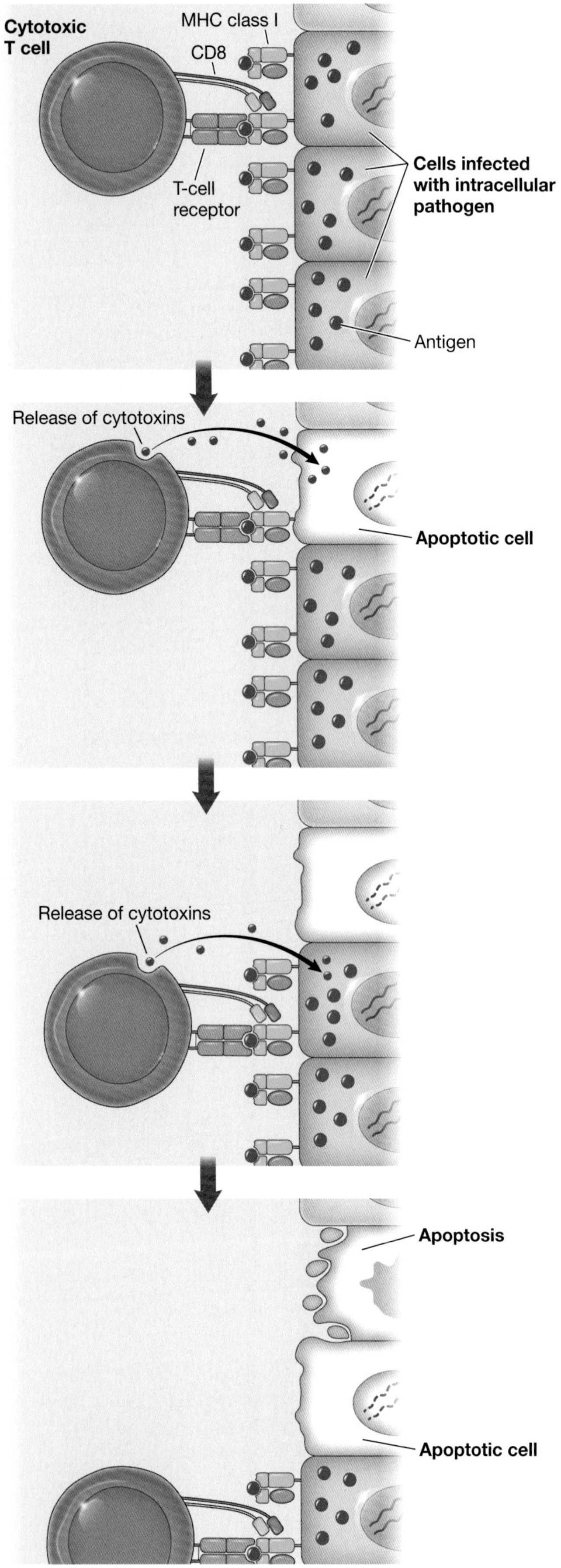

FIGURE 7.13 Cytotoxic T cells scan for intracellular pathogens Cytotoxic T cells monitor MHC class I molecules on the surface of cells to scan for the presence of the antigen their T-cell receptor can bind. Recognition of the receptor-specific antigen presented on the surface of a cell signals to the cytotoxic T cell the presence of an intracellular foreign antigen and drives activation of its effector function. The infected cell is destroyed by apoptosis. The same cytotoxic T cell can monitor neighboring cells for the presence of the same intracellular infection. (After P. Parham. 2014. *The Immune System*, 4th ed., W. W. Norton and Company.)

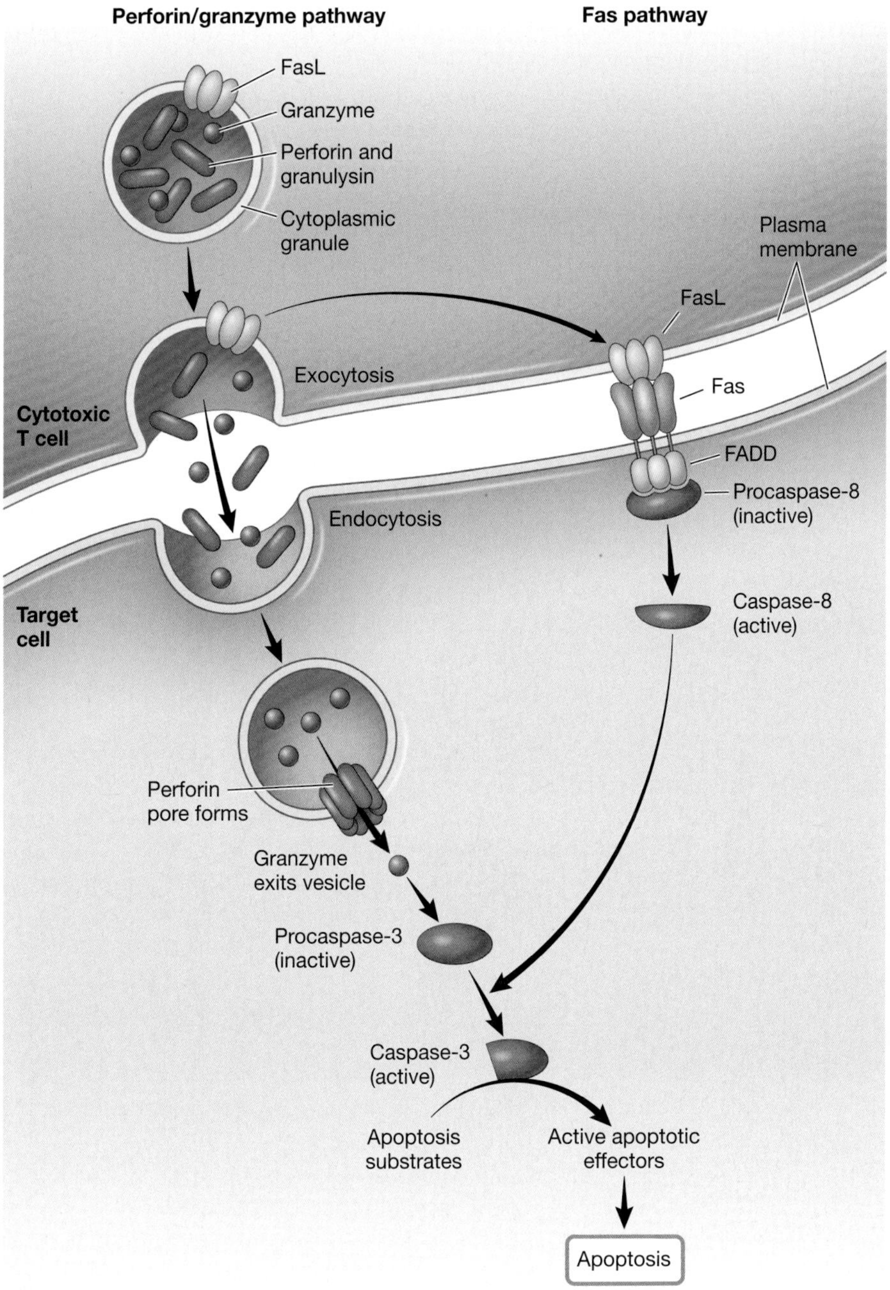

FIGURE 7.14 Cytotoxic T cells can induce apoptosis via two pathways Cytotoxic T cells release cytotoxins from cytoplasmic granules toward a target cell. The granules fuse with the T-cell plasma membrane during exocytosis in the T cell, and granule components are endocytosed by the target cell. In the first pathway, perforin and granulysin, two granule components, form pores in the endosomal membrane, allowing delivery of granzyme (another granule component) into the target cell. Granzyme activates caspase-3, inducing apoptosis. In the second pathway, the granule protein FasL, which is incorporated in the T-cell plasma membrane during granule exocytosis, interacts with Fas on the target cell. This interaction recruits FADD to the target cell membrane, activating the protease caspase-8, which in turn activates caspase-3 and induces apoptosis.

apoptosis Programmed cell death; a mechanism utilized to destroy a cell without releasing intracellular contents into the extracellular environment.

perforin Protein expressed by cytotoxic T cells that functions to perforate small holes in target cell membranes in order to deliver cytotoxins and induce apoptosis.

granulysin Protein expressed by cytotoxic T cells that functions to perforate small holes in target cell membranes in order promote pathogen lysis.

granzyme Serine protease expressed by cytotoxic T cells that is delivered to target cells to induce apoptosis.

Fas ligand (FasL) Cell-surface molecule expressed by cytotoxic T cells and natural killer cells that binds to Fas on target cells to induce apoptosis.

7.14 | What are the functions of helper T cells?

LEARNING OBJECTIVE

7.14.1 Compare and contrast the properties and functions of T_H1, T_H2, T_{FH}, and T_H17 helper T cells.

conjugate pair Cell-cell interaction of an effector T cell and a macrophage that the effector T cell will activate that is driven by proteins at the surface of both cells.

CD40 ligand (CD40L) Ligand expressed by helper T cells that must bind to CD40 on either macrophages or B cells in order to aid in proper macrophage or B-cell activation by the helper T cell.

Recall that there are various subsets of helper T cells (see Table 4.1). In this section, we'll focus on the T_H1, T_H2, T_{FH}, and T_H17 subsets. Each plays a distinct role in the adaptive immune response.

T_H1 helper T cells

The primary function of T_H1 helper T cells is to activate macrophages. Activated T_H1 cells leave the secondary lymphoid tissue and move to the site of infection. Macrophages there have begun to phagocytose pathogens, processing antigens via the MHC class II presentation pathway. T_H1 cells can bind to the MHC-peptide complex to form a strong immunological synapse with the presenting macrophage. This cell pair is known as a **conjugate pair**.

T_H1 helper T cells release cytokines to induce macrophage activation. This stimulation increases the macrophage's ability to engulf and destroy pathogens. Macrophage activation results in more efficient fusion of phagosomes with lysosomes to enhance digestion of phagocytosed particles. Activated macrophages also synthesize microbicidal molecules, including nitric oxide, reactive oxygen species, and proteases, that work to destroy engulfed pathogens.

Macrophages require two important signals for activation to occur, both supplied by T_H1 cells (**FIGURE 7.15**). The most important signal is provided by IFN-γ, secreted by T_H1 cells. The other necessary signal originates with the T-cell-surface molecule **CD40 ligand** (**CD40L**), which binds to CD40 on the macrophage surface.

Macrophage activation induces expression of more CD40 as well as TNF-α receptor, another cytokine capable of macrophage activation. Finally, macrophages produce TNF-α, which acts synergistically to enhance activation.

Unlike cytotoxic T cells, T_H1 helper T cells do not store the molecules necessary for macrophage activation but rather produce them upon interaction. The T_H1 cell of a conjugate pair initiates synthesis of molecules such as IFN-γ and CD40L. Contact with the macrophage must continue for this production to continue. The interaction allows the T_H1 cell to direct the production and location of CD40L on its surface as well as the direction to which the IFN-γ should be secreted.

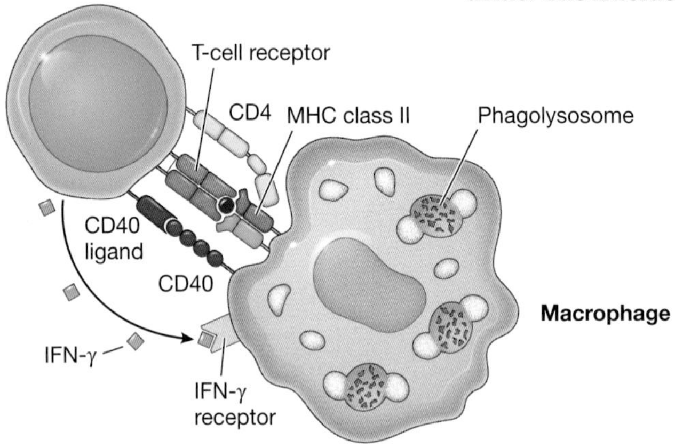

FIGURE 7.15 T_H1 helper T cells activate macrophages via the cytokine interferon-γ and CD40 ligand A T_H1 cell expresses a receptor that can recognize an MHC-peptide complex on the surface of a macrophage. The T_H1 cell binds to the macrophage and activates it through the interaction of CD40 ligand with CD40 on the macrophage surface and through the cytokine action of interferon-γ (IFN-γ).

T_H1 helper T cells are also critical in the host response to pathogens that reside in macrophages, such as mycobacteria. Intracellular pathogens in macrophages have evolved mechanisms to evade normal macrophage function. This intracellular niche also protects the organism from extracellular immune system components such as complement and antibodies. T_H1 cells and the IFN-γ and CD40L that they produce play a crucial role in activating the infected macrophage and enhancing its function. Chronically infected macrophages eventually lose the ability to become activated by this mechanism, so they can be destroyed by FasL produced by T_H1 cells.

T_H1 cells can also play a role in stimulating the production of opsonizing antibodies to extracellular pathogens. They produce the costimulatory signals required to activate B cells that recognize antigen. T_H1 cells promote the formation of opsonizing antibodies by inducing isotype switching of activated B cells to favor opsonizing antibodies.

T_H2 helper T cells

The main function of T_H2 cells is to activate B cells and granulocytes to combat multicellular parasites. T_H2 helper T cells release cytokines such as IL-5 to induce activation of these granulocytes. T_H2 helper T cells also secrete the cytokine IL-4, which activates B cells to express the immunoglobulin isotype IgE, used by granulocytes such as mast cells for pathogen recognition and the subsequent inflammatory response that works to expel pathogens.

T_H2 helper T cells can also play a role in macrophage activation. Some of the cytokines produced by T_H2 cells, including transforming growth factor-β (TGF-β), IL-4, IL-10, and IL-13, suppress macrophage activation. Thus, T_H2 cell action can suppress T_H1 cell responses.

follicular helper T cells (T_{FH}) A differentiated subset of T cells responsible for activating B cells in a germinal center in secondary lymphoid tissue.

primary focus Cluster of activated B cells and T cells that is induced in germinal centers and is important in an adaptive immune response.

Follicular helper (T_{FH}) T cells

Helper T cells that aid in the activation and differentiation of B cells in germinal centers of secondary lymphoid tissue are **follicular helper T cells (T_{FH})**. T_{FH} helper T cells are maintained in the follicle of secondary lymphoid tissues due to the expression of the cell-surface homing receptor CXCR5.

B cells that have migrated from a site of infection to a draining secondary lymphoid tissue will present antigens via MHC class II to T cells in the T-cell zone. If a B cell presents an antigen recognized by a T_{FH} cell, a conjugate pair is formed between the B cell and T cell. Upon B-cell activation and expansion of activated B cells and T cells within the secondary lymphoid tissue, a **primary focus** is formed, which is a cluster of proliferating activated T cells and B cells.

The T_{FH} cell first responds through the synthesis of CD40L, which binds to CD40 on the B-cell surface, prompting B-cell division. This interaction, along with the cytokine IL-4 secreted by the T_{FH} cell, promotes clonal expansion. Other cytokines produced by the T_{FH} cell, including IL-21, promote B-cell differentiation into plasma cells (**FIGURE 7.16**).

Antigen presentation in a B cell involves antigen binding to surface immunoglobulin and subsequent receptor-mediated endocytosis. Only B cells that have internalized a pathogen in this manner will process sufficient antigens to present to T cells. The processing may also provide additional epitopes for T-cell activation. Thus, a B cell and T cell may recognize the same pathogen via different epitopes. For example, a B cell will usually recognize surface molecules, and a T cell may recognize an internal protein antigen. The interaction between B cells and T_{FH} cells has provided a means to develop more potent vaccines (see Chapter 11).

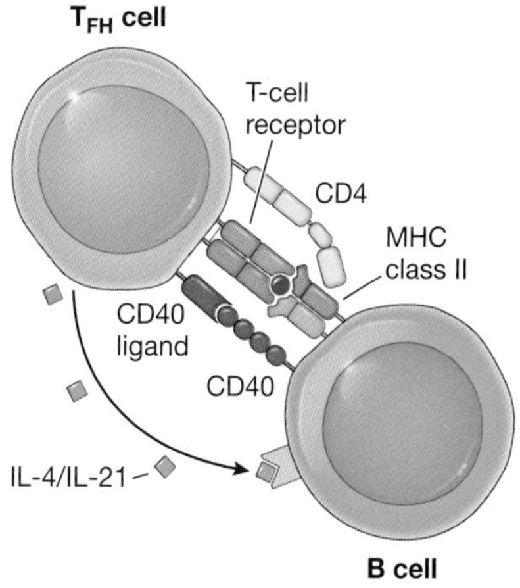

FIGURE 7.16 Follicular helper (T_{FH}) T cells activate B cells via the cytokines IL-4 and IL-21 and CD40 ligand A T_{FH} cell that expresses a receptor that can recognize an MHC-peptide complex on the surface of a B cell activates the B cell to which it is bound. The T_{FH} cell activates the B cell through the interaction of CD40 ligand with CD40 on the B-cell surface and through the cytokine action of IL-4 and IL-21, which drive B-cell proliferation and differentiation.

T_H17 helper T cells

T_H17 helper T cells are a recently identified subset of CD4 effector T cells. Their primary function is to stimulate neutrophils to clear extracellular bacteria and fungi. T_H17 cells are activated early in the adaptive immune response. Differentiation of CD4 T cells to the T_H17 subset is driven mainly by the cytokines TGF-β and IL-6. Upon differentiation, T_H17 T cells produce a variety of cytokines, but the major cytokine produced is IL-17, which induces expression of chemokines that recruit neutrophils to a site of infection. This allows for clearance of an extracellular bacterial or fungal infection.

CHECKPOINT QUESTION

1. Describe the interaction of a conjugate pair between a B cell and a follicular helper T cell.

7.15 | What are the functions of regulatory T cells?

LEARNING OBJECTIVE

7.15.1 Describe the properties and functions of regulatory T cells.

regulatory T cells (T_{regs}) Effector T cells that function to prevent autoimmunity and promote peripheral self-tolerance.

natural killer (NK) T cells Effector T cells that respond to glycolipid and perform functions that resemble both helper T cells (cytokine production) and cytotoxic T cells (apoptosis induction).

Regulatory T cells (T_{regs}) modulate the immune response, maintain tolerance to self-antigens, and prevent autoimmunity. They work to ensure that helper T cells and cytotoxic T cells are deactivated once they have served their purpose and destroyed a pathogen during an infection. Most regulatory T cells utilize the transcription factor Foxp3 to function in their role of immune response suppression, although a recently identified subset of regulatory T cells (Tr1) suppresses T-cell responses by secreting the anti-inflammatory cytokine IL-10.

A subset of T_{regs} develops to regulate effector CD4 and CD8 cells by suppressing activation. Some of these cells develop from autoreactive T cells in the thymus (known as thymus-derived or natural regulatory T cells). Others, known as peripheral or induced regulatory T cells, develop at the site of an immune response.

Regulatory T cells use a variety of different molecules to shut down effector T-cell function. Their mechanisms include:

- Utilizing immunosuppressive cytokines such as TGF-β, IL-35, and IL-10
- Utilizing micro RNAs to downregulate pathways necessary for activation of T-cell subsets
- Utilizing perforin and granzyme to bring about apoptosis
- Shutting down dendritic cell function to prevent further T-cell activation
- Disrupting metabolic function
- Expressing IL-2 receptors to sequester secreted IL-2 from activating neighboring T cells
- Expressing CTLA4 at their surface to suppress T-cell activation.

CHECKPOINT QUESTION

1. Explain the role of regulatory T cells in promoting peripheral tolerance.

7.16 | What are the functions of natural killer T cells?

LEARNING OBJECTIVE

7.16.1 Describe the properties and functions of natural killer T cells.

Natural killer (NK) T cells are an interesting subset that bears some resemblance to lymphocytes of the adaptive immune system and to NK cells of the innate immune system. Like other T cells, they develop in the thymus. But, instead of having a diverse T-cell receptor, NK T cells express a specific α:β T-cell receptor that recognizes a glycolipid presented by a CD1d molecule. NK T cells use this glycolipid–T-cell receptor interaction to combat various types of bacterial infections, including *Sphingomonas* and *Ehrlichia*. NK T cells can take on functions similar to those of helper T cells by secreting proinflammatory cytokines such as IL-6 and IFN-γ. They can also function in a manner similar to cytotoxic T cells by inducing apoptosis using Fas–Fas ligand interactions.

CHECKPOINT QUESTION

1. Why do natural killer T cells recognize only a specific subtype of bacterial infections?

Making Connections: T-Cell Differentiation and Effector Function

Activated T cells differentiate into a variety of effector T cells, each with a specific function in the cellular arm of the adaptive immune response. Some effector T cells work to combat intracellular infections, while others fight extracellular infections. One subset of effector T cells combats both types of infections, and yet another promotes peripheral tolerance. Differentiation of effector T cells depends on signals provided by the antigen-presenting cell, including specific cytokines and the MHC molecule presenting antigen.

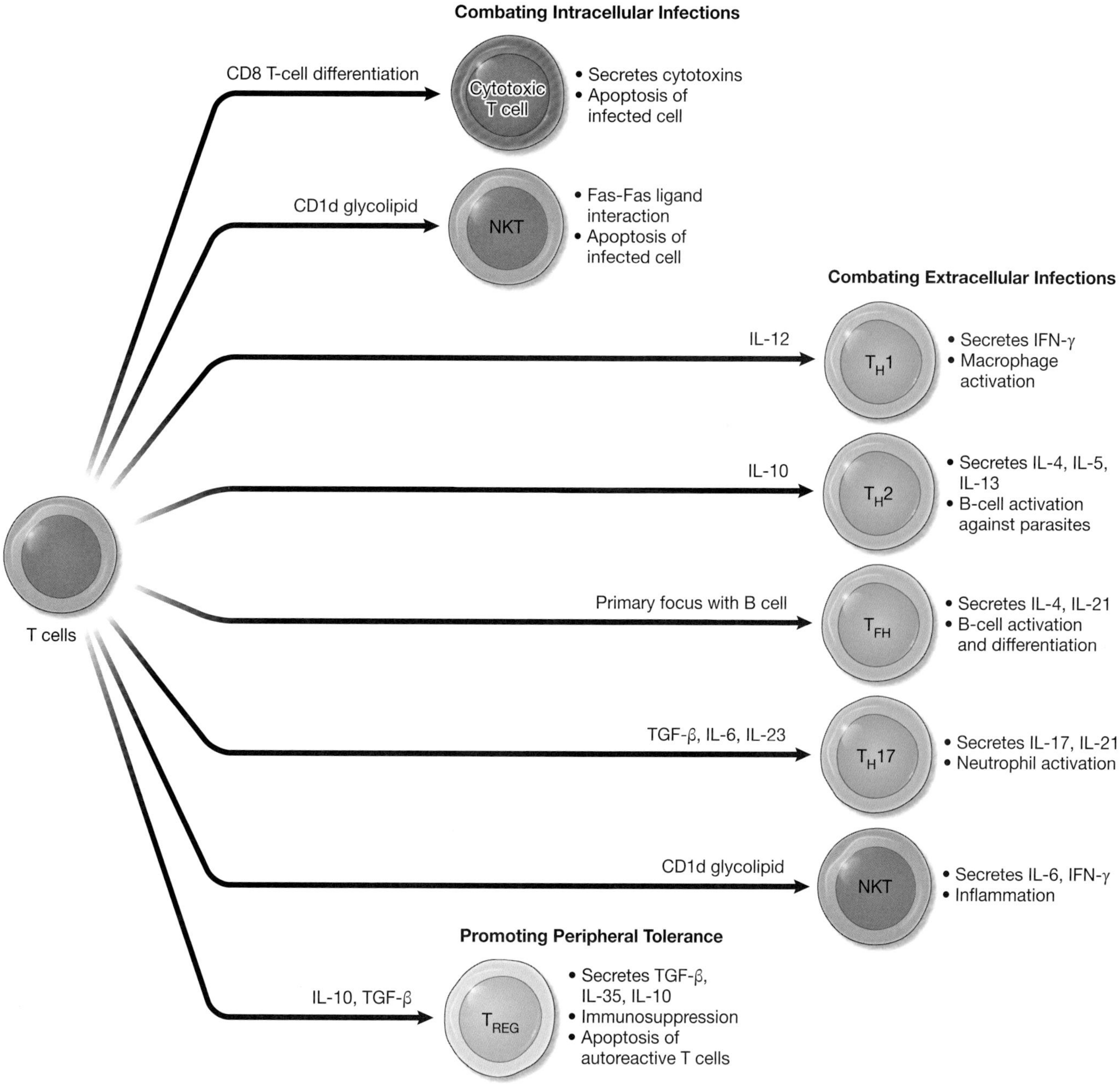

Upon activation, T cells undergo clonal expansion due to the action of IL-2 and differentiation based on interactions and/or cytokines provided by the antigen-presenting cell.

(Sections 1.10, 4.6, 4.7, 6.5, 7.12, 7.13, 7.14, 7.15, 7.16)

Each effector T cell is specialized in the type of infection it works to combat or the role it plays in peripheral tolerance. These effector functions are due to the cytokines or cytotoxins secreted by the effector T cell.

(Sections 4.5, 4.7, 5.8, 7.13, 7.14, 7.15, 7.16)

Summary

7.1 How do naïve T cells become effector T cells?

- Naïve T cells circulate through the circulatory and lymphatic systems until they enter a secondary lymphoid tissue. Here, a naïve T cell may engage with an antigen-presenting cell, activate, and become an effector T cell.
- Activated T cells expressing the CD4 coreceptor are destined to become helper T cells; those expressing the CD8 coreceptor will become cytotoxic T cells.

7.2 Which cells present antigen to T cells in secondary lymphoid tissue?

- The major antigen-presenting cells that function in T-cell activation are dendritic cells, macrophages, and B cells.

7.3 What is the role of dendritic cells in antigen presentation and T-cell activation?

- Dendritic cells play a primary role in T-cell activation. Immature dendritic cells function in phagocytosis and antigen processing, and mature dendritic cells function in secondary lymphoid tissue to present antigen to T cells.
- Dendritic cells are capable of processing a wide variety of foreign antigens, including bacteria, viruses, and pathogen products.

7.4 What is the role of macrophages in antigen presentation and T-cell activation?

- Macrophages are also involved in T-cell activation. Unlike dendritic cells, most macrophages reside in specific tissues, so just a limited number migrate to sites of infection.

7.5 What is the role of B cells in antigen presentation and T-cell activation?

- B cells, through the action of receptor-mediated endocytosis triggered by immunoglobulin engagement, can also process and present antigens to T cells.

7.6 How do T cells migrate into secondary lymphoid tissue?

- T cells migrate into secondary lymphoid tissues in a manner similar to neutrophil migration; T cells bind to cells of the secondary lymphoid tissue and migrate toward chemokines expressed in these tissues.

7.7 How are T cells activated?

- Naïve T cells require several interactions at the surface to promote activation: T-cell receptor binding to MHC-peptide complex, coreceptor interaction with the MHC molecule, and a costimulatory signal provided by CD28 on the T cell binding to B7 on the antigen-presenting cell.

7.8 What signal transduction process must occur for T-cell activation?

- The engagement of a T cell with an antigen-presenting cell induces a signaling cascade. The cascade begins through phosphorylation of cytoplasmic components of the receptor complex and ends in the activation of transcription factors that activate gene expression required for T-cell expansion.

7.9 What is the role of interleukin-2 in T-cell activation?

- Activation of T cells changes expression of many genes, especially the important cytokine interleukin-2, which can activate T cells in the area.

7.10 Why is costimulation important in T-cell activation?

- A lack of costimulatory signal during T-cell engagement causes the T cell to enter an inactive state known as anergy.

7.11 How do effector T cells utilize cytokines as part of the immune response?

- Many cytokines expressed by T cells bind to JAK receptors on target cell surfaces, which induces the activation of the transcription factor STAT.

7.12 What signaling process drives T-cell differentiation into various types of effector T cells?

- Cytokines produced during T-cell activation can induce T cells to differentiate into various types of effector T cells: T_H1 helper T cells, T_H2 helper T cells, T_{FH} helper T cells, T_H17 helper T cells, regulatory T cells, cytotoxic T cells, and natural killer T cells.
- Activated T cells no longer need a costimulatory signal in order to function.

7.13 How do cytotoxic T cells destroy target cells?

- Cytotoxic T cells monitor cells for intracellular infection. Upon binding to an infected target cell, a cytotoxic T cell destroys the infected cell by secreting proteins that induce apoptosis and by binding to apoptosis ligands such as Fas.

7.14 What are the functions of helper T cells?

- T_H1 helper T cells activate macrophages using CD40L and by secreting the cytokine IFN-γ.
- T_H2 helper T cells activate B cells and granulocytes and by secreting the cytokines IL-4, IL-5, and IL-6.
- T_{FH} helper T cells induce B-cell activation in germinal centers of secondary lymphoid tissue through the use of CD40L and the secretion of the cytokines IL-4 and IL-21.
- T_H17 helper T cells activate neutrophils through the secretion of IL-17.

7.15 What are the functions of regulatory T cells?

- Regulatory T cells downregulate the adaptive immune response and prevent overstimulation; they promote tolerance by inactivating self-reactive T cells.

7.16 What are the functions of natural killer T cells?

- Natural killer T cells function similarly to cytotoxic T cells; they also induce an inflammatory response through recognition of a glycolipid on the surface of certain bacteria.

Review Questions

7.1 Predict what would happen if an individual suffered from an inability to undergo T-cell priming.

7.2 Compare and contrast the action of the three different professional antigen-presenting cells.

7.3 Contrast the functions of immature and mature dendritic cells. Why do these two subsets of dendritic cells function in this manner?

7.4 How can macrophages that reside in secondary lymphoid tissue participate in antigen presentation to T cells for an infection residing outside of that secondary lymphoid tissue?

7.5 Why is the number of potential epitopes presented by B cells smaller compared to dendritic cells and macrophages?

7.6 Why do adaptive immune responses involving T cells require days to develop when innate immune responses only require hours?

7.7 Which of the following represents the interaction between the costimulatory molecule on professional antigen-presenting cells and the coreceptor the surface of naïve T cells?

A. DC-SIGN:ICAM-3
B. B7:CD28
C. ICAM-1:LFA-1
D. MHC class II: T-cell receptor
E. MHC class II:CD4

7.8 The process of T-cell activation begins at the immunological synapse and ends with the activation of transcription of genes important in T-cell differentiation and action. Describe the pathways activated upon stimulation of the T-cell receptor with a costimulatory signal, culminating in the activation of transcription factors.

7.9 Binding of ________ to its receptor induces T-cell proliferation and differentiation of activated T cells.

A. CD4
B. CD28
C. LFA-1
D. IL-2
E. IL-4

7.10 Explain the importance of T-cell anergy in peripheral tolerance.

7.11 Predict the consequence on cytokine signaling if an individual had a homozygous mutation in the protein kinase JAK.

7.12 If you had an isolated population of naïve T cells that you want to differentiate into T_H1 helper T cells, which of the following cytokines would you need to provide to the culture?

A. IL-10
B. IL-13
C. IL-17
D. IL-4
E. IFN-γ

7.13 Describe the action of cytotoxins during a cytotoxic T-cell response. How do cytotoxins ensure that they are causing minimal damage to surrounding tissue?

7.14 List the helper T cells and provide a brief description of the function of each.

7.15 Predict the consequence of an inability of an individual to produce the transcription factor Foxp3 in the context of the immune system.

7.16 Explain how NK T cells behave both like cytotoxic T cells and helper T cells.

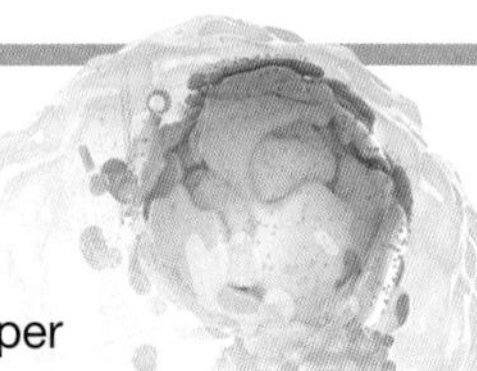

CASE STUDY REVISITED

Recurring Bacterial Infections

A 2-year-old girl is admitted to the emergency room presenting with a fever of 103°F, chills, and trouble breathing. Her parents report that she has had persistent coughing for nearly a week that she has started to produce green and bloody mucus. Based on these symptoms and the exam, a CT chest scan and pleural fluid culture are ordered. The CT chest scan shows infiltrates in the chest cavity and the pleural fluid culture comes back positive for bacterial infection. The initial diagnosis is bacterial pneumonia. Further history reveals that the child has suffered from three prior bacterial infections within the past month. Because of the recurring history of bacterial infections, the physician orders more laboratory tests to measure the white blood cell count and the presence of antibodies to *Haemophilus pneumoniae* (the causative agent of bacterial pneumonia). Laboratory test results show the patient's T cells (CD4 and CD8) and B cells to be at normal levels but indicate a very low concentration of antibodies to *H. pneumoniae*.

Think About...

Questions for individuals

1. Which cells are involved in the production of soluble immunoglobulins? (Hint: Don't consider only the cells directly involved but also the cells needed for this production to be activated.)
2. What cytokines are secreted by helper T cells to aid in the production of immunoglobulins to *H. pneumoniae*? Considering that the patient's current infection is associated with mucosal tissue, what soluble immunoglobulin would likely be made if everything in the patient were functioning normally?

Questions for student groups

1. Keeping in mind what you have learned about adaptive immune system function, what are some possible explanations for the girl's recurring bacterial infections? Consider the possible reasons in light of both overall immune system functioning and events at the molecular level.
2. How could you determine whether your possible explanations are correct? Assume that isolated cell populations from this patient can be harvested and cultured to provide some answers.
3. Why can a mutation in RAG1/RAG2 be ruled out?
4. Sequence analysis of the ZAP-70 gene reveals a mutation that would shift the reading frame of ZAP-70 mRNA, resulting in a truncated protein. Would these sequencing results be consistent with the patient's history and symptoms? Explain why or why not.

MAKING CONNECTIONS **Key Concepts**	**COVERAGE** **(Section Numbers)**
Comparing MHC class I and MHC class II processing and presentation in professional antigen-presenting cells	2.2, 2.3, 2.5, 2.8, 4.2, 4.6, 6.5, 7.2, 7.3, 7.4
Comparing the recognition of presented antigen via CD4 and CD8 T cells	1.9, 1.10, 2.8, 4.2, 4.7, 6.2, 6.4, 7.3, 7.4, 7.5
Signaling events and interactions required for proper T-cell activation	4.7, 5.8, 6.2, 6.6, 7.6, 7.7
Cytokines released and T-cell differentiation by antigen-presenting cells upon recognition of different pathogens	1.10, 4.6, 4.7, 6.5 7.6, 7.7, 7.8, 7.9, 7.12, 7.13, 7.14, 7.15, 7.16
Effector T-cell function	4.5, 4.7, 5.8, 7.13, 7.14, 7.15, 7.16

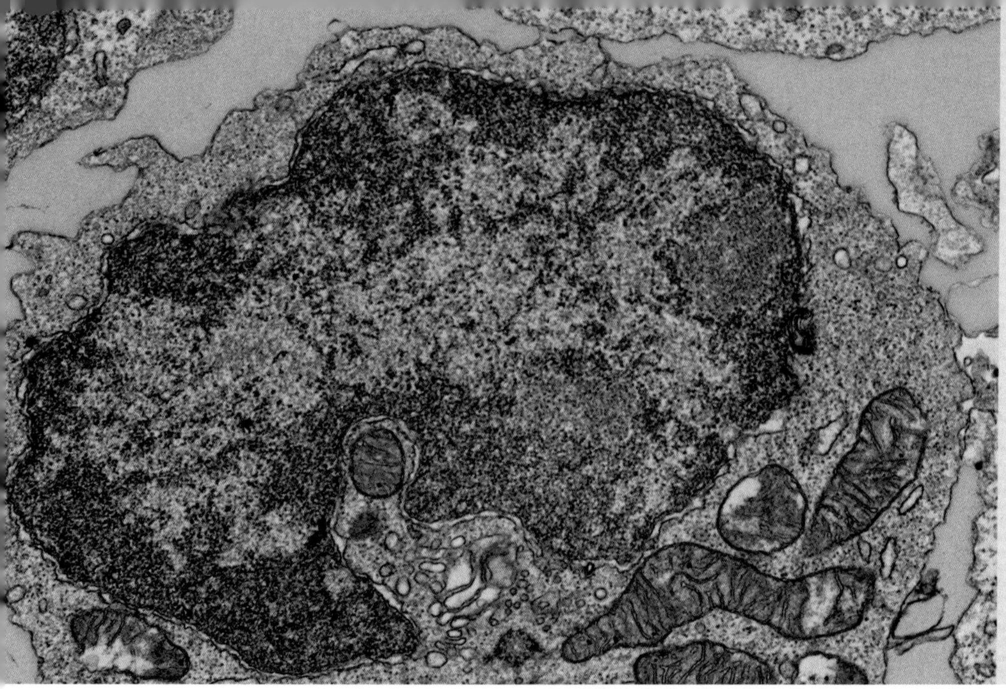

8

Development of B Lymphocytes

CASE STUDY: Maya's Story

Maya's childhood has been difficult. During her two years of life, Maya has been to the emergency room six times for various reasons, including painful ear infections and a bout with pneumonia. Her parents, Ana and Diego, are beside themselves and constantly worry about their next trip to the ER.

Unfortunately, Ana and Diego's fear of their next visit to the ER is realized sooner rather than later. Maya is sitting in the living room and acting very lethargic. Ana and Diego also notice that Maya seems to be having trouble catching her breath and is wheezing when she breathes. Ana remembers that labored breathing from the time Maya had pneumonia. Diego recognizes it too and tells Ana that he is going to get their coats and that she should put her shoes and Maya's shoes on for yet another ER trip.

In order to understand what Maya is experiencing, and to answer the case study questions at the end of the chapter, we will begin exploring the other arm of the adaptive immune system, namely the humoral arm of adaptive immunity. In Chapters 5, 6, and 7, we explored T-cell development, T-cell effector functions, and key molecular players, which are all at the center of the cellular arm of adaptive immunity.

B cells, the other lymphocyte of the adaptive immune system, work to provide humoral immunity. In a manner similar to that of T cells, B cells undergo a complex and intricate developmental program. The result is an organism with a diverse population of cells that are able to recognize a wide variety of antigens. Key

QUESTIONS Explored

8.1 Which B-cell populations will be our major focus?

8.2 Where and how do B cells develop in the fetus?

8.3 Which cell lineages begin B-cell development within the bone marrow?

8.4 What is the site of B-cell development in adults, and which developmental stages occur there?

8.5 What critical checkpoints are required for proper B-cell development?

8.6 What must occur before cells can pass beyond the pro-B cell stage?

8.7 What must occur for pre-B cells to become immature B cells?

8.8 How does negative selection of B cells occur in the bone marrow?

8.9 How does positive and negative selection of transitional B cells occur in the spleen?

8.10 How do mature B cells enter the circulation and become activated?

8.11 How do B-1 and marginal-zone B cells develop?

recombination events at the immunoglobulin genes of developing B cells must occur and critical checkpoints must be passed to ensure the production of functional immunoglobulin subunits. This developmental program also ensures that B cells do not target self-molecules within the organism. In this chapter, we explore the key events involved in B-cell development and the steps taken to ensure that the resulting B cell is capable of serving a useful role in the humoral adaptive immune system.

8.1 Which B-cell populations will be our major focus?

LEARNING OBJECTIVE

8.1.1 Identify the B-cell subtype that plays a primary role in surveying and combating infections through activation at secondary lymphoid tissues.

B-1 B cells B cells that reside in pleural and peritoneal cavities that are responsible for combating common pathogens of these tissues.

B-2 B cells B cells that reside in the circulatory system and secondary lymphoid tissues capable of mounting a typical humoral adaptive immune response; also known as *follicular B cells*.

The humoral arm of the adaptive immune response is populated by B cells produced during hematopoiesis. We will see later in this chapter and in Chapter 10 that there are several different subtypes of B cells that differ in their development and activities. These subtypes include B-1 B cells (also known as *CD5 B cells*), B-2 B cells (also known as *follicular B cells*), and marginal-zone B cells (**FIGURE 8.1** and **TABLE 8.1**).

B-1 B cells occupy and protect body cavities, including the peritoneal and pleural cavities, and their job is primarily to protect gut and lung tissue. In this text, we will focus most of our attention on **B-2 B cells**, which make up the majority of B cells responsible for surveying and combating infections

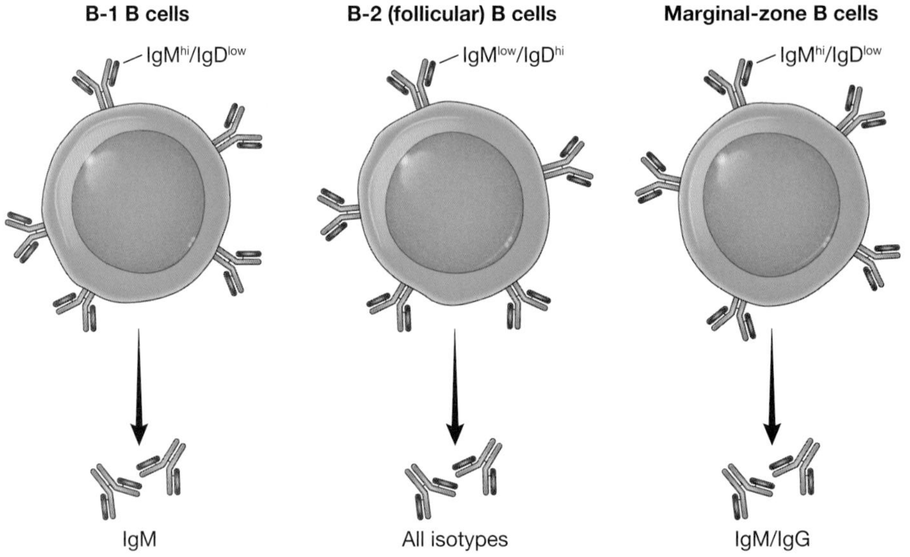

FIGURE 8.1 **Subsets of B cells** Three subsets of B cells are present in humans: B-1 B cells, B-2 (follicular) B cells, and marginal-zone B cells. B-1 B cells reside in pleural and peritoneal cavities and typically produce immunoglobulins that recognize patterns on pathogens that reside in these areas of the body and, as such, are limited in their diversity. Activated B-1 B cells primarily produce the immunoglobulin isotype IgM. B-2 (follicular) B cells are the major B lymphocyte present in humans and reside in most areas of the body. B-2 B cells express a wide diversity of surface immunoglobulins and, upon activation, they can differentiate to produce any of the immunoglobulin isotypes. Marginal-zone B cells reside in the spleen. They can produce the immunoglobulin isotypes IgM and IgG.

TABLE 8.1 | Comparison of B-Cell Subtypes and Their Activity

Characteristics	B-1 B Cells	B-2 (follicular) B Cells	Marginal-Zone B Cells
Major location	Pleural and peritoneal cavities	Secondary lymphoid tissue	Spleen marginal zone
Production of new B cells	Self-renewing of B-1 cells	Bone marrow	May be self-renewing
Diversity	Little diversity	High diversity	Somewhat diverse
Need for T-cell help	No	Yes	Unknown
Major isotypes produced	High levels of IgM	High levels of IgG	IgM, some IgG
Binding to carbohydrate antigens	Yes	Possible	Yes
Binding to protein antigens	Possible	Yes	Yes
Production of memory B cells	Little to none	Yes	Unknown
Presence of surface IgD	Little to none	On naïve B cells	Little to none

Source: Data from T. W. LeBien and T. F. Tedder. 2008. *Blood* 112: 1570–1580, and D. Allman and S. Pillai. 2008. *Curr Opin Immunol* 20: 149–157.

through activation at secondary lymphoid tissues. **Marginal-zone B cells** reside in the marginal zones of the spleen and primarily protect against bloodborne pathogens.

marginal-zone B cells B cells that reside in the marginal zone of the white pulp of the spleen and monitor for bloodborne pathogens.

CHECKPOINT QUESTION

1. Contrast B-1, B-2, and marginal-zone B cells in terms of their major location, diversity, and requirement for T-cell help.

8.2 | Where and how do B cells develop in the fetus?

LEARNING OBJECTIVE

8.2.1 Describe the anatomic locations and the processes by which B-cell development occurs in the fetus.

In adult mice and humans, B-cell development occurs primarily in the bone marrow and the spleen. However, since the presence of bone marrow occurs late in fetal development, B-cell development (and other hematopoiesis) must occur at a different location before bone marrow is present. Hematopoiesis in a fetal organism is important for development of a functioning immune system and is critical for oxygen delivery to the developing fetus.

As fetal growth occurs and new tissues form during the developmental stages, the location of hematopoiesis and the initiation and sites of B-cell development change. These changes are driven by where hematopoietic stem cells (HSCs) are produced. HSCs differentiate into erythroid cells, which are required for the delivery of oxygen to developing tissues. They can also produce myeloid and lymphoid progenitors. Lymphoid progenitors are precursors to B-cell production (as well as to production of T cells and NK cells; see Chapter 1).

Anatomy and timing during fetal development

Hematopoiesis begins in the yolk sac of the developing embryo at around 7 days after fertilization in mice and 21 days after fertilization in humans to produce oxygen-carrying erythroid cells (**FIGURE 8.2**). The appearance of fetal

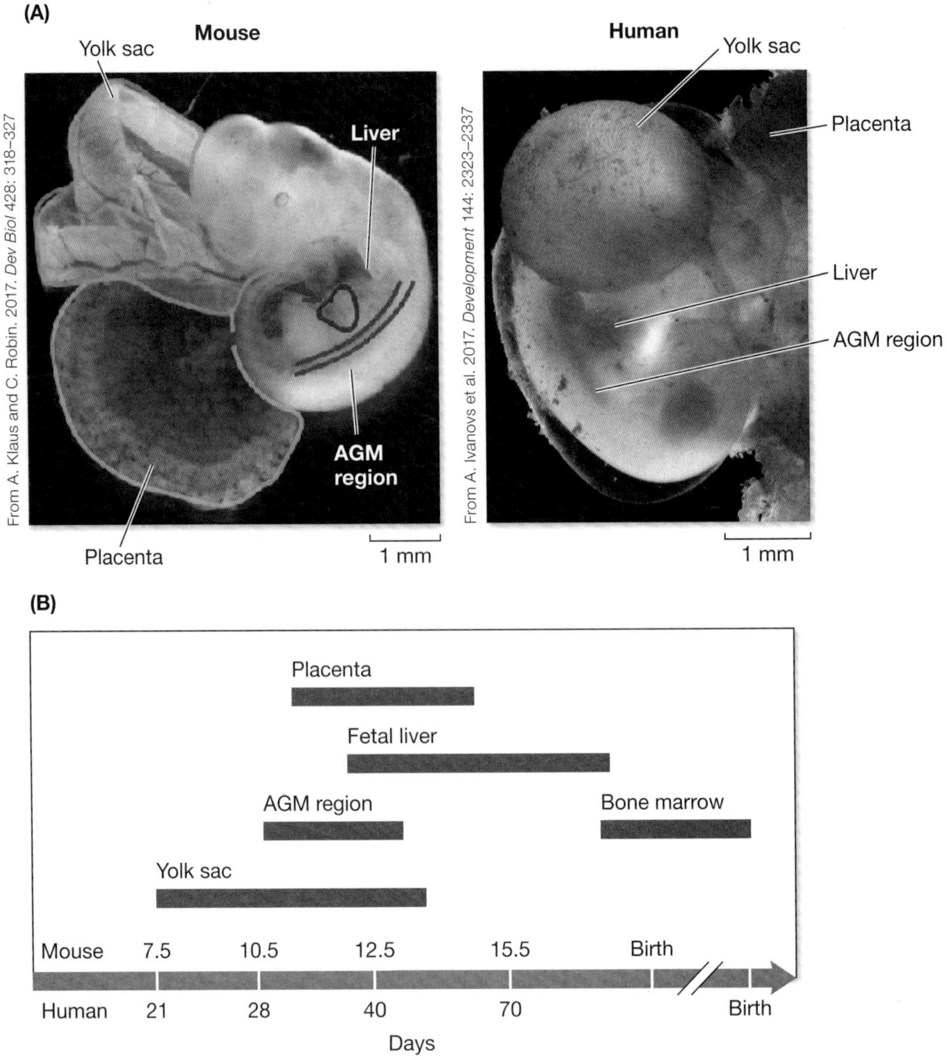

FIGURE 8.2 Locations of embryonic hematopoiesis in humans and mice (A) Hematopoiesis in both embryonic humans and mice occurs within the aorta-gonad-mesonephros (AGM) region, the yolk sac, the placenta, and fetal liver. (B) While the developmental timeline of humans and mice differs, the location and timing of hematopoiesis in both organisms corresponds within their individual developmental programs. Hematopoiesis begins in the yolk sac, expands to the placenta and AGM region, and expands further to the fetal liver as bones begin to develop. Once bone marrow forms, it becomes the primary location of hematopoiesis. (Data from J. Punt et al. 2018. *Kuby Immunology*, 8th ed., W. H. Freeman and Company: New York.)

HSCs occurs in the aorta-gonad-mesonephros (AGM) region a day later in mice, and on day 10, this region is the location of production of mature HSCs.

These HSCs contain the same capacity as do those in adult mice, as irradiated mice can replenish their blood cell population with the cells from HSCs isolated from fetal mice 10 days after gestation. After day 10 in mouse fetal development (day 28 in humans), HSCs can be found not only in the yolk sac and AGM but also in the placenta and fetal liver. The placenta continues to be the major location of HSC production until the liver takes over at day 13 in mice (day 40 in humans).

pre-B cell Developing B cell that has successfully undergone immunoglobulin heavy chain rearrangement and is in the process of rearranging its immunoglobulin light chain.

Within the liver at day 13, **pre-B cells** (also known as *precursor B cells*) can be observed. Pre-B cells are beginning their development into B cells, and by day 17 in mice, B cells capable of producing IgM at their cell surface can be detected. At approximately day 17 in mice (day 75 in humans), hematopoiesis begins to occur in bone marrow. The fetal liver serves as the location for B-cell development while the bone marrow continues to develop.

B-cell development in the fetal liver

HSCs derived from adult bone marrow in mice and humans do not actively proliferate until they receive activation signals. In contrast, HSCs produced during fetal development in the liver must continuously and rapidly divide to produce erythrocytes and populate the immune system.

B cells in mice and humans produced during hematopoiesis in the fetal liver are primarily B-1 B cells. Since the main threat of infection within the developing fetus is via gut or lung entry, it makes sense that fetal hematopoiesis focuses on the production of B-1 B cells to protect these locations from pathogen invasion.

B-1 B cells express antibodies that are cross-reactive; many of them recognize carbohydrate antigens shared by multiple microbes. They do not express terminal deoxynucleotidyl transferase (TdT) at high levels, and their V(D)J recombinase activity only uses a subset of variable V, D, and J gene segments. (TdT and V(D)J recombinase are key enzymes involved in immunoglobulin diversity that were discussed in Chapter 4.) Thus, B-1 B cells do not demonstrate as much immunoglobulin diversity as B-2 B cells. B cells produced during fetal development are therefore very similar in their antigen recognition to innate immune cells, as their immunoglobulins function in a similar manner to pathogen-associated molecular pattern (PAMP) receptors of neutrophils and macrophages (see Chapter 2). Thus, B-1 B cells serve as a bridge between the innate and adaptive immune systems.

During human fetal development, B-cell development switches location from the liver to the bone marrow. B cells that develop in the bone marrow in adult mice and humans are primarily destined to become B-2 B cells, following the developmental program described throughout the rest of this chapter. However, B-1 B cells in circulation within the pleural and peritoneal cavities are self-renewing—daughter B-1 B cells are produced from dividing B-1 B cells already in circulation.

CHECKPOINT QUESTIONS

1. Why are B-1 B cells the most prevalent type of B cells present during fetal development?
2. What is the key difference in diversity between B-1 and B-2 B cells? What drives this difference in diversity?

8.3 Which cell lineages begin B-cell development within the bone marrow?

LEARNING OBJECTIVE

8.3.1 Name and describe the five cell lineages that begin B-cell development in the bone marrow.

As we discovered in Chapter 1, hematopoiesis begins with a common HSC and is responsible for the production of all blood cells. As HSCs receive signals within the bone marrow, they begin the differentiation process into one of the many cell types of the circulatory system, including erythrocytes, innate immune cells, and lymphocytes. Because HSCs need to be both self-renewing and multipotent, what signals are required to drive cell division or differentiation? Furthermore, what drives HSC differentiation toward B-cell development, and what is the cell lineage of this differentiation?

stromal cells Cells within specific tissues that allow for intimate cell-to-cell contact; provide important signaling molecules to drive development and differentiation.

stem cell factor Factor responsible for activating the signals necessary for progression to lymphoid progenitor cell development.

multipotent progenitor cell Early precursor cell of hematopoiesis that is derived from differentiation of a hematopoietic stem cell in the bone marrow; has the ability to differentiate into a variety of blood cells given the right signal.

Hematopoietic stem cells

We know that HSCs can give rise to all blood cell types. Thus, by definition, they must be *multipotent* (programmable to differentiate into a number of different cell types). Furthermore, stem cells in general, including HSCs, must be *self-renewing* (able to divide and produce identical copies of the parent cell). They must provide an ample supply of undifferentiated cells that can be programmed to give rise to specific cell types.

HSCs, like most stem cells, also maintain their chromatin in a state that promotes transcriptional activation by having less-organized nucleosome structure and chromatin packing, as well as through acetylation and methylation of histones, which promotes transcription of genes located close to those modifications. HSCs also utilize several transcription factors to activate genes involved in differentiation of different blood cell types, depending on the signal provided to the differentiating stem cell. As HSCs are primed for B-cell differentiation, several transcription factors activate genes required for progression from a stem cell to a common lymphoid progenitor cell. These transcription factors include the proteins Ikaros, PU.1 (purine box factor 1), and E2A, which function either in HSC self-renewal (E2A) or in activating specific genes responsible for B-cell development (Ikaros, PU.1).

HSCs must also be able to interact with and receive signals from specialized cells in the bone marrow known as **stromal cells**. This requires cell-surface receptors that can receive these signals and interact with bone marrow stromal cells. One key cell-surface molecule expressed by HSCs is c-Kit, which binds to the signaling molecule **stem cell factor** (SCF). The interaction between c-Kit and SCF is essential in lymphocyte development and plays two roles: (1) keeping HSCs in contact with stromal cells to aid in the delivery of differentiation signals and (2) driving the differentiation of HSCs to the next progenitor, **multipotent progenitor cells**.

Multipotent progenitor cells

Multipotent progenitor cells derive from HSCs that have begun to differentiate. As with any stem cell, at this point, multipotent progenitor cells are no longer self-renewing. Instead, these cells begin to express several molecules that allow them to continue their differentiation protocol. Note that this protocol can be altered to allow the multipotent progenitor cell to differentiate into a number of different cell types.

Two key molecules expressed by multipotent progenitor cells are CD34 (a cell-surface receptor used to distinguish multipotent progenitor cells from other hematopoietic cells) and CXCR4 (a receptor for the stromal cell chemokine CXCL12). Stromal cells secrete CXCL12 to maintain a population of progenitor cells in the correct location as differentiation continues. Multipotent progenitor cells ultimately differentiate into lymphoid-primed multipotent progenitor cells.

Lymphoid-primed multipotent progenitor cells

Lymphoid-primed multipotent progenitor cells further continue the B-cell development pathway as they begin to express an important signaling molecule, fms-related tyrosine kinase 3 receptor (flt-3). This signaling molecule binds to its cell-surface ligand expressed by bone marrow stromal cells and, upon engagement, drives signaling that culminates in the expression of the IL-7 receptor in the lymphoid-primed multipotent progenitor cell.

IL-7 secreted by stromal cells in the bone marrow plays an important role in B-cell development. It promotes further differentiation through the early pro-B, pro-B, and pre-B stages of development. Lymphoid-primed multipotent progenitor cells destined to differentiate into lymphocytes begin to express key components needed for V(D)J recombination, including RAG1, RAG2, and TdT, at the end of this stage.

Early lymphoid progenitor cells

In Chapter 4 we learned that the RAG1 and RAG2 components of V(D)J recombinase are vital in driving the diversity seen in immunoglobulins and T-cell receptors. Thus, it stands to reason that cells expressing major components of V(D)J recombinase, including RAG1 and RAG2, are likely to develop into lymphocytes, since they are the only cells that utilize the V(D)J recombinase to further differentiate. The expression of V(D)J recombinase components at the end of the previous differentiation stage also marks the beginning of the early lymphoid progenitor cell stage. Cells that have differentiated into early lymphoid progenitor cells are destined to become lymphocytes. The type of lymphocyte they become is dictated in part by whether these cells remain in the bone marrow (B cells) or if they migrate to the thymus (T cells). Cells that remain in bone marrow continue to elevate expression of the IL-7 receptor, which promotes their further differentiation into common lymphoid progenitor cells.

Common lymphoid progenitor cells

Common lymphoid progenitor cells can further differentiate to become NK cells, innate lymphoid cells, T cells, or B cells. Cells that continue to bind to IL-7 expressed by stromal cells will become common lymphoid progenitor cells, progressing toward B-cell development. Signaling through the IL-7 receptor enhances survival of the common lymphoid progenitor cell and promotes chromatin remodeling. This allows for the exposure of the immunoglobulin genes, making them accessible as they recombine via the activity of V(D)J recombinase through the stages of B-cell development described in the next section.

CHECKPOINT QUESTIONS

1. Why do hematopoietic stem cells maintain a flexible chromatin structure and histone acetylation and methylation?
2. Which signal in differentiating progenitor cells is required to commit differentiation to B-cell development?

8.4 What is the site of B-cell development in adults, and which developmental stages occur there?

LEARNING OBJECTIVE

8.4.1 Identify the site of B-cell development in adults and describe the developmental stages that occur there.

We have seen that the site of B-cell development during gestation shifts as tissue development continues within the fetal mouse or human. As bone forms, the final shift to the bone marrow as the site of hematopoiesis occurs. In adults, the bone marrow serves not only as the site for B-cell development but also as the site for most hematopoiesis events (see Chapter 1), although some development and differentiation occurs elsewhere, such as T-cell development in the thymus. Bone marrow contains specialized cells that aid in the developmental process and signaling to drive differentiation events that produce the various cells of hematopoietic lineage. However, B cells complete their maturation process after leaving the bone marrow and migrating to the spleen (**FIGURE 8.3**).

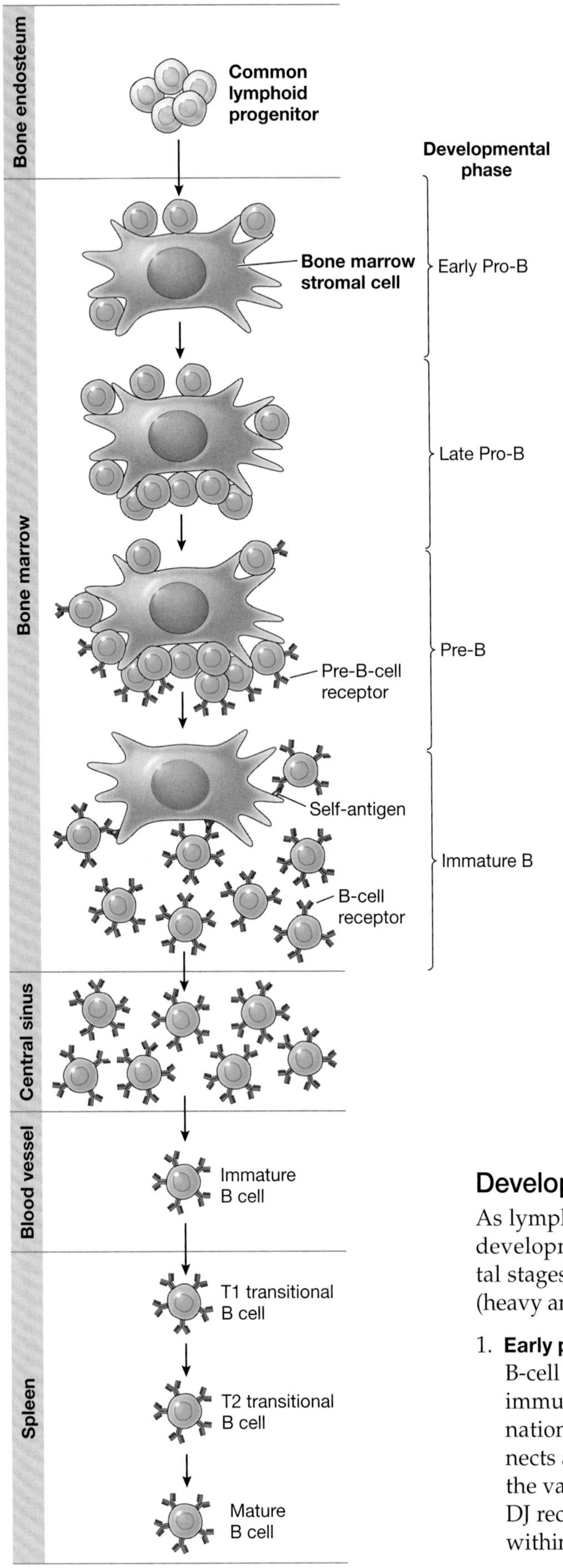

FIGURE 8.3 Location of B-cell development in adults B cells develop in the bone marrow, starting from a common lymphoid progenitor. Signals provided by bone marrow stromal cells initiate the differentiation of common lymphoid progenitor cells into B cells. During development, key events in immunoglobulin rearrangement occur during the early and late pro-B cell stages and the pre-B cell stage before the cell becomes an immature B cell. Within the bone marrow, further selection of the immature B cell occurs to promote central tolerance before the cell migrates to the spleen. Within the spleen, the immature B cell continues its developmental program through two transitional stages (T1 and T2) and ultimately becomes a mature B cell that can function in the humoral arm of the adaptive immune system.

B-cell development in bone marrow

During B-cell development in the bone marrow, precursor lymphoid progenitor cells require signals to prompt proper activation events to commit the lymphoid progenitor to the B-cell lineage. These signals start HSCs on the journey toward the production of erythrocytes, innate immune cells, and lymphocytes. Signals that drive commitment of cell differentiation are provided by bone marrow stromal cells. These cells bind to adhesion molecules on HSCs and differentiating cells and provide signaling molecules to further promote differentiation.

B cells begin their development as HSCs in the endosteum, which is located near the surface of the bone (**FIGURE 8.4**). These stem cells continue differentiation while in contact with stromal cells and receive cues that drive commitment to a particular B-cell lineage.

Cells that are differentiating into B cells undergo changes in cell-surface markers as a result of signaling events initiated by interaction with stromal cells. This change in cell-surface markers is analogous to changes seen in partially differentiated lymphoid progenitor cells destined for T-cell lineage commitment. In fact, these changes were significant in the discovery of the stages of B-cell development (see **KEY DISCOVERIES**).

Developmental stages in bone marrow

As lymphoid progenitor cells continue their journey through B-cell development, they will progress through important developmental stages (see Figure 8.3) to rearrange their immunoglobulin genes (heavy and light chains) and test for productive rearrangements:

1. **Early pro-B cells**: Lymphoid progenitor cells that have begun B-cell development initiate genetic recombination at the immunoglobulin heavy chain loci. As we have seen, recombination of the heavy chain requires recombination that connects a V, D, and J variable gene segment together to create the variable region. These recombination events begin with DJ recombination of the heavy chain immunoglobulin loci within early pro-B cells.

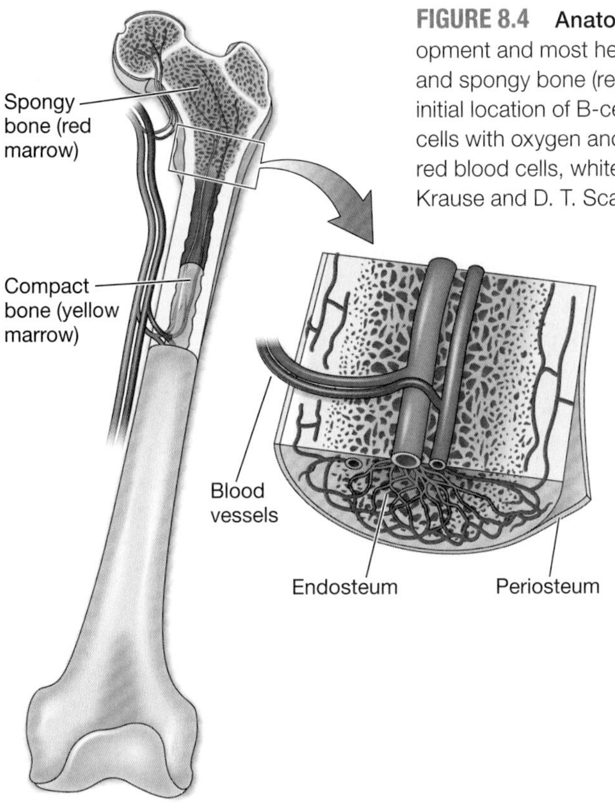

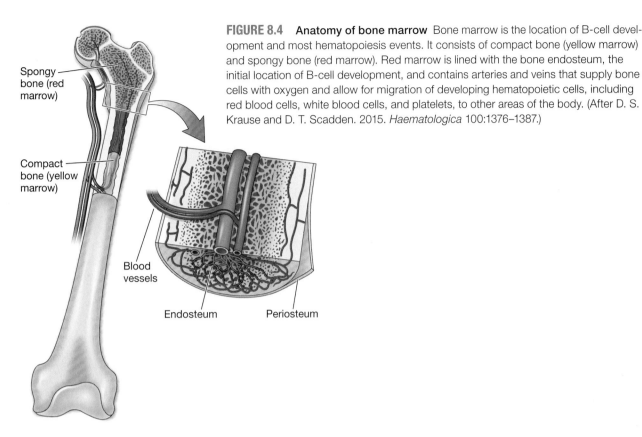

FIGURE 8.4 Anatomy of bone marrow Bone marrow is the location of B-cell development and most hematopoiesis events. It consists of compact bone (yellow marrow) and spongy bone (red marrow). Red marrow is lined with the bone endosteum, the initial location of B-cell development, and contains arteries and veins that supply bone cells with oxygen and allow for migration of developing hematopoietic cells, including red blood cells, white blood cells, and platelets, to other areas of the body. (After D. S. Krause and D. T. Scadden. 2015. *Haematologica* 100:1376–1387.)

2. **Pro-B cells**: After recombination of D and J variable gene segments of the heavy chain gene occurs, a V segment must be recombined with the newly formed DJ segment. These events occur in pro-B cells. At the end of the pro-B cell stage, developing cells will test the immunoglobulin heavy chain recombination events to determine whether these recombination events were productive or unproductive.
3. Pre-B cells: Once the developing cell tests for productive rearrangement of the immunoglobulin heavy chain locus and goes through expansion, the cell again initiates recombination events. This time, those events occur at the immunoglobulin light chain loci. V and J variable gene segments of the light chain loci rearrange, and at the end of the pre-B cell stage, another test determines whether the light chain locus has productively recombined through positive selection.
4. **Immature B cells**: Cells that have productively recombined both their heavy and light chains become immature B cells, which then undergo negative selection within the bone marrow. B cells that react to self-antigen go through a process known as **receptor editing** in an attempt to further rearrange their light chain and change the immunoglobulin specificity to no longer react with self-antigens. Immature B cells that continue to react with self-antigen are removed. Those that pass through all the developmental checkpoints migrate from the bone marrow as **transitional B cells** and travel to the spleen.

early pro-B cells Initial stage in B-cell development that occurs as common lymphoid progenitor cells receive B-cell developmental signals in bone marrow.

pro-B cell Developing B cell that is undergoing immunoglobulin heavy chain rearrangement.

immature B cells B cells that have completed initial development in bone marrow that must subsequently migrate to the spleen for further development.

receptor editing Mechanism utilized by B cells to attempt to rearrange the immunoglobulin light chain since, through initial somatic recombination events, they express a self-reactive immunoglobulin.

transitional B cells B cells that have finished development in the bone marrow but need to finish the developmental program within the spleen.

B-cell selection in the spleen

Transitional B cells exit the bone marrow and migrate to the spleen to complete their development and maturation. The spleen's main function in the body is

KEY DISCOVERIES

How did we learn of the stages involved in early B-cell development?

Article

R. R. Hardy, C. E. Carmack, S. A. Shinton, J. D. Kemp, and K. Hayakawa. 1991. Resolution and characterization of pro-B and pre-pro-B cell stages in normal mouse bone marrow. *Journal of Experimental Medicine.* 173: 1213–1225.

Background

Using transformed B-cell lines, the events and timing of immunoglobulin rearrangement had been well characterized as of the mid-1980s. However, these events had never been demonstrated in primary lymphoid tissue, limiting in vivo definition of the cells involved in each key stage of B-cell development. How were the stages of B-cell development, including the pre-B and pro-B cell stages, identified and characterized in bone marrow?

Early Research

Not only had earlier research established the likely order of rearrangement that must occur in developing B cells, but studies of bone marrow cells had established the presence of several cell-surface markers that were present on bone marrow cells that contained the cell-surface marker B220, which had been established as a B-cell-specific marker. These markers included heat-stable antigen (HSA), BP-1, and CD43, although CD43 was known to be absent from mature B cells. Since it was well established that cells changed their cell-surface marker profile during differentiation, including during hematopoiesis, it stood to reason that the B-cell-specific markers might be used as a "clock" to mark the stage at which a developing B cell was at in its differentiation program.

Think About...

1. Considering the research summary, state the hypothesis as it relates to expectations about cell-surface markers on developing B cells.
2. How would you measure the cell-surface markers on each B-cell population? How might you distinguish between cells in a complex population of non-clonal lineage?
3. How did the findings of this study allow us to define the stages of B-cell development and the cell-surface markers present during each developmental stage?

Article Summary

The researchers used flow cytometry and molecular biology to define the cell-surface markers associated with the early stages of B-cell development as well as later developmental stages, including immature and mature B cells.

To identify the cell-surface markers, bone marrow cells containing specific cell-surface markers were isolated into subpopulations via flow cytometry. A fluorescently labeled antibody was used to tag a specific cell-surface marker to separate B cells into subpopulations. The researchers also used molecular biology to determine whether immunoglobulin gene rearrangement was occurring in the subpopulations.

Conclusion

The researchers were able to isolate six different subpopulations of B cells, characterized by their cell-surface markers. Three of these populations contained the markers B220 and CD43, while the other three contained B220 but lacked CD43 (**FIGURE KD 8.1**).

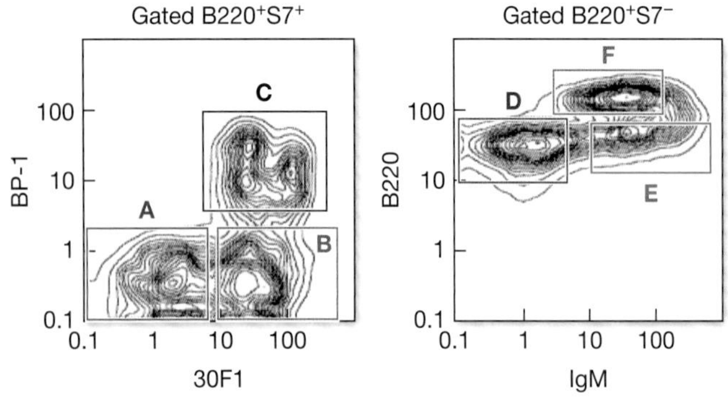

FIGURE KD 8.1 Identification of six different B-cell populations by flow cytometry Analysis of different B-lymphocyte populations that all contained the cell-surface marker B220 and either containing or lacking the second cell-surface marker S7 resolved six different subpopulations of B cells, three in the S7-positive population of developing B cells and three in the S7-negative population of developing B cells. The S7 marker is present on cells early in the developmental program, whereas the absence of S7 correlates with newly emerging B cells, suggesting that the six identified populations correlate with six distinct steps in the B-cell development program. (Data from © 1991 R. R. Hardy et al. originally published in *J Exp Med* 173: 1213–1225. doi.org/10.1084/jem.173.5.1213.)

These cells were categorized into different cell fractions: A through F. Interestingly, it was also known that fractions A, B, and C did not contain IgM at their cell surface, suggesting that cells that express both B220 and CD43 isolated in these experiments represented early stages in B-cell development. Indeed, polymerase chain reaction (PCR) analysis of immunoglobulin rearrangement demonstrated that fraction A did not have any immunoglobulin rearrangement, and fraction B began to show D-to-JH rearrangement. Fraction C did not show V(D)J recombination, although subsequent studies demonstrated that this does occur. Using these data, researchers were able to state that fraction A cells were pre-pro-B cells (bone marrow cells that have differentiated from common lymphoid progenitor cells and are committed to B-cell development), fraction B cells were early-pro-B cells, and fraction C were late-pro-B cells, or those that had just made it through the immunoglobulin checkpoint.

Analysis of the other three cell fractions (D, E, F) demonstrated the separation of B cells later in development. These studies were done via flow cytometry and measured the levels of IgM and IgD at the surface of the B-cell subpopulations. The low amount of IgM in fraction D cells allowed the researchers to determine that this population represented the small pre-B cells stage, and the presence of IgM on the surface of fraction E cells demonstrated that these cells were immature B cells. Fraction F cells were concluded to be mature B cells due to the levels of both IgM and IgD at their cell surface. Characterizing each stage in the development of the B220 cells fated to become B cells allowed the researchers to establish which cell-surface markers were representative of each stage in B-cell development and demonstrated that these cells were distinguishable due to cell-surface marker changes.

to filter and cleanse the blood. Equally important is its role in lymphocyte proliferation and the positive and negative selection of B cells.

Immature B cells in the spleen begin as **T1 transitional B cells**, which undergo negative selection, a process in which the affinity of the immunoglobulin is tested against self-antigens expressed in the spleen. T1 transitional B cells with immunoglobulins that recognize self-antigen with high affinity are removed from the B-cell population. T1 B cells that escape negative selection become **T2 transitional B cells**. The T2 B cells receive a survival signal, enabling them to transition into mature B cells, which enter the circulation to aid in humoral immunity.

T1 transitional B cell B cell that has migrated to the spleen and is being further tested via negative selection to promote peripheral tolerance.

T2 transitional B cell B cell that has migrated to the spleen and successfully passed through negative selection and is now receiving B-cell survival signals and final differentiation signals.

CHECKPOINT QUESTIONS

1. In B-cell development, what is the importance of stromal cells within the bone marrow?
2. List the four stages of B-cell development that occur in bone marrow, and briefly highlight the key processes that occur in each stage.

Making Connections: Production of B Cells

Hematopoiesis is the production of all types of blood cells. The location of these events, especially that of B-cell production, changes with the age of the organism. The production of B cells follows a strict program, moving through several developmental checkpoints before resulting in B cells that can act in the humoral arm of the adaptive immune system.

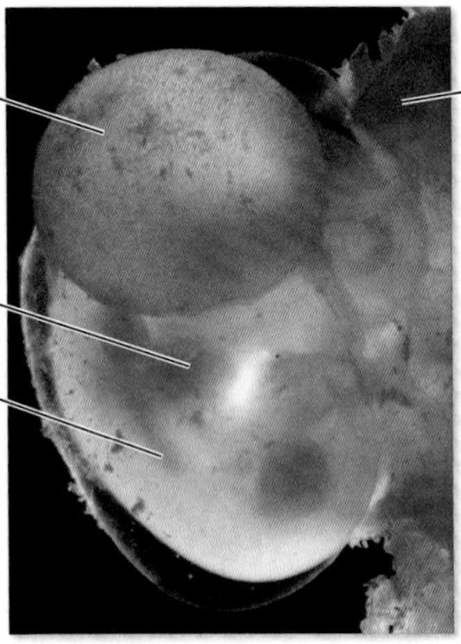

From A. Ivanovs et al. 2017. *Development* 144: 2323–2337.

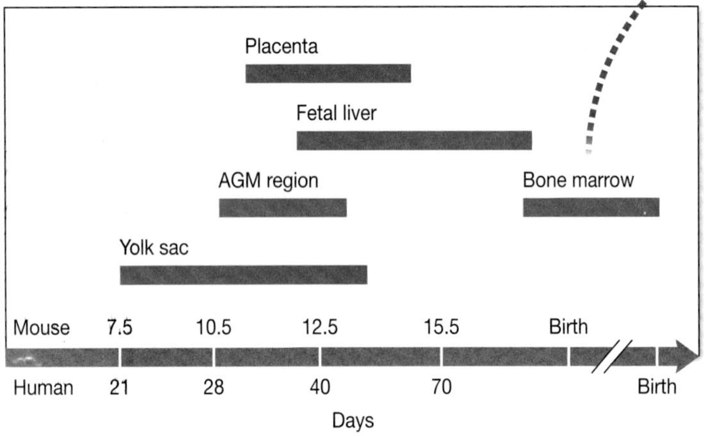

Data from J. Punt et al. 2018. *Kuby Immunology*, 8th ed., W. H. Freeman and Company: New York

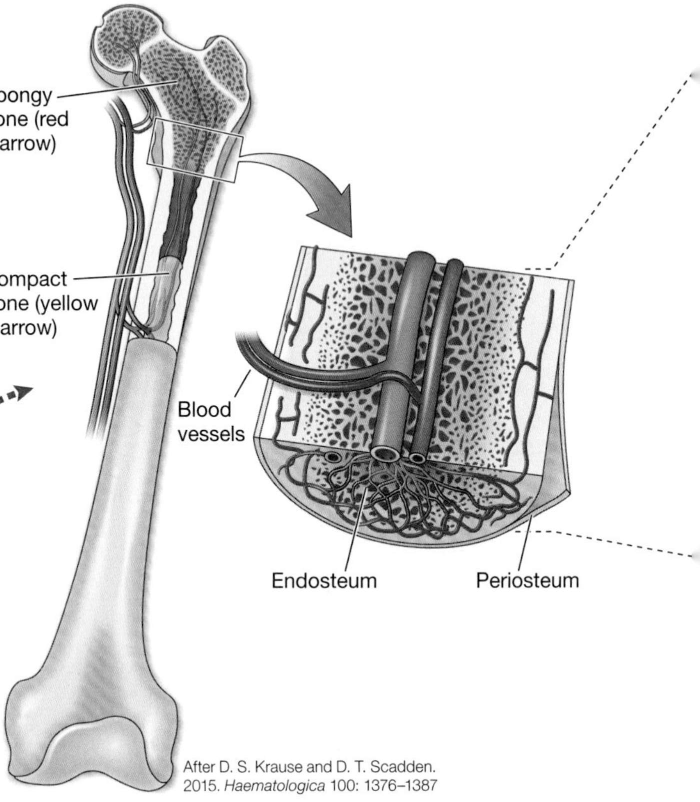

After D. S. Krause and D. T. Scadden. 2015. *Haematologica* 100: 1376–1387

B-cell development, a process that occurs during hematopoiesis, occurs in different parts of the human body depending on the stage of development. In the fetus, B-cell development occurs in the placenta, AGM region, yolk sac, and liver. In adults, it takes place in the bone marrow.

(Sections 1.9, 8.2, 8.2)

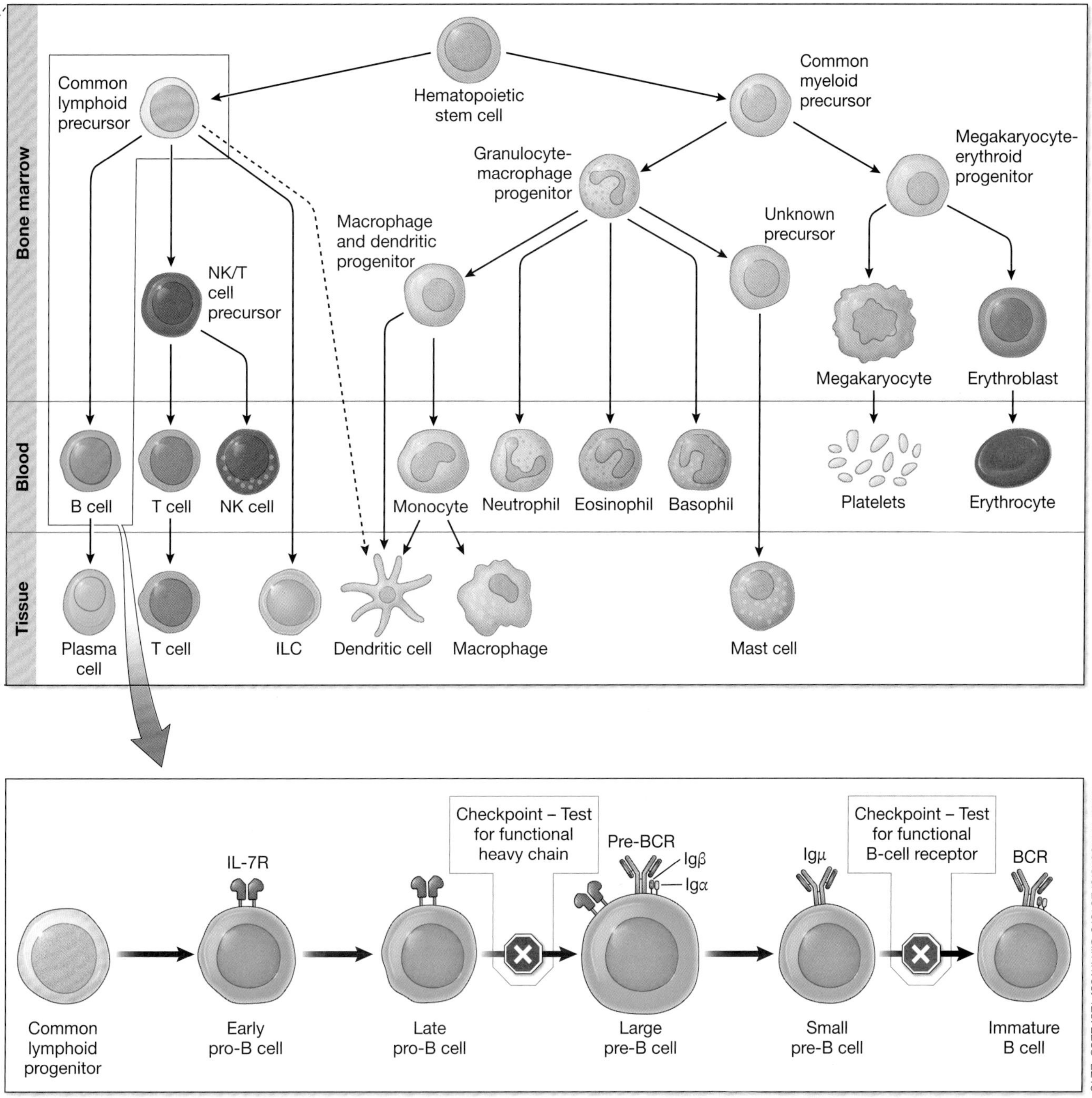

After F. Melchers. 2015. *J Clin Invest* 125: 2203–2210

B-cell development begins with hematopoietic stem cell differentiation into a common lymphoid progenitor. The cells then move through several different developmental stages moderated by key developmental checkpoints designed to verify the proper function of recombined immunoglobulin genes. The process results in several types of B cells capable of functioning in the humoral arm of adaptive immunity.

(Sections 1.6, 8.1, 8.3, 8.4, 8.5, 8.6, 8.7, 8.9, 8.11)

8.5 What critical checkpoints are required for proper B-cell development?

LEARNING OBJECTIVE

8.5.1 Explain the events that make up the two critical checkpoints in B-cell development.

Common lymphoid progenitor cells that receive signals committing them to B-cell development proceed through the early pro-B, pro-B, and pre-B cell stages on their way toward becoming immature B cells.

As discussed in Chapter 5, a major hurdle in proper T-cell development centers around the rearrangement of both chains of the T-cell receptor. In developing T cells, checkpoints test the rearrangement of these loci. Unproductive rearrangement of all loci of a T-cell receptor gene by V(D)J recombinase results in the removal of the cell by apoptosis. The mechanisms underlying B-cell development mirror those of T-cell development, and V(D)J recombinase and TdT function similarly in both processes. Thus, immunoglobulin locus recombination events can result in both productive and unproductive rearrangement of the loci.

These rearrangement events must be tested to prevent the production of B cells that cannot respond to antigen due to lack of a functional immunoglobulin at their cell surface. The development process includes checkpoints that test the rearrangement events of immunoglobulin heavy chain and light chain loci to determine whether these components will work properly in an expressed immunoglobulin. Even if the loci have been rearranged productively, because of the true randomness of variable-region formation of heavy and light chains, the antigen specificity of immunoglobulins could be toward self-antigens. Thus, B cells must be positively selected to test for functional immunoglobulin and negatively selected against if immunoglobulins recognize self-antigen.

B cells differ from T cells in their developmental checkpoint. T cells can produce two different types of T-cell receptors (αβ and γδ T-cell receptors), while B cells can produce only one functional type of immunoglobulin. (Remember, immunoglobulin isotypes are produced due to rearrangement of the heavy chain locus to alter the constant region used during heavy chain expression, not due to different genes for each isotype.) There are two critical checkpoints of the rearrangement events that occur during B-cell development (**FIGURE 8.5**):

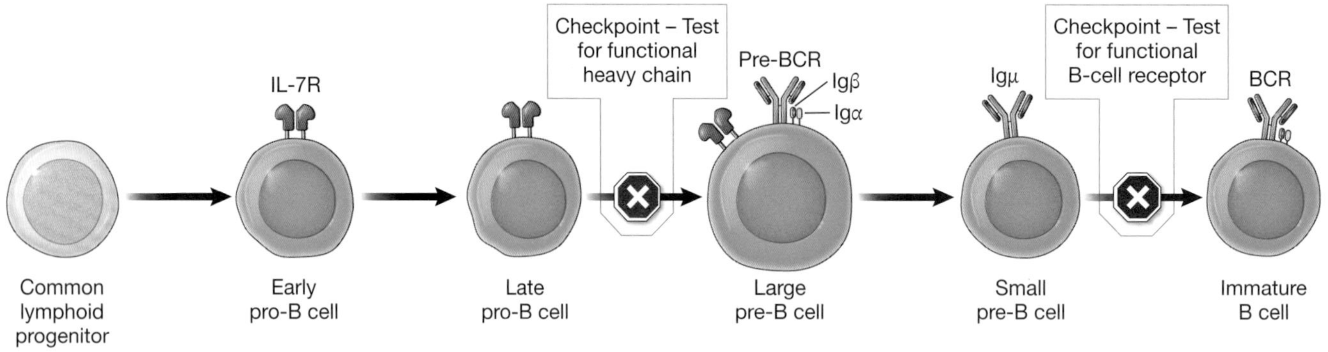

FIGURE 8.5 B-cell development checkpoints Development begins in the bone marrow when common lymphoid progenitor cells receive signals from stromal cells to differentiate into B cells. The first developmental stage is the pro-B stage in which somatic recombination occurs at the immunoglobulin heavy chain locus, first recombining a D and J segment during the early pro-B cell stage and then recombining a V segment with the DJ segment in the late pro-B cell stage. Next, developing B cells undergo a checkpoint to test for a functional heavy chain. Cells that pass this checkpoint become large pre-B cells. These cells undergo several cell divisions and become small pre-B cells, where somatic recombination of the immunoglobulin light chain locus occurs. The developing B cell reaches another checkpoint to test for a functional B-cell receptor (BCR). Immature B cells move forward via negative selection to ensure central tolerance before exiting the bone marrow. (After F. Mechers. 2015. *J Clin Invest* 125: 2203–2210.)

- *Checkpoint of the heavy chain of the immunoglobulin*: The first somatic recombination event to occur in developing B cells is the recombination of the immunoglobulin heavy chain locus. These recombination events occur first through the rearrangement of D and J gene segments of the heavy chain locus, followed by rearrangement of a V segment to the recombined DJ segment of the heavy chain locus. This checkpoint tests the rearrangement event of the heavy chain variable region. If the immunoglobulin heavy chain is recombined properly, the developing B cell continues with the recombination of the immunoglobulin light chain. Unproductive rearrangement of all immunoglobulin heavy chain loci commits the developing B cells to be removed by apoptosis.
- *Checkpoint of the light chain of the immunoglobulin*: The purpose of this checkpoint is to test for productive rearrangement of the immunoglobulin light chain locus. Immunoglobulin light chain rearrangement requires recombination of V and J variable segments of the light chain loci. Unproductive rearrangement of all of the immunoglobulin light chain loci commits the developing B cell to be removed by apoptosis.

surrogate light chain Protein complex formed when the proteins VpreB and λ5 are expressed and interact; takes the place of the immunoglobulin light chain in pro-B cells that need to test the rearrangement of the immunoglobulin heavy chain.

VpreB Protein component of the surrogate light chain that resembles the variable region of the immunoglobulin light chain; aids in testing for productive rearrangement of the immunoglobulin heavy chain.

λ5 Protein component of the surrogate light chain that resembles the constant region of the immunoglobulin light chain; aids in testing for productive rearrangement of the immunoglobulin heavy chain.

pre-B-cell receptor Receptor that contains rearranged immunoglobulin heavy chain, the surrogate light chain, and the signaling molecules Igα and Igβ; responsible for testing the immunoglobulin heavy chain for productive rearrangement.

CHECKPOINT QUESTIONS

1. What are the two critical checkpoints of B-cell development?
2. Why would you predict that developing B cells that can no longer productively recombine an immunoglobulin heavy chain or light chain locus would be removed via apoptosis?

8.6 What must occur before cells can pass beyond the pro-B cell stage?

LEARNING OBJECTIVE

8.6.1 Explain the processes of allelic exclusion of the immunoglobulin heavy chain locus and gene rearrangement after the heavy chain checkpoint.

Once a common lymphoid progenitor cell within the bone marrow has received signals driving activation of B-cell development, it begins the process of rearrangement of the variable regions at the heavy chain loci. During the early pro-B cell stage, the activities of RAG1, RAG2, and other V(D)J recombinase proteins rearrange immunoglobulin heavy chain loci by recombining a D variable segment with a J_H variable segment (see Figure 4.10).

As the developing B cell continues in the pro-B cell stage, it will recombine a V variable segment with the recombined DJ_H segment. Due to the activity of V(D)J recombinase, along with the removal and addition of P and N nucleotides by exonucleases and TdT, the rearrangement may be productive or unproductive. It is imperative that the immunoglobulin heavy chain has rearranged properly to ensure that the resultant B cell is capable of mounting an immune response and functioning in the humoral arm of adaptive immunity. To test the rearrangement of the heavy chain, the pro-B cell expresses not only the rearranged heavy chain but also a **surrogate light chain** to be able to test whether the rearranged heavy chain is properly expressed and capable of forming a functional immunoglobulin (**FIGURE 8.6**).

The surrogate light chain is composed of two proteins, **VpreB** and **λ5**. VpreB mimics the variable region of the immunoglobulin light chain, and λ5 mimics the constant region of the immunoglobulin light chain. The immunoglobulin must also form a complex with the immunoglobulin coreceptors Igα and Igβ to ensure that it can properly assemble into a receptor complex known as the **pre-B-cell receptor**. Interestingly, there is very little of this receptor at

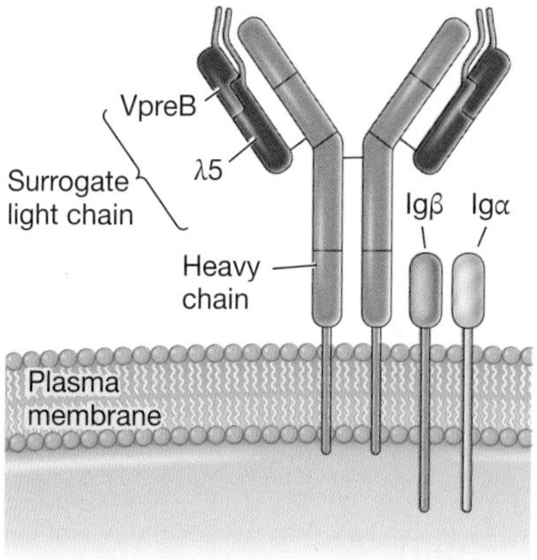

FIGURE 8.6 Pre-B-cell receptor
The pre-B-cell receptor forms after the pro-B cell stage of development. The complex includes the immunoglobulin heavy chain, produced from the recombined heavy chain locus, and the two surrogate light chain components VpreB and λ5. Formation of a functional pre-B-cell receptor allows the developing cell to move past the heavy chain checkpoint and continue to the pre-B cell stage.

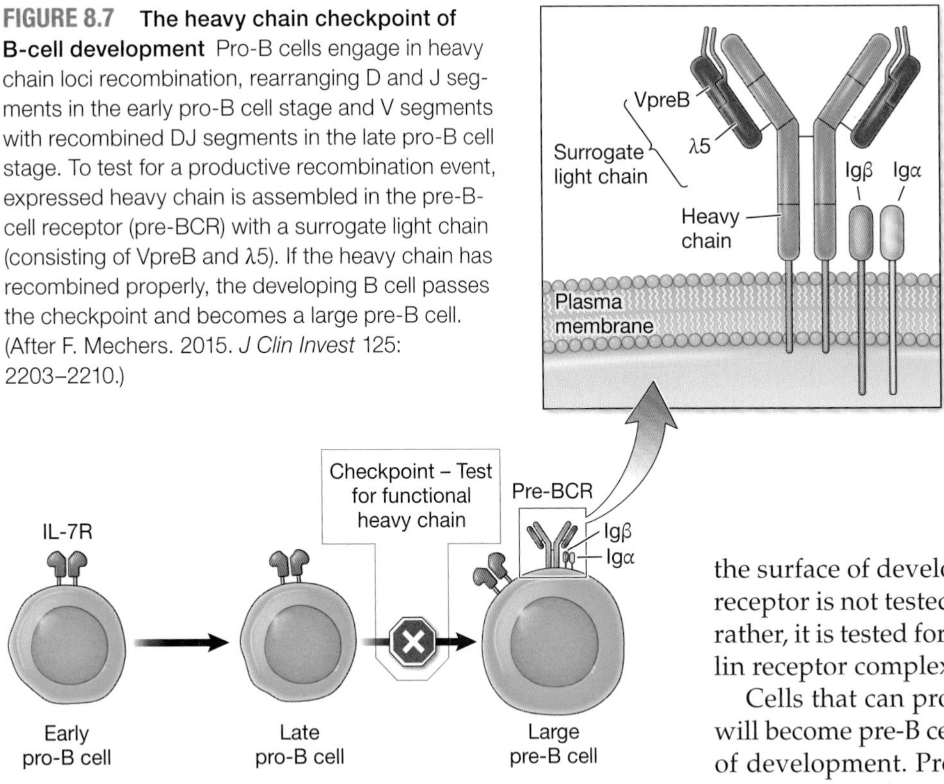

FIGURE 8.7 The heavy chain checkpoint of B-cell development Pro-B cells engage in heavy chain loci recombination, rearranging D and J segments in the early pro-B cell stage and V segments with recombined DJ segments in the late pro-B cell stage. To test for a productive recombination event, expressed heavy chain is assembled in the pre-B-cell receptor (pre-BCR) with a surrogate light chain (consisting of VpreB and λ5). If the heavy chain has recombined properly, the developing B cell passes the checkpoint and becomes a large pre-B cell. (After F. Mechers. 2015. *J Clin Invest* 125: 2203–2210.)

the surface of developing B cells, suggesting that the receptor is not tested for its ability to bind to antigen; rather, it is tested for assembly into an immunoglobulin receptor complex.

Cells that can properly form a pre-B-cell receptor will become pre-B cells and move on to the next stage of development. Pro-B cells that have exhausted all possibilities of immunoglobulin heavy chain rearrangement are removed from the developing B cell population via apoptosis (**FIGURE 8.7**).

Allelic exclusion of the immunoglobulin heavy chain locus

We have seen that a very important feature of an adaptive immune response is the singular specificity of each lymphocyte receptor for its antigen. In order for a T cell or B cell to recognize a single antigen via its cell-surface receptor, it must ensure that once a productive rearrangement of a receptor locus has occurred and been tested, it shuts down recombination events at the other locus to prevent synthesis of two different chains (with two different variable regions) within the same cell.

During T-cell development, this process of *allelic exclusion* occurs through the inactivation of V(D)J recombinase activity and rearrangement of chromatin structure; this inactivates the locus on the other chromosome, preventing polyspecificity toward different antigens. During B-cell development, once the immunoglobulin heavy chain rearrangement checkpoint is passed via the proper formation of the pre-B-cell receptor, developing B cells also engage in allelic exclusion. In these cells, the process occurs via a similar mechanism—expression of RAG1 and RAG2 is shut down while the developing B cell proliferates and produces more developing B cells with a properly rearranged immunoglobulin heavy chain. The chromatin at the immunoglobulin heavy chain locus is inactivated to prevent rearrangement events at that locus and to inactivate gene activity of the heavy chain locus that has not undergone somatic recombination.

The importance of allelic exclusion in B-cell function can easily be demonstrated if we consider the possible expression of two different immunoglobulin heavy chains within a single B cell. As a cell such as this assembles its immunoglobulins at the cell surface, there would be several possible antigen specificity combinations of the assembled immunoglobulins: those with two antigen-binding sites that bind an antigen of interest, those with one antigen-binding site specific for the antigen of interest, and those without an antigen-binding site specific for the antigen of interest (**FIGURE 8.8**).

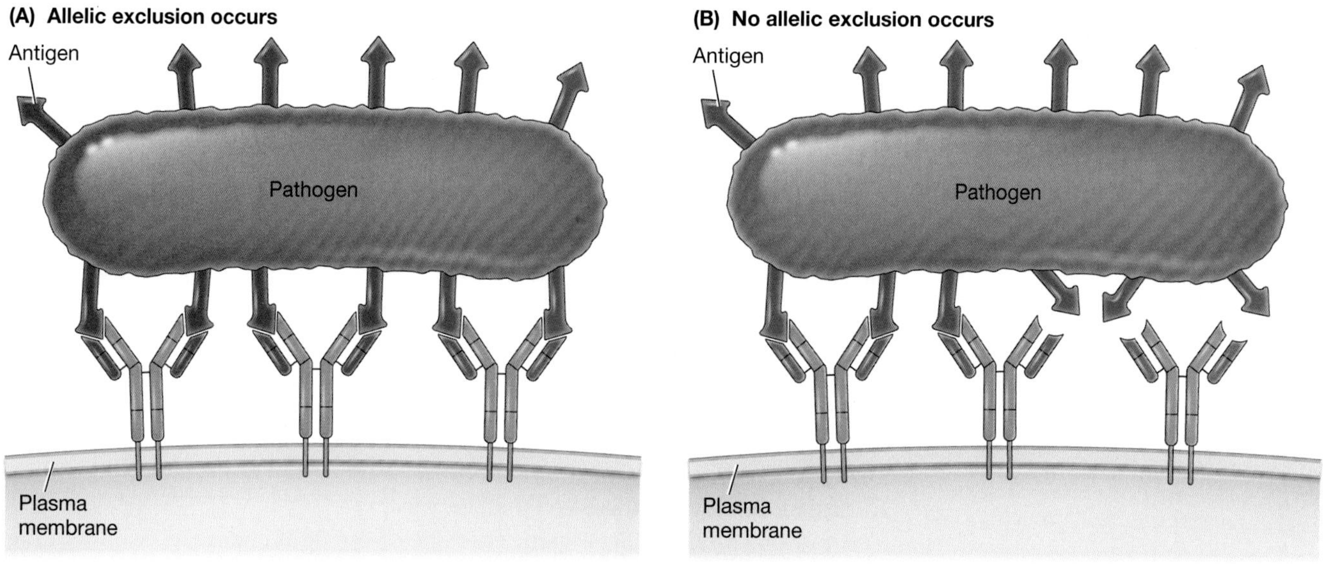

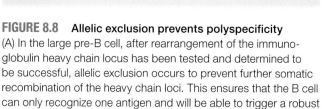

FIGURE 8.8 Allelic exclusion prevents polyspecificity (A) In the large pre-B cell, after rearrangement of the immunoglobulin heavy chain locus has been tested and determined to be successful, allelic exclusion occurs to prevent further somatic recombination of the heavy chain loci. This ensures that the B cell can only recognize one antigen and will be able to trigger a robust B-cell response when needed. (B) If allelic exclusion does not occur, somatic recombination of the heavy chain loci continues and could result in the B cell producing more than one heavy chain. Each heavy chain would drive different antigen specificity, and the ensuing polyspecificity of the B cell would dampen the humoral adaptive immune response. (After P. Parham. 2014. *The Immune System*, 4th ed., W. W. Norton and Company.)

Polyspecific immunoglobulins would complicate the ability to cluster immunoglobulins at the B-cell surface. Since clustering of immunoglobulin receptors plays an important role in activating a robust B-cell response, especially to multivalent antigens, lack of allelic exclusion would potentially inhibit a robust humoral response. The expression of two different immunoglobulin heavy chains complicates matters even further when we consider the initial response of a B cell, which is to produce IgM antibodies. IgM is produced as a pentameric protein, containing 10 heavy chains and 10 light chains. If a B cell was able to express two different immunoglobulin heavy chains, due to the structure of IgM, only about 0.1% of the assembled and secreted IgM would contain 10 appropriate heavy chains capable of binding to target antigen, thus limiting the robustness of the adaptive immune response triggered by B-cell activation.

Gene rearrangement after the heavy chain checkpoint

Developing B cells that can properly assemble a pre-B-cell receptor and that have gone through allelic exclusion and proliferation move to the next stage in B-cell development: they become pre-B cells. Before reactivating somatic recombination to drive light chain rearrangements, the pre-B cells are *large pre-B cells*. These cells engage in allelic exclusion and proliferate, increasing the number of developing B cells that have undergone heavy chain rearrangement. After proliferation, somatic recombination is again activated in *small pre-B cells* through the upregulation of expression of V(D)J recombinase and TdT. However, the location of recombination switches to the light chain loci.

CHECKPOINT QUESTIONS

1. What is the composition of the surrogate light chain, and how does it resemble the immunoglobulin light chain?
2. Why is allelic exclusion of an immunoglobulin heavy chain that has not recombined necessary to normal B-cell function?

8.7 What must occur for pre-B cells to become immature B cells?

LEARNING OBJECTIVE

8.7.1 List and describe the events that must occur for pre-B cells to become immature B cells.

Once the transition from pro-B to pre-B cell has occurred and the cell has proliferated (large pre-B cells) and reactivated somatic recombination by again expressing V(D)J recombinase and TdT, recombination events occur during the small pre-B cell stage at the immunoglobulin light chain loci. We have seen that there are two different types of immunoglobulin light chain genes—κ and λ light chains. Somatic recombination of the light chains begins at a random κ or λ light chain locus. Rearrangements of V and J variable segments at any of these loci are tested for the ability to form a functional immunoglobulin at the surface of the developing B cell (**FIGURE 8.9**).

An unproductive rearrangement at one of the light chain loci results in the lack of ability to form a functional immunoglobulin at the surface of the developing B cell. It also allows somatic recombination at the other light chain loci to continue. If rearrangement of all four immunoglobulin light chain loci is unproductive, the developing cell is removed from the cell population by apoptosis. However, due to the four different opportunities for a productive rearrangement during light chain recombination, many pre-B cells are capable of moving through the light chain checkpoint and becoming immature B cells.

CHECKPOINT QUESTION

1. Why would you predict that most pre-B cells are likely to move through the light chain checkpoint and develop into immature B cells?

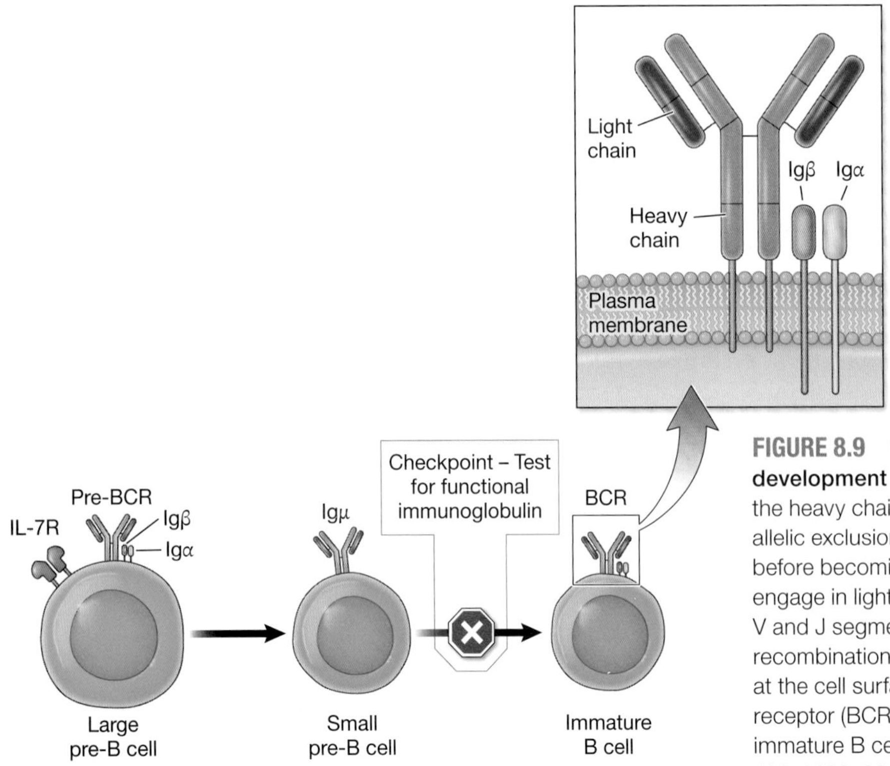

FIGURE 8.9 Immunoglobulin checkpoint of B-cell development Large pre-B cells that have passed the heavy chain checkpoint undergo cell division and allelic exclusion of the immunoglobulin heavy chain loci before becoming small pre-B cells. Small pre-B cells engage in light chain loci recombination, rearranging the V and J segments. To test for a productive light chain recombination event, the immunoglobulin is assembled at the cell surface and tested for functionality as a B-cell receptor (BCR). Cells that pass this checkpoint become immature B cells. (After F. Mechers. 2015. *J Clin Invest* 125: 2203–2210.)

Making Connections: B-Cell Development

Developmental stages require recombination events to produce a functional B-cell receptor. Each stage involves rearrangement of heavy and light chain loci and checkpoints to ensure that productive rearrangements have occurred.

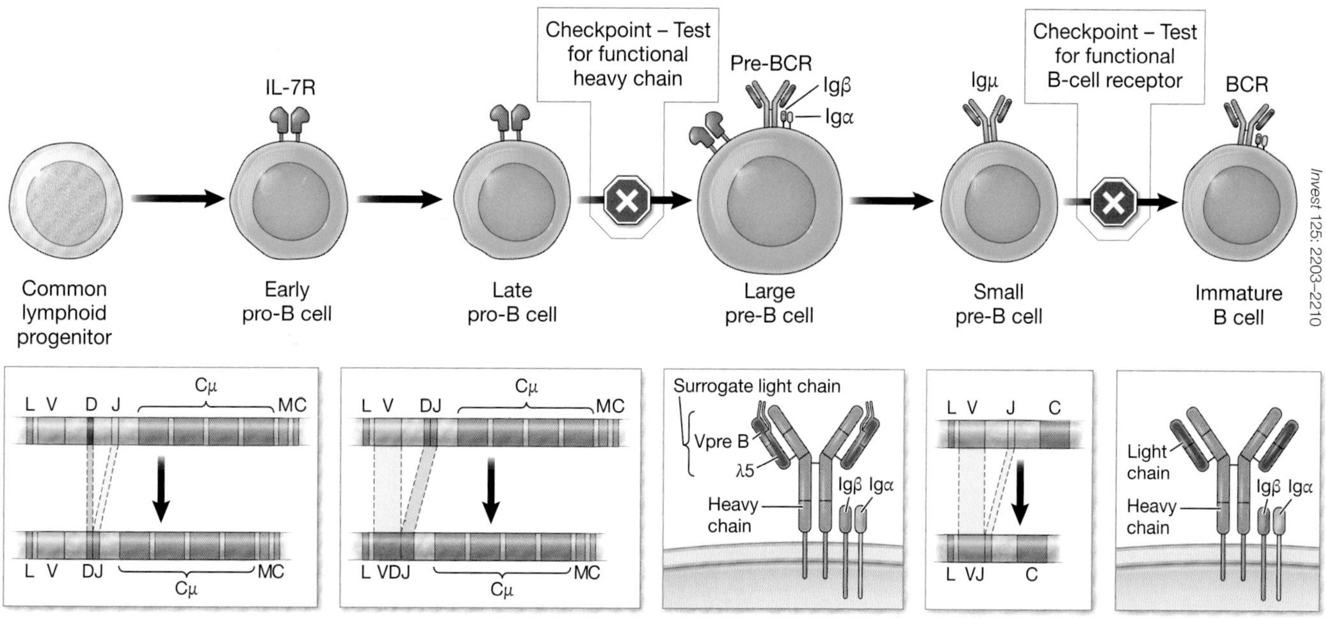

After F. Melchers. 2015. *J Clin Invest* 125: 2203–2210

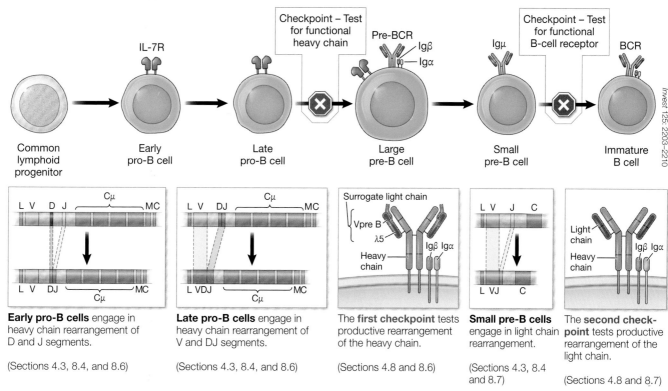

Early pro-B cells engage in heavy chain rearrangement of D and J segments.

(Sections 4.3, 8.4, and 8.6)

Late pro-B cells engage in heavy chain rearrangement of V and DJ segments.

(Sections 4.3, 8.4, and 8.6)

The **first checkpoint** tests productive rearrangement of the heavy chain.

(Sections 4.8 and 8.6)

Small pre-B cells engage in light chain rearrangement.

(Sections 4.3, 8.4 and 8.7)

The **second checkpoint** tests productive rearrangement of the light chain.

(Sections 4.8 and 8.7)

Early and late pro-B cells, after I.-L. Martensson et al. 2010. *FEBS Letters* 584: 2572–2579. Small pre-B cells, after W. Purves et al. 1995. *LIFE: The Science of Biology*, 4th ed., Sinauer Assoc., Sunderland, MA; T. Dörner and P. E. Lipsky. 2001. *Arthritis Rheum* 44: 2715-2727; P. Parham. 2014. *The Immune System*, 4th ed., W. W. Norton and Company, New York

8.8 How does negative selection of B cells occur in the bone marrow?

LEARNING OBJECTIVE

8.8.1 Summarize the three possible outcomes of negative selection of B cells in the bone marrow.

Positive selection of immature B cells occurs through testing the formation of a functional immunoglobulin at the surface of the developing B cell; this requires productive rearrangement at an immunoglobulin light chain locus. (Recall that heavy chain rearrangement was tested at the pro-B cell to pre-B cell checkpoint.) Since productive rearrangement of both heavy chain and light chain loci can form an antigen-binding site capable of recognizing self-antigens, a mechanism must exist to remove self-reactive B cells at this stage. Immature B cells within the bone marrow that express self-reactive immunoglobulins may have one of three fates: clonal deletion, receptor editing, or anergy (**FIGURE 8.10**).

Clonal deletion

Immature B cells with self-reactive antigen can be removed from the B-cell population through an immunoglobulin-mediated activation of apoptosis within the cell. This negative selection of self-reactive B cells within the bone marrow leads to **central tolerance** of the B-cell repertoire within an organism. The mechanism

central tolerance Negative selection processes that occur in primary lymphoid tissues that are responsible for the removal of self-reactive lymphocytes.

clonal deletion Removal of lymphocytes that express self-reactive receptors by apoptosis.

anergy Cellular state of non-responsiveness in which the lymphocyte is incapable of expansion and differentiation.

by which autoreactive immunoglobulins signal for **clonal deletion** is still being unraveled—recent evidence suggests that it may be related to the strength of the stimulation of the immunoglobulin receptor (see **EMERGING SCIENCE**).

Receptor editing

Self-reactive immature B cells in the bone marrow are capable of reactivating somatic recombination at the immunoglobulin light chain locus in an attempt to alter the antigen specificity. A new productive rearrangement of the light chain locus may alter the antigen specificity and convert the self-reactive B cell to one that is capable of recognizing a foreign antigen. If all receptor editing possibilities are exhausted, the developing B cell will die through clonal deletion.

Anergy

Self-reactive B cells within the bone marrow that can recognize soluble monovalent antigens may not be removed via clonal deletion or may not induce receptor editing. Instead, these cells become unresponsive to antigen and are put into an arrested state known as **anergy**. Anergic B cells still express immunoglobulin at their cell surface, although the isotype tends to be IgD rather than the IgM typically produced on the surface of a normal cell. Anergic B cells are incapable of responding to antigen and live for 2 to 10 days, compared to the approximately 80 days that normal B cells circulate.

Stringency of negative selection

The process of negative selection of self-reactive B cells is identical to the process of negative selection of self-reactive T cells. However, the level of stringency of negative selection of B cells in bone marrow differs from that of negative

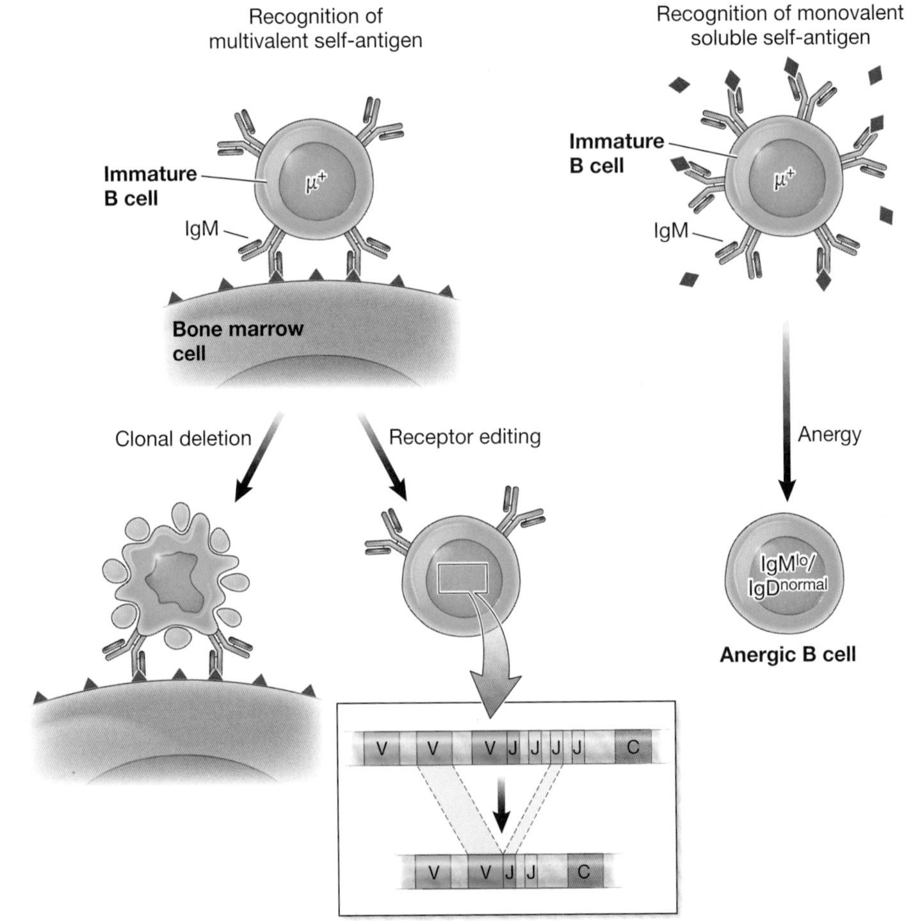

FIGURE 8.10 Fates of self-reactive immature B cells After testing for a functional B-cell receptor, immature B cells undergo negative selection to remove self-reactive cells. If the B cell expresses a receptor that recognizes a multivalent self-antigen in the bone marrow, one of two outcomes occurs: clonal deletion (the cell dies by apoptosis) or receptor editing, in which the cell reengages in immunoglobulin light chain recombination to attempt to produce a receptor that is no longer self-reactive. If the B-cell receptor recognizes a soluble monovalent self-antigen, the cell changes expression of the isotypes of receptors at the surface and expresses a small amount of IgM. This change in isotype expression induces a state of anergy (unresponsiveness), preventing the cell from responding to the antigen it recognizes.

EMERGING SCIENCE

How is the strength of immunoglobulin signaling that drives central tolerance regulated?

Article

F. Pala, H. Morbach, M. C. Castiello, J. N. Schickel, S. Scaramuzza, N. Chamberlain, B. Cassani, S. Glauzy, N. Romberg, F. Candotti, A. Aiuti, M. Mosticardo, A. Villa, and E. Meffre. 2015. Lentiviral-mediated gene therapy restores B cell tolerance in Wiskott-Aldrich syndrome patients. *The Journal of Clinical Investigation* 125: 3941–3951.

Background

Proper formation of an immunoglobulin receptor on the surface of a developing B cell is vital in verifying that the developing cell is capable of acting within the adaptive immune system. Signaling events in pre-B cells are activated upon passage through the immunoglobulin light chain checkpoint. However, the same receptor must be tested to determine whether the cell-surface immunoglobulin is self-reactive, and strong signals during negative selection can drive clonal deletion. This begs the question: How is signal strength fine-tuned to ensure that proper development of cells containing a functional immunoglobulin occurs as long as the immunoglobulin is not self-reactive? Interestingly, the Wiskott-Aldrich syndrome protein (WASp), a protein that connects extracellular signals to the actin cytoskeleton, has been implicated in the generation of both immunodeficiency and autoimmunity, and defects in T-cell peripheral tolerance in WASp-deficient T cells have been identified. Could such tolerance defects due to WASp be present in B cells as well?

The Study

Researchers hypothesized that WASp plays a role in promoting B-cell tolerance and clearance of autoreactive B cells. To test their hypothesis, researchers analyzed B-cell populations in healthy individuals compared to WAS patients. Their findings are summarized in **FIGURE ES 8.1**. They discovered that WAS patients contained fewer polyreactive antibodies compared to healthy individuals, suggesting that the lack of WASp increases central tolerance and clonal deletion.

(Continued)

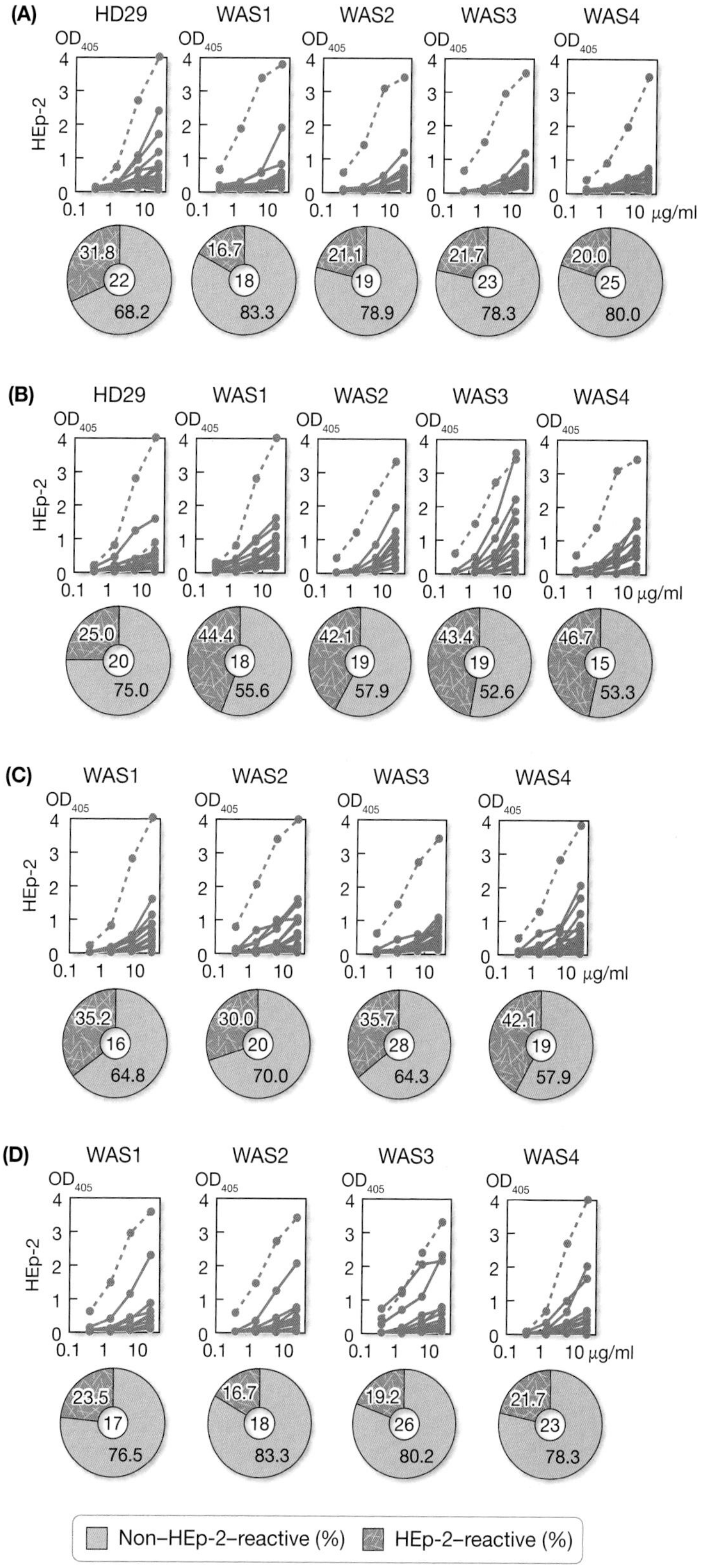

FIGURE ES 8.1 Absence of Wiskott-Aldrich syndrome protein (WASp) causes defects in B-cell central and peripheral tolerance In four different patients with Wiskott-Aldrich syndrome, the amount of self-reactive immunoglobulins from newly emerging B cells was lower than in normal individuals (A), whereas self-reactive mature B cells were more predominant in the same patients (B), suggesting defects in both central and peripheral tolerance in these individuals. Replacement of a wild-type form of WASp via gene therapy restored both central tolerance (C) and peripheral tolerance (D). (Data from F. Pala et al. 2015. *J Clin Investig* 125: 3941–3951. © 2015 American Society for Clinical Investigation.)

EMERGING SCIENCE (continued)

This increase in central tolerance in the absence of WASp correlated with an increase in B-cell activation through immunoglobulin signaling. Paradoxically, the researchers found a decrease in peripheral tolerance and identified more autoreactive antibodies produced by naïve B cells, which could be explained due to malfunction in the action of regulatory T cells. Interestingly, replacement of a wild-type WASp expression via gene therapy was able to restore normal B-cell development and central and peripheral tolerance.

Conclusion

This research provides evidence that signal strength through the immunoglobulin receptor plays an important role in activating central and peripheral tolerance. It also implicates the molecule WASp as a component responsible for regulating this strength. These findings suggest that WASp plays an important role in early B-cell development, as a large increase in central tolerance was seen upon removal of WASp. Although there was greater stringency in autoreactive cells during early B-cell development, studies on peripheral tolerance demonstrated that autoreactive B cells could still escape the bone marrow; they would then be unchecked by peripheral tolerance mechanisms, due to the lack of functioning regulatory T cells, which would aid in providing peripheral tolerance. These studies point to WASp as an important regulator of immunoglobulin signaling and suggest that it plays a role in tuning the strength of the signal that drives either further B-cell development or clonal deletion.

Think About...

1. What do you predict would happen if you were able to overexpress WASp in hematopoietic stem cells? Given these research findings, what might occur to B-cell development, central tolerance, and peripheral tolerance?
2. WASp has also been implicated in severe immunodeficiencies. Speculate as to how different mutations in the WASp gene can cause either immunodeficiency (lack of lymphocyte function) or increased immunoglobulin signaling (which appears to be higher lymphocyte function).
3. The researchers delivered a wild-type WASp expression gene via a lentivirus into hematopoietic stem cells before transplanting them into patients. Speculate as to why the researchers would use this tool for delivery of the wild-type gene into HSCs (consider looking at an overview of viral life cycles).

selection of T cells in the thymus. Negative selection of B cells is less stringent, likely due to B-cell activation being dependent on help from T cells. If a self-reactive B cell escapes negative selection in the bone marrow and enters the circulation, it still requires follicular helper T-cell activity to become activated. The follicular helper T cell would also have to be self-reactive to the same antigen, and since the stringency of negative selection of T cells in the thymus is high, self-reactive T cells tend to be limited within the circulation. Thus, circulation of self-reactive B cells is more tolerated as long as negative selection of T cells in the thymus is maintained at a stringent level.

Negative selection of T cells is further driven by inducing expression of specialized endocrine genes within the thymus using the transcription factor AIRE (see Chapter 5). However, there is no AIRE activity (or presence of an AIRE homolog) in bone marrow, suggesting that developing B cells are not exposed to endocrine-specific proteins as they develop in bone marrow. There may be self-reactive B cells that can recognize endocrine-specific proteins that were not displayed to the developing B cells; therefore, these B cells would escape negative selection and enter the circulation. Again, the self-reactive B cells would require activation help from a self-reactive T cell, so as long as there is high stringency of negative selection of T cells in the thymus, self-reactive B cells can be tolerated in the body.

Autoimmune diseases (discussed in Chapter 16) can be caused by the production of autoantibodies by self-reactive B cells, as occurs in systemic lupus erythematosus, rheumatoid arthritis, and type 1 diabetes. The autoantibodies recognize self-antigens and trigger humoral responses.

● **CHECKPOINT QUESTIONS**

1. Compare and contrast the three possible fates of negative selection of self-reactive immature B cells.
2. Why is a lower level of negative selection stringency during B-cell development tolerated in comparison to the very high stringency of negative selection that occurs during T-cell development?

8.9 How does positive and negative selection of transitional B cells occur in the spleen?

LEARNING OBJECTIVE

8.9.1 Describe how positive and negative selection of transitional B cells occurs in the spleen.

Approximately 25% of immature B cells exit the bone marrow primed to develop into mature B cells in the B-cell follicle of secondary lymphoid tissues such as lymph nodes. However, the other 75% must travel to the spleen, where they will complete their development into mature B cells. Once an immature B cell reaches the spleen, it undergoes two additional developmental stages before becoming a mature B cell. These two stages encompass a population of B cells known as *transitional B cells*, and the two stages reflect the progression of *T1 transitional B cells* to *T2 transitional B cells* on the way to becoming mature B cells (**FIGURE 8.11**). The transition from T1 to T2 to mature B cell takes 3 to 4 days.

Although the spleen has important physiological functions outside that of a secondary lymphoid tissue, because of its role in B-cell development, it stands

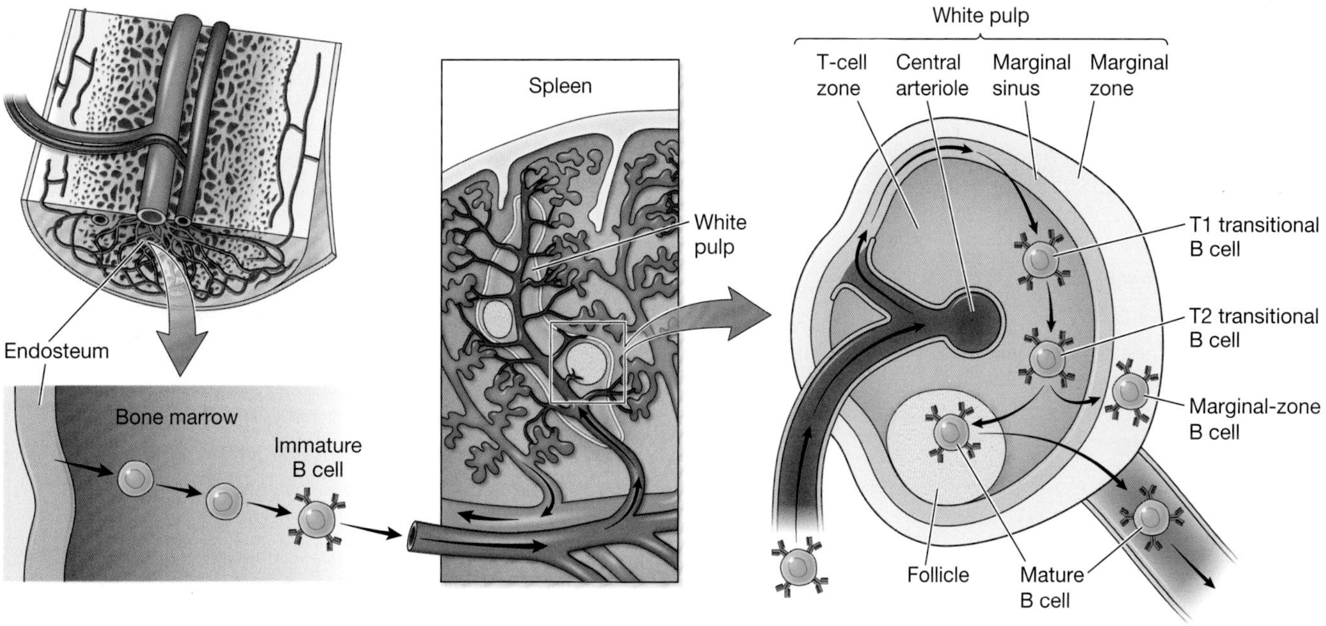

FIGURE 8.11 B-cell maturation in the spleen After passing negative selection in bone marrow, immature B cells migrate to the spleen, where they complete the maturation process. Immature B cells enter the spleen and become T1 transitional B cells, which migrate to the T-cell zone within the white pulp. T1 transitional B cells that pass negative selection in the spleen become T2 transitional B cells, which receive important survival signals and begin to express both initial isotypes of immunoglobulin (IgM and IgD) at their surface. After maturation, some T2 cells remain in the marginal zone and become marginal-zone B cells. However, most T2 cells become mature B cells within the follicle and reenter the circulation to function within the humoral arm of the adaptive immune system.

BAFF Stands for B-cell activation factor or B-cell survival factor. Survival factor for B cells secreted by cells such as follicular dendritic cells to promote survival of B cells capable of mounting a humoral adaptive immune response.

peripheral tolerance Negative selection/inactivation processes that occur in peripheral tissues to prevent the activation of self-reactive lymphocytes that have escaped negative selection in their primary lymphoid tissues.

to reason that its evolution is connected to that of the adaptive immune system (see **EVOLUTION AND IMMUNITY**).

T1 transitional B cells

T1 transitional B cells are characterized by several key cell-surface markers, including a high level of IgM, a low level of IgD, and the cell-surface markers CD24 and CD93. During this stage, BAFF-R, the receptor for the B-cell survival factor **BAFF**, is expressed. Note that expression of BAFF-R is relatively low in T1 cells compared to B cells later in development.

Immature B cells enter the spleen, transition into T1 cells, and migrate to the T-cell zone of the spleen. T1 cells are further tested for self-reactivity and go through a round of negative selection to further promote **peripheral tolerance**. Binding of a cell-surface immunoglobulin on a T1 cell activates cellular signaling through the immunoglobulin, resulting in a release of calcium and activation of an apoptotic signal to remove self-reactive B cells from the population (**FIGURE 8.12**). Approximately 55% to 75% of immature B cells that develop into T1 transitional B cells die through this mechanism of peripheral tolerance.

T2 transitional B cells

T2 transitional B cells are also characterized by key cell-surface markers, and they also express immunoglobulins at the surface. But the transition from the T1 to T2 stage marks the point at which we begin to see the initiation of alternative splicing of immunoglobulins and an increase in IgD expression at the cell surface. T2 cells express both IgM and IgD at the surface, although at this stage, IgM expression is still predominant. Furthermore, T2 cells express higher levels of BAFF-R at their cell surface as well as additional cell-surface molecules, including CD21 and CD23.

T2 cells within the spleen undergo positive selection in which signaling through the immunoglobulin stimulates survival signals via an increase in intracellular calcium and production of the second messenger diacylglycerol (see Figure 8.12). T2 cells that do not undergo positive selection within the spleen do not receive the survival signal and undergo apoptosis. Survival

EVOLUTION AND IMMUNITY

The Spleen

In Chapter 5, we learned that the evolution of the thymus coincides with that of T-cell development, due to its importance as the primary lymphoid tissue responsible for proper T-cell developmental cues. Since the spleen plays an important role in B-cell development, we would assume that the spleen evolved on a time frame similar to that of B-cell development, although the role of the spleen in circulation complicates this reasoning.

Recall that the white pulp of the spleen is the critical location of immune cell function and maturation, suggesting that white pulp evolution would be closely tied to adaptive immune system evolution. As we saw within lamprey (jawless fish), lymphocyte-like cells encode for multiple leucine-rich repeat (LRR) domains, termed *variable-lymphocyte receptors* (*VLRs*), and represent a system similar to an adaptive immune system. Within jawless fish, we can also observe the development of small foci of lymphoid cells in areas of the intestine responsible for hematopoiesis, suggesting that this location may be a functional equivalent of the spleen.

To further support this hypothesis, lymphocyte-like cells expressing VLRs are closely associated in these foci. Within adult cartilaginous fish, the white pulp primarily consists of B cells localized in a T-cell zone. The spleens of reptiles are more elaborately structured, but we do not see the germinal centers associated with the spleen in higher vertebrates, including birds. Finally, it is not until mammalian evolution that we observe the marginal zone, where marginal-zone B cells reside.

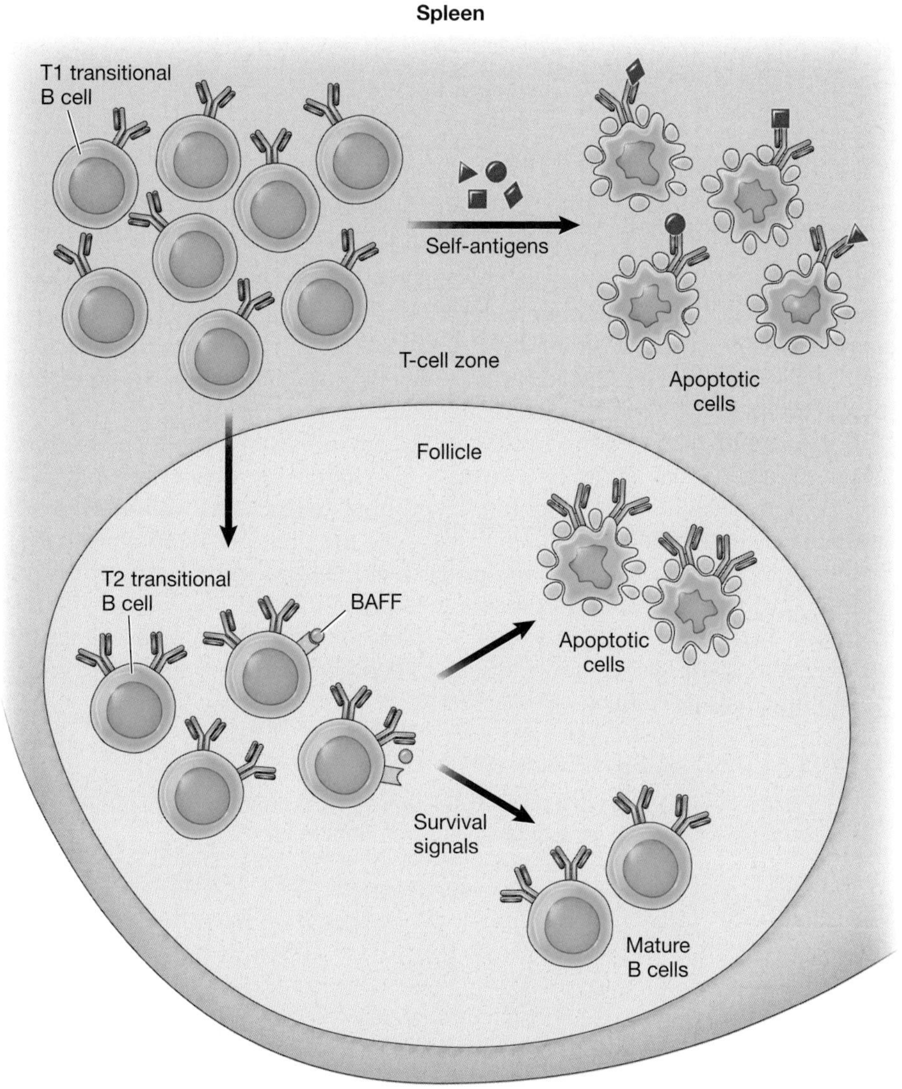

FIGURE 8.12 Selection of transitional B cells within the spleen Within the T-cell zone, self-antigens are presented and T1 transitional B cells undergo a second round of negative selection. T1 cells that recognize self-antigen are removed through clonal deletion to promote peripheral tolerance. T1 cells that pass negative selection migrate to the follicle and become T2 transitional B cells. Within the follicle, some T2 cells receive the important survival signal BAFF and finish the maturation process to become mature B cells. Those that do not receive the survival signal die by neglect.

signals are also dependent on BAFF-R on the cell surface, which drives the synthesis of anti-apoptotic factors and promotes survival of the developing B cell. Those that pass through positive selection within the spleen become mature B-2 B cells capable of circulating in the blood and lymphatic system, entering secondary lymphoid tissue, and becoming activated if their immunoglobulin receptor receives the proper signal.

CHECKPOINT QUESTIONS

1. Define the process of peripheral tolerance that occurs in B-cell development in the spleen.
2. Explain how T1 transitional B cells die by apoptosis while T2 cells survive when both receive signals through their immunoglobulin.

8.10 | How do mature B cells enter the circulation and become activated?

LEARNING OBJECTIVE

8.10.1 Outline the steps by which mature B cells enter the circulation and become activated.

naïve B cell Fully mature B cell that has not yet come into contact with antigen.

primary lymphoid follicle Location within a secondary lymphoid tissue where circulating B cells migrate to test for the presence of antigen their immunoglobulin is designed to recognize.

Once a B cell has matured in the spleen, it will enter the circulation where it can monitor antigens within secondary lymphoid tissues. Even though B cells exiting the spleen are considered mature B cells in terms of their development, these cells are also **naïve B cells** because they have not encountered antigen that their expressed immunoglobulin recognizes.

B cells travel through the circulatory system and can enter secondary lymphoid tissues such as a lymph node through a high endothelial venule (FIGURE 8.13). Upon entry into the secondary lymphoid tissue, the B cell migrates to a **primary lymphoid follicle**, where it scans for the presence of antigen that its immunoglobulin is designed to recognize. If antigen recognition occurs, the cell will go through an activation process (described in detail in Chapter 10). If it does not encounter antigen, it will leave the secondary lymphoid tissue via the efferent lymphatic system and circulate to another secondary lymphoid tissue to continue its surveillance.

CHECKPOINT QUESTION

1. Explain the migration of a naïve B cell through a secondary lymphoid tissue as it monitors for antigen via its immunoglobulin.

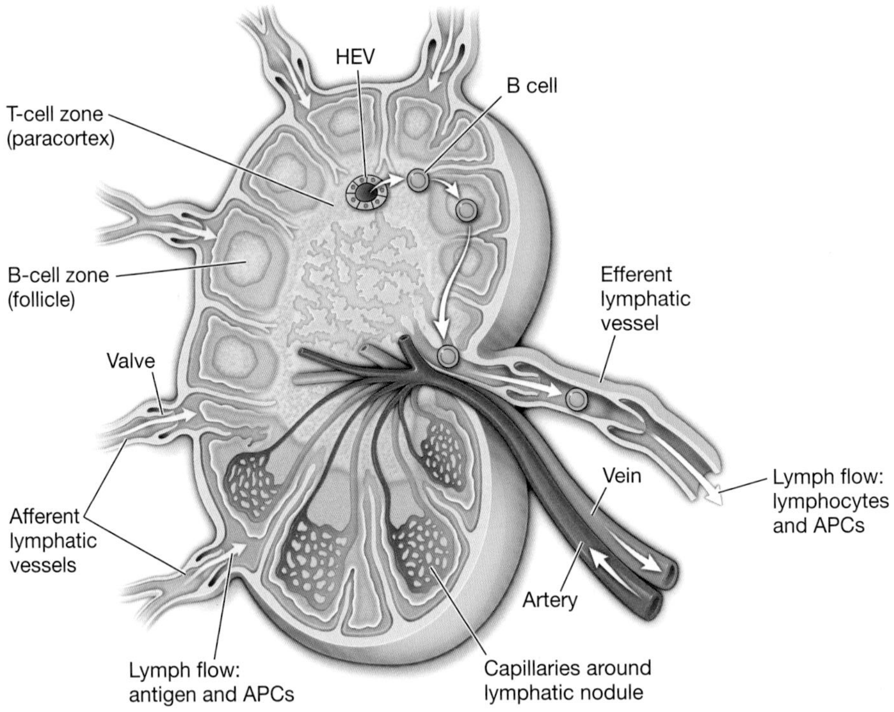

FIGURE 8.13 **Movement of B cells into and out of a lymph node** Mature B cells in circulation enter secondary lymphoid tissues via a high endothelial venule (HEV) and migrate to the T-cell zone. B cells activated here through the action of T cells activated by antigen-presenting cells (APCs) migrate to the follicle of the B-cell zone, where they may produce immunoglobulins to combat a recognized antigen. B cells that are not activated because they have not come in contact with the antigen their immunoglobulin recognizes exit via the efferent lymphatic vessels, reenter the circulation, and migrate to a different secondary lymphoid tissue.

8.11 How do B-1 and marginal-zone B cells develop?

LEARNING OBJECTIVE

8.11.1 Explain what is known about how B-1 and marginal-zone B cells develop.

The type of B-cell development addressed in the majority of this chapter was that of B-2 B cells (follicular B cells), since we know the most about these cells experimentally. However, other B-cell populations can develop and act at specialized locations within the body, including *B-1 B cells* and *marginal-zone B cells* (see Table 8.1).

B-1 B cells were discussed previously as resident B cells of the pleural and peritoneal cavities. Marginal-zone B cells reside in the outer zone of the white pulp of the spleen, where they function primarily to detect foreign antigens and pathogens located in the blood. Because of their distinct functions and locations within an organism, these cells develop differently than B-2 B cells.

B-1 B-cell development

We know that B-1 B cells patrol the pleural and peritoneal cavities of an organism and can make up approximately half of all of the B cells present in those locations. Unlike B-2 B cells, which undergo a large degree of somatic recombination at both the immunoglobulin heavy and light chain loci, B-1 B cells have a limited amount of immunoglobulin diversity. Most of the immunoglobulins expressed by B-1 B cells lack diversity due to limited V(D)J recombination and TdT activity; they commonly express immunoglobulins with antigen specificity toward common microbial carbohydrates. These immunoglobulins tend to be similar in action to receptors on innate immune cells that recognize PAMPs.

Interestingly, signals that drive negative selection of T-1 transitional B-2 cells, namely antigen recognition in the spleen, are required for positive selection of B-1 B cells. A strong affinity for self-antigens of the B-1 B cells seems to promote the activation of survival signals within developing cells. Another difference between the development of B-1 B cells and B-2 B cells is that B-1 B cell development appears to occur independently of the action of the B-cell survival factor BAFF.

A major debate surrounding the development of these two cell types focused on the question of whether both share a common progenitor cell or whether B-1 B cells arise from a different developmental lineage. Many pieces of evidence suggest that the development of B-1 and B-2 B cells occurs via different lineages. For instance, B-1 B cells are present much earlier in the development of an organism compared to B-2 B cells. (Recall that B-1 B cells are the predominant B cells present in the developing fetus, generated early on in the AGM region.)

B-1 B cells also have limited immunoglobulin diversity, in part due to their developing at a point earlier than when efficient TdT activity can occur, thus limiting the addition of N nucleotides to immunoglobulin variable segment junctions in B-1 B cells. The location of the different B-cell populations (B-1 in pleural and peritoneal cavities and B-2 in circulation and in secondary lymphoid tissues) also suggests independent lineage of the two populations.

Finally, regeneration of these two populations seems to occur independently. B-2 B cells regenerate through the developmental process described in this chapter, beginning with hematopoiesis and developmental stages in the bone marrow and transitional stages within the spleen, before they are capable of functioning. B-1 B cells seem to regenerate in the periphery of the organism. Irradiation of the bone marrow in a mouse depletes the B-2 B cell population but leaves the B-1 B cells fully functional. All of this evidence suggests the independent lineage of B-1 and B-2 B cells.

Making Connections: B-Cell Maturation

B-cell development requires the recombination of both the heavy chain and light chain immunoglobulin loci. While the developmental checkpoints test for immunoglobulin function, further steps are required to ensure that self-reactive B cells do not circulate. Events in the bone marrow and spleen drive the formation of mature B cells, which can circulate into secondary lymphoid tissue to scan for the presence of the antigen to which their immunoglobulin can bind.

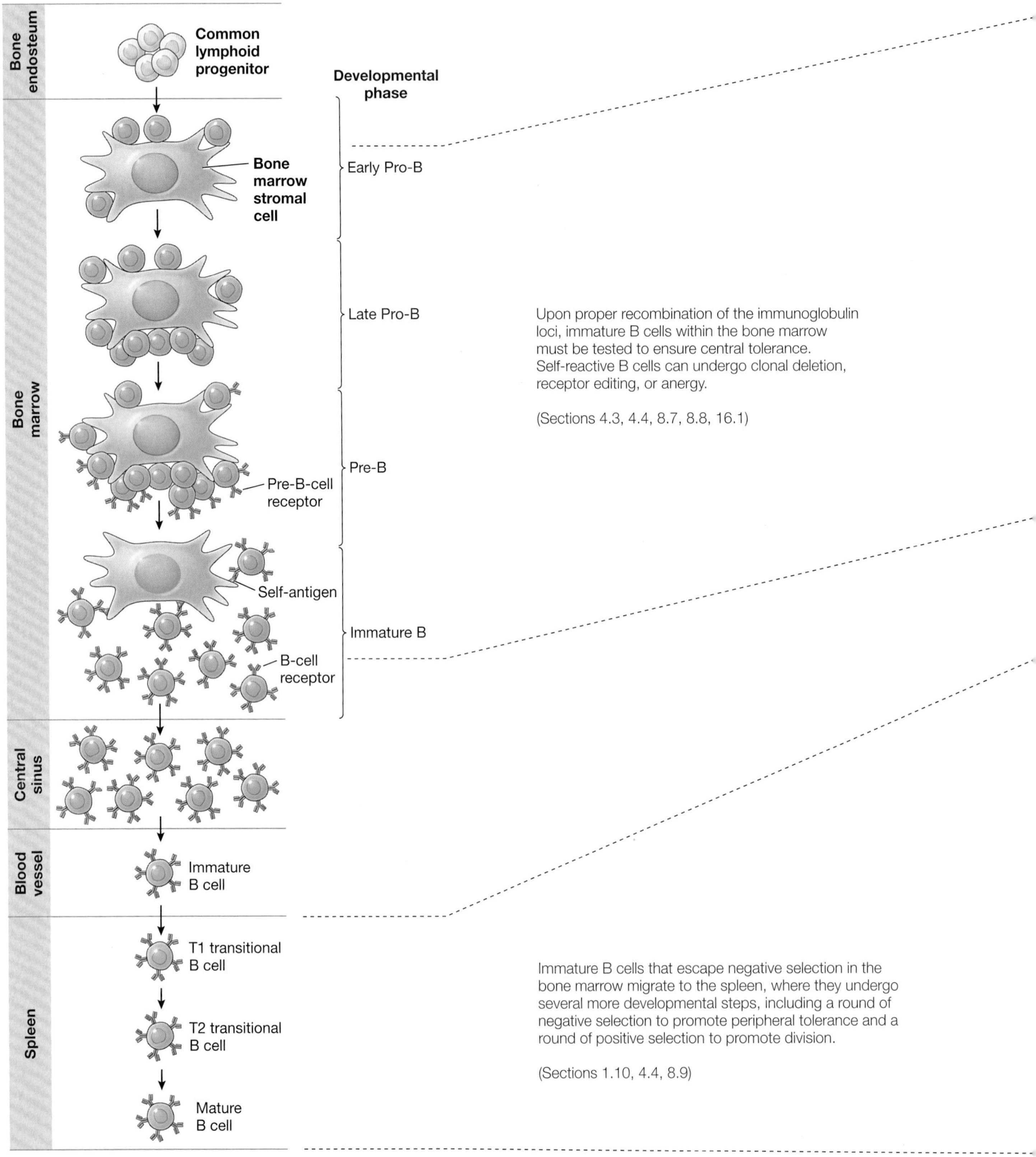

Upon proper recombination of the immunoglobulin loci, immature B cells within the bone marrow must be tested to ensure central tolerance. Self-reactive B cells can undergo clonal deletion, receptor editing, or anergy.

(Sections 4.3, 4.4, 8.7, 8.8, 16.1)

Immature B cells that escape negative selection in the bone marrow migrate to the spleen, where they undergo several more developmental steps, including a round of negative selection to promote peripheral tolerance and a round of positive selection to promote division.

(Sections 1.10, 4.4, 8.9)

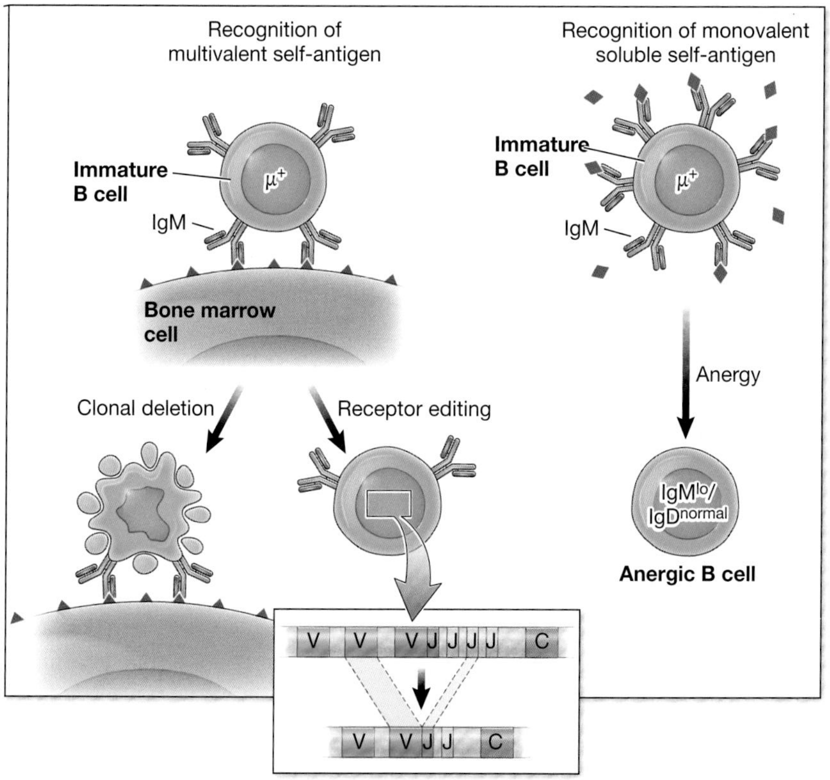

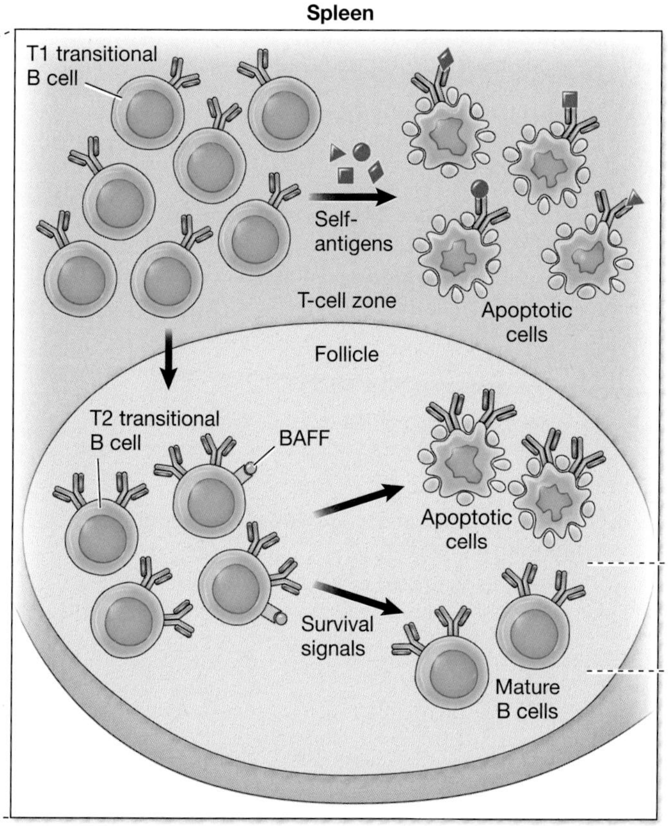

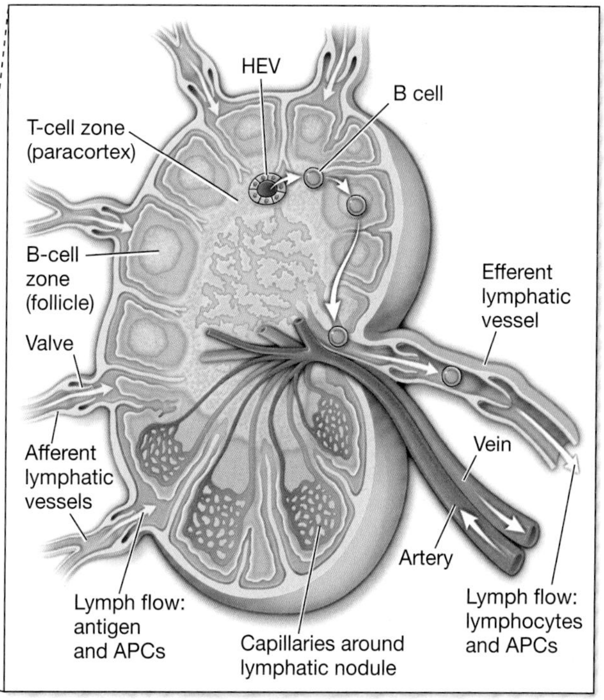

Mature B cells can circulate in the body and enter secondary lymphoid tissue to determine the presence of antigen their specific immunoglobulin can recognize.

(Sections 4.5, 4.8, 8.10, 9.4, 10.5)

Marginal-zone B-cell development

Marginal-zone B cells are so named due to their residence in the marginal zone of the white pulp in the spleen. They are characterized by a relatively high level of IgM at their surface, along with CD21, but have low levels of CD23 and IgD at their surface (see Table 8.1). They also express membrane-bound adhesion molecules and receptors of phospholipids that allow them to adhere to other cells. These receptors likely work to maintain the marginal-zone B cells in the spleen and to prevent them from leaving the secondary lymphoid tissue. Marginal-zone B cells can express immunoglobulins that recognize proteins or carbohydrates, but not surprisingly, these immunoglobulins seem to be specialized in recognizing antigens of bloodborne pathogens.

What leads to the development of marginal-zone B cells? Evidence suggests that marginal-zone B cells develop from the T2 transitional B-cell population, just as B-2 B cells do. Like B-2 B cells, marginal-zone B cells require signaling through BAFF-R. However, marginal-zone B cells also have similarities to developing B-1 B cells; for instance, both cell types require strong self-antigen recognition by their immunoglobulin to survive. Strong positive selection of T2 transitional B cells is a requirement for the development of marginal-zone B cells from this transitional B-cell population. Interestingly, marginal-zone B cells also require Notch signaling to differentiate.

CHECKPOINT QUESTION

1. Contrast the differences in B-1, B-2, and marginal-zone B cells with respect to their diversity and the role they play in the adaptive immune system.

Summary

8.1 Which B-cell populations will be our major focus?

- Three major types of B cells are produced: B-1 (CD5 B cells), B-2 (follicular B cells), and marginal-zone B cells. We focus most of our attention on B-2 B cells, which make up the majority of B cells responsible for combating infections through activation at secondary lymphoid tissues.

8.2 Where and how do B cells develop in the fetus?

- B-cell development in mice and humans follows similar patterns during fetal development, eventually developing in the liver, where B-1 B cells are produced to protect pleural and peritoneal cavities and in the bone marrow, where the majority of B cells are produced.

8.3 Which cell lineages begin B-cell development within the bone marrow?

- Hematopoietic stem cells must receive proper signals to transition to specific lymphoid progenitor cell lineages and ultimately become common lymphoid progenitor cells that can receive B-cell-specific differentiation signals.

8.4 What is the site of B-cell development in adults, and which developmental stages occur there?

- The bone marrow serves as the primary site of B-cell development in adult vertebrates; it is where B cells undergo developmental stages to produce functional immunoglobulins.

8.5 What critical checkpoints are required for proper B-cell development?

- Rearrangement of the heavy and light chain immunoglobulin loci occurs during specific B-cell developmental stages. The rearrangement of heavy chain and light chain loci is tested to ensure that the B cell is capable of functioning properly during a humoral immune response.

8.6 What must occur before cells can pass beyond the pro-B cell stage?

- Developing B cells require proper rearrangement of a heavy chain locus to transition from pro-B cells to pre-B cells.
- Upon passage through the heavy chain immunoglobulin checkpoint, developing B cells undergo allelic exclusion of the other heavy chain locus to ensure that they cannot develop into polyspecific B cells.

8.7 What must occur for pre-B cells to become immature B cells?

- Developing B cells require proper rearrangement of a light chain locus to transition from small pre-B cells to immature B cells.

8.8 How does negative selection of B cells occur in the bone marrow?

- Upon passage through both immunoglobulin rearrangement checkpoints, B cells are tested for self-reactivity. Self-reactive B cells are subject to one of three fates: clonal deletion, receptor editing, or anergy.

8.9 How does positive and negative selection of transitional B cells occur in the spleen?

- Development of a majority of B cells continues in the spleen, where immature B cells mature through a T1 and T2 transitional B-cell stage and undergo a second round of positive and negative selection.

8.10 How do mature B cells enter the circulation and become activated?

- Mature B cells can circulate and migrate into secondary lymphoid tissues, where they test for antigen recognition using their immunoglobulins and become activated upon antigen recognition.

8.11 How do B-1 and marginal-zone B cells develop?

- B-1 B cells are resident B cells of the pleural and peritoneal cavities, and marginal-zone B cells reside in the outer zone of the white pulp of the spleen. B-1 B cells develop much earlier in an organism's life compared to B-2 B cells. Resident B-1 B cells replenish the B-1 B cell population through self-replication rather than from a lymphoid progenitor in bone marrow. Marginal-zone B cells develop as an arm of the T2 transitional B-cell population.

Review Questions

8.1 Which B-cell subtype plays a primary role in combating infections through activation at secondary lymphoid tissues?

8.2 Explain why B-cell development in the fetus must occur in tissues such as the AGM and the liver.

8.3 Why are there several developmental stages of B-cell development within bone marrow?

8.4 What would you predict would be the fate of B-cell development in individuals whose bone marrow stromal cells are incapable of expressing stem cell factor (SCF)?

8.5 Suppose you had the means by which to label intracellular immunoglobulin heavy and light chains and separate cell populations via flow cytometry. If you were able to separate different B-cell developmental stages and analyze their heavy chain and light chain expression, what would your prediction be of the expression of immunoglobulin genes in each of the developmental stages? Draw plots demonstrating heavy chain levels (y-axis) and light chain levels (x-axis) in pro-B cells, pre-B cells, and immature B cells.

8.6 Why is allelic exclusion of unrearranged immunoglobulin heavy chain essential to normal B-cell function?

8.7 Why would you predict that a majority of pre-B cells ultimately become immature B cells and pass the pre-B cell checkpoint?

8.8 Explain the functional similarities and differences between clonal deletion and anergy in B-cell development.

8.9 What would you predict would be the fate of an individual who was asplenic (lacking a spleen) due to a genetic disorder in relation to functional B cells?

8.10 What is the function of the primary lymphoid follicle in a secondary lymphoid tissue?

8.11 Illustrate what you would predict flow cytometry data of B-1, B-2, and marginal-zone B cells would be if you looked at the three populations in regard to their IgM (y-axis) and IgD (x-axis) expression.

CASE STUDY REVISITED

Common Variable Immunodeficiency

A 2-year-old girl arrives at the emergency room with difficulty breathing. The child's parents mention that their daughter has had these symptoms before when she suffered from pneumonia. Further examination of her medical history reveals frequent and persistent bacterial infections. The physician orders an X-ray and confirms pneumonia in the patient. Because the child has had several visits to the ER due to bacterial infections, the physician orders a blood test to examine her white blood cell count and immunoglobulin levels. The blood test reveals a very low immunoglobulin level in the blood. In order to examine why the child has a low level of immunoglobulins even though she has a bacterial infection, the physician orders further genetic testing to examine genes involved in B-cell function. These genetic tests determine a deletion in the $\lambda 5$ gene, resulting in a lack of expression of $\lambda 5$. Because of the lack of expression of $\lambda 5$, the physician diagnoses the patient with common variable immunodeficiency.

Think About...

Questions for individuals

1. Define common variable immunodeficiency.
2. Why would common variable immunodeficiency lead to persistent bacterial infections?
3. Which protein complex is missing in this patient?

Questions for student groups

1. Which process within B-cell development is likely faulty due to the mutation identified in this case?
2. Would you expect to be able to detect any immature B cells within this patient? Why or why not?
3. Speculate on possible therapies that might be given to this patient.

MAKING CONNECTIONS

Key Concepts	COVERAGE (Section Numbers)
Locations where hematopoiesis occurs	1.9, 8.2, 8.3
B-cell developmental arm of hematopoiesis	1.6, 1.9, 8.2, 8.3
Stages of B-cell development and recombination events that occur during these stages	4.3, 4.8, 8.4, 8.5, 8.6, 8.7
Negative selection of B cells	4.3, 4.4, 8.7, 8.8, 16.1
B-cell maturation	1.10, 4.4, 4.5, 8.8, 8.9, 8.11
B-cell populations	1.8, 4.8, 8.2, 8.9, 8.11
B-cell activation	4.5, 4.8, 8.10, 9.4, 10.5

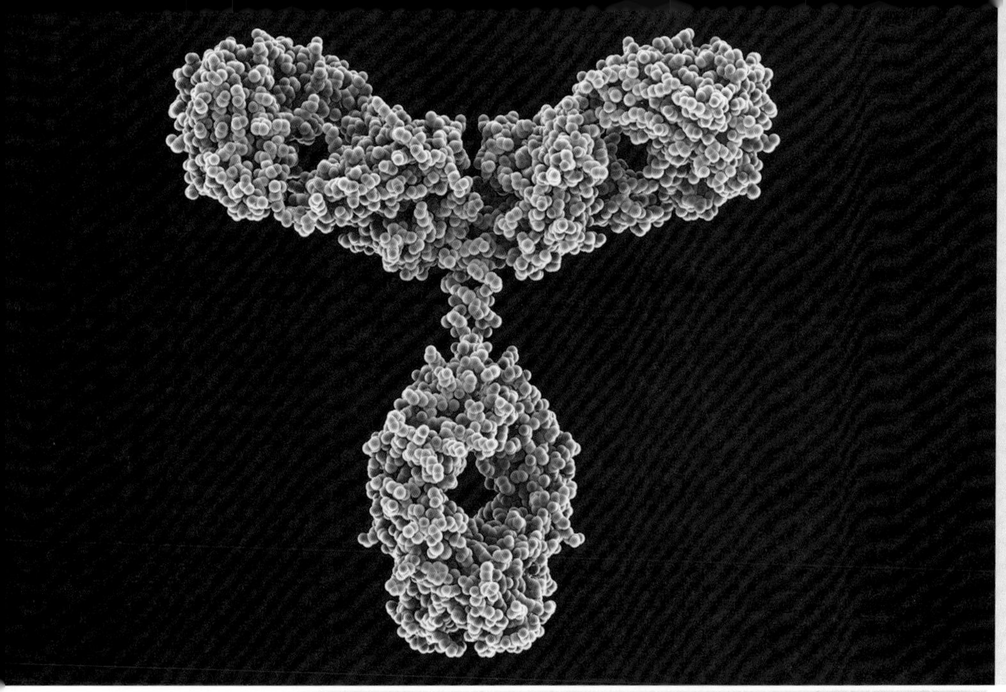

9

Immunoglobulins and B-Cell Diversity

CASE STUDY: Marshall's Story

Marshall is a typical 5-year-old boy; he is constantly on the go, and he sometimes gets into trouble or gets hurt. Troy remembers when he was his son's age; he thinks that Marshall is a "chip off the old block." What Troy doesn't remember from his childhood is the myriad of ear infections Marshall has suffered from over the past year. When Marshall recently visited an otolaryngologist (ear, nose, and throat doctor) to assess his recent ear infections, the doctor said that it didn't seem worthwhile to put tubes in Marshall's ears, since his eustachian tubes were likely developed enough to handle fluid buildup. This assessment didn't really satisfy Troy, since he wants Marshall to have a normal childhood without having to worry about constant earaches.

One morning, Marshall mentions that his ear hurts. "Not again," Troy mutters to himself as he goes to the medicine cabinet to get Marshall some ibuprofen. Troy says to Marshall, "Hey, bud, how bad does it hurt, scale of 1 to 10?" Marshall replies, "Fifty!" Troy knows that Marshall has a fairly high tolerance for pain, given the trouble he's gotten into in the past, so Troy knows that if Marshall says "fifty," he really means "fifty." Troy calls Marshall's pediatrician and asks to get an appointment for later that day.

In order to understand what Marshall is experiencing, and to answer the case study questions at the end of the chapter, we must examine the role of immunoglobulins in preventing infections. The humoral arm of the adaptive immune system relies on the action of B cells through the action of immunoglobulins, important players in the adaptive immune system. Immunoglobulins are proteins that serve as the B-cell receptor and also drive B-cell effector functions.

QUESTIONS Explored

9.1 What is the structure of immunoglobulins?

9.2 How does genetic recombination in immunoglobulins compare to recombination in T-cell receptors?

9.3 Why is alternative splicing important in antibody production?

9.4 What is the importance of somatic hypermutation and isotype switching in antibody maturation?

9.5 What is the nature of the different antibody isotypes?

The diversity of immunoglobulins determines B-cell diversity within an organism, since each B cell can produce only one type of immunoglobulin. The structure of immunoglobulins incorporates highly variable regions that aid in binding to antigen and constant regions that drive their effector functions. In this chapter we will explore the structure and composition of immunoglobulins, how they are produced, and mechanisms used to fine-tune their action within an immune response.

9.1 | What is the structure of immunoglobulins?

LEARNING OBJECTIVES

9.1.1 Describe the structure and composition of an immunoglobulin and the functions associated with its primary components.

9.1.2 Compare and contrast the structure of immunoglobulin and T-cell receptor genes.

Over the past few chapters, we have seen that the adaptive immune system is striking in its diversity and specialized in regard to the cells it utilizes to recognize and combat pathogens. We have focused our attention on the cell-mediated immune response, which relies on a repertoire of T cells to directly combat pathogens or to activate other immune system cells to assist in pathogen clearance.

The other arm of the adaptive immune response is humoral immunity, which employs the action of different, but equally specific, receptors known as *immunoglobulins*. There are many similarities between T-cell receptors and immunoglobulins:

- Both are very specific in the antigen they recognize.
- Their diversity is created through somatic recombination.
- Both are essential components of receptor complexes at the surface of adaptive immune cells required for activation of the immune response.

However, there are important differences, mainly due to their role in the adaptive immune response. Unlike T-cell receptors, which function solely to promote T-cell activation, immunoglobulins have additional effector functions. In addition to promoting B-cell activation, they also serve as soluble molecules that tag foreign particles and prevent the action of those particles through neutralization or participating in their removal. Their structure and ability to recognize a diverse array of antigens gives them a truly distinctive role during an adaptive immune response.

Important discoveries in immunoglobulin research

Immunoglobulins were first described by German bacteriologist Emil von Behring and Japanese bacteriologist Shibasabura Kitasato in 1890 when they reported that serum from animals immunized against diphtheria could be used to treat other animals infected with the same illness. Their findings suggested the presence of a molecule within the serum that was capable of treating illness and conferring immunity. Due to this discovery, von Behring was awarded the Nobel Prize in 1901.

In 1900, German biochemist Paul Ehrlich described the first theoretical structure of antibodies in his side-chain theory of the function of the immune system. He described an antibody as a branched molecule with multiple antigen-binding sites that was responsible for recognizing a specific antigen. This early model of immunoglobulin structure was refined by American biologist Gerald Edelman and British biochemist Rodney Porter, who independently confirmed the structure of immunoglobulins by determining their primary sequence (Edelman, see **KEY DISCOVERIES**) and their sensitivity to proteases such as papain (Porter, 1959). Because of their seminal work, Edelman and Porter shared the Nobel Prize in 1972. These studies provided the foundation for the determination of the three-dimensional structure of an immunoglobulin in 1977 by E. W. Silverton, Manuel Navia, and David Davies.

Immunoglobulin composition

For immunoglobulins to function efficiently in the adaptive immune system, they must be able to bind tightly to their cognate antigen and must link that bound antigen to other effector mechanisms of the immune system. Immunoglobulin structure enables these proteins to effectively perform both critical functions of humoral immunity.

There are five immunoglobulin isotypes—IgM, IgD, IgG, IgE, and IgA—and each has specific functions within the adaptive immune system (see Chapter 4). While the isotypes differ in their structure, composition, and function, they all contain a common repeating immunoglobulin domain that varies in number in the different isotypes.

KEY DISCOVERIES

How was the protein sequence of an entire immunoglobulin determined?

Article

G. M. Edelman, B. A. Cunningham, W. E. Gall, P. D. Gottlieb, U. Rutishauser, and M. J. Waxdal. 1969. The covalent structure of an entire gammaG immunoglobulin molecule. *Proceedings of the National Academy of Sciences USA* 63: 78–85.

Background

The structural features of immunoglobulins play critical roles in immunoglobulin binding to antigens and the effector functions of different immunoglobulins. Heavy and light chain subunits must interact with each other to provide the structural framework of an antigen-binding site and components that allow certain immunoglobulins to function as opsonins. Thus, knowledge of the structure of an intact immunoglobulin provides a window into the function of these important proteins of the humoral immune response. How was the structural composition of immunoglobulins determined?

Early Research

By the 1950s, it had been established that immunoglobulins play a critical role in a selective immune response. This selectivity had been traced to the variable regions of antibodies, and the physiological functions (many of which were unknown at the time) had been attributed to the constant regions. Immunoglobulins were known to be made up of light chains and heavy chains, but their arrangement had not been established. Furthermore, only portions of the amino acid sequence of immunoglobulin proteins had been established by protein sequence analysis.

Think About...

1. Provide examples of several experiments that could be conducted to determine the composition and structural arrangement of an immunoglobulin. These experiments do not have to reflect the time of this paper's publication, but be cognizant of what you must know to conduct your proposed experiments.
2. Some of your examples from Question 1 likely utilize state-of-the-art technology that the researchers in this article did not have at their disposal. How did the authors identify the sequences of the heavy and light chains and determine the position of disulfide bonds within each chain?
3. How did the findings of this study determine the composition of an entire immunoglobulin and, more importantly, its regions and how they are spatially arranged? What is the significance of identifying this arrangement in terms of immunoglobulin function?

Article Summary

The authors were center stage in identifying the structural determinants that drive immunoglobulin function. Studies had previously shown the diversity of immunoglobulins, the two major regions (variable and constant), and the general roles played by these regions. To ascertain how an immunoglobulin can form a functioning unit, the authors embarked on sequencing the entire immunoglobulin protein and determining how components of the heavy and light chains covalently interact with each other through disulfide bond formation.

To determine the protein sequence, researchers had to purify immunoglobulin and subject it to degradation to create manageable-size fragments that could be sequenced using the technology of the time. They used several different proteases (trypsin, chymotrypsin, and pepsin) and determined the sequence of the peptide fragments using

(Continued)

a technique known as Edman degradation. Each protease produced different peptide maps, which overlapped with each other, allowing the authors to piece together the entire sequence of the heavy and light chains. They were also able to determine which cysteine residues are involved in disulfide bond formation using a technique known as diagonal electrophoresis.

Conclusion

Using the techniques at their disposal, the authors reported the protein sequence of an entire immunoglobulin (a major feat at the time of this study). Their disulfide bond analysis allowed them to piece together a map of what an immunoglobulin looks like with respect to the interactions of heavy chains and light chains (**FIGURE KD 9.1**). This provided a unique perspective on the structural determinants of both the heavy and light chains and how they are arranged in relation to each other. Their studies demonstrated that the variable regions of the heavy and light chains are in close proximity and are able to constitute an antigen-binding site. Furthermore, they demonstrated that the constant regions of the heavy chain make up the Fc component of an immunoglobulin.

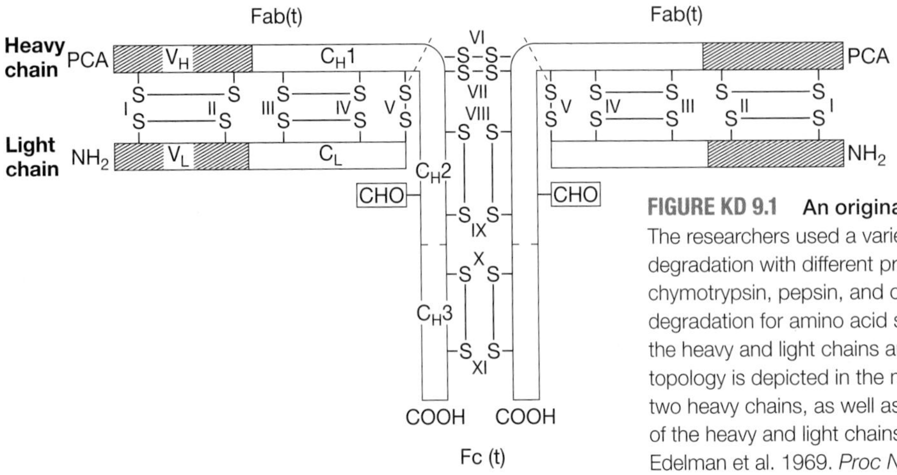

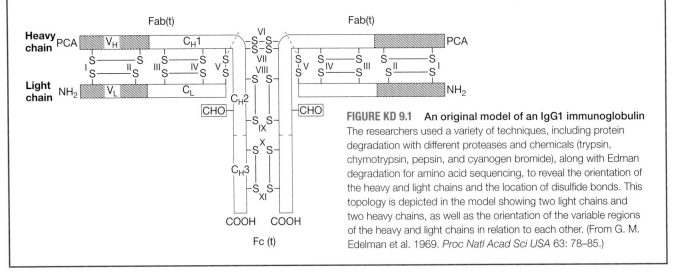

FIGURE KD 9.1 An original model of an IgG1 immunoglobulin The researchers used a variety of techniques, including protein degradation with different proteases and chemicals (trypsin, chymotrypsin, pepsin, and cyanogen bromide), along with Edman degradation for amino acid sequencing, to reveal the orientation of the heavy and light chains and the location of disulfide bonds. This topology is depicted in the model showing two light chains and two heavy chains, as well as the orientation of the variable regions of the heavy and light chains in relation to each other. (From G. M. Edelman et al. 1969. *Proc Natl Acad Sci USA* 63: 78–85.)

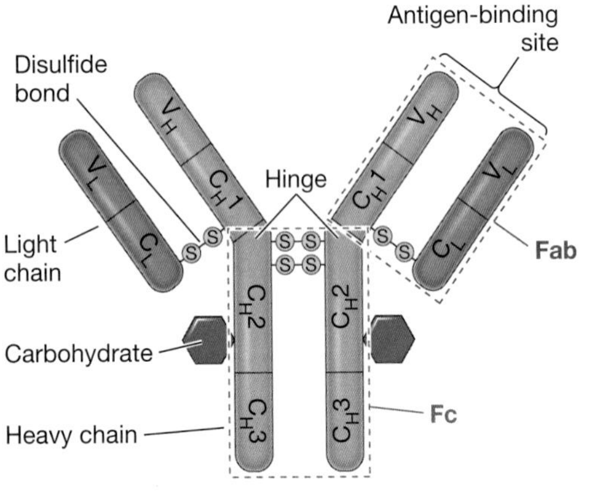

FIGURE 9.1 Structure of immunoglobulins Immunoglobulins consist of two identical heavy chains and two identical light chains. Heavy chains are covalently linked to each other and to the light chains via disulfide bonds. Heavy chains are glycosylated, which protects the hinge region from being degraded by extracellular proteases. The hinge region is susceptible to cleavage by a variety of proteases. If cleaved, it releases the Fc (fragment crystallizable) fragment, consisting of two or three of the heavy chain constant regions, and the Fab (fragment antigen binding) fragments, which consist of a constant and variable region of the heavy chain linked to a light chain.

Immunoglobulins are composed of four polypeptides: two identical **heavy chains** and two identical **light chains** (**FIGURE 9.1**). One IgG subclass is the exception: it can have two different heavy and light chains. Recall that there are two light chain immunoglobulin loci (λ and κ). Each light chain has a variable region and a constant region, and each heavy chain has one variable region and three or four constant regions, depending on the isotype.

The isotype is dictated by expression of a specific heavy chain, and there are five possibilities based on the expression of a different constant region (μ, δ, γ, ε, α heavy chain constant region loci, which encode for the five possible isotypes).

The two heavy chains are covalently linked via a disulfide bond between cysteines. Disulfide bonds also link each heavy chain to a light chain. Furthermore, immunoglobulins are glycosylated, especially on the heavy chains, at certain constant regions. The addition of these carbohydrates increases the interaction of the glycosylated domains with the environment and makes them more accessible to interact with other proteins, such as those of the complement system.

The interaction between each heavy and light chain allows the variable regions that are linked by a disulfide bond to be in close proximity, analogous to the proximity of the variable regions of the α and β chains of the T-cell receptor. Just as this interaction in the T-cell receptor produces a surface that serves as an antigen-binding site, the surface created upon interaction of the variable regions of the heavy chain and light chain serves this same purpose. Because each immunoglobulin consists of two heavy chains and two light chains, each immunoglobulin contains two antigen-binding sites within an immunoglobulin subunit (see Figure 9.1).

The flexible **hinge region** of the immunoglobulin lies within its constant region. This is a portion of the protein that does not fold into a stable structure. Because of its flexibility, the hinge region plays an important role in promoting binding to antigens, especially to those that contain repeating units of sequence or structure recognized by a single immunoglobulin. The constant regions of the light chains interact with the first constant region of the heavy chain, further extending the antigen-binding sites. These regions work in concert with the hinge region to aid in immunoglobulin–antigen interaction.

Following the critical discovery of the basic immunoglobulin structure, two important structural and functional elements were identified. Digestion of purified immunoglobulins with the protease papain causes cleavage of the hinge region, releasing two functionally significant fragments. The first is the **Fab** (fragment antigen binding), which contains a single antigen-binding site of the digested immunoglobulin. The other fragment is the **Fc** (fragment crystallizable), which contains the stem of a complete immunoglobulin molecule (see Figure 9.1). The Fc region plays an important role in binding to receptors of certain immune system cells, including macrophages, dendritic cells, and granulocytes such as mast cells. This view of an immunoglobulin allows us to appreciate how a single molecule is capable of connecting a foreign pathogen to an immune response. The Fab regions, which contain the antigen-binding sites, can bind to a foreign pathogen, and the attached Fc region can bind to receptors on other immune cells, thus linking the bound pathogen as a target for other immune cells.

B cells, which are responsible for expressing immunoglobulins, do so in two different forms—a membrane-bound form that serves as an antigen receptor for the B cell that expresses it, and a soluble form that functions as a targeted weapon against the foreign antigen it is designed to recognize (see Figure 4.4). B cells can express both the membrane-bound and soluble forms of immunoglobulins through alternative splicing of the transmembrane portion of the immunoglobulin protein product. This allows the immunoglobulin to be membrane-bound when the transmembrane region is expressed or to be soluble by removal of the transmembrane region during mRNA processing.

heavy chain Larger of the two subunits of immunoglobulins that contains a variable region for inclusion in an antigen-binding site and the Fc region for isotype effector function.

light chain Smaller of the two subunits of immunoglobulins that contains a variable region for inclusion in an antigen-binding site.

hinge region Flexible loop region located between the Fc and Fab regions of an immunoglobulin; provides flexibility to the Fab portion of an immunoglobulin for increased antigen access.

Fab Region of an immunoglobulin that contains the variable regions of both the heavy and light chains and the antigen-binding site.

Fc Region of an immunoglobulin that contains the isotype-specific heavy chain constant region of the immunoglobulin.

Immunoglobulin structure

Just as immunoglobulins share similarities with T-cell receptors at a functional level (binding to antigen using an antigen-binding site composed of variable regions from two different polypeptides), there are structural similarities as well. T-cell receptors consist of modular *immunoglobulin-like domains* within both the constant and variable regions of each receptor subunit. These domains facilitate protein interactions and have been evolutionarily maintained in several immune system proteins, including immunoglobulins, T-cell receptors, and major histocompatibility complex (MHC) molecules. The heavy and light chains of immunoglobulins are also composed of a modular array of these domains. A look at the three-dimensional structure of an immunoglobulin (**FIGURE 9.2**) reveals that the variable region domains are close together, forming an antigen-binding surface.

Arguably, the antigen-binding site of an immunoglobulin is its most important functional component (although we will see that the Fc region plays a critical role in regard to effector function). Each antigen-binding site is formed by the close proximity of variable regions of a heavy chain and light chain. We know that immunoglobulin diversity is a critical part of the diversity of the humoral adaptive immune response. Because that diversity results from changes introduced during somatic recombination of variable regions of the heavy and light chain genes, it stands to reason that the antigen-binding site of immunoglobulins is the most diverse in variability.

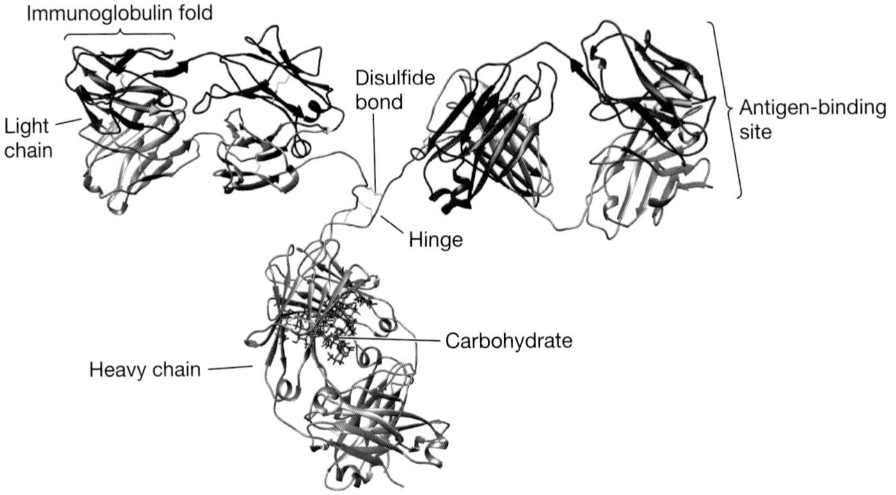

FIGURE 9.2 Three-dimensional structure of an immunoglobulin Immunoglobulins contain two heavy chains (gray) and two light chains (dark red/maroon). Each heavy chain has one variable region and three or four constant regions, each with a distinct immunoglobulin fold. Each light chain has one variable region and one constant region, each with a distinct immunoglobulin fold. Heavy chains are covalently linked to each other and to the light chains via disulfide bonds. The antigen-binding sites are located where a heavy chain and light chain variable region come together in three-dimensional space. (Data from L. J. Harris et al. 1997. *Biochem* 36: 1581–1597. © 1997 American Chemical Society; redrawn using Chimera and PDB file 1IGT.)

The variable regions of the light and heavy chains have locations within their amino acid sequence known as *hypervariable regions*, also called *complementarity-determining regions*. These hypervariable regions are three flexible loops of each variable region located at the junction sites of V, D, and J segments brought together during somatic recombination (**FIGURE 9.3**). Due to their location at these junction sites, the diversity of hypervariable regions is driven by both somatic recombination and junctional diversity (discussed in Chapter 4). While these three loops are distant from each other in the amino acid sequence of each immunoglobulin chain, they are in close proximity to

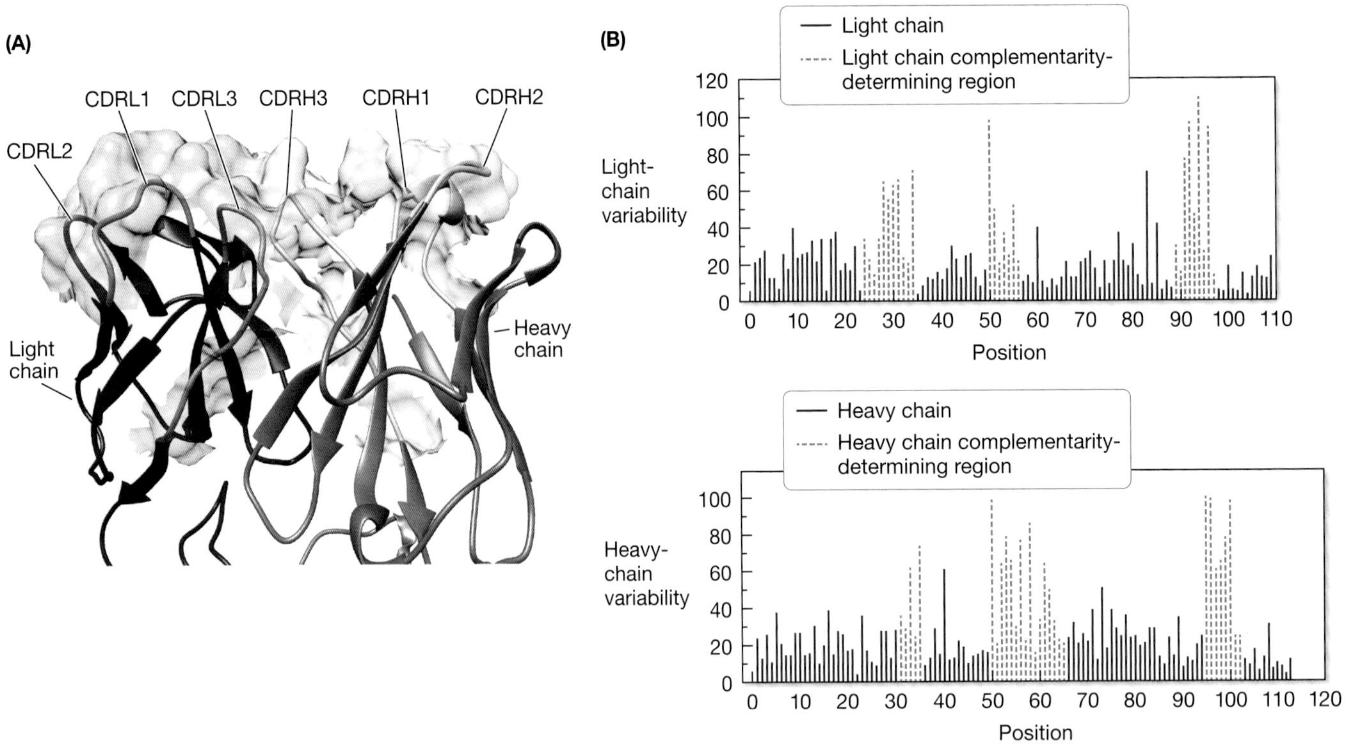

FIGURE 9.3 Hypervariable regions of immunoglobulins (A) Each light and heavy chain has three hypervariable loops, also known as complementarity-determining regions (CDR). Although these loops are distant from each other in the amino acid sequence of the chain, they are in close proximity in three-dimensional space and play a crucial role in forming the antigen-binding site. (B) A Kabat-Wu plot demonstrates that the level of highest variation in amino acid sequence can be seen at the three CDRs of the heavy chain and light chain. The variation in these loops allows for immunoglobulins with different CDRs to bind to unique antigens. (A, data from L. J. Harris et al. 1997. *Biochem* 36: 1581–1597. © 1997 American Chemical Society; redrawn using Chimera and PDB file 1IGT; B, after E. A. Kabat et al. 1991. *Sequences of Proteins of Immunological Interest*, Vol. 2, 5th ed. US Department of Health and Human Services, Public Health Service, National Institutes of Health. NIH Publication No. 91-3242.)

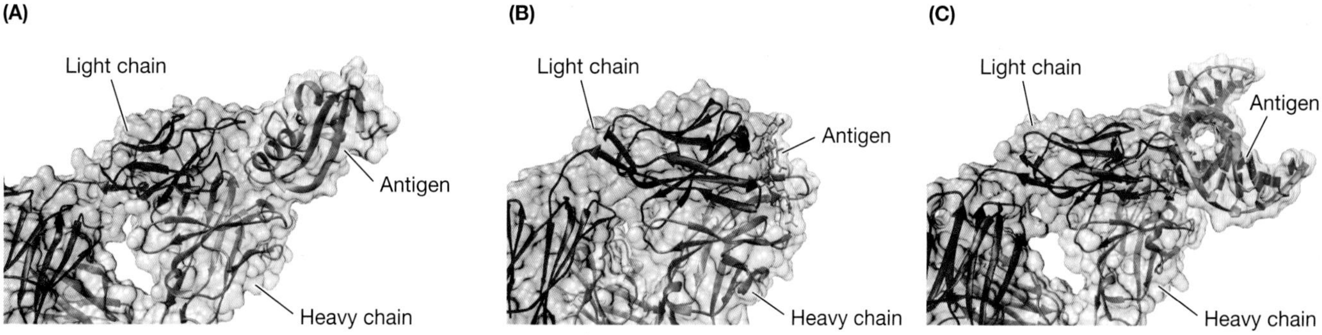

FIGURE 9.4 Immunoglobulin–antigen interactions The three-dimensional structure of immunoglobulins with different biological macromolecules, including protein, carbohydrate, and nucleic acid, shows the interface between the antigen-binding site and the antigen it binds. (A) An immunoglobulin binding to the vaccinia virus protein A33R. (B) An immunoglobulin binding to the carbohydrate N-acetyl-D-glucosamine. (C) An immunoglobulin binding to a photoactivated nucleic acid. (A, data from M. H. Matho et al. *PLOS Pathog* 11: e1005148–e1005148; redrawn using Chimera and PDB file 4LQF; B, data from C. Soliman et al. 2018. *J Biol Chem* 293: 5079–5089. doi.org/10.1074/jbc.RA117.001170, redrawn using Chimera and PDB file 6BE4; C, data from H. Yokoyama et al. 2013. *Acta Crystallogr D Biol Crystallogr* 69: 504–512; redrawn using Chimera and PDB file 3VW3.)

each other in three-dimensional space in the antigen-binding site. Since these regions are the most highly varied, differences within the loops lead to diversity in the types of antigens recognized.

Immunoglobulin epitopes

With a few exceptions, most T-cell receptors are only capable of binding to protein antigens. Immunoglobulins, however, are far more diverse in terms of the type of antigen they can bind. Given the right antigen-binding site, immunoglobulins can bind to any of the four major biomacromolecules (proteins, carbohydrates, lipids, and nucleic acids), especially those on a pathogen's surface. This ability is critical in fighting infection.

The antigen-binding site of an immunoglobulin varies dramatically in three-dimensional space, depending on the composition of the variable regions of the heavy and light chains. This surface variation means that there is an equal amount of variation in the types of molecular surfaces with which immunoglobulins can interact. Three-dimensional structural analysis demonstrates that the entire antigen-binding site plays a role in antigen recognition (**FIGURE 9.4**).

Each immunoglobulin recognizes a specific antigen and binds to that antigen at a specific location or structure called an *epitope*. An epitope usually consists of a short sequence or structural feature of the antigen (such as a short sequence of amino acids in protein antigens). Some antigens are recognized by more than one type of immunoglobulin or by more than one molecule of the same immunoglobulin, often seen when a pathogen has a repeated structure such as a polysaccharide at its surface.

Multivalent antigens are those that can bind to multiple copies of the same immunoglobulin because of a repeating epitope or that have multiple epitopes for multiple immunoglobulins (**FIGURE 9.5**). Multivalency of a single antigen promotes clustering of immunoglobulins on the antigen. This dramatically increases the number of immunoglobulins interacting with the antigen, contributing to efficient recognition and uptake. A multivalent antigen also promotes **avidity** with a B cell expressing multiple identical immunoglobulins that can bind the same antigen. Clustering of immunoglobulins is important in allowing for complement activation and in increasing efficiency of phagocytosis of pathogens by macrophages, neutrophils, and dendritic cells. Furthermore, the flexibility of the hinge region allows a single immunoglobulin to bind to multiple epitopes if an antigen is multivalent for that epitope, producing a higher affinity for that specific antigen.

multivalent antigen An antigen that either has more than one copy of an epitope or contains multiple epitopes for multiple immunoglobulins.

avidity The overall strength of binding of a protein complex to an antigen.

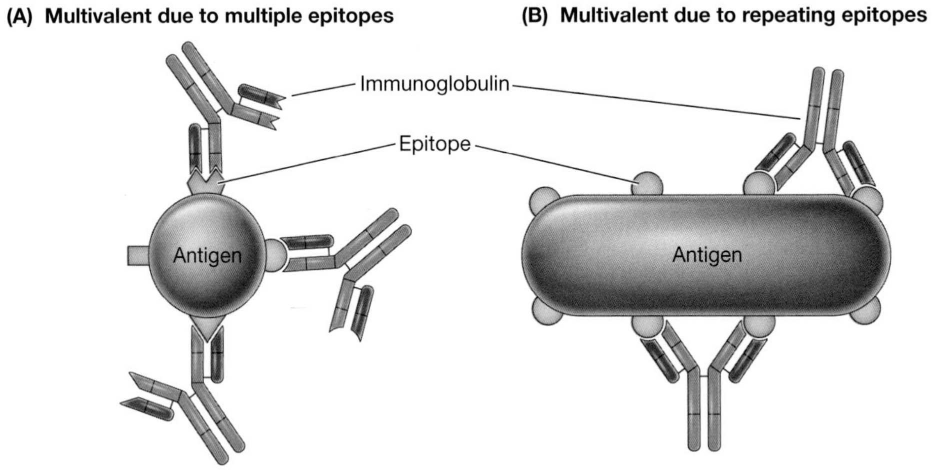

FIGURE 9.5 Multivalent antigens Immunoglobulins bind to antigens at locations known as epitopes. An antigen can be multivalent by having multiple epitopes, each allowing the binding of a different immunoglobulin (A), or by having a repeating epitope that allows binding of more than one immunoglobulin of the same type (B).

linear epitope Immunoglobulin epitope that incorporates structural determinants maintained by the linear sequence of the antigen.

conformational epitope Immunoglobulin epitope that incorporates structural determinants only maintained in the presence of native antigen structure.

Epitopes can also differ based on how they are recognized. Some immunoglobulins bind to **linear epitopes** (FIGURE 9.6). In this case, the immunoglobulin binds to antigen in a sequence-specific manner—the epitope is dictated by structural elements constituted by macromolecule residues in sequence (for instance, a short stretch of amino acids that are together in the sequence of the protein). Such epitopes are not typically lost when the three-dimensional structure of the antigen is destroyed, as commonly the linear sequence is still conserved.

Conversely, immunoglobulins can bind to epitopes created by the complete structure of the antigen. Such epitopes are **conformational epitopes** and may or may not be created by macromolecule residues that are close together in linear sequence. If conformational epitopes are created by residues far apart in linear sequence but close together in the native structure of the antigen, these epitopes are destroyed if the three-dimensional structure of the antigen is destroyed.

CHECKPOINT QUESTIONS

1. How many and which polypeptides make up a single immunoglobulin?
2. Define the Fab and Fc regions of an immunoglobulin.
3. Why are the hypervariable loops important in creating diversity in immunoglobulins?
4. Define linear and conformational epitopes.

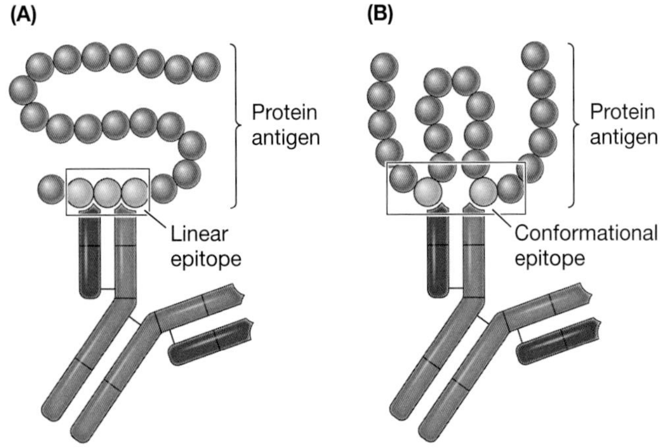

FIGURE 9.6 Linear and conformational epitopes (A) A linear epitope is one that is close in primary sequence (such as adjacent amino acids in the primary sequence of a protein). Linear epitopes are not destroyed if the three-dimensional structure of the antigen is destroyed, as long as the primary sequence is still intact. (B) A conformational epitope is dictated by the structure of the antigen. Conformational epitopes may or may not be driven by functional groups or atoms that are farther apart in the primary sequence, and they are always destroyed if the three-dimensional structure of the antigen is no longer intact. (After B. Forsström. 2014. Characterization of antibody specificity using peptide array technologies. [Dissertation] Kungliga Tekniska Högskolan, KTH, Royal Institute of Technology, School of Biotechnology, Solna, Sweden.)

Making Connections: Structure of the B-cell Receptor and Connection to Diversity

The B-cell receptor is required for the recognition of foreign antigen. The protein contains two copies of two different subunits, the heavy chain and the light chain. Developing B cells undergo somatic recombination of their immunoglobulin loci to produce a diverse population of B cells, capable of recognizing a wide variety of different antigens using their specific B-cell receptor. The structure of the B-cell receptor enables it to utilize this mode of generating diversity to recognize different antigens.

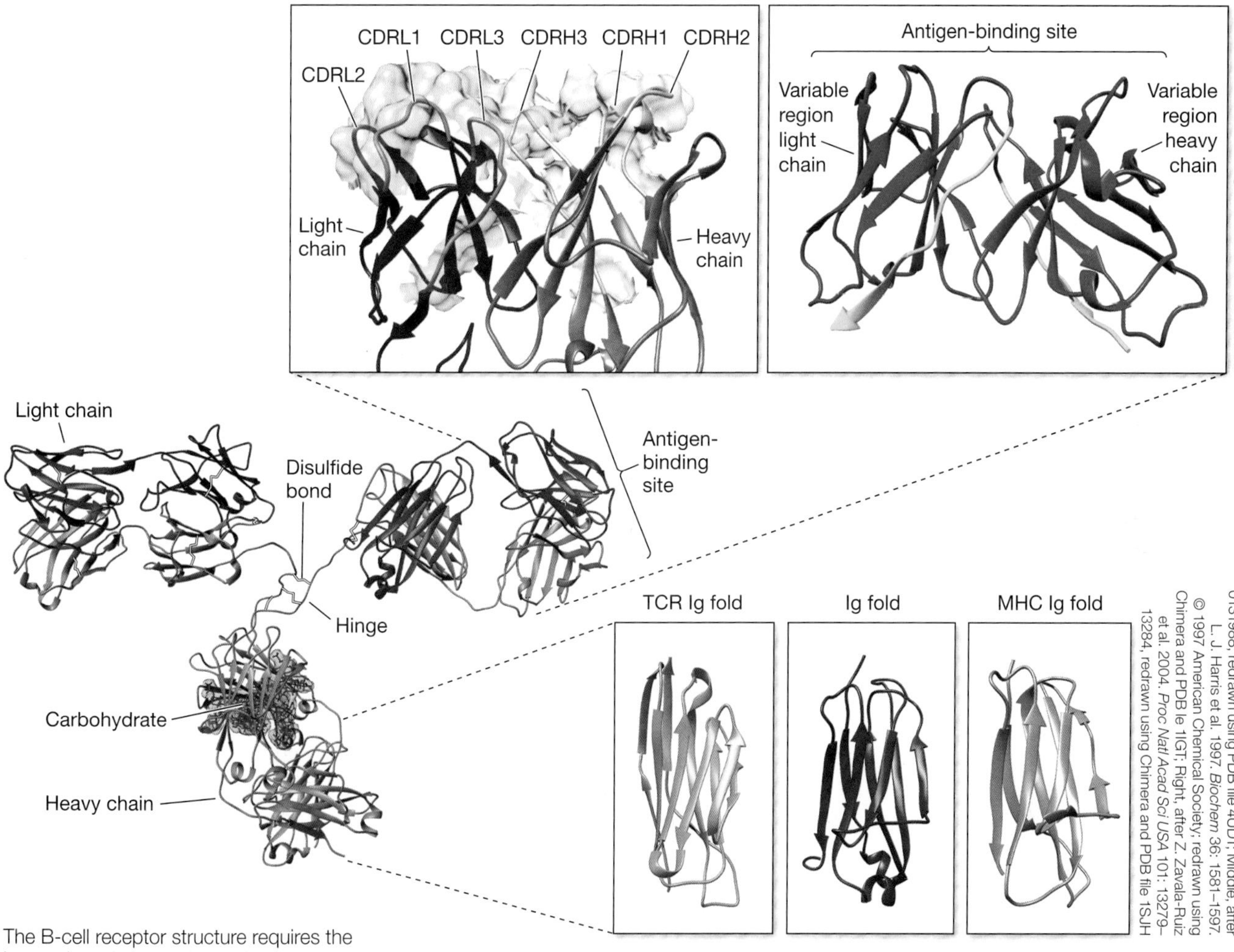

Both, data from L. J. Harris et al. 1997. *Biochem* 36: 1581–1597. © 1997 American Chemical Society; redrawn using Chimera and PDB file 1IGT

Data from L. J. Harris et al. 1997. *Biochem* 36: 1581–1597. © 1997 American Chemical Society; redrawn using Chimera and PDB file 1IGT

Left, after K. E. Rödström et al. 2015. *PLOS ONE* 10: e0131988/CC BY 4.0. doi:10.1371/journal.pone.0131988; redrawn using PDB file 4UDT; Middle, after L. J. Harris et al. 1997. *Biochem* 36: 1581–1597. © 1997 American Chemical Society; redrawn using Chimera and PDB le 1IGT; Right, after Z. Zavala-Ruiz et al. 2004. *Proc Natl Acad Sci USA* 101: 13279–13284, redrawn using Chimera and PDB file 1SJH

The B-cell receptor structure requires the interaction of two copies of the heavy chain with two copies of the light chain. The three-dimensional structure demonstrates the presence of two different immunoglobulin folds in the light chain and four different immunoglobulin folds in the heavy chain (although some isotypes may contain five). The immunoglobulin folds resemble those of T-cell receptors and MHC molecules.

(Sections 4.7, 4.8, 6.2, 6.4, 9.1, 10.2, 10.3)

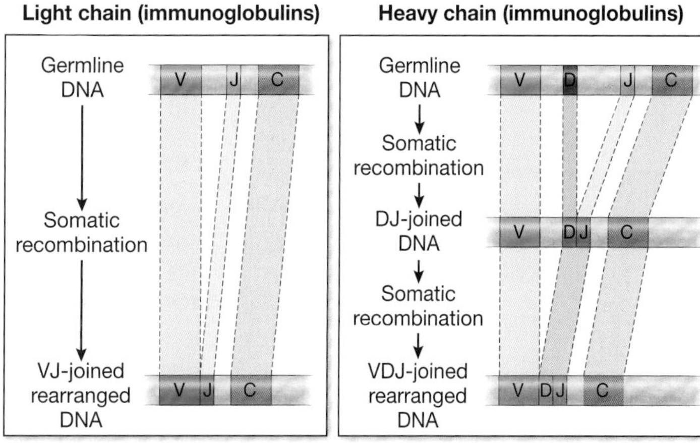

Both after W. Purves et al. 1995. *LIFE: The Science of Biology*, 4th ed., Sinauer Assoc., Sunderland, MA; T. Dörner and P. E. Lipsky. 2001. *Arthritis Rheum* 44: 2715–2727; P. Parham. 2014. *The Immune System*, 4th ed., W. W. Norton and Company, New York

The antigen-binding site of a B-cell receptor is located at the interface of the variable regions of the heavy chain and light chain. Each chain contributes a variable region, and each variable region has three hypervariable loops. The variability of this region is driven by somatic recombination.

(Sections 4.2, 4.8, 8.6, 9.1, 9.4)

9.2 How does genetic recombination in immunoglobulins compare to recombination in T-cell receptors?

LEARNING OBJECTIVE

9.2.1 Compare and contrast the mechanism of genetic recombination of immunoglobulins and T-cell receptors.

As discussed in Chapter 4, somatic recombination is at the very center of creating diversity of the adaptive immune system. The genes for the heavy and light chains of immunoglobulins contain many different V, D, and J segments (recall that the D segment is specific for the heavy chain loci). Due to the number of segments and junctional diversity, gene rearrangement can result in a very diverse population of B cells, each expressing an immunoglobulin that can recognize a different epitope.

Much of the genetic recombination that occurs at the immunoglobulin loci mirrors those that occur at the loci for T-cell receptors during T-cell recombination events. However, some events occur specifically in B cells to enable the expression of different immunoglobulin isotypes. The mechanisms driving isotype switching differ slightly from those of somatic recombination but still result in permanent alteration of the gene loci within B cells to express the correct isotype.

Similarities with T-cell receptor recombination

During B-cell development from the common lymphoid progenitor in the bone marrow, somatic recombination at the immunoglobulin loci occurs, beginning with rearrangement events at the heavy chain loci and continuing to rearrangement events at the light chain loci. Somatic recombination at these loci is identical to the process in T cells—the employment of V(D)J recombinase and the critical activities of proteins including RAG1, RAG2, exonucleases, and TdT promote recombination at signal sequences within the variable regions of the immunoglobulin loci and drive diversity. The events can produce productive or unproductive rearrangements, which are tested at specific developmental checkpoints (see Chapter 8). As in T-cell development, unproductive rearrangement events result in apoptosis of the developing B cell if they go unresolved. Productive rearrangements allow the developed B cell to express immunoglobulins at the cell surface, where they wait to engage their specific antigen.

Immunoglobulins at the B-cell surface resemble the T-cell receptor in that very little protein extends into the cell's cytoplasm. This lack of cytoplasmic tail on immunoglobulins requires that they interact with other cell-surface proteins to drive the signal transduction events needed for B-cell activation. Immunoglobulins interact with two cell-surface proteins, Igα and Igβ, which contain cytoplasmic domains that initiate signal transduction events upon antigen recognition (see Figure 4.19).

Recombination in isotype switching

Somatic recombination at the immunoglobulin loci happens during B-cell development within the bone marrow. This recombination occurs before the B cell ever comes in contact with antigen, similar to T-cell receptor recombination that occurs during development of naïve T cells in the thymus.

One important difference between rearrangement in T-cell receptor genes and immunoglobulin genes is that recombination of the T-cell receptor loci halts after the T cell has progressed through positive and negative selection in the thymus. However, recombination events can occur after the B cell has finished

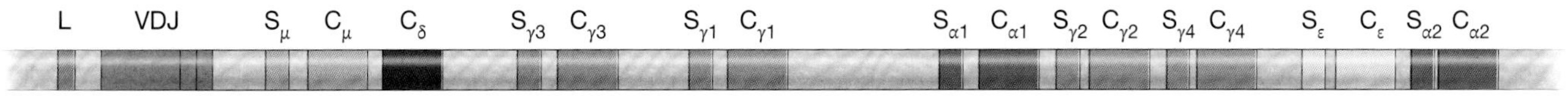

FIGURE 9.7 Structure of the immunoglobulin heavy chain constant region locus While the variable region of the heavy chain is subject to somatic recombination, there are several different functional units of constant regions, whose use in heavy chain production depends on the organization of the gene structure. Each constant region unit in the gene encodes a different isotype constant region, with the μ and δ constant regions functioning in naïve B cells. Each constant region also has a switch region (denoted S) on its 5′ end, which facilitates class switch recombination to drive expression of a different isotype in activated B cells. (After P. Parham. 2014. *The Immune System*, 4th ed., W. W. Norton and Company.)

development in the bone marrow. These events are not the same as those that allowed for formation of a functioning immunoglobulin. Instead, recombination in B cells that have left the bone marrow occurs within secondary lymphoid tissue and is referred to as **class switch recombination** (discussed in Section 9.4).

Immunoglobulin heavy chain genes have five constant regions (μ, δ, γ, ε, and α), which encode for the IgM, IgD, IgG, IgE, and IgA isotypes, respectively (**FIGURE 9.7**). Each constant region is preceded on its 5′ end by a *switch region* (denoted S) and followed by the heavy chain constant region subscript, which function during class switch recombination. B cells exiting the bone marrow leave with the ability to utilize either the μ or δ constant regions within their heavy chain.

B cells can mature further in secondary lymphoid tissue when they come into contact with the antigen their immunoglobulin recognizes. This maturation includes switching the isotype to enable the immunoglobulin to best combat its specific antigen.

class switch recombination Alteration of DNA at the immunoglobulin heavy chain locus within B cells, which promotes the incorporation of a different isotype constant region within the expressed immunoglobulin; also called *isotype switching*.

alternative splicing Maintenance or removal of specific RNA sequences that can promote the formation of different proteins with different properties or locations within cells.

CHECKPOINT QUESTIONS

1. What are the similarities and differences between somatic recombination in T cells and somatic recombination in B cells?
2. Which cell-surface proteins associate with immunoglobulins to form a signaling complex necessary for B-cell activation upon antigen recognition?
3. What process is used to recombine immunoglobulin heavy chain loci to drive isotype switching?

9.3 Why is alternative splicing important in antibody production?

LEARNING OBJECTIVE

9.3.1 Explain the role of alternative splicing in antibody production.

B cells use immunoglobulins to recognize pathogens and as a primary force in the humoral adaptive immune response. Upon proper development within the bone marrow, a naïve B cell circulating through the bloodstream and lymphatic system expresses two immunoglobulins at the cell surface: IgM and IgD. Circulating B cells enter secondary lymphoid tissue, where they attempt to engage foreign antigens with their cell-surface immunoglobulins. Antigen binding through its immunoglobulin induces clonal expansion of the B cell and differentiation of activated B cells into plasma cells. Antigen binding also sets several other processes within the plasma cell in motion, including production of soluble immunoglobulins. The production of IgM, IgD, and soluble immunoglobulins relies on **alternative splicing** mechanisms to differentially express these proteins that are key to the humoral adaptive immune response.

IgM and IgD production

Developing and naïve B cells capable of producing immunoglobulins make one of two isotypes: IgM and IgD. In fact, these are the only two isotypes that can be made simultaneously in a B cell. B cells that have gone through isotype switching only express one isotype. B cells that express IgM and IgD do so because of alternative mRNA splicing. The rearranged heavy chain locus of a naïve B cell that has not undergone isotype switching incorporates the variable region that has been recombined and the two closely located constant regions μ and δ.

During immunoglobulin transcription, a long primary RNA transcript is produced, which contains both the μ and δ constant regions as well as key components to allow for RNA processing and production of a functional membrane-bound immunoglobulin (**FIGURE 9.8**). Processing of this primary RNA transcript can include either the membrane-spanning region that is present right after the μ constant regions or the one present right after the δ constant regions. Close to the membrane-spanning region RNA segment is a polyadenylation site required for mRNA maturation. Polyadenylation at the specific site, along with alternative splicing maintaining either the μ constant regions or the δ constant regions, results in the production of mRNA that can express either IgM or IgD (see Figure 9.8).

Membrane-bound and soluble immunoglobulins

Naïve B cells and those still developing within bone marrow only express immunoglobulins (IgM and IgD) as membrane proteins at their cell surface. Naïve B cells then circulate to secondary lymphoid tissue to monitor for the antigen that their immunoglobulin can bind. Upon binding to antigen and activation of the engaged B cell, the B cell begins to express soluble immunoglobulins in addition to the membrane-bound immunoglobulins. All isotypes can be

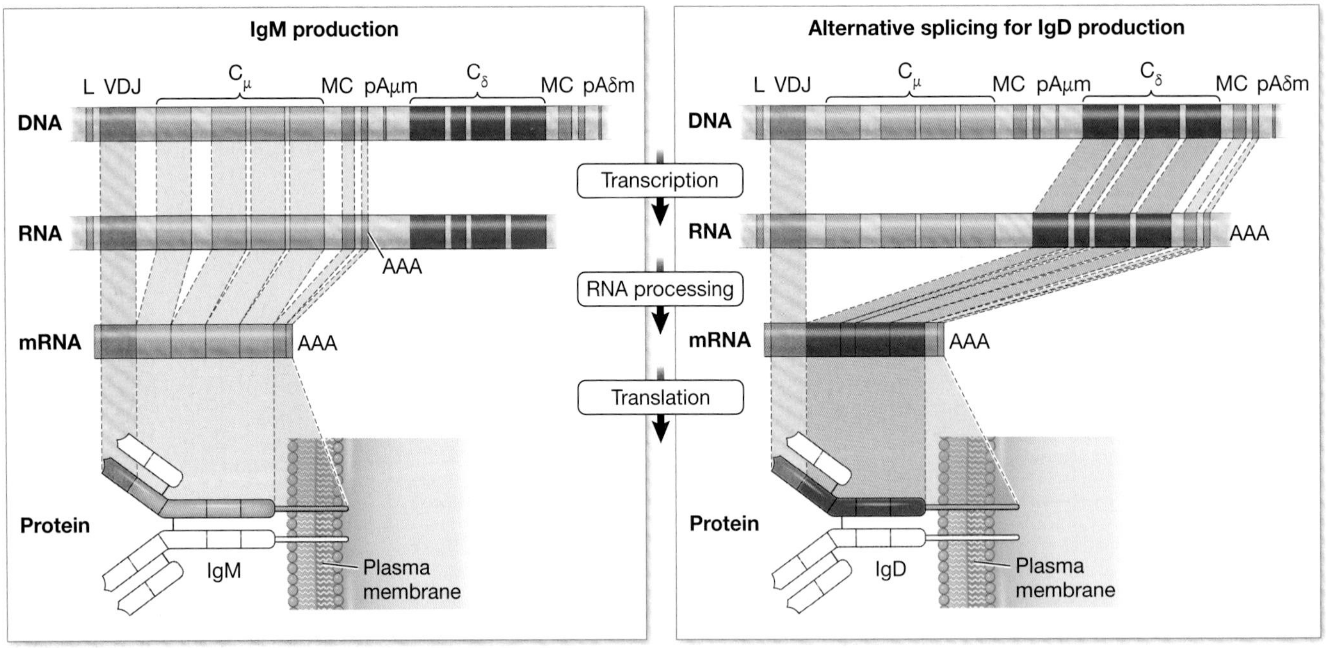

FIGURE 9.8 Alternative splicing drives synthesis of IgM or IgD The μ and δ constant regions are in close proximity on the immunoglobulin heavy chain locus. Naïve B cells can express both IgM and IgD, and the expression of each is dictated by alternative splicing. The RNA transcript of the heavy chain contains the constant regions for both IgM and IgD. Splicing removes the μ or the δ constant region from the transcript in the processing to mRNA. Each constant region also has a transmembrane domain (MC) and a site for polyadenylation (pA). The mature mRNA transcript contains the recombined VDJ region, either the μ or δ constant region (whichever is not removed by splicing), a transmembrane region, and is polyadenylated. Translation of the mature mRNA will synthesize either IgM (left) or IgD (right). (After P. Parham. 2014. *The Immune System*, 4th ed., W. W. Norton and Company.)

produced as both membrane-bound and soluble proteins. However, naïve B cells and memory B cells produce membrane-bound immunoglobulins, while terminally differentiated B cells (plasma cells) produce soluble immunoglobulins. The process is also driven by alternative RNA splicing (**FIGURE 9.9**).

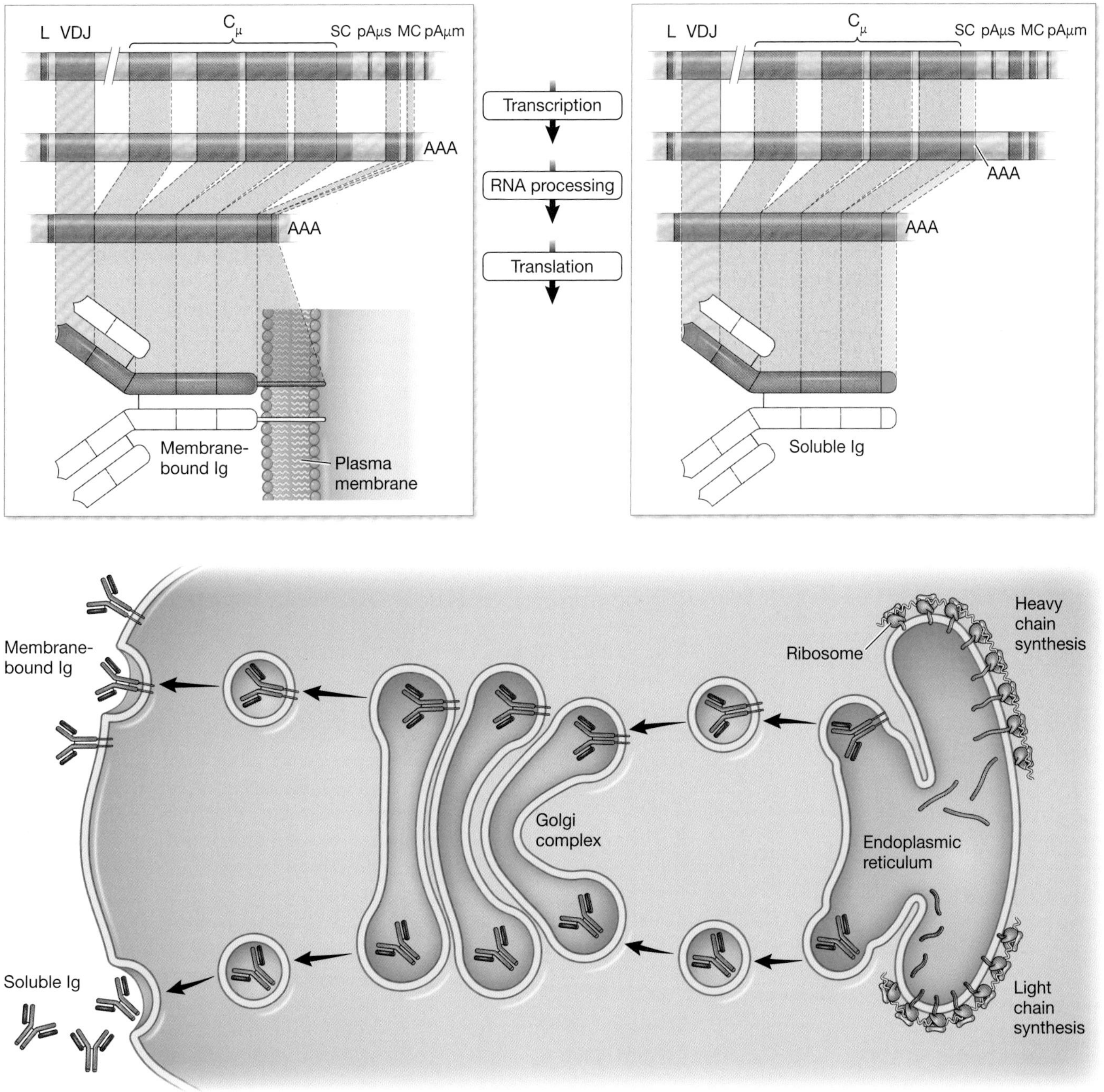

FIGURE 9.9 Alternative splicing drives synthesis of both membrane-bound and soluble immunoglobulins The heavy chain locus contains an alternative splice site that can incorporate a transmembrane region or remove it from the spliced mRNA. Incorporating the transmembrane region places the synthesized immunoglobulin within the endoplasmic reticulum (ER) membrane during co-translational import. As the immunoglobulin moves through the secretory pathway, it will always be maintained in the ER membrane, including when it reaches the plasma membrane. Removal of the transmembrane region from the mRNA during splicing still permits co-translational import of the immunoglobulin, but instead of being incorporated into the ER membrane, it is released into the ER lumen. During transit in the secretory pathway it will remain soluble until its vesicle fuses with the plasma membrane, where the immunoglobulin is secreted. (Top, after P. Parham. 2014. *The Immune System*, 4th ed., W. W. Norton and Company; bottom, after T. Mak et al. 2014. *Primer to the Immune Response*, 2nd ed. Academic Cell/Elsevier Inc., Burlington, MA.)

EVOLUTION AND IMMUNITY

Alternative Splicing and Immune System Diversity

An interesting example of receptor diversification is seen with the *Drosophila* (fruit fly) immunoglobulin superfamily member Down syndrome cell adhesion molecule (DSCAM). DSCAM in humans is involved in neural development. An abnormal abundance in a developing fetus can induce Down syndrome, thus the molecule name. This abundance is associated with the presence of a third copy of chromosome 21, a hallmark of Down syndrome.

DSCAM was identified in *Drosophila* as a protein involved in neuronal migration. However, its presence in hemolymph (the equivalent of blood in invertebrates) suggests that it may serve a role in pathogen recognition. The DSCAM protein product typically has 10 immunoglobulin domains. The DSCAM gene structure is interesting in that there are four different exons that can be encoded by multiple exon possibilities based on alternative splicing. Exon 4 has 12 possible alternative exons, exon 6 has 48 possible alternative exons, exon 9 has 33 possible alternative exons, and exon 10 has 2 possible alternative exons. These possibilities and alternative splicing events could theoretically produce 38,016 different DSCAM molecules. But does DSCAM play a role in immunity? A hint suggesting this to be the case was seen in a 2005 study by Schmucker and coauthors showing that hemocytes (invertebrate cells involved in phagocytosis) lacking DSCAM are less efficient at phagocytosis of *Escherichia coli* in vitro.

The primary RNA transcript of immunoglobulins contains a segment that encodes for a large hydrophobic sequence of amino acids required to span the plasma membrane. This RNA segment either can be maintained in the mature mRNA or can be removed during RNA splicing. Both mature mRNAs have a sequence required for co-translational transport into the secretory pathway of the B cell, where translation will begin at the endoplasmic reticulum (ER).

Maintenance of the segment that encodes for hydrophobic amino acids allows the immunoglobulin to be anchored in the plasma membrane during co-translational transport into the ER. The secretory pathway drives the movement of the membrane-bound immunoglobulin to the plasma membrane, where it can function as a membrane-bound receptor (see Figure 9.9).

Conversely, if the segment that encodes for hydrophobic amino acids is removed by RNA splicing, the co-translationally transported protein does not anchor in the ER membrane and instead is a soluble protein within the lumen of the ER. The secretory pathway promotes the secretion of the immunoglobulin from the B cell, and since it is not anchored to the membrane, it is secreted in a soluble form (see Figure 9.9).

Alternative splicing is a commonly used mechanism for making alternative forms of a protein, to change tissue-specific isoforms, to change localization in a cell, or to change binding specificity of the protein. The way in which alternative splicing promotes diversity of binding specificity in immune receptors can be seen in the DSCAM protein in fruit flies (see **EVOLUTION AND IMMUNITY**).

CHECKPOINT QUESTIONS

1. Why can naïve B cells express only IgM and IgD immunoglobulins?
2. What key component of an immunoglobulin is removed by alternative splicing to produce soluble, secreted immunoglobulins?

9.4 What is the importance of somatic hypermutation and isotype switching in antibody maturation?

LEARNING OBJECTIVES

9.4.1 Describe the mechanism and importance of somatic hypermutation in antibody maturation.

9.4.2 Describe the mechanism and importance of isotype switching in antibody maturation.

While IgM and IgD play important initial roles in B-cell function and in humoral adaptive immunity, the other immunoglobulin isotypes (IgA, IgG, and IgE) are also central to humoral adaptive immunity. These isotypes are not expressed in naïve B cells, only in B cells that have become activated by antigen binding and go through proper isotype switching.

A primary immune response that activates B cells can induce immunoglobulins to undergo **affinity maturation**. This increased **affinity** enhances the immunoglobulin's ability to recognize and bind to antigen quickly and efficiently and prompts the altered B cell to become activated more quickly, a hallmark of the secondary adaptive immune response and immunological memory.

One key enzyme, *activation-induced cytidine deaminase (AID)*, expressed in B cells is critical in driving both isotype switching and affinity maturation. Furthermore, active transcription of the immunoglobulin genes, which is driven during B-cell activation, is required for both processes to occur.

affinity maturation Process by which immunoglobulin genes gain higher affinity for antigen through mutagenesis of the immunoglobulin gene during B-cell activation.

affinity The strength of an interaction between an immunoglobulin antigen-binding site and its epitope.

activation-induced cytidine deaminase (AID) Enzyme within activated B cells that catalyzes the conversion of cytosine to uracil within specific sites in the immunoglobulin genes, driving somatic hypermutation and class switch recombination.

Somatic hypermutation

Recall that somatic recombination drives diversity of the immunoglobulin genes. This diversity, concentrated within the variable regions (especially the hypervariable loops) of the heavy and light chains, sets the stage for the potential production of well over one million different immunoglobulins.

After a B cell is activated in secondary lymphoid tissue, some of the B cells produced during clonal expansion will immediately begin to secrete antibodies that have the same antigen-binding site as the parental B cell that was originally activated. However, some B cells will engage in affinity maturation to further diversify the antigen-binding site. To carry out affinity maturation, B cells activate *somatic hypermutation*, a somewhat random mutagenic process that introduces point mutations within the immunoglobulin gene. These random mutations are focused on mutational hot spots within the hypervariable loops of the variable regions of the heavy and light chains.

The mutation frequency that occurs during somatic hypermutation (one mutation at every 10,000 base pairs per cell division) is much higher than the mutation frequency in other cells (one mutation at every 10,000,000,000 base pairs per cell division). This results in approximately one or two mutations per variable region per cell division, increasing the immunoglobulin diversity that can be produced after B-cell activation. This high mutation frequency may be driven by the presence of sequences of the immunoglobulin genes capable of preferentially recruiting the machinery required for somatic hypermutation (see **EMERGING SCIENCE**).

Activation-induced cytidine deaminase (**AID**) is the enzyme that plays a key role in this mutagenic process by deaminating the base cytosine, creating the base uracil (**FIGURE 9.10**). Recall that uracil is not present in DNA; one major

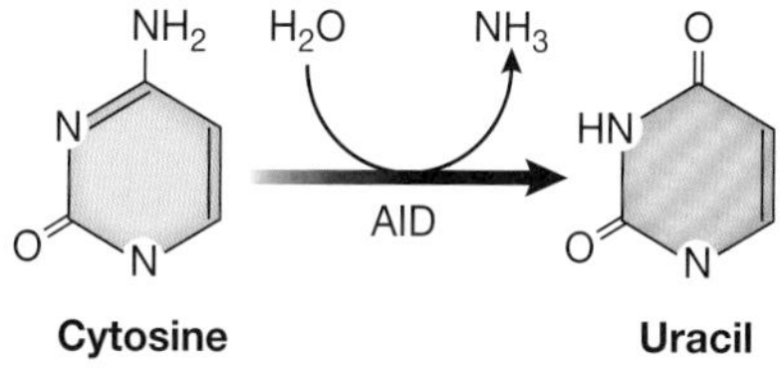

FIGURE 9.10 Activation-induced cytidine deaminase (AID) reaction AID is expressed in activated B cells and catalyzes the deamination of cytosine and the formation of uracil in DNA. (After T. Nishizawa and H. Suzuki. 2015. *BioMed Res Int.* Article ID 794378: 1–7.)

EMERGING SCIENCE

How does somatic hypermutation preferentially alter the variable regions of immunoglobulins within B cells?

Article

J-M. Buersteed, J. Alinikula, H. Arakawa, J. J. McDonald, and D. G. Schatz. 2014. Targeting of somatic hypermutation by immunoglobulin enhancer and enhancer-like sequences. *PLOS Biology* 12: e1001831.

Background

It had been well established by the beginning of the twenty-first century that activation-induced cytidine deaminase (AID) plays an essential role in B-cell maturation by driving somatic hypermutation and isotype switching. However, the mechanism underlying preferential targeting of AID to immunoglobulin loci had not been established. The mutation rate of immunoglobulin variable regions is much higher than the mutation rate at other regions of the genome within B cells by several orders of magnitude. Since AID activity requires single-stranded DNA, and since it can theoretically alter any cytosine base to uracil within DNA structure that it recognizes, it begged the question: How does AID specifically target immunoglobulin genes?

The Study

The mechanism by which AID preferentially mutates the immunoglobulin genes within B cells has been a highly debated topic for more than 20 years. While much evidence pointed to the presence of *cis*-acting regulatory elements (elements within the immunoglobulin gene responsible for regulating the mutation frequency), other findings served to question this hypothesis. Two important observations that called the *cis*-acting regulatory element hypothesis into question were (1) removing the putative elements from mice did not seem to abolish mutation frequency in immunoglobulin genes and (2) expression of AID in non-B-cell lines increased the mutation frequency in those lines.

In support of the *cis*-acting element hypothesis, a 10-kilobase region of the immunoglobulin locus, named the diversification activator (DIVAC), was shown to be necessary for mutation of a GFP transgene (a gene artificially introduced) in a chicken B-cell line. Knowing the presence of this 10-kilobase DIVAC, the researchers aimed to identify core sequence elements of the DIVAC that might be responsible for facilitating targeted mutation of immunoglobulin genes by AID.

The researchers refined the important sequence elements of the DIVAC to two key sequence regions they called the *cIgλE* and *3′ Core* regions. Further analysis of these regions demonstrated that they were functionally conserved in organisms other than chickens, including mice and humans. The researchers were also able to demonstrate that these same regions mirrored transcriptional enhancers (**FIGURE ES 9.1**).

Conclusion

The researchers identified key regulatory elements that function as transcriptional enhancers and, more importantly, as elements responsible for enhancing the specific targeting of hypermutation within a B-cell line. These regulatory elements are both structurally and functionally conserved in vertebrate organisms. Identification of these sequences may explain the means by which AID preferentially targets immunoglobulin loci during somatic hypermutation.

Think About...

1. Why do you think the researchers utilized a GFP transgene for their experiments rather than looking at mutation of the immunoglobulin gene within the B-cell line?
2. The researchers demonstrated that several transcription factor binding motifs (for NFκB, MEF2, PU.1, and IRF4) are necessary for the hypermutation they observed in their experiments. Design an experiment to test whether any or all of these binding motifs are sufficient for hypermutation.
3. Design an experiment to test whether AID can preferentially bind to DNA containing these regulatory elements. Would you need to use double-stranded or single-stranded DNA in your experiment? Why?

reason is cytosine's susceptibility to deamination by environmental factors. Deamination of cytosine to uracil activates DNA repair mechanisms that counteract the potential damage to DNA. These same repair mechanisms are activated based on the action of AID in B cells and have the potential to induce point mutations.

Introduction of cytosine-to-uracil changes by AID in activated B cells can result in a number of outcomes, resulting in mutation of the original genetic

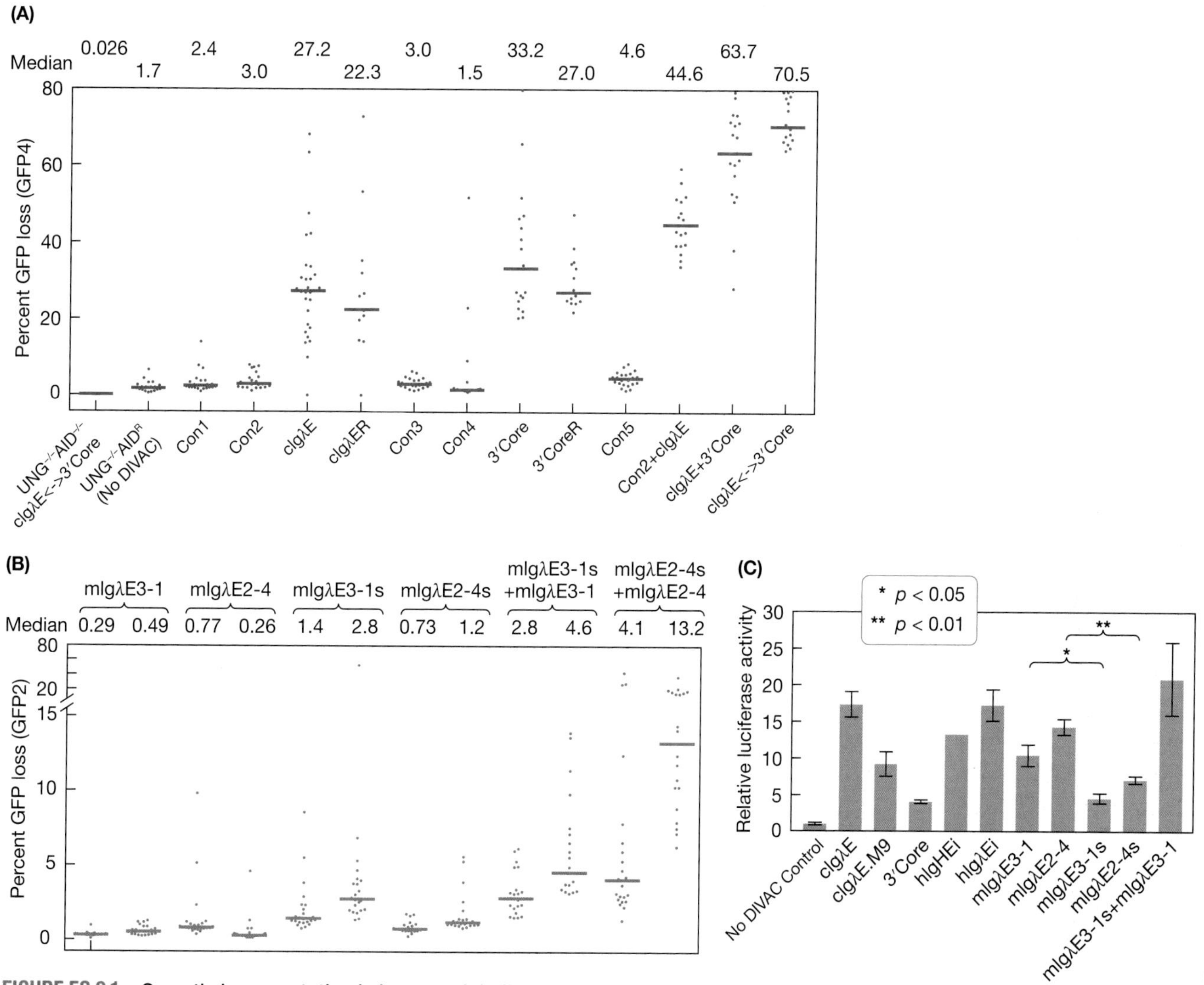

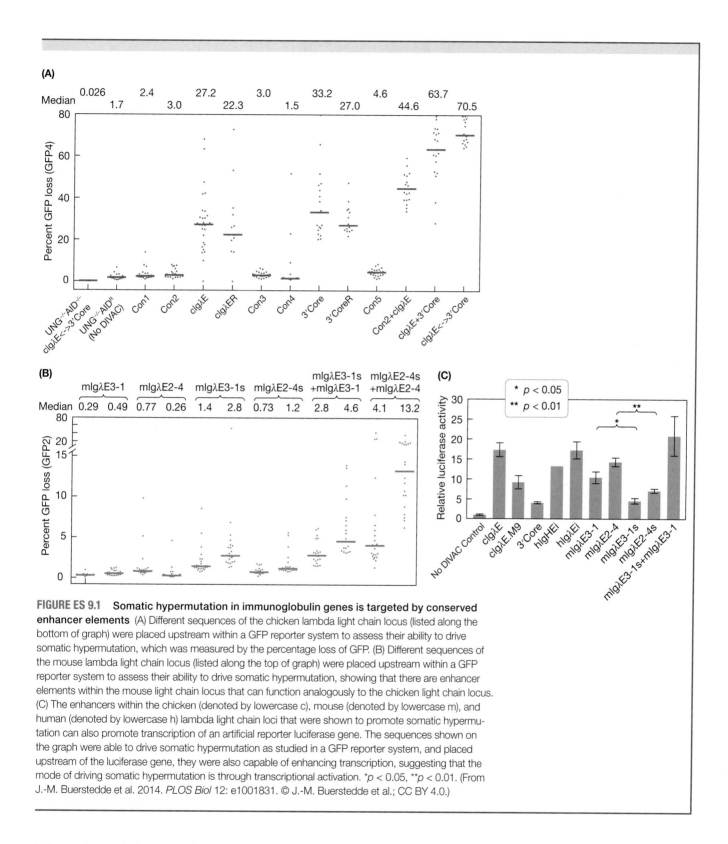

FIGURE ES 9.1 Somatic hypermutation in immunoglobulin genes is targeted by conserved enhancer elements (A) Different sequences of the chicken lambda light chain locus (listed along the bottom of graph) were placed upstream within a GFP reporter system to assess their ability to drive somatic hypermutation, which was measured by the percentage loss of GFP. (B) Different sequences of the mouse lambda light chain locus (listed along the top of graph) were placed upstream within a GFP reporter system to assess their ability to drive somatic hypermutation, showing that there are enhancer elements within the mouse light chain locus that can function analogously to the chicken light chain locus. (C) The enhancers within the chicken (denoted by lowercase c), mouse (denoted by lowercase m), and human (denoted by lowercase h) lambda light chain loci that were shown to promote somatic hypermutation can also promote transcription of an artificial reporter luciferase gene. The sequences shown on the graph were able to drive somatic hypermutation as studied in a GFP reporter system, and placed upstream of the luciferase gene, they were also capable of enhancing transcription, suggesting that the mode of driving somatic hypermutation is through transcriptional activation. $^{*}p < 0.05$, $^{**}p < 0.01$. (From J.-M. Buerstedde et al. 2014. *PLOS Biol* 12: e1001831. © J.-M. Buerstedde et al.; CC BY 4.0.)

information and changing the G-C base pair to a different pair (depending on the change induced by AID and the DNA repair mechanisms). Three possible mechanisms can repair the cytosine-to-uracil conversion, and all three have the ability to introduce mutation at the changed base pair (**FIGURE 9.11**):

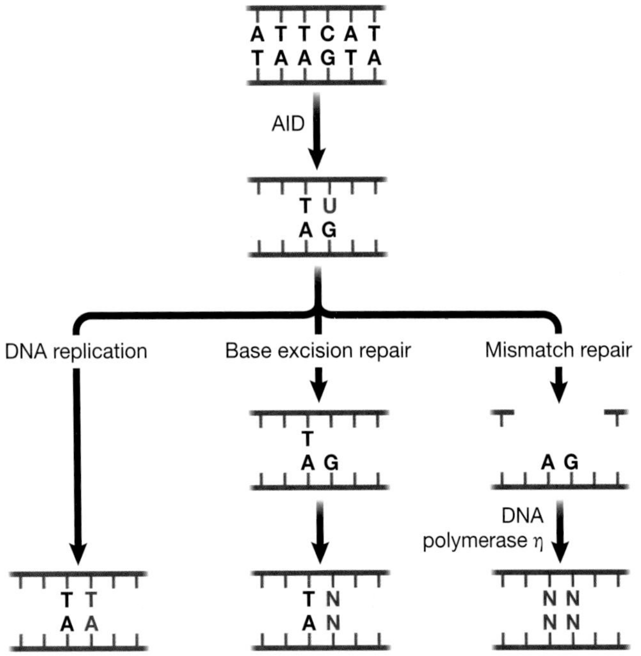

FIGURE 9.11 Deamination of cytosine to uracil in DNA can produce DNA mutations Due to the action of AID, the presence of the base uracil in DNA can activate several different repair mechanisms to remove the uracil and replace it with another base. In DNA replication, the replication machinery interprets the uracil as the base thymine and places the complementary base adenine on the newly synthesized strand. This results in the original cytosine base being synthesized to the base thymine. In base excision repair, the uracil base is removed from the DNA, creating an abasic site. Subsequent replication can result in the incorporation of any of the four bases (denoted N) in place of the original cytosine. In mismatch repair, the repair machinery recognizes the mismatched base pair and excises the uracil and several surrounding bases from the altered DNA. The action of the error-prone polymerase DNA polymerase η can result in mutation of any of the removed bases to another base (denoted N).

- *DNA replication*: The actively dividing B cell must replicate its DNA to continue progression through mitosis. A cytosine-to-uracil conversion prior to DNA replication can result in the replication machinery interpreting the presence of a uracil as a thymine base (the two bases look chemically similar). This misread by the DNA replication machinery will result in changing of the uracil base (from AID activity) to a thymine and the complementary base to an adenine, thus mutating the original C-G base pair to a T-A base pair.
- *Base excision repair*: A common mechanism employed by cells to repair cytosine-to-uracil conversions is base excision repair. This mechanism requires the activity of the enzyme uracil-DNA glycosylase, which cleaves the glycosidic bond between the uracil base and the deoxyribosephosphate backbone, creating an **abasic site**. An endonuclease then removes the abasic deoxyribosephosphate, allowing for short-patch base excision repair polymerases to replace the missing base. Because these polymerases are inherently error-prone, any of the four bases can theoretically be introduced into the original site containing a cytosine, resulting in a 75% chance of mutating the original cytosine to one of the other three bases and creating a point mutation at the original site.
- *Mismatch repair*: Because a guanine-uracil (G-U) is a nontraditional base pair, repair mechanisms that recognize the mismatched base pair can result in the excision of a larger stretch of DNA surrounding the uracil base. Error-prone polymerases, especially DNA polymerase η, can then fill in the excised gap. This mechanism not only can result in mutation at the original cytosine-to-uracil conversion site but can also result in point mutations in the surrounding excised DNA that was replicated by DNA polymerase η.

Isotype switching

Recombination events within B cells are not limited to somatic recombination at the variable regions of the heavy and light chain immunoglobulin loci. Activated B cells can also undergo class switch recombination, which allows them to produce immunoglobulins with effector functions suited for the tissue they are protecting and the pathogen they are combating. Isotype switching also requires the activities of AID and uracil-DNA glycosylase, but the mechanisms underlying class switch recombination are slightly different from those driving somatic hypermutation.

Located upstream of each of the constant regions (except for the δ constant region) are **switch regions** (also called *switch sequences*) that are rich in repetitive target sequences for AID activity. AID targets cytosine bases in both the donor switch region (region upstream of the current immunoglobulin constant region)

abasic site Site in DNA that does not contain a base connected to the sugar-phosphate backbone.

switch region DNA sequence upstream of immunoglobulin heavy chain constant regions that is targeted for recombination during class switch recombination; also called *switch sequence*.

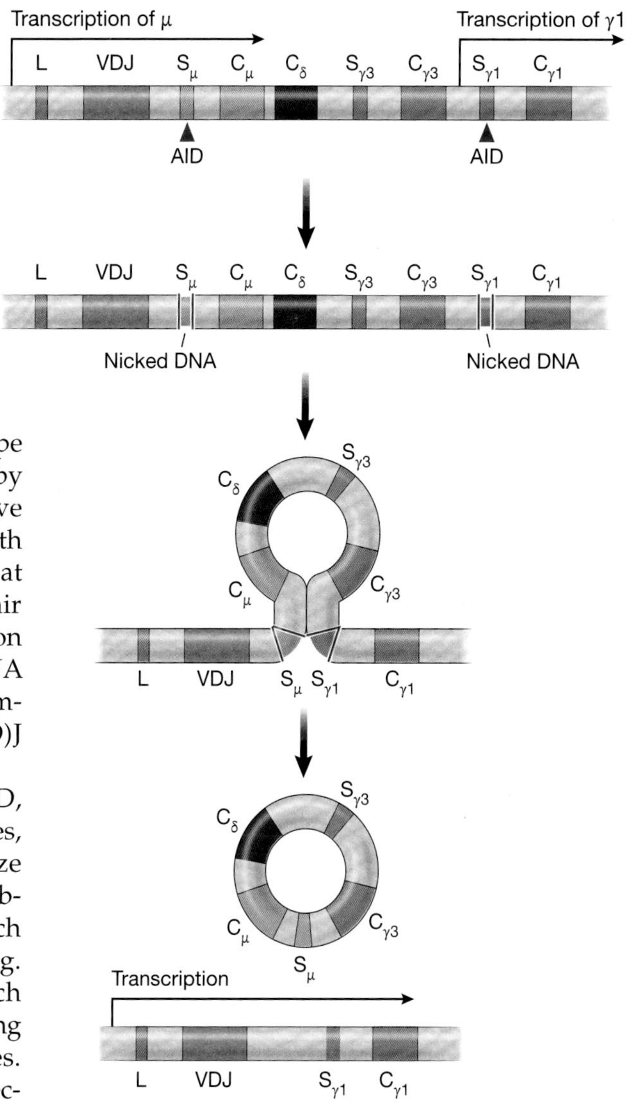

FIGURE 9.12 Class switch recombination Within activated B cells, transcription of different immunoglobulin heavy chain constant regions exposes the associated switch regions to cytosine deamination by AID. Repair mechanisms remove the uracil from the switch regions and create nicked DNA. Because of the sequence similarity among the switch regions and the breaks within the DNA, repair proteins facilitate recombination at the nicked switch regions. This results in recombination of the new constant region to a location for use during heavy chain transcription and removal of the intervening sequence in a manner similar to V(D)J recombination. (After F.-L. Meng et al. 2015. In *Molecular Biology of B Cells*, 2nd ed., pp. 345–362. Academic Press: London.)

and the acceptor switch region (region upstream of the isotype to be switched to) (**FIGURE 9.12**). Cytosine-to-uracil conversion by AID triggers the activity of uracil-DNA glycosylase to remove the uracil base, creating abasic sites in both switch regions. Both abasic sites are removed by an endonuclease, creating nicks at each of the switch regions. The nicks activate mismatch repair enzymes to create double-strand breaks. This results in ligation of the switch regions and excision of a circular piece of DNA between the two regions in a manner similar to somatic recombination (although this recombination is not driven by V(D)J recombinase).

Class switch recombination, driven by the action of AID, requires active transcription through the immunoglobulin genes, as AID requires a single-stranded DNA substrate to catalyze cytosine-to-uracil conversion. Transcription of the immunoglobulin genes is initiated upon activation within the B cell, which requires assistance from helper T cells and cytokine signaling. Importantly, specific cytokines can drive proper class switch recombination, and the cytokines that activate isotype switching are also secreted during different types of immune responses. Class switch recombination occurs in the germinal centers of secondary lymphoid tissues when helper T cells interact with a B cell presenting the appropriate MHC-peptide complex. The activated helper T cell secretes the cytokines needed to drive both B-cell activation and isotype switching. A summary of the action of these cytokines in isotype switching is shown in **TABLE 9.1**.

Because AID is involved in both somatic hypermutation and isotype switching, the loss of AID activity can limit the diversity of the immunoglobulin repertoire and the isotypes that can be produced. AID deficiency causes an immune disorder known as hyper IgM syndrome due to the inability of B cells to undergo class switch recombination. Furthermore, because AID promotes DNA changes, we might expect a connection to the development of certain types of cancer. In fact, AID activity is elevated in certain autoimmune diseases and oncogenesis.

TABLE 9.1 | Cytokines and the Isotype Switch They Initiate

Cytokine Signal	Isotype Produced by Activated B Cell
IFN-γ	IgG3, IgG2a
IL-4	IgG1, IgE
IL-5	IgA
TGF-β	IgA, IgG2b

Source: J. Stavnezer. 2000. *Curr Top Microbiol Immunol* 245: 127–168.

CHECKPOINT QUESTIONS

1. What three mechanisms repair the presence of uracil in DNA during somatic hypermutation?
2. What are the similarities and differences between somatic recombination and isotype switching?

Making Connections: Immunoglobulin Gene Structure, Recombination, and Expression

The heavy and light chain loci are arranged on chromosomes 2, 14, and 22. To produce a functional subunit, the immunoglobulin loci must undergo somatic recombination to construct a variable region. Unlike T-cell receptors, the effector function of immunoglobulins is dictated by the isotype. Isotypes are produced either through alternative splicing mechanisms or by class switch recombination.

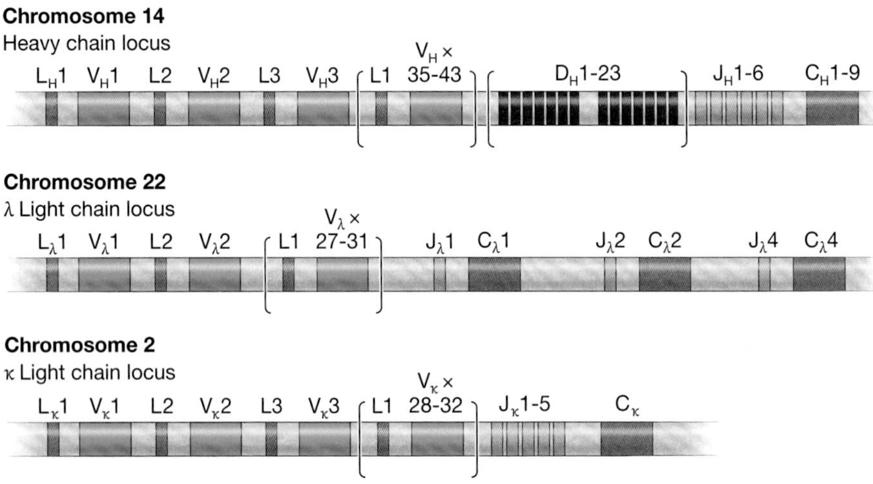

Data from S. Tonegawa. 1983. *Nature* 302: 575–581

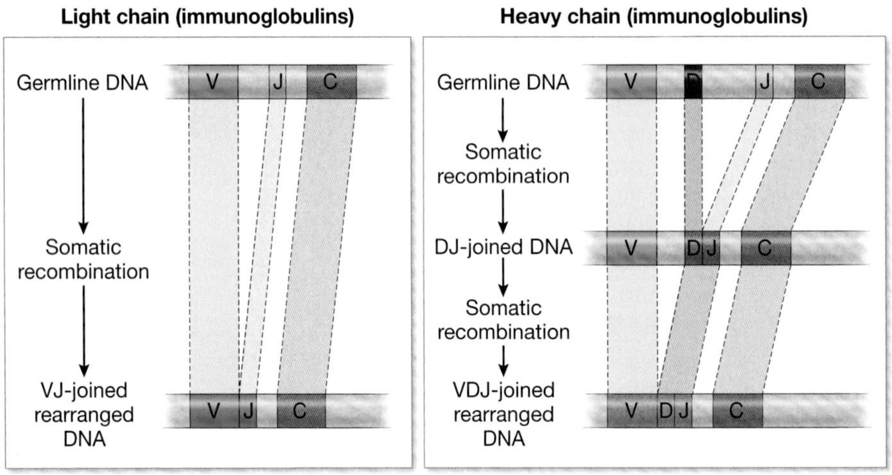

After W. Purves et al. 1995. *LIFE: The Science of Biology*, 4th ed., Sinauer Assoc., Sunderland, MA; T. Dörner and P. E. Lipsky. 2001. *Arthritis Rheum* 44: 2715–2727; P. Parham. 2014. *The Immune System*, 4th ed., W. W. Norton and Company, New York

In humans, immunoglobulin genes are located on three different chromosomes. The heavy chain locus is found on chromosome 14, whereas the two light chain loci are found on chromosome 22 (λ) and chromosome 2 (κ). During B-cell development, both the heavy and light chain loci undergo somatic recombination to produce a variable region that functions in antigen binding.

(Sections 4.3, 4.8, 8.6, 9.2)

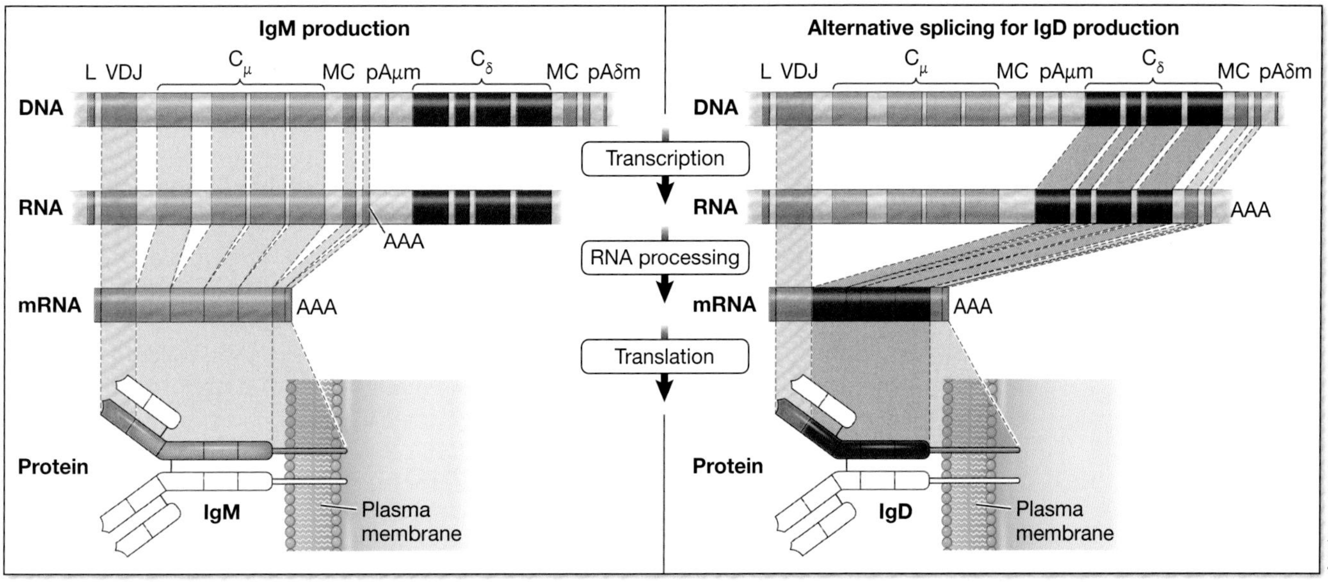

After P. Parham. 2014. *The Immune System*, 4th ed., W. W. Norton and Company

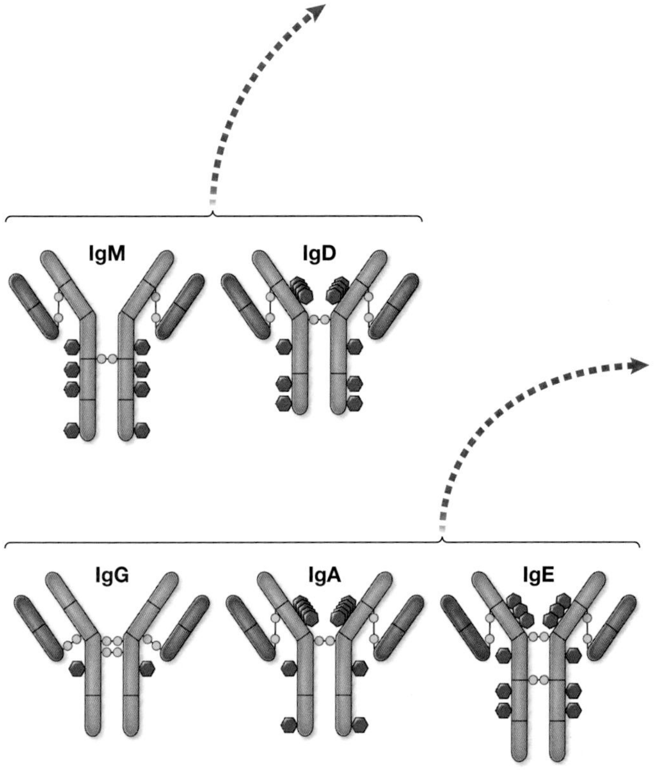

After T. Mak et al. 2014. *Primer to the Immune Response*, 2nd ed. Academic Cell/Elsevier Inc.: Burlington, MA

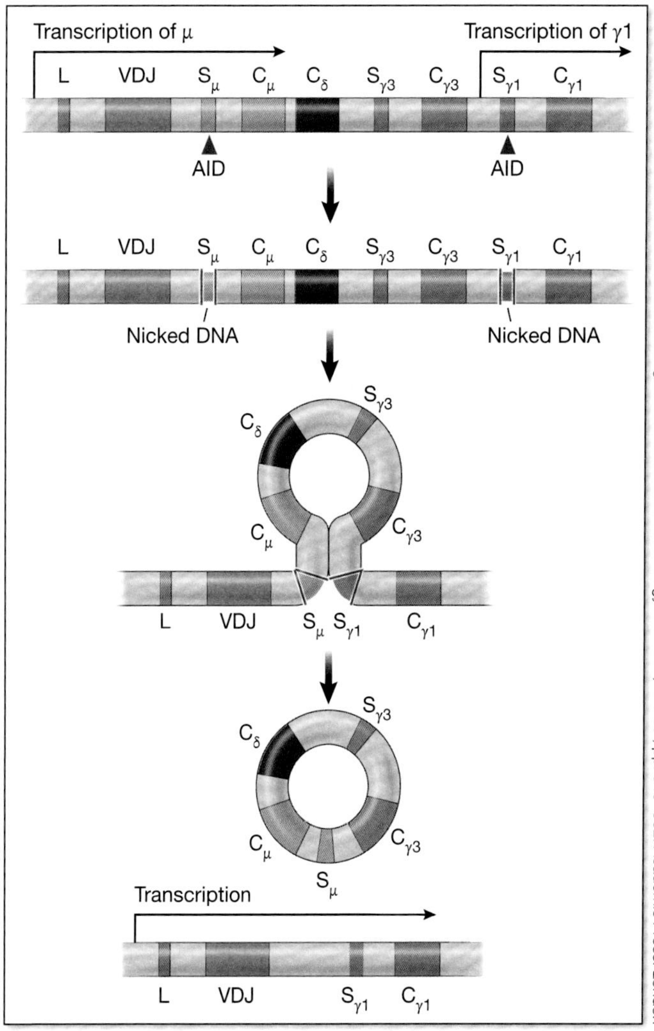

After F.-L. Meng et al. 2015. In *Molecular Biology of B Cells*, 2nd ed., pp. 345–362. Academic Press: London

While the effector function of T cells is at the level of the cell, the effector function of B cells is driven by the presence of specific immunoglobulin isotypes. Naïve B cells express IgM and IgD through alternative splicing mechanisms. Activated B cells may undergo class switch recombination to recombine the heavy chain constant region and produce a different immunoglobulin isotype.

(Sections 4.3, 4.8, 9.1, 9.2, 9.3, 9.4)

9.5 What is the nature of the different antibody isotypes?

LEARNING OBJECTIVE

9.5.1 Name the five antibody isotypes and explain the effector function(s) of each.

Isotype switching in activated B cells promotes the production of antibodies that induce an efficient immune response against the recognized pathogen within affected tissue. Antibody structure varies based on the composition of the heavy chain constant region loci and their physical structure when fully assembled (**FIGURE 9.13**).

Antibodies have three major effector functions: *neutralization*, *opsonization*, and *complement and immune cell activation*. Neutralizing antibodies directly block the action of pathogens and toxins and prevent their interaction with target cells. Opsonizing antibodies enhance pathogen recognition by other components of the immune system, including phagocytes. Finally, antibodies can prompt complement activation via the classical pathway (see Chapter 3) and can activate other immune cells, including natural killer (NK) cells and granulocytes.

Antibody isotypes promote these effector functions and interaction with other components of the immune system, including bridging with innate immune responses. The isotypes differ in their ability to be transported across barriers within the body. The key properties and effector functions of each immunoglobulin isotype are summarized in **TABLE 9.2** (and discussed further in Chapter 10).

IgM

The IgM isotype is one of the first expressed (along with IgD) in naïve B cells. Soluble IgM is present in a pentameric form through disulfide crosslinking of the heavy chains and through interaction with another soluble protein known

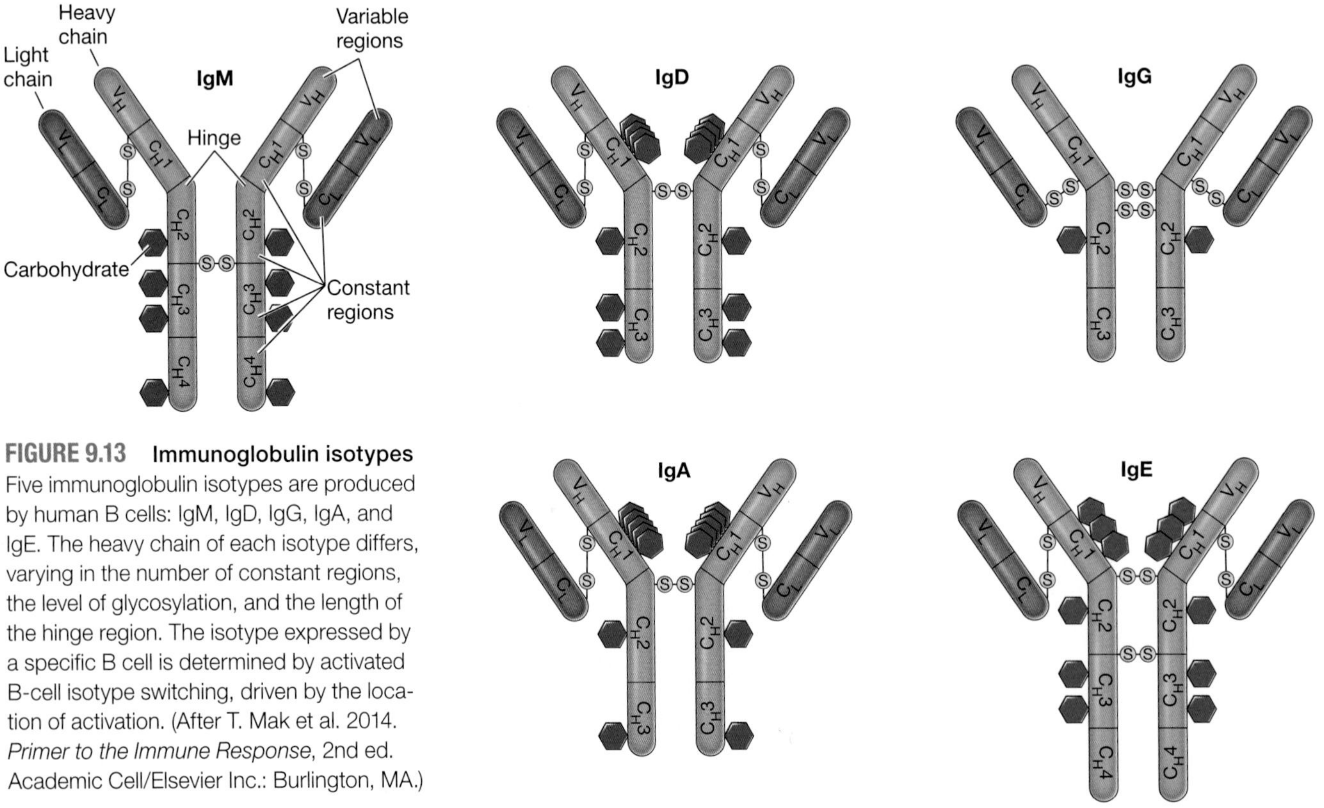

FIGURE 9.13 Immunoglobulin isotypes Five immunoglobulin isotypes are produced by human B cells: IgM, IgD, IgG, IgA, and IgE. The heavy chain of each isotype differs, varying in the number of constant regions, the level of glycosylation, and the length of the hinge region. The isotype expressed by a specific B cell is determined by activated B-cell isotype switching, driven by the location of activation. (After T. Mak et al. 2014. *Primer to the Immune Response*, 2nd ed. Academic Cell/Elsevier Inc.: Burlington, MA.)

TABLE 9.2 | Properties and Key Functions of Immunoglobulin Isotypes

		IgM	IgD	IgG1	IgG2	IgG3	IgG4	IgA	IgE
Properties	Mean serum level (mg/ml)	1.5	0.03	9	3	1	0.5	2.5	5×10^{-5}
	Transport across epithelium	+	–	–	–	–	–	+++ (dimeric)	–
	Transport across placenta	–	–	+++	+	++	++	–	–
	Extravascular diffusion	+	–	+++	+++	+++	+++	++ (monomer)	+
Effector functions	Neutralization	+	–	+++	+++	+++	+++	+++	–
	Opsonization	–	–	+++	+++	++	+	+	–
	NK cell sensitization	–	–	++	–	++	–	–	–
	Mast cell sensitization	–	+	+	–	+	–	–	+++
	Basophil sensitization	–	+	+	–	+	–	–	++
	Complement activation	+++	–	++	+	+++	–	+	–

Sources: H. W. Schroeder and L. Cavacini. 2010. *J Allergy Clin Immunol* 125: S41–S52, and K. E. Mostov. 1994. *Annu Rev Immunol* 12: 63–84.

as the **J chain** (**FIGURE 9.14**). Since soluble IgM is present in a pentameric form within the body, it can recognize antigen using its antigen-binding sites. Because IgM has 10 antigen-binding sites, it can have a high *avidity* (the combined strength of bond affinities) for antigen. However, since the IgM locus does not undergo somatic hypermutation, the binding *affinity* of each antigen-binding site tends to be lower than the affinity of class-switched immunoglobulins.

J chain Serum protein that interacts with IgM and IgA to provide additional quaternary structure.

The binding of IgM to an antigen allows it to neutralize antigens and serve as a trigger for the activation of complement via the classical pathway of activation—soluble IgM can serve as a substitute for C-reactive protein within the classical complement pathway. The structure of pentameric IgM sterically blocks the Fc components of each monomeric immunoglobulin. This prevents IgM from serving as an opsonin for phagocytic cells and other innate immune

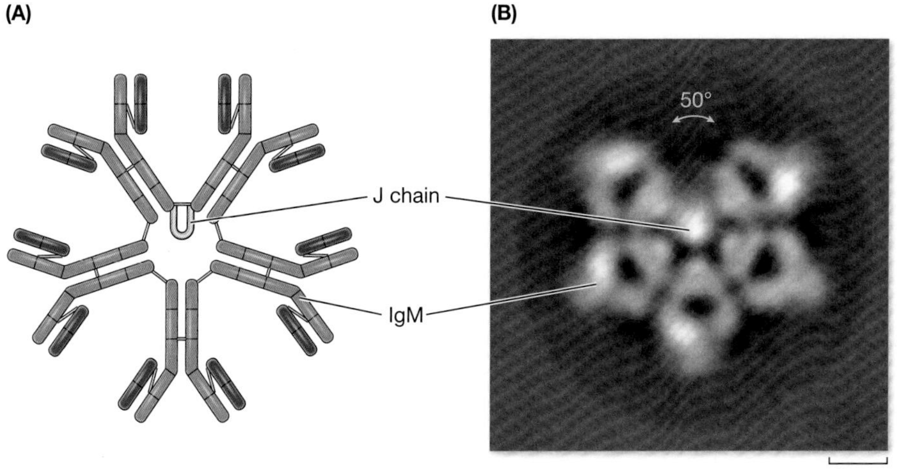

FIGURE 9.14 Structure of pentameric IgM (A) IgM assembles in a pentameric structure through disulfide bonds between the heavy chain regions and through association with the J chain protein. (B) A negative stain electron microscopy image shows that pentameric IgM bound to the J chain is not symmetrical; it has an opening of the pentameric structure of 50°. (B, from E. Hiramoto et al. 2018. *Sci Adv* 4: 1–9. © 2018 The Authors; CC BY 4.0.)

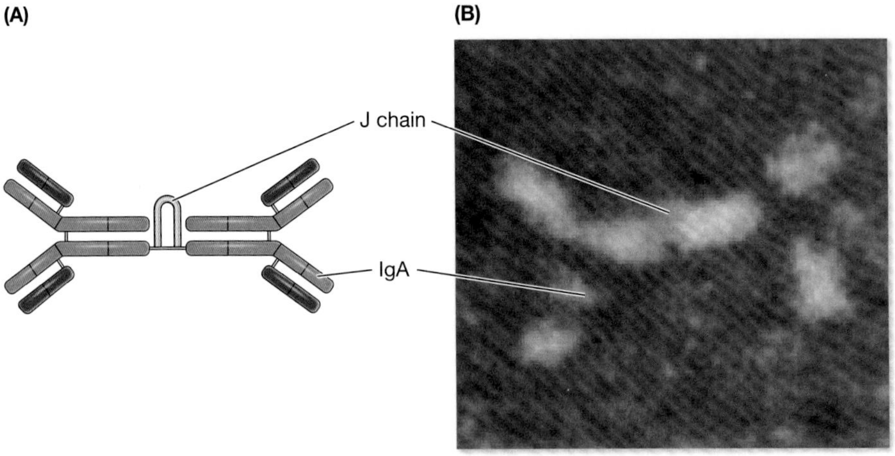

FIGURE 9.15 Structure of dimeric IgA (A) IgA often assembles in a dimeric structure through disulfide bonds between the heavy chain regions and through association with the J chain protein. (B) An electron microscopy image of purified IgA shows the dimeric structure of the molecule. (B, from E. A. Munn et al. 1971. *Nature* 231: 527–529.)

cells, as pentameric IgM cannot interact with Fc receptors at the surface of these cells. Interestingly, an Fc receptor for IgM has been identified, begging the question: What additional effector functions of IgM have not yet been discovered?

IgD

The IgD isotype is the other isotype expressed by naïve B cells due to alternative splicing with IgM. As with IgM, IgD is also involved in B-cell activation. The effector functions of IgD are far less defined compared to the other immunoglobulin isotypes, as it is not highly expressed in soluble form by activated B cells. However, IgD has been shown to play a role in mast cell and basophil activation in a manner similar to IgE.

IgA

Because mucosal surfaces (discussed further in Chapter 12) serve as entry points for a large number of pathogens, it is imperative that they are well protected by the immune system. IgA immunoglobulins play a major role in the protection of mucosal surfaces. IgA is present in both monomeric and dimeric forms, with the dimeric form being held together by J chain (**FIGURE 9.15**), the same polypeptide that aids in formation of pentameric IgM. IgA is efficiently transported across mucosal epithelial cells, where it works primarily as a neutralizing antibody. Because it is easily transported across mucosal epithelial cells, IgA is a major immunoglobulin present in breast milk. In infants who breastfeed, it can function in passive immunity.

One of 500 Caucasians have an immunodeficiency known as selective IgA deficiency. These individuals do not produce the IgA isotype, but surprisingly, 85% to 90% do not present any symptoms. This may likely be due to other isotypes, such as IgM and IgG, compensating for the loss of IgA production.

IgG

The IgG isotype is the most abundant immunoglobulin in serum. There are four subclasses (IgG1, IgG2, IgG3, IgG4), which differ slightly in the heavy chain constant regions and, more importantly, in the length of the hinge regions (**FIGURE 9.16**). The different hinge regions provide IgG flexibility between the Fc and Fab regions, and the different constant regions confer different effector functions. The high prevalence of IgG in serum is due to its primary role as a neutralizing and opsonizing antibody. IgG can serve as a substitute for C-reactive protein in the classical pathway for complement activation (similar to the action of IgM).

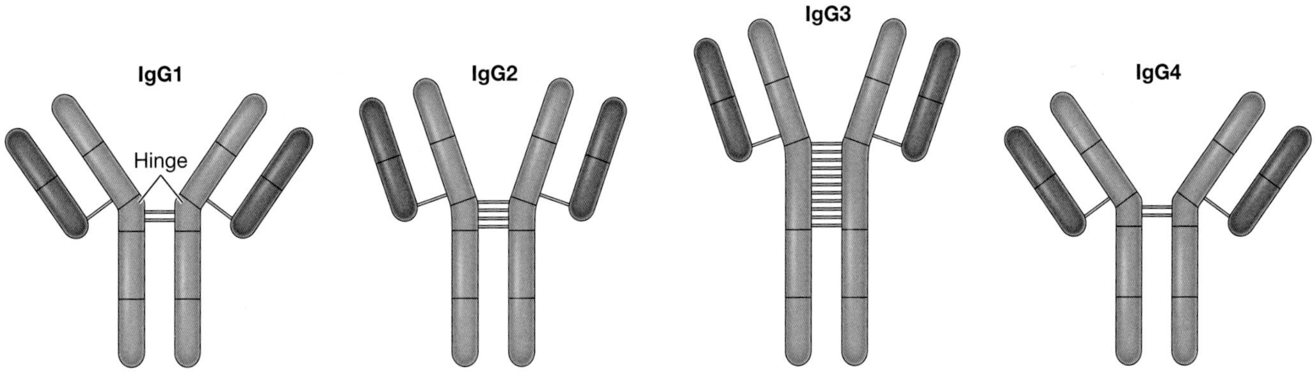

FIGURE 9.16 IgG subclasses The IgG isotype is diverse due to the presence of four different constant region components in the heavy chain locus. Class switch recombination allows activated B cells to produce one of four different IgG subclasses, which differ primarily in their hinge regions. Thus, the IgG subclasses differ in the flexibility of their Fab portions as they bind to antigen.

Interestingly, IgG4 can mix heavy and light chains. One heavy and light chain dimer of one IgG4 immunoglobulin can combine with a heavy and light chain dimer of another IgG4 immunoglobulin. This creates an IgG4 immunoglobulin that is bispecific—with two distinct antigen-binding sites, it can recognize two different antigens.

IgG is easily transported into extravascular sites and also serves an important protective role in a developing fetus, as it can be transported across the placental barrier from mother to fetus. Many innate immune cells, from phagocytes to NK cells, contain Fc receptors that recognize IgG, providing effector functions that bridge the adaptive and innate immune responses.

IgE

The IgE isotype is primarily involved with bridging the innate and adaptive immune response and combating inhaled or ingested pathogens, including parasites. IgE tightly associates with granulocytes such as mast cells, basophils, and eosinophils through interaction of its heavy chain constant region with Fc receptors on the surface of these cells. IgE engagement with its antigen on the surface of a granulocyte triggers degranulation and the release of inflammatory mediators, including histamine, to inhibit pathogen survival and expel it from the organism. IgE-mediated responses of granulocytes play an important role in the onset of certain allergic responses (discussed in Chapter 15).

CHECKPOINT QUESTIONS

1. Define neutralizing antibodies and opsonizing antibodies.
2. Why is IgM incapable of acting as an efficient opsonin?

Making Connections: Immunoglobulin Isotype Action

In humans, there are five immunoglobulin isotypes. These isotypes drive the effector functions of the humoral arm of the adaptive immune response. Each isotype has specific abilities that facilitate its role in protecting us from pathogen infection.

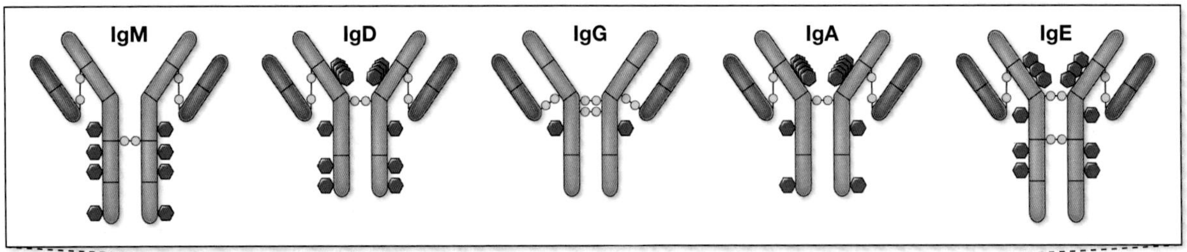

After T. Mak et al. 2014. *Primer to the Immune Response,* 2nd ed. Academic Cell/ Elsevier Inc.: Burlington, MA

The five immunoglobulin isotypes, IgM, IgD, IgG, IgA, and IgE, are produced by alternative splicing or class switch recombination.

(Sections 4.8, 9.1, 9.4, 9.5)

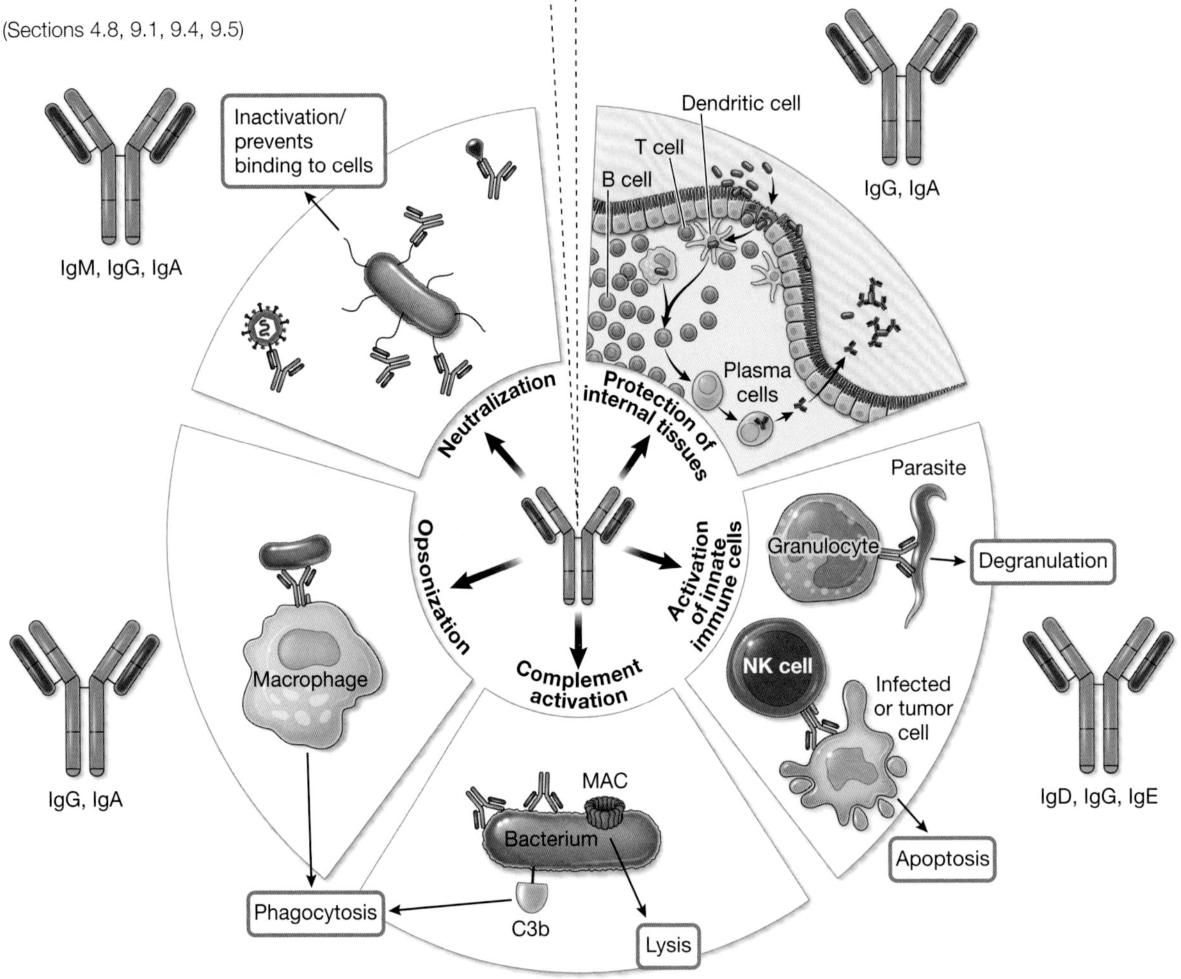

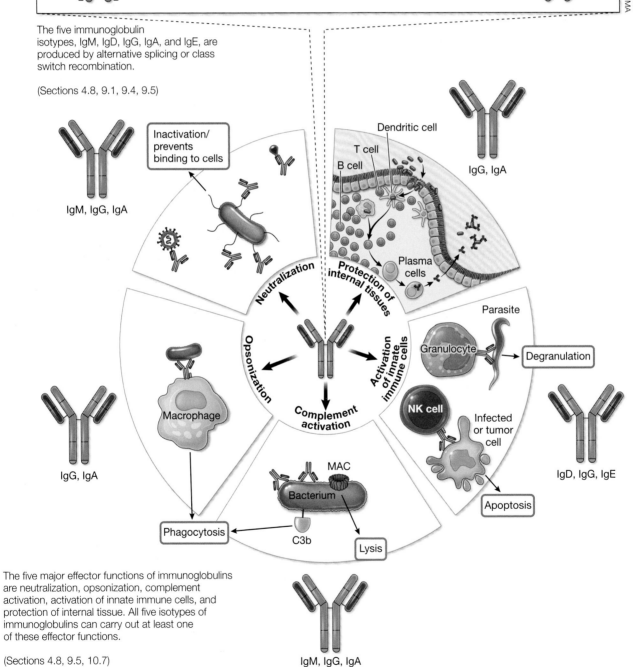

The five major effector functions of immunoglobulins are neutralization, opsonization, complement activation, activation of innate immune cells, and protection of internal tissue. All five isotypes of immunoglobulins can carry out at least one of these effector functions.

(Sections 4.8, 9.5, 10.7)

Summary

9.1 What is the structure of immunoglobulins?

- Immunoglobulins are made up of two heavy chains and two light chains and contain two antigen-binding sites composed of variable regions from the heavy and light chains.
- The three-dimensional structure of immunoglobulins promotes effector function and allows immunoglobulins to productively interact with a diverse array of antigens.
- Immunoglobulins recognize specific components of antigens known as epitopes. Epitopes are formed by structures close together in linear sequence or structures that are close together in three-dimensional space in the folded antigen.

9.2 How does genetic recombination in immunoglobulins compare to recombination in T-cell receptors?

- Somatic recombination of the immunoglobulin loci of B cells occurs via the same mechanism as T-cell receptor recombination, employing V(D)J recombinase and other key proteins involved in somatic recombination.
- Immunoglobulins at the surface of B cells interact with Igα and Igβ to form a complex capable of transducing the signal of antigen contact into the cytoplasm of the B cell.
- Immunoglobulin loci go through additional recombination after a B cell has come in contact with antigen. Class switch recombination provides a mechanism for a B cell to change the isotype of the expressed immunoglobulin to promote proper effector function.

9.3 Why is alternative splicing important in antibody production?

- The initial immunoglobulins expressed by naïve B cells are IgM and IgD.
- IgM and IgD are expressed together in naïve B cells due to alternative RNA splicing of the μ or δ constant regions of the immunoglobulin heavy chain.
- Expression of membrane-bound and soluble immunoglobulins after B-cell activation also occurs due to alternative RNA splicing, through maintenance or removal of the membrane-spanning domain.

9.4 What is the importance of somatic hypermutation and isotype switching in antibody maturation?

- Activated B cells can undergo affinity maturation and isotype switching to produce immunoglobulins with higher antigen affinity and more efficient effector functions.
- Somatic hypermutation occurs due to the activity of activation-induced cytidine deaminase (AID), which deaminates cytosine bases to uracil in sequence-specific regions within the variable regions of heavy and light chain loci. Repair mechanisms can mutate uracil to another base.
- Isotype switching also occurs due to the activity of AID acting at switch regions within the immunoglobulin heavy chain locus. Repair mechanisms recombine the DNA at the broken switch regions, resulting in expression of a new heavy chain constant region.

9.5 What is the nature of the different antibody isotypes?

- There are five different immunoglobulin isotypes (IgM, IgD, IgA, IgG, IgE). Each has specific effector functions that play a role in pathogen destruction and removal.
- Most antibodies function as either neutralizing antibodies, opsonizing antibodies, or antibodies that activate complement or immune cells as part of the humoral adaptive immune response.

Review Questions

9.1 Explain the roles of the Fab, Fc, and hypervariable loop regions of immunoglobulins in relation to their function.

9.2 Why does class switch recombination only occur in B cells? What is the functional significance of class switch recombination in B cells compared to the reason it is lacking in T cells?

9.3 **a.** Explain the importance of alternative splicing in immunoglobulin production.

b. Trace the production of a soluble immunoglobulin from the produced mRNA to its ultimate secretion.

9.4 Describe why AID activity promotes both somatic hypermutation and isotype switching.

9.5 Flow cytometry is a technique used to analyze surface molecules of various cells. You have the means to analyze populations of B cells, including their surface immunoglobulins, in individuals who have been exposed to a variety of different pathogens. Flow cytometry plots typically have two different cell-surface markers on each axis. For this question, draw four plots for each of the four different possible infections listed below. Plots should have a generic B-cell-specific marker on the *y*-axis, and each of the four plots should have one of the four different immunoglobulins, IgM, IgA, IgG, and IgE, on the *x*-axis. Presence of a B cell expressing a specific immunoglobulin should be represented as a circle in the upper-right quadrant of the plot, whereas if the B-cell population does not express the specific immunoglobulin, it should be represented as a circle in the upper-left quadrant of the plot.

A. Initial infection with a pathogen

B. Infection of a schistosome parasite

C. Infection of *Bacillus anthracis* in the bloodstream

D. Sinus mucosal infection of *Streptococcus pneumoniae*

CASE STUDY REVISITED

Hyper IgM Syndrome

A 5-year-old boy is taken to his primary care physician due to ear pain. A physical exam by the pediatrician reveals that the child has a fever and is suffering from otitis media (middle ear infection). A patient history reveals that he has suffered from five previous ear and sinus infections in the past year. Due to the recurring infections, the pediatrician orders further blood tests to examine white blood cells and immunoglobulin levels. The blood work reveals that his immunoglobulin serum levels are high in IgM and lacking in the immunoglobulin isotypes IgA, IgG, and IgE. Because of the lack of immunoglobulin isotypes, the pediatrician orders further genetic screening of the patient to examine any mutations associated with isotype switching. The genetic screening identifies a mutation in activation-induced cytidine deaminase (AID). Identification of this mutation allows the physician to diagnose the patient with hyper IgM syndrome.

Think About...

Questions for individuals

1. Define hyper IgM syndrome.
2. Why are this patient's symptoms consistent with hyper IgM syndrome?
3. Would you anticipate abnormal function of T cells in hyper IgM syndrome? Why or why not?

Questions for student groups

1. Why might there be a lack of other immunoglobulin isotypes in the patient's serum? If you were looking for defects of a particular protein, which protein would you suspect and why?
2. Would you anticipate somatic hypermutation in the IgM present in the patient? Why or why not?
3. Would you expect to observe IgD as an immunoglobulin present in the patient? How is IgD production different from IgA, IgG, and IgE production?

MAKING CONNECTIONS Key Concepts	COVERAGE (Section Numbers)
Immunoglobulin structure	4.2, 4.7, 4.8, 6.2, 6.4, 9.1, 10.2
Immunoglobulin junctional diversity	4.3, 4.8, 8.6, 9.1
Immunoglobulin antigen-binding sites	4.2, 4.8, 9.1, 9.4
Immunoglobulin alternative splicing	4.3, 4.8, 9.1, 9.3
Class switch recombination	4.3, 4.8, 9.1, 9.2, 9.4
Immunoglobulin isotype function	4.8, 9.1, 9.4, 9.5, 10.7

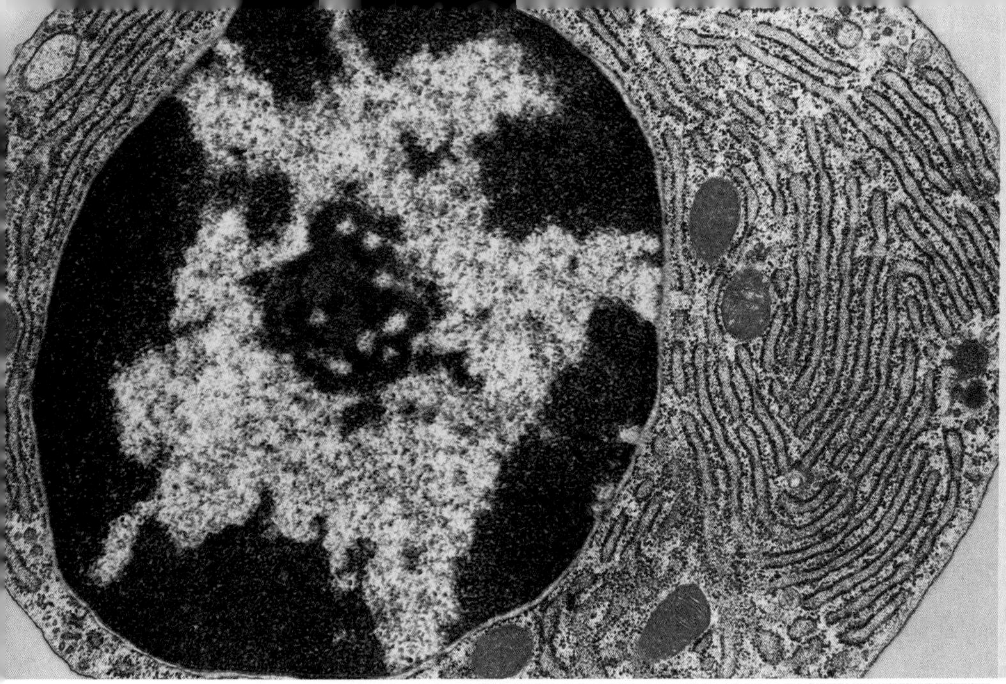

© Don W. Fawcett/Science Source

10

B-Cell-Mediated Adaptive Immunity

CASE STUDY: Marshall's Story Continued

Marshall's recent ear infections have been difficult. Troy wants to figure out what is going on with his son and understand the reason why he persistently suffers from ear infections. Most recently, Marshall was poked and prodded as his pediatrician took blood samples for testing. The tests revealed that Marshall is lacking in certain immunoglobulin isotypes. This didn't put Troy's mind at ease, however; he wants to hear how the pediatrician plans on treating Marshall, rather than hear a list of what Marshall doesn't have.

Marshall just finished his most recent regimen of antibiotics. Today is the day of Marshall's 2-week follow-up appointment with his pediatrician, and Troy recognizes that Marshall still seems to be in pain, suggesting that the antibiotics haven't helped. Troy hopes that Marshall's pediatrician has more news about Marshall's condition and a treatment plan, since it certainly seems that Marshall's ear infections aren't scheduled to stop anytime soon.

In order to understand what Marshall is experiencing, and to answer the case study questions at the end of the chapter, we will continue to explore the function of B cells in our adaptive immune system. Chapters 8 and 9 focused on B-cell development (driven by somatic recombination of immunoglobulin genes and leading to maturation of newly developed B cells) along with the structure of immunoglobulins, the primary players in B-cell effector function. To understand the means by which immunoglobulins act in a humoral immune response, we must explore the molecular mechanisms that drive clonal selection and expansion of B cells.

QUESTIONS Explored

10.1 Which signals and molecules are necessary for B-cell activation?

10.2 How are B cells activated in the absence or presence of T cells?

10.3 How do B cells migrate and behave in secondary lymphoid tissues?

10.4 How is immunoglobulin affinity maturation driven?

10.5 What are the properties and functions of the different antibody isotypes?

The signaling events that promote B-cell activation are remarkably similar to those that activate T cells. In fact, the same transcription factors activate both B-cell and T-cell clonal expansion. After B-cell activation, some differentiated B cells begin to secrete soluble immunoglobulins, while others continue to mature and take steps to refine immunoglobulin affinity for its antigen. Once secreted, immunoglobulins function in a variety of ways to orchestrate the steps involved in a humoral immune response.

10.1 Which signals and molecules are necessary for B-cell activation?

LEARNING OBJECTIVES

10.1.1 List the key molecules that are required for B-cell activation and outline the significant events of the signaling process.

10.1.2 Describe the processes of receptor crosslinking and costimulation in B-cell activation and explain why they are important.

Just as T cells express a unique T-cell receptor capable of responding to one specific antigen, B cells express an immunoglobulin on their surface that recognizes and responds to a specific antigen. Antigen recognition by a B cell signals the adaptive immune system to mount a humoral immune response to combat the foreign antigen. The signaling events drive B-cell activation, division, and differentiation in a manner similar to that of T-cell activation. These events include the activation of kinases within the B-cell cytoplasm and activation of transcription factors to drive gene expression of cytokines and other signaling molecules (**FIGURE 10.1**).

Antigen recognition is the first step of the activation process, and naïve B cells require further signals to drive clonal expansion. Proper activation of a naïve B cell occurs when membrane immunoglobulins cluster as a result of antigen interaction. In addition to signals from T_H cells or signaling events induced through receptors such as Toll-like receptors (TLRs), naïve B cells need a *costimulatory signal* to properly activate.

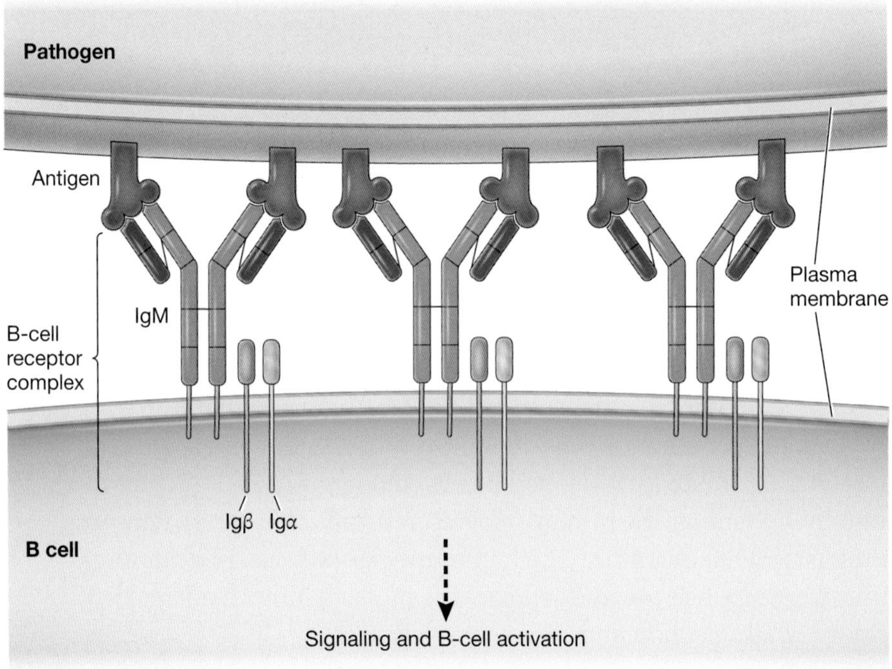

FIGURE 10.1 B-cell receptor engagement drives activation Each B cell expresses multiple copies of a receptor that recognizes a specific antigen. Antigen-receptor binding, along with signaling through the B-cell receptor complex (which includes Igα and Igβ), induces activation and clonal expansion. (After P. Parham. 2014. *The Immune System*, 4th ed., W. W. Norton and Company.)

Clustering of membrane immunoglobulins and the B-cell receptor complex

Recall that the immunoglobulin heavy and light chains have small cytoplasmic domains and require the action of Igα and Igβ to help relay signals to the B-cell cytoplasm. Igα and Igβ have immunoreceptor tyrosine-based activation motif (ITAM) sequences that play a role in B-cell activation similar to their role in the CD3 molecules involved in T-cell receptor signaling and activation. Immunoglobulin–antigen binding allows membrane immunoglobulins to cluster in a small area of the B-cell plasma membrane. This occurs due to the size of a typical antigen in comparison to the B cell. Immunoglobulin crosslinking (clustering) on the B-cell surface initiates signaling events that aid in activation. This process is driven by the action of proteins of the B-cell receptor complex, including Igα and Igβ (**FIGURE 10.2**). B-cell signaling events include the activation of kinases and phosphorylation of ITAMs located in the Igα and Igβ cytoplasmic domains.

The B-cell coreceptor

We know that naïve B cells require signaling through a coreceptor to properly activate and differentiate. The B-cell coreceptor is analogous to the T-cell coreceptor (where B7 on an antigen-presenting cell interacts with CD28 on the T-cell surface). The B-cell coreceptor is made up of three different polypeptides: CR2, CD19, and CD81 (**FIGURE 10.3**). CR2 is responsible for the binding required for the costimulatory signal. CR2 can bind to iC3b and C3d, both breakdown products of C3b fixed on the surface of a pathogen during complement activation. Both iC3b and C3d are generated after C3b fixation by the protein factor I; this cleavage event is facilitated by the interaction of CR1, a complement receptor on the B-cell surface. CD19 contains a cytoplasmic domain that plays a role in signaling events, and CD81 aids in the stabilization of the coreceptor complex. Together, the recognition of iC3b and C3d by CR2 and the activation of signaling cascades by the cytoplasmic domain of CR19 provide the costimulatory signal required for B-cell activation.

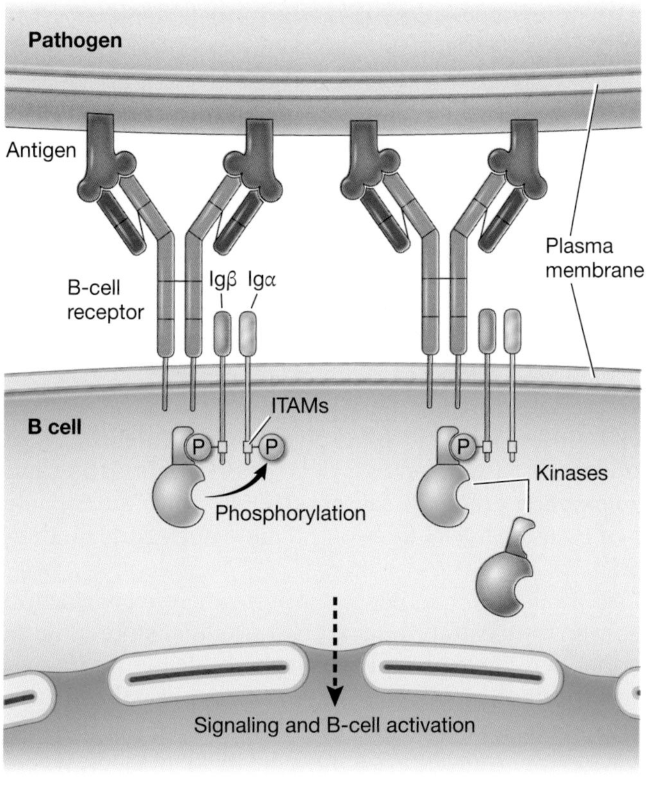

FIGURE 10.2 Clustering of B-cell receptor complexes drives activation Many antigens to which the receptor binds promote clustering of the receptor on the B-cell surface. Signaling through the B-cell receptor complex includes the action of kinases that phosphorylate ITAMs on the receptor complex proteins Igα and Igβ. This triggers signaling events that ultimately drive activation and clonal expansion. (Based in part on P. Parham. 2014. *The Immune System*, 4th ed., W. W. Norton and Company.)

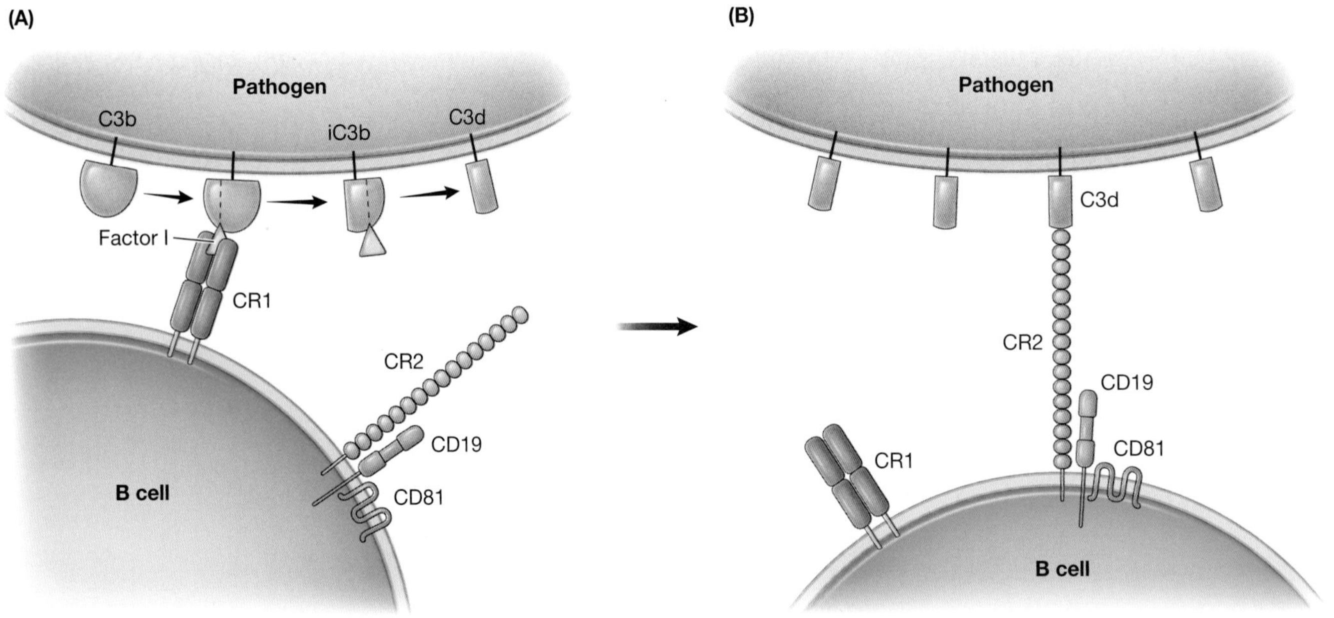

FIGURE 10.3 B-cell coreceptor binding (A) The B-cell coreceptor consists of three transmembrane proteins: CR2, CD19, and CD81. In order for CR2 to bind to C3d on the pathogen surface, C3b that was fixed on the pathogen surface during complement activation must be cleaved by the plasma protein factor I. The protein CR1 facilitates the cleavage of C3b by factor I to produce C3d. (B) Once C3d is present on the pathogen surface, it can bind to the CR2 component of the B-cell coreceptor. Binding of CR2 to C3d engages the coreceptor for B-cell activation. (After P. Parham. 2014. *The Immune System*, 4th ed., W. W. Norton and Company.)

Signaling at the immunoglobulin receptor complex

Clustering of membrane immunoglobulins triggers B-cell signaling events driven by the phosphorylation of ITAMs on Igα and Igβ. These phosphorylated tyrosines provide a binding site for signaling molecules that can only interact with tyrosine residues phosphorylated during receptor activation (see Figure 10.2).

Kinases that phosphorylate the ITAMs of Igα and Igβ include Blk, Lyn, and Fyn (**FIGURE 10.4**, Step 1). Phosphorylation allows for recruitment of Syk, a key kinase within B cells (Figure 10.4, Step 2). Recruitment of Syk to the phosphorylated ITAMs allows for Syk activation in a manner similar to the activation of ZAP-70 in T cells.

Upon activation, Syk induces three signal transduction pathways. These pathways lead to activation of transcription factors that initiate changes in gene expression. Signaling pathways are common in many types of cells, but their activation in B cells ultimately drives clonal expansion and differentiation. Syk initiates two of the three pathways through activation of the enzyme phospholipase C-γ (PLC-γ), which catalyzes the cleavage of phosphatidylinositol bisphosphate to diacylglycerol (DAG) and inositol triphosphate (IP_3) (Figure 10.4, Steps 3a and 3b). The other key pathway initiated by Syk is the phosphorylation and activation of **guanine-nucleotide exchange factors** (**GEFs**), which are capable of catalyzing exchange of guanine nucleotides on guanine nucleotide binding proteins (Figure 10.4, Step 3c).

IP_3 signals the opening of calcium channels, prompting an influx of calcium ions into the B-cell cytoplasm from extracellular and intracellular stores (Figure 10.4, Step 4a). Calcium ions activate a phosphatase called calcineurin (Figure 10.4, Step 5a), which in turn dephosphorylates the transcriptional activator NFAT (the same transcription factor activated in T-cell signaling) (Figure 10.4, Step 6a). This event allows cytoplasmic NFAT to relocate to the nucleus, where it can activate genes involved in clonal expansion and differentiation.

guanine-nucleotide exchange factor (GEF) Protein that aids in the activation of small GTP-binding proteins by promoting the exchange of bound GDP for GTP on the GTP-binding protein.

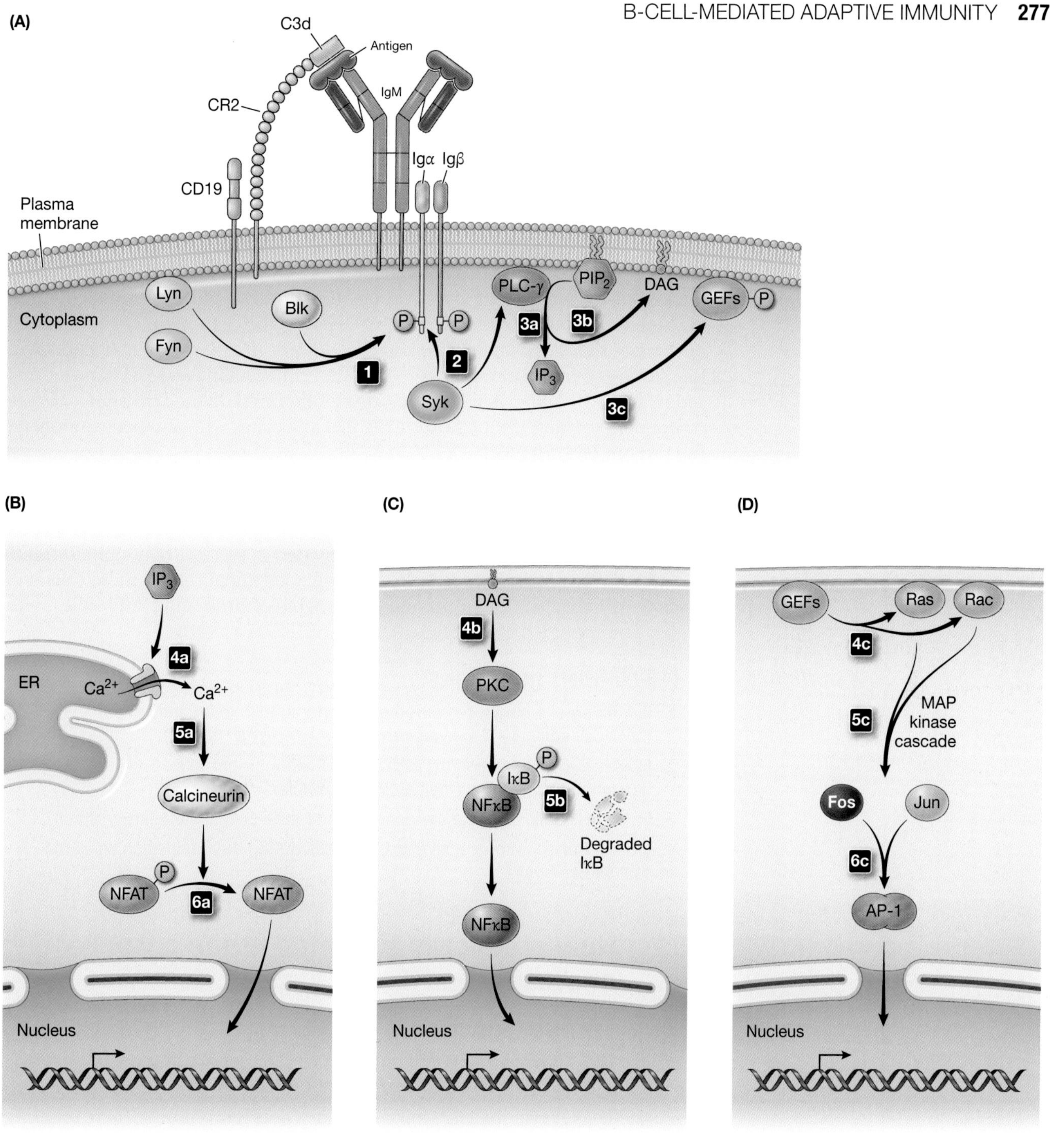

FIGURE 10.4 Signaling events that induce B-cell activation (A) Binding of antigen to membrane immunoglobulin along with a costimulatory signal activates several protein kinases, including Lyn, Fyn, and Blk. These kinases phosphorylate ITAMs on Igα and Igβ (Step 1). Phosphorylated ITAMs recruit the protein kinase Syk to the membrane, where it is activated (Step 2). Activated Syk phosphorylates phospholipase C-γ (PLC-γ), which catalyzes the cleavage of phosphatidylinositol bisphosphate (PIP_2) to inositol triphosphate (IP_3) and diacylglycerol (DAG) (Steps 3a and 3b). Activated Syk also phosphorylates and activates guanine-nucleotide exchange factors (GEFs) (Step 3c). (B) IP_3 leads to the activation of the transcription factor nuclear factor of activated T cells (NFAT). As a second messenger, IP_3 also activates calcium channels in the endoplasmic reticulum (ER) membrane, resulting in movement of calcium ions (Ca^{2+}) from the ER into the cytosol (Step 4a). Ca^{2+} activates the phosphatase calcineurin (Step 5a), which removes phosphates from the transcription factor (NFAT) (Step 6a). Dephosphorylated NFAT translocates to the nucleus and activates genes involved in B-cell activation. (C) DAG also leads to the activation of the transcription factor NFκB. It activates protein kinase C (PKC) (Step 4b), which phosphorylates IκB to target its destruction (Step 5b). Degradation of IκB releases the NFκB within the cytosol, allowing it to translocate into the nucleus and transcribe genes involved in B-cell activation. (D) GEFs lead to the activation of a third transcription factor, AP-1, involved in B-cell activation. GEFs activate the small GTPases Ras and Rac (Step 4c), triggering signaling through the MAP kinase cascade (Step 5c) and activating Fos. Fos interacts with Jun to form AP-1 (Step 6c). AP-1 translocates into the nucleus and transcribes genes involved in B-cell activation.

DAG, the other product of the PLC-γ reaction, is capable of initiating another signal transduction pathway through the activation of protein kinase C (PKC) (Figure 10.4, Step 4b), the same kinase activated during T-cell signaling. PKC activation drives activation of the transcription factor NFκB through destruction of the associated IκB protein (in a mechanism similar to TLR signaling and T-cell signaling) (Figure 10.4, Step 5b). Localization of NFκB to the nucleus of the B cell further promotes clonal expansion and differentiation.

Syk phosphorylates and activates GEFs, which are capable of catalyzing nucleotide exchange on guanine nucleotide binding proteins. Syk activation of these GEFs causes the nucleotide exchange and activation of the guanine nucleotide binding proteins Ras and Rac (Figure 10.4, Step 4c). Active Ras and Rac drive activation of a mitogen-activated protein kinase (MAP kinase) cascade (Figure 10.4, Step 5c). This cascade culminates with the activation of Fos, a component of transcription factor AP-1 that associates with another component, Jun (Figure 10.4, Step 6c).

The actions of the three transcription factors, NFAT, NFκB, and AP-1, all promote B-cell proliferation and differentiation. Some B cells differentiate into plasma cells and begin to produce soluble IgM in the initial rounds of a humoral immune response. Others are activated in secondary lymphoid tissue and migrate to a lymphoid follicle, forming a germinal center; these cells will eventually become either a plasma cell capable of producing antibodies with higher affinity or a memory cell.

CHECKPOINT QUESTIONS

1. Is binding of a single cell-surface immunoglobulin to an antigen sufficient for B-cell activation? Why or why not?
2. How is B-cell activation through a B-cell receptor complex similar to T-cell activation through the T-cell receptor complex?
3. How do B cells facilitate the generation of their own costimulatory signal?
4. How is transcriptional activation in B cells during their activation similar to that within T cells?

10.2 How are B cells activated in the absence or presence of T cells?

LEARNING OBJECTIVE

10.2.1 Explain how B cells are activated in response to thymus-dependent and thymus-independent antigens.

Mature B cells travel through the circulatory and lymphatic system, entering and exiting secondary lymphoid tissues. They become activated when they express the proper immunoglobulin capable of recognizing a foreign antigen. A B cell's entry into a lymphoid tissue and migration to the lymphoid follicle is driven by follicular dendritic cells. Follicular dendritic cells secrete the chemokine CXCL13 to signal B-cell migration to the lymphoid follicle.

Clonal expansion of the B cell is driven by the activation of signaling pathways that resemble those induced during T-cell activation. However, unlike T-cell activation, which requires engagement of a T-cell receptor with a major histocompatibility complex (MHC)-peptide complex, several different types of antigens can activate B-cell clonal expansion. There are two types of B-cell responses: T-independent (for thymus-independent) and T-dependent (for thymus-dependent). They differ based on the need for helper T-cell involvement (**FIGURE 10.5**):

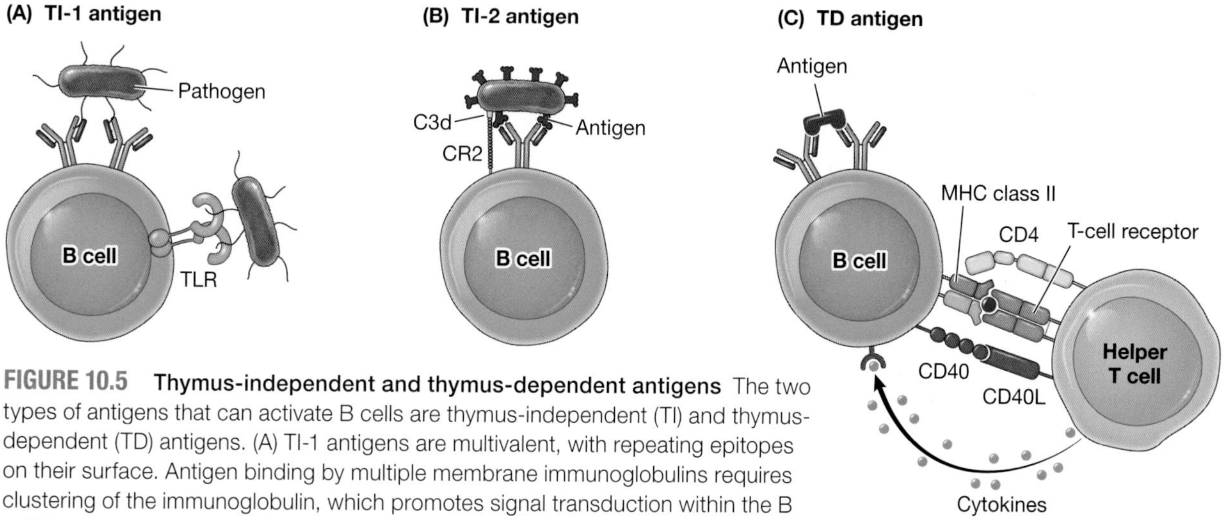

FIGURE 10.5 Thymus-independent and thymus-dependent antigens The two types of antigens that can activate B cells are thymus-independent (TI) and thymus-dependent (TD) antigens. (A) TI-1 antigens are multivalent, with repeating epitopes on their surface. Antigen binding by multiple membrane immunoglobulins requires clustering of the immunoglobulin, which promotes signal transduction within the B cell. The B cell achieves activation through signaling from the membrane immunoglobulins, along with binding and signaling through a pattern recognition receptor, such as a TLR at the B-cell surface. (B) TI-2 antigens also tend to be multivalent antigens that promote membrane immunoglobulin clustering and signaling. These antigens prompt B-cell activation because they become fixed with complement, driving the formation of C3d on their surface and allowing binding and signaling through the B-cell coreceptor. (C) TD antigens require engagement of the B-cell receptor with antigen, along with antigen processing via MHC class II processing. Processed antigen is displayed in MHC class II molecules to circulating CD4 helper T cells. Binding of the T-cell receptor from a helper T cell, along with cytokines produced by the helper T cell and signaling through the CD40/CD40 ligand complex, activates the bound B cell. (Based in part on J. Punt et al. 2018. *Kuby Immunology*, 8th ed., W. H. Freeman and Company: New York.)

- **T-independent (TI) response**: A subset of B cells can become activated without any help from T-cell activity. These include B-1 B cells and marginal-zone B cells (see Chapter 8). Some B cells express immunoglobulins that can recognize a pathogen through both its membrane immunoglobulin and a pattern recognition receptor such as TLRs (see Chapter 2). **T-independent (TI) antigens** are capable of activating a B cell independent of T-cell action. *TI-1 antigens* bind to an immunoglobulin and a pattern recognition receptor on the surface of the B cell. A *TI-2 antigen* can activate a B cell based on its ability to cluster immunoglobulins on the cell surface when enough antigen receptors bind with the antigen. Many times, TI-2 antigens occur as repetitive units on a pathogen surface, such as a repeating polysaccharide unit of a pathogen's cell wall. TI-2 antigens become fixed with complement during an innate immune response, promoting the formation of C3d on their surface and engagement with the B-cell coreceptor. When a sufficient number of membrane immunoglobulins are clustered at the cell surface, the engaged B cell activates.
- **T-dependent (TD) response**: Most B cells (B-2 B cells, see Chapter 8) require signals provided by helper T cells to properly activate when they bind with an antigen via their membrane immunoglobulin. **T-dependent (TD) antigens** are those that cannot stimulate B cells to produce antibodies without help from T cells. TD antigens are recognized by the B-cell receptor and internalized and processed via MHC class II processing. They are subsequently recognized by a T cell with a receptor capable of recognizing the same antigen (although not necessarily the same epitope). For a B cell that recognizes a TD antigen to become active, the T cell that engages

T-independent (TI) response B-cell response that can induce B-cell activation independent of aid from helper T-cell signaling.

T-independent (TI) antigen Antigen that is capable of activating a B cell in the absence of helper T-cell action.

T-dependent (TD) response B-cell response that requires aid from helper T cells to promote proper B-cell activation.

T-dependent (TD) antigen Antigen that can only activate a B cell with the aid of a helper T cell.

CD40 Cell-surface molecule on B cells or macrophages that must engage with CD40 ligand from a helper T cell to promote proper B-cell activation.

CD40 ligand (CD40L) Ligand expressed by helper T cells that must bind to CD40 on either macrophages or B cells in order to aid in proper macrophage or B-cell activation by a helper T cell.

with the B cell must provide other necessary signals. This occurs through cytokine signaling, using cytokines such as IL-4, and through cell-surface interaction between **CD40** on the B cell and **CD40 ligand** (**CD40L**) on the T cell. These signals act in concert with antigen recognition to promote B-cell activation and differentiation.

Thymus-independent antigens

B-1 and marginal-zone B cells express membrane immunoglobulins that can facilitate activation without assistance from T_H cells. These B cells activate in response to *TI-1* and *TI-2 antigens*, which differ based on the means by which they stimulate B-cell activation. Although we will focus on the major mechanism by which each of these antigens drives B-cell activation through action of the membrane immunoglobulin, note that B cells that recognize a TI antigen also require signaling through the B-cell coreceptor.

TI-1 ANTIGENS In addition to their individual immunoglobulins, B cells have other cell-surface receptors that recognize foreign molecules. B cells can express pattern recognition receptors, including those of the TLR family (see Chapter 2). Recognition of a pathogen-associated molecular pattern (PAMP) by a TLR in conjunction with engagement of the B cell's membrane immunoglobulin with antigen (and engagement of the B-cell coreceptor) can drive B-cell activation. Foreign molecules' interaction with both receptors can be sufficient to activate the B cell and induce division and differentiation into plasma cells.

One example of a TI-1 antigen is lipopolysaccharide (LPS). If a B cell expresses an LPS-specific immunoglobulin or recognizes another cell-surface molecule on a gram-negative bacterium, the binding of the membrane immunoglobulin with its ligand, along with the presence of LPS signaling through TLR4 recognition, can cause activation (**FIGURE 10.6**).

Another example is bacterial DNA, which is recognized by TLR9, a TLR expressed at a high level on B cells. Recognition of a bacterium via membrane immunoglobulin–antigen interaction results in phagocytosis of the bacterium and its degradation within the

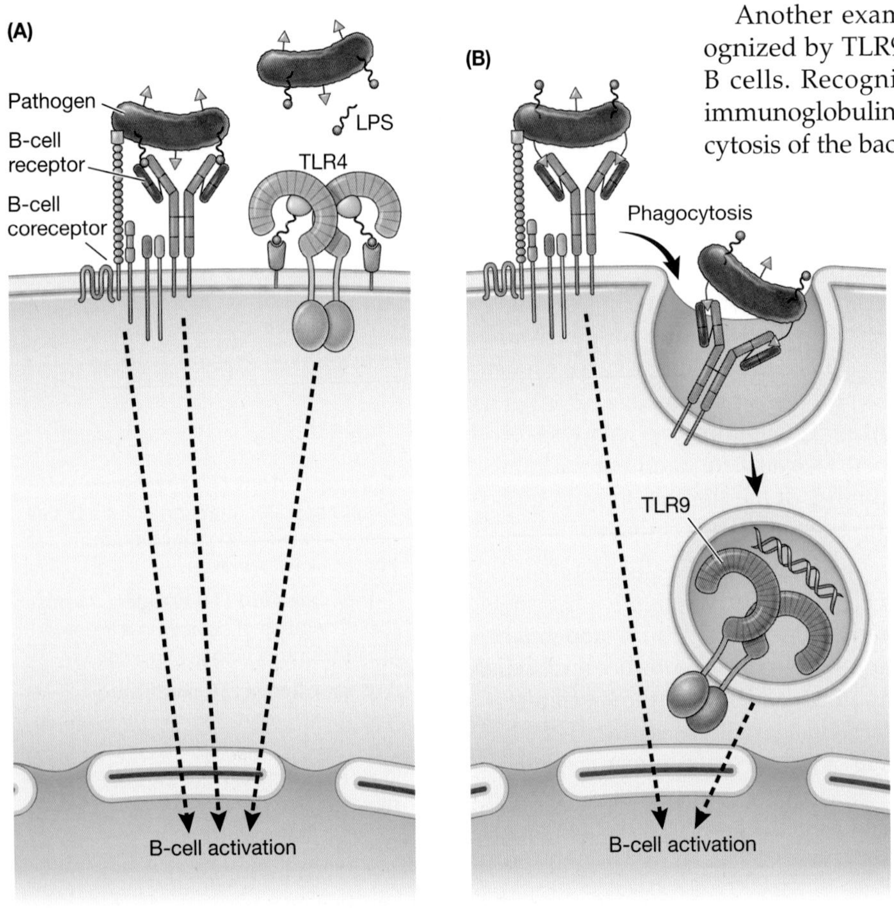

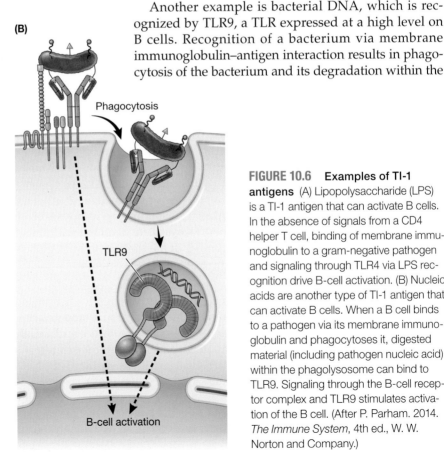

FIGURE 10.6 Examples of TI-1 antigens (A) Lipopolysaccharide (LPS) is a TI-1 antigen that can activate B cells. In the absence of signals from a CD4 helper T cell, binding of membrane immunoglobulin to a gram-negative pathogen and signaling through TLR4 via LPS recognition drive B-cell activation. (B) Nucleic acids are another type of TI-1 antigen that can activate B cells. When a B cell binds to a pathogen via its membrane immunoglobulin and phagocytoses it, digested material (including pathogen nucleic acid) within the phagolysosome can bind to TLR9. Signaling through the B-cell receptor complex and TLR9 stimulates activation of the B cell. (After P. Parham. 2014. *The Immune System*, 4th ed., W. W. Norton and Company.)

phagolysosome. Release of bacterial DNA through the degradation process leads to DNA recognition through TLR9 and activation of TLR9 signal transduction events (see Figure 10.6). Since the immunoglobulin has engaged its antigen, allowing for phagocytosis of the bacterium, the engagement of TLR9 with bacterial DNA serves as the necessary additional signal to prompt B-cell activation.

TI-2 ANTIGENS Some immunoglobulins can recognize cell-surface molecules that are repetitive in nature (for example, the repetitive polysaccharide component of a bacterial cell wall). Certain membrane immunoglobulins cluster in close proximity when recognition of this type of antigen occurs, and this clustering can trigger B-cell activation (**FIGURE 10.7**). Activation still requires costimulation of the B-cell coreceptor. Membrane immunoglobulin clustering plays an important role in both the activation of TI-2 antigen-specific B cells and activation of B-2 cells that require T cell help.

Many bacterial pathogens have a polysaccharide coat, and immunoglobulins that can recognize repeating polysaccharide structures can provide effective protection against these pathogens. This knowledge has been used in the creation of vaccines that promote the development and expansion of B cells with the ability to produce immunoglobulins that recognize polysaccharides. Unfortunately, children younger than 18 months and adults over age 65 have a hard time inducing TI-2 antigen-specific B cells. This has sparked the development of *conjugate vaccines*, which rely on helper T-cell function to promote the expansion of B cells that produce immunoglobulins that recognize polysaccharides (see Chapter 11).

Thymus-dependent antigens

The majority of B cells utilize signals from T cells to activate and differentiate. This process occurs after a B cell has entered a secondary lymphoid tissue where antigen is being presented (**FIGURE 10.8**). After B-cell migration into secondary lymphoid tissue, interaction between the B cell and a follicular dendritic cell expressing BAFF provides survival signals to the circulating B cells. Recall that BAFF stands for *B-cell activation factor* or *B-cell survival factor*. It is a cytokine in the tumor necrosis factor (TNF) family that serves as both an activating factor and a survival factor.

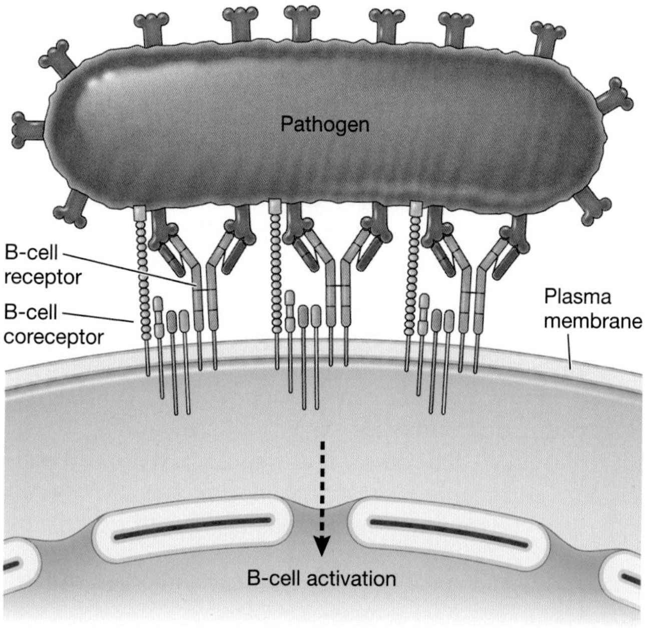

FIGURE 10.7 TI-2 antigens trigger B-cell receptor clustering TI-2 antigens typically contain repeating epitopes that can be recognized by a membrane immunoglobulin. Because of the repeating nature of epitopes present in the antigen and the small size of pathogens, membrane immunoglobulin binding to TI-2 antigen causes the immunoglobulins to cluster in close proximity within the plasma membrane. This clustering triggers signaling and B-cell activation. (After P. Parham. 2014. *The Immune System*, 4th ed., W. W. Norton and Company.)

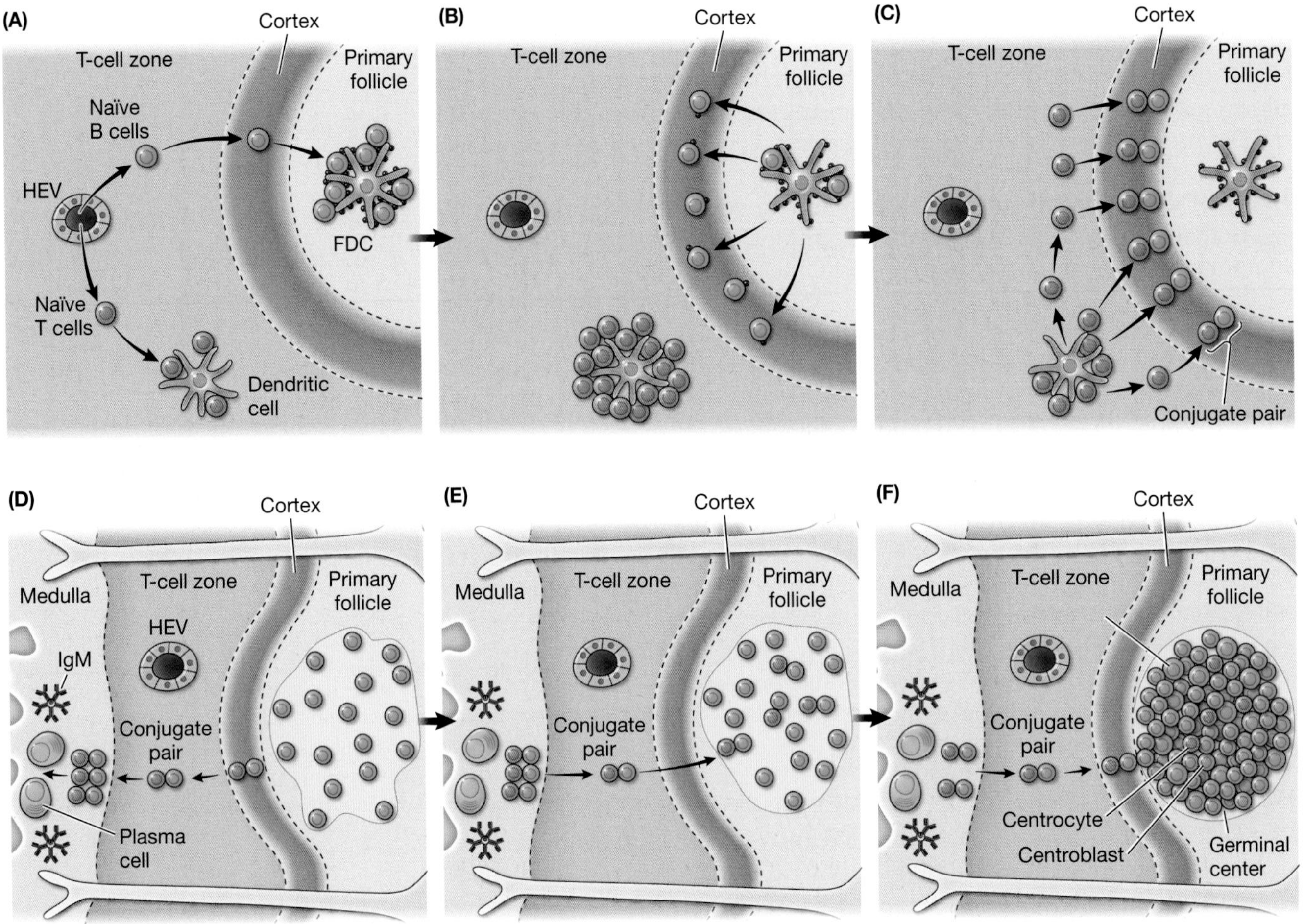

FIGURE 10.8 B-cell migration and activation in secondary lymphoid tissue (A) B cells enter secondary lymphoid tissue through high endothelial venules (HEVs) and migrate to a primary follicle, where they interact with follicular dendritic cells presenting antigen. T cells also enter secondary lymphoid tissue through HEVs. They migrate to the T-cell zone, where they interact with dendritic cells presenting antigen. (B) Activated B cells migrate to the cortex, while activated T cells in the T-cell zone undergo clonal expansion. (C) Activated B cells and T cells interact at the cortex, where they become a conjugate pair. (D) Some conjugate pairs migrate to the medulla, where the B cells receive signals prompting them to differentiate into plasma cells. Plasma cells begin to secrete IgM. (E) Some conjugate pairs migrate from the medulla to the primary follicle. (F) Conjugate pairs within the primary follicle provide signals to the B cells to differentiate into centroblasts. Centroblasts undergo somatic hypermutation, isotype switching, and a round of negative selection before becoming centrocytes. A centrocyte can differentiate into a plasma cell that can produce a different immunoglobulin isotype that has undergone somatic hypermutation, or it can differentiate into a memory cell to promote activation of a secondary adaptive immune response. (After P. Parham. 2014. *The Immune System*, 4th ed., W. W. Norton and Company.)

When a B cell interacts with an antigen via its membrane immunoglobulin, it engulfs the antigen via receptor-mediated endocytosis. The antigen is then processed and loaded on MHC class II molecules, which display peptides to CD4 T cells (see Chapter 4). Within the T-cell zone of the secondary lymphoid tissue, T cells have potentially become activated by dendritic cells presenting antigen to previously naïve T cells. This occurs if their T-cell receptor recognizes the MHC-peptide complex. The activated helper T cell expresses cytokines responsible for clonal expansion and differentiation of the T_H cell, as well as cytokines that aid in activating other immune system cells, including B cells. Signals from the T_H cell drive B-cell activation. Some of the activated B cells migrate out of the T-cell zone to become a *primary focus*, which is a cluster of activated B cells that will differentiate into plasma cells capable of expressing soluble IgM (see Figure 10.8).

TABLE 10.1 | Helper T-Cell Signals

Helper T-Cell Signal/Molecule	Action on B Cell
CD40 ligand	Binds to CD40 on B-cell surface; aids in B-cell activation
LFA-1	Binds to ICAM-1 on B-cell surface; aids in proper conjugate pair formation
IL-4	Aids in B-cell activation and proliferation
IL-5	Aids in B-cell differentiation
IL-6	Aids in B-cell differentiation

Other B cells migrate to the lymphoid follicle. Within the follicle, the area in which TD B-cell activation, proliferation, and differentiation occurs is known as the **germinal center**. Here, B cells become *centroblasts* and begin the processes of somatic hypermutation and isotype switching through the action of activation-induced cytidine deaminase. Once these processes have occurred in the germinal center, the centroblasts become **centrocytes**, which can undergo clonal selection (see Figure 10.8). The majority of centrocytes will become plasma cells that can express and secrete soluble immunoglobulins that have switched their isotype and have a higher affinity for their antigen. A small subset will become memory B cells capable of mounting an adaptive immune response upon subsequent exposure to the same antigen (discussed in Chapter 11).

germinal center Area in secondary lymphoid tissue where rapidly dividing B cells and T cells are activated and B cells undergo somatic hypermutation and isotype switching as they differentiate into centrocytes.

centrocyte Differentiating B cell within germinal centers that can undergo somatic hypermutation and isotype switching and differentiate into either a memory B cell or a plasma cell that can express mutated, isotype-specific immunoglobulins.

conjugate pair Cellular interaction between a helper T cell via its T-cell receptor and an antigen-presenting cell.

Signals provided by helper T cells

We know that the majority of B-cell activation requires signals provided by T_H cells (summarized in **TABLE 10.1**). So, how exactly do T cells aid in B-cell activation?

Membrane immunoglobulins bind antigen and allow for receptor-mediated endocytosis of the bound antigen. This endocytosis event enables the B cell to process the extracellular antigens and present them on MHC class II molecules. If the right T cell recognizes the presented antigen using its expressed T-cell receptor, it then engages the B cell and the two become a **conjugate pair**.

T-cell engagement and subsequent activation induces the expression of CD40 ligand on the T-cell surface, which binds to the CD40 molecule on the B cell. Signals through the conjugate pair allow the B cell to maintain a tight interaction with the T cell, as the CD40-CD40 ligand interaction drives expression of the adhesion molecule ICAM-1 on the B-cell surface. ICAM-1 binds tightly to LFA-1 on the surface of the T cell. All these cell-surface interactions tighten the conjugate pair interaction, prompting the T cell to secrete cytokines toward the engaged B cell, providing the signals needed for B-cell activation.

Some cytokines secreted by T cells, such as IL-4, aid in B-cell activation and proliferation. Others, such as IL-5 and IL-6, provide signals to some B cells that drive their differentiation into plasma cells. Another group of cytokines can induce B cells to switch isotypes to match the effector function needed during an infection.

CHECKPOINT QUESTIONS

1. Explain the interactions that must occur on the surface of a B cell that recognizes a thymus-dependent antigen for activation to take place.
2. Contrast the activation of B cells that recognize TI-1 antigens with those that recognize TI-2 antigens.
3. What signals are provided by helper T cells to activate B cells recognizing thymus-dependent antigens?

Making Connections: Thymus-Independent and Thymus-Dependent B-Cell Responses

B cells each express a receptor that recognizes a specific antigen. B-cell activation is triggered when the receptor engages its specific antigen. This activation may or may not require assistance from helper T cells. Thymus-independent antigens trigger B-cell activation without needing help from T cells, whereas thymus-dependent antigens require T-cell help. B cells reside in locations where they are most likely to come into contact with the specific antigen they recognize.

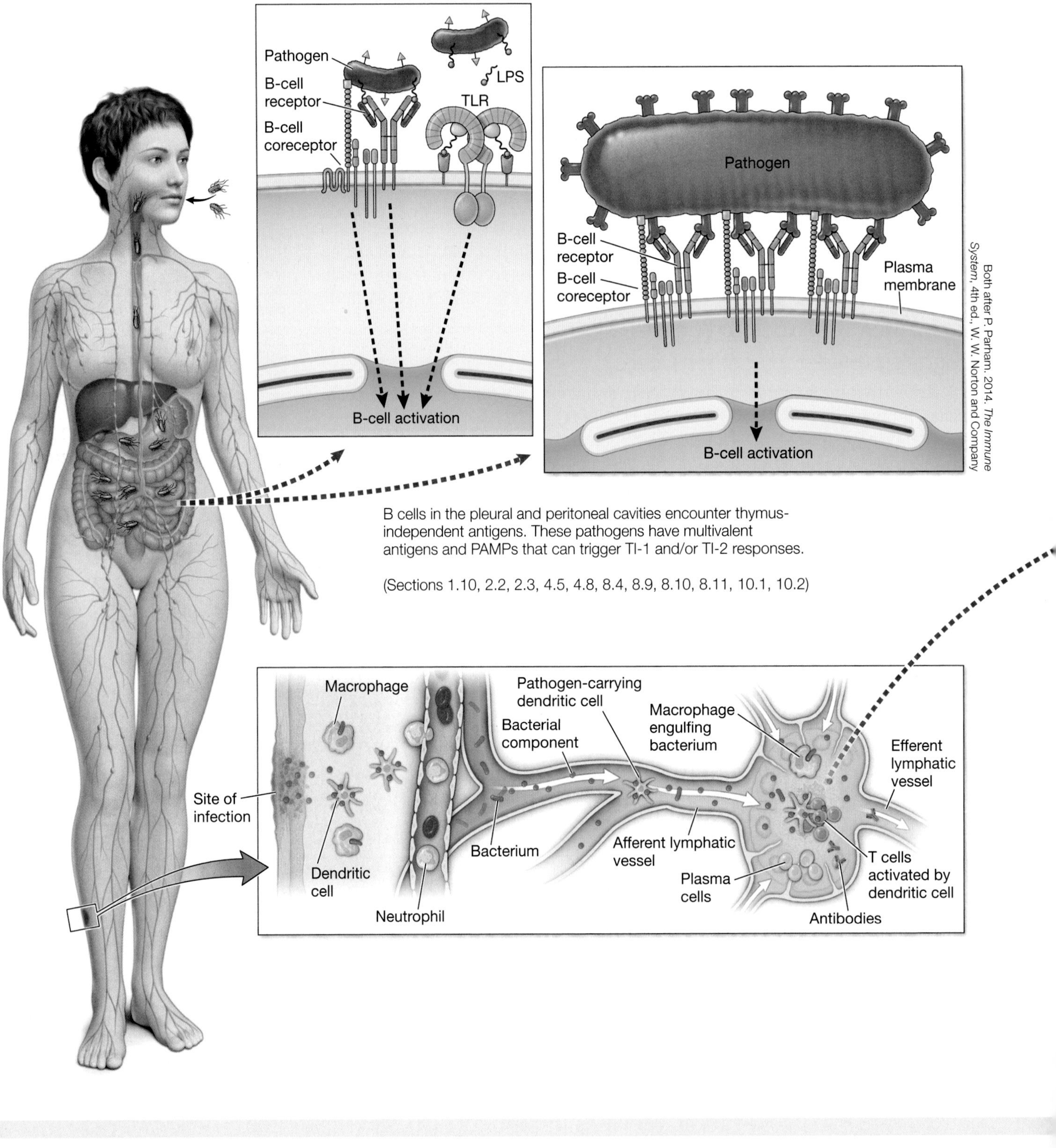

Both after P. Parham. 2014. *The Immune System*, 4th ed., W. W. Norton and Company

B cells in the pleural and peritoneal cavities encounter thymus-independent antigens. These pathogens have multivalent antigens and PAMPs that can trigger TI-1 and/or TI-2 responses.

(Sections 1.10, 2.2, 2.3, 4.5, 4.8, 8.4, 8.9, 8.10, 8.11, 10.1, 10.2)

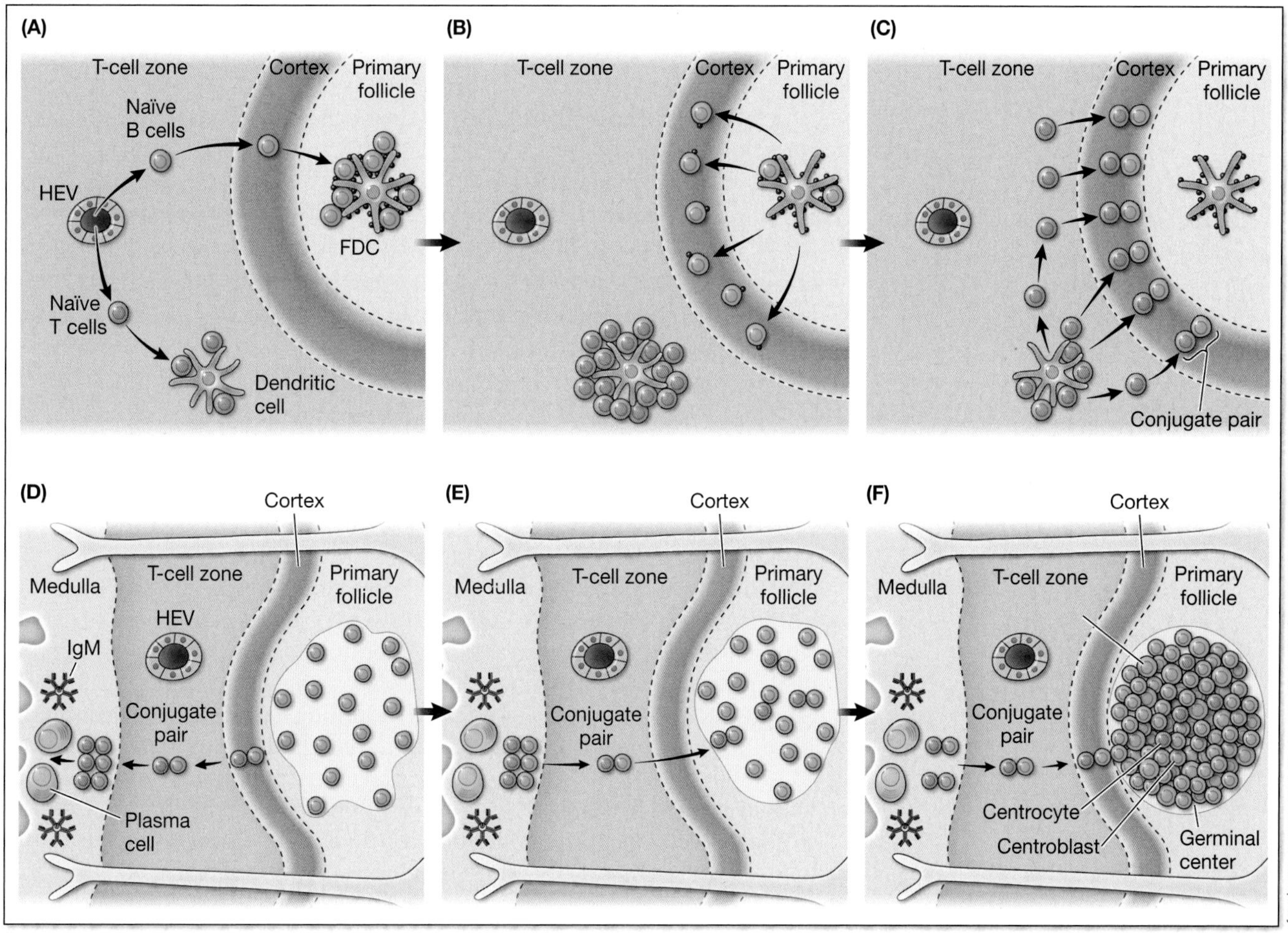

After P. Parham. 2014. *The Immune System*, 4th ed., W. W. Norton and Company

B cells in secondary lymphoid tissue encounter thymus-dependent antigens. These antigens require assistance from helper T cells to trigger B-cell activation.

(Sections 1.10, 2.2, 2.3, 4.5, 4.8, 8.4, 8.9, 8.10, 8.11, 9.2, 9.4, 9.5, 10.1, 10.2)

10.3 | How do B cells migrate and behave in secondary lymphoid tissues?

LEARNING OBJECTIVE

10.3.1 Describe how B cells migrate into secondary lymphoid tissues and their two possible fates once there.

B-cell activation is concentrated in secondary lymphoid tissue due to the diversity seen in immunoglobulins in naïve B cells. To efficiently activate an adaptive humoral immune response, B cells circulate in the circulatory and lymphatic systems and enter and exit secondary lymphoid tissues to monitor antigens. B cells migrate into secondary lymphoid tissue in the same manner that T cells migrate into lymphoid tissue.

Chemokines such as CCL19 and CCL21 drive B-cell movement into secondary lymphoid tissues. They also stimulate the migration of B cells into the T-cell zone of the lymphoid tissue, where B cells monitor for the presence of their antigens. In the absence of their specific antigen, B cells migrate to the B-cell zone through chemokine secretion by follicular dendritic cells. Here, B cells receive survival signals, including BAFF, before they exit the secondary lymphoid tissue and continue circulating.

If antigen recognition occurs, and the B cell forms a conjugate pair with a T cell in the T-cell zone of the secondary lymphoid tissue, two fates can occur (see Figure 10.8):

- The conjugate pair leaves the T-cell zone and migrates to the medulla, where B cells can differentiate into plasma cells.
- The conjugate pair first migrates to the medullary cords and then further to the lymphoid follicles, where B cells can differentiate into centrocytes.

The primary focus

Upon formation of a conjugate pair, the paired cells migrate into the medullary cords and form a **primary focus**, in which both cells are activated and undergo clonal expansion due to signaling events discussed earlier in this chapter. The B cell within the primary focus differentiates into a plasma cell and begins to express and secrete soluble IgM (**FIGURE 10.9**). IgM produced during clonal expansion in the primary focus can leave the lymphoid tissue through the efferent lymphatic system and circulate, where it binds to antigen and works to subdue it.

Although activation of B cells is vital in mounting a robust adaptive immune response, downregulation of B-cell proliferation is crucial to ensure that unregulated growth does not occur. One major player in downregulation is the transcriptional repressor **BLIMP-1** (B-lymphocyte-induced maturation protein), which alters B-cell function by inhibiting cell division and promoting the immunoglobulin expression and secretion required of plasma cells.

The secondary focus

Some conjugate pairs migrate to the lymphoid follicle, where signals from follicular dendritic cells and the T cell of the conjugate pair continue to drive B-cell activation and proliferation. The conjugate pair becomes a **secondary focus**, and signals provided to the B cell drive rapid division and differentiation into *centrocytes*. The rapidly dividing B cells and T cells make up the *germinal center* within the lymphoid follicle, and the rapid division of cells within the germinal center causes lymphoid tissue swelling, characteristic of an active adaptive immune response.

The signals not only drive B-cell division but also induce the expression of activation-induced cytidine deaminase (AID) to activate somatic hypermutation and isotype switching. We have previously discussed the mechanism of AID in the induction of somatic hypermutation and isotype switching (see Chapter 9).

primary focus Cluster of activated B cells and T cells that is induced in germinal centers and is important in an adaptive immune response.

BLIMP-1 B-cell transcription factor responsible for downregulating B-cell division and promoting plasma cell differentiation by inducing immunoglobulin expression.

secondary focus Conjugate pair of B cell and T cell that has migrated from the T-cell zone in a secondary lymphoid tissue to the lymphoid follicle for further B-cell differentiation.

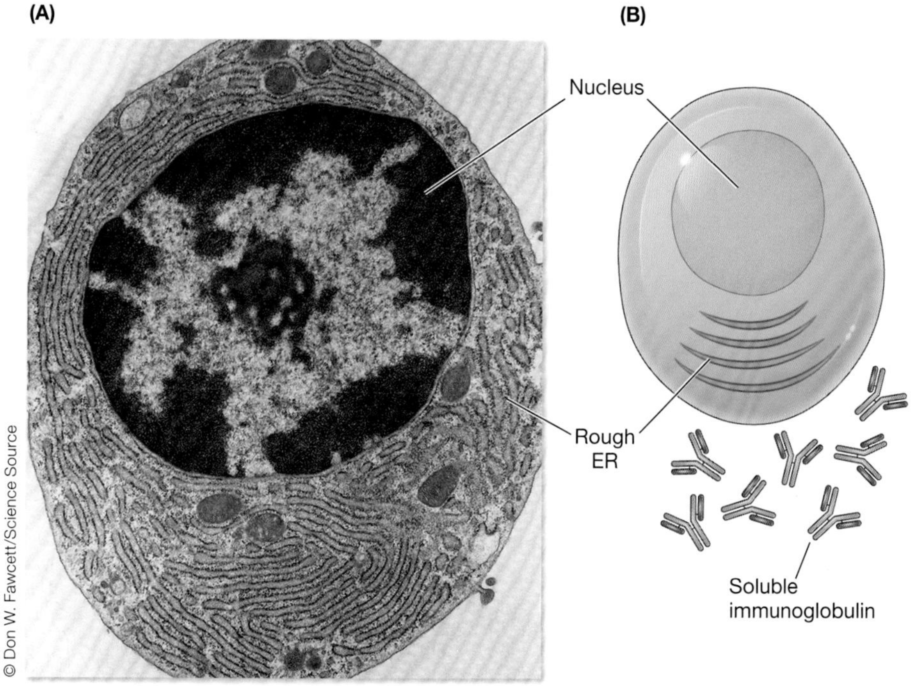

FIGURE 10.9 Plasma cells as immunoglobulin factories (A) An electron micrograph of a plasma cell shows the elaborate network of rough endoplasmic reticulum (ER) needed for the secretion of soluble immunoglobulins. (B) Plasma cells differentiate from B cells after activation; they specialize in the secretion of soluble immunoglobulins, which perform the effector functions of the humoral arm of the adaptive immune response. (After F. W. Alt et al. 2015. *Molecular Biology of B Cells*, 2nd ed., pp. 227–249. Academic Press: London.)

We will next explore the mechanisms that initiate these processes, the means by which proper isotype switching occurs, and the means by which B cells that have undergone somatic hypermutation continue to bind tightly to their antigen.

While some centrocytes undergo differentiation into plasma cells to express immunoglobulins, others differentiate into memory cells, which play a role in a secondary adaptive immune response. These long-lived memory B cells promote stronger and faster recognition of pathogens that were previously encountered.

CHECKPOINT QUESTION

1. Compare the differences seen in B-cell differentiation in a primary focus with those of a secondary focus.

10.4 How is immunoglobulin affinity maturation driven?

LEARNING OBJECTIVE

10.4.1 Explain the processes of affinity maturation and isotype switching that occur in germinal centers within secondary lymphoid tissues.

B-cell migration to the lymphoid follicle in a secondary lymphoid tissue and formation of a germinal center is crucial to inducing a robust humoral adaptive immune response. The centrocytes that differentiate as a result of B-cell activation within the germinal center are paramount to the formation of functional antibodies, tight antigen binding, and the formation of memory cells. We know that both somatic hypermutation (which drives affinity maturation) and isotype switching are initiated by the action of AID and DNA repair mechanisms that produce permanent changes in immunoglobulin genes (see Chapter 9). Random mutations in the variable regions can alter the antigen-binding site of the immunoglobulin to drive higher affinity for antigen. DNA recombination events can switch the constant region of the heavy chain used in immunoglobulin expression to promote immunoglobulin effector function. The B-cell maturation process within the germinal center requires another round of selection to ensure that the cell's immunoglobulin is still capable of recognizing antigen and does not recognize self-antigen.

Affinity maturation

A major differentiation event that occurs within the germinal centers is driven by the expression and action of AID. Somatic hypermutation, which is driven by the activity of AID through cytosine-to-uracil transitions, can alter the binding affinity of the originally expressed immunoglobulin on the activated B cell. This altered binding affinity only occurs if the deamination of cytosine bases and subsequent DNA repair mechanisms change the original cytosine base to another base, resulting in mutation of the original immunoglobulin gene. Since the activity of AID is random toward cytosine bases in the variable regions of the immunoglobulin genes (although isolated mutations are focused in the hypervariable regions, presumably due to the positive selection of these B-cell clones), the base changes can have a positive or negative effect on immunoglobulin affinity for its antigen. T cells, follicular dendritic cells, and macrophages within the germinal center, which all present antigens to centrocytes, promote additional rounds of positive and negative B-cell selection that have undergone somatic hypermutation (**FIGURE 10.10**).

POSITIVE SELECTION IN GERMINAL CENTERS Just as developing B cells undergoing somatic recombination need to test for proper immunoglobulin expression and function within bone marrow, the potential changes that occur in immunoglobulin genes during somatic hypermutation require additional rounds of testing to ensure proper function.

Centrocytes test the affinity of their immunoglobulins using antigens presented on the surface of follicular dendritic cells located in the germinal center. B cells expressing immunoglobulins with a high enough affinity to interact with antigens on the surface of a follicular dendritic cell can interact with that cell to gain survival signals, including BAFF. These centrocytes process the bound antigen and present it to helper T cells within the germinal center, where they gain further survival signals and ultimately inhibit apoptosis through the expression of Bcl-X_L. Because survival signals such as BAFF are only provided to B cells that can physically interact with follicular dendritic cells and helper T cells, and because these interactions are driven by proper immunoglobulin function, B cells with altered immunoglobulins that are no longer capable of interacting with antigen die by apoptosis (see Figure 10.10).

Only centrocytes with the highest affinity for antigen will successfully compete for interaction with follicular dendritic cells and helper T cells within the germinal center and receive the survival signals secreted by follicular dendritic cells and helper T cells. Thus, B cells that have undergone somatic hypermutation events that have increased the affinity of their immunoglobulin to antigen are best suited to interact with these cells and receive these survival signals.

This selective pressure favoring the tightest immunoglobulin–antigen interactions drives affinity maturation. This process allows centrocytes expressing immunoglobulins with higher affinity to antigen to differentiate into plasma cells that express more suitable immunoglobulins for antigen binding (compared to the original immunoglobulins). It is advantageous for our adaptive immune response to maintain these cells over the long term since their immunoglobulins are best suited to recognize the same pathogen in the future.

Affinity maturation can also promote differentiation into memory cells capable of mounting a quick and selective secondary immune response should the need arise (described in Chapter 11). Some centrocytes that have undergone affinity maturation and have successfully competed for survival signals from follicular dendritic cells and helper T cells may use those same survival signals

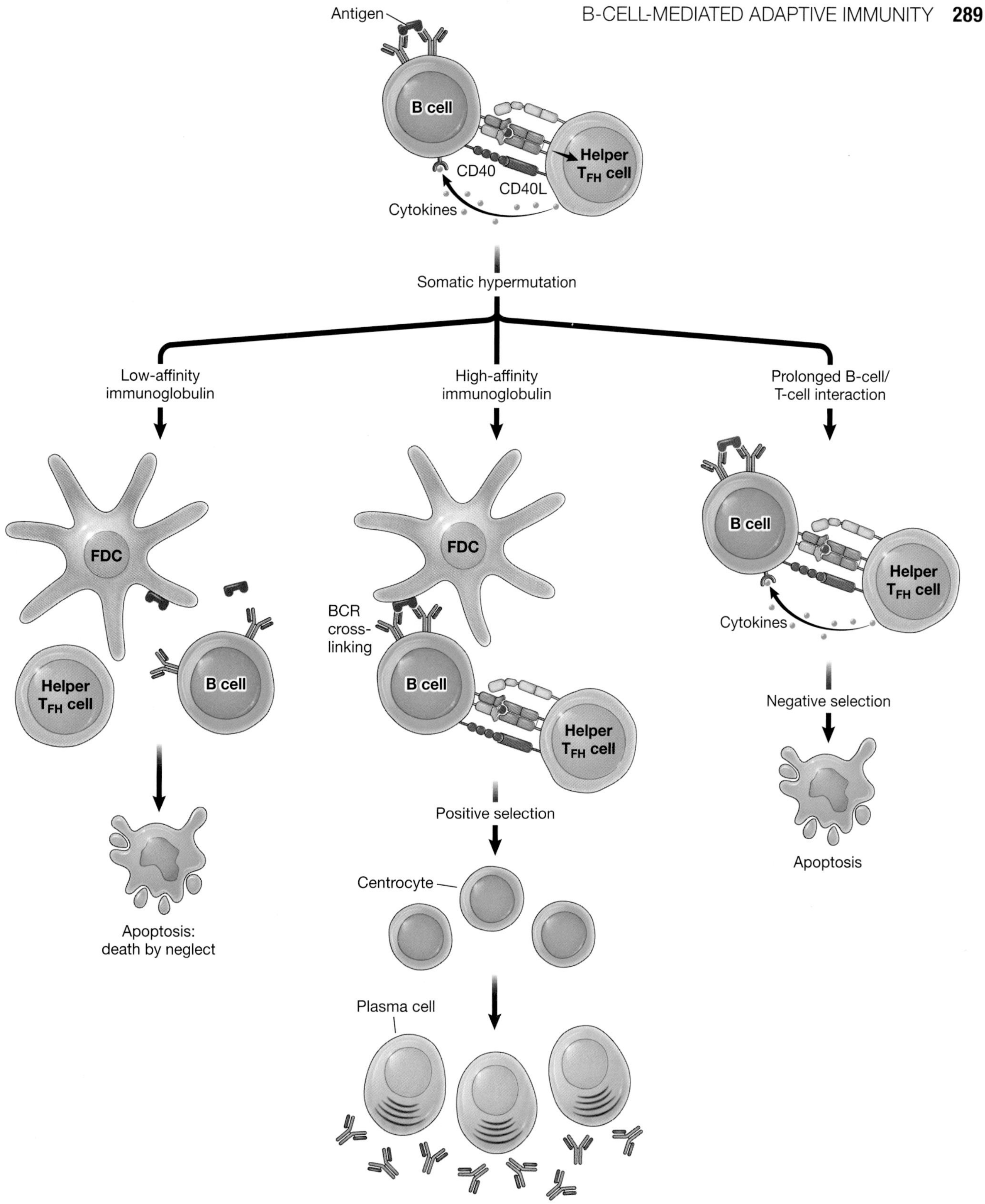

FIGURE 10.10 Positive and negative selection of B cells after somatic hypermutation B cells that interact with follicular helper (T_{FH}) T cells receive signals to activate somatic hypermutation. Mutation of the immunoglobulin loci can produce immunoglobulins with low affinity, high affinity, or self-reactive immunoglobulins. With low-affinity immunoglobulins, the B cell will no longer interact with follicular dendritic cells (FDC) or T_{FH} cells within the germinal center. The B cell does not receive survival signals and undergoes apoptosis due to neglect. With high-affinity immunoglobulins, the B cell interacts with FDCs and T_{FH} cells within the germinal center. Signals prevent apoptosis and promote differentiation into a centrocyte. Some centrocytes further differentiate into plasma cells that produce high-affinity soluble immunoglobulins. If somatic hypermutation produces immunoglobulins that are self-reactive, the B cell has a prolonged interaction with the T_{FH} cell within the germinal center. Negative signals from the T_{FH} cell drive negative selection of the potentially self-reactive B cell, resulting in apoptosis. (Based in part on P. Parham. 2014. *The Immune System*, 4th ed., W. W. Norton and Company.)

to differentiate to memory B cells. Recent evidence suggests that immune complexes in germinal centers allow for the generation of memory B cells through secretion of BAFF from follicular dendritic cells (see **EMERGING SCIENCE**).

EMERGING SCIENCE

How are centrocytes triggered to differentiate into memory B cells?

Article

S. Kang, A. B. Keener, S. Z. Jones, R. J. Benschop, A. Caro-Maldonado, J. C. Rathmell, S. H. Clarke, G. K. Matsushima, J. K. Whitmire, and B. J. Vilen. 2016. IgG-immune complexes promote B cell memory by inducing BAFF. *Journal of Immunology* 196: 196–206.

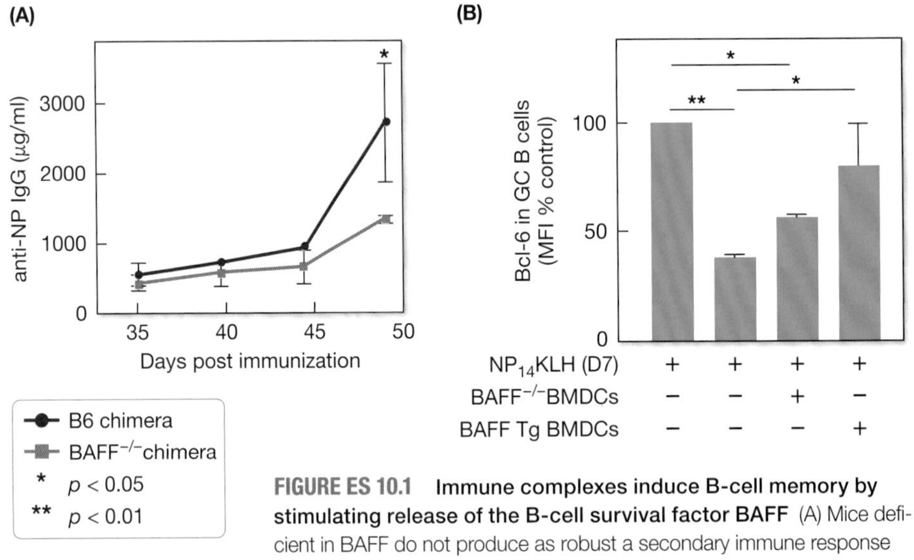

FIGURE ES 10.1 Immune complexes induce B-cell memory by stimulating release of the B-cell survival factor BAFF (A) Mice deficient in BAFF do not produce as robust a secondary immune response compared to wild-type mice when boosted with NP$_{14}$KLH. (B) Low Bcl-6 expression in CD16-deficient mice could be rescued when these mice were injected with BAFF-expressing bone marrow-derived dendritic cells (BMDCs). This rescue required the presence of BAFF, as BMDCs from BAFF-deficient mice could not rescue low Bcl-6 expression. n.s. = not significant, * $p < 0.05$, ** $p < 0.01$. (From S. Kang et al. 2016. *J Immunol* 196: 196–206. © 2016. The American Association of Immunologists, Inc.)

Background

We have seen that B cells that have undergone clonal selection and have formed a germinal center within a secondary lymphoid tissue have one of two major fates—plasma cells or memory B cells. Although it is well known that B cells within the germinal center are capable of differentiating into these two effector cells, the signals that steer the differentiation of activated centrocytes down each of these pathways are not fully established. Deciphering the roles of all of the cells involved in B-cell activation and the signals provided by follicular helper T cells and follicular dendritic cells would allow for a better understanding of the molecular signals that facilitate B-cell differentiation during activation and clonal expansion.

The Study

Researchers embarked on the identification of key signaling molecules involved in the generation of memory B cells. Since germinal centers in vivo have a variety of cells that play a role in B-cell activation and proper differentiation of centrocytes into plasma cells or memory cells, the researchers had to analyze signals that could be derived from the actions of each of these cells that had the potential to help drive memory B cells to a specific antigen.

A key hint in their studies came in the analysis of mice incapable of expressing BAFF. Mice lacking BAFF demonstrated a lower secondary immune response as well as defects in germinal center B cells and memory B cells. The researchers went on to demonstrate that the BAFF is secreted by dendritic cells in response to binding to IgG-immune complexes through the FcγR receptor CD16. They also demonstrated that secreted BAFF prompts memory B-cell formation through the activation of Bcl-6 expression (**FIGURE ES 10.1**). These data demonstrate a molecular mechanism that allows dendritic cells to promote the formation of memory B cells within a germinal center.

Think About...

1. How could you identify any additional molecules involved in the BAFF-dependent activation of Bcl-6 within B cells that are differentiating into memory B cells?
2. If you were able to express Bcl-6 constitutively in B cells in culture, do you think that they would differentiate into memory B cells within that culture? Why or why not?
3. Design an experiment that demonstrates that BAFF directly interacts with B cells to induce the expression of Bcl-6 rather than indirectly acting on another cell population within germinal centers. What are some important controls you would use in this experiment?

NEGATIVE SELECTION IN GERMINAL CENTERS Not only can somatic hypermutation increase the affinity of an immunoglobulin for its antigen (or not change the affinity if mutations are properly repaired), but a mutated immunoglobulin can also change the affinity of the immunoglobulin to that of recognizing self-antigen. This change could result in the development of B cells that can cause autoimmunity, so somatic hypermutation must also provide a means of negative selection to ensure that such self-reactivity does not occur.

Self-reactive B cells within the germinal center that arise due to somatic hypermutation and that interact with T_H cells for a prolonged period of time represent potentially autoreactive B cells, which are removed via apoptosis or are rendered inactive or anergic in a manner similar to the onset of anergy in developing B cells within bone marrow (see **KEY DISCOVERIES**). These mechanisms help ensure that B cells that are potentially autoreactive due to changes in immunoglobulin affinity (driven by somatic hypermutation) are not permitted to enter the circulation or become activated.

KEY DISCOVERIES

How do we know that negative selection of B cells occurs in germinal centers after somatic hypermutation?

Articles

S. Han, B. Zheng, J. Dal Porto, and G. Kelsoe. 1995. In situ studies of the primary immune response to (4-hydroxy-3-nitrophenyl) acetyl IV. Affinity-dependent, antigen-driven B cell apoptosis in germinal centers as a mechanism for maintaining self-tolerance. *Journal of Experimental Medicine* 182: 1635–1644.

B. Pulendran, G. Kannourakis, S. Nourl, K. G. C. Smith, and G. J. V. Nossal. 1995. Soluble antigen can cause enhanced apoptosis of germinal-centre B cells. *Nature* 375: 331–334.

K. M. Shokat and C. C. Goodnow. 1995. Antigen-induced B-cell death and elimination during germinal-centre immune responses. *Nature* 375: 334–338.

Background

We have seen that affinity maturation and isotype switching are key processes driven in germinal center B cells to aid in immunoglobulin effector function. Because somatic hypermutation is a random event, as random as V(D)J recombination during B-cell development, one can imagine that a mutation in immunoglobulins can alter the antigen specificity, resulting in an immunoglobulin capable of binding and responding to a self-antigen. Such autoimmunity is negatively selected for during V(D)J recombination, so it was reasonable to assume that another round of negative selection must occur in germinal centers to prevent an autoimmune disorder.

Early Research

At the time of these studies in 1995, it was known that somatic hypermutation and isotype switching, driven by AID activity, occurred within B cells in germinal centers. Because of the random nature of somatic hypermutation, it had been hypothesized that B cells that became self-reactive during the random process of somatic hypermutation would have to be selectively removed to prevent autoimmunity. However, the question remained: Does negative selection of B cells occur in germinal centers, and if so, what is the mechanism of autoreactive B-cell removal?

Think About...

1. Given the processes of negative selection of self-reactive T cells and B cells during their development in primary lymphoid tissue, what might be a hypothesis concerning the mechanism of negative selection of B cells in germinal centers?
2. Design several experiments that would aid in testing your hypothesis concerning the mechanism of negative selection of B cells.
3. If you determined that autoreactive B cells were eliminated from germinal centers via apoptosis, design experiments that could distinguish between apoptosis induction due to lack of a survival signal versus presence of a pro-apoptotic signal.

Articles Summary

Each of the three studies provided separate pieces of evidence that all reached the same conclusion—high-affinity recognition of antigen by B cells in germinal centers drives apoptosis within these potentially autoreactive B cells to

(*Continued*)

ensure that negative selection occurs after somatic hypermutation (FIGURE KD 10.1). Although the methodology of the three studies differs, each study was paramount in demonstrating the common mechanism of negative selection of B cells in germinal centers to prevent induction of autoimmunity that could result after somatic hypermutation of immunoglobulins.

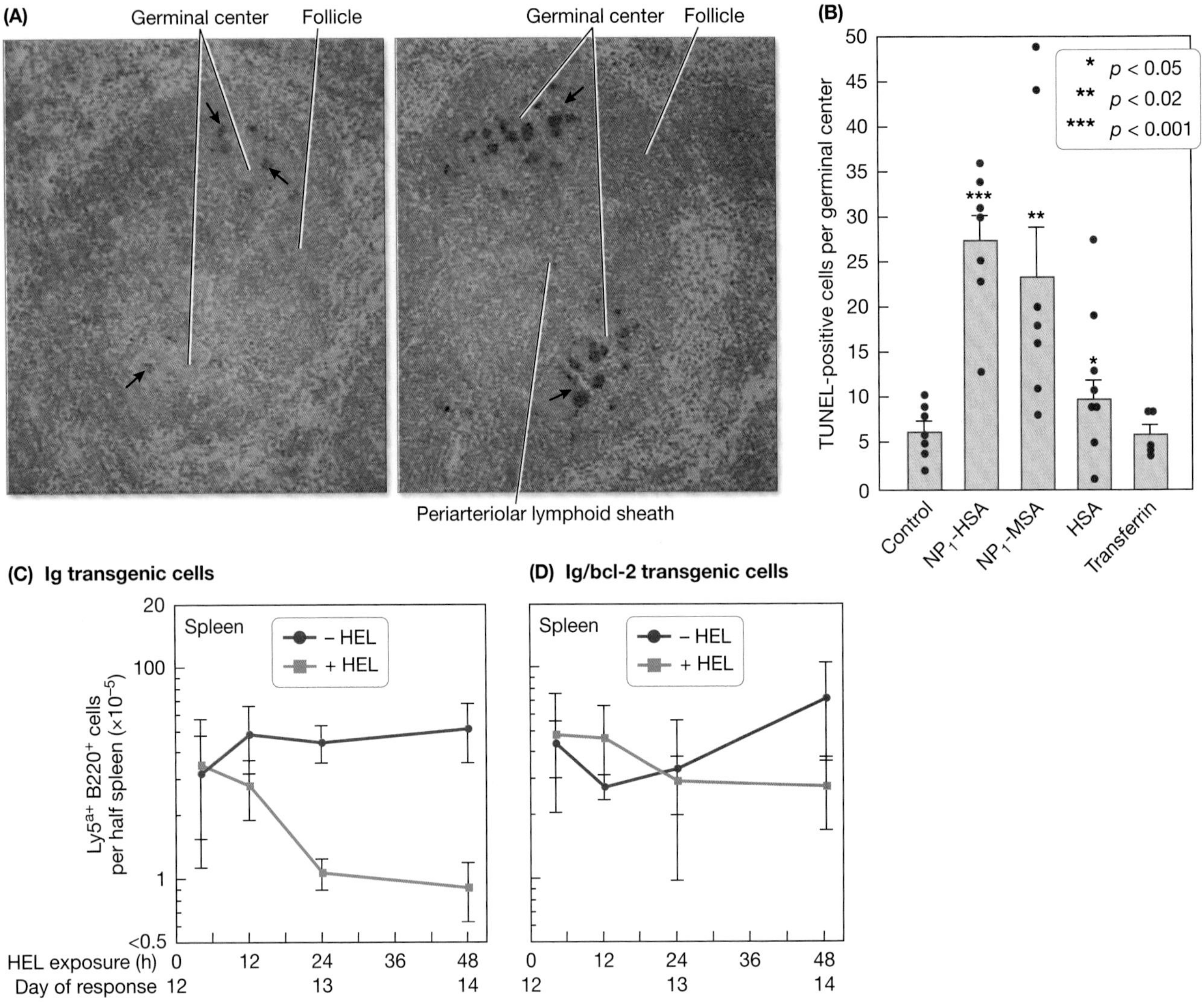

FIGURE KD 10.1 Self-peptides induce apoptosis in germinal centers (A) The spleens from mice injected with self-peptide (right) contain a higher number of apoptotic cells (arrows indicate stained apoptotic cells) within spleen germinal centers compared to mice injected with a control peptide (left). (B) Soluble antigen induces greater apoptosis in germinal center B cells. Injection of soluble NP_1-HSA or NP_1-MSA increases the number of apoptotic cells in germinal centers, as measured by TUNEL assay, in comparison to the irrelevant antigens HSA and transferrin. NP_1, (4-hydroxy-3-nitrophenyl)acetyl; HSA, human serum albumin; MSA, mouse serum albumin; * $p < 0.05$, ** $p < 0.02$, *** $p < 0.001$. (C) Self-antigen treatment lowers the number of B cells present in mice. Mice containing lysozyme-specific B cells (Ig transgenic cells) were injected or not injected with hen egg lysozyme (HEL), and the number of B cells were enumerated. Treatment of mice with HEL lowered the number of B cells isolated after 24 hours. (D) Loss of B cells after self-antigen treatment is driven by apoptosis. Addition of the anti-apoptosis gene *bcl-2* in the transgenic B cells prevented a reduction in the number of B cells after treatment with self-antigen, demonstrating that B-cell loss was driven by apoptosis. (A, © 1995 S. Han et al. Originally published in *J Exp Med* doi.org/10.1084/jem.182.6.1635; B, from B. Pulendran et al. 1995. *Nature* 375: 331–334; C,D, from K. M. Shokat and C. C. Goodnow. 1995. *Nature* 375: 334–338.)

Isotype switching

Not only does AID activity within centrocytes drive somatic hypermutation, but cytosine deamination in switch regions of immunoglobulin heavy chain loci can drive isotype switching. This allows activated centrocytes to express immunoglobulins with specific effector functions that contribute to an efficient humoral response that matches the need of the adaptive immune response.

Recall that AID activity requires single-stranded DNA as a substrate, and AID only acts on DNA regions that are actively undergoing transcription (or replication). How is proper isotype switching, which allows the correct immunoglobulin effector function to be selected in the switched immunoglobulin gene, driven to the proper regions of the immunoglobulin gene? The answer lies in cytokines secreted by helper T cells that assist in the activation and differentiation of centrocytes within the germinal center. Cytokines secreted by these T_H cells, including IL-4, IL-5, and TGF-β, are capable of activating transcription at specific switch regions within the heavy chain locus (**TABLE 10.2**). This directed transcription at certain switch regions allows AID to specifically target cytosine bases within these regions and direct proper isotype switching.

TABLE 10.2 | Cytokines Secreted by Helper T Cells

Cytokine	IgM	IgG1	IgG2	IgG3	IgG4	IgA	IgE
IL-4	–	++		–	++		++
IL-5						++	
IFN-γ	–	–		++			–
TGF-β	–			–		++	

Source: J. Stavnezer. 2000. Molecular processes that regulate class switching. *Curr Top Microbiol Immunol* 245: 127–168.

CHECKPOINT QUESTIONS

1. Why is it important that centrocytes undergo a second round of positive and negative selection within germinal centers?
2. How does the activity of helper T cells drive isotype switching?

10.5 | What are the properties and functions of the different antibody isotypes?

LEARNING OBJECTIVE

10.5.1 Describe the functions of the various antibody isotypes involved in the humoral adaptive immune response.

The main function of B cells is to produce soluble immunoglobulins with specific effector functions that target and remove pathogens. While higher eukaryotic organisms such as fish, birds, mice, and humans employ a slightly different repertoire of antibody isotypes to carry out this charge of the humoral adaptive immune response (see **EVOLUTION AND IMMUNITY**), they rely on several major immunoglobulin functions as mechanisms of fighting pathogen infection. These mechanisms include *neutralization*, *protection of internal tissues*, *activation of innate immune cells*, *complement activation*, and *clearance of small immune complexes* (**TABLE 10.3**).

Neutralizing antibodies

A large number of pathogens and toxic pathogen products interact with target cells during the course of an infection. For example, viral particles such as the influenza virus need to bind to the surface of a target cell using adhesion molecules to promote viral fusion with the cellular membrane and delivery of viral genomes and key viral proteins into the cell cytoplasm. Therefore, inhibiting this interaction is an efficient means of preventing toxin action and blocking pathogen entry into cells.

neutralizing antibody Antibody that can block pathogen or toxin interaction with a target cell.

Neutralizing antibodies are capable of interacting with both soluble toxins and surface molecules on pathogens to prevent them from interacting with target cells (**FIGURE 10.11**). These antibodies ultimately protect against pathogen infection (usually blocking the life cycle or infection cycle of the pathogen). Both IgG and IgA antibodies act as neutralizing antibodies against pathogens and toxins at various tissue surfaces.

TABLE 10.3 | Immunoglobulin Isotypes/Subtypes and Their Role in Major Antibody Functions

Effector Function	IgM	IgG	IgA	IgE
Neutralization	+	+++	+++	–
Protection of Internal Tissues	+	+	+++	–
Innate Immune Cell Activation	–	++	–	+++
Complement Activation	+++	+++	+	–
Small Immune Complex Clearance	–	+++	–	–

Source: H. W. Schroeder and L. Cavacini. 2010. *J Allergy Clin Immunol* 125: S41–S52.

EVOLUTION AND IMMUNITY

Immunoglobulins

The evolution of immunoglobulins in organisms occurred during the evolution of jawed vertebrates. The incorporation of immunoglobulins into the tree of life beginning with jawed vertebrates demonstrated a large diversity in the presence of different isotypes within organisms capable of expressing immunoglobulins. As we have seen, higher eukaryotic mammals express the immunoglobulin isotypes IgM, IgD, IgA, IgG, and IgE, which all bear different effector functions.

Interestingly, there are differences within immunoglobulin isotypes in organisms that have evolved B-cell signaling. Cartilaginous fish express three immunoglobulin isotypes—IgM, IgD, and IgNAR—with IgM playing the major immunoglobulin role and being present in both a monomeric and pentameric form within these species. With the evolution of teleost fish, we see the emergence of another immunoglobulin isotype, IgT/Z, which behaves differently at the protein level (monomer vs. multimer) depending on the tissue in which it resides.

We continue to see isotype divergence in amphibian species, with the presence of IgY, IgX, and IgF in addition to IgM and IgD, with IgX having a similar role as IgA in protection of mucosal surfaces. Within reptiles, we see four isotypes (IgM, IgD, IgY, and IgA-like), and thus the emergence of an IgA homolog. Avian species also express immunoglobulins, but only three isotypes: IgM, IgA, and IgY. With the diversity of different isotypes present within higher vertebrate species, we can gain an appreciation of the similarity of necessary effector functions that immunoglobulins possess to properly maintain humoral immunity.

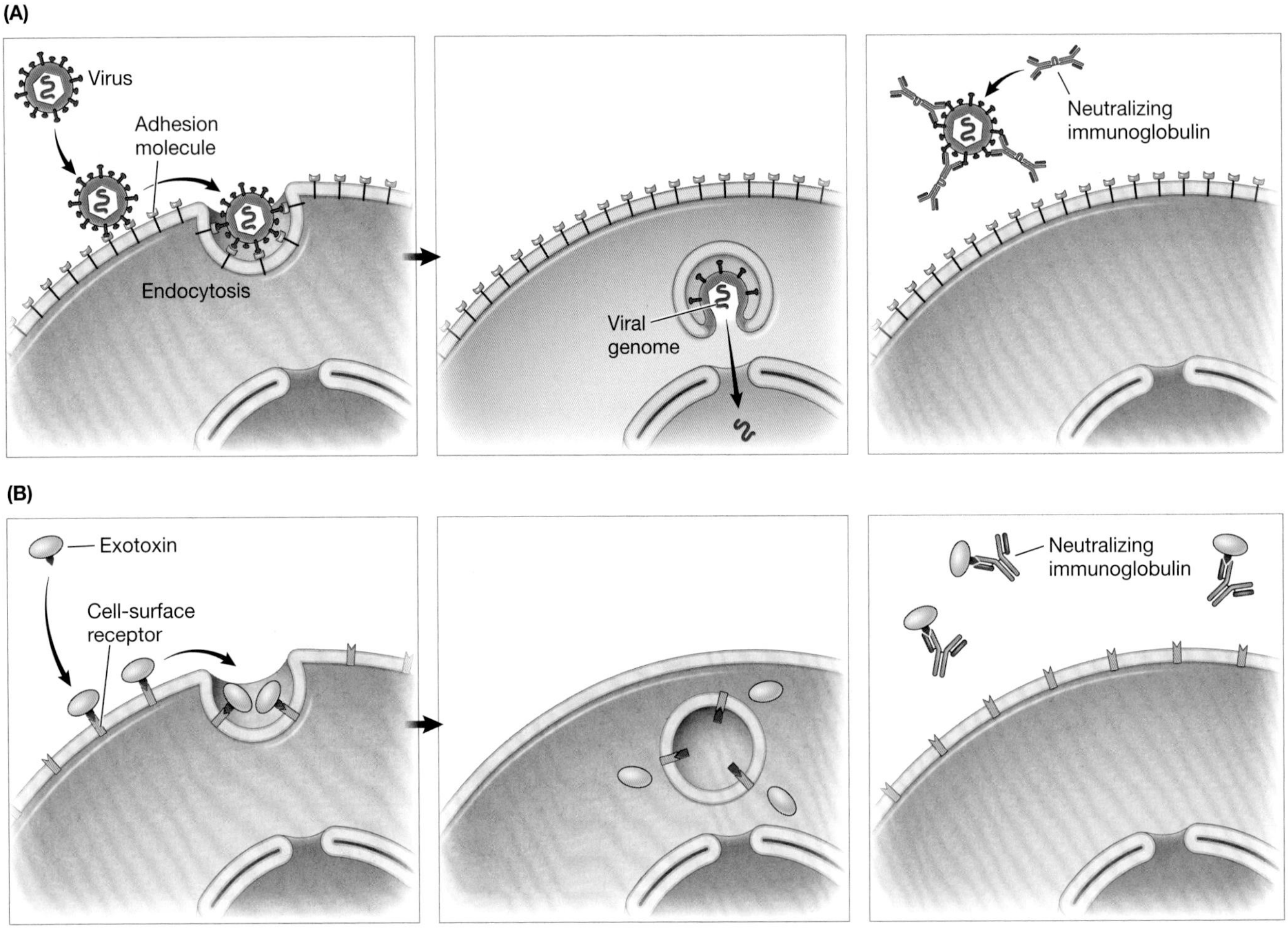

FIGURE 10.11 Action of neutralizing immunoglobulins (A) Viruses use surface proteins to bind to target cells within the host. Upon binding, the cells engulf the viral particles (endocytosis), allowing the viral genome to be incorporated into the cell. Neutralizing immunoglobulins bind to viral surface proteins and prevent their interaction with molecules in the cell membrane, blocking the life cycle of the virus. (B) Exotoxins, commonly secreted by pathogenic bacteria, disrupt normal cellular function. After binding to a target cell molecule, endocytosis delivers the toxin into the cytosol where it can exert its effect. Neutralizing immunoglobulins block exotoxin action by binding and preventing its interaction with the target cell molecule. (After P. Parham. 2014. *The Immune System*, 4th ed., W. W. Norton and Company.)

Making Connections: B-Cell Signaling and Activation

The activation of many B cells occurs in secondary lymphoid tissue with assistance from helper T cells functioning in the cellular arm of the adaptive immune response. Signals provided by the helper T cell can promote immediate B-cell activation and differentiation into plasma cells, or they can induce somatic hypermutation and isotype switching within the activated B cell to produce immunoglobulins of higher affinity and different effector function.

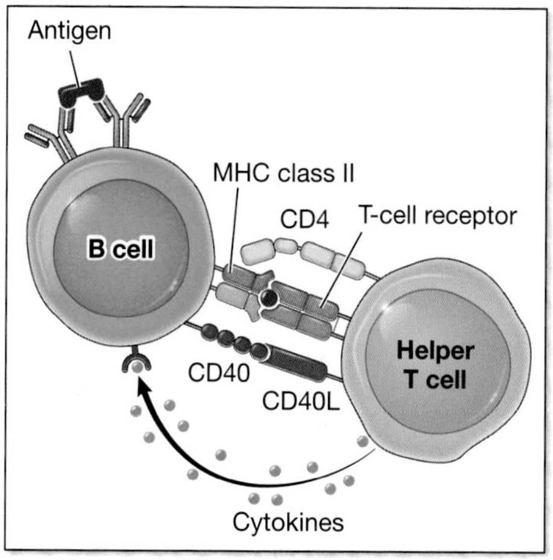

Based in part on J. Punt et al. 2018. *Kuby Immunology*, 8th ed., W. H. Freeman and Company: New York

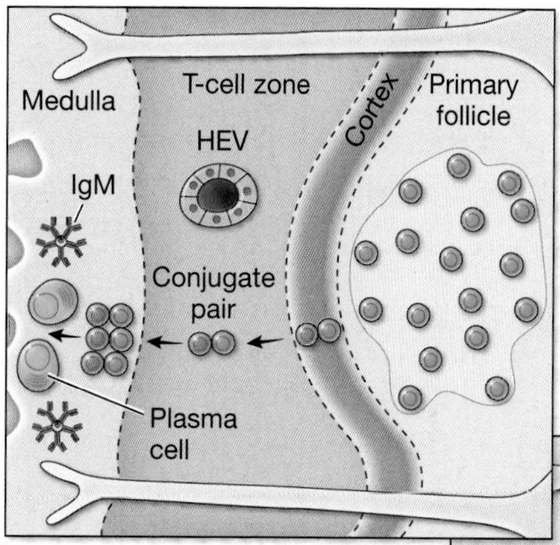

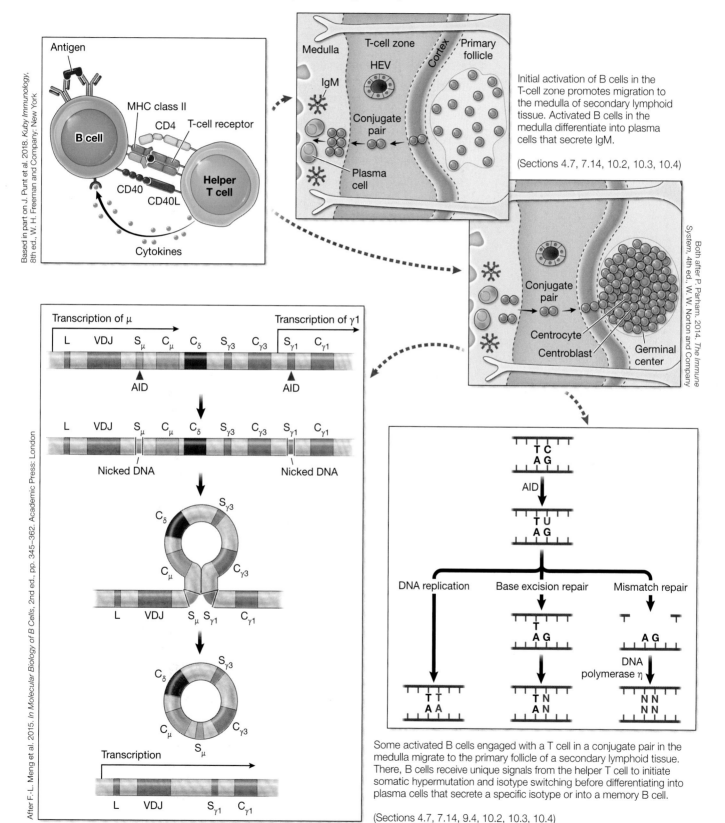

Initial activation of B cells in the T-cell zone promotes migration to the medulla of secondary lymphoid tissue. Activated B cells in the medulla differentiate into plasma cells that secrete IgM.

(Sections 4.7, 7.14, 10.2, 10.3, 10.4)

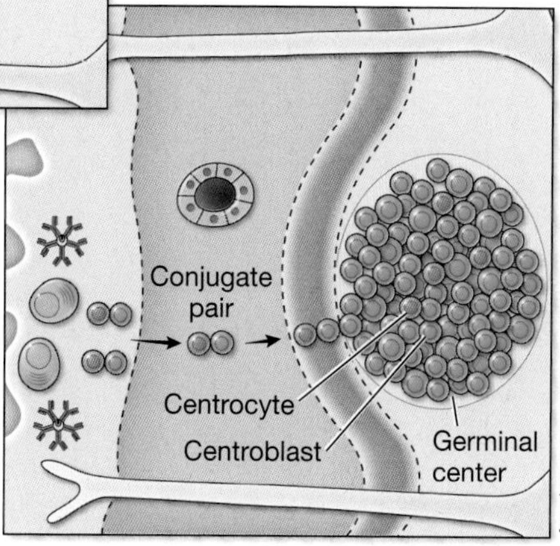

Both after P. Parham. 2014. *The Immune System*, 4th ed., W. W. Norton and Company

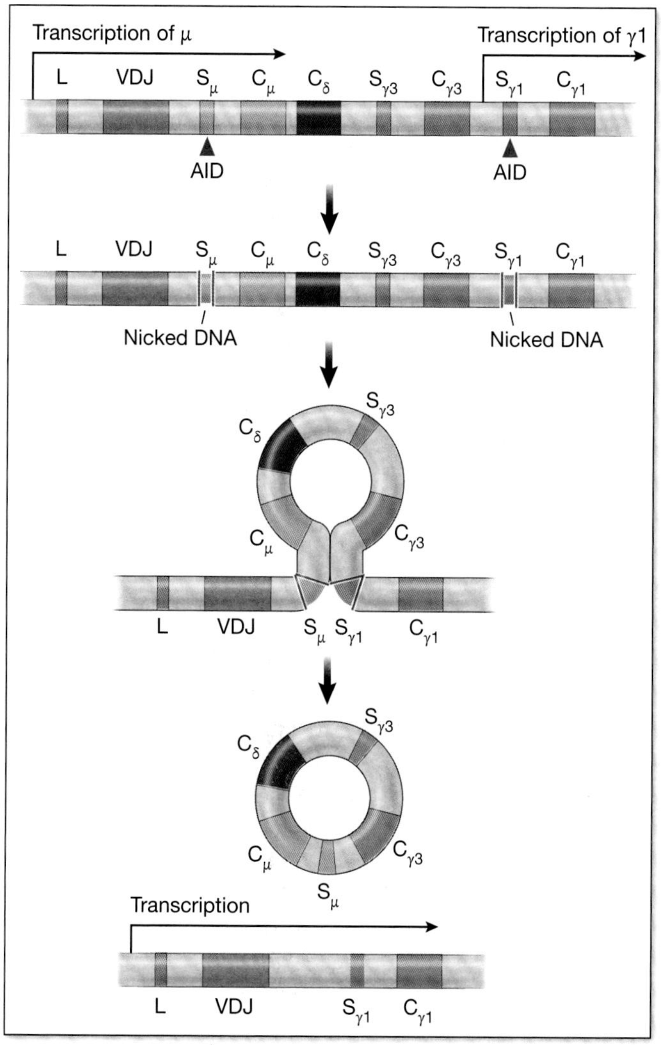

After F.-L. Meng et al. 2015. In *Molecular Biology of B Cells*, 2nd ed., pp. 345–362. Academic Press: London

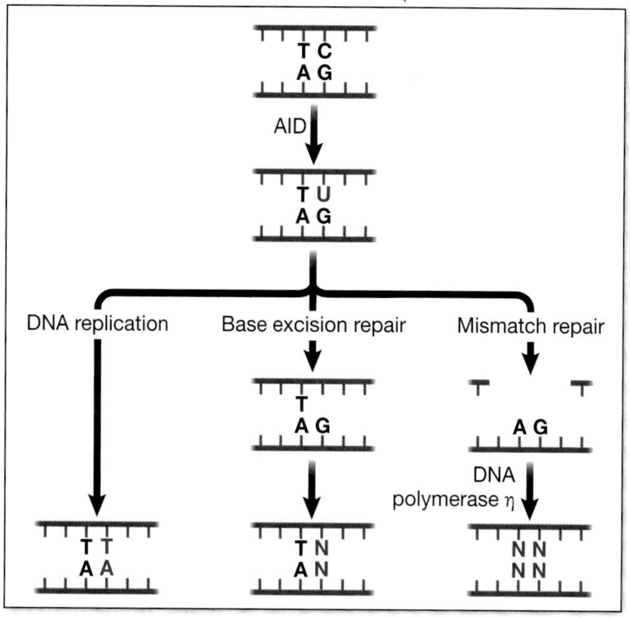

Some activated B cells engaged with a T cell in a conjugate pair in the medulla migrate to the primary follicle of a secondary lymphoid tissue. There, B cells receive unique signals from the helper T cell to initiate somatic hypermutation and isotype switching before differentiating into plasma cells that secrete a specific isotype or into a memory B cell.

(Sections 4.7, 7.14, 9.4, 10.2, 10.3, 10.4)

Fc receptors and innate immune cells

Fc receptors Receptors on the surface of innate immune cells capable of recognizing the Fc component (tail region) of an immunoglobulin.

FcγRI Fc receptor expressed by phagocytes capable of binding IgG that is opsonizing a foreign antigen to facilitate phagocytosis.

inflammatory mediator Molecules secreted by immune cells that induce inflammatory responses to activate innate immune cells and lymphocytes.

histamine Inflammatory mediator that is capable of increasing vascular permeability and inducing localized inflammatory responses in order to facilitate pathogen expulsion.

Recall that B-cell effector function is driven by immunoglobulin isotypes. Isotypes are determined by the constant region of the heavy chains, which is part of the Fc region of the immunoglobulin structure. This region drives effector function due to the presence of **Fc receptors** on the surface of various innate immune cells (**FIGURE 10.12**). Fc receptor binding to immunoglobulins enhances the function of the cells expressing these receptors.

PHAGOCYTIC CELLS Innate immune cells such as macrophages and neutrophils express Fc receptors capable of recognizing IgG that has opsonized the surface of pathogenic organisms. Receptors such as **FcγRI** allow these phagocytic cells to bind to the Fc portion of an immunoglobulin in a manner similar to the way in which complement receptors CR1 and CR2 recognize fixed C3b on the surface of pathogens. Binding of FcγRI receptors to opsonizing immunoglobulins on the pathogen surface facilitates phagocytic uptake by macrophages and neutrophils and removal of the tagged pathogen (see Figure 10.12).

GRANULOCYTES Several innate immune cells, including mast cells, basophils, and eosinophils, work to expel pathogens from the body by activating inflammatory responses and muscle contraction. These mechanisms facilitate the removal of pathogens via forceful expulsion through sneezing, coughing, vomiting, and diarrhea. The specific mechanism engaged depends on the location of the inflammatory response and the muscles involved. These expulsion mechanisms are triggered by the innate immune response via **inflammatory mediators**, such as **histamine**, secreted by granulocytes in response to the appropriate signals.

These granulocytes work in conjunction with the humoral adaptive immune response to generate the inflammatory responses required for pathogen expulsion. Cells such as mast cells express FcεRI receptors, which can bind to IgE immunoglobulins at the cell surface. IgE is typically a secreted, soluble immunoglobulin; however, most IgE in the body is associated tightly with granulocytes expressing FcεRI. Since these receptors bind to the IgE constant regions, a single granulocyte can bind to many different IgE immunoglobulins, each capable of binding a different antigen via its antigen-binding site. Engagement of a surface IgE that is engaged with an FcεRI receptor on a granulocyte

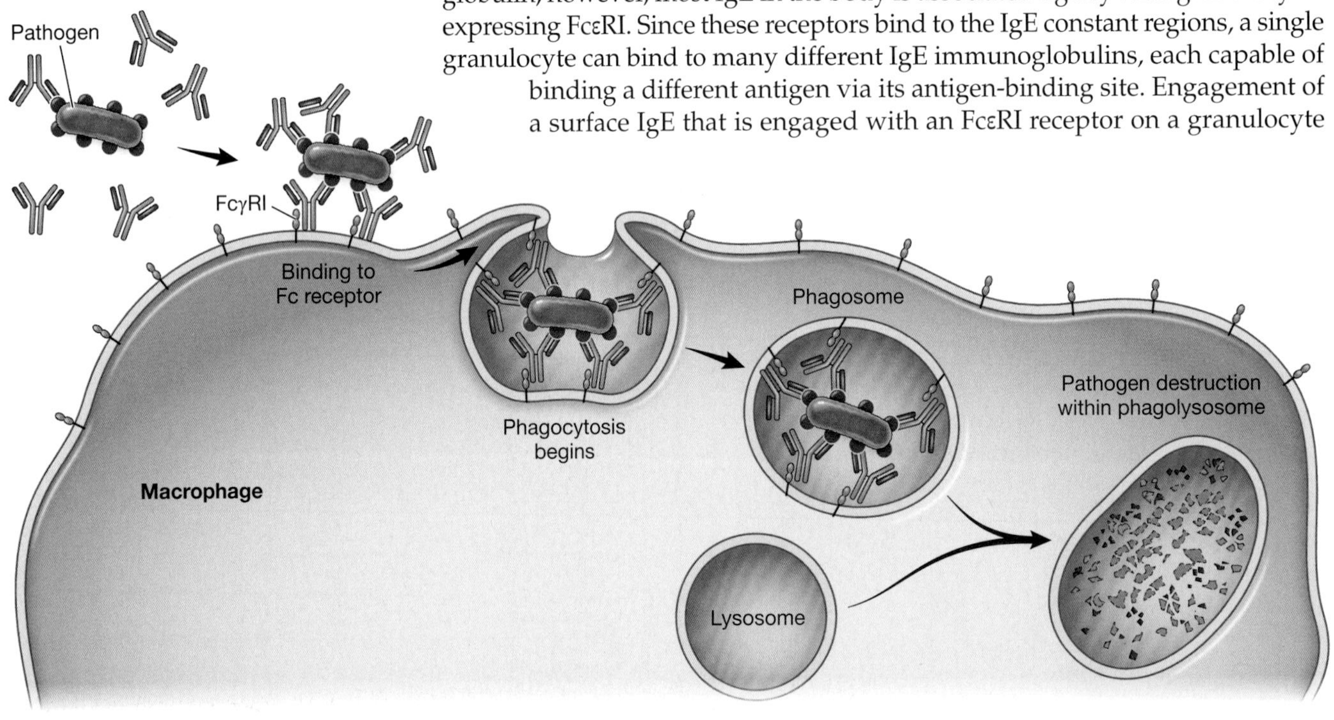

FIGURE 10.12 Fc receptors facilitate phagocytosis Fc receptors on the plasma membrane of professional phagocytic cells such as macrophages bind to the immunoglobulin heavy chain Fc constant region. When soluble immunoglobulin binds to the pathogen surface, it can act as an opsonin to bind to Fc receptors. This binding facilitates phagocytosis because of the clustering of Fc receptors at the cell surface. The pathogen is ultimately destroyed due to degradation within the phagolysosome.

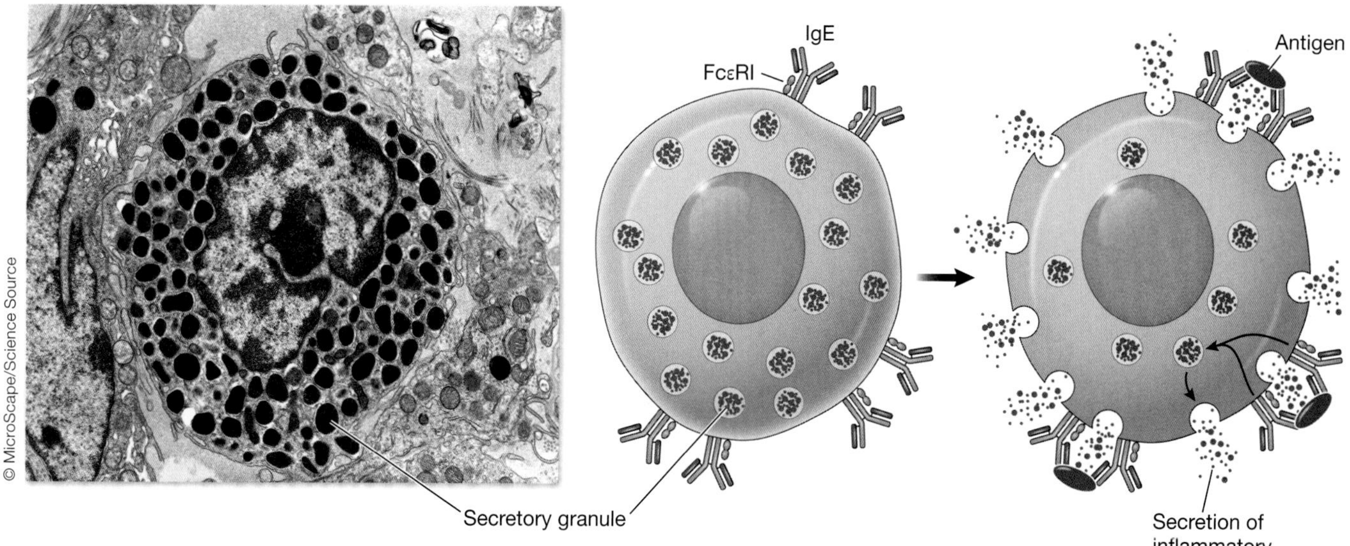

FIGURE 10.13 Granulocytes use Fc receptors for activation and degranulation Granulocytes such as mast cells, basophils, and eosinophils express the FcεRI receptor at their cell surface. FcεRI binds to the Fc portion of IgE and, upon binding of antigen to IgE, the granulocyte is activated. Activation results in the fusion of secretory granules with the plasma membrane and secretion of inflammatory mediators within the extracellular environment.

results in activation of the granulocyte and fusion of secretory granules with the plasma membrane. This results in the delivery of granulocyte "cargo" filled with inflammatory mediators into the extracellular environment (**FIGURE 10.13**).

NK CELLS We know that natural killer (NK) cells play an important role in recognition of intracellular infection. In one sense, NK cells act much like cytotoxic T cells in combating intracellular infections—both cell types employ mechanisms that induce infected cells to undergo apoptosis. A major difference between NK cells and cytotoxic T cells is the means by which they recognize an intracellular infection. While cytotoxic T cells employ their T-cell receptor to recognize peptides processed through the MHC class I pathway (responsible for displaying intracellular peptides), NK cells can use an Fc receptor, FcγRIII, to recognize cells that have been recognized by IgG immunoglobulins.

Virally infected cells have been hijacked by a pathogen that requires a host cell for propagation, which can occur through viral budding from the infected cell surface. Such budding events require the incorporation of viral proteins into the infected cell's plasma membrane, and thus, these viral proteins can serve as a novel antigen for soluble immunoglobulins such as IgG. Recognition of self-cells coated in IgG by FcγRIII on NK cells allows the NK cells to bind to the cells coated in IgG and target these cells for apoptosis through **antibody-dependent cell-mediated cytotoxicity** (**FIGURE 10.14**). Not only is such toxicity important in the body's natural ability to fight intracellular infections, but it has also been used in chemotherapy to treat certain cancers, including B-cell tumors (see Chapter 18).

Protection of internal tissues

Antibodies' effector functions depend on their efficient delivery to surfaces that may come into contact with pathogens or toxins. Antibodies such as IgM (made at the beginning of the humoral response) and IgG and IgA (made after isotype switching) are delivered across endothelial barriers from the bloodstream to extracellular spaces and mucosal surfaces. Once properly delivered, antibodies can protect surfaces as neutralizing antibodies, activate complement, or form small immune complexes.

antibody-dependent cell-mediated cytotoxicity NK cell-directed cytotoxic effector function toward cells that have been bound by antibodies on their cell surface.

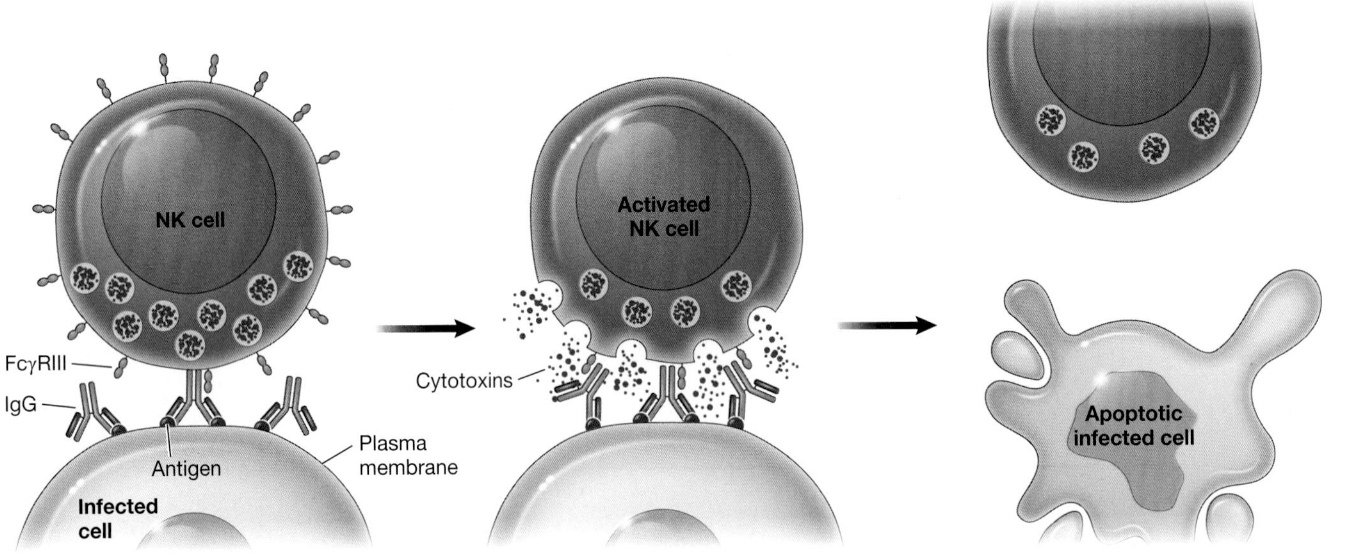

FIGURE 10.14 Antibody-dependent cell-mediated cytotoxicity Natural killer (NK) cells express the Fc receptor FcγRIII at their cell surface. In cells infected with an intracellular pathogen such as a virus, foreign antigen may be incorporated in the plasma membrane. Soluble IgG that recognizes the foreign antigen can coat the surface of the infected cell. When FcγRIII of an NK cell recognizes IgG at the surface of an infected cell, the NK cell is activated, triggering secretion of cytotoxins and apoptosis of the infected cell. The NK cell is released to continue to monitor for neighboring infected cells.

transcytosis Process of delivering immunoglobulins and other macromolecules from one membrane face of a cell that faces the bloodstream or mucosa-associated lymphoid tissue to the other face of a cell facing a different environment, such as extracellular or mucosal tissue.

FcRn Fc receptor expressed on endothelial cells that aids in protection and translocation of IgG from the bloodstream to an extracellular tissue environment; also called *neonatal Fc receptor*.

poly-Ig receptor Receptor on the surface of epithelial cells lining mucosal tissue that aids in the proper delivery of dimeric IgA (and IgM) to mucosal surfaces.

Antibodies produced during B-cell activation in secondary lymphoid tissues are transported with plasma through the circulatory and lymphatic systems. IgM, IgA, and IgG exit the lymph nodes, spleen, and other secondary lymphoid tissues via the efferent lymphatics and protect against bloodborne pathogens (septicemia).

The mechanism of antibody delivery into extracellular spaces is referred to generally as **transcytosis** (which describes the transport of various molecules from one surface of the cell to another). Immunoglobulins such as IgG and IgA are transported from the bloodstream by binding to one side of the endothelial cell layer and traveling to the other side of the endothelial cell so that it can be released, or exposed, to the extracellular environment. The specific mechanism is primarily driven by the type of receptor on the endothelial cell involved.

IgG is taken up by pinocytosis from the bloodstream, a process that normally facilitates degradation of proteins present in the plasma. However, when taken up by pinocytosis, IgG is not degraded; it is protected by binding to a specific receptor through its constant region. IgG immunoglobulins can bind to **FcRn** receptors (neonatal Fc receptors) in endothelial cells, which ultimately protects IgG from destruction in the lysosome and diverts IgG transport to the opposite side of the endothelial cell, facing the extracellular environment (**FIGURE 10.15**).

Dimeric IgA, the primary immunoglobulin that protects against infection at mucosal surfaces, is produced in the lamina propria associated with mucosa-associated lymphoid tissue (MALT). Antigens from pathogens are delivered to lymphoid tissues at mucosal surfaces in a different manner; rather than relying on inflammation and fluid drainage to the lymphoid tissue, specialized cells capture antigens from mucosal surfaces and deliver these antigens to MALT. This mechanism of antigen delivery prevents an inflammatory response and the leakage of mucosal fluids (normally kept isolated by a tight network of epithelial cells) into surrounding tissue. Thus, dimeric IgA must be efficiently delivered to the location of the pathogen via transcytosis, as it cannot be efficiently transported by fluid leakage back into the mucosal tissue. Epithelial cells lining mucosal surfaces express the **poly-Ig receptor**, which is capable of covalently binding to dimeric IgA via its associated J chain at the lamina propria and transporting it to the epithelial cell surface in contact with the mucosal surface (**FIGURE 10.16**). Because IgM is also associated with the J chain, the poly-Ig receptor can also deliver IgM to mucosal surfaces—this delivery of

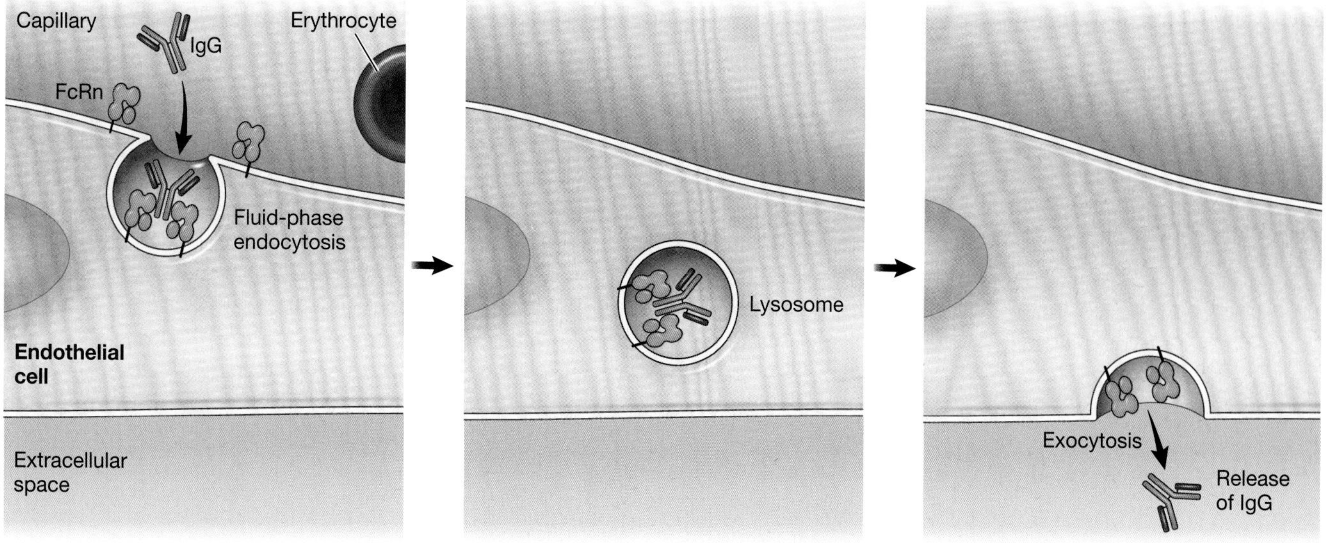

FIGURE 10.15 Transcytosis of IgG from circulation to extracellular spaces IgG in the bloodstream is taken up by endothelial cells lining the vessels via fluid-phase endocytosis. These endothelial cells also express FcRn, an Fc receptor that interacts with IgG. The binding of endocytosed IgG with FcRn is strengthened by the acidic environment of the lysosome, protecting IgG from degradation. Once the IgG and FcRn vesicle fuses with the plasma membrane of the endothelial cell a second time, the basic environment of the extracellular space lowers the affinity of FcRn for IgG, resulting in the release of IgG. (After P. Parham. 2014. *The Immune System*, 4th ed., W. W. Norton and Company.)

IgM at mucosal surfaces plays an important role in protecting against mucosal infections in individuals who suffer from selective IgA deficiency (mucosal immunity is discussed in Chapter 12).

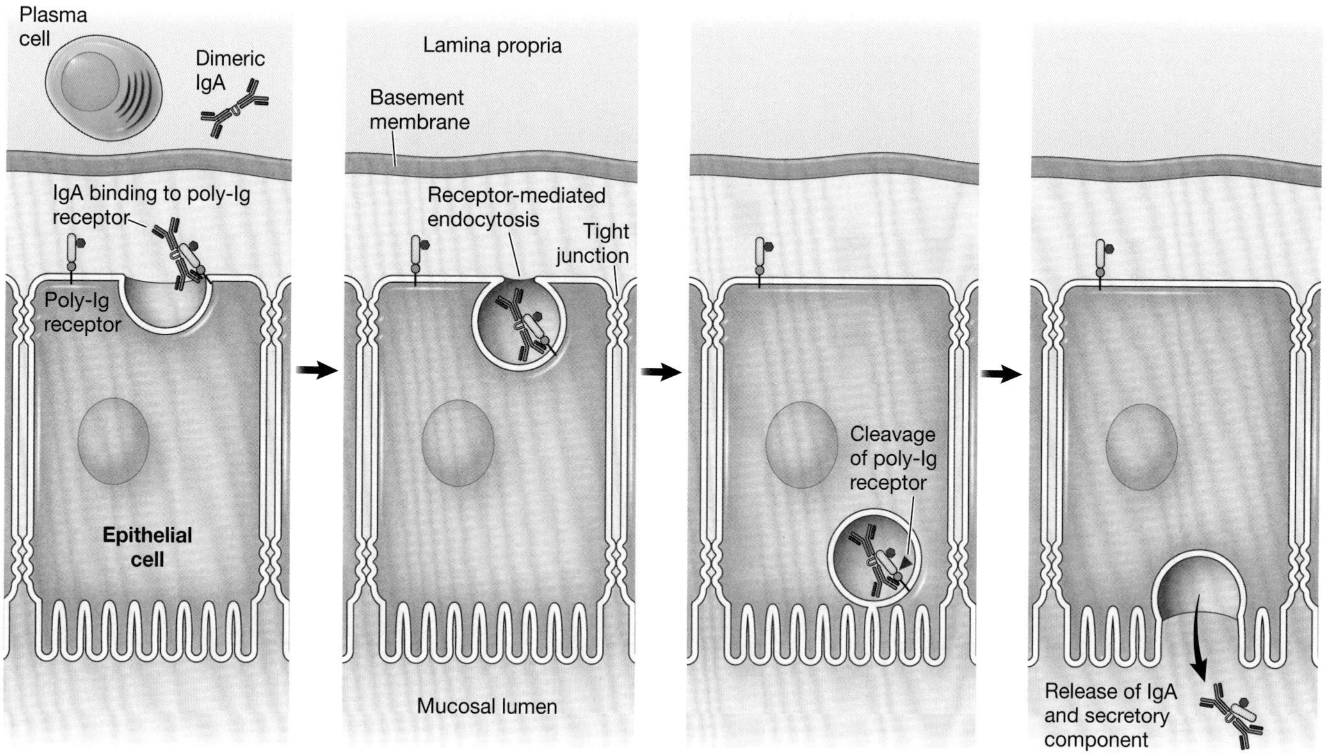

FIGURE 10.16 Transcytosis of IgA from mucosa-associated lymphoid tissue (MALT) to mucosal lumen Dimeric IgA, secreted by MALT plasma cells, binds to the poly-Ig receptor on the basolateral side of mucosal epithelial cells. The receptor and dimeric IgA are engulfed via receptor-mediated endocytosis. Digestion within the lysosomal vesicle degrades the poly-Ig receptor to its secretory component, and vesicle transport moves the dimeric IgA and secretory component to the luminal side of the epithelial cell. The vesicle fuses with the epithelial cell plasma membrane, resulting in release of the dimeric IgA in complex with the secretory component. (After P. Parham. 2014. *The Immune System*, 4th ed., W. W. Norton and Company.)

immune complexes Small complexes of soluble antigen bound with immunoglobulins and fixed complement on the surface marked for removal by phagocytic cells.

Complement activation

In Chapter 3 we saw that a major immune response capable of opsonizing and targeting pathogens for destruction is the complement pathway. The three pathways of complement activation (cleavage of C3 to C3a and C3b)—the alternative pathway, the lectin pathway, and the classical pathway—all require the formation of a C3 convertase capable of complement fixation on the surface of the foreign pathogen. Not only can immunoglobulins act as opsonins on the surface of foreign antigens, but they can also induce the classical pathway of complement activation.

We know that the classical pathway of complement activation begins with the binding of C-reactive protein to pathogens with phosphocholine on their surface (see Chapter 3). This binding causes the recruitment of C1 to the pathogen surface, which is associated with a protease capable of cleaving the C2 and C4 complement proteins. These cleavage events produce C4b and C2b, which associate to form the classical C3 convertase.

Both IgM and IgG are capable of working as a surrogate for C-reactive protein if they recognize a foreign antigen. IgM can bind to a pathogen surface if it contains the right antigen-binding site. However, because of its planar nature, IgM cannot engage with C1 and activate complement via the classical pathway until it changes from the planar form to its staple form. Even though IgM lacks the hinge region seen in IgG and IgA, the flexibility of its Fab arms promotes the formation of the staple form of IgM, allowing it to engage the foreign antigen and bind to C1. Binding of C1 by the staple form of IgM at a pathogen surface allows it to function in the activation of complement (**FIGURE 10.17**).

IgG is also capable of acting as a C-reactive protein surrogate. However, for IgG to interact with C1, at least two IgG molecules must bind in close proximity on the same surface to facilitate binding and activation of complement via the classical pathway (**FIGURE 10.18**). Once at least two IgG immunoglobulins have engaged with C1, the same mechanism of complement activation driven by C-reactive protein or the staple form of IgM results in C3b deposition on the foreign surface.

Small immune complexes

The ability to activate complement and fix C3b on the surface of a foreign molecule is not limited to a cellular pathogen. Binding of IgM or IgG to a soluble multivalent antigen within the bloodstream can also allow for C1 binding and C3b fixation to form **immune complexes** (see Figure 10.18). C3b fixation during the formation of immune complexes allows these soluble foreign particles to be recognized by phagocytes such as macrophages through their CR1 receptors.

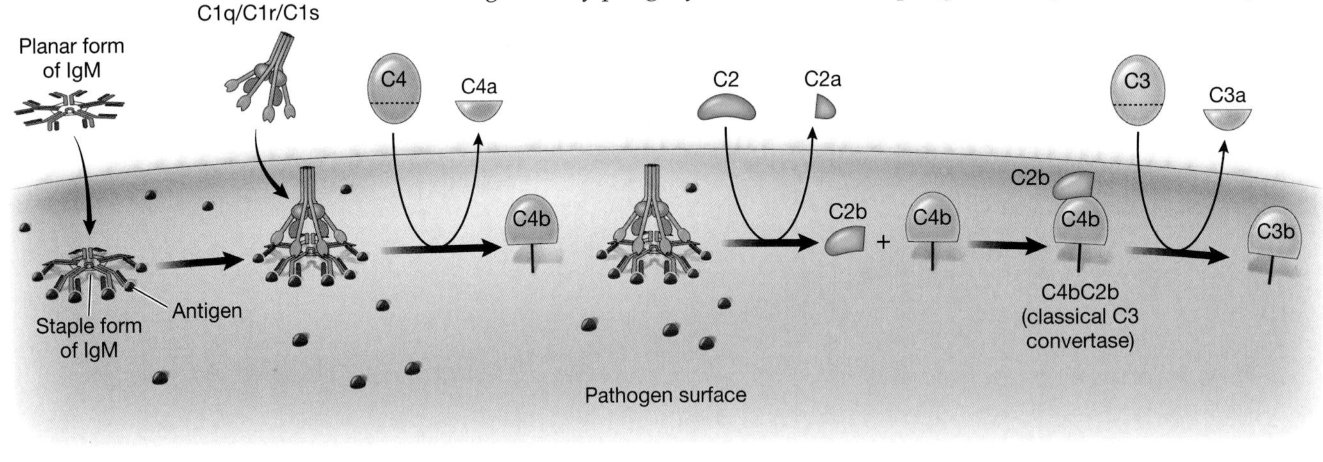

FIGURE 10.17 IgM action in the classical pathway of complement activation Pentameric IgM is a planar molecule when secreted by plasma cells. Flexibility of the hinge of IgM allows it to bind to repeating epitopes on a pathogen surface and convert to the staple form. The staple form IgM serves as a binding partner for C1q/C1r/C1s, which acts as a protease in the classical pathway of complement activation. C1q/C1r/C1s cleaves the complement proteins C2 and C4 to create the classical C3 convertase C4bC2b. The classical C3 convertase cleaves C3 into the anaphylatoxin C3a and the opsonin C3b.

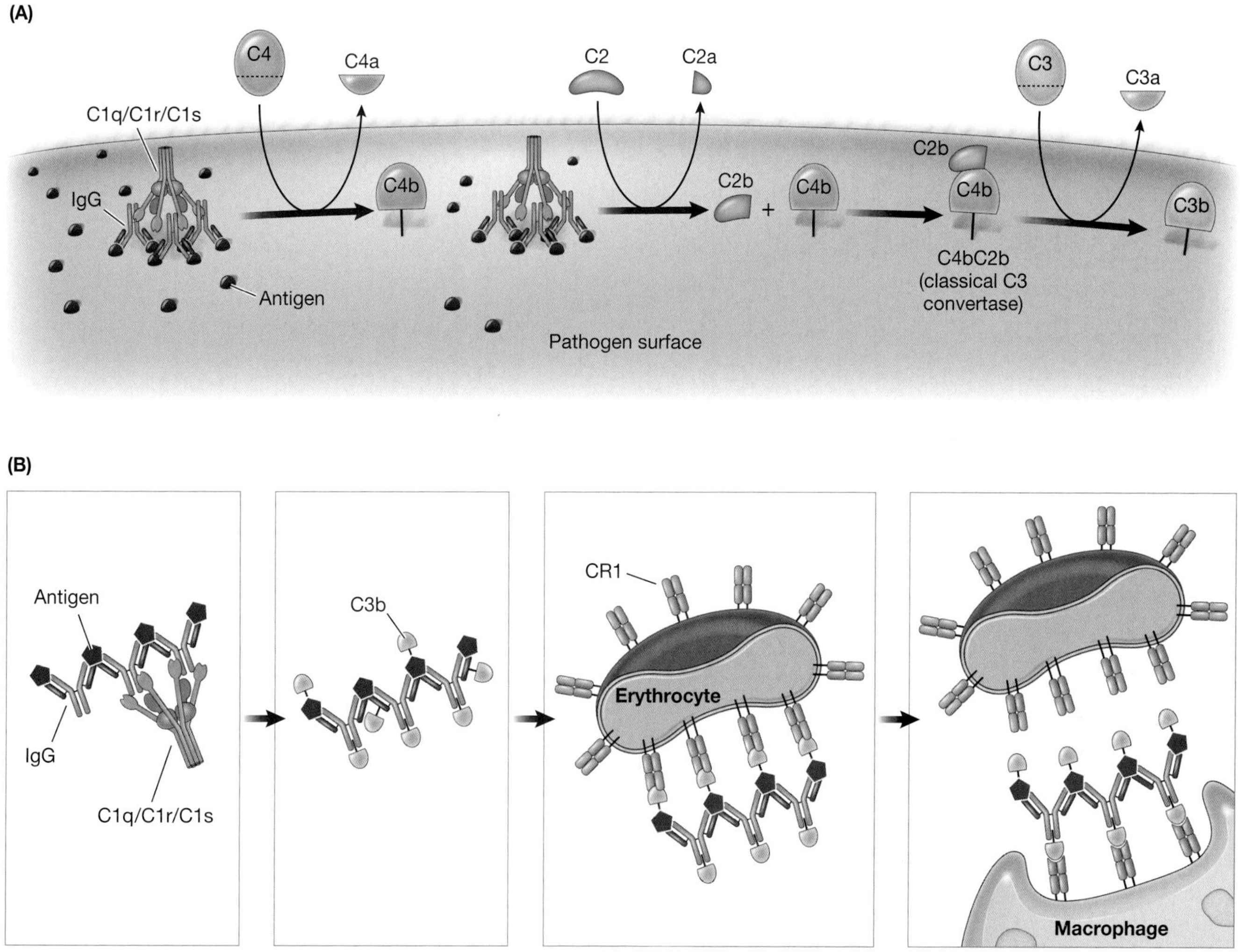

FIGURE 10.18 IgG action in the classical pathway of complement activation (A) IgG binds to antigen on a pathogen surface. Unlike IgM, at least two molecules of IgG must bind to antigen to allow it to bind to C1q/C1r/C1s. Binding to IgG allows C1q/C1r/C1s to create the classical C3 convertase C4bC2b. (B) IgG in the bloodstream binds to small foreign antigens and creates soluble immune complexes by fixing C3b to the surface of both the antigen and IgG via the classical pathway of complement activation. C3b on immune complexes binds to the complement receptor CR1 on erythrocytes, which then transport the immune complexes to macrophages. Macrophages use their CR1 receptors to bind to the immune complexes and destroy them via phagocytosis.

Erythrocytes also play an important role in immune complex clearance, as they express CR1 at their surface. The formation and presence of immune complexes within the bloodstream allows them to bind to red blood cells. As erythrocytes travel through the liver and spleen, they deliver immune complexes to macrophages residing in these tissues, resulting in efficient clearance.

Although immune complexes are removed by resident macrophages of the liver and spleen, persistence of immune complexes in the bloodstream can lead to their aggregation and precipitation within small blood vessels. Precipitated immune complexes can aggregate within the kidney glomeruli. Specialized cells are present at these locations to clear these complexes and repair any damage that may have occurred during immune complex aggregation.

CHECKPOINT QUESTIONS

1. Why would you predict that neutralizing antibodies typically recognize antigens at locations that are important in protein interactions?
2. How do IgM and IgG act as surrogates for C-reactive protein?

Making Connections: Immunoglobulin Isotype Actions—A Detailed Look

Each of the five isotypes of immunoglobulins in humans has specific effector functions. The five effector functions of immunoglobulins—neutralization, opsonization, complement activation, activation of innate immune cells, and protection of internal and mucosal tissues—are each carried out by specific isotypes made at discrete locations within the body.

Isotypes of immunoglobulins are determined by the Fc component of the immunoglobulin heavy chain. This Fc component of specific isotypes (e.g. IgG, IgE) bind to receptors at the surface of innate immune cells to promote their activation when bound to antigen. Other immunoglobulins (e.g. IgM, IgA) interact with the J chain to form complexes that contain multiple immunoglobulins to promote neutralization of pathogens and toxins. Still other immunoglobulins (e.g. IgM, IgG) have flexible hinges that promote complement activation.

(Sections 2.1, 2.7, 3.3, 4.8, 9.5, 10.5)

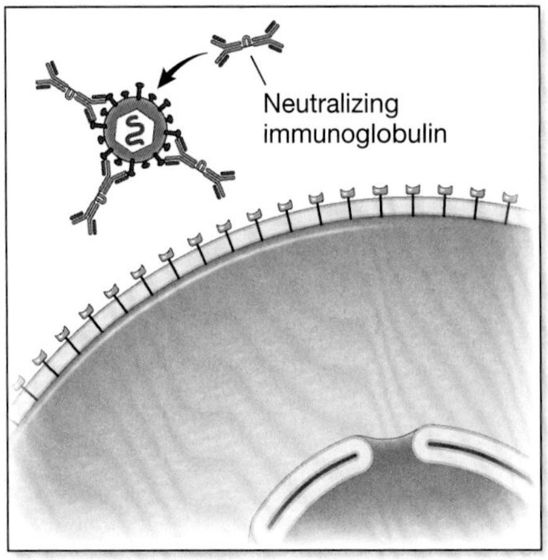

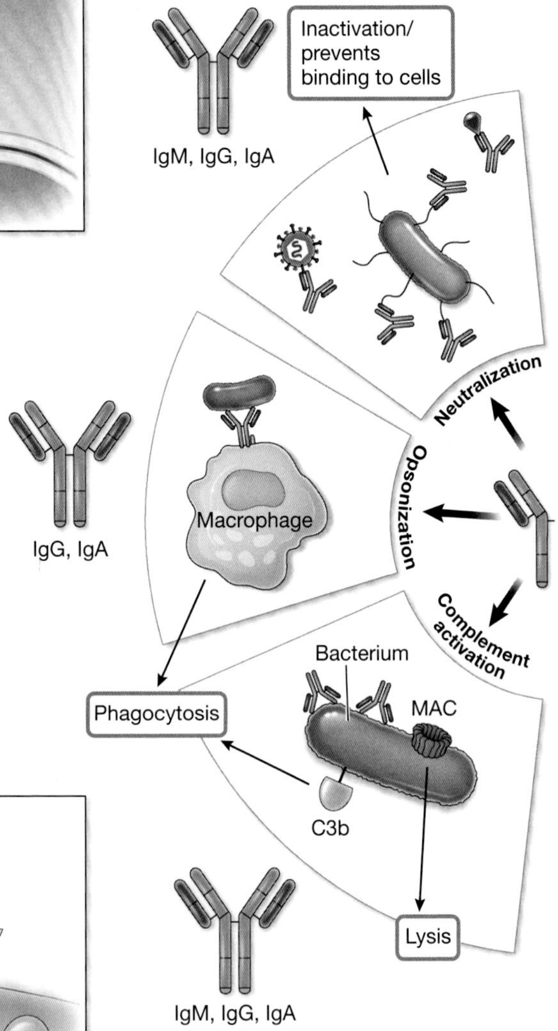

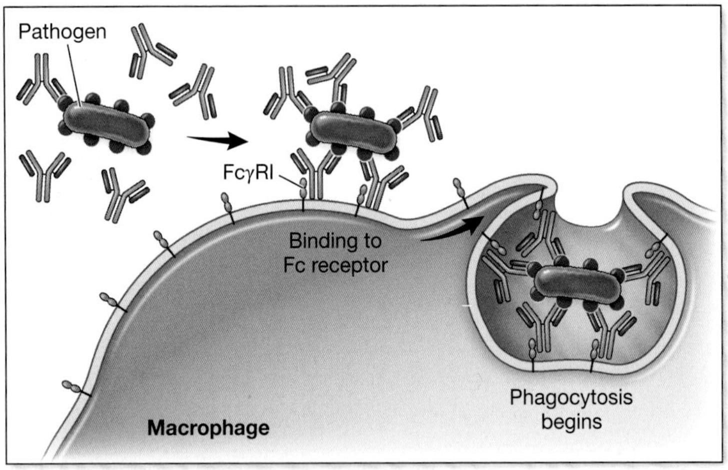

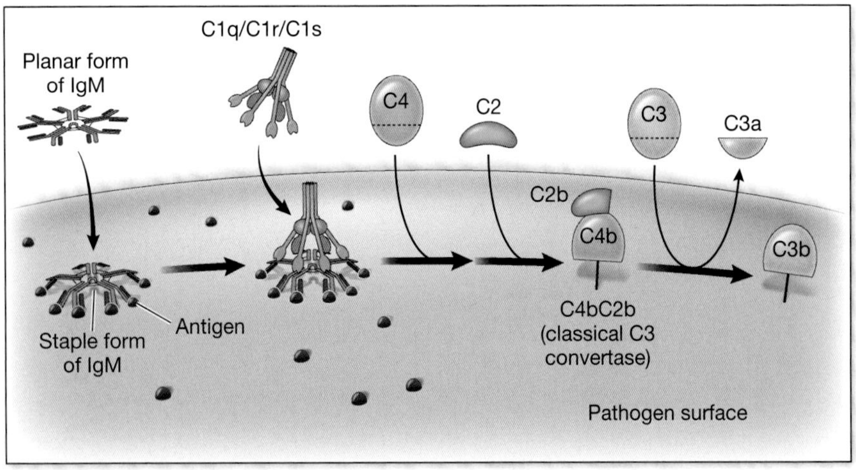

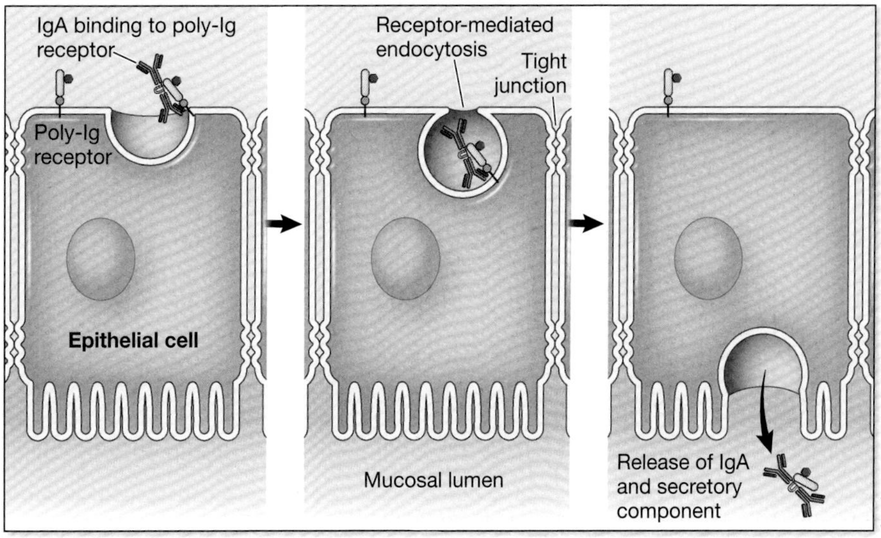

After P. Parham, 2014. *The Immune System*, 4th ed., W. W. Norton and Company

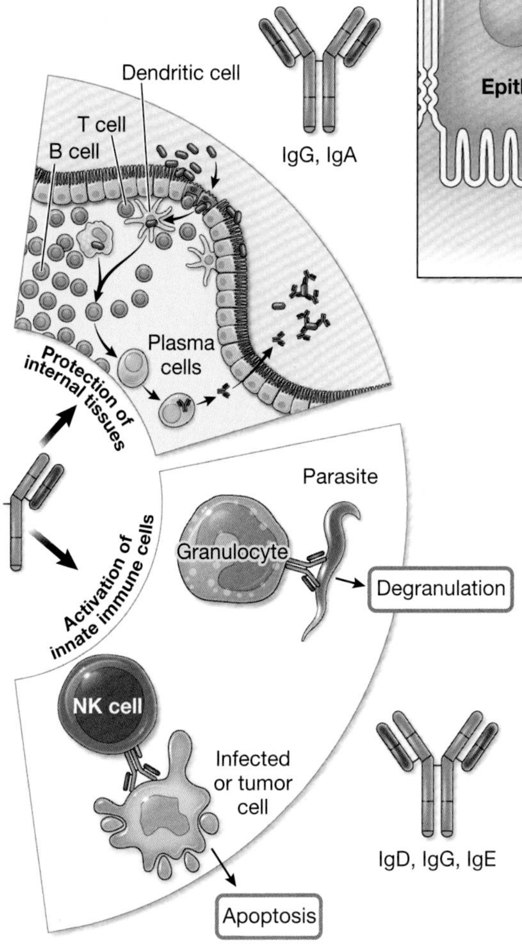

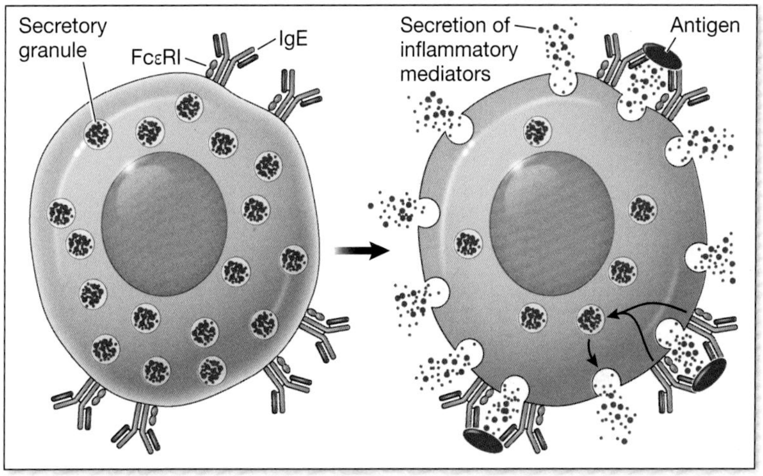

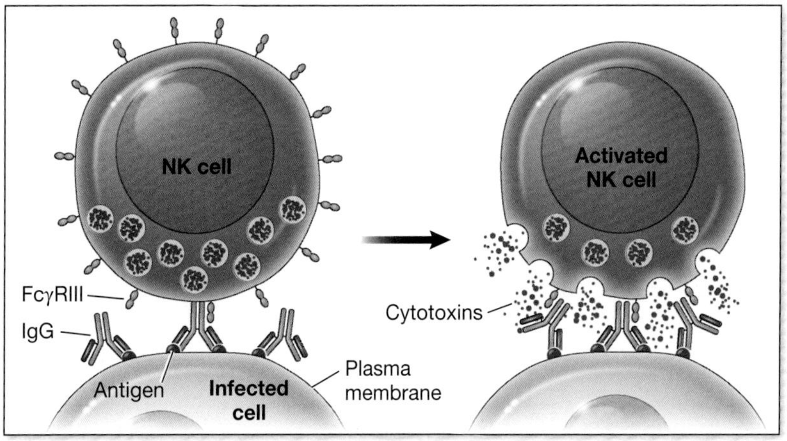

Summary

10.1 Which signals and molecules are required for B-cell activation?

- B cells can only become activated upon engagement of their cell-surface immunoglobulin with the antigen they are capable of recognizing.
- Upon antigen engagement, the immunoglobulin receptor complex activates signaling pathways that result in activation of three important transcription factors (NFAT, NFκB, AP-1) to promote B-cell clonal expansion and differentiation.
- Additional signals are required for proper B-cell activation—these signals are provided either by a helper T cell or by coordination of B-cell signaling events.
- Clustering of immunoglobulins at the B-cell plasma membrane during antigen interaction further promotes B-cell activation in conjunction with the action of the B-cell coreceptor, which recognizes complement breakdown products on the pathogen surface.

10.2 How are B cells activated in the absence or presence of T cells?

- Although most B cells require aid from helper T cell signals to properly activate upon binding antigen (T-dependent response), some B cells can become activated in a manner independent of helper T-cell action (T-independent response).

10.3 How do B cells migrate and behave in secondary lymphoid tissues?

- To monitor the presence of antigens, B cells migrate into secondary lymphoid tissues, where they test for the presence of the antigen their immunoglobulin can recognize in the T-cell zone.
- Activated B cells within secondary lymphoid tissue, which have formed a conjugate pair with a helper T cell, form a primary focus and begin clonal expansion and the production of IgM in medullary cords.

10.4 How is immunoglobulin affinity maturation driven?

- Some activated B cells within medullary cords migrate to a lymphoid follicle and become a germinal center, where they undergo clonal expansion and differentiation, as well as somatic hypermutation and isotype switching at their immunoglobulin loci.
- Due to the mutation that occurs during somatic hypermutation, B cells within the germinal center undergo another round of positive and negative selection to promote affinity maturation and limit self-recognition.

10.5 What are the properties and functions of the different antibody isotypes?

- Secreted immunoglobulins have a variety of functions within the humoral adaptive immune response, including neutralization, activation of innate immune cells, protection of internal tissues, complement activation, and clearance of small immune complexes.

Review Questions

10.1 Why are B-cell activation and T-cell activation similar in regard to the signaling cascades activated in both processes?

10.2 Why would you predict that a deficiency in factor I would cause problems in B-cell activation?

10.3 Briefly define the difference between thymus-dependent and thymus-independent B-cell responses.

10.4 Why would you suspect that thymus-independent antigens are typically non-proteinaceous?

10.5 What is the major difference in B cells activated in a primary focus versus B cells activated in a secondary focus?

10.6 Why do B cells in the germinal center need to go through additional selection (both positive and negative)?

10.7 The chemotherapeutic drug rituximab is a monoclonal antibody against the CD20 antigen on B cells and is an effective chemotherapeutic for B-cell tumors. Explain this drug's mechanism of action against B-cell tumors.

CASE STUDY REVISITED

Hyper IgM Syndrome Continued

A 5-year-old boy is taken to his primary care physician for a 2-week follow-up appointment for treatment of otitis media (middle ear infection). The patient is still complaining of ear pain, and a medical examination reveals that the antibiotic treatment was not successful in clearing the infection. Previous blood work had revealed that the patient's immunoglobulin serum levels were high in IgM and lacking in the immunoglobulin isotypes IgA, IgG, and IgE. The pediatrician had diagnosed the patient with hyper IgM syndrome. In the hope that his son's condition might be treated, the boy's father asks for further testing to determine why the patient is lacking certain immunoglobulin isotypes. The pediatrician decides to have flow cytometry analysis conducted on the patient's lymphocytes to determine the presence or absence of important cell-surface molecules. Analysis of T cells from the patient reveals that his T cells are incapable of expressing CD40 ligand.

Think About...

Questions for individuals

1. The case study in Chapter 9 also focused on hyper IgM syndrome. How do these two case studies differ?
2. Explain the connection between hyper IgM syndrome and the lack of isotype switching.

Questions for student groups

1. Why might a lack of CD40 ligand on the surface of T cells prevent isotype switching?
2. If this patient's helper T cells lack CD40 ligand, how might IgM still be present in the bloodstream?

MAKING CONNECTIONS Key Concepts	COVERAGE (Section Numbers)
B-cell responses to different antigens	1.10, 2.2, 2.3, 4.5, 4.8, 8.4, 8.9, 8.10, 8.11, 9.4, 9.5, 10.1, 10.2
B-cell activation	4.7, 7.14, 9.4, 10.2, 10.4
B-cell maturation during clonal expansion	9.4, 10.3, 10.4
B-cell action	9.5, 10.5
Immunoglobulin effector function	2.1, 2.7, 3.3, 4.8, 9.5, 10.5
Isotypes used for immunoglobulin effector functions	2.1, 2.7, 3.3, 4.8, 9.5, 10.5

11

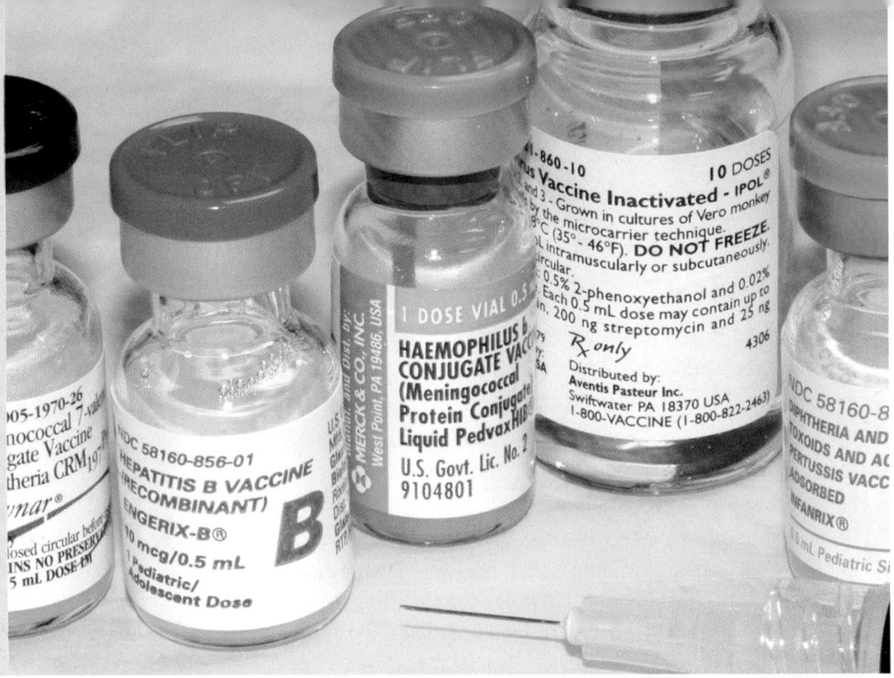

Immunological Memory and Vaccination

QUESTIONS Explored

11.1 How does immunological memory work to prevent and fight disease?

11.2 How does immunological memory develop?

11.3 How do vaccines prevent disease?

11.4 What are the different types of viral and bacterial vaccines?

11.5 What is the role of an adjuvant in a vaccine?

11.6 What key concerns affect vaccine development?

CASE STUDY: Aparna and Samir's Story

Aparna and Samir have been married for 2 years and are now hoping to have children. The past several home pregnancy tests were negative, but this month the couple remains hopeful. Aparna holds the test stick in her hand, anxious as the minutes tick by.

Samir looks at his wife. "Is it one line or two?"

Aparna looks at the stick, and a wave of excitement crashes over her. "It's two! It's positive! Samir, I'm so happy!"

Suddenly, Aparna recalls how difficult her mother's pregnancy was with her younger sister. Her sister came into this world greeted by a blood transfusion, intravenous fluids, and phototherapy. Aparna knows that her mother had received no prenatal care due to the family's low income. She is grateful that she and Samir have health insurance coverage and hopes this means that she and the baby will be well taken care of. Aparna plans to schedule an appointment with her OB/GYN and tells Samir that they both should be screened for blood type and any genetic markers that might indicate pregnancy complications or birth defects.

To understand what Aparna is concerned about and to answer the case study questions at the end of the chapter, we must understand the mechanisms of action of the primary and secondary adaptive immune responses. The adaptive immune response not only works to combat a pathogen the first time we encounter it but also produces cells that enable the immune system to "remember" the pathogen. This immunological memory allows for a more robust and quicker response to a subsequent infection by the same pathogen. Our understanding of the adaptive immune response has resulted in treatments that promote healthy pregnancy and childbirth. This understanding has also resulted in vaccines that protect against deadly pathogens.

11.1 How does immunological memory work to prevent and fight disease?

LEARNING OBJECTIVE

11.1.1 Differentiate between a primary and secondary immune response.

The adaptive immune system is a formidable weapon in targeting specific pathogens that are not controlled by the innate immune response. We have seen that the diversity of T-cell and B-cell receptors allows the adaptive immune system to recognize and combat a wide variety of pathogens. Clonal selection of lymphocytes to specific antigens allows the adaptive immune response to concentrate effector mechanisms to efficiently target and clear an infection. We have learned that secondary lymphoid tissues serve as a location for lymphocyte migration and antigen presentation to circulating lymphocytes. While this centralization increases the efficiency of the adaptive immune response, the engagement of a specific lymphocyte with its cognate antigen-presenting cell (APC) displaying the proper epitope still requires time. To reduce the length of time required to mount an effective adaptive immune response, *memory cells* that differentiated during a primary immune response "remember" a specific pathogen. These memory cells have undergone many of the steps required to become effector cells and are located where they need to be to recognize the pathogen.

Timeline of adaptive immune responses

The first time a pathogen infects an organism, it can take up to 14 days to resolve the infection. During this time, dendritic cells process antigens, migrate into a draining lymphoid tissue, and engage with T cells expressing a receptor capable of recognizing the MHC-peptide complex at the dendritic cell surface, and clonal selection occurs. Clonal selection of T cells leads to the activation of B cells, which then begin to secrete soluble IgM (**FIGURE 11.1**). Some of the

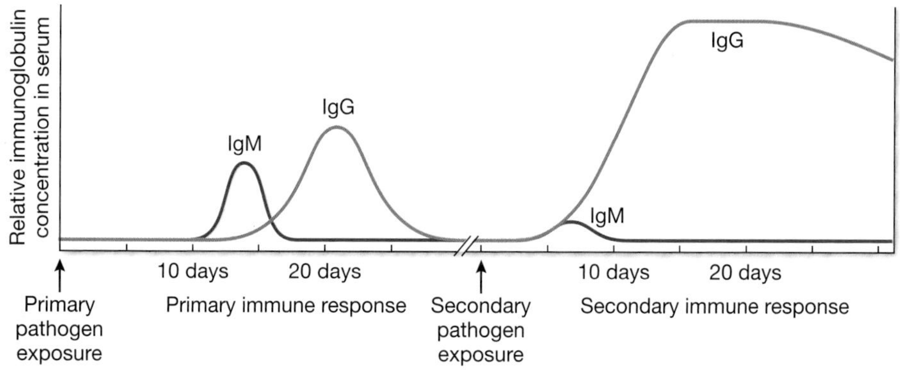

FIGURE 11.1 Immunoglobulin production during the primary and secondary immune responses Initial exposure to a pathogen elicits the activation of B cells, which initially produce IgM that recognizes epitopes present on the pathogen. A subset of B cells activated in the primary immune response migrates to lymphoid follicles, where they undergo somatic hypermutation and isotype switching. These B cells produce different immunoglobulin isotypes, which may include IgG. Upon secondary exposure to the same pathogen, the secondary immune response is triggered, activating memory B cells that have undergone somatic hypermutation and isotype switching during the primary immune response. These activated B cells produce the same isotype produced during the primary immune response but are activated much more rapidly. Some memory B cells with IgM on their surface that recognize the pathogen are also activated to produce IgM, although the number of IgM-producing memory B cells is low. (After N. Parker et al. 2016. *Microbiology*. Houston, Texas: OpenStax. CC BY 4.0. Access for free at: https://openstax.org/details/books/microbiology.)

activated B cells migrate to lymphoid follicles, where they ultimately produce other immunoglobulin isotypes. Thus, while the adaptive immune system can combat a wide variety of pathogens due to the expression of a diverse set of T-cell receptors and immunoglobulins, its efficiency is limited upon the first encounter with a pathogen because of the timing and the limited immunoglobulin isotypes produced during the primary response.

Although the primary adaptive immune response to a pathogen is slow due to the need to coordinate different arms of the immune system and the need for naïve lymphocytes to become activated, when the same pathogen is encountered a second (or any subsequent) time, the efficiency of the response is greatly improved (see Figure 11.1). While it takes about 2 weeks to clear a pathogen during a primary immune response, the secondary immune response can clear an infection much more rapidly—typically in 3 to 4 days. This faster rate is correlated with a much faster and more robust adaptive immune response. The more rapid secondary response is due to the differentiation and action of memory cells.

Memory cells

Memory cells are lymphocytes produced during a primary immune response. Memory cells are capable of quick activation to combat the same pathogen upon subsequent exposure. Furthermore, most memory cells produce immunoglobulins that have undergone both somatic hypermutation and isotype switching. During a secondary response, the activation of naïve B cells is limited to focus energy on the production of high-affinity immunoglobulins that can combat the same pathogen (see Figure 11.1).

Memory cells are an important product of clonal expansion (**FIGURE 11.2**). They have antigen receptors that are specific for the pathogen encountered during the initial exposure that resulted in their production. Memory cells not only respond more quickly to their specific antigen but also monitor for antigens in tissues other than secondary lymphoid tissues; memory B cells can even provide for better antigen recognition due to changes in antigen specificity (somatic hypermutation discussed in Chapter 9). Memory cells are paramount to the development and successful activity of vaccines, a triumph of immunology that has protected hundreds of millions of people from life-threatening diseases and eradicated certain diseases from the human population.

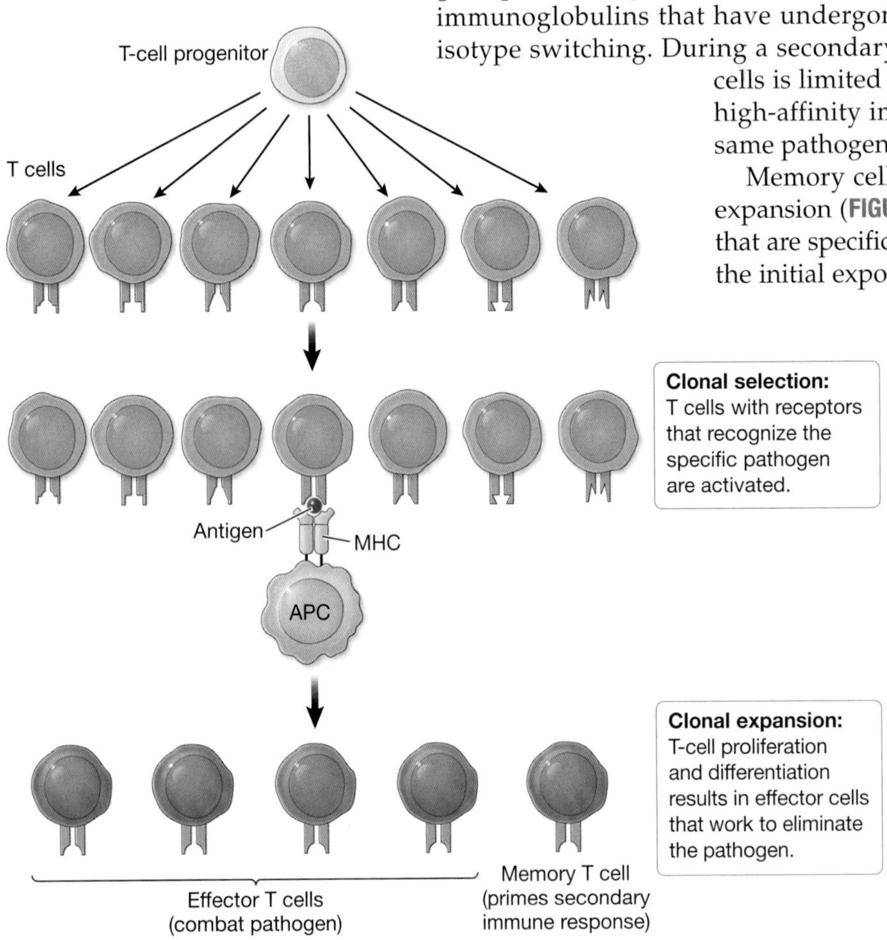

FIGURE 11.2 Production of memory cells Clonal selection occurs when a lymphocyte with the proper receptor recognizes the presence of a foreign antigen. Clonal expansion then occurs, in which the selected lymphocyte undergoes signaling events that drive division and differentiation. The majority of differentiated cells drive the effector function of the adaptive immune response. Some become memory cells that allow the adaptive immune system to respond more quickly to the presence of the same antigen upon subsequent exposure.

CHECKPOINT QUESTIONS

1. Compare and contrast the timing and robustness of a primary and secondary adaptive immune response.
2. Which cell type is responsible for activation of a quick secondary adaptive immune response?

11.2 | How does immunological memory develop?

LEARNING OBJECTIVES

11.2.1 Explain why immunological memory is a key part of the adaptive immune response.

11.2.2 Name the various types of memory T cells and explain their functions.

11.2.3 Describe the locations and activity of memory B cells.

Many of the previous chapters were devoted to explaining the development and actions of cells of the adaptive immune system. Lymphocyte development drives the production of a vast array of cells capable of recognizing a wide variety of antigens, and these cells subsequently circulate to determine if their antigen-specific receptor is capable of recognizing their cognate antigen. Clonal selection of antigen-specific lymphocytes drives expansion and differentiation of these cells into effector lymphocytes to efficiently clear the infection.

In this section, we focus on the differentiation of memory cells and explore the differences in the activation process and molecules expressed between memory cells and other effector lymphocytes produced during an adaptive immune response.

Clonal expansion and memory cells

The process of clonal selection and clonal expansion detailed in earlier chapters focused on the immediate clearance of the pathogen that the lymphocytes of the adaptive immune response recognized through their antigen-specific receptors. However, important products of clonal expansion during an adaptive immune response are the memory cells that differentiate from the naïve T and B cells activated upon clonal selection (see Figure 11.2). These long-lived memory cells are not directly involved in combating the pathogen whose antigens promoted their differentiation. Instead, they play a crucial role in the rapid adaptive immune response that ensues upon subsequent exposure to the same pathogen. Part of the speed and power of this response is due to the ability of some memory cells to monitor additional tissues beyond secondary lymphoid tissues, their ability to induce an adaptive immune response with a wider variety of costimulatory signals, and their enhanced affinity for antigen (in the case of memory B cells).

Memory T cells

When comparing the two major subsets of T cells—naïve T cells and memory T cells—there are several functional and molecular differences that distinguish the two groups. We know that naïve T-cell activation is driven primarily by dendritic cells that have migrated to a secondary lymphoid tissue after taking up and processing antigens. Naïve T cells develop in the thymus, enter the circulation, and migrate into secondary lymphoid tissues, where they may or may not come into contact with the MHC-peptide complex that their receptor can recognize. Their migration in the circulatory and lymphatic systems and travel that is limited to secondary lymphoid tissues enables an efficient primary adaptive immune response.

Memory T cells produced during a primary adaptive immune response can be divided into four subpopulations (**FIGURE 11.3**):

- **T memory stem cells (T_{SCM})**
- **Central memory T cells (T_{CM})**
- **Effector memory T cells (T_{EM})**
- **Resident memory T cells (T_{RM})**

T memory stem cells (T_{SCM}) Memory T cells that maintain pluripotency; they reside within the circulatory and lymphatic systems and secondary lymphoid tissues and are capable of further differentiating into central memory T cells or effector memory T cells upon subsequent activation.

central memory T cells (T_{CM}) Memory T cells that reside within the circulatory and lymphatic systems and in secondary lymphoid tissues; activation is analogous to naïve T-cell activation.

effector memory T cells (T_{EM}) Memory T cells that reside within the circulatory and lymphatic systems, secondary lymphoid tissues, and peripheral tissues; major role is monitoring for the presence of antigen in the lymphatic system and peripheral tissues.

resident memory T cells (T_{RM}) Memory T cells that reside within peripheral tissues; major role is the persistent monitoring of these tissues for repeat infection by the same pathogen.

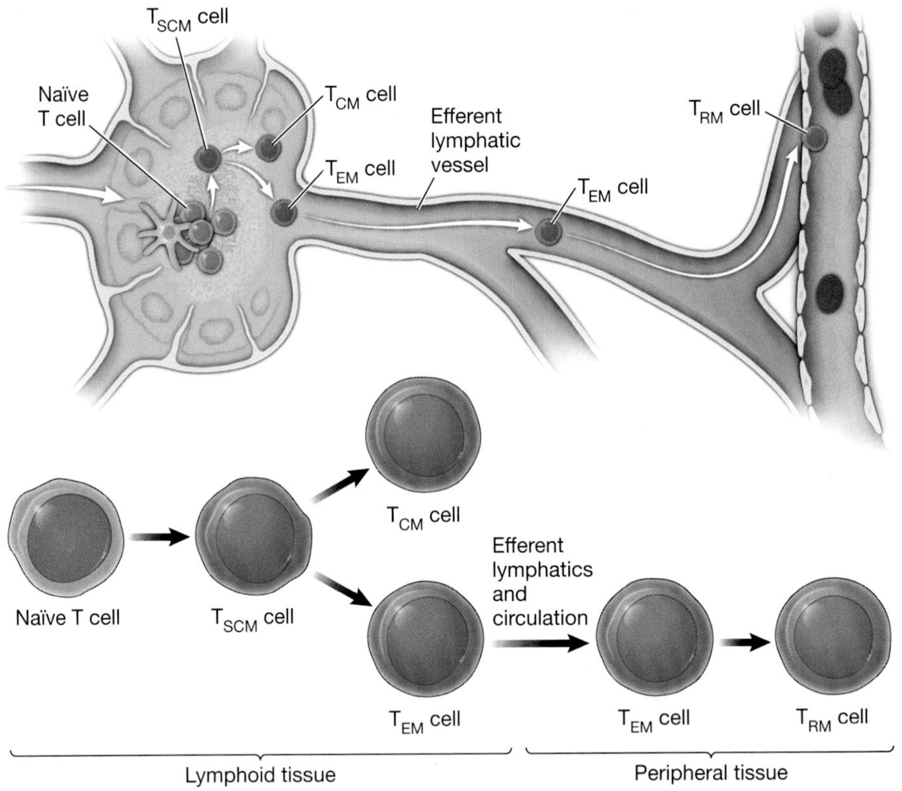

FIGURE 11.3 Memory T cell populations Upon activation of a naïve T cell with its cognate MHC-antigen complex, the cell will expand and differentiate into effector T cells and memory T cells. The first memory T cells produced during clonal expansion are T memory stem cells (T_{SCM} cells), which are maintained within the circulatory and lymphatic systems. T_{SCM} cells are pluripotent and can further differentiate into central memory T cells (T_{CM} cells) and effector memory T cells (T_{EM} cells), which also reside in the circulatory and lymphatic systems. T_{EM} cells can also migrate into peripheral tissue and produce resident memory T cells (T_{RM} cells), which remain in their tissue of origin. (After D. L. Farber et al. 2014. *Nat Rev* 14: 24–35.)

These memory T cells differ from naïve T cells in regard to the cell-surface molecules they express, the locations they populate, and the APC types that are capable of activating them. These subpopulations can be further defined based on their function, location, and cell-surface marker expression pattern.

T MEMORY STEM CELLS (T_{SCM}) The presence of memory T cells within both the circulatory system and peripheral tissues might suggest the production of a stem cell-like memory T cell capable of multipotency for memory T cells that can reside in these locations. Indeed, within the circulatory and lymphatic systems there resides a T-cell subset that has similar cell-surface markers as naïve T cells but has also gained the expression of cell-surface molecules more indicative of memory T cells. These include the beta chain of the IL-2 receptor (IL-2Rβ), which is typically expressed on memory T cells to prompt faster activation in the presence of IL-2. T memory stem cells (T_{SCM}) contain the potential for high survival, self-renewal, and multipotency for other memory T cells, all properties of stem cells, giving rise to their name. T_{SCM} cells are capable of further differentiation into two other types: central memory T cells and effector memory T cells.

CENTRAL MEMORY T CELLS (T_{CM}) Central memory T cells (T_{CM}) are usually located within the circulatory and lymphatic systems, migrating to and from secondary lymphoid tissues. Their mode of activation and migration is similar to that of naïve T cells; both are capable of entry into secondary lymphoid tissues, where they monitor antigen-presenting dendritic cells using their receptors to determine the presence or absence of specific antigens. T_{CM} cells are much longer-lived than their naïve T-cell and effector memory T-cell counterparts. It has been shown that T_{CM} cells can express a larger amount of IL-2, thus providing these cells with the ability to efficiently activate and differentiate into effector cells upon exposure to specific antigens. Upon recognition of an antigen by its T-cell receptor, activation of T_{CM} cells and subsequent differentiation into effector cells closely follows the activation of naïve T cells described in Chapter 7.

EFFECTOR MEMORY T CELLS (T_{EM}) Effector memory T cells (T_{EM}) are located within secondary lymphoid tissues and within the circulatory and lymphatic systems. However, T_{EM} cells also have the capacity to migrate into other peripheral tissues. These cells are aptly named due to their ability to quickly activate and differentiate into effector T cells in peripheral tissues without the need to migrate through secondary lymphoid tissues for antigen detection and activation. T_{EM} cells can do this because of their elevated ability to secrete effector cytokines, including IFN-γ and TNF-α. T_{EM} cells usually reside in peripheral tissues close to the original sites of infection responsible for their production. This allows these cells to protect locations from infections similar to those that caused their differentiation.

RESIDENT MEMORY T CELLS (T_{RM}) While effector memory T cells routinely enter and exit peripheral tissues, another subset of memory T cells maintains residence within a peripheral tissue. Once it has gained entry, it does not reenter the circulation. These resident memory T cells (T_{RM}) function similarly to T_{EM} cells in peripheral tissues, offering protection against recurring infection through quick activation and elevated effector function at potential sites of infection. It is speculated that T_{RM} cells play important roles in preventing persisting infections (including those of latent viruses such as herpes simplex virus) and in protecting against the establishment of new infections.

DIFFERENCES IN CELL-SURFACE MARKER EXPRESSION One of the classical distinguishing features between naïve T cells and memory T cells is the expression of a different isoform of the cell-surface T-cell signaling molecule CD45. CD45 can be expressed in two different isoforms: CD45RA and CD45RO. Naïve T cells express the CD45RA isoform, while memory T cells express the CD45RO isoform (**TABLE 11.1**). While CD45RA is part of the receptor signaling complex in naïve T cells and is required for activation, CD45RO in the receptor signaling complex of memory T cells is more efficient at signaling and initiating memory T-cell activation.

In addition, several other cell-surface markers can be used to differentiate naïve T cells and memory T cells. When these markers are used in conjunction with each other, the three major populations of T cells (naïve T cells, central memory T cells, and effector memory T cells) can be distinguished from one another.

One of these differentiating markers is the chemokine receptor CCR7, which we have seen is responsible for T-cell homing into secondary lymphoid tissues. Another is the molecule L-selectin (also known as CD62L), an important cell-surface molecule expressed by T cells for proper homing into secondary lymphoid tissue (see Chapter 7). L-selectin has a similar expression profile as CCR7 and is expressed more prominently by naïve and central memory T cells. It is not expressed by effector memory T cells. Finally, CD103, a member of the integrin alpha family of proteins, is expressed on resident memory T cells. It is not present on the surface of naïve, central memory, or effector memory T cells. CD103 is responsible for binding to E-cadherin on epithelial cells, suggesting that T_{RM} cells utilize this adhesion to maintain their presence in peripheral tissues.

TABLE 11.1 | Differences in Cell-Surface Molecule Expression

Cell Type	CD45RA	CD45RO	CCR7	L-selectin	CD103
Naïve T cell	+	–	+	+	–
Central memory T cell	–	+	+	+	–
Effector memory T cell	–	+	–	+/– (varies)	–
Resident memory T cell	–	+	–	+/– (varies)	+

Memory B cells

As we saw in Chapter 10, B cells are activated in secondary lymphoid tissues by CD4 helper T cells capable of recognizing the same antigen or different processed antigens from the same pathogen. This activation occurs in the T-cell zone, where activated B cells begin to produce IgM immunoglobulins to combat the pathogen.

Some of the conjugate pairs of T cells and B cells become a germinal center, where follicular helper T cells promote the activation of B cells, and these activated B cells undergo isotype switching and somatic hypermutation. The activated B cells within germinal centers can produce effector plasma cells responsible for secreting high-affinity, isotype-switched immunoglobulins. They can also differentiate into memory B cells, capable of reactivation and production of the same high-affinity, isotype-switched antibodies that were produced by the original germinal center B cell. Memory B cells localize in the spleen and migrate into other secondary lymphoid tissues using the circulatory and lymphatic systems.

Although this pathway of memory B-cell production is the most commonly described, a population of these cells has been shown to differentiate independent of germinal center activation. These memory B cells are not produced during follicular helper T-cell activation in a germinal center; rather, they are produced during CD4 helper T-cell activation in a primary focus. These cells express IgM upon reactivation, as they do not undergo somatic hypermutation and isotype switching during their production.

Suppression of naïve B cell activation in a secondary immune response

The production of memory B cells within the germinal center facilitates the generation of a cell population capable of producing high-affinity immunoglobulins of the correct isotype that can efficiently combat the pathogen responsible for activating the B cell. Thus, it stands to reason that these B cells would be the most useful upon subsequent exposure to the same pathogen, since the immunoglobulins produced have gained higher affinity and enhanced effector function due to somatic hypermutation and isotype switching.

To promote activation of memory B cells upon subsequent exposure to a specific pathogen, antibodies that bind to that pathogen can inhibit activation of naïve B cells that express immunoglobulins capable of recognizing that same pathogen. Many of the isotype-switched immunoglobulins produced in a germinal center are IgG. These soluble immunoglobulins neutralize and opsonize the pathogen whose antigens were responsible for B-cell activation. Persistence of these immunoglobulins and memory B cells capable of expressing them inhibits naïve B cell activation when they are bound to the pathogen. If a naïve B cell binds to a pathogen through its immunoglobulin when there is also soluble IgG bound to the pathogen surface, the Fc component of the bound IgG interacts with an inhibitory receptor, FcγRIIB1, on the surface of the naïve B cell. This interaction inhibits activation of the naïve B cell to limit the production of low-affinity IgM during a secondary immune response (which has limited effector function). This inhibition of naïve B-cell activation ensures that energy and resources are focused on the production of higher-affinity immunoglobulins during a secondary immune response (**FIGURE 11.4**).

This mechanism of inhibiting naïve B-cell activation is used medically in the prevention of hemolytic anemia of the newborn, which can occur when the mother is negative for an erythrocyte antigen known as Rhesus (Rh) factor but the father is Rh positive. If the developing fetus is Rh positive, red blood cells containing the antigen can drive activation of B cells capable of recognizing Rh when the mother is exposed to the fetal red blood cells at birth. This primary immune response produces low-affinity IgM but, more importantly, produces memory B cells capable of a more robust secondary immune response upon subsequent exposure to Rh.

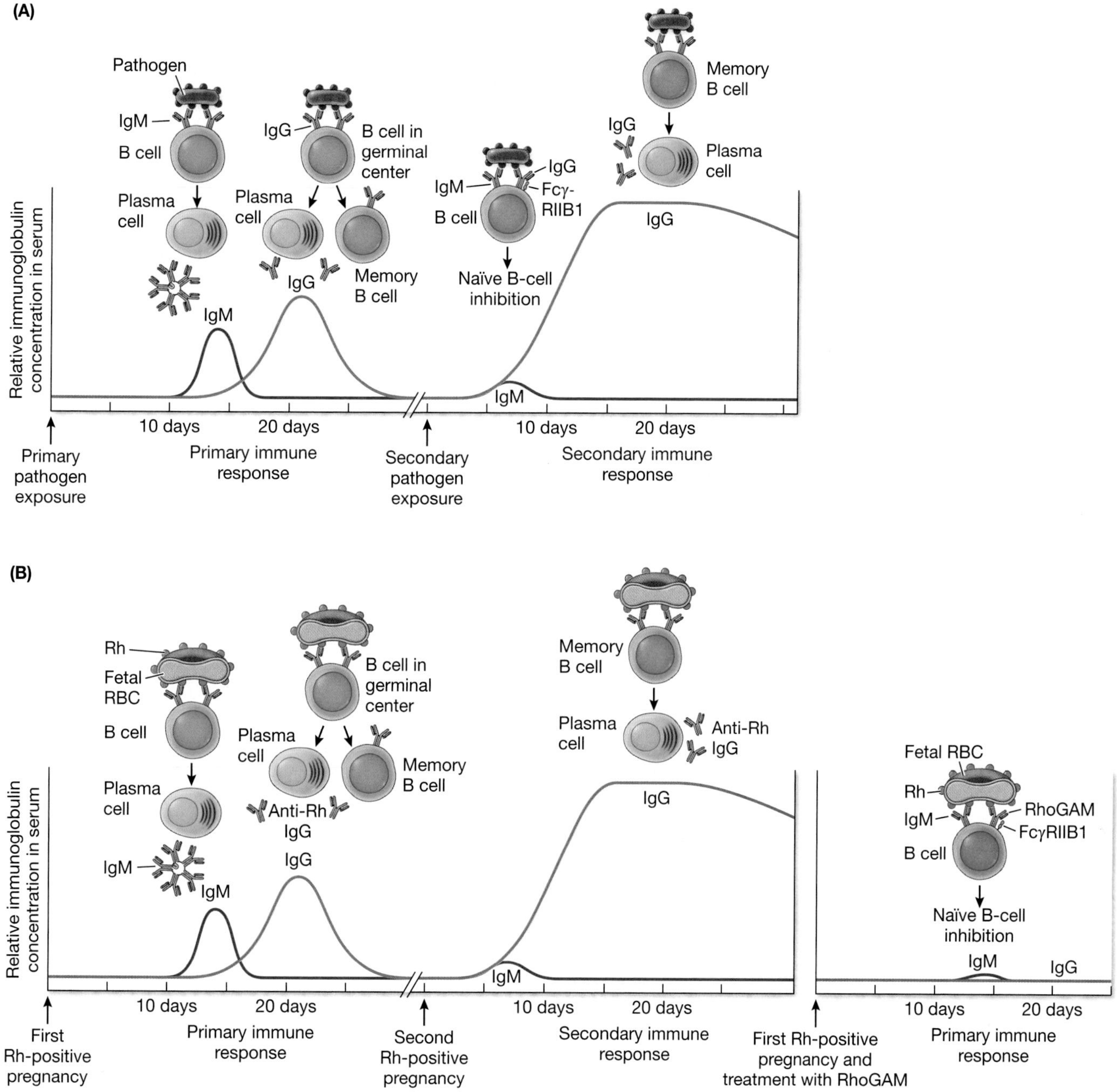

FIGURE 11.4 The secondary immune response blocks naïve B-cell activation (A) During a primary immune response, activated naïve B cells differentiate into plasma cells and secrete IgM that aids in pathogen recognition and destruction. Some B cells migrate to a lymphoid follicle, and germinal center B cells undergo somatic hypermutation and isotype switching. These activated B cells differentiate into plasma cells that secrete isotype-switched immunoglobulin (such as IgG) and memory B cells. Subsequent exposure to the same pathogen promotes activation of memory B cells, driving their differentiation into plasma cells that secrete isotype-switched immunoglobulin. The presence of high-affinity immunoglobulin that can bind to the same antigen as a naïve B cell results in recognition of the immunoglobulin by FcγRIIB1, an inhibitory receptor on the naïve B-cell surface, which prevents activation of the cell. (B) Left: A Rhesus factor (Rh)-negative female carrying a fetus that expresses Rh on his/her erythrocytes can initiate a B-cell response, whereby naïve B cells that come into contact with fetal red blood cells (RBCs) bind to Rh and activate to produce plasma cells expressing anti-Rh IgM. This activation results in the production of germinal center B cells that undergo isotype switching and produce higher-affinity anti-Rh IgG and memory B cells that produce the same immunoglobulin. Subsequent pregnancy with an Rh-positive fetus can activate the memory B cells. This activation results in production of high-affinity IgG that can cross the placental barrier and lysis of fetal RBCs, leading to fetal anemia. Right: To prevent activation of naïve B cells that express immunoglobulins that recognize Rh, females pregnant with an Rh-positive fetus are administered RhoGAM, an anti-Rh immunoglobulin. The immunoglobulin binds to the FcγRIIB1 inhibitory receptor on naïve B cells that recognize Rh, preventing their activation and the subsequent production of memory B cells. (Graphs after N. Parker et al. 2016. *Microbiology*. Houston, Texas: OpenStax. CC BY 4.0. Access for free at: https://openstax.org/details/books/microbiology.)

Making Connections: Primary and Secondary Adaptive Immune Responses

Lymphocytes of the adaptive immune system recognize pathogens using their specific receptor, a product of the recombination events that occurred during their development. Upon initial contact with a pathogen, the primary adaptive immune response is driven by the action of draining lymphoid tissues, where the majority of lymphocyte activation occurs. After activation, memory cells are produced that will promote a more robust secondary adaptive immune response upon subsequent exposure to the same pathogen.

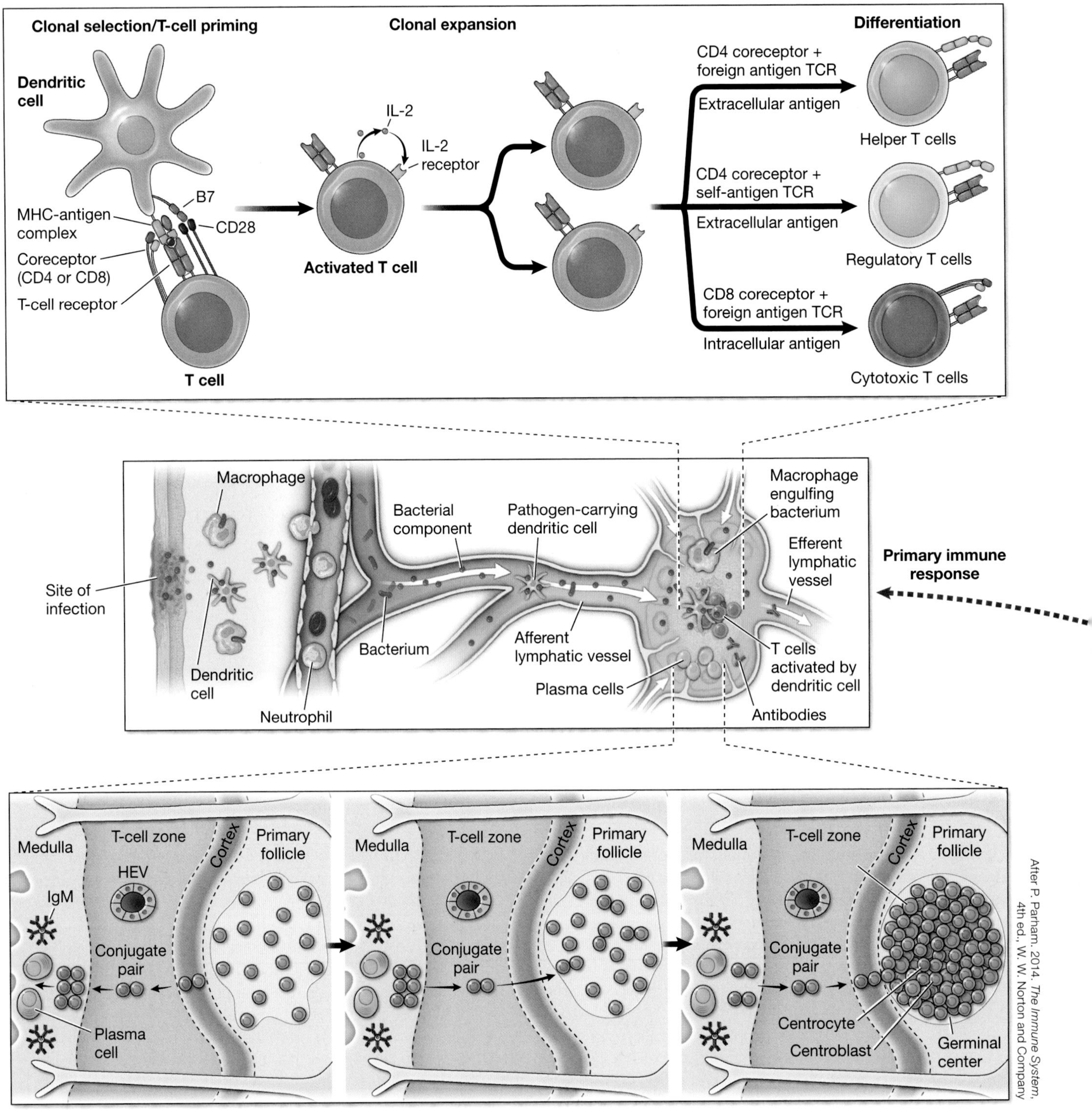

After P. Parham, 2014. *The Immune System*, 4th ed., W. W. Norton and Company

The primary adaptive immune response is concentrated in secondary draining lymphoid tissue. Inflammation promotes the migration of antigen-presenting dendritic cells into secondary lymphoid tissue, where T cells and B cells are activated.

(Sections 1.4, 1.8, 1.10, 7.1, 7.3, 7.7, 10.2, 10.3, 10.4)

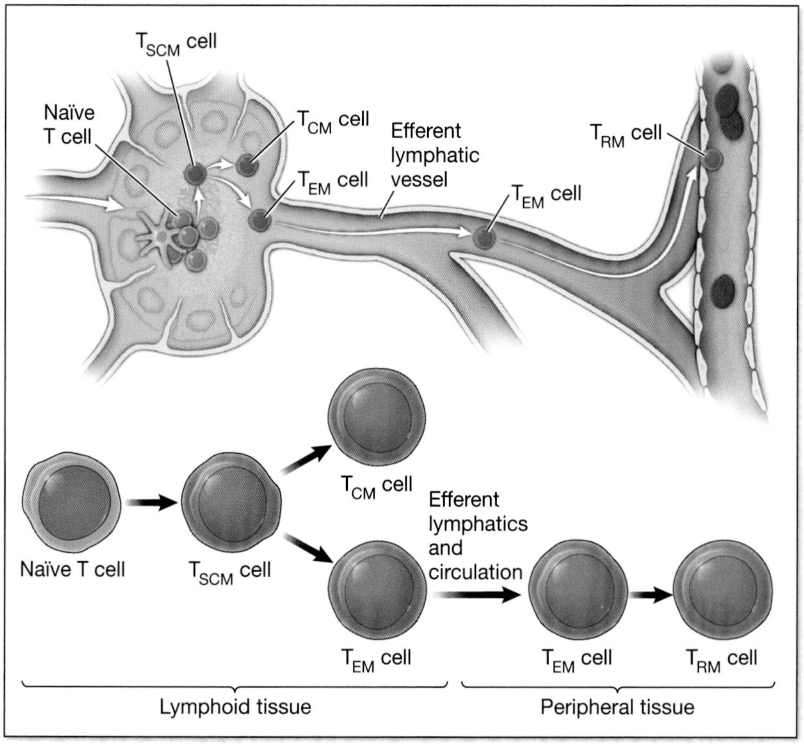

After D. L. Farber et al. 2014. *Nat Rev* 14: 24–35

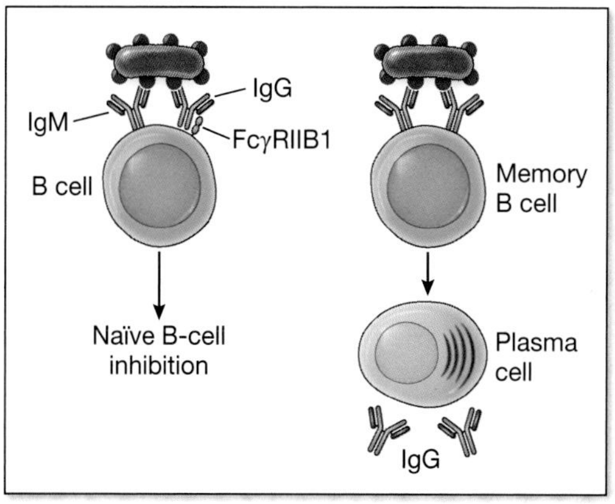

The secondary adaptive immune response is a result of activation of memory T and B cells produced during the primary adaptive immune response. The secondary response is faster because memory cells reside where they were produced. It is more robust because memory lymphocytes do not require costimulation, and memory B cells have undergone somatic hypermutation.

(Sections 1.8, 1.10, 11.1, 11.2)

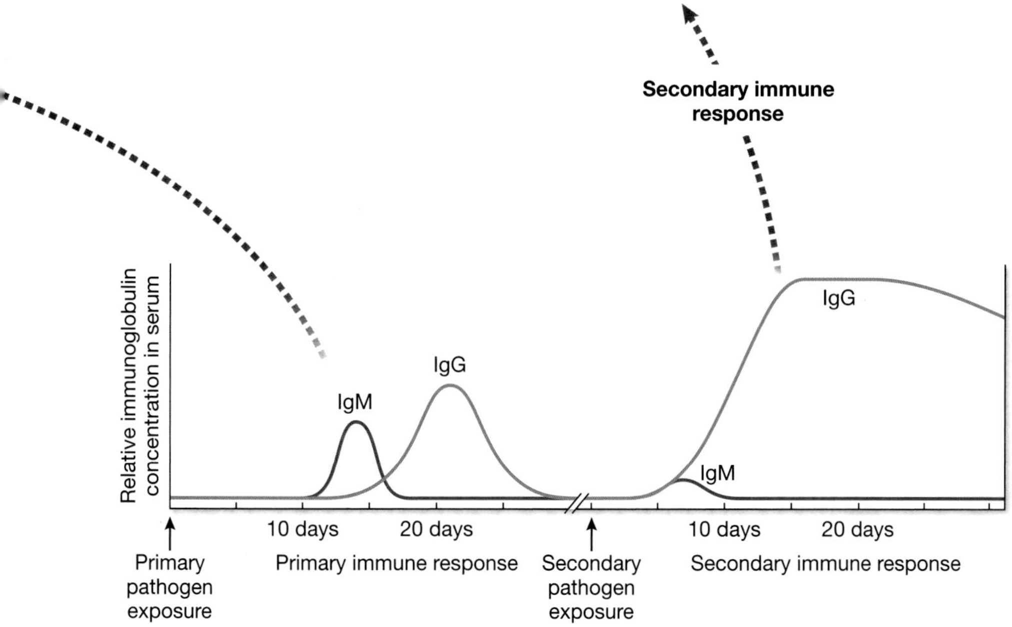

After N. Parker et al. 2016. *Microbiology*. Houston, Texas: OpenStax. CC BY 4.0. Access for free at: https://openstax.org/details/books/microbiology

A subsequent pregnancy with Rh incompatibility can drive the activation of these memory B cells and the production of high-affinity IgG immunoglobulins that can cross the placental barrier and bind to red blood cells in the developing fetus. These immunoglobulins can drive the clearance of fetal red blood cells containing Rh factor, resulting in fetal anemia (inadequate number or quality of red blood cells in the fetus).

To prevent this reaction, at-risk pregnant women are treated with the antibody RhoGAM® (Rh_0[D] immune globulin), which recognizes Rh on fetal red blood cells that have entered the mother's circulation. Because the antibody is bound to the red blood cells, any naïve B cells capable of recognizing the Rh-positive red blood cells will bind but will be inactivated due to negative signals driven by the RhoGAM immunoglobulin binding to FcγRIIB1 on the naïve B cells. This treatment effectively removes naïve B cells from the mother that are capable of being activated by Rh factor, thus protecting the Rh-positive red blood cells of the fetus from being destroyed or damaged by the mother's immune system (see Figure 11.4). RhoGAM also acts as an opsonin for fetal red blood cells that have moved into the mother's circulation to promote quick clearance before the Rh-positive cells can activate maternal B cells that express an immunoglobulin to Rh factor.

Persistence of memory cells

Lymphocytes require survival signals in order to persist. Naïve lymphocytes gain these survival signals when their T-cell receptors or membrane immunoglobulins engage with their cognate antigen. Circulating naïve lymphocytes that do not receive this survival signal are short-lived in circulation and are removed by apoptosis. However, memory lymphocytes do not seem to require survival signals through their T-cell receptors or immunoglobulins for long-term survival.

It is unclear what survival signals drive long-term persistence of memory lymphocytes, but we know that memory cells can survive for many years. It has been shown that secreted cytokines are capable of providing survival signals to subsets of memory lymphocytes. For example, IL-7 and IL-15 can promote survival of $CD8^+$ memory T cells without the need for T-cell receptor engagement. It has been suggested that these cytokines may come from other adaptive immune responses unrelated to the action of these memory lymphocytes. Further evidence has demonstrated that memory lymphocytes transcriptionally upregulate genes that aid in their utilizing survival signals for long-term persistence, including T-cell receptor and immunoglobulin transcription to promote cross-reactivity through these receptors as an additional survival signal.

CHECKPOINT QUESTIONS

1. Compare and contrast the locations of naïve, central memory, effector memory, and resident T cells.
2. What is the purpose of negative signaling in naïve B cells through engagement of the FcγRIIB1 receptor?

11.3 | How do vaccines prevent disease?

LEARNING OBJECTIVE

11.3.1 Explain the process by which vaccines prevent disease.

One of immunology's medical triumphs is the development and use of vaccines to protect populations from pathogenic disease. In the late 1700s, Edward Jenner, an English physician and scientist, created a vaccine for smallpox, which at the time was infecting 60% of the world population and resulting in 30% mortality.

Jenner had heard that people who had experienced a cowpox infection (a mild infection, common in farmers and milkmaids) did not seem to have a reaction when inoculated with smallpox material containing the virus (powdered dried scabs or pustule fluid), suggesting that the cowpox infection provided protection against the inoculated smallpox material. This process of inoculating an individual with smallpox material was known as **variolation**. He then postulated that a cowpox inoculation would protect people from smallpox. Jenner successfully demonstrated that cowpox inoculation was protective against smallpox, and thus was born the process of **vaccination**. In 1798, the results of his experiments were published and widespread use of the smallpox vaccine began. Prior to Jenner's discoveries, variolation reduced the number of infected persons by 15% to 30%. However, it also caused approximately 1% of those treated to actually contract the disease.

The world witnessed a recent triumph in the history of vaccination with the rapid development and production of multiple vaccines against the causative agent of COVID-19, SARS-CoV-2. The vaccines developed by Pfizer/BioNTech and Moderna are both mRNA vaccines (discussed in Section 11.4) that benefited from past studies aimed at the development of a vaccine against severe acute respiratory syndrome (SARS), which saw an outbreak between 2002 and 2004. The similarity of the coronaviruses that cause SARS and COVID-19, along with 30 years of research aimed at the development of mRNA vaccines, allowed the companies to develop, test, and produce mRNA COVID-19 vaccines in 11 months. The previous record for getting a vaccine to market was the mumps vaccine, developed in the 1960s, which took 4 years.

Although the process of vaccination is encountering some opposition in today's society (see the **CONTROVERSIAL TOPICS** box), most scientists herald vaccination as a triumph of immunology (see the **KEY DISCOVERIES** box). Furthermore, vaccination can benefit society as a whole because of the nature of

variolation Obsolete practice in which a small dose of a pathogen (smallpox virus) was introduced intranasally or through a small scratch under the skin to create a mild infection with the goal of providing immunity to a more serious infection.

vaccination Process of inoculating an individual with a harmless counterpart of a pathogen to mimic an infection to stimulate the adaptive immune response and the production of memory lymphocytes for fast, robust protection upon subsequent exposure to the pathogen.

CONTROVERSIAL TOPICS

Vaccines and Autism

The year was 1998, and a study led by Andrew Wakefield published in *The Lancet* initiated a cascade of events whose ripples are still being felt today. The published study focused on 12 British children and claimed to demonstrate a link between the measles, mumps, and rubella (MMR) vaccine and the onset of autism. This article was later retracted due to claims of conflict of interest and the inability of the scientific community to reproduce the findings. The conflict-of-interest claims were related to Wakefield's connection to two venture capital businesses, Immunospecifics Biotechnologies Ltd. and Carmel Healthcare Ltd., which stood to earn large profits from a reported connection between the MMR vaccine and autism. Furthermore, an investigation published in 2011 demonstrated that Wakefield's study was essentially fabricated (*British Journal of Medicine*, 2011. 342: c5347).

Since its publication, there have been large-scale studies, with extensive population data sets, which were reviewed by governmental agencies including the CDC, the American Academy of Pediatrics, and the Institute of Medicine of the National Academy of Sciences, and all reached the same conclusion—the MMR vaccine does not cause autism.

However, the debate rages on, championed by celebrities with their own anecdotes, the media, and online stories and accounts that fuel the beliefs of either side. Confounding the debate is the increased understanding of the autism spectrum and increasing numbers of children being diagnosed with some form of autism or related conditions. Furthermore, children are commonly diagnosed at approximately the same age as when they receive the MMR vaccine. Our current age of instant communication and social media influence contributes to polarizing the sides of this debate. Although many scientists will state with clear certainty and beyond reasonable doubt that vaccines do not cause autism, the debate continues, and will likely continue until the cause of autism is clearly determined.

Think About...

1. Which side of the vaccine-versus-autism debate do you relate to? What key pieces of information can you cite as supporting information?
2. What data might convince you to consider the other side's stance as more credible than your own?

KEY DISCOVERIES

How was vaccination developed against smallpox?

Article

Jenner, E. 1798. An inquiry into the causes and effects of the variole vaccinae, a disease discovered in some of the western counties of England, particularly Gloucestershire and known by the name of the cow-pox. London: Sampson Low, 1798.

Background

Smallpox was having a profound effect on the global population in the 1700s, infecting approximately 60% of the world, with an associated mortality rate of 30%. The massive impact of the disease warranted attempts to combat the infection and increase survival by protecting the population by any means necessary. However, at the time, pharmaceuticals did not yet exist and medical knowledge was limited. The pressing question was: How can the global population be protected from such a devastating disease?

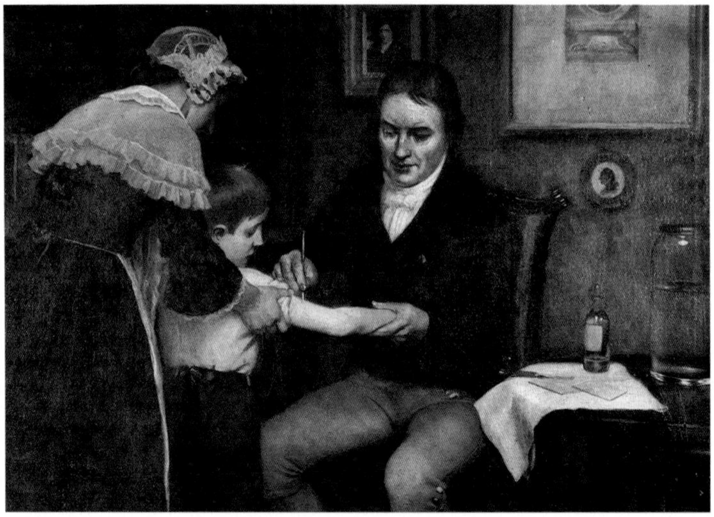

Early twentieth-century painting by Ernest Board depicts Dr. Edward Jenner performing his first vaccination in 1796.

Early Research

One of the standard means of protection against pathogens of the time was the practice of variolation: inoculation with a small amount of the disease-causing agent. Observations had been made that this could provide protection against subsequent infection from the same pathogen. However, this practice came with major risk—approximately 1% of those inoculated with the pathogen (by scratching the skin surface and applying infected material) would subsequently be infected with the pathogen. Although the practice of variolation brought with it the potential for protection against a pathogen, scientists and doctors of the time scrambled to find a safer alternative.

Think About...

1. Was variolation worth the risk of infection? How might a medical doctor of today wrestle with the conundrum of that time—deliberately infecting an individual with a pathogen to attempt to provide future protection? What are some of the ethical and moral questions associated with this situation?
2. If you lived in the eighteenth century and had to design a vaccine for smallpox (or any other pathogen), what questions might need to be answered? (Remember, scientists at the time had very limited technology compared to today!)
3. In the eighteenth century, what means might be available to test your vaccine?

Article Summary

Edward Jenner, often heralded as the father of immunology, utilized current knowledge and careful observations to attempt to design a protective measure against smallpox that bypassed variolation, thus reducing the risk of infection with the live virus. A major observation was that individuals who had been exposed to cowpox (including farmers and milkmaids) did not seem to have a reaction to smallpox variolation, suggesting that the cowpox exposure acted similarly to smallpox variolation and that cowpox exposure provided resistance to smallpox infection. Jenner postulated that cowpox blisters (and the pus of those blisters) might have a protective agent that could help prevent smallpox infection.

Jenner inoculated an 8-year-old boy with material he had isolated from a milkmaid, and although the boy experienced pain in his armpit, complained of a headache, and developed a fever, he remained protected from smallpox. Over the course of several months, the boy was inoculated with smallpox several times, never becoming infected with the smallpox virus. Jenner's findings on the vaccination of this child, and the other cases he documented in this seminal paper, gave birth to modern vaccination.

herd immunity, which is protection of unvaccinated individuals due to a large percentage of the population being vaccinated. This means the pathogen cannot infect enough people to start an epidemic.

The Centers for Disease Control and Prevention (CDC) provides a recommended immunization schedule for children from birth to 18 years of age. This schedule is endorsed by the American Academy of Family Physicians and the American Academy of Pediatrics, along with other medical organizations. The most recent version is shown in **FIGURE 11.5**.

herd immunity Protection of unvaccinated individuals within a population that is largely vaccinated against a specific pathogen, which prevents the spread of infection within the population.

Vaccine strategy

Vaccination takes advantage of the production of memory lymphocytes during a primary immune response. The strategy of vaccination is to induce a primary immune response by activating lymphocytes that will combat a specific pathogen without an actual infection by the pathogen.

As we have seen, this primary immune response can give rise to memory T and B cells and thus allow for protection upon subsequent contact with the pathogen. The production of memory T and B cells is the goal of vaccination. Vaccination provides immunity while removing the inherent danger associated

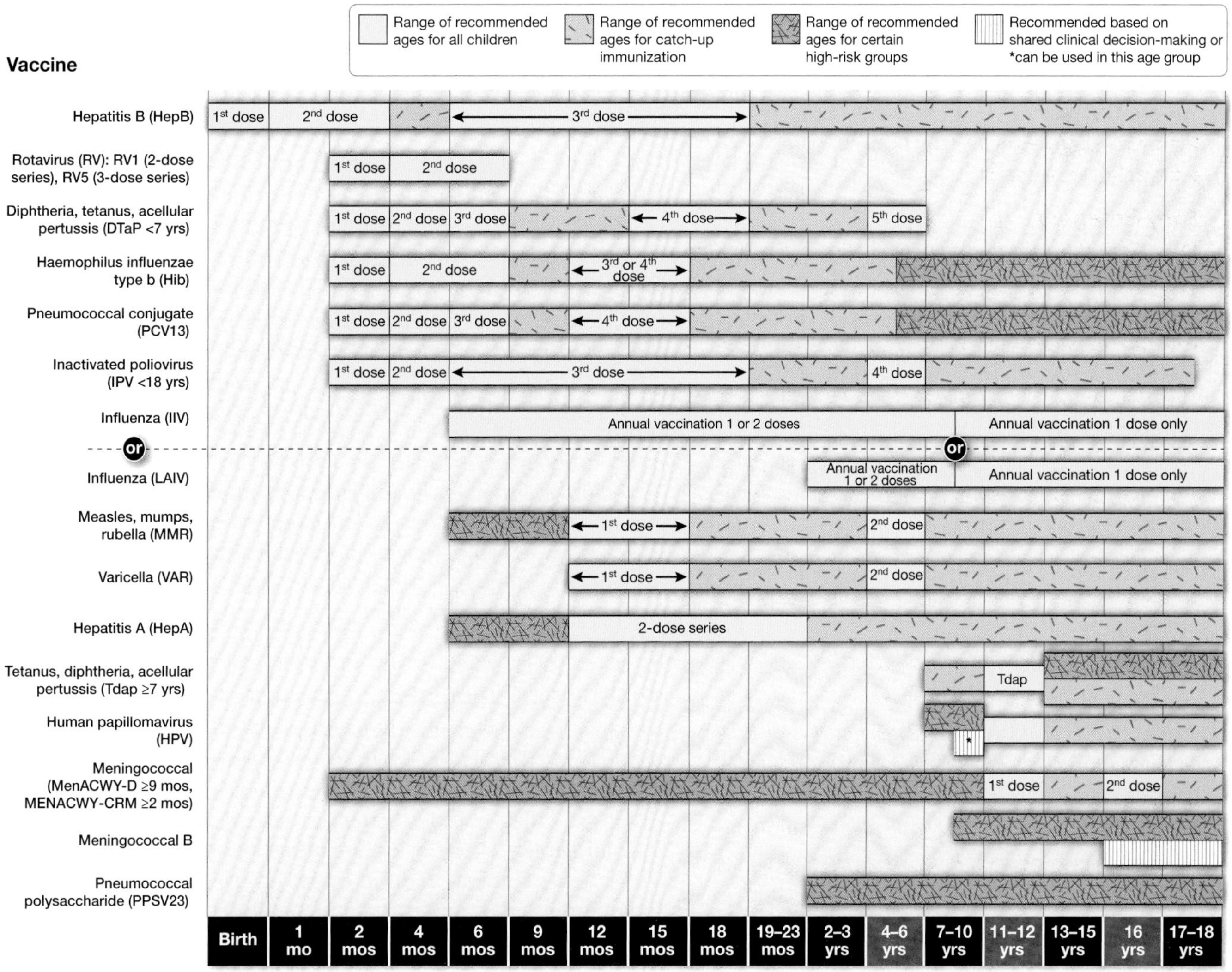

FIGURE 11.5 CDC-recommended child and adolescent immunization schedule for ages 18 years and younger, 2020 (Centers for Disease Control and Prevention. 2021. *Recommended Child and Adolescent Immunization Schedule for ages 18 years or younger*, US, 2021. 2/11/2021 update. US Dept. of Health and Human Services. www.cdc.gov/vaccines/schedules/downloads/child/0-18yrs-child-combined-schedule.pdf)

with pathogen contact, effectively tricking the adaptive immune system into mounting a primary immune response without the pathogen itself being present.

To activate these lymphocytes, antigens and epitopes from the pathogen must be utilized to allow for proper presentation to lymphocytes and to drive clonal selection and clonal expansion. Part of the strategy of vaccination is determining the right antigen that is capable of eliciting the desired primary immune response. We will see later in this chapter that there are several means by which antigens or epitopes can stimulate a proper primary immune response, resulting in an effective vaccine.

CHECKPOINT QUESTIONS

1. What were the key observations that allowed Jenner to hypothesize that cowpox inoculation would protect against smallpox infection?
2. How are Jenner's findings connected to our current knowledge of the adaptive immune system?

11.4 What are the different types of viral and bacterial vaccines?

LEARNING OBJECTIVE

11.4.1 Identify the various types of vaccines based on their components and actions.

inactivated vaccine Vaccine composed of chemically inactivated, irradiated, or heat-killed pathogen.

live attenuated vaccine Vaccine composed of a pathogen that has not been inactivated but is incapable of normal survival or pathogenesis in the host.

Edward Jenner was somewhat lucky in having access to a virus that was closely related to the smallpox virus, enabling its use as a vaccine. Most pathogens do not have closely related nonpathogenic counterparts, making vaccine design much trickier than simply selecting a different organism.

We have seen that the goal of vaccination is to stimulate the activation of a primary adaptive immune response and promote the production of memory cells capable of combating a specific pathogen. An effective vaccine must be capable of inducing this response without causing harm to the population being treated. Furthermore, since pathogens are recognized by different effector lymphocytes (e.g., viruses are typically recognized by CD8 T cells), vaccination must induce a primary immune response that promotes activation of the exact lymphocytes needed to combat the actual pathogen, if possible. In this section, we will explore the different types of vaccines that are used successfully for protection against a variety of pathogens. **TABLE 11.2** provides a summary of vaccine types that are currently available.

TABLE 11.2 | Types of Vaccines

Vaccine Type	Examples
Inactivated	Rabies Polio (Salk) Hepatitis A Influenza
Live attenuated	MMR (measles, mumps, rubella) Rotavirus Polio (Sabin) Varicella
Toxoid	Diphtheria Tetanus
Subunit	Pertussis Hepatitis B
Conjugate	Streptococcal pneumonia *Haemophilus influenzae*
Recombinant vector (in trials)	HIV Rabies Measles SARS-CoV-2 (virus that causes COVID-19)
DNA (in trials)	Influenza Herpes simplex virus
Messenger RNA	SARS-CoV-2 (virus that causes COVID-19)

Inactivated vaccines

Because there are so few nonpathogenic counterparts to human pathogens that can be used as a vaccine, most vaccines are composed of pathogens that have been killed or inactivated to destroy their ability to cause disease. Killing or inactivation of pathogens is typically done either through heat treatment (high temperatures can kill bacteria), irradiation (powerful irradiation can permanently damage the pathogen's DNA), or formalin treatment (formalin is a formaldehyde solution that crosslinks or covalently modifies proteins and DNA).

Since killing or inactivating pathogens via these methods theoretically still maintains many of the antigens that may be important in promoting a primary immune response, **inactivated vaccines** provide a simple means of inducing this response using a largely intact pathogen that cannot cause disease. While all three of these mechanisms effectively inactivate the pathogen, researchers have suggested that irradiation may produce a vaccine that best mimics the actual pathogen as this process does less damage to the associated antigens, allowing the inactivated virus or bacterium to closely mirror the live pathogen within the host.

Inactivated vaccines generally require a second dose as a booster because the first dose does not elicit an immune response that provides a high level of protection. An example of an inactivated vaccine that requires multiple doses is the polio vaccine. The CDC recommends four doses of the inactivated polio vaccine in children for full protection against the virus.

Live attenuated vaccines

While inactivated vaccines are a safe and effective means of protection against many pathogens, inactivation often requires a large amount of pathogen and must effectively destroy all of the pathogen to ensure safety. Another method of vaccination utilizes pathogen that has lost its ability to cause disease but does not kill the pathogen. These **live attenuated vaccines** behave similarly to inactivated vaccines—both contain many (if not all) of the possible antigens capable of inducing a protective adaptive immune response. One added benefit of a live attenuated vaccine is the ability for the pathogen within the vaccine to survive without causing disease. This survival allows the vaccine to more closely mimic an infection without causing negative reactions within the host (**FIGURE 11.6**). Because a live attenuated vaccine more closely mimics an infection, often a single immunization is enough to induce a robust primary immune response.

Because the vaccine mimics an infection, live attenuated vaccines tend to be better at inducing the correct type of adaptive immune response when compared to inactivated vaccines. Furthermore, they can activate the cellular arm of the adaptive immune response and induce production of secretory IgA in addition to IgG. Many viral vaccines are live attenuated vaccines, including those for measles, mumps, and rubella (MMR), chickenpox, and rotavirus (see Table 11.2). While live attenuated vaccines have demonstrated benefits, some have resulted in severe complications in

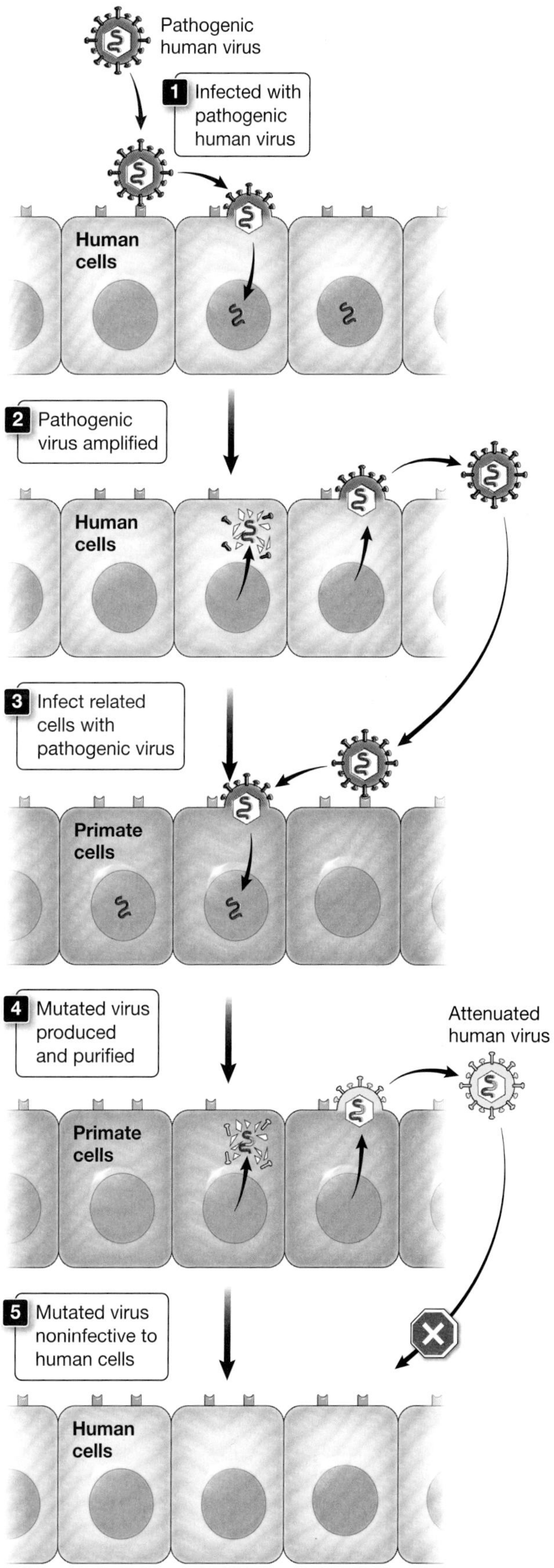

FIGURE 11.6 Production of attenuated vaccines First, a large amount of pathogenic virus is produced via infection of human cells grown in culture (Step 1). Through the life cycle of the virus, it is amplified and released into the cell culture media (Step 2). Amplified virus is used to infect a related cell line such as primate cells (Step 3). Viruses often undergo higher rates of mutation and may eventually mutate into a form that can infect the primate cells but not the human cell line. The mutated virus is produced and purified from primate cells (Step 4), but the once-pathogenic virus is incapable of infecting human cells, thus producing an attenuated vaccine incapable of producing disease in humans (Step 5). (After C. A. Janeway and P. Travers. 1997. *Immunobiology: The Immune System in Health and Disease*. Current Biology Limited/Garland Publishing: New York.)

toxoid Inactivated exotoxin employed in a vaccine to stimulate an adaptive immune response that involves production of neutralizing antibodies that block normal exotoxin activity.

immunocompromised individuals. Thus, the CDC recommends that immunocompromised individuals do not receive live attenuated vaccines such as the MMR, varicella (chickenpox), and rotavirus vaccines.

Toxoid vaccines

While the previously described vaccines utilize whole pathogens that have been inactivated or are impaired in pathogenesis, many vaccines also attempt to protect against the toxin products expressed by pathogens. This is especially true of bacterial pathogens, which commonly express exotoxins that are often responsible for the bacterial disease (see Chapter 13). Since the exotoxins are the causative agents of disease, activation of an adaptive immune response capable of neutralizing these toxins via immunoglobulin production is a common means to protect against the pathogenesis of these bacteria.

Diphtheria toxin and tetanus toxin are exotoxins produced by the pathogenic organisms *Corynebacterium diphtheriae* and *Clostridium tetani*, respectively. Protection against infections caused by these microorganisms is afforded by neutralizing these toxins. However, because these toxins are capable of inducing disease, they cannot be simply injected as a vaccine. Thus, just as pathogenic organisms can be inactivated, exotoxins can be inactivated as well. **Toxoids** are exotoxins inactivated by formalin treatment or heat used to elicit a primary adaptive immune response without the toxic activity of the exotoxin. If this response activates B cells capable of producing toxin-neutralizing immunoglobulins, toxoid vaccines can offer protection from the pathogenic effects of the bacteria (**FIGURE 11.7**).

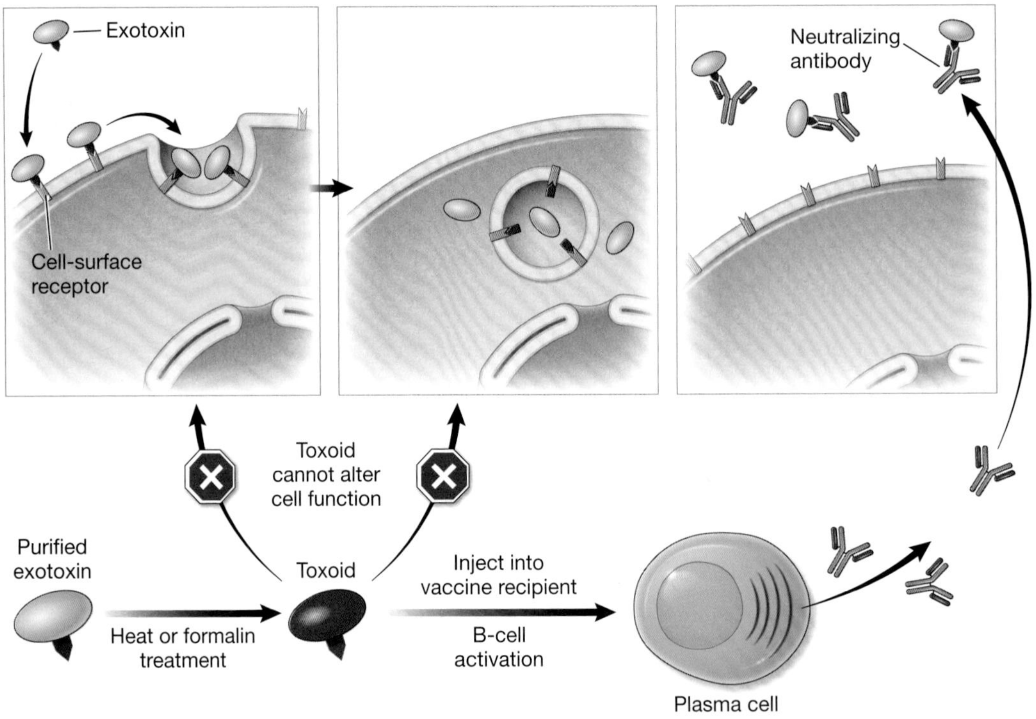

FIGURE 11.7 Toxoid generation for vaccine development Exotoxins secreted by pathogens target host cells and lead to altered cell function by binding to cell-surface receptors and delivering toxin components intracellularly. These toxin components affect normal cell function and can lead to disease. Purified exotoxins can be treated with heat or formalin to structurally or chemically alter them to create inactive toxoids. Inactive toxoids do not affect host cell function but can be recognized by the adaptive immune system. When recognized, they promote a primary immune response, activating B cells that express a receptor that recognizes the toxoid. The activated B cells differentiate into plasma cells, which produce neutralizing antibodies that prevent toxin action and entry into target cells.

Subunit vaccines

Just as neutralizing immunoglobulins are effective at halting the action of exotoxins, we also saw that neutralizing antibodies can protect cells from intracellular pathogens, especially viral infection. This is due to the action of the pathogen's adhesion molecules, which bind to target cells and gain entry. Neutralizing immunoglobulins that recognize these adhesion molecules can block infection by the intracellular pathogen. Thus, **subunit vaccines** represent one strategy used to raise a proper primary immune response and protect against cellular adhesion and entry into target cells (**FIGURE 11.8**). Since the adhesion molecules of viruses and bacteria are not directly responsible for generating disease symptoms, they can be employed to generate a primary adaptive immune response (similar to the strategy associated with toxoid vaccines). The resulting immune response activates B cells that can generate neutralizing antibodies to the specific cell adhesion molecules.

subunit vaccine Vaccine composed of a cell-surface antigen of a pathogen; designed to stimulate the production of neutralizing immunoglobulins that block pathogen interaction with target cells.

conjugate vaccine Vaccine composed of a weak antigen (typically incapable of mounting a robust adaptive immune response) coupled with a stronger antigen that can activate a protective primary immune response.

Conjugate vaccines

The term *conjugate* refers to something that is coupled or connected. A **conjugate vaccine** is a protective strategy based on coupling a weak antigen with a stronger antigen. These vaccines have multiple epitopes and thus are also considered *multivalent vaccines*.

Several pathogenic bacteria utilize a polysaccharide capsule as an evasive tactic that shields pathogen-associated molecular patterns (PAMPs) from

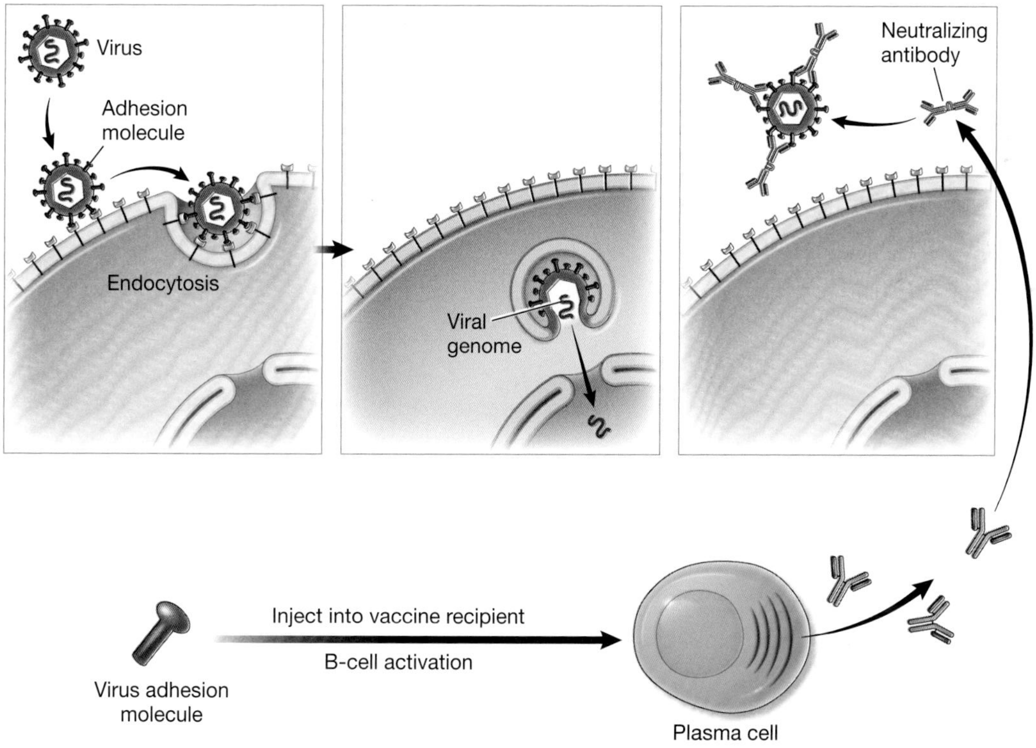

FIGURE 11.8 **Subunit vaccine development** Intracellular pathogens such as viruses bind to target cells via adhesion molecules to gain entry into the cell. During the normal viral life cycle, after binding to a target cell and endocytosis, the viral membrane fuses with the cell, and virus components (including the viral genome) enter the cytosol. Development of a subunit vaccine involves purification of pathogen adhesion molecules, which are then administered via injection to raise an adaptive immune response. The adhesion molecule activates B cells, driving differentiation of plasma cells and secretion of neutralizing antibodies that bind to the injected adhesion molecule and prevent the virus from binding to target cells.

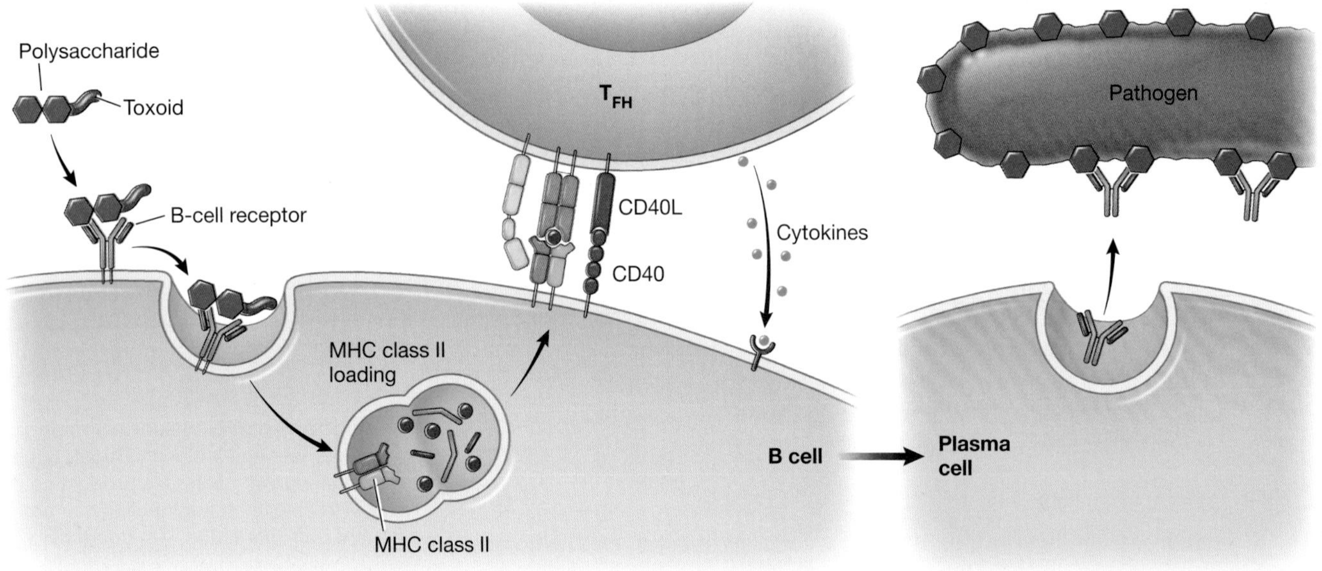

FIGURE 11.9 Conjugate vaccine development and action Since antigen presentation to T cells is largely limited to peptides, activation of B cells that express an immunoglobulin that can recognize a polysaccharide is limited. Conjugate vaccines are designed to overcome this limitation by linking the target polysaccharide to a toxoid. B cells expressing a receptor that binds to the polysaccharide phagocytose the conjugate material, which is degraded within the phagolysosome. Toxoid peptides are loaded onto MHC class II molecules and presented to helper T cells. A helper T cell that recognizes the MHC-toxoid complex activates the B cell expressing the immunoglobulin that binds to polysaccharide. The result is production of plasma cells that secrete the polysaccharide-specific immunoglobulin.

detection by the innate immune response (see Chapter 13). Healthy individuals are capable of raising an adaptive immune response that includes production of opsonizing antibodies that recognize these polysaccharide coats. However, in children younger than 18 months and in older adults, the adaptive immune response is less capable of efficiently activating B cells that produce these immunoglobulins. One reason for this reduced capability is that the generation of antibodies to polysaccharides is a T-cell-independent response (recall that T cells typically require protein antigen for activation). Young children and the elderly have a difficult time raising such a T-cell-independent response.

To counteract this difficulty, vaccines have been developed that link the polysaccharide coat (weak antigen) with a toxoid (strong antigen). This drives the activation of T cells that respond to the toxoid and B cells that respond to the polysaccharide. The activated B cells can then produce opsonizing antibodies to the polysaccharide coat of the bacterial pathogen (**FIGURE 11.9**).

Conjugate vaccines commonly require boosters to promote full protection against the pathogen. The conjugate vaccine against *Haemophilus influenzae* type b, a causative agent of bacterial meningitis, is an example of a conjugate vaccine that requires more than one dose.

Recombinant vector vaccines

Recombinant vector vaccines are one type of vaccine currently under development and undergoing clinical trials. The strategy underlying these vaccines is utilizing harmless bacteria or attenuated viruses to express a pathogenic antigen capable of eliciting an adaptive immune response. These vaccines take advantage of the ability to isolate a gene capable of encoding an antigen from a pathogen and placing the gene into a plasmid (small circular piece of DNA) or into a harmless or attenuated virus (**FIGURE 11.10**). The Janssen COVID-19 vaccine from Johnson and Johnson is an example of a recombinant vector vaccine that has been modified to express a surface protein of the SARS-CoV-2 virus.

FIGURE 11.10 Recombinant vector vaccine development These vaccines take advantage of recombinant DNA technology. After identification of a target antigen, the gene that encodes the antigen is amplified and placed into a nonpathogenic recombinant vector for delivery. The recombinant vector is injected into the vaccine recipient, infecting cells within the individual. While the recombinant vector is not pathogenic, it undergoes the life cycle of the vector, promoting insertion of the antigen gene into infected cells. Antigen is produced by infected cells and is either secreted for recognition by circulating B cells or processed via the MHC class I antigen processing and presentation pathway, where it can activate CD8 T cells with the correct receptor to recognize the MHC-antigen complex.

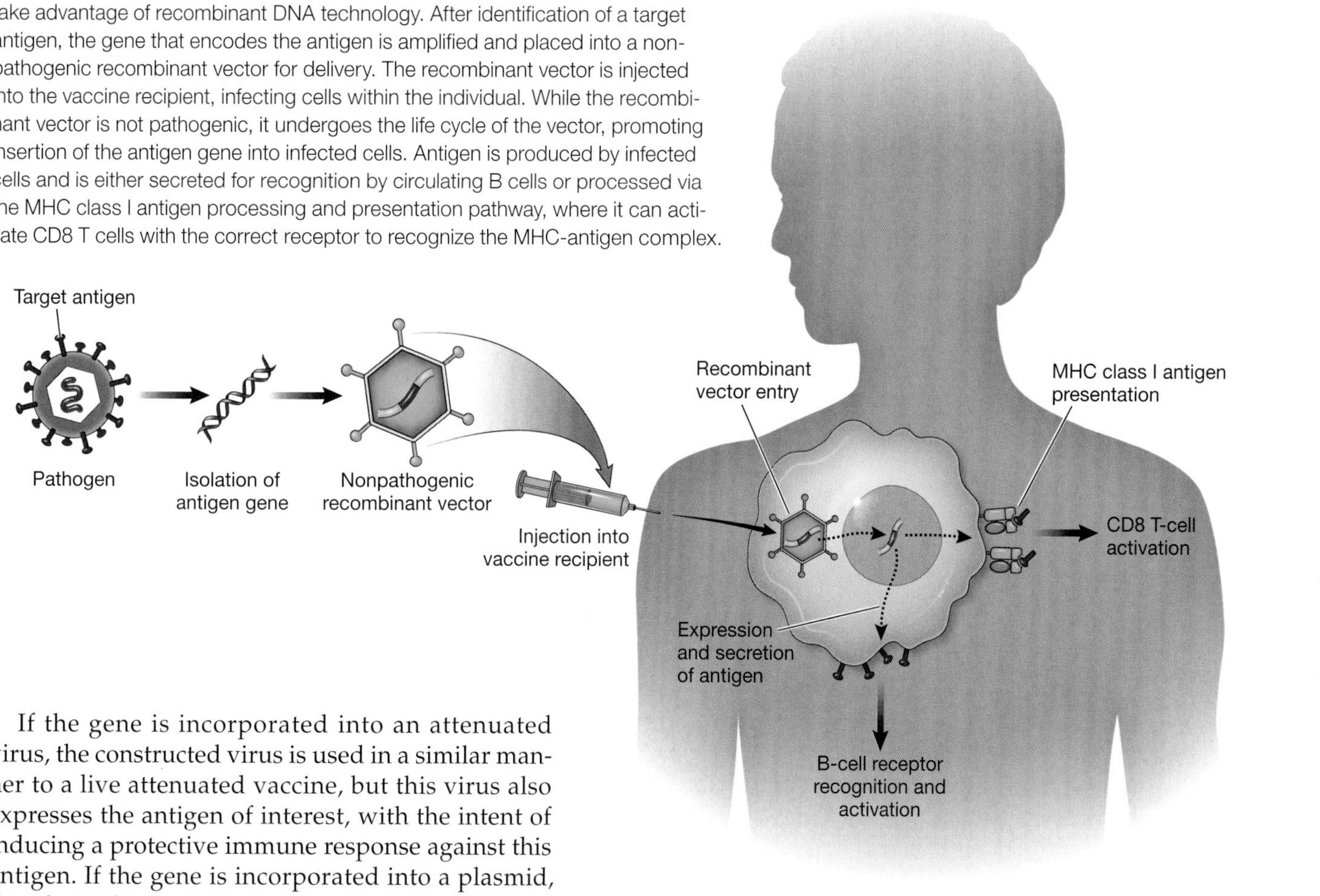

If the gene is incorporated into an attenuated virus, the constructed virus is used in a similar manner to a live attenuated vaccine, but this virus also expresses the antigen of interest, with the intent of inducing a protective immune response against this antigen. If the gene is incorporated into a plasmid, the plasmid is incorporated into harmless bacteria and the bacteria are used as the inoculated vaccine; the harmless bacteria express the antigen to which protection is to be mounted. Recombinant vector vaccines are currently in clinical trials for infections such as HIV, rabies, and measles.

DNA vaccines

DNA vaccines utilize a similar strategy to that of recombinant vector vaccines. However, instead of using a virus or bacteria as a carrier of the gene, the DNA vaccine is placed directly into the host in hopes of having host cells pick up the DNA and incorporate it into their genomic DNA (**FIGURE 11.11**). The DNA is typically mixed with microscopic gold particles, and the coated gold particles are then introduced with highly pressurized gas. Alternatively, the DNA vaccine can be mixed with a chemical that is easily taken up by the host cells. Vaccines currently under development attempt to incorporate DNA into target cells to express antigens within those host cells and activate CD8 T cells. DNA vaccines are currently being tested for influenza and herpes virus infections.

Messenger RNA vaccines

Messenger RNA (mRNA) vaccines rely on strategies similar to their DNA and recombinant vector vaccine counterparts. These vaccines are based on the central dogma of molecular biology, whereby mRNA serves as the code for protein translation. An mRNA molecule that encodes an antigen target is delivered into a target cell (**FIGURE 11.12**). The target cell takes up the mRNA to translate the antigen and induce an adaptive immune response.

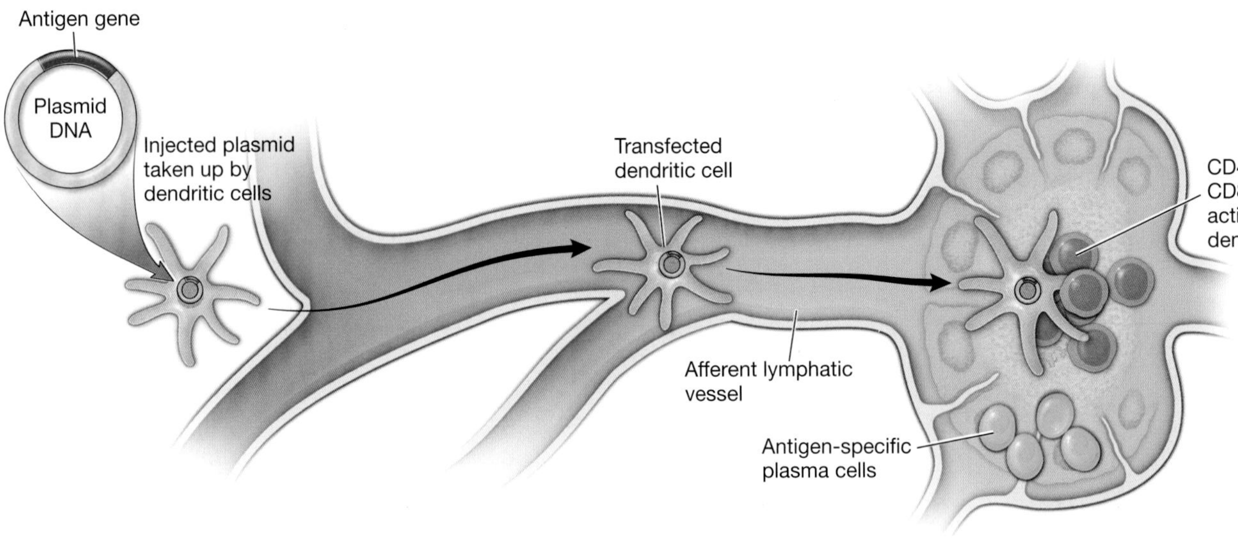

FIGURE 11.11 DNA vaccine development DNA vaccines are similar to recombinant vector vaccines, but rather than packaging an antigen gene in a recombinant vector, plasmid DNA containing an antigen gene is mixed with chemicals or particles easily taken up by cells. The mixture containing the DNA is then injected into an individual, where it is taken up by dendritic cells. The plasmid DNA within dendritic cells is used to encode antigen for both MHC class I and class II processing and presentation. Dendritic cells migrate to secondary lymphoid tissue, such as a lymph node, where they activate antigen-specific T cells and B cells, resulting in an adaptive immune response to the antigen. (After E. N. Gary and D. B. Weiner. 2020. *Curr Opin Immunol* 65: 21–27. CC BY 4.0. © The Authors.)

The mRNA can be mixed with cationic liposomes for delivery into target cells and may be packaged with protamines (positively charged nuclear proteins) for added stability.

One key issue associated with mRNA vaccines was discussed in Chapter 2—our innate immune system can recognize PAMPs as foreign entities. Injected mRNA may induce an innate immune response if it is recognized

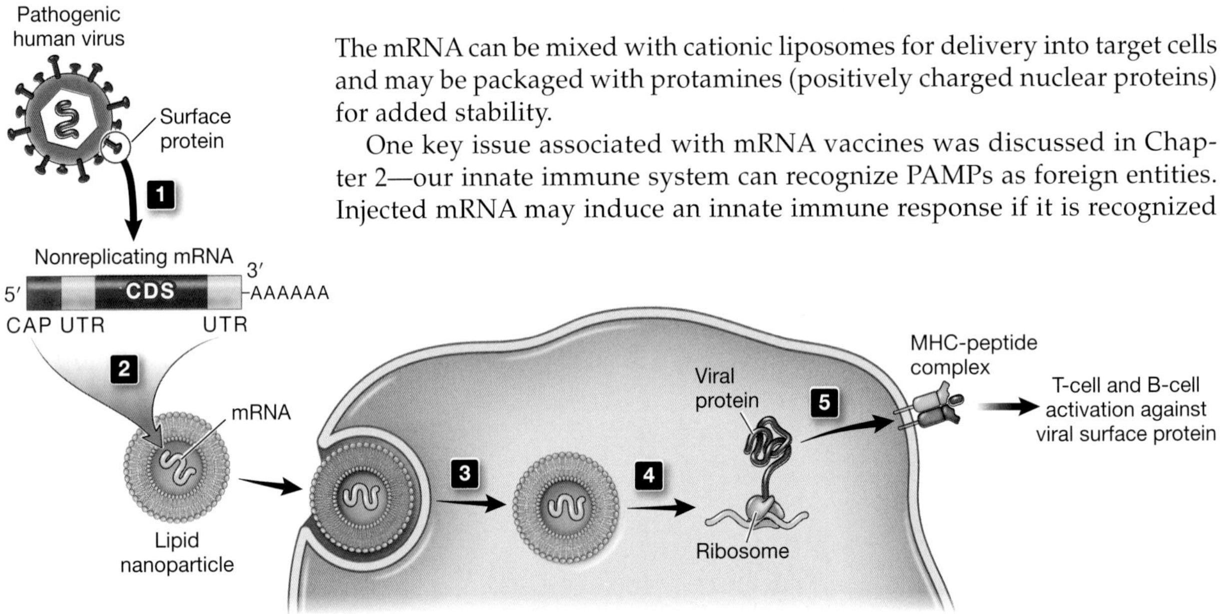

FIGURE 11.12 Production of mRNA vaccines The generation of an mRNA vaccine begins with the identification of a viral surface antigen, followed by the isolation and production of a nonreplicating mRNA that can encode this antigen (Step 1). The nonreplicating mRNA that is produced contains a protective cap and poly(A) tail, along with the untranslated regions (UTR) that flank the coding sequence (CDS). The synthesized mRNA is packaged in lipid nanoparticles (Step 2). After injection of the mRNA vaccine, it is taken up by cells and the mRNA is delivered to the cytosol (Step 3). The mRNA sequence is translated into the viral protein by the ribosome (Step 4). Cytosolic viral protein is processed and presented on MHC molecules on the surface of the cell (Step 5). This promotes T-cell and B-cell activation against the viral surface protein. (After N. A. C. Jackson et al. 2020. *npj Vaccines* 5: 11. CC BY 4.0. © The Authors.)

via pattern recognition receptors (PRRs) within the target cell. This innate immune response would target the cell with the mRNA molecule for destruction because it would trigger the same response our body uses to combat viral pathogens. To counter this possible recognition and prevent the mRNA-containing cell from being destroyed by the innate immune system, the mRNA can be purified and its nucleotides modified to limit recognition as a PAMP.

A recent success story has been the rapid development of two mRNA vaccines to combat the COVID-19 pandemic. These vaccines express a surface protein of the SARS-CoV-2 virus (see **EMERGING SCIENCE**). Because both the Pfizer/BioNTech and Moderna COVID-19 vaccines are approximately 50% effective after the first dose, it is recommended that two doses of this vaccine be administered (as the second dose raises the efficacy to greater than 90%).

CHECKPOINT QUESTION

1. Compare and contrast the strategies and actions of inactivated, live attenuated, toxoid, subunit, and conjugate vaccines.

EMERGING SCIENCE

Can messenger RNA be used for vaccination to a pathogen?

Article

Sahin, U., A. Muik, E. Derhovanessian, I. Vogler, L. M. Kranz, M. Vormehr, A. Baum, K. Pascal, J. Quandt, D. Maurus, S. Brachtendorf, V. Lorks, J. Sikorski, R. Hilker, D. Becker, A.-K. Eller, J. Grutzner, C. Boesler, C. Rosenbaum, M.-C. Kuhnle, U. Luxemburger, A. Kemmer-Bruck, D. Langer, M. Bexon, S. Bolte, K. Kariko, T. Palanche, B. Fischer, A. Schultz, P.-Y. Shi, C. Fontes-Garfias, J. L. Perez, K.A. Swanson, J. Loschko, I. L. Scully, M. Cutler, W. Kalina, C. A. Kyratsous, D. Cooper, P. R. Dormitzer, K. U. Jansen, and O. Tureci. 2020. COVID-19 vaccine BNT162b1 elicits human antibody and T_H1 T cell responses. *Nature* 586: 594–599.

Background

The COVID-19 pandemic caused by the SARS-CoV-2 virus took the world by storm, first identified in Wuhan, China, in December 2019 and rapidly infecting individuals worldwide. By the end of 2020, Johns Hopkins University reported more than 81 million COVID-19 cases globally, resulting in more than 1.7 million deaths. While mitigating factors, including lockdowns, travel restrictions, and requirements for social distancing and face coverings, were instituted in certain areas to lower exposure rates, it became increasingly apparent that true protection might not occur until the development of a safe and effective vaccine. Researchers raced to develop a vaccine, using both tried-and-true technology and the newer mRNA technology.

The Study

Researchers developed a vaccine that included an mRNA molecule that encodes for the receptor-binding domain of the viral spike protein, which is predicted to be a target for the generation of neutralizing antibodies. The mRNA molecule was made in vitro using modified nucleotides to protect the mRNA and minimize activation of an innate immune response and was packaged in lipid nanoparticles to aid in delivery to target cells. The researchers vaccinated 60 individuals with varying doses of the mRNA vaccine to test for the generation of a protective immune response. The vaccine induced a dose-dependent generation of immunoglobulins that recognized the spike protein receptor-binding domain that was capable of neutralizing virus infection (**FIGURE ES 11.1**). The researchers further demonstrated the activation of both CD4 and CD8 T cells due to the mRNA vaccine, with CD4 T cells preferentially differentiating into T_H1 helper T cells.

The researchers were able to successfully demonstrate the use of an mRNA vaccine to express the receptor-binding domain of the spike protein of SARS-CoV-2. The vaccine was shown to promote the production of immunoglobulins that bind to the spike protein and act as neutralizing antibodies. This research further paved the path for

(*Continued*)

EMERGING SCIENCE (continued)

the development of the first mRNA vaccine produced to combat the COVID-19 pandemic.

Think About...

1. Why did the researchers choose to generate the mRNA to express the virus spike protein? After looking into the biology of coronaviruses, are there any other potential targets the researchers could have chosen?
2. With your knowledge of induction of the adaptive immune response, which cells would be most beneficial in receiving the mRNA molecule for activation of the adaptive immune response and CD4 and CD8 T cells?
3. How would CD8 T cells be activated by an mRNA vaccine? How would CD4 T cells be activated by an mRNA vaccine?

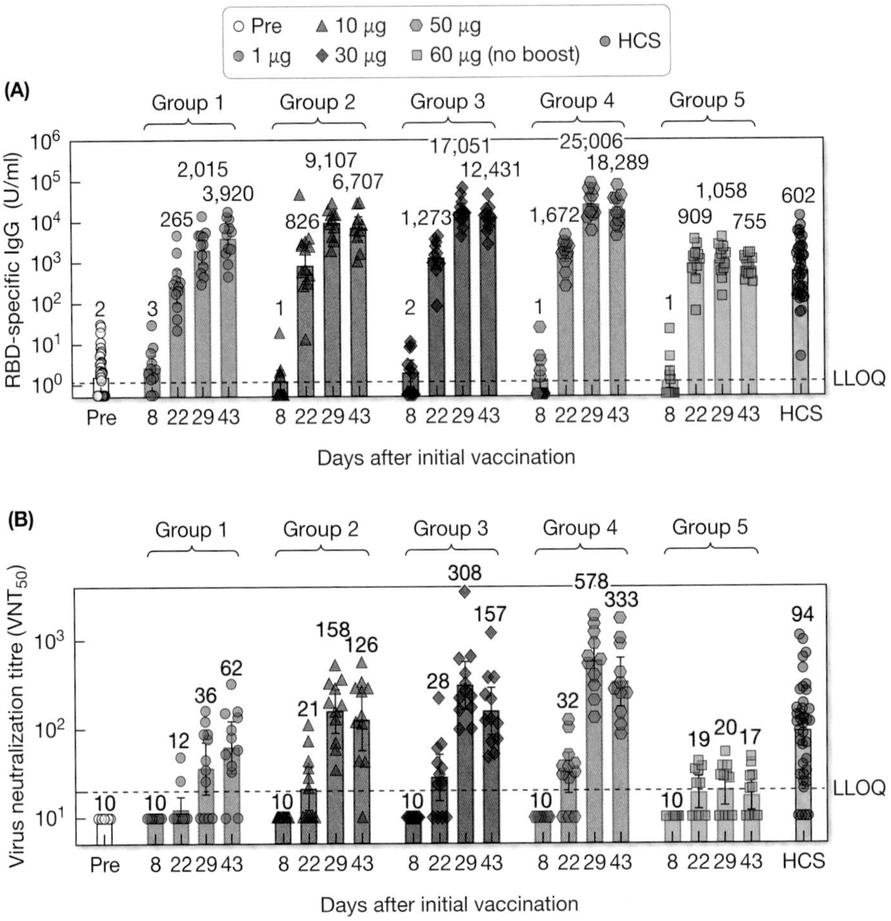

FIGURE ES 11.1 Effectiveness of an mRNA SARS-CoV-2 vaccine. (A) Injection of the mRNA vaccine of the SARS-CoV-2 spike protein receptor-binding domain into human participants induced the production of immunoglobulins specific to the receptor-binding domain in comparison with pre-injected control and to levels similar to that seen in serum from individuals who had recovered from COVID-19 (human convalescent serum [HCS]). (B) The same vaccine produced neutralizing antibodies against virus infection into Vero CCL81 cells grown in culture when compared to both pre-injection treatment and treatment with HCS. (From U. Sahin et al. 2020. *Nature* 586: 594–599. © The Authors.)

11.5 | What is the role of an adjuvant in a vaccine?

LEARNING OBJECTIVE

11.5.1 Explain how adjuvants in vaccines can make them more effective at preventing disease.

An effective vaccine is one that is capable of activating an adaptive immune response that will be protective against the pathogen that the vaccine is attempting to mimic. As long as the vaccine is indeed the first time an individual comes into contact with specific antigens from a potential pathogen, a primary adaptive immune response will be raised upon clonal selection and clonal expansion of the antigen-specific naïve T cells and B cells. However, recall that both naïve T and B cells require a costimulatory signal for proper activation—lack of this costimulatory signal during clonal selection and antigen-specific receptor engagement results in anergy of these naïve T and B cells (see Chapters 7 and 10). Thus, an effective vaccine not only must promote clonal selection of lymphocytes that will be protective against a pathogen but also must ensure that these lymphocytes receive a proper costimulatory signal so that the potentially protective lymphocytes are activated rather than removed from the population through anergy.

adjuvant Component of a vaccine used to enhance antigen-specific immune responses; important in stimulating the inflammatory response required for a complete adaptive immune response.

Inflammation revisited

One of the important immune responses of our innate immune system is the induction of an inflammatory response (see Chapters 1 and 2). This inflammatory response is important due to the secretion of cytokines that prompt complement activation and the activation of innate immune cells such as macrophages, neutrophils, and dendritic cells.

The activation of these cells not only increases their efficiency of action but also activates the expression of the costimulatory molecule B7 on dendritic cells, which is responsible for costimulation of the CD28 molecule on naïve T cells during antigen presentation. B-cell costimulation is also indirectly affected by the inflammatory response. Recall that B-cell costimulation utilizes the B-cell coreceptor, whereby CR2 on the B cell can bind to fixed complement components (namely iC3b and C3d) on a pathogen. Furthermore, B cells can be costimulated by the action of CD4 helper T cells: CD40L on a helper T cell can bind to CD40 at the B-cell surface and promote B-cell activation. Helper T cells will only express CD40L upon clonal selection and expansion, which in part is driven by the inflammatory response. Thus, vaccine development must employ a means to allow for clonal selection and expansion of protective lymphocytes while also inducing an inflammatory response to ensure proper lymphocyte activation.

Adjuvants

To work properly, a vaccine must be capable of activating an inflammatory response. This is especially true when the vaccine does not utilize a whole organism (is not an inactivated or live attenuated vaccine). Recall that the inflammatory response is typically driven by the recognition of PAMPs, and proteins do not typically belong to this family of molecules. To adequately induce an inflammatory response, substances known as **adjuvants** are added to the vaccine to strengthen the adaptive immune response by inducing inflammation.

To date, several substances have been approved for use as adjuvants in human vaccines. One that has been used safely for many decades is aluminum salts, or alum. Alum is thought to act by providing a reservoir of antigen at the injection site and promoting slow release to the immune system to help boost the immune response. Another approved adjuvant that functions in a similar manner is MF59, an oil-and-water emulsion thought to enhance the adaptive immune response through slow release of antigen. The other adjuvant approved for human use in the United States is AS04, which is a mixture of alum, and thus takes advantage of slow release of antigen, and a lipid molecule known as monophosphoryl lipid A, which is an agonist of TLR4 comparable to lipopolysaccharide (LPS). AS04 can directly induce an inflammatory response through the action of monophosphoryl lipid A binding to TLR4 and induction of inflammatory cytokine release.

While adjuvants work to slow the release of antigen to the immune system and prompt an inflammatory response, they are not particularly efficient in activating CD8 T cells, which are important in protecting against intracellular infections, especially viral infections. There have been recent advances in adjuvant production and antigen delivery systems aimed at enhancing the activation of

Making Connections: Vaccine Development and Targeted Pathogens

A properly developed vaccine results in a primary adaptive immune response that produces effector T cells and/or plasma cells capable of recognizing and destroying the targeted pathogen. That same primary adaptive immune response should produce memory T cells and/or memory B cells that will activate more quickly and more robustly should the vaccinated individual come into contact with the pathogen. Each vaccine strategy employs different techniques to identify and deliver antigens to the vaccinated individual, and some strategies are more useful at targeting specific pathogens.

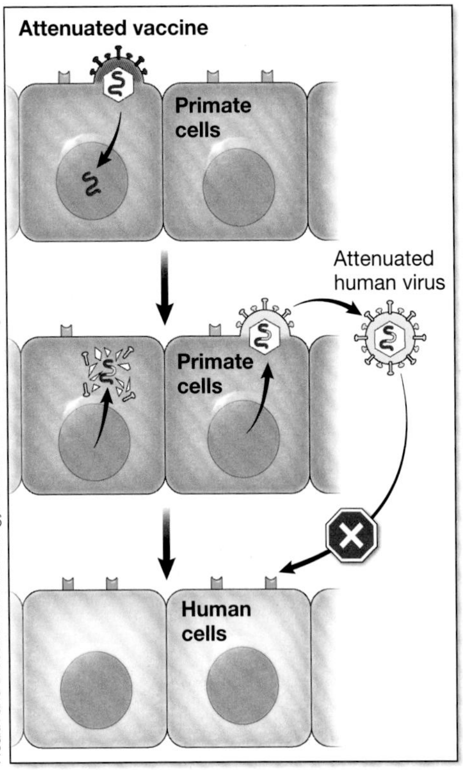

After C. A. Janeway and P. Travers. 1997. *Immunobiology: The Immune System in Health and Disease*. Current Biology Limited/Garland Publishing: New York

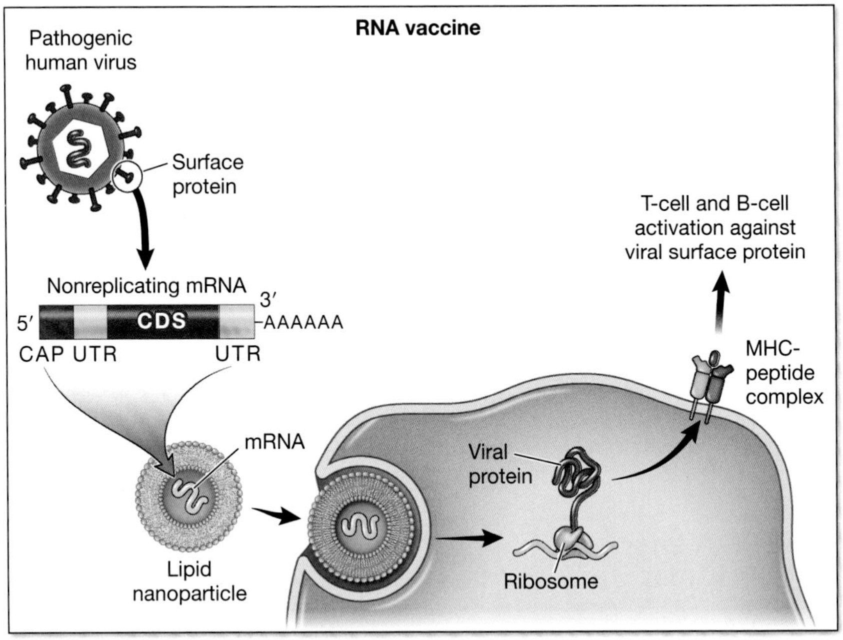

Targeted Pathogens				
Vaccine type	Bacteria	Viruses	Fungi	Protozoa
Toxoid	✓			
Conjugate	✓			
Attenuated	✓	✓		
RNA		✓		
Subunit	✓	✓	✓	✓
Recombinant vector	✓	✓	✓	✓
DNA	✓	✓	✓	✓

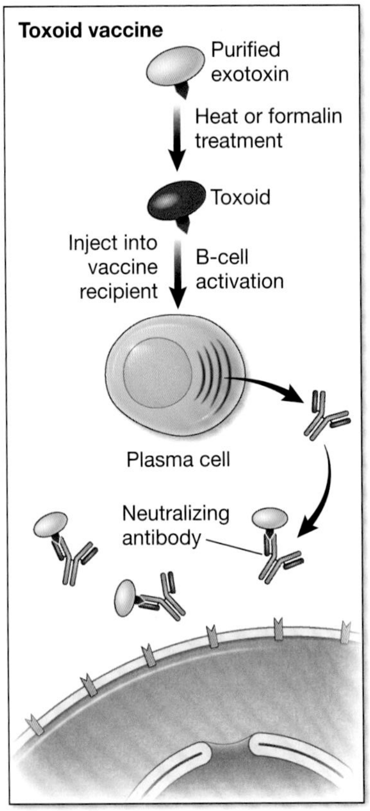

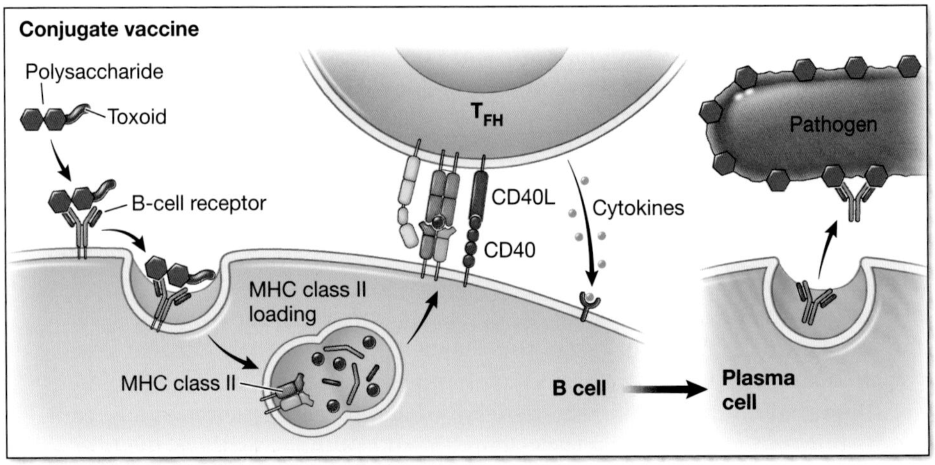

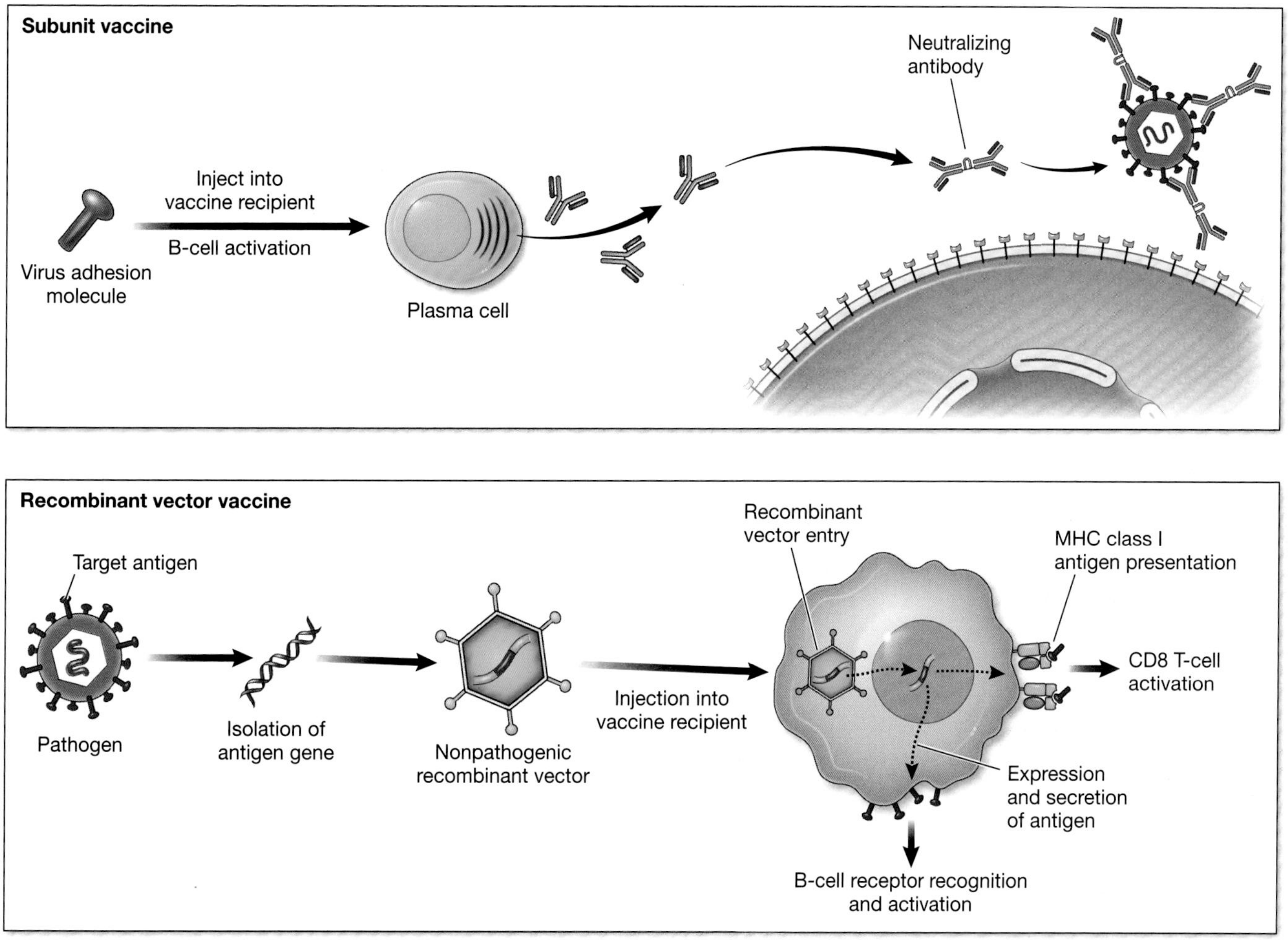

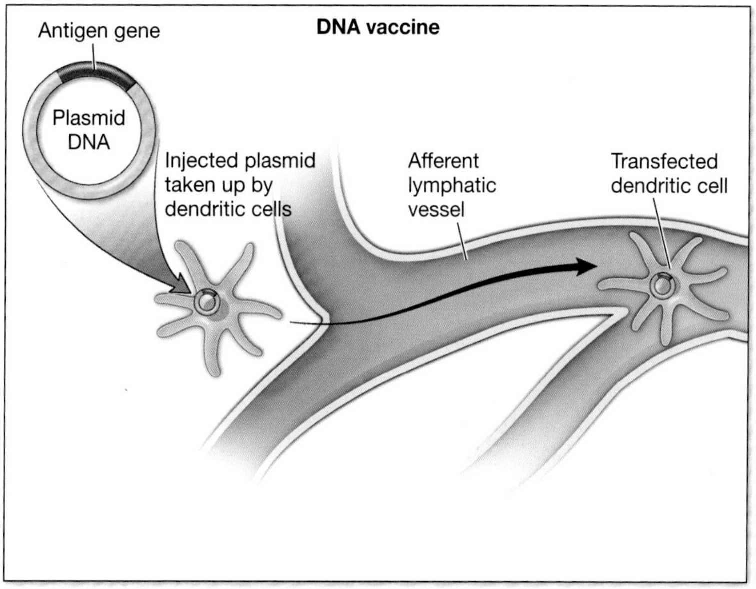

Some of the vaccine strategies known to produce a robust protective response are extremely effective against bacterial and viral pathogens. Attenuated vaccines, toxoid vaccines, conjugate vaccines, and subunit vaccines have all shown effective protection against these pathogens. Newer strategies, including recombinant vector vaccines and DNA vaccines, are less tested but show promise to produce a protective adaptive immune response to theoretically any pathogen, if a proper antigen is identified.

(Sections 1.3, 11.3, 11.4, 11.6, 12.2, 12.3)

CD8 T cells during a primary immune response. One such delivery system is the use of *immunostimulating complexes* (ISCOMs). These complexes are lipid carriers that encapsulate antigen, fuse with an APC, and deliver antigen into the APC cytoplasm. As we learned, cytoplasmic proteins are presented via MHC class I antigen presentation to CD8 T cells, and thus ISCOMs may be a viable option for use as an adjuvant to activate antigen-specific CD8 T cells.

Another additive currently in clinical trials is being used in DNA vaccines. To try to induce an inflammatory response, DNA vaccines that contain a gene for encoding antigen also sometimes include the common pathogen DNA motif CpG (cytosine-guanine motifs). Since these DNA motifs are common in pathogens, our innate immune response employs PRRs, namely TLR9, to detect these DNA motifs. Thus, the incorporation of CpG DNA motifs has been postulated to aid in a robust adaptive immune response due to activation of an inflammatory response and the activation of TLR9 signaling pathways.

CHECKPOINT QUESTION

1. Explain the functions and benefits of adjuvants in vaccines.

11.6 What key concerns affect vaccine development?

LEARNING OBJECTIVE

11.6.1 List important questions that must be addressed during vaccine development.

This chapter has focused on the creation of immunological memory and its employment during vaccination against pathogen infection. While we have explored several successful strategies for vaccinating against deadly and debilitating diseases, we have not explored many of the questions that must be considered when developing a vaccine. In this section, we will discuss these questions.

Is the vaccine safe?

An extremely important question that must be addressed during vaccine development is its safety for the population. Embedded in this broad question are considerations regarding the antigen itself: Can the antigen revert to being pathogenic? Can the antigen induce toxicity? Safety considerations include the safety of other chemicals within the preparation: Are the adjuvant components safe in humans? Was the vaccine developed in an organism that may induce an allergic response?

Often the question of safety must be addressed based on the specific type of vaccine and an analysis of the risks versus benefits the vaccine provides. For example, the polio vaccine is available as both an inactivated vaccine and a live attenuated vaccine. While the live attenuated polio vaccine is more effective than the inactivated version, the inactivated vaccine is safer because of the small risk of revertants (attenuated virus mutants that become pathogenic). Thus, the risk associated with the live attenuated vaccine must be weighed against the risk of the infection.

Another consideration related to vaccine safety concerns the pathogenesis of the disease. Sometimes the pathogenesis associated with the disease is caused not by the pathogen itself but rather by the immune response driven by the presence of the pathogen. For example, vaccination against respiratory syncytial virus (RSV) has largely been unsuccessful because the symptoms associated with RSV infection are driven by the immune response. RSV vaccination has caused more pronounced symptoms in vaccinated individuals, which is not the desired outcome of a vaccine.

Is the vaccine effective?

Any vaccine must be capable of inducing the proper immune response and adequate protection against the pathogen. Clinical trials are paramount in discovering the effectiveness of a vaccine—while animal models provide information about the potential effectiveness of a vaccine, protection in animals may not translate perfectly to humans.

Developing an effective vaccine often involves a long and difficult process. This is especially true in regard to viral vaccines when the pathogen is capable of quickly mutating and escaping memory lymphocyte activity. One example of the difficulties in developing effective vaccines is in the story of human immunodeficiency virus (HIV). HIV is capable of quickly mutating through its normal life cycle. This high mutation frequency can have a profoundly negative effect on generating a primary adaptive immune response that can produce memory lymphocytes that are useful upon exposure to the pathogen. In spite of research and development efforts, there are some pathogens whose life cycle and ability to evade the immune response have so far prevented the discovery of a vaccine, especially those that have a primarily intracellular life cycle and undergo latency, including herpes simplex virus, *Mycobacterium tuberculosis*, and *Plasmodium falciparum* (the causative agent of malaria).

What is the best mode of delivery?

While many vaccines are injected intramuscularly, sometimes it is most effective to deliver vaccines via different routes. For example, the live attenuated rotavirus vaccine is given orally to activate an adaptive immune response within mucosa-associated lymphoid tissue and to promote the activation of lymphocytes in locations where they most likely will come in contact with pathogen. In this case, protection is induced first within the tissues most likely to come into contact with the pathogen.

Who should receive the vaccine?

While it is easy to argue that vaccines against widespread, potentially pandemic, pathogens should be provided to as many people as possible (e.g., influenza vaccine, COVID vaccine), the question arises as to the distribution of vaccines when widespread pandemic is less likely under normal circumstances.

At present, vaccines against emerging infectious diseases, including Zika, Ebola, and dengue fever, are being developed. One can ask, who should receive these vaccines? It is easy to argue that those most likely to be in direct contact with these pathogens are best served by vaccination. However, the question becomes more complex, especially in regard to pathogens that have the potential to be used as bioterrorist weapons. Should everyone receive an anthrax vaccine given that it has successfully been used in bioterrorist attacks? Should a vaccine be recommended for first responders and those most likely to come into contact with the pathogen?

What is the vaccination schedule?

While some vaccines require administration of only a single dose, many vaccines require booster shots to provide a high level of protection. This is especially true with inactivated vaccines, which do not provide a high level of immunity after one dose. Boosters are also required when the level of immunity is lowered over time after the initial vaccination. For example, with the tetanus and diphtheria vaccines, boosters are recommended for adults every 10 years. When additional inoculation is required, it becomes increasingly challenging for people to comply with the vaccination schedule. The need for one or more booster shots is problematic in general but is especially concerning in parts of the world where individuals are less trusting of the medical profession and where they have to travel long distances to receive a vaccine.

How must the vaccine be stored?

The recent development of various COVID-19 vaccines highlighted an issue that is paramount to all vaccines: How must a vaccine be stored to maintain its efficacy? The first COVID-19 vaccine approved in multiple countries required storage at –70°C, a temperature that requires the use of ultra-low-temperature freezers (which are not widely available) or dry ice (which easily sublimates). A second vaccine required storage at normal refrigeration and freezer temperatures, so it could be shipped to areas in which ultra-low-temperature storage is not available. Vaccines requiring extreme cold for storage are more difficult to distribute to areas that do not have access to the equipment or electricity required for safe storage. Thus, the development process must lead to a global strategy to ensure the safest and most effective methods for distribution and storage.

What is the cost versus benefit of vaccine development?

Vaccine research and development is most often conducted by pharmaceutical companies, and undoubtedly, these companies engage in extensive cost/benefit analysis regarding all potential products. Questions about the utility of the vaccine, the number of individuals likely to be immunized, and the cost of developing, testing, and producing the vaccine all play a role in the development process. This cost/benefit issue can easily spark debate about whether a pharmaceutical company should best serve the population in general or its stakeholders when considering the development of a vaccine.

CHECKPOINT QUESTION

1. Summarize the important questions that need to be explored during vaccine development.

Summary

11.1 How does immunological memory work to prevent and fight disease?

- An adaptive immune response functions to eliminate a pathogen and develop immunological memory to more efficiently combat a subsequent infection with that pathogen.

11.2 How does immunological memory develop?

- Memory cells produced during an adaptive immune response include memory B cells, T memory stem cells, central memory T cells, effector memory T cells, and resident memory T cells.
- The different classes of memory T cells specialize in their localization and capacity to induce adaptive immune responses in tissues within the body.
- Once a primary adaptive immune response has been raised, the secondary adaptive immune response aims to prevent activation of naïve B cells that only express IgM—this is done through the action of inhibitory Fcγ receptors on the surface of naïve B cells.

11.3 How do vaccines prevent disease?

- Vaccination is aimed at inducing an effective primary adaptive immune response against a harmless subunit of a pathogen to produce memory cells to aid in combating the pathogen more quickly and robustly.

11.4 What are the different types of viral and bacterial vaccines?

- A variety of vaccination strategies have been employed as means to promote effective protection against different viral and bacterial pathogens; these include inactivated vaccines, live attenuated vaccines, toxoid vaccines, subunit vaccines, conjugate vaccines, recombinant vector vaccines, DNA vaccines, and mRNA vaccines.

11.5 What is the role of an adjuvant in a vaccine?

- Adjuvants are an important additive of vaccines required for inducing an appropriate and effective adaptive immune response. They work to increase the inflammatory response required to drive costimulatory signals needed for B-cell and T-cell activation.

11.6 What key concerns affect vaccine development?

- Many factors must be considered in the development and delivery of vaccines, including safety, efficacy, mode of delivery, cost, and the targeted population.

Review Questions

11.1 An individual who had been exposed to *Listeria monocytogenes* develops a protective adaptive immune response approximately 2 weeks after exposure. The individual contracts another *L. monocytogenes* infection 3 years later, and clearance of the infection after the second exposure takes approximately 2 weeks. Has this individual mounted a memory response to *L. monocytogenes*? Explain your answer.

11.2 If you were an immunologist attempting to isolate all memory T cells from peripheral blood, what marker would you use during flow cytometry analysis to enrich for your population of memory T cells?

11.3 Recently, measles outbreaks in the United States have become more common as vaccination safety has come into question. Infection of individuals who have been vaccinated against measles has also occurred during these outbreaks. Explain how these outbreaks demonstrate a failure to establish herd immunity in the population during a lull in vaccination.

11.4 Several different types of vaccines were discussed in this chapter. Some vaccines do not require adjuvants. Which type(s) of vaccine might not require an adjuvant? Explain your answer.

11.5 Describe an ideal adjuvant that would aid in the induction of a proper immune response to an intracellular virus such as herpes simplex virus.

11.6 We have explored some common questions related to vaccine development. Compose another question that should be addressed during this process and consider the answers that will support development of the vaccine.

CASE STUDY REVISITED

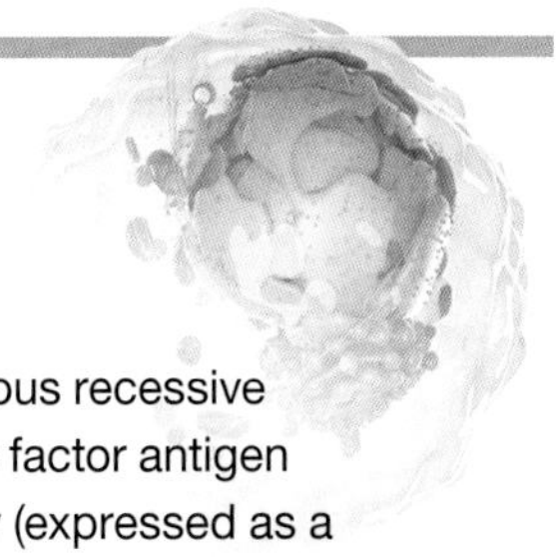

Potential for Hemolytic Anemia of a Newborn

A woman, age 24, and her husband visit her OB/GYN after a home pregnancy test shows a positive result. The doctor orders a blood test to confirm the pregnancy and determine the woman's blood type. Her test results reveal that she is blood type A and Rh-negative. The doctor recommends that the husband's blood be tested to determine any possible blood incompatibility between the mother and baby. Those results show that the husband is blood type AB and Rh-positive.

Think About...

Questions for individuals

1. What is the Rh factor?
2. Why is there a possible blood incompatibility between the mother and baby?

Questions for student groups

1. Being Rh-negative is a homozygous recessive trait in which a particular Rhesus factor antigen is lacking. What is the probability (expressed as a percentage) that the baby is Rh-positive if the father is heterozygous for the Rhesus factor antigen? What is the probability (expressed as a percentage) that the baby is Rh-positive if the father is homozygous for the Rhesus factor antigen?
2. The OB/GYN tells her patient that at 28 weeks, she will need a RhoGAM injection. Why would this recommendation be made?
3. What is the mechanism of action of RhoGAM in this treatment?

MAKING CONNECTIONS Key Concepts	COVERAGE (Section Numbers)
Primary adaptive immune responses—B-cell and T-cell activation	1.4, 1.8, 1.10, 7.1, 7.3, 7.7, 10.2, 10.3, 10.4
Secondary adaptive immune responses—Memory cell production and suppression of the primary immune response	11.1, 11.2
Vaccine development strategy	11.3, 11.4, 11.6
Vaccine action against pathogens	13.2, 13.3

12

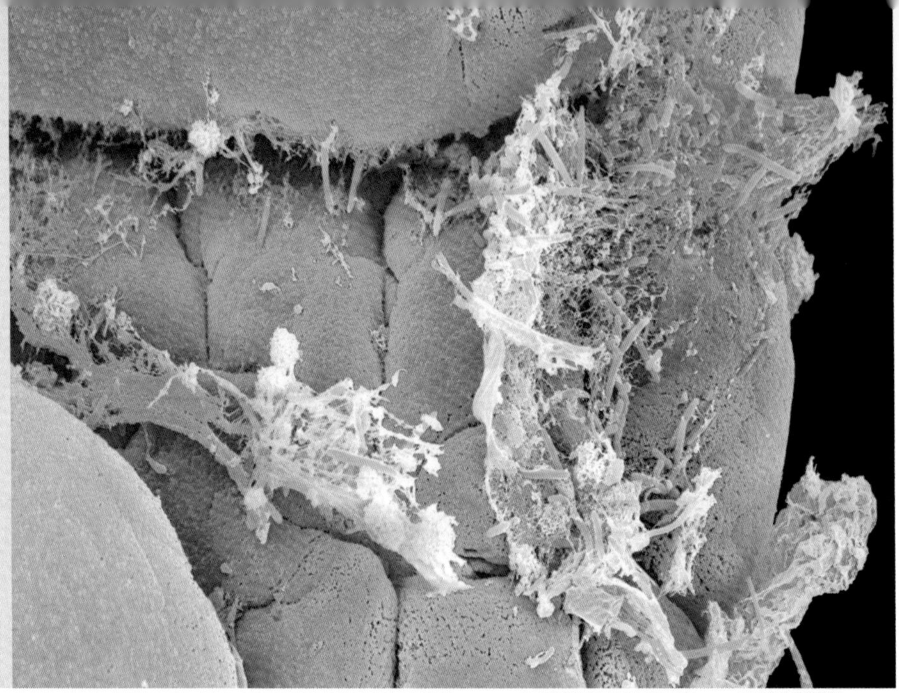

Mucosal Immunity

QUESTIONS Explored

CASE STUDY: Dylan's Story

It is a beautiful summer day and a great evening for a barbeque. Dylan has been brining the spare ribs overnight and all day. He's looking forward to showing off his family's recipe for the first time. He fires up the grill and adds his makeshift smoke box to slowly smoke the ribs. After a few hours, the ribs look perfect, and he asks his wife, Gabriella, to call the kids to the picnic table for dinner. After enjoying their meal, they clean up and toss a baseball around to enjoy the last bit of this summer day.

The following afternoon, Dylan is watching a baseball game on TV when he starts to feel nausea and abdominal pain. He dashes to the bathroom and vomits several times. Still feeling ill, he goes back to the couch to rest. Suddenly Gabriella rushes to the bathroom, and Dylan hears her vomiting. A short while later, his son yells from upstairs, "Mom! We don't feel good! Our stomachs really hurt!"

To understand what Dylan and his family are experiencing, and to answer the case study questions at the end of the chapter, we must understand the function of mucosal tissue, including its role in detecting and combating pathogens that come into contact with mucosal surfaces. Mucosal tissue is rich in lymphoid tissue because it is a potential entry point for pathogens. Lymphoid tissue associated with mucosal surfaces functions in a unique way to activate an immune response. In this chapter, we will compare this type of lymphoid tissue with draining lymphoid tissue (discussed in Chapter 2) and learn about the innate and adaptive immune responses employed at mucosal surfaces.

12.1 | What are mucosal surfaces?

LEARNING OBJECTIVES

12.1.1 Compare and contrast systemic and mucosal immunity.

12.1.2 Define *mucosa* and explain its structure and function.

12.1.3 Describe the characteristics of mucosae that help to prevent infection.

Mucosa, also called a mucous membrane, is a thin sheet of tissue that secretes mucus and covers or lines various body structures within the respiratory, gastrointestinal (GI), and genitourinary tracts. It consists of connective tissue covered by one or more layers of epithelial cells (**FIGURE 12.1**). The immune system must work to combat pathogens wherever they are present, and the *mucosae* (plural), also called mucosal surfaces, are an incredibly important front line that the immune system must monitor.

mucosa A mucous membrane; also referred to as a mucosal surface; plural is *mucosae*.

Mucosal surfaces play key physiological roles in respiration (gas exchange), food digestion, nutrient absorption, and reproduction, and so must be permeable to function properly. However, this permeability makes mucosal surfaces susceptible to a variety of pathogens, including airborne pathogens, those in contaminated food and water, and those responsible for sexually transmitted infections. Thus, the innate and adaptive immune systems have evolved to protect mucosal surfaces from the many pathogens that can access an organism through these permeable portals of entry.

Mucosal surfaces usually act as barriers between two different environments (for example, the mucosal lining of the small intestines separates the lumen of the intestines and the submucosal surfaces, including the blood vessels that transport absorbed nutrients).

In earlier chapters, we have discussed the delivery of antigens to the adaptive immune system based on an infection driving an inflammatory response and the action of draining lymphoid tissue. This response includes fluid leakage at the site of infection and the migration of specialized immune cells from the site of infection to secondary lymphoid tissue, where antigen presentation occurs. The innate and adaptive immune responses that protect most of our body (the bloodstream and tissues other than the mucosal surfaces discussed in this chapter) are referred to as systemic immunity.

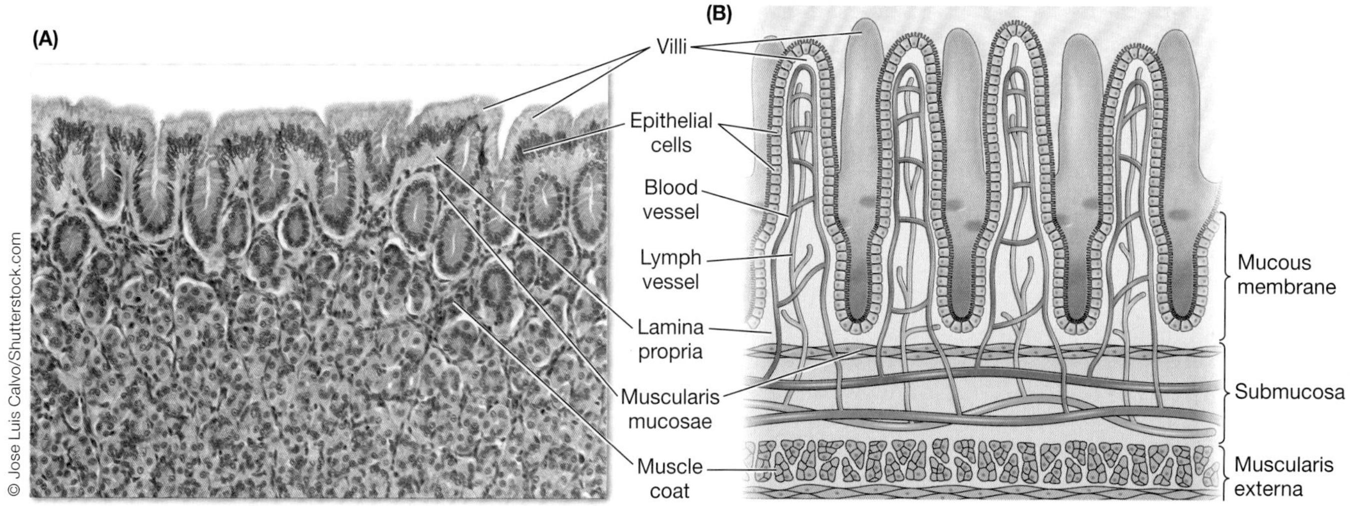

FIGURE 12.1 Structure of mucous membrane (A) Micrograph of mucous membrane. (B) The structure of mucous membranes facilitates their role in gas exchange, nutrient absorption, and protection of internal tissues. Mucous membranes typically contain a layer of epithelial cells organized in hairlike structures known as villi, which increases the surface area of the epithelial cell layer. Underneath the epithelial cell layer is an extracellular tissue known as the lamina propria. Mucous membranes are highly vascularized, containing a network of circulatory and lymphatic vessels that promote mucous membrane function and immune cell action.

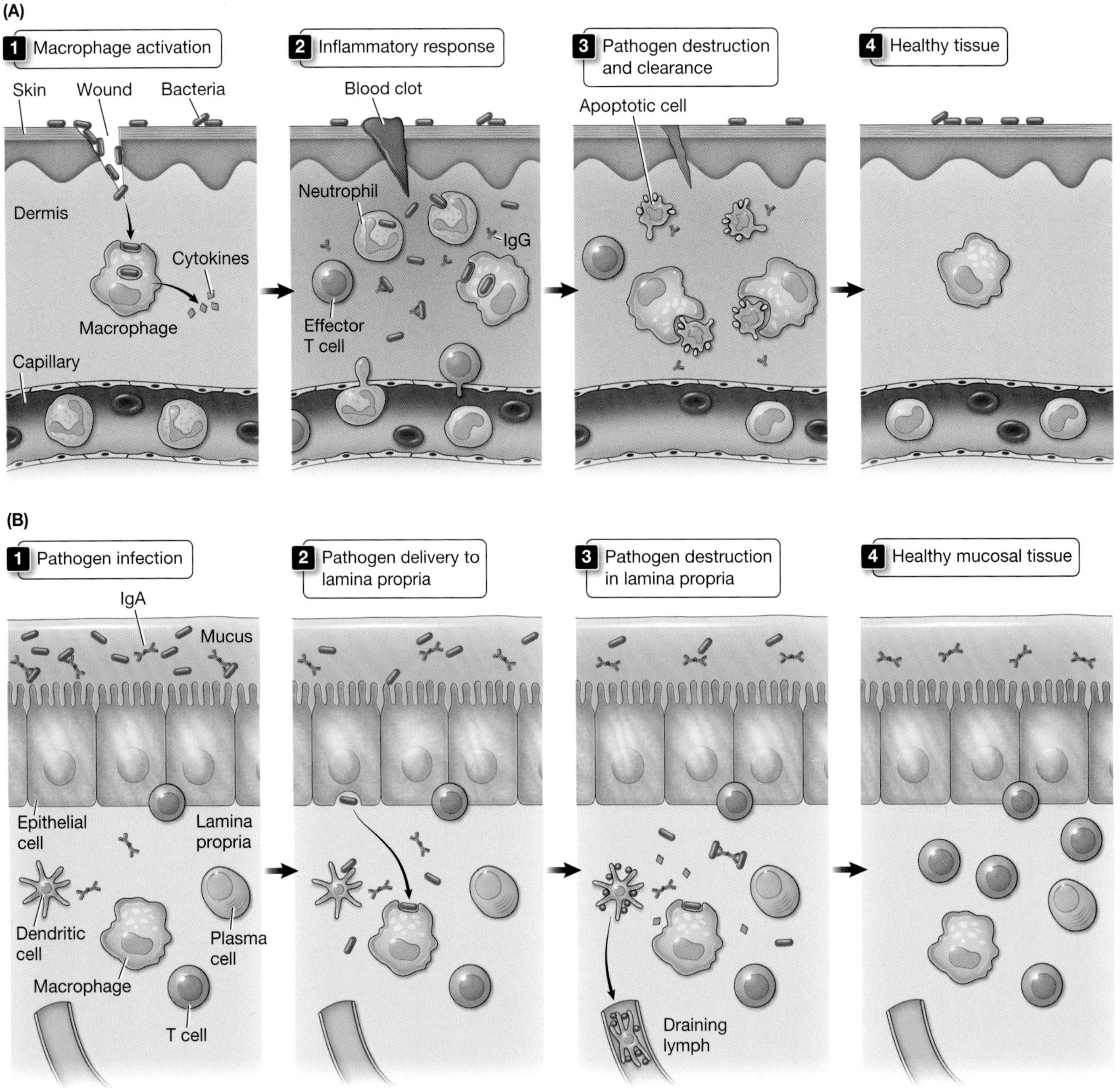

In mucosal tissue, an inflammatory response that would involve leakage of fluid between two compartments whose anatomy and physiology differ could cause complications and potential tissue damage. Increased permeability of mucosal tissue could allow increased movement of the microbiota into underlying tissues, activation of innate immune cells in the area, and an inflammatory response that can potentially damage cells lining the mucosal tissue. Thus, antigen delivery to lymphoid tissue associated with mucosal surfaces must rely on mechanisms that differ from inflammatory responses (**FIGURE 12.2**). *Mucosal immunity* relies on mechanisms similar to those we have seen in previous chapters. Its functions differ from systemic immunity in that they must also maintain the mucosal barrier to prevent contact between environments separated by mucosae. Interestingly, cells of the mucosae function not only as a barrier to pathogens but also as a sensor to their presence. This chapter focuses on the initial barriers that work to prevent pathogen infection at mucosal surfaces and the means by which an adaptive immune response can be stimulated, if necessary, to fight a pathogen that comes in contact with mucosae.

◀ **FIGURE 12.2 Comparison of a systemic immune response and an immune response in mucosal tissue** (A) In a systemic infection, a pathogen such as a bacterium gains entry to underlying tissue via an open wound. Resident macrophages can phagocytose these pathogens and activate cytokine expression through PRR activation after recognizing PAMPs on the pathogen (Step 1). Cytokine release drives an inflammatory response (Step 2), resulting in increased fluid flow to and from the wound. It also drives recruitment of other white blood cells such as neutrophils and effector T cells that may have been activated by an adaptive immune response. Immune cell action will result in pathogen destruction (Step 3). Neutrophils, which are short-lived, will die at the site of infection via apoptosis. Eventually, inflammation subsides, resulting in healthy uninfected tissue (Step 4). (B) Infection in mucosal tissue (Step 1) involves a different manner of delivery of antigen and pathogen to lymphoid tissue than occurs during an inflammatory response. An inflammatory response in mucosal tissue could damage underlying tissue due to the presence of degradative enzymes and low-pH fluid within the mucosal tissue. Endocytosis of the pathogen by epithelial cells (facilitated by immunoglobulins at the mucosal surface) delivers the pathogen to the underlying lamina propria (Step 2), where innate and adaptive immune cells reside. In the lamina propria, the pathogen is phagocytosed by resident innate immune cells such as macrophages and dendritic cells (Step 3). These cells secrete cytokines to activate other resident cells of the innate and adaptive immune response. Dendritic cells migrate to nearby mucosa-associated lymphoid tissue to further activate the adaptive immune response. Eventually the pathogen is cleared, resulting in healthy mucosal tissue (Step 4). (After P. Parham. 2014. *The Immune System*, 4th ed., W. W. Norton and Company.)

Mucus

While the structure and functions of mucosae differ based on their specific tissue type, all mucosal surfaces share some common characteristics. Mucosae are covered in **mucus**, a thick, viscous fluid that lubricates and protects the mucosal surfaces. Mucus contains several macromolecules, including proteoglycans and glycoproteins, that work in conjunction with peptides and enzymes to protect the mucosae from pathogen colonization. These enzymes and peptides, along with the ongoing secretion and flushing of mucus, function to prevent infection at mucosal surfaces.

mucus Thick, viscous secretion from epithelial cells of mucosal tissue that contains mucin and secreted defensins and immunoglobulins.

mucin Glycoprotein that protects mucosal tissue by increasing the viscosity of mucus, concentrating defensins at mucosal surfaces, and cross-linking secreted IgA at the mucosal surface.

Mucins

Mucins are an important family of glycoproteins that are secreted by epithelial cells (**FIGURE 12.3**). Mucins are large polypeptides (generally over 100 kD in molecular mass) with a high number of serine and threonine amino acids,

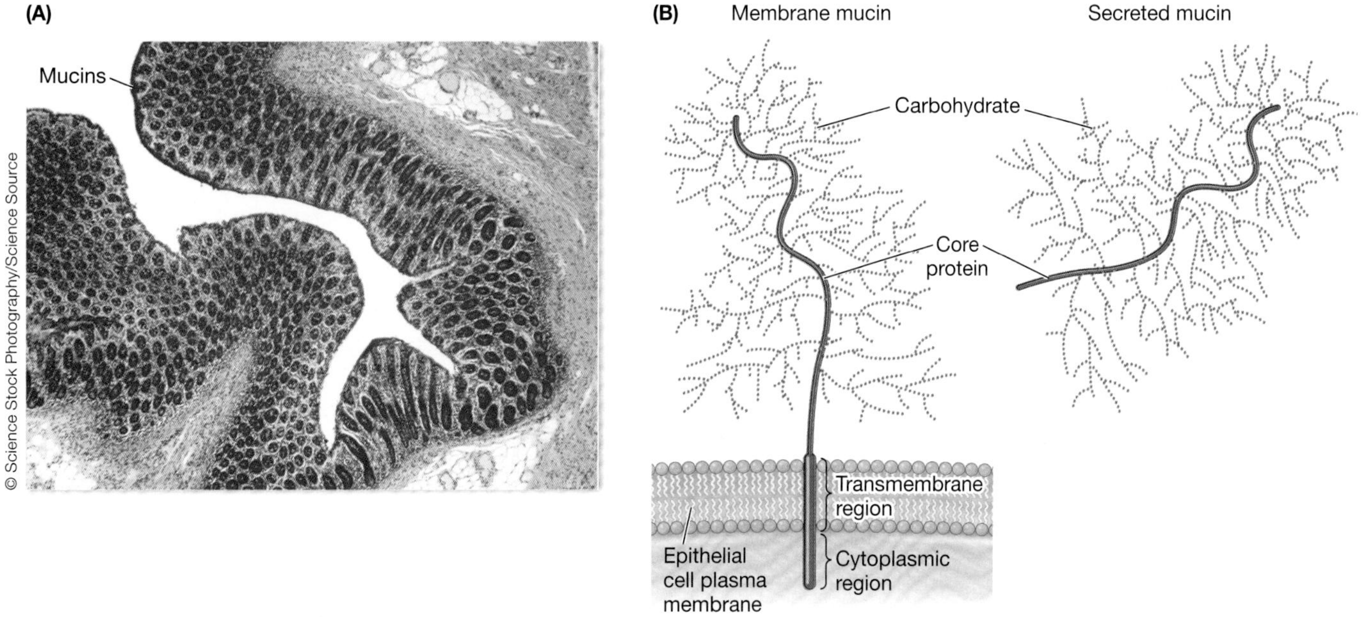

FIGURE 12.3 Mucins in mucosal tissue (A) Mucins can be detected histologically using a variety of stains, including mucin mucicarmine, which stains mucins on the surface of mucosal epithelial cells magenta. (B) Mucosal epithelial cells express both membrane mucins and secreted mucins, which are a component of mucus. Membrane mucins are composed of a core protein that contains a transmembrane region and carbohydrates linked to the core protein in the extracellular environment. Secreted mucins lack the transmembrane region in their core protein but still contain carbohydrates linked to the core protein. (B, after illustration by Jcastr07/CC BY-SA 4.0.)

which allows for *O*-linked glycosylation (addition of carbohydrates to serine and threonine residues) of mucins. The addition of polysaccharides and the presence of cysteine residues within the polypeptides greatly increase the molecular weight of these complexes and promote the viscosity of mucus. This viscosity makes it more difficult for pathogens to move through mucosal surfaces, an important defense against pathogens.

One of the sugar molecules added to mucins via glycosylation is sialic acid, which is negatively charged. This charge attracts positively charged defensins secreted by epithelial cells (see Chapter 1) to remain at the mucosal surface and increases their potential exposure to pathogens.

CHECKPOINT QUESTIONS

1. Why are mucins effective as a defense against infection at mucosal surfaces?
2. Compare and contrast the delivery of antigens to secondary lymphoid tissue when captured at an open wound and at a mucosal surface.

12.2 | What is mucosa-associated lymphoid tissue?

LEARNING OBJECTIVES

12.2.1 Define *mucosa-associated lymphoid tissue* (MALT) and explain where these tissues are found in the body.

mucosa-associated lymphoid tissue (MALT) Secondary lymphoid tissue that specializes in the uptake and clearance of pathogens at mucosal surfaces.

inductive compartment The area of mucosa-associated lymphoid tissue responsible for the clonal selection and expansion of T and B lymphocytes.

We have seen that the innate and adaptive immune systems work in tandem to prevent infection and disease. Secondary lymphoid tissue plays an important role to centralize the clonal selection and expansion of lymphocytes required to combat a foreign antigen, and this is also true of secondary lymphoid tissue associated with mucosal surfaces.

The main function of this type of secondary lymphoid tissue, known as **mucosa-associated lymphoid tissue (MALT)**, is to centralize activation of an adaptive immune response at mucosal surfaces (**FIGURE 12.4**). However, it is important that antigen delivery to the adaptive immune system in mucosal tissue occurs with minimal inflammation and tissue damage. Indeed, an inflammatory response associated with infection at mucosal surfaces tends to be the manifestation of symptoms associated with the infection. In this case, drainage to a secondary lymphoid tissue would cause more harm than good.

MALT lies beneath the mucosal epithelial cells and is sometimes referred to as the **inductive compartment** (**FIGURE 12.5**). This tissue is designed to promote antigen delivery to circulating lymphocytes and to produce lymphocytes that will ensure protection of the mucosal barriers. While the rest of the chapter will focus on MALT of the GI tract, keep in mind that many features of gut-associated lymphoid tissue are mirrored at other mucosal

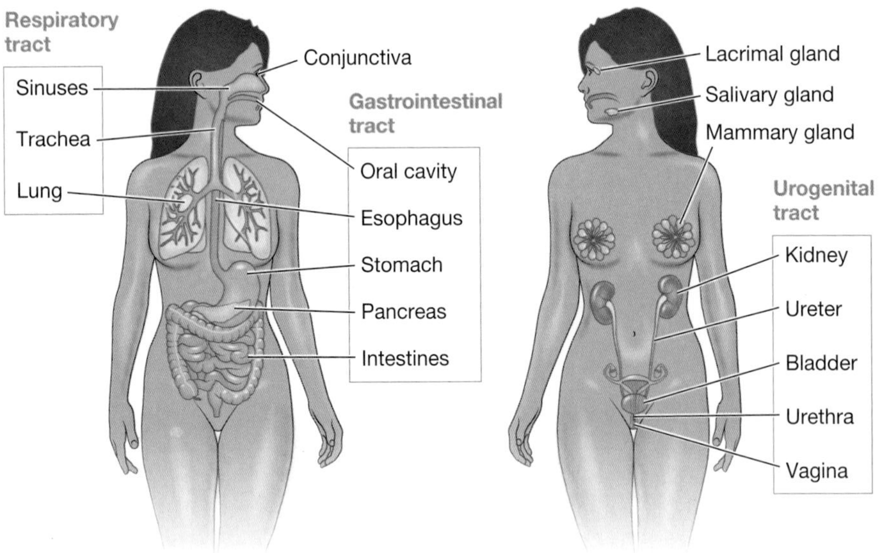

FIGURE 12.4 Mucosal tissues of the body Mucosal tissues include components of the respiratory tract (labeled in blue), gastrointestinal tract (labeled in green), and urogenital tract (labeled in purple). Mucosal tissues of the respiratory tract are found in the sinuses, trachea, and lungs. Those of the gastrointestinal tract are found in the oral cavity, esophagus, stomach, pancreas, and intestines. Mucosal tissues of the urogenital tract are found in the kidneys, ureters, bladder, urethra, and vagina. Other mucosal tissues are present in the lacrimal, salivary, and mammary glands.

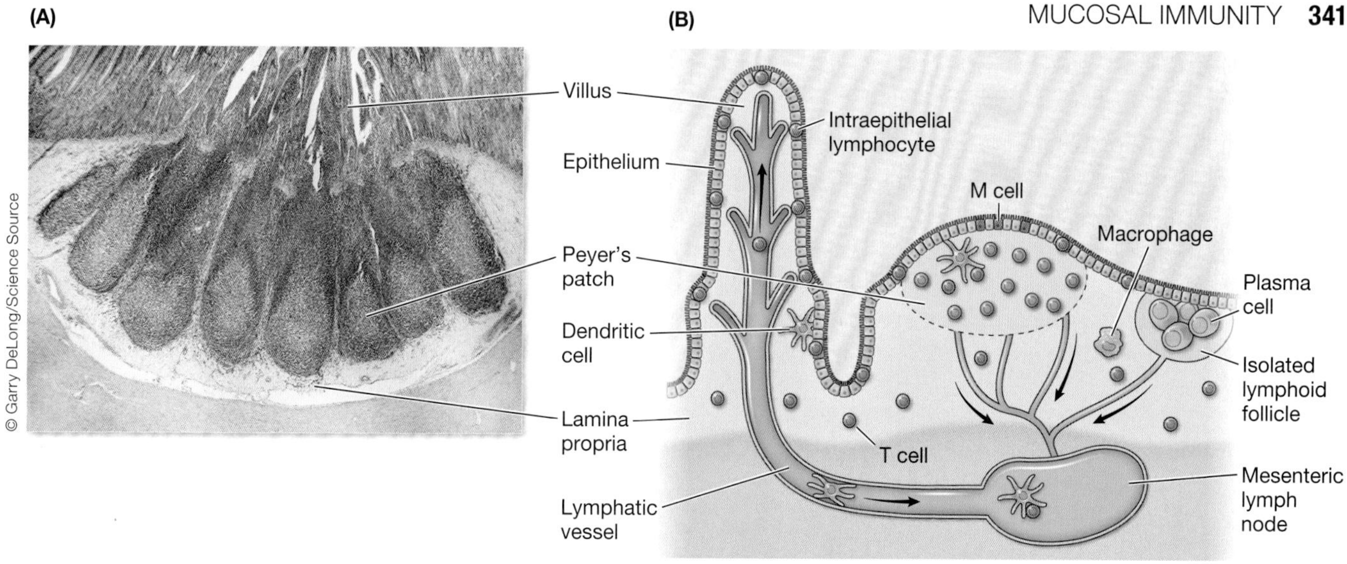

FIGURE 12.5 Mucosa-associated lymphoid tissue (MALT) (A) Micrograph of MALT of the small intestine. (B) MALT contains specialized cells and structures that enable pathogen delivery to lymphoid tissue and activation of an adaptive immune response. Epithelial cell barriers of mucosal tissue contain intraepithelial lymphocytes that protect the epithelial cell layer from intracellular infection. These barriers also contain M cells that deliver antigen to underlying Peyer's patches. Within Peyer's patches, lymphocytes are activated in response to antigen delivered by M cells. Within nearby isolated lymphoid follicles, B cells are activated to produce plasma cells. The lymphatic system connects mucosal tissue and Peyer's patches with mesenteric lymph nodes within the gut. (B, after R. Q. Wu et al. 2014. *Intl J Oral Sci* 6: 125–132.)

surfaces. MALT is also present in the respiratory tract (bronchus-associated lymphoid tissue, or BALT), in the eyes (conjunctival-associated lymphoid tissue, or CALT), and in the nasal cavity (nasal-associated lymphoid tissue, or NALT).

Gut-associated lymphoid tissue

A large amount of **gut-associated lymphoid tissue (GALT)** is present throughout the GI tract. Activated lymphocytes and innate immune cells such as macrophages reside in the connective tissue below the mucosal epithelial cells. This connective tissue layer is known as the **lamina propria**, sometimes referred to as the **effector compartment** because it contains the effector cells of the immune system. Along with the inductive compartment, mesenteric lymph nodes (see Figure 12.5) function as an extension to GALT to protect the gut mucosae.

The palatine and lingual tonsils and adenoids (also known as the pharyngeal tonsils) are types of GALT that function to protect the body from the multitude of microorganisms that enter the digestive and respiratory systems. Infection, especially in children, whose immune systems are not yet fully developed, can cause swelling of the tonsils. In the past, this condition usually resulted in tonsillectomy (surgical removal of the tonsils). Now that the importance of the tonsils as a secondary lymphoid tissue has been recognized, tonsillectomies are now recommended only in certain situations, such as when swelling has occurred more than seven times in one year and for pediatric sleep apnea, although there is some controversy surrounding this treatment (see the **CONTROVERSIAL TOPICS** box).

Due to the large amount of digested material and abundance of nutrients found in the small intestine, this organ is a location in which microorganisms (both good and bad) can thrive. Thus, the small intestine is rich in GALT and effector cells to defend against infection in this environment. **Peyer's patches**, also known as *aggregated lymphoid nodules*, are aggregations of GALT found in the ileum portion of the small intestine (see Figure 12.4). **Isolated lymphoid follicles** of the GI tract consist primarily of B cells (see Figure 12.5).

Microbiota in the gut

Recall from Chapter 1 that our *microbiota* consists of microbes that live in and on us, usually symbiotically, and compete with potential pathogens for space

gut-associated lymphoid tissue (GALT) Mucosa-associated lymphoid tissue that protects the gastrointestinal tract through adaptive immune response activation.

lamina propria Connective tissue that lies underneath the epithelial barrier of mucosal tissue.

effector compartment Connective tissue under the epithelial barrier that contains effector lymphocytes and innate immune cells.

Peyer's patch Specialized mucosa-associated lymphoid tissue of the small intestine.

isolated lymphoid follicle Follicle containing a germinal center and mainly B cells undergoing activation and differentiation.

CONTROVERSIAL TOPICS

Should tonsillectomy be used to treat sleep apnea?

From the early to mid-twentieth century, tonsillectomy was a commonplace surgery used as routine treatment for infected and inflamed tonsils (tonsillitis). As of the 1980s, the frequency of this procedure declined, and now it is only used in cases when other forms of treatment don't work or can't be used. We have learned that secondary lymphoid tissue is paramount to the protection of mucosal surfaces, and currently tonsillectomy is only recommended for repeated (at least seven) bouts of throat infection in a calendar year or when allergies to multiple antibiotics make treatment more problematic.

Tonsillectomies have become more common as a treatment for pediatric sleep apnea, which can be problematic or even fatal if untreated. Unfortunately, there has been only limited (in size and scope) research on the benefit of using this procedure to treat sleep apnea. The surgery is now performed on more than 500,000 children under age 15 each year, with the vast majority (90%) of procedures done to treat sleep apnea. This situation begs the questions:

- Is this truly the best way to treat a potentially fatal disease?
- Does the benefit of the surgery outweigh the potential risks? (Tonsillectomy can be fatal, most commonly due to uncontrollable hemorrhaging.)
- What is the long-term effect of this surgery on the patient's immune system?

Stories of children who suffer brain death due to tonsillectomy complications have been reported by the mainstream news media. In light of the availability of other forms of treatment for sleep apnea, including the use of continuous positive airway pressure (CPAP) machines and nutritional counseling to assist in weight loss (for obese patients), this topic is one that should be discussed and explored to ensure that the treatment truly fits the diagnosis. Parents should explore all available options and understand potential treatments and their risks to make an informed decision about the best treatment for a child suffering from sleep apnea.

Think About...

1. If you were a pediatrician treating an overweight child with sleep apnea, would you recommend a tonsillectomy? Why or why not? If not, what would your recommendations for treatment be?
2. Should childhood tonsillectomies for sleep apnea be banned in the United States? Why or why not? What type of evidence would be important to support your recommendation?

and nutrients. Gut microbiota aid in digestion and produce vitamins required for enzyme activity. They also produce metabolic products and antimicrobial agents that can prevent pathogen colonization. Thus, these organisms play a role in protecting mucosal surfaces, including those of the GI tract.

In order to thrive and colonize, ingested bacteria such as *Salmonella* traveling through the digestive tract must compete for space and nutrients with resident gut microbiota. The low pH of the stomach (1.5 to 3.5) and the presence of these microorganisms serve as two more lines of defense against invading bacteria and other pathogens. The concentration of symbiotic microorganisms increases along the GI tract, starting with up to 1000 bacteria per mL in the stomach and increasing to 10^5 to 10^8 bacteria per mL in the small intestine and 10^{10} to 10^{12} bacteria per mL in the colon.

Our microbiota may not just be involved in protecting us from pathogens we come into contact with at mucosal surfaces; it may also be responsible for the development of mucosal immunity. Animals raised in a germ-free environment have smaller Peyer's patches in their MALT and produce less soluble IgA, which is important in protecting mucosal tissues. In humans, changes in the normal microbiota of the gut can cause inflammation, which is seen in inflammatory bowel disease. Alterations of the microbiota of the respiratory tract have been implicated in diseases of the lung, including chronic obstructive pulmonary disease, asthma, and cystic fibrosis. The microbiota may also aid in the development of T_H17 helper T cells (which activate neutrophils), a predominant helper T cell present in the lamina propria of the intestines. While our microbiota plays a role in protecting us from infection, it is also likely involved in "educating" our mucosal immune system and response.

Although organisms of the microbiota generally do not cause disease, they are the primary cause of infections associated with mucosal surfaces, especially within the gut. Bacterial strains that include *Escherichia coli*, *Helicobacter* species, and *Shigella* species can cross the epithelial barrier, causing opportunistic infections. GI disease can also occur if we ingest pathogenic bacteria such as *Salmonella*, pathogenic strains of *E. coli* such as enteropathogenic *E. coli* (or EPEC), viruses such as rotavirus, and parasitic helminths such as hookworms or roundworms.

microfold cells (M cells) Cells that deliver antigens from the lumen of mucosal tissue to dendritic cells in mucosa-associated lymphoid tissue.

transcytosis Transport of a soluble molecule from one side of a cellular barrier to the other.

Preventing an inappropriate immune response

How does the body prevent the activation of immune responses against the microbiota? After all, our microbiota includes microorganisms that contain pathogen-associated molecular patterns (PAMPs) that are recognized by a variety of receptors (see Chapter 2). Recognition of a PAMP by a receptor typically induces immune responses that include inflammation. So what prevents mucosal tissues from being in a constant state of inflammation? And how does the immune system distinguish good microorganisms (microbiota) from bad microorganisms (pathogens)?

A simple model exists to explain this ability to distinguish beneficial microorganisms from those with the potential to cause harm by limiting the interaction of the microbiota with epithelial cells and cells of the immune system. The presence of mucus, mucins, soluble immunoglobulins such as IgA, and antimicrobial peptides at mucosal surfaces limits the interactions of the microbiota with intestinal epithelial cells. Additionally, the tight junctions holding these cells together prevent migration of microorganisms to underlying tissue.

Another model, first proposed by immunologist Polly Matzinger in 1994, suggests that the immune system primarily responds to cells that have been injured or stressed. This model, known as the Danger model, is based on the idea that the normal microbiota does not promote stress or injury of mucosal tissue. It hypothesizes that only pathogens capable of causing disease (or microbiota that may have gained access to underlying tissue, causing an opportunistic infection) induce an immune response by driving changes due to stress or injury. The Danger model suggests that stressed or injured cells release danger signals as a result of damage caused by the pathogen. Research has revealed a variety of danger signals, called damage-associated molecular patterns (DAMPs), that can induce an immune response. DAMPs include nucleic acids, heat shock proteins, cytokines, and plasma membrane proteins. Many DAMPs are recognized by pattern recognition receptors (PRRs) that recognize PAMPs. Thus, while good and bad microorganisms may be distinguished by location, the Danger model suggests that only the bad microorganisms cause damage and promote an immune response within mucosal tissue.

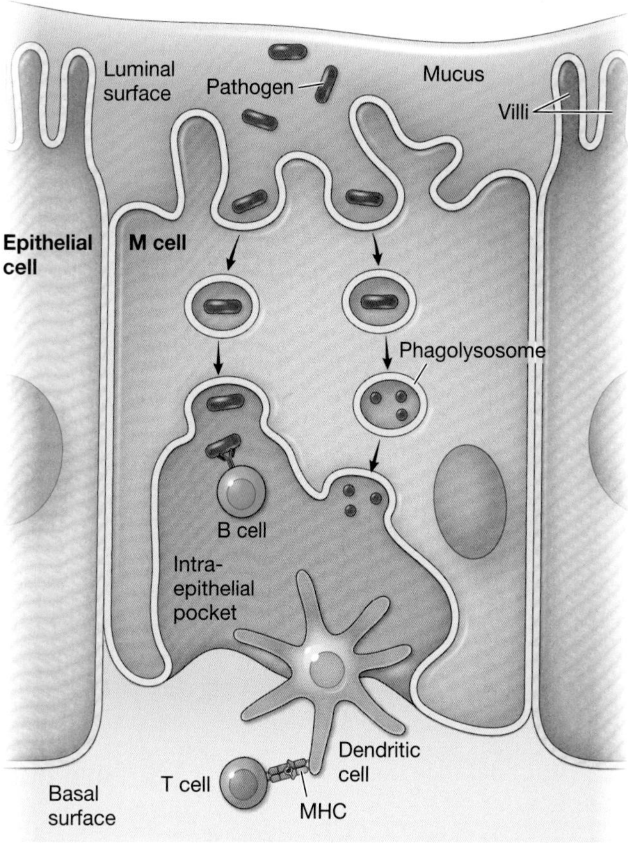

FIGURE 12.6 Transcytosis of pathogens by M cells M cells embedded in the epithelial cell layer of mucous membranes deliver pathogens and antigens to underlying lymphoid tissue through transcytosis. Pathogens are phagocytosed by the M cell on the luminal side of the epithelial cell layer and transported via vesicles to the basal side, toward the intraepithelial pocket. Some of the phagocytosed pathogen is delivered intact, and some is digested via the phagolysosome into antigen components before transcytosis. B cells on the basal side bind to antigen via their receptor for antigen processing and presentation. Dendritic cells on the basal side process pathogens and antigens for presentation to nearby T cells. (After P. Parham. 2014. *The Immune System*, 4th ed., W. W. Norton and Company.)

M cells and mucosal immunity

If a goal of an immune response at mucosal surfaces is to limit inflammation, and MALT does not function as draining lymphoid tissue, how then are antigens and pathogens delivered to the underlying MALT? One mechanism of antigen and pathogen delivery is the action of **microfold cells** (**M cells**), which are specialized intestinal epithelial cells that deliver antigens to GALT through **transcytosis** (**FIGURE 12.6**). The luminal face of an M cell binds microorganisms. Upon binding to a microbe, the M cell engulfs it

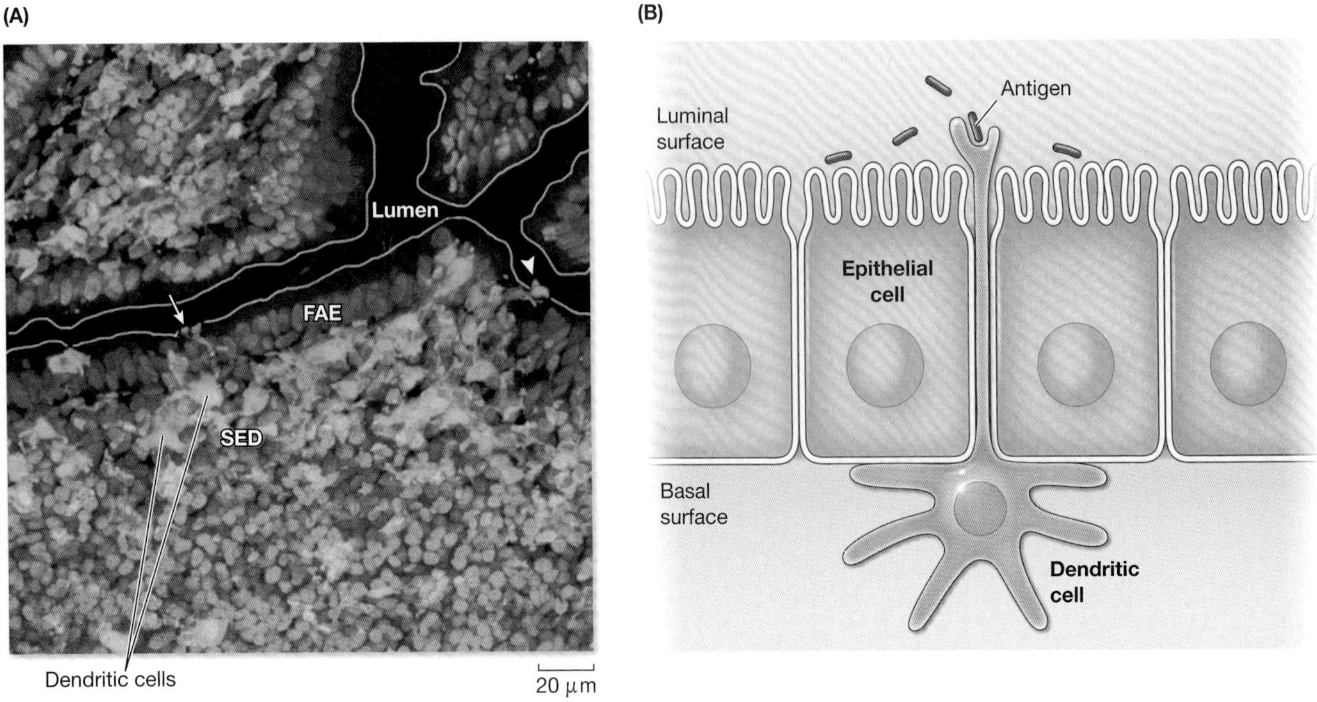

FIGURE 12.7 Sampling of lumen antigens by dendritic cells (A) Micrograph of dendritic cells (green) extending dendrites from the subepithelial dome (SED) of a Peyer's patch across the follicle-associated epithelium (FAE) into the lumen (arrow). The arrowhead depicts migration of a dendritic cell into the lumen. (B) Dendritic cells extend dendrites between epithelial cells of mucous membranes to capture antigen on the luminal side of an epithelial cell barrier for processing and presentation in underlying mucosa-associated lymphoid tissue. (B, after M. Kool et al. 2012. *F1000 Biol Reports* 4: 6.)

intraepithelial pocket Invagination of apical surface of M cells promoting efficient transport of antigens through the M cells to dendritic cells of the lymphoid tissue.

and transports it to its apical side, which faces the GALT. The apical surface of M cells has an extensive fold, creating an **intraepithelial pocket**, which provides a location for delivery of antigens to dendritic cells and T cells located in the adjacent MALT.

Dendritic cells

Antigens are also delivered to MALT through the action of dendritic cells located in the lamina propria. During an infection, dendritic cells can extend projections through the mucosal epithelial barrier and bind to pathogens within the gut lumen (**FIGURE 12.7**). These dendritic cells take up the pathogen, process it, and move into the MALT or into an adjacent mesenteric lymph node, where the antigen is presented to T cells via mechanisms discussed in earlier chapters.

Mucosal epithelial cells

While mucosal epithelial cells create a physical barrier between the luminal and basal sides of mucosal tissue, they also recognize pathogens using a number of mechanisms. They express a variety of Toll-like receptors (TLRs, see Chapter 2), which are capable of recognizing PAMPs such as lipopolysaccharide (LPS) and bacterial flagella. PAMP recognition by the epithelial cell triggers a signaling cascade that activates the transcription factor NFκB and the production of inflammatory cytokines such as IL-1 and IL-6, antimicrobial peptides such as defensins, and chemokines to recruit neutrophils, eosinophils, dendritic cells, and T cells (**FIGURE 12.8**). Although these cytokines are inflammatory cytokines, the high turnover rate of mucosal epithelial cells controls the inflammatory response, preventing damage to mucosae.

While mucosal epithelial cells express a variety of TLRs to recognize PAMPs, they are also capable of activating NFκB by detecting bacterial cell wall components using intracellular proteins of the **NOD** (nucleotide-binding oligomerization domain) family. Two members of the NOD family, NOD1 and NOD2, are present in mucosal epithelial cells and can recognize the presence of pathogens at mucosal surfaces. NOD1 binds to the tripeptide motif of the peptidoglycan component of bacterial cell walls, and NOD2 binds to the muramyl peptide motif. NOD1 or NOD 2 binding to these intracellular bacterial components results in the activation of a signaling cascade:

NOD Stands for nucleotide-binding oligomerization domain; an epithelial protein responsible for the detection of intracellular microorganism components and activation of cytokine secretion to activate effector mechanisms within mucosa-associated lymphoid tissue.

1. NOD1 or NOD2 binding causes activation of the protein kinase RIP2/RICK.
2. RIP2/RICK activates the protein kinase TAK1 through phosphorylation.
3. TAK1 activates the IκB-kinase (IKK) complex.
4. IKK phosphorylates the IκB/NFκB complex.
5. IκB is targeted for destruction by the proteasome.
6. Released cytosolic NFκB is free to be transported to the nucleus.
7. NFκB activates transcription of the same cytokines, chemokines, and antimicrobial peptides that are produced in TLR signaling (see Figure 12.8).

The NOD family of proteins is actively involved in maintaining homeostasis of the small intestine—mutations of these proteins have been linked to the onset of Crohn's disease and irritable bowel syndrome. Although the mechanism driving these diseases is unclear, there seems to be a link between NOD function and the induction of autophagy, a process that is important in the clearance of bacterial infection (see the **EMERGING SCIENCE** box).

CHECKPOINT QUESTIONS

1. Explain the process of antigen delivery to mucosa-associated lymphoid tissue driven by M cells.
2. Would you predict that an organism that is deficient in the production of NOD2 would be more or less susceptible to infection in mucosal tissue? Why?

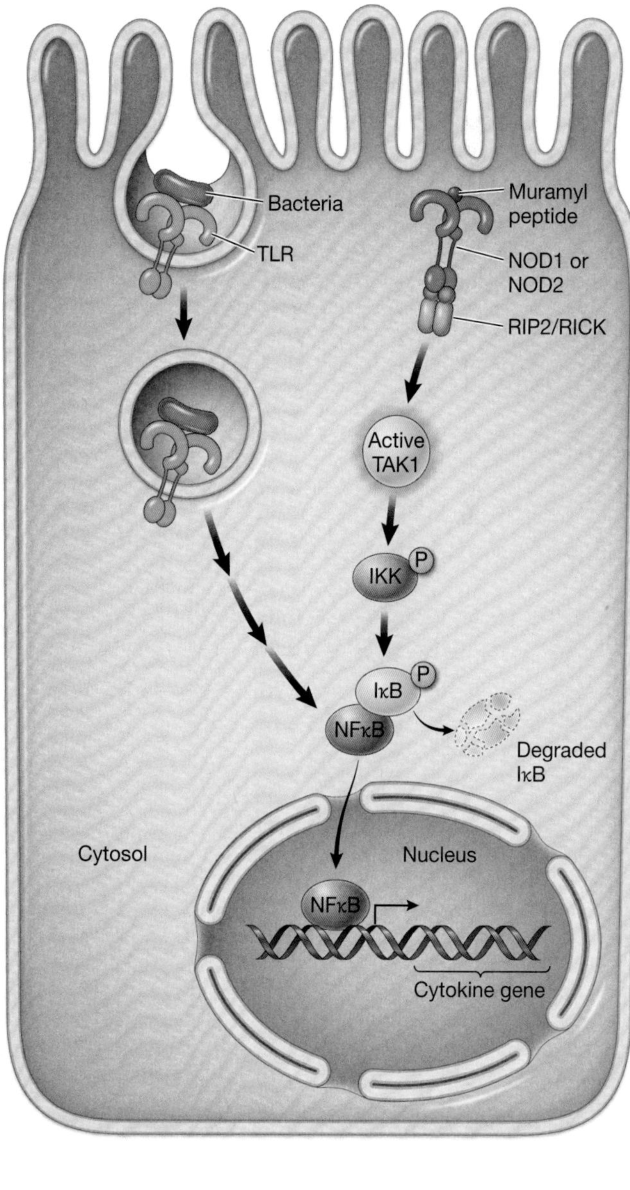

FIGURE 12.8 Detection of intracellular bacteria by mucosal epithelial cells Mucosal epithelial cells express TLRs that bind to bacterial PAMPs and activate cytokine production. These epithelial cells also express NOD1 and NOD2, which bind to bacterial cell wall breakdown products such as muramyl peptide. Binding activates the kinase RIP2/RICK, which phosphorylates and activates TAK1. Active TAK1 phosphorylates IκB kinase (IKK), which phosphorylates IκB. Phosphorylated IκB is targeted for degradation, releasing the transcription factor NFκB. Free NFκB translocates to the nucleus, where it activates cytokine gene transcription.

EMERGING SCIENCE

Why does mutation of the NOD2 protein predispose individuals to Crohn's disease?

Article

L. H. Travassos, L. A. M. Carneiro, M. Ramjeet, S. Hussey, Y. G. Kim, and J. G. Magalhaes. 2010. NOD1 and NOD2 direct autophagy by recruiting ATG16L1 to the plasma membrane at the site of bacterial entry. *Nature Immunology* 11: 55–62.

Background

Crohn's disease is a chronic inflammatory disorder of the small intestines. Its symptoms can be debilitating and include abdominal pain, severe diarrhea, and weight loss. Several polymorphisms and mutations have been identified that genetically predispose certain individuals to the disease. One of the first mutations and the most common mutation identified is a frameshift mutation in NOD2, a protein known to be involved in bacterial sensing by mucosal epithelial cells. The mechanism by which this mutation causes Crohn's disease was not fully understood until further mutations were identified that predisposed certain individuals to the disease, including the protein ATG16L1, which is known to be a key player in autophagy (the process of intracellular digestion). Researchers wanted to determine if there was a functional link between the NOD2 protein and ATG16L1 in epithelial cells.

The Study

Researchers hypothesized that there was a functional link between the ability for NOD2 to detect intracellular bacteria and the induction of autophagy of those bacteria through the action of ATG16L1. The researchers first wanted to demonstrate that the NOD function was connected to the induction of autophagy, and indeed, treatment of cells with NOD agonists was capable of inducing autophagosome formation. These agonists were ineffective at autophagosome formation in NOD-deficient cells (**FIGURE ES 12.1**). Their studies also suggested that the induction of autophagy driven by NOD was independent of the typical NOD signaling pathway used to signal the presence of intracellular bacterial components (thus, it was independent of RIP2/RICK and NFκB).

Because ATG16L1 was also known to be linked to Crohn's disease pathology, the researchers wanted to know if there was a functional link between NOD and ATG16L1. They were able to demonstrate that NOD and ATG16L1 interact with each other and, importantly, that NOD causes ATG16L1

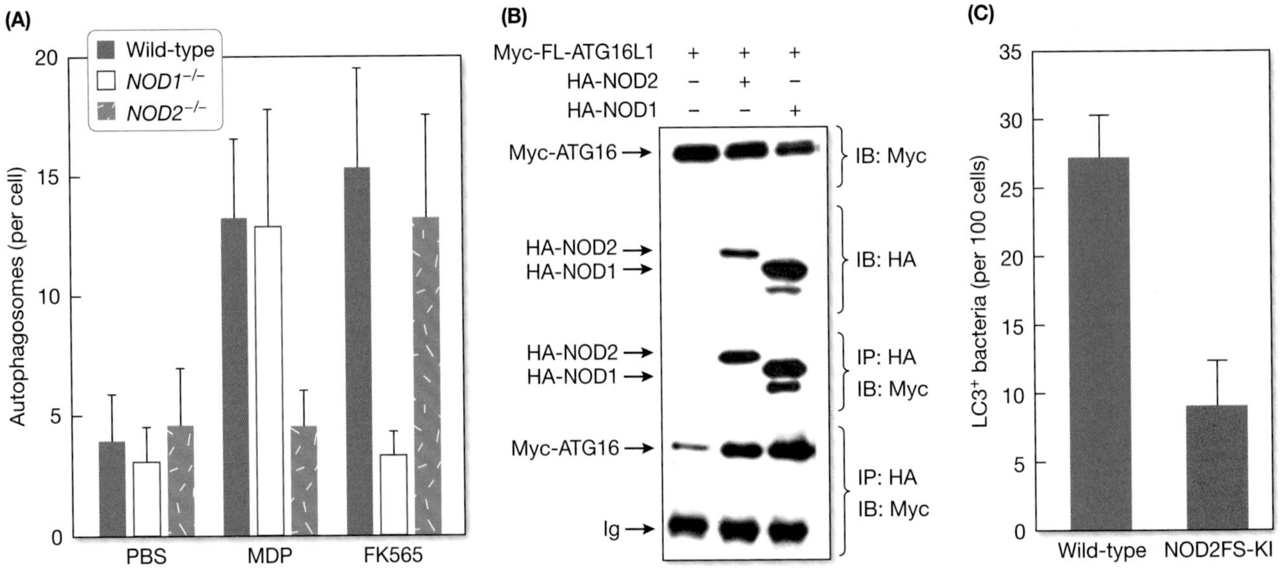

FIGURE ES 12.1 NOD1 and NOD2 aid in autophagosome formation through the recruitment of ATG16L1 (A) Autophagosomes in macrophages from wild-type mice and mice deficient in either NOD1 or NOD2 were quantified after injection of mice with a NOD1 agonist (FK565) or a NOD2 agonist (muramyl dipeptide; MDP). Phosphate-buffered saline (PBS) was used as a negative control of autophagosome induction. (B) The interaction of NOD proteins with ATG16L1 was tested using co-immunoprecipitation assays. HEK293 cells were transfected with a Myc-tagged ATG16L1 and HA-tagged forms of NOD1 or NOD2. Presence of each protein was assessed via Western blot. NOD proteins were immunoprecipitated, and the presence of Myc-tagged ATG16L1 in the immunoprecipitated proteins was assessed via Western blot. (C) *Shigella flexneri* in autophagosomes (based on LC3 detection) was quantified in HeLa cells expressing either wild-type NOD2 or a common Crohn's disease NOD2 mutant, NOD2FS-KI. (From L. H. Travassos et al. 2010. *Nat Immunol* 11: 55–62. © 2009 Nature Publishing Group.)

to localize to autophagosomes engulfing intracellular bacteria. The most common NOD mutant failed to allow ATG16L1 recruitment to autophagosomes to take place. These data demonstrate a functional link between the NOD family and the ability to clear intracellular bacteria by autophagy. The findings suggest that an inability to induce autophagy may be a causative factor in the onset of Crohn's disease.

Think About...

1. How would you use knowledge of ATG16L1 mutations and other polymorphisms known to predispose to Crohn's disease to further support this study?
2. Autophagy is a process used by all cells in different ways. Why do you suspect or predict that somatic mutations in ATG16L1 and/or the NOD family will only manifest as an inflammatory bowel disease?
3. Knowing the functional link between NOD, ATG16L1, and a lack of induction of autophagy in individuals suffering from Crohn's disease, hypothesize how you might manipulate this pathway (with agonists or antagonists) to alleviate Crohn's disease symptoms.

12.3 | Which immune effector cells protect mucosal surfaces?

LEARNING OBJECTIVE

12.3.1 Name the innate immune cells that play a role in mucosal immunity and describe their functions.

Both innate and adaptive immune cells are critical in the protection of mucosal surfaces, using effector mechanisms described in earlier chapters. The effector mechanisms of immune cells at mucosal surfaces must also function in a way that minimizes inflammation to prevent damage to tissue that is separated from the lumen of the mucosae. In this section, we take a closer look at the innate and adaptive immune cells that protect mucosal surfaces and the effector mechanisms employed by the immune system to prevent infection at mucosal surfaces.

Innate immune cells

The workings of the innate immune system provide an efficient means to remove the threat of a large number of disease-causing pathogens. Innate immune system effector mechanisms quickly identify and target many of the pathogens we encounter. However, many of the mechanisms discussed in Chapter 2 were related to the actions of inflammatory cytokines as part of a robust innate immune response to threatening pathogens. Since an inflammatory response is potentially harmful at mucosal surfaces, in these locations, the innate immune response must be efficient at pathogen clearance while at the same time preventing inflammation that can damage underlying tissue. Intestinal macrophages, gut dendritic cells, and innate lymphoid cells function in the innate immune response to combat mucosal infection without inducing the typical inflammatory response seen with open wound infections.

INTESTINAL MACROPHAGES Intestinal macrophages populate the lamina propria and remove microorganisms (via phagocytosis) that can potentially invade the underlying tissue. While gut macrophages are as efficient as their counterparts in other body regions at destroying foreign antigens, they lack the ability to produce and secrete inflammatory cytokines such as TNF-α, IL-1, and IL-6. This differentiation ensures that mucosal surfaces are protected from damage. Furthermore, intestinal macrophages are not able to activate naïve T cells since they do not express the costimulatory molecule B7 and do not secrete cytokines required for T-cell clonal expansion, although the cytokines

Making Connections: Draining and Mucosa-Associated Lymphoid Tissue

When a pathogen causes a local infection, the innate immune system produces an inflammatory response accompanied by migration of innate immune cells to the site of infection and drainage of fluid to nearby lymphoid tissue to activate an adaptive immune response. Conversely, when the immune system recognizes a pathogen within mucosal tissue, inflammation is limited to prevent leakage of mucosal proteins and enzymes into underlying tissue. Instead, mucosal epithelial cells, M cells, and dendritic cells aid in pathogen detection and work to deliver pathogens and antigens to mucosa-associated lymphoid tissue. Mucosal tissue contains a wide array of innate and adaptive immune cells capable of mounting a rapid immune response.

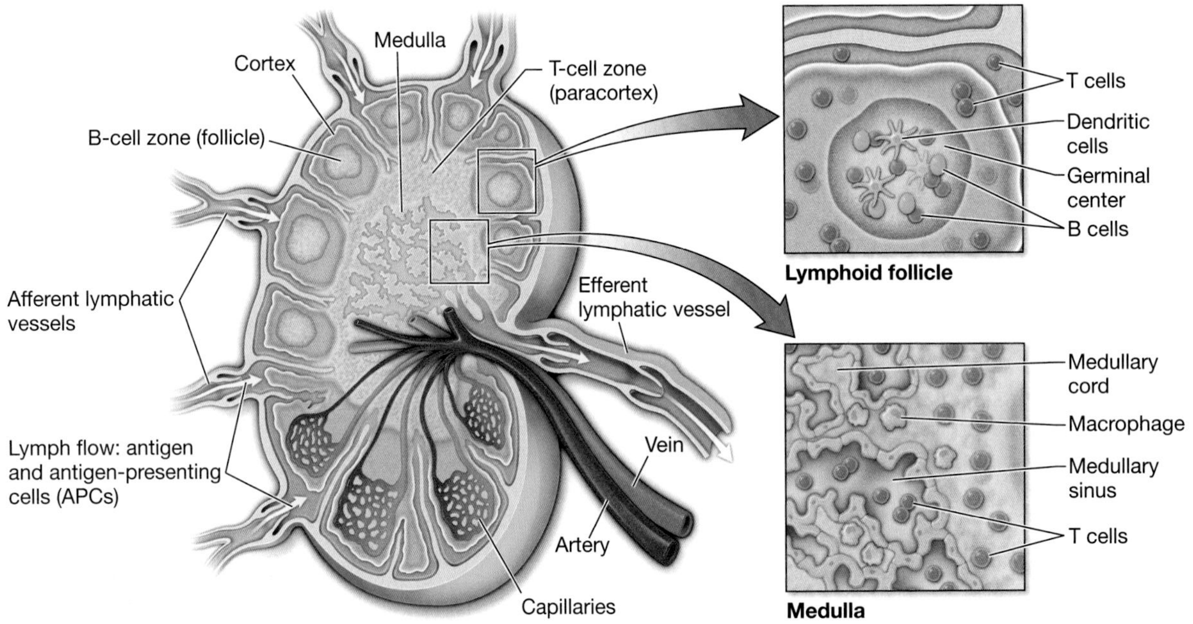

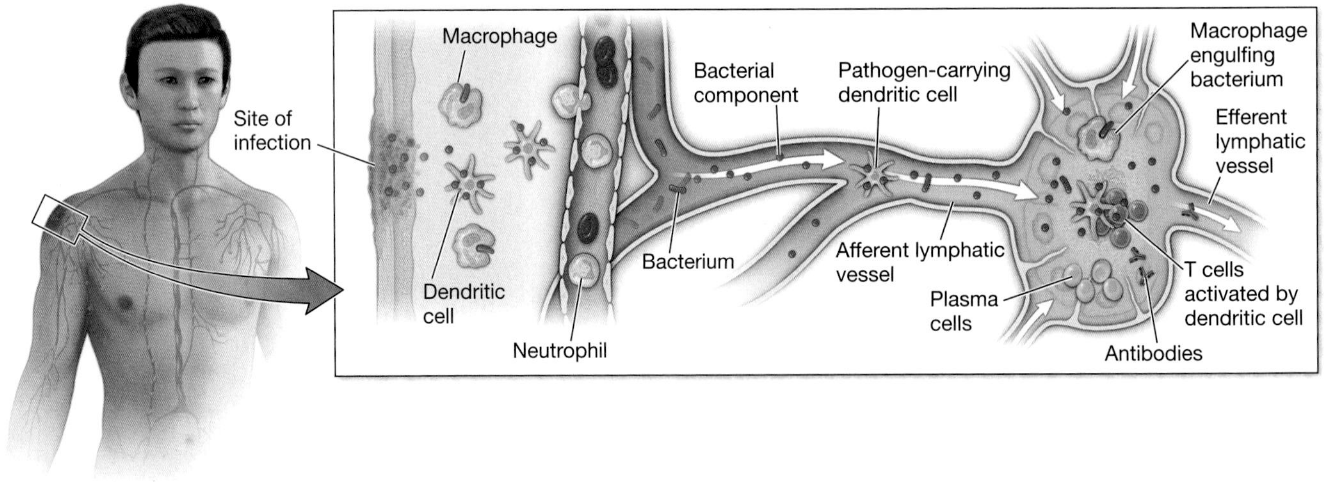

Local infection prompts an inflammatory response driven by the action of innate immune cells at a site of infection and the release of inflammatory cytokines. This promotes movement of pathogen and antigen to a draining lymphoid tissue to activate the adaptive immune response.

(Sections 1.6, 1.9, 2.2, 2.4, 2.5, 2.8, 4.3, 7.2, 7.3, 7.6, 10.3)

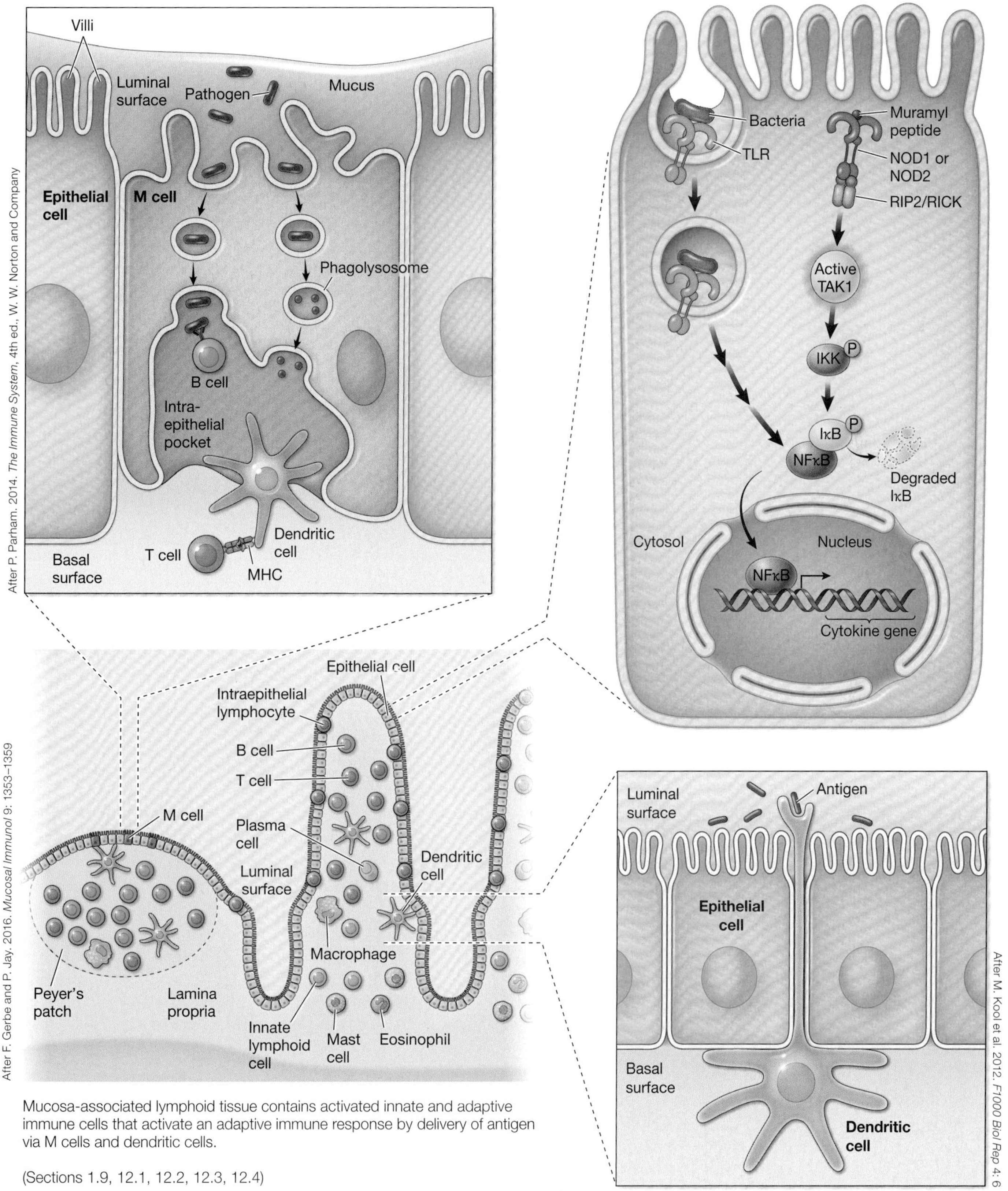

Mucosa-associated lymphoid tissue contains activated innate and adaptive immune cells that activate an adaptive immune response by delivery of antigen via M cells and dendritic cells.

(Sections 1.9, 12.1, 12.2, 12.3, 12.4)

oral tolerance Immune system tolerance to certain substances, such as food, that are ingested orally and are not harmful to the body.

they do secrete help support the maintenance of T_H17 helper T cells. Thus, intestinal macrophages work primarily in an innate immune response to recognize and phagocytose organisms rather than to induce an adaptive immune or inflammatory response. Another important feature of intestinal macrophages is their ability to aid in repair of epithelial cell damage by secreting the epithelial cell proliferation factor PGE2.

How do intestinal macrophages differ from other macrophages in the body? The answer lies in their inability to induce the activation of the transcription factor NFκB, which we know to be vital in the activation of cytokine production. To limit NFκB activation, macrophages that have differentiated in the small intestine lose the ability to produce many cell-surface receptors and signaling molecules that activate this transcription factor upon ligand interaction, including receptors for the Fc portion of immunoglobulins, the LPS coreceptors CD14 and MD2, and signaling molecules that are important in the TLR pathway such as MyD88 and TRAF6. While these receptors help "normal" macrophages engulf opsonized microorganisms (or, in the case of CD14 and MD2, help them recognize gram-negative organisms), engagement of these receptors leads to inflammatory cytokine production through activation of NFκB. Thus, the absence of these receptors in intestinal macrophages means that an inflammatory response does not occur or is limited.

GUT DENDRITIC CELLS Gut dendritic cells play a dual role in protecting the GI mucosae. Their first role is to provide **oral tolerance** to the vast number of foreign antigens that the body encounters during the process of digestion. Since most digested material presents no disease threat (although allergies to certain substances can induce an immune response; see Chapter 15), dendritic cells that take up these foreign molecules must be prevented from inducing an immune response.

How does this occur? In healthy mucosae, a subset of gut dendritic cells that reside underneath M cells is responsible for the uptake of digested food material transcytosed by M cells. Once the dendritic cells have taken up digested material, they activate antigen-specific T cells to expand and differentiate into regulatory T cells (T_{regs}). The T_{regs} then utilize their effector mechanisms to prevent the (abnormal) expansion and activation of T cells that would produce an adaptive immune response to harmless food antigens.

Gut dendritic cells also play an important role in recognizing both commensal organisms and foreign pathogens. While commensal organisms normally aid in digestion and prevent pathogen colonization, they can cause disease if they are left unchecked and permitted to invade the surrounding tissue. Thus, dendritic cells continually uptake the small number of commensal organisms that cross the epithelial barrier and enter the MALT. There, the dendritic cells process commensal organism antigens, most often via MHC class II presentation.

Commensal organism antigens presented by gut dendritic cells activate antigen-specific CD4 T cells, driving the activation of antigen-specific B cells. The activated B cells expand and differentiate into plasma cells and secrete antigen-specific dimeric IgA (after isotype switching occurs) that is capable of binding and neutralizing commensal organisms. Under normal conditions in a healthy organism, binding of dimeric IgA to commensal organisms in the gut lumen has no effect other than preventing them from invading the surrounding tissue.

As previously described, during an infection, gut dendritic cells have the ability to send projections through the epithelial barrier to capture pathogenic microorganisms and process them in the same manner as commensal organism antigens are processed. These dendritic cells can activate antigen-specific CD4 T cells in the MALT or in a nearby mesenteric lymph node to activate antigen-specific B cells and promote the production of neutralizing antibodies.

Innate lymphoid cells

Innate lymphoid cells (ILCs) are abundant in mucosal tissue. These cells are a recently defined cell lineage derived from the common lymphoid progenitor during hematopoiesis. ILCs are similar to T cells in that they secrete cytokines to promote the function of other innate and adaptive immune cells. However, they differ from T cells in that their activation is not induced by engaging a specific antigen. Instead, ILCs bind to cytokines present upon tissue damage or to microbial products.

Three major types of ILCs present in mucosal tissue are ILC1, ILC2, and ILC3:

- ILC1s activate macrophages and dendritic cells through the secretion of IFN-γ. The activated macrophages and dendritic cells remove pathogens through phagocytosis and present antigens to T cells in the vicinity.
- ILC2s respond to proteins secreted during helminth infection, such as IL-33 and thymic stromal lymphopoietin (TSLP). They secrete cytokines such as IL-5, which promotes activation of resident granulocytes and expulsion of the helminth.
- ILC3s secrete the cytokines IL-22 and IL-17 when triggered by cytokines such as IL-1β. IL-17 secreted by ILC3s drives activation of neutrophils to promote pathogen clearance. IL-22 prompts epithelial cells to secrete antimicrobial peptides and produce mucus.

Migration of effector lymphocytes

CD4 helper T cells and B cells play an important role in the protection of mucosal surfaces. Naïve B and T cells, after undergoing positive and negative selection in the bone marrow, spleen, and thymus, migrate through the circulatory system and into MALT and mesenteric lymph nodes via the mechanisms described in earlier chapters:

1. Cells of these secondary lymphoid tissues secrete chemokines such as CCL19 and CCL2, which bind to the cell-surface receptor CCR7 on the surface of naïve lymphocytes.
2. Naïve B cells and T cells leave the circulation through a high endothelial venule and enter the MALT (**FIGURE 12.9**).
3. Naïve B cells and T cells that do not encounter antigen within these tissues exit via the efferent lymphatic system and migrate to a different secondary lymphoid tissue to monitor for their specific antigen.
4. If the naïve B cells or T cells engage an antigen that their receptor can recognize, they are retained in the MALT, where they expand and differentiate into effector lymphocytes, whose action will be discussed in the following section.

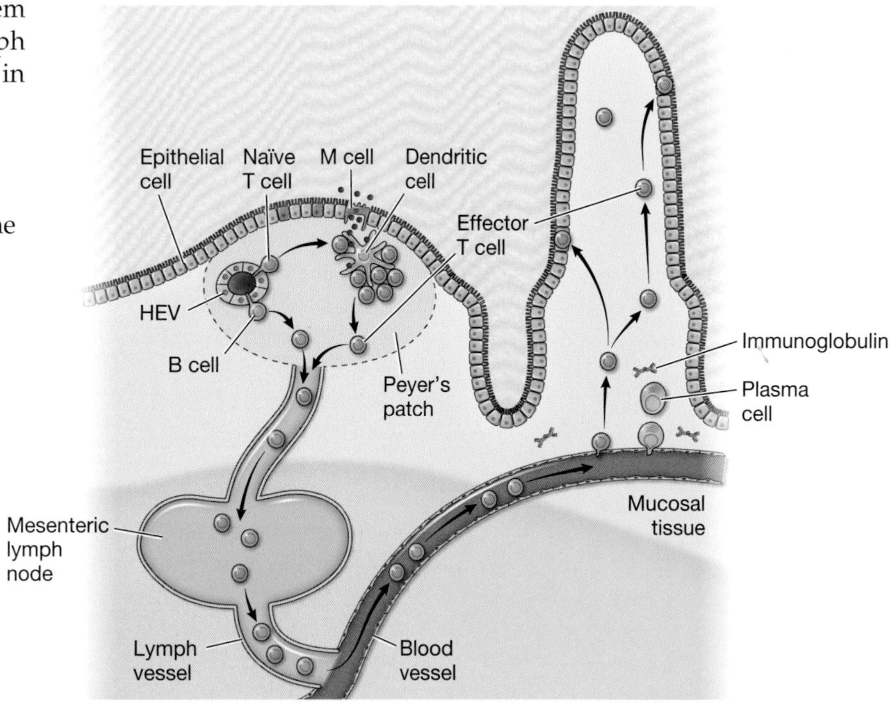

FIGURE 12.9 Migration of lymphocytes in mucosa-associated lymphoid tissue (MALT) Naïve T cells and B cells enter MALT such as a Peyer's patch from a high endothelial venule (HEV). T cells interact with dendritic cells via their receptor, and activated T cells differentiate into effector T cells, including cytotoxic T cells and T_H2 helper T cells. B cells and effector T cells travel through the lymphatic system, through a mesenteric lymph node, and through the circulatory system to nearby mucosal tissue. Here, B cells secrete immunoglobulins. Effector T cells migrate into the lamina propria of nearby mucosal tissue. (After A. Mowat. 2003. *Nat Rev Immunol* 3: 331–341.)

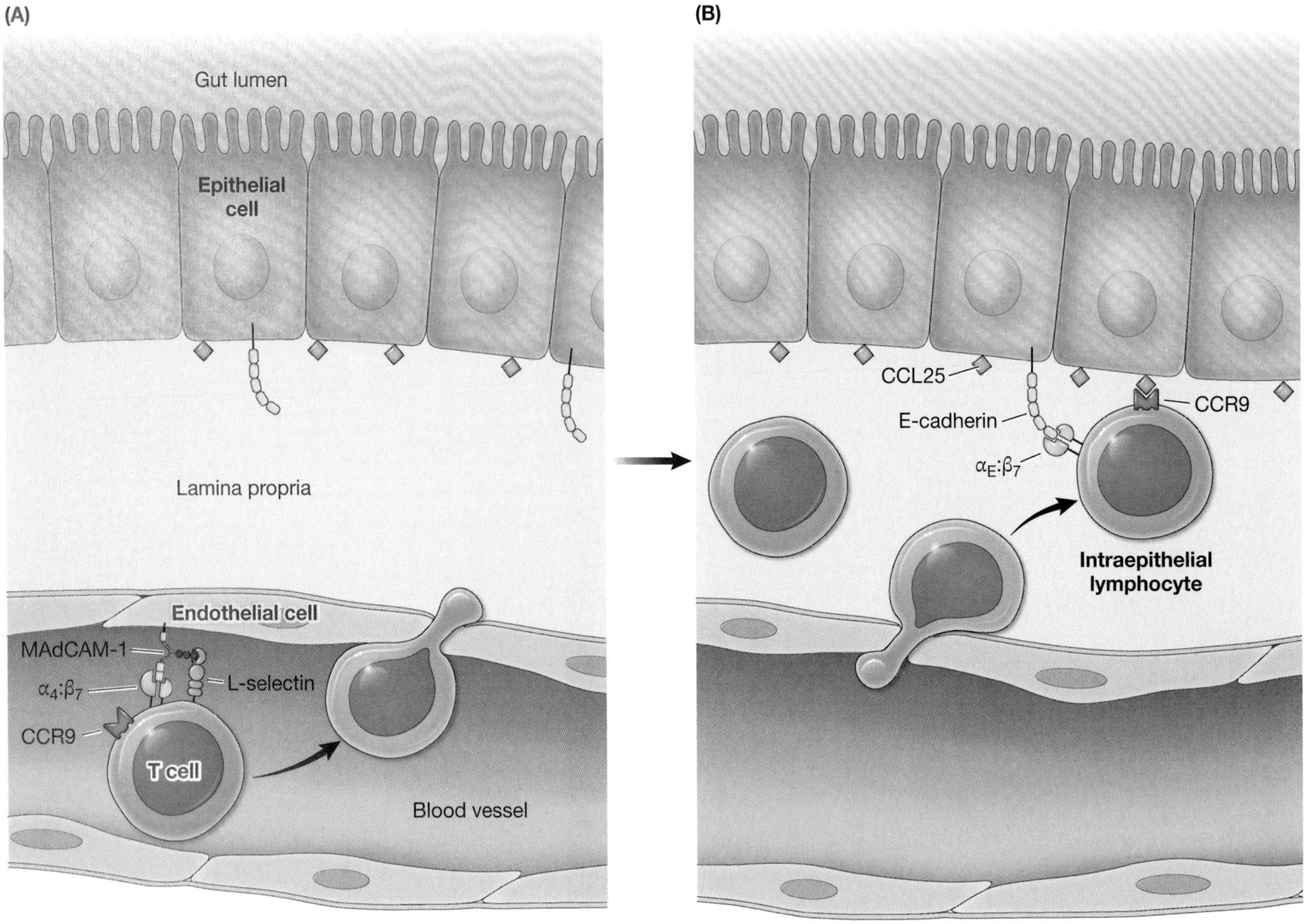

FIGURE 12.10 Extravasation and migration of T cells to mucosal tissue (A) Circulating T cells destined for entry into mucosal tissue express cell-surface receptors including L-selectin and the integrin α_4:β_7, which bind to the endothelial cell receptor MAdCAM-1. This tight binding prompts the T cell to migrate out of the bloodstream and into the underlying mucosal tissue. (B) T cells destined to become intraepithelial T cells switch the expression of their cell-surface integrin to α_E:β_7, which binds to the epithelial cell receptor E-cadherin. T-cell migration to the epithelial cell layer is further promoted by the chemokine CCL25, which binds to the T-cell receptor CCR9. (After N. Iijima and A. Iwasaki. 2015. *Trends Immunol* 36: 556–564.)

intraepithelial lymphocyte Cytotoxic T cells embedded in the epithelial cell barrier of mucosal tissue.

During activation, the lymphocytes develop an enhanced ability to migrate into other MALT. They inactivate expression of CCR7 and the cell-adhesion molecule L-selectin (required for entry into most secondary lymphoid tissues) and begin to express adhesion molecules specific to mucosal homing. One example of this change in receptor expression is the increase in the cell-adhesion molecule integrin α_4:β_7.

The integrin α_4:β_7 can bind to another cell-adhesion molecule, the mucosal addressin MAdCAM-1, promoting tight binding between the effector lymphocyte and the tissue it is engineered to protect. Expression of the chemokine receptor CCR9 facilitates migration into the mucosal tissue, as it binds to the chemokine CCL25 secreted by mucosal epithelial cells (**FIGURE 12.10**). There may also be an increase in the integrin α_E:β_7 in cells destined to become **intraepithelial lymphocytes**, which are T cells that become part of the epithelial barrier. Effector cells destined to become intraepithelial lymphocytes utilize their integrin to bind to the epithelial cell adhesion protein E-cadherin. Thus, a lymphocyte activated in MALT is uniquely equipped to further migrate and protect only mucosal tissue due to these changes in cell-surface receptors and adhesion molecules.

CHECKPOINT QUESTIONS

1. How do dendritic cells provide oral tolerance for an organism?
2. Explain the importance of changes of the cell-surface receptor profile of lymphocytes activated in mucosa-associated lymphoid tissue.

12.4 How do lymphocytes protect mucosal surfaces?

LEARNING OBJECTIVES

12.4.1 Name the effector lymphocytes involved in mucosal immunity and describe the mechanisms they use to fight infection.

12.4.2 Explain the role of antibodies in mucosal immunity.

12.4.3 Describe how the body combats an intestinal helminth infection.

Mucosal surfaces require constant protection, and we have seen that different types of cells are involved in this task. Changes in cell-surface molecules and inhibition of transcription factor activation are used to prevent inflammation and to guide antigen-specific lymphocytes to mucosal surfaces. Lymphocytes work to neutralize pathogens in a variety of ways, including the action of helper T cells, intraepithelial T cells, and plasma cells that produce dimeric IgA.

Effector lymphocytes in a healthy GI tract

A healthy mucosal surface is populated with effector lymphocytes and cells of the innate immune system, including intestinal macrophages and gut dendritic cells (**FIGURE 12.11**). Also present are mast cells, which provide protection via IgE-mediated activation, especially against parasite infections. In addition to these important innate immune cells, healthy gut mucosae also contain a predominance of effector lymphocytes standing ready to mount an adaptive immune response should the need arise. The presence of these effector lymphocytes in healthy mucosal tissue is in stark contrast to the rest of the body, which activates lymphocytes only when an infection is present.

EFFECTOR T CELLS Healthy mucosal tissue is populated with a large number and variety of effector T cells. These T cells are heterogeneous in that they each express a different receptor, allowing the population to recognize a multitude of different microorganisms and pathogens. The T-cell receptors expressed can be either $\alpha:\beta$ or $\gamma:\delta$ receptors, and the T cells can have either the CD4 or CD8 coreceptor present on their surface.

Most effector T cells with the CD4 coreceptor reside in the lamina propria, while the majority of those expressing the CD8 coreceptor are intraepithelial lymphocytes embedded within the epithelial layer, representing about 10% of the cells in this layer. Several types of CD4 effector T cells are present in the lamina propria, including T_H2 and T_{FH} helper T cells. To drive plasma cells to produce the immunoglobulins IgE and IgA, another type of CD4

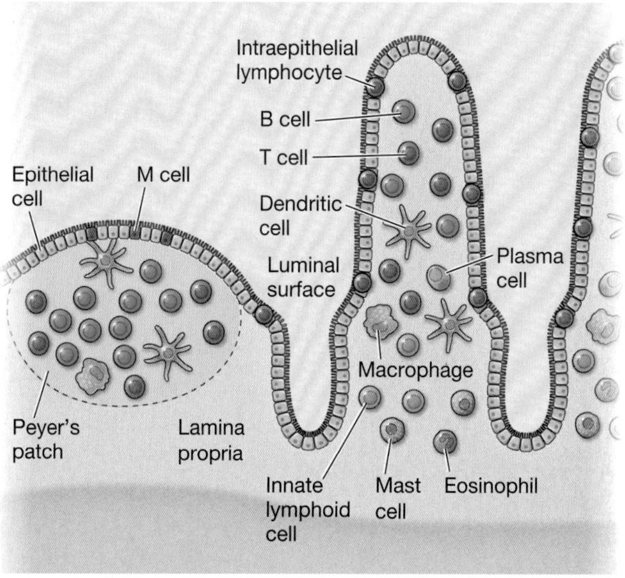

FIGURE 12.11 Immune cells of healthy mucosal tissue To combat the many pathogens in contact with mucosal surfaces, many different types of immune cells reside within the lamina propria of mucosal tissue. These include activated T cells within the lamina propria, intraepithelial lymphocytes within the epithelial cell layer, and plasma cells. In addition, many types of innate immune cells reside in mucosal tissue, including macrophages and dendritic cells (capable of phagocytosis and antigen presentation), innate lymphoid cells (capable of macrophage activation, mucus production, and secretion of antimicrobial peptides), and mast cells and eosinophils (both capable of parasite and helminth expulsion). (After F. Gerbe and P. Jay. 2016. *Mucosal Immunol* 9: 1353–1359.)

effector T cell is needed: T_H17 helper T cells. As discussed in Chapter 7, T_H17 helper T cells secrete IL-17 to activate nearby neutrophils to aid in pathogen destruction. Intraepithelial lymphocytes not only fight intracellular infection, but because of their unique adhesive properties, they also interact with epithelial cells to help maintain the epithelial barrier.

PLASMA CELLS Plasma cells are also present in high numbers within the lamina propria. These cells are responsible for secreting the large amount of immunoglobulin required to neutralize the many pathogens that can be present at mucosal surfaces. In fact, the plasma cells in mucosal tissue represent approximately 80% of the 75 billion (7.5×10^{10}) total plasma cells in an adult human body. Immunoglobulins produced by plasma cells in the lamina propria are transported to the mucosal surface via transcytosis (described in Chapter 10).

While some of the plasma cells in the lamina propria secrete pentameric IgM to act as a neutralizing antibody at the mucosal surface, many cells have differentiated and gone through isotype switching to produce IgG, IgE, or, most commonly, IgA (for a historical look at IgA isotype switching in MALT, see the **KEY DISCOVERIES** box).

Recall that the driving factor of proper isotype switching is the action of specific cytokines that activate transcription and facilitate recombination events at the proper region of the immunoglobulin heavy chain locus. Isotype switching to IgA is facilitated mainly by T_{FH} helper T cells secreting the cytokine TGF-β. To further drive isotype switching to IgA, gut dendritic cells secrete a variety of other cytokines and effector molecules, including IL-4 and IL-10, and the soluble gas nitric oxide, which is produced by dendritic cells by the action of the enzyme inducible nitric oxide synthase (iNOS). Nitric oxide facilitates expression of the TGF-β receptor on the surface of B cells, allowing them to respond more efficiently to the cytokine and preferentially switch to produce IgA.

Antibody action

One key effector mechanism of lymphocytes engaged in protecting mucosal surfaces is the production of secretory IgA. After it is secreted by plasma cells in MALT, IgA, along with IgM, can be transported by epithelial cells to neutralize pathogens at the mucosal surface (see Figure 10.16). Immunoglobulins present at the mucosal surface remain there due to the viscosity of mucus and through disulfide bond formation with mucin through cysteine residues within the immunoglobulin and those that link mucin molecules.

Recall that the structure of IgM is pentameric, and the Fc portion is located in the interior portion of the planar structure. IgA is typically dimeric at mucosal surfaces and is held together by the same J chain that holds pentameric IgM together (**FIGURE 12.12**). While IgM is capable of activating complement via the classical pathway, the Fc portion of dimeric IgA cannot bind the C1 protease that cleaves C2 and C4. Thus, IgA at mucosal surfaces serves primarily as a neutralizing antibody, as it is incapable of activating complement via the classical pathway.

IgA has two subclasses: IgA1 and IgA2. Both subclasses can be made as monomeric IgA (in nonmucosal tissues) and dimeric IgA. The major difference between the two subclasses lies in the hinge region connecting the Fab and Fc regions of the immunoglobulin heavy chain. IgA1 has a longer, more flexible hinge region, which provides more flexibility and capacity to bind to multiple repeated epitopes on a single pathogen (see Figure 12.12). However, this longer hinge region makes IgA1 more susceptible to protease cleavage (analogous to the differences between IgG subclasses). In fact, several pathogenic organisms, including *Streptococcus pneumoniae*, *Haemophilus influenzae*, and *Neisseria gonorrhoeae*, employ synthesis and secretion of proteases to combat the action of IgA1 through proteolytic cleavage at the hinge region. IgA2 functions similarly to IgA1 and is typically made in locations where bacterial populations, and likely protease expression, are high.

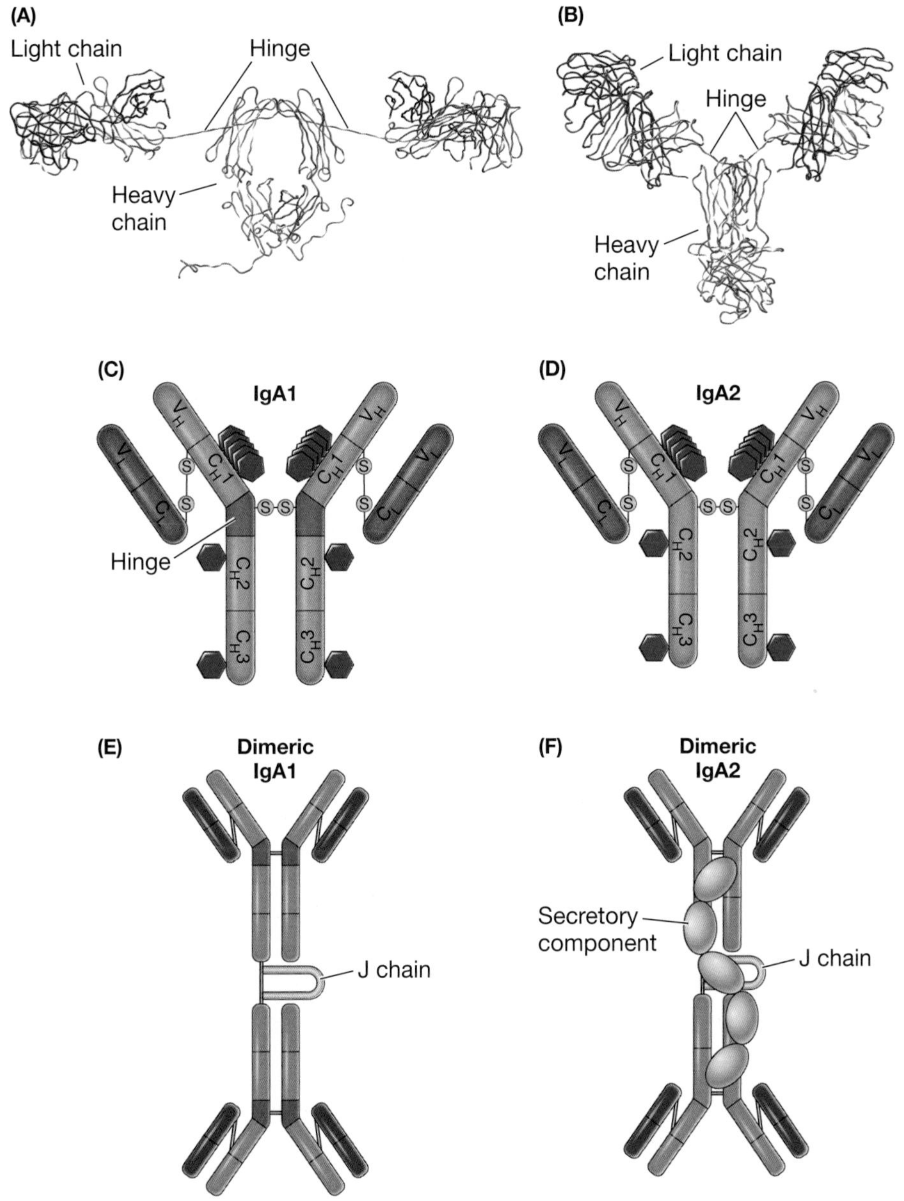

FIGURE 12.12 The two subclasses of the immunoglobulin IgA IgA has two subclasses: IgA1 and IgA2. (A) The structure of IgA1 reveals its long hinge region, which provides flexibility in the antigen-binding site to bind antigens that may be spaced differently on a pathogen surface. (B) The structure of IgA2 reveals a shorter hinge region than that of IgA1. Thus, IgA2 has fewer flexible antigen-binding sites but is more resistant to protease cleavage. (C) The IgA1 subclass heavy chain has one variable region and three constant regions. (D) The IgA2 subclass heavy chain has one variable region and three constant regions. (E) Two IgA1 proteins can dimerize with the aid of the J chain to form dimeric IgA1. (F) Two IgA2 proteins can dimerize with the aid of the J chain and interact with the secretory component to form dimeric IgA2. (A, data from A. Almogren et al. 2006. *J Mol Biol* 356: 413–431; redrawn using Chimera and PDB file 2ESG. B, data from A. Bonner et al. 2009. *J Biol Chem* 284: 5077–5087; redrawn using Chimera and PDB file 3CM9. C–F, based in part on T. Mak et al. 2014. *Primer to the Immune Response*, 2nd ed. Academic Cell/Elsevier Inc.: Burlington, MA.)

Because of the role IgA plays in protecting mucosal tissues, individuals with the immunodeficiency selective IgA deficiency are susceptible to infections at mucosal surfaces, including sinus infections, respiratory infections (including bronchitis and pneumonia), GI infections, and chronic diarrhea. Interestingly, many individuals with selective IgA deficiency have mild illness and are never diagnosed due to their mild symptoms. This immunodeficiency is further explored in Chapter 14.

Effector T-cell action against helminth infection

Helminths are parasitic worms that can colonize the intestine and even endanger the life of the host. Unlike commensal organisms such as certain bacteria that aid in digestion and prevent pathogen colonization, helminths are always harmful to the host.

The immune system has developed a variety of mechanisms to combat helminth infection. As with any immune response in mucosal tissue, the goal is to fight infection while minimizing an inflammatory response. Indeed, immune responses to helminths that stimulate the release of inflammatory cytokines tend to be more harmful than helpful to the host organism. To minimize the release of inflammatory mediators, CD4 T cells activated in response to helminth infections tend to be T_{FH} cells and T_H2 cells. These helper T cells tend to activate the production of both IgE-producing plasma cells and cytokines that increase the likelihood of shedding helminths that are not attached to mucosal surfaces or located within cells.

KEY DISCOVERIES

How do we know that MALT is a prominent location for isotype switching to IgA?

Article

E. C. Butcher, R. V. Rouse, R. L. Coffman, C. N. Nottenburg, R. R. Hardy, and I. L. Weissman. 1982. Surface phenotype of Peyer's patch germinal center cells: Implications for the role of germinal centers in B cell differentiation. *Journal of Immunology* 129: 2698–2707.

Background

We have seen that a major immunoglobulin present in organisms is secretory IgA, mainly because it is the immunoglobulin responsible for neutralizing pathogens over the large surface area of mucosal surfaces. We also have seen that a B cell's ability to produce IgA is based on active class switch recombination, whereby a B cell originally activated and producing IgM recombines the immunoglobulin heavy chain locus to produce IgA. One might suspect that lymphoid tissue associated with protecting mucosal surfaces would be the center for producing activated plasma cells capable of secreting IgA for mucosal protection.

Early Research

At the time of this study, it was hypothesized, based on a large amount of evidence, that germinal centers are the location of B-cell differentiation into plasma cells or memory cells by the action of helper T-cell activity. However, due to the inability to definitively detect germinal center lymphoid cells because of a lack of a specific marker, this hypothesis was difficult to test. The question remained: Do germinal centers indeed represent a location where B-cell activation and isotype switching occur?

Think About...

1. If your prediction was that germinal centers mainly contain T cells and B cells, design a series of experiments to test your hypothesis. How would you definitively demonstrate the presence of these two cell populations? What would be your source of germinal centers?
2. If your source of germinal centers was from gut-associated lymphoid tissue, which cell-surface immunoglobulins would you predict to be present on B cells located in these germinal centers (if B cells were present in the isolated germinal centers)?
3. Design experiments to test your prediction of the types of cell-surface immunoglobulins present on B cells isolated from mucosa-associated lymphoid tissue germinal centers.

Article Summary

The researchers who conducted this study made important discoveries concerning the population of cells present in isolated Peyer's patches. Using the affinity of peanut agglutinin (PNA) to binding of germinal center cells, these scientists were at the forefront of research employing two-laser flow cytometry to demonstrate that the majority of cells isolated from Peyer's patches are B cells and that these cells primarily express IgA at their surface (**TABLES**). This seminal research paper described the cell population of germinal centers and demonstrated that germinal centers of Peyer's patches are the locations of isotype switching to IgA of activated B cells within mucosa-associated lymphoid tissues.

Non-immunoglobulin Surface Antigens of Peyer's Patch Lymphocytes

Antibody	Specificity	Cell Source	% positive by FACS
RA3-3A1	B220	BALB/c	99
53-7.3	Lyt-1	BALB/c	3
53-6.7	Lyt-2	BALB.K	1
49-h4	ThB	BALB/c	4, 6
11-5.2	Ia.2 (I-A^k)	BALB.K	98, 100
11-4.1	H-$2K^K$	BALB.K	97

Source: Data from E. C. Butcher et al. 1982. *J Immunol* 129: 2698–2707.

Surface Immunoglobulin Determinants of Peyer's Patch Lymphocytes

Antibody	Specificity	Cell Source	% positive by FACS
Rα Kappa	κ	BALB/c	90
H10.4.22 (αDelta)	Igh-5a	BALB/c	0
Rα Mu	μ	BALB/c	17
Rα Gamma	γ	BALB/c	15
Rα Alpha	α	BALB/c	68
71-14	α	BALB/c	69, 38

Source: Data from E. C. Butcher et al. 1982. *J Immunol* 129: 2698–2707.

Making Connections: Immune Responses of Mucosa-Associated Lymphoid Tissue

Healthy mucosal tissue is filled with innate and adaptive immune cells that act quickly when a pathogen is detected. Plasma cells within mucosal tissue secrete IgA and IgE isotypes, as these effectively promote pathogen clearance and neutralization.

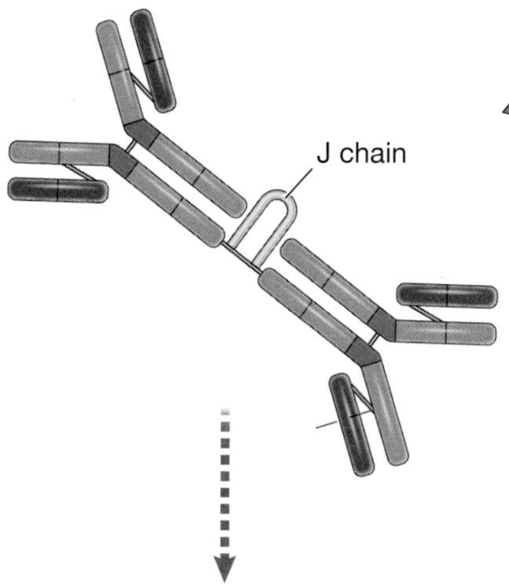

Based in part on T. Mak et al. 2014. *Primer to the Immune Response*, 2nd ed. Academic Cell/Elsevier Inc.: Burlington, MA

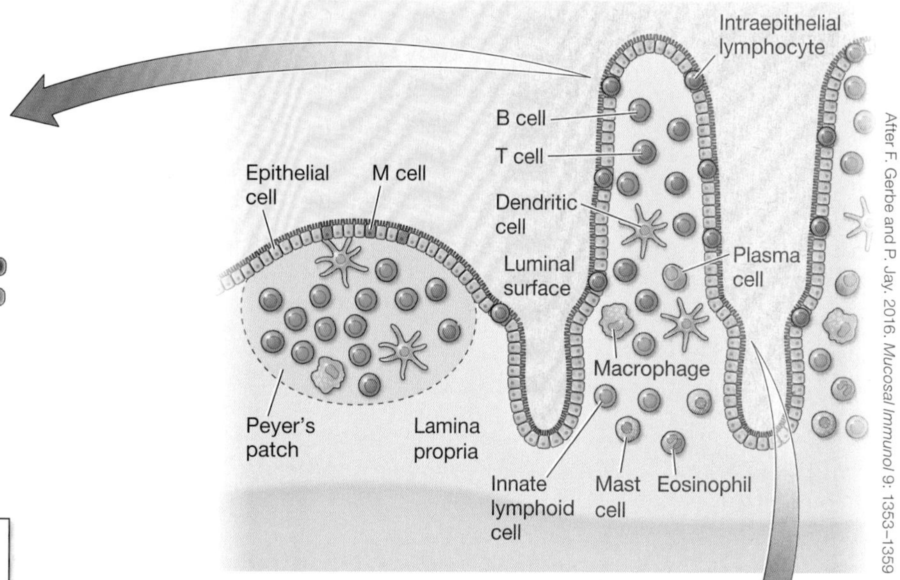

After F. Gerbe and P. Jay. 2016. *Mucosal Immunol* 9: 1353–1359

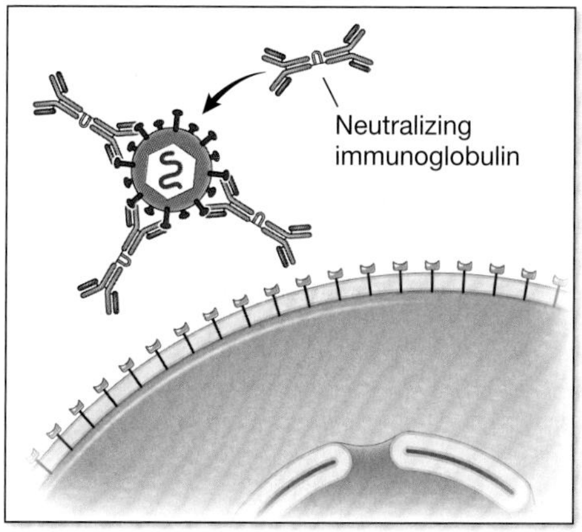

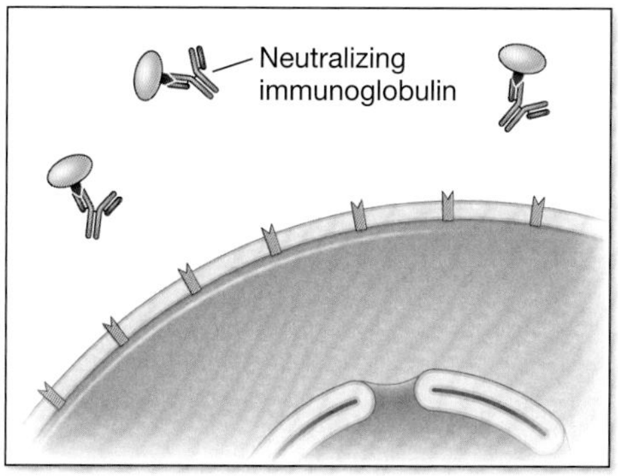

After P. Parham. 2014. *The Immune System*, 4th ed., W. W. Norton and Company

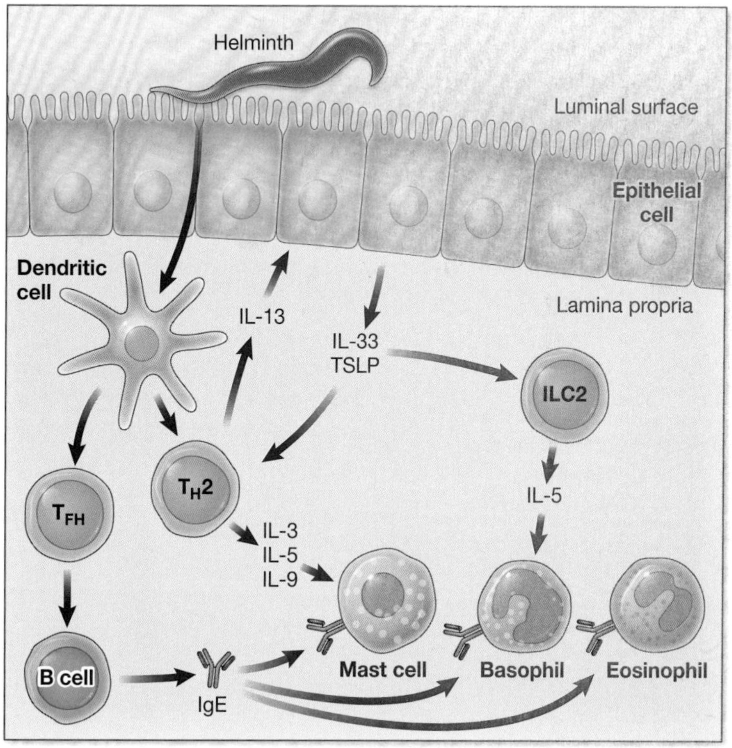

Healthy mucosa contains a wide variety of innate immune cells, including macrophages, dendritic cells, innate lymphoid cells, mast cells, eosinophils, and basophils. These cells promote pathogen destruction or the induction of an adaptive immune response. As part of the adaptive immune response, granulocytes and plasma cells that express IgA and IgE are activated. Effector T cells are also present in mucosal tissue; these cells work to support the activity of the innate immune cells.

(Sections 1.5, 7.14, 10.5, 12.3, 12.4)

Many immune responses associated with mucosal tissue are designed to limit inflammation and tissue damage and to neutralize or expel the pathogen. The primary immunoglobulins present in mucosal tissue include IgA, which works to neutralize pathogens and pathogen toxins, and IgE, which promotes granulocyte activity. Degranulation of granulocytes results in pathogen expulsion through sneezing, coughing, vomiting, or diarrhea.

(Sections 1.5, 10.5, 12.3, 12.4)

How does the mucosal tissue drive activation and differentiation of T_H2 cells? The process begins with pathways discussed earlier in this chapter based on epithelial cell action and detection of foreign microorganisms through PAMP receptors such as TLRs and the action of NOD. When a helminth is recognized, these pathways are triggered within epithelial cells, driving activation of NFκB and secretion of two pro-T_H2 cytokines, IL-33 and TSLP. These two cytokines also activate resident ILC2s to drive the activation of granulocytes.

Through the action of gut dendritic cells that have processed helminth antigens, CD4 T cells differentiate into T_H2 cells and T_{FH} helper T cells to drive activation of B cells and promote isotype switching to IgE. IgE works in conjunction with mast cells, eosinophils, and basophils to promote expulsion of a pathogen through effector molecules that stimulate muscle contraction within the tissue.

Cytokines produced by T_H2 cells can promote the recruitment of mast cells (IL-3, IL-9) and eosinophils (IL-5) and increase the production and turnover of epithelial cells within the intestine (IL-13) to promote epithelial sloughing and release of the helminths from the small intestine (**FIGURE 12.13**). A T_H2 cell response is desired during helminth infections for all of these reasons. Furthermore, since T_H1 cell responses tend to promote an inflammatory response due to the release of IFN-γ and activation of macrophages, the mucosal response to helminth infection mirrors that of antigen delivery by minimizing the

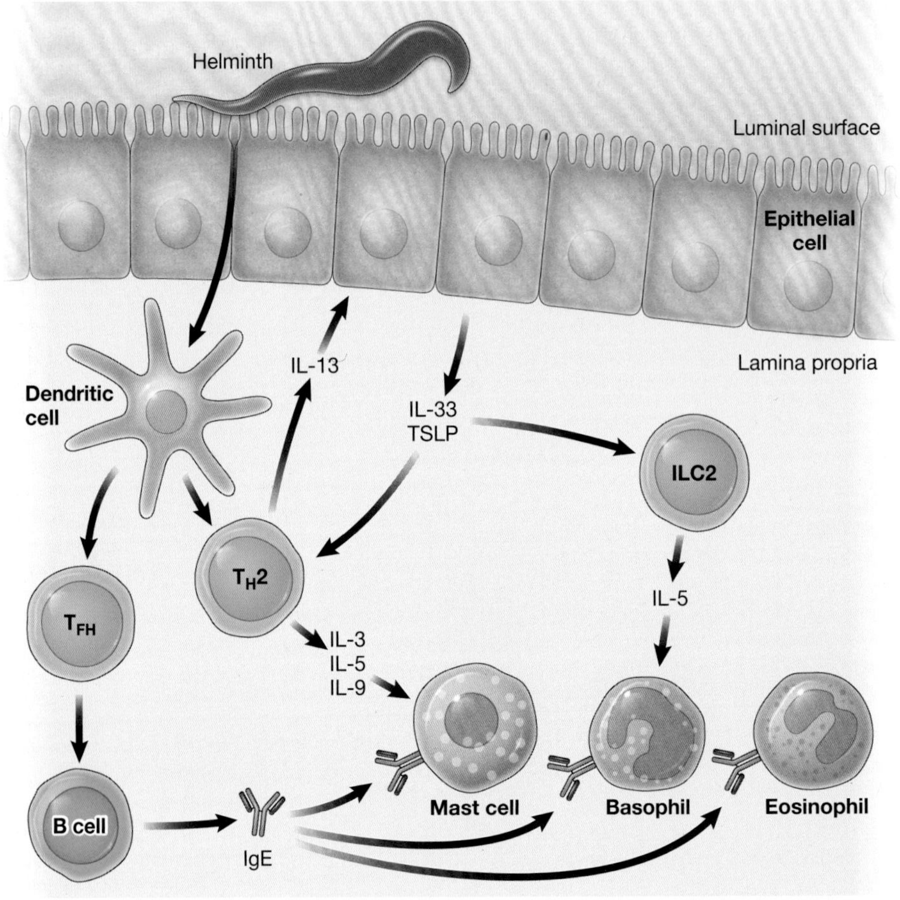

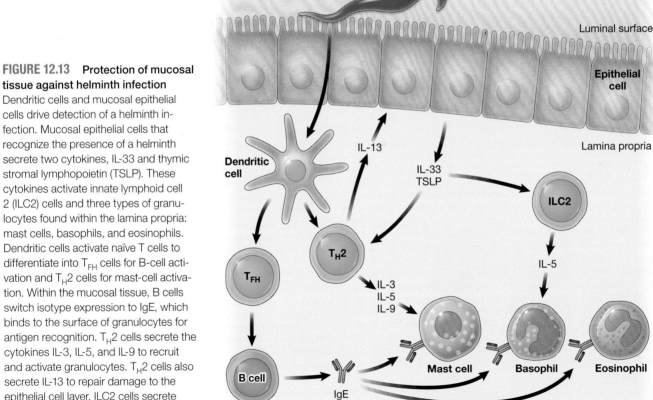

FIGURE 12.13 Protection of mucosal tissue against helminth infection Dendritic cells and mucosal epithelial cells drive detection of a helminth infection. Mucosal epithelial cells that recognize the presence of a helminth secrete two cytokines, IL-33 and thymic stromal lymphopoietin (TSLP). These cytokines activate innate lymphoid cell 2 (ILC2) cells and three types of granulocytes found within the lamina propria: mast cells, basophils, and eosinophils. Dendritic cells activate naïve T cells to differentiate into T_{FH} cells for B-cell activation and T_H2 cells for mast-cell activation. Within the mucosal tissue, B cells switch isotype expression to IgE, which binds to the surface of granulocytes for antigen recognition. T_H2 cells secrete the cytokines IL-3, IL-5, and IL-9 to recruit and activate granulocytes. T_H2 cells also secrete IL-13 to repair damage to the epithelial cell layer. ILC2 cells secrete the cytokine IL-5 to activate basophils.

inflammatory response and limiting T_H1 cell activation. IFN-γ also acts counter to IL-13 and prolongs the stability of epithelial cells, which results in a more hospitable environment for helminths attempting to survive in the small intestine.

CHECKPOINT QUESTIONS

1. How are B cells activated in mucosa-associated lymphoid tissue signaled to preferentially isotype switch to IgA?
2. Why is the activation of T_H2 helper T cells the best means of protecting against helminth infections in the GI tract compared to T_H1 cells?

Summary

12.1 What are mucosal surfaces?

- Mucosal surfaces carry out important physiological functions; they require a unique set of immune system defense mechanisms to prevent an inflammatory response.
- First lines of defense of mucosal surfaces include the viscosity of mucus, which is aided by the action of mucins and defensins secreted to aid in pathogen lysis.

12.2 What is mucosa-associated lymphoid tissue?

- Mucosal surfaces are protected by mucosa-associated lymphoid tissue (MALT), which functions as other secondary lymphoid tissue does in systemic immunity but minimizes the need for inflammation for antigen delivery.
- Delivery of antigen to MALT is aided by the action of both M cells and dendritic cells that minimize the need for fluid leakage for delivery to lymphoid tissue.
- Mucosal epithelial cells also play a role in protecting mucosal surfaces through the action of Toll-like receptors (TLRs) and intracellular NOD proteins, which are both involved in pathogen recognition.

12.3 Which immune effector cells protect mucosal surfaces?

- Intestinal macrophages protect mucosal surfaces primarily by acting to phagocytose pathogens and limit their role in inducing an inflammatory response.
- Gut dendritic cells aid in the induction of oral tolerance and protection of mucosal surfaces by efficiently delivering antigens to lymphocytes that migrate to gut-associated lymphoid tissue.

12.4 How do lymphocytes protect mucosal surfaces?

- Mucosal surfaces are also protected by effector lymphocytes that remain at the ready in the area; these include CD4 and CD8 T cells and plasma cells.
- Most effector lymphocytes remain associated with the lamina propria of mucosal tissue, although some CD8 T cells intersperse with the epithelial layers to protect mucosal surfaces as intraepithelial lymphocytes.
- Plasma cells that protect mucosal surfaces secrete isotype switched antibodies, primarily dimeric IgA, to aid in pathogen neutralization.
- Effector T cells protect against helminth infection by attempting to drive differentiation into T_H2 helper T cells, which are the most efficient in helminth clearance due to their effector mechanisms.

Review Questions

12.1 If mucin were isolated from mucosal tissue, what types of molecules might be present in the isolated mucin?

12.2 In the laboratory, a transgenic mouse has been created that has a homozygous deletion for RICK. Assuming that mice of this type have no developmental abnormalities and can survive, predict the effect of this mutation when these mice are challenged with *Salmonella*.

12.3 Your lab partner has isolated macrophages from a mouse to study the effects of diet changes on the transcriptional profile of intestinal macrophages. He shows you flow cytometry data he used to analyze the macrophage population, and he is excited that the cells are positive for the LPS coreceptor CD14. You tell your lab partner that he does not have a population of intestinal macrophages. Explain why.

12.4 The evolutionary arms race between vertebrate organisms and their immune responses and pathogens is a constant struggle. You are leading a research team studying a newly identified helminth that appears to have evolved virulence mechanisms capable of driving an inflammatory response in the small intestine and preventing efficient clearance by the host organism. Which types of virulence factors may be present in the genome of the newly identified helminth?

CASE STUDY REVISITED

Trichinosis

A family of four arrives at the hospital emergency department with symptoms of nausea, abdominal pain, and vomiting. The patients are a male, age 32, female, age 31, boy, age 6, and a girl, age 4. In reporting what food they have consumed in the last 24 hours, the father tells the ER doctor about the spare ribs he cooked on the grill the night before. The doctor suspects food poisoning or a parasitic infection. She orders a blood test and requests that stool samples from each patient be brought to the lab within 24 hours. The blood test results show that they all have an increased number of eosinophils in their blood, which may indicate a parasitic infection. The stool cultures do not reveal a bacterial infection, but those for two of the patients come back positive for trichinella (a roundworm). The doctor makes a diagnosis of trichinosis and prescribes albendazole for all members of the family.

Think About...

Questions for individuals

1. What are some common causes of trichinosis?
2. What are some common symptoms of trichinosis?

Questions for student groups

1. If you were to run an ELISA (enzyme-linked immunosorbent assay that detects a specific molecule using immunoglobulins) on cytokines within the intestinal mucosal tissue of these patients, which cytokines would you predict to be elevated?
2. Trichinosis, if caught early, is treated with an antiparasitic medication such as albendazole. Look up the mechanism of action of albendazole. How does it act as an antiparasitic? Do you predict that it might have any side effects in humans given the target of the drug?

MAKING CONNECTIONS **Key Concepts**	**COVERAGE** **(Section Numbers)**
Review of draining lymphoid tissue action in lymphocyte activation	1.6, 1.9, 2.2, 2.4, 2.5, 2.8, 4.3, 7.2, 7.3, 7.6, 10.3
Action of mucosa-associated lymphoid tissue in antigen delivery and lymphocyte activation	1.9, 12.1, 12.2, 12.3, 12.4
Innate immune cells of mucosal tissue	1.5, 12.3, 12.4
Lymphocyte action in mucosal tissue	1.5, 7.14, 10.5, 12.3, 12.4
IgA and IgE action in mucosal tissue	1.5, 10.5, 12.3, 12.4

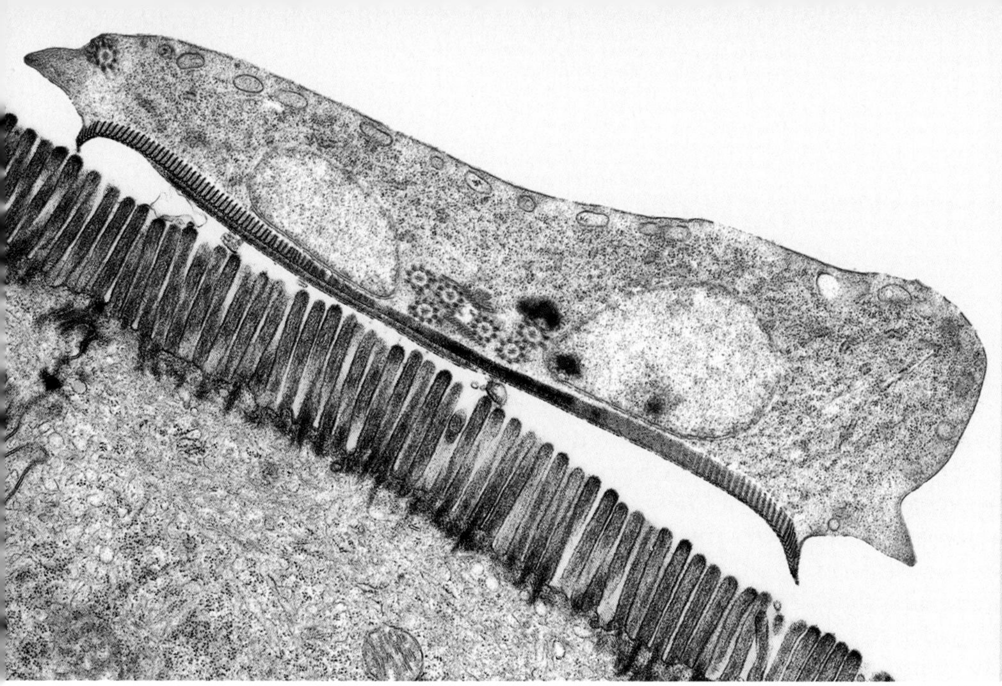

13

Pathogen Evasion of the Immune System

● CASE STUDY: Sabrina's Story

On a beautiful spring day Sabrina is outside gardening. It's her favorite thing to do in the springtime—she enjoys growing vegetables and planting flowers that will provide color all summer long. Like many gardeners, she isn't a fan of clearing weeds, but she works diligently to prepare her garden for the plants and seedlings she purchased yesterday at the farmer's market. While trying to dig out a stubborn weed, her hand slips and she cuts herself. Sabrina goes inside and runs her hand under water to clean off the dirt before bandaging the wound. Before long, the pain subsides and she is back outside taking advantage of the last few hours of sunlight.

The following morning, Sabrina wakes up in a cold sweat. She has chills and cannot seem to get warm. She is also having a problem catching her breath. At first, Sabrina figures she might have been outside in the sun too long the day before. But she feels a nagging pain in her hand and is worried that she may not have done a good enough job cleaning the wound. She calls her mother and asks her to drive her to the emergency room to get checked out.

To understand what Sabrina is experiencing, and to answer the case study questions at the end of the chapter, we must revisit the mechanisms our body uses to combat pathogens and the mechanisms pathogens use to evade our immune responses. In this chapter, we will explore the evolutionary arms race between our immune system, which is an efficient weapon against many pathogens, and the many pathogens that have evolved elaborate mechanisms to survive immune system defenses.

QUESTIONS Explored

13.1 What are the primary ways that pathogens evade immune system defenses?

13.2 How does genetic variation allow pathogens to evade the immune system?

13.3 How do pathogens hide from the immune system?

13.4 How can pathogens downregulate the immune system?

13.5 How do superantigens prevent a focused immune response?

13.1 | What are the primary ways that pathogens evade immune system defenses?

LEARNING OBJECTIVE

13.1.1 Name at least four tactics used by pathogens to evade the immune system.

In previous chapters, we have seen that an "orchestra" made up of a diverse range of cells and substances comes into play as part of the body's immune system defenses against foreign antigens. We have learned that elegant and elaborate mechanisms are used to combat pathogens and prevent disease. In this chapter, we will examine the means by which pathogens evade the immune system or reduce the effectiveness of the immune response. **TABLE 13.1** provides an overview, and the specifics of each mechanism are discussed in the following sections.

Tactics used by pathogens to evade the body's defenses include (**FIGURE 13.1**):

TABLE 13.1 | Mechanisms Used by Pathogens to Evade the Immune System

Immune response	Mechanism	Examples of pathogens employing this mechanism
Pathogen recognition of surface antigens	Genetic variation of surface antigens	*Salmonella* species *Vibrio cholerae*
	Mutation and recombination	Influenza virus
	Gene conversion of surface antigens	*Neisseria* species *Trypanosoma* species
	Latency	Herpesvirus
Phagocytosis	Escaping the phagosome	*Listeria monocytogenes*
	Preventing phagosome fusion with lysosome	*Mycobacterium tuberculosis*
	Survival in the phagolysosome	*Cryptococcus neoformans*
	Inhibiting phagocytosis	*Yersinia pestis*
Antimicrobial peptides/proteins	Blocking antimicrobial peptide/protein action	*Escherichia coli* *Salmonella* species *Pseudomonas aeruginosa* *Treponema denticola*
Cytokine signaling/inflammation	Disrupting Toll-like receptor recognition	*Salmonella* species *Porphyromonas gingivalis*
	Disrupting cytokine signaling	*Staphylococcus aureus* *Shigella flexneri* *Yersinia pestis* *Leishmania* species
Complement activation	Inhibiting complement fixation	*Haemophilus influenzae* *Streptococcus pneumoniae* *Escherichia coli* *Neisseria meningitidis* *Staphylococcus aureus* Herpesvirus Human cytomegalovirus
Antigen processing and presentation	Blocking MHC class I presentation	Human cytomegalovirus
T-cell action	Inhibiting T-cell activation	*Helicobacter pylori* *Neisseria gonorrhoeae*
	Activating nonspecific T-cell response by superantigens	*Staphylococcus aureus*
B-cell action	Degrading immunoglobulins	*Staphylococcus* species *Haemophilus influenzae* *Neisseria* species
	Blocking IgA by superantigens	*Staphylococcus aureus*

- *Genetic variation*: Because many of the tools employed by our immune system involve proteins that bind to molecules on the pathogen surface, pathogens may have variations in genetic information that cause changes to their surface molecules. These differences may result from the presence of different genes in pathogen variants, from active genetic recombination events, or from high rates of mutation during the pathogen life cycle.

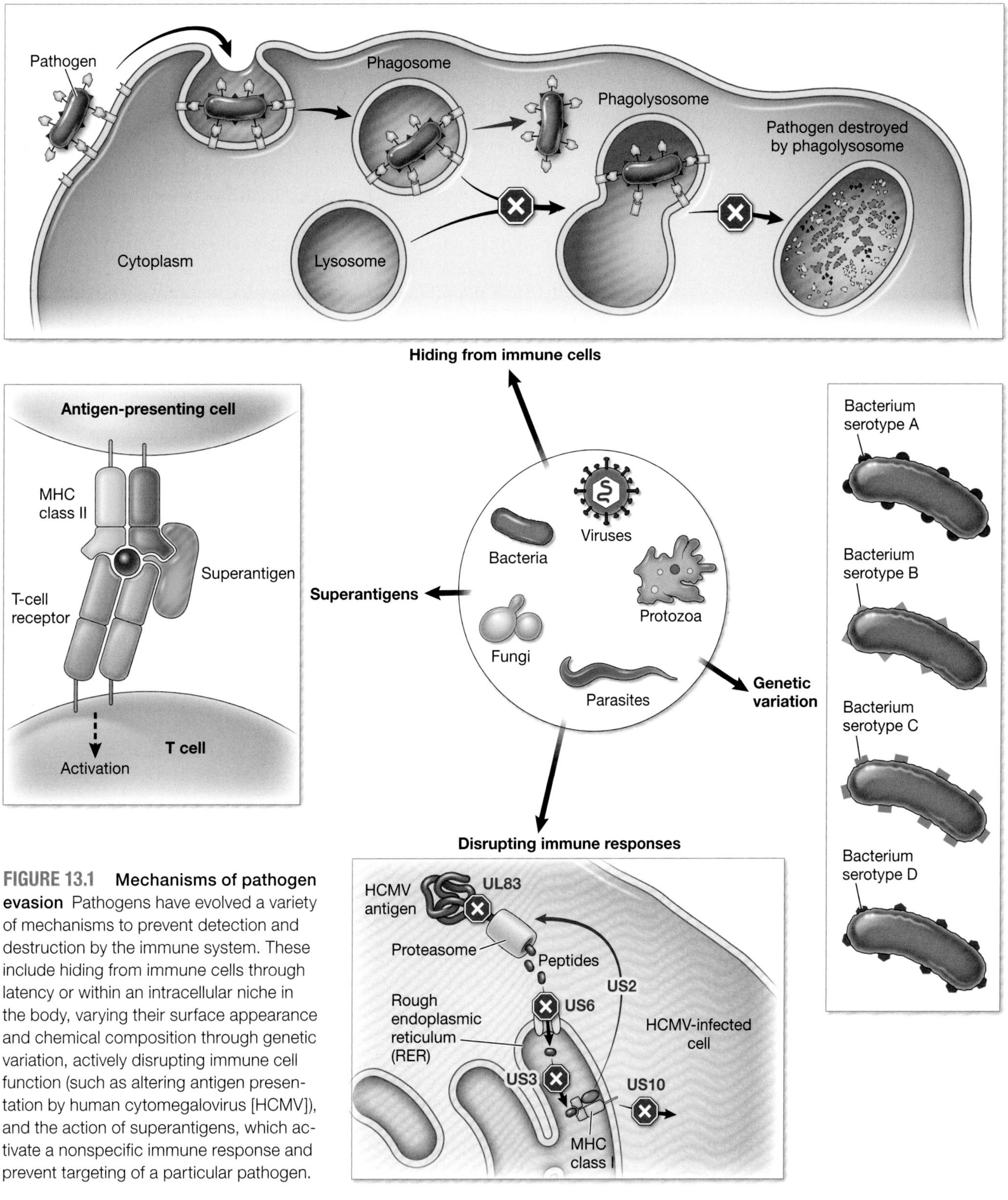

FIGURE 13.1 Mechanisms of pathogen evasion Pathogens have evolved a variety of mechanisms to prevent detection and destruction by the immune system. These include hiding from immune cells through latency or within an intracellular niche in the body, varying their surface appearance and chemical composition through genetic variation, actively disrupting immune cell function (such as altering antigen presentation by human cytomegalovirus [HCMV]), and the action of superantigens, which activate a nonspecific immune response and prevent targeting of a particular pathogen.

- ***Avoiding recognition***: If a pathogen can avoid being recognized by the immune system, it can evade destruction. Some pathogens use macromolecules that mimic host self-molecules to prevent recognition. Others avoid recognition by residing in locations that can conceal them, such as inside a cell or organelle, where extracellular components cannot reach them. To prevent recognition, extracellular pathogens usually mask their outer surface, while intracellular pathogens usually escape detection by living within an organism's cells.
- ***Altering the host immune response***: Some pathogens avoid destruction by producing substances that alter the immune response. Because these products (*toxins* and *superantigens*) affect immune system functioning, they can produce symptoms commonly seen during an infection.

Vertebrates have evolved amazing innate and adaptive immune responses to survive the onslaught of a wide variety of pathogens on a daily basis. However, the ongoing survival of multicellular organisms and pathogens has created an "evolutionary arms race" in which the key to the organism's survival is getting rid of the pathogen, and the key to pathogen survival is to overcome the mechanisms of the innate and adaptive immune systems.

We generally think of pathogens as a nuisance, potentially detrimental, and sometimes even life-threatening. In many cases, their ability to cause disease is due to mechanisms that have evolved to ensure their survival when faced with the threat of the immune system (see Figure 13.1). Some mechanisms make it more difficult for a pathogen to be detected. These include making a coating that renders normal recognition mechanisms of the immune system ineffective or residing in a location that is inaccessible to the immune system. Some pathogens employ a high mutation rate to alter their recognition by the immune system without affecting pathogenesis. Others express virulence factors that actively target specific immune response pathways to prevent efficient clearance.

CHECKPOINT QUESTIONS

1. List the means by which a pathogen can survive within a host organism through evasion of the immune system.
2. One important immune system mechanism is phagocytosis. Predict how a pathogen would escape phagocytosis (i.e., which steps of phagocytosis a pathogen could block or evade).

13.2 How does genetic variation allow pathogens to evade the immune system?

LEARNING OBJECTIVE

13.2.1 Describe how bacteria and viruses use genetic variation to promote their survival.

The immune system uses many strategies to distinguish between self and nonself. Innate immune cells, B cells, and T cells all employ receptors that can recognize foreign pathogens. Pathogen recognition via these receptors often relies on detection of pathogen surface molecules or the use of opsonins such as antibodies to mark a pathogen as foreign.

Because the immune system recognizes pathogen surface molecules using receptors or antibodies, by changing their surface, pathogens can avoid detection. How exactly do pathogens evade the immune system through surface molecule variation? The following sections describe several genetic variation mechanisms used by pathogens to avoid detection by the host immune system. This genetic variation can be as simple as the presence of different genes responsible for encoding different surface molecules and as complex as gene rearrangement events that induce expression of different cell-surface markers. While we have used certain pathogens as examples, keep in mind that many other pathogenic organisms employ one or more of these mechanisms to evade detection by the immune system.

serotype Members of the same species of a pathogen that differ in surface molecules.

hemagglutinin Adhesion molecule expressed by the influenza virus in order to bind to target cells.

neuraminidase Virus surface enzyme involved in the processing and formation of infectious viral particles.

Bacterial serotypes

While the immune system employs many mechanisms to combat bacterial pathogens, one very important tactic is the production of opsonizing antibodies that recognize surface antigens, bind to them, and tag the bacteria as foreign. Some bacteria have developed a mechanism designed to evade this immune response—they express different surface antigens without changing the bacterial genus and species. Indeed, many species have multiple **serotypes**, which differ on their surface and thus are not recognized by the same immunoglobulins (**FIGURE 13.2**). Examples of species with multiple serotypes include *Salmonella* (over 2600 serotypes) and *Vibrio cholerae* (over 200 serotypes). Because of these differences in surface molecules, organisms that produce opsonizing antibodies that recognize bacteria of one serotype must mount an independent immune response to recognize a different serotype. This provides a pathogen with a different serotype a chance to survive within the host.

Viral mutation and recombination

Immune responses that combat viral infection tend to take the form of neutralizing antibodies that target key viral surface molecules required for interaction with target cells and membrane fusion. A typical example is the influenza virus, which has two important surface proteins: **hemagglutinin**, which is responsible for viral fusion with sialic acid sugars on target cells, and **neuraminidase**, which is responsible for the release of mature viral particles from the host cell. In fact, these two viral proteins are used to characterize and name various influenza virus serotypes (e.g., H1N1). Neutralizing

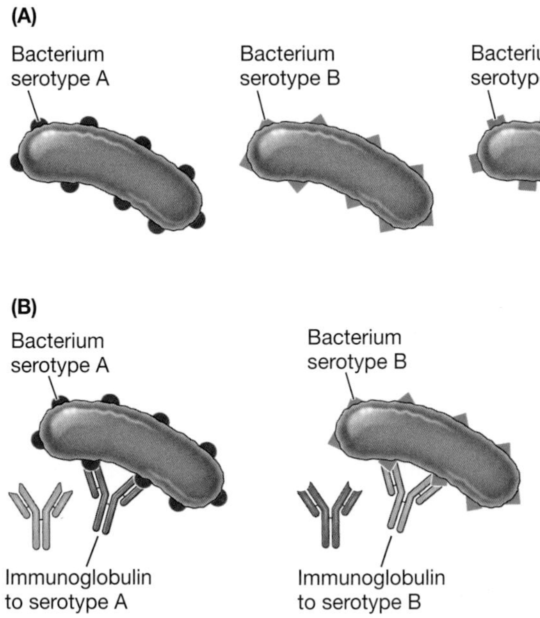

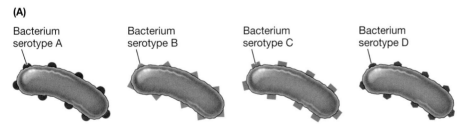

FIGURE 13.2 Bacteria of different serotypes evade recognition (A) Bacteria of the same species may exist as different serotypes, meaning that their surfaces differ chemically. (B) Bacterial serotypes recognized by an immunoglobulin that binds to a specific serotype epitope are not recognized by immunoglobulins that recognize different serotype epitopes. The unique epitope provides a mechanism for each bacterial serotype to evade recognition in individuals who have come into contact with different serotypes.

antigenic drift Random mutation of a pathogen that changes epitopes of a previous adaptive immune response and renders lymphocytes from that immune response nonreactive to the changed pathogen.

epidemic Localized spread of infectious disease within a population.

antibodies that protect against influenza typically target these two viral surface molecules. Thus, just as bacterial serotypes protect different variants of bacterial species, it stands to reason that changes to these viral surface molecules would offer protection if the changes prevent previously raised antibodies from binding.

How can a virus change its surface molecules to evade the action of neutralizing antibodies? Many viruses, including the influenza virus, have an RNA genome, making them prone to errors during genome replication because the enzyme responsible for replication typically does not contain proofreading activity. During genome replication, errors can be introduced that change the surface molecules, effectively preventing the immune system from recognizing the new variant of a virus. Such a change is referred to as **antigenic drift** (**FIGURE 13.3**). The virus responsible for COVID-19, SARS-CoV-2, underwent antigenic drift as it spread through the human population. Changes occurred in the spike protein of the virus, which binds to target cells. To date, three virus variants with mutations in the genome have been identified (B.1.1.7, B.1.135, and P1 found in the United Kingdom, South African, and Brazil), which allow the virus to become more transmissible.

New viral variants that arise due to antigenic drift can lead to local outbreaks, or **epidemics**, of viral infection, as populations will have not mounted

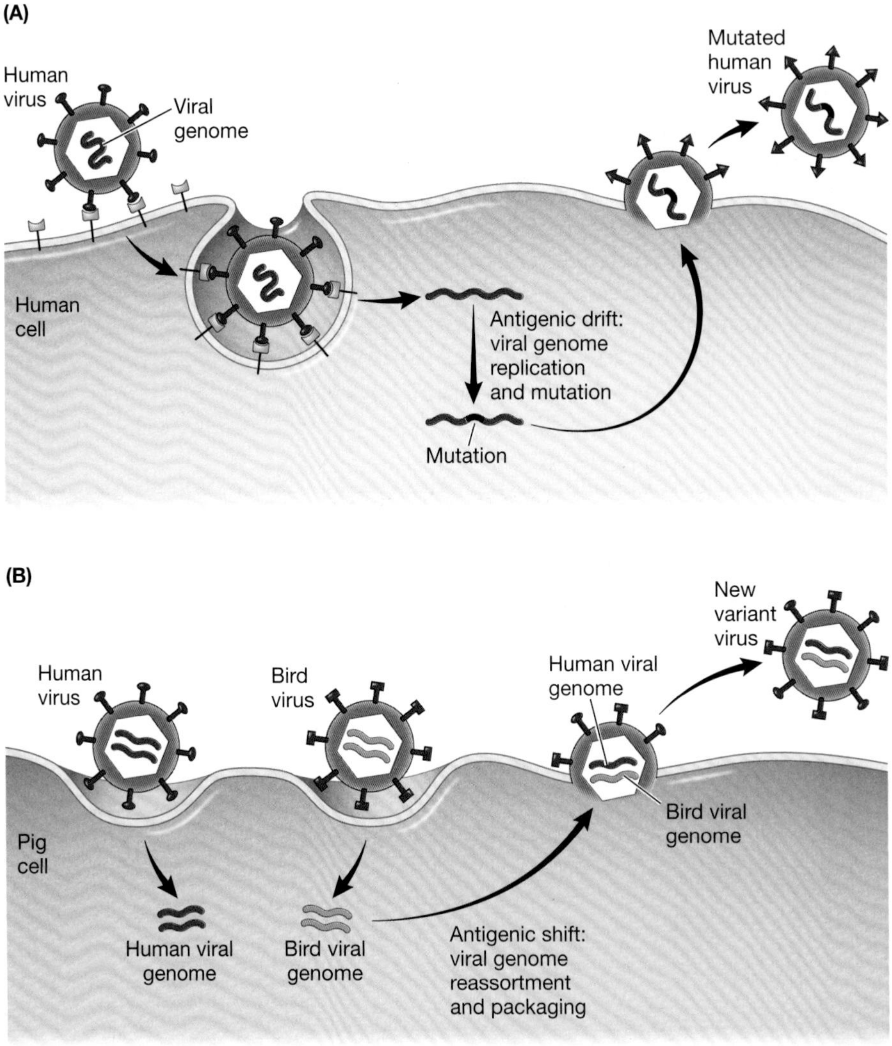

FIGURE 13.3 Antigenic drift versus antigenic shift (A) Antigenic drift occurs in viruses that are highly mutable. After infecting a target cell, viral genome replication results in random changes within the viral genome. These random mutations may occur in genes responsible for producing viral proteins, resulting in alterations in viral surface proteins that adhere to target cells. If these mutations do not affect the ability to bind cells, the altered virus is still infective, but may not be recognized by the adaptive immune response that was raised against the original virus. (B) Antigenic shift occurs in viruses that can exchange genetic information. Infection of a common cell, such as a pig cell, with two closely related virus variants that are pathogenic in different species, such as a human virus and a bird virus, leads to the genetic information of both virus variants being present in the same cell. Viral replication and packaging of the two viral genomes might result in reassortment of the genome, whereby genetic components of both the human and bird viruses are packaged in a new variant virus.

an immune response to the newly emerged virus. Infections that result in the spread of disease across continents or worldwide are known as **pandemics** (see **CONTROVERSIAL TOPICS**).

pandemic Spread of infectious disease over a large area, such as across continents or globally.

While random mutation and antigenic drift result in new viral variants that can escape previously mounted immune responses, variants can also emerge via another means. If close variants of viral strains infect the same cell, their genetic components and surface molecules create new viral particles that

CONTROVERSIAL TOPICS

Is it possible to safely study dangerous pathogens?

This chapter focuses on the molecular actions of various pathogens that evade or subvert the immune system. Our knowledge of these molecular actions arises from the meticulous science conducted by dedicated researchers aimed at understanding these pathogens. Research is conducted either to learn more about the nature of a pathogen and its evolution or in the interest of improving human health. However, science still has much to uncover regarding the mysteries of pathogen action.

The emergence of new pathogens and recent pandemics of SARS-CoV-2, Ebola, and Zika virus cause us to take stock of how much we know, and how much we need to learn, about pathogen action. In the wake of elevated global terrorism, we as a society must be concerned about the misuse of important information. The sad fact is that this knowledge can be twisted and used to harm and not help people. This raises an important question for today's society and scientific community: What type and how much research should be conducted on dangerous pathogens that can be used as bioterrorism agents?

And there is a key follow-up question within this debate: How do we disseminate research-based knowledge that we as a society deem worthy of funding and attention, given that dissemination is a cornerstone of the scientific process but can ultimately provide information to those who wish us harm? As a society, we must grapple with these questions and consider the implications of our actions, as all decisions bring inherent risks.

Leading scientists are on both sides of this debate, making the issue even more confusing and contentious in the public eye. How are citizens, who may or may not be relatively well versed in science, to grapple with these issues? The reporting of accidental mishandling by laboratories of potentially dangerous pathogens such as smallpox and anthrax only fuels the flames of this debate.

One group of scientists and legal and bioethics experts, the Cambridge Working Group, was created in 2014 to address the issues surrounding research on potential pandemic pathogens. Their stance has been to exercise extreme caution in such research and to ensure that proper regulation is in place to allow the research to be conducted in as safe a manner as possible in light of both public health and scientific goals. The group argues that the creation of potential pandemic pathogens should stop until all the risks and benefits have been weighed, and until all safer experiments have been determined to be substandard in regard to the useful information that can be gained compared to experiments utilizing the potential pandemic pathogen. The list of supporting scientists is impressive, featuring many prominent experts in various fields.

On the other side of the debate, a group named Scientists for Science argues that scientists are capable of conducting research on potential pandemic pathogens safely, and that in the interest of science, this research will bring about important discoveries about microbial pathogenesis and human health. The group argues that it is often difficult to analyze the risk involved in such research, as this information builds over time as each experiment unravels the mystery of a given pathogen. Their list of supporting scientists is equally impressive.

These questions must be discussed across all levels within our society, as all individuals will ultimately be affected by any decisions made. Scientists must play an important role by informing the public based on data they can analyze and risks they can predict. The public must allow their voice to be heard through the election of policymakers who will eventually turn the winning side of the debate into law.

Think About...

1. Both groups on opposite sides of the debate about potential pandemic pathogen research have prominent scientists endorsing their recommendations. What actions would you suggest to an "average" citizen trying to choose a side in this debate?
2. What are some likely benefits of conducting scientific research on potential pandemic pathogens? What are some risks of conducting this type of scientific research?

antigenic shift Recombination event of closely related pathogens that promote the formation of an altered pathogen capable of escape from any prior adaptive immune response raised due to lack of specificity to the newly formed pathogen.

antigenic variation Process of changing a recognition target of the immune system through gene conversion or rearrangement.

contain components of both viruses. For example, if a human influenza virus and bird influenza virus were to infect the same target cell, viral replication would result in the production of a virus containing components from both the human and bird strains. Such exchange between viral variants is called **antigenic shift** (see Figure 13.3). This process can cause worldwide pandemics due to the novel nature of the newly recombined pathogen.

Pathogens that originate in an animal reservoir can cause pandemics. Viruses within animal reservoirs can mutate prior to entry into the human population, or they can mutate to become more transmissible once they begin to infect humans. SARS-CoV-2 is an example of such a virus. SARS-CoV-2 is genetically related to viruses seen in bats, and although its origin has not yet been determined, its similarity to coronaviruses in bat species suggests that it originated in an animal reservoir. Through evolution, SARS-CoV-2 acquired mutations in genes that encode proteins that interact with target cells. These mutations facilitated transmission of SARS-CoV-2 to humans because they allowed the virus to bind to human cells. Thus, as humans come in closer contact with animal species that were previously separated by geographic and natural boundaries, the chance of new emerging pathogens capable of causing pandemics increases.

Gene conversion

One of the most fundamental mechanisms used by the adaptive immune system is the genetic recombination of immunoglobulin and T-cell receptor genes during lymphocyte development to promote the generation of a diverse population of cells expressing receptors with different ligands. This process causes a change in the surface molecules expressed in these lymphocytes. Since recombination is used by our T cells and B cells to alter the composition of cell-surface proteins, could pathogens use similar mechanisms to change the way that pathogens "look" to the immune system? Indeed, some pathogens utilize gene conversion or rearrangement, causing **antigenic variation**, to survive when the selective pressure of the immune system is great. Antigenic variation can occur via three different mechanisms (**FIGURE 13.4**):

- Having multiple copies of a gene that all vary and contain their own on/off switch
- Having a single expression locus and many silent gene copies that can occupy the expression locus via gene rearrangement when necessary
- Having a highly variable region in a gene product that can change when necessary.

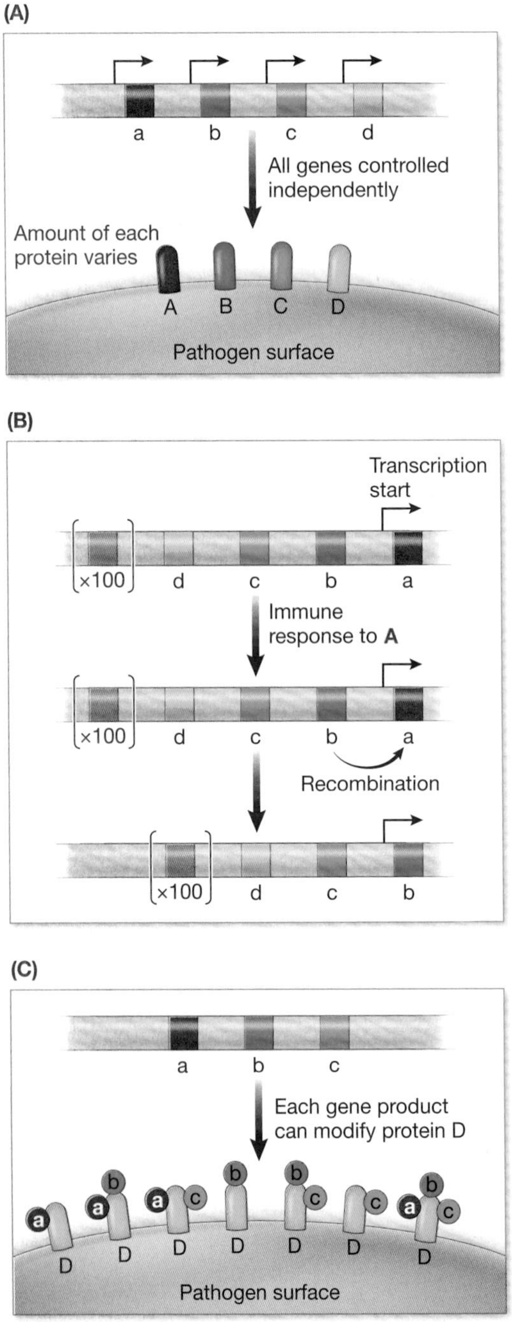

FIGURE 13.4 Mechanisms of genetic variation in pathogens (A) In one type of genetic variation, individual genes are independently regulated. Each gene of a cell-surface protein might be regulated to vary the amount of protein at the surface. As long as each of the protein variants can function in the same manner, the amount of each protein at the cell surface will vary, resulting in proteins that function but are not recognized by the immune system. (B) Genetic variation in which a pathogen recombines gene variants into an active transcription site. If the immune response drives recognition of the first gene product in the active transcription site, the pathogen recombines its genome and replaces the gene with a variant that functions similarly but is not recognized by the immune system. (C) Genetic variation in which different pathogen enzymes modify a cell-surface protein independently. Modification by various pathogen genes varies or masks the epitopes on the cell-surface protein, resulting in protein variants that may not be recognized by the immune system. (After P. Parham. 2014. *The Immune System*, 4th ed., W. W. Norton and Company.)

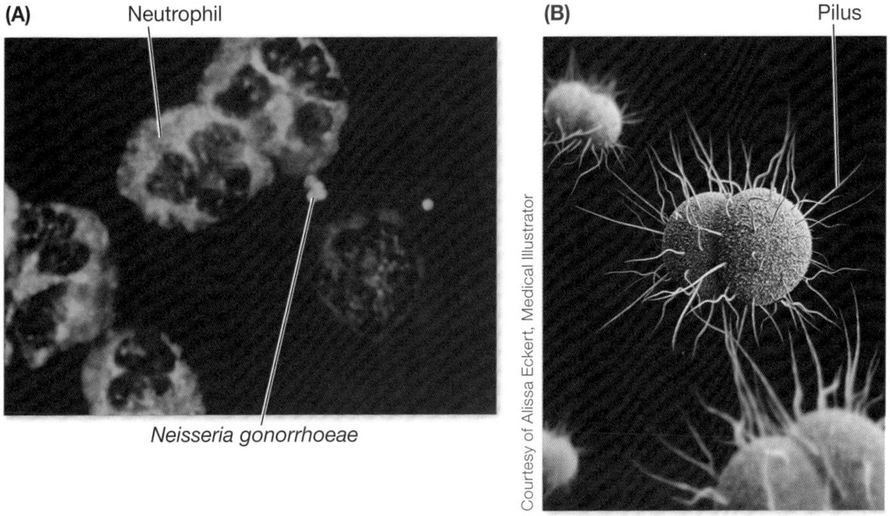

FIGURE 13.5 ***Neisseria gonorrhoeae*** (A) The gram-negative bacteria *Neisseria gonorrhoeae* in the presence of neutrophils. (B) *Neisseria gonorrhoeae* expresses pili to adhere to target cells.

A good example of a pathogen that uses all of these mechanisms of antigenic variation is the *Neisseria* species of bacteria (**FIGURE 13.5**), the causative agent of gonorrhea and meningitis. *Neisseria* utilize multiple gene copies of an outer membrane protein called Opa, each under the control of their own expression switch, to vary which Opa proteins are expressed within the host. These copies also vary the expression of Opa to evade surface recognition molecules such as immunoglobulins of the host immune system.

Neisseria species also engage in the rearrangement of multiple silent genes into a single expression locus. The *Neisseria* pilin gene, *pilE*, which expresses the pilus that plays an important role in host cell attachment, is encoded at a single locus. However, *Neisseria* have many silent pilin gene variants, *pilS*, that can recombine into the *pilE* locus, thus creating different pili on the cell surface depending on which gene is located at the *pilE* locus and expressed by the bacteria.

Neisseria bacteria also have the ability to vary the sugars associated with the ends of their lipooligosaccharide (LOS, analogous to lipopolysaccharide [LPS]). To change which sugars are attached, the bacteria rely on the activity of gene products capable of adding different sugars as terminal sugars on LOS. LOS of *Neisseria* is vital in the interaction between the pathogen and human cells, but by varying the attached sugars, *Neisseria* can maintain the function of LOS without being recognized by the immune system.

Another example of a pathogen that employs gene conversion to hide from the immune system is the *Trypanosoma* species, parasitic protozoa that cause a variety of diseases, including sleeping sickness and Chagas disease. Trypanosomes express a variant surface glycoprotein (VSG) from a single locus. The VSG coats the protozoan, preventing it from being targeted by complement fixation on its surface and shielding surface antigens from immunoglobulin action. However, VSG can induce a robust adaptive immune response in its host. How then do trypanosomes continue to use the VSG to serve as a protective barrier and also prevent recognition of the VSG by the host immune system? Trypanosomes contain approximately 1000 VSG genes, although only one is located in the expression locus. Through recombination events, they can change which VSG gene is located in the expression locus and effectively alter the VSG coat appearance due to the presence of a region that is highly varied among the genes.

CHECKPOINT QUESTIONS

1. Explain the difference between antigenic drift and antigenic shift.
2. What mechanisms can be used by bacterial pathogens to achieve antigenic variation?

13.3 How do pathogens hide from the immune system?

LEARNING OBJECTIVES

13.3.1 Explain how viruses evade the immune system by entering a dormant state.

13.3.2 Explain how pathogen niches within the host enable evasion of immune system defenses.

In both the innate and adaptive immune response, recognizing a pathogen as a foreign entity is paramount to destruction of the pathogen. Just as manipulation of cell-surface molecules is an effective means of avoiding detection, the niche of the pathogen is also very important. Residing in a location that is not accessible to innate immune cells or soluble immunoglobulins provides pathogens one more mechanism for evasion.

Viral latency

Viruses often hijack the machinery of host cells to allow for viral genome replication by integrating their viral genomes into host cell chromosomes. To destroy the virus, the immune system must recognize the intracellular viral infection and destroy the infected host cell. Recognition frequently occurs because the virus is rapidly replicating and producing viral proteins, which can be processed by cytosolic antigen presentation and presented via MHC class I molecules on the surface of the host cell. Eventually, this presentation results in targeting of the infected cell by CD8 cytotoxic T cells and, in some cases, activation of virus-specific B cells capable of producing neutralizing antibodies. In host cells that are rapidly producing virus, the presentation of viral particles is more efficient, and clearing the infection becomes easier for the adaptive immune response through the action of cytotoxic T cells. Thus, one strategy used by viruses is to limit antigen presentation within the host cell. This minimizes the potential activation of virus-specific CD8 T cells.

Viruses such as the herpes simplex virus and human immunodeficiency virus (HIV) (**FIGURE 13.6**) can incorporate their viral genomes into target cells and undergo a state of **latency**, or a dormant state. In this dormant state, the viral genome is present but the host cells lower their production of viral particles (**FIGURE 13.7**). Reactivation of viral genome and protein expression is pathogen specific, but often takes advantage of the cell that the virus infects. For instance, we will see later in this book that HIV primarily infects CD4 T cells and utilizes mechanisms of T-cell

latency Period of dormancy of a pathogen while within a host.

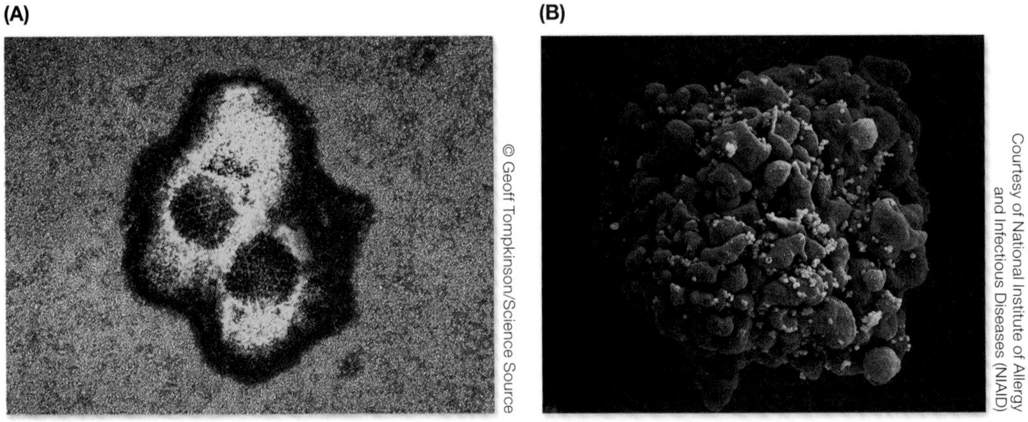

FIGURE 13.6 Viral pathogens (A) Micrograph of herpesvirus. (B) Scanning electron micrograph image of human immunodeficiency virus (yellow) infecting a T cell (red).

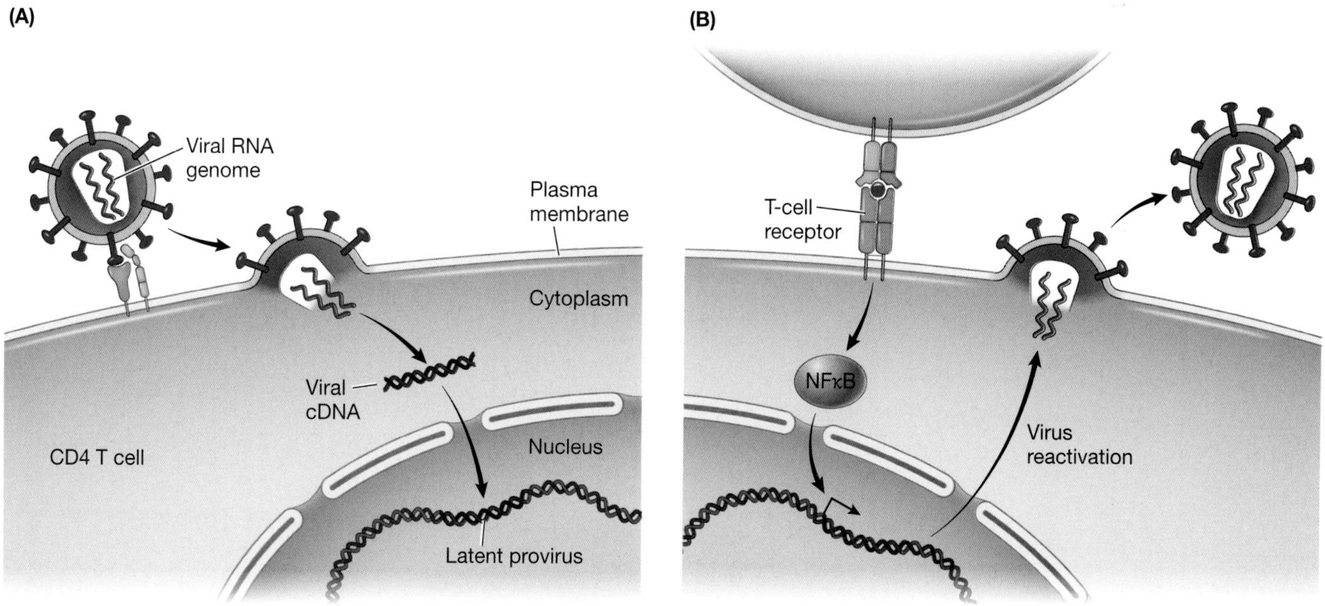

FIGURE 13.7 Viral latency in HIV infection (A) Human immunodeficiency virus (HIV) is an RNA virus that infects CD4 T cells. Upon infection, the RNA genome of the virus present in the cytoplasm is converted to a cDNA genome. The cDNA genome travels to the nucleus, where it is integrated into the cell genome and enters a latent state. (B) Upon activation of an infected T cell, the transcription factor NFκB reactivates the HIV viral genome, resulting in the production of both viral RNA genome and RNA used for translation of viral proteins. Both the RNA genome and translated proteins are assembled into new virus.

activation to also activate its own viral genome and protein expression. Upon reactivation, viral particles are produced and can infect neighboring cells, allowing for viral persistence. The latency strategy permits the viral genome to remain incorporated in the host cell genome and undetected by the immune system until activation of viral expression can occur and allow for viral replication.

Pathogen niches

Detection by the immune system (innate or adaptive immune response) is the first step in mounting a successful immune response. We have just seen that viruses can avoid detection by reducing production of viral particles and existing in a dormant state. Other pathogens utilize a variety of tactics to avoid detection.

Many of the innate and adaptive immune responses target bacteria that are extracellular. Phagocytosis, opsonization, and neutralization of extracellular bacteria by soluble immunoglobulins can all be efficient means of clearing an infection. One might suspect that pathogenic bacteria could resist clearance by avoiding the extracellular environment that makes them susceptible to these mechanisms. Indeed, some pathogenic organisms do reside in an intracellular environment to avoid detection. Often, these intracellular pathogens are efficiently phagocytosed by innate immune cells such as neutrophils and macrophages, but they execute survival strategy mechanisms to avoid destruction by the phagocyte.

If we consider the process of phagocytosis (see Chapter 2), we can envision several ways a pathogen could avoid destruction (**FIGURE 13.8**):

- Escaping the phagosome and surviving within the cytosol of the phagocyte
- Preventing phagosome–lysosome fusion into a phagolysosome
- Surviving in the phagolysosome without being digested by the phagocyte.

In the following sections, we'll examine some ways in which pathogenic organisms maintain an intracellular niche within a host organism and avoid destruction by phagocytosis.

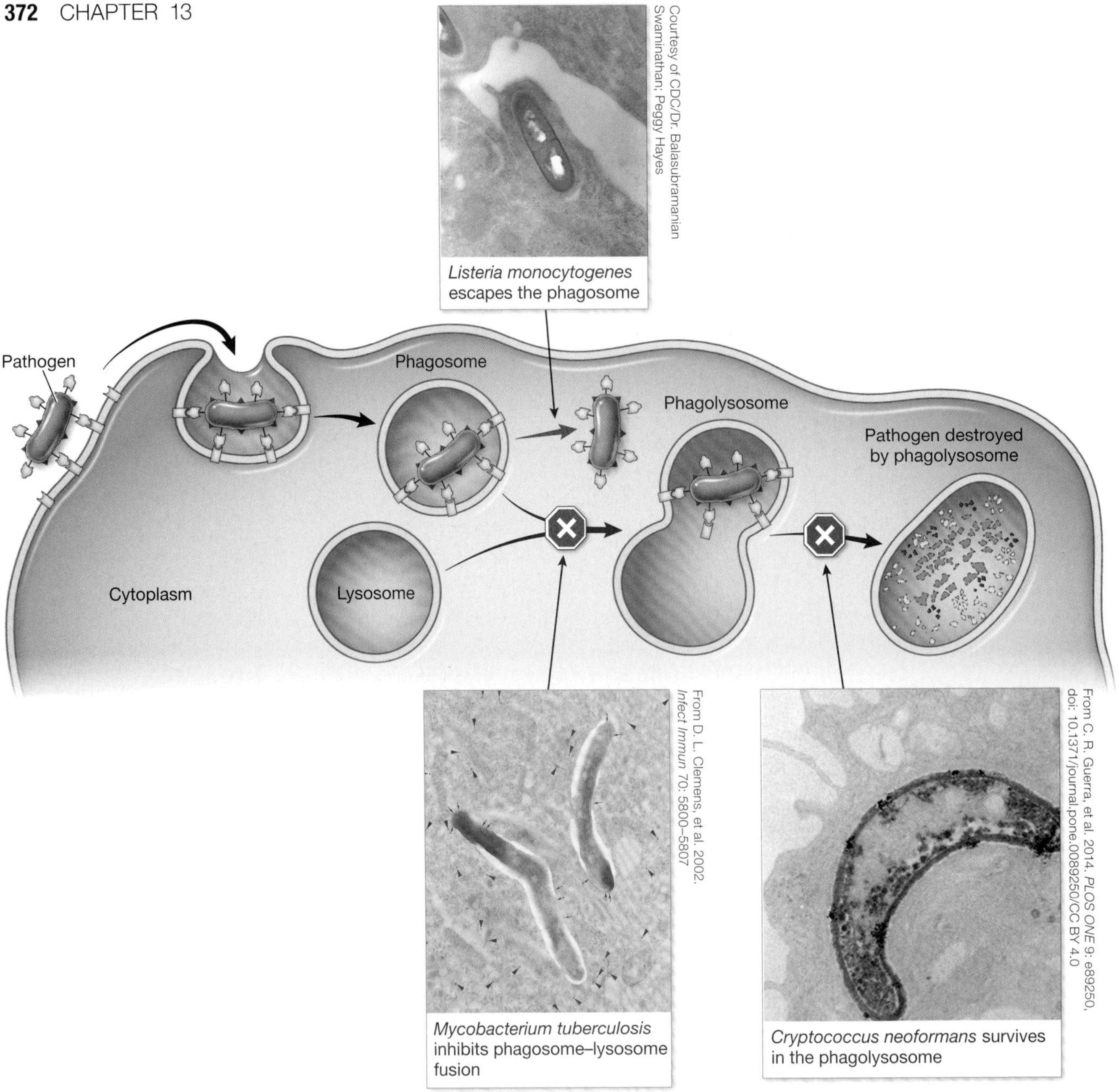

FIGURE 13.8 Pathogen mechanisms of evading destruction by phagocytosis Three examples of pathogens that use a variety of mechanisms to avoid destruction by phagocytosis are *Listeria monocytogenes*, *Mycobacterium tuberculosis*, and *Cryptococcus neoformans*. *Listeria monocytogenes* expresses proteins that allow the bacteria to escape the phagosome shortly after phagocytosis so it can remain in the cytoplasm of the cell. *Mycobacterium tuberculosis* (arrows in micrograph) survives within phagocytes by inhibiting fusion of phagosomes and lysosomes, protecting the pathogen in the phagosomal compartment. The phagosome is separate from the cytoplasm, which is labeled with gold beads (arrowheads in micrograph). *Cryptococcus neoformans* survives within the phagolysosome, avoiding digestion and destruction in the organelle.

cadherin A transmembrane calcium-binding protein family that is important in cell-to-cell adhesion interactions.

internalin Pathogen adhesion molecule involved in target-cell binding and activation of phagocytosis events within the target cell.

ESCAPING THE PHAGOSOME Several species of pathogenic bacteria live within the cytosol of host cells after they have been engulfed and, subsequently, escape the phagosome. One example is *Listeria monocytogenes*, which primarily infects intestinal epithelial cells, inducing these cells to uptake the bacteria and ultimately escaping the endocytic vesicle. *Listeria* cells bind to epithelial cell **cadherins** through a bacterial surface **internalin** molecule. This binding induces intracellular events within the intestinal epithelial cell that result in rearrangement of the actin cytoskeleton and formation of a phagocytic cup to trick the epithelial cell into phagocytosing the bacterium. Once inside the intestinal epithelial cell, *Listeria* uses a pore-forming toxin known as listeriolysin O (LLO) to create holes in the phagosome membrane. The membrane ruptures, allowing the bacteria to escape into the cytosol of the epithelial cell (see Figure 13.8). The intracellular *Listeria* is then protected from the action of phagocytic cells in the area and from neutralizing antibodies present at the mucosal surface.

BLOCKING PHAGOSOME–LYSOSOME FUSION Recall that efficient destruction of phagocytosed material requires both uptake and fusion of the phagosome with the lysosome of the phagocytic cell, since the lysosome contains all of the digestive enzymes required for breakdown of endocytosed material. Thus, if a pathogen can prevent phagosome–lysosome fusion, it can reside in an endocytic vesicle without being destroyed by the phagocyte. This blocking action not only protects the internalized pathogen but also limits presentation of pathogen-specific epitopes to T cells via class II presentation, as this presentation requires efficient digestion within the lysosome before loading of MHC class II molecules can occur.

One well-studied pathogen that uses this method of protection is *Mycobacterium tuberculosis*. Upon phagocytosis by macrophages, the bacteria use several different mechanisms to prevent fusion of the phagosome with lysosomes of the phagocyte (see Figure 13.8). Some of these mechanisms target proteins responsible for membrane trafficking and fusion.

M. tuberculosis secretes protein phosphatases such as PtpA that can dephosphorylate and inactivate host cell proteins necessary for vesicular trafficking. The bacteria can further block phagosome maturation as they block acidification of the lumen, or inside, of the phagosome, a process that requires the action of proton pumps in the phagosomal membrane. Under normal conditions, the proton pumps actively transport protons into the phagosomal lumen through the action of ATP hydrolysis to lower the pH to the level required for lysosomal enzyme activity (about 4.5). The bacteria block the action of these proton pumps by preventing their assembly at the phagosomal membrane surface, thus preventing acidification of the phagosomal lumen. These strategies support the intracellular (and intraphagosomal) survival of *M. tuberculosis* within macrophages.

SURVIVING IN THE PHAGOLYSOSOME A few pathogens can survive the acidic and destructive environment of the phagolysosome and use that vesicle as a protective niche, thereby avoiding destruction by the immune system. These pathogens are able to survive the action of degradation pathways within the phagolysosome and replicate efficiently in an acidic environment (see Figure 13.8).

One such pathogen is *Cryptococcus neoformans*, a fungus capable of causing meningoencephalitis in immunocompromised individuals. To prevent destruction by lysosomal mechanisms, *C. neoformans* employs a polysaccharide coat (capsule) to inhibit the action of reactive oxygen species produced within the lysosome. It has also been suggested that the fungus can inhibit the activity of lysosomal proteases, including cathepsins, although how this occurs has not yet been determined. Interestingly, *C. neoformans* grows optimally at a pH around 5, which is very close to the acidic environment of the lysosome. During *C. neoformans* infection, the phagolysosome pH eventually increases, presumably due to membrane damage and permeability within the phagolysosome. This membrane permeability is driven by cryptococcal growth, but the exact mechanism has not yet been determined.

CHECKPOINT QUESTIONS

1. List the advantages and disadvantages associated with a pathogen surviving within an intracellular niche.
2. Why are mechanisms that inhibit the process of phagocytosis so important when employed by intracellular pathogens?

13.4 How can pathogens downregulate the immune system?

LEARNING OBJECTIVE

13.4.1 List and describe six strategies used by pathogens to block or inhibit normal immune function.

toxin Product of a living organism that can cause disease within an organism.

endotoxin Component of a pathogenic organism capable of eliciting a disease state.

exotoxin Secreted product of a pathogenic organism capable of eliciting a disease state.

In the evolutionary arms race between multicellular organisms and pathogens, higher eukaryotic organisms have evolved the elegant mechanisms of a multilayered immune response, which we have discussed in earlier chapters. The complexity and redundancy of the immune response allow higher eukaryotic organisms to survive the constant onslaught of pathogens in the environment seeking to survive and replicate within a host organism.

However, pathogens have a distinct advantage—the speed at which they can evolve (see **EMERGING SCIENCE**). Pathogens can evolve much more rapidly because typically they are unicellular, their life cycle or division time is much faster, and they are capable of faster recombination events, including the uptake or transfer of genetic elements with other organisms in the surrounding environment. Just as higher eukaryotic species have elaborate mechanisms to thwart pathogen invasion and disease, pathogens have evolved mechanisms to prevent immune defenses from working properly, increasing their chance of survival.

Pathogens have devised a variety of means to subvert each type of immune response. We have seen that some pathogens rely on evasion strategies to avoid detection by innate or adaptive immune responses. Others actively target immune response pathways to disrupt mechanisms designed to detect and destroy foreign invaders. Whether facing an innate or adaptive response, the overall objective of pathogens is to prevent destruction by the host and promote their own survival (**FIGURE 13.9**). In the following sections, we discuss pathogens that target various mechanisms of the immune response. You may wish to review these mechanisms in previous chapters to appreciate how pathogens have evolved to subvert these systems.

Toxins

A bacterial **toxin** is a poisonous substance with the ability to alter normal immune system behavior or cellular activity. Toxins are categorized into two groups: endotoxins and exotoxins. An **endotoxin** is a molecule present as a component of the pathogenic organism. LPS is an endotoxin that is a component of the outer membrane of gram-negative bacteria. An **exotoxin** is a product secreted by a pathogen, usually a protein, that alters normal cellular machinery. Some toxins, such as anthrax toxin, alter signaling pathways within target cells. Others, such as toxic-shock syndrome 1 from *Staphylococcus*, are *superantigens*, which induce a nonspecific immune response that makes it difficult to destroy the pathogen.

Blocking antimicrobial peptides and proteins

A key first line of defense of the innate immune response is the activity of antimicrobial proteins and peptides such as lysozyme, which cleaves the polysaccharide component of bacterial cell walls, and defensins, which aid in pathogen lysis by creating pores in plasma membranes. Inhibition of these first-line players is of course beneficial for pathogen survival, and pathogens have evolved mechanisms for this purpose. For instance, several lysozyme inhibitors have been identified in gram-negative bacteria.

Escherichia coli expresses a protein named Ivy that is capable of binding to lysozyme with high affinity and blocking its proteolytic activity. A more broadly conserved lysozyme inhibitor family has been identified in gram-negative organisms, including *E. coli*, *Salmonella* species, and *Pseudomonas aeruginosa*. One type

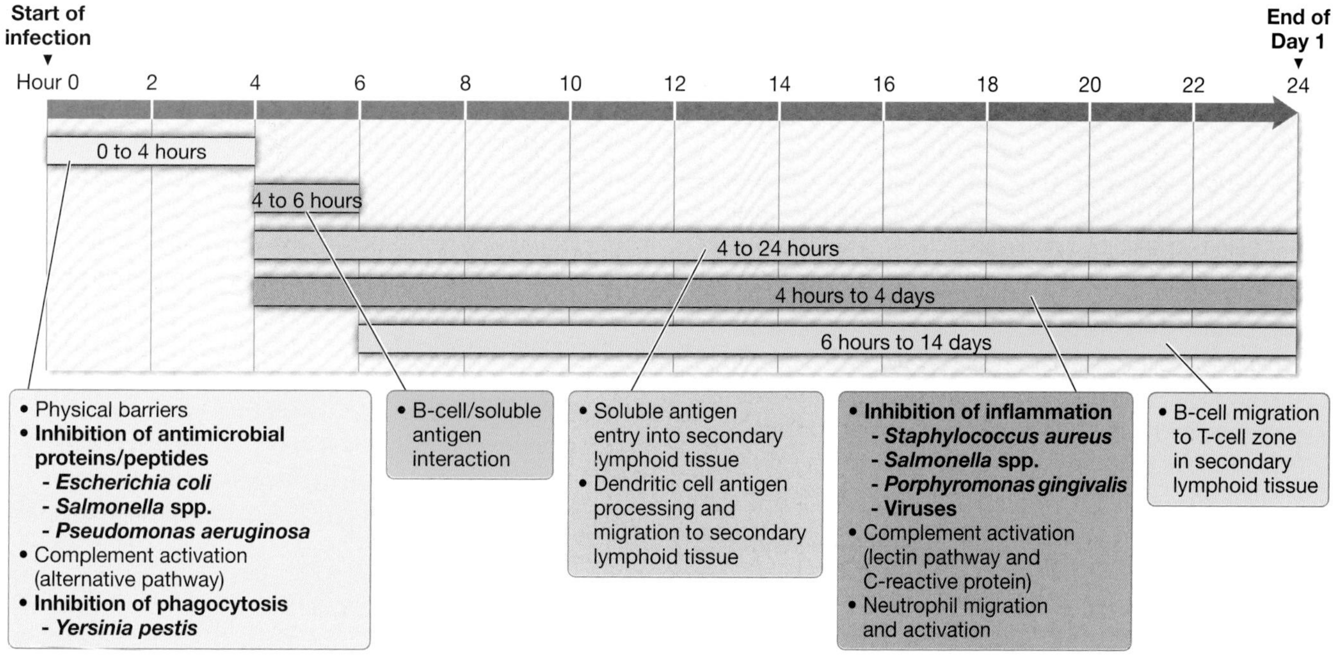

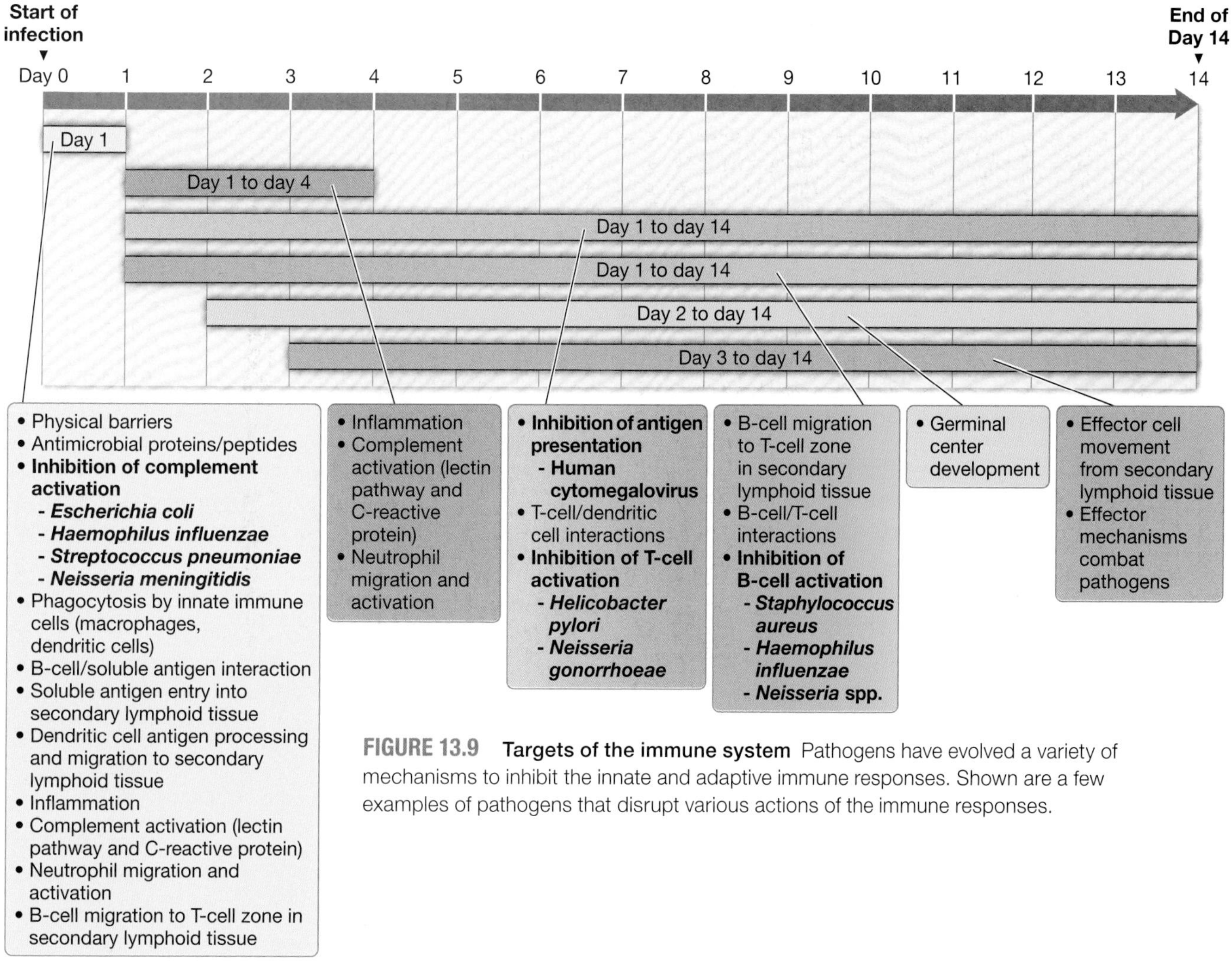

FIGURE 13.9 Targets of the immune system Pathogens have evolved a variety of mechanisms to inhibit the innate and adaptive immune responses. Shown are a few examples of pathogens that disrupt various actions of the immune responses.

How do bacteria evolve as a function of time and space in the presence of the selective pressure of antibiotics?

Article

M. Baym, T. D. Lieberman, E. D. Kelsie, R. Chait, R. Gross, I. Yelin, and R. Kishony. 2016. Spatiotemporal microbial evolution on antibiotic landscapes. *Science* 353: 1147–1151.

Background

Because of the presence of emerging and increasingly antibiotic-resistant bacteria, studies have been undertaken to understand the means by which antibiotic resistance evolves. The understanding of how this evolution occurs has previously been based on studies conducted in well-controlled environments lacking real-life selective pressure driven by spatially distinct concentrations of antibiotic as a selective pressure agent.

Models suggested that changes in the spatial structure of a selective pressure can alter the means of selection: survival is not dictated by competition but rather by being the first to evolve to the selective pressure. However, a method to study the process of evolution in a spatiotemporal manner based on selective pressure had not been designed until this study.

The Study

Researchers devised a means to allow for an increased selective pressure (increasing antibiotic concentration) in a soft agar matrix. Bacteria were plated in an area where no antibiotic was present, where they utilized nutrients until they were depleted. Upon nutrient depletion, the bacteria were capable of migration through the soft agar media until they interacted with the first level of antibiotic (**FIGURE ES 13.1**).

This experimental design allowed the researchers to observe over time the evolution of bacteria capable of surviving the increasing selective pressure of increasing antibiotic concentration. Using these techniques, the researchers were also able to isolate distinct mutations that explained the mechanism driving the evolutionary adaptation due to the increased selective pressure. They were also able to follow the additive changes driving adaptation to increasing antibiotic concentrations.

Think About...

1. If you designed this experiment using the same antibiotic but different pathogens, would you expect to see similar or different mutations driving increased antibiotic resistance?
2. Could you alter the experiment to study the evolution of antibiotic resistance to nonmotile bacteria? How would you conduct such an experiment?
3. Is this experimental protocol a good model for explaining antibiotic resistance based on treatment of bacterial infections within the population? Why or why not?

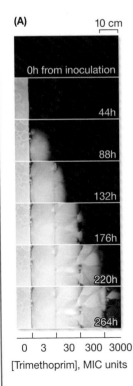

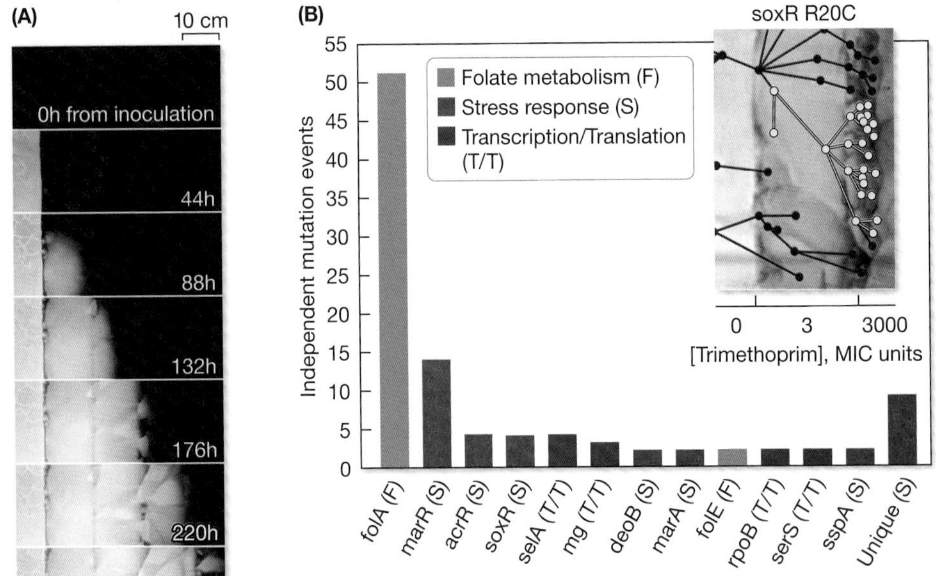

FIGURE ES 13.1 Selective pressure in the presence of antibiotic promotes mutants capable of growth. (A) *E. coli* were seeded on a microbial evolution and growth arena (MEGA) plate containing increasing concentrations of the antibiotic trimethoprim. Over time, mutant *E. coli* that was resistant to higher doses of trimethoprim could be isolated. (B) Sequence analysis of individual mutants isolated from the MEGA plate demonstrated the isolation of independent mutants of different functional categories, including genes involved in folate metabolism, the stress response, and transcription/translation. MIC, minimum inhibitory concentration. (From M. Baym et al. 2016. *Science* 353: 1147–1151; doi: 10.1126/science.aag0822. Reprinted with permission from AAAS.)

of inhibitor, named PliC (for periplasmic lysozyme inhibitor of c-type lysozymes), resides in the periplasmic space of gram-negative bacteria. A second type, called MliC (for membrane-bound lysozyme inhibitor of c-type lysozymes), resides in the membrane of gram-negative organisms. These inhibitors are important survival factors because inhibition of lysozyme is a key strategy in subverting the host organism's immune response.

We know that defensins play an important protective role against infection. Thus, blockage of defensin action is an effective strategy used by a number of pathogens. *P. aeruginosa* (**FIGURE 13.10**) utilizes rhamnolipids (a type of glycolipid) to block the expression of defensins, likely due to the inhibition of signaling pathways required for the activation of defensin synthesis. Another pathogen with resistance to defensins, *Treponema denticola*, an oral pathogen, utilizes both a unique outer membrane to prevent defensin binding and efflux pumps to transport cytoplasmic peptides, including host defensins, out of the bacteria and prevent their action.

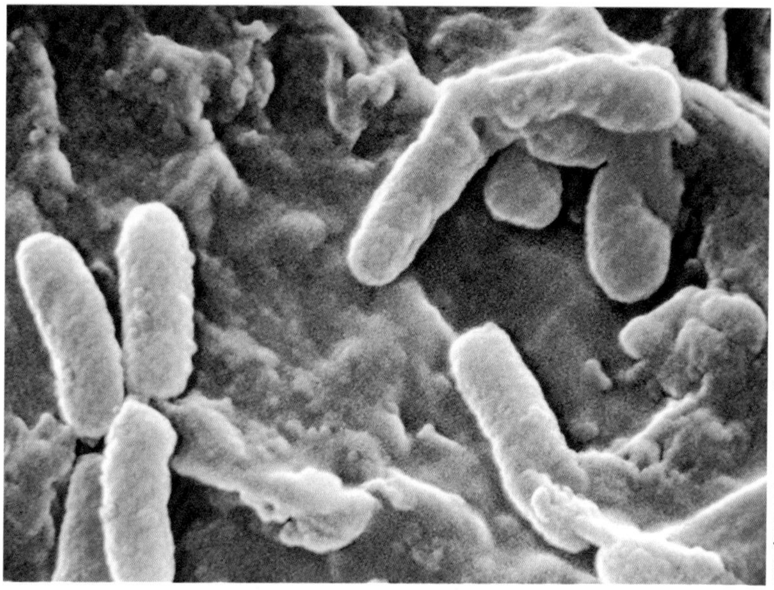

FIGURE 13.10 Micrograph of *Pseudomonas aeruginosa*

Disrupting phagocytosis

Phagocytosis is a major component of the innate immune response, whereby neutrophils and macrophages engulf and destroy extracellular pathogens, and dendritic cells process extracellular pathogen antigens via MHC class II presentation to prompt a proper adaptive immune response. The process of phagocytosis requires membrane remodeling, primarily driven by the action of the actin cytoskeleton.

A wide range of extracellular pathogens have evolved mechanisms to inhibit the actin cytoskeleton. A good example of such a pathogen is *Yersinia pestis*, the causative agent of plague. *Y. pestis* utilizes a type III secretion system, which functions as a molecular syringe, to inject several bacterial proteins into the cytoplasm of target eukaryotic cells (**FIGURE 13.11**).

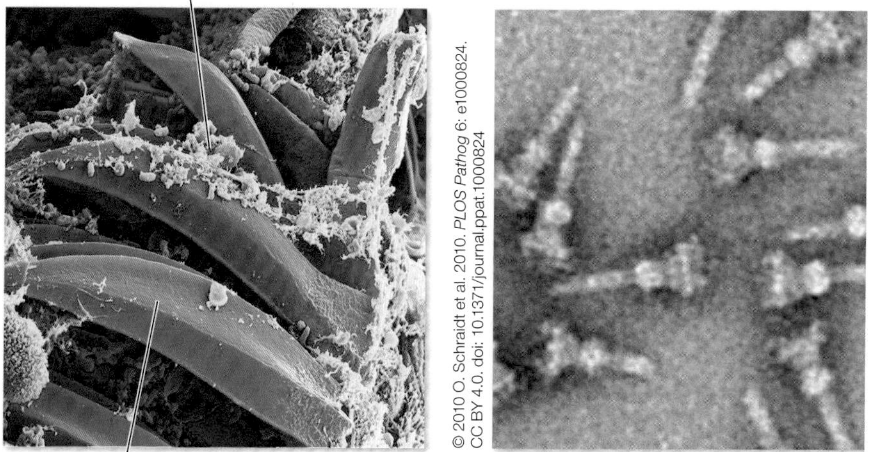

FIGURE 13.11 ***Yersinia pestis*, the causative agent of the plague** (A) Micrograph of *Yersinia pestis* (yellow) within the proventriculus (purple) of a flea's digestive system. (B) *Yersinia pestis*, as well as other pathogenic bacteria, use a type III secretion system to inject bacterial toxins into target cells. Shown is an electron micrograph of purified type III secretion systems from *Salmonella*.

virulence factor Product of a pathogen that supports pathogen survival and aids disease pathogenesis.

Several of these effector proteins target phagocytosis mechanisms, including YopH, a protein tyrosine phosphatase that prevents integrin action. (See **KEY DISCOVERIES** to learn more about the identification of YopH.) YopE is a protein that inactivates GTP-binding proteins required for actin rearrangements during phagocytosis, and YpkA is a protein kinase that disrupts the actin cytoskeleton. Together, these bacterial effector proteins work synergistically to inhibit phagocytosis.

Disrupting cytokine signaling

Cytokines play an important role in innate and adaptive immune cell activation, induction of an inflammatory response, and activation of complement pathways. Cytokine activity requires expression by the activated cell, cytokine secretion, and binding of the cytokine to its target receptor where it can activate signaling within the target cell. Several bacterial **virulence factors**, or proteins involved in the survival and pathogenesis of an organism, have been shown to

KEY DISCOVERIES

How did we learn about the action of *Yersinia* effector proteins and the ability of YopH to dephosphorylate phosphotyrosine?

Article

K. Guan and J. E. Dixon. 1990. Protein tyrosine phosphatase activity of an essential virulence determinant in *Yersinia*. *Science* 249: 553–556.

Background

The late 1980s and early 1990s were the high point in understanding phosphorylation in signaling cascades and the presence of enzymes responsible for the reversible addition and removal of phosphate from serine, threonine, and tyrosine residues. The removal of phosphate from phosphorylated residues was determined to be catalyzed by protein phosphatases. The questions at the time of these discoveries remained:

- How do protein tyrosine phosphatases catalyze removal of phosphate from phosphotyrosine?
- Which organisms employ such phosphatase activity in their natural cell biology?

Early Research

Protein tyrosine phosphatases were beginning to be identified and cloned, and a cysteine residue was shown to be important in the catalytic activity of the identified and cloned protein tyrosine phosphatases. These phosphatases were demonstrated to play essential roles in signaling events, including those of the adaptive immune system. Identification of these enzymes was of paramount importance to understanding their role in cell biological processes and their mechanism of action.

Think About...

1. How would you go about identifying unknown proteins to have protein tyrosine phosphatase activity if you had important sequence information of several known protein tyrosine phosphatases?
2. If you isolated a potential protein tyrosine phosphatase using bioinformatics approaches, what experiments could you do to test for protein tyrosine phosphatase activity?
3. Imagine the isolation of a protein tyrosine phosphatase from an organism that was known to not carry out tyrosine phosphorylation. Speculate on its role in the function of the organism and how and why it may have acquired such an enzyme.

Article Summary

Researchers utilized known protein tyrosine phosphatase sequences to analyze protein databases for proteins containing similar sequences, especially centered around the cysteine residue known for catalytic activity. These analyses revealed a bacterial pathogen, *Yersinia*, to contain and express a protein similar to protein tyrosine phosphatases (**FIGURE KD 13.1**).

This discovery was of interest for several reasons, including the fact that bacteria are known to not contain phosphotyrosine, and that the identified protein, YopH, was known to be involved in the virulence of *Yersinia*. The researchers wanted to determine whether YopH contains protein tyrosine phosphatase activity. To answer this question, they expressed YopH in *E. coli* and tested the phosphatase activity of both the wild-type protein and a mutant target the expression of cytokines. *Y. pestis* expresses an effector protein, YopJ, that blocks the activation of the transcription factor NFκB. Recall that NFκB is required to activate the transcription of a variety of immune genes, including those that express cytokines. *Shigella flexneri* employs a similar mechanism of NFκB inhibition to block cytokine expression using the bacterial effector protein OspG. Viruses can also block cytokine expression, and a number of viral mechanisms have been identified that block NFκB activation to subvert the action of antiviral interferon cytokines.

Another strategy pathogens use to block cytokine action is preventing its ability to bind to its target receptor—this blocks downstream signaling cascades required for proper response to the presence of the cytokine. Viruses have been shown to block chemokine action using a similar strategy. Viral proteins can bind to the chemokine and prevent it from acting as a proper signal for cell migration, or they can bind to the chemokine receptor and prevent detection of chemokine presence. Bacterial pathogens have also been shown to

YopH with the predicted essential cysteine residue mutated to an alanine or serine. Indeed, wild-type YopH was able to dephosphorylate phosphotyrosine, but mutation of the essential cysteine obliterated phosphatase activity. This study was the first to identify a bacterial protein capable of dephosphorylating phosphotyrosine and suggested a potential role of YopH in host cells during *Yersinia* pathogenesis.

(A)

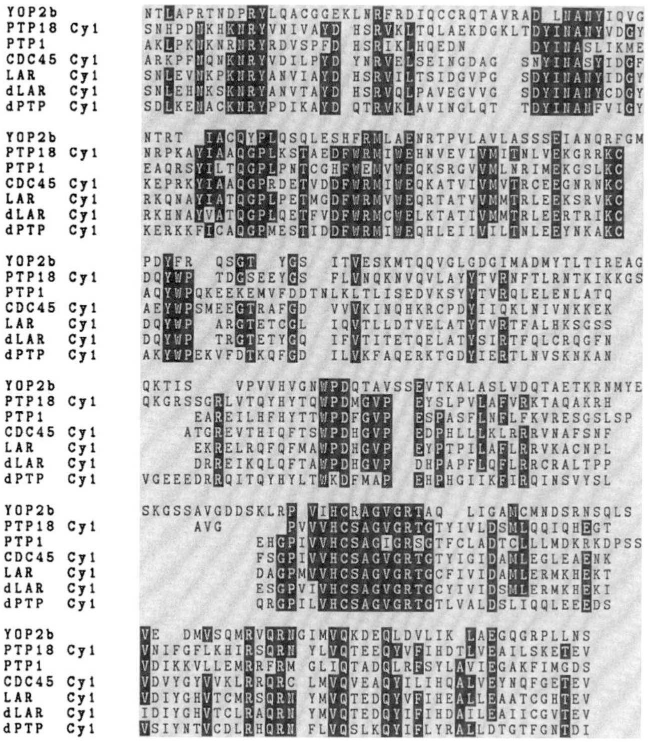

(B)

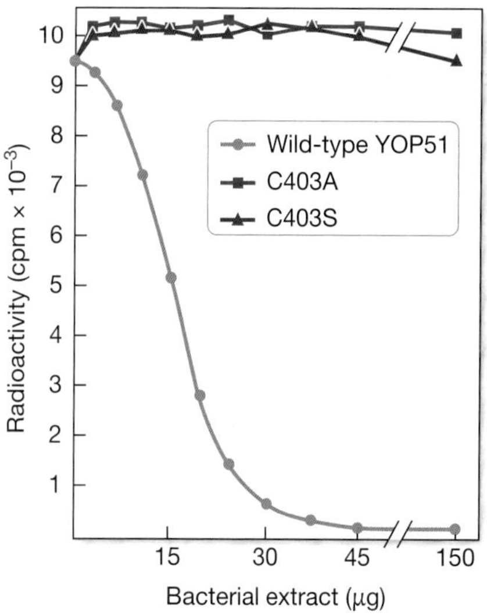

FIGURE KD 13.1 ***Yersinia pestis*** **expresses a protein tyrosine phosphatase.** (A) Sequence alignment of the *Y. pestis* YOP2b with other protein tyrosine phosphatases reveals key phosphatase determinants in the *Y. pestis* toxin. (B) Recombinant expression of the *Yersinia enterolitica* YOP51, which bears 99% identity to YOP2b, demonstrates that YOP51 is a protein tyrosine phosphatase. Radioactive artificial substrate was incubated in the presence of wild-type YOP51 (blue circles) or in the presence of mutants in YOP51 predicted to disrupt phosphatase activity, C403A (red squares), and C403S (purple triangles). (From K. L. Guan and J. E. Dixon. 1990. *Science*: 249: 553–556; doi: 10.1126/science.2166336. Reprinted with permission from AAAS.)

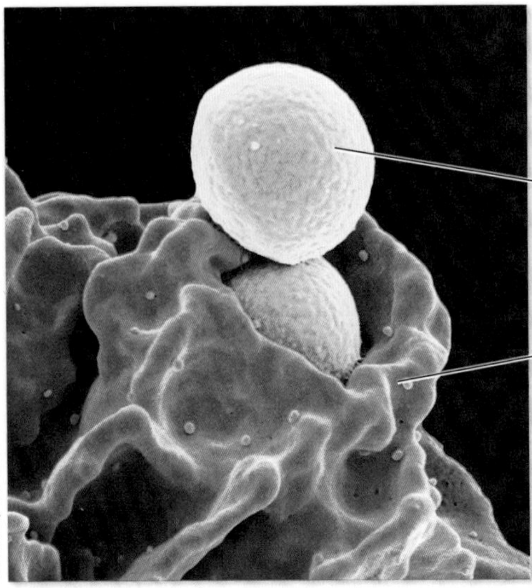

FIGURE 13.12 Micrograph of *Staphylococcus aureus* (yellow) and a neutrophil (purple) A neutrophil can be seen phagocytosing a bacterium.

block cytokine signaling. *Staphylococcus aureus* (FIGURE 13.12) produces a virulence factor, protein A, that binds to the TNF-α receptor and prevents proper signaling of the inflammatory cytokine in the wake of a staphylococcal infection.

Eukaryotic parasites are also capable of disrupting cytokine signaling. *Leishmania* is a parasite that can suppress the secretion of cytokines. This parasite causes leishmaniasis, resulting in skin sores and potentially life-threatening damage to internal organs, including the spleen, liver, and bone marrow. The *Leishmania* surface molecule lipophosphoglycan (LPG) inhibits the production of important inflammatory cytokines, such as TNF-α, and cytokines involved in T-cell activation, such as IL-12. LPG inhibits the production of these cytokines by preventing activation of protein kinase C (PKC), thus preventing activation of transcription factors needed for cytokine expression. *Leishmania* prevents the production of both an inflammatory response and T-cell activation by inhibiting signaling pathways required for secretion of these important cytokines of the innate and adaptive immune responses.

Disrupting detection by Toll-like receptors

We have seen that a wide variety of Toll-like receptors (TLRs) are responsible for detecting pathogen-associated molecular patterns, stimulating inflammatory responses, activating innate immune cells, and inducing the lectin complement pathway. Preventing TLR signaling or blocking detection by TLRs allows pathogens to persist without a proper immune response. Thus, some pathogens have evolved mechanisms to block TLR signaling and detection.

As previously discussed, TLR4 is responsible for the recognition of LPS from gram-negative bacteria. To prevent detection, some bacteria have evolved mechanisms to alter the structure of LPS, namely the lipid A portion of LPS (FIGURE 13.13). *Salmonella* can modify its lipid A moiety using lipid modification enzymes, including the removal of the third fatty acyl unit, which renders the modified LPS much less potent as a TLR4 ligand. Some organisms also have the ability to utilize different lipid A structures on their LPS. The oral pathogen *Porphyromonas gingivalis* has heterogeneity within the lipid A component of LPS, and each varied LPS is capable of acting differently on TLR signaling pathways; some can act as agonists and some can act as antagonists, depending on the lipid A structure.

Disrupting the complement system

The complement cascade is an extensive pathway utilized by the innate immune system to aid in the opsonization and destruction of extracellular pathogens. The activation of C3 by a C3 convertase, regardless of the mechanism of activation (alternative, classical, or lectin pathway), allows C3b to be deposited on the pathogen surface, which acts as a tag for complement receptors on phagocytic cells such as macrophages and neutrophils. The complement cascade is also capable of lysis of extracellular pathogens through the formation of the membrane-attack complex when C5 is processed on the surface of the pathogen. Blockage of the complement system, either by preventing complement fixation

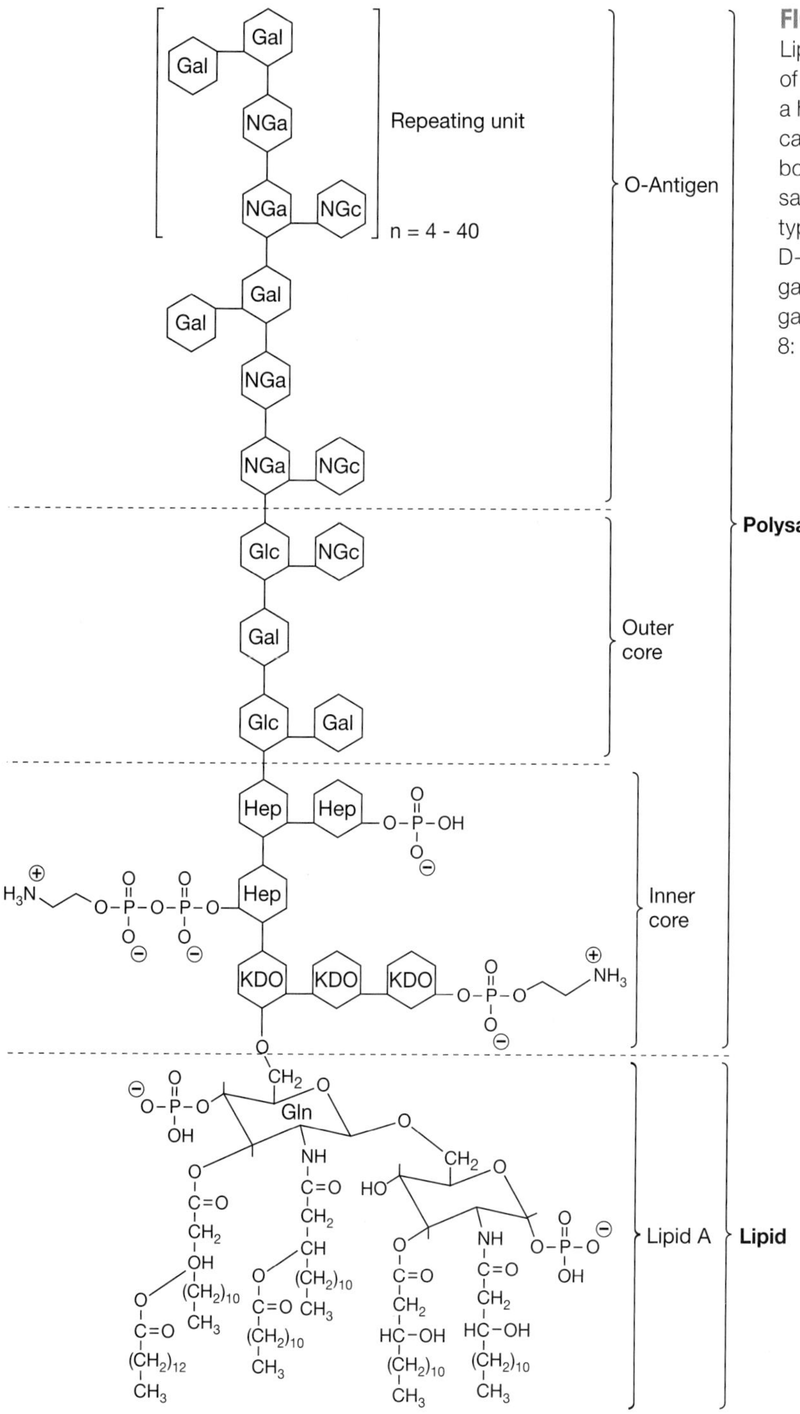

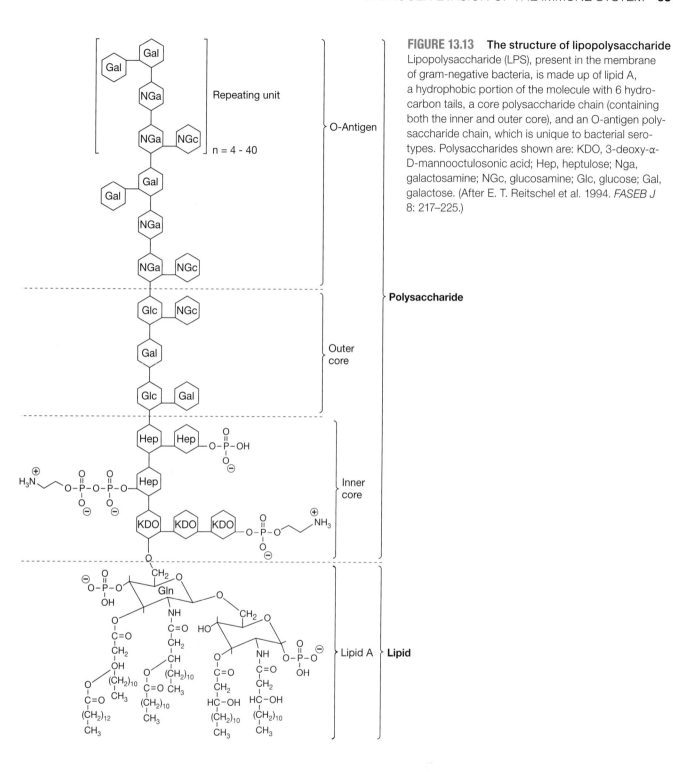

FIGURE 13.13 The structure of lipopolysaccharide Lipopolysaccharide (LPS), present in the membrane of gram-negative bacteria, is made up of lipid A, a hydrophobic portion of the molecule with 6 hydrocarbon tails, a core polysaccharide chain (containing both the inner and outer core), and an O-antigen polysaccharide chain, which is unique to bacterial serotypes. Polysaccharides shown are: KDO, 3-deoxy-α-D-mannooctulosonic acid; Hep, heptulose; Nga, galactosamine; NGc, glucosamine; Glc, glucose; Gal, galactose. (After E. T. Reitschel et al. 1994. *FASEB J* 8: 217–225.)

or formation of the membrane-attack complex, is an effective strategy for pathogens seeking to escape from an innate immune response. Indeed, many pathogens have evolved mechanisms to avoid the effects of the complement arm of the innate immune response.

A simple means for a pathogen to block complement opsonization is to modify the surface to prevent efficient C3 interaction. Several pathogens use a polysaccharide coat to prevent C3 fixation on their surface: *Streptococcus pneumoniae* (**FIGURE 13.14**), *Haemophilus influenzae*, *E. coli*, and *Neisseria meningitidis* bacteria all utilize a polysaccharide capsule to avoid destruction.

FIGURE 13.14 Micrograph of ***Streptococcus pneumoniae*** (purple) being attacked by a white blood cell (pink)

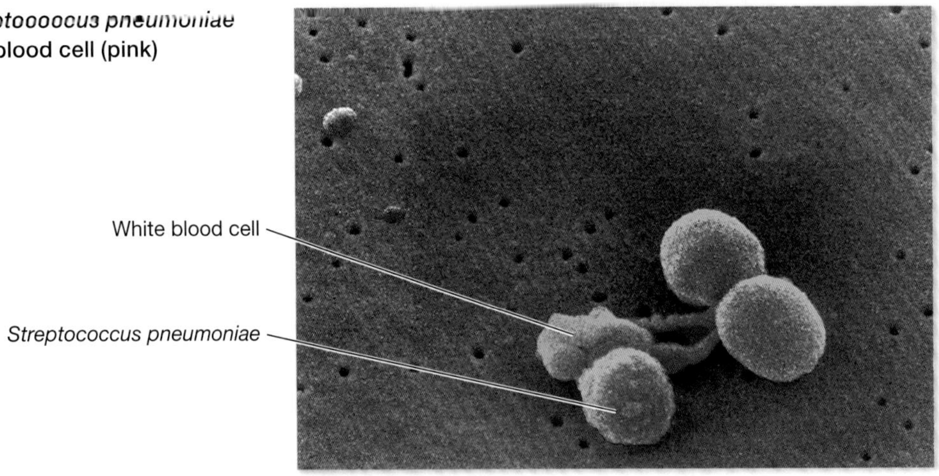

Courtesy of CDC/Janice Haney Carr

Some pathogens have evolved pathways to actively inhibit complement activation. We learned in Chapter 3 that complement fixation is inhibited on the surface of self-noninfected cells due to the action of complement inhibitors such as DAF and MCP. Certain viruses, including human cytomegalovirus, induce the cellular expression of these inhibitors to prevent lysis of the infected cell and allow continuation of the viral life cycle. Other viruses incorporate these inhibitors as viral coat proteins to prevent complement fixation of virions.

While certain viruses take advantage of host inhibitors of complement, other pathogens make use of inhibitors that block complement activation. Herpes simplex virus expresses a protein called glycoprotein C-1 that binds and inhibits C3b. Other viruses express proteins related to DAF and MCP to actively block complement activation. Certain bacteria can also inhibit complement activation: *S. aureus* expresses a protein that inhibits the action of C3 convertase, thus limiting the activation of complement and the fixation of C3b on its surface.

Disrupting the adaptive immune system

We have explored in great detail the complex mechanisms of the adaptive immune system that work to target specific pathogens, clear infection, and develop memory of the presence of a pathogen so that any subsequent infection can be dealt with swiftly. Cellular immunity and humoral immunity are driven by the action of T and B lymphocytes and rely on proper antigen presentation to T cells via MHC class I or class II pathways. The arsenal employed by the adaptive immune response—cellular and humoral immunity—is a potent force that pathogens must overcome in order to survive.

CD8 cytotoxic T cells target many viruses, as these lymphocytes are important in detecting and combating intracellular pathogens. As we have seen, CD8 T-cell activation requires display of a specific antigen to a CD8 T cell with the correct receptor through MHC class I antigen presentation. Recall that this presentation displays antigens that are cytoplasmic in nature, which are degraded by the proteasome into peptides that are then transported into the endoplasmic reticulum and loaded onto MHC class I molecules. Blockage of any of these steps reduces the capacity of the infected cell to properly display intracellular antigen peptides, potentially protecting a virus that has infected the cell. This strategy is employed by many viruses. One elegant example of a virus that subverts MHC class I presentation via multiple mechanisms is human cytomegalovirus, which expresses several different proteins to block class I antigen presentation (**FIGURE 13.15**). Human cytomegalovirus uses viral proteins to shut

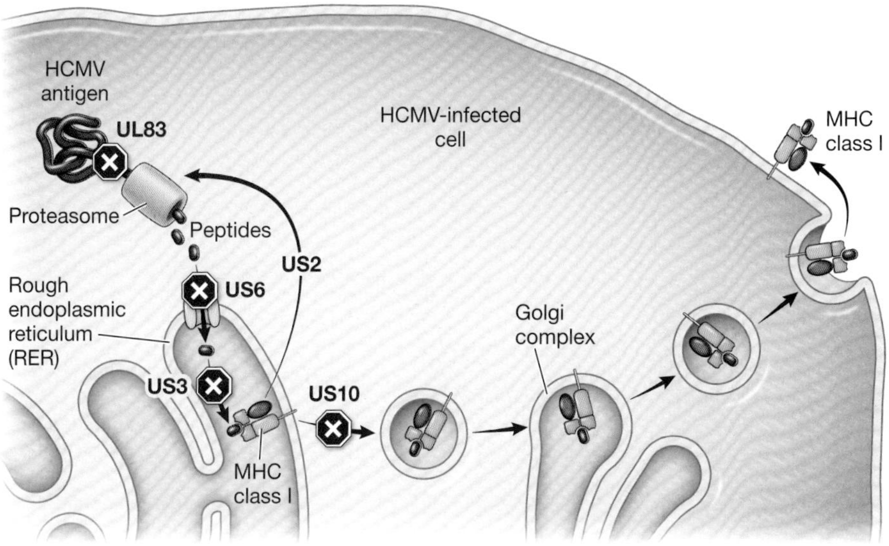

FIGURE 13.15 Mechanisms of inhibition of MHC class I presentation by human cytomegalovirus The genome of human cytomegalovirus (HCMV) contains genes that express proteins (UL83, US2, US3, US6, and US10) that block MHC class I antigen processing and presentation at several steps in the pathway. This blocking limits the number of MHC class I molecules presenting antigen at the surface of an infected cell. The decreased antigen presentation serves to hide HCMV, preventing detection by the adaptive immune system.

down MHC class I antigen presentation at almost every step, making this virus extremely efficient at blocking CD8 T-cell activation.

Bacterial pathogens have also evolved mechanisms that inhibit various aspects of the adaptive immune response. Unlike viruses, which tend to inhibit antigen presentation, pathogenic bacteria usually block lymphocyte activation or lymphocyte effector mechanisms. For instance, *Helicobacter pylori* expresses the toxin VacA, which inhibits T-cell proliferation by blocking the expression of IL-2 through inhibition of the transcription factor NFAT, thus limiting the action of IL-2 on T cells engaged in clonal selection. *Neisseria gonorrhoeae* blocks the proliferation of T cells using the Opa toxins it expresses, which interact with a cell-surface adhesion molecule, CEACAM1, and block signaling through this adhesion molecule.

B-cell activation and the production of antibodies also aid in the detection and neutralization of bacterial pathogens. Thus, a good strategy for bacterial pathogens would be to target antibodies and prevent their action. Since many bacteria interact with mucosal surfaces, a widely employed subversion strategy is the degradation of IgA by expression of IgA proteases (recall that IgA is a major protective immunoglobulin of mucosal surfaces). Such IgA proteases are expressed by *Staphylococcus*, *H. influenzae*, and *Neisseria*, and secretion of IgA proteases by these bacteria causes digestion of IgA immunoglobulins at mucosal surfaces and a disruption of the humoral immune response, allowing pathogen escape and survival.

CHECKPOINT QUESTIONS

1. Explain why many viruses express proteins that block MHC class I peptide presentation.
2. Several examples of mechanisms described in this section demonstrate pathogen blockage of transcription factors. Why is this an effective strategy for blocking innate or adaptive immune responses?

Making Connections: Pathogen Evasion of the Innate Immune Response

The innate immune system employs a variety of mechanisms to recognize and destroy pathogens, including inflammation, phagocytosis, complement fixation, and apoptosis of intracellularly infected cells. This selective pressure has driven the evolution of a variety of mechanisms used by pathogens to evade recognition and destruction by the innate immune system.

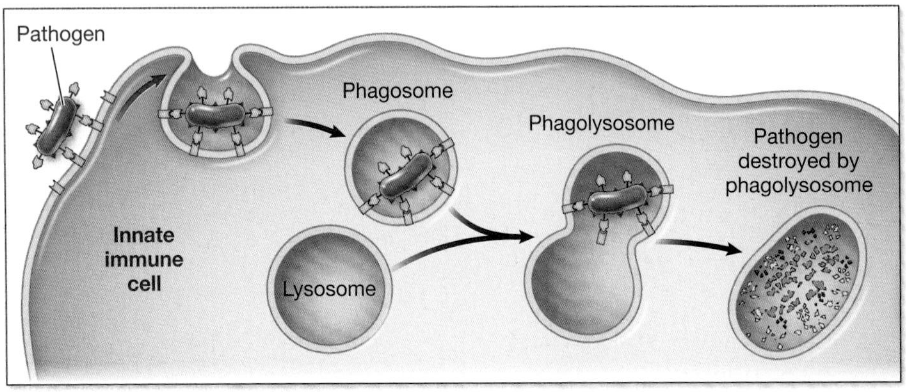

Phagocytosis evaded by

- Genetic variation (*Salmonella, V. cholerae*)
- Hiding from immune cells (*L. monocytogenes, M. tuberculosis*)
- Disrupting immune responses (*Y. pestis*)

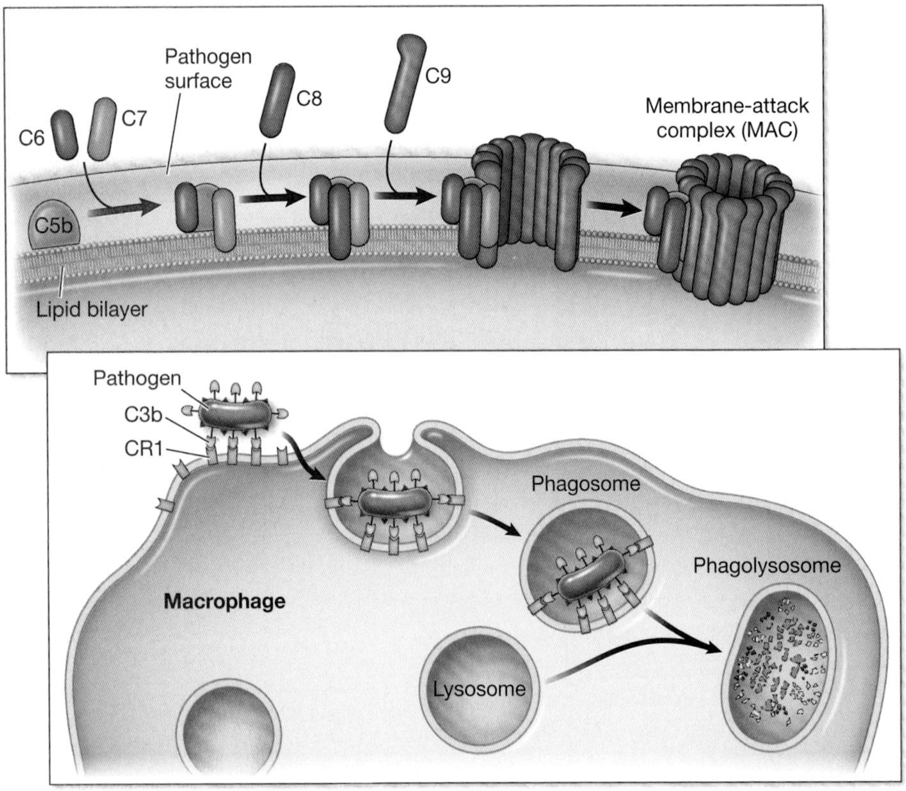

After P. Parham. 2014. *The Immune System*, 4th ed., W. W. Norton and Company

Complement activation evaded by

- Genetic variation (*S. pneumoniae, E. coli*)
- Hiding from immune cells (*L. monocytogenes, M. tuberculosis*)
- Disrupting immune responses (*S. aureus*)

The innate immune responses are designed to recognize and destroy pathogens through the action of proteins, receptors, and cytokines that facilitate the binding of pathogen patterns and the activation of innate immune cells. Pathogens use a variety of evasion strategies to counteract these mechanisms.

(Sections 2.1, 2.3, 2.2, 2.4, 2.5, 2.7, 3.1, 3.2, 13.1, 13.2, 13.3, 13.4)

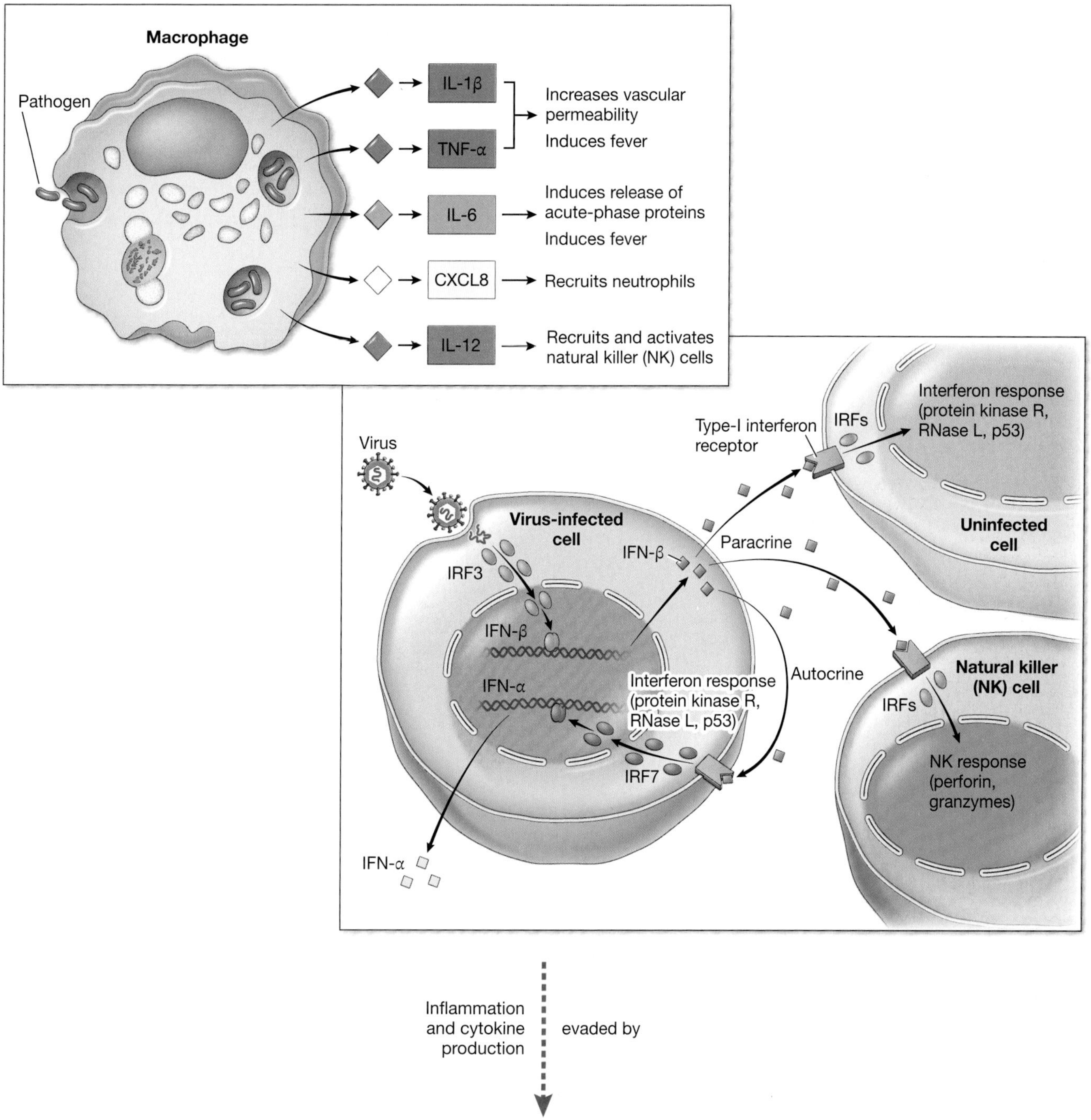

- Genetic variation (*Salmonella, P. gingivalis*)
- Hiding from immune cells (Herpesvirus, HIV)
- Disrupting immune responses (*Y. pestis*)

13.5 How do superantigens prevent a proper focused immune response?

LEARNING OBJECTIVE

13.5.1 Define *superantigen* and explain how these substances interfere with the immune response.

superantigen Toxin produced by a pathogen that is capable of general suppression of the immune system by nonspecific activation of T lymphocytes.

Certain bacteria alter cellular immunity through the expression of **superantigens**, which are toxins capable of activating a nonspecific lymphocyte response or blocking lymphocyte effector function. Superantigens are a family of proteins made by pathogens that have a profound effect on the immune system. Rather than targeting a specific molecule in a specific pathway, they target an entire arm of the immune response to suppress that response. In this section, we discuss examples of pathogens that suppress the immune response using different agonist or antagonist mechanisms. In each case, the outcome is an organism-level suppression of the immune system.

Staphylococcal enterotoxin

One example of a superantigen is staphylococcal enterotoxin. This toxin acts as a superantigen because of its ability to bind to both MHC class II molecules and T-cell receptors. Normally, the interaction of an MHC class II molecule on the surface of an antigen-presenting cell and a T-cell receptor is antigen specific. Recall that a T-cell receptor can only engage with the correct MHC molecule presenting the right antigen epitope. However, staphylococcal enterotoxin removes this specificity and allows any MHC class II molecule presenting any peptide to interact with any T-cell receptor (**FIGURE 13.16**). Because of this engagement, a robust, but nonspecific, T-cell response is engaged, resulting in a release of inflammatory cytokines, including IL-2 and TNF-α. The large-scale release of these cytokines can cause septic shock due to the effects they have within the bloodstream, causing a rapid loss of blood pressure as the vasculature permits fluid efflux.

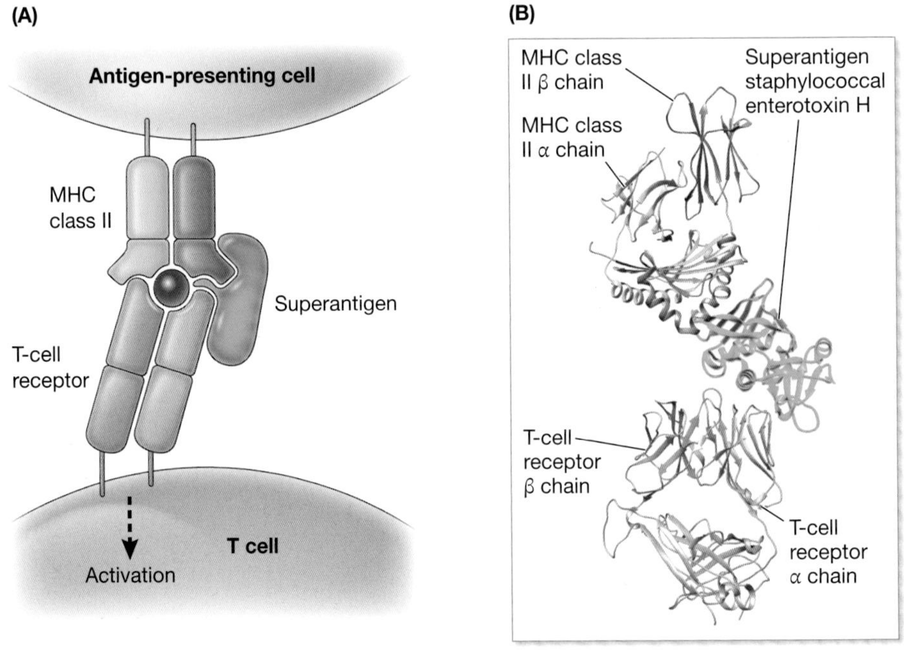

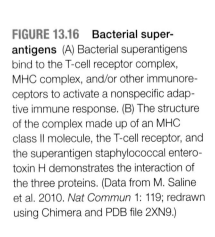

FIGURE 13.16 Bacterial superantigens (A) Bacterial superantigens bind to the T-cell receptor complex, MHC complex, and/or other immunoreceptors to activate a nonspecific adaptive immune response. (B) The structure of the complex made up of an MHC class II molecule, the T-cell receptor, and the superantigen staphylococcal enterotoxin H demonstrates the interaction of the three proteins. (Data from M. Saline et al. 2010. *Nat Commun* 1: 119; redrawn using Chimera and PDB file 2XN9.)

Staphylococcal enterotoxin causes further nonspecific signaling through the T-cell receptor by interacting with CD28, the costimulatory molecule required for T-cell activation. The superantigen functions to produce the required signal through the costimulatory receptor CD28 and allow for nonspecific activation of the T cell rather than inducing anergy of the T cell.

The other profound effect of staphylococcal enterotoxin action is the lack of specificity of the T-cell response. The lack of clonal selection before clonal expansion prevents a proper adaptive immune response from being raised. Furthermore, activated T cells with bound superantigen are removed from the population through apoptosis, ultimately lowering the number of T cells in circulation.

Staphylococcal superantigen-like protein

Another example of a superantigen-like molecule that acts to suppress the adaptive immune system is the staphylococcal superantigen-like protein 7 (SSLP7). SSLP7 is capable of blocking the action of both IgA and the complement pathway. SSLP7 can bind to secretory IgA, which, as discussed earlier, is an important weapon of humoral immunity against mucosal pathogens. Binding of SSLP7 to secretory IgA prevents its ability to interact with FcαRI receptors, which are receptors on the surface of phagocytic cells that, when bound to secretory IgA in complex with antigen, initiate phagocytosis (**FIGURE 13.17**). The SSLP7–secretory IgA interaction prevents efficient phagocytosis, as the IgA is incapable of interacting with its cognate Fc receptor. SSLP7 also has the ability to block the action of the complement protein C5, an important precursor protein for the formation of the membrane-attack complex during complement activation and an aid in pathogen lysis.

CHECKPOINT QUESTION

1. Would you characterize the staphylococcal enterotoxin superantigen as an agonist or antagonist? Why?

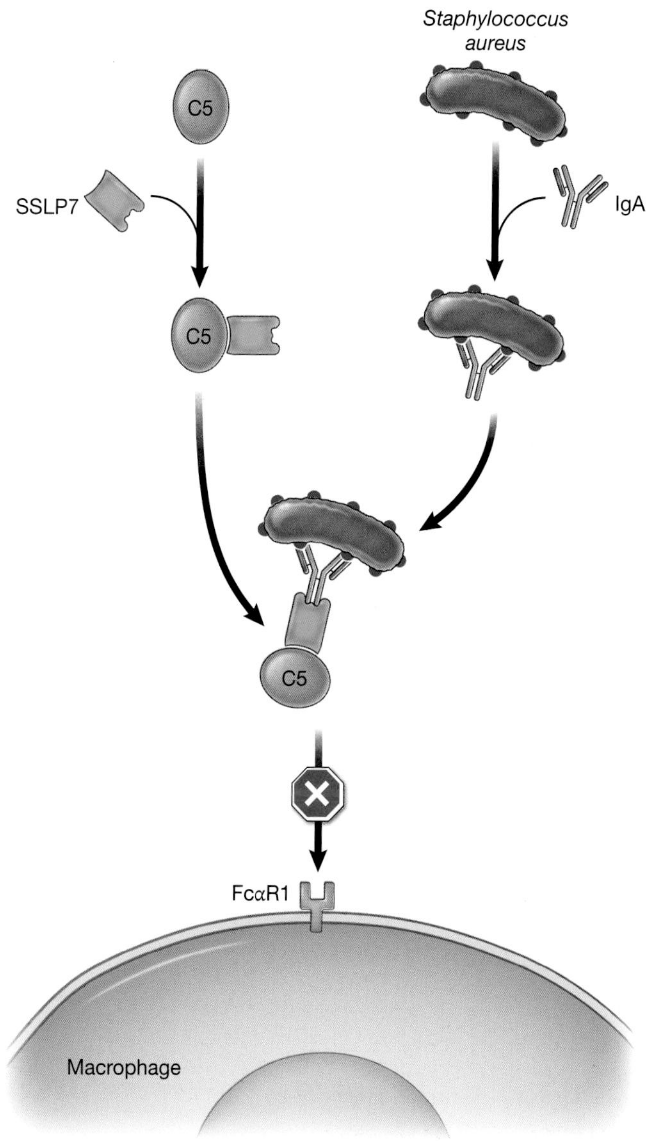

FIGURE 13.17 Mechanism of action of staphylococcal superantigen-like protein 7 Macrophages express the receptor FcαR1 that binds to the Fc component of IgA on the surface of a bacterium to drive phagocytosis. *Staphylococcus aureus* secretes staphylococcal superantigen-like protein 7 (SSLP7), which binds to the Fc component of IgA on the surface of *S. aureus*, and the complement protein C5. This interaction prevents IgA on the bacterium surface from interacting with the Fc receptor on the macrophage surface, preventing phagocytosis. (After P. Parham. 2014. *The Immune System*, 4th ed., W. W. Norton and Company.)

Making Connections: Pathogen Evasion of the Adaptive Immune Response

The adaptive immune system uses specific receptors expressed by B cells and T cells to recognize pathogens. Differentiation of clonally selected lymphocytes drives their effector functions in order to destroy those pathogens. Selective pressure from the adaptive immune response has driven the evolution of a variety of mechanisms used by pathogens to evade recognition and destruction by the adaptive immune system.

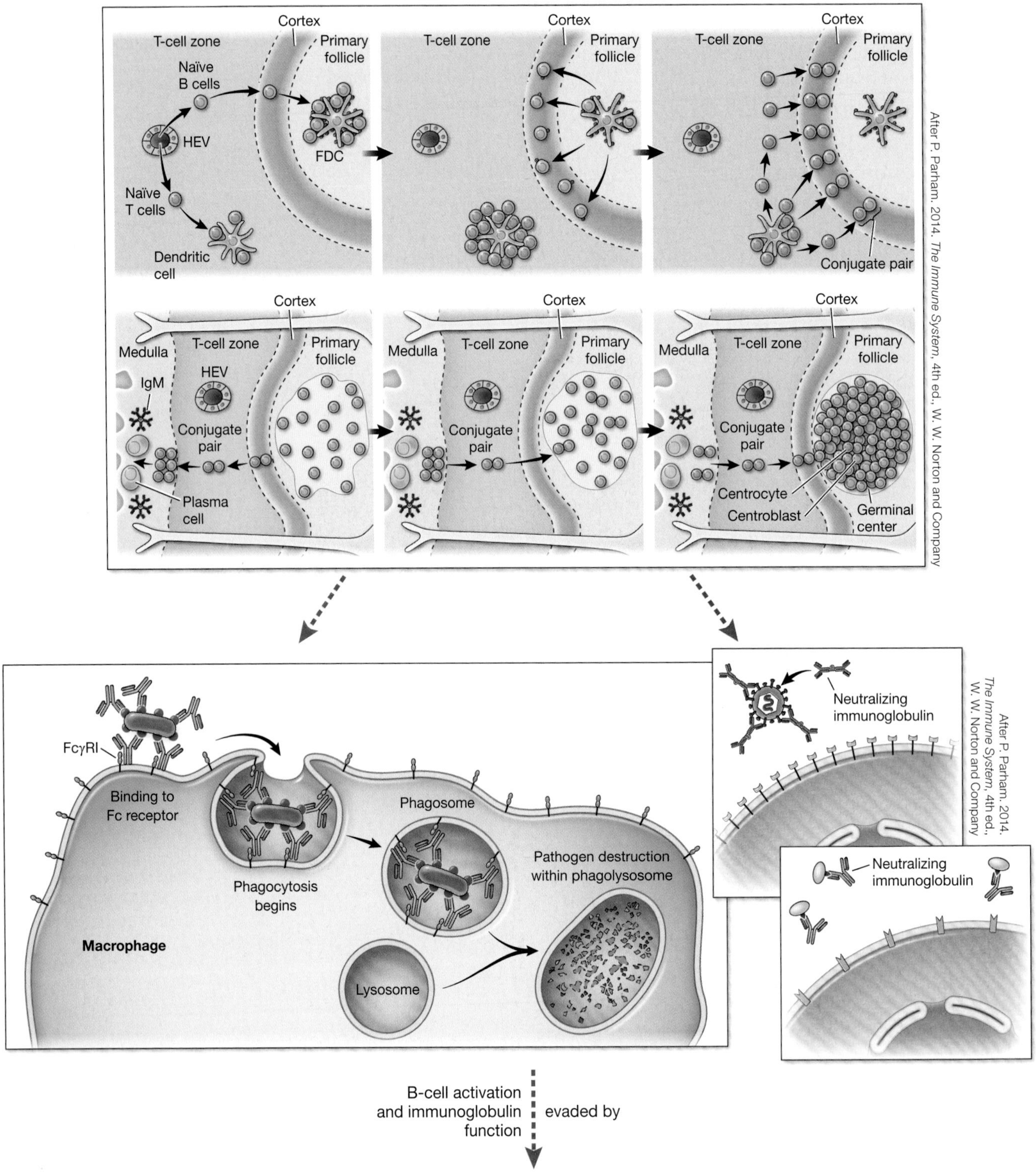

- Genetic variation (*S. pneumoniae, E. coli*)
- Hiding from immune cells (*L. monocytogenes*)
- Disrupting immune responses (*Staphylococcus, Neisseria*)
- Superantigens (*S. aureus*)

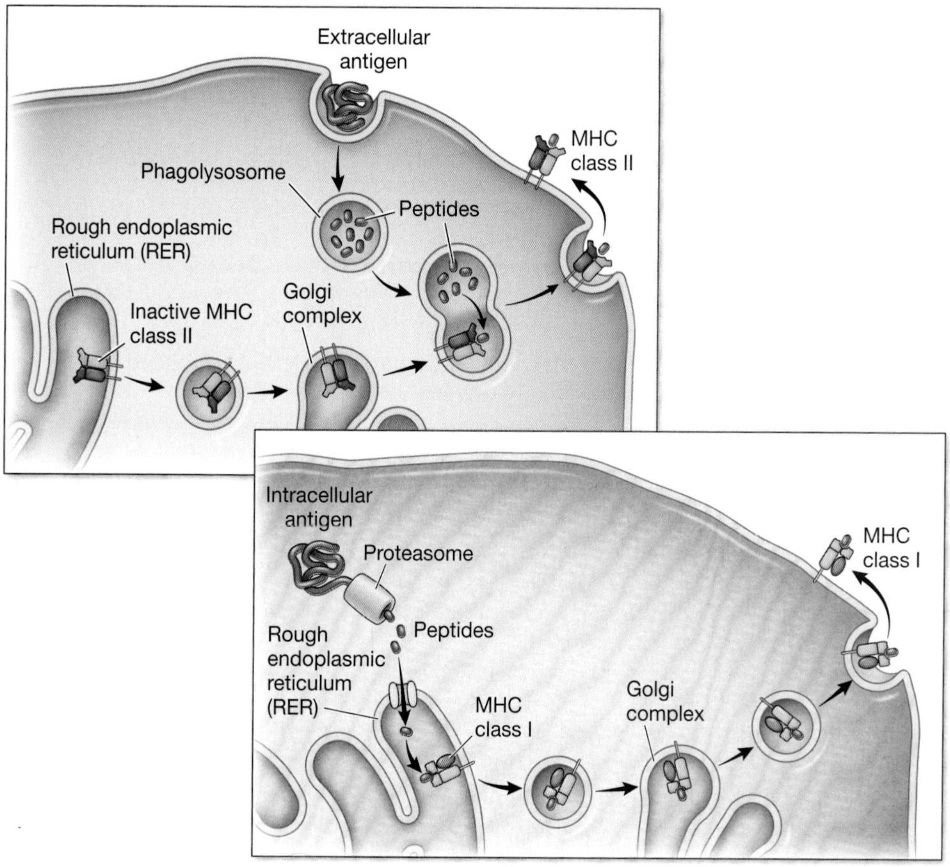

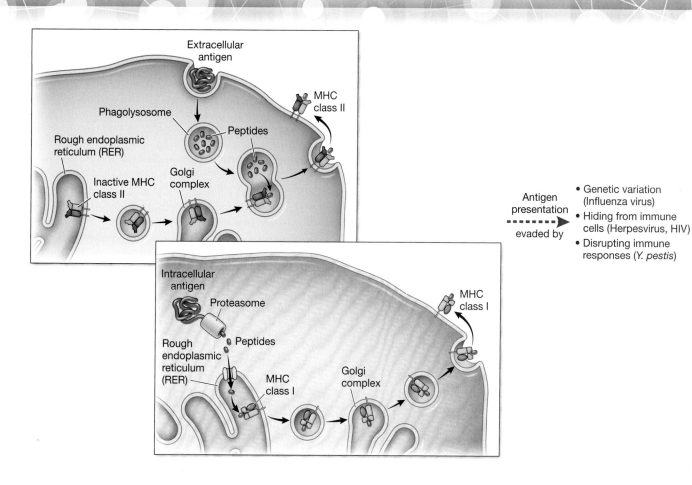

Antigen presentation evaded by

- Genetic variation (Influenza virus)
- Hiding from immune cells (Herpesvirus, HIV)
- Disrupting immune responses (*Y. pestis*)

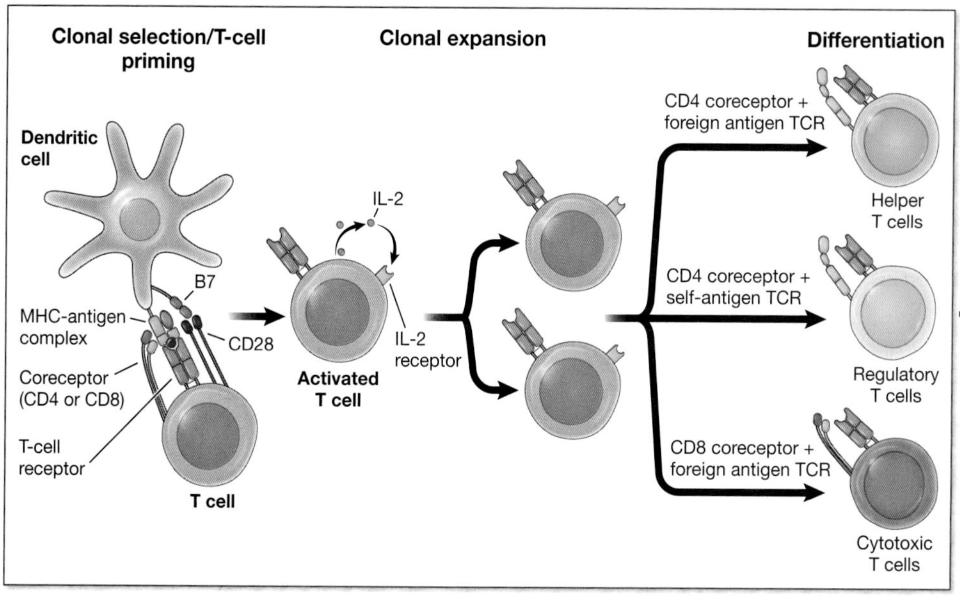

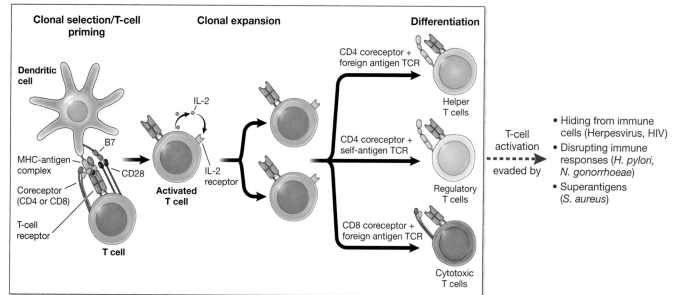

T-cell activation evaded by

- Hiding from immune cells (Herpesvirus, HIV)
- Disrupting immune responses (*H. pylori, N. gonorrhoeae*)
- Superantigens (*S. aureus*)

The adaptive immune responses are designed to recognize and destroy pathogens through the action of B cells and T cells and their effector functions. Pathogens use a variety of evasion strategies to counteract these mechanisms.

(Sections 4.2, 4.6, 4.7, 4.8, 6.1, 6.6, 7.1, 7.7, 7.9, 10.1, 10.5, 13.1, 13.2, 13.3, 13.4, 13.5)

Summary

13.1 What are the primary ways that pathogens evade immune system defenses?

- Pathogens use a variety of mechanisms to subvert or evade the immune system, including genetic variation, hiding from immune system cells, blocking or disrupting the immune response, downregulating immune system activity, and using superantigens to induce a nonspecific immune response.

13.2 How does genetic variation allow pathogens to evade the immune system?

- Pathogens can use genetic variation to change their cell-surface molecules to avoid detection by the immune system. Genetic variation can be driven by the presence of serotypes, mutation of cell-surface molecules, or the expression of different genes that encode cell-surface molecules.

13.3 How do pathogens hide from the immune system?

- Pathogens can hide from detection by including a dormant state in their life cycle or by residing in an intracellular niche that cannot be detected by normal immune cell action.
- Bacterial pathogens can reside intracellularly by inducing phagocytosis and then halting the phagocytic process before they are digested by the phagocyte.

13.4 How can pathogens downregulate the immune system?

- Pathogens have evolved a variety of mechanisms that work to disrupt immune system function, including inhibiting antimicrobial peptides and proteins, disrupting phagocytosis, disrupting cytokine signaling, blocking detection by TLRs, inhibiting the complement system, and disrupting various actions of the adaptive immune system.

13.5 How do superantigens prevent a focused immune response?

- Certain pathogens can express superantigens that induce a nonspecific lymphocyte response. These superantigens can prompt a systemic activation of lymphocytes, causing clonal expansion of a wide number of lymphocytes and the inability of the adaptive immune system to target the pathogen.

Review Questions

13.1 You have identified a novel virus that is causing an epidemic in North America. The virus appears to induce a robust inflammatory response, during which a large titer of virus can be detected within the patient's blood. During the viral infection cycle, after a period of 2 weeks, the viral titer decreases dramatically. However, upon secondary infection by an independent pathogen, the titer of the original virus increases as well. Your group hypothesizes that the virus has an ability to undergo a state of dormancy and reemerge upon subsequent pathogen infection that induces another inflammatory response. Which of the five general mechanisms of evasion of the immune system would categorize this state of dormancy?

13.2 Avian influenza virus (H5N1), commonly known as bird flu, has been feared as a potentially infectious agent in humans, and concern arose that it could cause a major pandemic. Cases of H5N1 occasionally infect humans, but they are rare and only occur under certain conditions. Would human cases of H5N1 be an example of antigenic drift or antigenic shift? Why?

13.3 You have identified a novel bacterial pathogen that survives inside of macrophages during its infection cycle. A fellow graduate student within your lab is convinced that the bacteria are cytoplasmic, but you hypothesize that the bacteria are living within the phagosome of macrophages. Design one or two simple experiments to test your hypothesis.

13.4 You have isolated a novel extracellular pathogen that survives well within a mouse host. You are trying to predict the presence of virulence factors necessary for its survival. Make some predictions about the types of virulence factors you expect the pathogen to employ for survival.

13.5 You have isolated a novel protein that utilizes a molecular mechanism analogous to a C3 convertase (it cleaves soluble C3 into C3a and C3b), and you characterize it as a superantigen. If the mode of action is correct (i.e., it acts as a C3 convertase), is this mode of action one of an agonist or an antagonist? Is this properly characterized as a superantigen?

CASE STUDY REVISITED

Septic Shock

A woman, age 32, comes to the hospital emergency department complaining of chills and difficulty breathing. Initial observation shows that she has low blood pressure and a fever of 102°F. The exam by the ER physician reveals that the woman has a cut on her right hand that is reddened, suggesting an infection at the wound site. Blood culture demonstrates the presence of *Staphylococcus aureus*. ELISA (**e**nzyme-**l**inked **i**mmuno**s**orbent **a**ssay) testing confirms the presence of TNF-α in the bloodstream. The presence of TNF-α in the blood, combined with low blood pressure and fever, allow the doctor to diagnose septic shock.

Think About...

Questions for individuals

1. What are some potential causes for a patient presenting with low blood pressure?
2. What are the actions of the inflammatory cytokine TNF-α?

Questions for student groups

1. What would you predict as the causative agent of the septic shock?
2. Would you predict that the patient would be more susceptible to other pathogen infections while experiencing septic shock? Why or why not?

MAKING CONNECTIONS Key Concepts	**COVERAGE** (Section Numbers)
Phagocytosis and mechanisms of pathogen evasion	2.1, 2.2, 13.1, 13.2, 13.3, 13.4
Inflammation and mechanisms of pathogen evasion	2.3, 2.4, 2.5, 2.7, 13.1, 13.2, 13.3, 13.4
Complement activation and mechanisms of pathogen evasion	2.1, 3.1, 3.2, 13.1,
Antigen presentation and mechanisms of pathogen evasion	4.2, 4.6, 6.1, 6.6, 13.1, 13.2, 13.3, 13.4, 13.5
T-cell activation and mechanisms of pathogen evasion	4.2, 4.7, 7.1, 7.7, 7.9, 13.1, 13.2, 13.3, 13.4, 13.5
B-cell function and mechanisms of pathogen evasion	4.2, 4.8, 10.1, 10.5, 13.1, 13.2, 13.3, 13.4, 13.5

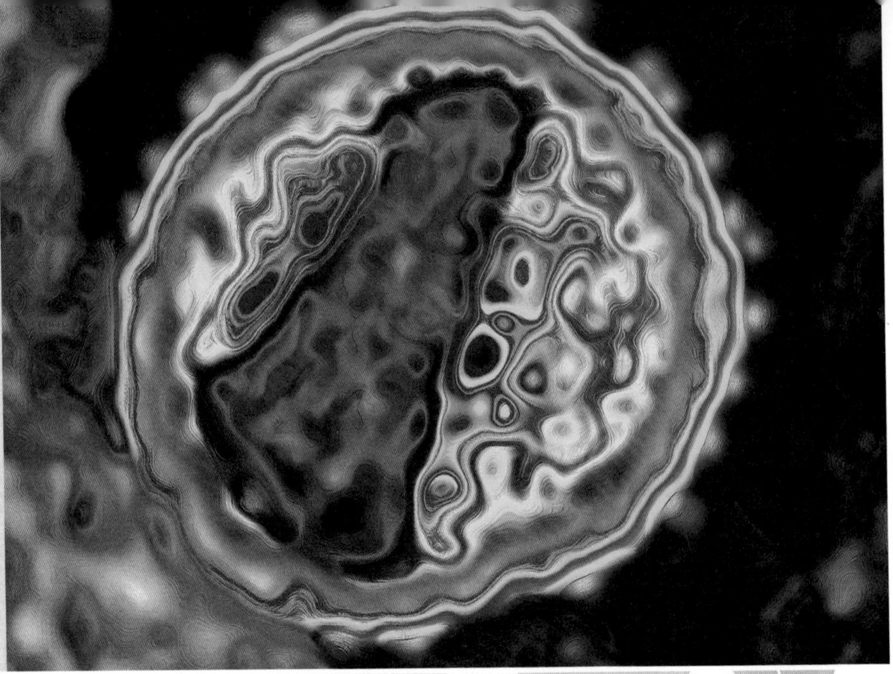

© James Cavallini/Science Source

14

Immunodeficiencies

QUESTIONS Explored

14.1 What is immunodeficiency?

14.2 How does inherited immunodeficiency differ from acquired immunodeficiency?

14.3 How does a deficiency in the innate immune system lead to disease?

14.4 How does a combined deficiency in lymphocyte development or action lead to disease?

14.5 How does a deficiency in T-cell action lead to disease?

14.6 How does a deficiency in B-cell action lead to disease?

14.7 How does HIV cause AIDS?

CASE STUDY: Donte's Story

Life as new parents has been challenging for Kayla and Justice—sleepless nights, constant trips to the store to purchase that forgotten item, and learning on the fly how to care for their son, Donte. Donte seems to be developing normally. However, they notice that he suffers from chronic diarrhea and cries much of the time, and they plan to speak with their pediatrician about this. One day, Donte wakes up crying. When Kayla enters his room she sees a red rash all over his body and realizes this is probably something much worse than an upset stomach. At 6 months of age, Donte has not yet been vaccinated against measles, and Kayla fears the worst. Kayla and Justice decide to take him to the emergency room.

To understand what Donte is experiencing, and to answer the case study questions at the end of this chapter, we must understand the problems that occur when immune system function is impaired. In earlier chapters we have seen how the innate and adaptive immune systems work to combat pathogens. What happens, then, if our immune system cannot function properly? An inability of an arm of the immune system to function is referred to as an *immunodeficiency*, which can be caused by a variety of factors. The severity of the problem associated with an immunodeficiency depends on which part of the immune response is impaired.

Some immunodeficiencies are inherited, while others are induced by environmental factors. Commonly, the lack of an immune cell or an immune cell function results in increased likelihood of infection with pathogens that are normally combated by that cell or cell function. We will explore examples of immunodeficiencies of the innate and adaptive immune systems and the problems that result.

14.1 | What is immunodeficiency?

LEARNING OBJECTIVE

14.1.1 Compare and contrast primary immunodeficiencies, secondary immunodeficiencies, and acquired immunodeficiencies.

Because of the importance of the various types of immune responses in protecting our bodies from infection, one can imagine that malfunction of any of these immune processes would have deleterious effects. This chapter focuses on what happens when things go wrong—specifically, the health consequences for individuals who cannot properly mount one or more of the immune responses we have examined in detail in earlier chapters.

The inability to establish an effective immune response is known as **immunodeficiency**. This condition can be classified into several types:

- *Primary immunodeficiencies* are disorders that occur because some part of the immune system is missing or defective due to genetic disorders.
- *Secondary immunodeficiencies* are caused by non-inherited factors such as malnutrition, age, infection, disease, drugs, or exposure to toxins.
- *Combined immunodeficiencies* involve impairments in both B-cell and T-cell function due to inherited mutations.

We have seen throughout this text the many mechanisms used by the innate and adaptive immune responses to defend against infection and disease. These mechanisms include the action of specialized cells of hematopoietic origin and that of plasma proteins (e.g., complement) and cytokines to efficiently tag, target, and destroy pathogens. Proper development and action of proteins and cells of the immune system promote efficient recognition and clearance of pathogens; thus, immunodeficiencies can result in persistent infections and infections with opportunistic pathogens. These immunodeficiencies can manifest due to genetic abnormalities or may be caused by pathogens or environmental factors that prevent a proper immune system response.

immunodeficiency A disorder in an innate or adaptive immune response caused by a lack of molecules required for inducing the response or a lack of cells required for the response due to improper development.

inherited immunodeficiency A disorder in an innate or adaptive immune response caused by mutations within genes in the patient that prevent the proper responses from occurring.

CHECKPOINT QUESTION

1. Differentiate between the three different types of immunodeficiencies and their causes.

14.2 | How does inherited immunodeficiency differ from acquired immunodeficiency?

LEARNING OBJECTIVE

14.2.1 Differentiate between inherited and acquired immunodeficiencies.

While all immunodeficiencies prevent a critical immune response pathway from functioning properly, the onset of such problems can occur via two major means. **Inherited immunodeficiency** is caused by mutations in genes responsible for the action of specific immune response molecules or in genes that play a role in the development of immune cells. Remember from your study of genetics that there are several modes of inheritance for a disease caused by a single gene. If there are no chromosomal abnormalities, all somatic cells have two copies of every nuclear chromosome and, thus, two copies of every gene. Genes can either be on the autosomal chromosomes (22 pairs in each somatic cell) or on the sex chromosomes (X or Y). An autosomal dominant disease or an X-linked dominant disease caused by a single gene is due to a dominant allele and only requires one copy of that allele to be present for the disease to manifest. An autosomal recessive disease is due to two recessive alleles and requires both copies of that allele to be present for the disease to manifest. In an X linked recessive disease, males with the recessive allele will manifest the disease, but females will require two copies of the recessive gene to manifest the disease.

These mutations can produce one of three outcomes. In dominant immunodeficiencies, a gene product's activity is altered, and the presence of the altered gene product is sufficient to drive the disease phenotype. A second outcome is that a dysfunctional gene product may result, preventing its action in the immune system. A third outcome is that the mutation can change the expression of a gene due to changes in its promoter region or affecting its transcriptional activity. The nature of the mutation dictates the severity of the disorder. It also determines whether a single arm of the immune response is blocked (usually caused by mutation of a gene whose protein product plays an important role in that immune response) or whether the immunodeficiency is extremely severe, completely blocking the presence of immune cells (**FIGURE 14.1**).

Primary immunodeficiencies affect approximately 1 in 1200 people in the United States. The relative prevalence of primary immunodeficiencies within the population varies due to the severity of loss of a particular immune cell type and each individual's susceptibility to infection. All immunodeficiencies cause problems with combating pathogens, but some are more tolerable within the population than others.

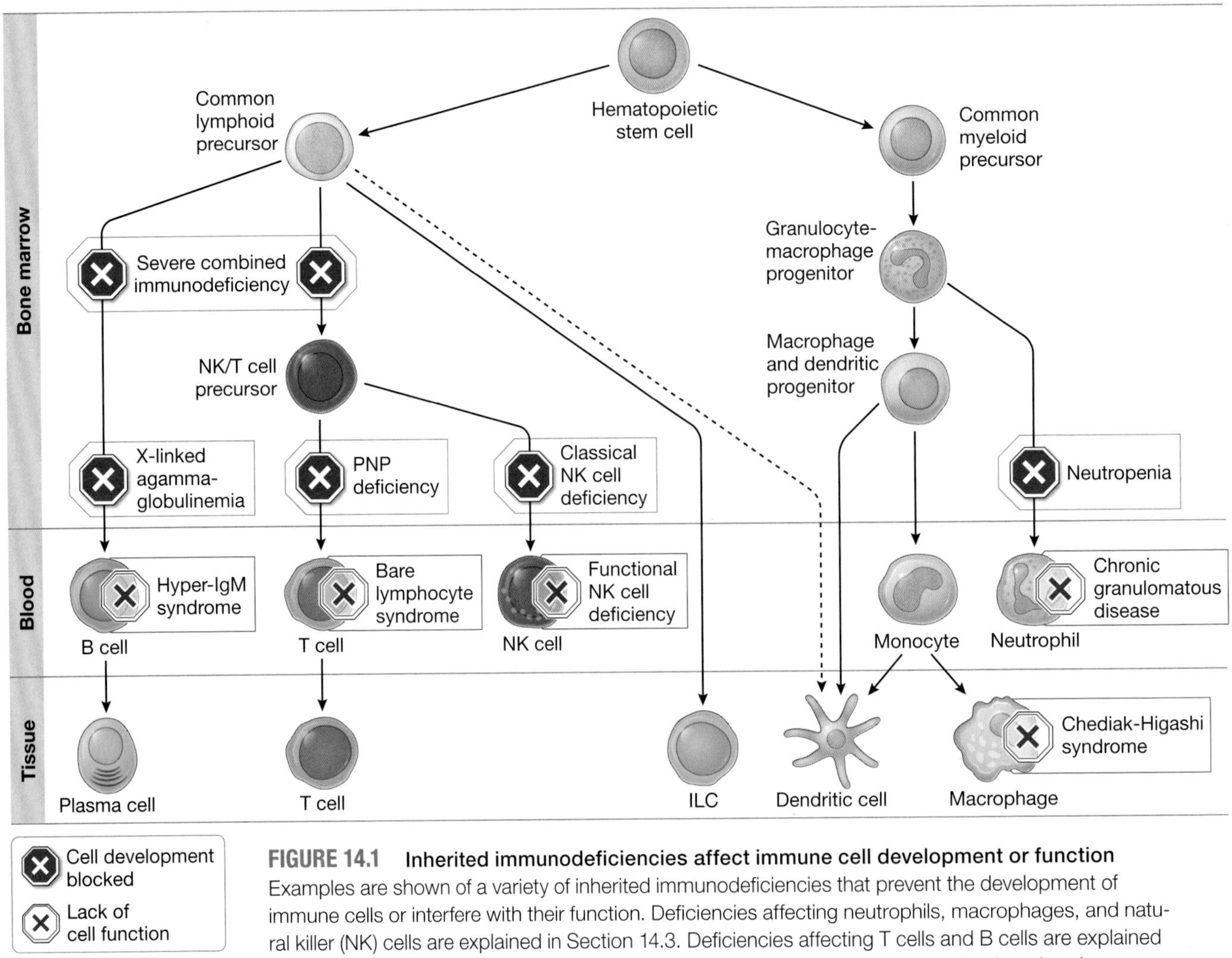

FIGURE 14.1 Inherited immunodeficiencies affect immune cell development or function
Examples are shown of a variety of inherited immunodeficiencies that prevent the development of immune cells or interfere with their function. Deficiencies affecting neutrophils, macrophages, and natural killer (NK) cells are explained in Section 14.3. Deficiencies affecting T cells and B cells are explained in Sections 14.4, 14.5, and 14.6. ILC, innate lymphoid cell; PNP, purine nucleoside phosphorylase.

Among the more than 200 inherited immunodeficiencies that have been identified, B-cell immunodeficiencies are the most prevalent, far outnumbering other primary immunodeficiencies. Combined immunodeficiencies have a higher prevalence than T-cell immunodeficiencies, and T-cell immunodeficiencies are more prevalent than those affecting the innate immune system. The lower prevalence of T-cell and innate immune system immunodeficiencies is likely due to the significant morbidity associated with the loss of T-cell or innate immune cell function compared to that of B-cell immunodeficiencies.

While inherited immunodeficiencies are caused by genetic abnormalities, **acquired immunodeficiency** is driven by the suppression of the immune system by external factors (**FIGURE 14.2**), including immunosuppressive agents, aging, malnutrition, or infectious agents such as the human immunodeficiency virus (HIV). Although the causes differ, both types ultimately prevent normal immune function and can lead to persistent and opportunistic infections that may be fatal.

acquired immunodeficiency
Immunodeficiency caused by a pathogen or environmental factor that has altered or blocked an innate or adaptive immune response, preventing normal immune system function.

Inherited immunodeficiency

Inherited immunodeficiencies are caused by genetic mutations that affect the immune response. Malfunction of a gene product involved in immune response function can result either in actions that go awry or in the absence of necessary

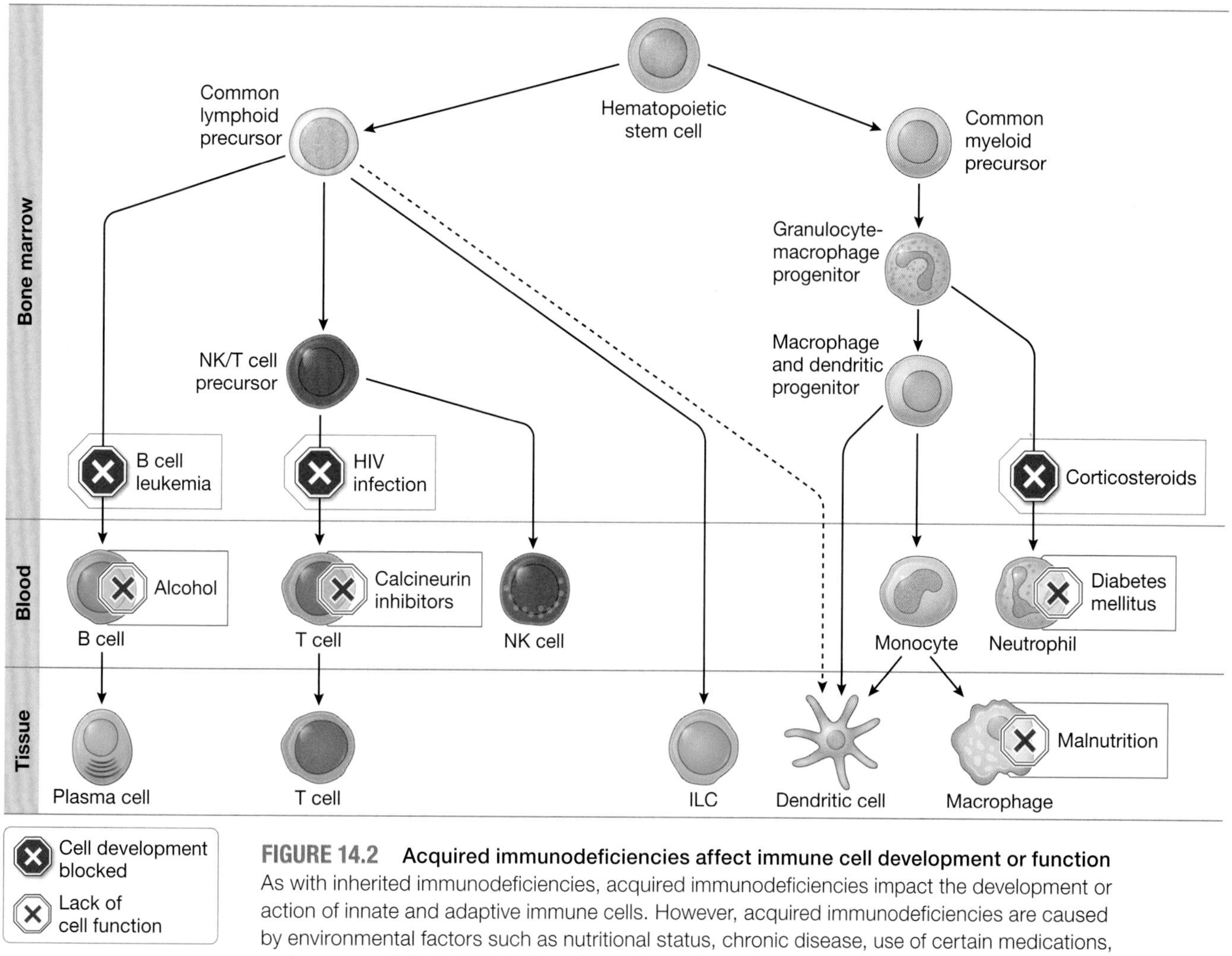

FIGURE 14.2 Acquired immunodeficiencies affect immune cell development or function
As with inherited immunodeficiencies, acquired immunodeficiencies impact the development or action of innate and adaptive immune cells. However, acquired immunodeficiencies are caused by environmental factors such as nutritional status, chronic disease, use of certain medications, and overuse of alcohol. Note that, although not shown here, some environmental factors may impact more than one immune cell type. ILC, innate lymphoid cell.

functions. For instance, an inherited immunodeficiency in the C3 component of complement can result in an inability to fix complement on the surface of pathogens, preventing pathogen destruction and clearance. Individuals who suffer from this type of immunodeficiency are susceptible to recurring infection by gram-negative bacteria due to the importance of complement fixation on gram-negative bacteria for efficient recognition and destruction.

One can imagine that abnormalities in genes associated with any of the pathways important in proper immune cell function can lead to a variety of immunodeficiency disorders. Indeed, many inherited immunodeficiencies are related to the improper functioning of an immune response protein (**TABLE 14.1**).

Another cause of inherited immunodeficiency is mutation in a gene responsible for the proper development of immune cells. Lack of development of a specific cell type (such as occurs with severe combined immunodeficiency or

TABLE 14.1 | Examples of Inherited Immunodeficiencies

Immunodeficiency	Mutation	Symptom
Immunodeficiencies in First Defenses of the Innate Immune System		
Crohn's disease	NOD Defensins	Gastrointestinal (GI) inflammation Susceptibility to GI infection
Immunodeficiencies in the Complement System		
Complement proteins (e.g., C3)	Varies (see Table 14.2)	Varies (see Table 14.2)
Hereditary angioneurotic edema	C1INH	Edema and swelling Asphyxiation
Immunodeficiencies in Phagocyte Function		
Leukocyte adhesion deficiency	Common subunit in LFA-1, CR3, CR4	Recurring infections with encapsulated bacteria
Chediak-Higashi syndrome	LYST	Recurring infections with encapsulated bacteria
Chronic granulomatous disease	NADPH oxidase	Recurring infections with encapsulated bacteria
Immunodeficiencies in Natural Killer Cell Function		
Classical NK cell deficiency	GATA2 MCM4	Lack of NK cells in circulation Persistent viral infection
Functional NK cell deficiency	CD16	Lack of ADCC Persistent viral infection
Severe Combined Immunodeficiencies		
Ommen syndrome	RAG1/2	Lack of T-cell and B-cell development
Adenosine deaminase deficiency	Adenosine deaminase	Lack of T-cell and B-cell development
T-cell or B-cell Immunodeficiencies		
Cytokine receptor and JAK3 deficiency	Common gamma cytokine receptor JAK3	Lack of T-cell development
PNP deficiency	Purine nucleoside phosphorylase	Lack of T-cell development
Bare lymphocyte syndrome	TAP Transcription factor for MHC class II expression	Lack of peptide presentation
X-linked agammaglobulinemia	Btk	Lack of B-cell development
Hyper-IgM syndrome	CD40 AID	Lack of production of immunoglobulin isotypes
Selective IgA deficiency	Lack of mucosal IgA	Recurring mucosal infections (ear, bronchial) Susceptibility to allergic asthma

Sources: M. S. Lim and K. S. J. Elenitoba-Johnson. 2004. *J Mol Diagnostics* 6: 59–83; D. J. Unsworth. 2008. *J Clin Pathol* 61: 1013–1017; R. Caruso et al. 2014. *Immunity* 41: 898–908; J. S. Orange. 2013. *J Allergy Clin Immunol* 132: 515–526.

X-linked agammaglobulinemia) leads to an improper immune response. This is not because a mechanism has been altered but rather because that mechanism is not present. Examples of developmental defects that lead to inherited immunodeficiencies are shown in Table 14.1.

Because inherited immunodeficiencies are driven by genetic defects related to at least one arm of the immune system, the most common treatment is bone marrow transplantation to replace the defective cell or process of the hematopoietic system (discussed in detail in Chapter 17). While bone marrow transplantation is the common means for treating immunodeficiencies, recent advances have been made in the use of gene therapy (addition of a normally functioning gene into a diseased individual). Some gene therapy trials have worked extremely well in the treatment of certain combined immunodeficiencies. For example, in the treatment of X-linked severe combined immunodeficiency, gene therapy replaced a cytokine receptor needed for lymphocyte development. However, other trials have resulted in the onset of leukemia in individuals receiving treatment. Because of the promise of gene therapy, research has been focused on reducing the possibility of onset of certain cancers that have occurred during trials.

Acquired immunodeficiency

As with inherited immunodeficiencies, acquired immunodeficiencies also prevent actions related to a normal immune response or lead to a lack of necessary immune cells. Common consequences also occur, namely higher risk for infection with both typical and opportunistic pathogens. However, acquired immunodeficiency is brought about due to environmental factors, which can include pathogens with the ability to inactivate normal immune cell function. Environmental factors that drive acquired immunodeficiencies usually have an immunosuppressive impact, preventing an arm of the immune system from properly combating pathogens. These factors include immunosuppressive drugs taken by individuals with an autoimmune disease or those who have undergone organ transplants. Other factors include malnutrition, aging, environmental toxins, chemotherapies, radiation, and, notably, HIV, the causative agent of acquired immunodeficiency syndrome (AIDS). Individuals with acquired immunodeficiencies may develop certain cancers, especially those of the hematopoietic system.

CHECKPOINT QUESTION

1. Compare and contrast inherited immunodeficiency and acquired immunodeficiency.

14.3 How does a deficiency in the innate immune system lead to disease?

LEARNING OBJECTIVE

14.3.1 Provide examples of inherited immunodeficiencies of the innate immune system.

In this section, we focus on examples of inherited immunodeficiencies that affect the many immune response pathways that are important in preventing pathogen infection. (Keep in mind that environmental factors can also block the same pathways and result in acquired immunodeficiency.)

Immunodeficiencies in first defenses

Recall that the skin, mucosal surfaces, and epithelia serve as physical barriers to prevent pathogen colonization and infection. Epithelial cells utilize intracellular molecules such as nucleotide-binding oligomerization domain-containing proteins (NODs) to recognize pathogen components. Peptides and enzymes such

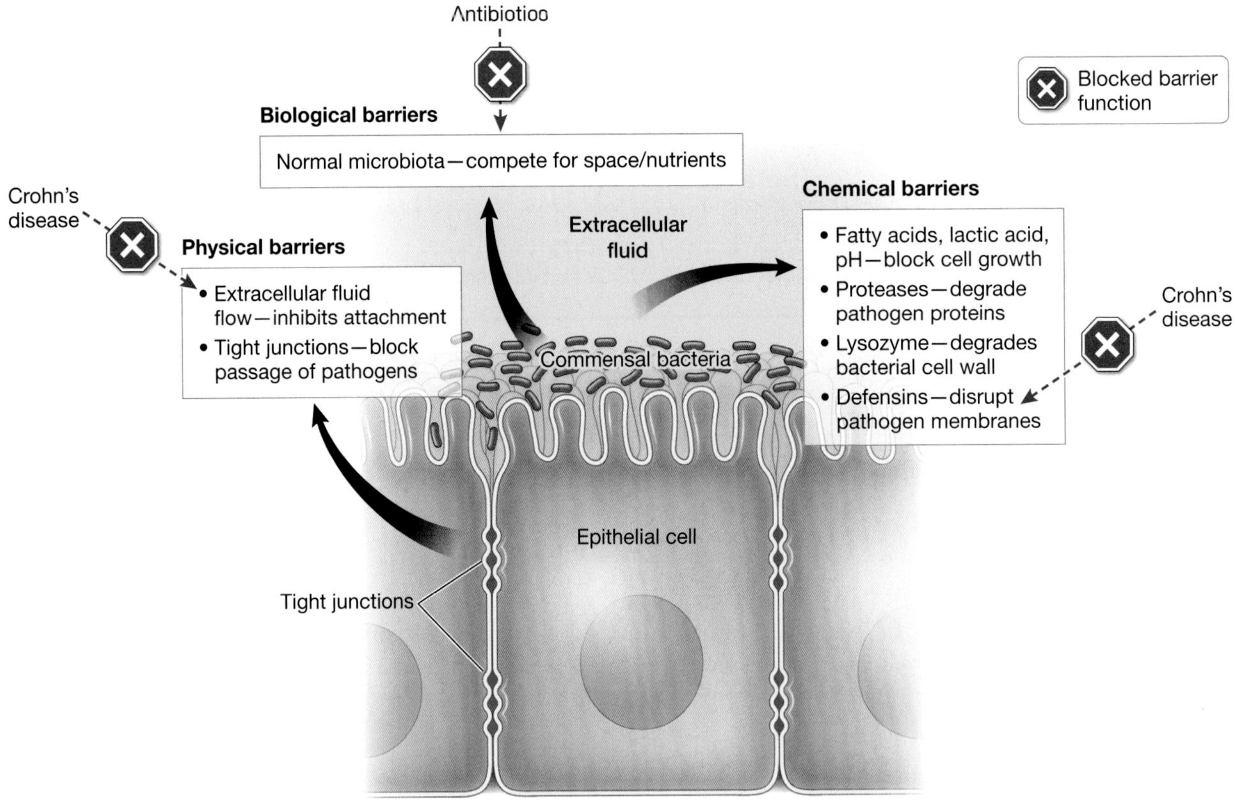

FIGURE 14.3 Immunodeficiencies that affect physical and chemical barriers The action of epithelial cells at mucosal surfaces works to fight infection. Conditions such as Crohn's disease or taking antibiotics impact several of the body's physical and chemical barriers, including extracellular fluid flow, the normal microbiota associated with mucosal tissue, and defensin action.

Crohn's disease Chronic inflammatory bowel disease.

as defensins enhance the protective nature of these barriers. Thus, defects in our protective barriers, in molecules associated with pathogen recognition in epithelial cells, or in peptides and enzymes that enhance pathogen-fighting activities at the surfaces can lead to increased susceptibility to infections (**FIGURE 14.3**).

One disease associated with inflammation and malfunction of mucosal surfaces is **Crohn's disease**. Recent evidence suggests that Crohn's disease has many similarities with immunodeficiencies. In Chapter 12, we saw that a defect in NOD, a protein responsible for detecting pathogen components within epithelial cells, could lead to Crohn's disease and cause chronic inflammation and infection within the small intestine. Defects in the defensin family of proteins have also been associated with the onset of Crohn's disease.

Immunodeficiencies of the complement system

We know that the complement system is an important arm of the innate immune system, utilizing plasma proteins to activate pathogen opsonization through cleavage of the complement protein C3. In order to activate cleavage of C3, the complement system relies on several different mechanisms that produce a different C3 convertase (**FIGURE 14.4**). The activation of C3 cleavage is tightly controlled to ensure that there is plenty of C3 present in plasma when needed and to prevent improper C3 fixation on non-foreign cells (**TABLE 14.2**).

Improper activation or regulation of C3 cleavage can lead to immunodeficiencies that render individuals highly susceptible to infection with pathogens that are normally cleared by a functioning complement system, including

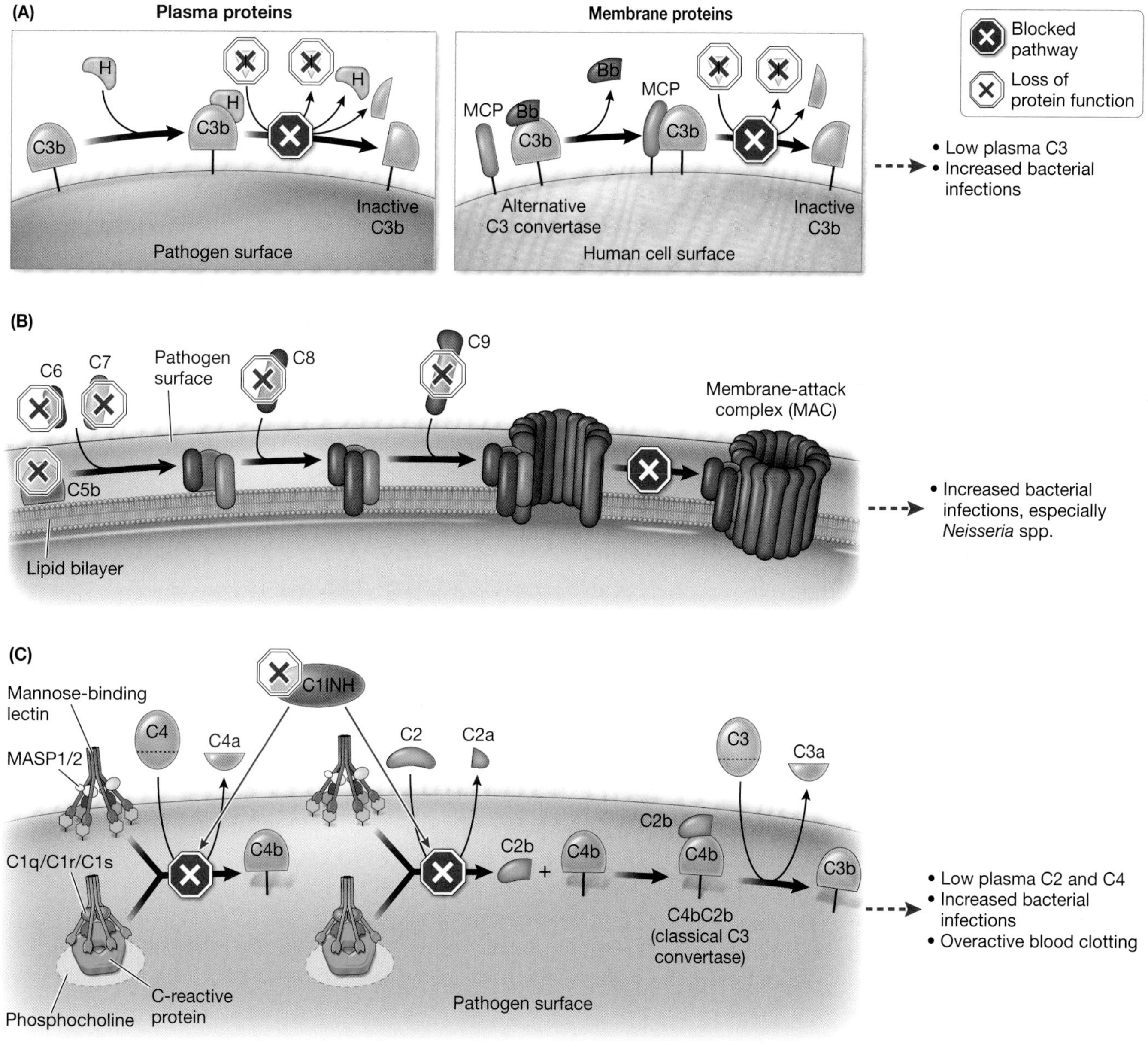

FIGURE 14.4 Immunodeficiencies of complement fixation (A) Loss of factor I, which inhibits complement fixation, leads to nonspecific cleavage of C3 in plasma and increased fixation of C3 on the surface of both pathogens and human cells. This decreases the amount of plasma C3 available for fixation on pathogen surfaces. The result is increased susceptibility to bacterial infections. (B) Loss of any of the proteins involved in formation of the membrane-attack complex results in the inability to perforate pathogen membranes. This leads to increased bacterial infections, especially with bacteria such as *Neisseria* that are sensitive to action of the membrane-attack complex. (C) In hereditary angioneurotic edema (HANE), loss of C1INH, an inhibitor of the C1 protease, leads to nonspecific cleavage of C2 and C4 in plasma and a low level of these two complement proteins. This results in an inability to fix complement via the lectin and classical pathways and increased bacterial infections. Since C1INH also acts as an inhibitor of blood-clotting proteases, loss of C1INH increases nonspecific vasodilation, swelling, and edema. (A,B after P. Parham. 2014. *The Immune System*, 4th ed., W. W. Norton and Company.)

encapsulated bacteria and *Neisseria* species of bacteria. These immunodeficiencies can either prevent proper C3 cleavage or cause depletion in C3 due to inefficient action of naturally occurring inhibitors that work to limit C3 activation when an infection is not present. For example, in Chapter 3, we learned about the regulatory function of factor I in preventing nonspecific activation of C3. Individuals lacking functional factor I have an extremely low plasma level of C3 and are incapable of efficiently fixing C3b on the surface of pathogens.

TABLE 14.2 | Proteins Involved in Complement Activation

Complement Protein Mutated	Symptoms
C1, C2, C4	Improper clearance of immune complexes
C3	Recurring infection with encapsulated bacteria
C5-C9 (membrane-attack complex proteins)	Susceptibility to *Neisseria*
Factor D, Factor P	Recurring infection with encapsulated bacteria
Factor I	Recurring infection with encapsulated bacteria
DAF, CD59	Paroxysmal nocturnal hemoglobinuria
C1INH	Hereditary angioneurotic edema

Source: D. J. Unsworth. 2008. *J Clin Pathol* 61: 1013–1017.

In addition to C3 activation, the complement system utilizes the action of complement proteins in the formation of a membrane-attack complex. Complement proteins C5, C6, C7, C8, and C9 coordinate to form pores in pathogen membranes and threaten pathogen survival. Bacterial pathogens, especially *Neisseria* species, are susceptible to action of the membrane-attack complex. Disrupting the action of any of these complement proteins can lead to a higher risk of infection with *Neisseria*.

One player in the complement system that has not yet been discussed is the regulatory protein C1 inhibitor (C1INH). This inhibitor blocks the activities of C1 proteases involved in the classical pathway of complement activation by covalently binding to proteases and inactivating the proteolytic activity of these enzymes. In the immunodeficiency disease **hereditary angioneurotic edema** (**HANE**), individuals have low plasma levels of both C2 and C4 and overproduction of the C2 fragment C2a due to overactive C1 proteases.

Along with low levels of C2 and C4, individuals suffering from HANE also have overactive blood clotting pathways. C1INH is capable of regulating proteases involved in clotting pathways due to the similarities blood clotting proteases have with C1 proteases of the complement pathways. Bradykinin is a protein of the blood clotting cascade that is activated to a high degree in individuals suffering from HANE. Like C2a, bradykinin is a potent vasodilator. The swelling and edema seen in individuals with HANE is caused by overproduction of C2a and overactive bradykinin. This condition can be fatal if the swelling results in airway constriction.

Immunodeficiencies in phagocyte function

We know that phagocytic cells play an important role in the innate immune response. These cells migrate to the site of infection and engulf pathogens they find there. Through the action of NADPH oxidase and the respiratory burst driven by neutrophil granules, along with the action of lysosomal proteins, phagocytosed pathogens are efficiently destroyed, limiting symptoms and the spread of infection.

Since phagocytes play an important role in the innate immune response, developmental defects or lack of these cells can lead to recurring bacterial infections (**FIGURE 14.5**). **Neutropenia** is an abnormally low level of neutrophils in the blood. Two forms of neutropenia caused by inherited immunodeficiency are cyclic and congenital neutropenia. Cyclic neutropenia results in changing levels of neutrophils in circulation, from normal to low and back to normal again due to fluctuations in neutrophil production in the bone marrow. Congenital neutropenia results in a lack of neutrophils because these cells fail to develop properly during hematopoiesis. The onset of both cyclic and congenital neutropenia is tied to mutation in the neutrophil elastase gene, although the means by which neutrophil elastase malfunction leads to improper neutrophil development is still unclear.

hereditary angioneurotic edema (HANE) Disease caused by the lack of production of the protein C1 inhibitor (C1INH), causing the overactivity of blood-clotting proteases and buildup of fluids in areas where blood-clotting activation occurs.

neutropenia Abnormally low levels of neutrophils in the blood.

LEUKOCYTE ADHESION DEFICIENCY Improper migration or function of professional phagocytes can prevent proper pathogen clearance, resulting in recurring or chronic infection. Phagocytic cells such as neutrophils must be able to efficiently exit the circulatory system at sites of infection to target pathogens for destruction. This migration is driven by adhesion of the leukocytes to

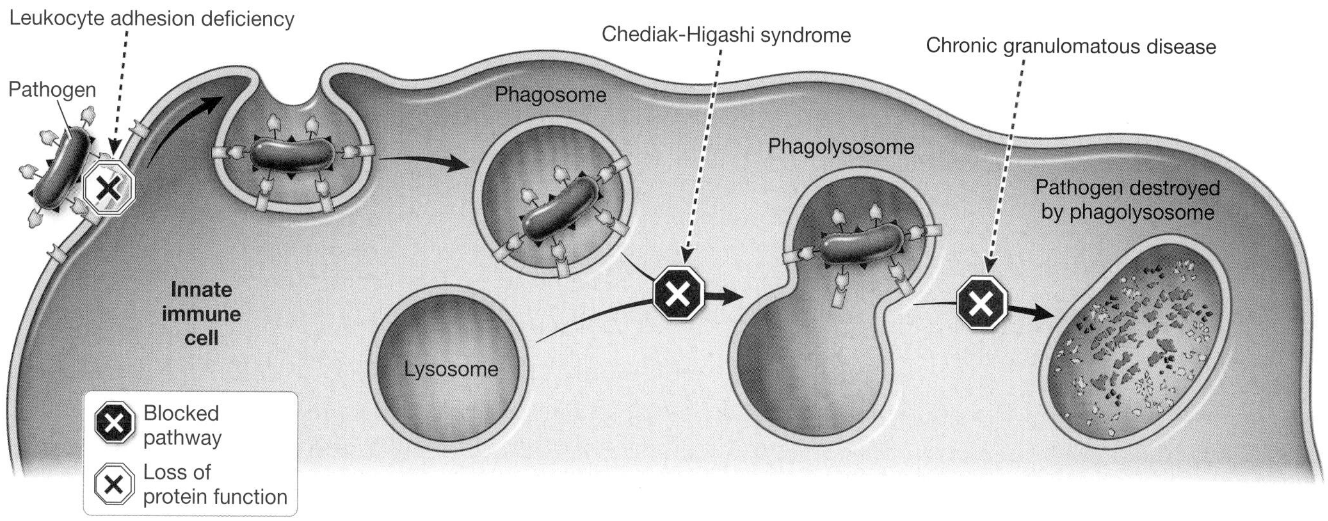

FIGURE 14.5 Immunodeficiencies of phagocyte function The loss of important steps of phagocytosis leads to an increased susceptibility to infection by extracellular pathogens. Leukocyte adhesion deficiency is an inability to properly bind to pathogens to initiate phagocytosis. This occurs due to a lack of cell-surface adhesion molecules on the phagocyte. Chediak-Higashi syndrome is caused by phagosomes' inability to fuse with lysosomes to digest phagocytosed material. Chronic granulomatous disease is caused by phagocytes' inability to destroy pathogens within the phagolysosome. This is seen more commonly when neutrophils' granule components do not work properly to facilitate phagolysosome action.

endothelial cells adjacent to the site of infection, which typically have altered surfaces due to the inflammatory response driven by the infection.

Tight binding of phagocytic cells to endothelial cell surfaces is driven by the leukocyte receptors LFA-1, CR3, and CR4. In one form of **leukocyte adhesion deficiency**, a common subunit of these receptors is mutated, namely the β-2 integrin subunit (also known as CD18), preventing phagocytes from efficient migration to sites of infection. Furthermore, since the subunit is important in CR3 and CR4 activity, these phagocytes are unable to recognize fixed C3b on the pathogen surface, preventing clearance by phagocytosis. Thus, individuals suffering from this condition experience recurring infections with extracellular pathogens due to inefficient migration to and recognition of these pathogens, even if they have been properly opsonized with complement.

CHEDIAK-HIGASHI SYNDROME In addition to being able to efficiently migrate to and engulf microorganisms, phagocytic cells must also efficiently process the engulfed material. They do this by promoting fusion of the phagosome with the lysosome and digesting the engulfed material using lysosomal enzymes or destroying the engulfed material through the action of enzymes within granules (as is the case with neutrophils). Since these mechanisms are crucial to the roles phagocytic cells play in the innate immune system, abnormalities in phagocytosis are certain to inhibit the innate immune system's capacity to clear pathogens. Problems with normal phagolysosome formation and function can be seen in **Chediak-Higashi syndrome**, a rare autosomal recessive disorder driven by a mutation in the *LYST* gene, whose protein product is responsible for proper vesicular transport and formation of a phagolysosome. Individuals with Chediak-Higashi syndrome are prone to bacterial infections, including those caused by *Staphylococcus aureus* and *Streptococcus* species.

CHRONIC GRANULOMATOUS DISEASE **Chronic granulomatous disease** is another example in which improper phagocytic cell function manifests as

leukocyte adhesion deficiency Immunodeficiency caused by a mutation in a phagocyte cell-surface molecule, preventing migration of phagocytes to a site of infection and preventing proper recognition of pathogens opsonized with C3b by phagocytic cells.

Chediak-Higashi syndrome Immunodeficiency that prevents proper phagosome–lysosome fusion, and thus prevents the action of phagocytic cells in the innate immune system.

chronic granulomatous disease Immunodeficiency that prevents the proper action of neutrophils in the innate immune system, causing persistent infection and the formation of granulomas due to inefficient clearance of neutrophils by macrophages.

an immunodeficiency. This condition is caused by mutation in genes responsible for the function of NADPH oxidase. Since this enzyme is responsible for initiating the respiratory burst mechanism, malfunction of NADPH oxidase results in inefficient pathogen clearance and the formation of granulomas (masses of immune cells that form due to chronic inflammation). Within the granuloma, macrophages attempt to rid neutrophils that have not efficiently cleared the infection. Individuals with chronic granulomatous disease commonly have recurring infections that include pneumonia, skin abscesses, and bacterial and fungal infections caused by organisms such as *S. aureus*, *Listeria* species, and *Candida* species.

Immunodeficiencies in natural killer cell function

We have seen that natural killer (NK) cells are important mediators of the innate immune system and act to combat intracellular infections, especially viral infections. By monitoring cell surfaces, NK cells detect differences at the cell surface using activating and inhibitory receptors typically induced when a virus attempts to hide from other arms of the immune system, including detection by cytotoxic T cells (**FIGURE 14.6**).

Some viruses may reduce MHC class I presentation to lower cytoplasmic epitopes displayed on the surface of the infected cell; this also lowers the amount of viral epitopes displayed. Through the action of their inhibitory and activating receptors, NK cells monitor cell surfaces for discrepancies in MHC class I levels as an indirect means of detecting a virus-infected cell. NK cells are especially suited to target herpesvirus-infected cells, which use mechanisms to downregulate MHC class I molecules at the cell surface.

Because NK cells play a vital role in fighting viral infections, impaired function or lack of NK cells can result in increased susceptibility to these pathogens. NK cell deficiency occurs in two forms: classical and functional.

classical NK cell deficiency Immunodeficiency that prevents the proper development of natural killer cells.

Classical NK cell deficiency is characterized by a lack of NK cells in the affected individual. This disease is extremely rare; it has been described

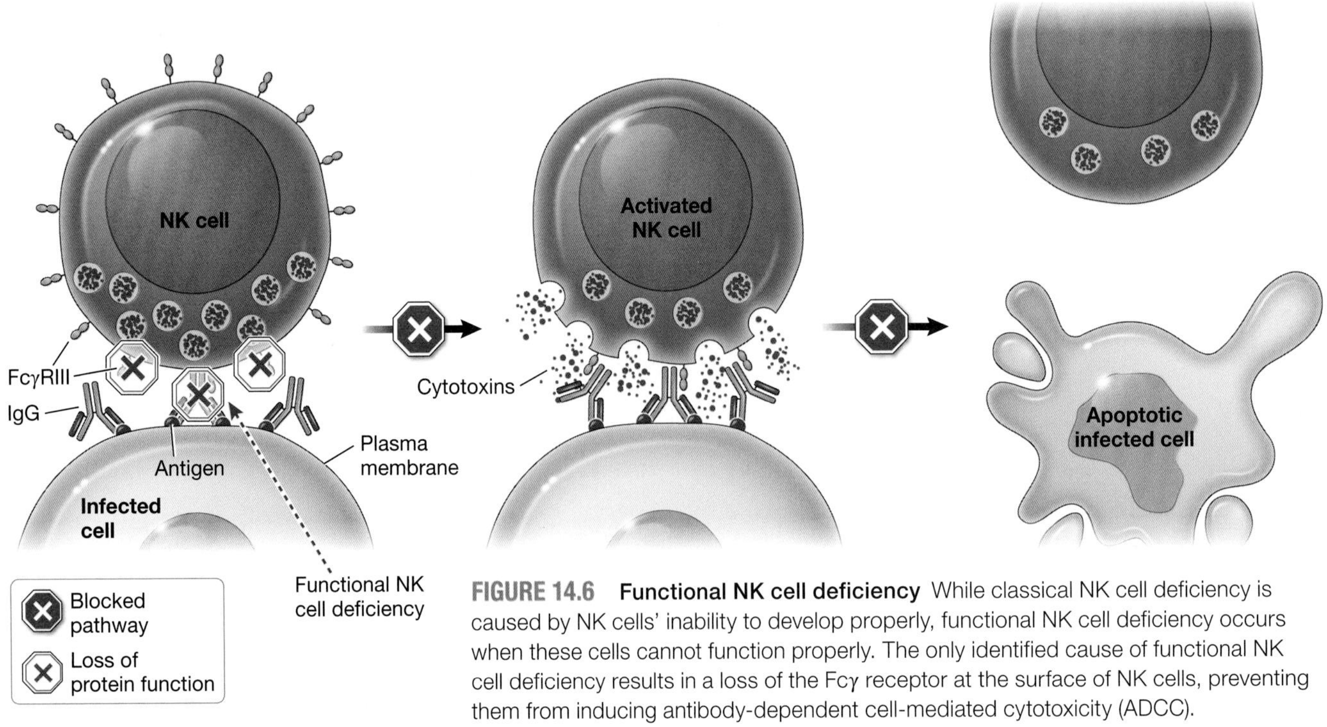

FIGURE 14.6 Functional NK cell deficiency While classical NK cell deficiency is caused by NK cells' inability to develop properly, functional NK cell deficiency occurs when these cells cannot function properly. The only identified cause of functional NK cell deficiency results in a loss of the Fcγ receptor at the surface of NK cells, preventing them from inducing antibody-dependent cell-mediated cytotoxicity (ADCC).

in fewer than 30 individuals. Two causes have been identified. The first is a haploinsufficiency (presence of one functional gene that cannot provide enough protein product for normal function) in the *GATA2* gene. This gene encodes a hematopoietic transcription factor responsible for proper hematopoiesis, so it is easy to see why improper *GATA2* activity would prevent normal production of NK cells.

The second mutation is an autosomal recessive mutation in the *MCM4* gene, which is an important player in DNA replication. The current hypotheses regarding the mechanism of NK cell deficiency in individuals who suffer from this *MCM4* mutation are either that NK cell development is specifically altered or that NK cell survival is decreased (both hypotheses are possible).

Functional NK cell deficiency is caused by improper NK cell activity. The only mutation that has been identified to date as a cause of functional NK deficiency is a mutation in the gene responsible for encoding for CD16. CD16 is an Fcγ receptor on the surface of NK cells responsible for recognizing IgG antibody that is bound to a cell through antigen recognition. Recognition of this antibody by CD16 induces antibody-dependent cell-mediated cytotoxicity (ADCC), resulting in destruction of the tagged cell. Without the presence of this Fc receptor, NK cells cannot function properly in ADCC, and infected cells are not cleared by the innate immune system.

functional NK cell deficiency Immunodeficiency that prevents the proper activity of natural killer cells and results in persistent intracellular (typically viral) infection.

severe combined immunodeficiency (SCID) Immunodeficiency that specifically affects the adaptive immune system (either B cells, T cells, or both).

Omenn syndrome Combined immunodeficiency that results from mutation in genes that encode proteins in the V(D)J recombinase, resulting in the inability of B cells and T cells to develop properly.

graft-versus-host disease Autoreactive disease induced in individuals who have undergone bone marrow transplantation, where mature T cells and B cells that matured in a donor recognize recipient antigens as foreign.

CHECKPOINT QUESTIONS

1. Why does a defect in a negative regulator of complement activation cause problems in innate immune responses?
2. Compare and contrast classical and functional NK cell deficiency as related to innate immune response immunodeficiencies.

14.4 How does a combined deficiency in lymphocyte development or action lead to disease?

LEARNING OBJECTIVE

14.4.1 Provide examples of inherited immunodeficiencies that alter lymphocyte development or action.

Abnormalities in immune cell development and function can lead to adaptive immune response immunodeficiencies. Some of the most debilitating are known as **severe combined immunodeficiencies (SCIDs)**.

SCIDs linked to T-cell and B-cell development

Recall from Chapter 4 that T cells and B cells develop a diverse repertoire of cell-surface receptors through somatic recombination and V(D)J recombinase. Since the action of V(D)J recombinase is vital for positive selection of T cells and B cells during their development in primary lymphoid tissues, malfunction in lymphocyte-specific components of V(D)J recombinase, including *RAG1* and *RAG2*, leads to impaired lymphocyte development and, thus, immunodeficiency. A loss of both *RAG1* and *RAG2* activity results in an absence of properly developed B cells and T cells and severe combined immunodeficiency.

Partial loss of V(D)J recombinase activity through missense mutations in at least one of the *RAG1* or *RAG2* alleles can result in an immunodeficiency known as **Omenn syndrome**, which can lead to recurrent infections with opportunistic pathogens, or can manifest with a similar phenotype to **graft-versus-host disease.** This autoimmune disease (discussed in more detail in Chapter 17) is

Making Connections: Immunodeficiencies of the Innate Immune System

The innate immune system constitutes our first lines of defense against pathogens. Thus, loss of the ability to produce innate immune cells or proteins leads to infections with pathogens that are normally handled efficiently by the innate immune system.

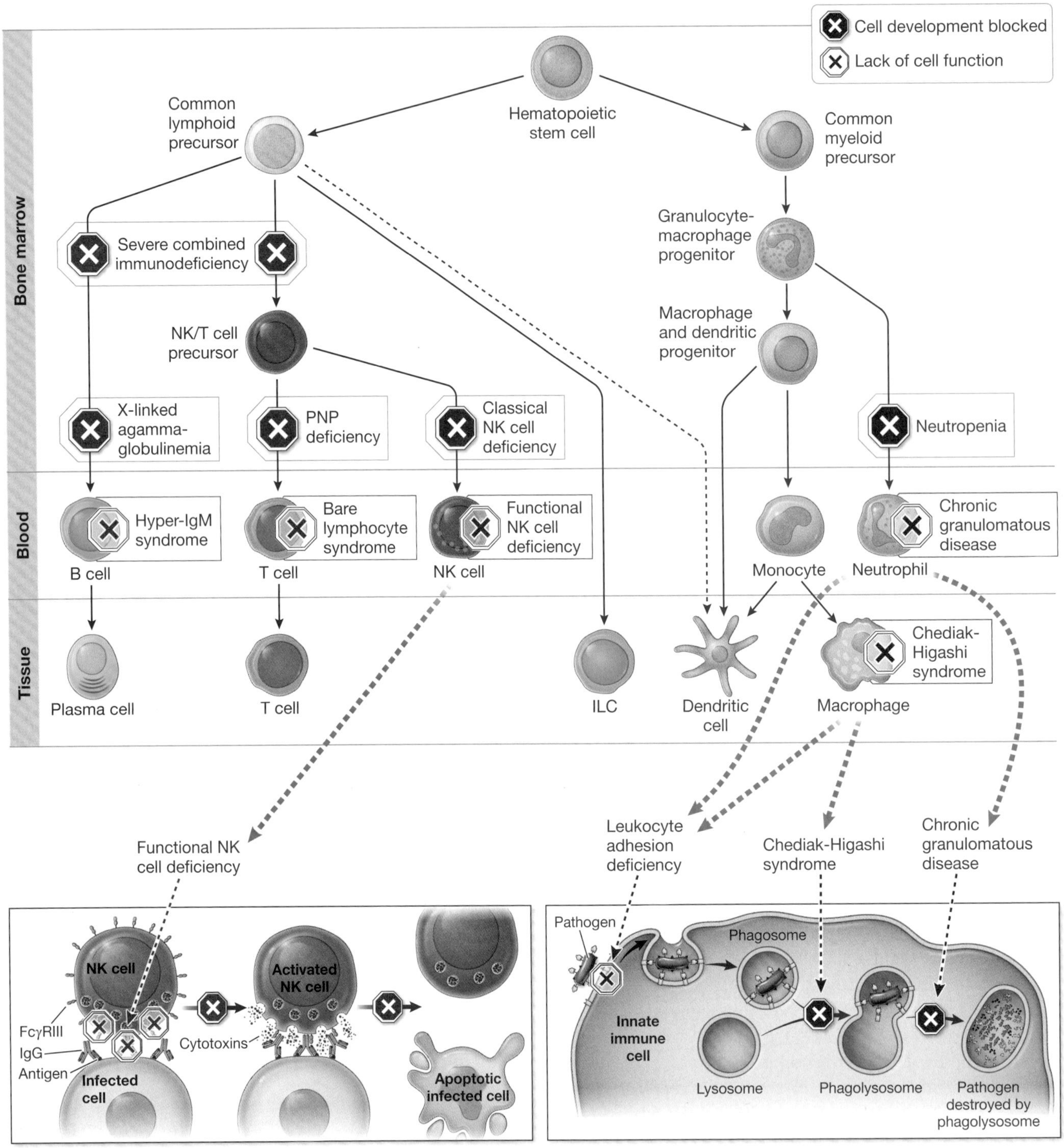

Hematopoiesis produces all cells of the circulatory system, including leukocytes. Disrupting the production of white blood cells results in susceptibility to infection. Innate immune cells function in pathogen recognition and destruction through the action of a variety of cellular proteins, and their loss can inhibit the action of innate immune cells.

(Sections 1.4, 1.5, 2.1, 2.2, 2.5, 2.6, 2.7, 3.1, 3.2, 3.4, 10.5, 14.1, 14.2, 14.3) caused by the presence of self-reactive T cells that attack normal tissues. The autoimmune activity occurs due to a limited production of diverse T cells that results from decreased V(D)J recombinase activity.

Adenosine deaminase deficiency is an autosomal recessive disorder in the gene that encodes the enzyme adenosine deaminase, causing lack of both B cells and T cells. This enzyme is involved in the purine salvage pathway and is responsible for converting adenosine and deoxyadenosine into inosine and deoxyinosine. Loss of adenosine deaminase activity causes an accumulation of dATP in cells. Accumulation of dATP in developing lymphocytes causes inhibition of ribonucleotide reductase, another key cell division enzyme, which works to convert ribonucleotides to deoxyribonucleotides. This inhibition causes lack of production of the monomers required for DNA synthesis, and DNA must be synthesized in rapidly dividing cells such as B cells and T cells produced in primary lymphoid tissue. The result is that lymphocytes cannot complete the S phase of the cell cycle. Disruption in the cell cycle prevents proper lymphocyte production, resulting in a lack of circulating B cells and T cells. Both *RAG1/RAG2* deficiency and adenosine deaminase deficiency cause SCID during infancy and must be treated with bone marrow transplantation.

adenosine deaminase deficiency Immunodeficiency caused by the lack of the enzyme adenosine deaminase, which is required for the salvage pathways of purine nucleotides.

JAK3 deficiency Immunodeficiency in the gene that encodes JAK3, a kinase responsible for proper cytokine signaling, resulting in improper T-cell development.

CHECKPOINT QUESTION

1. Why does disruption of *RAG1* and *RAG2* cause immunodeficiency involving lack of both B cells and T cells?

14.5 | How does a deficiency in T-cell action lead to disease?

LEARNING OBJECTIVE

14.5.1 Provide examples of inherited immunodeficiencies that alter T-cell development or action.

Immunodeficiencies linked to T-cell development

We have seen that T cells carry out a multitude of essential functions in adaptive immunity. As you might expect, a deficiency in T-cell development or function has a profoundly negative impact on the adaptive immune response (**FIGURE 14.7**).

T cells function as initiators of adaptive immune responses (as with specific CD4 helper T-cell subclasses) or as weapons of the cellular arm of the adaptive immune response (as with CD8 cytotoxic T cells). Normal T-cell development occurs due to the productive rearrangement of the T-cell receptor loci and the mechanisms that ensure central tolerance (preventing self-reactive T cells from entering the lymphocyte population). Thus, the cellular arm of adaptive immunity is tightly controlled at the level of development. But what happens when T cells do not develop properly?

MUTATION IN CYTOKINE RECEPTORS AND JAK3 DEFICIENCY One genetic defect that prevents T-cell production is an X-linked recessive mutation in the common gamma chain component of the cytokine receptors that recognize cytokines such as IL-2, IL-4, IL-7, IL-9, and IL-15. This mutation prevents proper signal transduction utilizing these cytokines, resulting in impaired development of T cells within the thymus. Specifically, the mutation prevents activation of the kinase JAK3, which is required for downstream signaling events within developing thymocytes. Therefore, **JAK3 deficiency** prevents proper T-cell development.

FIGURE 14.7 Immunodeficiencies of T-cell development Mutations in T-cell development create a lack of T cells in circulation. Omenn syndrome is caused by a lack of RAG1 and RAG2, the proteins responsible for V(D)J recombination, which prevents T cells from moving past initial developmental checkpoints. Adenosine deaminase (ADA) deficiency and purine nucleoside phosphorylase (PNP) deficiency inhibit DNA synthesis, preventing the cell division required for thymocyte production. (After C. A. Janeway Jr. 2001. *Immunology: The Immune System in Health and Disease*, 5th ed. Garland Science/Norton: New York.)

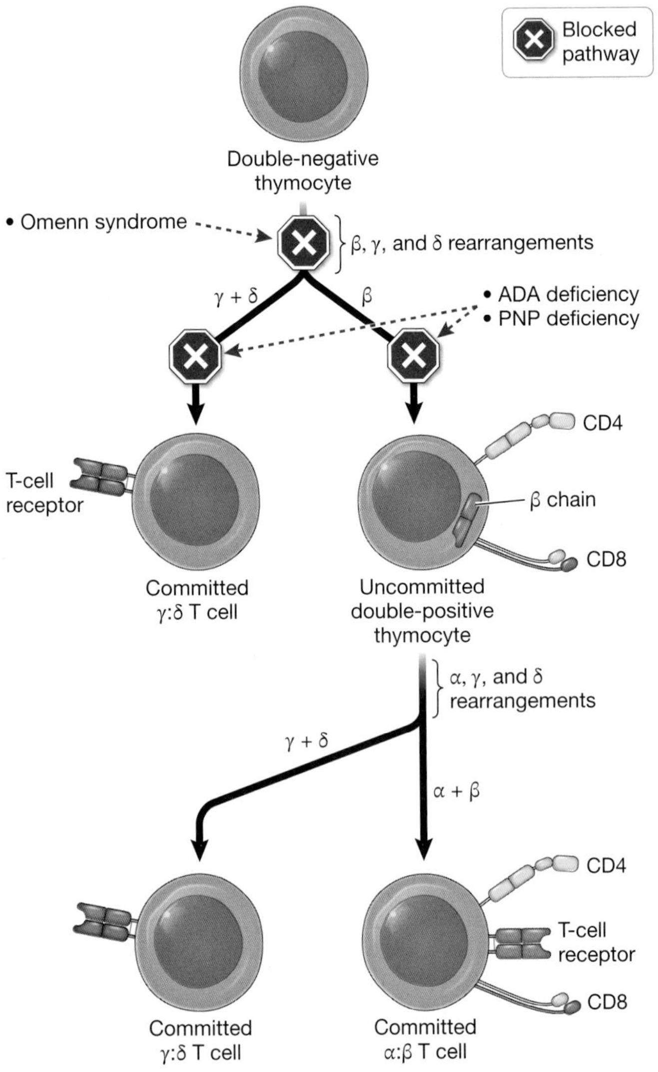

PNP deficiency Immunodeficiency caused by a mutation in the gene encoding purine nucleoside phosphorylase, a protein required for purine salvage pathways.

PNP DEFICIENCY **PNP deficiency** is a deficiency in the enzyme purine nucleoside phosphorylase (PNP), which lowers the number of T cells within the white blood cell population. This enzyme is also involved in the purine salvage pathway (similarly to adenosine deaminase). However, mutation in the PNP gene seems to specifically affect T cells, causing a buildup of dGTP and leading to apoptosis, limiting their numbers in the organism.

DIGEORGE SYNDROME We have explored the importance of the thymus as a primary lymphoid tissue in T-cell development. Absent or impaired thymus function prevents the proper development of T cells, resulting in loss of that arm of the adaptive immune response. DiGeorge syndrome (also called *22q11.2 deletion syndrome*) is caused by the absence of a small portion of chromosome 22. Individuals with this disease often have heart abnormalities, developmental delays, and learning problems. In some patients, the disease also negatively impacts thymus development and function. Individuals with improper thymus development have lower numbers of T cells or a lack of T cells, leading to increased bacterial, viral, and fungal infections. The absence of proper T-cell development (which includes negative selection of self-reactive T cells) can also result in autoimmune conditions.

Immunodeficiencies linked to T-cell function

T-cell activation requires the T-cell receptor to recognize a peptide presented on the surface of a cell via MHC molecules. Thus, if proper MHC-peptide presentation doesn't occur, the T cell cannot be activated (**FIGURE 14.8**).

bare lymphocyte syndrome (BLS) Immunodeficiency caused by the inability of cells to properly load MHC complexes, thus preventing proper presentation and activation of T cells.

BARE LYMPHOCYTE SYNDROME Deficiencies in antigen presentation can lead to **bare lymphocyte syndrome** (**BLS**). One type of BLS, caused by mutations in TAP genes, prevents proper activation of CD8 T cells, which are activated by MHC class I-peptide complexes. Recall that TAP works to transport peptide fragments generated by the proteasome across the endoplasmic reticulum (ER) membrane, where they can be loaded onto MHC class I molecules via a peptide-loading complex. If TAP is not functioning properly, then MHC class I molecules cannot be loaded with proteasomal peptides and cannot translocate to the cell surface.

Another type of BLS prevents proper activation of CD4 T cells, which are activated by MHC class II-peptide complexes. A mutation in transcription factors responsible for the expression of MHC class II molecules prevents expression of these molecules, so antigen-presenting cells capable of phagocytosing extracellular pathogens cannot present pathogen peptides on their surface and thus cannot activate CD4 T cells.

OTHER MUTATIONS Recognition of an MHC-peptide complex via a T-cell receptor induces signaling events that activate the engaged T cell. The action of key signaling molecules, including components of the coreceptor such as the CD3 complex and the protein kinase ZAP-70, is required for T-cell activation and clonal expansion. Mutation in ZAP-70 or the CD3 complex prevents proper T-cell activation. Since these proteins are also important in T-cell development,

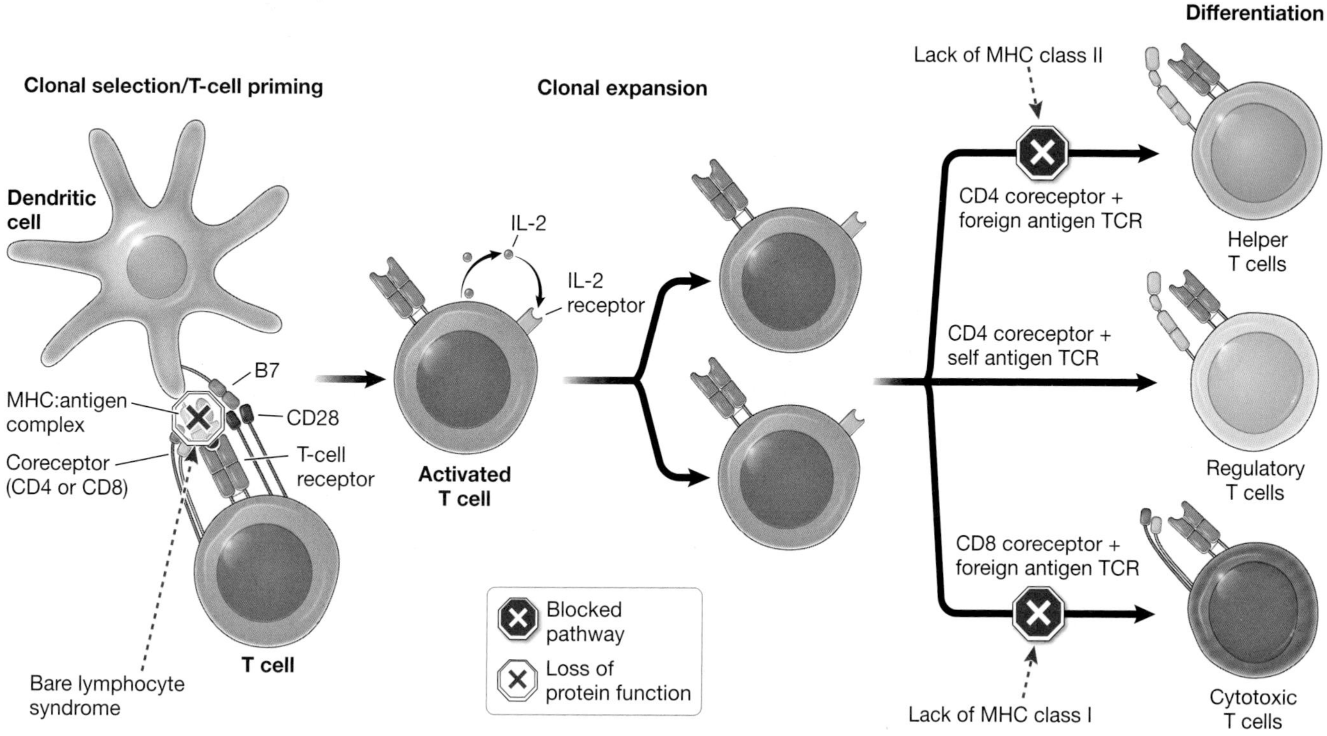

FIGURE 14.8 Bare lymphocyte syndrome Lack of MHC molecules at the surface of antigen-presenting cells results in bare lymphocyte syndrome, which prevents proper activation of subsets of T cells due to an inability to properly signal through the T-cell receptor (TCR). Loss of MHC class I molecules prevents proper activation of CD8 T cells, and loss of MHC class II molecules prevents proper activation of CD4 T cells.

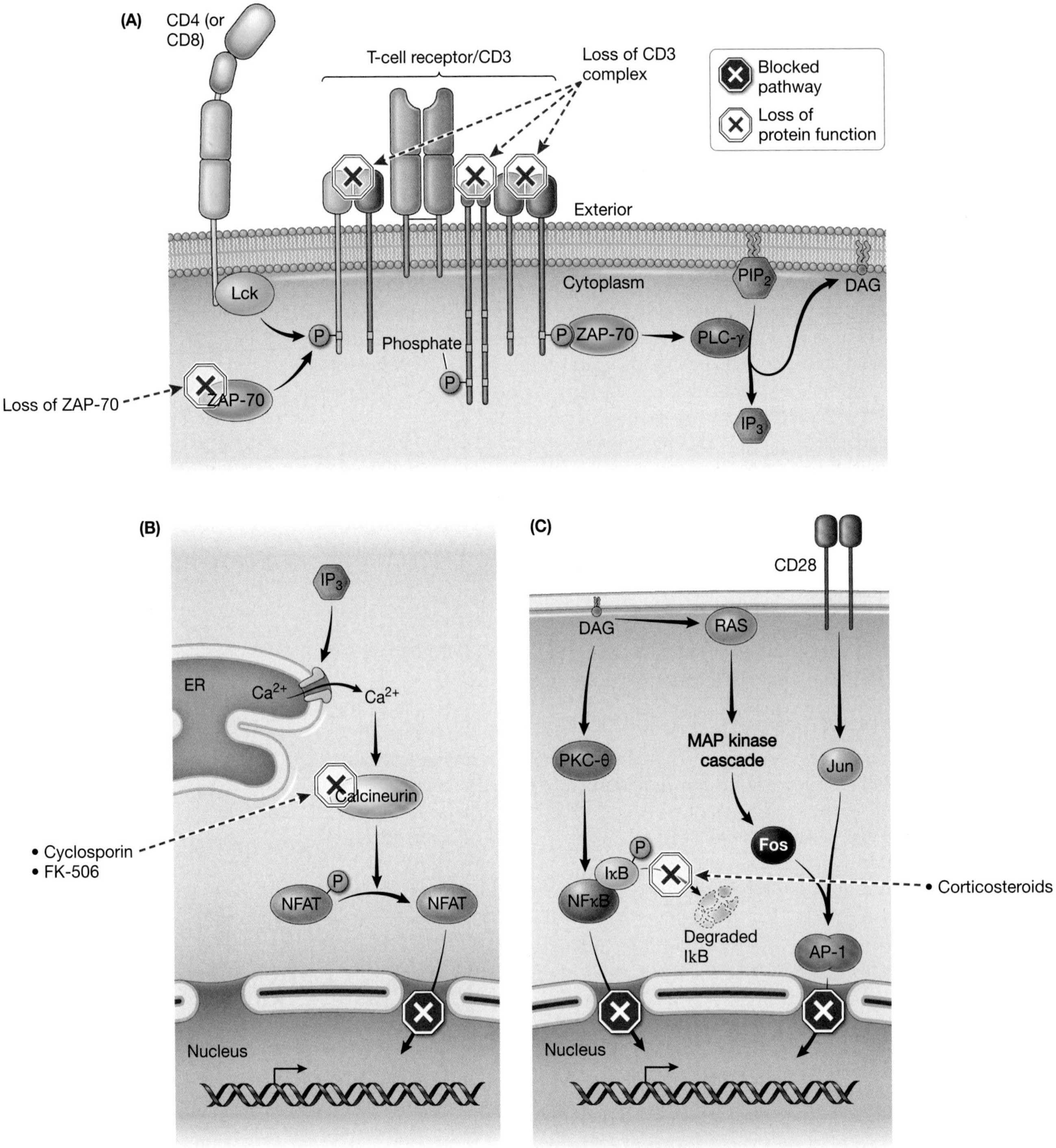

FIGURE 14.9 Immunodeficiencies of T-cell activation (A) Loss of key signaling molecules such as ZAP-70 or proteins of the CD3 complex prevent intracellular transduction of T-cell-receptor engagement. Loss of these signaling events prevents activation of the transcription factors required for T-cell activation and clonal expansion. (B) Immunosuppressants such as cyclosporin and FK-506 prevent activation of the transcription factor NFAT and proper T-cell activation. (C) Immunosuppressants such as corticosteroids increase the expression of IκB and block activation of NFκB. Loss of NFκB activation impairs the ability of T cells to undergo clonal expansion.

mutation in ZAP-70 or the CD3 complex can result in a decreased number of T cells in circulation and can manifest similarly to immunodeficiencies in T-cell development (**FIGURE 14.9**).

● CHECKPOINT QUESTION

1. Provide an example of a mutation that can cause an immunodeficiency due to a lack of proper T-cell development and a lack of proper T-cell signaling.

14.6 How does a deficiency in B-cell action lead to disease?

LEARNING OBJECTIVE

14.6.1 Provide examples of inherited immunodeficiencies that alter B-cell development or action.

Just as some immunodeficiencies specifically affect the function and development of T cells, others can alter the function and development of B cells. In this section, we will see that these disorders mirror the developmental and functional disorders that affect T cells.

Immunodeficiencies linked to B-cell development

Certain immunodeficiencies specifically affect B-cell development, and the best characterized is **X-linked agammaglobulinemia**, an X-linked recessive disease that prevents B-cell survival. Individuals with this condition (primarily males) have undetectable levels of plasma immunoglobulin. This immunodeficiency is caused by mutation on the gene *Btk* on the X chromosome, which encodes for Bruton's tyrosine kinase (BTK), an important factor in providing a survival signal during development. B cells lacking BTK cannot develop past the pre-B-cell stage. Since mutation of this gene prevents proper B-cell development within bone marrow, no immature B cells can leave the bone marrow to undergo normal activation and differentiation into plasma cells, thus eliminating immunoglobulins from plasma.

Immunodeficiencies linked to B-cell function

Some of the more common immunodeficiencies that affect B-cell function block the production of one or several immunoglobulin isotypes. One such immunodeficiency is **hyper-IgM syndrome**, in which affected individuals are capable of producing IgM but have low levels of IgA and IgG. Hyper-IgM syndrome is caused by a mutation in CD40 ligand. Recall that CD40 ligand, expressed by activated T cells, is required for driving centrocytes to undergo somatic hypermutation and isotype switching when it engages CD40 at the B-cell surface. Loss of this signal prevents B cells from receiving signals provided by T cells to activate these two processes, thus preventing the production of isotypes other than IgM (**FIGURE 14.10**).

Another cause of hyper-IgM syndrome is a defect in a component of the kinase IKK. Without functional IKK, the signaling required for isotype switching downstream of CD40 cannot be activated. Recall that IKK activates NFκB by phosphorylating IκB and signaling for its degradation. This frees cytoplasmic NFκB to migrate to the nucleus and function as a transcription factor to activate genes required for isotype switching (see Figure 14.10). Thus, nonfunctional IKK prevents NFκB activation and subsequent activation of isotype switching.

Hyper-IgM syndrome can also be caused by mutations in activation-induced cytidine deaminase (AID) or uracil-DNA glycosylase. Recall from Chapter 9 that AID randomly deaminates cytosine bases at transcriptionally active genes, resulting in the presence of uracil bases within those genes. Uracil-DNA glycosylase can excise the uracil base, and DNA repair enzymes can replace the abasic site (via somatic hypermutation) or can drive recombination events at switch regions (via isotype switching). Since AID and uracil-DNA glycosylase are necessary for proper isotype switching in B cells, loss of either of their activities prevents B cells from producing isotype-switched immunoglobulins.

Certain B-cell immunodeficiencies affect protection of mucosal surfaces. Dimeric IgA at mucosal surfaces works to neutralize pathogens and protect those surfaces from infection. Loss of this important neutralizing antibody at mucosal surfaces, as seen in **selective IgA deficiency**, can lead to recurring infections such as ear infections, bronchitis, and pneumonia. Individuals with

X-linked agammaglobulinemia X-linked immunodeficiency preventing proper B-cell development past the pre-B-cell stage; caused by a mutation in the *Btk* gene, which encodes for Bruton's tyrosine kinase, a protein required for proper B-cell development.

hyper-IgM syndrome Immunodeficiency in B-cell function that prevents proper isotype switching in activated B cells, resulting in the lack of immunoglobulin isotypes except for IgM.

selective IgA deficiency Immunodeficiency that causes an inability to produce IgA immunoglobulins without impacting the production of other immunoglobulin isotypes.

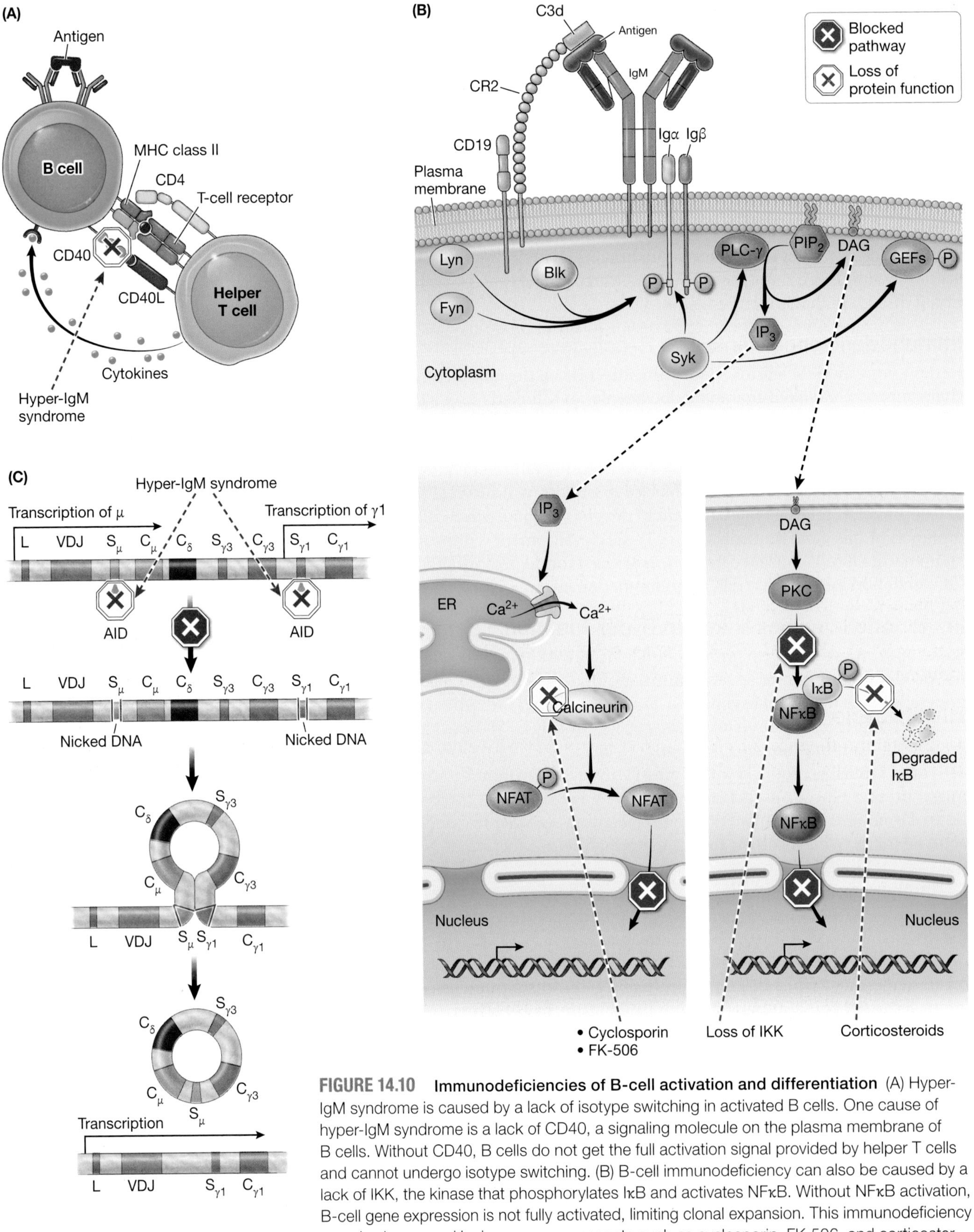

FIGURE 14.10 Immunodeficiencies of B-cell activation and differentiation (A) Hyper-IgM syndrome is caused by a lack of isotype switching in activated B cells. One cause of hyper-IgM syndrome is a lack of CD40, a signaling molecule on the plasma membrane of B cells. Without CD40, B cells do not get the full activation signal provided by helper T cells and cannot undergo isotype switching. (B) B-cell immunodeficiency can also be caused by a lack of IKK, the kinase that phosphorylates IκB and activates NFκB. Without NFκB activation, B-cell gene expression is not fully activated, limiting clonal expansion. This immunodeficiency can also be caused by immunosuppressants such as cyclosporin, FK-506, and corticosteroids, which can block transcription factor activation. (C) Hyper-IgM syndrome can also be caused by the lack of activation-induced cytidine deaminase (AID). Lack of AID prevents conversion of cytosine to uracil bases in switch regions of the immunoglobulin heavy chains, which is required for class switch recombination. (A, after T. W. Mak et al. 2005. *The Immune Response: Basic and Clinical Principles*, p.125. Elsevier: Burlington, MA. C, after F.-L. Meng et al. 2015. In *Molecular Biology of B Cells*, pp. 345–362. Academic Press: London.)

selective IgA deficiency may also suffer from allergic asthma (see Chapter 15) or autoimmune disorders (see Chapter 16). Interestingly, some individuals diagnosed as lacking IgA do not suffer associated symptoms, presumably because their bodies compensate by using other immunoglobulins in place of IgA. At present, this variance in how selective IgA deficiency affects different individuals and its mechanism of onset are not well understood.

CHECKPOINT QUESTIONS

1. X-linked agammaglobulinemia is primarily seen in sons of asymptomatic female carriers of a mutation in *Btk*. Explain why.
2. Would you anticipate that IgM immunoglobulins in an individual suffering from hyper-IgM syndrome would be capable of undergoing affinity maturation? Why or why not?

14.7 | How does HIV cause AIDS?

LEARNING OBJECTIVE

14.7.1 Describe the life cycle of HIV and its role in the onset of AIDS, and define how certain HIV therapies aim to alter the normal HIV life cycle.

Up to this point, this chapter has focused on the effects of genetic mutations that alter innate and adaptive immune responses. Pathogens are also capable of suppressing the immune system in their quest to survive within their host. The best-characterized example of a pathogen that can induce immunodeficiency within a host is the **human immunodeficiency virus** (**HIV**). If untreated, HIV can induce acquired immunodeficiency syndrome (AIDS; see **KEY DISCOVERIES**).

human immunodeficiency virus (HIV) Retrovirus that infects CD4 T cells and macrophages and is capable of causing the disease acquired immunodeficiency syndrome (AIDS).

retrovirus Virus that has an RNA genome that must be reverse transcribed to a DNA copy and incorporated into the target cell's genome, before viral particles can further be produced.

HIV life cycle

To understand the means by which HIV induces AIDS, we must explore the HIV life cycle and the means by which the virus interacts with and shuts down components of the immune system.

HIV is an RNA **retrovirus**, meaning that its genome is packaged in mature viral particles as an RNA molecule (**FIGURE 14.11**). The RNA genome within a

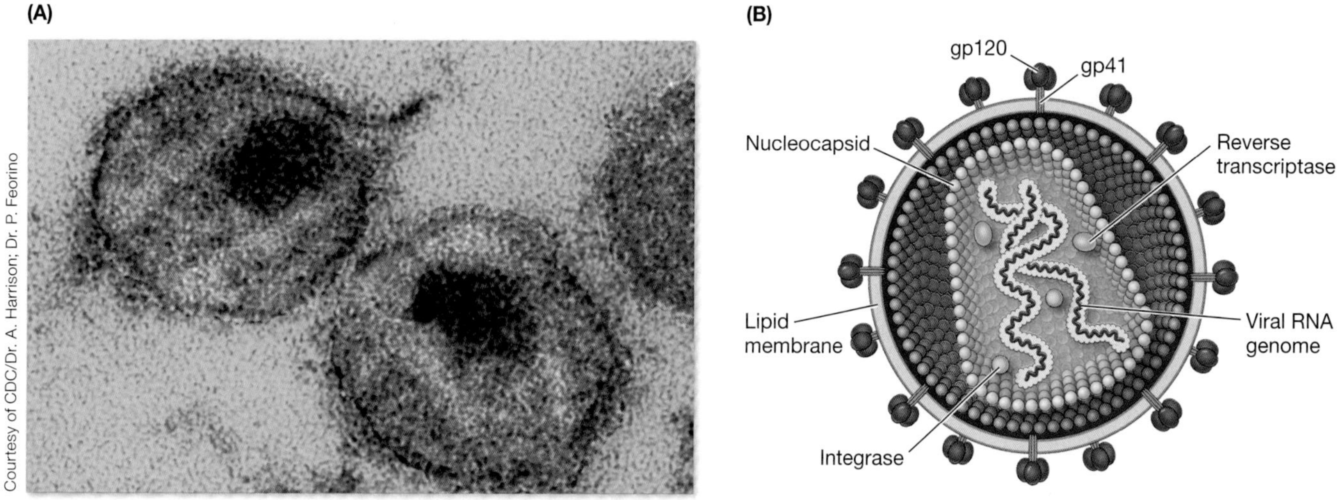

FIGURE 14.11 Human immunodeficiency virus (A) Micrograph of human immunodeficiency virus (HIV), the causative agent of acquired immunodeficiency syndrome (AIDS). (B) HIV is an RNA retrovirus, meaning its genome is present as an RNA molecule. Each viral particle has a nucleocapsid core, which contains the genome along with key proteins required for the HIV life cycle, including reverse transcriptase and integrase. The core is surrounded by a lipid membrane containing two glycoproteins—gp120 and gp41—required for infection in target cells.

Making Connections: Immunodeficiencies of the Adaptive Immune System

The adaptive immune system constitutes the humoral and cellular arms of our specific immune defenses. Loss of the ability to produce lymphocytes leads to severe problems associated with pathogen infection. When lymphocytes lose their ability to function properly, the result is an increased susceptibility to infection.

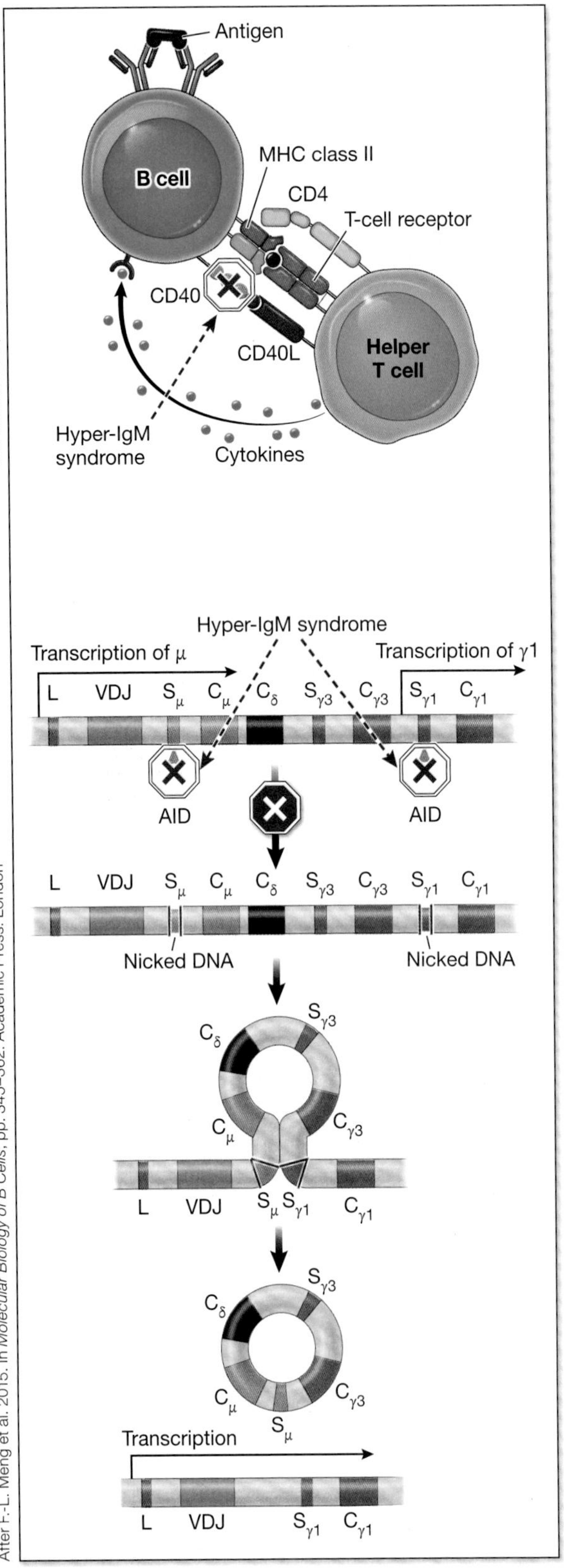

After T. W. Mak et al. 2005. *The Immune Response: Basic and Clinical Principles*, p.125. Elsevier: Burlington, MA

After F.-L. Meng et al. 2015. In *Molecular Biology of B Cells*, pp. 345–362. Academic Press: London

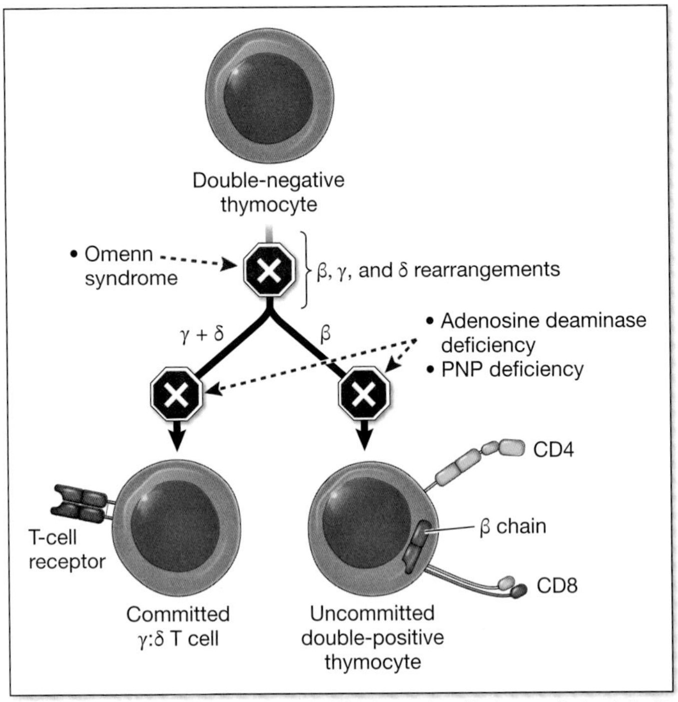

After C. A. Janeway Jr. 2001. *Immunology: The Immune System in Health and Disease*, 5th ed. Garland Science/Norton: New York

Hematopoiesis produces all cells of the circulatory system, including lymphocytes. Disrupting the production of B cells and T cells results in severe combined immunodeficiency.

(Sections 1.5, 4.4, 5.4, 6.6, 7.1, 7.8, 9.4, 10.2, 14.1, 14.2, 14.4, 14.5)

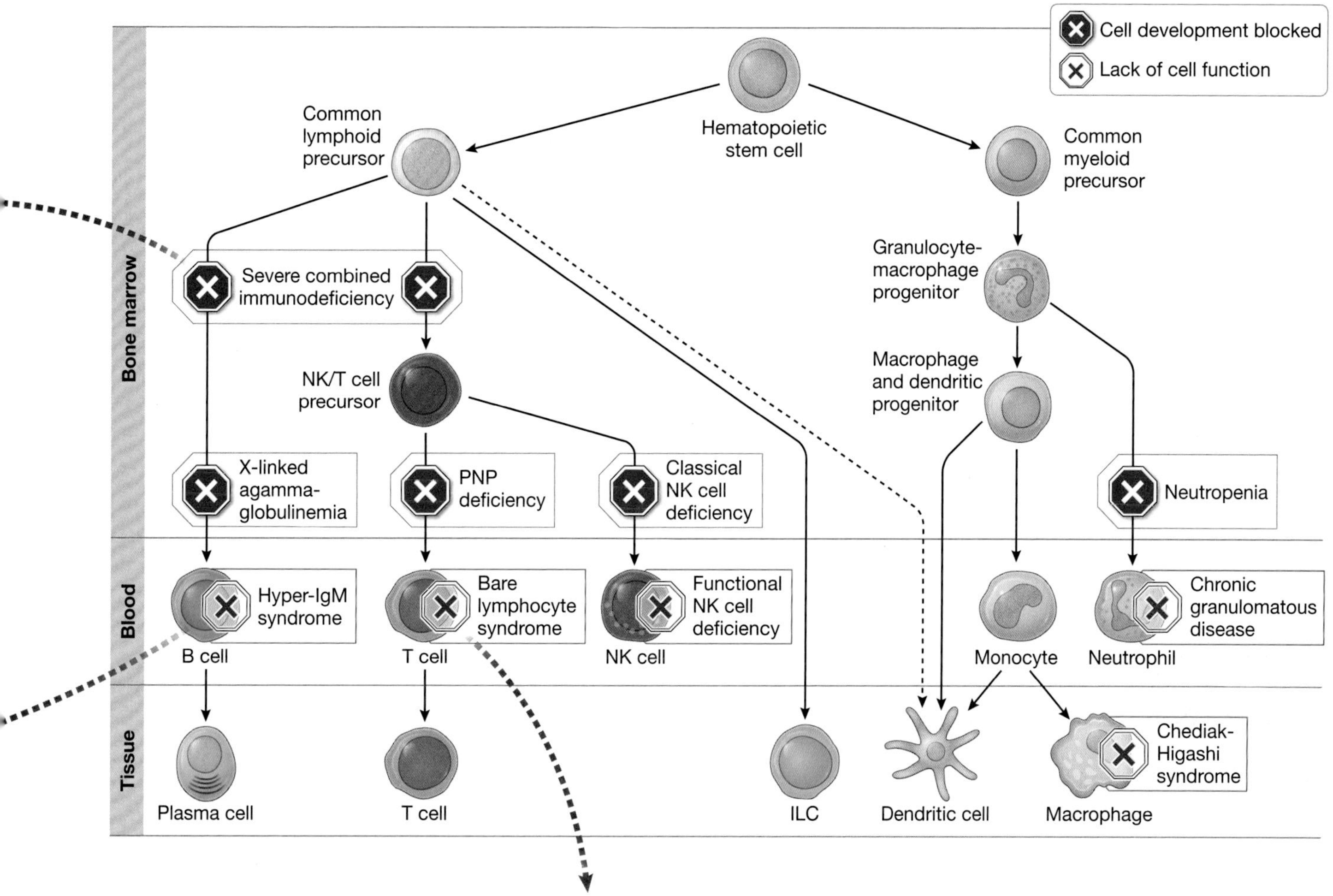

Cell development blocked
Lack of cell function
Bone marrow
Blood
Tissue
Hematopoietic stem cell
Common lymphoid precursor
Common myeloid precursor
Severe combined immunodeficiency
NK/T cell precursor
Granulocyte-macrophage progenitor
Macrophage and dendritic progenitor
X-linked agammaglobulinemia
PNP deficiency
Classical NK cell deficiency
Neutropenia
Hyper-IgM syndrome
Bare lymphocyte syndrome
Functional NK cell deficiency
Chronic granulomatous disease
B cell
T cell
NK cell
Monocyte
Neutrophil
Chediak-Higashi syndrome
Plasma cell
T cell
ILC
Dendritic cell
Macrophage

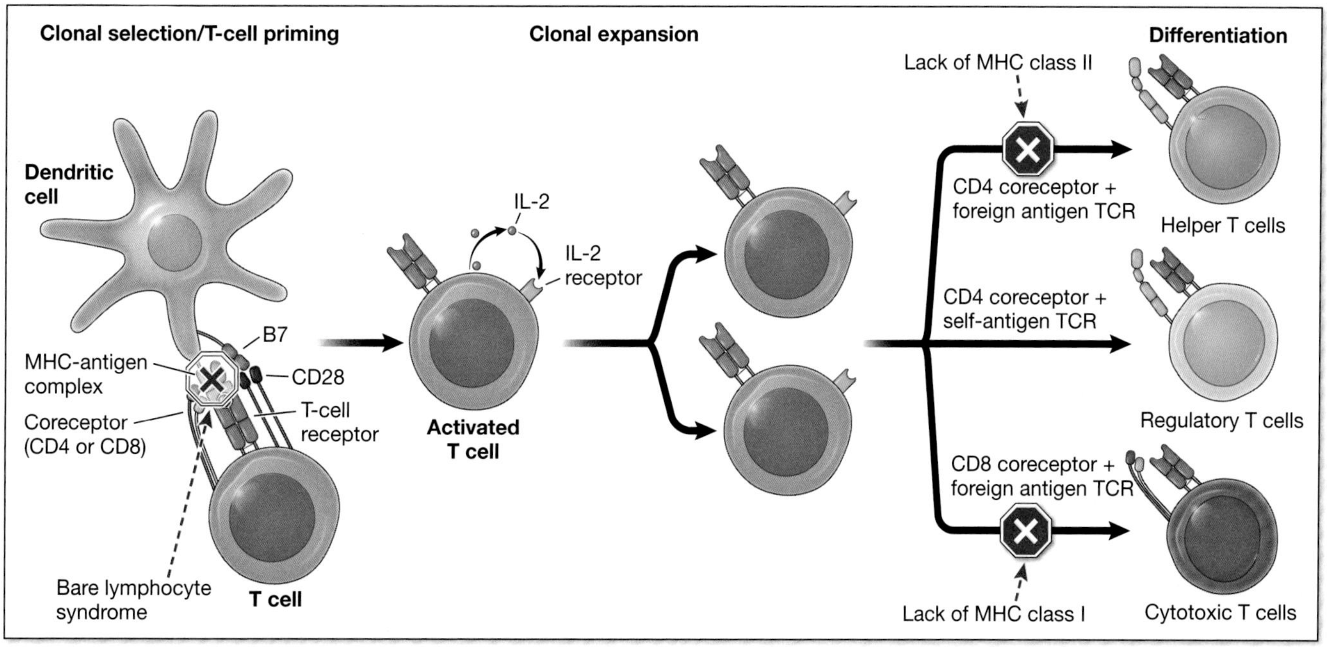

Clonal selection/T-cell priming
Clonal expansion
Differentiation
Dendritic cell
B7
MHC-antigen complex
CD28
Coreceptor (CD4 or CD8)
T-cell receptor
Bare lymphocyte syndrome
T cell
IL-2
IL-2 receptor
Activated T cell
Lack of MHC class II
CD4 coreceptor + foreign antigen TCR
Helper T cells
CD4 coreceptor + self-antigen TCR
Regulatory T cells
CD8 coreceptor + foreign antigen TCR
Cytotoxic T cells
Lack of MHC class I

KEY DISCOVERIES

How was the causative agent of AIDS identified?

Articles

R. C. Gallo, P. S. Sarin, E. P. Gelmann, M. Robert-Guroff, E. Richardson, V. S. Kalyanaraman, D. Mann, G. D. Sidhu, R. E. Stahl, S. Zolla-Pazner, J. Leibowitch, and M. Popovic. 1983. Isolation of human T-cell leukemia virus in acquired immune deficiency syndrome (AIDS). *Science* 220: 865–867.

F. Barre-Sinoussi, J. C. Chermann, F. Rey, M. T. Nugeyre, S. Chamaret, J. Gruest, C. Dauguet, C. Axler-Blin, F. Vezinet-Brun, C. Rouzioux, W. Rozenbaum, and L. Montagnier. 1983. Isolation of T-lymphotropic retrovirus from a patient at risk for acquired immune deficiency syndrome (AIDS). *Science* 220: 868–871.

Background

AIDS had first been clinically observed in the early 1980s in the United States, although the first probable case likely dates back to the 1920s in what is now the Democratic Republic of the Congo. The disease was characterized clinically in individuals who suffered from rare opportunistic infections, including pneumocystis pneumonia and Kaposi sarcoma. The disease had been prevalent in the gay community and was first termed gay-related immune deficiency (GRID) by researchers in the field. The Centers for Disease Control and Prevention (CDC) observed that the disease was seen in hemophiliacs, Haitians, homosexuals, and heroin addicts, and coined the term 4H disease. The CDC soon recognized that the disease was not limited to these populations and in 1982 introduced the term acquired immunodeficiency syndrome (AIDS). Although the disease had been categorized as an acquired immunodeficiency, the question remained: What must be acquired to drive onset of the immunodeficiency?

Early Research

In the early 1980s, while the disease was showing global prevalence and infection was on the rise, the causative agent had yet to be discovered. It was known that individuals affected by AIDS suffered from issues associated with impaired T-cell function and a diminished number of CD4 T cells, suggesting that the causative agent was specifically targeting this cell subpopulation. Other T-cell-specific viruses such as human T-cell lymphotropic virus (HTLV) had been described prior to these studies.

Think About...

1. What would you hypothesize might be a possible causative agent of AIDS? What types of pathogens could potentially cause depletion of CD4 T cells?
2. What experiments could you use to determine if T cells in AIDS patients were infected by your hypothesized pathogen?
3. What experiments could you use to determine if the infectious agent were responsible for the onset of AIDS (assuming you could do similar experiments in a model system or in isolated cell culture)?

Article Summary

Researchers utilized the knowledge that AIDS patients had dysfunction in T-cell population subtypes as the basis for searching for a potential causative agent. With knowledge of the presence of other T-cell viruses, specifically HTLV, they searched for the presence of HTLV markers, including proteins p19 and p24, and found that AIDS patients were positive for the protein or positive for presence of antibodies to these proteins. This finding suggested that an HTLV family member was present in AIDS patients. Viral particles could also be seen in these patients via electron microscopy (**FIGURE KD 14.1**). These results demonstrated the presence of a T-cell-specific virus connected with AIDS, establishing the foundational principles that would drive later HIV research.

reverse transcriptase HIV enzyme responsible for creating a cDNA genome copy using the RNA genome from a viral particle that has just infected a target cell.

HIV protease HIV enzyme responsible for proteolysis of HIV proteins needed for proper assembly and formation of a virion.

viral particle is enclosed in a nucleocapsid with other proteins required for the initial stages of the life cycle, including **reverse transcriptase**, **HIV protease**, and **integrase**. The nucleocapsid is surrounded by a membrane formed upon viral budding from an infected cell. Within the membrane are **gp120** and **gp41**, two viral glycoproteins involved in binding viral particles to target cells and fusing the viral membrane with the target cell membrane. **TABLE 14.3** summarizes the important players in the HIV life cycle.

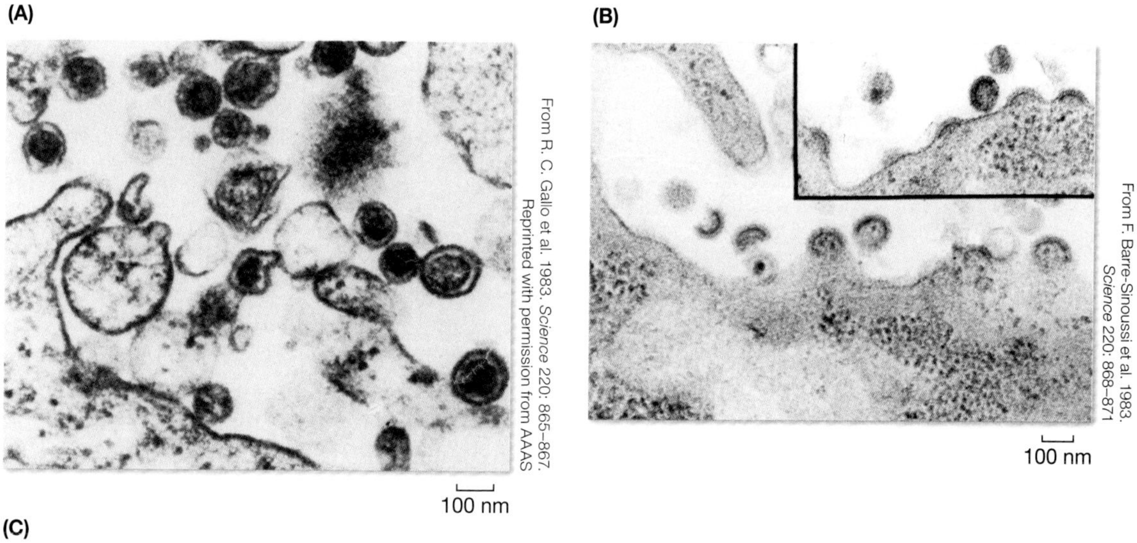

(C)

Characteristics of the cultured T cells	Source of T cells		
	AIDS patient (EP)	Normal cord blood T cells (C183)	Cord blood (C183) T cells infected with $HTLV_{EP}$
HTLV expression			
p24 (ng/mg)	40	0	500
p19 (percent positive cells)	9	0	37
Reverse transcriptase activity (pmole/ml extract)	81	0	21
Virus particles (electron microscopy)	+	N. D.	+

FIGURE KD 14.1 Discovery of HIV. (A) Electron micrograph of T cells from a patient with acquired immunodeficiency syndrome (AIDS). (B) Electron micrograph of cord lymphocytes producing HIV. Isolated human T-cell leukemia virus (HTLV, now known as HIV) is measured by the presence of key proteins, including p24, p19, and reverse transcriptase, in a patient with AIDS. The virus was isolated and used to infect normal cord blood cells to verify that HIV is the transmissible agent of AIDS. (C, from R. C. Gallo et al. 1983. *Science* 220: 865–867. Reprinted with permission from AAAS.)

TABLE 14.3 | Human Immunodeficiency Virus (HIV) Proteins

HIV Protein	Function in Life Cycle
gp120	Recognition of CD4, CCR5, CXCR4 receptors
gp41	Viral fusion with target cell membrane
Reverse transcriptase	Synthesis of cDNA copy of viral RNA genome
Integrase	Integration of viral cDNA genome into target cell genome
Tat	Enhancing transcription of viral RNA
Rev	Inducing synthesis of viral protein translation
Protease	Processing of nucleocapsid proteins for viral packaging

Source: J.A. Levy. 1993. *Microbiol Rev* 57: 183–289.

integrase HIV enzyme responsible for incorporating the cDNA genome of HIV into the target cell genome.

gp120 Glycoprotein of 120 kilodaltons expressed on the surface of the HIV virus; responsible for binding of viral particles to CD4 on CD4 T cells and macrophages.

gp41 Glycoprotein of 41 kilodaltons expressed on the surface of the HIV virus; responsible for efficient viral fusion with the target cell membrane.

The HIV replication cycle, like most viral replication cycles, requires interaction and hijacking of target cell processes for viral production and budding (**FIGURE 14.12**):

1. ***Viral binding and fusion.*** Infectious virus, transmitted through exposure to infected body fluid (via sexual activity or contact with contaminated blood or other body fluids), binds to a target cell. HIV is capable of interacting with CD4 at the surface of CD4 T cells (T-tropic) or macrophages (M-tropic) using its glycoprotein gp120. Efficient virus binding to target cells requires the action of a coreceptor; M-tropic HIV utilizes the coreceptor CCR5, whereas T-tropic strains utilize the coreceptor CXCR4. Some individuals have a deletion mutation in CCR5 (the CCR5-Δ32 mutation) that can render them resistant to HIV infection. The CCR5-Δ32 mutation has been used in stem-cell transplantation in the removal of HIV from an individual suffering from acute myeloid leukemia (see **EMERGING SCIENCE**). Efficient binding of HIV to a target cell engages gp41 to promote fusion of the viral membrane with the target cell membrane and delivery of the genome and nucleocapsid proteins into the target cell cytosol.
2. ***Reverse transcription and cDNA genome synthesis.*** Upon delivery of the genome and nucleocapsid proteins into the target cell cytoplasm, the HIV protein reverse transcriptase produces a double-stranded cDNA copy of the RNA genome. It is important to note that the reverse transcriptase does not have proofreading capabilities as it synthesizes cDNA, causing the reverse transcriptase to be highly mutagenic. The high mutation rate of reverse transcriptase as it synthesizes the cDNA copy of the HIV genome is problematic as researchers and physicians develop strategies to combat the virus and attempt to develop an effective HIV vaccine.
3. ***Genome integration.*** Upon synthesis of the cDNA copy of the viral genome, it is transported to the nucleus of the target cell, where the viral protein integrase incorporates the viral genome into the target cell DNA. Once the cDNA viral genome is inserted into the target cell DNA, the virus remains latent until the infected cell becomes activated during normal immune response events.
4. ***Viral activation and transcription.*** Recall that upon activation of a CD4 T cell during an adaptive immune response, the induction of certain transcription factors, including NFκB, drives clonal expansion of the CD4 T cell. In HIV-infected CD4 T cells, induction of signaling, which activates NFκB, also activates viral transcription. Viral mRNA is produced, spliced, and transported to the cytoplasm, where it is used to encode translation of the viral proteins Tat and Rev. Tat is responsible for further induction of viral RNA transcription and translation of proteins required for viral assembly, whereas Rev assists in the production of full-length viral RNA genomes.
5. ***Viral packaging and budding.*** During viral transcription and translation, proteins that need to be incorporated into the nucleocapsid of new viral particles (including reverse transcriptase and integrase) are synthesized. During this process, other important viral proteins are made, including gp160 and protease. The gp160 glycoprotein is processed in the secretory pathway into the two viral envelope glycoproteins gp120 and gp41, which are then transported to the plasma membrane to be incorporated into the viral membrane upon budding. The protease is responsible for inducing maturation of other viral nucleocapsid proteins. The last step is packaging of the nucleocapsid and budding of the virus from the infected cell. Upon budding, the infectious virus can contact another cell and continue the life cycle.

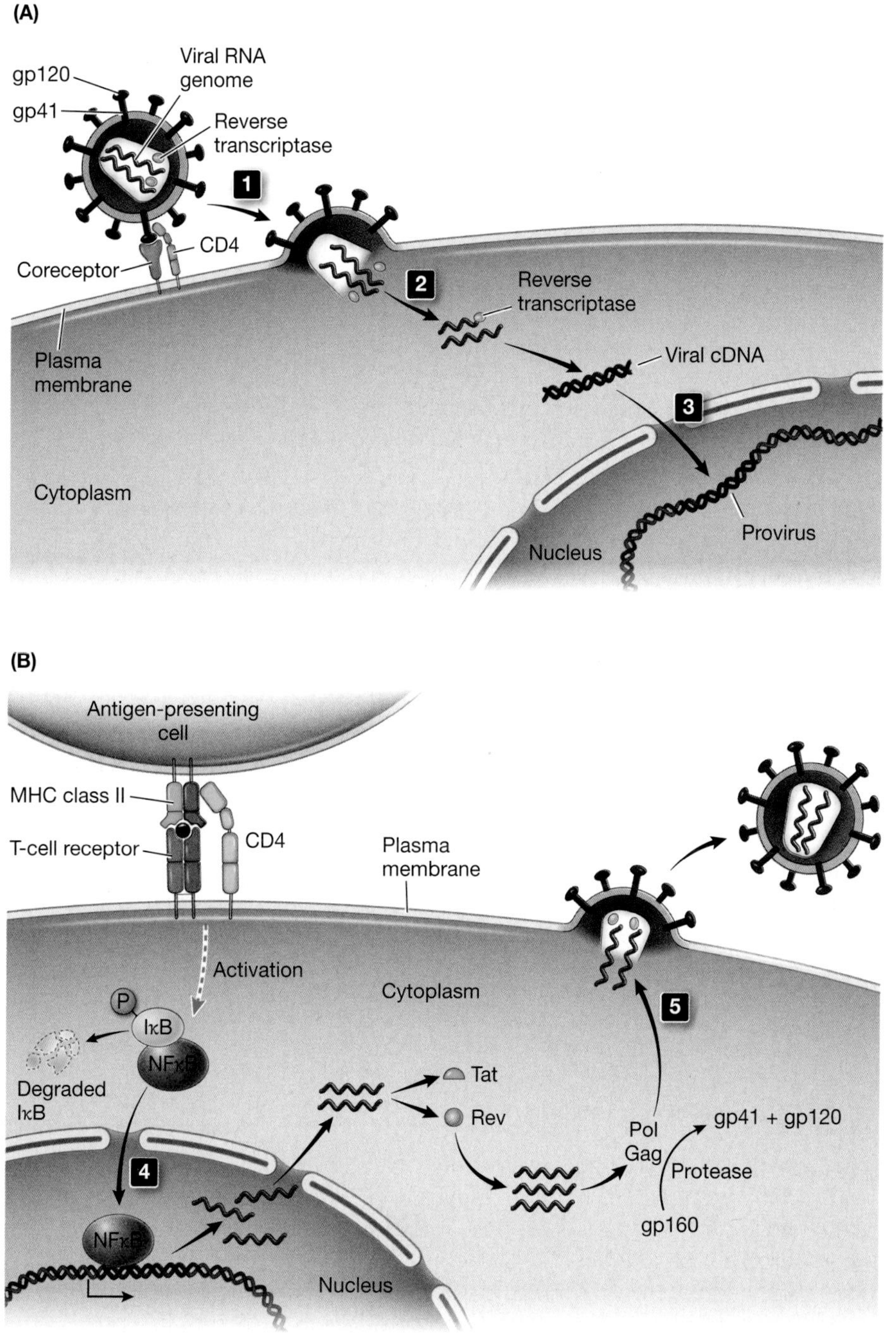

FIGURE 14.12 **HIV life cycle** (A) The HIV life cycle begins when the virus, via the viral coat protein gp120, binds to a target cell such as a T cell via the plasma membrane protein CD4 and a coreceptor. The viral membrane fuses with the target cell membrane, facilitated by the viral coat protein gp41, releasing the nucleocapsid components into the cytoplasm, including the viral genome and reverse transcriptase (Step 1). Reverse transcriptase then makes a cDNA copy of the viral genome (Step 2). The viral cDNA copy is transported to the nucleus and integrated into the host cell genome via the viral protein integrase (Step 3), where the viral genome becomes a latent provirus. (B) Activation of an infected T cell results in activation of the transcription factor NFκB, which initiates transcription of viral RNA (Step 4). Viral RNA serves as a messenger for viral protein translation, beginning with the transcription of two viral proteins, Tat and Rev. Tat increases viral RNA transcription, whereas Rev increases transcription of full-length viral RNA genomes. Viral RNA transcripts are translated into key packaging proteins, including Gag, Pol, and gp160, which is cleaved by the viral protease into gp120 and gp41. Viral proteins and genomes are packaged into a newly formed nucleocapsid and new virus buds from the infected cell (Step 5).

EMERGING SCIENCE

Can HIV be cured through stem-cell transplantation?

Article

G. Hutter, D. Nowak, M. Mossner, S. Ganepola, A. Mubig, K. Allers, T. Schneider, J. Hofmann, C. Kucherer, O. Blau, I. W. Blau, W. K. Hofmann, and E. Thiel. 2009. Long-term control of HIV by *CCR5* delta32/delta32 stem-cell transplantation. *New England Journal of Medicine* 360: 692–698.

Background

As of 2009, it was known that HIV entry into T cells required CD4 binding in combination with the binding of a coreceptor (CCR5 or CXCR4). Furthermore, it had been shown that homozygous deletion of 32 bp within CCR5 confers resistance to HIV infectivity. The most widely accepted treatment for HIV infection at the time was highly active antiretroviral therapy (HAART). The HIV-positive patient in this study, known as the Berlin patient, presented with acute myeloid leukemia. The patient was being treated for HIV infection with HAART and, to treat the leukemia, received an allogeneic stem-cell transplantation from a donor patient carrying the CCR5-Δ32 deletion.

The Study

Upon stem-cell transplantation of the HIV-positive individual with bone marrow from a CCR5-Δ32 donor, the researchers attempted to verify that the recipient seroconverted to have a CCR5-Δ32 phenotype and also studied the HIV viral titer in the patient. While the patient was undergoing HAART treatment before the transplantation, stem-cell transplantation from a CCR5-Δ32 donor demonstrated a loss of viral titer and an increase in the number of circulating CD4 T cells within the recipient (**FIGURE ES 14.1**). This study further solidified the role of the CCR5 receptor in HIV pathogenesis and may suggest another possible point of intervention in individuals infected with HIV.

Think About...

1. What is your hypothesis of the mechanism by which stem-cell transplantation from a donor homozygous for the CCR5-Δ32 deletion confers HIV resistance in an HIV-positive patient?
2. What experiments do the authors use to demonstrate loss of viral load and control of the HIV infection after stem-cell transplantation?
3. Do you think the stem-cell transplantation used in this study is a viable option as a global treatment of HIV? Why or why not?

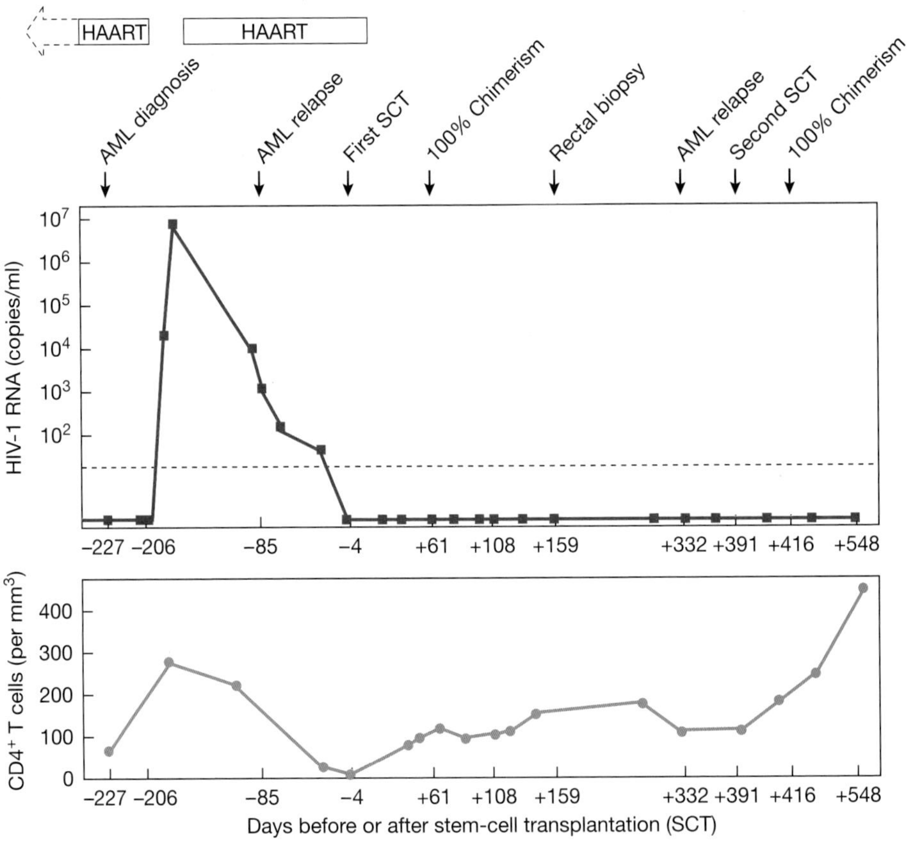

FIGURE ES 14.1 Stem-cell transplantation using CCR5 Δ32/Δ32 cells leads to loss of HIV infection. A patient diagnosed with acute myeloid leukemia (AML) was also HIV positive. Doctors treated the AML using stem-cell transplantation from a donor who was homozygous for the CCR5 Δ32/Δ32 mutation in hopes of curing the HIV infection. Upon stem-cell transplantation, the patient's viral load decreased to undetectable, and remained undetectable after antiviral treatment was stopped. Furthermore, with the absence of HIV infection, the patient's CD4 T-cell levels increased. Dotted line in top graph represents the limit of HIV-mRNA detection. HAART = highly active antiretroviral therapy. (From *The New England Journal of Medicine*, Gero Hütter, M.D., et al. Long-Term Control of HIV by CCR5 Delta32/Delta32 Stem-Cell Transplantation, 360: 692–698. © 2009 Massachusetts Medical Society. Reprinted with permission from Massachusetts Medical Society.)

Acquired immunodeficiency syndrome

After initial infection with HIV and initial detection of viral load and HIV-specific antibodies, the virus enters a state of clinical latency, and budding virus appears to be kept in check by the host immune system. If left untreated, the clinical latency phase can last from 3 to 20 years, averaging approximately 8 years. During this time, the number of CD4 T cells begins to decline to levels incapable of sustaining a properly functioning adaptive immune system (**FIGURE 14.13**).

The onset of AIDS is clinically connected to the point at which CD4 T cells are depleted to an extremely low number (2×10^5 CD4 T cells/mL) or when an HIV-positive individual sustains an opportunistic infection associated with HIV infection. Surprisingly, although the scientific data support HIV as the causative agent of AIDS, there are individuals who deny this connection (see **CONTROVERSIAL TOPICS**).

Individuals with an HIV infection that progresses to AIDS are extremely prone to opportunistic infections. Common infections are pneumonia and respiratory tract infection with the opportunistic fungi *Candida albicans*. Many other infections have been reported, including *Toxoplasma gondii* (cause of toxoplasmosis), *Mycobacterium tuberculosis* (cause of tuberculosis), *Cryptococcus neoformans* (cause of bacterial meningitis), *Cryptosporidium* (cause of chronic diarrhea), and *Pneumocystis jirovecii* (cause of pneumonia).

Patients with AIDS are also prone to viral-induced cancers, including Kaposi sarcoma and central nervous system lymphomas (caused by Epstein–Barr virus). Both of these cancers can be induced by opportunistic infection with viruses of the herpesvirus family.

Current treatments

Due to the life cycle and the high mutation rate of HIV, an effective vaccine has not yet been developed in spite of a significant amount of research. Much of the challenge arises from the high mutation rate of reverse transcriptase from HIV—this mutation rate has resulted in a highly variable range of HIV strains within the human population, making it difficult to develop a vaccine that can prevent infection with all of the variant HIV strains. Furthermore, vaccines are generally aimed at developing immune responses that will allow an individual to survive exposure to a given disease, but to date, there are no known cases of individuals surviving AIDS. The viral latency associated with HIV infection also complicates the efficacy of an HIV vaccine. While an effective HIV

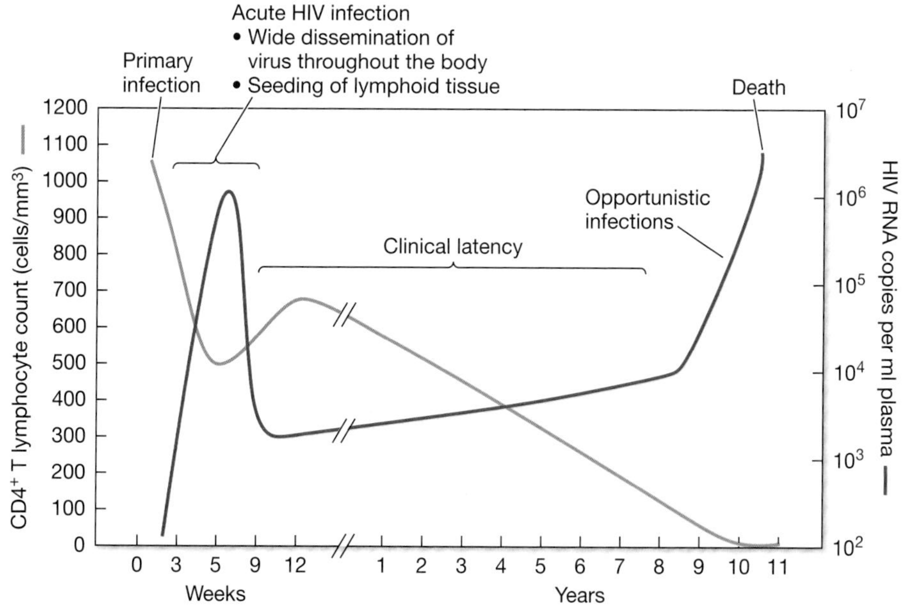

FIGURE 14.13 The course of HIV progression to AIDS in humans After initial infection with HIV, viral genome titer increases dramatically, with an initial decrease in CD4 T cells, until viral latency occurs. Over the course of progression, during clinical latency, viral titer remains largely unchanged, but CD4 T cells continue to decrease in number. Eventually, if untreated, CD4 T-cell levels are decreased to an extent that allows onset of infection by opportunistic pathogens until the patient eventually succumbs. Antiretroviral therapy (ART) can prevent the onset of AIDS. It has been estimated that from 1995 to 2015, ART prevented more than 9.5 million people with HIV from developing AIDS. (From Sigve/Wikimedia. Hiv-timecourse copy.svg/ CC0 1.0.)

CONTROVERSIAL TOPICS

Is HIV truly the causative agent of AIDS?

Research from the past 30 years has overwhelmingly demonstrated that HIV infection can induce the onset of AIDS. Surprisingly, there are individuals (including some in the scientific community) who subscribe to HIV/AIDS denialism—the belief that HIV does not cause AIDS. Some people deny the existence of HIV, while others believe it exists but is incapable of causing AIDS.

After the discovery of a T-cell-specific virus by two research groups in patients suffering from AIDS and reclassification of the genetically indistinguishable viruses isolated in these two studies to HIV, in 1987 the scientist Peter Duesberg published an article in *Cancer Research* claiming that HIV was not the causative agent of AIDS. He proposed that the disease was induced by recreational drug use or the use of antiretroviral therapy, thus fueling HIV/AIDS denialism claims.

Many of the claims associated with HIV/AIDS denialism are inconsistent or scattered in their reason for the onset of AIDS and/or their denial of HIV as the causative agent. In addition to Duesberg's propositions, other denialists claim that HIV testing is inaccurate and that HIV does not fulfill Koch's postulates for the microbial causation of a disease.

Although there are individuals who deny the HIV–AIDS connection, including a few prominent scientists, the scientific community and the published record overwhelmingly demonstrate the connection. Abundant evidence exists that HIV is indeed a real virus and does fulfill Koch's postulates if not held to extremely stringent standards (since Koch's postulates were written to test the causation of bacterial, not viral, infection to disease onset).

While discussion and debate about scientific findings are worthwhile and an important component of the scientific method (promoting testing and retesting of hypotheses to verify results), it is irresponsible to incite debate based on rhetoric alone without supporting scientific evidence.

Think About...

1. How could you convince an HIV denialist that HIV is the causative agent of AIDS? Which key pieces of evidence would you cite as support for your point of view?
2. What data might convince you to consider the denialism stance as more credible than your own?

vaccine has not yet been developed, there is hope for the future. Several studies have demonstrated that monoclonal antibodies can be protective neutralizing antibodies against HIV. Furthermore, there are HIV-infected individuals who have remained asymptomatic (without onset of AIDS) after years of infection, suggesting that they are capable of keeping the viral load and infection at bay.

Since a safe and effective vaccine has yet to be developed, the current treatment options employed in HIV infections are antiretroviral therapies. These pharmaceuticals work to halt the normal HIV life cycle. Drugs against reverse transcriptase (nucleoside analog and nonnucleoside reverse transcriptase inhibitors) and the HIV protease are the most commonly prescribed pharmaceuticals for HIV-positive individuals. Integrase and fusion inhibitors are also used in treatment.

The challenges associated with the development of an effective vaccine also cause complications in antiretroviral therapy—antiretrovirals constitute a selective pressure against the HIV virus, and the high mutation rate of reverse transcriptase can promote the production of virus that has undergone spontaneous mutation, rendering the antiretroviral therapy useless against the mutant HIV. To prevent the occurrence of spontaneous mutations capable of evading a single antiretroviral therapy, a common approach is the use of highly active antiretroviral therapy (HAART), which employs several different antiretroviral agents within a cocktail of typically three compounds. The combination therapy is usually much more effective at inhibiting the HIV life cycle than a single pharmaceutical agent because the random mutagenesis caused by reverse transcriptase typically cannot overcome the selective pressure of the three different compounds simultaneously.

Combined with HAART as a long-term therapy, another avenue being explored is the use of CRISPR (clustered regularly interspaced short palindromic repeats) gene-editing technology to remove the HIV provirus from infected cells.

Researchers demonstrated that primates infected with simian immunodeficiency virus (SIV) could have SIV proviral DNA removed using the CRISPR gene-editing technique to remove the latent virus from the DNA genome of infected cells. This therapy might prove useful in HIV treatment in humans and could provide a means to a cure by removing the HIV genome from the target cell genome.

CHECKPOINT QUESTIONS

1. List and describe the stages of the HIV life cycle.
2. Explain why the activity of HIV reverse transcriptase complicates both vaccine design and antiretroviral therapy.

Summary

14.1 What is immunodeficiency?

- The three general types of immunodeficiencies are primary immunodeficiencies (genetic defects in one arm of the immune system), secondary immunodeficiencies (blockage of an immune system mechanism due to environmental factors or pathogen infection), and combined immunodeficiencies (lack of both B-cell and T-cell function).

14.2 How does inherited immunodeficiency differ from acquired immunodeficiency?

- Inherited immunodeficiencies are genetic defects in an immune response, whereas acquired immunodeficiencies are caused by an external agent preventing the action of an immune response.

14.3 How does a deficiency in the innate immune system lead to disease?

- Immunodeficiencies that affect the innate immune system can prevent the action of first lines of defense, complement activation, phagocyte function, and natural killer cell function.

14.4 How does a combined deficiency in lymphocyte development or action lead to disease?

- Immunodeficiencies that prevent lymphocyte development or action can cause the most severe combined immunodeficiencies, since lymphocytes are required for the induction of an adaptive immune response.

14.5 How does a deficiency in T-cell action lead to disease?

- Immunodeficiencies that prevent T-cell action or T-cell development can cause some of the severe immunodeficiencies, since T cells are responsible for activating many different cell types during an adaptive immune response.

14.6 How does a deficiency in B-cell action lead to disease?

- Immunodeficiencies that prevent B-cell action or B-cell development can prevent immunoglobulin production or can prevent proper isotype switching and somatic hypermutation.

14.7 How does HIV cause AIDS?

- HIV is a retrovirus capable of inducing the onset of acquired immunodeficiency syndrome (AIDS) due to its life cycle and the target cells it infects.
- HIV therapies currently in use function to alter the activities of specific HIV proteins and processes required for the life cycle and infectivity of HIV.

Review Questions

14.1 Define each of the following immunodeficiencies as a primary immunodeficiency, secondary immunodeficiency, or combined immunodeficiency:

A. AIDS

B. Bare lymphocyte syndrome

C. Chediak-Higashi syndrome

14.2 You have been presented with a case of an individual suffering from recurring infections. You suspect an immunodeficiency but are unsure if it is an inherited immunodeficiency or an acquired immunodeficiency. Describe some basic tests that you could conduct to distinguish between the two possibilities.

14.3 Many immunodeficiencies associated with abnormal complement activation result in higher susceptibility to infections by encapsulated bacteria. Explain why.

14.4 You have isolated white blood cells from both a normal patient and a patient suffering from Omenn syndrome. You decide to use flow cytometry to quantify the number of T cells and B cells present in each white blood cell sample. Illustrate what you would expect the flow cytometry pattern to be for each sample in regard to their CD4 (*y*-axis) and IgM (*x*-axis) expression.

14.5 Explain why bare lymphocyte syndrome, typically caused by an abnormality in MHC-peptide presentation,

(*Continued*)

Review Questions (*continued*)

leads to immunodeficiency in T-cell function. Would you predict that T cells in an individual with bare lymphocyte syndrome would develop properly? Why or why not?

14.6 A couple have a son who suffers from X-linked agammaglobulinemia. The father has normal levels of IgA and IgG in his plasma. The parents recently found out that they will have another child. What is the probability that their second child will suffer from X-linked agammaglobulinemia if the child will be (a) male or (b) female?

14.7 A typical nucleoside analog reverse transcriptase inhibitor is the compound zidovudine (also known as azidothymidine, or AZT). Zidovudine has an azido group at the 3′ carbon of the deoxyribose sugar. Why is this compound an inhibitor of reverse transcriptase?`

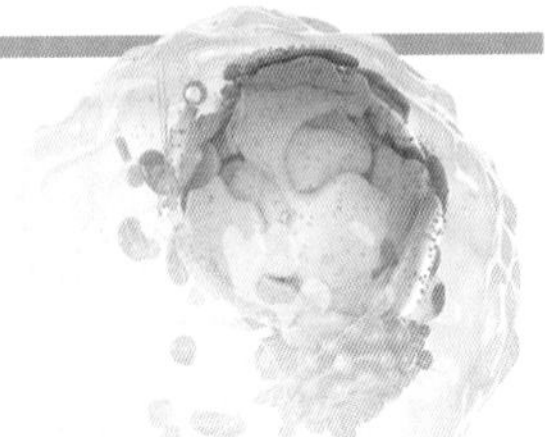

CASE STUDY REVISITED

Severe Combined Immunodeficiency

A male infant, age 6 months, is brought to the hospital emergency department with a red rash all over his body. The parents state that he has suffered from chronic diarrhea and that the rash manifested that morning. Physical examination reveals many swollen lymph nodes. A complete blood panel shows that the baby lacks circulating B and T cells. Suspecting Omenn syndrome, the medical team orders sequence analysis for the *RAG1* and *RAG2* genes; however, sequence analysis does not reveal mutation in either gene. Further sequence analysis reveals that there is a missense mutation in the adenosine deaminase gene, and the diagnosis is severe combined immunodeficiency syndrome. An MHC crossmatch is ordered for bone marrow transplantation.

Think About...

Questions for individuals

1. What is Omenn syndrome, and how is this disease related to the lack of circulating B and T cells?
2. Investigate the missense mutation in the adenosine deaminase gene and its connection to severe combined immunodeficiency. How is this mutation inherited? What does this inheritance mean in regard to the genetics of the parents (who do not manifest severe combined immunodeficiency)?

Questions for student groups

1. Why was the first suspicion in this case a mutation in *RAG1* or *RAG2*? Was *RAG1* and *RAG2* sequence analysis warranted?
2. Why does mutation in the adenosine deaminase gene, and thus malfunction of a purine salvage pathway, prevent B cells and T cells from circulating in the bloodstream?
3. Why is a bone marrow transplant warranted in this case?

MAKING CONNECTIONS Key Concepts	COVERAGE (Section Numbers)
Review of innate immune system cells and function	1.4, 1.5, 2.1, 2.2, 2.5, 2.6, 2.7, 3.1, 3.2, 3.4, 10.5
Immunodeficiencies of the innate immune system	14.1, 14.2, 14.3
Review of adaptive immune system cells and function	1.5, 4.4, 5.4, 6.6, 7.1, 7.8, 9.4, 10.2
Immunodeficiencies of the adaptive immune system	14.1, 14.2, 14.4, 14.5

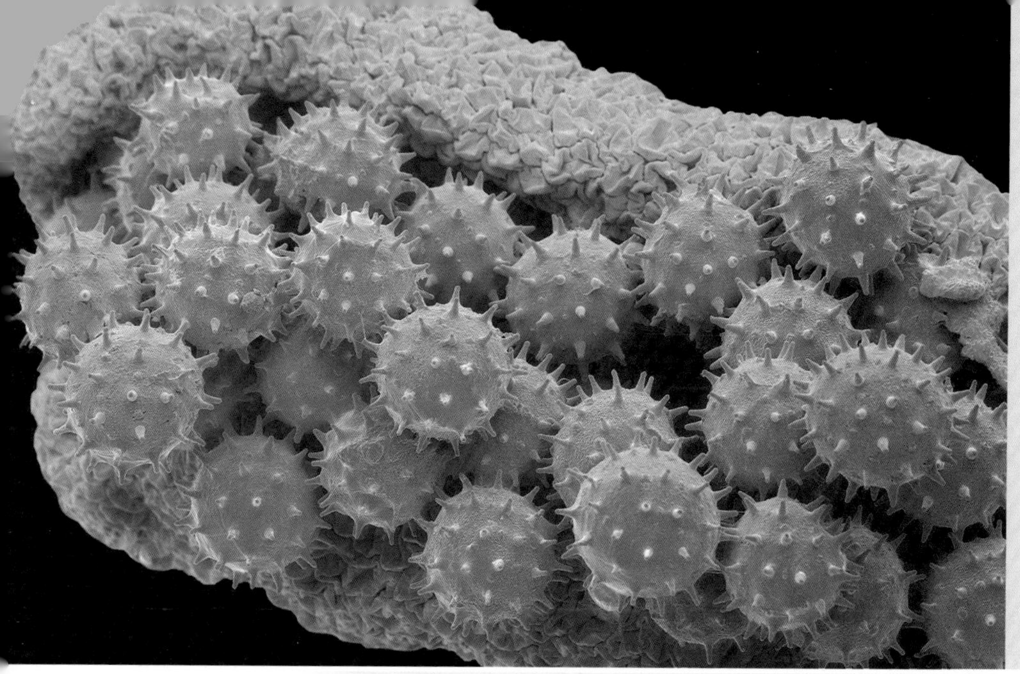

15

© Science Photo Library/Alamy Stock Photo

Allergies and Hypersensitivity Reactions

CASE STUDY: Aleksandar's Story

Aleksandar is looking forward to having a pleasant dinner with his wife, Viktoria. The couple is from Bulgaria, and moussaka made with ground pork and potatoes is their traditional Sunday dinner. They open a bottle of wine and get ready to sit down to enjoy their meal. Since it is such a beautiful day, they decide to eat on their back deck. Aleksandar goes out to the deck to open the umbrella to shade the table. As he opens the umbrella, his wife calls out, "Be careful—I saw some hornets flying around the umbrella yesterday!"

Unfortunately, Viktoria's warning comes a bit too late. As Aleksandar opens the umbrella, he is greeted by several hornets that have been building a nest in the umbrella. He tries to avoid the insects, but they defend their new construction, and Aleksandar is stung on his hand and his neck before he retreats inside. It's the first time he's been stung by a hornet. He doesn't recall ever having been stung by a bee, either. When Aleksandar rushes into the house, Viktoria suggests that they eat indoors and have the nest dealt with by professionals.

Shortly after dinner, Viktoria notices a rash developing on Aleksandar's neck. Aleksandar says that he is having difficulty swallowing and breathing, and his airway feels like it is constricting. He starts to panic, and the couple leaves immediately for the emergency room.

To understand what Aleksandar is experiencing and to answer the case study questions at the end of the chapter, we must explore the various hypersensitivity reactions that can occur. We have seen the means by which the immune system recognizes and destroys pathogens. These responses depend on the immune system's ability to recognize foreign antigens through the action of cellular receptors.

QUESTIONS Explored

15.1 What are the different types of hypersensitivity responses?

15.2 How does type I hypersensitivity occur?

15.3 What factors are responsible for type II hypersensitivity?

15.4 How does type III hypersensitivity occur?

15.5 What is delayed-type (type IV) hypersensitivity?

What happens, though, if the immune system recognizes and responds to foreign antigens that are not pathogenic? Individuals may have hypersensitivity reactions (also called allergic reactions) to innocuous material. In this chapter, we will explore the various types of hypersensitivity reactions and see how these responses are identical to the ways in which the body reacts to foreign pathogens.

15.1 What are the different types of hypersensitivity responses?

LEARNING OBJECTIVES

15.1.1 Compare and contrast the four types of hypersensitivity reactions.

15.1.2 Describe the role of inflammatory mediators in the onset of hypersensitivity reactions.

Throughout this book we have seen that the innate and adaptive immune responses employ various mechanisms to recognize and combat foreign pathogens to prevent disease. Recognition of foreign molecules by PAMP receptors, T-cell receptors, and immunoglobulins drives induction of these mechanisms, including inflammation and activation of the humoral and cellular arms of the adaptive immune system.

While these mechanisms are designed to destroy potentially harmful pathogens, activation of these processes in the presence of foreign, but normally innocuous, substances can lead to reactions that range from mildly annoying to life-threatening. Stimulation of an immune response to harmless foreign material is referred to as an **allergy**, a type of **hypersensitivity reaction**. These hypersensitivity reactions can include immune responses to innocuous foreign substances or to self-antigens, such as those seen in autoimmune diseases (see Chapter 16). Hypersensitivity reactions can be caused by activation of either an innate or adaptive immune response, and thus differ in their symptoms and in the time it takes for the reaction to occur.

Allergy and hypersensitivity

Every day, we come into contact with a multitude of foreign materials that are not pathogenic. However, the immune system may or may not respond to these normally harmless materials as it would respond to a dangerous pathogen.

Among the most common types of hypersensitivity reactions are allergic reactions, which affect approximately 30% of adults and 40% of children worldwide. These reactions, known as *type I hypersensitivity* reactions, are driven by the recognition of **allergens** (molecules that can induce type I hypersensitivity) and the release of molecules that drive the allergic response. Recognition of an allergen induces a hypersensitivity response typically through the release of inflammatory mediators.

Hypersensitivity reactions vary and are based on the immune response pathway involved in recognizing the foreign material (**TABLE 15.1**). The major types of hypersensitivity reactions are:

- *Type I hypersensitivity*: These reactions, which include allergic reactions, are driven by the action of granulocytes such as mast cells, eosinophils, and basophils in response to a foreign molecule capable of binding to IgE associated with these granulocytes. This recognition causes release of inflammatory mediators from the granulocytes.
- *Type II hypersensitivity*: These reactions are caused by the production of immunoglobulins, such as IgG and IgM, which recognize cell-surface molecules and target cells containing these molecules via the complement pathway or antibody-dependent cell-mediated cytotoxicity.

allergy Type I hypersensitivity reaction driven by the activation of granulocytes that recognize an allergen via associated IgE antibodies and induce signaling and degranulation through their Fcε receptor.

hypersensitivity reaction Reaction caused by an immune response driven by the recognition of an innocuous material within an organism.

allergen Molecule with the ability to induce type I hypersensitivity when recognized by IgE immunoglobulins in complex with granulocytes.

TABLE 15.1 | Types of Hypersensitivity Reactions

Hypersensitivity	Antigen	Immune Recognition Molecule	Effector Mechanism	Examples of Symptoms
Type I	Soluble allergen	IgE	Mast cell, basophil, and eosinophil degranulation	Many, depending on allergen entry (see Table 15.3)
Type II	Modified cell-surface molecule	IgG/IgM	Complement activation Antibody-dependent cell-mediated cytotoxicity	Blood transfusion reactions, including hemolytic anemia
Type III	Soluble antigen	IgG/IgM	Immune complex formation	Serum sickness Arthus reaction
Type IV	Soluble or cell-associated antigen	T-cell receptor	Macrophage activation Eosinophil activation Cytotoxicity	Contact dermatitis Granulomas

- *Type III hypersensitivity*: These reactions are caused by the action of IgG recognizing soluble antigens and forming immune complexes, resulting in the unnecessary activation of complement and activation of phagocytic cells.
- *Type IV hypersensitivity*: These reactions are driven by the activation of T cells by a foreign, but harmless, antigen. Because these reactions require T-cell activation, and thus require time to occur, they are often referred to as *delayed-type hypersensitivity* reactions. The response induced during delayed-type hypersensitivity is dictated by the type of T cells that are activated during the reaction.

Inflammatory mediators

As we will see in the next section, a major mechanism involved in producing an allergic response (type I hypersensitivity) is the recognition of an allergen by IgE immunoglobulins associated with the surface of granulocytes. This recognition activates the granulocytes and causes the release of inflammatory mediators contained within the granules of these cells (**FIGURE 15.1**). Some of these inflammatory mediators include **histamine**, **leukotrienes**, **prostaglandins**, **heparin**, and the cytokine TNF-α.

We know that inflammatory mediators such as TNF-α work to increase vascular permeability. This allows migration of other immune cells (such as neutrophils

histamine Inflammatory mediator produced and secreted by activated granulocytes such as mast cells capable of both increasing vascular permeability and inducing smooth muscle contraction; histamine release from granulocytes is a primary factor in inducing type I hypersensitivity.

leukotrienes Inflammatory mediators derived from arachidonic acid that can induce smooth muscle contraction, increase vascular permeability, and aid in the production of other inflammatory cytokines.

prostaglandin Inflammatory mediator secreted by activated granulocytes capable of inducing vasodilation, fever, and pain at the site of release.

heparin Inflammatory mediator secreted by activated granulocytes that aids in increasing vascular permeability.

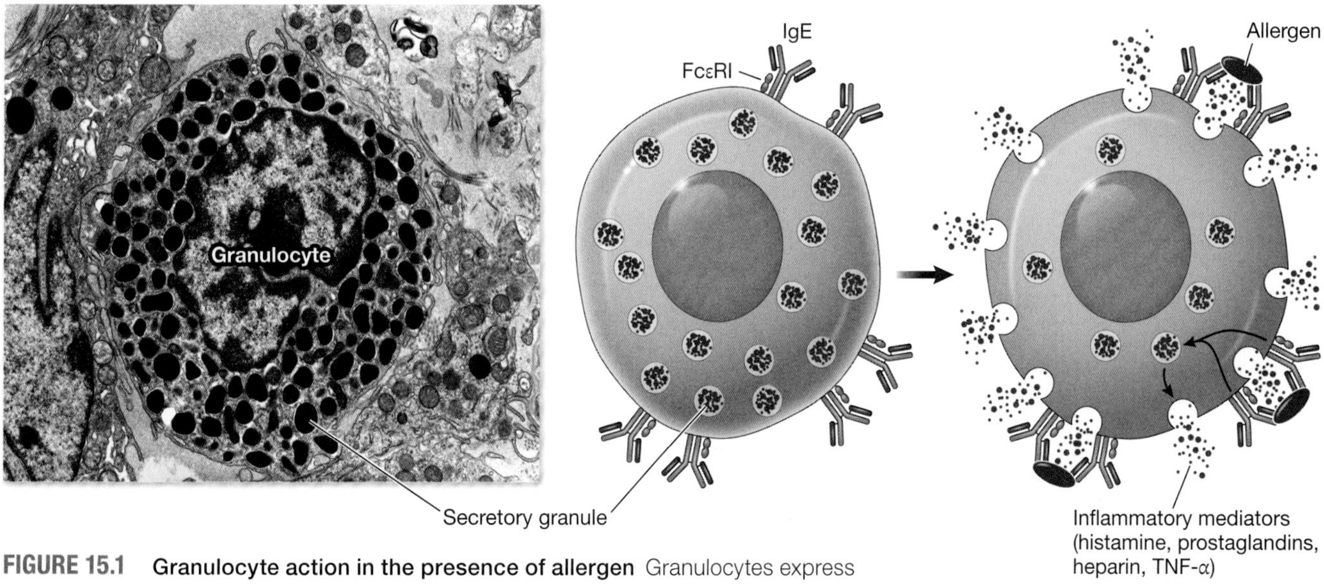

FIGURE 15.1 Granulocyte action in the presence of allergen Granulocytes express the cell-surface protein FcεRI, which binds IgE produced by plasma cells. When the bound IgE binds to allergen and clusters at the cell surface, the result is degranulation of the granulocyte and secretion of inflammatory mediators, including histamine, prostaglandins, heparin, and TNF-α.

and effector cells of the adaptive immune system) to the site of infection and potential migration of dendritic cells to draining secondary lymphoid tissue to activate the adaptive immune response. Inflammatory mediators also cause localized temperature increase to aid in the immune response and inhibit pathogen division and survival. Inflammatory mediators such as histamine and heparin, secreted by granulocytes such as mast cells in response to binding of an antigen to associated IgE immunoglobulins, behave in a similar manner to TNF-α.

● CHECKPOINT QUESTION

1. Differentiate between the four types of hypersensitivity reactions, focusing on the immune response activated during each type.

15.2 | How does type I hypersensitivity occur?

LEARNING OBJECTIVES

15.2.1 Explain the normal action of granulocytes and how granulocyte action leads to type I hypersensitivity.

15.2.2 Explain how allergy manifestations are driven by granulocyte action due to IgE recognition of an innocuous allergen.

A type I hypersensitivity reaction is driven by the presence of an allergen within an individual whose immune system recognizes this allergen as a potentially harmful foreign molecule. This recognition is driven by the presence of allergen-binding IgE in close contact with Fcε receptors on the surface of granulocytes. Binding of IgE to its antigen causes degranulation of these granulocytes and the release of inflammatory mediators. In this section, we will explore the action of different granulocytes and how their activation can drive the many different allergic reactions we see in our population today.

Allergies are one of the fastest-growing childhood health problems and are especially prevalent in developed nations. It has been suggested that increased exposure to allergens such as dust mites due to children spending more time indoors may be a contributing factor. The *hygiene hypothesis* speculates that the preponderance of allergies in the world, especially in developed countries, is caused by the lack of exposure to microorganisms and parasites (both pathogenic and symbiotic) early in childhood, thus limiting the priming of the immune system. This lack of exposure to a large number of antigens may increase the likelihood of allergic reactions later in life (see the **CONTROVERSIAL TOPICS** box).

Granulocytes mediating inflammation

As we will see later in this section, type I hypersensitivity reactions occur when allergens bind IgE and signal the release of inflammatory mediators from granules contained within cells associated with the IgE immunoglobulins that utilize Fcε receptors to bind IgE (see Chapter 10 to review the action of Fcε receptors). To appreciate the many means by which allergens can induce type I hypersensitivity, let's take a closer look at the granulocytes responsible for causing these hypersensitivity reactions.

MAST CELLS We have seen that an effector function of B cells that aids in the expulsion of parasites and pathogens is the action of mast cells in conjunction with associated IgE. These mast cells are filled with granules standing at the ready (**FIGURE 15.2**), waiting for detection of a foreign pathogen. Recognition causes the granules to fuse with the mast cell's plasma membrane and to release inflammatory mediators (**TABLE 15.2**).

TABLE 15.2 | Inflammatory Mediators Secreted by Mast Cells

Inflammatory Mediator	Activity
Histamine, heparin	Increase vascular permeability Smooth muscle contraction
TNF-α	Inflammation
IL-4	T_H2 activation
IL-3, IL-5, GM-CSF	Eosinophil activation
Prostaglandins	Increase vascular permeability Smooth muscle contraction

Source: E. Z. M. da Silva et al. 2014. *J Histochem Cytochem* 62: 698–738.

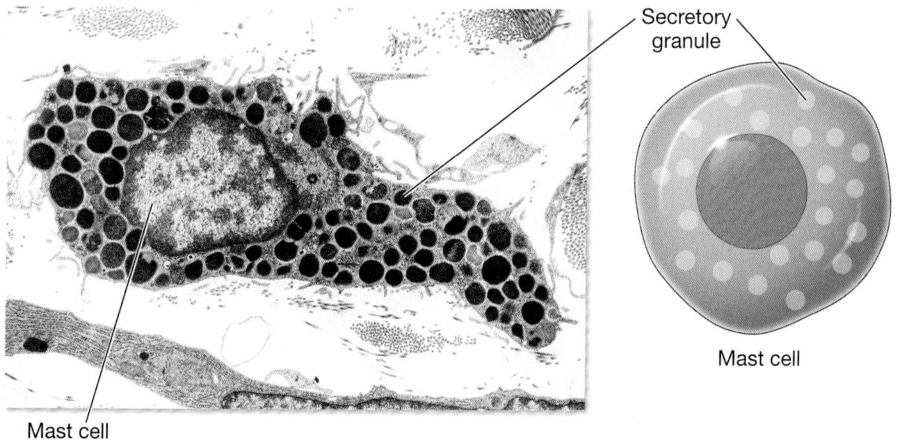

FIGURE 15.2 Mast cells Mast cells are an important granulocyte of the innate immune system. These cells are filled with secretory granules containing inflammatory mediators. Recognition of antigen (or allergen) results in mast cell degranulation and inflammation.

CONTROVERSIAL TOPICS

Why has the prevalence of allergies and asthma increased so dramatically in developed countries?

The incidence of allergic reactions (type I hypersensitivity) has increased dramatically in developed countries, and the number of people affected has steadily risen from the late twentieth century to today. The Centers for Disease Control and Prevention reported that in U.S. children younger than 18 years of age, the incidence of food allergies increased from 5.6% in 2014 to 6.1% in 2017. Over the same time period and in the same age group, the incidence of skin allergies increased from 11.8% to 12.5%.

In an effort to explain the rapid increase specifically in developed countries (the incidence of allergy in developing nations remained relatively constant), scientists hypothesized the causation. In the late 1980s, scientists proposed the *hygiene hypothesis*, which postulates that a lower level of microorganism infection during childhood shifts the balance of immune responses, leading to increased incidence of allergy. There were several pieces of supporting evidence for this hypothesis:

- Asthma incidence seemed to increase more dramatically due to development of a nation and cleanliness of air quality within the home.
- Exposure of children or pregnant women to farm environments, grain products, and cow's milk seemed to lower incidence of allergies to grain, milk, and pollen.

Biologically, the hygiene hypothesis is rooted in the development of T cells and a shift in the balance in T-cell responses that are favored during early childhood. Many scientists have focused their research on studying the shift in balance in children of developed nations toward T_H2 responses, which favor the production of IgE, the hallmark antibody in allergy. It has been postulated that exposure to parasites early in childhood and methods of food handling and cooking may also contribute to allergy. Perhaps exposure to parasites in early childhood (more common in developing nations) prompts the adaptive immune system to properly focus T_H2 responses toward combating those pathogens rather than leaving a "void" to be filled by unnecessary responses to innocuous material.

Other studies focus on a connection between pathogen exposure and the proper development of a broad set of regulatory T cells that may prevent autoimmune disorders. Data supporting the hygiene hypothesis not only supports the connection of cleanliness and increased incidence of allergy, it also supports the connection between these same environments and increased incidence in autoimmune diseases such as type 1 diabetes.

While the hygiene hypothesis seems to demonstrate a clear connection between the level of industrialization and increased water and food sanitation, lack of pathogen exposure, and the increased incidence of allergy and autoimmunity, alternative hypotheses have been proposed. One hypothesis known as the *old friends hypothesis* is a refinement of the hygiene hypothesis. It suggests that it is not lack of exposure to certain pathogens that increases the likelihood of allergies but rather lack of exposure to normal microbiota. Recall that the microbiota is important not just in protecting us from colonization by potential pathogens but also in ensuring the normal development and function of our immune system. Perhaps a lack of exposure to certain microbial species that aid in immune system development may alter the way in which an immune system reacts to innocuous material.

(Continued)

CONTROVERSIAL TOPICS *(continued)*

Currently, some studies are focusing on early or late exposure to solid foods, the types of solid foods and milks and formulas introduced to infants, and breastfeeding. Hopefully immunologists will soon be able to delineate the connection so progress can be made in reducing the incidence of conditions that range from mildly annoying to life-threatening.

Think About...

1. Do you think the data support the claims of the hygiene hypothesis? Why or why not?
2. How would scientists begin to test causation of higher allergy or autoimmunity risk (keep in mind that many current studies are epidemiological studies)?
3. What data might convince you to change your opinion regarding the hygiene hypothesis?

The action of mast cells upon degranulation is extensive (**FIGURE 15.3**). The molecules secreted during degranulation can affect a variety of cell types in the area of action. These effects include an increase in vasodilation and vascular permeability: when these actions are directed toward a pathogen infection, they promote migration of immune cells to the site of infection and migration of dendritic cells to draining lymphoid tissues to activate an adaptive immune response.

Mast cell degranulation can also activate smooth muscle fiber contraction to drive expulsion of the pathogen. The purpose of this muscle contraction is to prompt coughing, sneezing, vomiting, or diarrhea to expel the pathogen from the mucosal tissue (in the respiratory tract or gastrointestinal tract).

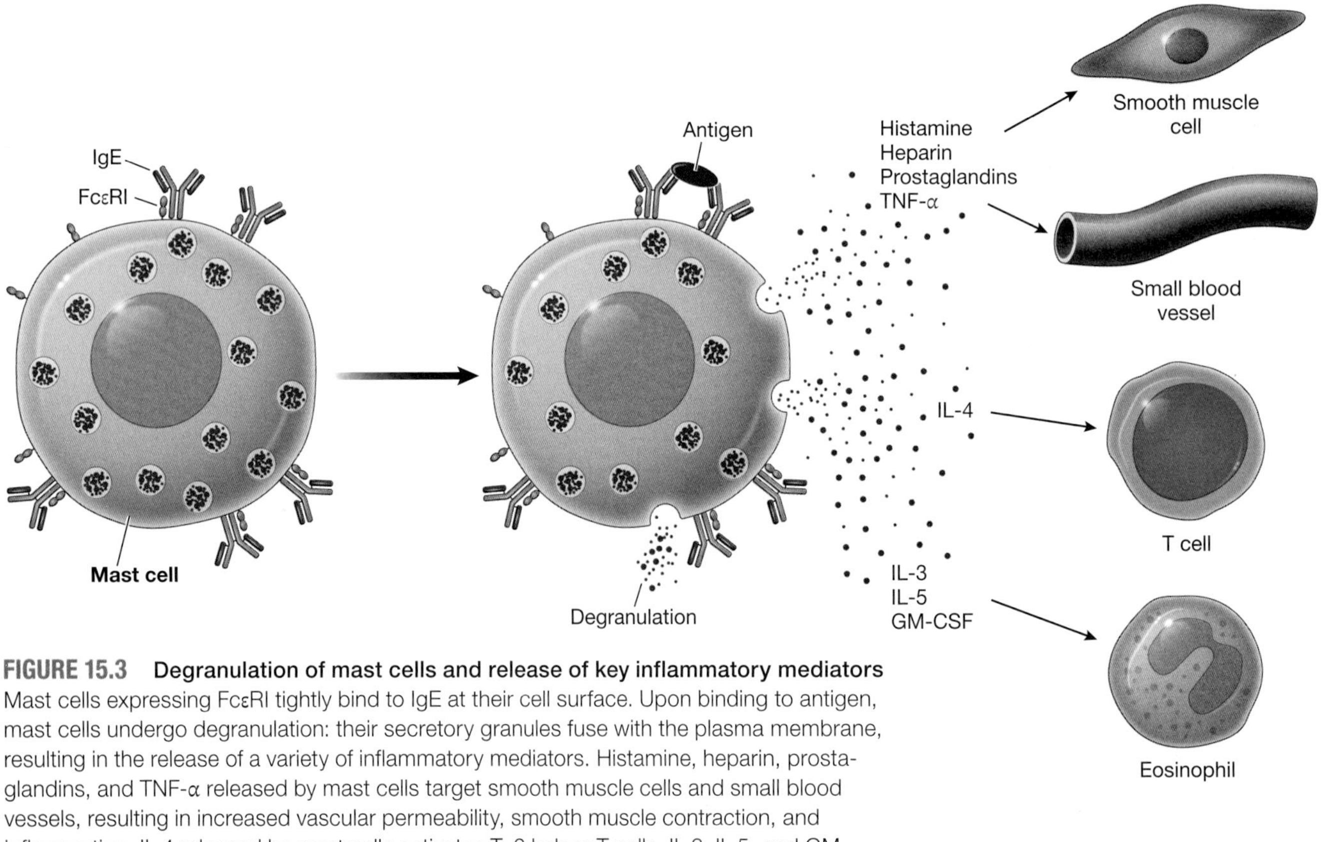

FIGURE 15.3 Degranulation of mast cells and release of key inflammatory mediators Mast cells expressing FcεRI tightly bind to IgE at their cell surface. Upon binding to antigen, mast cells undergo degranulation: their secretory granules fuse with the plasma membrane, resulting in the release of a variety of inflammatory mediators. Histamine, heparin, prostaglandins, and TNF-α released by mast cells target smooth muscle cells and small blood vessels, resulting in increased vascular permeability, smooth muscle contraction, and inflammation. IL-4 released by mast cells activates T_H2 helper T cells. IL-3, IL-5, and GM-CSF released by mast cells activate nearby eosinophils to increase the release of other inflammatory mediators.

While normal mast cell activation plays an important role in the innate immune response, especially in pathogen expulsion, activation of mast cells due to allergen recognition can be a nuisance or even life-threatening. Degranulation is driven by allergen recognition by IgE, and depending on the location of the activated mast cells and the mode of entry of the allergen, the induced response can cause a variety of different type I hypersensitivity reactions (discussed in the section *IgE and type I hypersensitivity*).

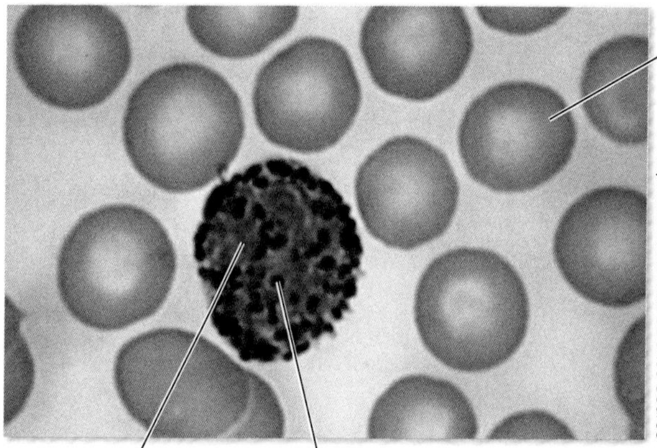

FIGURE 15.4 **Light micrograph of a basophil** A basophil filled with secretory granules can be seen with surrounding red blood cells.

BASOPHILS Basophils are a type of granulocyte that work to promote pathogen expulsion. They are particularly important in our immune response to parasites, especially helminths (parasitic worms). Basophils are so named because their granules are filled with proteins that are easily stained with basic dyes (**FIGURE 15.4**). Because they represent a very small proportion of white blood cells (approximately 0.5%), our understanding of basophil action is still evolving. We do know that activation of basophil degranulation closely mirrors that of mast cell degranulation through recognition of antigen via associated IgE.

EOSINOPHILS Eosinophils are yet another type of granulocyte capable of binding to IgE at their surface and releasing a variety of immune system effector proteins in response to antigens, including parasitic helminths. Eosinophils are named based on their ability to be stained with acid dyes, such as the histological stain eosin (**FIGURE 15.5**).

In response to IgE signaling, these white blood cells are capable of secreting inflammatory mediators such as histamine and prostaglandins and toxic molecules, including major basic protein, eosinophil peroxidase, eosinophil cationic protein, and eosinophil-derived neurotoxin. Major basic protein can further induce mast cell and basophil degranulation. Eosinophil peroxidase creates reactive oxygen species through the reaction of hydrogen peroxide and halide ions such as chloride. Eosinophil cationic protein and eosinophil-derived neurotoxins are both ribonucleases, enzymes that cleave RNA substrates and are thought to combat viral genomes.

IgE and type I hypersensitivity

Why do we consider the degranulation of mast cells, basophils, and eosinophils when discussing the concept of hypersensitivity? The answer lies in what prompts activation of all three cell types, namely IgE at the cell surface tightly bound with the Fc receptor FcεRI. While antigen recognition by IgE associated with these granulocytes drives inflammation and vasodilation to aid in pathogen removal, recognition of an allergen by an IgE immunoglobulin can induce a variety of type I hypersensitivity reactions (see **KEY DISCOVERIES** box). The specific nature of the reaction depends on the allergen's mode of entry and the location of the granulocytes recognizing the allergen (**FIGURE 15.6**).

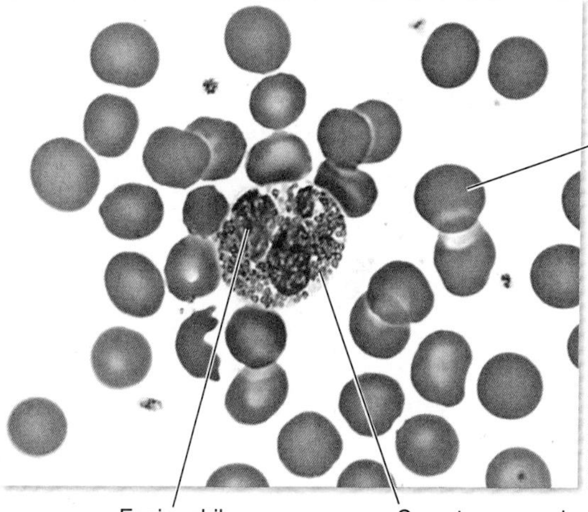

FIGURE 15.5 **Light micrograph of an eosinophil** An eosinophil filled with secretory granules can be seen with surrounding red blood cells.

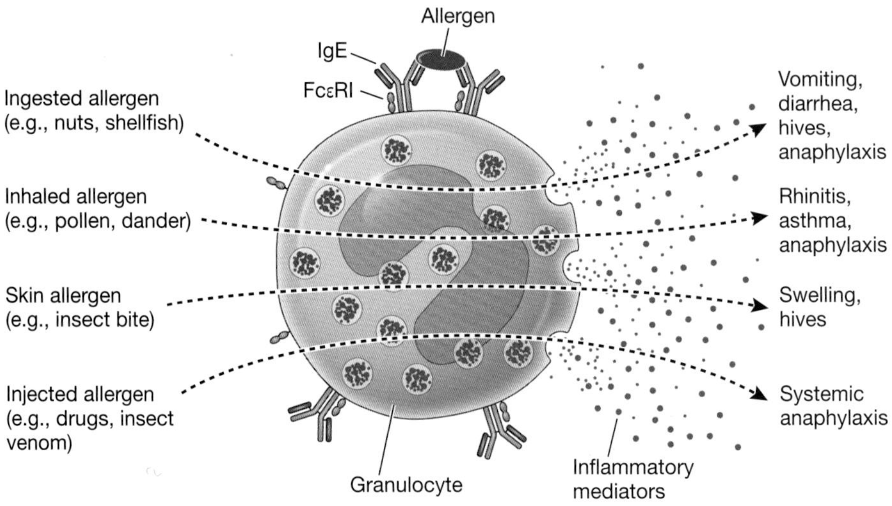

FIGURE 15.6 Type I hypersensitivity reactions Type I allergic reactions activate granulocytes and result in the release of inflammatory mediators where contact with the allergen occurs. Ingestion of a food allergen can result in vomiting, diarrhea, hives, and/or anaphylaxis. Inhalation of an allergen can produce reactions that include rhinitis and an asthma attack. Skin contact with or inhalation of an allergen can result in localized swelling and hives. Injection of an allergen into the bloodstream can result in systemic anaphylaxis.

Type I hypersensitivity reactions

While all type I hypersensitivity reactions are driven by allergens recognized by granulocyte-associated IgE, these reactions vary in level of severity and the tissues affected (**TABLE 15.3**). Possible routes of entry include skin contact, inhalation, ingestion, and injection. Most allergens are small peptides, proteins, or drugs that can induce allergic responses. The responses triggered by these allergens drive inflammation at points of contact or can induce granulocyte activation that results in smooth muscle contraction to facilitate removal of the antigen from the body.

Because type I hypersensitivity is driven by the action of IgE in association with granulocytes, an individual who suffers from an allergy must first become sensitized to a specific allergen. The person's first exposure to an allergen elicits a primary adaptive immune response in which antigen-presenting cells phagocytose and process the antigen via MHC class II presentation, resulting in the

TABLE 15.3 | Type I Hypersensitivity Reactions

Reaction	Possible Allergen	Allergen Entry Route	Response
Wheal-and-flare	Insect bite or sting Allergens used in skin testing for allergies	Skin contact	Localized swelling Hives
Food allergy	Nuts Shellfish Fruit	Ingestion	Vomiting Diarrhea Hives Increased vascular permeability
Rhinitis	Pollen Dust mite Pet dander	Inhalation	Nasal mucosa irritation and edema
Asthma	Pollen Dust mite	Inhalation	Bronchial constriction Inflammation of respiratory tract
Systemic anaphylaxis	Drugs Insect venom Food allergens	Injection Ingestion (food allergens) Inhalation	Edema Increased vascular permeability Occluded trachea Death

activation of allergen-specific helper T cells. Recall that these helper T cells can induce activation of antigen-specific B cells. This sensitization (first exposure) to allergen exposure drives clonal selection and expansion of allergen-specific B cells; a subset of B cells differentiates into centrocytes and undergoes isotype switching to express IgE immunoglobulins.

KEY DISCOVERIES

How was IgE identified as the important molecule involved in type I hypersensitivity?

Article

K. Ishizaka and T. Ishizaka. 1967. Identification of γE-antibodies as a carrier of reaginic activity. *Journal of Immunology* 99: 1187–1198.

Background

While allergic reactions had been identified very early in the history of immunology, the molecule responsible for inducing the allergic response remained elusive. Classic experiments attempting to separate molecules involved in the reaction were designed to answer the question: Which immune molecule is responsible for allergic reactions?

Early Research

In the 1920s, studies by Prausnitz and Kustner found that a nonallergic individual could demonstrate an "allergic" wheal-and-flare reaction when serum from an allergic individual was injected into the nonallergic individual, followed by introduction of the allergen to the same site. These wheal-and-flare reactions were termed P-K reactions and were likely driven by immunoglobulins present in the allergic individual's serum.

Think About...

1. How could you identify the type of immunoglobulin that might be responsible for the allergic reaction?
2. If your serum contained different types of immunoglobulins, how might you purify different immunoglobulins from the serum sample?
3. How could you test that your purified immunoglobulin was solely responsible for the allergic reaction you observed?

Article Summary

Using serum from an individual allergic to ragweed pollen as the source of immunoglobulins, the researchers used gel filtration and ion-exchange chromatography to purify the immunoglobulins responsible for the P-K reaction. When they tested the purified antibodies for the capacity to induce a ragweed-pollen–specific P-K reaction, they found that the presence of IgE immunoglobulins strongly correlated with the induction of these reactions in nonallergic subjects (**TABLE**). Furthermore, researchers were able to show that when IgE immunoglobulins were removed from the serum that was originally capable of inducing a P-K reaction, the reaction no longer occurred. These findings were paramount in identifying IgE immunoglobulin as an important effector molecule in the induction of allergic responses.

Distribution of Skin-Sensitizing Activity and Immunoglobulins Following Ion-Exchange Chromatography

Serum	Fraction	Minimum dose	IgE present	IgG present	IgA present
P	A	for P-K > 4	–	+++	–
	B	> 4	–	+++	–
	C	> 4	–	+++	–
	D	1	–	+++	–
	E	0.5	±	++	–
	F	0.1	+	++	–
	G	0.2	+	++	–
	H	0.5	±	+	–
	I	> 4	-	+	-
	J	> 4	–	–	–
	K	> 4	–	–	–
A	A	0.002	++	++	–
	B	0.0006	+++	+	±
	C	0.0014	++	+	±
	D	0.005	++	+	+
	E	0.017	++	–	+
	F	0.13	+	–	+

Source: K. Ishizaka, T. Ishizaka, and M. M. Hornbrook. 1967. *J Immunol* 98: 490–500.

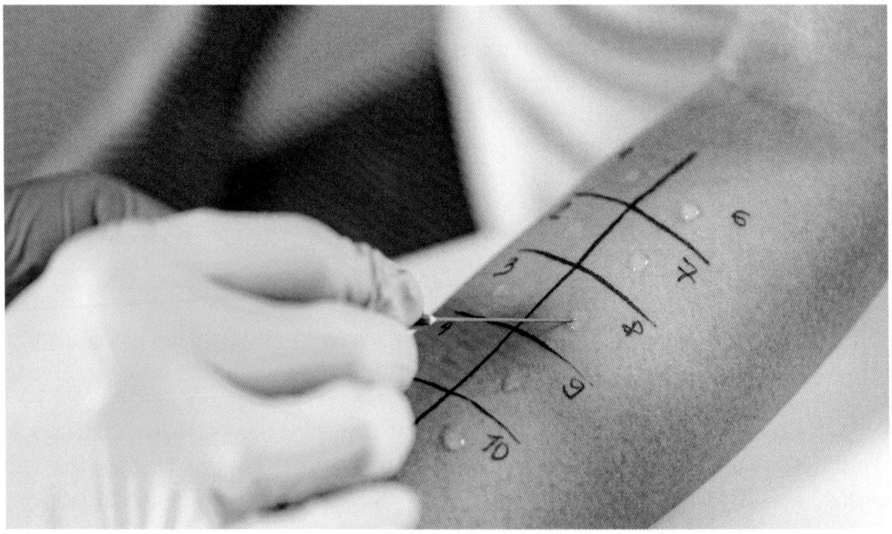

FIGURE 15.7 Skin prick test for common allergens A skin prick allergy test involves applying a small amount of liquid containing a common allergen such as pollen, mold, pet dander, dust mites, and foods to the forearm. A small needle penetrates the skin and introduces a small amount of liquid containing the allergen underneath the skin. Reactions can occur within 15 minutes of exposure to the allergen.

In a patient with allergies, the memory cells produced following initial exposure to allergen have IgE on their plasma membrane with a high affinity for the allergen. Thus, sensitization "primes" the adaptive immune response: memory cells are at the ready to produce higher-affinity immunoglobulins at a faster rate during a secondary adaptive immune response. Remember that a secondary immune response is a faster and more robust response due to the production of memory cells and the processes of somatic hypermutation and isotype switching. Sensitization to an allergen drives the formation of allergen-specific memory cells that can respond more quickly and more effectively upon subsequent exposure to that allergen.

While all type I hypersensitivity reactions are driven by IgE-induced degranulation of mast cells, basophils, and eosinophils, the entry point of the allergen dictates the type of allergic reaction that ensues.

wheal-and-flare reaction Localized inflammation that occurs due to subcutaneous presence of an allergen capable of inducing type I hypersensitivity; presents as a raised lesion surrounded by reddish, inflamed tissue.

urticaria Hive formation due to an inflammatory response during a hypersensitivity reaction.

WHEAL-AND-FLARE REACTION The **wheal-and-flare reaction** occurs when a small amount of allergen is injected into the skin via an insect bite or sting or as part of testing to diagnose specific allergies (**FIGURE 15.7**). Upon injection with an allergen to which a person is allergic, an inflammatory response occurs at the site within minutes of exposure (**FIGURE 15.8**). This typically results in a *wheal*, a raised skin lesion containing fluid, surrounded by a red, itchy area, the *flare*. This initial inflammatory response is an immediate reaction. More widespread swelling can occur as a late-phase reaction.

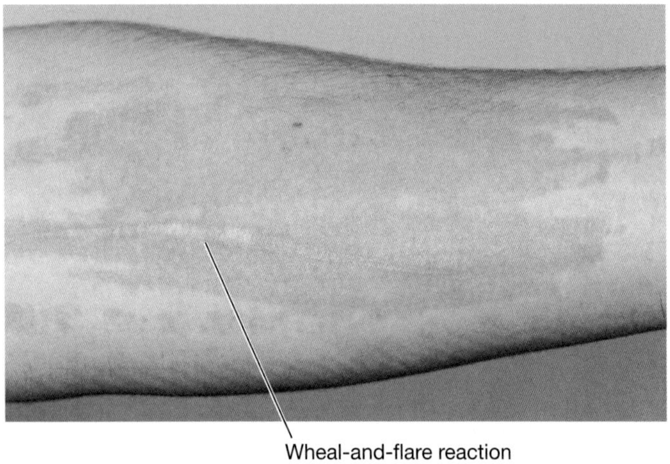

FIGURE 15.8 Wheal-and-flare reaction Exposure to an allergen via the skin results in immediate swelling (wheal) and redness (flare) due to a type I hypersensitivity response that drives localized inflammation.

OTHER SKIN REACTIONS The skin reaction known as **urticaria** is commonly called *hives*. Hives are itchy, raised, red welts on the skin that erupt due to contact with an allergen. They can also occur during systemic anaphylaxis if allergen within the bloodstream is transported to the skin and activates skin-associated granulocytes. Contact with an allergen can also cause a localized rash known as *allergic eczema*. Because skin contact with an allergen can induce an inflammatory response at the site of contact due to degranulation of mast cells, another possible result is *angioedema* (swelling of the tissue beneath the skin surface).

SYSTEMIC ANAPHYLAXIS Allergen presence in the bloodstream can lead to widespread activation of blood-vessel–associated mast cells throughout the body. This causes degranulation and release of histamine and other inflammatory mediators, causing anaphylactic shock symptoms very similar to the action of TNF-α released in the bloodstream during a systemic infection (see Chapter 2). Inflammatory mediators such as histamine in the bloodstream can lead to body-wide vasodilation, causing fluid leakage, loss of blood pressure, and organ failure. Fluid movement into extracellular tissues can cause connective tissue swelling and lead to trachea constriction and death.

FOOD ALLERGY Ingestion of food can lead to allergic reactions due to the many different proteins associated with the ingested material. Allergic reactions to milk, eggs, peanuts, tree nuts, wheat, soy, fish, and shellfish account for about 90% of identified food allergies. Most of these allergies are reactions to proteins present in these foods. A common reaction to food allergens is activation of granulocytes within the gastrointestinal (GI) tract, which can induce vomiting or diarrhea, depending on the location of the release of the inflammatory mediators from the activated granulocytes. While reactions within the GI tract are the most common type associated with ingestion of food allergens, if these allergens are absorbed and transported through the bloodstream, the potential arises for a rash or hives to develop or a systemic anaphylaxis reaction as previously described. Because of the risk of anaphylaxis reaction and the size of the population affected (the Centers for Disease Control and Prevention [CDC] estimates that 4% to 6% of children in the United States suffer from food allergies), there has been an increased focus on research, treatment, and education (see **EMERGING SCIENCE** box).

RHINITIS AND ASTHMA We have seen that the type I hypersensitivity reaction is connected to the location of allergen contact. The same is true with allergens that are inhaled, such as pollens. Inhalation of allergens can lead to activation of granulocytes associated with respiratory mucosal tissue.

Rhinitis is inflammation of the nasal mucous membrane. The release of inflammatory mediators in respiratory mucosa leads to swelling and obstruction of nasal passages and can lead to increased mucus production. While mild rhinitis symptoms are usually not serious, if an allergen triggers activation of granulocytes within the bronchial tubes or lungs, the inflammatory response can lead to airway constriction due to contraction of smooth muscle associated with the respiratory tract and the production of excess mucus.

Asthma is a chronic respiratory disorder that involves inflammation of the airways; its symptoms include wheezing, coughing, and difficulty breathing. Allergens are a common trigger for asthma attacks.

Genetic predisposition to type I hypersensitivity

Residents of developed nations seem to have an increased predisposition to type I hypersensitivity to some of the allergens discussed in this chapter; these individuals are said to be **atopic**. Atopic individuals seem to express more IgE immunoglobulins that are capable of recognizing innocuous allergens.

There are genetic polymorphisms within these populations that appear to shift the balance of IgE signaling within individuals who are predisposed to type I hypersensitivity. For example, there are polymorphisms in MHC class II genes, T-cell receptor genes, IgE Fcε receptor genes, and IL-4 signaling genes that can make a person more susceptible to allergy. The identification of polymorphisms that promote predisposition to allergy will undoubtedly shed light on the susceptibility of certain individuals to type I hypersensitivity. However, since epigenetic factors and environmental factors such as air pollution,

atopic Having a predisposition to mount a type I hypersensitivity reaction to common allergens.

EMERGING SCIENCE

Is the type of sugar attached to IgE immunoglobulins responsible for driving peanut allergies?

Article

K. T. C. Shade, M. E. Conroy, N. Washburn, M. Kitaoka, D. J. Huynh, E. Laprise, S. U. Patil, W. G. Shreffler, and R. M. Anthony. 2020. Sialylation of immunoglobulin E is a determinant of allergic pathogenicity. *Nature* 582: 265–270.

Background

Increasing prevalence of nut allergies has been observed in developed countries, but the factors responsible for the increase in these allergies is unknown. It is well established that IgE responses are a driving force in allergic reactions; however, total and allergen-specific IgE do not always correlate well with the onset of allergies. Because glycosylation plays an important role in the function of IgG, the researchers sought to determine if there is correlation with different sugars attached to IgE and the onset of peanut allergies and hypersensitivity responses.

The Study

Researchers isolated IgE from individuals who reported peanut allergies and others who did not report allergies to peanuts. They found that the IgE isolated from patients with peanut allergies was more reactive when it was preloaded and crosslinked on the surface of mast cells compared to IgE from individuals who did not have peanut allergies. Analysis of the sugars attached to the IgE immunoglobulins isolated from patients with peanut allergies revealed a higher amount of sialic acid attached compared to IgE isolated from nonallergic counterparts. The researchers employed a cutaneous anaphylaxis assay in mice by sensitizing mice intradermally in their ears with ovalbumin-specific immunoglobulins that either contained sialic acid naturally, had their sialic acid removed, or had their sialic acid removed and reattached. The researchers found a cutaneous anaphylaxis response in mice containing IgE with sialic acid attached, either naturally or reattached after removal (**FIGURE ES 15.1**). The study reveals a link between allergic responses and the presence of sialic acid as a sugar attached to IgE, suggesting the role of sialic acid in IgE activity when attached to granulocytes.

Think About...

1. Can the findings in this study be used to predict whether an individual might be predisposed to peanut allergies? If so, how?
2. Would researchers need to be concerned about naturally occurring sialic acid? What are some examples of where sialic acid glycosylation is found?
3. With this knowledge, what are some possible therapeutic interventions that can be used to prevent a peanut allergy? What might some of the challenges be with these therapeutic interventions?

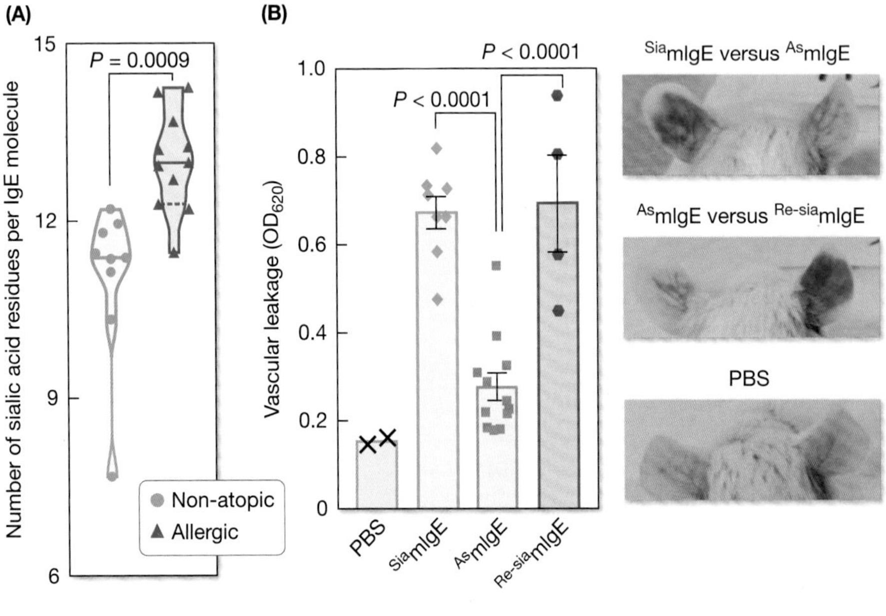

FIGURE ES 15.1 Correlation between the presence of sialic acid on IgE and the manifestation of peanut allergy responses (A) IgE isolated from patients with peanut allergy contains significantly more sialic acid attached to purified IgE compared to IgE isolated from patients who lack a peanut allergy. (B) Mice were injected intradermally in their ears with either phosphate-buffered saline (PBS) or different ovalbumin-specific IgE immunoglobulins, either containing sialic acid (^{Sia}mIgE), IgE with the sialic acid removed (^{As}mIgE), or IgE that had removed and reattached sialic acid ($^{Re\text{-}sia}$mIgE). IgE containing sialic acid caused an increase in vascular leakage and response to antigen in the presence of Evans Blue dye, which measures plasma extravasation at the injection site. (From K. T. C. Shade et al. 2020. *Nature* 582: 265–270. © 2020, The Authors.)

exposure to tobacco smoke, and consumption of processed foods are also involved, the puzzle of fully understanding the role of these polymorphisms in the onset of allergy remains complex.

Diagnostic tests and treatments

Several tests are used to diagnose specific allergies. The most common tests used by allergists are the skin prick test and intradermal skin test, in which potential allergens are introduced underneath the skin. If an allergen induces a wheal-and-flare response (swelling and redness), it may indicate that the individual is allergic to that allergen. A blood test that can detect allergen-specific IgE is used to test for the presence of immunoglobulins capable of reacting with a specific allergen. Another test applies a potential allergen using a patch that adheres to the patient's skin to determine if the applied allergen results in hives.

There are a wide range of available treatments for allergy symptoms, and other treatments focus on preventing a serious or life-threatening allergic reaction from occurring. Several pharmaceutical agents can be used to suppress type I hypersensitivity. Corticosteroids are used in topical creams to treat hives and skin reactions associated with allergen exposure. They are used in oral and inhaled forms to treat rhinitis and asthma. Corticosteroids work to suppress type I hypersensitivity because they suppress the action of the inflammatory mediators associated with the hypersensitivity response. Antihistamines, as their name implies, work to suppress type I hypersensitivity by suppressing the inflammatory action of histamine secreted by activated granulocytes. Epinephrine is used when individuals can suffer from systemic anaphylaxis in the presence of an allergen that has entered the circulation.

Immunotherapy has also been used to reduce type I hypersensitivity. Immunotherapy involves the administration of small amounts of allergen orally or by injection. Treatment with small amounts of allergen can be effective by desensitizing an individual in the presence of the allergen. This prevents an allergic reaction and instead induces a normal immune response and development of different immunoglobulins (other than IgE) to the antigen that do not induce granulocyte activation.

CHECKPOINT QUESTIONS

1. Why is granulocyte activation driven by IgE signaling different from B-cell activation driven by IgE signaling?
2. Why is systemic anaphylaxis driven by inflammatory mediators during an allergic reaction similar to shock induced by TNF-α secretion?

15.3 What factors are responsible for type II hypersensitivity?

LEARNING OBJECTIVES

15.3.1 Describe the type of allergen responsible for the onset of type II hypersensitivity reactions.

15.3.2 Define the immunoglobulin actions that drive type II hypersensitivity.

A type II hypersensitivity reaction is a sensitivity to innocuous material driven by the action of IgM and IgG immunoglobulins that recognize chemically active cell-surface molecules. Upon recognizing certain cell-surface molecules, these immunoglobulins can induce targeting of cells that are mistakenly identified as foreign. The means by which IgM and IgG act in type II hypersensitivity reactions are as opsonins for phagocytosis, activation of complement on the surface of the recognized cell, and activation of antibody-dependent cell-mediated cytotoxicity (**FIGURE 15.9**).

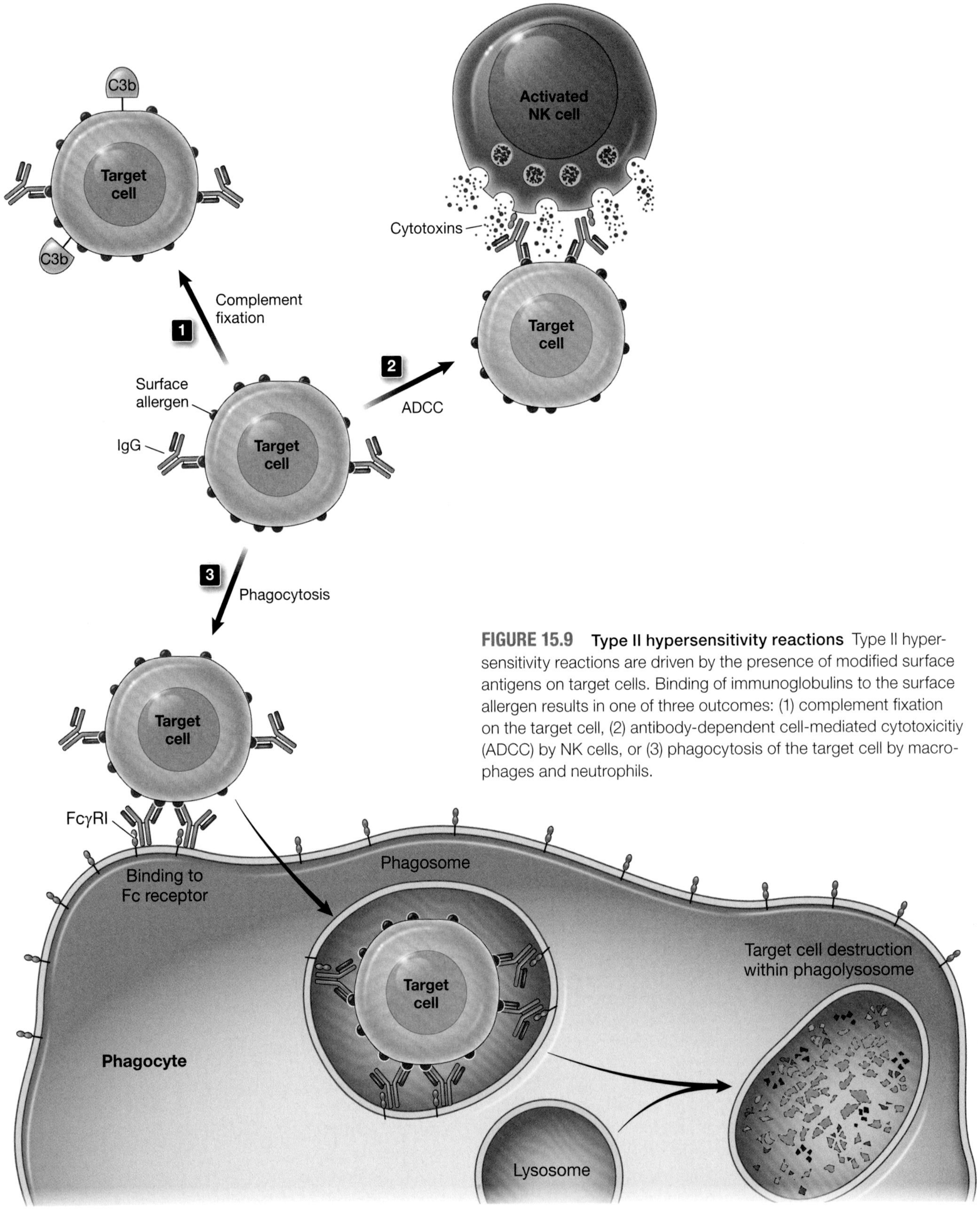

FIGURE 15.9 **Type II hypersensitivity reactions** Type II hypersensitivity reactions are driven by the presence of modified surface antigens on target cells. Binding of immunoglobulins to the surface allergen results in one of three outcomes: (1) complement fixation on the target cell, (2) antibody-dependent cell-mediated cytotoxicitiy (ADCC) by NK cells, or (3) phagocytosis of the target cell by macrophages and neutrophils.

Making Connections: Mucosal Immunity and Type I Hypersensitivity

An important component of mucosal immunity is granulocyte activation and the action of IgE bound to receptors on these cells, which include mast cells, basophils, and eosinophils. Degranulation of these cells in mucosal tissue prompts the expulsion of pathogens from mucosal tissue. However, if granulocytes are bound to IgE that recognizes an allergen, they can induce an inflammatory response where they bind antigen, resulting in a type I hypersensitivity response.

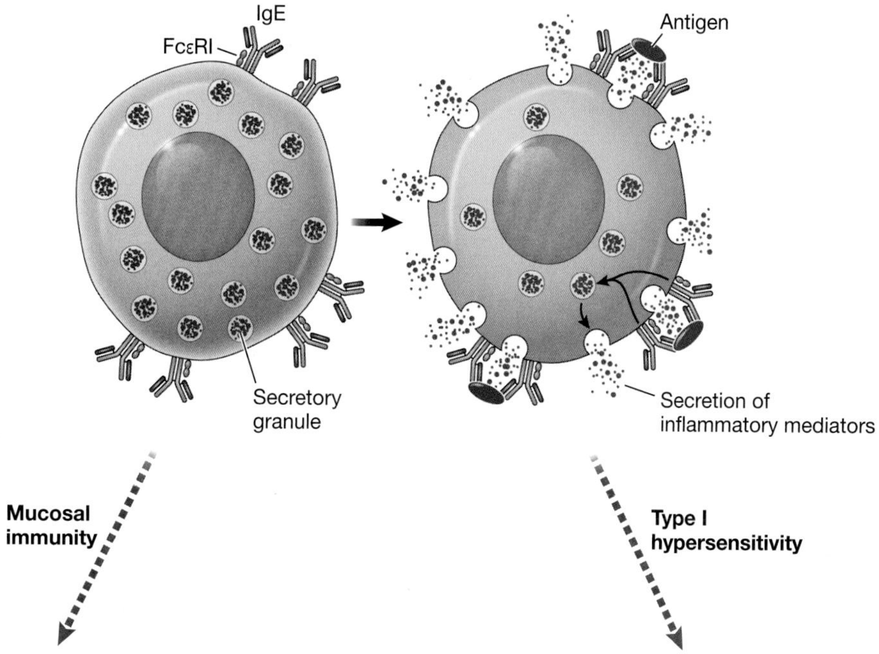

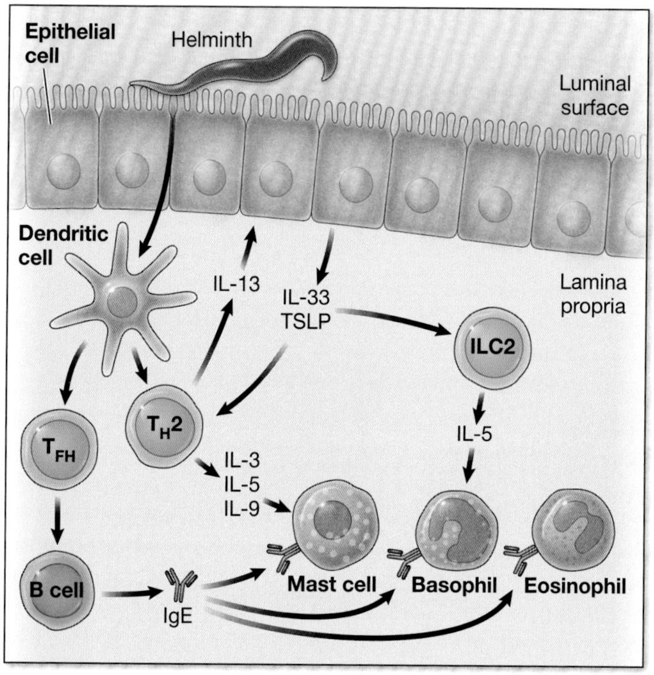

Mucosal immunity, especially to pathogens such as parasites, requires smooth muscle contraction to expel the pathogen from the site of infection. This coordinated immune response involves the action of mucosal epithelial cells, plasma cells expressing IgE, and granulocytes secreting inflammatory mediators.

(Sections 1.4, 1.5, 1.9, 10.4, 10.5, 12.1, 12.2, 12.3, 12.4)

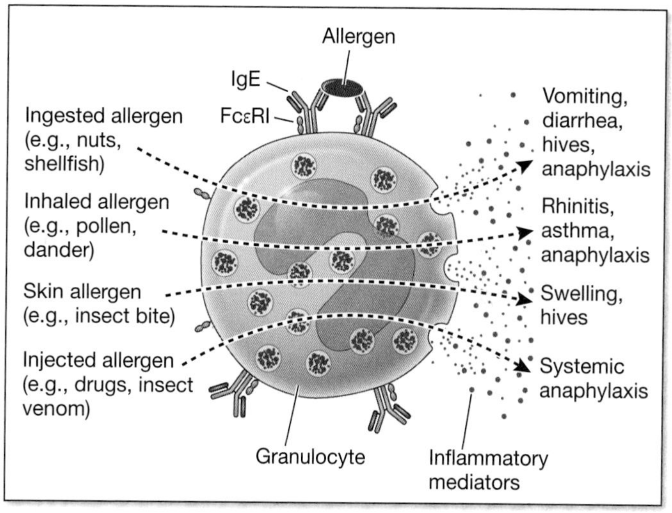

Binding of allergen to IgE at the granulocyte surface induces the release of inflammatory mediators at the site of recognition. This type I hypersensitivity reaction manifests based on the location of granulocyte activation.

(Sections 1.5, 10.5, 15.1, 15.2)

Opsonization revisited

Recall that soluble immunoglobulins with accessible Fc domains, such as IgG, can serve as opsonins for phagocytic cells that express Fc receptors capable of recognizing those immunoglobulin isotypes. For example, phagocytic cells such as macrophages express FcγRI receptors to recognize foreign material opsonized by IgG immunoglobulins (see Chapter 10). Recognition by these Fc receptors induces clustering of those Fc receptors and induction of phagocytosis. Since recognition of opsonizing immunoglobulins by phagocytic cells is independent of the antigen to which they are bound, IgG can serve as an opsonin of cells harboring these immunoglobulins for phagocytic cells such as macrophages.

Complement activation revisited

IgM and IgG can also induce the classical pathway of complement activation. IgM (in its staple form) or two molecules of IgG are capable of binding to C1, which through the activity of an associated protease causes activation through cleavage of the C2 and C4 complement proteins and formation of the classical C3 convertase responsible for cleaving C3 into C3a and C3b. Fixation of C3b on the cell surface promotes phagocytosis due to recognition of C3b through cell receptors such as CR1 and can induce formation of the complement membrane-attack complex (MAC), leading to lysis of the targeted cell. Thus, recognition of a harmless cell-surface molecule by IgM or IgG can induce complement activation and fixation of C3b or formation of an MAC on cells bearing these harmless surface molecules.

Antibody-dependent cell-mediated cytotoxicity

We also saw in Chapter 10 that immunoglobulins such as IgG can serve as a trigger for antibody-dependent cell-mediated cytotoxicity, a mechanism by which an antibody induces destruction of the tagged cell through the action of NK cells. NK cells, using FcγRIII receptors, recognize a bound IgG immunoglobulin and target the cell by releasing cytotoxic molecules and inducing apoptosis. Just as in the cases of phagocytic cell activation and complement activation, cells bound to immunoglobulin molecules at their surface can also cause hypersensitivity due to the activation of NK cells as part of an innate immune response.

Type II hypersensitivity reactions

Recognition of harmless cell-surface molecules by immunoglobulins can induce a variety of effector mechanisms, all of which result in destruction of the cell through phagocytosis, lysis through action of the MAC, or destruction by NK cells. Because many type II hypersensitivity reactions are driven by the action of IgG, and since IgG is prevalent in blood, common examples of type II reactions involve complications within the circulatory system.

BLOOD TRANSFUSION REACTIONS Since MHC complexes are not present on red blood cells, blood transfusions are relatively safe and not likely to activate an immune response in the recipient. However, a type II hypersensitivity reaction can occur due to an improper blood transfusion. Erythrocytes are rich in surface proteins, glycoproteins, and glycolipids, and these molecules are encoded by different alleles in the population. Because these molecules differ from person to person, it is possible that one individual may have immune cells that will recognize blood from another individual as foreign based on these cell-surface differences. Transfusion of such blood can potentially induce a type II hypersensitivity reaction.

Antigens on the surface of red blood cells (RBCs) are actually carbohydrate components of lipids in the RBC membranes and are responsible for the A, B, AB, and O blood types we see in the human population (**FIGURE 15.10**). Individuals possessing immunoglobulins to certain blood type glycolipid antigens

(A)

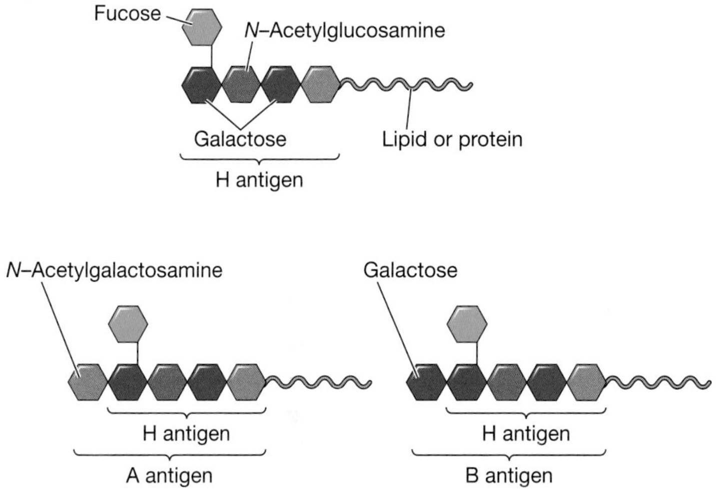

(B)

Blood-Group Phenotype	Antigens on Erythrocytes	Serum Antibodies
A	A	Anti-B
B	B	Anti-A
AB	A and B	None
O	H	Anti-A and anti-B

FIGURE 15.10 ABO blood groups (A) ABO blood groups are determined based on the presence of oligosaccharides on proteins or lipids on erythrocyte plasma membranes. The H antigen is present in all individuals as a core oligosaccharide of five sugars that is added to plasma membrane proteins or lipids. Individuals who are blood type A add *N*-acetylgalactosamine to the H antigen, and individuals who are blood type B add galactose to the H antigen. (B) The presence of blood group antigens determines an individual's blood type. Furthermore, the presence of these blood group antigens determines which soluble immunoglobulins are present in those individuals due to tolerance, preventing B cells that recognize self-blood types from being produced. Individuals who are blood type A produce immunoglobulins to B antigen. Those who are blood type B produce immunoglobulins to A antigen. Individuals who are blood type O (meaning they cannot make A or B antigen) produce immunoglobulins to both A and B antigens. Those who are blood type AB do not produce immunoglobulins to any blood group antigens. (A after G. Litwack 2017. *Human Biochemistry*. Academic Press)

can mark blood cells bearing those glycolipids as "foreign" and drive a type II hypersensitivity reaction.

The three glycolipid antigens on blood cells are H antigen, A antigen, and B antigen. The presence of different antigens is driven by the presence or absence of two glycosyltransferases capable of altering the H antigen (the precursor to A and B antigen) with a different carbohydrate. Individuals with type O blood only have H antigen glycolipids, as they lack the two glycosyltransferases that can modify H antigen. Individuals with type A blood express a glycosyltransferase that adds the terminal carbohydrate *N*-acetylgalactosamine to the H antigen lipid. Individuals with type B blood express a glycosyltransferase that adds the terminal carbohydrate galactose to the H antigen lipid. Individuals with type AB blood express both glycosyltransferases and thus can make both A and B antigen glycolipids.

Also contributing to differences in blood type is the presence of the Rhesus factor (Rh) antigen, of which there are 50 different antigens. The Rh D antigen is the most important and the one most considered in blood typing and in relation to hemolytic anemia. An individual who is blood-type positive, such as AB+, is positive for Rh D antigen.

Individuals who are blood type A produce antibodies to the B antigen, and vice versa. Individuals who are blood type AB do not express immunoglobulins to either A or B antigens. Individuals who are blood type O produce immunoglobulins that can recognize both A and B antigens. Proper blood typing ensures that an individual in need of a blood transfusion does not receive donor blood that would prompt a type II hypersensitivity reaction.

When considering the production of antibodies to A or B antigens in an individual who does not express these antigens, an interesting question arises: How does an individual who does not express a specific antigen create immunoglobulins to an antigen their body has never seen? For instance, a person who is blood type O expresses immunoglobulins to both A and B antigens, but that individual has never come in contact with those antigens (and thus could not raise a specific adaptive immune response to those antigens). The current hypothesis is that the A and B antigens contain carbohydrates similar to those seen on common microorganisms, including influenza virus and gram-negative organisms such as *E. coli*, and it is these microorganisms that provide the antigen for the production of immunoglobulins capable of recognizing blood-type antigens. All individuals are exposed to these microorganisms but due to negative selection, individuals with similar blood-type antigens do not have B cells capable of producing immunoglobulins to that antigen. For example, an individual who is blood type B, which makes immunoglobulins capable of recognizing A antigen, does not make immunoglobulins to B antigen because B cells undergoing development are negatively selected if they are capable of recognizing B antigen. Production of the A antigen antibodies in this individual occurs when that person is exposed to the microorganism that bears antigens similar to the A antigen carbohydrate.

HEMOLYTIC DISEASE OF THE NEWBORN Another possible type II hypersensitivity reaction that can arise is also due to blood incompatibility, but in this case the reaction is caused by recognition of the RBC antigen Rh antigen. This type II reaction can occur during pregnancy when an Rh-negative mother is pregnant with an Rh-positive fetus, since the father was Rh-positive. During a first such pregnancy, there are not usually complications since there is little exposure of the mother to fetal blood due to the placental barrier. Furthermore, exposure to Rh antigen would mount a primary immune response and the production of IgM immunoglobulins, which are also incapable of passing across the placental barrier.

During delivery, the mother's exposure to fetal cells is increased, allowing the mother's immune system to potentially respond to the Rh antigen and produce memory to these foreign molecules. This response would eventually drive the formation of B cells that express Rh-specific IgG, which is capable of crossing the placental barrier. Upon a subsequent Rh-positive pregnancy, the Rh-specific IgG production would result in the opsonization of fetal RBCs and destruction of those cells because of the induced type II hypersensitivity response. The loss of fetal RBCs could cause severe anemia (deficiency of RBCs or of hemoglobin in the blood) or even death of the developing fetus.

DRUG-INDUCED ANEMIA Certain antibiotics, such as penicillin, are capable of inducing hemolytic anemia because they can covalently modify self-antigens by attaching to proteins on the surface of RBCs. These covalently modified antigens may be seen as foreign by the individual's immune system and cause the production of immunoglobulins that induce type II hypersensitivity and reduction of RBCs.

Diagnostic tests and treatments

Since many common type II hypersensitivity reactions are driven by incompatibilities with antigen present on the surface of RBCs, the most common means used to prevent such reactions is blood typing and **crossmatching**. Blood typing identifies a person's blood type and Rh factor. Crossmatching is a test used to check for harmful interactions between a potential donor and recipient of blood, tissue, or an organ, especially testing for HLA compatibility.

In Chapter 11, we explored Rh-antigen incompatibility and the use of RhoGAM as a therapeutic strategy to prevent fetal hemolytic anemia. RhoGAM is a solution of antibodies that prevents activation of a primary immune response to Rh antigen. First, RhoGAM functions as an opsonin on fetal RBCs that contain the Rh antigen. Binding of RhoGAM immunoglobulins to Rh-positive fetal RBCs works to clear these cells before they are recognized by the mother's immune

crossmatching Blood test used before blood transfusion or tissue/organ transplantation to check for harmful interactions between donor blood and that of a potential recipient.

system. In Chapter 11 we also explored the action of the inhibitory Fcγ receptor on naïve B cells that, when engaged with an IgG immunoglobulin, prevent their activation. The use of RhoGAM as a treatment during an Rh-positive pregnancy acts to suppress activation of naïve B cells capable of recognizing Rh antigen. When a naïve B cell binds to Rh antigen via the immunoglobulin and also comes into contact with RhoGAM antibodies via an inhibitory Fcγ receptor, activation of that naïve B cell is shut down. Thus, RhoGAM treatment provides two separate therapeutic interventions against a type II hypersensitivity response.

CHECKPOINT QUESTIONS

1. Describe the three ways by which IgM or IgG can induce type II hypersensitivity.
2. Explain why a person who is blood type AB can receive blood from a donor with blood type O, but a person who is blood type O cannot receive blood from a donor with blood type AB.

15.4 | How does type III hypersensitivity occur?

LEARNING OBJECTIVE

15.4.1 Explain how type III hypersensitivity reactions are driven by inefficient clearance of immune complexes.

As with type II hypersensitivity, type III hypersensitivity is induced by the action of IgG or IgM immunoglobulins. In this case, the immunoglobulins recognize soluble antigens and prompt the formation of immune complexes (antigen–antibody complexes), a typical process in the presence of soluble antigen. While the formation of immune complexes is a normal part of the immune response, inefficient clearance can result in hypersensitivity reactions due to abnormal activation of inflammatory responses and cells that attempt to clear the immune complexes.

Immune complexes revisited

Recall that immune complexes can form when immunoglobulins such as IgG bind to soluble antigen, drive the fixation of complement on the antigen surface, and promote the clearance of these molecular complexes via phagocytosis (see Chapter 10). The effectiveness of the clearance process depends on the size of the immune complexes—larger complexes can fix complement more efficiently and are more easily removed through normal clearance mechanisms, especially phagocytosis by cells expressing complement receptors and Fc receptors. However, small immune complexes are more difficult to opsonize with complement and are not efficiently cleared, which can lead to type III hypersensitivity (**FIGURE 15.11**).

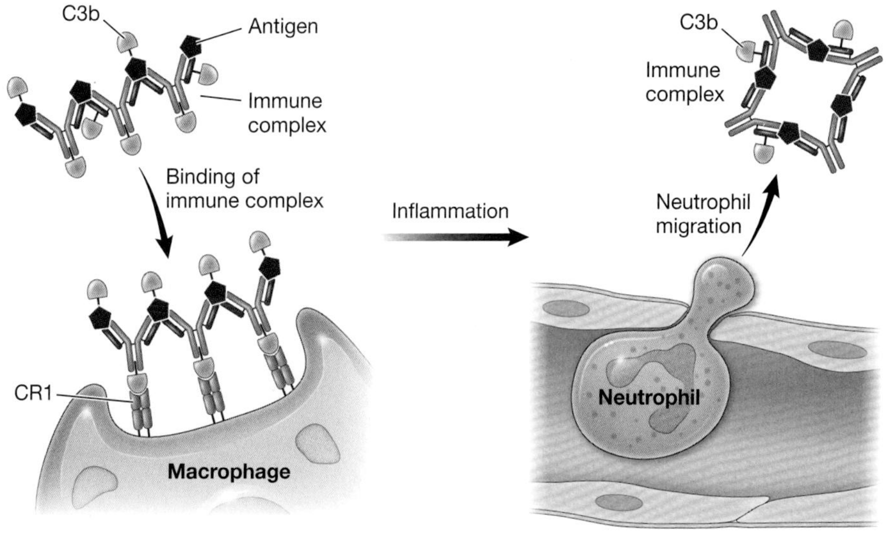

FIGURE 15.11 Type III hypersensitivity reactions Type III hypersensitivity reactions occur when antigens bound to immunoglobulins cause complement activation and the formation of immune complexes. If these immune complexes are not efficiently cleared, macrophages bind to them via CR1 receptors, causing an inflammatory response and the recruitment of neutrophils to the site of inflammation.

Type III hypersensitivity reactions

When immune complexes form but are inefficient in the fixation of complement and thus are inefficiently cleared, they remain present in the bloodstream and can be deposited in blood vessels and tissues. Factors that influence whether immune complexes drive type III hypersensitivity include:

- The affinity of a particular antigen to a specific location or tissue
- Disruption of normal phagocytosis
- The chemical nature of the antigen (influences the efficiency of engulfment by phagocytic cells)

Deposited immune complexes can bind to neutrophils and macrophages, which then secrete inflammatory cytokines at the site of deposition. The ensuing inflammation allows the immune complexes to migrate into the tissue, due to increased permeability of the vasculature, and induce complement activation. This complement activation can lead to localized inflammatory responses due to the production of C3a and C5a and further recruitment of innate immune cells to the site. The inflammatory response can cause vasculitis (inflammation of the blood vessels), glomerulonephritis (inflammation of kidney glomeruli), or arthritis (inflammation of the joints), depending on where the immune complexes were deposited.

SERUM SICKNESS **Serum sickness** is a classic example of a type III hypersensitivity reaction. During the early twentieth century, diphtheria was treated by using serum from horses that had been inoculated with the diphtheria bacterium *Corynebacterium diphtheriae*. While the horse serum was capable of clearing the bacterial infection, it was also capable of inducing a type III hypersensitivity reaction characterized by fever, vasculitis, glomerulonephritis, and arthritis (**FIGURE 15.12**).

serum sickness A type III hypersensitivity reaction driven by the formation of immune complexes that contain immunoglobulins from another organism.

While horse serum is rarely used in modern medicine, the incidence of serum sickness has increased in the population due to the use of mouse monoclonal antibodies as immunotherapeutics to treat a variety of diseases, including cancers. This side effect, which typically manifests 7 to 10 days after exposure to the antigen, is sometimes minimized by genetic engineering to make the immunoglobulin used for the immunotherapy as similar as possible to human immunoglobulins.

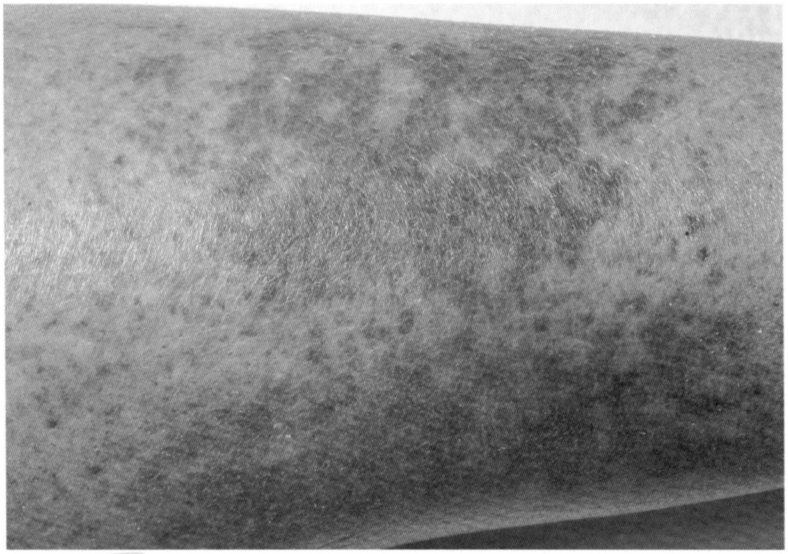

FIGURE 15.12 **Serum sickness** Serum sickness occurs due to a type III hypersensitivity reaction. The injection of a foreign serum, protein, or drug results in the formation of small immune complexes where immunoglobulins bind to the foreign antigen. Often, serum sickness manifests as vasculitis, which is due to the inflammatory response produced by the activation of macrophages by small immune complexes.

ARTHUS REACTION An *Arthus reaction* is a localized type III hypersensitivity response driven by the formation of immune complexes at the site of deposition of a large amount of antigen, followed by an inflammatory response, and peaks 4 to 10 hours after exposure to the antigen. This reaction results in complement fixation on the immune complexes and production of the anaphylatoxins C3a and C5a. Recall that these two cleaved complement subunits increase vascular permeability, cause degranulation of cells such as mast cells, and promote an inflammatory response. The symptoms associated with an Arthus reaction include the responses driven by C3a and C5a, including localized inflammation, swelling, and bleeding (**FIGURE 15.13**).

Autoantigens and immune complexes

Chronic type III hypersensitivity reactions can be induced by the presence of immune complexes containing **autoantigens** (antigens recognized during autoimmune responses). Some autoantigens induce the formation of immune complexes that are not easily cleared and result in type III hypersensitivity reactions. We will explore autoimmune disorders in detail in Chapter 16. Examples of those that result from type III hypersensitivity are systemic lupus erythematosus (its characteristic butterfly rash is due to type III hypersensitivity) and rheumatoid arthritis (induced by type III hypersensitivity of the joints).

Diagnostic tests and treatment

Other than attempting to observe circulating immune complexes (not a definitive test for type III hypersensitivity), not many diagnostic tools can be employed to test for this hypersensitivity. Treatments typically focus on anti-inflammatory pharmaceuticals, including the use of corticosteroids (for similar purposes as in treating type I hypersensitivity reactions).

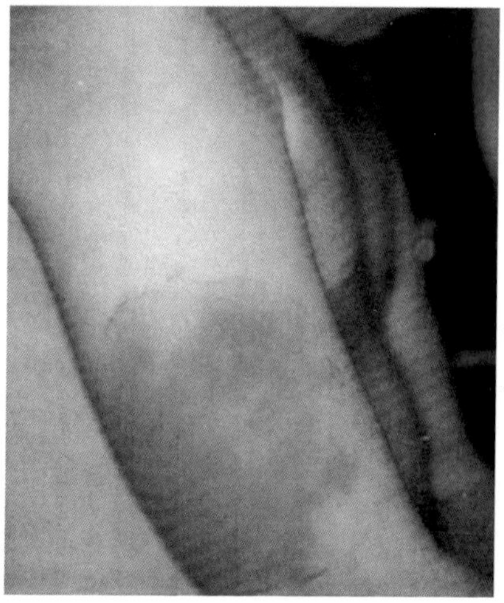

From A. Kroger. 2011. *Contraindications and Precautions to Vaccination*. National Center for Immunization and Respiratory Diseases. DHHS/CDC

FIGURE 15.13 Arthus reaction The Arthus reaction is a type III hypersensitivity reaction caused by localized inflammation and leakage of blood vessels at the injection site. The reaction produces a rash due to vasculitis at the injection site and results in localized tissue necrosis.

• CHECKPOINT QUESTION

1. Contrast normal immune complex clearance with the lack of clearance seen during type III hypersensitivity.

15.5 What is delayed-type (type IV) hypersensitivity?

LEARNING OBJECTIVES

15.5.1 Describe the allergen responsible for the onset of type IV hypersensitivity reactions.

15.5.2 Differentiate type IV hypersensitivity from the other three types in regard to the amount of time required to induce a reaction.

Hypersensitivity reactions types I through III are induced by abnormal responses of the humoral arm of the adaptive immune system. In contrast, type IV hypersensitivity is induced by the cellular arm of the adaptive immune system. Type IV is also known as **delayed-type hypersensitivity** because of the time required for symptoms to develop. Most often, type IV hypersensitivity is an inflammatory response to intracellular pathogens. In all cases, it is driven by T-cell activation and effector mechanisms.

Phases of type IV hypersensitivity

Type IV reactions require the processing and presentation of antigens via antigen-presenting cells (APCs), such as macrophages, on MHC molecules. These are presented to T cells, usually CD4 T cells that will develop into T_H1 helper T cells, but the activated cells can also be CD4 T cells that will develop into T_H17 or T_H2 helper T cells or that can be presented via MHC class I molecules to CD8 T cells (**FIGURE 15.14**). Type IV requires two phases to induce the hypersensitivity reaction, which is another reason for the delayed onset of symptoms.

SENSITIZATION PHASE The first phase of type IV hypersensitivity is the *sensitization phase*. As in any adaptive immune response, APCs must present antigen to T cells via MHC molecules to activate the T cell bearing the receptor that can recognize the MHC-antigen complex. This activation typically occurs due to the action of localized APCs, not in the context of a secondary lymphoid tissue. Often the APCs are macrophages, but resident endothelial cells can also serve as APCs during this phase (see Figure 15.14).

autoantigens Self-antigens recognized as foreign during an autoimmune response.

delayed-type hypersensitivity Allergic reaction driven by the activation of an intracellular antigen processed and presented to T cells; commonly the effector T cells activated during a delayed-type hypersensitivity reaction are helper T cells that activate macrophages and an inflammatory response through the secretion of IFN-γ.

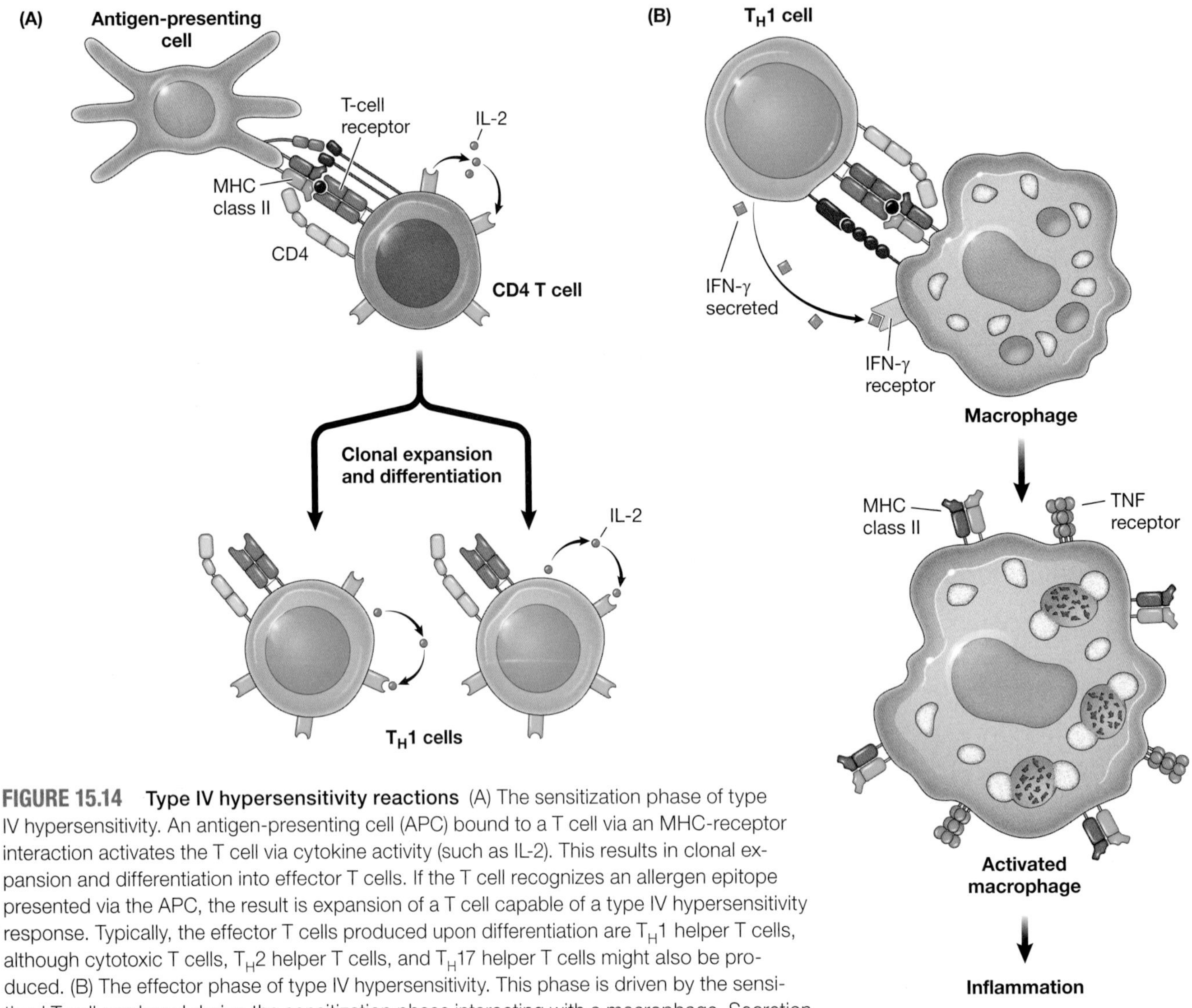

FIGURE 15.14 Type IV hypersensitivity reactions (A) The sensitization phase of type IV hypersensitivity. An antigen-presenting cell (APC) bound to a T cell via an MHC-receptor interaction activates the T cell via cytokine activity (such as IL-2). This results in clonal expansion and differentiation into effector T cells. If the T cell recognizes an allergen epitope presented via the APC, the result is expansion of a T cell capable of a type IV hypersensitivity response. Typically, the effector T cells produced upon differentiation are T_H1 helper T cells, although cytotoxic T cells, T_H2 helper T cells, and T_H17 helper T cells might also be produced. (B) The effector phase of type IV hypersensitivity. This phase is driven by the sensitized T cell produced during the sensitization phase interacting with a macrophage. Secretion of IFN-γ by the helper T cell results in activation of the macrophage and the hypersensitivity response. The macrophages produce an inflammatory response at the site of activation.

EFFECTOR PHASE After a T cell has been sensitized to a particular antigen, subsequent exposure to that antigen drives the differentiation of that T cell into an effector T cell, beginning the *effector phase* of delayed-type hypersensitivity. Since the most common T cell activated during a type IV reaction is a T_H1 helper T cell, activation causes secretion of cytokines such as IFN-γ and CXCL8 and the activation and recruitment of macrophages to the site of T-cell activation. Since this multi-step process requires approximately 48 to 72 hours after exposure to the antigen, the symptoms associated with delayed-type hypersensitivity do not appear until this maximal activation.

Type IV hypersensitivity reactions

Type IV hypersensitivity reactions are often observed in the context of intracellular infections. Intracellular bacteria, including mycobacterial species, viruses (including herpes simplex virus and smallpox), and intracellular fungi (including *Candida albicans*), can all induce type IV hypersensitivity reactions. Delayed-type hypersensitivity often drives macrophage activation, resulting in enhanced macrophage phagocytosis and lytic activities. If the antigen is not easily removed, prolonged macrophage activation and cellular destruction can be observed and may result in the formation of granulomas due to the extensive macrophage activity and resultant tissue damage.

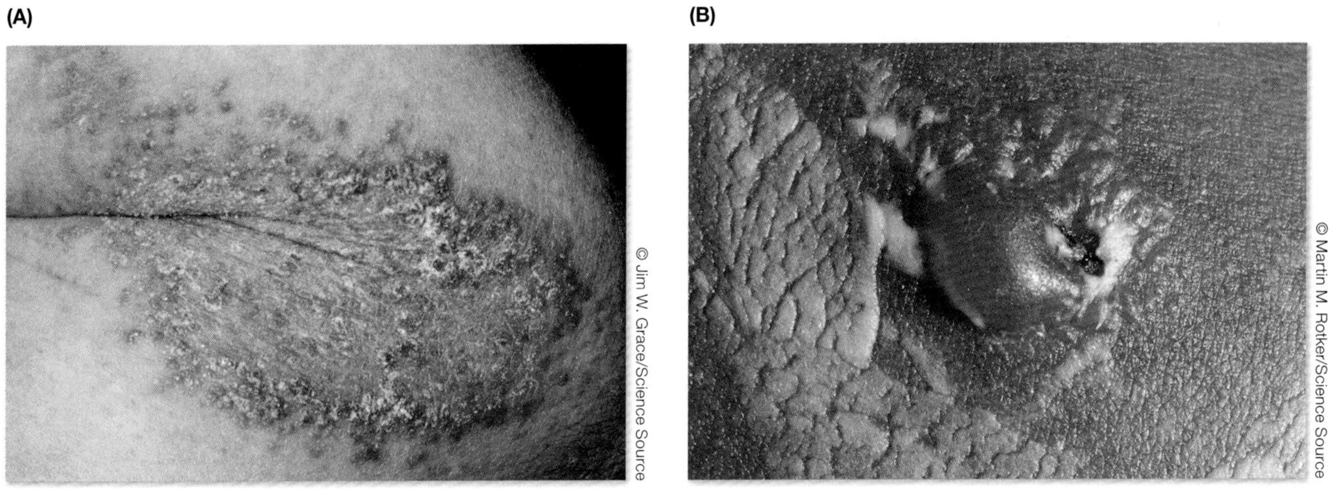

FIGURE 15.15 Photographs of type IV hypersensitivity reactions (A) Contact dermatitis occurs in individuals who are allergic to poison ivy. (B) The tuberculin test is a type IV hypersensitivity reaction that occurs in individuals who have previously been exposed to *Mycobacterium tuberculosis* antigens.

While type IV hypersensitivity is most frequently associated with intracellular infections, certain chemicals can also induce this response. Such is the case with *contact dermatitis* caused by reactive chemicals, including those seen in poison oak and poison ivy (**FIGURE 15.15**). These compounds can covalently modify skin proteins, which are then processed and presented to T cells. Activation of antigen-specific T cells induces the localized inflammation associated with this condition.

Diagnostic tests and treatments

Type IV hypersensitivity can be diagnosed in a manner similar to the skin tests used for allergy testing. However, type IV hypersensitivity tests require monitoring over a longer period of time (usually 48 to 72 hours) to determine if an inflammatory response has occurred. One classic test used to determine delayed-type hypersensitivity is the tuberculin test, used to test prior exposure to *Mycobacterium tuberculosis*.

TUBERCULIN TEST The tuberculin test is used as a diagnostic tool to observe prior exposure to *M. tuberculosis* antigen. A small amount of *M. tuberculosis* antigen is applied subcutaneously, and the site of application is monitored. An inflammatory reaction that arises in 48 to 72 hours (see Figure 15.15) suggests that the patient has had prior exposure and sensitization to the *M. tuberculosis* antigen. This test can be complicated by several factors. A positive test result can occur due to current infection with the bacteria, prior infection and clearance of a tuberculosis (TB) infection (since these individuals will have memory T cells capable of a type IV hypersensitivity reaction), and vaccination (for the same reason as those who have cleared a tuberculosis infection). Thus, a positive tuberculin test often requires further investigation to determine whether the positive result is due to an active infection. Since European countries commonly vaccinate to prevent TB infection, individuals from these countries often have false-positive results due to the presence of memory T cells that recognize the *M. tuberculosis* antigen used in the test.

CHECKPOINT QUESTION

1. Why does delayed-type hypersensitivity often not surface until 48 to 72 hours after antigen exposure?

Making Connections: Allergic Reactions—When the Immune Response Goes Wrong

The immune response employs many different weapons to recognize and destroy pathogens that can cause us harm. However, if immune responses are aimed at innocuous materials we come into contact with (allergens), it results in effects that vary from being a nuisance to being life-threatening.

Type I Hypersensitivity

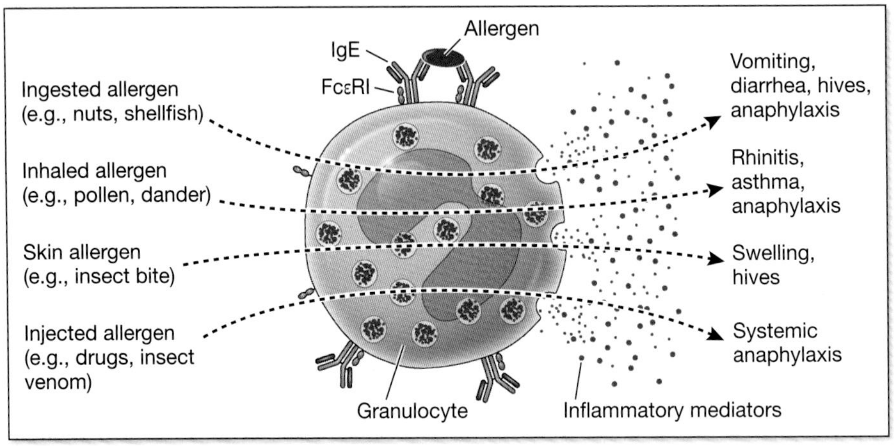

Type I hypersensitivity is caused by the degranulation of granulocytes upon binding of an allergen to IgE at their surface. This results in release of inflammatory mediators at the site of recognition and induces a variety of effects depending on where this recognition occurs.

(Sections 1.5, 10.5, 15.1, 15.2)

Type II Hypersensitivity

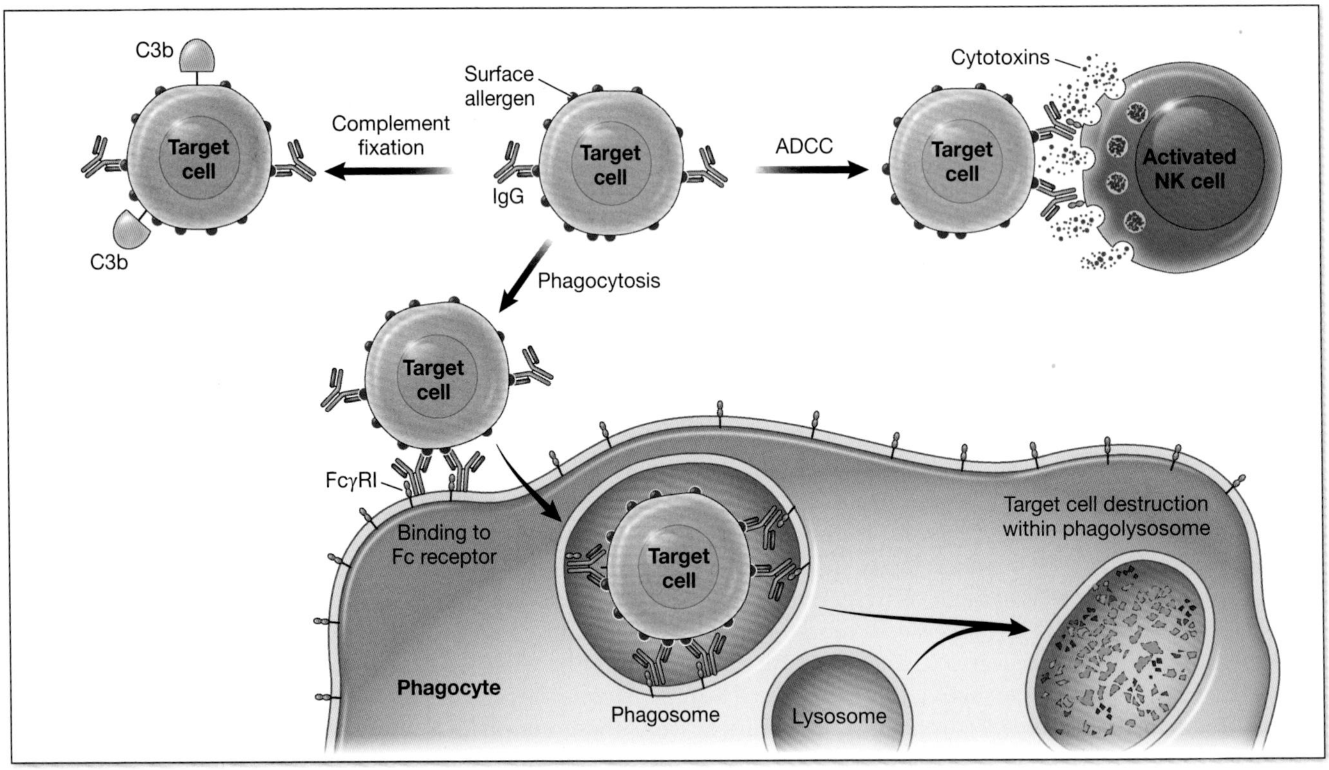

Type II hypersensitivity is caused by the production of immunoglobulins, either IgM or IgG, to modified surface molecules on cells. Binding of these immunoglobulins to modified antigens results in phagocytosis of the opsonized cell, targeting the cell for destruction by NK cells, and complement activation.

(Sections 1.5, 2.1, 2.7, 3.1, 3.3, 10.5, 15.3)

Type III Hypersensitivity

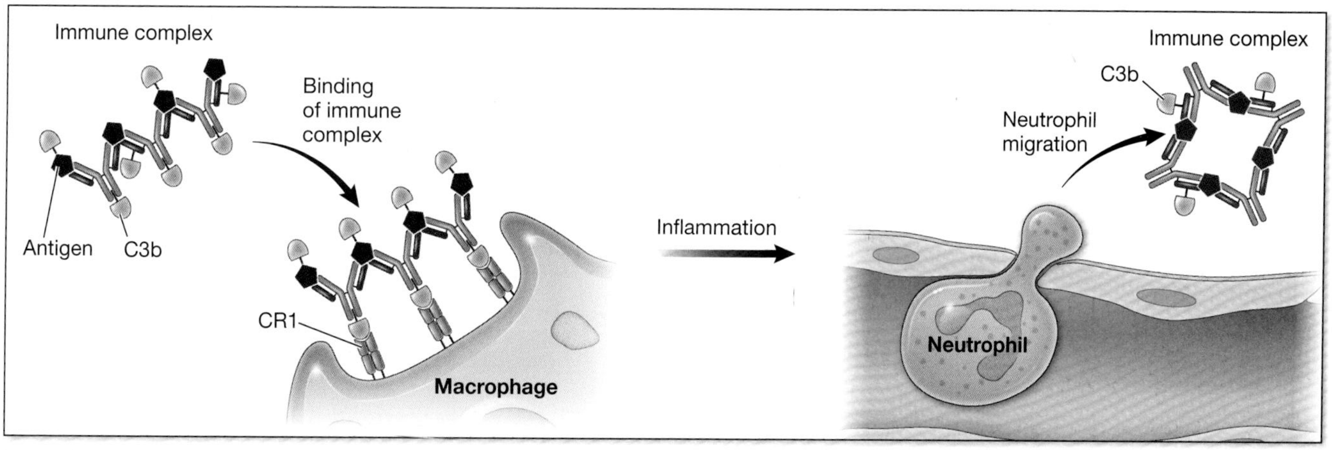

Type III hypersensitivity is caused by the formation of small immune complexes to allergens through immunoglobulin binding to the allergen and complement fixation. Recognition of immune complexes by macrophages induces their activation and an ensuing inflammatory response, resulting in vascular permeability and an influx of neutrophils to the site of recognition.

(Sections 1.5, 2.5, 2.6, 10.5, 15.4)

Type IV Hypersensitivity

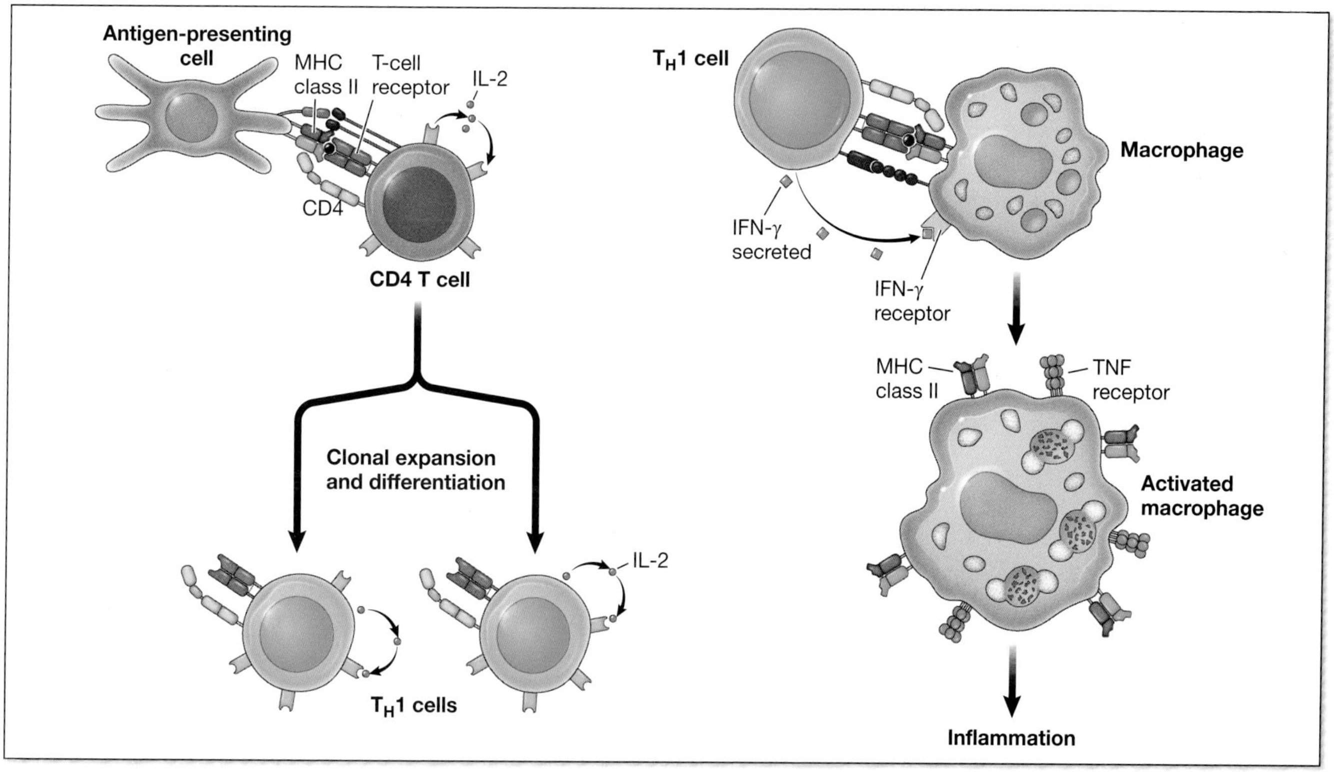

Type IV hypersensitivity is caused by the activation of T cells due to recognition of allergen presented by an antigen-presenting cell. Activation of the T cell results in their effector function, which varies based on the type of effector T cell produced. Most commonly, type IV hypersensitivity reactions result in activation of T_H1 helper T cells, which secrete IFN-γ.

(Sections 1.5, 2.5, 4.2, 4.7, 7.7, 7.14, 15.5)

Summary

15.1 What are the different types of hypersensitivity responses?

- There are four types of hypersensitivity reactions: type I, type II, type III, and type IV. These differ based on the immune response responsible for inducing the response.

15.2 How does type I hypersensitivity occur?

- Type I hypersensitivity reactions are driven by allergen recognition by IgE bound to granulocytes including mast cells, basophils, and eosinophils.
- The type of allergy driven by a type I hypersensitivity reaction manifests primarily based on the location of contact with the allergen.

15.3 What factors are responsible for type II hypersensitivity?

- Type II hypersensitivity reactions are driven by the action of soluble immunoglobulins, especially IgM and IgG, reacting with cell-surface molecules and inducing complement fixation or destruction of the cells by antibody-dependent cell-mediated cytotoxicity.

15.4 How does type III hypersensitivity occur?

- Type III hypersensitivity reactions are caused by inefficient clearance of soluble antigens bound as small immune complexes in the bloodstream; these immune complexes can induce localized inflammation at the site of accumulation.

15.5 What is delayed-type (type IV) hypersensitivity?

- Unlike types I through III reactions, type IV hypersensitivity is an induction of the cellular arm of the adaptive immune response, typically caused by intracellular pathogens, such as *Mycobacterium tuberculosis*, at locations outside of secondary lymphoid tissue.
- Because type IV hypersensitivity reactions involve activation of T cells, which requires time, they are also referred to as delayed-type hypersensitivity reactions.

Review Questions

15.1 You are a physician, and a patient comes to your office with hives. You suspect that the patient is suffering from an allergic reaction but want to rule out a type IV hypersensitivity reaction. What types of tests would help you distinguish between a type I and a type IV hypersensitivity reaction?

15.2 Postulate why immunotherapy can be an effective treatment for type I hypersensitivity.

15.3 A patient comes into the emergency room suffering from multiple gunshot wounds and significant loss of blood. While you have been successful in slowing the loss of blood, you realize that a blood transfusion is needed. However, you do not know the patient's blood type and do not have time to carry out a crossmatch test. Which type of blood should be given and why?

15.4 A patient goes to an urgent care clinic due to pain and a rash on a large portion of the body. A medical history reveals that the patient is undergoing treatment with the monoclonal antibody Humira® as a therapy for rheumatoid arthritis. You suspect that the rash is a vasculitis reaction due to serum sickness. Explain why.

15.5 Why would a tuberculin test produce a positive result in an individual who has recovered from a previous tuberculosis infection?

CASE STUDY REVISITED

Systemic Anaphylaxis

A 29-year-old man is brought to the emergency room with a rash and difficulty breathing and swallowing. The man has a weak, rapid pulse and low blood pressure. His wife reports that he was stung by hornets about 2 hours earlier. She also reports that her husband does not recall having been stung by a bee or hornet before this incident. The ER physician diagnoses systemic anaphylaxis and immediately treats the patient with epinephrine, intravenous cortisone, and oxygen to help him breathe.

Think About...

Questions for individuals

1. What is the purpose of cortisone treatment for this patient?
2. What is the purpose of epinephrine treatment for this patient?
3. What components of hornet venom may cause an allergic reaction?

Questions for student groups

1. Why is it significant that the patient does not recall having been stung by a bee or hornet before this incident?
2. If this patient were stung again, would you predict that the allergic response would be stronger or weaker than what he is currently experiencing? Why?
3. A common prescription for individuals who are allergic to bee or hornet stings is an autoinjector containing epinephrine. Why would this be prescribed for an individual rather than having epinephrine administered at a hospital?

MAKING CONNECTIONS Key Concepts	COVERAGE (Section Numbers)
Comparison of granulocyte action during mucosal immunity versus type I hypersensitivity	1.4, 1.5, 1.9, 10.4, 10.5, 12.1, 12.2, 12.3, 12.4, 15.1, 15.2
Type I hypersensitivity and IgE action	1.5, 10.5, 15.1, 15.2
Type II hypersensitivity and IgG action	1.5, 2.1, 2.7, 3.1, 3.3, 10.5, 15.3
Type III hypersensitivity and immune complex clearance	1.5, 2.5, 2.6, 10.5, 15.4
Type IV hypersensitivity and T-cell action	1.5, 2.5, 4.2, 4.7, 7.7, 7.14, 15.5

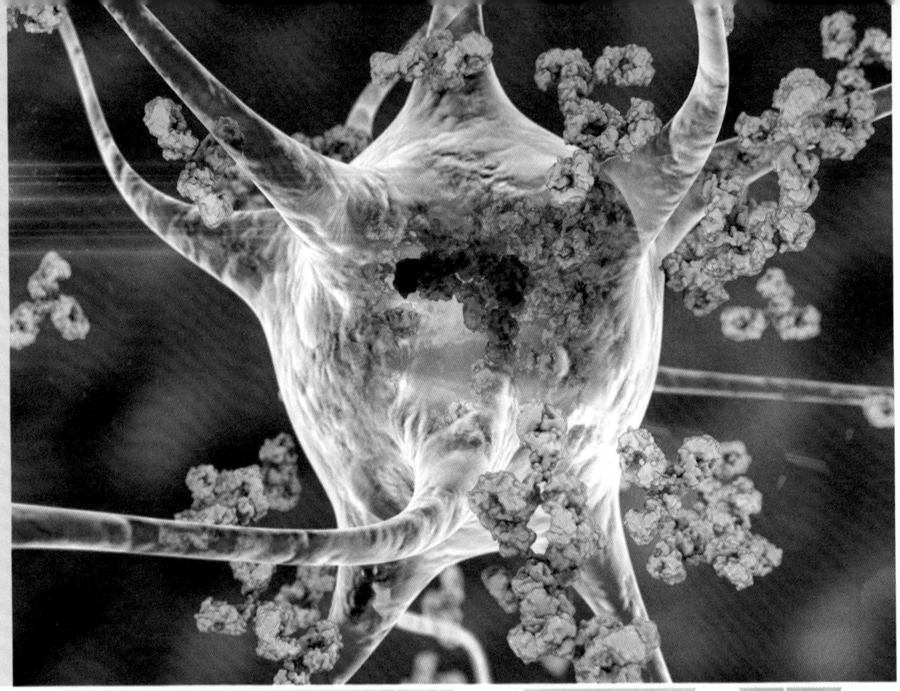

16 Autoimmune Diseases

QUESTIONS Explored

16.1 What is the relationship between self-tolerance and autoimmunity?

16.2 How do genetic and environmental factors affect the progression of autoimmune disease?

16.3 Why do many autoimmune diseases target endocrine glands?

16.4 What are the similarities between hypersensitivity and autoimmune diseases?

16.5 How do autoantibodies promote the progression of autoimmune disease?

CASE STUDY: Kai's Story

"It's pizza night, everyone!" Derrick comes in the front door holding the familiar boxes associated with Friday nights. He is greeted by his wife, Olivia, who takes the boxes and carries them out to the kitchen.

"Dinner, Kai!" Olivia calls out to their son, who is in his bedroom finishing up his homework.

Kai comes downstairs, thinking that pizza night used to be so much more fun when he didn't end up feeling crummy later in the evening. However, he didn't want to upset his parents, especially when this was their Friday night tradition and he enjoyed catching up on the week with them. During dinner, they chat about how school is going for Kai and the events of his parents' work week. Before long, they are clearing the table and moving to the living room to watch a movie.

A few hours later, Kai gets up and runs to the bathroom. He cries out, "Mom, I don't feel good." Olivia rushes to the bathroom, and Kai tells her that his stomach hurts and that he has thrown up and has diarrhea. Olivia realizes that this has occurred on a regular basis over the past few months and almost always on Friday nights. Tonight's episode seems to be the worst so far, and she decides to take Kai to the hospital ER.

To understand Kai's symptoms, and to answer the case study questions at the end of this chapter, we will explore problems that occur when tolerance to self-antigens is lost. Cells of the adaptive immune system undergo somatic recombination events to provide a diverse array of receptors at their surfaces. During this developmental process, these cells are tested to determine whether they recognize self-antigen. But what happens if cells that recognize self-antigen aren't removed during this process? A consequence of the loss of tolerance within the adaptive immune system is autoimmunity—the immune system recognizes self-tissue as "foreign" and destroys that tissue.

In this chapter, we will see remarkable similarities between autoimmune diseases and the hypersensitivity responses discussed in Chapter 15. While the antigens recognized in hypersensitivity reactions (foreign innocuous antigens) differ from those recognized in autoimmune diseases (self-antigens), in both cases, recognition prompts activation of adaptive immune responses that can be damaging and life-threatening.

autoimmunity Adaptive immune response raised against a self-antigen due to lack of central or peripheral tolerance.

16.1 What is the relationship between self-tolerance and autoimmunity?

LEARNING OBJECTIVE

16.1.1 Explain how loss of self-tolerance during lymphocyte development or circulation can lead to autoimmunity.

We have seen that the immune system employs many mechanisms to avoid self-recognition and targeting of normal cells. Because the immune system uses powerful tools to recognize and combat pathogens, it is critical to ensure that these defense mechanisms are aimed only at foreign molecules and organisms rather than at self-molecules or cells.

Paramount to the immune response are the activities that drive self-tolerance, including central and peripheral tolerance. As we saw in Chapters 5 and 8, central tolerance is driven by the mechanism of negative selection in primary lymphoid tissues, and as discussed in Chapter 7, peripheral tolerance occurs because of the process of anergy and the action of regulatory T cells. Failure of these mechanisms can lead to the targeting and destruction of tissues through **autoimmunity**, the misidentification of self-molecules and cells as foreign and the subsequent targeting of those molecules and cells by the immune response.

Negative selection and self-tolerance revisited

The diversity of the adaptive immune system is driven by the recombination events that occur at the T-cell receptor and immunoglobulin loci during T-cell and B-cell development. The action of enzymes involved in recombination, including V(D)J recombinase and terminal deoxynucleotidyl transferase (TdT), provides an elaborate mechanism for increasing lymphocyte diversity, but receptors created through these recombination events may recognize self-molecules. While the process of positive selection in bone marrow and the thymus ensures proper B-cell and T-cell function if they encounter their cognate antigen in secondary lymphoid tissue, it is the process of negative selection (see Chapters 5 and 8) that drives *central tolerance*. Central tolerance ensures that properly assembled T-cell receptors and immunoglobulins do not recognize self-antigens, as developing T cells and B cells that express receptors that recognize self-antigens are removed by apoptosis due to negative selection (**FIGURE 16.1**).

Complementing central tolerance are mechanisms that drive *peripheral tolerance*, including the activities of regulatory T cells (see Chapter 7), the requirement of helper T cells to drive B-cell activation, and activation of anergy of T cells and B cells that do not receive a costimulatory signal. Regulatory T cells suppress effector T-cell action using a variety of different mechanisms, including secreting immunosuppressive cytokines such as IL-10 and TGF-β. Helper T cells, including T_H2 and T_{FH} helper T cells, drive activation of B cells and

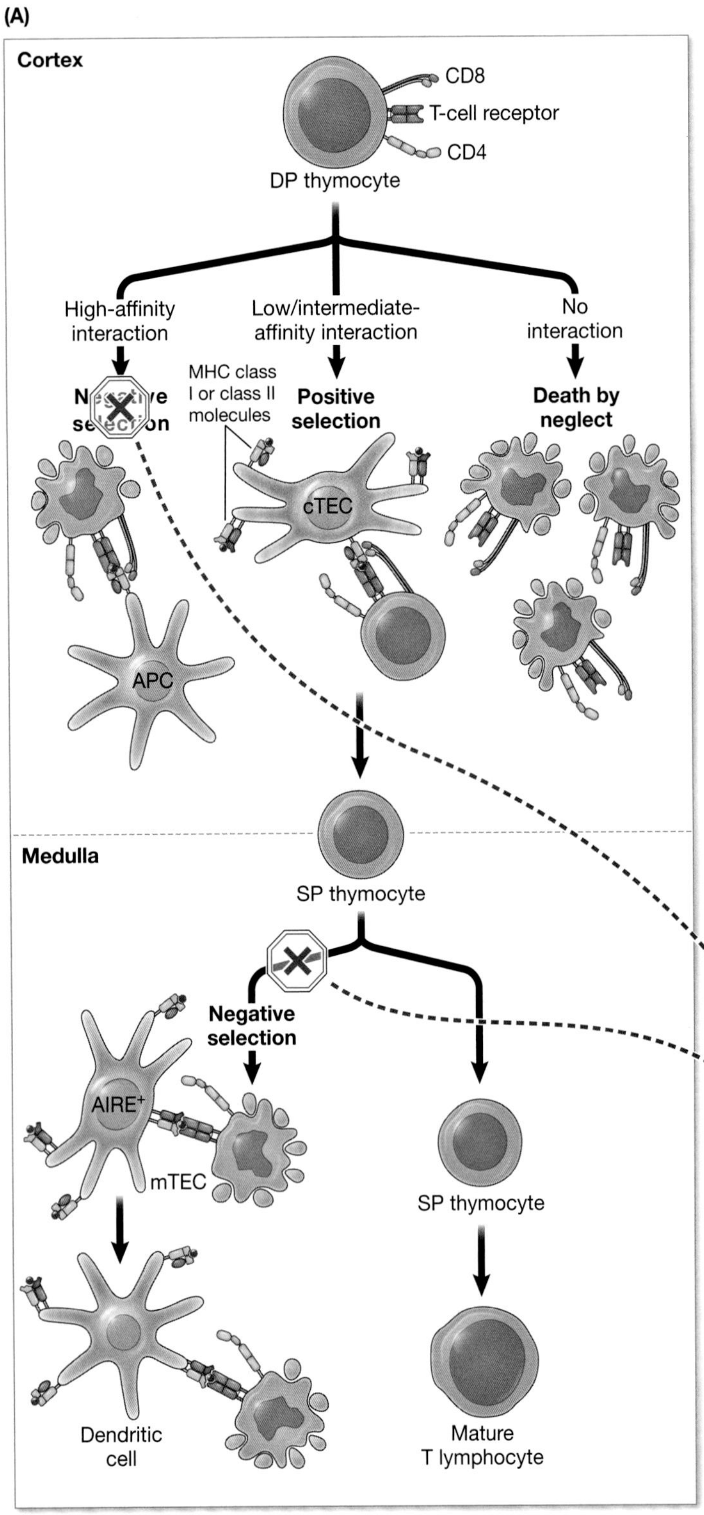

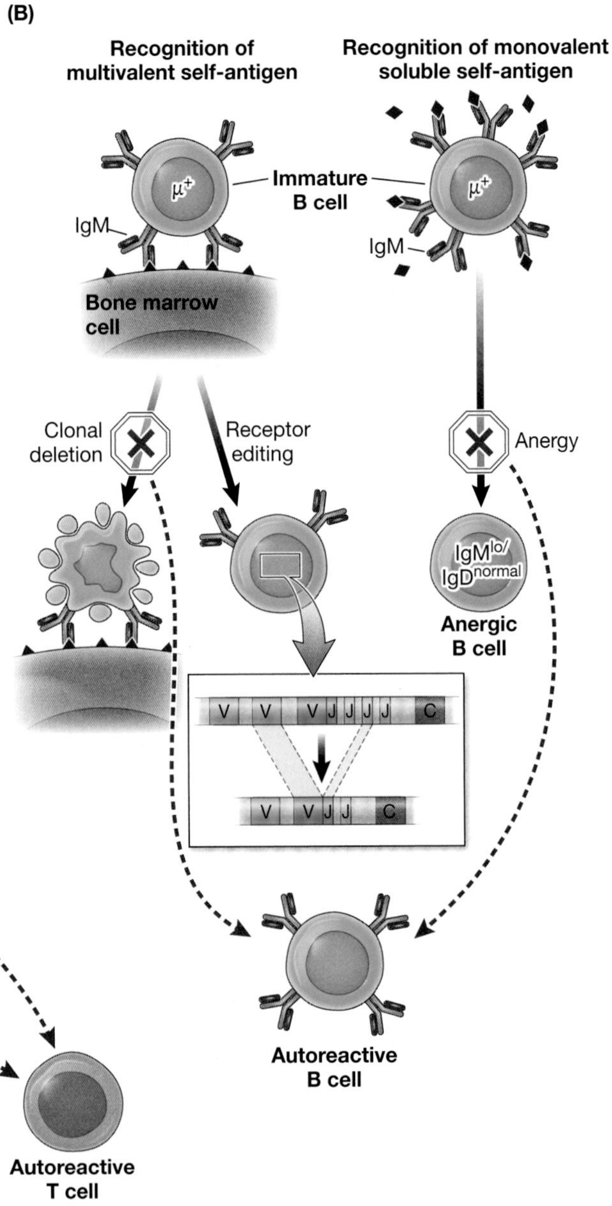

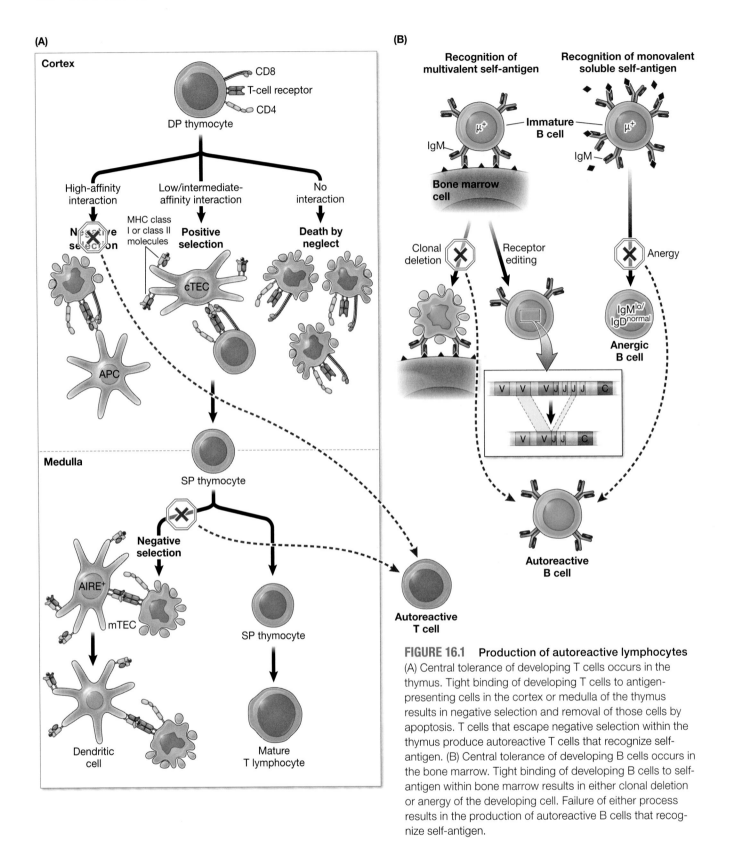

FIGURE 16.1 Production of autoreactive lymphocytes (A) Central tolerance of developing T cells occurs in the thymus. Tight binding of developing T cells to antigen-presenting cells in the cortex or medulla of the thymus results in negative selection and removal of those cells by apoptosis. T cells that escape negative selection within the thymus produce autoreactive T cells that recognize self-antigen. (B) Central tolerance of developing B cells occurs in the bone marrow. Tight binding of developing B cells to self-antigen within bone marrow results in either clonal deletion or anergy of the developing cell. Failure of either process results in the production of autoreactive B cells that recognize self-antigen.

promote isotype switching and somatic hypermutation. By limiting the presence of autoreactive T cells in circulation, the activation of autoreactive B cells (that require activation by T cells that recognize the same antigen) is prevented.

A costimulatory signal is required to activate T cells and B cells in secondary lymphoid tissue. This costimulatory signal is provided by other arms of the immune system, including the inflammatory response or complement activation. Lack of this costimulatory signal creates an unresponsive state in T cells and B cells, which ensures that their activation is connected to the activation of other immune system responses.

These processes prevent self-reactivity (also known as autoreactivity) in most individuals, but autoimmune diseases can result when the processes fail. While symptoms of many autoimmune conditions are specific to the disease, a common thread of all autoimmune disorders is failure of tolerance and induction of an inappropriate adaptive immune response.

TABLE 16.1 | Organ-Specific Autoimmune Diseases

Autoimmune Disease	Antigen/Mutated Gene	Effector Mechanism
Goodpasture syndrome	Renal and lung basement membrane	Autoantibodies
Graves disease	TSH receptor	Agonist autoantibodies
Hashimoto disease	Thyroid proteins	Autoantibodies, T_H1 cells
Insulin-dependent (type 1) diabetes mellitus	Pancreatic β cells	Autoantibodies
Myasthenia gravis	Acetylcholine receptors	Antagonistic autoantibodies

Source: R. H. Scofield. 2004. *Lancet* 363: 1544–1546.

Autoimmunity

When central or peripheral tolerance fails to prevent negative selection or inactivation of the adaptive immune system, autoimmunity arises. The nature of each autoimmune disease differs because each condition is caused by the targeting of different tissues by the immune system. However, because the failure of central or peripheral tolerance mechanisms drives the onset of autoimmunity, all autoimmune diseases are caused by self-reactivity by the cellular or humoral arms of the adaptive immune system.

Many autoimmune diseases target specific organs, and others are *systemic autoimmune diseases*, affecting multiple body systems. We will discuss examples of both organ-specific and systemic autoimmune diseases. Some of the most common are summarized in **TABLES 16.1** and **16.2**.

● CHECKPOINT QUESTION

1. Why are autoimmune diseases malfunctions of either the cellular or humoral arms of the adaptive immune response?

TABLE 16.2 | Systemic Autoimmune Diseases

Autoimmune Disease	Antigen/Mutated Gene	Effector Mechanism
Multiple sclerosis (MS)	Nervous tissue	Autoantibodies, T_H1 and cytotoxic T cells
Rheumatoid arthritis (RA)	Connective tissue	Autoantibodies, immune complexes
Systemic lupus erythematosus (SLE)	DNA, nucleosomal proteins	Autoantibodies, immune complexes
Immune dysregulation, polyendocrinopathy, enteropathy, X-linked (IPEX) syndrome	*FOXP3* mutation	Loss of T_{reg} cells
Autoimmune polyendocrinopathy-candidiasis-ectodermal dystrophy (APECED)	*AIRE* mutation	Loss of central tolerance

Source: R. H. Scofield. 2004. *Lancet* 363: 1544–1546.

16.2 How do genetic and environmental factors affect the progression of autoimmune disease?

LEARNING OBJECTIVES

16.2.1 Discuss how mutations in genes responsible for self-tolerance induce autoimmune disease, and provide specific examples.

16.2.2 Discuss how environmental factors influence autoimmune disease, and provide specific examples.

autoantibody Antibody capable of recognizing a self-antigen and that is responsible for the induction of autoimmunity in certain diseases.

Both genetic and environmental factors can alter negative selection and regulation of the adaptive immune system and induce autoimmunity. In Chapters 5 and 7 we saw that negative selection and tight regulation of T cells is important for controlling the adaptive immune system and that helper T cells activate a variety of other immune cells, including B cells. The production of autoreactive T cells can result in autoimmune diseases due to activation of these cells and the subsequent activation of other immune cells. B cells are often activated in autoimmune diseases as many manifest due to the production of **autoantibodies** (self-reactive antibodies).

While failure of central and peripheral tolerance leads to a variety of autoimmune diseases due to production of autoreactive lymphocytes, others are caused by factors that affect the function of autoreactive lymphocytes. Some individuals have genetic predispositions for certain types of autoimmunity due to particular MHC polymorphisms that are capable of presenting self-antigens to autoreactive T cells. Others are at risk of developing autoimmune diseases due to environmental factors such as chemical exposure and physical trauma.

Failure to ensure tolerance due to improper T-cell action can occur in a variety of ways (**FIGURE 16.2**):

- *Failure of negative selection of T cells.* Since self-reactivity is limited through the action of negative selection and central tolerance, failure to target and remove self-reactive T cells from the white blood cell population during their development can induce autoreactivity.
- *Failure of regulating T-cell activation.* T-cell activation requires recognition of an MHC-peptide complex by a T-cell receptor and a costimulatory signal. Loss of the requirement for a costimulatory signal can limit peripheral tolerance.
- *Failure of action of regulatory T cells.* Recall that peripheral tolerance is also driven by the action of regulatory T cells. These T cells recognize self-antigens and instead of inducing an immune response, they inhibit the action of other T cells that recognize the same antigen. Thus, a lack of activity of regulatory T cells can prevent proper peripheral tolerance.

Failure of negative selection of T cells

Recall from Chapter 5 that during T-cell development, positive selection ensures that developed T cells have functioning T-cell receptor components capable of binding to MHC molecules in complex with peptide with moderate affinity. Negative selection ensures that T cells that bind to MHC-peptide complexes too tightly within the thymus are removed by apoptosis, as they represent potentially self-reactive T cells.

During the process of negative selection, it is important that T cells are exposed to a variety of possible self-antigens, since after they leave the thymus, they may become exposed to those same self-antigens within the body. Furthermore, the vast repertoire of peptides presented to T cells must span

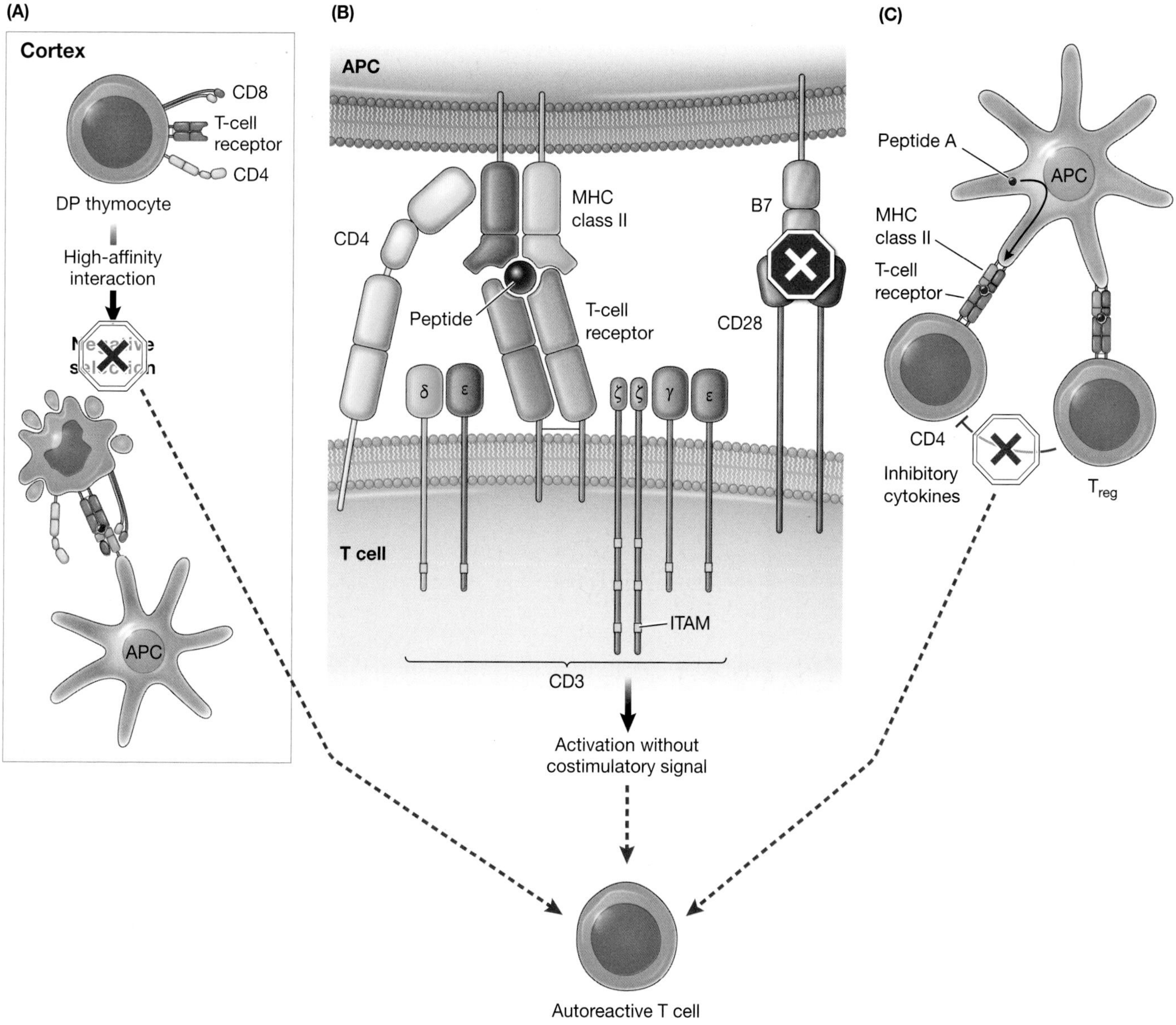

FIGURE 16.2 Causes of autoreactive T-cell production (A) Central tolerance of T cells occurs when the T-cell receptor of a developing thymocyte is tested for affinity to self-antigen. A developing thymocyte that expresses a T-cell receptor with a high affinity for self-antigen is normally removed by apoptosis. Loss of this step in negative selection can result in the production of autoreactive T cells. (B) Proper T-cell activation requires the T-cell receptor to engage with an MHC-peptide complex on the surface of an antigen-presenting cell (APC). The APC must also express the ligand B7 for a proper costimulatory signal through the T-cell surface protein CD28. This results in the phosphorylation of immunoreceptor tyrosine-based activation motifs (ITAMs). If a T cell can activate without a costimulatory signal, it may be able to produce an autoreactive T cell. (C) Peripheral tolerance is assisted by the action of regulatory T (T_{reg}) cells. When a T_{reg} cell binds to the same APC as a T cell recognizing the same peptide, it secretes cytokines that prevent activation of the T cell. Loss of action of T_{reg} cells can result in the activation of autoreactive T cells.

tissue-specific proteins to ensure that T cells exiting the thymus will not target cells displaying these tissue-specific proteins (a problem often seen in autoimmune diseases).

To drive the expression of tissue-specific proteins within resident thymic antigen-presenting cells (APCs) to developing T cells, the transcription factor AIRE (autoimmune regulator) is activated, thus improving the efficiency of negative selection. A lack of the transcription factor AIRE results in incomplete negative selection and possible autoimmunity to certain tissues. For example,

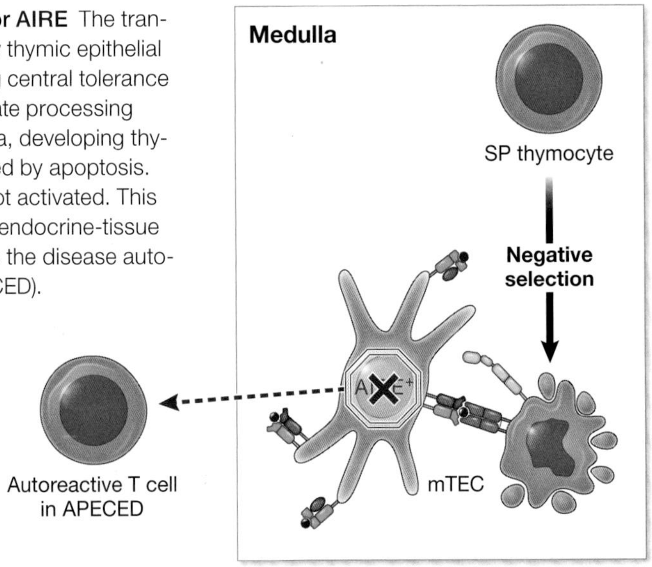

FIGURE 16.3 **Autoimmunity due to loss of the transcription factor AIRE** The transcription factor autoimmune regulator (AIRE) is expressed in medullary thymic epithelial cells (mTECs) in the thymus. AIRE plays an important role in promoting central tolerance due to its ability to activate endocrine tissue-specific genes and facilitate processing and presentation of the endocrine-specific proteins. Within the medulla, developing thymocytes that tightly bind to mTECs presenting self-antigen are removed by apoptosis. If mTECs cannot express AIRE, endocrine tissue-specific genes are not activated. This might allow developing thymocytes expressing a T-cell receptor to an endocrine-tissue antigen to escape negative selection. Lack of AIRE expression causes the disease autoimmune polyendocrinopathy-candidiasis-ectodermal dystrophy (APECED).

autoimmune polyendocrinopathy-candidiasis-ectodermal dystrophy (APECED) Autoimmune disease caused by a lack of the AIRE transcription factor, resulting in a lower amount of negative selection of tissue-specific self-reactive T cells and thus an increase in tissue-specific autoimmunity. Also known as *autoimmune polyendocrine syndrome type 1* (*APS-1*).

autoimmune polyendocrinopathy-candidiasis-ectodermal dystrophy (**APECED**; also known as *autoimmune polyendocrine syndrome type 1*, or *APS-1*) is caused by absence of the expression of AIRE. Since autoreactive T cells are produced due to this disease, autoreactive B cells are also commonly induced by the cognate autoreactive T cells. The result is autoimmune reactions from both the cellular and the humoral arms of the adaptive immune system, leading to the malfunction of many endocrine glands expressing tissue-specific genes (**FIGURE 16.3**).

Failure of peripheral tolerance of T cells

Recall that both central and peripheral tolerance play a role in preventing self-reactivity by the adaptive immune system. Peripheral tolerance is maintained through several mechanisms, including making sure that T cells are not activated unless required.

Inflammatory mediators induce the expression of the T-cell costimulatory molecule B7 on the surface of APCs (**FIGURE 16.4**). Expression of B7 on the cell surface provides the activating signal to T cells when it engages with CD28 on the surface of T cells that recognize an MHC-peptide complex presented by the APC using its T-cell receptor. Lack of this costimulatory signal aids in peripheral tolerance because without the signal, the T cell undergoes anergy. Using this mechanism of peripheral tolerance ensures that any self-reactive T cell that escaped negative selection in the thymus is prevented from activating and causing tissue damage through an autoimmune response.

While CD28 at the cell surface contributes to proper T-cell activation during antigen recognition, we also saw that CTLA4 (also expressed by T cells to bind to B7) works to downregulate T-cell activation and is used during the normal inhibitory function of regulatory T cells. If there is alteration in the requirement for the costimulatory signal or in the inhibition of T cells, peripheral tolerance will be limited, and a higher incidence of autoimmune disease is likely.

Several genetic polymorphisms in CTLA4 have been linked to autoimmune diseases. Mutations in CTLA4 have been connected to insulin-dependent (type 1) diabetes mellitus, Graves disease, Hashimoto disease, and systemic lupus erythematosus (SLE). Most of the autoimmune diseases connected with mutations in CTLA4 result in a loss of CTLA4 function and, thus, overactivation of T cells.

The mechanisms of CTLA4 function and regulation of T-cell activation are complex, as CTLA4 is expressed as both a membrane-bound and a soluble form by alternative splicing (analogous to cell-surface and soluble immunoglobulins).

Both forms play important inhibitory roles in T-cell activation. In some diseases, such as SLE, production of the soluble form of CTLA4 is decreased, demonstrating the inhibitory role of soluble CTLA4 in T-cell signaling. A significant source of soluble CTLA4 comes from regulatory T cells, which are key mediators of peripheral tolerance (see Chapter 7).

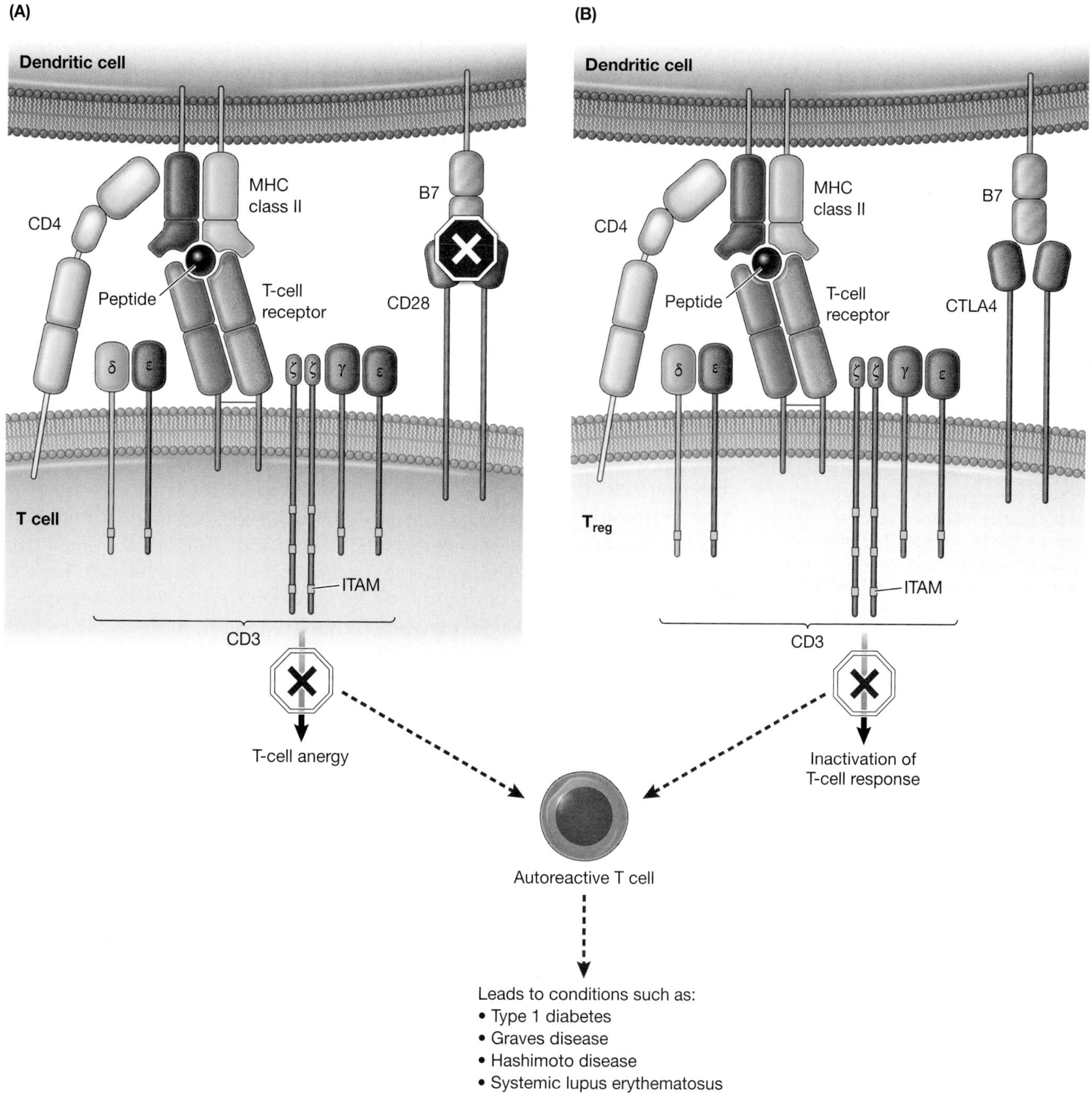

FIGURE 16.4 Autoimmunity due to abnormal T-cell signaling (A) A T cell is activated when its receptor binds to an MHC-peptide complex on the surface of an antigen-presenting cell (APC). This activation requires a costimulatory signal in which B7 on the APC binds to the T-cell plasma membrane protein CD28 and results in the phosphorylation of immunoreceptor tyrosine-based activation motifs (ITAMs). If a costimulatory signal is not present, the T cell bound to the APC does not activate and, instead, undergoes anergy. Loss of the costimulatory signal can result in loss of T-cell anergy and activation of autoreactive T cells. (B) T cells use the cell-surface protein CTLA4 to dampen activation and as a means to shut down clonal expansion. Regulatory T cells use signaling through CTLA4 to promote peripheral tolerance. Loss of the ability to shut down T-cell activation through CTLA4 results in the inability to inactivate autoreactive T cells. Abnormal T-cell signaling is seen in autoimmune diseases such as type 1 diabetes, Graves disease, Hashimoto disease, and systemic lupus erythematosus (SLE).

immune dysregulation, polyendocrinopathy, enteropathy, X-linked (IPEX) syndrome
Autoimmune disease characterized by a defective *FOXP3* gene and a lack of production of regulatory T cells and lowered peripheral tolerance.

Guillain–Barré syndrome
Autoimmune disease resulting in the demyelination of peripheral nerves and muscle weakness.

The loss of membrane-bound CTLA4 also prevents normal regulatory T-cell function. Regulatory T cells (T_{reg}) work to inhibit the activation of naïve T cells that recognize the same self-antigen as the T_{reg} cell (**FIGURE 16.5**). When a T_{reg} cell and an autoreactive naïve T cell come in contact with the same APC, the T_{reg} cell, through signaling driven by CTLA4, secretes immunosuppressive cytokines such as IL-4, IL-10, and TGF-β to prevent activation of the autoreactive T cell binding the same APC. Since CTLA4 is important in T_{reg}-cell function, mutations in CTLA4 can not only produce overactive T cells but can also inhibit the proper peripheral tolerance function of T_{reg} cells.

Improper production of T_{reg} cells is likely to result in problems in peripheral tolerance. While T_{reg} cells follow the same initial steps of development as other T-cell subpopulations (see Chapter 5), they are not subjected to the same mechanisms of negative selection as are other T-cell types upon self-antigen recognition. Instead, they develop into T_{reg} cells in a microenvironment of the thymus. All T_{reg} cells express the transcriptional regulator FOXP3 to aid in their function in peripheral tolerance.

One disorder associated with the lack of T_{reg}-cell activity is **immune dysregulation, polyendocrinopathy, enteropathy, X-linked (IPEX) syndrome**. This disease results from a mutation in *FOXP3*, which leads to a lack of functioning T_{reg} cells and causes autoimmune responses in a variety of tissues, including the gastrointestinal tract, thyroid, and pancreas. Malfunction of T_{reg} cells has also been implicated in the onset of other autoimmune diseases, including type 1 diabetes, **Guillain–Barré syndrome**, multiple sclerosis (MS), and myasthenia gravis.

Predisposition to autoimmunity based on sex

While in theory, autoimmune disease can occur in both men and women, approximately 80% of individuals with autoimmune disease are women. The reason for this preponderance of autoimmune disease in women is unknown, but several hypotheses are supported by research findings. One hypothesis suggests that women are more susceptible to autoimmune disease because they have two X chromosomes, whereas men have one X chromosome. In women, one of the two X chromosomes is inactivated in every cell, and which one is inactivated is determined on a cell-to-cell basis. Since the two X chromosomes differ (one from the mother and one from the father), it is possible that many of the genes on the X chromosomes are different alleles and potentially express polymorphic antigens. This increases the likelihood of autoimmunity when the immune system reacts to one of the two polymorphic antigens from an X chromosome as a foreign molecule.

Another hypothesis suggests that increased levels of certain hormones may increase the likelihood of autoimmune responses. Studies have demonstrated an increased risk of autoimmunity in females going through puberty compared to males, likely due to differences in hormone levels. Higher levels of estrogen have been associated with SLE and rheumatoid arthritis (RA). During pregnancy, increased levels of hormones, including estriol, progesterone, and prolactin, may impact the onset of autoimmune responses. Postpartum and menopausal hormonal changes have been associated with an increased incidence of RA.

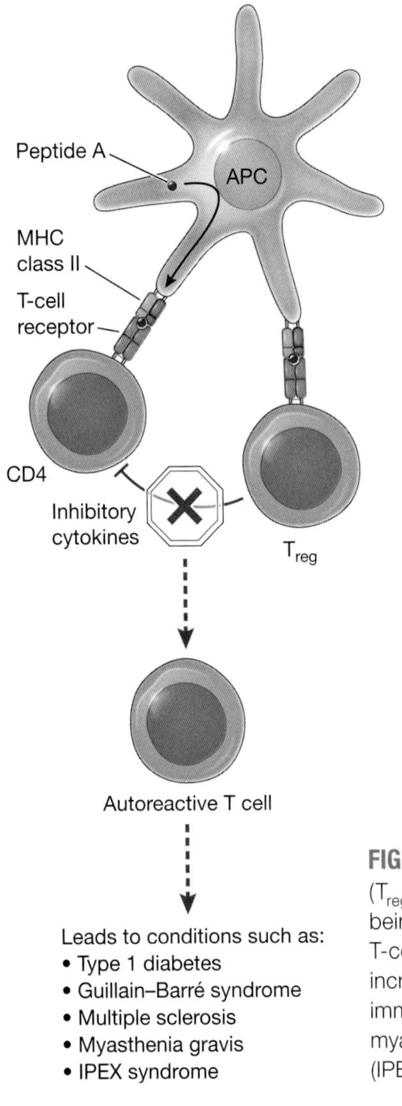

FIGURE 16.5 Autoimmunity due to abnormal regulatory T-cell function Regulatory T (T_{reg}) cells play a role in peripheral tolerance. They work to prevent autoreactive T cells from being activated by antigen-presenting cells presenting self-antigen that the autoreactive T-cell receptor can bind. Loss of T_{reg} cells can result in loss of peripheral tolerance and increases the potential activation of autoreactive T cells. Loss of T_{reg} cells is seen in autoimmune diseases such as type 1 diabetes, Guillain–Barré syndrome, multiple sclerosis, myasthenia gravis, and immune dysregulation, polyendocrinopathy, enteropathy, X-linked (IPEX) syndrome.

HLA isotypes and predisposition to autoimmunity

Recall that a proper immune response from the cellular arm of adaptive immunity requires proper T-cell function and presentation of antigen via APCs on MHC (HLA in humans) molecules at their cell surface. Because of the role played by MHC molecules in antigen presentation and T-cell activation and the intricacies of the immunological synapse created when a T-cell receptor interacts with an MHC-peptide complex, polymorphisms within the MHC loci may predispose some individuals to autoimmune responses.

As we learned in Chapter 6, HLA loci are highly polymorphic, and thus there are many alleles of the different HLA isotypes within the human population. HLA standard nomenclature is based on the gene associated with the HLA allotype, the allele group of the HLA allotype, and the specific isotype expressed by that gene (**FIGURE 16.6**). Classic genetic analyses, along with current powerful genotyping tools including the Immunochip (a microarray capable of looking at single nucleotide polymorphisms), have revealed HLA alleles that predispose individuals to certain autoimmune diseases (**TABLE 16.3**). While having these alleles does not guarantee onset of an autoimmune disease, these studies have revealed that certain HLA alleles are associated with a higher incidence of specific autoimmune diseases, thus demonstrating a connection between certain HLA polymorphisms and onset of autoimmunity.

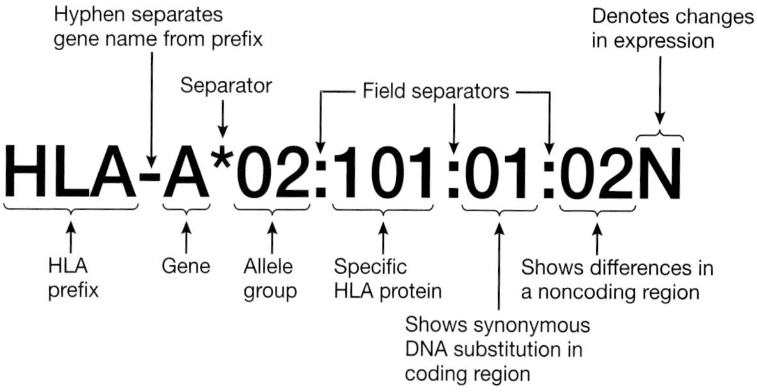

FIGURE 16.6 Standard HLA nomenclature Each HLA allele name has up to four unique digits separated by colons. The number before the first colon describes the antigen carried by the allotype. The second number describes the HLA subtype. The third number distinguishes alleles that vary within the coding region by silent mutations. The fourth number distinguishes alleles that differ in their noncoding regions. The letter suffix denotes where the HLA protein is expressed: N for null, L for low surface expression, S for secreted, C for cytoplasmic, A for aberrant expression, or Q for questionable. (After J. Robinson et al. 2015. *Nucleic Acids Res* 43: D423–431; J. Robinson et al. 2000. *Tissue Antigens* 55: 280–287.)

Failure of negative selection and peripheral tolerance of B cells

While central tolerance and peripheral tolerance of self-reactive T cells play a critical role in preventing autoimmunity, these processes also prevent the production of autoreactive B cells. B cells developing in the bone marrow assess the self-reactivity of newly recombined immunoglobulins and activate receptor

TABLE 16.3 | MHC Polymorphisms Linked to Autoimmune Disease

Disease	Classic Associated Locus	Fine-Mapping Locus
Rheumatoid arthritis (RA)	HLA-DRB1 AA 70–74	HLA-DRB1: AA 11,13,71,74 HLA-DRB1: AA 11,57,74 HLA-DRB1: AA 11,71,74
Celiac disease	HLA-DQ locus (DQ2, DQ8)	HLA-DQA1: AA 25,47 HLA-DQB1: AA 57,74
Systemic lupus erythematosus (SLE)	HLA-DRB1 (*0301 and *1501)	HLA-DRB1: AA 11,12,26 HLA-DRB1*1501-HLA-DQB1*0602 HLA-DRB1*0301
Insulin-dependent (type 1) diabetes mellitus	HLA-DRB1, DQB1, heterozygosity for HLA-DR3/HLA-DR4	HLA-DQB1: AA 57 HLA-DRB1: AA 13 HLA-DRB1: AA 71
Multiple sclerosis (MS)	HLA-DRB1*1501	HLA-DRB1*1501
Graves disease	HLA-DRB1*0301-DQA*0501-DQB*0201 (DR3-DQ2 haplotype)	HLA-DPB1: AA 35,55 HLA-DPB1*0501

Source: V. Matzaraki et al., 2017. *Genome Biol* 18: 76.

Goodpasture syndrome Autoimmune disease caused by the production of antibodies that recognize collagen expressed in blood vessels and the basement membrane of tissues, resulting in the onset of vasculitis.

celiac disease Autoimmune disease caused by the production of an adaptive immune response against the protein gluten (found in wheat, barley, and rye), resulting in targeting of small intestinal epithelial cells and the destruction of microvilli.

editing (if they recognize a multivalent antigen in the bone marrow) or induce anergy (if they recognize a monovalent antigen in the bone marrow). If receptor editing fails to produce a functioning immunoglobulin that is not self-reactive, the B cell will undergo apoptosis and be removed by negative selection. Apoptosis is activated by a protein known as Bim. Loss of Bim results in loss of central tolerance and the production of autoreactive B cells.

Anergic B cells are incapable of being activated and are removed quickly from circulation because of their short half-life. The activation of anergy within B cells that recognize monovalent self-antigens is driven by signaling phosphatases including SHIP-1 and SHP-1. Furthermore, Bim plays a role in the quick removal of anergic B cells, as it promotes apoptosis of anergic B cells that cannot receive and respond to survival signals. Loss of SHIP-1, SHP-1, or Bim prevents the onset of anergy and allows B cells that recognize self-antigen to enter the circulation.

SLE is the primary autoimmune disease associated with loss of central tolerance of B cells due to a loss of Bim, SHIP-1, or SHP-1. This disease is characterized by production of autoantibodies and the targeting of a variety of peripheral tissues in autoimmune and inflammatory responses due to the action of autoreactive B cells.

Environmental factors that trigger autoimmunity

While genetic predisposition accounts for the occurrence of some autoimmune conditions, environmental factors also influence the predisposition to or onset of autoimmune disease. These environmental factors can be noninfectious or infectious agents capable of inducing autoimmune responses. As with HLA polymorphisms, exposure to these environmental agents does not guarantee an autoimmune reaction but rather is linked to an increased risk for such a disorder. Environmental factors can also exacerbate symptoms associated with autoimmune diseases.

NONINFECTIOUS AGENTS ASSOCIATED WITH AUTOIMMUNITY Cigarette smoke is a noninfectious environmental agent that can increase the severity of symptoms associated with **Goodpasture syndrome**, a disease that causes vasculitis (inflammation of blood vessels) due to autoimmunity to those vessels. Because of the high degree of vascularization of the kidneys, most patients with this condition experience inflammation of the kidneys. Those who are smokers also suffer from pulmonary hemorrhaging because of the damage done to lung tissue by the cigarette smoke, thus allowing autoantibodies to access underlying pulmonary blood vessels and induce complement activation on those cells. Cigarette smoke has also been linked with an increased risk of onset of RA.

Environmental stressors have also been implicated in worsening the symptoms of certain autoimmune diseases. SLE symptoms can be triggered by both emotional stress and exposure to UV light. Oxidative stress, or an overabundance of reactive oxygen species that cannot be detoxified by naturally occurring enzymatic or reducing mechanisms, is capable of increasing the severity of RA symptoms.

Another example can be seen in individuals who have **celiac disease**, an autoimmune disorder that causes gastrointestinal inflammation due to autoimmunity to gluten, a protein found in wheat, rye, and barley. Ingestion of gluten causes an immune response to small intestinal epithelial cells, leading to damage to the microvilli necessary for proper nutrient absorption in the small intestine. While individuals with celiac disease are genetically predisposed to have an autoimmune response, the response does not occur until the person consumes gluten.

Another form of gluten sensitivity known as nonceliac gluten sensitivity (NCGS) is not related to an autoimmune reaction. There has been some debate surrounding NCGS as a clinical diagnosis, and further research is needed (see the **CONTROVERSIAL TOPICS** box).

CONTROVERSIAL TOPICS

Do we really need all those gluten-free foods found in grocery stores?

The answer to this question may be more complex than you think. Over the past decade, there has been an increasing preponderance of food products labeled "gluten-free," including foods that a reasonable person would not expect to contain gluten (such as peanut butter, dried cranberries, and canned mushrooms) since gluten is a protein specific to wheat, rye, and barley. These gluten-free labels have increased in number largely due to the fact that a higher percentage of the population now chooses to eat a gluten-free diet. According to a 2017 study published in *Mayo Clinic Proceedings*, the population of individuals eating a gluten-free diet tripled from 0.5% to 1.7% of the U.S. population between 2009 and 2014, whereas the number of confirmed celiac disease cases (the autoimmune disease known to be caused by a reaction to gluten), has remained unchanged at approximately 0.7% of the population.

Individuals without celiac disease who avoid gluten in their diets may be perceived as following a fad diet since they don't suffer from the autoimmune disease or a wheat allergy. While some who claim that they experience fewer intestinal issues after switching to a gluten-free diet may be experiencing a placebo effect, there may be a significant portion of the population who are actually sensitive to gluten or to some other antigen present in gluten-containing foods.

Individuals with non-celiac gluten sensitivity (NCGS) suffer from intestinal symptoms associated with the ingestion of gluten-containing foods yet lack a positive celiac disease diagnosis. The pathogenesis of NCGS is not well understood, and it may be caused by gluten or by another protein present in gluten-containing foods. Because the actual cause of NCGS has yet to be discovered, it remains a controversial issue, one that some researchers question as a true clinical diagnosis. Until a confirmed causative agent of NCGS is discovered, the controversy about gluten sensitivity is likely to remain a topic of debate.

Think About...

1. How might a scientist attempt to discover a cause of NCGS?
2. The estimated number of individuals affected by NCGS worldwide is between 0.5% and 13%. Why do you think there is such a wide disparity between the lower and upper estimates? Do you think this wide range further strengthens or weakens the case of NCGS being an accepted clinical diagnosis?
3. Do you think that NCGS is truly a clinical illness or is related to the popularity of the gluten-free diet? What evidence from an opposing viewpoint might sway your opinion?

INFECTIOUS AGENTS ASSOCIATED WITH AUTOIMMUNITY Various pathogens are capable of inducing autoimmune responses. Recall from Chapter 13 that pathogens utilize a variety of mechanisms to evade detection and destruction by the immune system. A strategy used by some pathogens, such as *Streptococcus pyogenes*, is expressing molecules that are very similar to proteins in the body, a process known as **molecular mimicry**. By mimicking our own proteins, pathogens take advantage of central and peripheral tolerance mechanisms of the adaptive immune system. If our adaptive immune system recognizes these molecules as foreign and a proper adaptive immune response is raised, the resulting cells may also recognize our closely related proteins as antigens.

molecular mimicry Mechanism used by pathogens to produce molecules similar or identical to host proteins in an attempt to evade the adaptive immune response by taking advantage of the processes of central and peripheral tolerance.

An example of molecular mimicry inducing the onset of autoimmunity is rheumatic fever, which causes inflammation of heart tissue, blood vessels, and joints. *Streptococcus pyogenes* infections can cause rheumatic fever due to the production of antibodies that bind to *S. pyogenes* antigens such as the carbohydrate *N*-acetylglucosamine. The antibodies also recognize heart muscle proteins such as cardiac myosin and drive complement activation on the surface of heart muscle. In this case, while the adaptive immune response is acting as intended to get rid of the infectious agent, because of the mechanisms used by the pathogen to evade the immune system, the response raised against the pathogen can also target self-antigens, mistaking them for foreign molecules.

Molecular mimicry is also thought to promote autoimmunity in Guillain–Barré syndrome. Infection with the gram-negative bacteria *Campylobacter jejuni* drives production of immunoglobulins that can cross-react with epitopes on gangliosides, which are lipids that are prevalent in the plasma membrane of cells of the nervous system. Production of these antibodies that recognize both *C. jejuni* antigens and gangliosides can lead to immunoglobulins binding to nerve cells and the targeting of those cells for destruction by an autoimmune response.

CHECKPOINT QUESTIONS

1. Why are many genetic predispositions to autoimmune disease connected to T-cell function?
2. When is a normal adaptive immune response to a pathogen capable of inducing an autoimmune response?

16.3 Why do many autoimmune diseases target endocrine glands?

LEARNING OBJECTIVE

16.3.1 Explain why the anatomy and physiology of endocrine tissues make them susceptible to autoimmune diseases.

While we will see in Section 16.5 that some autoimmune diseases are systemic, inducing autoimmune reactions throughout the body, many autoimmune diseases target specific endocrine tissues. Endocrine glands are highly specialized in their function and commonly express tissue-specific proteins. Due to their physiological roles, they are highly vascularized, making them susceptible to autoimmune responses in individuals who are predisposed to problems with central or peripheral tolerance.

Endocrine gland structure and function

Endocrine glands are highly vascularized tissues (**FIGURE 16.7**) that secrete hormones and regulate a wide range of physiological processes such as metabolism. Hormones are delivered via the bloodstream to tissues throughout the body to regulate these biological processes. To carry out its normal function, endocrine glands must express specialized hormones that induce specific tissue responses.

The unique activity of each endocrine gland, along with the extensive vasculature associated with endocrine tissue, makes them highly susceptible to autoimmune diseases that target those tissues due to the accessibility of immune cells and effector molecules such as immunoglobulins. Let's look at two examples of endocrine glands and their normal physiological functions to better understand their susceptibility to autoimmune diseases.

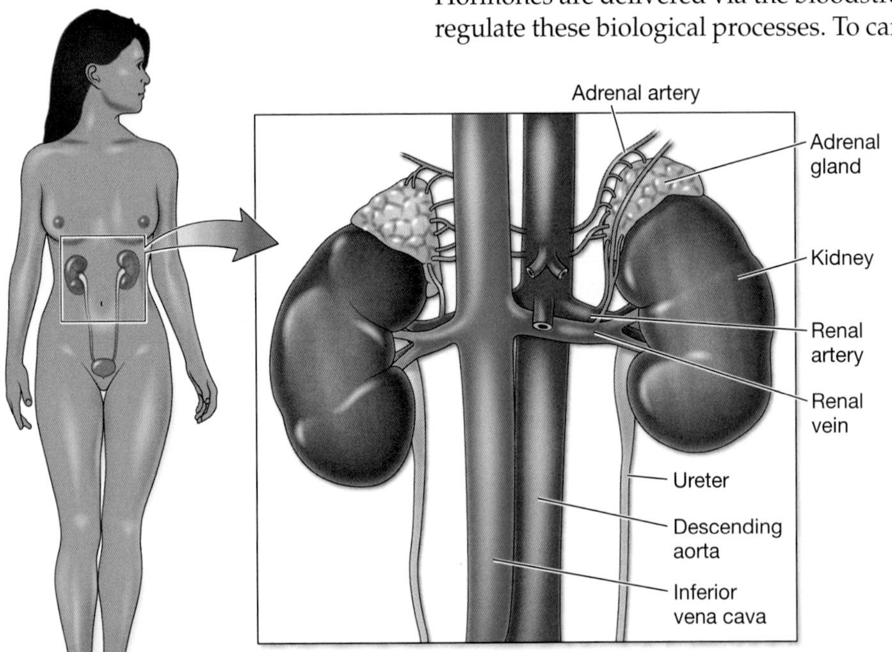

FIGURE 16.7 Adrenal glands as an example of endocrine tissue An adrenal gland is located above each kidney. These glands secrete hormones such as adrenaline. Because hormones secreted by the adrenal glands must be transported throughout the body, they are highly vascularized.

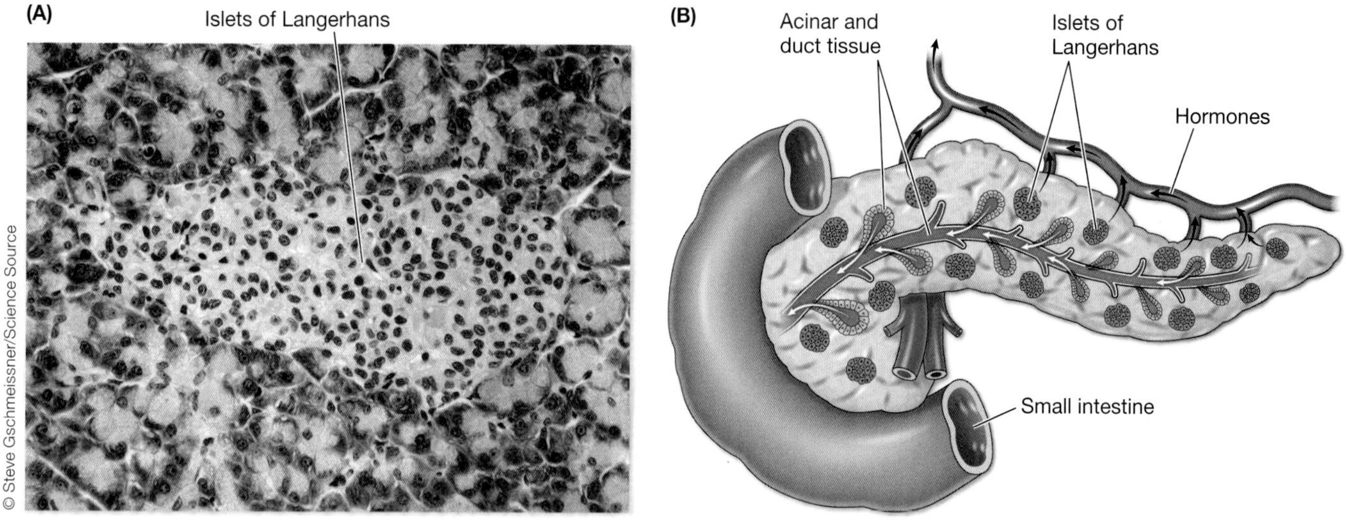

FIGURE 16.8 Pancreas The pancreas is an endocrine tissue responsible for regulating blood glucose levels. (A) Groups of cells in the islets of Langerhans contain both alpha and beta cells. (B) The pancreas secretes the hormone glucagon from alpha cells when blood glucose levels are low and the hormone insulin from beta cells when blood glucose levels are high. It is connected with the circulatory system so these hormones can circulate in the bloodstream.

PANCREAS The pancreas is an important endocrine gland that regulates blood glucose levels and metabolism. Within the pancreas are cell clusters known as islets of Langerhans that contain several types of specialized cells, including pancreatic α cells and pancreatic β cells (**FIGURE 16.8**). Pancreatic α cells are responsible for the secretion of the hormone glucagon, and pancreatic β cells secrete the hormone insulin. These hormones are released into the bloodstream based on blood glucose levels and induce either glucose uptake by peripheral tissues (when insulin is secreted) or the activation of glucose production by the liver (when glucagon is secreted). They also regulate metabolic processes within peripheral tissues. Because of its specialized protein production and the extensive vasculature of the pancreas, pancreatic cells are targeted in the autoimmune disease insulin-dependent (type 1) diabetes mellitus.

THYROID The thyroid is another important endocrine gland; it regulates the body's basal metabolic rate. The thyroid performs this function by secreting two hormones: triiodothyronine (T_3) and tetraiodothyronine (thyroxine, T_4) (**FIGURE 16.9**). These two hormones are produced from the thyroid-specific protein precursor thyroglobulin, which is stored in thyroid follicles and iodinated on tyrosine residues through the action of the enzyme thyroid peroxidase. Stored thyroglobulin remains in thyroid follicles until the pituitary gland releases thyroid-stimulating hormone (TSH), which signals the thyroid to release T_3 and T_4. Released TSH binds to the TSH receptor on the surface of thyroid epithelial cells. This binding prompts the uptake of thyroglobulin from the follicle into thyroid epithelial cells, proteolytic processing of thyroglobulin into T_3 and T_4, and release of T_3 and T_4 into the bloodstream, which increase metabolism in target tissues. Because the thyroid is highly vascularized and expresses tissue-specific proteins, including thyroglobulin, the TSH receptor, and thyroid peroxidase, it is the targeted tissue in Graves disease and Hashimoto disease.

Autoimmune diseases of the endocrine glands

We have seen that endocrine glands can be prime targets for autoimmune responses when central or peripheral tolerance goes awry. If tissue-specific proteins are not efficiently displayed to developing lymphocytes, central tolerance

FIGURE 16.9 Action of the thyroid gland The thyroid gland is responsible for regulating basal metabolic rate. This endocrine gland secretes two important hormones, T_3 and T_4, which are made from the precursor protein thyroglobulin expressed in thyroid epithelial cells and stored in the thyroid follicle. Thyroid epithelial cells transport iodine from the bloodstream into the thyroid follicle, where it is attached to thyroglobulin. Iodinated thyroglobulin is endocytosed by thyroid epithelial cells, where it is cleaved into the mature T_3 and T_4 hormones. These hormones are stored in secretory vesicles in thyroid epithelial cells until signaled to be released. This signal comes from the pituitary gland, which secretes thyroid-stimulating hormone (TSH). TSH binds to the TSH receptor on the surface of thyroid epithelial cells, stimulating fusion of secretory vesicles containing T_3 and T_4 with the plasma membrane and release of the hormones into the bloodstream.

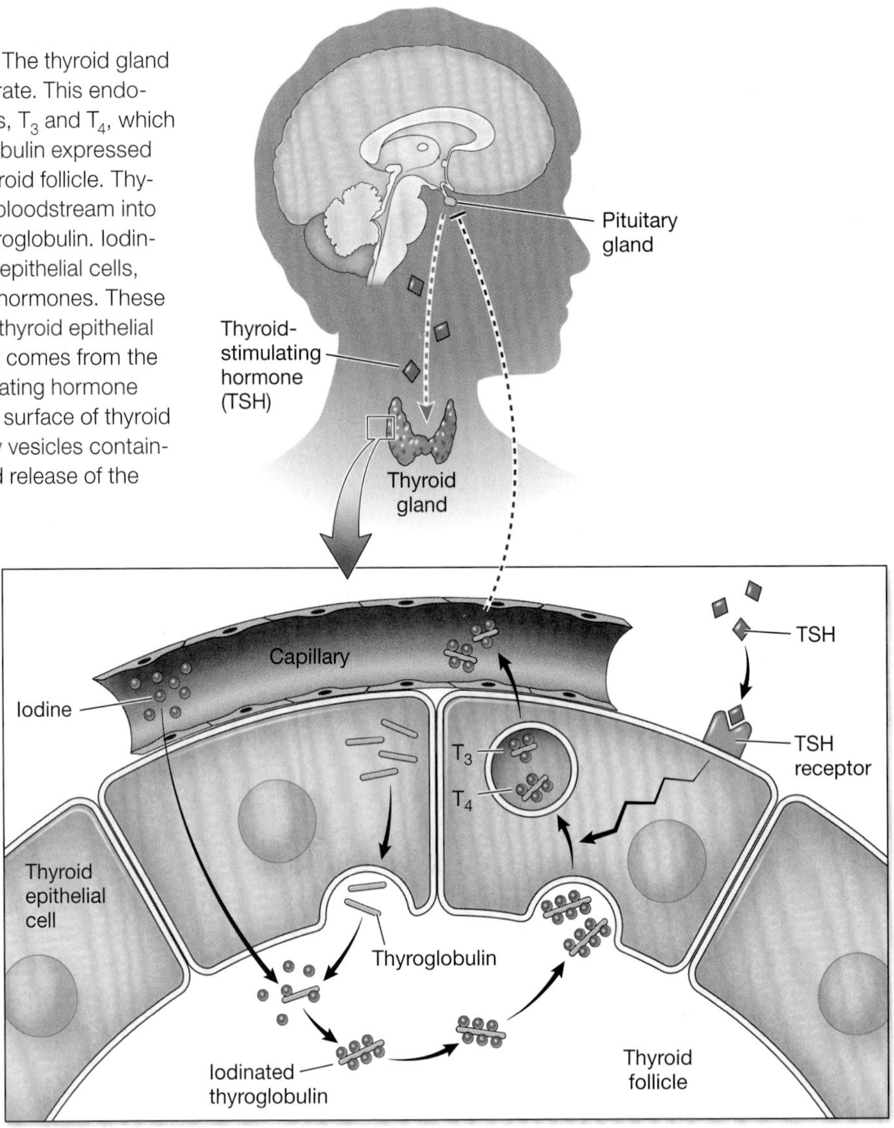

to these proteins can be lost, resulting in production of autoreactive lymphocytes that can target the endocrine glands. The pancreas and thyroid are both targets of organ-specific autoimmune diseases.

TYPE 1 DIABETES MELLITUS **Insulin-dependent (type 1) diabetes mellitus** is characterized by the destruction of pancreatic β cells in the islets of Langerhans (**FIGURE 16.10**). The exact means by which these cells are targeted by an autoimmune response is not known, as more than 50 genes have been identified as genetic factors. In addition, several environmental factors have also been identified as predisposing an individual to the onset of the disease.

Loss of pancreatic β cells is driven by an adaptive autoimmune response that leads to the production of autoantibodies to pancreatic β cell-specific proteins, including insulin, glutamate decarboxylase, and the phosphatase-related protein IA-2. The presence of these autoantibodies can be used as markers for the onset of the disease in combination with other clinical diagnoses, including hyperglycemia and a decrease in endogenous insulin. Because of the loss of pancreatic β cells and the reduced or nonexistent capacity to produce insulin, the typical treatment involves insulin injections or an insulin pump that delivers porcine, bovine,

insulin-dependent (type 1) diabetes mellitus Autoimmune disease caused by the targeting of pancreatic β cells by the immune system, resulting in their destruction and the inability of the pancreas to produce insulin.

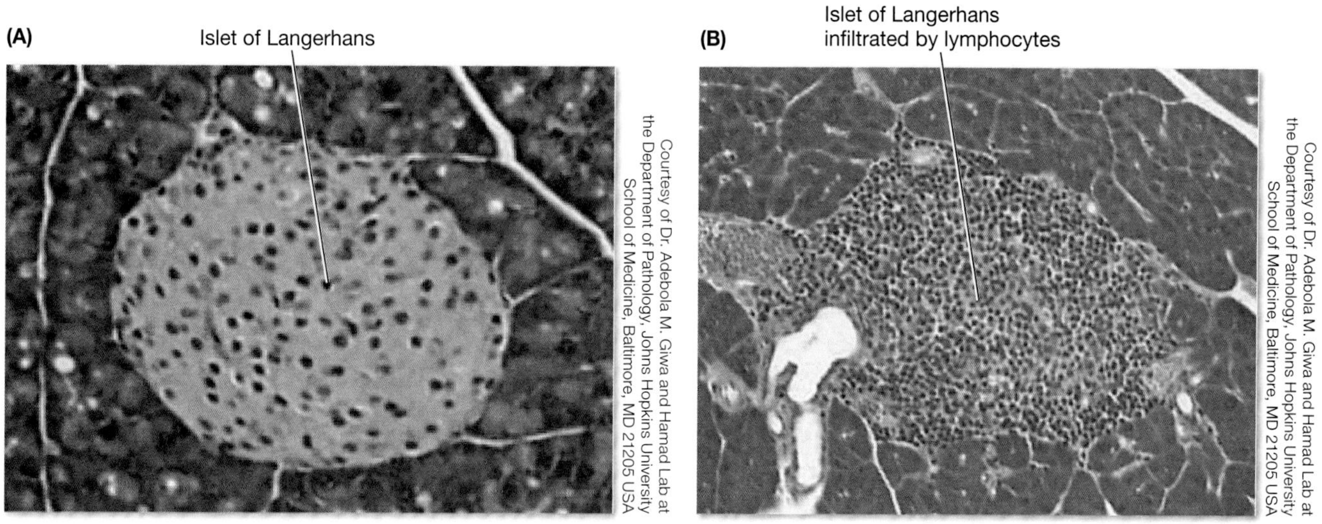

FIGURE 16.10 Type 1 diabetes Type 1 diabetes is an autoimmune response against cells of pancreatic islets of Langerhans. (A) A normal islet of Langerhans in the pancreas has an even distribution of alpha and beta cells, as seen with hematoxylin-and-eosin staining. (B) An islet of Langerhans in the pancreas of a patient with type 1 diabetes, visualized with hematoxylin-and-eosin staining, shows evidence of a large number of infiltrating cells, likely lymphocytes, inducing an autoimmune response.

or recombinant human insulin. The insulin helps to regulate metabolism, and immunosuppressive drugs may be used to minimize the autoimmune response.

GRAVES DISEASE **Graves disease** is an autoimmune disease caused by the production of thyroid-specific autoantibodies that bind to the TSH receptor on the surface of thyroid epithelial cells. These autoantibodies act as **agonist** antibodies upon binding to the TSH receptor, mimicking the binding of TSH to the TSH receptor. This interaction stimulates the release of T_3 and T_4 from the thyroid, increasing basal metabolic rate and resulting in **hyperthyroidism**. Individuals with Graves disease experience symptoms that include heat intolerance, weight loss, excessive sweating, and warm skin. Patients are prescribed drugs (methimazole or carbimazole) that prevent thyroid function by blocking iodine uptake by the thyroid and thus preventing maturation of thyroglobulin into T_3 and T_4. Often, Graves disease patients must undergo irradiation with radioactive iodine and surgery to remove the hyperactive thyroid, followed by thyroid hormone treatment to compensate for the loss of natural thyroid hormone production.

HASHIMOTO DISEASE **Hashimoto disease**, also known as Hashimoto thyroiditis, targets the thyroid but more closely resembles the pathology of type 1 diabetes. Individuals with Hashimoto disease develop an autoimmune response that results in the production of thyroid-specific autoantibodies, including those that can recognize thyroid peroxidase, thyroglobulin, and TSH receptor. These autoantibodies are capable of mediating antibody-dependent cell-mediated cytotoxicity of thyroid cells, resulting in destruction of the thyroid and causing **hypothyroidism**. Common symptoms include fatigue, weight gain, hair loss, dry skin, and sensitivity to cold. Treatment usually involves prescribed thyroid hormones to aid in regulation of basal metabolism. Dietary and lifestyle changes can assist in managing this disease.

Graves disease Autoimmune disease caused by the production of agonist antibodies against the thyroid-stimulating hormone receptor, resulting in hyperthyroidism due to elevated secretion of thyroid hormone.

agonist Molecule that can bind to a receptor and mimic a ligand, driving receptor activation.

hyperthyroidism Condition in which the thyroid is overactive and produces too much thyroid hormone.

Hashimoto disease Autoimmune disease caused by the production of antibodies against thyroid-specific antigens, resulting in the destruction of the thyroid and hypothyroidism. Also known as *Hashimoto thyroiditis*.

hypothyroidism Condition in which the thyroid is underactive and does not produce adequate amounts of thyroid hormone.

• CHECKPOINT QUESTION

1. Why are endocrine glands highly susceptible to autoimmune reactivity?

Making Connections: Autoimmunity

Developing lymphocytes engage in recombination of receptor genes to increase diversity. However, this process may result in the production of lymphocyte receptors that recognize self-antigen. There are mechanisms that prevent the activation and circulation of self-reactive lymphocytes. When these mechanisms fail, the result is autoimmune disease.

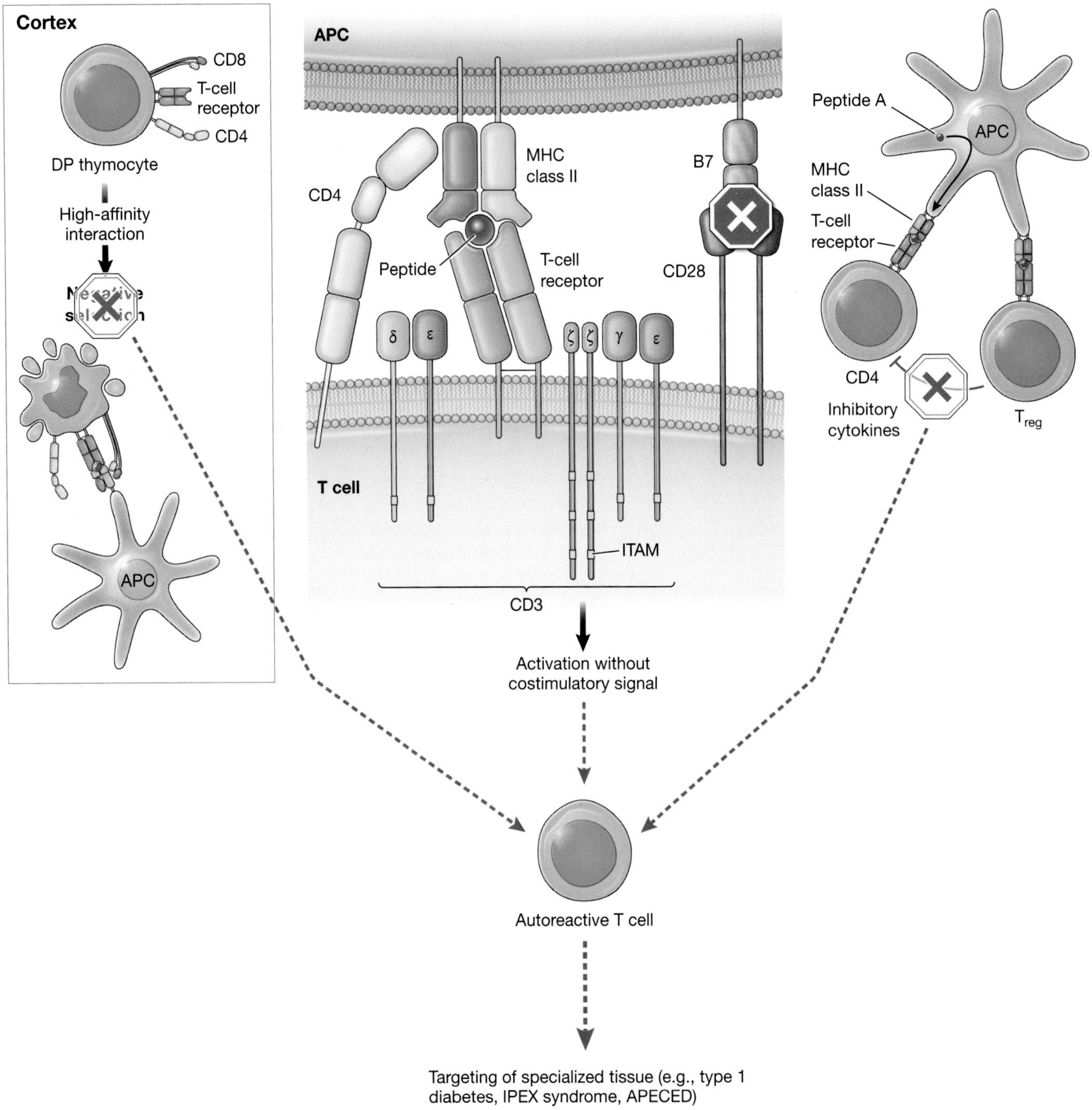

Problems associated with central and peripheral tolerance of autoreactive T cells lead to a variety of autoimmune diseases. These diseases are associated with the production of autoreactive T cells, which target highly specialized cells or endocrine tissues.

(Sections 1.3. 1.8, 4.5, 5.7, 7.10, 7.15, 16.1, 16.2, 16.3)

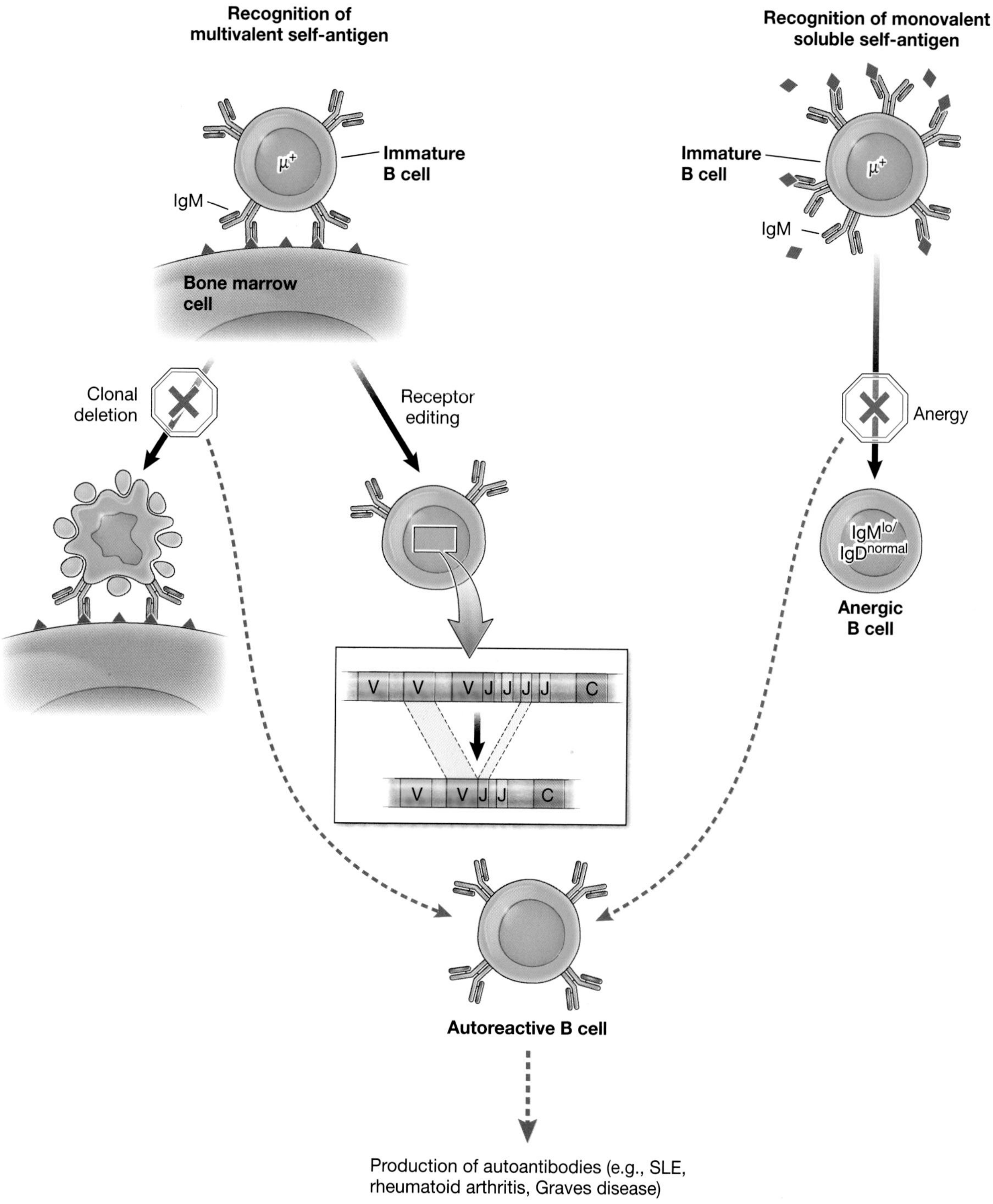

Problems associated with central and peripheral tolerance of autoreactive B cells result in the production of autoantibodies that recognize self-antigen.

(Sections 1.3, 1.8, 4.5, 8.8, 8.9, 10.5, 16.1, 16.4, 16.5)

16.4 What are the similarities between hypersensitivity and autoimmune diseases?

LEARNING OBJECTIVE

16.4.1 Compare the onset of hypersensitivity reactions to autoimmune diseases.

We learned in Chapter 15 that there are four different types of hypersensitivity reactions whereby the immune system responds to innocuous foreign antigens. A variety of responses ensue, depending on the entry point of the antigen and the location where it is recognized by the immune system. Although hypersensitivity reactions and autoimmune diseases differ in the antigens that they recognize (hypersensitivity reactions are caused by innocuous foreign particles, and autoimmune diseases are caused by recognition of self-antigens), there are similarities between the two.

Hypersensitivity revisited

Recall that the four types of hypersensitivity differ in terms of subsequent reactions and the antigens that induce those reactions:

- Type I hypersensitivity is driven by recognition of allergens by IgE on the surface of granulocytes and degranulation, resulting in inflammation at the site of recognition.
- Type II hypersensitivity is driven by the recognition of cell-surface molecules and target cells containing these molecules via the complement pathway or antibody-dependent cell-mediated cytotoxicity.
- Type III hypersensitivity is driven by the action of IgG or IgM recognizing soluble antigens and forming small immune complexes.
- Type IV hypersensitivity is driven by the activation of T cells by an innocuous antigen, resulting in IFN-γ release and activation of macrophages.

Although hypersensitivity reactions and autoimmune disease fundamentally differ in the antigens recognized, the resulting immune responses bear striking similarities. Indeed, autoimmune diseases mirror the responses seen in type II, type III, and type IV hypersensitivity responses.

Autoimmune diseases as hypersensitivity reactions

In this section, we will compare several autoimmune diseases with the three hypersensitivity reactions that they resemble to highlight some of the similarities in immune system reactions. We will also see that some autoimmune diseases can be caused by reactions that mimic multiple hypersensitivity reactions.

CONDITIONS RELATED TO TYPE II HYPERSENSITIVITY Type II hypersensitivity reactions are caused by the recognition of cell-surface molecules by either IgG or IgM, which act as targets for either complement activation or antibody-dependent cell-mediated cytotoxicity of the tagged cell. Several autoimmune diseases result in the destruction of cells associated with the affected tissue, including some discussed earlier in this chapter.

Recall that Goodpasture syndrome is caused by the destruction of the basement membranes of the kidneys and lungs, resulting in extensive tissue damage and vasculitis of these tissues. The destruction of the basement membranes is driven by recognition of self-antigens by autoantibodies and the activation of complement on the membrane surfaces within the kidney and lungs.

We have seen that Hashimoto disease results in hypothyroidism due to the destruction of thyroid epithelial cells, thus preventing the release of thyroid hormones T_3 and T_4. Hashimoto disease is caused by the recognition of thyroid-specific proteins by autoantibodies and subsequent destruction of thyroid epithelial cells by antibody-dependent cell-mediated cytotoxicity. Thus, the process that drives the onset of Hashimoto disease is similar to type II hypersensitivity.

Another autoimmune disease that mimics the reaction of type II hypersensitivity is **myasthenia gravis**. This disease affects neuromuscular junctions due to the production of autoantibodies that recognize the acetylcholine receptor at those sites. These autoantibodies can induce the symptoms of myasthenia gravis in two ways. First, they are **antagonist** antibodies that, when bound to the acetylcholine receptor, prevent normal receptor function and proper neuromuscular signaling. Second, because these antibodies recognize a cell-surface protein, they can induce complement activation or antibody-dependent cell-mediated cytotoxicity. The neuromuscular junction is damaged due to cellular destruction.

myasthenia gravis Autoimmune disease caused by the production of antagonist antibodies to the acetylcholine receptor, resulting in improper synaptic transmission between nerves and muscles leading to muscle weakness and loss of function.

antagonist Molecule that can bind to a receptor and prevent receptor function.

CONDITIONS RELATED TO TYPE III HYPERSENSITIVITY Type III hypersensitivity reactions are caused by the production of small immune complexes created by IgG or IgM recognizing antigens that are not efficiently cleared. Deposition of these immune complexes at locations within the bloodstream causes aberrant complement activation and localized inflammation at the site. Two systemic autoimmune diseases, RA and SLE, have symptoms of inflammation due to the presence of autoantibodies that recognize soluble self-antigens and the deposition of the resultant immune complexes in specific tissues. Inflammation of blood vessels (vasculitis, as in SLE) and inflammation of joints (arthritis, as in RA) occurs due to the inefficient clearance of immune complexes.

CONDITIONS RELATED TO TYPE IV HYPERSENSITIVITY Type IV hypersensitivity reactions, or delayed-type hypersensitivity, are driven by the activation of antigen-specific T cells responding to intracellular antigens that have been processed and presented and the subsequent targeting of the involved cells. Two autoimmune diseases—type 1 diabetes and celiac disease—are examples of disorders that result from the destruction of tissue-specific cells due to a T-cell response.

Recall that insulin-dependent (type 1) diabetes mellitus is caused by the destruction of tissue-specific cells, namely pancreatic β cells. Due to the complex factors involved in the onset of this disease, the exact cause of the destruction of pancreatic β cells is unknown. However, the production of autoantibodies to intracellular proteins of pancreatic β cells, including insulin and glutamate dehydrogenase, suggests that the activation of T cells capable of recognizing these intracellular antigens bears similarities to type IV hypersensitivity reactions.

Celiac disease is caused by the uptake of gluten, a group of digestion-resistant proteins present in various cereal grains, in the digestive tract. After uptake in the digestive tract, gluten is delivered to the underlying lamina propria either through transcytosis by small intestinal epithelial cells or between junctions of epithelial cells. Here, gluten is taken up by APCs, including dendritic cells and macrophages. Within the APCs, gluten peptides are modified (through deamidation) by the enzyme tissue transglutaminase (an autoantigen used as a marker for the identification of individuals affected by celiac disease). These peptides can be presented via MHC class II molecules to CD4 T cells. CD4 T cells that recognize MHC-gluten peptide complexes induce celiac disease through an inflammatory response that targets the epithelial cells and destroys the microvilli. Because this autoimmune disease occurs through a modified intracellular antigen that is presented to T cells, it resembles a type IV hypersensitivity reaction, which involves activation of T cells by modified intracellular allergens presented on MHC molecules.

CHECKPOINT QUESTION

1. Summarize the mechanisms of type II, type III, and type IV hypersensitivity reactions and provide an example of an autoimmune disease that resembles these reactions.

Making Connections: Similarities Between Hypersensitivity and Autoimmunity

Hypersensitivity reactions are driven by an adaptive immune response to innocuous allergens. This inappropriate response activates effector mechanisms that mimic those used to combat pathogens. Autoimmune diseases are similar to hypersensitivity reactions; they differ only in regard to the antigen recognized. In autoimmune diseases, the adaptive immune response targets self-antigens, not innocuous foreign material.

Comparing Type II Hypersensitivity and Autoimmunity

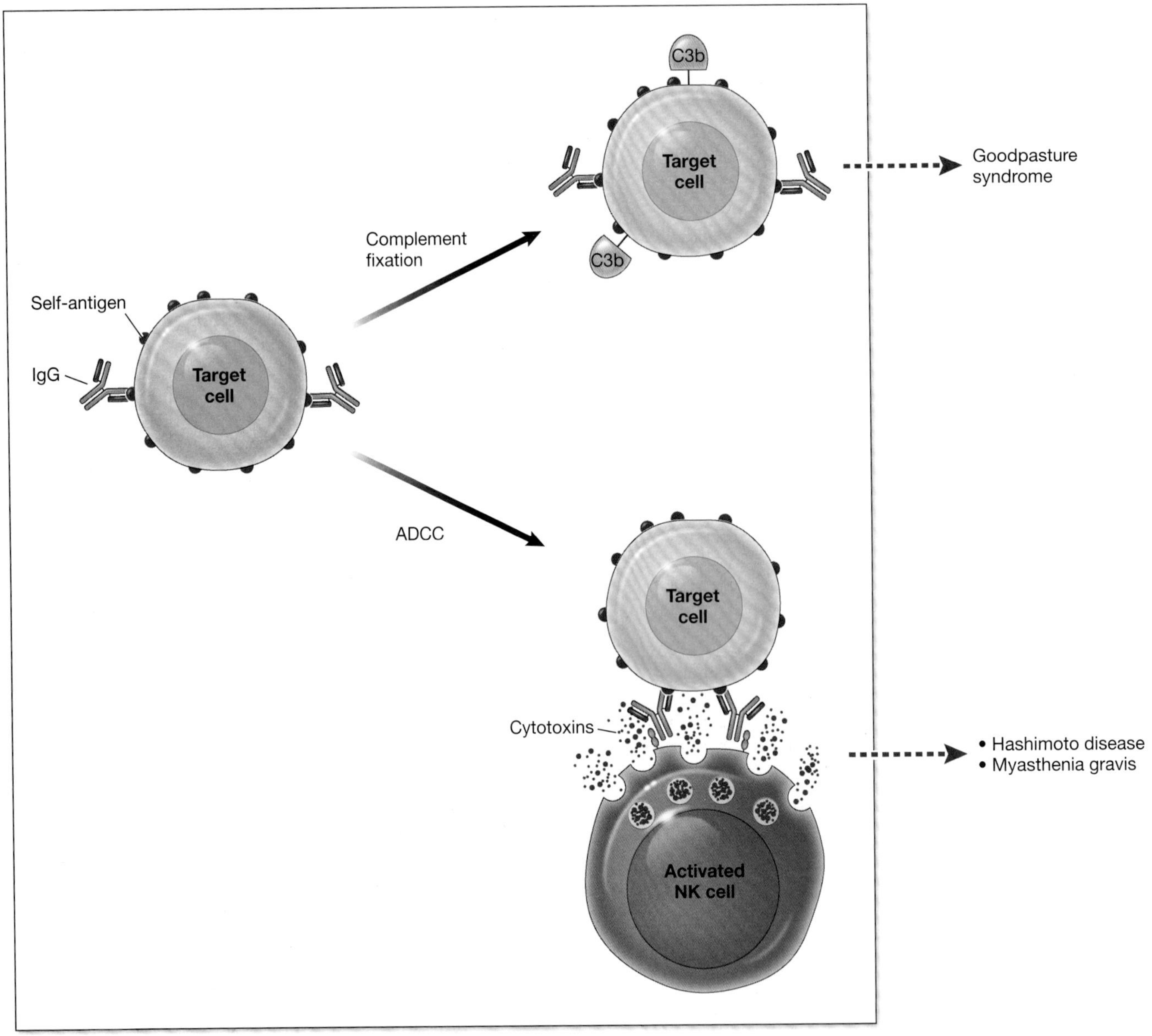

Type II hypersensitivity reactions are induced by recognition of modified surface antigens by immunoglobulins. This recognition is followed by effector mechanisms such as antibody-dependent cell-mediated cytotoxicity (ADCC) or complement fixation. Autoantibodies that recognize self-antigen can trigger similar responses.

(Sections 1.5, 2.7, 3.1, 3.3, 10.5, 15.3, 16.4)

Comparing Type III Hypersensitivity and Autoimmunity

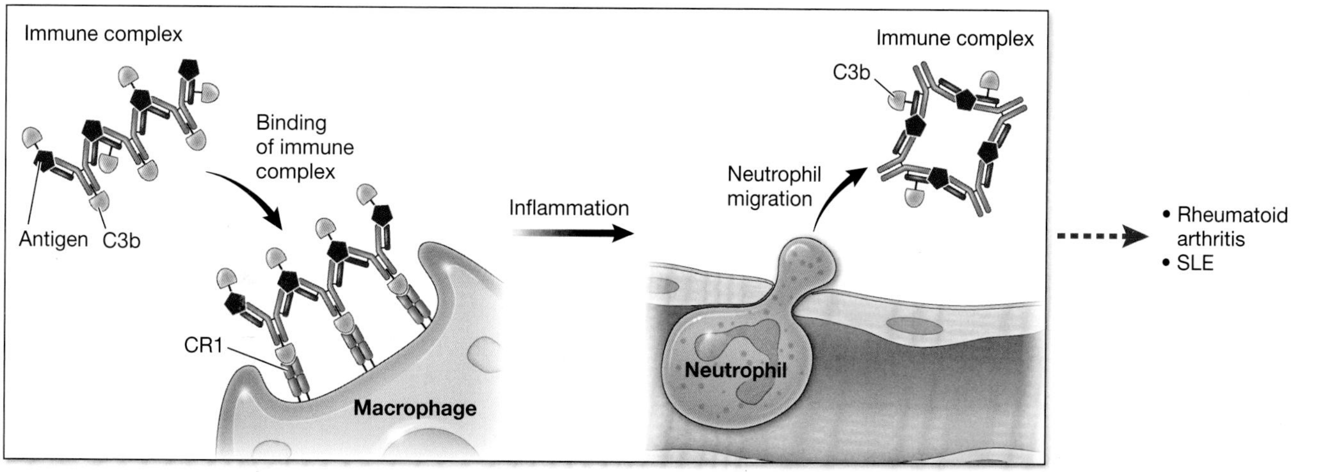

Type III hypersensitivity reactions are induced by production of small immune complexes that are not efficiently cleared from the bloodstream. Macrophages that bind these immune complexes activate an inflammatory response and infiltration of neutrophils. Autoantibodies that recognize soluble antigen produced in autoimmune diseases such as RA and systemic lupus erythematosus (SLE) can result in formation of immune complexes and induction of inflammation.

(Sections 1.5, 2.5, 2.6, 10.5, 15.4, 16.4)

Comparing Type IV Hypersensitivity and Autoimmunity

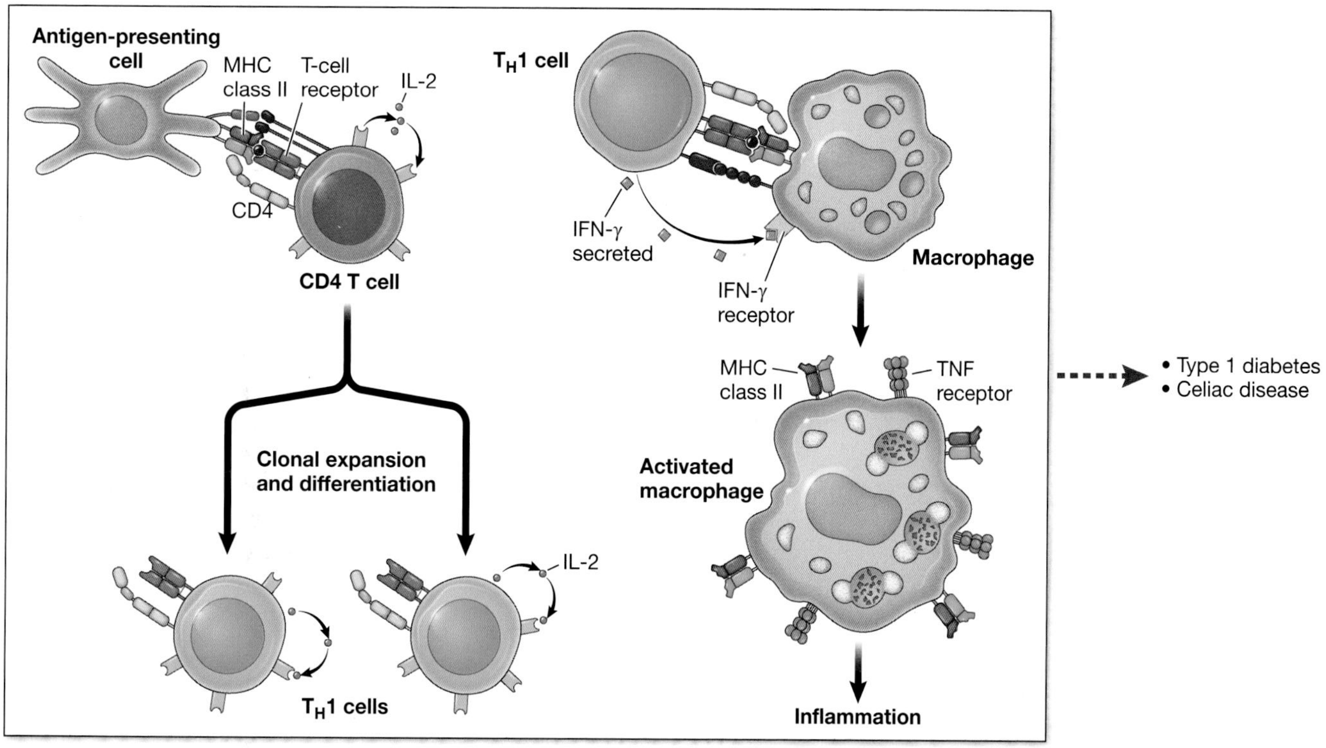

Type IV hypersensitivity reactions are induced by activation and clonal expansion of T cells that recognize modified antigens. Following this recognition, the clonally expanded T cells activate macrophages and induce inflammation. Autoantibodies that recognize self-antigen can trigger similar responses. Autoreactive T cells are activated in diseases such as type 1 diabetes and celiac disease, producing immune responses similar to type IV hypersensitivity.

(Sections 1.5, 2.5, 4.2, 4.7, 7.7, 7.14, 15.5, 16.4)

16.5 How do autoantibodies promote the progression of autoimmune disease?

LEARNING OBJECTIVE

16.5.1 Explain why many autoimmune diseases are caused by the production and action of autoantibodies.

Taking another look at Table 16.1, notice that autoantibodies are the immune effector responsible, at least in part, for the onset of each autoimmune disease. We have seen many examples of autoimmune diseases driven by the presence of these autoantibodies and the resulting reactions in the body. In this section, we will explore the role of autoantibodies in the onset of certain autoimmune diseases and learn how these autoantibodies induce either localized or systemic responses.

Organ-specific autoimmunity

We have seen that autoantibodies to cell-surface receptors can have possible opposing effects, acting as either agonists or antagonists (**FIGURE 16.11**). In Graves disease, autoantibodies that bind to the TSH receptor are capable of acting as an agonist of that receptor, mimicking the presence of TSH and inducing hyperthyroidism through the elevated secretion of T_3 and T_4.

On the other end of the spectrum are the antagonistic autoantibodies produced in myasthenia gravis. These autoantibodies also bind to a cell-surface molecule, specifically acetylcholine receptors, but act to prevent the activity and signaling of these receptors. These two autoimmune diseases demonstrate

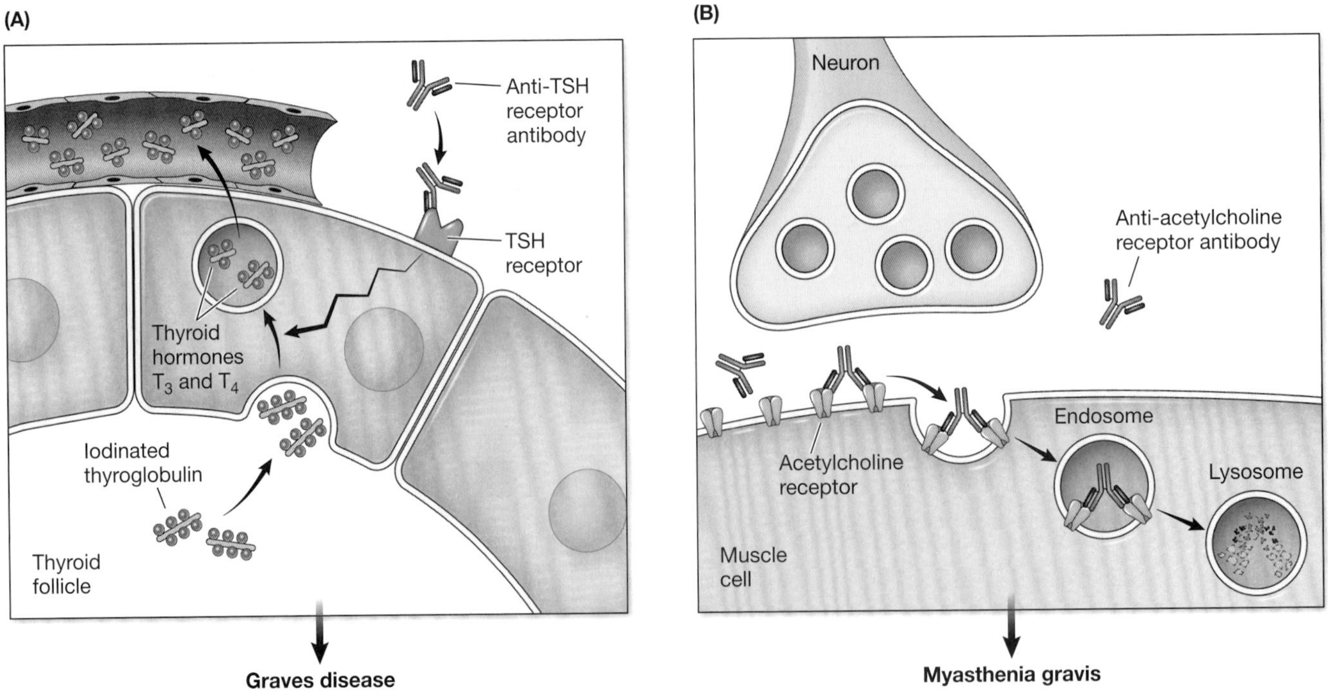

FIGURE 16.11 Agonist versus antagonist immunoglobulins in autoimmune diseases (A) Graves disease is caused by the production of agonist immunoglobulins that bind to the TSH receptor on thyroid epithelial cells. Binding of anti-TSH receptor antibodies to the TSH receptor induces release of excessive amounts of the hormones T_3 and T_4, causing Graves disease and resultant hyperthyroidism and increased metabolic rate. (B) Myasthenia gravis is caused by the production of antagonist immunoglobulins that bind to the acetylcholine receptor at the junction between neurons and muscle cells. Binding of anti-acetylcholine receptor antibodies to the acetylcholine receptor leads to endocytosis of the acetylcholine receptor and its destruction in the lysosome. This causes improper signaling at neuromuscular junctions and the skeletal muscle weakness of myasthenia gravis.

the potential to produce autoantibodies that have very different activities upon recognition of cell-surface molecules.

In addition to the production of autoantibodies that can alter the activity of a cell-surface molecule, tissue-specific autoantibodies can induce type II or type III hypersensitivity-like reactions. We saw in diseases such as RA, SLE, Goodpasture syndrome, and Hashimoto disease that the autoimmune responses caused by self-antigen recognition bear striking similarity to the hypersensitivity reactions discussed in Chapter 15. Thus, while the antigen prompting various nonproductive adaptive immune responses may differ, the immune response utilizes the same antibody effector mechanisms (complement activation, antibody-dependent cell-mediated cytotoxicity, inflammation, and immune complex clearance) required for the clearance of pathogens.

systemic lupus erythematosus (SLE) Systemic autoimmune disease characterized by the production of a variety of autoantibodies, including antibodies to nucleic acids, lipids, and nucleosomal proteins, resulting in the production of small immune complexes; mimics type III hypersensitivity.

Systemic autoimmunity

Certain autoimmune diseases act systemically, negatively affecting many organs and systems within the body based on the autoimmune response raised during the disease state. Because of the nature of an autoimmune response, for a disease to cause systemic symptoms, the self-antigen recognized during the autoimmune response must be a common antigen present throughout the body (rather than a specialized antigen as with organ-specific autoimmune diseases). Here, we consider three systemic autoimmune diseases—SLE, RA, and MS—as examples of diseases in which common cellular components are recognized and that involve systemic (vs. localized) effects and symptoms.

SYSTEMIC LUPUS ERYTHEMATOSUS **Systemic lupus erythematosus (SLE)** is an autoimmune disease that can stimulate an immune response to almost every cell present in an affected individual. According to the Centers for Disease Control and Prevention (CDC), women between the ages of 15 and 44 are most at risk of developing this disease. The autoantibodies produced in SLE are capable of recognizing very common cellular components, including nucleic acids, nucleoproteins responsible for DNA packaging, phospholipids, and other cell-surface molecules. The production of these autoantibodies, especially to phospholipids and cell-surface molecules, leads to inflammation and targeted destruction of cells possessing these common surface molecules, resulting in systemic immune responses.

Tissue destruction causes the release of the intracellular autoantigens such as nucleic acids and nucleoprotein components, resulting in formation of soluble immune complexes and causing reactions similar to type III hypersensitivity reactions, including joint inflammation, blood vessel damage, vasculitis, and kidney damage, due to the deposition of immune complexes in these locations (**FIGURE 16.12**). Vasculitis caused during an SLE

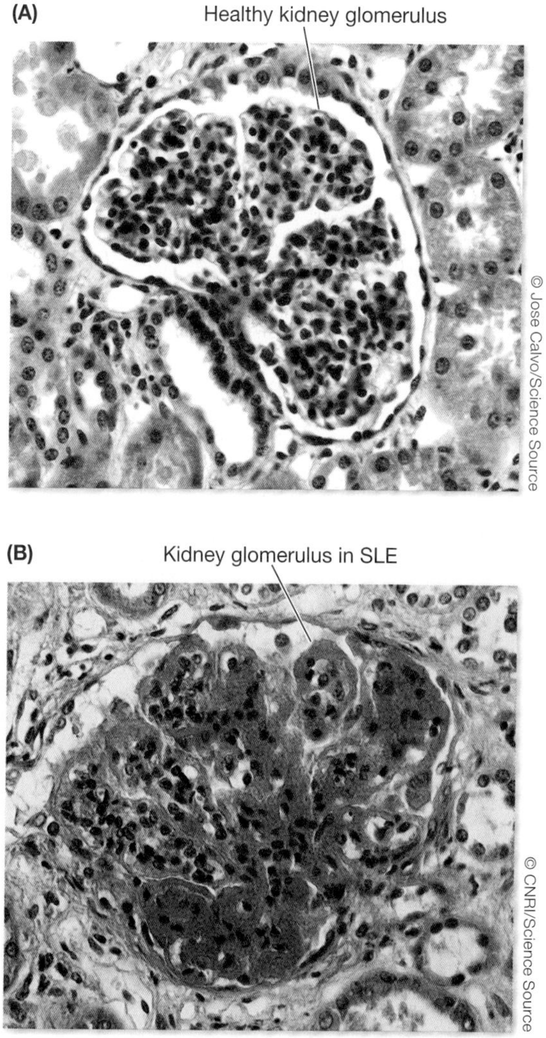

FIGURE 16.12 Kidney damage in systemic lupus erythematosus (SLE) (A) Glomeruli are tufts of blood vessels within the nephrons of the kidney that filter the blood. (B) Damage to glomeruli occurs in SLE due to the deposition of immune complexes that have not been properly removed from the bloodstream.

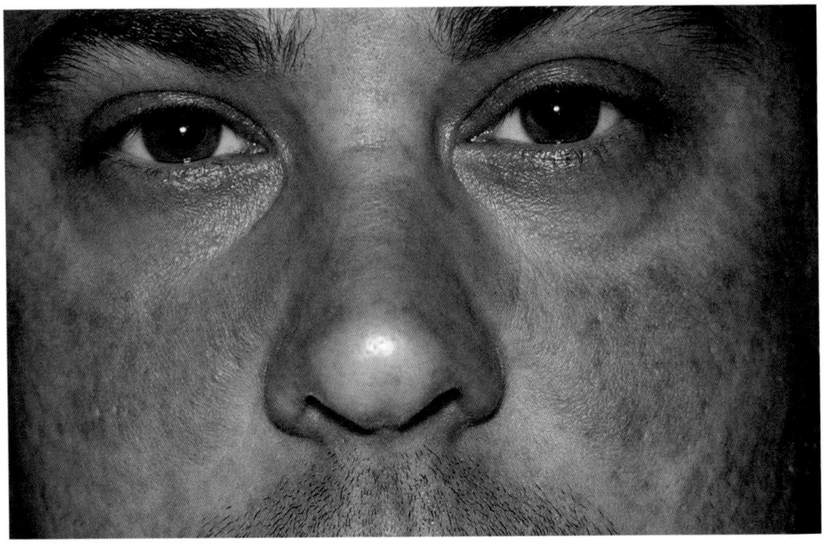

FIGURE 16.13 Facial butterfly rash of systemic lupus erythematosus (SLE) About 50% of SLE patients experience the telltale butterfly rash on the cheeks and nose. The rash appears red and may be smooth or may have a scaly or bumpy texture. Its onset is unpredictable. In some individuals it arises due to sun exposure, while in others it appears spontaneously. Some patients report that appearance of the rash is indicative of the impending onset of other symptoms.

reaction is what creates the characteristic butterfly rash commonly seen on the face of an individual experiencing a flare of SLE symptoms (**FIGURE 16.13**).

The most common treatments include anti-inflammatory medications such as nonsteroidal anti-inflammatory drugs (NSAIDs) and immunosuppressive drugs such as corticosteroids to limit the autoimmune response. More recently, immunotherapies have also been used. For example, belimumab, a human monoclonal antibody that inhibits B-cell activating factor (BAFF) and thus B-cell survival, was approved by the Food and Drug Administration (FDA) in 2011 as a therapy to treat SLE, although its effectiveness in treating severe cases has not been demonstrated.

RHEUMATOID ARTHRITIS A systemic autoimmune disease that affects 1% to 3% of the U.S. population is **rheumatoid arthritis** (**RA**). In this disease, which occurs primarily in older adults (over age 60), an autoimmune response causes chronic inflammation of the joints (**FIGURE 16.14**). A large proportion of those with this condition produce **rheumatoid factor**, anti-immunoglobulin antibodies capable of recognizing the Fc region of IgG or IgM (see **KEY DISCOVERIES** box for the study that discovered rheumatoid factor). These autoantibodies form immune complexes containing soluble IgG or IgM, and the immune complexes deposit in

rheumatoid arthritis (RA) Systemic autoimmune disease characterized by the inflammation of joints.

rheumatoid factor Autoantibodies capable of recognizing the Fc region of IgG molecules.

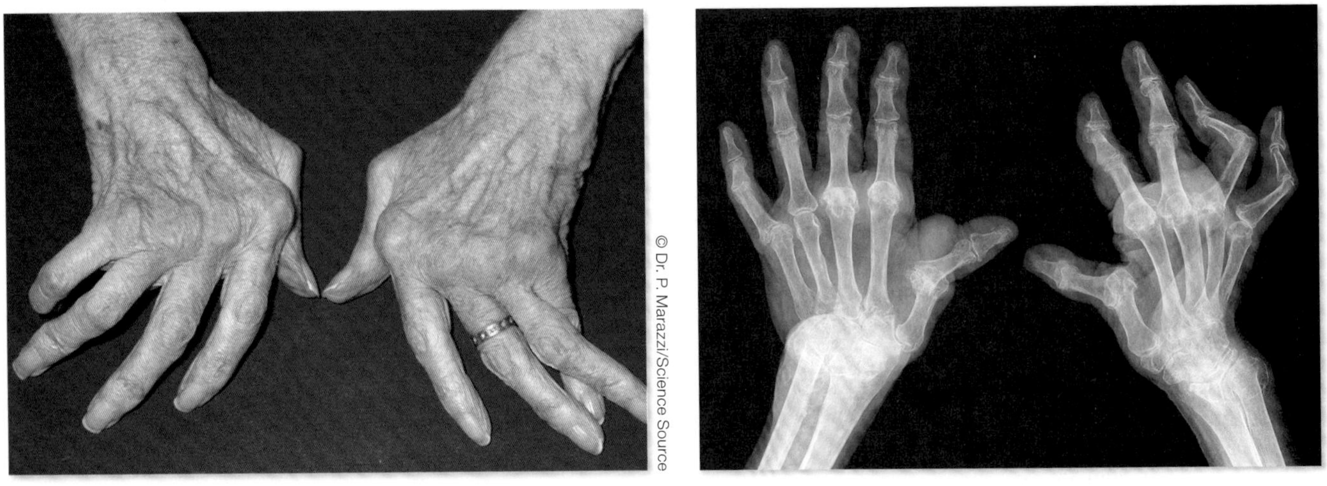

FIGURE 16.14 Effect of rheumatoid arthritis on the hands (A) Rheumatoid arthritis is an autoimmune disease that results in joint inflammation, often resulting in disfiguration of the fingers and hands. (B) Abnormal joint morphology can be seen via radiography.

KEY DISCOVERIES

How was the production of rheumatoid factor in rheumatoid arthritis discovered?

Article

Franklin, E. C., H. R. Holman, H. J. Muller-Eberhard, and H. G. Kunkel. 1957. An unusual protein component of high molecular weight in the serum of certain patients with rheumatoid arthritis. *Journal of Experimental Medicine* 105: 425–438.

Background

Pioneer immunologist Paul Ehrlich was responsible for recognizing that the power of the immune response requires an equally powerful means to prevent self-recognition. In his studies, Ehrlich coined the term *horror autotoxicus*, meaning the horror of self-toxicity. He recognized that while we have the capacity to combat many different pathogens, we also must have the capacity to keep our weapons of the immune system in check to prevent self-reactivity. However, we know that certain diseases arise due to self-reactivity, many of which result from the production of autoantibodies. We have seen that rheumatoid factor is an autoantibody that recognizes self-IgG immunoglobulins and is often present in RA. This leads to the question: How was rheumatoid factor discovered?

Early Research

Researchers had known that normal human serum contained a high-molecular-weight protein complex that could be observed by ultracentrifugation. When normal human serum was analyzed by ultracentrifugation, this high-molecular-weight complex migrated with a sedimentation coefficient of 19*S*. Analysis of proteins in this complex suggested that the major protein components were α2-globulin and γ-globulin and revealed that other proteins might also be part of this complex. Normal human serum had never been demonstrated to have protein complexes with sedimentation coefficients greater than 19*S*, but researchers were interested in the composition of serum from patients with RA.

Think About...

1. Describe a typical ultracentrifugation experiment separating large molecules based on their sedimentation coefficient. What are some means by which you can determine the composition of different complexes you isolated? How could you detect proteins that might be a part of these complexes?
2. If you isolated a complex that sedimented at a different size in samples from a patient with RA compared to a patient without RA (control), what are some questions you would want to answer about the composition of that complex? What experiments would you conduct to answer these questions?

Article Summary

Researchers separated components in normal serum and in serum from patients with RA. As they expected, normal human serum contained a complex that sedimented at a sedimentation coefficient of 19*S*. Interestingly, in 14 of the sera from 31 RA patients, a larger complex with a sedimentation coefficient of approximately 22*S* was observed. The researchers, who were aware of the fact that sera from RA patients could facilitate γ-globulin precipitation and sheep cell agglutination, tested whether the 22*S* complex also had this activity. They observed that the 22*S* complex contained factors capable of driving the same γ-globulin precipitation and sheep cell agglutination that could be observed from sera from patients with RA (**TABLE**). The researchers hypothesize at the end of their report the possibility that the 22*S* component represents a soluble antigen–antibody complex. These findings were among the first to demonstrate that Ehrlich's *horror autotoxicus* was plausible and was a likely driving force for a disease state in humans.

Relationship of Amount of 22*S* Component to γ-Globulin Precipitation and Sheep Cell Agglutination Activity

	γ-globulin precipitation					Sheep cell agglutination						
	Dilution					Dilution						
Amount of 22*S* Material	Neg.	0–25	50	100	>150	<32	64	128	256	512	1024	2048
None	6	8	2			1	2		1	1		
<75 mg percent		3	4					1	2	2		
75–125 mg percent			2	1				1			2	
125–175 mg percent			1	1	1						2	1
>300 mg percent					1					1		

Source: E. C. Franklin et al. 1957. *J Exp Med* 105: 425–438. © 1957 by The Rockefeller Institute for Medical Research New York.

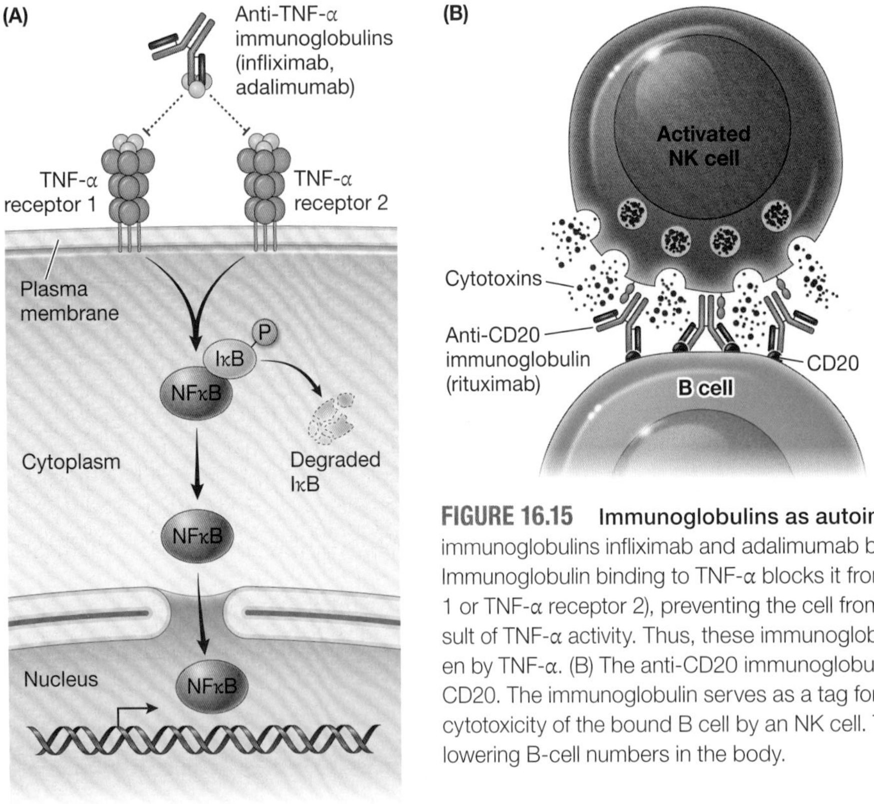

FIGURE 16.15 Immunoglobulins as autoimmune disease therapy (A) The anti-TNF-α immunoglobulins infliximab and adalimumab bind to TNF-α during an inflammatory response. Immunoglobulin binding to TNF-α blocks it from interacting with its receptor (TNF-α receptor 1 or TNF-α receptor 2), preventing the cell from activating genes controlled by NFκB as a result of TNF-α activity. Thus, these immunoglobulins dampen the inflammatory response driven by TNF-α. (B) The anti-CD20 immunoglobulin rituximab binds to the B-cell surface protein CD20. The immunoglobulin serves as a tag for activating antibody-dependent cell-mediated cytotoxicity of the bound B cell by an NK cell. Thus, rituximab dampens autoimmunity by lowering B-cell numbers in the body.

joints and drive inflammation through a type III hypersensitivity-like response. RA treatment mirrors that of SLE treatment since both cause similar symptoms, albeit at different locations and for different reasons. NSAIDs are used to limit the inflammatory response, and corticosteroids are used as immunosuppressive agents. Several immunotherapies have been developed to treat RA, including antibodies such as infliximab and adalimumab, which block the action of the inflammatory cytokine TNF-α. The antibody rituximab recognizes CD20 at the surface of B cells and works as an immunosuppressive agent via targeted B-cell destruction through antibody-dependent cell-mediated cytotoxicity (**FIGURE 16.15**).

MULTIPLE SCLEROSIS Another systemic autoimmune disease prevalent in the population is **multiple sclerosis** (**MS**), which affects more than 400,000 people in the United States. Women are more likely than men to develop the disease, which is usually diagnosed between the ages of 20 and 50. Affected individuals experience muscle weakness, vision problems, and problems with coordination, due to the targeting and destruction of myelin-producing cells of the nervous system and impaired neuronal function (**FIGURE 16.16**).

Autoantibodies recognize common antigens present in all myelin-producing cells, including myelin basic protein and proteolipid protein. Although the exact mechanism of MS onset is unclear, it has been suggested that T_H1 helper T cells play an important role. Unfortunately, there are limited medications for MS treatment, although much interest has been focused on the use of immunotherapy to treat the disease. Several monoclonal antibodies have been approved to treat some types of multiple sclerosis (see **EMERGING SCIENCE** box).

multiple sclerosis (MS) Autoimmune disease caused by demyelination of the central nervous system and white blood cell migration into the brain.

CHECKPOINT QUESTION

1. Contrast the autoantigens responsible for the onset of tissue-specific autoimmune diseases and those responsible for the onset of systemic autoimmune diseases.

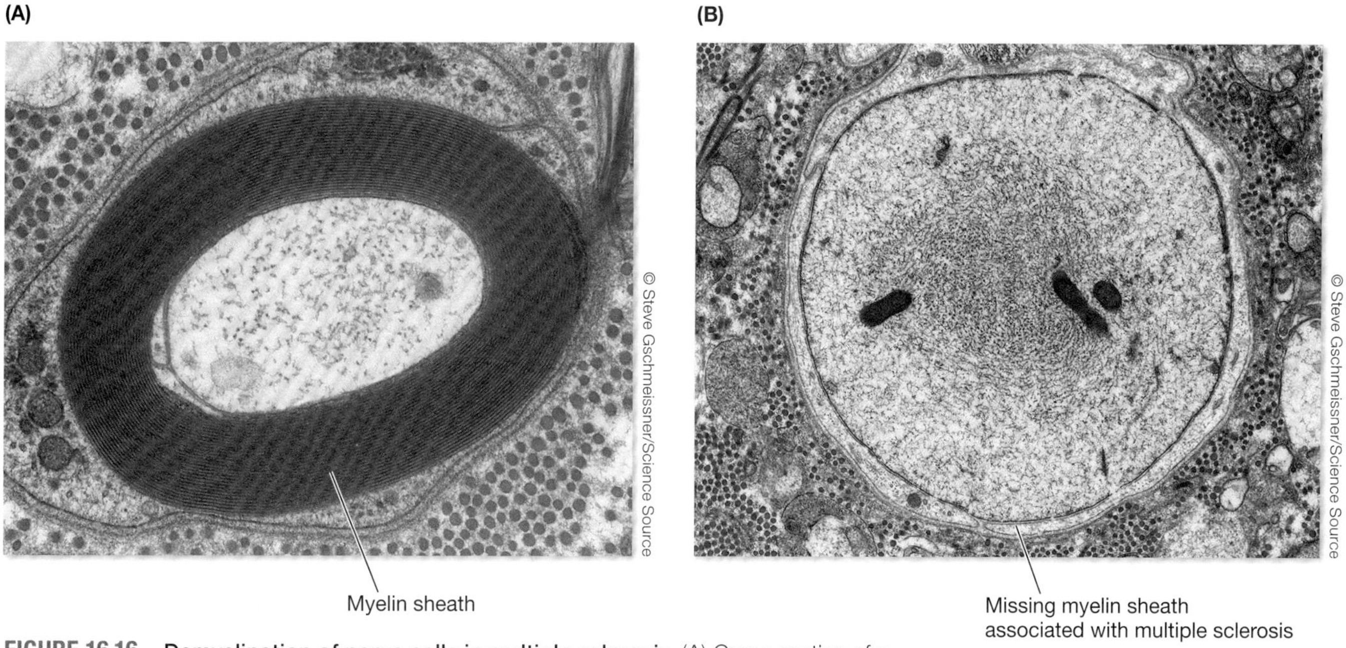

FIGURE 16.16 Demyelination of nerve cells in multiple sclerosis (A) Cross section of a normal nerve cell seen via transmission electron microscopy shows a prominent myelin sheath, which acts as an insulator of the cell for electrical signaling. (B) Multiple sclerosis results in the loss of the myelin sheath, leading to impaired signaling to neighboring nerve cells.

EMERGING SCIENCE

How might monoclonal antibodies be used to combat multiple sclerosis (MS)?

Article

Kappos, L., D. Li, P. A. Calabresi, P. O'Connor, A. Bar-Or, F. Barkhof, M. Yin, D. Leppert, R. Glanzman, J. Tinbergen, and S. L. Hauser. 2011. Ocrelizumab in relapsing-remitting multiple sclerosis: A phase 2, randomised, placebo-controlled, multicentre trial. *Lancet* 378: 1779–1787.

Background

While the pathophysiology of demyelination of nerve cells and inflammation of the brain is known to be causative to the onset of MS, therapeutics to the disease are limited. B cells have been implicated as a contributing factor in disease onset, as both antigen-presenting cells and producers of autoantibodies, suggesting that inhibition of B-cell action might be an effective therapeutic strategy. Drugs such as rituximab, which inhibit B cells by binding to CD20 and inducing antibody-dependent cell-mediated cytotoxicity, had been effectively used in other therapies, including some leukemias, lymphomas, and rheumatoid arthritis. Researchers wanted to test whether a humanized version of an anti-CD20 antibody, ocrelizumab, might be an effective MS treatment.

The Study

Researchers placed 218 patients age 18 to 55 who had been diagnosed with relapsing–remitting multiple sclerosis in one of four treatment groups (placebo, low-dose ocrelizumab, high-dose ocrelizumab, or interferon beta-1a as a positive control). During the treatment protocol, individuals were monitored for MS lesions and relapse using brain MRI scans and for MS phenotypes as measured by an Expanded Disability Status Scale. The study found statistically significant differences in the treatment groups. There was a reduction in brain lesions in both the low-dose (by 89%) and high-dose (by 96%) ocrelizumab recipients. Individuals in these two groups remained lesion-free at a higher rate compared to both the placebo group and the interferon beta-1a group (77% low-dose and 88% high-dose). In other words, humanized anti-CD20 immunoglobulin ocrelizumab lowers brain lesions in patients with multiple sclerosis (**TABLE**). This study demonstrated the effectiveness of B-cell therapy relative to the MS treatment that was being used at the time and paved the way for further safety and efficacy studies of B-cell therapy for MS.

(*Continued*)

EMERGING SCIENCE (continued)

Think About...

1. This study presents findings of a phase 2 trial. What phases of therapy discovery are required for getting a drug to market in the United States, and what must occur in each of these phases? This drug was approved in 2017 by the U.S. Food and Drug Administration. What studies must have followed this phase 2 trial?
2. What other controls (besides the placebo and interferon beta-1a) might have been employed in this study? Would these controls be positive or negative controls?
3. What other data were presented in the article related to the study being a phase 2 study?

Onset of Brain Lesions in Patients with Multiple Sclerosis After Treatment

	Placebo (n = 54)	Ocrelizumab 600 mg (n = 55)	Ocrelizumab 2000 mg (n = 55)	Interferon beta-1a (n = 54)
Number of T1 lesions over weeks 12, 16, 20, 24				
n (%)	54 (100%)	51 (93%)	52 (95%)	52 (96%)
Mean (SD)	5.5 (12.5)	0.6 (1.5)	0.2 (0.7)	6.9 (16.0)
Median (min–max)	1.6 (0–79)	0.0 (0–7)	0.0 (0–3)	1.0 (0–78)
95% CI	0.8–2.6	–	–	0.0–2.0
P value (ocrelizumab vs. placebo)	–	<0.0001	<0.0001	–
P value (ocrelizumab vs. interferon beta-1a)	–	<0.0001	<0.0001	–
Number of new lesions over weeks 4, 8, 12, 16, 20, 24				
Mean (SD)	6.6 (14.2)	0.8 (2.0)	0.8 (2.2)	7.2 (16.3)
Median (min–max)	2.2 (0–93)	0.0 (0–11)	0.0 (0–14)	1.0 (0–95)
95% CI	1.0–4.0	0.0–0.0	0.0–0.0	0.0–2.0
P value vs. placebo	–	<0.0001	<0.0001	0.9

Source: L. Kappos et al. 2011. *Lancet* 378: 1779–1787. © 2011, with permission from Elsevier.

Summary

16.1 What is the relationship between self-tolerance and autoimmunity?

- Autoimmune diseases are induced due to a lack of central or peripheral tolerance of the adaptive immune system.

16.2 How do genetic and environmental factors affect the progression of autoimmune diseases?

- The failure of central and peripheral tolerance, which drives the onset of autoimmune disease, can be caused by both genetic and environmental factors.

16.3 Why do many autoimmune diseases target endocrine glands?

- Organ-specific autoimmune diseases commonly target endocrine glands for two main reasons: these glands are highly specialized and synthesize tissue-specific proteins and are also highly vascularized.

16.4 What are the similarities between hypersensitivity and autoimmune diseases?

- Although the type of antigen differs between hypersensitivity and autoimmune diseases, the ensuing reactions overlap. Certain autoimmune diseases bear symptoms caused by reactions that mirror hypersensitivity reactions.

16.5 How do autoantibodies promote the progression of autoimmune diseases?

- Autoantibodies are a common product of autoimmune diseases and can be found in cases of both organ-specific and systemic disease.

Review Questions

16.1 Explain the connection between central tolerance, peripheral tolerance, and autoimmune diseases.

16.2 You are attempting to create a new mouse model that can be used to study the importance of regulatory T cells in peripheral tolerance and the onset of autoimmune reactions. Provide an experiment that would aid in the development of the mouse model that will allow you to study the importance of regulatory T cells.

16.3 You have developed an antibody to the human TSH receptor capable of binding to the extracellular domain of the receptor and blocking TSH binding, not through binding site competition but by binding site occlusion through steric hindrance. Since the antibody does not seem to bind to the same site as TSH to the TSH receptor, would you predict that this antibody would act as an agonist to TSH receptor function? Why or why not?

16.4 You are part of a scientific team that has identified a novel autoimmune disease. This disease causes destruction of adrenal gland cells responsible for the production of adrenaline. Your team wants to determine what type of autoimmune reaction is responsible for the destruction of the adrenal gland cells and how it relates to hypersensitivity reactions. Can you rule out any hypersensitivity reactions with the limited amount of information you have? How would you distinguish which type of reaction related to hypersensitivity can be causing the adrenal gland cell death?

16.5 Why might you predict that there are so few examples of systemic autoimmune diseases that result from type II hypersensitivity-type reactions?

CASE STUDY REVISITED

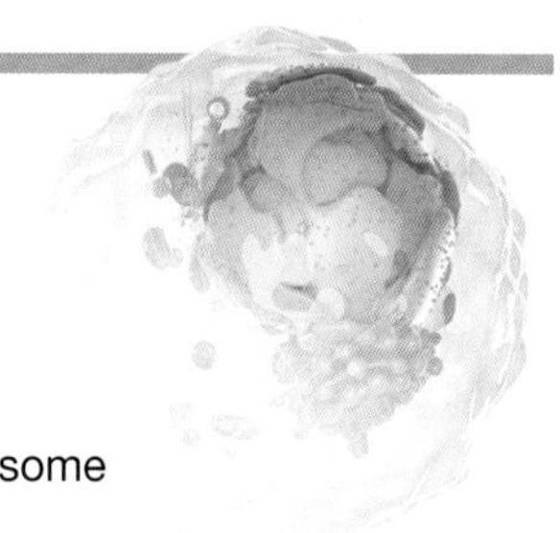

Celiac Disease

A boy, age 10, is brought to the hospital emergency department with symptoms of nausea, vomiting, and diarrhea. His mother tells the doctor that her son has experienced repeated bouts of diarrhea over several months, often after consuming pizza for dinner. She tells the doctor that she and her husband never feel ill after eating the same pizza. The doctor suspects celiac disease, an autoimmune disease in which the lining of the small intestine becomes damaged due to inflammation caused by an immune response to gluten, a protein found in wheat, barley, and rye. The doctor orders a blood test, which is positive for tissue transglutaminase antibodies. Because of the positive result, the doctor also orders a biopsy of the boy's small intestine, which confirms the diagnosis of celiac disease.

Think About...

Questions for individuals

1. Define *celiac disease*. What are some common symptoms?
2. Why would a biopsy of the small intestine reveal celiac disease? How would the boy's tissue sample differ from normal small intestine tissue?

Questions for student groups

1. Since an inflammatory response to gluten is typical in celiac disease, would you predict that there was an elevated or reduced activity of the transcription factor NFκB? Why?
2. Would you predict that individuals with celiac disease would be more or less likely to suffer from intestinal microbial infections? Why?

MAKING CONNECTIONS Key Concepts	**COVERAGE** (Section Numbers)
T-cell development and signaling as it relates to tolerance and onset of autoimmunity	1.3, 1.8, 4.5, 5.7, 7.10, 7.15, 16.1, 16.2, 16.3
B-cell negative selection as it relates to tolerance and onset of autoimmunity	1.3, 1.8, 4.5, 8.8, 8.9, 10.5, 16.1, 16.4, 16.5
Comparing type II hypersensitivity and autoimmunity	1.5, 2.7, 3.1, 3.3, 10.5, 15.3, 16.4
Comparing type III hypersensitivity and autoimmunity	1.5, 2.5, 2.6, 10.5, 15.4, 16.4
Comparing type IV hypersensitivity and autoimmunity	1.5, 2.5, 4.2, 4.7, 7.7, 7.14, 15.5, 16.4

17

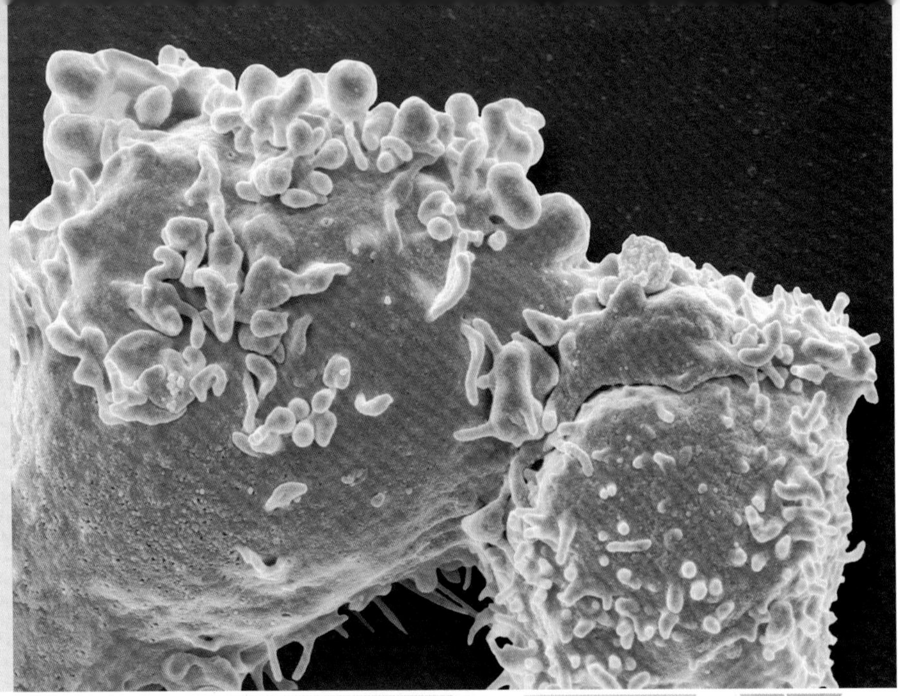

Transplantation and Immune Responses

QUESTIONS Explored

- **17.1** What is the history of organ and tissue transplantation?
- **17.2** Which medical conditions result in the need for tissue or blood cell transplantation?
- **17.3** Which types of alloreactions can occur as a result of transplantation?
- **17.4** What mechanisms drive solid organ transplant alloreactions?
- **17.5** What factors contribute to a successful transplant?
- **17.6** How are transplant rejections suppressed pharmacologically?
- **17.7** How is graft-versus-host disease used to treat other diseases?

CASE STUDY: Sabrina's Story

Sabrina is getting ready for another day at the lab bench. She has been working on her dissertation research for the past few years and is feverishly putting in hours to finish collecting data for her first publication. Before heading to the lab, Sabrina tells her husband, Sven, that her head is really hurting. "Again?" he replies. "Isn't that your third or fourth migraine this month? Maybe you need to consider taking it a bit easier or seeing a doctor."

Sabrina says, "Well, I certainly can't take it easier since I'm on the brink of getting this paper out. I guess I'd better make an appointment."

Sabrina sees her doctor later that day. At the physician's office, the nurse checks her vital signs (pulse, temperature, respiration rate, and blood pressure) before Sabrina enters the exam room. The doctor asks Sabrina how she is feeling and why she is visiting today. Sabrina tells her doctor about the migraines this month and that they seem to occur about the same number of times each month. She reports that she has been putting in long hours in the lab and just assumed the migraines were due to stress or fatigue.

The doctor is concerned about Sabrina's high blood pressure and asks if there is a family history of hypertension. Sabrina isn't sure about her family history but recalls that her grandfather died from a heart attack. The doctor prescribes a calcium channel blocker to lower Sabrina's blood pressure.

The following year, Sabrina wakes up one day feeling a bit under the weather. She's well enough to take on another day, and after all, she's on the last chapter of her dissertation and has to keep pushing to completion. She walks downstairs to start working, feels a massive wave of nausea, and has to run to the bathroom. Sven hears her vomiting and asks what he can do to help. The wave of nausea gives way to an urge to urinate, but Sabrina finds that she can't even though her bladder feels full. She decides that she needs to see her physician as soon as possible.

To understand what Sabrina is experiencing and to answer the case study questions at the end of the chapter, we must explore the connection between transplantation medicine and the workings of the immune system. We have seen that the adaptive immune system specializes in recognizing many foreign antigens to prevent the onset of diseases caused by pathogen challenge. However, there are times when it is medically necessary to introduce foreign tissue into an individual due to organ failure or replacement of the hematopoietic system. In transplantation medicine, the goal is maintaining the tissue or hematopoietic system in the recipient. But the remarkable mechanisms of the adaptive immune system present a challenge in this regard. This chapter focuses on how the immune system reacts to transplanted tissue and the interventions used to prevent transplant rejection.

transfusion Transfer of whole blood or blood components from a donor to a recipient.

17.1 What is the history of organ and tissue transplantation?

LEARNING OBJECTIVE

17.1.1 Describe key historical findings in transplantation medicine and define the common types of organ and tissue transplantation.

The field of immunology has been instrumental in modern medicine's ongoing quest to develop treatments for many types of infections and disease. Organ and tissue transplantation is one critical medical procedure that requires an in-depth understanding of the actions of the immune system. To better appreciate the advances achieved in transplantation medicine, we'll take a brief look at the history of blood transfusion, tissue transplantation, and bone marrow transplantation (**FIGURE 17.1**).

Blood transfusion

Blood **transfusion** is one of the simplest procedures that involves the transfer of cells from one individual (donor) to another (recipient). This process may transfuse blood cells or blood-cell components such as platelets. The first recorded successful blood transfusion into a human was performed in the seventeenth century by French physician Jean-Baptiste Denys (also spelled Denis), who introduced sheep's blood into a 15-year-old boy who had been previously treated with leeches. Unfortunately, later transfusions performed by Denys resulted in the death of the recipient, likely due to the ABO hypersensitivity reaction summarized in Chapter 15. Another seventeenth-century physician, Richard Lower, conducted similar blood transfusions in England using sheep's blood. However, because of the controversy surrounding blood transfusion at that time, both France and England made blood transfusions illegal.

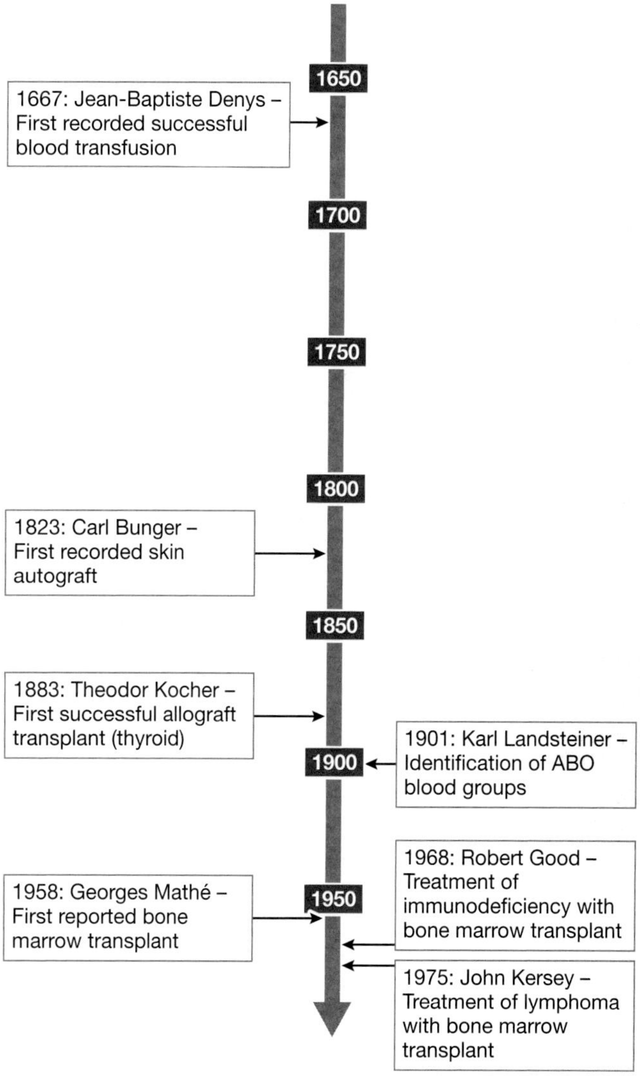

FIGURE 17.1 Important events in the history of transplantation

autograft Transplantation of tissue from one area on an individual to a different area on the same individual.

isograft Tissue transplantation between genetically identical individuals.

allograft Tissue transplantation between genetically distinct individuals.

xenograft Tissue transplantation between different species.

alloantigens Antigens that differ between individuals of the same species.

transplantation rejection Rejection of transplanted donor tissue or an organ by the recipient's immune system.

The procedure remained risky until the early twentieth century, when the A, B, and O blood types were identified by Karl Landsteiner, an American biologist born in Austria. This discovery resulted in the knowledge that different blood types were incompatible and could not be mixed because the immune response caused destruction of donor red blood cells by recipient antibodies. The discovery of blood-type compatibility was a major breakthrough, leading to the use of blood transfusion during surgery and to treat blood loss due to trauma.

Solid organ transplantation

Solid organ transplantation involves the transfer of a donor organ into a patient whose own organ has failed to function properly or has been damaged irreparably. These transplants can be performed in various ways:

- An **autograft** uses an individual's own tissue.
- An **isograft** involves the transfer of tissue between two genetically identical individuals (identical twins).
- An **allograft** involves the transfer of tissue between two genetically distinct individuals.
- A **xenograft** involves the transfer of tissue between individuals of different species.

Because of the specificity of the immune system and its ability to recognize and target foreign molecules, **alloantigens** (antigens that differ between individuals) can result in immune responses. These alloreactions are caused by the positive and negative selection of lymphocytes. Recall that positive selection of lymphocytes tests the production of a functioning lymphocyte receptor, and negative selection tests whether these receptors recognize self-antigens. Negative selection promotes tolerance to self-MHC molecules, but if a transplanted tissue contains different MHC molecules, these molecules might be recognized as foreign, leading to activation of an adaptive immune response. These responses account for many of the complications associated with solid organ transplantation and are more common when tissue is transferred between two genetically distinct individuals.

The history of tissue transplantation may actually date back thousands of years. The first documented case was a skin autograft performed in the early 1800s by German surgeon Carl Bunger, who transferred skin from a patient's inner thigh to the patient's nose to repair damage done by syphilis. A major breakthrough occurred in 1883, when Swiss surgeon Theodor Kocher successfully conducted an allograft thyroid transplant. Since this historic procedure, various organs, including arteries, veins, kidneys, livers, lungs, and hearts, have been successfully transplanted.

The major complication associated with allograft transplants is **transplantation rejection**, which occurs when the recipient's immune system targets donor tissue alloantigens. With the advent of immunosuppressive drugs (discussed in Section 17.6 in this chapter) and sensitive HLA typing and crossmatch testing, the rates of successful allograft transplantation and survival after transplantation have increased. Stem cell applications, including the use of stem cells to regenerate tissue and artificially grow genetically matched organs in vitro, and artificial organ transplantation are now being researched as alternatives to current methods of allograft transplantation (see **EMERGING SCIENCE**).

Hematopoietic stem cell transplantation

Transplantation of the hematopoietic stem cell system is another major medical breakthrough of the twentieth century. The first reported bone marrow transplant was performed in 1958 by Georges Mathé, a French oncologist and immunologist. The transplantation was done to treat patients who had been exposed

EMERGING SCIENCE

Can bioartificial tissue substitute for donor tissue in transplantation medicine?

Articles

Macchiarini, P., P. Jungebluth, T. Go, M. A. Asnaghi, L. E. Rees, T. A. Cogan, A. Dodson, J. Martorell, S. Bellini, P. P. Parnigotto, S. D. Dickinson, A. P. Hollander, S. Mantero, M. T. Conconi, and M. A. Birchall. 2008. Clinical transplantation of a tissue-engineered airway. *Lancet* 372: 2023–2030.

Gonfiotti, A., M. O. Jaus, D. Barale, S. Baiguera, C. Comin, F. Lavorini, G. Fontana, O. Sibila, G. Rombola, P. Jungebluth, and P. Macchiarini. 2014. The first tissue-engineered airway transplantation: 5-year follow-up results. *Lancet* 383: 238–244.

Background

Prior to this medical breakthrough, the treatment of airway defects was problematic and could be done only with the removal and repair of a small piece of the trachea (30% of the total trachea length in children or 6 cm of the trachea in adults). Furthermore, graft development to this point had not been successful. Researchers hypothesized that since bioengineered tissue had been used for human organ replacement of the bladder and skin, a bioengineered trachea could provide suitable treatment for a patient with a tracheal disorder.

The Study

Researchers isolated a trachea from a deceased donor and then decellularized the donor trachea using an enzymatic treatment to remove any donor cells from the removed trachea. The resulting matrix left after decellularization was used as a medium for the introduction of epithelial cells and chondrocytes isolated from the recipient in a bioreactor (**FIGURE ES 17.1**). Upon recellularization, the bioengineered trachea was implanted into the patient, resulting in complete opening of the obstructed left bronchus. A 5-year follow-up demonstrated continued functionality of the bioengineered trachea. These studies demonstrated the creation of a bioengineered trachea that was successfully implanted with recipient epithelial cells and chondrocytes into a patient suffering from chronic tracheitis (infection of the trachea) and bronchomalacia (weakening of the bronchial tubes), which continued to be functional 5 years after transplantation.

Think About...

1. Why did the researchers first decellularize the donor trachea matrix?
2. Three-dimensional printing of artificial tissues is also being actively pursued as a means to create tissues for implantation. What might some potential complications be when considering 3D printing in terms of the recipient's immune response to these artificial materials?
3. What other tissues might be similarly treated by this method of decellularization before transplantation?

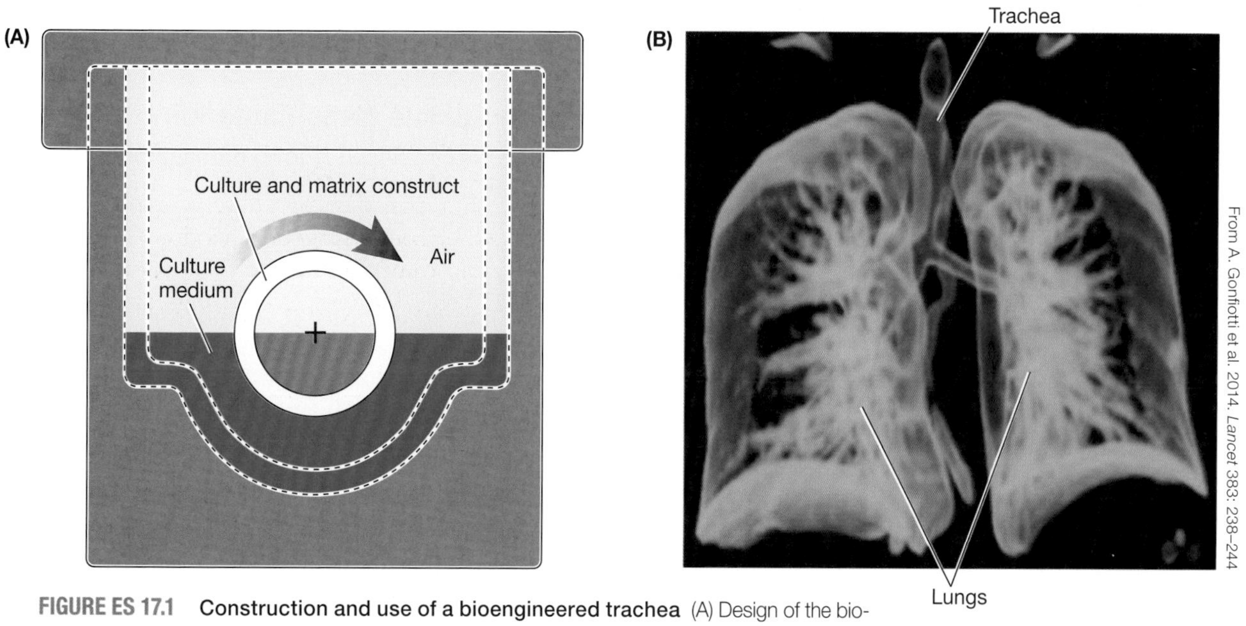

FIGURE ES 17.1 Construction and use of a bioengineered trachea (A) Design of the bioreactor used to recellularize a trachea isolated from a deceased donor. Cells from the trachea were first removed by enzymatic treatment. The decellularized trachea was then placed in the bioreactor with recipient chondrocytes and epithelial cells used to attach to the trachea. (B) Three-dimensional reconstruction image of a computed tomography (CT) scan of the recipient's lungs 5 years after implantation of the bioengineered trachea, displaying normal tracheal and lung function. (A, from P. Macchiarini et al. 2008. *Lancet* 372: 2023–2030. © 2008 Elsevier Ltd.)

to radioactivity due to a nuclear power plant accident. From the 1950s through the 1970s, American physician and hematologist Donnall Thomas and his team finetuned the process of bone marrow transplantation.

Other pioneers of this procedure were American physicians Robert Good, who in 1968 successfully treated an immunodeficiency with a bone marrow transplant, and John Kersey, who in 1975 was the first to successfully use a bone marrow transplant to treat lymphoma. While most bone marrow transplants are used to treat immunological disorders or cancers associated with the hematopoietic system, the use of these transplants to treat other diseases, including HIV, is being explored (see Emerging Science in Chapter 14).

CHECKPOINT QUESTION

1. Differentiate between autograft, isograft, and allograft transplants.

17.2 Which medical conditions result in the need for tissue or blood cell transplantation?

LEARNING OBJECTIVE

17.2.1 Compare and contrast disease states that are treated by solid organ transplantation or hematopoietic stem cell transplantation.

Physicians and researchers who played important roles in the breakthroughs of blood transfusion and tissue and bone marrow transplantation have been awarded the Nobel Prize in Physiology or Medicine, demonstrating the importance of these therapies in contemporary medicine. While blood transfusion is primarily used to replenish blood cells and/or platelets, transplanted cells, tissues, and organs are used to replace the function of failing or diseased tissue for many reasons, including those described in the following sections.

Solid organ and tissue transplantation

Tissues and organs are replaced due to damage or malfunction. While not every organ and tissue has been or can be successfully transplanted, the medical field has used this procedure to repair or replace many organs and tissues (**TABLE 17.1**). In some cases, the donor tissue or organ can be obtained from a living person (as with a kidney); in other cases, it comes from a person who is deceased (as with a cornea). The wait time for a transplantable organ averages 3.5 years due to the shortage of organs available for transplant. Recent dialogue has been opened worldwide in regard to increasing the number of potential organ donors and the regulation and legalization of human organ trade (see **CONTROVERSIAL TOPICS**).

TABLE 17.1 | Transplanted Tissues/Organs and Reasons for Transplant

Transplanted Tissue or Organ	Medical Condition Requiring Transplant
Skin	Burns Extensive necrotic skin infection
Cornea	Scarring due to infection or physical damage Chemical damage
Blood vessels	Coronary artery bypass Diabetes
Bone	Situations in which tissue reconstruction is required (e.g., dental implants, knee replacement)
Heart valves	Heart valve disease Congenital defects (e.g., valve stenosis, malformed heart valve)
Heart	Heart failure Coronary artery disease
Kidney	Kidney failure End-stage renal disease
Lungs	Chronic obstructive pulmonary disease (COPD) Pulmonary fibrosis
Liver	Liver failure due to viral hepatitis or cirrhosis
Pancreas	Type 1 diabetes
Stomach	Digestive system failure Stomach cancer
Intestines	Intestinal failure
Testes	Testicular cancer Congenital absence of testes

Source: www.organdonor.gov/learn/what-can-be-donated

Bone marrow transplantation

Hematopoietic stem cell transplantation is used to restore proper function of the bone marrow. If a subpopulation of cells produced in bone marrow fails to function or develop properly, then replacement of the stem cell system responsible for the production of those malfunctioning or improperly developed cells can resolve the issue. Under these circumstances, the patient's bone marrow is

CONTROVERSIAL TOPICS

Should organ trade be legalized and regulated?

Transplantation medicine has taken extremely large strides over the past century, pushing boundaries and providing individuals with organ failure a chance for prolonged life. However, complications have arisen, especially in light of the huge demand for organs and the lack of supply.

According to the U.S. Department of Health and Human Services, there are currently over 116,000 Americans in need of an organ transplant, with 20 people on the waiting list dying each day. In 2016 there were just over 33,000 organ transplant surgeries, showing that the demand for life-saving organ transplants far outweighs the availability of donor organs. This demand for donor organs has resulted in societal issues, including human organ trafficking and black market organ sales. Until 2013, organ trade had been illegal in all countries except Iran. Organ sales were legal in India until 1994 and in the Philippines until 2008, when all organ sales were banned in these countries.

The demand has driven illegal sales of organs, where on the black market kidneys can fetch more than $150,000, although the donor would receive only approximately 3% of this amount, with the rest going to the individuals engaged in the illegal sales. A major medical concern with black market organs is the potential for infection in both the donor and recipient, since the surgeries may be performed in substandard conditions.

Furthermore, since the selling of organs can potentially provide income to individuals in debt, those most commonly targeted for black market organs are poor. Because of safety concerns associated with black market organs, Australia and Singapore have legalized monetary compensation for living organ donors. In 2014, Robert Truog, director of the Center for Bioethics at Harvard Medical School, wrote a letter to President Obama, Attorney General Holder, and Congressional leaders pleading for support for projects to study compensation for live kidney donors due to high demand and the relative safety of kidney donation when the surgery is done properly. Compensating live donors, he argues, would improve the safety of donors and recipients alike. With the many social, medical, and ethical issues associated with this topic, it will likely continue to be a heated debate into the next decade.

Think About...

1. Do you think financial compensation for organ donation should be legalized or not? What evidence or reasoning supports your views? What evidence or reasoning from the opposing viewpoint might sway your view?
2. How might legalized compensation for organ donation be regulated? Would you propose that the federal government regulate the practice or another entity? If organ sales were legal, should donors be taxed on the sale of an organ? Why or why not?
3. What types of laws might need to be enacted to protect the rights of a donor selling an organ legally?
4. What considerations or information would you recommend a potential donor examine before deciding to donate?

destroyed through chemotherapy and irradiation prior to the transplant. Bone marrow transplant is also performed due to uncontrolled growth (cancer) of a subset of the hematopoietic cells. Disorders that may be treated with bone marrow transplantation are shown in **TABLE 17.2**.

HEMATOPOIETIC CELL GENETIC ABNORMALITIES Recall from Chapter 1 that the hematopoietic system has three lineage arms: the myeloid, lymphoid, and erythroid lineages. While we have focused much of our attention on the lymphoid and myeloid arms, the erythroid lineage (responsible for the production of erythrocytes and platelet-producing megakaryocytes) can also be affected by genetic abnormalities, as with thalassemia and sickle cell anemia, both of which alter hemoglobin function. Genetic abnormalities affecting the erythroid lineage can potentially be treated with bone marrow transplantation.

Recall from Chapter 14 that numerous immunodeficiencies affect many facets of the immune system. While some immunodeficiencies affect the activity of soluble molecules responsible for immune responses, such as the complement system, many alter the presence or activity of hematopoietic system cells. Because abnormalities that affect blood cell development and function are driven

TABLE 17.2 | Conditions Treated with Bone Marrow Transplant

Medical Condition Requiring Transplant	Affected Hematopoietic Cells Replaced
Leukemias, including: Acute myeloid leukemia (AML) Acute lymphoblastic leukemia (ALL) Chronic myeloid leukemia (CML)	B and/or T lymphocytes
Lymphomas: Hodgkin Non-Hodgkin	B and/or T lymphocytes
Red blood cell disorders: Thalassemia Sickle cell anemia Fanconi anemia (failure to produce red blood cells)	Erythrocytes
Severe-combined immunodeficiency	B and T lymphocytes
Wiskott-Aldrich syndrome	T lymphocytes and platelets

Source: R. L. Soutar and D. J. King. 1995. *Brit Med J* 310: 31–36.

by mutation in critical genes, bone marrow transplantation can effectively replenish the hematopoietic system with stem cells that are not genetically altered and thus are capable of producing normally functioning red or white blood cells.

HEMATOPOIETIC CELL CANCER Because of the need to produce large quantities of cells and quickly replenish the many different cell types produced by hematopoiesis, cells developing in the bone marrow are prone to uncontrolled cell division and becoming cancerous. Recall that during their development, B and T lymphocytes must undergo recombination events at their receptor loci to properly function. The recombination events that occur during development, while aimed at rearranging the lymphocyte receptor loci, are inherently mutagenic to the lymphocyte genome as the process relies on many of the same actions as non-homologous end-joining DNA repair mechanisms. Improper recombination events can fuse lymphocyte receptor loci with genes involved in cellular division and alter the cell-division control mechanism of those lymphocytes (**FIGURE 17.2**). There are two main types of hematopoietic cell cancers associated with abnormal lymphocyte development and division: leukemias and lymphomas. Leukemias are cancers that begin within the bone marrow due to abnormal recombination events that prevent proper development of lymphocytes. Several leukemias, including acute myeloid leukemia and acute lymphoblastic leukemia, result from genetic changes that occur during hematopoiesis development and differentiation. In these situations, an undifferentiated precursor is "frozen" in an uncontrolled dividing state. The other cancers that affect lymphocytes of the hematopoietic system are lymphomas. Lymphomas are caused by lymphocytes that undergo rapid and uncontrolled division in lymphatic tissue such as lymph nodes. These cancers result from genetic changes that occur during the clonal expansion phase of activation of the immune response, producing lymphocytes that can no longer stop dividing. The genetic alterations that occur to induce leukemia and lymphoma often require removal of the affected individual's hematopoietic system and replacement with donor bone marrow.

• CHECKPOINT QUESTION

1. Compare how removal of a diseased or damaged organ before transplanting a new one is analogous to chemotherapy and irradiation done prior to a bone marrow transplant.

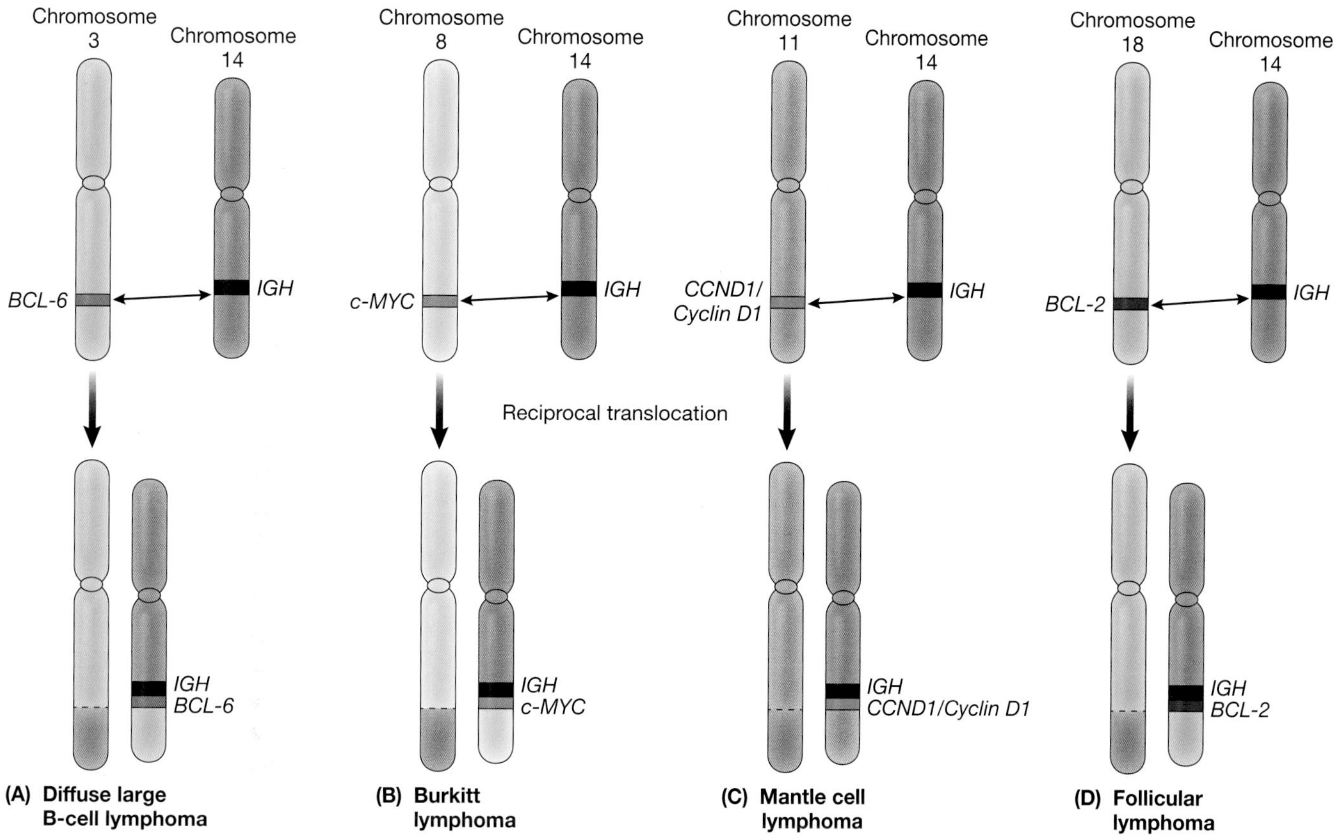

FIGURE 17.2 Translocations that occur during B-cell development result in cancer During B-cell development, recombination occurs at both the heavy chain and light chain immunoglobulin loci. The immunoglobulin heavy chain locus (IGH) is located on chromosome 14. If, during recombination, a reciprocal translocation (swapping of chromosomal ends) occurs between chromosome 14 at the heavy chain locus and another chromosome containing a proto-oncogene (gene that positively regulates cell division), the result is cancer. (A) Diffuse large B-cell lymphoma results from a reciprocal translocation between chromosome 14 at the heavy chain locus and chromosome 3 at the BCL-6 locus. BCL-6 is a transcription factor involved in IL-4 responses. (B) Burkitt lymphoma results from a reciprocal translocation between chromosome 14 at the heavy chain locus and chromosome 8 at the c-MYC locus. Myc is a transcription factor that activates genes involved in proliferation. (C) Mantle cell lymphoma results from a reciprocal translocation between chromosome 14 at the heavy chain locus and chromosome 11 at the CCND1/Cyclin D1 locus. Cyclin D1 regulates kinases involved in the cell cycle. (D) Follicular lymphoma results from a reciprocal translocation between chromosome 14 at the heavy chain locus and chromosome 18 at the BCL-2 locus. Bcl-2 is an inhibitor of apoptosis. (After J. Zheng. 2013. *Oncol Rep* 30: 2011–2019.)

17.3 Which types of alloreactions can occur as a result of transplantation?

LEARNING OBJECTIVE

17.3.1 Name and explain the types of alloreactions that can occur as a result of solid organ transplantation and hematopoietic stem cell transplantation.

While organ and bone marrow transplantation can effectively treat extremely debilitating diseases and organ failure, potentially life-threatening complications can occur due to the immune system's capacity to recognize the difference between self- and nonself antigens. Solid organs, when transplanted into a recipient, can present genetically distinct antigens on polymorphic HLA molecules. While we saw in Chapter 6 that the population benefits from the presence of polymorphic HLA molecules due to the ability to present a large number of different antigens from disease-causing pathogens, these same HLA polymorphisms can cause challenges to the success of a transplant. As we saw in Chapter 4, the recipient's immune system may identify the donor organ as

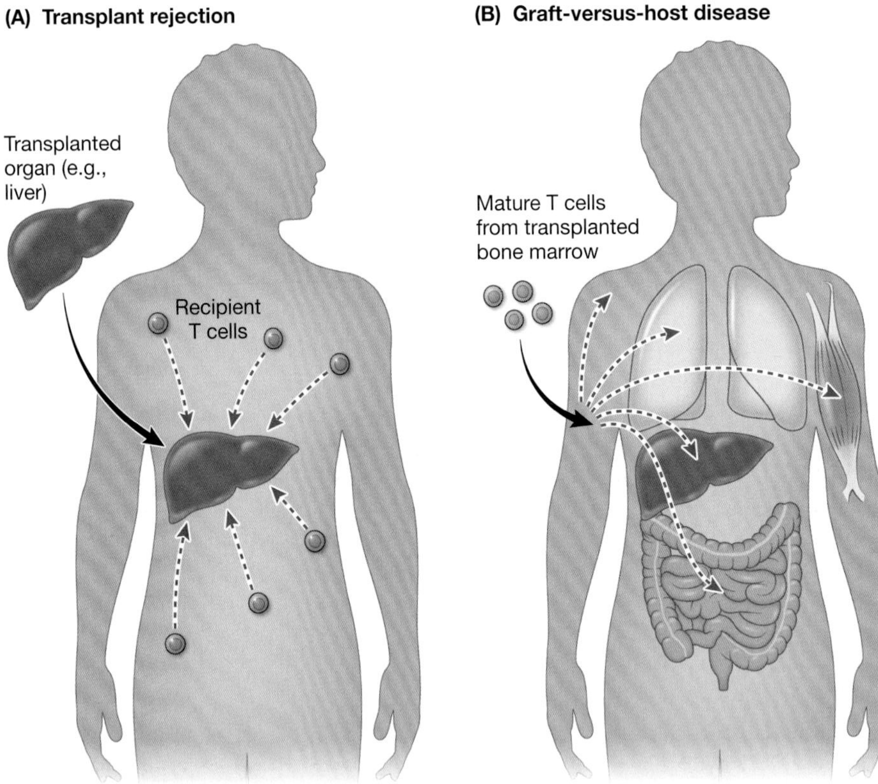

FIGURE 17.3 Alloreactions involving transplanted tissue (A) Transplantation of a solid organ or tissue, such as the liver, can result in an alloreaction in which the recipient's adaptive immune system recognizes the transplanted tissue as foreign, inducing a response that results in organ/tissue rejection. (B) Transplantation of bone marrow cells can result in an alloreaction in which immune system cells of the transplanted bone marrow target tissues of the recipient. Common targets include the liver, gastrointestinal tract, skin, lungs, and the neuromuscular system. This immune response is known as graft-versus-host disease.

alloreaction Immune system reaction in response to an allograft transplant.

graft-versus-host disease Alloreaction triggered by a bone marrow transplant in which the newly transplanted bone marrow develops and recognizes the recipient's tissues and organs as foreign.

foreign, which can lead to transplant rejection or destruction of the transplanted organ by the recipient's immune system (**FIGURE 17.3**). The speed at which rejection can occur depends on differences between the donor and recipient blood type and MHC molecules.

In contrast, the major **alloreaction** that can occur during a hematopoietic stem cell transplant is **graft-versus-host disease**, in which the newly transplanted hematopoietic system recognizes the recipient's tissues and organs as nonself and targets them in an immune response (see Figure 17.3). Although the complications associated with bone marrow transplantation differ from those related to solid organ transplants, the driving force behind both alloreactions is the same, namely recognition of tissue or organs as nonself by the immune system.

Solid organ rejection

We have seen that our innate and adaptive immune responses typically require specific activating signals. What activates an immune response to transplanted tissue? The patient's failing or injured organ is commonly inflamed due to immune system reactions. The damage and inflammation that occur within an individual in need of an organ transplant further complicate the difficult transplant surgery. Furthermore, the donor organ is also typically inflamed due to lack of blood supply to the organ while it was being harvested or for even longer if it was isolated from a cadaver. This inflammation primes the donated organ as a target for the immune system if the cells of the immune system recognize the organ as nonself.

HYPERACUTE REJECTION Recall from Chapter 15 that the presence of antibodies to A and B blood antigens in an individual can lead to a hypersensitivity reaction and result in hemolytic anemia if mismatched blood is given during a blood transfusion. This reaction is a type II hypersensitivity reaction, caused by the presence of soluble immunoglobulins to cell-surface molecules.

It is hypothesized that these antibodies arise due to similarity of the A and B antigens to bacterial cell-surface polysaccharides, since individuals who do not express one of these antigens should not produce immunoglobulins when they have not come in contact with the antigen due to mechanisms of tolerance during B-cell development. The presence of these preexisting antibodies is a problem in blood transfusions and, potentially, in organ transplantation, as these antigens are also expressed on blood vessel endothelial cells.

hyperacute rejection Rejection of a transplanted organ due to destruction of the vasculature of the organ through complement fixation driven by the presence of preexisting blood antigen or HLA antibodies.

Further complicating matters is the possible presence of preexisting antibodies to HLA molecules, which may have been produced if the individual had been pregnant (and raised the anti-HLA antibodies to polymorphisms given to the fetus from the father), had received previous blood transfusions, or had undergone prior transplant surgery. The recognition of either blood antigens or HLA molecules on vascular endothelial cells by preexisting immunoglobulins can lead to destruction of the blood vessels through complement fixation via the classical pathway of complement activation and the action of the C1 protease (C1q/C1r/C1s). This can prevent proper oxygenation of the transplanted organ, resulting in **hyperacute rejection** of the organ (**FIGURE 17.4**). Hyperacute rejection mirrors a type II hypersensitivity reaction recognizing blood group antigens or HLA molecules on the surface of the transplanted organ. Because the recipient's antibodies are preexisting, hyperacute rejection can occur even during the surgery itself. The only means known to prevent this type of rejection is through blood type matching and typing and crossmatching of HLA molecules (see Section 17.5).

ACUTE REJECTION While proper blood typing and HLA crossmatching prevent hyperacute rejection, the inflammation resulting from solid organ transplantation can activate donor dendritic cells. Activated donor dendritic cells increase antigen presentation and produce costimulatory molecules such as B7 required for the activation of T cells. In a similar manner to infection with a pathogen, these donor dendritic cells migrate to secondary lymphoid tissue, where they present antigens on donor MHC molecules. Recipient circulating T cells infiltrate the secondary lymphoid tissue, where they survey the surface of the donor dendritic cells. If a recipient T cell recognizes a donor dendritic cell MHC–peptide complex via its T-cell receptor, it becomes activated (as we saw in Chapter 7) and can use effector mechanisms against cells expressing the same donor MHC–peptide complexes (the transplanted tissue). Because it requires T-cell activation through donor antigen processing and presentation, acute rejection mimics type IV (delayed-type) hypersensitivity reactions.

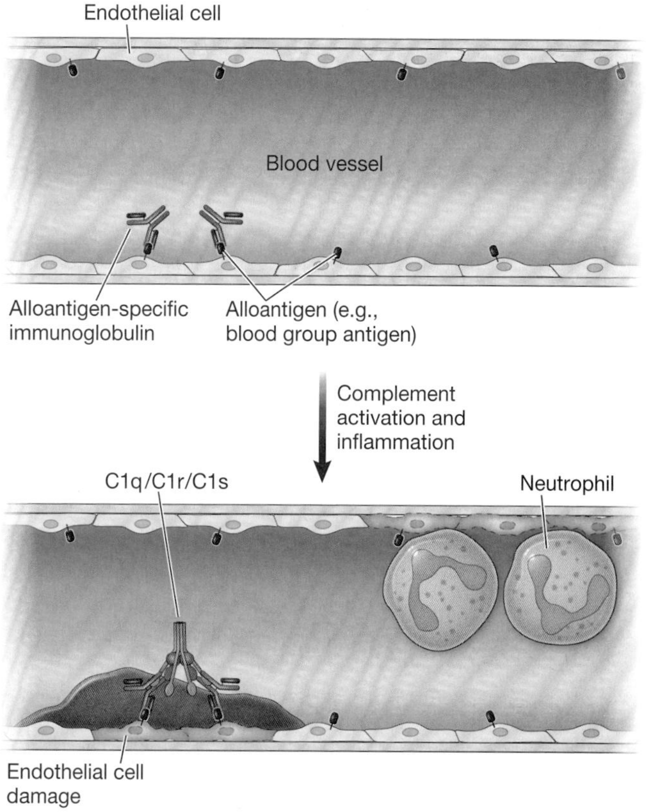

FIGURE 17.4 Hyperacute rejection of transplanted tissue Hyperacute rejection occurs when recipient immunoglobulins bind to surface antigens on the transplanted tissue. Recipient immunoglobulins that can trigger a hyperacute rejection include those that bind to blood group antigens. Binding of these immunoglobulins to surface antigens on the transplanted tissue results in complement activation via the classical pathway by the action of the protease C1q/C1r/C1s of the endothelial cells and triggers an inflammatory response. These responses promote neutrophil recruitment to the transplanted tissue and damage to endothelial cells. (After A. K. Abbas and A. H. Lichtman. 2004. *Basic Immunology Functions and Disorders of the Immune System*, 2nd ed. Saunders: Philadelphia.)

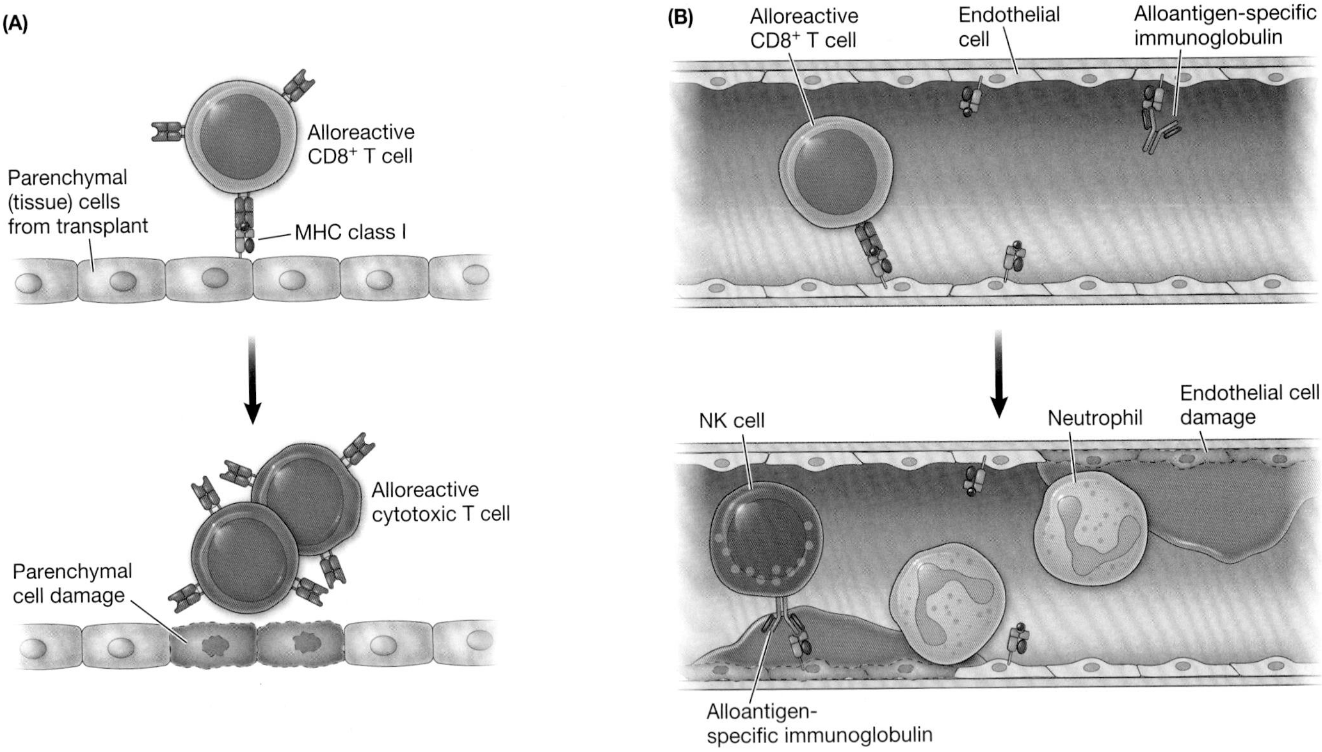

FIGURE 17.5 Acute rejection of transplanted tissue Acute rejection occurs when alloreactive lymphocytes recognize antigen presented on the donor organ. (A) One outcome of an acute rejection occurs when alloreactive CD8 T cells recognize an MHC class I–antigen complex on the surface of parenchymal (tissue) cells. This recognition promotes cytotoxic T-cell actions against the parenchymal cells and causes apoptosis of the parenchymal cell. (B) Another outcome of an acute rejection occurs in the blood vessel of the tissue when CD8 T cells recognize an MHC class I–antigen complex on the endothelial cell surface. Acute rejection can also occur when alloreactive antibodies are produced due to tissue damage resulting from the activation of alloreactive B cells at a nearby draining lymphoid tissue. Inflammation and recruitment of neutrophils within the blood vessel result in activation and recruitment of CD8 T cells, driving apoptosis of recognized endothelial cells. The presence of alloreactive immunoglobulins produced during B-cell activation promotes endothelial cell damage and apoptosis due to the action of circulating natural killer (NK) cells and the induction of antibody-dependent cell-mediated cytotoxicity. (After A. K. Abbas and A. H. Lichtman. 2004. *Basic Immunology Functions and Disorders of the Immune System*, 2nd ed. Saunders: Philadelphia.)

acute rejection Rejection of a transplanted organ due to recognition of donor MHC–peptide complexes presented by donor dendritic cells to recipient T cells and the ensuing effector mechanisms upon activation of those T cells.

chronic rejection Rejection of a transplanted organ due to immune reactions targeting the vasculature of the transplanted organ, resulting in loss of oxygen and ultimately organ failure.

Effector mechanisms can include cytotoxic T-cell responses toward donor cells, activation of area macrophages, and activation of B cells that express immunoglobulins that recognize donor antigens, driving the fixation of complement and antibody-dependent cell-mediated cytotoxicity by natural killer (NK) cells toward donor cells (**FIGURE 17.5**). The ensuing rejection driven by these mechanisms is referred to as **acute rejection**. Acute rejection reactions require approximately the same amount of time as a primary adaptive immune response. They can be prevented by proper HLA typing and crossmatching and treated using immunosuppressive therapies.

CHRONIC REJECTION While hyperacute and acute rejection of a transplanted organ can happen within hours or days of the transplant surgery, **chronic rejection** can occur months or years later. Quite often, chronic rejection is the culprit causing a transplanted organ to ultimately fail. In 2019 data from the Scientific Registry of Transplant Recipients, the 10-year graft failure percentage when a donated kidney was from a living donor was approximately 30%, demonstrating that approximately one-third of transplanted kidneys succumb to failure within 10 years due to chronic rejection.

Chronic rejection occurs as a result of immune responses that target the vasculature of the transplanted organ. It is driven by effector mechanisms that resemble those that drive acute rejection. Production of antibodies that recognize vasculature antigens and cell-mediated effector mechanisms such as T-cell responses result in a chronic state of inflammation. While chronic rejection is also caused by reactions to alloantigens, it requires processing and presentation of donor antigens by recipient antigen-presenting cells (APCs). Because it requires T-cell activation through donor antigen processing and presentation, chronic rejection mimics type IV (delayed-type) hypersensitivity reactions. The inflammation causes increased swelling of the vasculature, preventing efficient blood flow through the transplanted organ and eventually causing organ failure (**FIGURE 17.6**). Although therapies such as immunosuppressive drugs may prolong survival following organ transplantation, these treatments do not seem to be able to prevent chronic rejection. Ultimately, the patient may need to undergo another organ transplant due to failure of the transplanted organ.

Graft-versus-host disease

While organ rejection is the most common alloreaction seen in solid organ transplantation, a different alloreaction can occur with hematopoietic stem cell transplantation. As noted earlier in the chapter, individuals who undergo bone marrow transplants are first subjected to chemotherapy and irradiation to rid the body of rapidly dividing cells associated with the bone marrow. While this treatment accomplishes its goal, it also affects other rapidly dividing cells, including those of the skin, intestines, and liver, causing damage and inflammation in these tissues and organs. Dendritic cells associated with those tissues become activated and migrate to secondary lymphoid tissue, where they present antigens to mature T cells present in the transplanted bone marrow. These mature T cells have developed and matured in the donor's thymus and have gone through negative selection in the presence of donor dendritic cells; as a result, they may be alloreactive to cells that differ in the HLA molecules, resulting in graft-versus-host disease

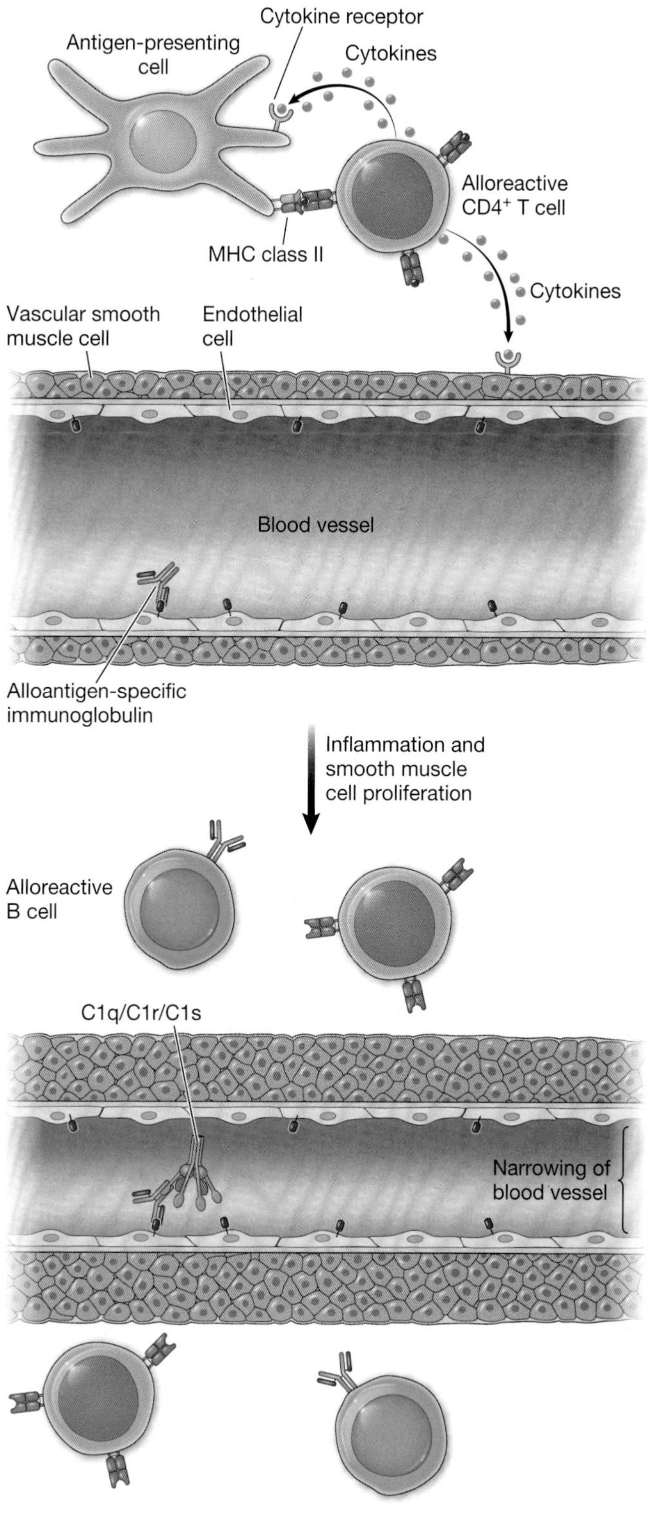

FIGURE 17.6 **Chronic rejection of transplanted tissues**
Chronic rejection mirrors a delayed-type hypersensitivity reaction, whereby antigen-presenting cells process antigens in the donor tissue and present them to alloreactive CD4 T cells via MHC class II complexes. Activated alloreactive T cells can prompt activation of alloreactive B cells to secrete immunoglobulins that recognize antigens within the donor tissue. Furthermore, alloreactive T cells can secrete cytokines to induce proliferation of smooth muscle cells surrounding the donor tissue blood vessels. Alloreactive immunoglobulins induce activation of complement via the classical pathway and the action of the C1q/C1r/C1s protease and an inflammatory response, driving further immune cell action within the donor tissue. Smooth muscle cell proliferation leads to narrowing of blood vessels within the transplanted tissue. This causes ischemia (decreased blood supply) and decreased oxygen delivery to the transplanted tissue. (After A. K. Abbas and A. H. Lichtman. 2004. *Basic Immunology Functions and Disorders of the Immune System,* 2nd ed. Saunders: Philadelphia.)

minor histocompatibility antigen Alloantigens that differ by the presented peptide rather than the MHC molecule presenting the peptide.

direct pathway of allorecognition Pathway of presentation of alloantigens in which donor dendritic cells transplanted with the organ or tissue migrate to a secondary lymphoid tissue and present antigens to recipient T cells.

targeting the inflamed tissues. Because bone marrow transplants are commonly used to treat immune system disorders, it is important that proper HLA typing and crossmatching occurs before the procedure.

Like chronic rejection of an organ transplant, there can be a chronic state of graft-versus-host disease, which mimics an autoimmune disease. The newly developing lymphocytes target a number of tissues within the recipient, including the skin, liver, eyes, mouth, lungs, and gastrointestinal tract. Although the mechanism of onset of chronic graft-versus-host disease is not known, it is thought to be connected to complications of immature hematopoietic cells maturing within the recipient. Individuals with chronic graft-versus-host disease are usually treated with immunosuppressive drugs, which suppress the action of the transplanted bone marrow cells and thus counteract the action of the transplanted tissue.

The likelihood of preventing graft-versus-host disease is increased through proper HLA matching between donor and recipient. Graft-versus-host disease can still occur in individuals with closely matched HLA loci, especially males who receive bone marrow transplants from their identical sisters. Mature T cells from the sister's donated bone marrow react to antigens present from the processing of proteins expressed from Y chromosome genes known as H-Y antigens. H-Y antigens can be presented on both class I and class II HLA molecules. These antigens are called **minor histocompatibility antigens** because the antigenic determinant is driven by the peptide bound to the HLA molecule more than the presenting HLA molecule. While H-Y antigens are one example of minor histocompatibility antigens, these antigens can also be derived from other expressed loci that are polymorphic in the population.

CHECKPOINT QUESTION

1. State the timing and explain the causes of hyperacute, acute, and chronic rejection.

17.4 What mechanisms drive solid organ transplant alloreactions?

LEARNING OBJECTIVE

17.4.1 Differentiate between the direct, indirect, and semidirect pathways of allorecognition.

In the last section we saw that two of the main concerns with solid organ transplantation are acute and chronic rejection, since hyperacute rejection can typically be avoided through proper blood typing and ensuring that there are no preexisting anti-HLA antibodies present in the recipient. These two types of rejection occur because tolerance for circulating T cells in a recipient is established to self-MHC molecules of the recipient. The recipient's T cells have not been exposed to donor MHC molecules during their development, and so circulating T cells within the recipient may recognize donor MHC molecules as foreign. The speed at which acute or chronic rejection occurs differs, with acute rejection potentially occurring within 1 to 2 weeks of the transplant and chronic rejection occurring within months or years. The difference in timing is due to the nature of antigen presentation that occurs and the cells involved.

Direct pathway of allorecognition

Acute rejection of solid organ transplants is driven by the **direct pathway of allorecognition** (**FIGURE 17.7**). During the transplant, donor white blood cells,

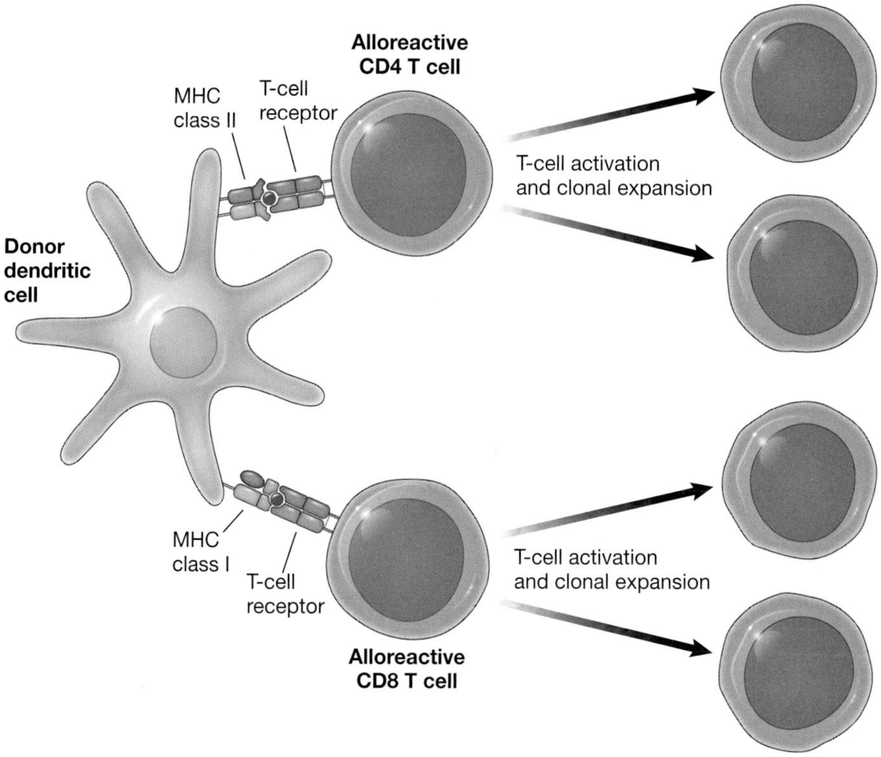

FIGURE 17.7 Direct pathway of allorecognition The direct pathway is driven by the action of donor dendritic cells in transplanted tissue. The inflammation present due to the transplant promotes migration of donor dendritic cells to draining lymphoid tissue, where alloreactive CD4 and CD8 T cells within secondary lymphoid tissue that recognize alloantigens bind to the donor dendritic cell. The interaction between the MHC–alloantigen complex and the T-cell receptor promotes activation and clonal expansion of the alloreactive T cells. (After T. Šarić et al. 2008. *Cells Tissues Organs* 188: 78–90.)

including dendritic cells, are introduced into the recipient. As mentioned earlier, there is inflammation within the organ, which prompts activation of the donor dendritic cells, driving expression of costimulatory receptors such as B7 and an increase in antigen presentation in the dendritic cells. The inflammation also increases vascular permeability and allows the dendritic cell to migrate to a nearby secondary lymphoid tissue to present processed antigens to circulating T cells.

Because dendritic cells express both HLA class I and class II molecules, the donor dendritic cells that drive the direct pathway present alloantigens via donor MHC molecules to recipient T cells containing either CD4 or CD8 coreceptors, depending on the HLA molecule used during presentation. If the recipient T cell expresses a receptor capable of binding to the donor MHC–peptide complex, it will activate (as described in Chapter 7), undergoing clonal expansion and differentiation into effector T cells that will target cells containing the donor MHC–peptide complex. These effector mechanisms can include production of cytotoxic T cells (discussed in Chapter 7) that ultimately target the transplanted organ.

Indirect pathway of allorecognition

While during the direct pathway of allorecognition donor dendritic cells present donor MHC–peptide complexes to recipient T cells, chronic rejection of a

Making Connections: Alloreactions of Transplantations

Transplanted tissues and hematopoietic stem cells illustrate a modern marvel of medicine. However, adaptive immune responses pose a challenge associated with acceptance of the transplanted cells or tissue. The various adaptive immune responses to transplanted cells or tissue mirror responses discussed in Chapters 7 and 10.

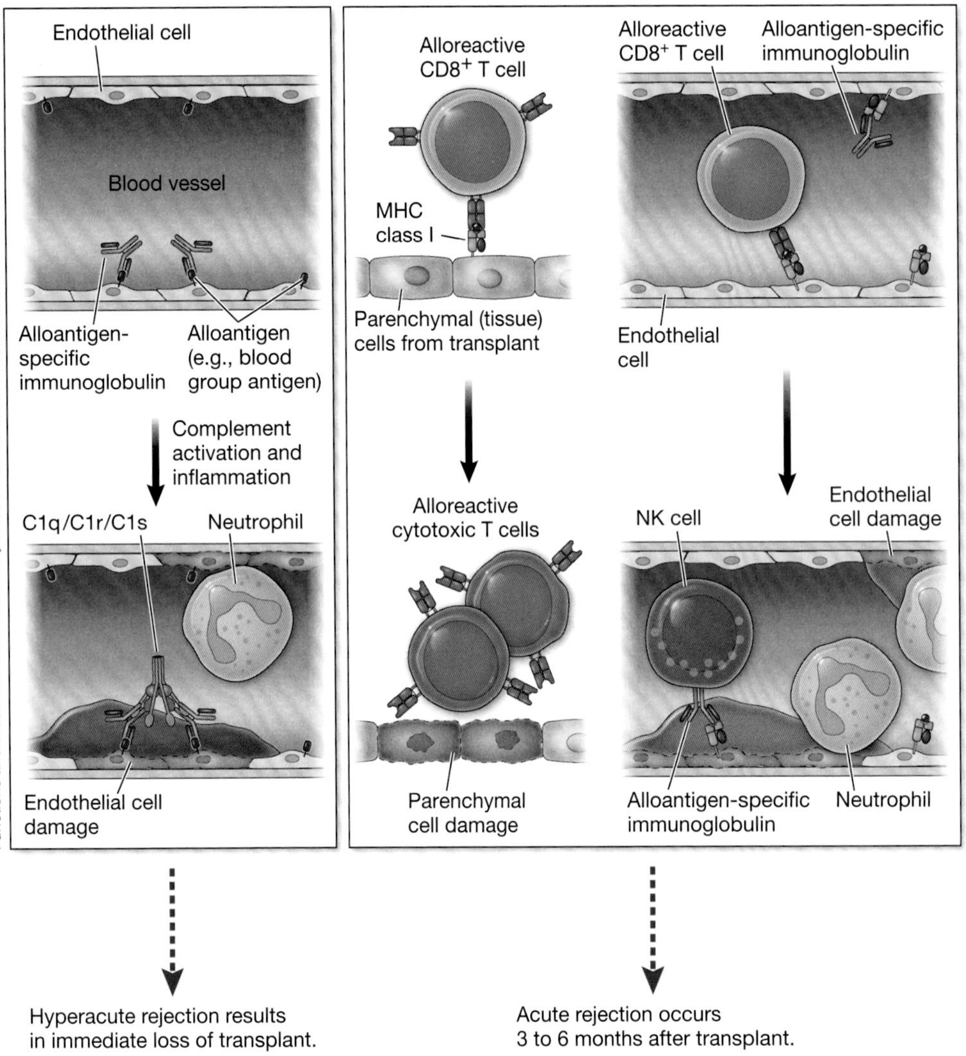

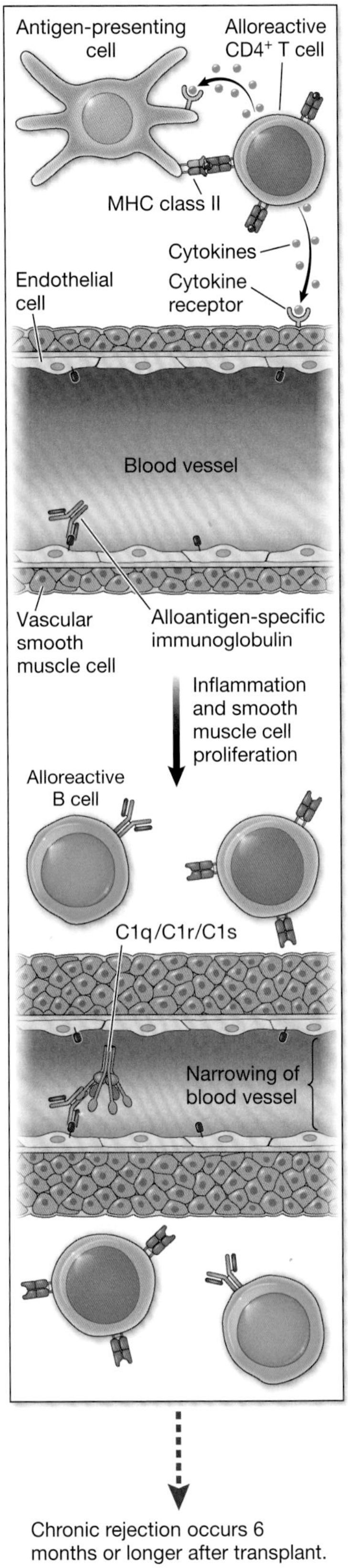

Solid organ transplants are often met with adaptive immune responses driven by the presence of immunoglobulins in circulation or the activation of alloreactive T cells within the recipient. These reactions can lead to immediate rejection of the transplanted tissue or rejection over a longer period of time if untreated.

(Sections 1.4, 1.8, 3.1, 3.3, 4.2, 4.5, 4.6, 4.7, 4.8, 5.7, 6.5, 6.7, 7.1, 10.5, 17.1, 17.3)

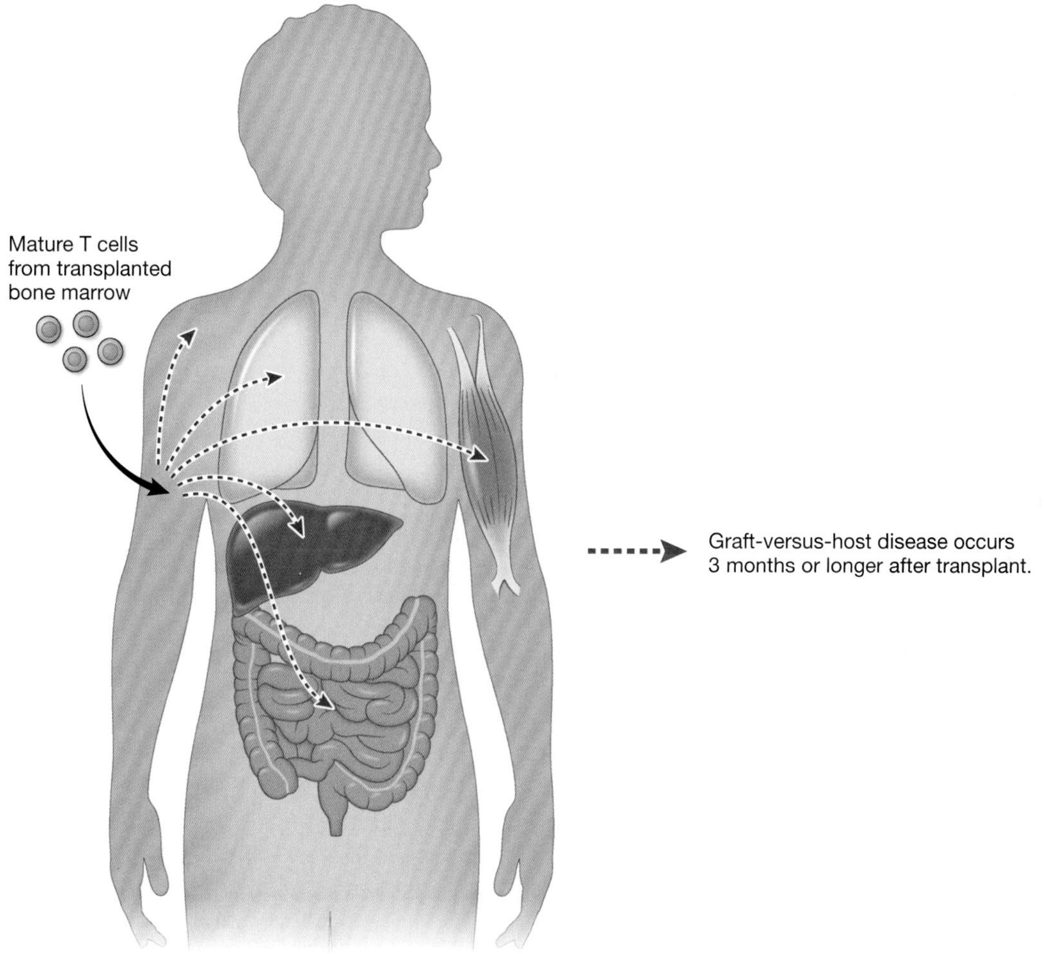
Mature T cells
from transplanted
bone marrow
Graft-versus-host disease occurs
3 months or longer after transplant.

Hematopoietic stem cell transplantation is often met with an adaptive immune response in which immune cells of the transplanted bone marrow react with recipient tissues, resulting in graft-versus-host disease

(Sections 1.5, 4.2, 4.5, 4.6, 4.7, 4.8, 5.7, 6.5, 6.7, 7.1 17.1, 17.3)

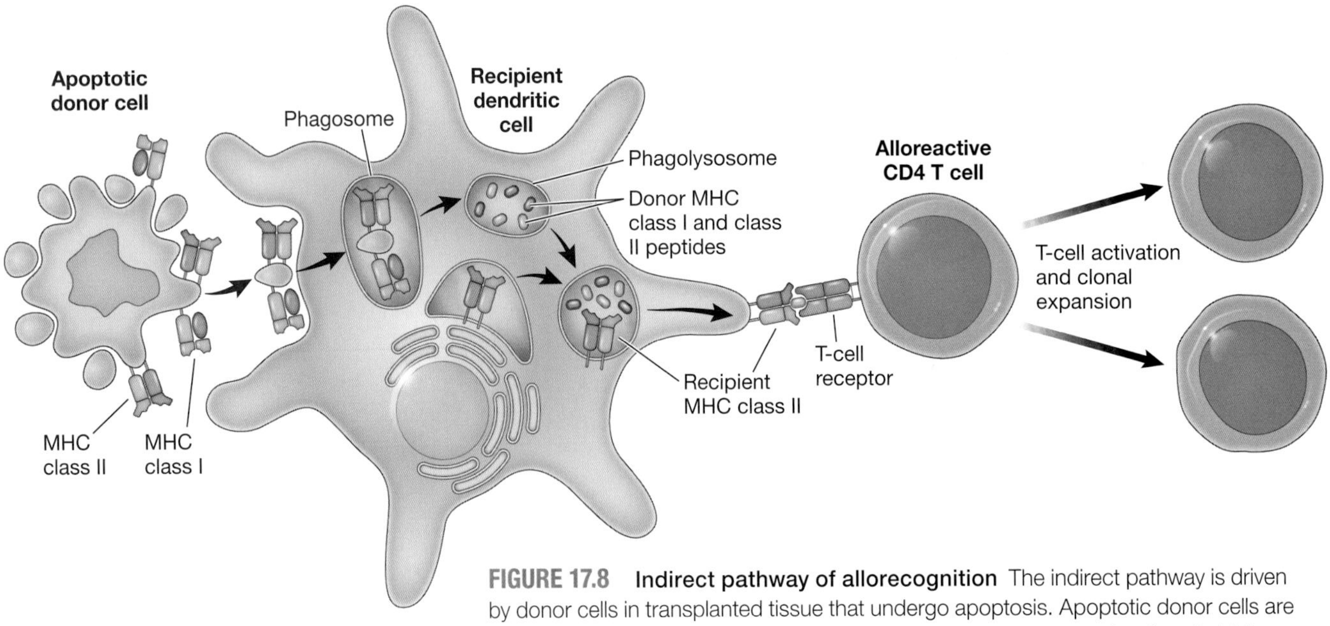

FIGURE 17.8 Indirect pathway of allorecognition The indirect pathway is driven by donor cells in transplanted tissue that undergo apoptosis. Apoptotic donor cells are phagocytosed by recipient dendritic cells, which migrate to secondary lymphoid tissue. Within the recipient dendritic cells, donor MHC class I and class II molecules are digested into peptides in the phagolysosome and loaded onto MHC class II molecules of recipient dendritic cells. These MHC class II molecules present donor MHC peptides to CD4 T cells. T cells with a receptor that recognizes allogeneic MHC peptides undergo activation and clonal expansion. (After T. Šarić et al. 2008. *Cells Tissues Organs* 188: 78–90.)

indirect pathway of allorecognition Pathway of presentation of alloantigens in which donor cells die by apoptosis in secondary lymphoid tissue, where they are phagocytosed by recipient dendritic cells, which then process donor cell antigens and present them to recipient T cells.

solid organ transplant is primarily driven by the **indirect pathway of allorecognition** (**FIGURE 17.8**). The initial stages of the direct and indirect pathways of allorecognition are identical: both pathways involve the migration of dendritic cells into a draining secondary lymphoid tissue. However, the indirect pathway diverges from the direct pathway at this point.

During the process, apoptosis of donor cells triggers initiation of the indirect pathway. These donor cells die by apoptosis due to inflammation and ischemia of the transplanted tissue. The resulting cell debris is phagocytosed by recipient dendritic cells located in the inflamed transplanted tissue or by recipient dendritic cells and recipient B cells in the secondary lymphoid tissue. The recipient dendritic cells and B cells that have engulfed the cell debris of the donor cells process donor antigens mainly through the MHC class II presentation pathway, since the cell debris is taken up via phagocytosis and is presented in a similar manner to antigens from extracellular pathogens. This pathway can present any donor antigens, including HLA allotypes, from the donor cells.

Recipient cells that have engulfed donor cell debris present to recipient T cells, mainly containing CD4 coreceptors. If the recipient T cell binds to the recipient dendritic cell MHC–peptide complex, it is activated and undergoes clonal expansion and differentiation. This pathway of allorecognition is the primary mechanism driving the production of anti-HLA antibodies to donor HLA allotypes, as the HLA molecules from donor cells are processed via the class II presentation pathway and peptides are displayed to recipient CD4 T cells (**FIGURE 17.9**). Upon activation, these recipient T cells may become helper T cells capable of activating B cells within the same secondary lymphoid tissue.

If the presenting B cell is displaying donor HLA allotype peptides via its MHC class II molecules, the activated B cell will become a plasma cell. This plasma cell can then produce anti-HLA allotype immunoglobulins or immunoglobulins to other alloantigens, causing an increase in soluble antibodies capable of recognizing antigens on the transplanted organ and resulting in increased targeting of the transplanted organ as nonself.

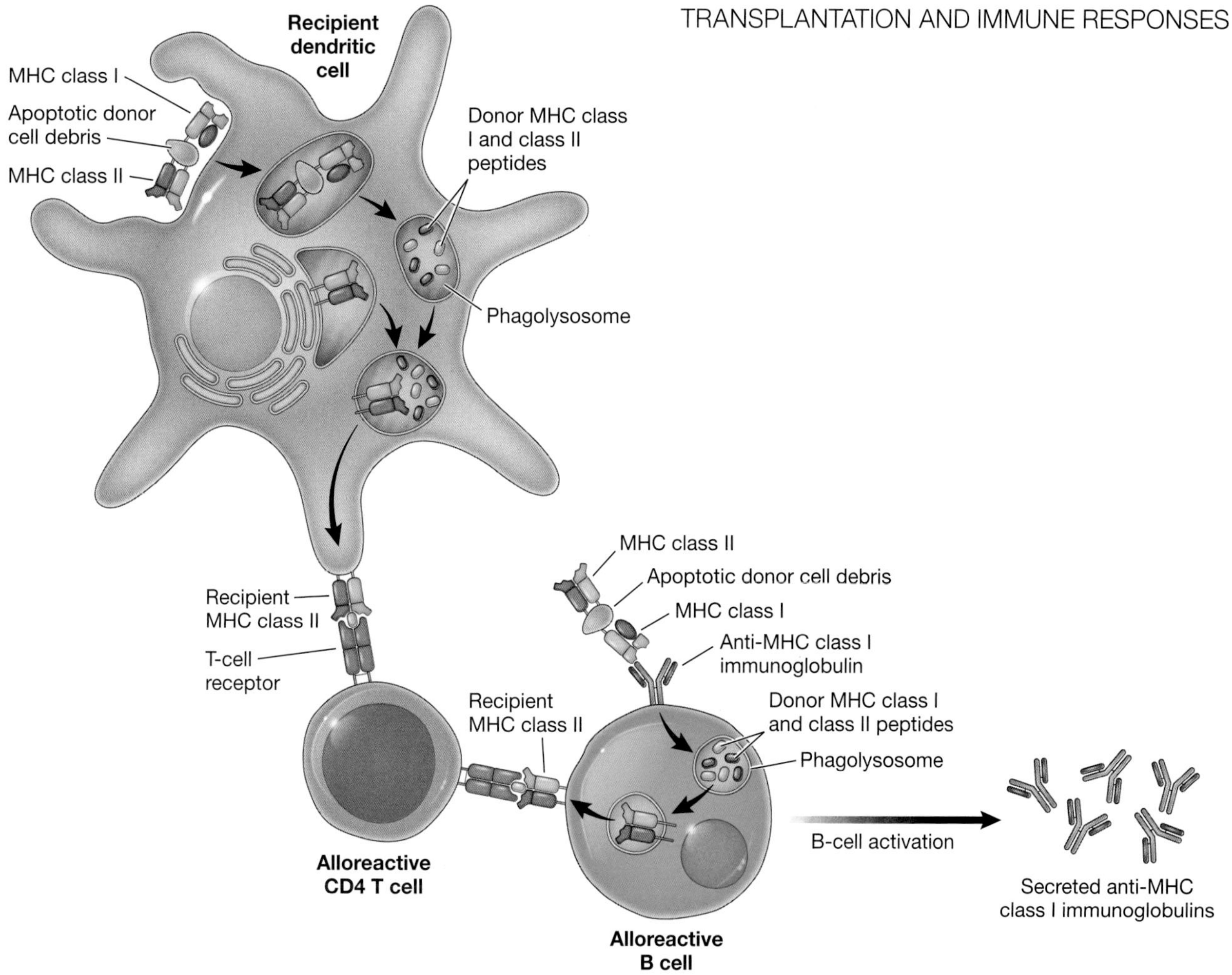

FIGURE 17.9 Activation of alloreactive B cells that recognize donor MHC molecules The indirect pathway of allorecognition is driven by phagocytosis of apoptotic donor cell debris that contains donor MHC molecules. These MHC class I and class II molecules are degraded in the phagolysosome and loaded onto MHC class II molecules of the recipient dendritic cell that phagocytosed the donor cell debris. Alloreactive CD4 T cells that recognize MHC class I and class II peptides presented on an MHC class II molecule will be activated. Alloreactive B cells that make an immunoglobulin that binds to donor MHC molecules bind to donor cell debris containing the MHC molecule and phagocytose the donor cell debris. MHC class I and class II molecules are digested within the phagolysosome of the B cell, and peptides produced after digestion are loaded and presented via MHC class II molecules on the surface of the B cell. If the alloreactive T cell recognizes the same MHC-peptide complex presented by both the recipient dendritic cell via the indirect pathway of allorecognition and an MHC-peptide complex presented by a B cell, the B cell forms a conjugate pair with the alloreactive T cell and is activated, resulting in secretion of alloreactive anti-MHC class I immunoglobulins. (After P. Parham. 2014. *The Immune System*, 4th ed., W. W. Norton and Company.)

Semidirect pathway of allorecognition

A third pathway has been proposed based on experimental evidence in animal transplantation models. The **semidirect pathway of allorecognition** involves the exchange of intact donor MHC class I molecules with recipient dendritic cells. This occurs via cell-to-cell interaction and exchange of plasma membrane material or by the action of exosomes (extracellular vesicles) that contain donor MHC class I molecules. This model has been difficult to support in transplantation in humans. However, in vitro studies have demonstrated exchange of MHC class I molecules between cultured dendritic cells and within mouse models of transplanted tissues. Thus, the semidirect pathway involves the recognition of intact donor MHC molecules (indicative of the direct pathway) being presented by a recipient dendritic cell (indicative of the indirect pathway).

semidirect pathway of allorecognition Pathway of presentation of alloantigens in which intact donor MHC molecules are transferred to recipient dendritic cells via cell-to-cell interaction or vesicle exchange, followed by presentation of those intact donor MHC molecules by the recipient dendritic cells.

CHECKPOINT QUESTION

1. Compare and contrast the direct, indirect, and semidirect pathways of allorecognition.

17.5 What factors contribute to a successful transplant?

LEARNING OBJECTIVE

17.5.1 Explain the importance of HLA typing and crossmatching in organ and tissue transplantation medicine.

We have seen that organ rejection can cause the same problems that created the need for the transplant to begin with, namely organ failure. Graft-versus-host disease can be problematic in a bone marrow transplant because it can mimic an autoimmune disease and target tissues and organs including the skin, gastrointestinal tract, and liver. Preventing or minimizing either of these potential reactions is paramount to the success of an organ or bone marrow transplant and a key goal in transplantation medicine.

As we have seen, solid organ rejection and graft-versus-host disease can be caused when the recipient's immune system detects nonself antigens. The most intense transplant reactions are driven by blood type reactions or differences between HLA molecules (for a glimpse of the discovery of allograft rejection, see **KEY DISCOVERIES**). Thus, minimizing differences between donor and recipient HLA molecules is a primary goal of transplant medicine.

HLA matching

The most successful transplants are autografts and isografts, demonstrating that shared genetic identity greatly increases the chances that a transplant will be accepted by the recipient's immune system. Autograft transplants are most common for blood transfusions, bone grafts, and skin grafts, although they have also been used with in vitro chemotherapy to treat leukemia.

While autografts and isografts provide shared genetic identity (the donor and recipient are the same individual or are genetically related), these transplants are not always viable as options. Since most transplants are allografts, *HLA typing* is performed to determine similarities between polymorphic HLA molecules of the potential donor and the recipient, especially HLA-A and HLA-B class I molecules and HLA-DR class II molecules. The importance of HLA typing for transplantation depends in part on the immunological privilege of the tissue, which refers to how susceptible a tissue is to an immune response due to inflammation. Some tissues, such as the cornea, are said to have immunological privilege because they are less susceptible to immune responses due to inflammation. Other tissues, such as the kidney, are much more susceptible to immune responses driven by inflammation.

HLA typing and crossmatching are conducted using DNA sequencing techniques to determine the sequence similarities between donor and recipient HLA molecules, along with testing serum levels for the presence of preexisting HLA immunoglobulins in cases of previous pregnancy or transplant. HLA matching is performed prior to both bone marrow and solid organ transplantation. However, because organ transplant recipients are also commonly treated with immunosuppressive agents, HLA mismatches between donor and recipient can be tolerated to a greater degree with solid organ transplants.

CHECKPOINT QUESTION

1. Why are parents or siblings regarded as potential organ donors for a patient?

KEY DISCOVERIES

How was allograft rejection first connected to immune response reactions?

Article

Medawar, P. B. 1945. A second study of the behaviour and fate of skin homografts in rabbits: A report to the War Wounds Committee of the Medical Research Council. *Journal of Anatomy* 79: 157–176.

Background

Peter Medawar was a pioneer in skin graft research. His work on the subject began during World War II, and his first published work involved autograft transplants of skin of soldiers wounded during the war. Medawar's work focused on methods for safe removal and reintroduction of the skin graft on wounded individuals. His research expanded into skin graft transplants, where he used rabbits as a model system for skin graft replacements and performed both autograft and allograft transplants. Medawar was surprised to observe that allograft skin graft transplants in rabbits were not well tolerated and were rejected by the recipient. The question remained: Why were these allograft transplants rejected?

Early Research

Medawar had shown that unlike autograft transplants, allograft transplants (what he termed *homograft* transplants because they were transplants between the same species of animal) resulted in the "homograft problem," whereby the graft did not survive. In this initial study, Medawar observed that graft survival depended on three phenomena: the size of the skin graft, whether the recipient had previously had a skin graft from the same donor, and an unanalyzed element of genetic diversity. He also observed that the reaction that killed the graft was specific to the graft and not the recipient's native skin.

Think About...

1. What would your hypothesis be about the mechanism of homograft rejection given these initial observations?
2. What experiments would you conduct to test this hypothesis? What controls could you use to demonstrate that the reactions observed were definitively due to allograft rejection and not due to other reactions in the animal?
3. How might you have been able to determine if an immune reaction was responsible for the allograft rejection? Consider that the date of this publication was 1945. Had an immunosuppressive drug been discovered by this point in time? If you could observe different cells within the allograft via microscopy, what immune cells might you try to detect?

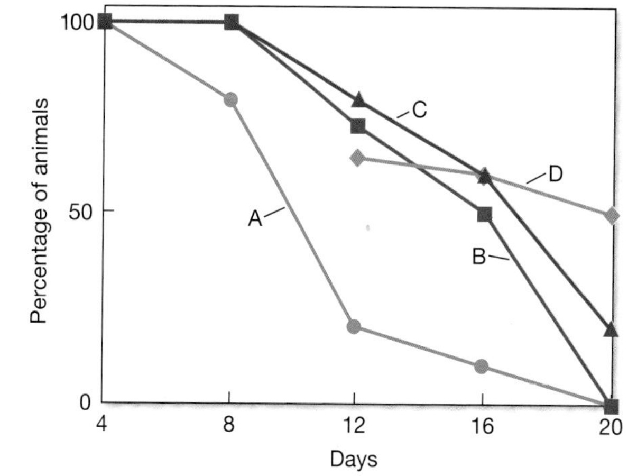

FIGURE KD 17.1 Survival of skin grafts on rabbits over time Each line represents a different skin graft size and transplant technique. Those transplanted by open style promote growth of the graft into the surrounding wound area. Those transplanted by fitted style transplant a graft the same size as the wound. The dosage refers to the size of the graft (high dosage = 0.36–0.44 g; medium dosage = 0.045–0.055 g; low dosage = 0.006 g). Curve A = high dosage, open style; Curve B = medium dosage, open style; Curve C = low dosage, open style; Curve D = medium dosage, fitted style. (After P. B. Medawar. 1945. *J Anat* 79: 157–176.)

Article Summary

Medawar systematically studied the survival of different skin grafts on rabbits, conducting experiments that varied the mechanism of the graft and the size of the skin graft. Medawar performed "fitted style" grafts, in which the size of the graft exactly matched the wound made on the recipient animal, and "open style" grafts, in which the size of the graft was smaller than the wound to allow for cellular growth from the graft into the wound. Within his studies, he also performed transplantations of different graft sizes.

Medawar conducted his experiments in 201 rabbits in this study and 181 rabbits in the study he had previously published, resulting in 802 different allograft transplants. Of these transplants, which were performed in genetically diverse rabbits, 708 of the allografts died on the recipient, regardless of the mechanism he used or the size of the graft (**FIGURE KD 17.1**). Medawar concluded that the genetic diversity between donor and recipient was the driving factor in the allograft rejection, and he hypothesized that an immune response was responsible for the death of the skin graft. This study pioneered the discovery of the genetic diversity factors responsible for allograft reactions and was essential to providing a means for successful tissue transplants.

17.6 | How are transplant rejections suppressed pharmacologically?

LEARNING OBJECTIVE

17.6.1 Describe the types of pharmacological agents used as immunosuppressive drugs and provide examples.

immunosuppressive Substance capable of inhibiting an innate or adaptive immune response.

While HLA matching increases the chances of a successful transplant and is vital prior to bone marrow transplantation, the inflammatory response within the transplanted organ and the likely adaptive immune response following surgery pose potential complications. To prevent or reduce these reactions, **immunosuppressive** drugs are administered after organ transplants to minimize targeting of the organ by the immune system and to prolong organ acceptance. Common immunosuppressive drugs used for this purpose are corticosteroids, cytotoxic drugs, and T-cell activation inhibitors (**TABLE 17.3**), which all block immune responses that may occur as a result of the transplant (**FIGURE 17.10**).

While immunosuppressive drugs are used in transplantation cases to limit the activation of the immune response, they are not specific toward white blood cells that will target transplanted tissue. Since the same mechanisms that activate lymphocytes that target transplanted tissue are used to activate lymphocytes that target pathogens, immunosuppressive drugs used during transplantation cases make the individual more susceptible to infection. Typically, transplant patients are kept on an elevated dose of immunosuppressive drugs immediately after the surgery in the induction phase of immunosuppression. This higher dose of immunosuppressive drugs is given to ensure that the immune system of the recipient does not target the transplanted tissue. However, during this phase, the patient is more susceptible to infection because of the inactivation of the immune system, including infection of opportunistic pathogens. After approximately 6 months of the higher dose of immunosuppressive drugs, the maintenance phase of immunosuppression begins, and the dose is lowered. Individuals in the

TABLE 17.3 | Examples of Immunosuppressive Drugs

Immunosuppressive Drug	Cellular Target	Effect on the Immune Response
Hydrocortisone	Glucocorticoid receptor	Production of IκB and lowering of inflammatory response
Prednisone	Glucocorticoid receptor	Production of IκB and lowering of inflammatory response
Azathioprine	Phosphoribosyl pyrophosphate amidotransferase	Inhibits purine biosynthesis in rapidly dividing cells, inhibiting mitosis
Mycophenalate motefil	Inosine monophosphate dehydrogenase	Inhibits guanine biosynthesis in rapidly dividing cells, inhibiting mitosis
Cyclophosphamide	DNA alkylation and crossklinking	Inhibits DNA replication in rapidly dividing cells, inhibiting mitosis
Methotrexate	Dihydrofolate reductase	Inhibits thymidine biosynthesis in rapidly dividing cells, inhibiting mitosis
Cyclosporin A	Cyclophilin	Inhibits calcineurin, blocking NFAT activation
FK-506	FK-binding proteins	Inhibits calcineurin, blocking NFAT activation
Rapamycin	mTOR	Inhibits G1 to S progression of the cell cycle and clonal expansion of T cells
Anti-CD3 antibodies	CD3	Opsonin for phagocytosis of T cells
Basiliximab	IL-2 receptor	Blocks IL-2 signaling and T-cell activation
Belatacept	B7	Causes T-cell anergy

Source: P. F. Halloran. 2004. *NEJM* 351: 2715–2729.

maintenance phase of immunosuppressive drug treatment are still susceptible to infection, but the infection risk is lower. Because the transplanted tissue will be rejected by the recipient's immune system without limiting its activation, transplant patients are maintained on immunosuppressive drugs throughout their life and cannot be completely weaned off of these medications.

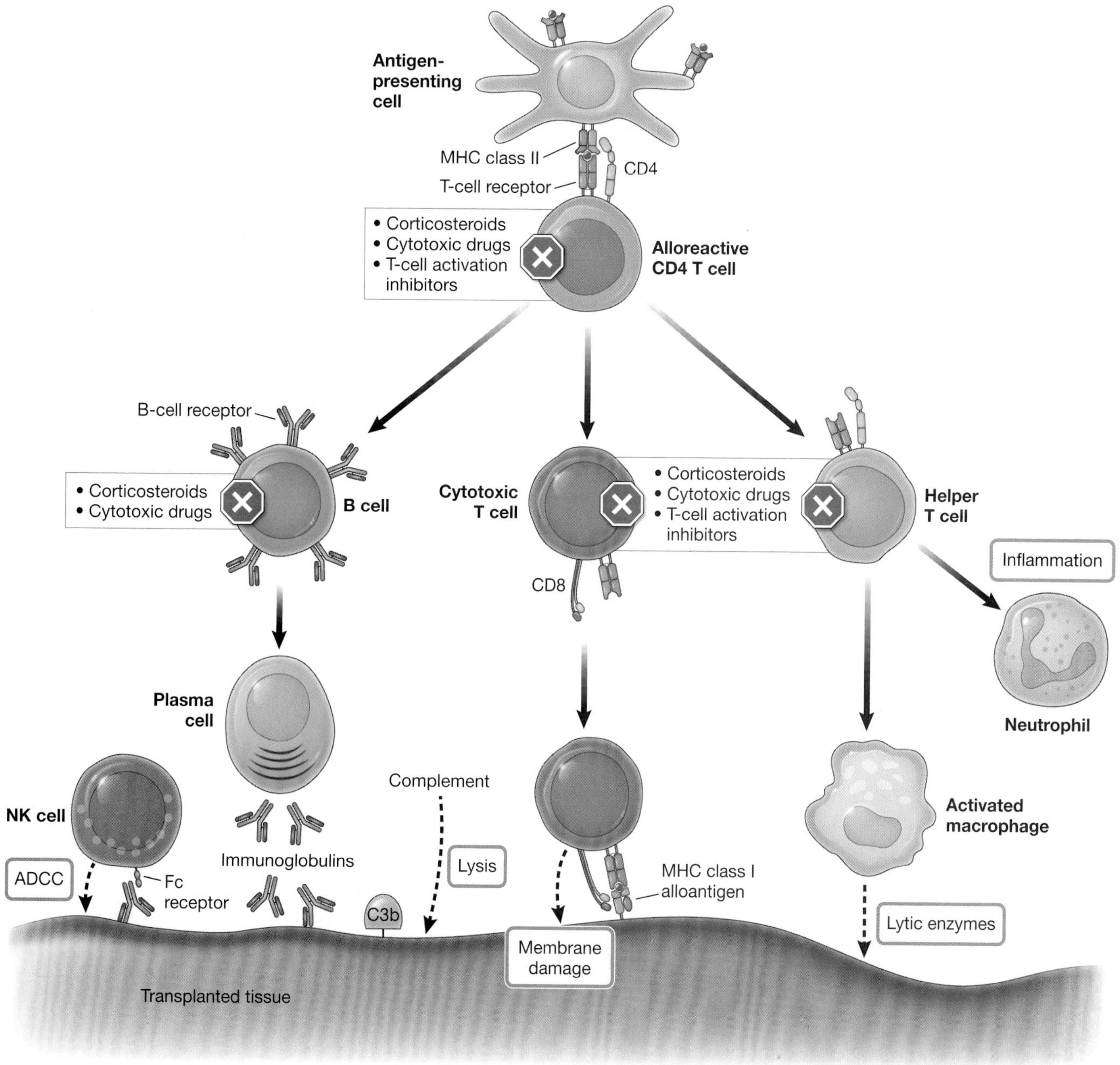

FIGURE 17.10 Alloreactions against solid organ transplants and common agents used to prevent transplant rejection Transplant rejection is driven by alloreactive lymphocytes activated by the presence of alloantigens from the transplanted tissue. Chronic rejection begins with donor antigens presented by antigen-presenting cells, resulting in the activation of alloreactive CD4 T cells. Depending on which helper T cell is produced after differentiation of the activated CD4 T cell, a number of different innate and adaptive immune responses ensue. Production of helper T cells results in the activation of neutrophils, macrophages, and/or B cells, resulting in inflammation, secretion of lytic enzymes, and production of alloreactive immunoglobulins. Alloreactive immunoglobulins produced by plasma cells can activate complement or induce antibody-dependent cell-mediated cytotoxicity (ADCC) by natural killer (NK) cells. Cytotoxic CD8 T cells might also be activated by helper T cells. Once activated, they target cells of the transplanted tissue for apoptosis. Immunosuppressive drugs, including corticosteroids, cytotoxic drugs, and T-cell activation inhibitors, can block activation of cells responsible for transplant rejection. (After J. Punt et al. 2018. *Kuby Immunology*, 8th ed., W. H. Freeman and Company: New York.)

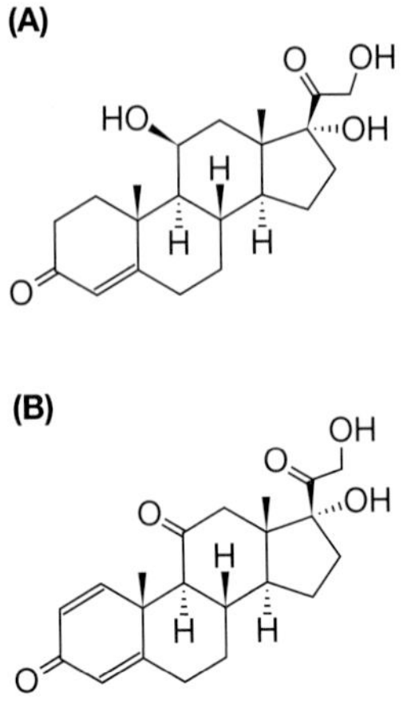

FIGURE 17.11 Chemical structure of two corticosteroids (A) Hydrocortisone. (B) Prednisone.

Corticosteroids

Cortisol is a naturally occurring steroid hormone produced by the adrenal glands. Natural or synthetic forms of this hormone used as a pharmaceutical are known as hydrocortisone (**FIGURE 17.11**). Because of its anti-inflammatory properties, hydrocortisone can be prescribed to reduce inflammation. However, in transplant cases, prednisone, a more potent synthetic corticosteroid, is typically used as an anti-inflammatory agent in concert with other immunosuppressive drugs to prevent organ rejection.

Corticosteroids such as hydrocortisone and prednisone inhibit inflammation by crossing plasma membranes and binding to intracellular steroid hormone receptors, which, in the absence of hormone, remain inactive through binding to heat shock protein 90 (Hsp90) (**FIGURE 17.12**). Binding of steroid hormone (in this case, hydrocortisone or prednisone) induces release of Hsp90 from the receptor, which is now free to traverse the nuclear envelope and control transcription of genes responsible for inhibiting the inflammatory response.

One important gene whose transcription is increased in the presence of corticosteroids is the NFκB inhibitor IκB. Increased expression of IκB causes decreased activity of the transcription factor NFκB, which plays an important role in the activation of the inflammatory response (as we saw in Chapter 2). Lack of NFκB activity causes decreased expression of inflammatory genes, including IL-1. Furthermore, lack of NFκB activity prevents the cell-surface alterations necessary for proper lymphocyte homing to the site of inflammation. Thus, in transplant cases, corticosteroid treatment works to limit rejection by preventing lymphocytes from homing to the transplanted organ and damaging transplanted tissue.

cyclosporin A Immunosuppressive drug that blocks T-cell activation by binding to cyclophilins and inhibiting calcineurin activation.

FK-506 Immunosuppressive drug that blocks T-cell activation by binding to FK-binding proteins and inhibiting calcineurin activation.

cyclophilin A peptidyl-prolyl isomerase that, in complex with cyclosporin A, blocks calcineurin activity and T-cell activation.

FK-binding protein A peptidyl-prolyl isomerase, that, in a similar manner to cyclophilin, binds to FK-506 and blocks calcineurin activity and T-cell activation.

Cytotoxic drugs

One of the features of hematopoietic cells, especially white blood cells, is their rapid and continued proliferation to replenish the supply of cells turned over during development and those at the end of their natural life cycle. In light of this characteristic, drugs that inhibit cell proliferation are used as another type of immunosuppressive agent in transplant cases. These cytotoxic drugs block proliferation not only of white blood cells but also of any rapidly dividing cells in the body.

Azathioprine is converted within cells to a nucleotide analog that inhibits purine biosynthesis by inhibiting the enzyme phosphoribosyl pyrophosphate

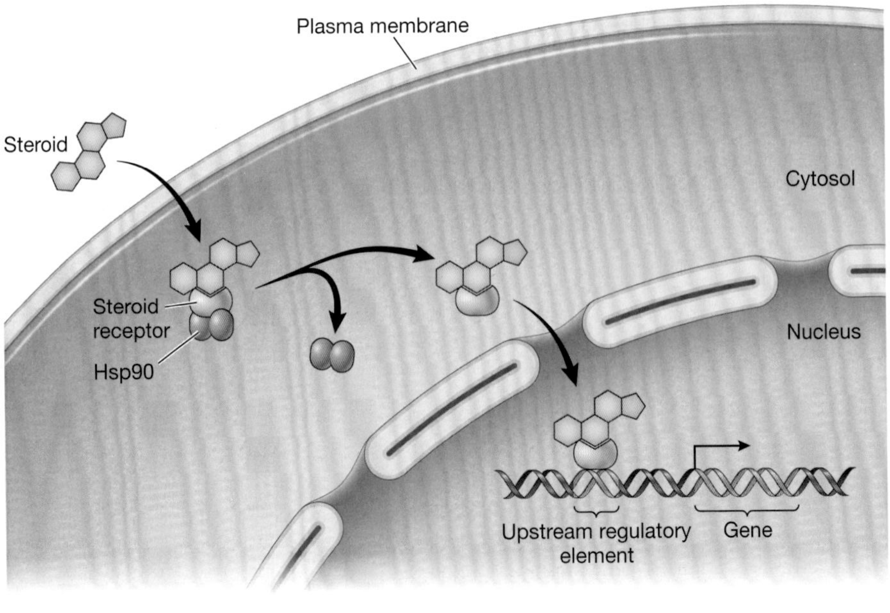

FIGURE 17.12 Activation of gene expression by steroid hormones Steroid hormones cross plasma membranes and interact with steroid hormone receptors in the cytosol. These receptors are kept inactive by heat shock protein 90 (Hsp90) within the cytosol until a steroid hormone binds to the receptor. Steroid hormone binding to its receptor causes release of Hsp90, allowing the hormone and its receptor to transit into the nucleus, where the steroid hormone receptor acts as a transcriptional activator. The receptor activates expression of genes that contain an upstream regulatory element recognized by the receptor. (After S. LeVay and J. Baldwin. 2011. *Human Sexuality*, 4th ed., Web Topic 5.1. Oxford University Press/Sinauer: Sunderland, MA.)

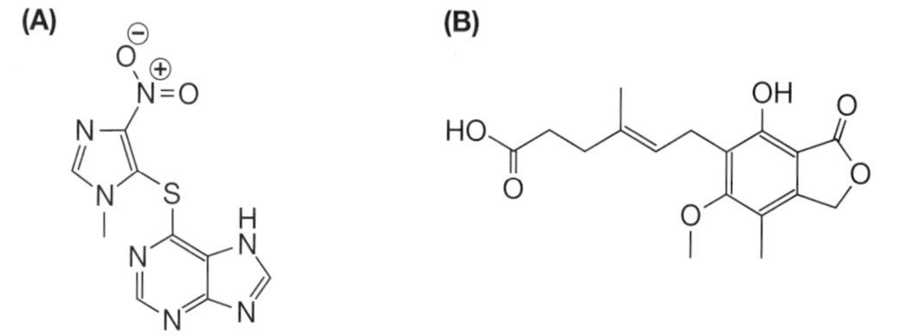

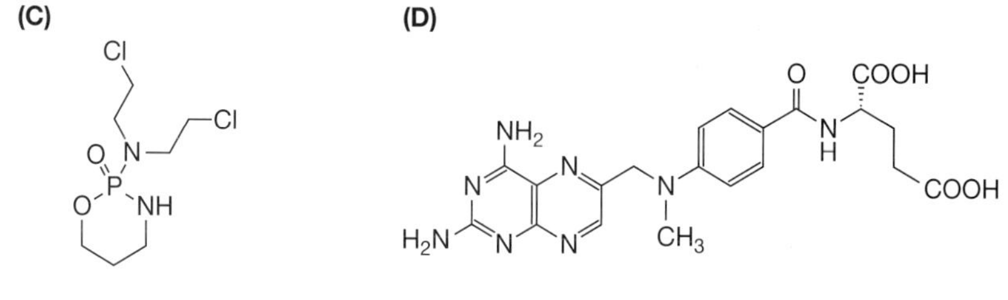

FIGURE 17.13 Chemical structures of cytotoxic drugs (A) Azathioprine is a purine analog that inhibits synthesis of DNA and RNA. (B) Mycophenolic acid inhibits synthesis of guanine-containing nucleotides. (C) Cyclophosphamide irreversibly cross-links DNA at guanine residues, leading to apoptosis. (D) Methotrexate inhibits synthesis of thymine-containing nucleotides.

amidotransferase, and thus prevents DNA replication (**FIGURE 17.13**). This drug, along with corticosteroids, aided the first kidney transplant between unrelated patients. Several other drugs are also used to prevent the division of white blood cells. These include mycophenalate motefil, which, when converted in vivo to mycophenolic acid, inhibits guanine biosynthesis by inhibiting the enzyme inosine monophosphate dehydrogenase; cyclophosphamide, which alkylates and crosslinks DNA molecules, preventing DNA replication; and methotrexate, which inhibits the enzyme dihydrofolate reductase, an enzyme critical in thymidine biosynthesis.

T-cell activation inhibitors

Another family of pharmaceutical immunosuppressive agents works more specifically by directly inhibiting T-cell activation. Recall that during T-cell activation, several transcription factors are activated, including NFAT (**FIGURE 17.14**). NFAT is activated in T cells by calcineurin, a calcium-dependent phosphatase that has increased activity in T cells with proper receptor signaling because of the rise in intracellular calcium levels.

Two compounds are used to inhibit the activation of calcineurin and thus inhibit NFAT activation. These compounds—**cyclosporin A** and **FK-506**—inhibit NFAT activation via similar mechanisms. Cyclosporin A binds to a peptidyl-prolyl isomerase **cyclophilin**, and this complex binds to calcineurin and blocks its phosphatase activity. FK-506 blocks calcineurin by binding to **FK-binding proteins**, another family of peptidyl-prolyl isomerases that, when in complex with FK-506, block calcineurin phosphatase activity.

A third pharmaceutical agent, rapamycin, is another immunosuppressant that is capable of blocking T-cell proliferation by blocking IL-2 signaling. This drug is an inhibitor of the protein mechanistic target of rapamycin, (mTOR), a kinase that promotes the transition of the G1 phase to S phase in the cell cycle. Recall that clonally selected T cells signal activation of clonal expansion and differentiation through the release of various cytokines, including IL-2. IL-2 functions to activate signaling cascades responsible for the rapid expansion of the clonally selected T cell. These signaling cascades include the action of mTOR to promote progression of the cell cycle within clonally selected T cells. Rapamycin prevents T-cell clonal expansion by blocking progression of the cell cycle of T cells triggered by IL-2.

Various immunological therapies employing monoclonal antibodies can also be used to limit T-cell activity in transplant patients (**FIGURE 17.15**). One class of antibodies, antithymocyte globulin and anti-CD3 antibodies, works to remove T cells from circulation by binding to surface molecules on T cells and facilitating

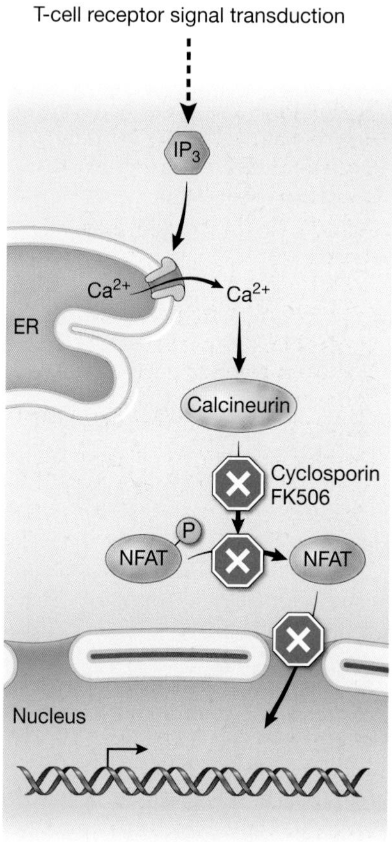

FIGURE 17.14 T-cell activation inhibitors as immunosuppressant The compounds cyclosporin and FK-506 inhibit T-cell activation by preventing activation of the transcription factor NFAT. NFAT requires activation through the action of the phosphatase calcineurin, which is activated in T cells via IP_3 and an influx of calcium ions from the endoplasmic reticulum (ER) into the cytosol. Cyclosporin and FK-506 block calcineurin action and thus prevent T-cell activation by preventing NFAT activation.

phagocytosis by professional phagocytes using their Fc receptors to recognize opsonized T cells. Another monoclonal antibody used to block T-cell activation is basiliximab, which binds to the high-affinity IL-2 receptor and blocks IL-2 signaling and T-cell proliferation.

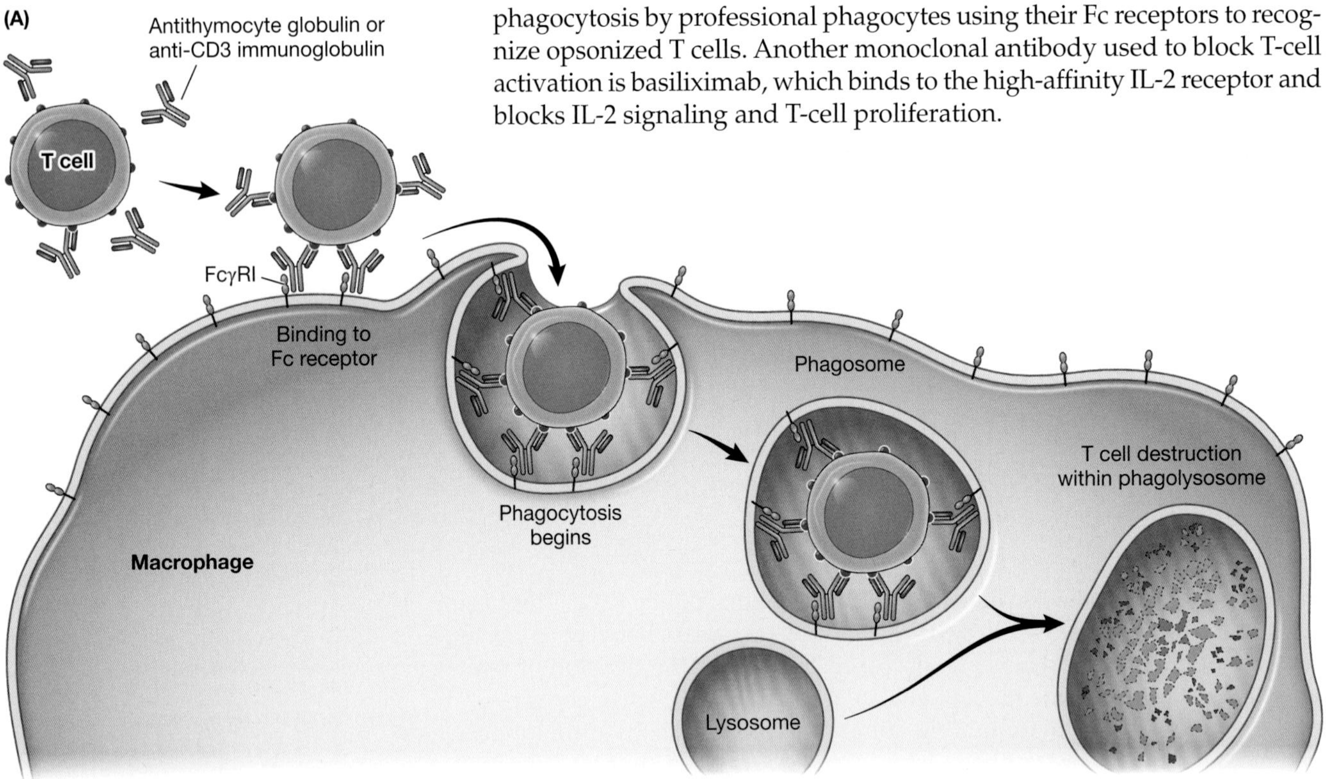

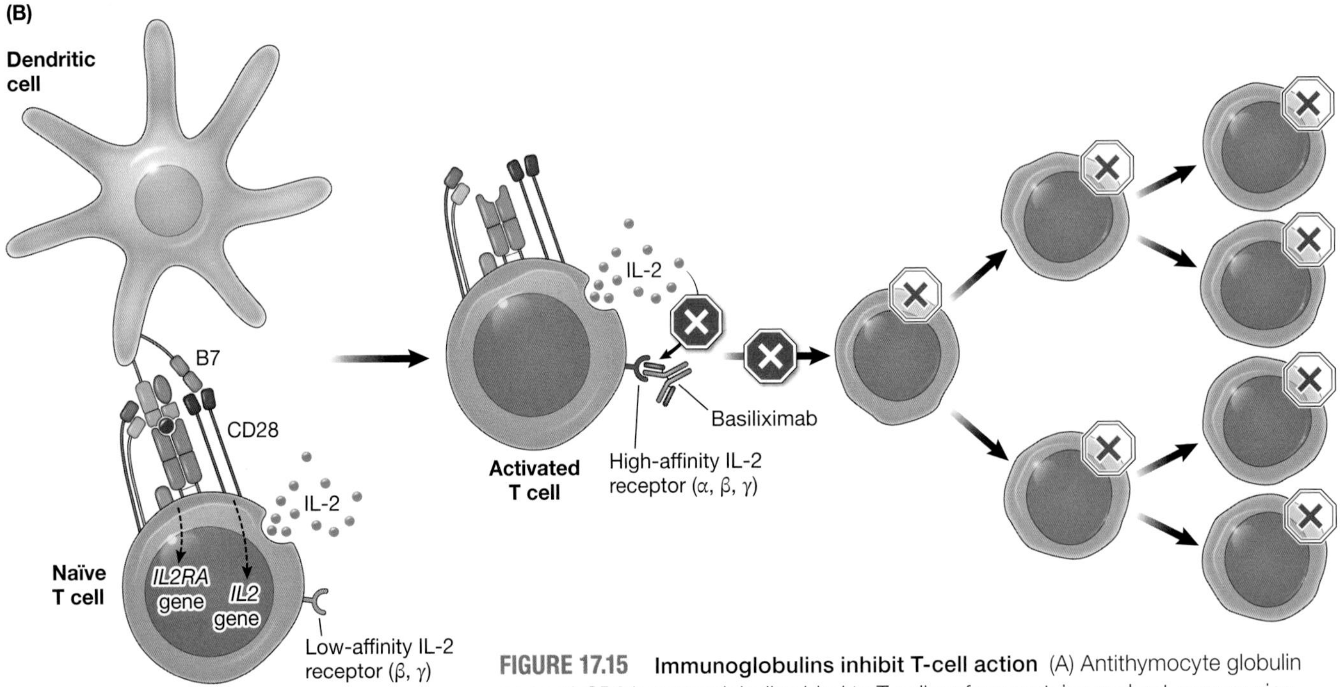

FIGURE 17.15 Immunoglobulins inhibit T-cell action (A) Antithymocyte globulin and anti-CD3 immunoglobulins bind to T-cell surface proteins and act as opsonins. Macrophages bind to opsonizing antibodies on the surface of T cells via their Fc receptors and phagocytose the T cells, resulting in their destruction. (B) The immunoglobulin basiliximab binds to the high-affinity IL-2 receptor. Binding of basiliximab to the IL-2 receptor prevents IL-2 binding and clonal expansion of the T cell.

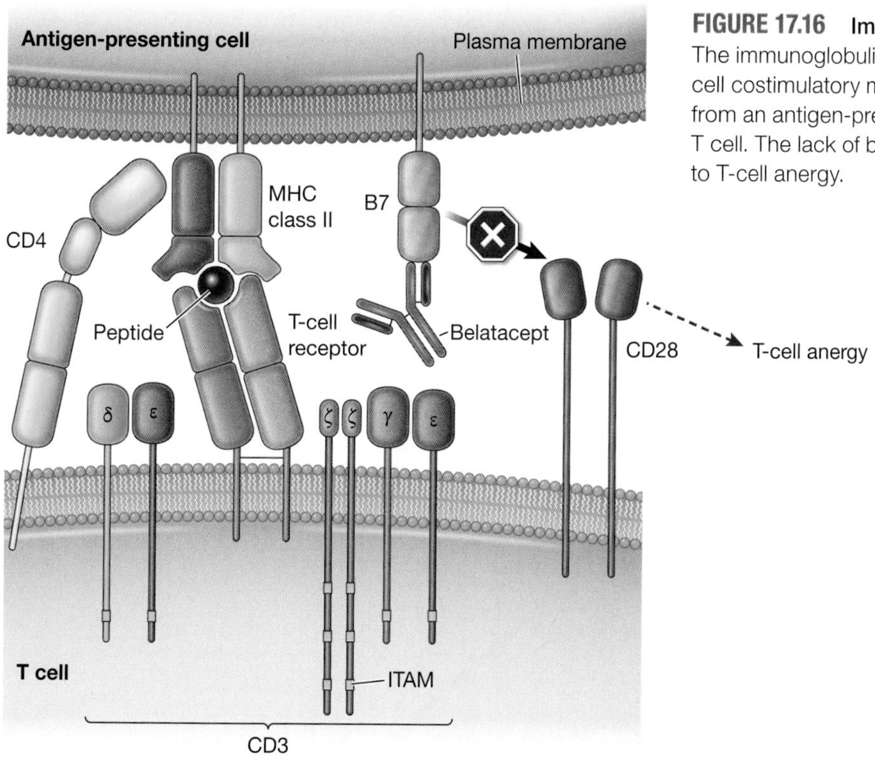

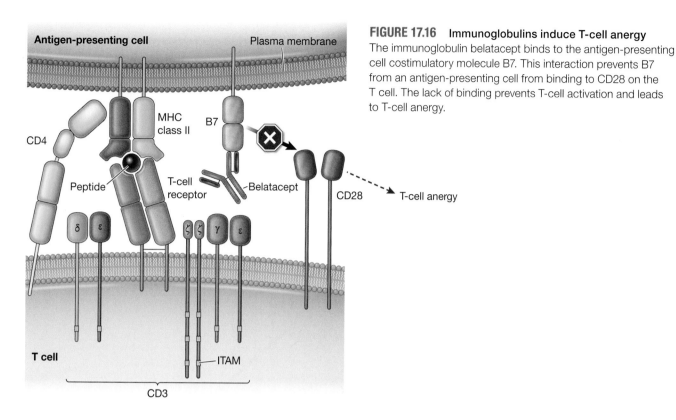

FIGURE 17.16 Immunoglobulins induce T-cell anergy The immunoglobulin belatacept binds to the antigen-presenting cell costimulatory molecule B7. This interaction prevents B7 from an antigen-presenting cell from binding to CD28 on the T cell. The lack of binding prevents T-cell activation and leads to T-cell anergy.

Belatacept is another example of an inhibitor of T-cell activation: it is a fusion protein consisting of the extracellular domain of CTLA4 and the IgG heavy chain. Recall that T cells require both T-cell receptor signaling and a costimulatory signal that uses the protein CD28 and the APC protein B7 (**FIGURE 17.16**). Recall from Chapter 7 that another T-cell membrane protein, CTLA4, is capable of binding B7 on APCs with a much higher affinity to downregulate activated T cells. Belatacept binds to B7 on the APC surface and prevents CD28 on T cells from binding to its costimulatory receptor. The lack of costimulatory signal causes the T cell to undergo anergy, preventing its activation.

CHECKPOINT QUESTION

1. Explain how cyclosporin A and FK-506 function to block T-cell activation.

17.7 How is graft-versus-host disease used to treat other diseases?

LEARNING OBJECTIVE

17.7.1 Explain how the graft-versus-leukemia effect works to prevent recurrence of abnormal lymphocyte growth.

We have seen that graft-versus-host disease occurs with bone marrow transplants due to mature donor T cells recognizing recipient tissue as nonself. While graft-versus-host disease can be problematic to individuals receiving a bone marrow transplant, the alloreaction can sometimes be beneficial in the treatment of certain diseases, including leukemia.

Making Connections: Preventing Transplant Alloreactions

Since alloreactions of transplanted tissues occur due to action of the adaptive immune system, many therapies used to prevent alloreactions prevent adaptive immune response action or clonal expansion. These therapies include HLA matching, along with the use of immunosuppressive drugs to limit transplant rejection.

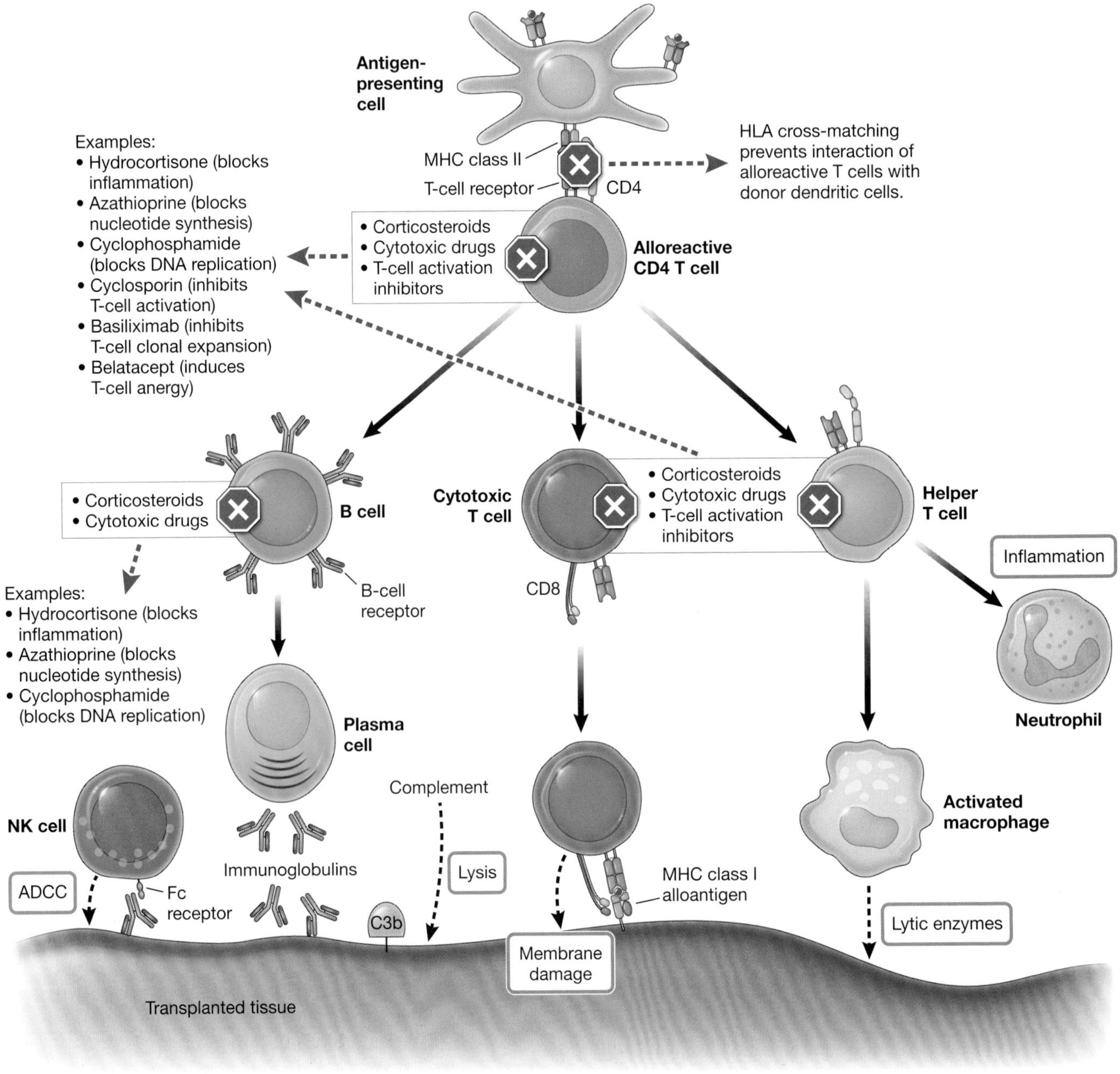

After J. Punt et al. 2018. *Kuby Immunology*, 8th ed., W. H. Freeman and Company: New York

Treatments used to combat alloreactions of transplanted tissue include corticosteroids, which block inflammation, cytotoxic drugs, which prevent cell division, and activation inhibitors. These treatments take advantage of the requirement of clonal expansion for a robust immune response and inhibition of signaling events that occur to drive adaptive immune responses.

(Sections 4.2, 4.5, 4.6, 4.7, 5.7, 6.5, 6.7, 7.1, 7.3, 7.7, 7.8, 7.9, 7.10, 10.5, 17.1, 17.3, 17.4, 17.5, 17.6)

Graft-versus-leukemia effect

A major driving force of graft-versus-host disease is the presence in transplanted bone marrow of mature T cells that react to the recipient's tissue as nonself. We might then predict that removing mature T cells from bone marrow before the transplant would minimize graft-versus-host disease. Indeed, this is the case. However, these transplants tend to have an increased risk of graft rejection and recurrence of cancer. Mature T cells in transplanted bone marrow assist in the destruction of uncontrolled leukemia cells that have survived chemotherapy and irradiation. This phenomenon is known as the **graft-versus-leukemia effect**.

graft-versus-leukemia effect Targeting of residual leukemia cells within a recipient (not destroyed by prior chemotherapy and irradiation) by mature T cells present in the transplanted bone marrow.

This effect is supplemented by the action of NK cells when the recipient's cells express fewer inhibitory receptors to NK cells than the donor's cells. Donor NK cells that develop shortly after the transplant target recipient cells that express fewer inhibitory receptors. Such a reaction supports the targeting of leukemia cells. Eventually, as the bone marrow transplant matures in the recipient, the NK cells that develop are re-educated to recognize the inhibitory receptor levels of both the donor and recipient.

Solid organ transplant tolerance

It has been observed that transplantation between dizygotic (nonidentical) twins is sometimes better tolerated than other allogeneic transplantations, likely due to mixing of blood circulation during development and, thus, chimerism of the hematopoietic system. Such blood chimerism can also be observed in individuals who have undergone a previous organ transplant, presumably due to the presence of donor blood that the recipient has tolerated. These two findings suggest that if donor hematopoietic stem cells are present during a solid organ transplant, the recipient may be more tolerant of the transplant, and organ rejection may be prevented. Research is being actively pursued to aid in chimerism establishment and maintenance in transplant patients. Success has been reported in several animal models, including canines and nonhuman primates. The ultimate goal is to develop a protocol for inducing tolerance via chimerism in transplant patients.

• CHECKPOINT QUESTION

1. Explain the graft-versus-leukemia effect and how it pertains to the recurrence of leukemia in patients when mature T cells are removed from bone marrow before transplant.

Summary

17.1 What is the history of organ and tissue transplantation?

- The three main types of transplantation performed are blood transfusions, solid organ transplantations, and hematopoietic stem cell transplantations.
- Transplantations can be characterized by the source of the tissue or organ being transplanted: autografts use an individual's own tissue, isografts use tissue from a genetically identical donor, allografts use tissue from a genetically distinct donor, and xenografts use tissue from a different species.

17.2 What medical conditions result in the need for tissue or blood cell transplantation?

- Solid organ transplantation and hematopoietic stem cell transplantation can be used to treat a variety of disorders and diseases, including organ failure, developmental abnormalities, and cancer.

17.3 Which types of alloreactions can occur as a result of transplantation?

- Several reactions can occur as a result of transplantation and put the recipient at risk, including organ rejection and graft-versus-host disease.
- Organ rejection can occur very quickly or after a long period of time, depending on the genetic similarities between donor and recipient.

17.4 What mechanisms drive solid organ transplant alloreactions?

- Solid organ transplant alloreactions can occur via the direct pathway of allorecognition, in which donor dendritic cells activate recipient T cells; via the indirect pathway, in which donor antigens are presented by recipient APCs to recipient T cells;

(Continued)

Summary *(continued)*

and via the semidirect pathway, in which intact donor MHC molecules are transferred to recipient dendritic cells by cell–cell interaction or by extracellular vesicles.

17.5 What factors contribute to a successful transplant?

- Alloreactions are minimized through the use of HLA typing and matching to prevent recognition of donor tissue as foreign by the recipient and to prevent graft-versus-host disease.

17.6 How are transplant rejections suppressed pharmacologically?

- Various pharmacological agents that act to suppress inflammation, prevent cellular growth, and inhibit T-cell activation are used to minimize alloreactions in transplantations.

17.7 How is graft-versus-host disease used to treat other diseases?

- The graft-versus-leukemia effect employs the graft-versus-host disease alloreaction to target and destroy recipient leukemia cells.

Review Questions

17.1 Is a kidney transplant between nonidentical twins an example of an autograft, isograft, or allograft? Explain.

17.2 Why is a typical protocol of chemotherapy and irradiation used to treat the patient prior to a bone marrow transplant?

17.3 A patient undergoing a kidney transplant is on the operating table. Immediately after the transplant, the kidney becomes spotted and blue in color, suggesting that the organ is not receiving adequate oxygen and is in failure. If the kidney has come from a healthy donor, why might the surgeons suspect that the patient is experiencing hyperacute rejection?

17.4 You are developing a method capable of destroying dendritic cells within an organ prior to transplantation surgery. Will this technique prevent the direct pathway to allorecognition, the indirect pathway to allorecognition, or both? Explain.

17.5 Explain why, in transplantation medicine, much of the focus during HLA matching is on HLA-A, HLA-B, HLA-C, and HLA-DR alleles. (You may need to refer back to Chapter 6 for help in answering.)

17.6 After solid organ transplantation surgery, a combination pharmaceutical therapy of corticosteroids and T-cell inactivation inhibitor (such as FK-506) is typically prescribed. Explain why both drugs are prescribed.

17.7 A leukemia patient has undergone chemotherapy and irradiation treatment prior to bone marrow transplantation, as a matched donor had been identified. A few weeks after the bone marrow transplant, the patient is given an infusion of isolated T cells from the donor. Explain why this second infusion of T cells was used.

● CASE STUDY REVISITED

Kidney Failure

A 26-year-old woman visits her physician due to nausea and vomiting and a decreased ability to urinate. She was diagnosed the year before with hypertension and placed on calcium channel blockers to lower her blood pressure. During her exam, the patient's blood pressure is found to be high, in spite of the fact she reports taking the medication as prescribed. The physician orders blood tests for blood urea nitrogen and creatinine levels to test for kidney function. The tests reveal very high levels of blood urea nitrogen (40 mg/dL; normal range 7–20 mg/dL) and creatinine (3.02 mg/dL; normal range 0.59–1.04 mg/dL). Because of the high blood urea nitrogen and creatinine levels, the nausea, and the inability to urinate, the doctor diagnoses kidney failure, orders dialysis, and places the patient on the transplant waiting list.

Think About...

Questions for individuals

1. Explain why urea and creatinine levels are measured via blood test to evaluate kidney function.
2. Describe why hypertension might lead to kidney failure.
3. Conduct research online to determine if treatment with calcium channel blockers can lead to kidney damage or kidney failure.

Questions for student groups

1. During the initial screening for transplant donors, the physician asks the woman about prior pregnancies and previous transplants, and orders a test to detect anti-HLA antibodies. Why are these actions taken?
2. Why would the physician ask the patient for a list of both immediate and extended family members after placing her on the transplant waiting list?
3. The physician finds the presence of anti-HLA antibodies in the patient's bloodstream. Why might this make her husband and children less likely donor candidates?

MAKING CONNECTIONS Key Concepts	**COVERAGE** (Section Numbers)
Comparing alloreactions of solid organ transplantation with normal innate and adaptive immune responses	1.4, 1.8, 3.1, 3.3, 4.2, 4.5, 4.6, 4.7, 4.8, 5.7, 6.5, 6.7, 7.1, 10.5, 17.1, 17.3
Comparing alloreactions of hematopoietic stem cell transplantation with normal adaptive immune responses	1.5, 4.2, 4.5, 4.6, 4.7, 4.8, 5.7, 6.5, 6.7, 7.1, 17.1, 17.3
The importance and complications of HLA matching in transplantation	4.2, 4.5, 4.6, 4.7, 5.7, 6.5, 6.7, 7.3, 17.1, 17.3, 17.4, 17.5
Blocking innate and adaptive immune responses to limit transplantation alloreactions	4.2, 4.7, 7.1, 7.3, 7.7, 7.8, 7.9, 7.10, 10.5, 17.1, 17.3, 17.4, 17.6

18

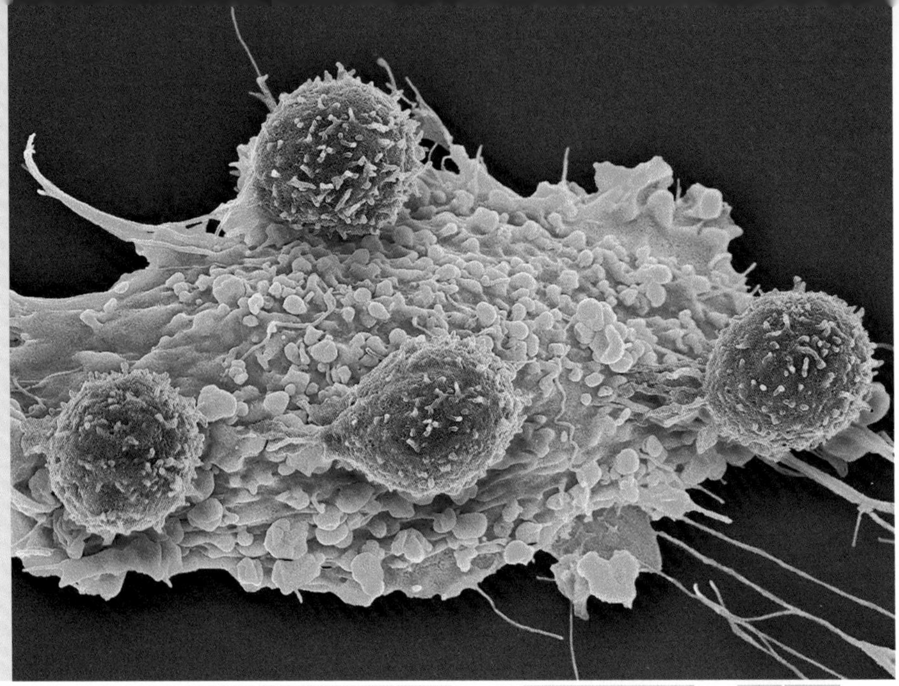

Cancer and the Immune System

QUESTIONS Explored

18.1 What causes cancer?

18.2 What is the importance of proto-oncogenes and tumor suppressors in cancer development?

18.3 How do cancer cells evade destruction by the immune system?

18.4 How does the immune system destroy cancer cells?

18.5 How can the immune system be manipulated to promote tumor destruction?

CASE STUDY: Connor's Story

"Mom! I don't feel good enough to go to school! It's my stomach again." Another Monday morning, and this seems to be a recurring theme with Heather's 8-year-old son, Connor. His complaints have increased recently. Heather goes into Connor's room and he tells her, "I feel really tired, and my stomach was doing somersaults all night."

Weeks ago, Heather had searched some medical websites to try to figure out what might be going on based on Connor's symptoms, but she found nothing conclusive. Some common possibilities included drug allergies (Connor was not taking any drugs) and viral infection. Heather decides that it's time to take Connor to his pediatrician.

To understand what Connor is experiencing and to answer the case study questions at the end of the chapter, we will explore the interaction of the immune system with cancer cells. Cancer occurs due to mutations within cells that affect the ability to control the cell cycle and cell division. Two key gene families—proto-oncogenes and tumor suppressor genes—that work to promote progression through the cell cycle (or as the "gatekeepers" of the cell cycle) are targets for mutations that can lead to the onset of cancer. Does our immune system recognize or target these cancer cells? We will explore the battle that ensues between cancer cells, which aim to continue to survive and divide, and immune mechanisms, which aim to recognize and target harmful pathogens and cells.

18.1 | What causes cancer?

LEARNING OBJECTIVE

18.1.1 Describe the hallmarks of cancer and the driving forces behind mutation of normal cells and the formation of tumor cells.

Cancer is a group of diseases driven by abnormal or uncontrolled cell growth. It has devastating global effects: in 2020, there were 19.2 million new reported cases of cancer, affecting 0.2% of the world population. In that year, cancer killed 9.9 million people. The medical community has focused much attention and financial resources on the study of the disease and on developing treatments for various types of cancers. Our understanding of the immune response has driven the creation of cancer therapies that enable the immune system to effectively target and destroy cancer cells.

cancer Group of diseases characterized by the uncontrolled growth of abnormal cells.

tumor Abnormal growth of cells within an organism.

benign Denoting a tumor that is noncancerous and localized.

malignant Denoting a tumor composed of cells that can invade surrounding tissue and spread to other regions of the body (metastasize).

metastasis Process of cancer-cell migration through the circulatory system to a secondary site within the body.

leukemia Cancer of circulating hematopoietic cells.

lymphoma Cancer of lymphoid tissue.

myeloma Cancer of the bone marrow.

How cancer develops

To understand how the immune system can target cancer cells, we must first appreciate the underlying forces behind the development of cancer and the complexities associated with attempting to target and eliminate cancer cells using chemotherapy or immune-based therapies. The driving force behind the onset of cancer is DNA mutation, which can be caused by a variety of mechanisms and ultimately results in cells' inability to control their growth. Later in this section we will explore the causative agents that drive DNA mutation and the onset of cancer.

Specific traits promote the onset of tumor growth; these are referred to as the hallmarks of cancer:

- *Self-sufficiency in growth signals*—Cancer cells are capable of growth in the absence of growth factors.
- *Insensitivity to anti-growth signals*—Inhibitory signals that normally halt growth are inactive toward cancer cells.
- *Evading apoptosis*—Cancer cells are insensitive to signals promoting their own programmed cell death.
- *Limitless replicative potential*—Cancer cells have infinite potential to divide.
- *Sustained angiogenesis*—Cancer cells gain access to nutrients and oxygen through increased blood vessel growth.
- *Tissue invasion and metastasis*—Cancer cells leave a primary site of growth and migrate to other organs and tissues.

Abnormal cell growth causes the formation of a **tumor**, which can be either **benign** or **malignant** (**FIGURE 18.1**). Benign tumors are more easily managed by surgery, as they are encapsulated and localized and have not spread to other areas of the body. Because they are encapsulated, with defined boundaries, surgeons can determine the borders of benign tumors and remove them without a high risk of leaving cancerous cells behind. Quite often, benign tumors are also observed without any further treatment because they may either be resorbed by the body or may never enlarge and lead to cancer. Malignant tumors are more difficult to treat because they often do not have distinct boundaries, and cancerous cells may have spread to other areas of the body, a process known as **metastasis**. Cancer can cause uncontrolled growth of various types of cells. For example, cancers associated with cells and tissues of the immune system include **leukemia**, which affects circulating hematopoietic cells; **lymphoma**, which affects lymphoid tissue; and **myeloma**, which affects the bone marrow.

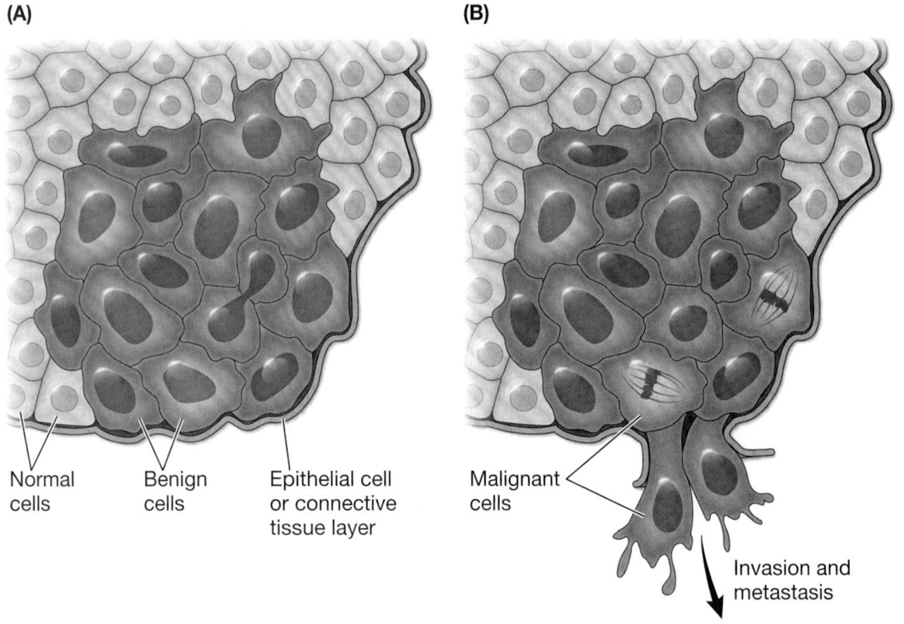

FIGURE 18.1 Benign and malignant tumors (A) A benign tumor is a cluster of abnormal cells that does not invade surrounding tissues. Cells within the tumor have mutated to allow them to divide more rapidly but have limited potential to grow and divide uncontrollably. In the case of a benign tumor, abnormal cells cannot spread past a barrier created by epithelial cells or connective tissue. (B) A malignant tumor is a cluster of cancerous cells with the ability to invade past the epithelial cell or connective tissue layer and metastasize (spread) to other tissues of the body. A malignant tumor contains mutated cells with the ability to grow and divide rapidly and uncontrollably.

proto-oncogene Gene that encodes a protein involved in the process of cell division.

tumor suppressor gene Gene that encodes a protein involved in regulating cell division.

angiogenesis Growth of new blood vessels.

Malignant tumors pose a serious threat because they consist of cells that can migrate to other tissues and organs and affect their function through blood occlusion and by depriving tissues of nutrients. How, then, does a malignant tumor develop? There are defined steps that promote progression of cancer (**FIGURE 18.2**):

- *DNA mutation*—Because cellular division is a tightly regulated and coordinated process, key players in this process must change before a cell can divide uncontrollably. These players change at the genetic level through mutation of their coding sequences that either activates them and prevents them from being shut down (these genes positively control the cell cycle and are called **proto-oncogenes**) or prevents their ability to shut down the cell cycle (these genes negatively control the cell cycle and are called **tumor suppressor genes**).
- *Angiogenesis*—Cellular division demands resources and energy. Tumors that are not connected to the circulatory system are self-limiting and cannot grow beyond 1 to 2 millimeters in size. Cells not connected to the circulatory system are not provided nutrients for growth, are limited in their oxygen supply (which can only diffuse 1 to 2 millimeters), and cannot get rid of metabolic waste products. **Angiogenesis** (the growth of new blood vessels) enables tumor growth beyond this size and allows connection to the circulatory system, which provides access for migration to secondary sites within the body.
- *Metastasis*—One regulatory signal that prevents abnormal cell division is cell-to-cell contact and anchorage of cells to the extracellular matrix. Some cells can divide only if they are attached to an extracellular matrix (referred to as *anchorage dependence*) until they come into contact with a neighboring cell. Under normal conditions, when a dividing cell population contacts neighboring cells, inhibitory signals inhibit cell division since the division is no longer necessary to "fill a gap" in the tissue. Likewise, cells requiring anchorage dependence to divide will no longer divide if contact with the extracellular matrix is lost due to similar inhibitory signals. However, during the mutations that drive the onset of cancer, some genes that provide this negative signal are mutated as well, preventing these cells from shutting down cell division upon cell-to-cell

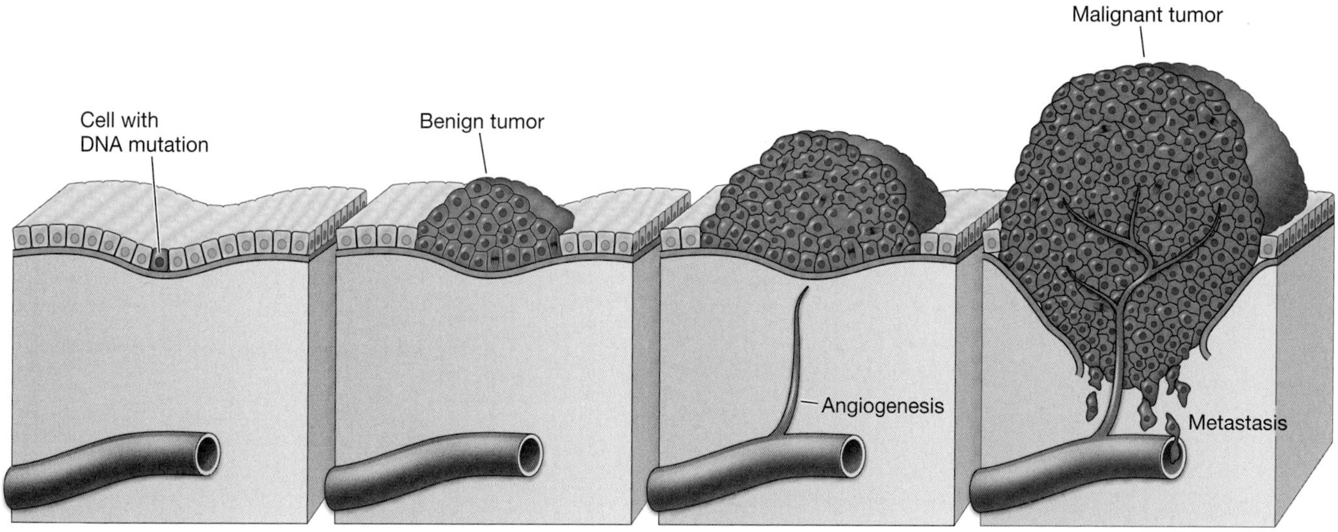

FIGURE 18.2 **Onset of cancer** Cancerous growth begins due to mutations of DNA in genes that regulate cell division. Mutation of genes involved in controlling cell division and growth leads to formation of a benign tumor. Continued changes to cells within the benign tumor may result in angiogenesis (new blood vessel growth) to promote delivery of nutrients to the rapidly dividing cells. Eventually, the tumor may become malignant (cancerous) and invade neighboring tissues and/or metastasize to other locations in the body.

contact or loss of connection with the extracellular matrix. Malignant tumor cells begin to over-express proteases that promote destruction of the extracellular matrix, providing an easier path for these cells to migrate into the circulatory system and travel to secondary sites.

Cancer cells and DNA mutation

Many cells in our body can undergo cellular division when provided with the right external signals. During the cell cycle, a cell reaches critical checkpoints to ensure that the mechanisms required for cell division, including replication of the parent cell's genetic information, have not malfunctioned before passing on the newly replicated genetic information to the daughter cells. The proteins involved in the intricate steps of cell division are tightly regulated to ensure that cell growth and division do not occur if the cell has not properly replicated its genetic information or was damaged during the replication process. Either outcome prevents the daughter cell from being viable or causes the daughter cell to be deleterious in the future.

The genes that encode proteins involved in the cell cycle fall into two major categories: proto-oncogenes and tumor suppressor genes. Because these two gene families encode proteins responsible for the process and regulation of cell division, malfunction of these gene products can result in uncontrolled division, leading to cancer. Both genetic and environmental factors influence DNA mutation and the progression of cancer.

GENETIC PREDISPOSITION TO CANCER Although many causative agents for the onset of cancer are environmental factors, individuals can have a genetic predisposition to cancer if they have inherited a mutated form of a tumor suppressor gene that encodes a protein responsible for regulating the cell cycle. Inherited genetic mutations play a role in 5% to 10% of all cancers, and specific mutations have been identified in more than 50 hereditary cancer syndromes. Typically, this occurs due to passage of a mutation of one of the two copies of a tumor suppressor gene, thus only requiring the other gene to be

mutagen Substance or agent capable of inducing DNA damage and increasing the rate of DNA mutation.

mutated to prevent its regulatory function. Two such examples are *p53* and *BRCA2*. The *p53* gene encodes a protein responsible for noting the presence of DNA damage and preventing further progression of mitosis in cells where DNA damage has occurred. This prevents the damaged DNA from passing to the daughter cells that would be produced during mitosis. Mutation of both copies of the *p53* gene is strongly correlated with the onset of cancer. Mutation of one copy of *p53* occurs in diseases such as Li-Fraumeni syndrome, an autosomal dominant disease that predisposes affected individuals to cancer development. Since one of the copies of the *p53* gene has already been mutated in individuals with Li-Fraumeni syndrome, only mutation of the wild-type *p53* gene is required to prevent its function and promote cancer progression.

A similar example is seen with the *BRCA2* gene, whose protein product is responsible for DNA repair during DNA replication. Single-stranded breaks in DNA are repaired during DNA replication using homologous recombination repair mechanisms. The protein product of *BRCA2* is responsible for recognizing these single-stranded breaks and facilitating the action of homologous recombination repair proteins to act and repair these breaks. If homologous repair mechanisms do not function properly, other repair mechanisms, including error-prone repair mechanisms, compensate for the loss of homologous recombination repair. Just as with the *p53* gene, mutation of one copy of *BRCA2* can predispose an individual to cancer, since only the wild-type *BRCA2* gene would need to be mutated to prevent *BRCA2* function in those carrying the first mutation. Nonfunctional *BRCA2* within an individual increases unrepaired DNA, thereby increasing the likelihood of mutation within those cells.

ENVIRONMENTAL FACTORS THAT FACILITATE CANCER Environmental factors are the most common culprit leading to cancer. Since genetic predisposition accounts for only 10% or less of all cancers, environmental factors are responsible for the onset of cancer in 90% to 95% of cases. **Mutagens** are environmental factors (such as toxic chemicals) that can induce DNA damage and increase mutations. Mutagens increase the rate of mutation since repair mechanisms that remove damaged DNA may also induce the formation of a mutation in the repaired DNA. While the mutations do not automatically cause cancer, if they occur in the genes responsible for cellular division, they can lead to cancer. In general, three groups of environmental factors can act as mutagens: physical agents, chemical mutagens, and certain pathogens.

Physical Agents Both ultraviolet light and ionizing radiation are capable of damaging DNA. UV light can induce the formation of pyrimidine dimers (**FIGURE 18.3**), which interfere with the action of DNA polymerase during DNA replication because of the distortion they cause in the DNA double helix. Ionizing radiation can cause DNA strands to break, either directly or through the formation of reactive hydroxyl radicals. These strand breaks increase the possibility of unintended recombination events and chromosomal translocations due to non-homologous end-joining DNA repair mechanisms (the same mechanism used to fuse excised hairpins during V(D)J recombination).

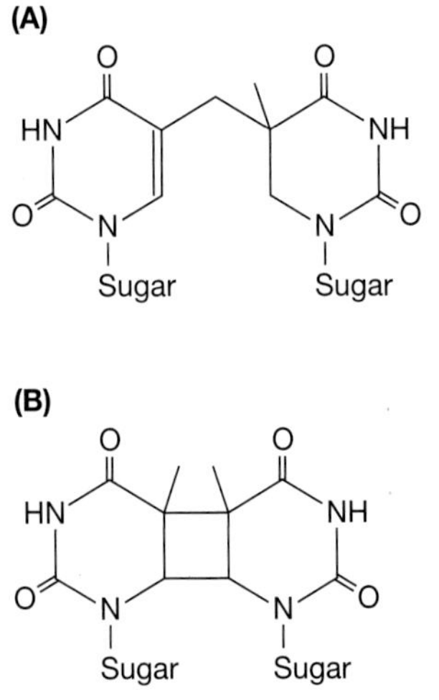

FIGURE 18.3 Pyrimidine dimers (A) Chemical structure of a 6-4 thymine dimer. (B) Chemical structure of a cyclobutane thymine dimer.

Chemicals Chemicals induce DNA mutations most commonly due to single-base alterations (**FIGURE 18.4**). Chemicals such as nitrous acid can induce oxidative deamination, causing cytosine bases to be chemically modified to uracil. Recall that the uracil change by activation-induced cytidine deaminase (AID) enzymatically modifies cytosine to uracil during somatic hypermutation; the change catalyzed by nitrous acid is mutagenic for similar reasons.

Reactive oxygen species, which can be present at high concentrations in metabolically active cells or created by exogenous chemicals such as benzene

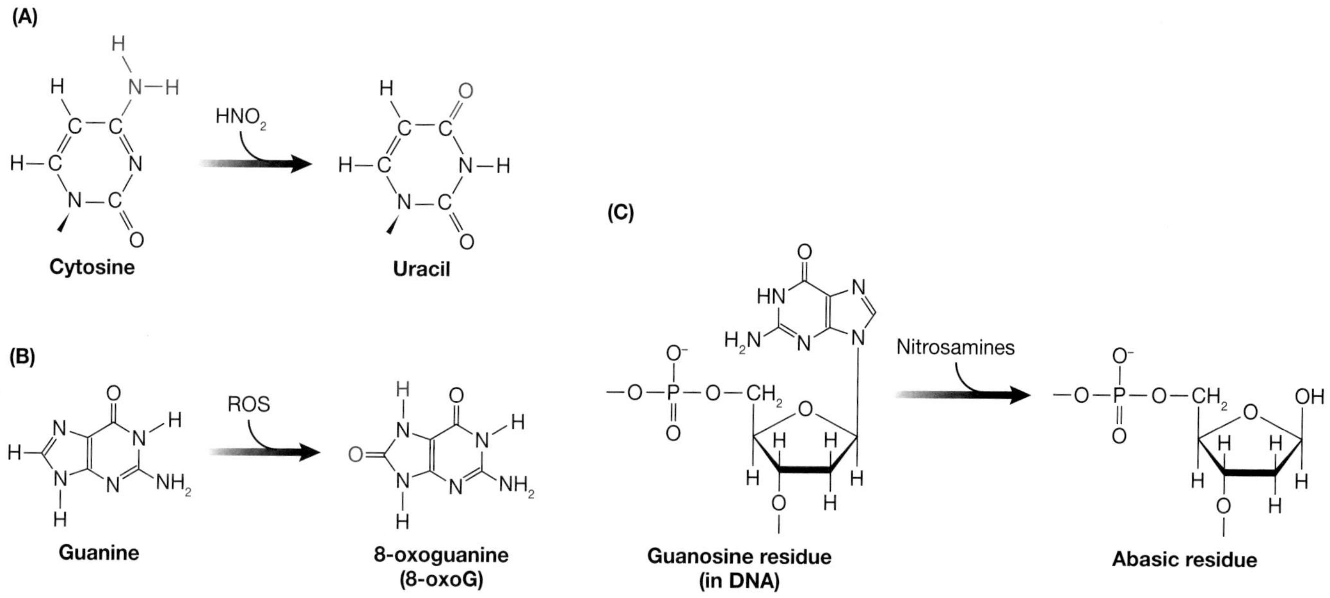

FIGURE 18.4 Chemical modification of nucleotides (A) Nitrous acid (HNO_2) reacts with cytosine, catalyzing its deamination to uracil. (B) Reactive oxygen species (ROS) oxidize guanine bases in DNA, converting them to 8-oxoguanine. (C) Nitrosamines, such as those found in tobacco smoke, react with nucleic acids and destabilize the glycosidic bond between the base and ribose sugar. This creates an abasic site within the DNA, which is ultimately repaired by DNA repair mechanisms.

present in tobacco smoke, can cause a wide variety of base changes, potentially resulting in mutation. For instance, reactive oxygen species can alter guanine bases to form 8-oxoguanine, which is capable of base pairing with adenine bases. Thus, conversion of guanine to 8-oxoguanine can alter the base on the complementary strand during replication from the wild-type cytosine to adenine. Alkylating agents, such as nitrosamines present in tobacco, can destabilize the glycosidic bond between the nitrogenous base and the ribose sugar within nucleotides, creating an abasic site, which can ultimately be changed during DNA repair, in a similar manner to the alterations seen during somatic hypermutation. Intercalating agents, such as ethidium bromide, can also cause mutations. These agents insert between base pairs of DNA, resulting in alterations of the structure of the DNA double helix and possibly resulting in frameshift mutations during DNA replication.

Pathogens Certain pathogens are known to cause abnormal cell growth (**TABLE 18.1**). Approximately 10% to 20% of cancers are caused by infection. Most cancer-causing pathogens are viruses or bacteria. In fact, the discovery of oncogenes was facilitated by the discovery of a chicken virus, known as sarcoma virus, which causes tumor formation (see **KEY DISCOVERIES**). Typically, these pathogens induce cancer by altering normal proteins that play a role in regulating cell division. For instance, human papillomavirus (HPV) expresses two proteins, E6 and E7, which inactivate two tumor suppressor genes, p53 and Rb. By inactivating these proteins, HPV promotes uncontrolled cell growth and can cause cancer; cervical cancer is the most common type associated with this virus. While certain pathogens are capable of inducing cancer, it should be noted that cancer is not transmissible from person to person; only the pathogen itself can be transmitted.

TABLE 18.1 | Pathogens That Can Cause Cancer

Pathogen	Cancer Caused by Infection
Human papillomavirus (HPV)	Primarily cervical cancer Cancers of the genitals, testes, anus, rectum, and oropharynx
Epstein–Barr virus	Burkitt lymphoma Nasopharyngeal cancer
Hepatitis B and C viruses	Liver cancer
Human herpesvirus 8	Kaposi sarcoma
Human T-lymphotrophic virus type 1 (HTLV-1)	Adult T-cell leukemia and lymphoma
Merkel cell polyomavirus	Merkel cell carcinoma
Helicobacter pylori	Stomach cancer
Chlamydia trachomatis	Cervical cancer

Sources: P.S. Moore and Y. Chang. 2010. *Nat Rev Cancer* 10: 878–889; A. Gagnaire et al. 2017. *Nat Rev Microbiol* 15: 109–128.

KEY DISCOVERIES

How was the first oncogenic virus discovered?

Article

Rous, P. 1911. A sarcoma of the fowl transmissible by an agent separable from the tumor cells. *Journal of Experimental Medicine* 13: 397–411.

Background

We have learned that mutation of proto-oncogenes and tumor suppressor genes is the driving force behind the development of uncontrolled cell growth and tumor formation. The story of the identification of oncogenes is connected to the identification of a virus capable of inducing tumor formation in chickens. How did the identification of a tumor-causing chicken virus lead to the identification of oncogenic viruses and eventually oncogenes?

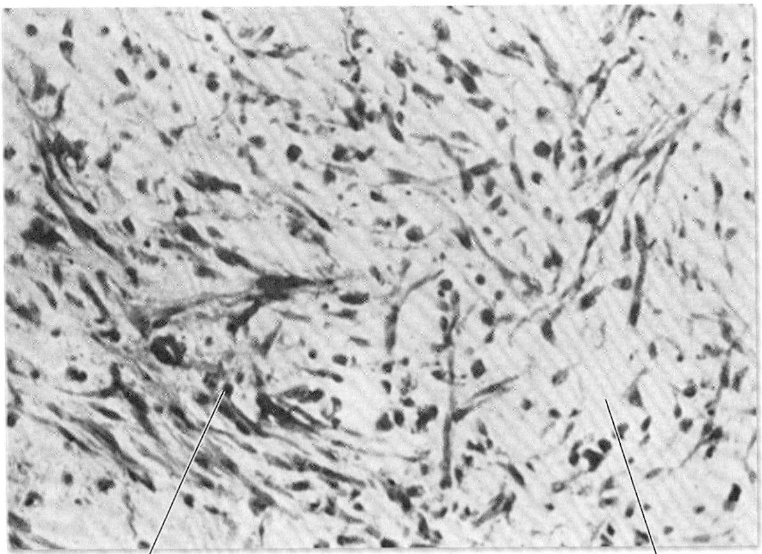

FIGURE KD 18.1 Cancerous growth induced by a cell-free filtrate isolated from a tumor from a different animal A cell-free filtrate was isolated from the tumor of a chicken and injected into another animal, resulting in the onset of a sarcoma in the injected animal.

Early Research

In the early twentieth century, American medical researcher Peyton Rous became aware of a transmissible sarcoma within chickens and undertook studies to determine the causative agent. Previous studies had never identified transmissible cancers in other organisms, including rats, mice, and dogs, so this newly discovered transmissible cancer was a curiosity unseen by the scientific community at the time.

Think About...

1. Knowing that the sarcoma in chickens was transmissible, what might you predict as factors capable of inducing this cancer?
2. How would you test your prediction of the causative agent? What experiments would you need to employ to test your hypothesis?
3. Would you predict that this chicken sarcoma could be transmitted to other organisms? How would your experimental findings inform your answer to this question?

Article Summary

Rous created cell-free filtrates from tumors isolated from chickens to test the transmissibility of sarcoma between animals. The cell-free filtrates were created through centrifugation and filtration through a Berkefeld filter capable of filtering out bacterial species. After injecting this cell-free filtrate into other chickens, similar sarcomas would eventually form (**FIGURE KD 18.1**). Rous discovered that this cell-free filtrate could not induce sarcoma formation in other birds.

Much later, in the 1970s, the transmissible agent was identified and named Rous sarcoma virus. Cutting-edge technology at the time led to the discovery of the *v-src* gene expressed in Rous sarcoma virus, a protein kinase hallmarked as the first identified oncogene. This discovery was driven by Rous's initial findings of the transmissible sarcoma from chickens passed via cell-free filtrates.

INFLAMMATION AND CANCER A hallmark of systemic infections by pathogens is the onset of an inflammatory response. This response is aided by inflammatory cytokines secreted by innate immune cells recognizing the pathogens using pattern recognition receptors. While we have focused on the role of inflammation in pathogen infection, inflammation also plays an important role in wound repair and tissue regeneration by promoting cell proliferation. While functions of the inflammatory response are crucial in pathogen destruction and wound healing, they use mechanisms that can promote mutation. Macrophages and neutrophils employ reactive oxygen species to target and destroy pathogens, including those that have the capacity to alter DNA bases such as conversion of guanine to 8-oxoguanine. Neutrophils that undergo degranulation can release these reactive oxygen species into the extracellular environment, where they can act on nearby cells, including cells within the individual. Chronic inflammation, caused by exposure to cancer-inducing pathogens such as *Helicobacter pylori* or by an underlying condition such as irritable bowel syndrome, can induce the onset of cancer due to the mutagenic properties of inflammation.

● CHECKPOINT QUESTION

1. Why is DNA mutation required for the onset of cancer?

18.2 What is the importance of proto-oncogenes and tumor suppressors in cancer development?

LEARNING OBJECTIVE

18.2.1 Explain the role of proto-oncogenes and tumor suppressor genes in controlling cell division and how mutation of these genes plays a role in the development of cancer.

We have seen that DNA mutation is the key driving force in the onset of cancer. While the mutations that occur are random, to induce the onset of cancer, they must specifically alter the action of proteins that drive or control the process of cell division. Because of the complexities and redundancy in the regulation and process of cell division, a single mutation of both copies of a gene within somatic cells typically doesn't lead to tumor formation. Indeed, the development of cancer requires multiple mutations of different genes involved in regulating the cell cycle.

oncogene Mutant form of a proto-oncogene that promotes uncontrolled cell division.

The two gene classes that can be mutated and cause uncontrolled cell division are proto-oncogenes and tumor suppressor genes. While mutation in both gene families can result in uncontrolled cell division, the mutations differ dramatically in how they can affect the function of their gene products and drive cancer formation (**FIGURE 18.5**).

Proto-oncogenes

Many factors are required for the initiation and progression of cell division. Cells must be provided a signal, typically from a growth factor, to initiate the process. These growth factors must bind to a receptor on the target cell to initiate signal transduction cascades and the transcription of genes required for cell growth, DNA replication, and mitosis. All genes involved in initiating or driving the process of cell growth and mitosis are referred to as *proto-oncogenes*. Mutation of a proto-oncogene that promotes a gain of function such as hyperactivity or an inability to be regulated can drive uncontrolled cell division. Such mutations need to alter only one of the two copies of the proto-oncogene since a gain-of-function mutation on one copy is sufficient for a proto-oncogene to promote uncontrolled cell division. When a proto-oncogene is mutated in such a way, the gene is referred to as an **oncogene**.

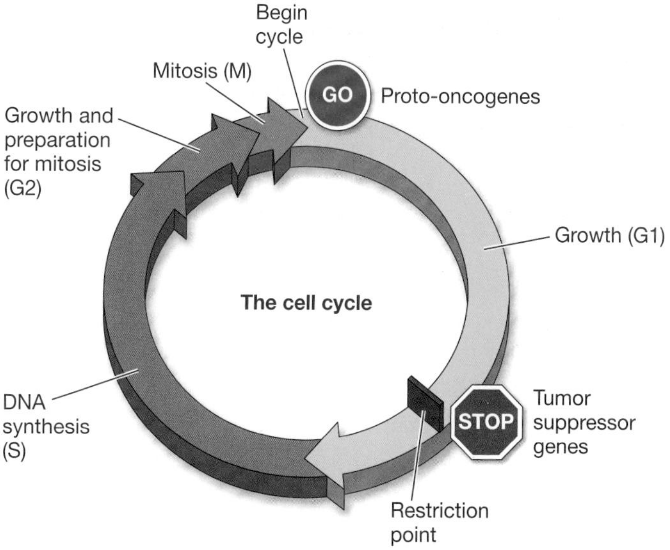

FIGURE 18.5 Proto-oncogenes and tumor suppressor genes control the cell cycle Proto-oncogenes promote entry into the cell cycle by activating pathways needed to progress through the G1 phase and for DNA synthesis (S phase). Mutation of proto-oncogenes that increases their activity within a cell results in uncontrolled cell division. Tumor suppressor genes function at checkpoints within the cell cycle to prevent damaged or mutated cells from continuing through the cycle. Mutation of tumor suppressor genes that decrease or lose their activity results in uncontrolled cell division. (After T. Sandal. 2002. *The Oncologist* 7: 73–81. © 2002 AlphaMed Press.)

Tumor suppressor genes

While proto-oncogenes are responsible for initiation and progression of cellular division, *tumor suppressor genes* are the gatekeepers of cell division: their gene products prevent the progression of the cell cycle. Typically, the gene products of tumor suppressor genes regulate the cell cycle by monitoring the state of the cell undergoing division. They check for DNA damage, as with *p53*, or they prevent progression through a stage until the cell has received the proper signal. Because tumor suppressor genes produce negative regulators of cell growth and division, both copies of the tumor suppressor gene in somatic cells must be mutated to create a loss of function to the extent that the gene no longer functions as a gatekeeper of the process. This requirement of loss of function of both copies of tumor suppressor genes for the onset of uncontrolled cell division is referred to as the *two-hit hypothesis* or *Knudsen hypothesis*.

● CHECKPOINT QUESTION

1. Explain why proto-oncogenes require a gain-of-function mutation to induce cancer while tumor suppressor genes require a loss-of-function mutation to induce cancer.

18.3 | How do cancer cells evade destruction by the immune system?

LEARNING OBJECTIVE

18.3.1 Describe mechanisms used by cancer cells to avoid the innate and adaptive immune response, and provide examples of each.

We have seen that mutation changes the activities of gene products of proto-oncogenes and tumor suppressors, leading to cancer. These changes alter the amino acid sequence of gene products, creating a new antigen that the immune system can recognize as nonself, since these antigens will be processed and presented on MHC class I molecules. This recognition of altered antigens within cancer cells allows the immune system to target the mutated cells using effector mechanisms discussed in earlier chapters.

Some tumor cells increase expression of certain antigens (in comparison to normal cells), and these are known as **tumor-associated antigens**. Others express antigens not expressed by normal cells, usually due to mutation. These are known as **tumor-specific antigens**. Both antigen types play a role in the immune system's ability to recognize and target tumor cells. In fact, we will see that the immune system is relatively efficient in targeting and destroying mutated cells that may have become cancerous.

If mutated cells can be recognized as nonself by the immune system, how can cancer cells survive, given the power and efficiency of the adaptive immune response? Obviously, some cancerous cells manage to avoid destruction by the immune system. We will see that tumors that progress to cancer have the ability to both evade and manipulate the immune response.

Evading the immune response

In Chapter 13, we saw that certain pathogens can survive within a host by evading immune responses that would normally target and eliminate them. Because of the selective pressure of targeting and destroying mutated cancer cells by the immune response, some cancer cells undergoing uncontrolled cell division may mutate genes and gain a selective advantage that allows them to evade the immune response in a manner similar to that used by pathogens. Two

tumor-associated antigen Antigen expressed at higher levels by tumor cells in comparison to normal cells.

tumor-specific antigen Antigen expressed only by tumor cells, typically a mutated protein.

such mechanisms are cleavage of NKG2D receptor ligands from the cell surface and downregulation of MHC class I molecules from the cell surface.

Transformed epithelial cells typically increase the expression of the NKG2D ligand MIC on their surface. Recall that the NKG2D receptor is an activating receptor on natural killer (NK) cells. Recognition of NKG2D ligands on the surface of host cells shifts the balance of inhibition versus activation of NK cells toward the selective targeting and destruction of cells expressing the NKG2D ligand.

Other cells that express NKG2D to detect host cell changes are cytotoxic T cells and γ:δ T cells, which target MIC-expressing cells. How, then, do cancerous epithelial cells, which express a ligand for an activating receptor to these cell types, avoid destruction by these cells? Some cancerous epithelial cells mutate to overexpress proteases that can remove MIC from their cell surface and create a soluble form of the extracellular domain of MIC (**FIGURE 18.6**). Since the extracellular domain is still capable of binding to NKG2D, it can still interact with the receptor on the surface of NK cells, cytotoxic T cells, and γ:δ T cells. However, these cells are not recognizing the cell that expressed MIC; they are just recognizing the soluble form of MIC. Binding of MIC to the NKG2D receptor induces endocytosis of NKG2D from the cell surface, reducing the amount of NKG2D available to recognize MIC on the surface of abnormal cells and allowing cancer cells to survive the effector mechanisms of NK cells, cytotoxic T cells, and γ:δ T cells.

Because tumor cells express mutated antigens to the adaptive immune system, especially cytotoxic T cells, it stands to reason that tumor cells that can escape recognition by cytotoxic T cells would evade these cells through alteration of MHC class I processing and presentation due to the selective pressure of destruction by these cells. Indeed, mutations in antigen presentation, such as producing a defective TAP transporter or defective MHC class I molecules, result in a lower amount of MHC class I on the surface of the tumor cells and less antigen presentation to cytotoxic T cells (**FIGURE 18.7**).

One of the functions of NK cells is to monitor levels of MHC class I molecules on the surface of host cells. A reduced number of these molecules at the cell surface signals to NK cells that

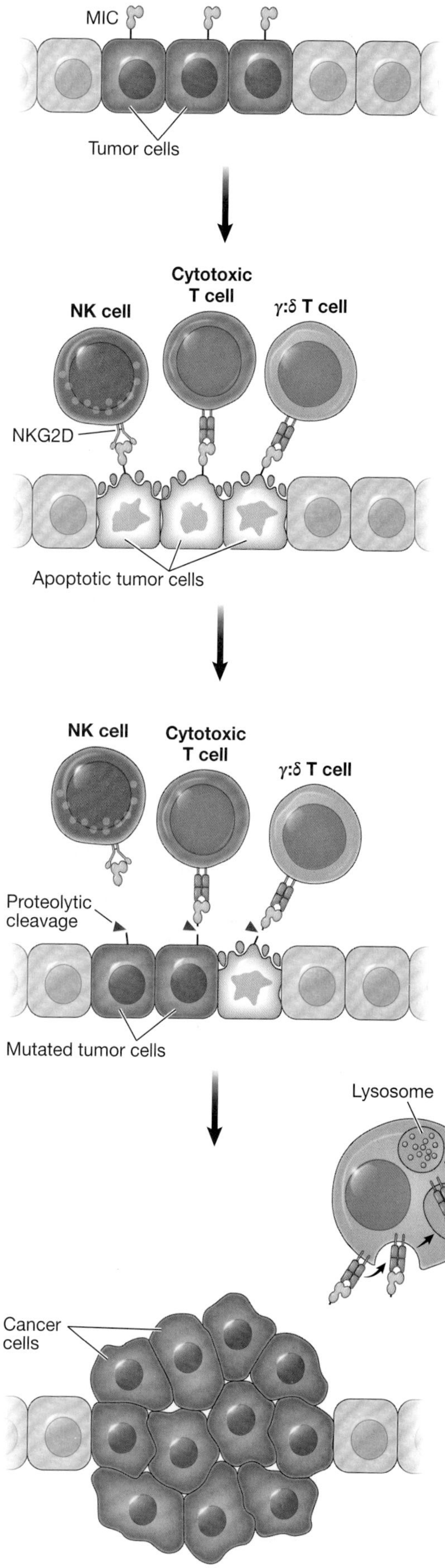

FIGURE 18.6 Tumor cells expressing soluble MIC escape the immune system Immune system cells, including natural killer (NK) cells, γ:δ T cells, and cytotoxic T cells, recognize the presence of cell-surface MHC class I chain related (MIC) proteins on cells that express MIC due to viral infection or induction of uncontrolled division. NK cells bind MIC using an activating receptor NKG2D, and T cells interact with MIC via their T-cell receptors. MIC recognition by immune cells targets the cell expressing MIC for apoptosis via the effector mechanisms of NK cells and cytotoxic T cells. Mutated cells can escape recognition using a protease that releases MIC from the cell surface, producing a soluble extracellular form of MIC. The soluble MIC still binds to receptors on the surface of mutated cells, but the effector cell does not target the mutated cell for destruction, allowing it to continue to divide uncontrollably. (After P. Parham. 2014. *The Immune System*, 4th ed., W. W. Norton and Company.)

alterations have occurred and that the cell should be targeted for destruction, since this lowering is associated with viral infection. Some NK cells can target these altered cells, and the graft-versus-leukemia effect described in Chapter 17 takes advantage of this phenomenon. However, tumor cells can alter how they are recognized by NK cells, and the downregulation of MHC class I presentation at the cell surface is often accompanied by the downregulation of activating ligands on the surface as well.

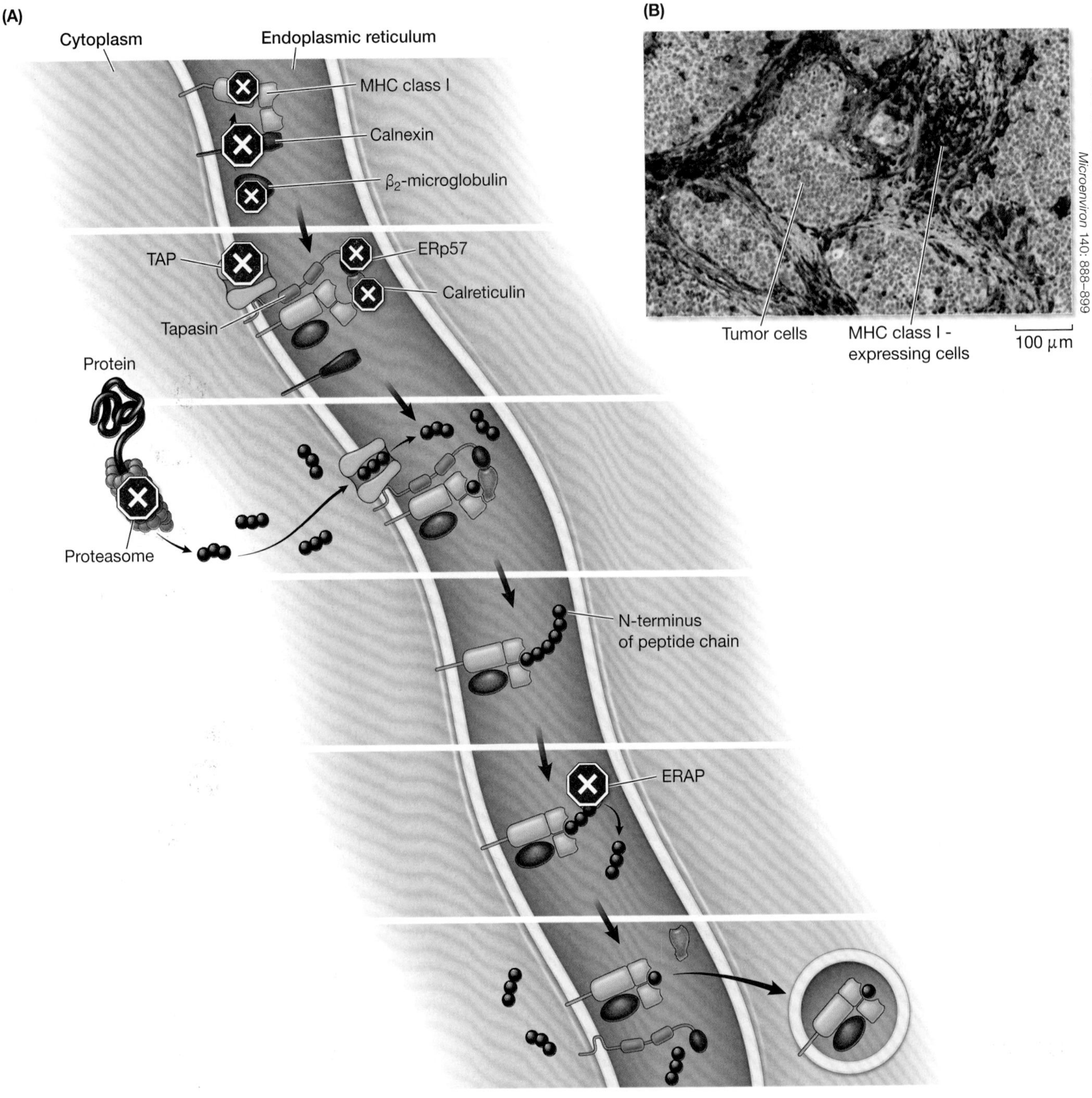

FIGURE 18.7 Mechanisms tumor cells employ to lower MHC class I presentation (A) Many of the proteins involved in MHC class I antigen processing and presentation have been identified as proteins that are altered in various cancers. Red octagons with an X in them represent loss-of-function mutations identified in certain cancers. (B) Micrograph of tumor cells lacking expression of MHC class I molecules. Cells expressing MHC class I molecules are stained brown.

Manipulating the immune response

While some tumor cells are not destroyed because they escape detection by innate and adaptive immune cells, others avoid destruction because they alter the activity of the immune cells responsible for recognizing and destroying them. This manipulation can include the lack of a costimulatory signal for T cells, alteration of apoptotic signals on the surface of tumor cells, and secretion of anti-inflammatory cytokines by tumor cells (**FIGURE 18.8**).

Recall that proper activation of T cells requires not only recognition of an MHC-peptide complex on the surface of a cell by a T-cell receptor but also a costimulatory signal B7 on the antigen-presenting cell binding to CD28 on the T-cell surface. Lack of the costimulatory signal induces anergy of the T cell as

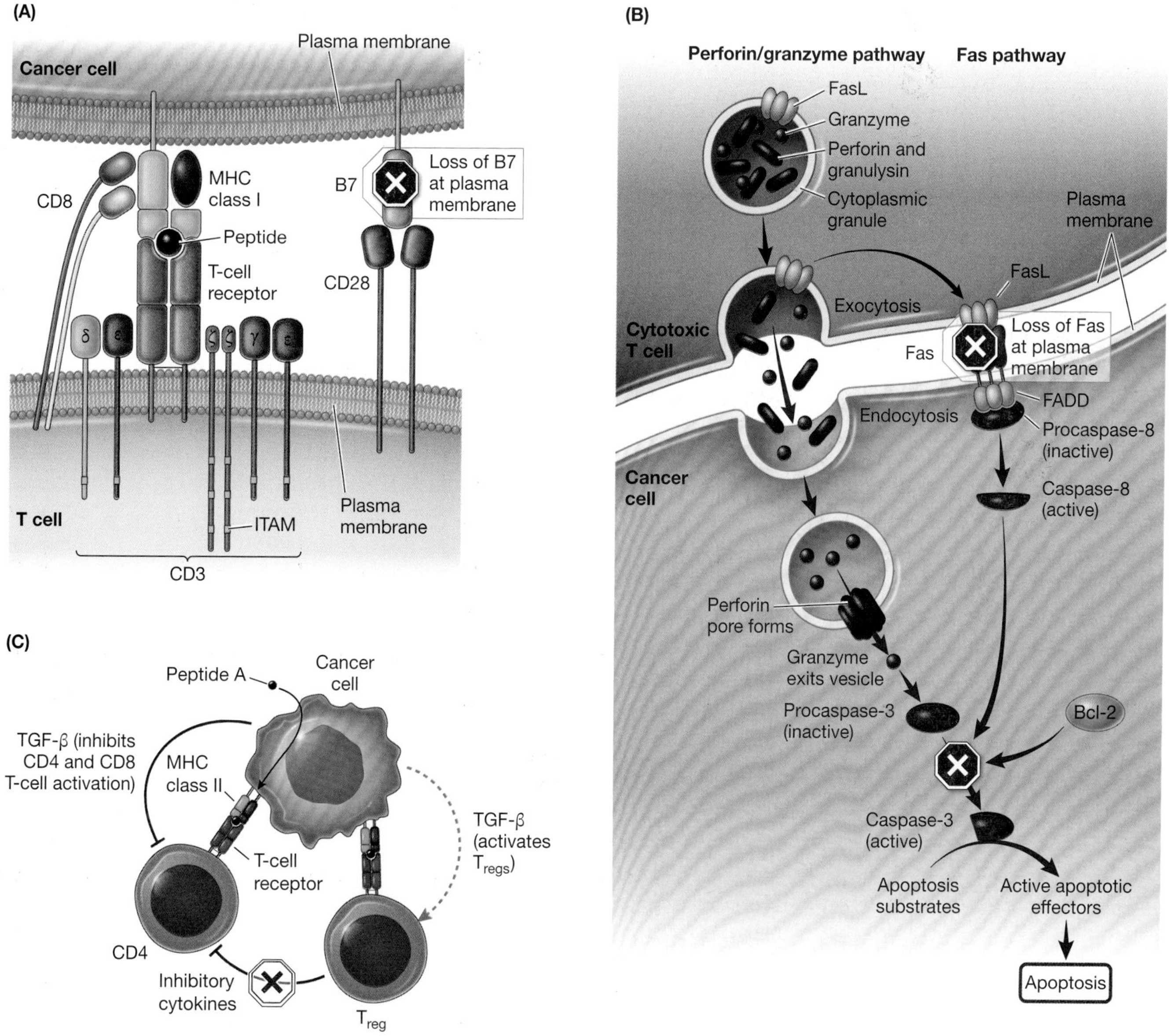

FIGURE 18.8 Cancer cells downregulate pathways that promote their destruction (A) Cancer cells often do not express the costimulatory molecule B7. T cells that recognize a cancer cell that lacks B7 at the surface undergo anergy rather than activation. (B) Cancer cells escape apoptosis using a variety of mechanisms. One way is by limiting the amount of Fas at their plasma membrane. A lack of Fas at the surface of a cancer cell prevents activation of apoptosis signaled by Fas ligand (FasL) from a cytotoxic T cell. (C) Cancer cells prevent activation of T cells that recognize cell-surface antigens presented by MHC class I or class II molecules by secreting the anti-inflammatory cytokine TGF-β, which activates regulatory T cells and inhibits activation of nearby helper or cytotoxic T cells.

a protective mechanism to prevent the activation of autoreactive T cells that escaped negative selection in the thymus. Tumor cells do not tend to express B7 on their surface and, as a result, T cells that recognize tumor cell MHC-peptide complexes commonly undergo anergy due to the lack of costimulatory signal.

The uncontrolled cell division associated with the onset of cancer produces tumor cells with a higher mutation frequency, allowing them to alter molecules that normally aid in the recognition and destruction of abnormal cells by the immune system. One effector mechanism used by NK cells and cytotoxic T cells is the induction of apoptosis of target cells. Thus, if tumor cells mutate molecules involved in the induction of their own apoptosis, they can evade such signals from NK cells and cytotoxic T cells. Some tumor cells can mutate their surface molecules to downregulate pro-apoptotic receptors such as Fas or upregulate anti-apoptotic factors such as Bcl-2 to prevent programmed cell death driven by the immune response.

We have seen that many immune system responses are driven by the secretion of cytokines to either promote, in the case of inflammatory cytokines, or downregulate, in the case of anti-inflammatory cytokines, the immune response. If tumor cells were capable of secreting anti-inflammatory cytokines, they could potentially inactivate immune cells that might otherwise target them for destruction. Some tumor cells upregulate secretion of the anti-inflammatory cytokine TGF-β, which can both recruit and activate regulatory T cells and inhibit other cells such as cytotoxic T cells and T_H1 helper T cells, ultimately creating an environment less susceptible to activating a T-cell response and promoting survival of the tumor cells.

Inflammation and suppression of the immune response

Earlier in this chapter we explored how inflammation can promote mutagenesis in cells during the process of wound repair as it drives cell proliferation. Chronic inflammation in tissue, caused either by underlying conditions such as irritable bowel syndrome or by infection, correlates with the onset of cancer in affected tissues. Since a normal inflammatory response aids in the targeting of infected or altered cells by prompting an adaptive immune response, why does chronic inflammation paradoxically promote the formation of cancer rather than target these cells?

The answer to this question may lie in the production of cells derived from our hematopoietic system that phenotypically look like cells explored earlier in this textbook, namely neutrophils and macrophages, but that behave much differently during the normal course of an infection. The progression of cancer strongly correlates with the production of cells known as myeloid-derived suppressor cells (MDSCs). Two main types of MDSCs have been described: PMN-MDSCs (for polymorphonuclear MDSCs), which phenotypically resemble neutrophils, and M-MDSCs (for monocytic MDSCs), which phenotypically resemble monocytes, the precursors of macrophages. While these cells mirror the morphology of their immune cell counterparts, their activities are much different. Rather than promoting immune responses and targeting pathogens and infections as neutrophils and macrophages do, MDSCs function as immune suppressors. Their immunosuppressive function is driven by production of anti-inflammatory cytokines and elevated levels of the enzyme arginase, reactive oxygen species, and nitric oxide. Thus, rather than functioning in the classical arm of innate immunity, MDSCs may prevent targeting of cancer cells by immune responses because of their immunosuppressive activity.

CHECKPOINT QUESTION

1. Explain how tumor cells can be similar to virus-infected cells in their evasion of the immune response.

18.4 How does the immune system destroy cancer cells?

LEARNING OBJECTIVE

18.4.1 Describe mechanisms used by the innate and adaptive immune systems to target and destroy cancer cells.

immunosurveillance Function of the immune system to target and destroy mutated and tumor cells.

Cancer tends to be an age-related disease, usually occurring later in life. Exceptions are usually cancers associated with hematopoietic cells such as leukemia, in which abnormal lymphocyte development can cause abnormal growth of cells arrested in an early developmental stage. Since cancer commonly affects older individuals, it is logical to assume that younger, stronger immune systems have effectively targeted and destroyed mutated tumor cells before they spread and cause cancer.

Individuals who have undergone an organ transplant and treatment with immunosuppressive drugs (see Chapter 17) are more prone to the onset of tumor formation. Furthermore, rare forms of cancer occur in individuals who are immunocompromised, such as in those infected with HIV. For example, cancer-causing viruses such as human herpesvirus 8 (HHV8) and Epstein–Barr virus can induce cancer formation in these individuals. These findings suggest that the immune system is capable of recognizing and destroying tumor cells or cells that have mutated and may become tumor cells, a function referred to as **immunosurveillance**. We have seen that tumor cells and virus-infected cells bear similarities in their evasion of the immune system, likely because immunosurveillance of tumor cells relies on the same effector mechanisms used to detect and destroy virus-infected cells.

Tumor antigens

For tumor cells to escape the control of the cell cycle, they must mutate gene products that control the cycle and drive cell division. These mutations cause changes in the amino acid sequence of proteins involved in these processes and can create new epitopes that are nonself epitopes not presented to T cells during negative selection in the thymus. These tumor-specific antigens serve as markers of cells that have been modified and no longer act as normal cells.

Some tumor cells change the expression profile of specific proteins to overcome the control of the cell cycle. These tumor-associated antigens can serve as a means for the immune system to distinguish self versus nonself cells and can also be used therapeutically (discussed in Section 18.5). Before we explore the use of the immune system as a therapeutic agent, let's look at immune system mechanisms used to recognize and destroy tumor cells.

Innate immune responses to cancer

Mutagenesis in tumor cells promotes the survival of tumor cells by limiting their recognition by the immune system. However, these changes can be a driving force to induce innate immune responses. We have seen that tumor cells often limit MHC class I processing and presentation due to selective pressure, allowing these cells to avoid the adaptive immune response that might be raised to tumor-specific antigens, just as virus-infected cells are capable of lowering cell-surface presentation of MHC class I molecules to evade recognition by cytotoxic T cells. To combat the reduced MHC class I expression seen in these cells, NK cells express an inhibitory receptor known as NKG2A, capable of monitoring levels of MHC class I at cell surfaces. This inhibitory receptor monitors MHC class I levels by binding to a specific MHC class I molecule, HLA-E, which binds to and presents the leader sequence of MHC class I molecules HLA-A, HLA-B, and HLA-C during their normal processing

Making Connections: Mechanisms Used by Cancer Cells to Evade Immune Responses

Cancer cells can mimic pathogens by gaining mutations that provide evasion mechanisms against immune responses. By evading immune responses that would normally target them for destruction, cancer cells can continue to propagate and potentially spread to other areas of the body.

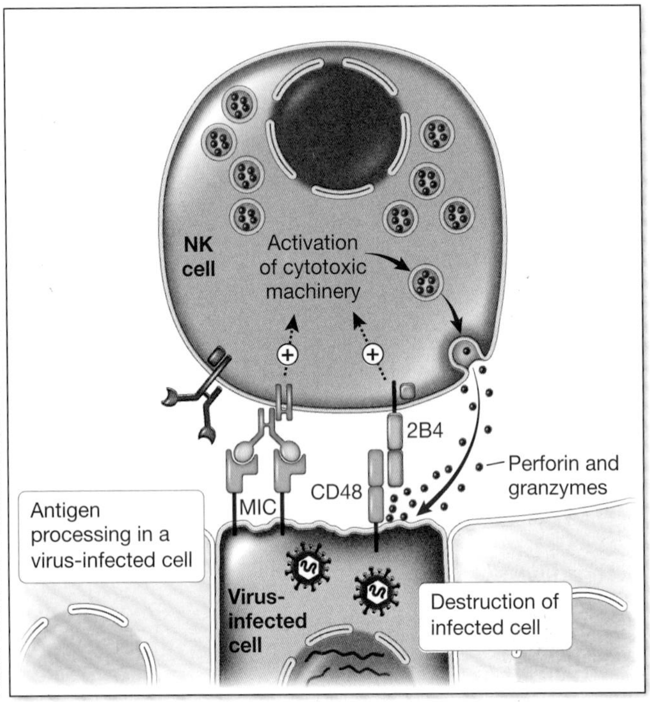

After P. Parham. 2014. *The Immune System*, 4th ed., W. W. Norton and Company

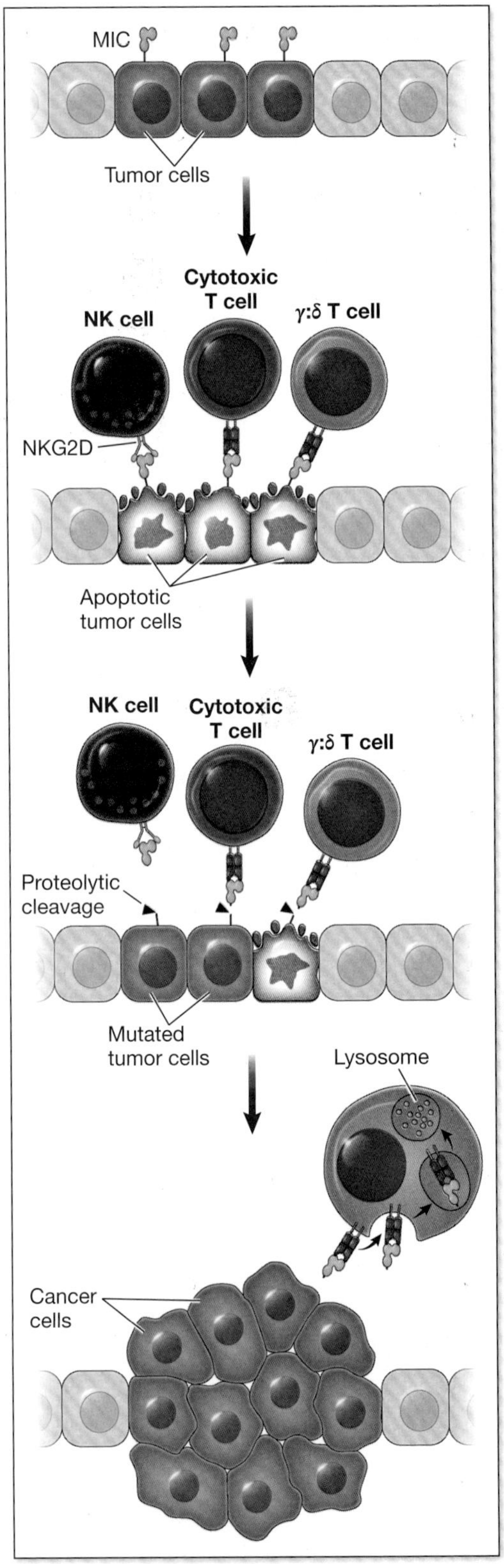

After P. Parham. 2014. *The Immune System*, 4th ed., W. W. Norton and Company

Natural killer (NK) cells monitor changes at the surface of cells in our body. The presence of MIC at the cell surface is an activating signal for NK cells, γ:δ T cells, and cytotoxic T cells. Cancer cells can bypass responses from these cells by producing a protease that cleaves MIC from the cell surface.

(Sections 2.7, 7.13 13.3, 13.4, 18.2, 18.3)

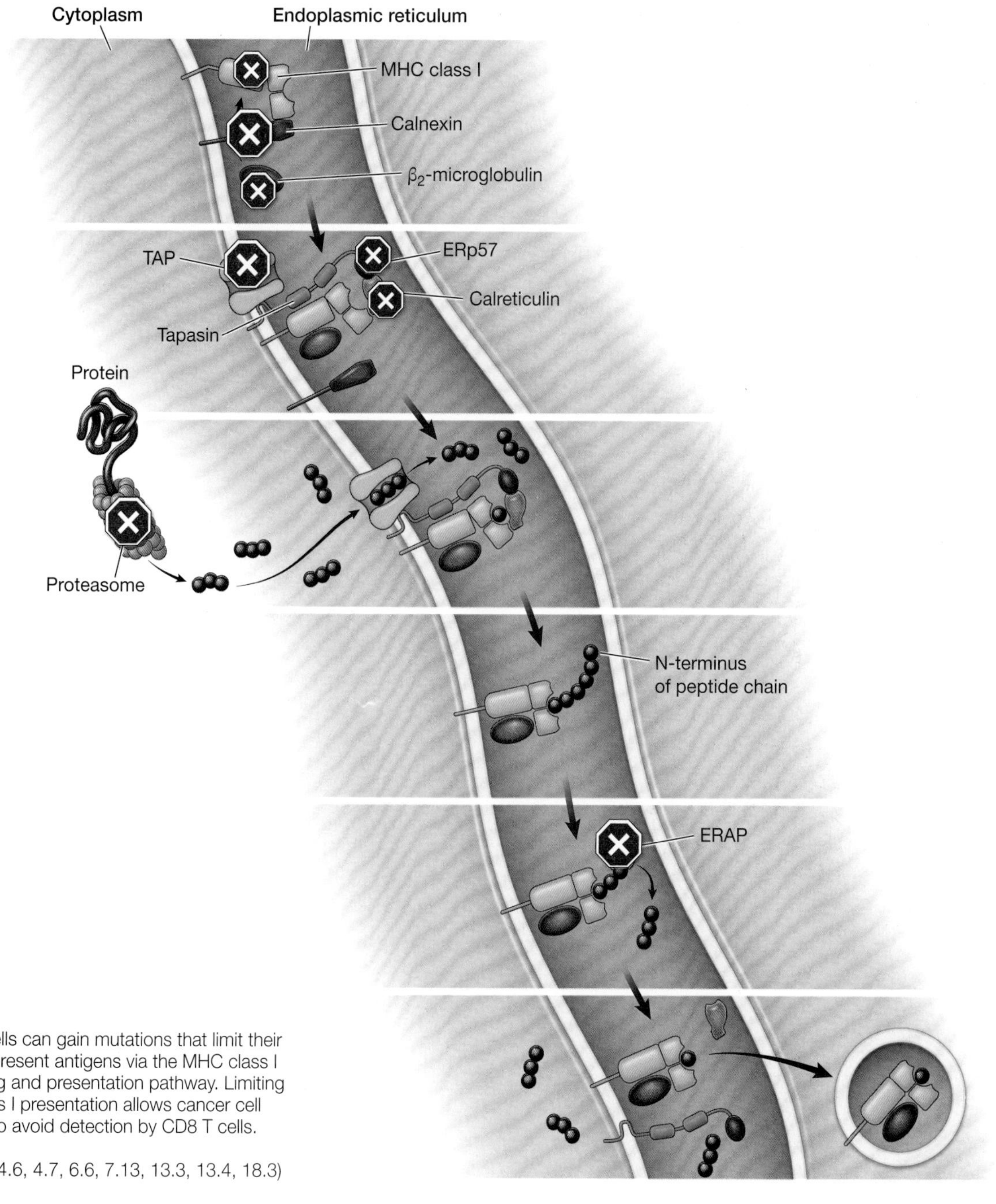

Cancer cells can gain mutations that limit their ability to present antigens via the MHC class I processing and presentation pathway. Limiting MHC class I presentation allows cancer cell antigens to avoid detection by CD8 T cells.

(Sections 4.6, 4.7, 6.6, 7.13, 13.3, 13.4, 18.3)

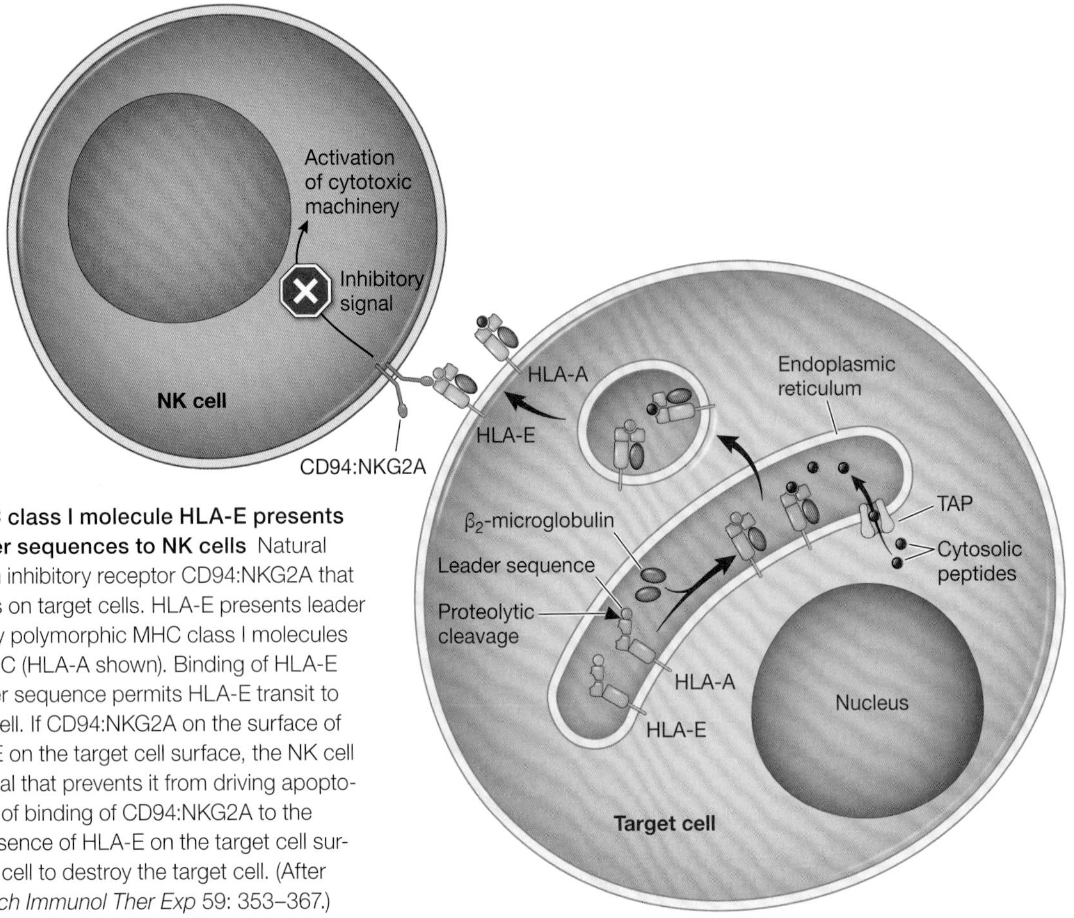

FIGURE 18.9 The MHC class I molecule HLA-E presents other MHC class I leader sequences to NK cells Natural killer (NK) cells express an inhibitory receptor CD94:NKG2A that binds to HLA-E molecules on target cells. HLA-E presents leader sequences from the highly polymorphic MHC class I molecules HLA-A, HLA-B, and HLC-C (HLA-A shown). Binding of HLA-E with an MHC class I leader sequence permits HLA-E transit to the surface of the target cell. If CD94:NKG2A on the surface of an NK cell binds to HLA-E on the target cell surface, the NK cell receives an inhibitory signal that prevents it from driving apoptosis in the target cell. Lack of binding of CD94:NKG2A to the target cell signifies the absence of HLA-E on the target cell surface and prompts the NK cell to destroy the target cell. (After M. Iwaszko et al. 2011. *Arch Immunol Ther Exp* 59: 353–367.)

in the endoplasmic reticulum (**FIGURE 18.9**). In normal cells, HLA-A, HLA-B, and HLA-C are processed through the endoplasmic reticulum, where their leader sequence is cleaved via the same process as any protein entering the endoplasmic reticulum. HLA-E binds to these leader sequences and, once bound, is transported to the cell surface, where it is recognized by NKG2A on NK cells. This recognition sends an inhibitory signal to the NK cells, which recognize the cells as normal and do not attack them.

Further promoting the recognition of HLA-A, HLA-B, and HLA-C molecules by NK cells are inhibitory receptors of the killer cell immunoglobulin-like receptor (KIR) family. KIR receptors on the surface of NK cells recognize proper levels of these three MHC class I molecules on the surface of cells and, when bound, suppress the cytotoxic function of the NK cell.

However, in tumor cells, lowered expression of HLA-A, HLA-B, and HLA-C molecules reduces these three molecules at the cell surface as well as the amount of HLA-E, since less leader sequence is present. The loss of HLA-A, HLA-B, HLA-C, and HLA-E from the cell surface decreases the inhibitory signal sent to NK cells, causing the NK cells to become activated and target the cells with a lower level of these class I molecules.

Adaptive immune responses to cancer

While the innate immune response can eliminate some tumor cells, the adaptive immune system also targets tumor cells and plays a role in immunosurveillance. DNA mutation in tumor cells can cause amino acid mutations or more drastic recombination events creating fused genes, allowing these tumor cells to bypass cell cycle control (**TABLE 18.2**). These tumor-specific antigens represent

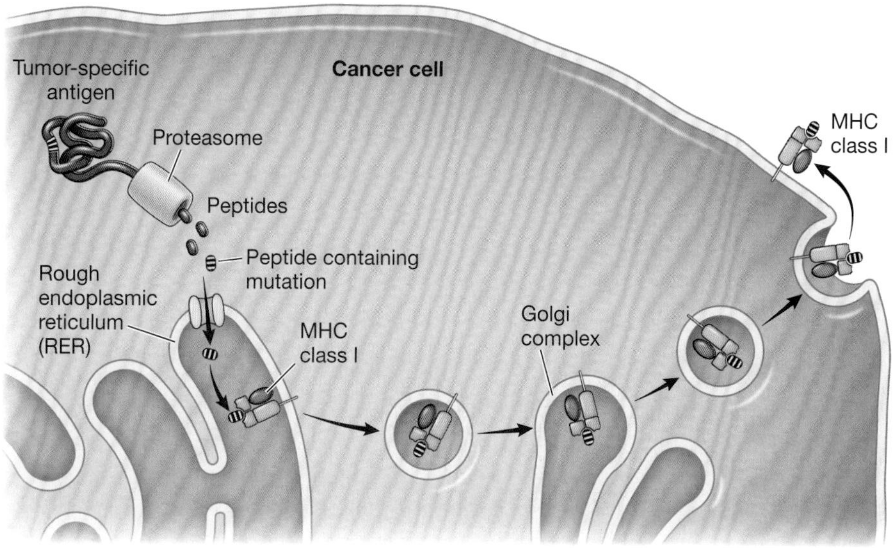

FIGURE 18.10 Presentation of tumor-specific antigens by cancer cells Cancer is caused by gene mutations that alter the function of proteins involved in cell cycle regulation or progression. These mutations can alter the amino acid sequence of proteins, creating tumor-specific antigens within cancer cells. Since all cytoplasmic proteins are processed and presented by cells using MHC class I presentation, tumor-specific antigens with altered amino acid sequences represent novel antigens that are potential targets of the cellular arm of the adaptive immune system.

novel epitopes that can be presented on MHC class I molecules (**FIGURE 18.10**). Since these mutations are not normally expressed, CD8 T cells that have gone through negative selection have not been subjected to these epitopes during development, and some may express a receptor that is capable of recognizing MHC class I molecules presenting these tumor-specific epitopes. Recognition of MHC-tumor-specific antigen peptide complexes by these CD8 T cells results in tumor cell destruction through apoptosis via granule activation from cytotoxic T cells as discussed in Chapter 7.

CHECKPOINT QUESTION

1. Explain how the inhibitory NK receptor NKG2A is capable of monitoring MHC class I levels on cell surfaces.

TABLE 18.2 | Tumor-Specific Antigens

Antigen	Cancer	Epitope[a]
BCR-Abl fusion protein	Chronic myeloid leukemia	SS**KALQRPV** GFKQSS**KAL** ATGFKQSS**KALQRPVAS**
Elongation factor 2	Lung squamous cell carcinoma	ETVSE**Q**SNV
Myosin class I	Melanoma	**KINKNPKYK**
p53	Squamous cell carcinoma	VVP**C**EPPEV
K-ras	Pancreatic adenocarcinoma	VVVGA**V**GVG
SIRT2	Melanoma	KIFSEVT**L**K
Triosephosphate isomerase	Melanoma	GELIG**I**LNAAKVPAD

Source: Cancer Antigenic Peptide Database, https://caped.icp.ucl.ac.be/

[a] Red amino acids denote mutated amino acids in the epitope compared to the wild-type protein.

18.5 How can the immune system be manipulated to promote tumor destruction?

LEARNING OBJECTIVE

18.5.1 Provide examples of immunological tools used to recognize and destroy cancer cells and explain how each carries out these functions.

The traditional therapy used for cancer treatment is "slash, burn, and poison"—surgery, radiation, and chemotherapy. A major complication in cancer treatment is the fact that, although cancer is driven by DNA mutation, the mutated cells are often very similar to normal host cells. Surgery, radiation therapy, and chemotherapy are used to remove tumor cells, damaging the cellular DNA beyond repair and preventing cell division by blocking processes within the cell cycle.

We have seen that the immune system is capable of immunosurveillance of tumor cells and can be efficient in targeting these cells. We have also seen that the immune system can be manipulated to specifically target tumor cells (see Emerging Science in Chapter 2). In this section, we will discuss advances in the field of immunology that take advantage of immunosurveillance of tumor cells or use immune system effector mechanisms to target tumor cells for destruction.

Vaccination

In Chapter 11, we explored the role of vaccination in priming the adaptive immune response against foreign pathogens. The activation of a primary adaptive immune response to innocuous material produces memory cells capable of targeting pathogens upon a subsequent encounter. Since the adaptive immune response can be trained to combat specific pathogens, might it also be used to target and destroy tumor cells since these cells have differences due to mutation? Immunologists are exploring vaccination as a means to amplify adaptive immune cells with effector mechanisms aimed at tumor cells and to produce memory cells capable of preventing tumor progression before it begins.

We have seen that some pathogens are capable of inducing cancer. One such pathogen is human papillomavirus. HPV infection is strongly correlated with cervical cancer and genital warts, in part due to the expression of oncogenes E6 and E7 by the viral genome. A recently developed HPV vaccine has shown protection against not only viral infection but also the cervical lesions that can develop into cancer. The Centers for Disease Control and Prevention now recommends that children aged 11 to 12, young women through the age of 26, and young men through the age of 21 be vaccinated against HPV to prevent viral infection and lower the incidence of cervical cancer (**FIGURE 18.11**). Because HPV is sexually transmitted, the vaccination recommendation has met with some controversy (see **CONTROVERSIAL TOPICS**).

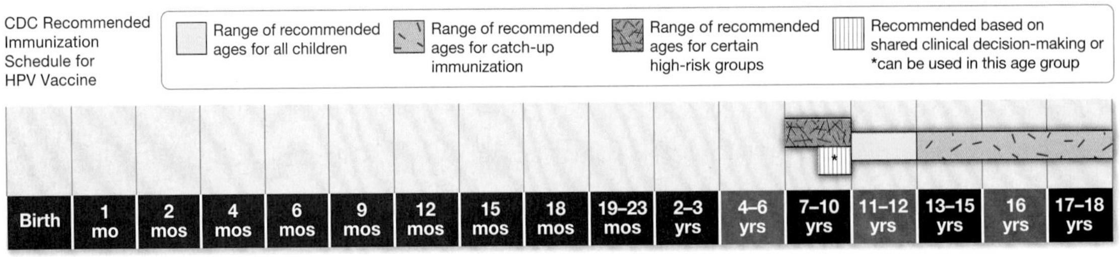

FIGURE 18.11 Centers for Disease Control and Prevention (CDC) recommended adolescent immunization schedule for the human papillomavirus (HPV) vaccine, 2020 (Centers for Disease Control and Prevention. 2021. *Recommended Child and Adolescent Immunization Schedule for ages 18 years or younger, United States, 2021.* 2/11/2021 update. U.S. Dept. of Health and Human Services. https://www.cdc.gov/vaccines/schedules/downloads/child/0-18yrs-child-combined-schedule.pdf.)

CONTROVERSIAL TOPICS

Should preteens receive the HPV vaccine?

We have seen in this chapter that human papillomavirus (HPV) infection is connected to the onset of various forms of cancer, mainly cervical cancer, due to its capacity to express oncogenes that halt the action of tumor suppressor gene products. Because HPV is regarded as a causative agent of cervical cancer, the Centers for Disease Control and Prevention (CDC) and the World Health Organization have recommended HPV vaccination in preteen children to prevent the cancers correlated with HPV infection. The CDC estimates that HPV vaccination may prevent up to 70% of cervical cancers, highlighting the importance of this strategy in preventing not only a sexually transmitted infection (STI) but also a potentially deadly disease.

Since 2006, when it was first commercially available, the HPV vaccine has been included in routine vaccinations in 71 countries and has been shown to be safe and effective. However, as discussed in the Controversial Topics box in Chapter 11, vaccination is often a hot-button topic, and the HPV vaccine has not escaped controversy. Much of the debate stems from the fact that HPV is sexually transmitted. Certain religious and conservative groups have taken the stance that the vaccination may lead to early sexual activity in preteens and teens because it would provide a false sense of security against STIs. These groups, along with some parent organizations, oppose mandatory HPV vaccination prior to entering the public school system, arguing that HPV is transmissible only through sexual activity and not through daily contact with other students.

The American Academy of Pediatrics opposes the viewpoint that sexual activity among teens will increase due to administration of the HPV vaccine. In 2014, the American Cancer Society and the CDC established the National HPV Vaccination Roundtable, which represents more than 70 public, private, and voluntary organizations, including the Academic Pediatric Association, the American Academy of Family Physicians, and the National Cancer Institute. The Roundtable's goal is to promote vaccination to reduce illness and death caused by HPV-associated cancers.

Thus, although HPV has been shown to be strongly associated with cervical cancer, and the vaccine has been tested for both safety and efficacy, the stigma associated with discussing teen sexual activity and STIs has made the HPV vaccine a more controversial medical topic in contrast to other life-saving vaccines with no connection to sexual activity.

Think About...

1. Do you support mandatory HPV vaccination of preteens entering the public school system? Why or why not? What data from the opposing argument might make you reconsider your stance on this issue?
2. What are some strategies for discussing HPV infection and vaccination with preteen children that would separate it from the discussion of sexual activity?
3. Would you predict that an increase in HPV vaccination would correlate with an increase in sexual activity? Would you predict that it would correlate with an increase or decrease in the use of contraception? Explain your answers.

Other strategies have been aimed at inducing an adaptive immune response specifically to tumor cells (**FIGURE 18.12**). These strategies include:

- Whole tumor cell vaccines, in which autologous or allogeneic tumor cells are injected as a means of antigen presentation
- Peptide/protein-based vaccines aimed at activating adaptive immune responses toward tumor-specific or tumor-associated antigens
- Use of mRNA vaccines (the same strategy used for vaccination against SARS-CoV-2) to deliver and express tumor-associated or tumor-specific antigens
- Use of dendritic cells loaded with tumor-specific or tumor-associated epitopes.

While all of these strategies can, in theory, drive an adaptive immune response to tumor cells, the development of such vaccines has met with many technical challenges. Currently only two cancer vaccines have been approved by the Food and Drug Administration (FDA): sipuleucil-T, which targets metastatic prostate cancer, and talimogene laherparepvec, which targets metastatic melanoma.

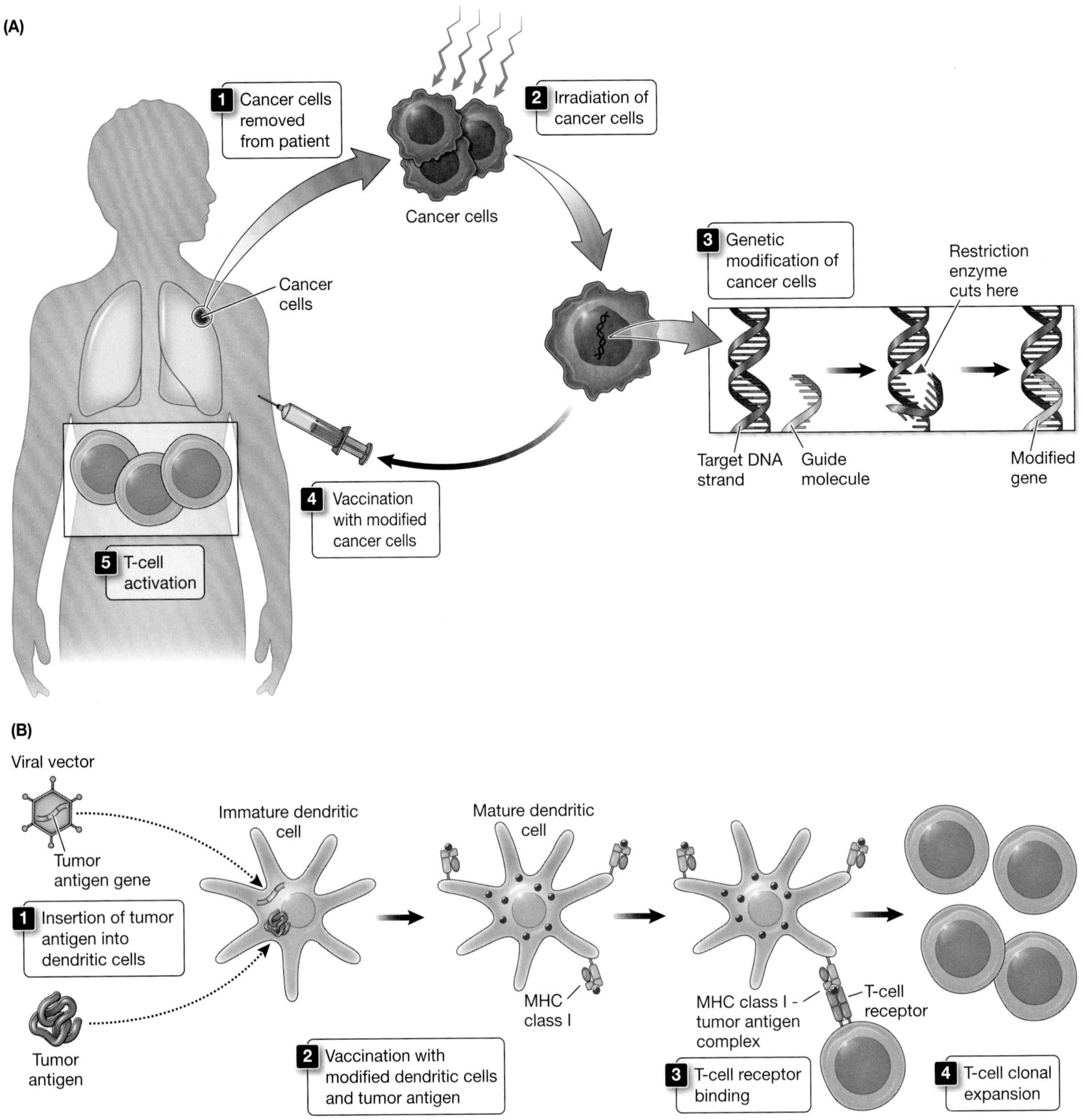

FIGURE 18.12 Cancer vaccine strategies (A) One strategy involves removal of cancer cells, modifying those cells to be antigenic and no longer cancer-causing, and reinjecting the modified cells into the patient. The process begins with removal of cancer cells from the patient (Step 1). The cells are irradiated to destroy their ability to cause cancer (Step 2). The cells' DNA is modified to aid in tumor antigen presentation (Step 3). The modified cells are injected into the patient (Step 4), where they may activate tumor antigen-specific T cells (Step 5). (B) Another strategy is introduction of tumor antigens into dendritic cells. Introduction is achieved by a viral vector containing the gene for a tumor antigen or direct insertion of tumor antigen into immature dendritic cells (Step 1). The dendritic cells mature to process and present tumor antigens and are injected into a patient (Step 2). Mature dendritic cells in lymphoid tissue may bind to tumor antigen-specific T cells (Step 3) before they undergo clonal expansion and differentiation (Step 4). (A, based in part on LUNGevity.org. 2021. *What is a Therapeutic Cancer Vaccine?* Updated March 11, 2021. LUNGevity Foundation, Chicago, IL. https://lungevity.org/for-patients-caregivers/lung-cancer-101/treatment-options/immunotherapy. DNA editing graphic based on M. Corones. 2015. Data Dive blog, April 27, 2015. Graphic by W. Foo, Reuters, http://blogs.reuters.com/data-dive/2015/04/27/how-find-and-replace-for-human-dna-works/; B, after T. Ireland. 2018. Science Focus blog, Dec. 13, 2018. Home of *BBC Science Focus Magazine*. https://www.sciencefocus.com/the-human-body/is-a-cancer-vaccine-on-the-horizon/)

Increase in costimulatory signals

Many tumor cells are not efficiently targeted by the adaptive immune system because they do not express high levels of the costimulatory molecule B7 on their surface. The low level of B7 can bind to CD28 or CTLA4. Because CTLA4 can bind to B7 with much higher affinity, it tends to compete for any B7 present on the surface of tumor cells, resulting in downregulation of T cells that can recognize tumor cells.

This problem can be counteracted by using antibodies that recognize CTLA4 on the surface of T cells to block the binding of CTLA4 with B7 on tumor cells. This enhances the probability of B7 interacting with T-cell CD28 to provide the proper costimulatory signal to the T cell that can recognize and target the tumor cells. Anti-CTLA4 antibodies have been approved for use in treating melanoma. This therapy may also be helpful in the treatment of breast cancer, which has been shown to have high levels of CTLA4.

Cytokines

We have seen many instances in which cytokines activate cells of the innate and adaptive immune systems. These cytokines promote cellular division and differentiation into effector cells of the immune response to fight infection. Thus, it stands to reason that cytokine therapy might be a useful strategy for activating immune cells to promote targeting and destruction of cancer cells.

One cytokine that can promote activation of the adaptive immune response is IL-2, which is responsible for activating T cells during clonal expansion and differentiation. IL-2, at high doses, has been shown to have therapeutic effects against cancer, likely due to the activation of CD8 T cells that recognize tumor-specific antigens. Activation of these T cells promotes differentiation of cytotoxic T cells that recognize cancer cells, prompting their pro-apoptotic activity toward tumor cells. IL-2 treatment also suppresses production of the immunosuppressive receptor programmed cell death protein 1 (PD-1), which shuts down T-cell activation by binding to its ligand PD-L1. Interestingly, PD-L1 has been shown to be expressed at higher levels in certain cancers, suggesting that this is one mechanism used by cancer cells to evade the immune response by driving immunosuppression through the action of PD-1. IL-2 treatment would counteract this strategy by limiting the amount of PD-1 expressed by immune cells. One complication of IL-2 treatment is its ability to also activate regulatory T cells (T_{regs}), which work to suppress the immune response (see Chapter 7).

The interferon family of cytokines is also promising in cancer therapy (interferon-α was the first cancer immunotherapy approved by the FDA [in 1986]). Recall from Chapter 2 that type I interferons are important activators of antiviral responses and subsequent activation of NK cells. The type I interferon IFN-α can also increase the efficiency of antigen processing and presentation of dendritic cells by increasing MHC class II expression and increasing the presence of the costimulatory molecule B7 at the surface of dendritic cells. These dendritic cells can then process and present tumor-specific antigens to CD4 and CD8 T cells. Another interferon, IFN-γ, can activate CD8 T cells and NK cells, making it promising for cancer treatment.

Granulocyte–macrophage colony-stimulating factor (GM-CSF) is another cytokine with antitumor activity. It promotes proliferation and differentiation of cells of the myeloid lineage of hematopoiesis, including dendritic cells and neutrophils. Production and differentiation of these dendritic cells promotes antigen processing and presentation of tumor-specific antigens and activation of adaptive immune responses against tumor cells. Because GM-CSF facilitates proliferation of neutrophils, it has also been used as a treatment for neutropenia (low neutrophil count) caused by chemotherapy.

Making Connections: Targeting the Immune Response to Combat Cancer

While cancer cells might gain mutations that make them less susceptible to targeting by the immune system, the alterations that occur make them susceptible to treatment that employs an immune response. The immune response can be activated through expression of tumor-specific antigens, by changes in cell-surface molecules, through vaccination, or through adoptive transfer of altered T cells.

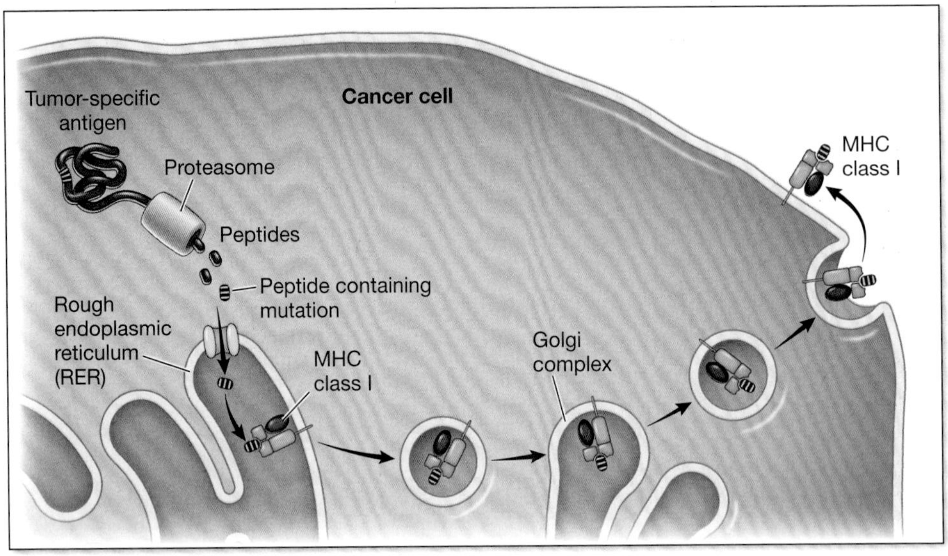

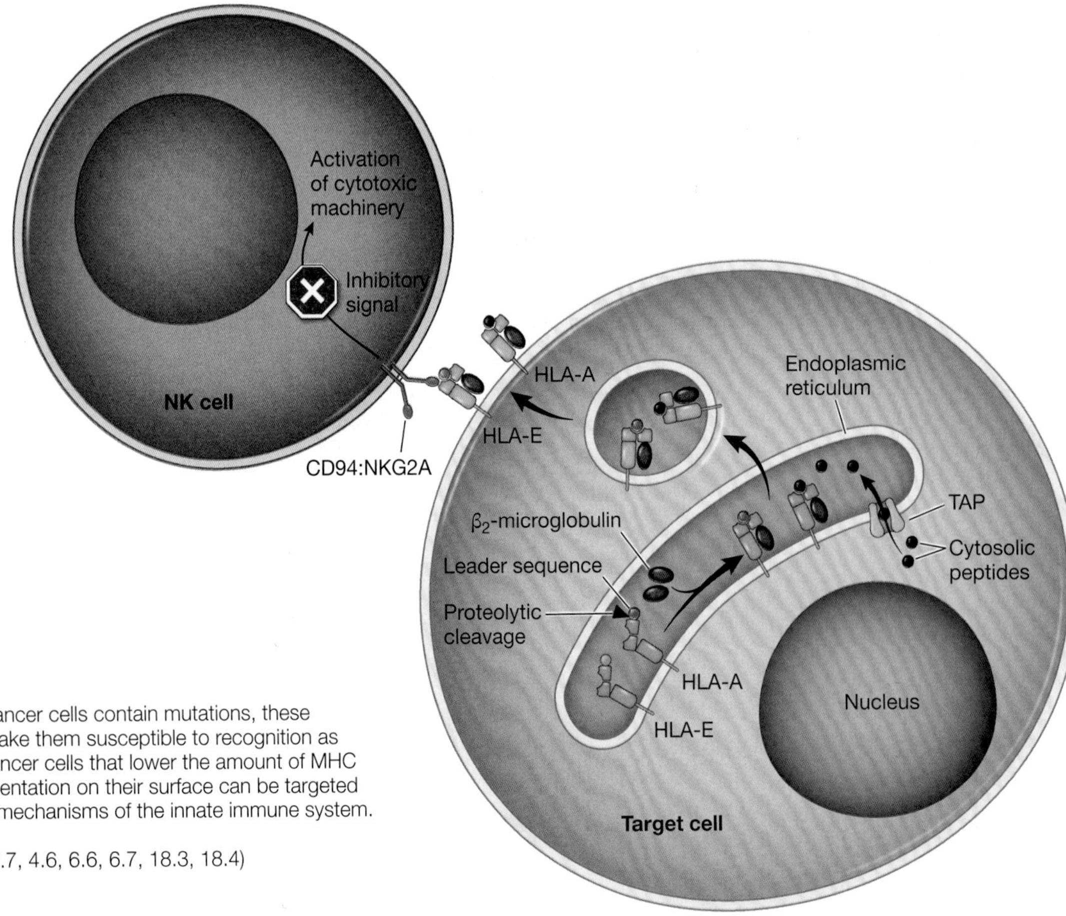

After M. Iwaszko et al. 2011. *Arch Immunol Ther Exp* 59: 353–367

Because cancer cells contain mutations, these changes make them susceptible to recognition as nonself. Cancer cells that lower the amount of MHC class I presentation on their surface can be targeted by various mechanisms of the innate immune system.

(Sections 2.7, 4.6, 6.6, 6.7, 18.3, 18.4)

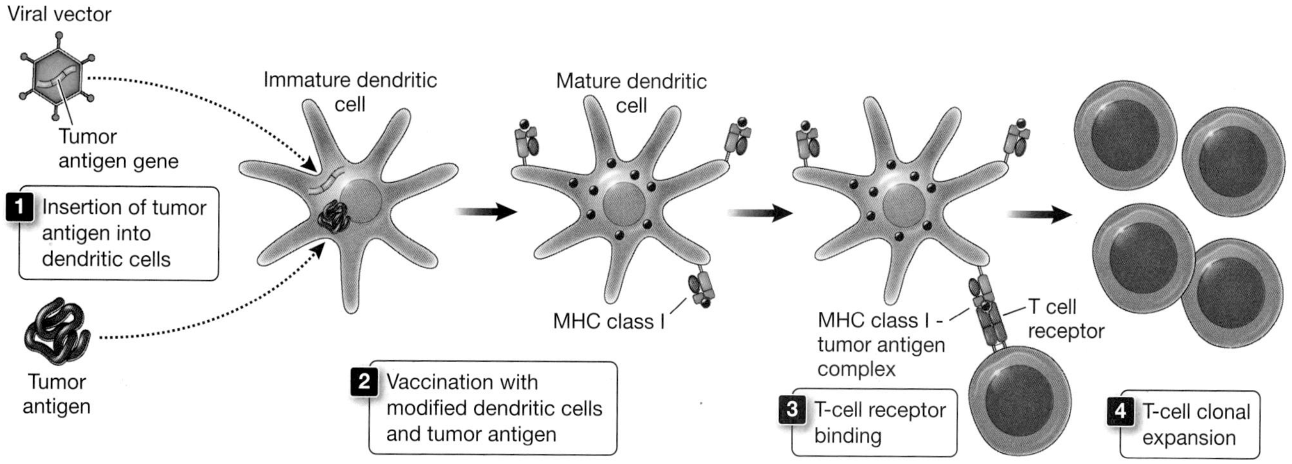

After T. Ireland. 2018. Science Focus blog, Dec. 13, 2018. Home of BBC *Science Focus Magazine*. https://www.sciencefocus.com/the-human-body/is-a-cancer-vaccine-on-the-horizon/

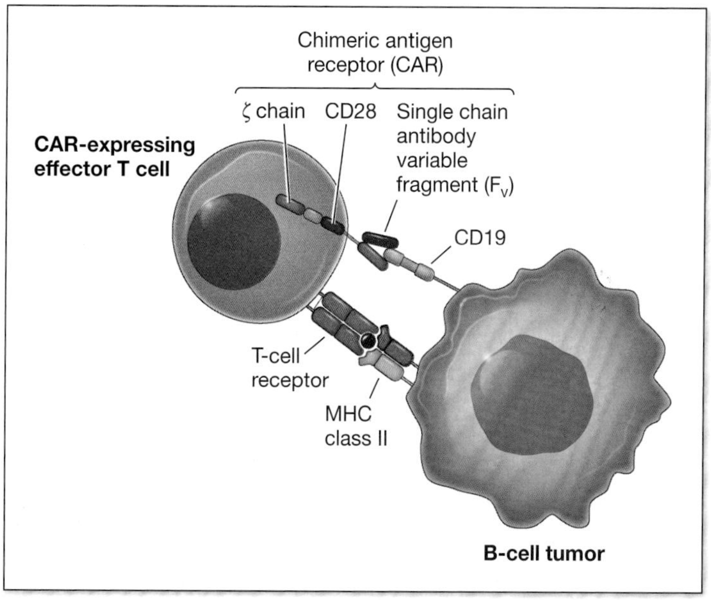

Vaccination and adoptive transfer using gene therapy are both attractive models for the treatment of cancer. Adoptive transfer has been successfully employed against B-cell leukemia using chimeric antigen receptor (CAR) T therapy.

(Sections 7.3, 7.7, 7.13, 11.3, 11.4, 18.4, 18.5)

Another cytokine with antitumor activity is IL-12, which activates production of IFN-γ in CD8 T cells and NK cells. As explained earlier in this section, IFN-γ increases the cytotoxic activity of CD8 T cells and NK cells, both of which recognize intracellular infections and mutated cells such as cancer cells. While this cytokine might seem to have potential for being able to activate cells that recognize and destroy cancer cells, it has shown limited effectiveness against many cancers and can cause systemic toxicity.

Monoclonal antibodies

Monoclonal antibodies can be used in the diagnosis and treatment of cancer. The high binding specificity of monoclonal antibodies, along with their capacity to invoke effector function of the humoral arm of adaptive immunity, makes them useful in a variety of applications.

For cancer diagnosis, a monoclonal antibody that can bind to a tumor-specific or tumor-associated antigen can be fused with a tracer molecule such as radioactive iodine and used to detect the presence of a tumor that expresses the tumor-specific antigen.

Monoclonal antibodies specific to tumor-specific or tumor-associated antigens can also be used to treat tumor cells via antibody-dependent cell-mediated cytotoxicity (see Chapter 10), whereby NK cells can recognize cells marked with immunoglobulin through their Fc receptor and target the immunoglobulin-bound cell for apoptosis. Several monoclonal antibodies have been approved for cancer treatment using this strategy (**TABLE 18.3**). One example is the use of anti-HER2 antibodies for the treatment of breast cancer. (HER2 is a tumor-associated antigen in many breast cancers.)

Another approved monoclonal antibody therapy is the artificial bispecific monoclonal antibody known as a bispecific T-cell engager. These antibodies can bind to a tumor-specific or tumor-associated antigen with one antigen binding site and the CD3 receptor of T cells, bringing a tumor cell in close proximity to a T cell. The cell–cell interaction allows a cytotoxic T cell engaged with the tumor cell via the bispecific antibody to drive apoptosis of the tumor cell. One example that has been approved for cancer treatment is a bispecific antibody that binds both T cells and CD19 on the surface of B cells for the treatment of B-cell cancers. (CD19 is a tumor-associated antigen on B cells.)

Another family of immunoglobulins used in cancer treatment works to prevent suppression of the adaptive immune response. This therapy, known as checkpoint therapy, uses immunoglobulins that are referred to as immune checkpoint inhibitors because they block the action of key checkpoint proteins that shut down the adaptive immune response. Recall from earlier in this chapter that both CTLA4 and PD-1 can suppress T-cell activation. Furthermore,

TABLE 18.3 | Monoclonal Antibodies Used in Cancer Treatment

Antibody	Antigen	Cancer Treated	Conjugate to Antibody
Alemtuzumab	CD52	Chronic lymphocytic leukemia	N/A
Trastuzumab	HER2	Breast cancer	N/A
Rituximab	CD20	Non-Hodgkin lymphoma	N/A
Ibritumomab tiuxetan	CD20	Non-Hodgkin lymphoma	Yttrium-90
Brentuximab vedotin	CD30	Hodgkin lymphoma	MMAE
Ado-trastuzumab emtansine	HER2	Breast cancer	DM1
Blinatumomab	CD19	Acute lymphoblastic leukemia	N/A

Source: M. S. Castelli et al. 2019. *Pharmacol Res Perspect* 7: e00535.

cancer cells sometimes increase the amount of CTLA4 they produce, or they produce the ligand of PD-1. Thus, an immunoglobulin to CTLA4 or PD-1 prevents these two proteins from functioning as an immune checkpoint protein by blocking the interaction of the proteins with their ligands. Immune checkpoint inhibitors function analogously to neutralizing antibodies, as they prevent the interaction of proteins required for a subsequent response.

Other monoclonal antibodies have been conjugated with cytotoxic drugs or radioactive elements to target chemotherapy or radiation treatment to cancer cells. The binding of the monoclonal antibodies to their tumor-specific antigen concentrates the conjugated cytotoxic drug or radioactive element at the tumor to enhance the effects of the drug or radioactivity.

Gene therapy

Recently, the use of modified T cells through gene therapy in a process called CAR-T (for *chimeric antigen receptor T cell*) therapy has been explored as a means to train extracted T cells to recognize certain types of leukemia. CAR-T therapy employs a gene that encodes a transmembrane protein on the surface of the modified T cells. The transmembrane protein contains an extracellular domain that specifically binds to a target cancer cell and an intracellular domain that aids in T-cell activation. In the drug Kymriah™, which was approved for the treatment of acute lymphoblastic leukemia, extracted T cells are infected with a retrovirus that contains DNA sequence for the production of a chimeric T-cell receptor that can recognize the B-cell surface protein CD19 (see **EMERGING SCIENCE**). To promote activation through this chimeric receptor, it also contains the intracellular region of CD28 (the signaling molecule that acts as a costimulatory signal for T-cell activation) and the ζ chain of CD3. The chimeric T cells recognize the leukemic cells using a receptor that recognizes an MHC-peptide complex on the surface of the B cell, along with the chimeric receptor binding to CD19 on the surface of the B cell. This interaction prompts activation of the chimeric T cell and allows it to target the B cell for destruction (**FIGURE 18.13**). While this gene therapy technique is still in its infancy, it is likely that it will be further developed and used to treat other forms of cancer.

CHECKPOINT QUESTION

1. Explain how monoclonal antibodies can be employed in cancer treatment.

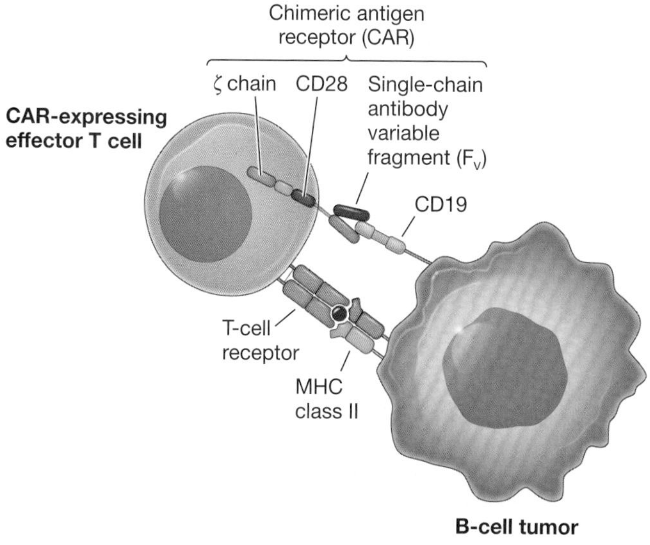

FIGURE 18.13 Treatment of B-cell tumors with modified T cells expressing a chimeric antigen receptor (CAR) A modified T cell engineered to produce a CAR containing a single-chain antibody variable fragment that binds to CD19 on the surface of a B cell can target B cells in an adaptive immune response. Binding of the CAR to CD19 promotes signaling through the CAR's cytoplasmic domain, which contains the cytoplasmic domain of the costimulatory molecule CD28 and the ζ chain of the CD3 complex. This interaction can act as a surrogate costimulatory signal to activate a T cell bound to an MHC-peptide complex on the B-cell tumor.

EMERGING SCIENCE

Can gene therapy be used to train the immune system to target cancer cells?

Article

Kalos, M., B. L. Levine, D. L. Porter, S. Katz, S. A. Grupp, A. Bagg, and C. H. June. 2011. T cells with chimeric antigen receptors have potent antitumor effects and can establish memory in patients with advanced leukemia. *Science Translational Medicine* 3: 95ra73.

Background

While we have seen that the immune system may be capable of targeting and destroying tumor cells, the response is limited to nonself tumor-specific antigens because of central tolerance driven during T-cell development. The researchers speculated that T cells genetically modified to express chimeric antigen receptors (CARs) composed of an antibody-binding domain with an intracellular domain capable of T-cell activation would aid in targeting tumor cells in an MHC-independent fashion.

The Study

Researchers genetically modified T cells to create CART19 cells, which express an anti-CD19 binding domain (an antigen restricted to B cells) with both the CD3 ζ chain and another costimulatory domain known as 4-1BB. They infused these genetically modified cells into three patients to test their antitumor effects and capacity to grow in vivo. The researchers were able to show that these cells were capable of expansion and persistence for 6 months in vivo (**FIGURE ES 18.1**).

Importantly, they showed that treatment of two patients with CART19 cell infusion was able to eliminate the clonal population in these patients, demonstrating that the genetically modified T cells were capable of eliminating the leukemic B-cell population. These studies were paramount in demonstrating the development of genetically modified T-cell populations capable of mounting specific adaptive immune responses to leukemia cells. The significant findings were instrumental in the development of the Food and Drug Administration (FDA)-approved therapy Kymriah™.

Think About...

1. Why do you think the researchers used an antibody's antigen-binding site to CD19 as the extracellular domain of the CAR rather than a T-cell receptor capable of binding to CD19 peptide-MHC complexes?
2. What steps would you need to take to generate a chimeric antigen receptor? What information would you need to generate this receptor, and what tools would you have to employ to construct this receptor?
3. How might this therapy be expanded to treat other cancers? What knowledge would be needed to employ this new therapy?

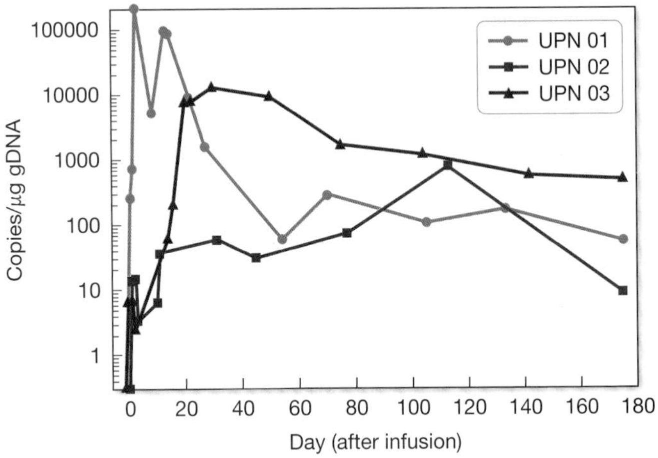

FIGURE ES 18.1 **Persistence of CART19 cells in the circulation and their ability to target B-cell leukemia cells** The amount of CAR19 was quantified in the blood of three patients (UPN 01, UPN 02, UPN 03) using quantitative polymerase chain reaction (qPCR). (M. Kalos et al., 2011. *Sci Translation Med* 3: 95ra73.)

Characterization of Tumors in Leukemia Patients Treated with CART19 Therapy

Patient (UPN)	Tissue	Time point	Chronic lymphocytic leukemia heavy chain clone reads	Clone frequency (% productives)
01	PB	Pre-infusion	184,256	99.713
01	PB	Day −1	407,592	99.758
01	PB	Day +28	0	0
01	PB	Day +176	0	0
01	BM	Day +28	0	0
01	BM	Day +176	0	0
03	PB	Pre-infusion	19,948,508	90.367
03	PB	Day −1	1,231,018	88.860
03	PB	Day +31	0	0
03	PB	Day +176	0	0
03	BM	Day +31	0	0
03	BM	Day +176	0	0

Source: M. Kalos et al., 2011. *Sci Transl Med* 3: 95ra73.
PB, peripheral blood; BM, bone marrow

Summary

18.1 What causes cancer?

- Cancer is driven by DNA mutation of specific genes that alter a cell's capacity to regulate growth and division.
- In order for DNA mutation to cause cancer, it must promote uncontrolled cell growth, prevent the action of inhibitory growth signals, promote the evasion of apoptosis, provide a continued supply of nutrients, promote limitless division, and drive tissue invasion and spread.
- While cancer is driven by DNA mutation, both genetic and environmental factors can promote its onset.

18.2 What is the importance of proto-oncogenes and tumor suppressor genes in cancer development?

- Proto-oncogenes produce products responsible for inducing cell division, and tumor suppressor genes produce products responsible for controlling cell division. Mutation of proto-oncogenes and tumor suppressor genes is the primary cause of the onset of cancer.

18.3 How do cancer cells evade destruction by the immune system?

- Cancer cells avoid destruction by immune system effector mechanisms by avoiding detection or manipulating the immune response that would normally target these cells.

18.4 How does the immune system destroy cancer cells?

- The innate immune system can recognize cancer cells in the same way it recognizes virus-infected cells that have altered antigen presentation at the surface.
- The adaptive immune system can recognize cancer cells by recognizing tumor-associated or tumor-specific antigens expressed by these cells.

18.5 How can the immune system be manipulated to promote tissue destruction?

- The field of immunology has provided novel means to detect and treat cancer using strategies that include vaccination against tumor cells, increasing the costimulatory signal provided to T cells that recognize tumor cells, using cytokine therapy to activate cells of the innate and adaptive immune systems, employing monoclonal antibodies to detect and specifically target tumor cells, and gene therapy to use modified T cells to target and destroy tumor cells.

Review Questions

18.1 You work for the Environmental Protection Agency and have been conducting field tests in a community adjacent to a chemical plant due to increased incidence of cancer within this community. You have isolated high levels of a chemical byproduct in water samples taken from the site. What experiments would you do to determine if the chemical byproduct is a mutagen that could cause the increased occurrence of cancer?

18.2 In your research lab, you are working with a cell line to identify novel genes involved in the cell cycle. You randomly mutate your cell line and isolate cells that are capable of dividing in the absence of growth factor, suggesting that the newly mutated cell line has lost the requirement of an external signal to undergo cell division. You map the mutation to a gene whose function had been unknown. What experiment(s) might you do to determine if the mutated gene is a proto-oncogene or a tumor suppressor gene?

18.3 Would you predict that treatment with inflammatory cytokines would increase or decrease the likelihood of cancer targeting by the immune system? Explain.

18.4 Why are adaptive immune responses to tumor cells typically driven by CD8 T-cell responses?

18.5 Most monoclonal antibodies used in cancer treatment recognize cell-surface molecules. Explain why.

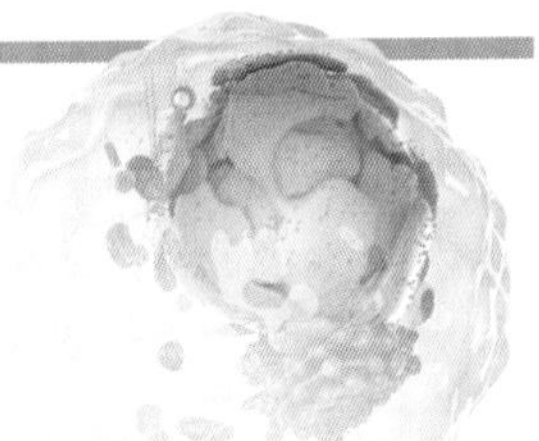

CASE STUDY REVISITED

Acute Lymphoblastic Leukemia

An 8-year-old boy complaining of frequent nausea and fatigue is brought to his pediatrician. With the patient's medical history, the nausea is not likely due to food poisoning because the boy's parents ate the same foods and did not experience nausea. The pediatrician runs tests to determine if common mutations in genes associated with Crohn's disease are present and also orders a colonoscopy. After initial testing, the pediatrician rules out gastrointestinal issues and takes a blood sample for further tests. The boy's white blood cell (WBC) count is extremely high (60,000 WBC/mL) even though there is no sign of infection. Genotyping by fluorescence in situ hybridization reveals translocations of multiple chromosomes, indicative of pre-B-cell acute lymphoblastic leukemia. The patient is referred to an oncologist, who suggests induction chemotherapy, starting with cytarabine, for cancer treatment.

Think About...

Questions for individuals

1. Look up the chemical structure of cytarabine. What would you propose as its mode of action as chemotherapy?
2. What stage of B-cell development is likely altered in the patient's acute lymphoblastic leukemia?
3. What are some common translocations that occur in acute lymphoblastic leukemia?

Questions for student groups

1. Would bone marrow irradiation and transplant be warranted in this case? Why or why not? Why wasn't surgery suggested as a course of action?
2. Given what you discovered about cytarabine, would you predict that other cell types might be affected by this drug?
3. Would the monoclonal antibody rituximab be useful as a therapeutic in this case? What would you need to know about the leukemia cells to make this determination?

MAKING CONNECTIONS Key Concepts	COVERAGE (Section Numbers)
Evasion of the innate immune system by cancer cells	2.7, 7.13, 13.3, 13.4, 18.2, 18.3
Evasion of the adaptive immune system by cancer cells	4.6, 4.7, 6.6, 7.13, 13.3, 13.4, 18.3
Targeting cancer cells that have altered antigen processing and presentation	2.7, 4.6, 6.6, 6.7, 18.3, 18.4
Manipulation of the adaptive immune response to target cancer cells	7.3, 7.7, 7.13, 11.3, 11.4, 18.4, 18.5

Glossary

Numbers in brackets indicate the chapter in which the term is defined.

A

abasic site Site in DNA that does not contain a base connected to the sugar-phosphate backbone. [9]

acquired immunodeficiency Immunodeficiency caused by a pathogen that has altered or blocked an innate or adaptive immune response, preventing normal immune system function. [14]

activation-induced cytidine deaminase (AID) Enzyme within activated B cells that catalyzes the conversion of cytosine to uracil within specific sites in the immunoglobulin genes, driving somatic hypermutation and class switch recombination. [9]

acute rejection Rejection of a transplanted organ due to recognition of donor MHC–peptide complexes presented by donor dendritic cells to recipient T cells and the ensuing effector mechanisms upon activation of those T cells. [17]

adaptive immunity Mechanisms used by the immune system that target specific products of pathogens through the use of T cells and B cells. [1]

adenosine deaminase deficiency Immunodeficiency caused by the lack of the enzyme adenosine deaminase, which is required for the salvage pathways of purine nucleotides. [14]

adhesion molecule Receptor on the surfaces of two interacting cells that aids in cellular contact. [2]

adjuvant Component of a vaccine used to enhance antigen-specific immune responses; important in stimulating the inflammatory response required for a complete adaptive immune response. [11]

affinity The strength of an interaction between an immunoglobulin antigen-binding site and its epitope. [9]

affinity maturation Process by which immunoglobulin genes gain higher affinity for antigen through mutagenesis of the immunoglobulin gene during B-cell activation. [4, 9]

agonist Molecule that can bind to a receptor and mimic a ligand, driving receptor activation. [16]

agranulocyte White blood cell that is devoid of granules. Refers to lymphocytes (B cells, T cells, NK cells, innate lymphoid cells) and monocytes (macrophages, dendritic cells). [1]

allele Alternative form of a gene located at the same location on a chromosome. [4]

allelic exclusion Inactivation of somatic recombination and active chromatin remodeling in order to prevent further recombination of specific loci of adaptive immune receptors. [5]

allergen Innocuous foreign molecule that can elicit a hypersensitivity response. Molecule with the ability to induce type I hypersensitivity when recognized by IgE immunoglobulins in complex with granulocytes. [1, 15]

allergy Hypersensitivity reaction to an innocuous foreign antigen. Type I hypersensitivity reaction driven by the activation of granulocytes that recognize an allergen via associated IgE antibodies and induce signaling and degranulation through their Fcε receptor. [1, 15]

alloantigens Antigens that differ between individuals of the same species. [17]

allogeneic Molecules from a source within the same species that are genetically different, as from a different individual. [6]

allograft Tissue transplantation between genetically distinct individuals. [17]

alloreaction Immune system reaction in response to an allograft transplant. [17]

allotype Protein product of a particular allele of a gene. [6]

alternative C3 convertase C3 convertase produced when the opsonin C3b binds to factor B and bound factor B is cleaved by factor D to produce C3bBb at the pathogen surface. [3]

alternative C5 convertase C5 convertase produced when C3b binds to the alternative C3 convertase C3bBb; responsible for cleaving C5 into C5a and C5b. [3]

alternative pathway First pathway of complement activation to be activated that relies on complement system proteins always present in the plasma for the activation of C3. [3]

alternative splicing Maintenance or removal of specific RNA sequences that can promote the formation of different proteins with different properties or locations within cells. [9]

anaphylatoxin Molecule capable of activating an inflammatory response by triggering degranulation of cells capable of inducing inflammation. [3]

anchor residues Residues of a peptide bound to an MHC molecule that are essential in interacting with the peptide-binding groove; typically need to have specific properties in all peptides that can bind to that particular MHC molecule. [6]

anergy Cellular state of nonresponsiveness in which the lymphocyte is incapable of expansion and differentiation. [7, 8]

angiogenesis Growth of new blood vessels. [18]

antagonist Molecule that can bind to a receptor and prevent receptor function. [16]

antibodies Soluble immunoglobulins synthesized by B cells to bind to antigens; act to neutralize pathogens and toxins, aid in complement activation, and aid as a tag to promote phagocytosis. [1]

antibody-dependent cell-mediated cytotoxicity NK cell-directed cytotoxic effector function toward cells that have been bound by antibodies on their cell surface. [10]

antigen Molecule recognized by a T-cell receptor, immunoglobulin, or antibody. [1]

antigen-presenting cells Phagocytic cells of the innate immune system that process engulfed materials and present them to T cells and B cells. [1]

antigenic drift Random mutation of a pathogen that changes epitopes of a previous adaptive immune response and renders lymphocytes from that immune response nonreactive to the changed pathogen. [13]

antigenic shift Recombination event of closely related pathogens that promote the formation of an altered pathogen capable of escape from any prior adaptive immune response raised due to lack of specificity to the newly formed pathogen. [13]

antigenic variation Process of changing a recognition target of the immune system through gene conversion or rearrangement. [13]

apoptosis Programmed cell death; a mechanism utilized to destroy a cell without releasing intracellular contents into the extracellular environment. Results in the formation of cellular debris removed via phagocytic cells such as macrophages. [1, 7]

atopic Having a predisposition to mount a type I hypersensitivity reaction to common allergens. [15]

autoantibody Antibody capable of recognizing a self-antigen and that is responsible for the induction of autoimmunity in certain diseases. [16]

autoantigens Self-antigens recognized as foreign during an autoimmune response. [15]

autocrine Signaling mechanism whereby the same cell producing a signaling molecule responds to that signaling molecule. [7]

autograft Transplantation of tissue from one area on an individual to a different area on the same individual. [17]

autoimmune polyendocrinopathy-candidiasis-ectodermal dystrophy (APECED) Autoimmune disease caused by a lack of the AIRE transcription factor, resulting in a lower amount of negative selection of tissue-specific self-reactive T cells and thus an increase in tissue-specific autoimmunity. Also known as *autoimmune polyendocrine syndrome type 1* (*APS-1*). [16]

autoimmunity Adaptive immune response raised against a self-antigen due to lack of central or peripheral tolerance. [16]

autologous Self-molecules within an individual that are not recognized by lymphocytes due to immune tolerance. [6]

avidity The overall strength of binding of a protein complex to an antigen. [9]

B

B cell Lymphocyte that is the primary cell involved in the humoral adaptive immune response; recognizes specific antigens using cell surface immunoglobulins and differentiates into plasma cells or centrocytes when activated. [1]

B-1 B cells B cells that reside in pleural and peritoneal cavities that are responsible for combating common pathogens of these tissues. [8]

B-2 B cells B cells that reside in the circulatory system and secondary lymphoid tissues capable of mounting a typical humoral adaptive immune response. Also known as *follicular B cells*. [8]

BAFF Stands for *B-cell activation factor* or *B-cell survival factor*. Survival factor for B cells secreted by cells such as follicular dendritic cells to promote survival of B cells capable of mounting a humoral adaptive immune response. [8]

balancing selection Selective pressure that promotes heterozygosity of a gene within a population. [6]

bare lymphocyte syndrome (BLS) Immunodeficiency caused by the inability of cells to properly load MHC complexes, thus preventing proper presentation and activation of T cells. [14]

basophil Granulocyte responsible for the clearance of parasites. [1]

benign Denoting a tumor that is noncancerous and localized. [18]

BLIMP-1 B-cell transcription factor responsible for downregulating B-cell division and promoting plasma cell differentiation by inducing immunoglobulin expression. [10]

bone marrow Primary lymphoid tissue that serves as the location of hematopoiesis and the maturation of B cells. [1]

C

C-reactive protein An acute-phase response protein that functions to initiate activation of the classical complement pathway. [3]

C3 convertase Protease produced during activation of complement that is capable of cleaving the complement protein C3 into C3a and C3b. [3]

cadherin A transmembrane calcium-binding protein family that is important in cell-to-cell adhesion interactions. [13]

calnexin Membrane-bound ER chaperone that aids in folding of proteins in the secretory pathway. [6]

calreticulin ER chaperone involved in the folding of proteins in the secretory pathway; functions in the peptide-loading complex to promote proper folding of MHC class I molecules loaded with peptide. [6]

cancer Group of diseases characterized by the uncontrolled growth of abnormal cells. [18]

CD4 helper T cells Activated and differentiated T cells that function to activate either B cells or macrophages. [7]

CD4 T-cell coreceptor involved in selection and binding to MHC class II molecules. [4]

CD40 Cell-surface molecule on B cells or macrophages that must engage with CD40 ligand from a helper T cell to promote proper B-cell activation. [10]

CD40 ligand (CD40L) Ligand expressed by helper T cells that must bind to CD40 on either macrophages or B cells in order to aid in proper macrophage or B-cell activation by the helper T cell. [7, 10]

CD8 T-cell coreceptor involved in selection and binding to MHC class I molecules. [4]

CD8 cytotoxic T cells Activated and differentiated T cells that function to target and destroy cells infected intracellularly. [7]

celiac disease Autoimmune disease caused by the production of an adaptive immune response against the protein gluten (found in wheat, barley, and rye), resulting in targeting of small intestinal epithelial cells and the destruction of microvilli. [16]

central memory T cells (T_{CM}) Memory T cells that reside within the circulatory and lymphatic systems and in secondary lymphoid tissues; activation is analogous to naïve T-cell activation. [11]

central tolerance Negative selection processes that occur in primary lymphoid tissues that are responsible for the removal of self-reactive lymphocytes. [5, 8]

centrocyte Differentiating B cell within germinal centers that can undergo somatic hypermutation and isotype switching and differentiate into either a memory B cell or a plasma cell that can express mutated, isotype-specific immunoglobulins. [10]

Chediak-Higashi syndrome Immunodeficiency that prevents proper phagosome–lysosome fusion, and thus prevents the action of phagocytic cells in the innate immune system. [14]

chemokine Soluble secreted protein that acts as an attractant molecule to promote cell migration to a specific area. [2]

chronic granulomatous disease Immunodeficiency that prevents the proper action of neutrophils in the innate immune system, causing persistent infection and the formation of granulomas due to inefficient clearance of neutrophils by macrophages. [14]

chronic rejection Rejection of a transplanted organ due to immune reactions targeting the vasculature of the transplanted organ, resulting in loss of oxygen and ultimately organ failure. [17]

class switch recombination Alteration of DNA at the immunoglobulin heavy chain locus within B cells, which promotes the incorporation of a different isotype constant region within the expressed immunoglobulin. Also called *isotype switching*. [9]

class-II associated invariant chain peptide (CLIP) Peptide produced upon digestion of the invariant chain that remains bound to MHC class II molecules until exchange with extracellular peptides generated during phagocytosis. [6]

classical C3 convertase C3 convertase produced when C4b binds to C2a (both products of the lectin and classical pathways of complement activation). [3]

classical NK cell deficiency Immunodeficiency that prevents the proper development of natural killer cells. [14]

classical pathway Third pathway of complement activation to be activated that relies on the use of C-reactive protein or an antibody to initiate complement activation. [3]

clonal deletion Removal of lymphocytes that express self-reactive receptors by apoptosis. [8]

clonal expansion Proliferation and differentiation of a T cell or B cell that has engaged its receptor with its antigen. [1]

clonal selection Process by which a specific T cell or B cell recognizes its antigen through action of its receptor. [1]

cluster of differentiation (CD) System used in the nomenclature of different cell-surface molecules in white blood cells. [2]

coding joint Segment of DNA joined together during V(D)J recombination that encodes for the variable region of a T-cell receptor or immunoglobulin. [4]

complement System of soluble plasma proteins that act to opsonize and lyse pathogens. Also known as *complement system*. [2]

complement activation Proteolytic processing of complement proteins to promote function in opsonization and pathogen destruction. [3]

complement component 3 (C3) Protein component of the complement system that is cleaved during complement activation to produce C3a (a chemokine) and C3b (an opsonin). [2, 3]

complement fixation Covalent attachment of a complement protein product to the surface of a pathogen. [3]

complement receptor Cell-surface receptor family that binds to complement proteins fixed on the surface of a pathogen, allowing for recognition and phagocytosis. [3]

complement receptor 1 (CR1) Cell-surface receptor expressed by macrophages that binds to C3b fixed on the surface of a pathogen, allowing for enhanced recognition and phagocytosis. [3]

complement receptor 3 (CR3) Cell-surface receptor expressed by macrophages and neutrophils that binds to iC3b fixed on the surface of a pathogen, allowing for enhanced recognition and phagocytosis. [3]

complement receptor 4 (CR4) Cell-surface receptor expressed by macrophages and neutrophils that binds to iC3b fixed on the surface of a pathogen, allowing for enhanced recognition and phagocytosis. [3]

complement system System of soluble plasma proteins that acts to opsonize and lyse pathogens. Also known as *complement*. [3]

conformational epitope Immunoglobulin epitope that incorporates structural determinants only maintained in the presence of native antigen structure. [9]

conjugate pair Cell-cell interaction of an effector T cell and a macrophage that the effector T cell will activate that is driven by proteins at the surface of both cells. Cellular interaction between a helper T cell via its T-cell receptor and an antigen-presenting cell. [7, 10]

conjugate vaccine Vaccine composed of a weak antigen (typically incapable of mounting a robust adaptive immune response) coupled with a stronger antigen that can activate a protective primary immune response. [11]

constant region Region of a T-cell receptor, immunoglobulin, or antibody where the amino acid sequence is similar among molecules. [4]

costimulatory signal Signal required during activation of lymphocytes in addition to antigen engagement with its receptor. [7]

Crohn's disease Chronic inflammatory bowel disease. [14]

cross-presentation Mechanism of peptide presentation whereby phagocytic cells transport endocytosed proteins into the cytosol for presentation via the MHC class I pathway. [6]

crossmatching Blood test used before blood transfusion or tissue/organ transplantation to check for harmful interactions between donor blood and that of a potential recipient. [15]

cyclophilin A peptidyl-prolyl isomerase that, in complex with cyclosporin A, blocks calcineurin activity and T-cell activation. [17]

cyclosporin A Immunosuppressive drug that blocks T-cell activation by binding to cyclophilins and inhibiting calcineurin activation. [17]

cytokine Soluble protein product secreted by cells to aid the immune system by increasing inflammation, chemotaxis to sites of infection, or immune cell signaling and differentiation. [1, 2]

cytosolic diversion Process by which endocytosed proteins are transported into the cytosol for cross-presentation. [6]

cytosolic innate receptor Intracellular receptor capable of recognizing PAMPs that may be present in the cytosol of cells. [2]

cytotoxic T cell CD8-positive T cell that has been activated by recognition of a specific epitope; targets cells with intracellular infections through the actions of cytokines and cytotoxins. [1]

cytotoxins Protein products produced by immune cells that act to target and kill intracellularly infected cells. [7]

D

decay-accelerating factor (DAF) Cell-surface protein expressed by human cells that inhibits complement activation by inactivating the alternative C3 convertase on the surface of human cells. [3]

defensin Antimicrobial peptide that is capable of disrupting cellular membranes by inserting into them. [1]

delayed-type hypersensitivity Allergic reaction driven by the activation of an intracellular antigen processed and presented to T cells; commonly the effector T cells activated during a delayed-type hypersensitivity reaction are helper T cells that activate macrophages and an inflammatory response through the secretion of IFN-γ. [15]

dendritic cell White blood cell that specializes in phagocytosis of foreign pathogens and presentation of antigen epitopes to the adaptive immune system. [1]

diapedesis Mechanism by which a leukocyte squeezes between two endothelial cells. [2]

direct pathway of allorecognition Pathway of presentation of alloantigens in which donor dendritic cells transplanted with the organ or tissue migrate to a secondary lymphoid tissue and present antigens to recipient T cells. [17]

directional selection Selective pressure that promotes maintenance of a single allele within a population. [6]

DN1 Double-negative 1 thymocytes; thymocytes that have entered the thymus and initiated somatic recombination at the T-cell receptor loci. [5]

DN2 Double-negative 2 thymocytes; thymocytes that have committed to T-cell development. [5]

DN3 Double-negative 3 thymocytes; thymocytes that have finished β-chain rearrangement and are active in β-chain allelic exclusion and proliferation to engage in α-chain rearrangement. [5]

DN4 Double-negative 4 thymocytes; thymocytes that are active in α-chain rearrangement. [5]

double-negative thymocyte Developing thymocyte within the thymus that does not express either T-cell coreceptor (CD4 or CD8). [5]

double-positive thymocyte Developing thymocyte within the thymus that expresses both T-cell coreceptors (CD4 and CD8). [5]

E

early pro-B cells Initial stage in B-cell development that occurs as common lymphoid progenitor cells receive B-cell developmental signals in bone marrow. [8]

effector cell Immune system cell that has been activated and changes its function to specialize in pathogen recognition and destruction. [1]

effector compartment Connective tissue under the epithelial barrier that contains effector lymphocytes and innate immune cells. [12]

effector memory T cells (T_{EM}) Memory T cells that reside within the circulatory and lymphatic systems, secondary lymphoid tissues, and peripheral tissues; major role is monitoring for the presence of antigen in the lymphatic system and peripheral tissues. [11]

endocrine Signaling mechanism whereby cells in one area respond to a signaling molecule produced in a different area; typically these signaling molecules are transported via the bloodstream to the responding cell. [7]

endoplasmic reticulum aminopeptidase (ERAP) Protease located in the ER lumen that trims the N-terminus of peptides bound to MHC class I molecules to bind peptides more tightly to the peptide-binding groove. [6]

endotoxin Component of a pathogenic organism capable of eliciting a disease state. [13]

eosinophil Granulocyte that specializes in clearance of parasitic infections. [1]

epidemic Localized spread of infectious disease within a population. [13]

epitope Region of an antigen that is recognized by an immunoglobulin, antibody, or T-cell receptor. [1]

erythrocyte Red blood cell responsible for transporting oxygen from the lungs to peripheral tissues. [1]

exonuclease Enzyme capable of removing nucleotides from the ends of nucleic acids. [4]

exotoxin Secreted product of a pathogenic organism capable of eliciting a disease state. [13]

extracellular pathogen Pathogen that resides within an organism without being enclosed by a host's cellular membrane. [1]

extravasation Process of migration of neutrophils from the circulation to a site of infection. [2]

F

Fab Region of an immunoglobulin that contains the variable regions of both the heavy and light chains and the antigen-binding site. [9]

factor B Protease zymogen that functions in the complement cascade; cleavage of factor B by factor D causes its activation and ability to function as a subunit of C3 convertase. [3]

factor D Protease of the complement system that functions to activate factor B through proteolytic cleavage. [3]

factor H Plasma protein that enhances the cleavage of C3b to iC3b by factor I. [3]

factor I Serine protease that inactivates C3b through its cleavage into a smaller fragment known as iC3b, which cannot function as a component of a C3 convertase. [3]

factor P Plasma protein that enhances the activity of the alternative C3 convertase to aid in complement activation and fixation. Also known as *properdin*. [3]

Fas ligand (FasL) Cell-surface molecule expressed by cytotoxic T cells and natural killer cells that binds to Fas on target cells to induce apoptosis. [7]

Fc Region of an immunoglobulin that contains the isotype-specific heavy chain constant region of the immunoglobulin. [9]

Fc receptor Receptor on the surface of an innate immune cell capable of recognizing the Fc component (tail region) of an immunoglobulin. [2, 10]

FcRn Fc receptor expressed on endothelial cells that aids in protection and translocation of IgG from the bloodstream to an extracellular tissue environment. Also called *neonatal Fc receptor*. [10]

FcγRI Fc receptor expressed by phagocytes capable of binding IgG that is opsonizing a foreign antigen to facilitate phagocytosis. [10]

FK-506 Immunosuppressive drug that blocks T-cell activation by binding to FK-binding proteins and inhibiting calcineurin activation. [17]

FK-binding protein A peptidyl-prolyl isomerase, that, in a similar manner to cyclophilin, binds to FK-506 and blocks calcineurin activity and T-cell activation. [17]

follicular helper T cells (T_{FH}) A differentiated subset of T cells responsible for activating B cells in a germinal center in secondary lymphoid tissue. [7]

functional NK cell deficiency Immunodeficiency that prevents the proper activity of natural killer cells and results in persistent intracellular (typically viral) infection. [14]

G

gene family Genes present within an organism that express isotypes of the same biochemical function; these genes are related evolutionarily as they are formed through duplication events from a single original gene. [6]

gene rearrangement Recombination event in a gene that changes the original genetic allele and promotes alternate function of the gene. [4]

genetic polymorphism Presence of multiple alleles within a population that encode for a particular gene product/isotype. [6]

germinal center Area in secondary lymphoid tissue where rapidly dividing B cells and T cells are activated and B cells undergo somatic hypermutation and isotype switching as they differentiate into centrocytes. [10]

glycoprotein Cell-adhesion molecule containing specific carbohydrates that binds to selectin. [2]

Goodpasture syndrome Autoimmune disease caused by the production of antibodies that recognize collagen expressed in blood vessels and the basement membrane of tissues, resulting in the onset of vasculitis. [16]

gp120 Glycoprotein of 120 kilodaltons expressed on the surface of the HIV virus; responsible for binding of viral particles to CD4 on CD4 T cells and macrophages. [14]

gp41 Glycoprotein of 41 kilodaltons expressed on the surface of the HIV virus; responsible for efficient viral fusion with the target cell membrane. [14]

graft-versus-host disease Autoreactive disease induced in individuals who have undergone bone marrow transplantation, where mature T cells and B cells that matured in a donor recognize recipient antigens as foreign. Alloreaction triggered by a bone marrow transplant in which the newly transplanted bone marrow develops and recognizes the recipient's tissues and organs as foreign. [14, 17]

graft-versus-leukemia effect Targeting of residual leukemia cells within a recipient (not destroyed by prior chemotherapy and irradiation) by mature T cells present in the transplanted bone marrow. [17]

granulocyte White blood cell that contain granules. Refers to neutrophils, basophils, eosinophils, and mast cells. [1]

granuloma Inflammatory mass that contains fused macrophages that have engulfed more neutrophils or antigen than they are capable of digesting. [2]

granulysin Protein expressed by cytotoxic T cells that functions to perforate small holes in target cell membranes in order promote pathogen lysis. [7]

granzyme Enzymes produced by cytotoxic innate and adaptive immune cells that promote the induction of apoptosis to limit the presence of an intracellular infection. Serine protease expressed by cytotoxic T cells that is delivered to target cells to induce apoptosis. [2, 7]

Graves disease Autoimmune disease caused by the production of agonist antibodies against the thyroid-stimulating hormone receptor, resulting in hyperthyroidism due to elevated secretion of thyroid hormone. [16]

guanine-nucleotide exchange factor (GEF) Protein that aids in the activation of small GTP-binding proteins by promoting the exchange of bound GDP for GTP on the GTP-binding protein. [10]

Guillain–Barré syndrome Autoimmune disease resulting in the demyelination of peripheral nerves and muscle weakness. [16]

gut-associated lymphoid tissue (GALT) Mucosa-associated lymphoid tissue that protects the gastrointestinal tract through adaptive immune response activation. [12]

H

haplotype Expressed isotypes of all alleles of a gene family within an individual. [6]

Hashimoto disease Autoimmune disease caused by the production of antibodies against thyroid-specific antigens, resulting in the destruction of the thyroid and hypothyroidism. Also known as *Hashimoto thyroiditis*. [16]

heavy chain Larger of the two subunits of immunoglobulins that contains a variable region for inclusion in an antigen-binding site and the Fc region for isotype effector function. [9]

hemagglutinin Adhesion molecule expressed by the influenza virus in order to bind to target cells. [13]

hematopoiesis Development of cells of the circulatory system from a common stem cell ancestor. [1]

heparin Inflammatory mediator secreted by activated granulocytes that aids in increasing vascular permeability. [15]

herd immunity Protection of unvaccinated individuals within a population that is largely vaccinated against a specific pathogen, which prevents the spread of infection within the population. [11]

hereditary angioneurotic edema (HANE) Disease caused by the lack of production of the protein C1 inhibitor (C1INH), causing the overactivity of blood-clotting proteases and buildup of fluids in areas where blood-clotting activation occurs. [14]

heterozygosity The presence of two different alleles of a gene within an individual's genome or the genome of a population. [6]

high endothelial venule (HEV) Specialized blood vessel in secondary lymphoid tissues containing large endothelial cells that express lymphocyte adhesion molecules. [7]

hinge region Flexible loop region located between the Fc and Fab regions of an immunoglobulin; provides flexibility to the Fab portion of an immunoglobulin for increased antigen access. [9]

histamine Inflammatory mediator that is capable of increasing vascular permeability and inducing localized inflammatory responses in order to facilitate pathogen expulsion. Inflammatory mediator produced and secreted by activated granulocytes such as mast cells capable of both increasing vascular permeability and inducing smooth muscle contraction; histamine release from granulocytes is a primary factor in inducing type I hypersensitivity. [10, 15]

HIV protease HIV enzyme responsible for proteolysis of HIV proteins needed for proper assembly and formation of a virion. [14]

HLA-DM MHC class II molecule that functions in promoting peptide exchange of CLIP with extracellular peptides generated by phagocytosis. [6]

human immunodeficiency virus (HIV) Retrovirus that infects CD4 T cells and macrophages and is capable of causing the disease acquired immunodeficiency syndrome (AIDS). [14]

human leukocyte antigen (HLA) Family of immune proteins in humans that includes cell-surface proteins responsible for the presentation of peptide antigens of T cells. Referred to as MHC proteins in other vertebrates. [4]

hyper-IgM syndrome Immunodeficiency in B-cell function that prevents proper isotype switching in activated B cells, resulting in the lack of immunoglobulin isotypes except for IgM. [14]

hyperacute rejection Rejection of a transplanted organ due to destruction of the vasculature of the organ through complement fixation driven by the presence of preexisting blood antigen or HLA antibodies. [17]

hypersensitivity reaction Reaction caused by an immune response driven by the recognition of an innocuous material within an organism. [15]

hyperthyroidism Condition in which the thyroid is overactive and produces too much thyroid hormone. [16]

hypervariable region Loops of the variable regions of the T-cell receptor subunits or the heavy and light chains of immunoglobulins that have the most diverse differences; these regions contribute significantly to the engagement of the T-cell receptor or an immunoglobulin with its antigen. [6]

hypothyroidism Condition in which the thyroid is underactive and does not produce adequate amounts of thyroid hormone. [16]

I

immature B cells B cells that have completed initial development in bone marrow that must subsequently migrate to the spleen for further development. [8]

immune complexes Small complexes of soluble antigen bound with immunoglobulins and fixed complement on the surface marked for removal by phagocytic cells. [10]

immune dysregulation, polyendocrinopathy, enteropathy, X-linked (IPEX) syndrome Autoimmune disease characterized by a defective *FOXP3* gene and a lack of production of regulatory T cells and lowered peripheral tolerance. [16]

immune response The body's response to the presence of foreign particles or microorganisms (antigens). [1]

immunity State of being resistant to infection by a specific pathogen. [1]

immunization Process used to prevent disease by exposing an individual to a nonpathogenic product of a pathogenic organism to promote mounting of a primary immune response and immunological memory. [1]

immunodeficiency A disorder in an innate or adaptive immune response due to the lack of molecules required for inducing the response or a lack of cells required for the response due to improper development. [14]

immunoglobulins Proteins made by B cells that act as specific antigen receptors for B-cell recognition, activation, and effector function. [1]

immunological memory Production of memory T cells and memory B cells during clonal expansion, which act in a faster, more robust adaptive immune response upon subsequent contact with the same antigen. [1]

immunological synapse Connection between a T cell and an antigen-presenting cell that aids in the clustering of T-cell receptor signaling molecules and activation of the T cell. [7]

immunology Branch of biomedical science that studies immunity and the mechanisms of the immune response. [1]

immunoproteasome Proteasome complexes with 11S subunit that promote digestion of proteins into peptides of 8 to 10 amino acids in length for better presentation on MHC class I molecules. [6]

immunoreceptor tyrosine-based activation motifs (ITAMs) Cytoplasmic amino acid sequences containing tyrosine that are targets of phosphorylation and subsequent activation of signaling events. [7]

immunosuppressive Substance capable of inhibiting an innate or adaptive immune response. [17]

immunosurveillance Function of the immune system to target and destroy mutated and tumor cells. [18]

inactivated vaccine Vaccine composed of chemically inactivated, irradiated, or heat-killed pathogen. [11]

indirect pathway of allorecognition Pathway of presentation of alloantigens in which donor cells die by apoptosis in secondary lymphoid tissue, where they are phagocytosed by recipient dendritic cells, which then process donor cell antigens and present them to recipient T cells. [17]

inductive compartment The area of mucosa-associated lymphoid tissue responsible for the clonal selection and expansion of T and B lymphocytes. [12]

inflammasome Complex of proteins that interact to produce a means to efficiently activate caspase-1 and allow efficient activation of IL-1. [2]

inflammation Response triggered by cytokines that results in vasodilation, redness, and swelling; this response aids the innate immune response by promoting immune cell migration to a site of infection and immune cell action at an infection site. [1]

inflammatory mediator Molecules secreted by immune cells that induce inflammatory responses to activate innate immune cells and lymphocytes. [10]

inherited immunodeficiency A disorder in an innate or adaptive immune response caused by mutations within genes in the patient that prevent the proper responses from occurring. [14]

innate immunity Mechanisms used by the immune system that target pathogens based on common components and in a relatively nonspecific manner. [1]

innate lymphoid cell (ILC) White blood cell of the innate immune system that works to activate other innate immune cells, including macrophages and dendritic cells, at a site of infection. [1]

insulin-dependent (type 1) diabetes mellitus Autoimmune disease caused by the targeting of pancreatic β cells by the immune system, resulting in their destruction and the inability of the pancreas to produce insulin. [16]

integrase HIV enzyme responsible for incorporating the cDNA genome of HIV into the target cell genome. [14]

integrin Cell-surface glycoprotein receptor important in mediating cell-to-cell adhesion interactions. [2]

intercellular adhesion molecule (ICAM) Cell-adhesion molecule that mediates cell-to-cell contact, commonly with integrins. [2]

interferon (IFN) Cytokine that activates cells important in targeting viral infection. [2]

interleukin (IL) Cytokine secreted by white blood cells. [2]

internalin Pathogen adhesion molecule involved in target-cell binding and activation of phagocytosis events within the target cell. [13]

intracellular pathogen Pathogen that resides within a cell of the infected organism. [1]

intraepithelial lymphocyte Cytotoxic T cells embedded in the epithelial cell barrier of mucosal tissue. [12]

intraepithelial pocket Invagination of apical surface of M cells promoting efficient transport of antigens through the M cells to dendritic cells of the lymphoid tissue. [12]

invariant chain Protein in the secretory pathway that binds to MHC class II molecules and blocks the peptide-binding groove from binding to intracellular peptides transported into the ER via TAP; is proteolytically cleaved in the MHC compartment to CLIP. [6]

isograft Tissue transplantation between genetically identical individuals.

isolated lymphoid follicle Follicle containing a germinal center and mainly B cells undergoing activation and differentiation. [12, 17]

isotype switching Mechanism used by activated B cells to incorporate a different heavy chain constant region into the functioning immunoglobulin gene to refine the activity of the immunoglobulin based on need. [4]

isotype Various protein products that have functional and structural similarities within an organism. [6]

isotypes Different protein versions of a gene family that have related functions. [4]

J

J chain Serum protein that interacts with IgM and IgA to provide additional quaternary structure. [9]

JAK3 deficiency Immunodeficiency in the gene that encodes JAK3, a kinase responsible for proper cytokine signaling, resulting in improper T-cell development. [14]

L

λ5 Protein component of the surrogate light chain that resembles the constant region of the immunoglobulin light chain; aids in testing for productive rearrangement of the immunoglobulin heavy chain. [8]

lamina propria Connective tissue that lies underneath the epithelial barrier of mucosal tissue. [12]

latency Period of dormancy of a pathogen while within a host. [13]

lectin Molecule capable of binding to carbohydrates. [2]

lectin pathway Second pathway of complement activation to be activated that relies on the use of mannose-binding lectin to initiate complement activation. [3]

leukemia Cancer of circulating hematopoietic cells. [18]

leukocyte White blood cell. Includes granulocytes, agranulocytes, and cells involved in blood clotting. [1]

leukocyte adhesion deficiency Immunodeficiency caused by a mutation in a phagocyte cell-surface molecule, preventing migration of phagocytes to a site of infection and preventing proper recognition of pathogens opsonized with C3b by phagocytic cells. [14]

leukotrienes Inflammatory mediators derived from arachidonic acid that can induce smooth muscle contraction, increase vascular permeability, and aid in the production of other inflammatory cytokines. [15]

light chain Smaller of the two subunits of immunoglobulins that contains a variable region for inclusion in an antigen-binding site. [9]

lineage commitment Process by which developing double-positive thymocytes switch to the expression of a single coreceptor and become single-positive thymocytes. [5]

linear epitope Immunoglobulin epitope that incorporates structural determinants maintained by the linear sequence of the antigen. [9]

lipopolysaccharide (LPS) Endotoxin that is a major component of cell walls in gram-negative bacteria. [2]

live attenuated vaccine Vaccine composed of a pathogen that has not been inactivated but is incapable of normal survival or pathogenesis in the host. [11]

lymph Fluid of the lymphatic system that allows for plasma and lymphocyte movement to and from lymphoid tissue. [1]

lymph node Secondary lymphoid tissue that contains anatomy specialized in antigen presentation to T cells and B cells and lymphocyte activation and proliferation. [1]

lymphatic system Circulatory system of lymph and lymphoid tissue that allows for lymphocyte migration to and from these tissues. [1]

lymphocyte Cells of the lymphoid arm of hematopoiesis, including the cells of the adaptive immune system (T cells and B cells) and some cells of the innate immune system (innate lymphoid cells and NK cells). [1]

lymphoid follicle Location within a lymph node where activated T cells migrate to promote activation of B cells during an infection. [1]

lymphoid tissue Specialized tissue of the lymphatic system that functions in the development and activation of T cells and B cells. [1]

lymphoma Cancer of lymphoid tissue. [18]

lysosome Organelle in eukaryotic cells that facilitates digestion of internalized material due to the presence of digestive enzymes and toxic substances. [2]

lysozyme Enzyme present in saliva, tears, and respiratory tract fluid that cleaves the polysaccharide of bacterial cell walls. [2]

M

M cell Cells associated with mucosa-associated lymphoid tissue that specialize in pathogen and antigen delivery to those tissues. [1]

macrophage Agranulocyte that specializes in phagocytosis of foreign antigens and apoptotic cells. [1]

major histocompatibility complex (MHC) Family of immune proteins in mice and other higher vertebrates that includes cell-surface proteins responsible for the presentation of peptide antigens to T cells. Referred to as HLA proteins in humans. [4]

malignant Denoting a tumor composed of cells that can invade surrounding tissue and spread to other regions of the body (metastasize). [18]

mannose-binding lectin (MBL) Acute-phase response protein that functions to initiate the lectin pathway of complement activation. [3]

marginal-zone B cells B cells that reside in the marginal zone of the white pulp of the spleen and monitor for bloodborne pathogens. [8]

mast cell Granulocyte that specializes in induction of an inflammatory response; often closely associated with IgE isotype antibodies. [1]

megakaryocyte White blood cell that specializes in the production of platelets. [1]

membrane cofactor protein (MCP) Cell-surface protein that inhibits complement activation by enhancing C3b cleavage by factor I on the surface of human cells. [3]

membrane-attack complex (MAC) Complex of complement proteins C5, C6, C7, C8, and C9 that work in concert to form holes in bacterial and eukaryotic membranes. [2, 3]

memory cells Differentiated long-lived T cells and B cells that drive secondary immune responses. [1]

metastasis Process of cancer-cell migration through the circulatory system to a secondary site within the body. [18]

MHC class I A major histocompatibility protein consisting of two polypeptides, an α chain transmembrane protein and soluble β_2-microglobulin; displays peptides produced through intracellular processing of antigens. [4]

MHC class II A major histocompatibility protein consisting of two transmembrane polypeptides, an α chain and a β chain; displays peptides produced through extracellular processing of antigens. [4]

MHC restriction Selection process driven by a recombined and expressed cell-surface T-cell receptor that can only interact with a specific self-MHC molecule. [5]

microbiota Microbial species that reside in symbiosis with an organism on or in tissues such as the skin and intestines. [1]

microfold cells (M cells) Cells that deliver antigens from the lumen of mucosal tissue to dendritic cells in mucosa-associated lymphoid tissue. [12]

minor histocompatibility antigen Alloantigens that differ by the presented peptide rather than the MHC molecule presenting the peptide. [17]

molecular mimicry Mechanism used by pathogens to produce molecules similar or identical to host proteins in an attempt to evade the adaptive immune response by taking advantage of the processes of central and peripheral tolerance. [16]

monocyte Circulating white blood cell that migrates to tissue and differentiates into a macrophage or a dendritic cell. [1]

mucin Glycoprotein that protects mucosal tissue by increasing the viscosity of mucus, concentrating defensins at mucosal surfaces, and cross-linking secreted IgA at the mucosal surface. [12]

mucosa A mucous membrane. Also referred to as a *mucosal surface*; plural is *mucosae*. [12]

mucosa-associated lymphoid tissue (MALT) Secondary lymphoid tissue that specializes in the uptake and clearance of pathogens at mucosal surfaces. [1, 12]

mucus Thick, viscous secretion from epithelial cells of mucosal tissue that contains mucin and secreted defensins and immunoglobulins. [12]

multiple sclerosis (MS) Autoimmune disease caused by demyelination of the central nervous system and white blood cell migration into the brain. [16]

multipotent progenitor cell Early precursor cell of hematopoiesis that is derived from differentiation of a hematopoietic stem cell in the bone marrow; has the ability to differentiate into a variety of blood cells given the right signal. [8]

multivalent antigen An antigen that either has more than one copy of an epitope or contains multiple epitopes for multiple immunoglobulins. [9]

mutagen Substance or agent capable of inducing DNA damage and increasing the rate of DNA mutation. [18]

myasthenia gravis Autoimmune disease caused by the production of antagonist antibodies to the acetylcholine receptor, resulting in improper synaptic transmission between nerves and muscles leading to muscle weakness and loss of function. [16]

myeloma Cancer of the bone marrow. [18]

N

N nucleotides Nucleotides added to the ends of variable segments involved in the formation of a coding joint; these are added to the ends by terminal deoxynucleotidyl transferase (TdT). [4]

naïve B cell Fully mature B cell that has not yet come into contact with antigen. [8]

natural killer (NK) cell White blood cell of the innate immune system that works to clear intracellular infections, including those caused by viruses. [1]

natural killer (NK) T cells Effector T cells that respond to glycolipid and perform functions that resemble both helper T cells (cytokine production) and cytotoxic T cells (apoptosis induction). [7]

negative selection Process by which lymphocytes are removed from the body if their receptors recognize self-molecules. [4]

neuraminidase Virus surface enzyme involved in the processing and formation of infectious viral particles. [13]

neutralizing antibody Antibody that can block pathogen or toxin interaction with a target cell. [10]

neutropenia Abnormally low levels of neutrophils in the blood. [14]

neutrophil Granulocyte that migrates to sites of infection and phagocytoses pathogens located there. [1]

NOD Stands for nucleotide-binding oligomerization domain; an epithelial protein responsible for the detection of intracellular microorganism components and activation of cytokine secretion to activate effector mechanisms within mucosa-associated lymphoid tissue. [12]

nuclear factor κB (NFκB) Transcription factor that activates gene expression during both innate and adaptive immune responses; inactivated by binding to IκB in the cytoplasm. [2]

O

Omenn syndrome Combined immunodeficiency that results from mutation in genes that encode proteins in the V(D)J recombinase, resulting in the inability of B cells and T cells to develop properly. [14]

oncogene Mutant form of a proto-oncogene that promotes uncontrolled cell division. [18]

opportunistic pathogen Organism with the potential to cause disease in an immunocompromised host. [1]

opsonin Protein product on a pathogen surface that marks the pathogen for phagocytosis by neutrophils and macrophages. [2]

opsonin receptor Protein capable of recognizing an opsonin on a foreign pathogen or molecule. [2]

opsonization Process of marking a pathogen or foreign molecule as one that needs to be targeted for removal or destruction by phagocytosis. [3]

oral tolerance Immune system tolerance to certain substances, such as food, that are ingested orally and are not harmful to the body. [12]

P

P nucleotides Nucleotides present in the overhangs left when V(D)J recombinase opens up the hairpin formed during V(D)J recombination. [4]

pandemic Spread of infectious disease over a large area, such as across continents or globally. [13]

paracrine Signaling mechanism whereby cells adjacent to a cell producing a signaling molecule respond to that signaling molecule. [7]

pathogen Organism with the potential to cause disease. [1]

pathogen-associated molecular pattern (PAMP) Molecules commonly seen on the surfaces of microorganisms that have been conserved evolutionarily but are absent on the surface of eukaryotic cells; used by innate immune cells to recognize foreign microorganisms. [2]

pattern recognition receptor (PRR) Proteins on the surface of innate immune cells that are capable of recognizing molecules known as PAMPs. [2]

peptide-binding groove Domain within an MHC molecule that binds to a peptide to present the bound peptide to a T-cell receptor. [6]

peptide-binding motif Sequence of amino acids of a peptide that are essential for interaction with an MHC molecule. [6]

peptide-loading complex Protein complex within the ER that assists in binding intracellular peptides to the peptide-binding groove of MHC class I molecules. [6]

perforin Protein expressed by cytotoxic T cells that functions to perforate small holes in target cell membranes in order to deliver cytotoxins and induce apoptosis. [2, 7]

peripheral tolerance Negative selection/inactivation processes that occur in peripheral tissues to prevent the activation of self-reactive lymphocytes that have escaped negative selection in their primary lymphoid tissues. [5, 8]

Peyer's patch Specialized mucosa-associated lymphoid tissue of the small intestine. [12]

phagocytosis Process of engulfing and killing cells through receptor binding and endocytosis. [1]

phagolysosome Membrane-bound organelle that results from fusion of a lysosome with a phagosome; specializes in digestion of phagocytosed material. [2]

phagosome Membrane-bound vesicle created during the internalization of a pathogen during phagocytosis. [2]

plasma cell B cell that has been activated by antigen and specializes in the secretion of antibodies into plasma. [1]

platelet Nonnucleated product of megakaryocytes that functions in blood clotting. [1]

PNP deficiency Immunodeficiency caused by a mutation in the gene encoding purine nucleoside phosphorylase, a protein required for purine salvage pathways. [14]

poly-Ig receptor Receptor on the surface of epithelial cells lining mucosal tissue that aids in the proper delivery of dimeric IgA (and IgM) to mucosal surfaces. [10]

polymorphic Genes that encode for the same protein; represented in a large number of different alleles in the population. [4]

positive selection Process by which lymphocytes are tested to determine if their receptor is functioning properly. [4]

pre-B cell Developing B cell that has successfully undergone immunoglobulin heavy chain rearrangement and is in the process of rearranging its immunoglobulin light chain. [8]

pre-B-cell receptor Receptor that contains rearranged immunoglobulin heavy chain, the surrogate light chain, and the signaling molecules Igα and Igβ; responsible for testing the immunoglobulin heavy chain for productive rearrangement. [8]

pre-T cell Thymocyte that has successfully passed the β-chain checkpoint; pre-T cells engage in β-chain allelic exclusion and proliferate to produce cells that can further engage in T-cell development through α-chain rearrangement. [5]

pre-Tα chain Protein expressed in thymocytes that have finished β-chain rearrangement; acts as a surrogate α chain to test proper expression and action of the recombined β chain. [5]

primary focus Cluster of activated B cells and T cells that is induced in germinal centers and is important in an adaptive immune response. [7, 10]

primary immune response Adaptive immune response generated upon initial exposure to a pathogen. [4]

primary lymphoid follicle Location within a secondary lymphoid tissue where circulating B cells migrate to test for the presence of antigen their immunoglobulin is designed to recognize. [8]

primary lymphoid tissue Lymphoid tissue that serves as a location for lymphocyte development. [1]

pro-B cell Developing B cell that is undergoing immunoglobulin heavy chain rearrangement. [8]

productive rearrangement Genetic recombination at the immunoglobulin and T-cell receptor loci that creates a gene that can produce a functional subunit. [6]

progenitor cell Precursor cell that can differentiate into several specialized blood cell types during hematopoiesis. [1]

promiscuous binding specificity Ability of a single MHC molecule to bind to a variety of peptides in its peptide-binding groove. [6]

properdin Plasma protein that enhances the activity of the alternative C3 convertase to aid in complement activation and fixation. Also known as *factor P*. [3]

prostaglandin Inflammatory mediator secreted by activated granulocytes capable of inducing vasodilation, fever, and pain at the site of release. [15]

protease inhibitor Molecule that is capable of preventing the enzymatic activity of proteases. [2]

proteasome Multi-subunit enzyme complex that cleaves intracellular cytosolic proteins into peptides for presentation on MHC class I molecules or removal of damaged or abnormal proteins. [6]

proto-oncogene Gene that encodes a protein involved in the process of cell division. [18]

R

receptor Molecule (typically a protein) that binds to another molecule to sense the environment, interact with other cells, and respond to stimuli. [1]

receptor editing Mechanism utilized by B cells to attempt to rearrange the immunoglobulin light chain since, through initial somatic recombination events, they express a self-reactive immunoglobulin. [8]

recombination activating gene 1 (*RAG1*) Gene activated in developing lymphocytes that functions in the recombination of variable regions in T-cell receptors and immunoglobulins as part of the V(D)J recombinase enzyme complex. [4]

recombination activating gene 2 (*RAG2*) Gene activated in developing lymphocytes that functions in the recombination of variable regions in T-cell receptors and immunoglobulins as part of the V(D)J recombinase enzyme complex. [4]

recombination signal sequences (RSSs) Sequence motifs that flank variable region gene segments and serve as recognition sites of enzymes involved in V(D)J recombination. [4]

regulatory T cells (T_{regs}) Effector T cells that function to prevent autoimmunity and promote peripheral self-tolerance. [7]

resident memory T cells (T_{RM}) Memory T cells that reside within peripheral tissues; major role is the persistent monitoring of these tissues for repeat infection by the same pathogen. [11]

respiratory burst Process of oxygen consumption due to the action of NADPH oxidase. [2]

retrovirus Virus that has an RNA genome that must be reverse transcribed to a DNA copy and incorporated into the target cell's genome, before viral particles can further be produced. [14]

reverse transcriptase HIV enzyme responsible for creating a cDNA genome copy using the RNA genome from a viral particle that has just infected a target cell. [14]

rheumatoid arthritis (RA) Systemic autoimmune disease characterized by the inflammation of joints. [16]

rheumatoid factor Autoantibodies capable of recognizing the Fc region of IgG molecules. [16]

S

scavenger receptor Innate immune cell receptor that binds to negatively charged molecules at the surface of microorganisms. [2]

secondary focus Conjugate pair of B cell and T cell that has migrated from the T-cell zone in a secondary lymphoid tissue to the lymphoid follicle for further B-cell differentiation. [10]

secondary immune response Adaptive immune response generated upon second and subsequent exposure to a pathogen. [4]

secondary lymphoid tissue Lymphoid tissue that serves as a location for activating lymphocytes. [1]

selectin Cell-adhesion molecule that binds to specific carbohydrates on another cell-adhesion molecule. [2]

selective IgA deficiency Immunodeficiency that causes an inability to produce IgA immunoglobulins without impacting the production of other immunoglobulin isotypes. [14]

self-antigen Constituent of normal tissue that stimulates an immune response. Also known as autoantigen. [1]

semidirect pathway of allorecognition Pathway of presentation of alloantigens in which intact donor MHC molecules are transferred to recipient dendritic cells via cell-to-cell interaction or vesicle exchange, followed by presentation of those intact donor MHC molecules by the recipient dendritic cells. [17]

serotype Members of the same species of a pathogen that differ in surface molecules. [13]

serum sickness A type III hypersensitivity reaction driven by the formation of immune complexes that contain immunoglobulins from another organism. [15]

severe combined immunodeficiency (SCID) Immunodeficiency that specifically affects the adaptive immune system (either B cells, T cells, or both). [14]

signal joint Circle of DNA originally located between two variable segments in the germline DNA that is removed upon V(D)J recombination of variable segments. [4]

single-positive thymocyte Developing thymocyte that has undergone lineage commitment and expresses only a single coreceptor. [5]

somatic hypermutation Preferential action of activation-induced cytosine deaminase (AID) at the variable regions of immunoglobulin genes to further enhance affinity for its epitope. [4]

somatic recombination DNA recombination (gene rearrangement) that occurs in developing T cells and B cells, also known as V(D)J recombination. [4]

spleen Secondary lymphoid tissue important in red blood cell filtration and monitoring pathogen presence in the bloodstream. [1]

stem cell factor Factor responsible for activating the signals necessary for progression to lymphoid progenitor cell development. [8]

stromal cells Cells within specific tissues that allow for intimate cell-to-cell contact; provide important signaling molecules to drive development and differentiation. [8]

subunit vaccine Vaccine composed of a cell-surface antigen of a pathogen; designed to stimulate the production of neutralizing immunoglobulins that block pathogen interaction with target cells. [11]

superantigen Toxin produced by a pathogen that is capable of general suppression of the immune system by nonspecific activation of T lymphocytes. [13]

surrogate light chain Protein complex formed when the proteins VpreB and λ5 are expressed and interact; takes the place of the immunoglobulin light chain in pro-B cells that need to test the rearrangement of the immunoglobulin heavy chain. [8]

switch region DNA sequence upstream of immunoglobulin heavy chain constant regions that is targeted for recombination during class switch recombination; also called *switch sequence.* [9]

systemic infection Infection that is present in the bloodstream and widespread throughout the body. [2]

systemic lupus erythematosus (SLE) Systemic autoimmune disease characterized by the production of a variety of autoantibodies, including antibodies to nucleic acids, lipids, and nucleosomal proteins, resulting in the production of small immune complexes; mimics type III hypersensitivity. [16]

T

T cell Lymphocyte that is the primary cell involved in the cell-mediated adaptive immune response; once activated, functions to activate other immune cells or target intracellular infections. [1]

T memory stem cells (T_{SCM}) Memory T cells that maintain pluripotency; they reside within the circulatory and lymphatic systems and secondary lymphoid tissues and are capable of further differentiating into central memory T cells or effector memory T cells upon subsequent activation. [11]

T-cell priming Process whereby effector T cells are produced to drive the cellular arm of the adaptive immune system; also called *T-cell activation.* [7]

T-dependent (TD) antigen Antigen that can only activate a B cell with the aid of a helper T cell. [10]

T-dependent (TD) response B-cell response that requires aid from helper T cells to promote proper B-cell activation. [10]

T-independent (TI) antigen Antigen that is capable of activating a B cell in the absence of helper T-cell action. [10]

T-independent (TI) response B-cell response that can induce B-cell activation independent of aid from helper T-cell signaling. [10]

T1 transitional B cell B cell that has migrated to the spleen and is being further tested via negative selection to promote peripheral tolerance. [8]

T2 transitional B cell B cell that has migrated to the spleen and successfully passed through negative selection and is now receiving B-cell survival signals and final differentiation signals. [8]

tapasin ER protein responsible for peptide loading and peptide exchange on MHC class I molecules. [6]

terminal deoxynucleotidyl transferase (TdT) Enzyme involved in the addition of nucleotides to the ends of recombining variable segments during lymphocyte development. [4]

T_H1 helper T cells Effector T cells that function to aid in activation of macrophages. [7]

T_H17 helper T cells Effector T cells that function to aid in neutrophil activation. [7]

T_H2 helper T cells Effector T cells that function to aid in activation of B cells. [7]

thymocyte Developing immature T cells present in the thymus. [5]

thymus Primary lymphoid tissue that serves as the location for T cell development. [1]

tolerance Inactivation of immune response against self-antigens. [1]

Toll-like receptor (TLR) Innate immune cell receptor family capable of recognizing a variety of different microbial products and activating key signaling cascades to induce an innate immune response. [2]

toxin Product of a living organism that can cause disease within an organism. [13]

toxoid Inactivated exotoxin employed in a vaccine to stimulate an adaptive immune response that involves production of neutralizing antibodies that block normal exotoxin activity. [11]

transcytosis Process of delivering immunoglobulins and other macromolecules from one membrane face of a cell that faces the bloodstream or mucosa-associated lymphoid tissue to the other face of a cell facing a different environment, such as extracellular or mucosal tissue. Transport of a soluble molecule from one side of a cellular barrier to the other. [10, 12]

transfusion Transfer of whole blood or blood components from a donor to a recipient. [17]

transitional B cells B cells that have finished development in the bone marrow but need to finish the developmental program within the spleen. [8]

transplantation rejection Rejection of transplanted donor tissue or an organ by the recipient's immune system. [17]

transporter associated with antigen processing (TAP) Protein complex embedded in the ER membrane; responsible for transporting peptides generated by the proteasome into the ER for MHC class I loading. [6]

tumor Abnormal growth of cells within an organism. [18]

tumor suppressor gene Gene that encodes a protein involved in regulating cell division. [18]

tumor-associated antigen Antigen expressed at higher levels by tumor cells in comparison to normal cells. [18]

tumor-specific antigen Antigen expressed only by tumor cells, typically a mutated protein. [18]

U

unproductive rearrangement Genetic recombination at the immunoglobulin and T-cell receptor loci that creates a gene that cannot produce a functional subunit. [6]

upregulation Increase in production of proteins within a cell due to activation of gene expression. [7]

urticaria Hive formation due to an inflammatory response during a hypersensitivity reaction. [15]

V

V(D)J recombinase Enzyme complex that functions in developing lymphocytes to recombine segments of the variable regions of T-cell receptors and immunoglobulins and promote the formation of a functional gene product. [4]

vaccination Process of inoculating an individual with a harmless counterpart of a pathogen to mimic an infection to stimulate the adaptive immune response and the production of memory lymphocytes for fast, robust protection upon subsequent exposure to the pathogen. [11]

vaccine Nonpathogenic product from a pathogen formulated to be administered to stimulate a primary immune response and provide protection from infection by that specific pathogen. [1]

variable region Region of a T-cell receptor, immunoglobulin, or antibody where the amino acid sequence is different among molecules; the antigen-binding site of T-cell receptors, immunoglobulins, and antibodies is created by the interaction of variable regions of different receptor subunits (e.g., the variable regions of the T-cell receptor α and β chains interact to form a domain containing the antigen-binding site). [4]

variolation Obsolete practice in which a small dose of a pathogen (smallpox virus) was introduced intranasally or through a small scratch under the skin to create a mild infection with the goal of providing immunity to a more serious infection. [11]

virulence factor Product of a pathogen that supports pathogen survival and aids disease pathogenesis. [13]

VpreB Protein component of the surrogate light chain that resembles the variable region of the immunoglobulin light chain; aids in testing for productive rearrangement of the immunoglobulin heavy chain. [8]

W

wheal-and-flare reaction Localized inflammation that occurs due to subcutaneous presence of an allergen capable of inducing type I hypersensitivity; presents as a raised lesion surrounded by reddish, inflamed tissue. [15]

X

X-linked agammaglobulinemia X-linked immunodeficiency preventing proper B-cell development past the pre-B-cell stage; caused by a mutation in the *Btk* gene, which encodes for Bruton's tyrosine kinase, a protein required for proper B-cell development. [14]

xenograft Tissue transplantation between different species. [17]

References

Chapter 1

Bajenoff, M., J. G. Egen, H. Qi, et al. 2007. Highways, byways and breadcrumbs: Directing lymphocyte traffic in the lymph node. *Trends in Immunology* 28: 346–352.

Becker, A. J., McCulloch, E. A., and Till, J. E. 1963. Cytological demonstration of the clonal nature of spleen colonies derived from transplanted mouse marrow cells. *Nature* 197: 452–454.

Belkaid, Y., and T. Hand. 2014. Role of the microbiota in immunity and inflammation. *Cell* 157: 121–141.

Boehm, T., I. Hess, and J. B. Swann. 2012. Evolution of lymphoid tissues. *Trends in Immunology* 33: 315–321.

Burnet, F. M. 1959. *The Clonal Selection Theory of Acquired Immunity*. Cambridge University Press, Cambridge, England.

Catron, D. M., A. A. Itano, K. A. Pape, et al. 2004. Visualizing the first 50 hr of the primary immune response to a soluble antigen. *Immunity* 21: 341–347.

Cichocki, F., E. Sitnicka, and Y. T. Bryceson. 2014. NK cell development and function—Plasticity and redundancy unleashed. *Seminars in Immunology* 26: 114–126.

Cullen, S. M., A. Mayle, L. Rossi, and M. A. Goodell. 2014. Hematopoietic stem cell development: An epigenetic journey. *Current Topics in Developmental Biology* 107: 39–75.

Della Chiesa, M., E. Marcenaro, S. Sivori, et al. 2014. Human NK cell response to pathogens. *Seminars in Immunology* 26: 152–160.

Deriu, E., J. Z. Liu, M. Pezeshki, et al. 2013. Probiotic bacteria reduce *Salmonella typhimurium* intestinal colonization by competing for iron. *Cell Host Microbe* 14: 26–37.

Ehrlich, P., and C. Bolduan. 1906. *A General Review of the Recent Work in Immunity*. John Wiley & Sons, New York.

Finlay, B. B., and G. McFadden. 2006. Anti-immunology: Evasion of the host immune system by bacterial and viral pathogens. *Cell* 124: 767–782.

McSorley, S. J. 2014. Immunity to intestinal pathogens: lessons learned from *Salmonella*. *Immunological Reviews* 260: 168–182.

Mebius, R. E., and G. Kraal. 2005. Structure and function of the spleen. *Nature Reviews Immunology* 5: 606–616.

Metchnikoff, E. 1905. *Immunity in the Infectious Diseases*. Macmillan, New York.

Mildner, A., and S. Jung. 2014. Development and function of dendritic cell subsets. *Immunity* 40: 642–656.

Munoz, M. A., M. Biro, and W. Weninger. 2014. T cell migration in intact lymph nodes in vivo. *Current Opinion in Cell Biology* 30: 17–34.

Nauseef, W. M., and N. Borregaard. 2014. Neutrophils at work. *Nature Immunology* 15: 602–611.

Pabst, R. 2007. Plasticity and heterogeneity of lymphoid organs: What are the criteria to call a lymphoid organ primary, secondary, or tertiary? *Immunology Letters* 112: 1–8.

Paul, W. E., 2012. *Fundamental Immunology*, 7th ed. Lippincott Williams & Wilkins, Philadelphia.

Chapter 2

Abram, C. L., and C. A. Lowell. 2009. The ins and outs of leukocyte integrin signaling. *Annual Review of Immunology* 27: 339–362.

Arend, W. P., G. Palmer, and C. Gabay. 2008. IL-1, IL-18, and IL-33 families of cytokines. *Immunological Reviews* 223: 20–38.

Bell, J. K., Botos, I., Hall, et al. 2005. The molecular structure of toll-like receptor 3 ligand binding domain. *Proceedings of the National Academy of Sciences USA* 102: 10976–10980.

Beutler, B., and E. T. Rietschel. 2003. Innate immune sensing and its roots: The story of endotoxin. *Nature Reviews Immunology* 3: 169–176.

Bryant, C. E., M. Symmons, and N. J. Gay. 2015. Toll-like receptor signaling through macromolecular protein complexes. *Molecular Immunology* 63: 162–165.

Della Chiesa, M., E. Marcenaro, S. Sivori, et al. 2014. Human NK cell response to pathogens. *Seminars in Immunology* 26: 152–160.

Dunkelberger, J. R., and W. C. Song. 2010. Complement and its role in innate and adaptive immune responses. *Cell Research* 20: 34–50.

Fickenscher, H., S. Hor, H. Kupers, et al. 2002. The interleukin-10 family of cytokines. *Trends in Immunology* 23: 89–96.

Froy, O. 2005. Regulation of mammalian defensin expression by Toll-like receptor-dependent and independent signaling pathways. *Cellular Microbiology* 7: 1387–1397.

Gee, K., C. Guzzo, N. F. Che Mat, et al. 2009. The IL-12 family of cytokines in infection, inflammation, and autoimmune disorders. *Inflammation & Allergy Drug Targets* 8: 40–52.

Glaser, R., J. Harder, H. Lange, et al. 2005. Antimicrobial psoriasin (S100A7) protects human skin from *Escherichia coli* infection. *Nature Immunology* 6: 57–64.

Jones, L. L., and D. A. Vignali. 2011. Molecular interactions within the IL-6/IL-12 cytokine/receptor superfamily. *Immunologic Research* 51: 5–14.

Kawai, T., and S. Akira. 2010. The role of pattern recognition receptors in innate immunity: Update on Toll-like receptors. *Nature Immunology* 11: 373–384.

Kemper, C., and D. E. Hourcade. 2008. Properdin: New roles in pattern recognition and target clearance. *Molecular Immunology* 45: 4048–4056.

Kerrigan, A. M., and G. D. Brown. 2009. C-type lectins and phagocytosis. *Immunobiology* 214: 562–575.

Lemaitre, B. 2004. The road to Toll. *Nature Reviews Immunology* 4: 521–527.

Lionakis, M. S., and M. G. Netea. 2013. Candida and host determinants of susceptibility to invasive candidiasis. *PLOS Pathogens* 9: e1003079.

Litvack, M. L., and N. Palaniyar. 2010. Soluble innate immune pattern-recognition proteins for clearing dying cells and cellular components: Implications on exacerbating or resolving inflammation. *Innate Immunity* 16: 191–200.

Mansour, S. C., O. M. Pena, and R. E. W. Hancock. 2014. Host defense peptides: Front-line immunomodulators. *Trends in Immunology* 35: 443–450.

Markiewski, M. M., B. Nilsson, K. N. Ekdahl, et al. 2007. Complement and coagulation: Strangers or partners in crime? *Trends in Immunology* 28: 184–192.

Medzhitov, R. 2007. Recognition of microorganisms and activation of the immune response. *Nature* 449: 819–826.

Medzhitov, R. 2013. Pattern recognition theory and the launch of modern innate immunity. *Journal of Immunology* 191: 4473–4474.

Michalek, M., C. Gelhaus, O. Hecht, et al. 2009. The human antimicrobial protein psoriasin acts by permeabilization of bacterial membranes. *Developmental and Comparative Immunology* 33: 740–746.

Micheau, M., and J. Tschopp. 2003. Induction of TNF receptor 1-mediated apoptosis via two sequential signaling complexes. *Cell* 114: 181–190.

Nauseef, W. M., and N. Borredaard. 2014. Neutrophils at work. *Nature Immunology* 15: 602–611.

Netea, M. G., C. Wijmenga, and L. A. J. O'Neill. 2012. Genetic variation in Toll-like receptors and disease susceptibility. *Nature Immunology* 13: 535–542.

Palm, N. W., and R. Medzhitov. 2009. Pattern recognition receptors and control of adaptive immunity. *Immunology Review* 227: 221–233.

Parham, P. 2014. *The Immune System*, 4th ed., W. W. Norton and Company.

Pegram, H. J., D. M. Andrews, M. J. Smyth, et al. 2011. Activating and inhibitory receptors of natural killer cells. *Immunology and Cell Biology* 89: 216–224.

Potlorak, A., X. He, I. Smirnova, et al. 1998. Defective LPS Signaling in C3H/HeJ and

C57BL/10ScCr Mice: Mutations in Tlr4 Gene. *Science* 282: 2085–2088.

Punt, J., J. A. Owen, S. A. Stranford, et al. 2018. *Kuby Immunology*, 8th ed., W. H. Freeman and Company: New York.

Raman, D., T. Sobolik-Delmaire, and A. Richmond. 2011. Chemokines in health and disease. *Experimental Cell Research* 317: 5755–5789.

Rot, A., and U. H. von Andrian. 2004. Chemokines in innate and adaptive host defense: Basic chemokinese grammar for immune cells. *Annual Review of Immunology* 22: 891–928.

Sadler, A. J., and B. R. G. Williams. 2008. Interferon-inducible antiviral effectors. *Nature Reviews Immunology* 8: 559–568.

Salzman, N. H., K. Hung, D. Haribhai, et al. 2010. Enteric defensins are essential regulators of intestinal microbial ecology. *Nature Immunology* 11: 76–83.

Schroder, K., and J. Tschopp. 2010. The inflammasomes. *Cell* 140: 821–832.

Takeuchi, O., and S. Akira. 2010. Pattern recognition receptors and inflammation. *Cell* 140: 805–820.

Tesar, B. M. and D. R. Goldstein. 2007. Toll-like receptors and their role in transplantation. *Frontiers in Bioscience-Landmark.* 12: 4221–4238.

Tonn, T., D. Schwabe, H. G. Klingemann, et al. 2013. Treatment of patients with advanced cancer with the natural killer cell line NK-92. *Cytotherapy* 15: 1563–1570.

Vercammen, E. et al. 2008. Sensing of Viral Infection and Activation of Innate Immunity by Toll-Like Receptor 3. *Clin Microbiol Rev* 21: 13–25.

Vijay, K. 2018. Toll-like receptors in immunity and inflammatory diseases: Past, present, and future. *International Immunopharmacology* 59: 391–412.

Walczak, H. 2011. TNF and ubiquitin at the crossroads of gene activation, cell death, inflammation and cancer. *Immunological Reviews* 244: 9–28.

Chapter 3

Dunkelberger, J. R., and W. C. Song. 2010. Complement and its role in innate and adaptive immune responses. *Cell Research* 20: 34–50.

Gros, P., F. J. Milder, and B. J. Janssen. 2008. Complement driven by conformational changes. *Nature Reviews Immunology* 8: 48–58.

Jarczak, J., E. M. Kosciuczuk, P. Lisowski, et al. 2013. Defensins: Natural components of human innate immunity. *Human Immunology* 74: 1069–1079.

Jones, L. L., and D. A. Vignali. 2011. Molecular interactions within the IL-6/IL-12 cytokine/receptor superfamily. *Immunologic Research* 51: 5–14.

Kemper, C., and D. E. Hourcade. 2008. Properdin: New roles in pattern recognition and target clearance. *Molecular Immunology* 45: 4048–4056.

Kemper, C., and J. P. Atkinson. 2007. T-cell regulation: With complements from innate immunity. *Nature Reviews Immunology* 7: 9–18.

Kerrigan, A. M., and G. D. Brown. 2009. C-type lectins and phagocytosis. *Immunobiology* 214: 562–575.

Mansour, S. C., O. M. Pena, and R. E. W. Hancock. 2014. Host defense peptides: Front-line immunomodulators. *Trends in Immunology* 35: 443–450.

Markiewski, M. M., B. Nilsson, K. N. Ekdahl, et al. 2007. Complement and coagulation: Strangers or partners in crime? *Trends in Immunology* 28: 184–192.

Mayilyan, K. R. 2012. Complement genetics, deficiencies, and disease associations. *Protein & Cell* 3: 487–496.

Menny, A., M. Serna, C. M. Boyd, et al. 2018. CryoEM reveals how the complement membrane attack complex ruptures lipid bilayers. *Nature Communications* 9: 5316.

Müller-Eberhard, H. J., and O. Gotze. 1972. C3 proactivator convertase and its mode of action. *The Journal of Experimental Medicine* 135: 1003–1008.

Parham. P. 2014. *The Immune System*, 4th ed., W. W. Norton and Company.

Peterson, S. L., H. X. Nguyen, O. A. Mendez, et al. 2017. Complement Protein C3 Suppresses Axon Growth and Promotes Neuron Loss. *Scientific Reports* 7: 12904.

Podack, E., R, J. Tschoop, and H. J. Muller-Eberhard. 1982. Molecular organization of C9 within the membrane attack complex of complement. Induction of circular C9 polymerization by the C5b-8 assembly. *Journal of Experimental Medicine* 156: 268.

Punt, J., J. A. Owen, S. A. Stranford, et al. 2018. *Kuby Immunology*, 8th ed., W. H. Freeman and Company: New York.

Ramadan, M. A., A. K. Shrive, D. Holden, et al. 2002. The three-dimensional structure of calcium-depleted human C-reactive protein from perfectly twinned crystals. *Acta Crystallographica Section D: Structural Biology* 58: 992–1001.

Sunyer, J. O., H. Boshra, G. Lorenzo, et al. 2003. Evolution of complement as an effector system in innate and adaptive immunity. *Immunologic Research* 27: 549–564.

Ward, P. A. 2009. Functions of C5a receptors. *Journal of Molecular Medicine* 87: 375–378.

Chapter 4

Boehm, T., and T. H. Rabbitts. 1989. The human T cell receptor genes are targets for chromosomal abnormalities in T cell tumors. *FASEB Journal* 3: 2344–2359.

Calame, K., J. Rogers, P. Early, et al. 1980. Mouse Cmu heavy chain immunoglobulin gene segment contains three intervening sequences separating domains. *Nature* 284: 452–455.

Capone, M., F. Watrin, C. Fernex, et al. 1993. TCR beta and TCR alpha gene enhancers confer tissue- and stage-specificity on V(D)J recombination events. *EMBO Journal* 12: 4335–4346.

Cesari, I. M., and M. Weigert. 1973. Mouse lambda-chain sequences. *Proceedings of the National Academy of Sciences USA* 70: 2112–2116.

Chien, Y. H., N. R. Gascoigne, J. Kavaler, et al. 1984. Somatic recombination in a murine T-cell receptor gene. *Nature* 309: 322–326.

Chien, Y., D. M. Becker, T. Lindsten, et al. 1984. A third type of murine T-cell receptor gene. *Nature* 312: 31–35.

Chun, J. J., D. G. Schatz, M. A. Oettinger, et al. 1991. The recombination activating gene-1 (RAG-1) transcript is present in the murine central nervous system. *Cell* 64: 189–200.

Clatworthy, A. E., M. A. Valencia-Burton, J. E. Haber, and M. A. Oettinger. 2005. The MRE11-RAD50-XRS2 complex, in addition to other non-homologous end-joining factors, is required for V(D)J joining in yeast. *Journal of Biological Chemistry* 280: 20247–20252.

Cooper, M. D., and M. N. Alder. 2006. The evolution of adaptive immune systems. *Cell* 124: 815–822.

Davis, M. M., K. Calame, P. W. Early, D. L. Livant, et al. 1980. An immunoglobulin heavy-chain gene is formed by at least two recombinational events. *Nature* 283: 733–739.

Davis, M. M., Y. H. Chien, N. R. Gasciogne, and S. M. Hedrick. 1984. A murine T cell receptor gene complex: Isolation, structure, and rearrangement. *Immunological Reviews* 81: 235–258.

Dörner, T. and Lipsky, P. E. 2001. Immunoglobulin variable-region gene usage in systemic autoimmune diseases. *Arthritis & Rheumatism* 44: 2715–2727.

Early, P. W., M. M. Davis, D. B. Kaback, et al. 1979. Immunoglobulin heavy chain gene organization in mice: Analysis of a myeloma genomic clone containing variable and alpha constant regions. *Proceedings of the National Academy of Sciences USA* 76: 857–861.

Early, P., H. Huang, M. Davis, et al. 1980. An immunoglobulin heavy chain variable region gene is generated from three segments of DNA: VH, D, and JH. *Cell* 19: 981–992.

Edelman, G. M., B. A. Cunningham, W. E. Gall, et al. 1969. The covalent structure of an entire gammaG immunoglobulin molecule. *Proceedings of the National Academy of Sciences USA* 63: 78–85.

Fahrer, A. M., J. F. Bazan, P. Papathanasiou, et al. 2001. A genomic view of immunology. *Nature* 409: 836–838.

Flajnik, M. F., and M. Kasahara. 2004. Origin and evolution of the adaptive immune system: Genetic events and selective pressures. *Nature Reviews Genetics* 11: 47–59.

Fugmann, S. D. 2001. *RAG1* and *RAG2* in V(D)J recombination and transposition. *Immunologic Research* 23:23–39.

Garboczi, D. N., P. Ghosh, U. Utz, et al. 1996. Structure of the complex between human T-cell receptor, viral peptide and HLA-A2. Nature 384: 134–141. doi: 10.1038/384134a0.

Gascoigne, N. R., Y. Chien, D. M. Becker, et al. 1984. Genomic organization and sequence of T-cell receptor beta-chain constant- and joining-region genes. *Nature* 310: 387–391.

Gilfillan, S., A. Dierich, M. Lemeur, et al. 1993. Mice lacking TdT: Mature animals with an immature lymphocyte repertoire. *Science* 261: 1175–1178.

Harris, L. J., S. B., Larson, K. W. Hasel, and A. McPherson. 1997. Refined structure of an intact IgG2a monoclonal antibody. Biochemistry 36: 1581–1597.

Hedrick, S. M., D. I. Cohen, E. A. Nielsen, and M. M. Davis. 1984. Isolation of cDNA clones encoding T cell-specific membrane-associated proteins. *Nature* 308: 149–153.

Hedrick, S. M., E. A. Nielsen, J. Kavaler, et al. 1984. Sequence relationships between putative T-cell receptor polypeptides and immunoglobulins. *Nature* 308: 153–158.

Heidelberger, M., and F. E. Kendall. 1937. The molecular weight of antibodies. *Journal of Experimental Medicine* 65: 393–414.

Horton, R., L. Wilming, V. Rand, et al. 2004. Gene map of the extended human MHC. *Nature Reviews Genetics* 5: 889–899.

Hozumi, N., and S. Tonegawa. 1976. Evidence for somatic rearrangement of immunoglobulin genes coding for variable and constant regions. *Proceedings of the National Academy of Sciences USA* 73: 3628–3632.

International Human Genome Sequencing Consortium. 2004. Finishing the euchromatic sequence of the human genome. *Nature* 431: 931–945.

Kavaler, J., M. M. Davis, and Y. Chien. 1984. Localization of a T-cell receptor diversity-region element. *Nature* 310: 421–423.

Koralov, S. B., T. I. Novobrantseva, J. Konigsmann, et al. 2006. Antibody repertoires generated by VH replacement and direct VH to JH joining. *Immunity* 25: 43–53.

Koretsky, G. A., and P. S. Myung. 2001. Positive and negative regulation of T-cell activation by adaptor proteins. *Nature Reviews Immunology* 1: 95–107.

Kurosaki, T. 2011. Regulation of BCR signaling. *Molecular Immunology* 48: 1287–1291.

La Gruta, N. L., P. G. Thomas, A. I. Webb, et al. 2008. Epitope-specific TCRbeta repertoire diversity imparts no functional advantage on the $CD8^+$ T cell response to cognate viral peptides. *Proceedings of the National Academy of Sciences USA* 105: 2034–2039.

Li, X. C., and M. Raghavan. 2010. Structure and function of major histocompatibility complex class I antigens. *Current Opinions in Organ Transplantation* 15: 499–504.

Nisonoff, A., F. C. Wissler, and L. N. Lipman. 1960. Properties of the major component of a peptic digest of rabbit antibody. *Science* 132: 1770–1771.

Oettinger, M. A., D. G. Schatz, C. Gorka, and D. Baltimore. 1990. RAG-1 and RAG-2, adjacent genes that synergistically activate V(D)J recombination. *Science* 248: 1517–1523.

Parham. P. 2014. *The Immune System*, 4th ed., W. W. Norton and Company.

Paul, W. E. 2012. *Fundamental Immunology*, 7th ed. Lippincott Williams & Wilkins, Philadelphia.

Punt, J., J. A. Owen, S. A. Stranford, et al. 2018. *Kuby Immunology*, 8th ed., W. H. Freeman and Company: New York.

Purves, W. 1995. LIFE-The Science of Biology, 4th ed., Sinauer Associates, Sunderland, MA.

Ranjaragan, S., and R. A. Mariuzza, 2014. T cell receptor bias for MHC: co-evolution or co-receptors? *Cellular and Molecular Life Sciences* 71: 3059–3068.

Raphael, I., S. Nalawade, T. N. Eager, and T. G. Forsthuber. 2015. T cell subsets and their signature cytokines in autoimmune and inflammatory diseases. *Cytokine* 74: 5–17.

Rivera-Munoz, P., L. Malivert, S. Derdouch, et al. 2007. DNA repair and the immune system: From V(D)J recombination to aging lymphocytes. *European Journal of Immunology* 37(Suppl 1): S71–S82.

Rock, K. L., I. A. York, and A. L. Goldberg. 2004. Post-proteasomal antigen processing for major histocompatibility complex class I presentation. *Nature Immunology* 5: 670–677.

Rodstrom, K. E. J., Regenthal, P., and Lindkvist-Petersson, K. 2015. Structure of Staphylococcal Enterotoxin E in Complex with Tcr Defines the Role of Tcr Loop Positioning in Superantigen Recognition. *PLOS One* 10: 01318.

Saito, H., D. M. Kranz, Y. Takagaki, et al. 1984. Complete primary structure of a heterodimeric T-cell receptor deduced from cDNA sequence. *Nature* 309: 757–762.

Saito, H., D. M. Kranz, Y. Takagaki, et al. 1984. A third rearranged and expressed gene in a clone of cytotoxic T lymphocytes. *Nature* 312: 36–40.

Schatz, D. G., and D. Baltimore. 1988. Stable expression of immunoglobulin gene V(D)J recombinase activity by gene transfer into 3T3 fibroblasts. *Cell* 53: 107–115.

Schatz, D. G., and Y. Ji. 2011. Recombination centres and the orchestration of V(D)J recombination. *Nature Reviews Immunology* 11: 251–263.

Schatz, D. G., M. A. Oettinger, and D. Baltimore. 1989. The V(D)J recombination activating gene, RAG-1. *Cell* 59: 1035–1048.

Siegrist, C.-A., Vaccine Immunology, Chapter 2 in Plotkin, S. A., Orenstein, W. A., and Offit, P. A. *Vaccines*, 6th ed., Elsevier Inc.

Smith-Garvin, J. E., G. A. Koretzky, and M. S. Jordan. 2009. T cell activation. *Annual Reviews in Immunology* 27: 591–619.

Soulas-Sprauel, P., P. Rivera-Munoz, L. Malvert, et al. 2007. V(D)J and immunoglobulin class switch recombinations: A paradigm to study the regulation of DNA end-joining. *Oncogene* 26: 7780–7791.

Stern, L. J., J. H. Brown, T. S. Jardetzky, et al. 1994. Crystal structure of the human class II MHC protein HLA-DR1 complexed with an influenza virus peptide. *Nature* 368: 215–221.

The MHC sequencing consortium. 1999. Complete sequence and gene map of a human major histocompatibility complex. *Nature* 401: 921–923.

Tonegawa, S., C. Brack, N. Hozumi, and R. Schuller. 1977. Cloning of an immunoglobulin variable region gene from mouse embryo. *Proceedings of the National Academy of Sciences USA* 74: 3518–3522.

Vaure, C. and Liu, Y. 2014. A comparative review of toll-like receptor 4 expression and functionality in different animal species. *Frontiers in Immunology* 5: 316.

Walls, A. C., Y-J. Park, M. A. Tortorici, et al. 2020. Structure, function, and antigenicity of the SARS-CoV-2 spike glycoprotein. *Cell* 180: 1–12.

Wearsch, P. A., and P. Cresswell. 2008. The quality control of MHC class I peptide loading. *Current Opinion in Cell Biology* 20: 624–631.

Weigert, M. G., I. M. Cesari, S. J. Yonkovich, and M. Cohen. 1970. Variability in the lambda light chain sequences of mouse antibody. *Nature* 228: 1045–1047.

Wu, H., Haist, V., Baumgärtner, W. and Shugart, K. 2010. Sustained viral load and late death in Rag2-/- mice after influenza A virus infection. *Virology Journal* 7: 72.

Yin, Y., Wang, X. X., Mariuzza, R. A. 2012. Crystal structure of a complete ternary complex of T-cell receptor, peptide-MHC, and CD4. *Proceedings of the National Academy of Sciences USA* 109: 5405–5410.

Chapter 5

Alam, S. M., G. M. Davies, C. M. Lin, et al. 1999. Qualitative and quantitative differences in T-cell receptor binding of agonist and antagonist ligands. *Immunity* 10: 227–237.

Anderson, G., S. Baik, J. E. Cowan, et al. 2014. Mechanisms of thymus medulla development and function. *Current Topics in Microbiology and Immunology* 373: 49–67.

Anderson, M. S., E. S. Venanzi, L. Klein, et al. 2002. Projection of an immunological self-shadow within the thymus by the aire protein. *Science* 298: 1395–1401.

Baldwin, T. A., K. A. Hogquist, and S. C. Jameson. 2004. The fourth way? Harnessing aggressive tendencies in the thymus. *Journal of Immunology* 173: 6515–6520.

Baldwin, T. A., M. M. Sandau, S. C. Jameson, and K. A. Hogquist. 2006. The timing of TCR alpha expression critically influences T-cell development and selection. *Journal of Experimental Medicine* 202: 111–121.

Cabaniols, J.-P., N. Fazilieau, A. Casrouge, et al. 2001. Most α/β T-cell receptor-diversity is due to terminal deoxynucleotidyl transferase. *Journal of Experimental Medicine* 194: 1385–1390.

Carpenter, A. C., and R. Bosselut. 2010. Decision checkpoints in the thymus. *Nature Immunology* 11: 666–673.

Ceredig, R., and T. Rolink. 2002. A positive look at double-negative thymocytes. *Nature Reviews Immunology* 2: 888–897.

Ellmeier, W., L. Haust, and R. Tschismarov. 2013. Transcriptional control of CD4 and CD8 coreceptor expression during T-cell development. *Cellular and Molecular Life Sciences* 70: 4537–4553.

Elmore, S. A., D. Dixon, J. R. Hailey, et al. 2016. Recommendations from the INHAND Apoptosis/Necrosis Working Group. *Toxicology and Pathology* 44: 173–88.

Gascoigne, N. R. 2010. $CD8^+$ thymocyte differentiation: T cell two-step. *Nature Immunology* 11: 189–190.

Ge, Q., and Y. Zhao. 2013. Evolution of thymus organogenesis. *Developmental and Comparative Immunology* 39: 85–90.

Germain, R. 2002. T-cell development and the CD4-CD8 lineage decision. *Nature Reviews Immunology* 2: 309–322.

Hogquist, K. A., and S. C. Jameson. 2014. The self-obsession of T cells: how TCR signaling thresholds affect fate 'decisions' and effector function. *Nature Immunology* 15: 815–823.

Hogquist, K. A., M. A. Gavin, and M. J. Bevan. 1993. Positive selection of $CD8^+$ T cells induced by major histocompatibility complex binding peptides in fetal thymic organ culture. *Journal of Experimental Medicine* 177: 1469–1473.

Hogquist, K. A., S. C. Jameson, and M. J. Bevan. 1996. Strong agonist ligands for the T-cell receptor do not mediate positive selection of functional $CD8^+$ T cells. *Immunity* 3: 79–86.

Hogquist, K. A., S. C. Jameson, W. R. Heath, et al. 1994. T-cell receptor antagonist peptides induce positive selection. *Cell* 76: 17–27.

Hogquist, K. A., T. A. Baldwin, and S. C. Jameson. 2006. Central tolerance: Learning self-control in the thymus. *Nature Reviews Immunology* 5: 772–782.

Janeway, Jr., C. A. 2001. *Immunology: The Immune System in Health and Disease*, 5th ed. Garland Science/Norton: New York.

Kisielow, P., H. Bluthmann, U. D. Staerz, et al. 1988. Tolerance in T-cell receptor transgenic mice involves deletion of nonmature $CD4^+8^+$ thymocytes. *Nature* 333: 742–746.

Klein, L., B. Kyewski, P. M. Allen, and K. A. Hogquist. 2014. Positive and negative selection of the T-cell repertoire: what thymocytes see (and don't see). *Nature Reviews Immunology* 14: 377–391.

Klein, L., M. Hinterberger, G. Wimsberger, and B. Kyewski. 2009. Antigen presentation in the thymus for positive selection and central tolerance induction. *Nature Reviews Immunology* 9: 833–844.

Kyewski, B., and P. Peterson. 2010. Aire, master of many trades. *Cell* 140: 24–26.

Lefranc, M.-P. IMGT®, the international ImMunoGeneTics database. 2001. *Nucleic Acids Research* 29: 207–209.

Ma, D., Y. Wei, and F. Liu. 2013. Regulatory mechanisms of thymus and T-cell development. *Developmental and Comparative Immunology* 39: 91–102.

Marrack, P., J. McCormack, and J. Kappler. 1989. Presentation of antigen, foreign major histocompatibility complex proteins and self by thymus cortical epithelium. *Nature* 338: 503–506.

Mathis, D., and C. Benoist. 2009. Aire. *Annual Review of Immunology* 27: 287–312.

McNeil, L. K., T. K. Starr, and K. A. Hogquist. 2006. A requirement for sustained ERK signaling during thymocyte positive selection in vivo. *Proceedings of the National Academy of Sciences USA* 102: 13574–13579.

Morris, G. P., and P. M. Allen. 2012. How the TCR balances sensitivity and specificity for the recognition of self and pathogens. *Nature Immunology* 13: 121–128.

Mouchess, M. L., and M. Anderson. 2014. Central tolerance induction. *Current Topics in Microbiology and Immunology* 373: 69–86.

Naito, T., and I. Taniuchi. 2013. Roles of repressive epigenetic machinery in lineage decision of T cells. *Immunology* 139: 151–157.

Nowell, C., and F. Radtke. 2017. Notch as a tumour suppressor. *Nature Reviews Cancer* 17: 145–159.

Parkin, J., and B. Cohen, 2001. An overview of the immune system. *The Lancet* 357: 1777–1789.

Page, D. M., L. P. Kane, J. P. Allison, and S. M. Hedrick. 1993. Two signals are required for negative selection of $CD4^+CD8^+$ thymocytes. *Journal of Immunology* 151: 1868–1880.

Passos, G. A., C. A. Speck-Hernandez, A. F. Assis, and D. A. Mendes-da-Cruz. 2017. Update on *Aire* and thymic negative selection. *Immunology* 153: 10–20.

Peterson, P., T. Org, and A. Rebane. 2008. Transcriptional regulation by AIRE: molecular mechanisms of central tolerance. *Nature Reviews Immunology* 8: 948–957.

Pui, J. C., D. Allman, L. Xu, et al. 1999. Notch1 expression in early lymphopoiesis influences B versus T lineage determination. *Immunity* 11: 299–308.

Punt, J., J. A. Owen, S. A. Stranford, et al. 2018. *Kuby Immunology*, 8th ed., W. H. Freeman and Company: New York.

Qin, C., Zhou, L., Hu, Z., et al. 2020. Dysregulation of immune response in patients with coronavirus 2019 (COVID-19) in Wuhan, China. *Clinical Infectious Diseases* 71: 762–768.

Sambandam, A., I. Maillard, V. P. Zediak, et al. 2006. Notch signaling controls the generation and differentiation of early T lineage progenitors. *Nature Immunology* 6: 663–670.

Schmitt, T. M., and J. C. Zuniga-Pflucker. 2002. Induction of T-cell development from hematopoietic progenitor cells by delta-like-1 in vitro. *Immunity* 17: 749–756.

Starr, T. K., S. C. Jameson, and K. A. Hogquist. 2003. Positive and negative selection of T cells. *Annual Review of Immunology* 21: 139–176.

Teh, H. S., P. Kisielow, B. Scott, et al. 1988. Thymic major histocompatibility complex antigens and the alpha beta T-cell receptor determine the CD4/CD8 phenotype of T cells. *Nature* 335: 229–233.

Uematsu, Y., S. Ryser, Z. Dembic, et al. 1988. In transgenic mice the introduced functional T-cell receptor beta gene prevents expression of endogenous beta genes. *Cell* 52: 831–841.

Venanzi, E. S., C. Benoist, and D. Mathis. 2004. Good riddance: thymocyte clonal deletion prevents autoimmunity. *Current Opinion in Immunology* 16: 197–202.

von Boehmer, H., H. S. Teh, and P. Kisielow. 1989. The thymus selects the useful, neglects the useless and destroys the harmful. *Immunology Today* 10: 57–61.

Wilson, A., H. R. MacDonald, and F. Radtke. 2001. Notch1-deficient common lymphoid precursors adopt a B cell fate in the thymus. *Journal of Experimental Medicine* 194: 1003–1012.

Zuniga-Pflucker, J. C. 2004. T-cell development made simple. *Nature Reviews Immunology* 4: 67–72.

Chapter 6

Allison, T. J., C. C. Winter, J.-J. Fournié, et al. 2001. Structure of a human γδ T-cell antigen receptor. *Nature* 411: 820–824.

Capone, M., F. Watrin, C. Fernex, et al. 1993. TCR beta and TCR alpha gene enhancers confer tissue- and stage-specificity on V(D)J recombination events. *EMBO Journal* 12: 4335–4346.

Chang, H.-C., K. Tan, J. Ouyang, et al. 2005. Structural and mutational analyses of a CD8αβ heterodimer and comparison with the CD8αα homodimer. *Immunity* 23: 661–671.

Chien, Y. H., N. R. Gascoigne, J. Kavaler, et al. 1984. Somatic recombination in a murine T-cell receptor gene. *Nature* 309: 322–326.

Chien, Y., D. M. Becker, T. Lindsten, et al. 1984. A third type of murine T-cell receptor gene. *Nature* 312: 31–35.

Davis, M. M., Y. H. Chien, N. R. Gasciogne, and S. M. Hedrick. 1984. A murine T cell receptor gene complex: isolation, structure, and rearrangement. *Immunological Reviews* 81: 235–258.

Doytchinova, I. A., and D. R. Flower. 2005. In silico identification of supertypes for class II MHCs. *Journal of Immunology* 174: 7085–7095.

Hedrick, S. M., D. I. Cohen, E. A. Nielsen, and M. M. Davis. 1984. Isolation of cDNA clones encoding T cell-specific membrane-associated proteins. *Nature* 308: 149–153.

Horton, R., L. Wilming, V. Rand, et al. 2004. Gene map of the extended human MHC. *Nature Reviews Genetics* 5: 889–899.

Huppa, J. B., and M. M. Davis. 2013. The interdisciplinary science of T-cell recognition. *Advances in Immunology* 119: 1–50.

International Human Genome Sequencing Consortium. 2004. Finishing the euchromatic sequence of the human genome. *Nature* 431: 931–945.

Ito, Y., M. Hashimoto, K. Hirota, et al. 2014. Detection of T cell responses to a ubiquitous cellular protein in autoimmune disease. *Science* 346: 363–368.

Jusforgues-Saklani, H., M. Uhl, N. Blachere, et al. 2008. Antigen persistence is required for dendritic cell licensing and $CD8^+$ T cell cross-priming. *Journal of Immunology* 181: 3067–3076.

Koretsky, G. A., and P. S. Myung. 2001. Positive and negative regulation of T-cell activation by adaptor proteins. *Nature Reviews Immunology* 1: 95–107.

Kuhns, M. S., and H. B. Badgandi. 2012. Piecing together the family portrait of TCR-CD3 complexes. *Immunological Reviews* 250: 120–243.

Leahy, D. J. 1995. A structural view of CD4 and CD8. *FASEB Journal* 9: 17–25.

Leone, P., E. C. Shin, F. Perosa, et al. 2013. MHC class I antigen processing and presenting machinery: organization, function, and defects in tumor cells. *Journal of the National Cancer Institute* 105: 1172–1187.

Li, Y., Y. Yin, and R. A. Mariuzza. 2013. Structural and biophysical insights into the role of CD4 and CD8 in T cell activation. *Frontiers in Immunology* 4: 1–11.

Natarajan, K., A. C. McShan, J. Jiang, et al. 2017. An allosteric site in the T-cell receptor Cβ domain plays a critical signalling role. *Nature Communications* 8: 15260.

Niu, L., H. Cheng, S. Zhang, et al. 2013. Structural basis for the differential classification of HLA-A*6802 and HLA-A*6801 into the A2 and A3 supertypes. *Molecular Immunology* 55: 381–392.

Parham, P. 2014. *The Immune System*, 4th ed., W. W. Norton and Company.

Paul, W. E. 2012. *Fundamental Immunology*, 7th ed. Lippincott Williams & Wilkins, Philadelphia.

Punt, J., J. A. Owen, S. A. Stranford, et al. 2018. *Kuby Immunology*, 8th ed., W. H. Freeman and Company: New York.

Rammensee, H.-G., J. Bachmann, N. N. Emmerich, et al. 1999. SYFPEITHI: database for MHC ligands and peptide motifs. *Immunogenetics* 50: 213–219.

Ranjaragan, S., and R. A. Mariuzza. 2014. T cell receptor bias for MHC: co-evolution or co-receptors? *Cellular and Molecular Life Sciences* 71: 3059–3068.

Robinson, J., A. Malik, P. Parham, et al. 2000. IMGT/HLA—a sequence database for the human major histocompatibility complex. *Tissue Antigens* 55: 280–287.

Robinson, J., J. A. Halliwell, J. H. Hayhurst, et al. 2015. The IPD and IMGT/HLA database: allele variant databases. *Nucleic Acids Research.* 43: D423–D431.

Roche, P. A., and K. Furuta. 2015. The ins and outs of MHC class II-mediated antigen processing and presentation. *Nature Reviews Immunology*. 15: 203–216.

Rock, K. L., I. A. York, and A. L. Goldberg. 2004. Post-proteasomal antigen processing for major histocompatibility complex class I presentation. *Nature Immunology* 5: 670–677.

Rödström, K. E, P. Regenthal, and K. Lindkvist-Petersson. 2015. Structure of Staphylococcal Enterotoxin E in Complex with TCR Defines the Role of TCR Loop Positioning in Superantigen Recognition. *PLOS ONE* 10: e0131988.

Saini, S. K., H. Schuster, V. R. Ramnarayan, et al. 2015. Dipeptides catalyze rapid peptide exchange on MHC class I molecules. *Proceedings of the National Academy of Sciences USA* 112: 202–207.

Saito, H., D. M. Kranz, Y. Takagaki, et al. 1984. Complete primary structure of a heterodimeric T-cell receptor deduced from cDNA sequence. *Nature* 309: 757–762.

Saito, H., D. M. Kranz, Y. Takagaki, et al. 1984. A third rearranged and expressed gene in a clone of cytotoxic T lymphocytes. *Nature* 312: 36–40.

Smith-Garvin, J. E., G. A. Koretzky, and M. S. Jordan. 2009. T cell activation. *Annual Reviews in Immunology* 27: 591–619.

Snell, G. D., and G. F. Higgins. 1951. Alleles at the histocompatibility-2 locus in the mouse as determined by tumor transplantation. *Genetics* 36: 306–310.

Trowsdale, J., and J. C. Knight. 2013. Major histocompatibility genomics and human disease. *Annual Review of Genomics and Human Genetics* 14: 301–323.

van Kasteren, S. I., H. Overkleeft, H. Ovaa, and J. Neefjes. 2014. Chemical biology of antigen presentation by MHC molecules. *Current Opinion in Immunology* 26: 21–31.

Wallaert, A., Durinck, K., Taghon, T. et al. 2017. T-ALL and thymocytes: A message of noncoding RNAs. *Journal of Hematology and Oncology* 10: 66.

Wang, R., K. Natarajan, and D. H. Margulies. 2009. Structural basis of the CD8αβ/MHC class I interaction: Focused recognition orients CD8β to a T cell proximal position. *Journal of Immunology* 183: 2554–2564.

Wearsch, P. A., and P. Cresswell. 2008. The quality control of MHC class I peptide loading. *Current Opinion in Cell Biology* 20: 624–631.

Wills, C., and D. R. Green. 1995. A genetic herd-immunity model for the maintenance of MHC polymorphism. *Immunological Reviews* 143: 263–292.

Yin, Y., X. X. Wang, and R. A. Mariuzza. 2012. Crystal structure of a complete ternary complex of T-cell receptor, peptide-MHC, and CD4. *Proceedings of the National Academy of Sciences USA* 109: 5045–5410.

Zanker, D., and W. Chen. 2014. Standard and immunoproteasomes show similar peptide degradation specificities. *European Journal of Immunology* 44: 3500–3503.

Zavala-Ruiz, Z., I. Strug, B. D. Walker, B. D., et al. A hairpin turn in a class II MHC-bound peptide orients residues outside the binding groove for T cell recognition. *Proceedings of the National Academy of Sciences USA* 101: 13279–13284.

Chapter 7

Bedoya, S. K., B. Lam, K. Lau, and J. Larkin III. 2013. Th17 cells in immunity and autoimmunity. *Clinical and Developmental Immunology* 2013: 986789.

Bernard, D., J. D. Hansen, L. D. Pasquier, et al. 2007. Costimulatory receptors in jawed vertebrates: Conserved CD28, odd CTLA4 and multiple BTLAs. *Developmental and Comparative Immunology* 31: 255–271.

Bour-Jordan, H., J. H. Esensten, M. Martinez-Llordella, et al. 2011. Intrinsic and extrinsic control of peripheral T-cell tolerance by costimulatory molecules of the CD28/B7 family. *Immunological Reviews* 241: 180–205.

Chan, A. C., M. Iwashima, C. W. Turck, and A. Weiss, (1992) ZAP-70: A 70 kd protein-tyrosine kinase that associates with the TCR zeta chain. *Cell* 71: 649–662.

Crotty, S. 2011. Follicular helper CD4 T cells (TFH). *Annual Review of Immunology* 29: 621–663.

Gratz, I. K., M. D. Rosenblum, and A. K. Abbas. 2013. The life of regulatory T cells. *Annals of the New York Academy of Sciences* 1283: 8–12.

Hebeisen, M., L. Baitsch, D. Presotto, et al. 2013. SHP-1 phosphatase activity counteracts increased T-cell receptor affinity. *Journal of Clinical Investigation* 123: 1044–1056.

Kaer, L. V. 2007. NKT cells: T lymphocytes with innate effector functions. *Current Opinion in Immunology* 19: 354–364.

Kaiko, G. E., J. C. Horvat, K. W. Beagley, and P. M. Hansbro. 2007. Immunological decision making: How does the immune system decide to mount a helper-T cell response? *Immunology* 123: 326–338.

Kapsenberg, M. L. 2003. Dendritic cell control of pathogen-driven T-cell polarization. *Nature Reviews Immunology* 3: 984–993.

Linsley, P., and S. Nadler. 2009. The clinical utility of inhibiting CD28-mediated costimulation. *Immunological Reviews* 229: 307–321.

McGhee, J. 2005. The world of T_H1/T_H2 subsets: First proof. *Journal of Immunology* 175: 3–4.

Moreau, H. D., and P. Bousso. 2014. Visualizing how T cells collect activation signals in vivo. *Current Opinion in Immunology* 26: 56–62.

Mosmann, T., H. Cherwinski, M. Bond, et al. 1986. Two types of murine helper T cell clone. I. Definition according to profiles of lymphokine activities and secreted proteins. *Journal of Immunology* 136: 2348–2357.

Nakayamada, S., H. Takahashi, Y. Kanno, and J. J. O'Shea. 2012. Helper T cell diversity and plasticity. *Current Opinion in Immunology* 24: 297–302.

Navarro, M. N., and D. A. Cantrell. 2014. Serine-threonine kinases in TCR signaling. *Nature Immunology* 15: 808–814.

Olivar, R., A. Luque, M. Naranjo-Gomez. 2013. The α7β0 isoform of the complement regulator C4b-binding protein induces a semimature, anti-inflammatory state in dendritic cells. *Journal of Immunology* 190: 2857–2872.

Parham, P. 2014. *The Immune System*, 4th ed., W. W. Norton and Company.

Pepper, M., and M. K. Jenkins. 2011. Origins of CD4(+) effector and central memory T cells. *Nature Immunology* 12: 467–471.

Reiner, S. 2008. Inducing the T cell fates required for immunity. *Immunological Research* 42: 160–165.

Rodriguez-Fernandez, J. L., L. Riol-Blanco, C. Delgado-Martin, and C. Escribano-Diaz. 2009. The dendritic cell side of the immunological synapse: Exploring terra incognita. *Discovery Medicine* 8: 108–112.

Rudd, C., A. Taylor, and H. Schneider. 2009. CD28 and CTLA-4 coreceptor expression and signal transduction. *Immunological Reviews* 229: 12–26.

Saito, T., T. Yokosuka, and A. Hashimoto-Tane. 2010. Dynamic regulation of T-cell activation and costimulation through TCR-microclusters. *FEBS Letters* 584: 4865–4871.

Sawant, D. V., and Vignali, D. A. A. 2014. Once a T_{reg}, always a T_{reg}? *Immunological Reviews* 259: 173–191.

Sharpe, A. 2009. Mechanisms of costimulation. *Immunological Reviews* 229: 5–11.

Smith-Garvin, J., G. Koretsky, and M. Jordan. T-cell activation. *Annual Review of Immunology* 27: 591–619.

Tellier, J., and S. L. Nutt. 2013. The unique features of follicular T cell subsets. *Cellular and Molecular Life Sciences* 70: 4771–4784.

Walunas, T. L., D. J. Lenschow, C. Y. Bakker, et al. 1994. CTLA-4 can function as a negative regulator of T-cell activation. *Immunity* 1: 405–413.

Wan, Y. Y. 2010. Multi-tasking of helper T cells. *Immunology* 130: 166–171.

Wiedemann, A., D. Depoil, M. Faroudi, and S. Valitutti, 2006. Cytotoxic T lymphocytes kill multiple targets simultaneously via spatiotemporal uncoupling of lytic and stimulatory synapses. *Proceedings of the National Academy of Sciences USA* 103: 10985–10990.

Wing, K., T. Yamaguchi, and S. Sakaguchi. 2011. Cell-autonomous and -non-autonomous roles of CTLA-4 in immune regulation. *Trends in Immunology* 32: 428–433.

Chapter 8

Allman, D., and S. Pillai. 2008. Peripheral B cell subsets. *Current Opinion in Immunology* 20: 149–157.

Boehm, T., I. Hess, and J. B. Swann. 2012. Evolution of lymphoid tissues. *Trends in Immunology* 33: 315–321.

Cambier, J. C., S. B. Gauld, K. T. Merrell, and B. J. Vilen. 2007. B-cell anergy: From transgenic models to naturally occurring anergic B cells? *Nature Reviews Immunology* 7: 633–643.

Carsetti, R., M. M. Rosado, and H. Wardmann. 2004. Peripheral development of B cells in mouse and man. *Immunological Reviews* 197: 179–191.

Casola, S. 2007. Control of peripheral B-cell development. *Current Opinion in Immunology* 19: 143–149.

Dorshkind, K., and E. Montecino-Rodriguez. 2007. Fetal B-cell lymphopoiesis and the emergence of B-1 cell potential. *Nature Reviews Immunology* 7: 213–219.

Dzierzak, E., and N. A. Speck. 2008. Of lineage and legacy: the development of mammalian hematopoietic stem cells. *Nature Immunology* 9: 129–136.

Fuxa, M., and J. A. Skok. 2007. Transcriptional regulation in early B cell development. *Current Opinion in Immunology* 19: 129–136.

Hardy, R. R., C. E. Carmack, S. A. Shinton, et al. 1991. Resolution and characterization of pro-B and pre-pro-B cell stages in normal mouse bone marrow. *Journal of Experimental Medicine* 173: 1213–1225.

Hoek, K. L., P. Antony, J. Lowe, et al. 2006. Transitional B cell fate is associated with developmental stage-specific regulation of diacylglycerol and calcium signaling upon B cell receptor engagement. *Journal of Immunology* 177: 5405–5413.

Ivanovs, A., S. Rybtsov, E. S. Ng, et al. 2017. Human haematopoietic stem cell development: from the embryo to the dish. *Development* 144: 2323–2337.

Klaus, A., and C. Robin. 2017. Embryonic hematopoiesis under microscopic observation. *Developmental Biology* 428: 318-327.

Krause, D. S., and D. T. Scadden. 2015. A hostel for the hostile: the bone marrow niche in hematologic neoplasms. *Haematologica* 100: 1376–1387.

Kurosaki, T., H. Shinohara, and Y. Baba. 2010. B cell signaling and fate decision. *Annual Review of Immunology* 28: 21–55.

Mackay, F., and P. Schneider. 2009. Cracking the BAFF code. *Nature Reviews Immunology* 9: 491–502.

Mikkola, H. K., and. S. H. Orkin. 2006. The journey of developing hematopoietic stem cells. *Development* 133: 3733–3744.

Monroe, J. G., and K. Dorshkind. 2007. Fate decisions regulating bone marrow and peripheral B lymphocyte development. *Advances in Immunology* 95: 1–50.

Montecino-Rodriguez, E., and K. Dorshkind. 2012. B-1 B cell development in the fetus and adult. *Immunity* 36: 13–21.

Nagasawa, T. 2006. Microenvironmental niches in the bone marrow required for B-cell development. *Nature Reviews Immunology* 6: 107–116.

Nemazee, D. 2006. Receptor editing in lymphocyte development and central tolerance. *Nature Reviews Immunology* 6: 728–740.

Pala, F. et al., 2015. *The Journal of Clinical Investigation* 125: 3941–3951.

Palis, J., R. J. Chan, A. Koniski, et al. 2001. Spatial and temporal emergence of high proliferative potential hematopoietic precursors during murine embryogenesis. *Proceedings of the National Academy of Sciences USA* 98: 4528–4533.

Parham, P. 2014. *The Immune System*, 4th ed., W. W. Norton and Company.

Perez-Vera, P., A. Reyes-Leon, and E. M. Fuentes-Panana. 2011. Signaling proteins and transcription factors in normal and malignant early B cell development. *Bone Marrow Research* 2011: 502751.

Pieper, K., B. Grimbacher, and H. Eibel. 2013. B-cell biology and development. *Journal of Allergy and Clinical Immunology* 131: 959–971.

Pillai, S., and A. Cariappa. 2009. The follicular versus marginal zone B lymphocyte cell fate decision. *Nature Reviews Immunology* 9: 767–777.

Punt, J., J. A. Owen, S. A. Stranford, et al. 2018. *Kuby Immunology*, 8th ed., W. H. Freeman and Company: New York.

Rajewsky, K., and H. von Boehmer. 2008. Lymphocyte development: Overview. *Current Opinion in Immunology* 20: 127–130.

Ramirez, J., K. Lukin, and J. Hagman. 2010. From hematopoietic progenitors to B cells: Mechanisms of lineage restriction and commitment. *Current Opinion in Immunology* 22: 177–184.

Srivastava, B., W. J. Quinn 3rd, K. Hazard, et al. 2005. Characterization of marginal zone B cell precursors. *Journal of Experimental Medicine* 202: 1225–1234.

Tokoyoda, K., T. Egawa, T. Sugiyama, et al. 2004. Cellular niches controlling B lymphocyte behavior within bone marrow during development. *Immunity* 20: 707–718.

Tung, J. W., and L. A. Herzenberg. 2007. Unraveling B-1 progenitors. *Current Opinion in Immunology* 19: 150–155.

von Boehmer, H., and F. Melchers. 2010. Checkpoints in lymphocyte development and autoimmune disease. *Nature Immunology* 11: 14–20.

Whitlock, C. A., and O. N. Witte. 1982. Long-term culture of B lymphocytes and their precursors from murine bone marrow. *Proceedings of the National Academy of Sciences USA* 79: 3608–3612.

Yarkoni, Y., A. Getahun, and J. C. Cambier. 2010. Molecular underpinning of B-cell anergy. *Immunological Reviews* 237: 249–263.

Yin, T., and L. Li. 2006. The stem cell niches in bone. *Journal of Clinical Investigation* 116: 1195–1201.

Chapter 9

Berek, C., G. M. Griffiths, and C. Milstein. 1985. Molecular events during maturation of the immune response to oxazolone. *Nature* 316: 412–418.

Buerstedde, J.-M, J. Alinikula, H. Arakawa, et al. 2014. Targeting of somatic hypermutation by immunoglobulin enhancer and enhancer-like sequences. *PLOS Biology* 12: e1001831.

Burnet, F. M. 1957. A modification of Jerne's theory of antibody production using the concept of clonal selection. *Australian Journal of Science* 20: 67–69.

Catron, D. M., A. A. Itano, K. A. Pape, et al. 2004. Visualizing the first 50 hr of the primary immune response to soluble antigen. *Immunity* 21: 341–347.

Chandra, V., A. Bortnick, and C. Murre. 2015. AID targeting: Old mysteries and new challenges. *Trends in Immunology* 36: 527–535.

Edelman, G. M., B. A. Cunningham, W. E. Gall, et al. 1969. The covalent structure of an entire gammaG immunoglobulin molecule. *Proceedings of the National Academy of Sciences USA* 63: 78–85.

Edholm, E. S., E. Bengten, and M. Wilson. 2011. Insights into the function of IgD. *Developmental and Comparative Immunology* 35: 1309–1316.

Eisen, H. N., and G. W. Siskind. 1964. Variations in affinities of antibodies during the immune response. *Biochemistry* 3: 996–1008.

Flajnik, M. F., and M. Kasahara. 2010. Origin and evolution of the adaptive immune system: Genetic events and selective pressures. *Nature Reviews Genetics* 11: 47–59.

Forsström, B. 2014. *Characterization of antibody specificity using peptide array technologies.* (Dissertation) Kungliga Tekniska Högskolan, KTH, Royal Institute of Technology, School of Biotechnology, Solna, Sweden.

Hannum, L. G., A. M. Haberman, S. M. Anderson, and M. J. Shlomchik. 2000. Germinal center initiation, variable gene region hypermutation, and mutant B cell selection without detectable immune complexes on follicular dendritic cells. *Journal of Experimental Medicine* 192: 931–942.

Harris, L. J., Larson, S. B., Hasel, K. W., et al. 1997. Refined structure of an intact IgG2a monoclonal antibody. *Biochemistry* 36: 1581–1597.

Heidelberger, M., and F. E. Kendall. 1936. Quantitative studies on antibody purification: I. The dissociation of precipitates formed by Pneumococcus specific polysaccharides and homologous antibodies. *Journal of Experimental Medicine* 64: 161–172.

Heidelberger, M., and K. O. Pedersen. 1937. The molecular weight of antibodies. *Journal of Experimental Medicine* 65: 393–414.

Hiramoto, E., A. Tsutsumi, R. Suzuki, et al. 2018. The ImG pentamer is an asymmetric pentagon with an open groove that binds the AIM protein. *Science Advances* 4: 1–9.

Jacob, J., J. Przylepa, C. Miller, and G. Kelsoe. 1993. In situ studies of the primary immune response to (4-hydroxy-3-nitrophenyl)acetyl. III. The kinetics of V region mutation and selection in germinal center B cells. *Journal of Experimental Medicine* 178: 1293–1307.

Kabat, E. A., T. T. Wu, H. M. Perry, et al. 1991. *Sequences of Proteins of Immunological Interest*, Vol 2, 5th ed. U. S. Department of Health and Human Services, Public Health Service, National Institutes of Health. NIH Publication No. 91–3242.

Koshland, M. E. 1975. Structure and function of the J chain. *Advances in Immunology* 20: 41–69.

Kubagawa, H., S. Oka, Y. Kubagawa, et al. 2014. The long elusive IgM Fc receptor, FcμR. *Journal of Clinical Immunology* 34: S35–45.

Kumar, R., L. J. DiMenna, J. Chaudhuri, and T. Evans. 2014. Biological function of activation-induced cytidine deaminase (AID). *Biomedical Journal* 37: 269–283.

LeBien, T. W., and T. F. Tedder. 2008. B lymphocytes: How they develop and function. *Blood* 112: 1570–1580.

Mak, T., et al. 2014. *Primer to the Immune Response*, 2nd ed., Academic Cell/Elsevier Inc.: Burlington, MA.

Matho, M. H., A., Schlossman, X., Meng, et al. 2015. Structural and functional characterization of anti-a33 antibodies reveal a potent cross-species orthopoxviruses neutralizer. *PLOS Pathog* 11: e1005148–e1005148.

Meng, F.-L. et al. 2015. The Mechanism of IgH Class Switch Recombination. In *Molecular Biology of B Cells*, 2nd ed., pp. 345–362. Academic Press: London.

Mostov, K. E. 1994. Transepithelial transport of immunoglobulins. *Annual Review of Immunology* 12: 63–84.

Munn, E. A., A. Feinstein, and A. J. Munro. 1971. Electron microscope examination of free IgA molecules and of their complexes with antigen. *Nature* 231: 527–529.

Muramatsu, M., H. Nagaoka, R. Shinkura, et al. 2007. Discovery of activation-induced cytidine deaminase, the engraver of antibody memory. *Advances in Immunology* 94: 1–36.

Nimmerjahn, F., and J. V. Ravetch. 2010. Antibody-mediated modulation of immune responses. *Immunological Reviews* 236: 265–275.

Nishizawa, T., and H. Suzuki. 2015. Gastric carcinogenesis and underlying molecular mechanisms: *Helicobacter pylori* and novel targeted therapy. *BioMed Research International*. 1–7.

Nisonoff, A., F. C. Wissler, and L. N. Lipman. 1960. Properties of the major component of a peptic digest of rabbit antibody. *Science* 132: 1770–1771.

Parham, P. 2014. *The Immune System*, 4th ed., W. W. Norton and Company.

Raff, M. C., M. Feldmann, and S. De Petris. 1973. Monospecificity of bone marrow-derived lymphocytes. *Journal of Experimental Medicine* 137: 1024–1030.

Schroeder, H. W. and Cavacini, L. 2010. Structure and function of immunoglobulins. *Journal of Allergy and Clinical Immunology* 125: S41-S52.

Soliman, C., Walduck, A. K., Yuriev, E., et al. 2018. Structural basis for antibody targeting of the broadly expressed microbial polysaccharide poly-N-acetylglucosamine. *Journal of Biology and Chemistry* 293: 5079–5089.

Stavnezer, J., and C. E. Schrader. 2014. IgH chain class switch recombination: Mechanism and regulation. *Journal of Immunology* 193: 5370–5378.

Tiselius, A. 1937. Electrophoresis of serum globulin. I. *Biochemical Journal* 31: 313–317.

Tiselius, A., and E. A. Kabat. 1938. Electrophoresis of immune serum. *Science* 87: 416–417.

Tiselius, A., and E. A. Kabat. 1939. An electrophoretic study of immune sera and purified antibody preparations. *Journal of Experimental Medicine* 69: 119–131.

Tong, P., and D. R. Wesemann. 2015. Molecular mechanisms of IgE class switch recombination. *Current Topics in Microbiology and Immunology* 388: 21–37.

Wiegert, M. G., I. M. Cesari, S. J. Yonkovich, and M. Cohn. 1970. Variability in the lambda light chain sequences of mouse antibody. *Nature* 228: 1045–1047.

Yokoyama, H., Mizutani, R., Satow, Y. 2013. Structure of a double-stranded DNA (6–4) photoproduct in complex with the 64M-5 antibody Fab. *Acta Crystallogr D Biol Crystallogr* 69: 504–512.

Chapter 10

Ada, G. L., and P. Byrt. 1969. Specific inactivation of antigen-reactive cells with 125I-labelled antigen. *Nature* 222: 1291–1292.

Allen, C. D., T. Okada, and J. G. Cyster. 2007. Germinal-center organization and cellular dynamics. *Immunity* 27: 190–202.

Allen, C. D., T. Okada, H. L. Tang, and J. G. Cyster. 2007. Imaging of germinal center selection events during affinity maturation. *Science* 315: 528–531.

Alt, F. W. et al. 2015. *B Cell Memory and Plasma Cell Development. Molecular Biology of B Cells*, 2nd ed. London: Academic Press.

Dal Porto, J. M., S. B. Gauld, K. T. Merrell, et al. 2004. B cell antigen receptor signaling 101. *Molecular Immunology* 41: 599–613.

Fleire, S. J., J. P. Goldman, Y. R. Carrasco, et al. 2006. B cell ligand discrimination through a spreading and contraction response. *Science* 312: 738–741.

Germain, R. N., M. Bajenoff, F. Castellino, et al. 2008. Making friends in out-of-the-way places: How cells of the immune system get together and how they conduct their business as revealed by intravital imaging. *Immunological Reviews* 221: 163–181.

Han, S., B. Zheng, J. Dal Porto, et al. 1995. In situ studies of the primary immune response to (4-hydroxy-3-nitrophenyl)acetyl. IV. Affinity-dependent, antigen-driven B cell apoptosis in germinal centers as a mechanism for maintaining self-tolerance. *Journal of Experimental Medicine* 182: 1635–1644.

Harwood, N. E., and F. D. Batista. 2009. The antigen expressway: Follicular conduits carry antigens to B cells. *Immunity* 30: 177–179.

Kang, S., A. B. Keener, S. Z. Jones, et al. 2016. IgG-Immune Complexes Promote B Cell Memory by Inducing BAFF. *The Journal of Immunology* 196: 196–206.

Kohler, G., and C. Milstein. 1975. Continuous cultures of fused cells secreting antibody of predefined specificity. *Nature* 256: 495–497.

Mitchell, G. F., and J. F. Miller. 1968. Cell to cell interaction in the immune response. II. The source of hemolysin-forming cells in irradiated mice given bone marrow and thymus or thoracic duct lymphocytes. *Journal of Experimental Medicine* 128: 821–837.

Pape, K. A., D. M. Catron, A. A. Itano, and M. K. Jenkins. 2007. The humoral immune response is initiated in lymph nodes by B cells that acquire soluble antigen directly in the follicles. *Immunity* 26: 491–502.

Parham, P. 2014. *The Immune System*, 4th ed., W. W. Norton and Company.

Parra, D., F. Takizawa, and J. O. Sunyer. 2013. Evolution of B cell immunity. *Annual Review of Animal Bioscience* 1: 65–97.

Pulendran, B., G. Kannourakis, S. Nouri, et al. 1995. Soluble antigen can cause enhanced apoptosis of germinal-centre B cells. *Nature* 375: 331–334.

Punt, J., J. A. Owen, S. A. Stranford, et al. 2018. *Kuby Immunology*, 8th ed., W. H. Freeman and Company: New York.

Schneider, P. 2005. The role of APRIL and BAFF in lymphocyte activation. *Current Opinion in Immunology* 17: 282–289.

Schroeder, H. W. and L. Cavacini. 2010. Structure and function of immunoglobulins. *Journal of Allergy and Clinical Immunology* 125: S41–S52.

Shokat, K. M., and C. C. Goodnow. 1995. Antigen-induced B-cell death and elimination during germinal-centre immune responses. *Nature* 375: 334–338.

Stavnezer, J. and C. E. Schrader. 2014. IgH chain class switch recombination: Mechanism and regulation. *Journal of Immunology* 193: 5370–5378.

Chapter 11

Castellino, F., G. Galli, G. D. Giudice, and R. Rappuoli. 2009. Generating memory with vaccination. *European Journal of Immunology* 39: 2100–2105.

Centers for Disease Control and Prevention. 2021. *Recommended Child and Adolescent Immunization Schedule for ages 18 years or younger, United States, 2021*. 2/11/2021 update. U.S. Dept of Health and Human Services. https://www.cdc.gov/vaccines/schedules/downloads/child/0-18yrs-child-combined-schedule.pdf.

Datta, S. K., S. Okamoto, T. Hayashi, et al. 2006. Vaccination with irradiated *Listeria* induces protective T cell immunity. *Immunity* 25: 143–152.

Farber, D. L., N. A. Yudanin, and N. P. Restifo. 2014. Human memory T cells: Generation, compartmentalization and homeostasis. *Nature Reviews Immunology* 14: 24–35.

Gary, E. N., and D. B. Weiner. 2020. DNA vaccines: Prime time is now. *Current Opinion in Immunology* 65: 21–27.

Gebhardt, T., S. N. Mueller, W. R. Heath, and F. R. Carbone. 2013. Peripheral tissue surveillance and residency by memory T cells. *Trends in Immunology* 34: 27–32.

Jackson, N. A. C., K. E. Kester, D. Casimiro, et al. 2020. The promise of mRNA vaccines: a biotech and industrial perspective. *npj Vaccines* 5: 11.

Janeway. C. A., and P. Travers. 1997. *Immunobiology: The immune System in Health and Disease*. Current Biology Limited/Garland Publishing: New York.

MacLeod, M. K. L., J. W. Kappler, and P. Marrack. 2010. Memory CD4 T cells: Generation, reactivation, and re-assignment. *Immunology* 130: 10–15.

Mahnke, Y. D., T. M. Brodie, F. Sallusto, et al. 2013. The who's who of T-cell differentiation: Human memory T-cell subsets. *European Journal of Immunology* 43: 2797–2809.

Nolz, J. C., G. R. Starbeck-Miller, and J. T. Harty. 2011. Naïve, effector and memory CD8 T-cell trafficking: Parallels and distinctions. *Immunotherapy* 3: 1223–1233.

Pardi, N., M. J. Hogan, F. W. Porter, and D. Weissman. 2018. mRNA vaccines—a new era in vaccinology. *Nature Reviews Drug Discovery* 17: 261–279.

Parker, N., M. Schneegurt, A.-H. Thi Tu, et al. 2016. *Microbiology*. Houston, Texas: OpenStax. https://openstax.org/details/books/microbiology.

Pulendran, B., and R. Ahmed. 2011. Immunological mechanisms of vaccination. *Nature Immunology* 12: 509–517.

Sadanand, S. 2011. Vaccination: The present and the future. *Yale Journal of Biology and Medicine* 84: 353–359.

Sahin, U., A. Muik, Evelyna D. et al. 2020. COVID-19 vaccine BNT162b1 elicits human antibody and T H 1 T cell responses. *Nature* 586: 594–599.

Sallusto, F., A. Lanzavecchia, K. Araki, and R. Ahmed. 2010. From vaccines to memory and back. *Immunity* 33: 451–463.

Takemori, T., T. Kaji, Y. Takahashi, et al. 2014. Generation of memory B cells inside and outside germinal centers. *European Journal of Immunology* 44: 1258–1264.

Taylor, J. J., and M. K. Jenkins. 2011. $CD4^+$ memory T cell survival. *Current Opinion in Immunology* 23: 319–323.

Chapter 12

Abeles, S. R., and D. T. Pride. 2014. Molecular bases and role of viruses in the human microbiome. *Journal of Molecular Biology* 426: 3892–3906.

Almogren, A., P. B. Furtado, Z. Sun, et al. 2006. Purification, properties and extended solution structure of the complex formed between human immunoglobulin A1 and human serum albumin by scattering and ultracentrifugation. *Journal of Molecular Biology* 356: 413–431.

Avila, M., D. M. Ojcius, and O. Yilmaz. 2009. The oral microbiota: Living with a permanent guest. *DNA and Cell Biology* 28: 405–411.

Bonner, A., A. Almogren, P. B. Furtado, et al. (2009) The nonplanar secretory IgA2 and near planar secretory IgA1 solution structures rationalize their different mucosal immune responses. *Journal of Biological Chemistry* 284: 5077–5087.

Butcher, E. C., R. V., Rouse, R L. Coffman, et al. 1982. Surface phenotype of Peyer's patch germinal center cells: Implications for the role of germinal centers in B cell differentiation. *Journal of Immunology* 129: 2698–2707.

Cresci, G. A., and E. Bawden. 2015. Gut microbiome: What we do and don't know. *Nutrition in Clinical Practice* 30: 734–746.

Davis, B. K., H. Wen, and J. P. Ting. 2011. The inflammasome NLRs in immunity, inflammation, and associated diseases. *Annual Reviews of Immunology* 29: 707–735.

Elinav, E., T. Strowig, J. Henao-Meija, and R. A. Flavell. 2011. Regulation of the antimicrobial response by NLR proteins. *Immunity* 34: 665–679.

Froy, O. 2005. Regulation of mammalian defensin expression by Toll-like receptor-dependent and independent signaling pathways. *Cellular Microbiology* 7: 1387–1397.

Fukata, M., A. S. Vamadevan, and M. T. Abreu. 2009. Toll-like receptors (TLRs) and Nod-like receptors (NLRs) in inflammatory disorders. *Seminars in Immunology* 21: 242–253.

Gerbe, F., and P. Jay 2016. Intestinal tuft cells: epithelial sentinels linking luminal cues to the immune system. *Mucosal Immunology* 9: 1353–1359.

Kool, M., et al. 2012. Cellular networks controlling Th2 polarization in allergy and immunity. *F1000 Biology Reports* 4: 6.

Lelouard, H., M. Fallet, B. de Bovis, et al. 2012. Peyer's patch dendritic cells sample antigens by extending dendrites through M cell-specific transcellular pores. *Gastroenterology* 142: 592–601.

Mak, T. et al. 2014. *Primer to the Immune Response*, 2nd ed., Academic Cell/Elsevier Inc.: Burlington, MA.

McDermott, A. J., and G. B. Huffnagle. 2013. The microbiome and regulation of mucosal immunity. *Immunology* 142: 24–31.

McGhee, J. R. 2005. Peyer's patch germinal centers: The elusive switch site for IgA. *Journal of Immunology* 175: 1361–1362.

McSorley, S. J. 2014. Immunity to intestinal pathogens: Lessons learned from *Salmonella*. *Immunological Reviews* 260: 168–182.

Metzinger, P. 2003. Tolerance, danger, and the extended family. *Annual Review of Immunology* 12: 991–1045.

Mowat, A. 2003. Anatomical basis of tolerance and immunity to intestinal antigens. *Nature Reviews Immunology* 3: 331–341.

Neish, A. S. 2014. Mucosal immunity and the microbiome. *Annals of the American Thoracic Society* 11: S28–S32.

Panda, S. K., and M. Colonna. 2019. Innate lymphoid cells in mucosal immunity. *Frontiers in Immunology* 10:861.

Parham, P. 2014. *The Immune System*, 4th ed., W. W. Norton and Company.

Salzman, N. H. 2011. Microbiota–immune system interaction: An uneasy alliance. *Current Opinion in Microbiology* 14: 99–105.

Salzman, N. H., K. Hung, D. Haribhai, et al. 2010. Enteric defensins are essential regulators of intestinal microbial ecology. *Nature Immunology* 11: 76–83.

Travassos, L. H., L. A. M. Carneiro, M. Ramjeet, et al. 2010. Nod1 and Nod2 direct autophagy by recruiting ATG16L1 to the plasma membrane at the site of bacterial entry. *Nature Immunology* 11: 55–62.

Wu, R. Q. et al. 2014. The mucosal immune system in the oral cavity: An orchestra of T cell diversity. *International Journal of Oral Science* 6: 125–132.

Chapter 13

Alcami, A., and U. H. Koszinowski. 2000. Viral mechanisms of immune evasion. *Trends in Microbiology* 8: 410–418.

Baym, M., T. D. Lieberman, E. D. Kelsic, et al. 2016. Spatiotemporal microbial evolution on antibiotic landscapes. *Science* 353: 1147–1151. doi: 10.1126/science.aag0822.

Bliska, J. B., K. Guan, J. E. Dixon, and S. Falkow. 1991. Tyrosine phosphate hydrolysis of host proteins by an essential *Yersinia* virulence determinant. *Proceedings of the National Academy of Sciences USA* 88: 1187–1191.

Callewaert, L., A. Aertsen, D. Deckers, et al. 2008. A new family of lysozyme inhibitors contributing to lysozyme tolerance in gram-negative bacteria. *PLOS Pathogens* 4: e1000019.

Clemens, D. L., B.-Y. Lee, and M. A. Horwitz. 2002. The Mycobacterium tuberculosis phagosome in human macrophages is isolated from the host cell cytoplasm. *Infection and Immunity* 70: 5800–5807.

DeLeon-Rodriguez, C. M., and A. Casadevall. 2016. *Cryptococcus neoformans*: Tripping on acid in the phagolysosome. *Frontiers in Microbiology* 7: 164.

Finlay, B. B., and G. McFadden. 2006. Anti-immunology: Evasion of the host immune system by bacterial and viral pathogens. *Cell* 124: 767–782.

Guan, K. L., and J. E. Dixon. 1990. Protein tyrosine phosphatase activity of an essential virulence determinant in Yersinia. *Science*: 249: 553–556. doi: 10.1126/science.2166336.

Guerra, C. R., S. H. Seabra, W. de Souza, and S. Rozental. 2014. *PLOS One* 9: e89250. doi: 10.1371/journal.pone.0089250.

Hmama, Z., S. Pena-Diaz, S. Joseph, and Y. Av-Gay. 2015. Immunoevasion and immunosuppression of the macrophage by *Mycobacterium tuberculosis*. *Immunological Reviews* 264: 220–232.

Juris, S. J., A. E. Rudolph, D. Huddler, et al. 2000. A distinctive role for the *Yersinia* protein kinase: Actin binding, kinase activation, and cytoskeleton disruption. *Proceedings of the National Academy of Sciences USA* 97: 9431–9436.

Juris, S. J., F. Shao, and J. E. Dixon. 2002. *Yersinia* effectors target mammalian signaling pathways. *Cellular Microbiology* 4: 201–211.

Knodler, L. A., J. Celli, and B. B. Finlay. 2001. Pathogenic trickery: Deception of host cell processes. *Nature Reviews Molecular Cell Biology* 2: 578–588.

Langley, R., B. Wines, N. Willoughby, et al. 2005. The staphylococcal superantigen-like protein 7 binds IgA and complement C5 and inhibits IgA-FcαRI binding and serum killing of bacteria. *Journal of Immunology* 174: 2926–2933.

Lima, T. A., and M. B. Lodoen. 2019. Mechanisms of human innate immune evasion by *Toxoplasma gondii*. *Frontiers in Cellular and Infection Microbiology* 9: 103.

Lorenzo, M. E., H. L. Ploegh, and R. S. Tirabassi. 2001. Viral immune evasion strategies and the underlying cell biology. *Seminars in Immunology* 13: 1–9.

Luganini, A., M. E. Terlizzi, and G. Gribaudo. 2016. Bioactive molecules released from cells infected with the human cytomegalovirus. *Frontiers in Microbiology* 7: 715.

McSorley, S. J. 2014. Immunity to intestinal pathogens: Lessons learned from *Salmonella*. *Immunological Reviews* 260: 168–182.

Merrell, D. S., and S. Falkow. 2004. Frontal and stealth attach strategies in microbial pathogenesis. *Nature* 430: 250–256.

Saline, M., K. E. J. Rodstrom, G. Fischer, G., et al. 2010. The Structure of superantigen complexed with Tcr and Mhc reveals novel insights into superantigenic T cell activation. *Nature Communications* 1: 119.

Schraidt, O., M. D. Lefebre, M. J. Brunner, et al. 2010 Topology and organization of the *Salmonella typhimurium* Type III secretion needle complex components. *PLOS Pathogens*.

Soulat, D., and C. Bogdan. 2017. Function of macrophage and parasite phosphatases in leishmaniasis. *Frontiers in Immunology* 8: 1838.

Chapter 14

Barre-Sinoussi, F., J. C. Chermann, F. Rey, et al. 1983. Isolation of a T-lymphotropic retrovirus from a patient at risk for acquired immune deficiency syndrome (AIDS). *Science* 220: 868–871. doi: 10.1126/science.6189183.

Brenchley, J. M., and D. C. Douek. 2008. HIV infection and the gastrointestinal immune system. *Mucosal Immunity* 1: 23–30.

Buckley, R. H. 2004. Molecular defects in human severe combined immunodeficiency and approaches to immune reconstitution. *Annual Review of Immunology* 22: 625–655.

Caruso, R., N. Warner, N. Inohara, and G. Nunez. 2014. NOD1 and NOD2: Signaling, host defense, and inflammatory disease. *Immunity* 41: 898–908.

Chasela, C. S., M. G. Hudgens, D. J. Jamieson, et al. 2010. Maternal or infant antiretroviral drugs to reduce HIV-1 transmission. *New England Journal of Medicine* 362: 2271–2281.

Chinen, J., and W. T. Shearer. 2009. Secondary immunodeficiencies, including HIV infection. *Journal of Allergy and Clinical Immunology* 125: S195–S203.

Cichocki, F., E. Sitnicka, and Y. T. Bryceson. 2014. NK cell development and function—plasticity and redundancy unleashed. *Seminars in Immunology* 26: 114–126.

Douek, D. C., M. Roederer, and R. A. Koup. 2009. Emerging concepts in the immunopathogenesis of AIDS. *Annual Reviews in Medicine* 60: 471–484.

Fischer, A. 2007. Human primary immunodeficiency diseases. *Immunity* 27: 835–845.

Fried, A. J., and F. A. Bonilla. 2009. Pathogenesis, diagnosis, and management of primary antibody deficiencies and infections. *Clinical Microbiology Reviews* 22: 396–414.

Gallo, R. C., P. S. Sarin, E. P. Gelmann, et al. 1983. Isolation of human T-cell leukemia virus in acquired immune deficiency syndrome (AIDS). *Science* 220: 865–867. doi: 10.1126/science.6601823.

Granich, R., S. Crowley, M. Vitoria, et al. 2010. Highly active antiretroviral treatment for the prevention of HIV transmission. *Journal of the International AIDS Society* 13: 1.

Hladik, F., and T. J. Hope. 2009. HIV infection of the genital mucosa in women. *Current HIV/AIDS Reports* 6: 20–28.

Hütter, G., D. Nowak, M. Mossner, et al. 2009. Long-term control of HIV by CCR5 Delta32/Delta32 stem-cell transplantation. *New England Journal of Medicine* 360: 692–698.

Lim, M. S., and K. S. J. Elenitoba-Johnson. 2004. The molecular pathology of primary immunodeficiencies. *Journal of Molecular Diagnostics* 6: 59–83.

Moore, J. P., S. G. Kitchen, P. Pugach, and J. A. Zack. 2004. The CCR5 and CXCR4 coreceptors—central to understanding the transmission and pathogenesis of human immunodeficiency virus type 1 infection. *AIDS Research and Human Retroviruses* 20: 111–126.

Notarangeleo, L. D. 2010. Primary immunodeficiencies. *Journal of Allergy and Clinical Immunology* 125: S182–S194.

O'Shea, J. J., M. Husa, D. Li, et al. 2004. Jak3 and the pathogenesis of severe combined immunodeficiency. *Molecular Immunology* 41: 727–737.

Orange, J. S. 2013. Natural killer cell deficiency. *Journal of Allergy and Clinical Immunology* 132: 515–526.

Piacentini, L., M. Biasin, C. Fenizia, and M. Clerici. 2009. Genetic correlates of protection against HIV infection: The ally within. *Journal of Internal Medicine* 265: 110–124.

Unsworth, D. J. 2008. Complement deficiency and disease. *Journal of Clinical Pathology* 61: 1013–1017.

Chapter 15

Abramson, J., and I. Pecht. 2007. Regulation of the mast cell response to the type 1 Fc epsilon receptor. *Immunological Reviews* 217: 231–254.

Adam, J., W. J. Pichler, and D. Yerly. 2011. Delayed drug hypersensitivity: Models of T-cell stimulation. *British Journal of Clinical Pharmacology* 71: 701–707.

Cavani, A., and A. De Luca. 2010. Allergic contact dermatitis: Novel mechanisms and therapeutic perspectives. *Current Drug Metabolism* 11: 228–233.

da Silva, E. Z. M., M. C. Jamur, and C. Oliver. 2014, Mast cell function: A new vision of an old cell. *Journal of Histochemistry and Cytochemistry* 62: 698–738.

Galli, S. J., S. Nakae, and M. Tsai. 2005. Mast cells in the development of adaptive immune responses. *Nature Immunology* 6: 135–142.

Gladman, A. C. 2006. Toxicodendron dermatitis: Poison ivy, oak, and sumac. *Wilderness and Environmental Medicine* 17: 120–128.

Holgate, S. T., R. Djukanovic, T. Casale, and J. Bousquet. 2005. Anti-immunoglobulin E treatment with omalizumab in allergic diseases: An update on anti-inflammatory activity and clinical efficacy. *Clinical and Experimental Allergy* 35: 408–416.

Ishizaka, K., T. Ishizaka, and M. M. Hornbrook. 1967. Allergen-binding activity of γE, γG and γA antibodies in sera from atopic patients in vitro measurements of reaginic antibody. *Journal of Immunology* 98: 490–501.

Karasuyama, H., K. Mukai, K. Obata, et al. 2011. Nonredundant roles of basophils in immunity. *Annual Review of Immunology* 29: 45–69.

Kroger, A. *Contraindications and precautions to vaccination*. National Center for Immunization and Respiratory Diseases. North Carolina Statewide Immunization Conference August 11, 2011.

Litwack, G. 2017. *Human Biochemistry*. Academic Press.

Madore, A. M., and C. Laprise. 2010. Immunological and genetic aspects of asthma and allergy. *Journal of Asthma and Allergy* 3: 107–121.

Minai-Fleminger, Y., and F. Levi-Schaffer. 2009. Mast cells and eosinophils: The two key effector cells in allergic inflammation. *Inflammation Research* 58: 631–638.

Misevic, G. 2018. ABO Blood groups system. *Asia-Pacific Journal of Blood Types and Genes* 2: 71–84.

Navines-Ferrer, A., E. Serrano-Candelas, G. J. Molina-Molina, and M. Martin. 2016. IgE-related chronic diseases and anti-IgE based treatments. *Journal of Immunological Research* 2016: 8163803.

Okada, H., C. Kuhn, H. Feillet, and J. F. Bach. 2010. The "hygiene hypothesis" for autoimmune and allergic diseases: An update. *Clinical and Experimental Immunology* 160: 1–9.

Picard, M., and V. R. Galvao. 2017. Current knowledge and management of hypersensitivity reactions to monoclonal antibodies. *Journal of Allergy and Clinical Immunology Practice* 5: 600–609.

Rosenwasser, L. J. 2011. Mechanisms of IgE inflammation. *Current Allergy and Asthma Reports* 11: 178–183.

Rothenberg, M. E., and S. P. Hogan. 2006. The eosinophil. *Annual Review of Immunology* 24: 147–174.

Sampson, H. A. 2016. Food allergy: Past, present and future. *Allergology International* 65: 363–369.

Shade, K.-T. C., M. E. Conroy, N. Washburn, et al. 2020. Sialylation of immunoglobulin E is a determinant of allergic pathogenicity. *Nature* 582: 265–270.

Sicherer, S. H., and H. A. Sampson. 2009. Food allergy: Recent advances in pathophysiology and treatment. *Annual Review of Medicine* 60: 261–277.

Vercelli, D. 2008. Discovering susceptibility genes for asthma and allergy. *Nature Reviews Immunology* 8: 169–182.

Yamamoto, F. 2004. Review: ABO blood group system—ABH oligosaccharide antigens, anti-A and anti-B, A and B glycosyltransferases, and ABO genes. *Immunohematology* 20: 3–22.

Chapter 16

Anderson, M. S., E. S. Venanzi, Z. Chen., et al. 2005. The cellular mechanism of AIRE control of T cell tolerance. *Immunity* 23: 227–239.

Angum, F., T. Khan, J. Kaler, L. Siddiqui, and A. Hussain. 2020. The prevalence of autoimmune disorders in woman: A narrative review. *Cureus* 12: e8094.

Bour-Jordan, H., J. H. Esensten, M. Martinez-Llordella, et al. 2011. Intrinsic and extrinsic control of peripheral T-cell tolerance by costimulatory molecules of the CD28/B7 family. *Immunological Reviews* 241: 180–205.

Coronel-Restrepo, N., I. Posso-Osorio, J. Naranjo-Escobar, and G. J. Tobon. 2017. Autoimmune diseases and their relation with immunological, neurological and endocrinological axes. *Autoimmunity Reviews* 16: 684–692.

Costa, V. S., T. C. Mattana, and M. E. da Silva. 2010. Unregulated IL-23/IL-17 immune response in autoimmune diseases. *Diabetes Research and Clinical Practice* 88: 222–226.

Danikowski, K. M., S. Jayaraman, and B. S. Prabhakar. 2017. Regulatory T cells in multiple sclerosis and myasthenia gravis. *Journal of Neuroinflammation* 14: 117.

Di Giuseppe, D., A. Discacciati, N. Orsini, and A. Wolk. 2014. Cigarette smoking and risk of rheumatoid arthritis: A dose-response meta-analysis. *Arthritis Research & Therapy* 16: R61.

Diny, N. L., N. R. Rose, and D. Cihakova. 2017. Eosinophils in autoimmune diseases. *Frontiers in Immunology* 8: 484.

Ferretti, C., and A. La Cava. 2016. Adaptive immune regulation in autoimmune diabetes. *Autoimmunity Reviews* 15: 236–241.

Franklin, E. C., H. R. Holman, H. J. Müller-Eberhard, and H. G. Kunkel. 1957. An unusual protein component of high molecular weight in the serum of certain patients with rheumatoid arthritis. *Journal of Experimental Medicine* 105: 425–438.

Gratz, I. K., M. D. Rosenblum, and A. K. Abbas. 2013. The life of regulatory T cells. *Annals of the New York Academy of Sciences* 1283: 8–12.

Hafler, D. A., J. M. Slavik, D. E. Anderson, et al. 2005. Multiple sclerosis. *Immunological Reviews* 204: 208–231.

Kappos, L., D. Li, P. A. Calabresi, et al. 2011. Ocrelizumab in relapsing-remitting multiple sclerosis: a phase 2, randomised, placebo-controlled, multicentre trial. *Lancet* 378: 1779–1787.

Kochi, Y. 2016. Genetics of autoimmune diseases: Perspectives from genome-wide association studies. *International Immunology* 28: 155–161.

Linsley, P. S., and S. G. Nadler. 2009. The clinical utility of inhibiting CD28-mediated costimulation. *Immunological Reviews* 229: 307–321.

Matzaraki, V., V. Kumar, C. Wijmenga, and A. Zhernakova. 2017. The MHC locus and genetic susceptibility to autoimmune and infectious diseases. *Genome Biology* 18: 76.

Quinonez-Flores, C. M., S. A. Gonzalez-Chavez, D. Del Rio Najera, and C. Pacheco-Tena. 2016. Oxidative stress relevance in the pathogenesis of the rheumatoid arthritis: A systematic review. *BioMed Research International* 2016: 6097417.

Rioux, J. D., and A. K. Abbas. 2005. Paths to understanding the genetic basis of autoimmune diseases. *Nature* 435: 584–589.

Rivellese, F., A. Nerviani, F. W. Rossi, et al. 2017. Mast cells in rheumatoid arthritis: Friends or foes? *Autoimmunity Reviews* 16: 557–563.

Robinson J., J. A. Halliwell, J. H. Hayhurst, et al. 2015. The IPD and IMGT/HLA database: Allele variant databases. *Nucleic Acids Research* 43: D423–431.

Robinson, J., A. Malik, P. Parham, et al. 2000. IMGT/HLA - a sequence database for the human major histocompatibility complex. *Tissue Antigens* 55: 280–287.

Rudd, C. E., A. Taylor, and H. Schneider. 2009. CD28 and CTLA-4 coreceptor expression and signal transduction. *Immunological Reviews* 229: 12–26.

Sawant, D. V., and D. A. A. Vignali. 2014. Once a T_{reg}, always a T_{reg}? *Immunological Reviews* 259: 173–191.

Scofield, R. H. 2004. Autoantibodies as predictors of disease. *Lancet* 363: 1544–1546.

Steward-Tharp, S. M., Y. J. Song, R. M. Siegel, and J. J. O'Shea. 2010. New insights into T cell biology and T cell-directed therapy for autoimmunity, inflammation, and immunosuppression. *Annals of the New York Academy of Sciences* 1183: 123–148.

Thomas, R. 2010. The balancing act of autoimmunity: Central and peripheral tolerance versus infection control. *International Reviews of Immunology* 29: 211–233.

Tsokos, G. C., and N. R. Rose. 2017. Immune cell signaling in autoimmune diseases. *Clinical Immunology* 181: 1–8.

Veldhoen, M. 2009. The role of T helper subsets in autoimmunity and allergy. *Current Opinion in Immunology* 21: 606–611.

von Boehmer, H., and F. Melchers. 2010. Checkpoints in lymphocyte development and autoimmune disease. *Nature Immunology* 11: 14–20.

Ward, F. J., L. N. Dahal, S. K. Wijesekera, et al. 2013. The soluble isoform of CTLA-4 as a regulator of T-cell responses. *European Journal of Immunology* 43: 1274–1285.

Wing, K., and S. Sakaguchi. 2010. Regulatory T cells exert checks and balances on self tolerance and autoimmunity. *Nature Immunology* 11: 7–13.

Wing, K., T. Yamaguchi, and S. Sakaguchi. 2011. Cell-autonomous and -non-autonomous roles of CTLA-4 in immune regulation. *Trends in Immunology* 32: 428–433.

Zharkova, O., T. Celhar, P. D. Cravens, et al. 2017. Pathways leading to an immunological disease: Systemic lupus erythematosus. *Rheumatology* 56: i55–i66.

Chapter 17

Abdelnoor, A. M., R. Ajib, M. Chakhtoura, et al. 2009. Influence of HLA disparity, immunosuppressive regimen used, and type of kidney allograft on production of anti-HLA class-I antibodies after transplant and occurrence of rejection. *Immunopharmacology and Immunotoxicology* 31: 83–87.

Chinen, J., and R. H. Buckley. 2010. Transplantation immunology: Solid organ and bone marrow. *Journal of Allergy and Clinical Immunology* 125: S324–335.

Hart, A., J. M. Smith, M. A. Skeans, et al. 2021. OPTN/SRTR 2019 Annual Data Report: Kidney. *American Journal of Transplantation* 21: 21–137.

Hickey, M. J., N. M. Valenzuela, and E. F. Reef. 2016. Alloantibody generation and effector function following sensitization to human leukocyte antigen. *Frontiers in Immunology* 7: 30. doi: 10.3389/fimmun.2016.00030.

Issa, F., A. Schiopu, and K. J. Wood. 2010. Role of T cells in graft rejection and transplantation tolerance. *Expert Review of Clinical Immunology* 6: 155–169.

Ravindra, K., J. Leventhal, D. Song, and S. T. Ildstad. 2012. Chimerism and tolerance in solid organ transplantation. *Journal of Clinical and Cellular Immunology* S9: 003.

Sayegh, M. H., and C. B. Carpenter. 2004. Transplantation 50 years later—progress, challenges, and promises. *New England Journal of Medicine* 351: 2761–2766.

Turka, L. A., and R. I. Lechler. 2009. Towards the identification of biomarkers of transplantation tolerance. *Nature Reviews Immunology* 9: 521–526.

Turka, L. A., K. Wood, and J. A. Bluestone. 2010. Bringing transplantation tolerance into the clinic: Lessons from the ITN and RISET for the Establishment of Tolerance consortia. *Current Opinion in Organ Transplantation* 15: 441–448.

Waldmann, H., and S. Cobbold. 2004. Exploiting tolerance processes in transplantation. *Science* 305: 209–212.

Chapter 18

Aldrich, J. F., D. B. Lowe, M. H. Shearer, et al. 2010. Vaccines and immunotherapeutics for the treatment of malignant disease. *Clinical and Developmental Immunology* 2010: 697158.

Allison, J. P., A. A. Hurwitz, and D. R. Leach. 1995. Manipulation of costimulatory signals to enhance antitumor T-cell responses. *Current Opinion in Immunology* 7: 682–686.

Boon, T., P. G. Coulie, and B. Van den Eydne. 1997. Tumor antigens recognized by T cells. *Immunology Today* 18: 267–268.

Cancer Antigenic Peptide Database, https://caped.icp.ucl.ac.be/.

Castelli, M. S., P. McGonigle, and P. J. Honrby. 2019. The pharmacology and therapeutic applications of monoclonal antibodies. *Pharmacology Research and Perspectives* 7: e00535.

Centers for Disease Control and Prevention. 2021. Recommended Child and Adolescent Immunization Schedule for ages 18 years or younger, United States, 2021. 2/11/2021 update. U.S. Dept of Health and Human Services.

Chen, L., S. M. Park, A. V. Tumanov, et al. 2010. CD95 promotes tumour growth. *Nature* 465: 492–496.

Cho, H. J., Y. K. Oh, and Y. B. Kim. 2011. Advances in human papilloma virus vaccines: A patent review. *Expert Opinion on Therapeutic Patents* 21: 295–309.

Chulpanova, D. S., K. V. Kiateva, A. R. Green, et al. 2020. Molecular aspects and future perspectives of cytokine-based anti-cancer immunotherapy. *Frontiers in Cellular and Developmental Biology* 8: 402. doi: 10.3389/fcell.2020.00402.

Corones, M. 2015. How "find and replace" for human DNA works. Data Dive Blog. April 27, 2015. http://blogs.reuters.com/data-dive/2015/04/27/how-find-and-replace-for-human-dna-works/.

Gagnaire, A., B. Nadel, D. Raoult, et al. 2017. Collateral damage: insights into bacterial mechanisms that predispose host cells to cancer. *Nature Reviews Microbiology* 15: 109–128.

Greten, F. R., and S. I. Grivennikov. 2019. Inflammation and cancer: Triggers, mechanisms, and consequences. *Immunity* 51: 27–41.

Hanahan, D., and R. A. Weinberg. Hallmarks of cancer: The next generation. *Cell* 144: 646–674.

Houghton, A. N., J. S. Gold, and N. E. Blachere. 2001. Immunity against cancer: Lessons learned from melanoma. *Current Opinion in Immunology* 13: 134–140.

Ireland, T. 2018. Is a cancer vaccine on the horizon? Science Focus blog. Home of BBC Science Focus Magazine. https://www.sciencefocus.com/the-human-body/is-a-cancer-vaccine-on-the-horizon/.

Iwaszko, M., and Bogunia-Kubik, K. 2011. Clinical significance of the HLA-E and CD94/NKG2 interaction. *Archivum Immunologiae et Therapiae Experimentalis* 59: 353.

Leone, P., E.-C. Shin, F. Perosa, et al. 2013. MHC class I antigen processing and presenting machinery: Organization, function, and defects in tumor cells. *Journal of the National Cancer Institute* 105: 1172–1187.

Lesterhuis, W. J., J. B. Haanen, and C. J. Punt. 2011. Cancer immunotherapy—revisited. *Nature Reviews in Drug Discovery* 10: 591–600.

Moore, P. S., and Y. Chang. 2013. Why do viruses cause cancer? Highlights of the first century of human tumour virology. *Nature Reviews Cancer* 10: 878–889.

Parham, P. 2014. *The Immune System*, 4th ed., W. W. Norton and Company.

Perea, F., M. Bernal, A. Sánchez-Palencia, et al. 2016. The absence of HLA class I expression in non-small cell lung cancer correlates with the tumor tissue structure and the pattern of T cell infiltration. *Tumor Immunology and Microenvironment* 140: 888–899.

Rous, P. 1911. A sarcoma of the fowl transmissible by an agent separable from the tumor cells. *Journal of Experimental Medicine* 13: 397–411. doi: 10.1084/jem.13.4.397.

Sandal, T. 2002. Molecular aspects of the mammalian cell cycle and cancer. *The Oncologist* 7: 73–81.

Sharma, P., K. Wagner, J. D. Wolchok, and J. P. Allison. 2011. Novel cancer immunotherapy agents with survival benefit: Recent successes and next steps. *Nature Reviews in Cancer* 11: 805–812.

Tagliamonte, M., A. Petrizzo, M. L. Tornesello, et al. 2014. Antigen-specific vaccines for cancer treatment. *Human Vaccines & Immunotherapeutics* 10: 3332–3346.

Veglia, F., M. Perego, and D. Gabrilovich. 2018. Myeloid-derived suppressor cells coming of age. *Nature Immunology* 19: 108–119.

Vesely, M. D., M. H. Kershaw, R. D. Schreiber, and M. J. Smyth. 2011. Natural innate and adaptive immunity to cancer. *Annual Review of Immunology* 29: 235–271.

Weinberg, R. A. 1996. How cancer arises. *Scientific American* 275: 62–70.

Yip, K. W., and J. C. Reed. 2008. Bcl-2 family proteins and cancer. *Oncogene* 27: 6398–6406.

Index

Note: Page numbers in *italics* indicate figures, tables, or definitions of key terms.

B

C

D

E

F

G

H

J

K

L

M

N

O

P

R

T

About the Book

Editors: Jason Noe and Joan Kalkut
Project Editor: Martha Lorantos
Development Editors: Kerry O'Neill and Eric Sinkins
Permissions Supervisor: Michele Beckta
Copy Editor: Wendy Walker
Production Manager: Joan Gemme
Photo Researcher: Mark Siddall
Book Design and Production: Meg Britton Clark
Cover Design: Meg Britton Clark
Illustration Program: Dragonfly Media Group
Indexer: Linda Mamassian
Cover and Book Manufacturer: LSC Communications